Handbook of
Heat Transfer
Applications

OTHER McGRAW-HILL REFERENCE BOOKS OF INTEREST

HANDBOOKS

American Institute of Physics • AMERICAN INSTITUTE OF PHYSICS HANDBOOK

Baumeister and Avallone • MARKS' STANDARD HANDBOOK FOR MECHANICAL ENGINEERS

Brady and Clauser • MATERIALS HANDBOOK

Burlington and May • HANDBOOK OF PROBABILITY AND STATISTICS WITH TABLES

Chopey and Hicks • HANDBOOK OF CHEMICAL ENGINEERING CALCULATIONS

Condon and Odishaw • HANDBOOK OF PHYSICS

Considine • PROCESS INSTRUMENTS AND CONTROLS HANDBOOK

Dean • LANGE'S HANDBOOK OF CHEMISTRY

Fink and Beaty • STANDARD HANDBOOK FOR ELECTRICAL ENGINEERS

Fink and Christiansen • ELECTRONICS ENGINEERS' HANDBOOK

Grant • HACKH'S CHEMICAL DICTIONARY

Harris • HANDBOOK OF NOISE CONTROL

Harris and Crede • SHOCK AND VIBRATION HANDBOOK

Hicks • STANDARD HANDBOOK OF ENGINEERING CALCULATIONS

Hopp and Hennig • HANDBOOK OF APPLIED CHEMISTRY

Juran • QUALITY CONTROL HANDBOOK

McLellan and Shand • GLASS ENGINEERING HANDBOOK

Maynard • INDUSTRIAL ENGINEERING HANDBOOK

Perry • ENGINEERING MANUAL

Perry and Green • PERRY'S CHEMICAL ENGINEERS' HANDBOOK

Rohsenow, Hartnett, and Ganić • HANDBOOK OF HEAT TRANSFER FUNDAMENTALS

Rosaler and Rice • STANDARD HANDBOOK OF PLANT ENGINEERING

Tuma • ENGINEERING MATHEMATICS HANDBOOK

Tuma • HANDBOOK OF PHYSICAL CALCULATIONS

Turner and Malloy • THERMAL INSULATION HANDBOOK

Turner and Malloy • HANDBOOK OF THERMAL INSULATION DESIGN ECONOMICS

ENCYCLOPEDIAS

CONCISE ENCYCLOPEDIA OF SCIENCE AND TECHNOLOGY
ENCYCLOPEDIA OF ELECTRONICS AND COMPUTERS
ENCYCLOPEDIA OF ENERGY
ENCYCLOPEDIA OF ENGINEERING
ENCYCLOPEDIA OF PHYSICS

DICTIONARIES

DICTIONARY OF SCIENTIFIC AND TECHNICAL TERMS
DICTIONARY OF MECHANICAL AND DESIGN ENGINEERING
DICTIONARY OF COMPUTERS

Handbook of
Heat Transfer
Applications

Second Edition

Editors

Warren M. Rohsenow

Department of Mechanical Engineering
Massachusetts Institute of Technology

James P. Hartnett

Energy Resources Center
University of Illinois at Chicago

Ejup N. Ganić

Faculty of Mechanical Engineering
University of Sarajevo, Yugoslavia

McGraw-Hill Book Company

New York   St. Louis   San Francisco   Auckland   Bogotá   Hamburg
Johannesburg   London   Madrid   Mexico   Montreal   New Delhi
Panama   Paris   São Paulo   Singapore   Sydney   Tokyo   Toronto

06061444

Library of Congress Cataloging in Publication Data
Main entry under title:

Handbook of heat transfer applications.

Rev. ed. of part of: Handbook of heat transfer /
edited by Warren M. Rohsenow, J.P. Hartnett. c1973.
Companion volume to Handbook of heat transfer
fundamentals.
Includes bibliographies and index.
1.Heat—Transmission—Handbooks, manuals, etc.
2.Mass transfer—Handbooks, manuals, etc.
I.Rohsenow, Warren M.   II.Hartnett, J. P. (James P.)
III.Ganić, Ejup N.   IV.Handbook of heat transfer.
V.Handbook of heat transfer fundamentals.
QC320.4.H36   1985   621.402′2   85-10997
ISBN 0-07-053553-1

$\mathcal{D}$
621.4022
HAN

1234567890   KGP/KGP   898765

ISBN 0-07-053553-1

The editors for this book were Harold B. Crawford, Richard K.
Mickey, and Peggy Lamb, the designer was Mark E. Safran, and the
production supervisor was Thomas G. Kowalczyk. It was set in Bas-
kerville by University Graphics, Inc.

Printed and bound by The Kingsport Press.

Contents

Detailed contents of each chapter appear on the opening page of the
chapter. A detailed alphabetical index follows Chapter 12.

Contributors

BARTZ, J. A., *Electric Power Research Institute, Palo Alto, California* (Chapter 10. Cooling Towers and Cooling Ponds)

BERGLES, ARTHUR E., *Department of Mechanical Engineering, Iowa State University* (Chapter 3. Techniques to Augment Heat Transfer)

BOTTERILL, J. S. M., *Chemical Engineering Department, University of Birmingham, England* (Chapter 6. Heat Transfer in Fluidized and Packed Beds)

CHENG, PING, *Department of Mechanical Engineering, University of Hawaii at Manoa* (Chapter 11. Geothermal Heat Transfer)

CHIANG, H. D., *Department of Mechanical Engineering, University of Minnesota* (Chapter 12. Measurement of Temperature and Heat Transfer)

CHO, Y. I., *Jet Propulsion Laboratory, California Institute of Technology, Pasadena, California* (Chapter 2. Nonnewtonian Fluids)

CLARK, JOHN A., *Department of Mechanical Engineering, University of Michigan* (Chapter 8. Thermal Energy Storage)

GABOR, J. D., *Reactor Analysis and Safety Division, Argonne National Laboratory* (Chapter 6. Heat Transfer in Fluidized and Packed Beds)

GOLDSTEIN, R. J., *Department of Mechanical Engineering, University of Minnesota* (Chapter 12. Measurement of Temperature and Heat Transfer)

HARTNETT, J. P., *Energy Resources Center, University of Illinois at Chicago* (Chapter 1. Mass Transfer Cooling; Chapter 2. Nonnewtonian Fluids)

KREITH, FRANK, *Solar Energy Research Institute, Golden, Colorado* (Chapter 7. Solar Energy)

KUSUDA, T., *Thermal Analysis Group, National Bureau of Standards, Washington, D.C.* (Chapter 9. Heat Transfer in Buildings)

MAULBETSCH, J. S., *Electric Power Research Institute, Palo Alto, California* (Chapter 10. Cooling Towers and Cooling Ponds)

MUELLER, A. C., *Consultant, Church Hill, Maryland* (Chapter 4. Heat Exchangers: Part 1. Heat Exchanger Basic Thermal Design Methods, Part 2. Process Heat Exchangers)

RABL, A., *Center for Energy and Environmental Sciences, Princeton University*   (Chapter 7. Solar Energy)

SHAH, RAMESH K., *General Motors Corporation, Harrison Radiator Division, Lockport, New York*   (Chapter 4. Heat Exchangers: Part 1. Heat Exchanger Basic Thermal Design Methods, Part 3. Compact Heat Exchangers)

TIEN, C. L., *Department of Mechanical Engineering, University of California, Berkeley*   (Chapter 5. Heat Pipes)

Preface

Introduction

Heat transfer plays an important role in practically every industrial and environmental process as well as in the vital areas of energy production and conversion. In the generation of electrical power, whether by nuclear fission, the combustion of fossil fuels, magneto-hydrodynamic processes, or the use of geothermal energy sources, numerous heat transfer problems must be solved. These problems involve conduction, convection, and radiation processes and relate to the design of components such as boilers, condensers, and turbines. Engineers are constantly confronted with the need to maximize or minimize heat transfer rates and to maintain the integrity of materials under conditions of extreme temperature.

In the preparation of the second edition of the *Handbook of Heat Transfer* it was the editors' goal to provide the information needed by engineers to deal with the heat transfer problems encountered in their daily work. The major objective of the editors was to prepare a handbook which contains information essential for practicing engineers, consultants, research engineers, university professors, students, and technicians involved with heat transfer technology. Since the publication of the first edition of the *Handbook of Heat Transfer* there have been many developments in the field, both in fundamentals and in applications. Consequently, to achieve our goal it was necessary to expand the first edition of the *Handbook of Heat Transfer* into two separate handbooks: *Handbook of Heat Transfer Fundamentals* and *Handbook of Heat Transfer Applications*.

Coverage

The *Handbook of Heat Transfer Applications* provides broad coverage of the practical aspects of the field, building on several chapters in the first edition and adding nine completely new chapters. In particular, the chapter Mass Transfer Cooling offers more detailed information on film cooling, while the chapter Techniques to Augment Heat Transfer has been completely updated to reflect new developments. The section Heat Exchangers has been greatly expanded to cover basic design methods with special attention given to both compact and process heat exchangers.

Seven of the new chapters cover recent developments in the critical field of energy. These include Heat Transfer in Fluidized and Packed Beds; Solar Energy; Geothermal

Heat Transfer; Cooling Ponds and Cooling Towers, Thermal Energy Storage; Heat Pipes; and Heat Transfer in Buildings.

An important chapter on Measurement of Temperature and Heat Transfer has been added to the second edition. Rounding out the coverage is a chapter dealing with Non-newtonian Fluids, a topic of increasing interest in the chemical, pharmaceutical, and food industries.

It is assumed that the user of the *Handbook of Heat Transfer Applications* has a knowledge of the basic information appearing in the companion text, the *Handbook of Heat Transfer Fundamentals*. Taken together, these two handbooks provide the most comprehensive coverage available of the science and art of heat transfer.

Units

It is recognized at this time that the English Engineering System of units cannot be completely replaced by the International System (SI). Transition from the English System of units to SI will proceed at a rational pace to accommodate the needs of the profession, industry, and the public. The transition period will be long and complex, and duality of units probably will be demanded for at least one or two decades. Both SI and English units have been incorporated in this edition to the maximum extent possible, with the goal of making the Handbook useful throughout the world. Each numerical result, table, figure, and equation in the handbook is given in both systems of units, wherever presentation in dimensionless form is not given. In a few cases some tables are presented in one system of units, mostly to save space, and conversion factors are printed at the end of such tables for the reader's convenience.

Nomenclature

An attempt has been made by the editors to use a unified nomenclature throughout the Handbook. Given the breadth of the technical coverage, some exceptions will be found. However one symbol has only one meaning within any given section. Each symbol is defined at the end of each section of the Handbook. Both SI and English units are given for each symbol in the nomenclature list.

Index

An added feature in this edition is a comprehensive alphabetical index designed to provide quick reference to information. Taken together with the Table of Contents, this edition now provides quick and easy access to any topic in the book.

Acknowledgments

The editors owe a great deal to the dedication with which the authors of the second edition made their expertise available. Their cooperation on the content and length of their manuscripts and in incorporating all of the above-mentioned specifications coupled with the high quality of their work has resulted in a Handbook which we believe will fulfill the needs of the engineering community for many years to come. We also wish to thank the professional staff at McGraw-Hill Book Company who were involved with the production of the Handbook at various stages of the project for their outstanding cooperation and continued support. The outstanding editorial work of Richard K. Mickey and Peggy Lamb is gratefully acknowledged. Finally, thanks are also due to the staff of the Energy Resources Center at the University of Illinois at Chicago, especially Dr. E. Y. Kwack and J. Wiet who provided proofreading assistance and organizational help throughout the editorial process.

Closing Remarks

The Handbook is ultimately the responsibility of the editors. Meticulous care has been exercised to minimize errors, but it is impossible in a work of this magnitude to achieve an error-free publication. Accordingly the editors would appreciate being informed of any errors so that they may be eliminated from subsequent printings. The editors would also appreciate suggestions from readers on possible improvements in the usefulness of the Handbook so that they may be included in future editions.

W. M. ROHSENOW
J. P. HARTNETT
E. N. GANIĆ

ABOUT THE EDITORS

Warren M. Rohsenow, D.Eng., is professor of mechanical engineering and director of the Heat Transfer Laboratory at the Massachusetts Institute of Technology. For his outstanding work in the field of heat transfer, Dr. Rohsenow received the Max Jacob Memorial Award and was elected to the National Academy of Engineering.

James P. Hartnett, Ph.D., is professor of mechanical engineering at the University of Illinois at Chicago and director of the Energy Resources Center. For his contributions to the field of heat transfer, Dr. Hartnett has received the ASME Memorial Award and the Luikov Medal of the International Centre of Heat and Mass Transfer.

Ejup N. Ganić, Sc.D., is professor of mechanical engineering at the University of Sarajevo, Yugoslavia. From 1977 to 1982 he was a consultant to the Argonne National Laboratory and, until 1982, he was a tenured associate professor at the University of Illinois at Chicago.

CHAPTER 1

Mass Transfer Cooling

By J. P. Hartnett

University of Illinois, Chicago

A. INTRODUCTION

The term *mass transfer cooling* includes transpiration cooling, film cooling with a liquid, and film cooling with a gas as well as various ablation schemes. A pictorial representation of various forms of mass transfer cooling is given in Fig. 1. With the exception of film cooling with a gas all of the systems are physically similar. The major difference in the methods shown in Fig. 1*b* through *d* is that the mass transfer distribution may be independently controlled for the systems in Fig. 1*b* and *c* while for the other systems the mass transfer rate is set by the thermodynamics of the system.

In light of these considerations, Sec. B of this chapter will deal with transpiration cooling and then follow with a brief note on the applicability of these results to liquid film cooling and ablation. Throughout Sec. B the effects of suction will also be discussed. The chapter will conclude with Sec. C dealing with gaseous film cooling.

B. TRANSPIRATION COOLING

1. Forced-Convection Laminar Flow

a. The Flat Plate with Constant Properties

The system of differential equations describing the physical situation where a constant-property gas flows in laminar motion over a porous flat plate with mass addition into the boundary layer (Fig. 1b) is

$$\text{Continuity} \qquad \frac{\partial u}{\partial x} + \frac{\partial v}{\partial y} = 0 \tag{1}$$

$$\text{Momentum} \qquad \rho u \frac{\partial u}{\partial x} + \rho v \frac{\partial u}{\partial y} = \mu \frac{\partial^2 u}{\partial y^2} \tag{2}$$

$$\text{Energy} \qquad \rho c_p u \frac{\partial T}{\partial x} + \rho c_p v \frac{\partial T}{\partial y} = k \frac{\partial^2 T}{\partial y^2} + \mu \left(\frac{\partial u}{\partial y}\right)^2 \tag{3}$$

$$\text{Species} \qquad \rho u \frac{\partial Y}{\partial x} + \rho v \frac{\partial Y}{\partial y} = \rho D_{12} \frac{\partial^2 Y}{\partial y^2} \tag{4}$$

Strictly speaking, in a constant-property flow there is no need for the species equation. However, it can be assumed that the injected gas, although having the same transport and thermodynamic properties as the free-stream gas, is given special identification (e.g., it may be an isotope of the free-stream gas or it may be "tagged" with a radioactive tracer). The resulting mass fraction profile and the value of the mass fraction at the wall may be of value in actual binary flows when the free-stream and the secondary gas are not markedly different in physical properties (e.g., nitrogen injected into air).

This system of equations requires seven boundary conditions. The following boundary conditions lead to a set of ordinary differential equations:

$$\left.\begin{array}{l} u = 0 \\ T = T_w \\ v \sim u_\infty/\sqrt{\text{Re}_x} \end{array}\right\} \text{ at } y = 0 \tag{5}$$

$$\left.\begin{array}{l} u = u_\infty \\ T = T_\infty \\ Y = 0 \end{array}\right\} \text{ at } y \to \infty \tag{6}$$

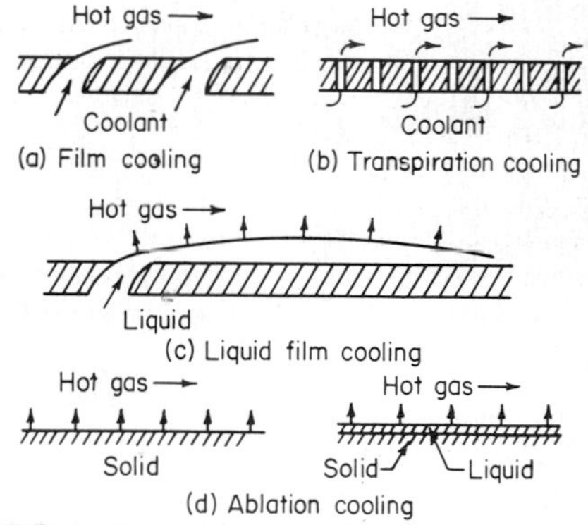

FIG. 1 Various mass transfer cooling schemes.

The next and last boundary condition states that there is no net flow of boundary-layer fluid into the plate surface. This fixes the value of the mass fraction at the surface $Y_{y=0}$

$$v = -\frac{D_{12}}{1 - Y} \frac{\partial Y}{\partial y} \qquad \text{at } y = 0 \tag{7}$$

Note that the resulting solution is restricted to the case where the distribution of the injected mass varies as $x^{-1/2}$. This blowing distribution is selected since it gives rise to a system of ordinary differential equations. For this physical system the resulting velocity distribution, local skin friction coefficient, and local dimensionless heat transfer coefficients are shown in Figs. 2 to 4 [1].

For mass transfer to a laminar boundary on a flat plate the following conclusions may be drawn from Fig. 2:

1. Mass addition to a zero–pressure gradient boundary layer results in an S-shaped velocity profile. Since this is known to be an unstable profile, mass transfer is destabilizing (i.e., mass addition to a laminar boundary-layer flow on a flat plate may cause the boundary layer to become turbulent). This conclusion does not apply to flows with favorable pressure gradient.

   Conversely, removal of mass from a zero–pressure gradient boundary layer (i.e., suction) is stabilizing.

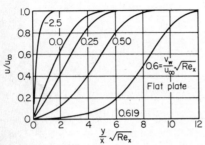

FIG. 2 Dimensionless velocity distribution u/u_∞ for constant-property laminar flow over a flat plate for various values of the blowing parameter $(v_w/u_\infty)\sqrt{Re_x}$ [1].

2. The maximum value of the dimensionless blowing parameter $(\rho_w v_w/\rho_\infty u_\infty)\sqrt{Re_x}$ is 0.619. Beyond this value the boundary-layer equations do not describe the flow.

   The local shearing stress may be determined from Fig. 3 and the equation

$$\tau_w = \frac{c_f}{2}\rho u_\infty^2 \qquad (8)$$

The local heat transfer rate is calculated as

$$q_w'' = h(T_w - T_{aw}) \qquad (9)$$

The local heat transfer coefficient is given on Fig. 4 as a function of the blowing or suction rate. The adiabatic wall temperature T_{aw} is determined from the relation

$$T_{aw} = T_\infty + r\frac{u_\infty^2}{2c_p} = T_\infty\left(1 + r\frac{\gamma - 1}{2}Ma_\infty^2\right) \qquad (10)*$$

In Eq. (10) the temperatures are absolute values in °R or K. The recovery factor r is given in Fig. 5. Note that the recovery temperature approaches the free-stream temperature for low-velocity flow and is equal to it in the limiting case of zero Mach number.

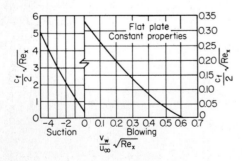

FIG. 3 Local skin friction coefficient for constant-property laminar flow over a flat plate for various values of the blowing parameter $(v_w/u_\infty)\sqrt{Re_x}$ [1].

*The adiabatic wall temperature is also called the recovery temperature. These two terms are used interchangeably throughout the chapter.

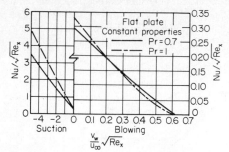

FIG. 4 Local dimensionless heat transfer coefficient $Nu/\sqrt{Re_x}$ for constant-property laminar flow over a flat plate for various values of the blowing parameter $(v_w/u_\infty)\sqrt{Re_x}$ [1].

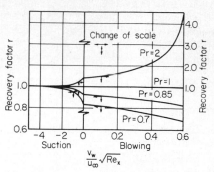

FIG. 5 Recovery factor for constant-property laminar flow over a flat plate for various values of the blowing parameter $(v_w/u_\infty)\sqrt{Re_x}$ [1].

b. The Flat Plate with Variable Properties Including Foreign Gas Injection into an Air Boundary Layer

The effect of variable physical properties is taken into account by the use of the Eckert reference method [2]. First the heat transfer coefficient, the skin friction coefficient, and the recovery factor are determined for a solid surface exposed to the same free-stream conditions and held at the same surface temperature as the mass transfer–cooled plate. The reference temperature T^* is first calculated from

$$T^* = T_\infty + 0.5(T_w - T_\infty) + 0.22(T_{aw0} - T_\infty) \qquad (11)$$

where

$$T_{aw0} = T_\infty + r_0^* \frac{u_\infty^2}{2c_p^*} \qquad (12)$$

and

$$r_0^* = \sqrt{Pr^*} \qquad (13)$$

The physical properties of the free-stream gas are known as functions of temperature and pressure and it is assumed that the wall temperature and the free-stream velocity and temperature are prescribed. The Prandtl number Pr^* and the specific heat c_p^* are to be evaluated at the reference temperature T^*. An initial estimate of these two properties is made, leading to a value for T_{aw0} and for T^*. New values of c_p^* and Pr^* and T^* may now be determined since T^* is known. The calculation is repeated until a consistent set of values of c_p^*, Pr^*, and T^* is achieved. The local skin friction coefficient c_{f0} and local Stanton number St_0 are then calculated from

$$\frac{c_{f0}}{2} = \frac{0.332}{\sqrt{u_\infty x/\nu^*}} \qquad (14)$$

and

$$St_0 = \frac{c_{f0}}{2}(Pr^*)^{-2/3} \qquad (15)$$

The reference temperature method has been shown to be valid for air, nitrogen, carbon dioxide, and hydrogen [3, 4].

The local skin friction coefficient c_f and the local Stanton number St in the presence of mass transfer with air as the free-stream gas for various injectant gases are shown in

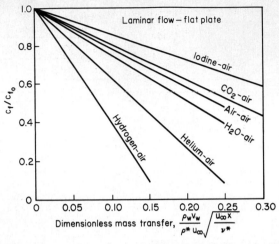

FIG. 6 Normalized skin friction coefficient c_f/c_{f0} for transpiration cooling in a laminar boundary layer on a flat plate as a function of the dimensionless mass transfer for foreign gas injection into an air boundary layer [5].

Figs. 6 and 7 in a normalized form c_f/c_{f0} and St/St_0 as a function of the dimensionless mass transfer rate

$$\frac{\rho_w v_w}{\rho^* u_\infty} \sqrt{\frac{u_\infty x}{\nu^*}}$$

where ρ^* and ν^* are density and kinematic viscosity of the free-stream gas evaluated at

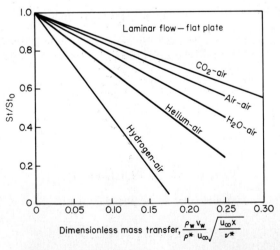

FIG. 7 Normalized Stanton number St/St_0 for transpiration cooling in a laminar boundary layer on a flat plate as a function of the dimensionless mass transfer for foreign gas injection into an air boundary layer [5].

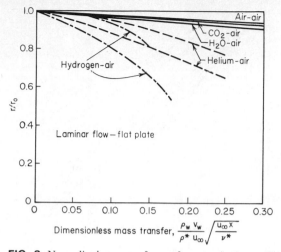

FIG. 8 Normalized recovery factors for transpiration cooling in a laminar boundary layer on a flat plate as a function of the dimensionless mass transfer [5].

T^* [5]. The normalized recovery factor r/r_0 is given on Fig. 8 [5]. The local shearing stress and the local heat transfer rate are then calculated:

$$\tau_w = \frac{c_{f0}}{2} \frac{c_f}{c_{f0}} \rho^* u_\infty^2 \tag{16}$$

$$q_w'' = \mathrm{St}_0 \frac{\mathrm{St}}{\mathrm{St}_0} \rho^* c_p^* u_\infty (T_w - T_{aw}) \tag{17}$$

The adiabatic wall temperature T_{aw} is calculated from

$$T_{aw} = T_\infty + \frac{r}{r_0} r_0^* \frac{u_\infty^2}{2c_p^*} \tag{18}$$

For other coolant gases injected into an air boundary layer it is recommended that the following approximate formulas be used if the flow is laminar:

$$\frac{c_f}{c_{f0}} = 1 - 2.08 \left(\frac{M_2}{M_1}\right)^{1/3} \frac{G}{\rho^* u_\infty} \sqrt{\frac{u_\infty x}{\nu^*}} \tag{19}$$

$$\frac{\mathrm{St}}{\mathrm{St}_0} = 1 - 1.90 \left(\frac{M_2}{M_1}\right)^{1/3} \frac{G}{\rho^* u_\infty} \sqrt{\frac{u_\infty x}{\nu^*}} \tag{20}$$

$$\frac{r}{r_0} = 1 \qquad \text{(or alternatively Fig. 11)} \tag{21}$$

Here M_1 and M_2 are the molecular weights of the injected gas and the free stream, respectively.

c. The Flat Plate with Variable Properties for Free-Stream Gases Other than Air

For free-stream gases other than air there are analytical predictions of normalized skin friction coefficients, Stanton numbers, and recovery factors for carbon dioxide, hydrogen,

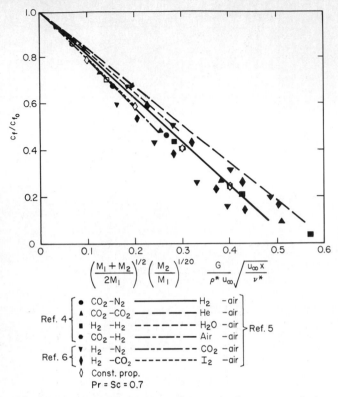

FIG. 9 Normalized skin friction coefficient c_f/c_{f_0} for mass transfer into foreign gas-free streams on a flat plate as a function of the dimensionless mass transfer rate [4].

and nitrogen free streams. These results together with those reported above for an air free stream are shown on Figs. 9 to 11. Table 1 summarizes the range of variables covered in these studies [4]. With the exception of M_1, all properties in the function plotted along the abscissa are those of the free-stream gas.

The wall mass fraction values which may be of value in the case of ablation are shown in Figs. 12 and 13 [4, 5].

The calculation procedure outlined in this section will yield reasonable estimates of the heat transfer provided that the wall temperature is considerably lower than the recovery temperature, a condition encountered in most flight applications. However, if the wall temperature is close to the recovery temperature, and if the coolant gas is much lighter than the free-stream gas, the influence of diffusion thermo becomes important.

Influence of Diffusion Thermo.   In those cases where the secondary gas differs from the main-stream gas there exists a mass flow due to the temperature gradient, the so-called thermal diffusion or Soret effect, and a heat flow due to the concentration gradient, the diffusion thermo or Dufour effect. The former is not of critical importance, but the latter is very important when the wall temperature is close to the recovery value (or equiv-

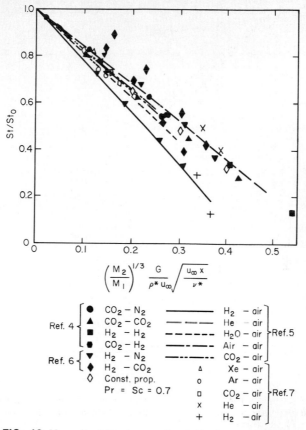

FIG. 10 Normalized Stanton number St/St_0 for mass transfer into foreign gas-free streams on a flat plate as a function of the dimensionless mass transfer rate [4].

alently if the wall temperature is close to the free-stream value under low-velocity conditions). Under these conditions adiabatic wall temperature is defined as the temperature where the conduction heat flow is equal and opposite to the diffusion thermo:

$$k_w \left(\frac{\partial T}{\partial y} \right)_w = \rho_w v_w \left(\frac{RT}{M} \frac{k_T}{Y} \right)_w \tag{22}$$

Calculations of this adiabatic wall temperature have been carried out for low-velocity conditions for an air free stream for a number of injected gases [7]. The results are shown on Fig. 14. In general, if a lightweight gas is injected, the recovery or adiabatic wall temperature is greater than the free-stream temperature because of the diffusion-thermo effect, and conversely, if a heavy coolant is used, the recovery temperature is lower than the free-stream value. For the gases shown, the heat transfer may still be calculated from Eq. (9) where the adiabatic wall temperature is determined from Fig. 14. For high velocities caution is in order if the wall temperature is close to recovery conditions, especially if lightweight coolants are to be used.

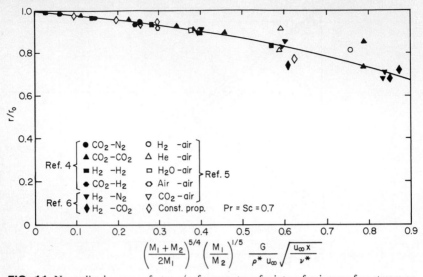

FIG. 11 Normalized recovery factor r/r_0 for mass transfer into a foreign gas-free stream on a flat plate as a function of the mass transfer rate [4].

TABLE 1 Gas Combinations and Flow Parameters Applicable to Figs. 9–11

Ref	Gas pair	Ma_∞	T_∞, K	T_w/T_∞
4	CO_2-N_2	0, 4, 8, 12	218	2, 6
	CO_2-H_2	0, 4, 8, 12	218	2, 6
	CO_2-CO_2	0, 4, 8, 12	218	2, 6
	H_2-H_2	0, 4, 8, 12	218	2, 6
6	H_2-N_2	0, 4, 8, 12	218	2, 6
	H_2-N_2	4	555	1, 2, 3
	H_2-N_2	0, 4	1110	0.5
	H_2-CO_2	0, 4, 8, 12	218	2, 6
	H_2-CO_2	4, 8	555	1, 2, 3
	H_2-CO_2	0, 4	1110	0.5
5	H_2-air	0	218	0.5, 2, 6
	H_2-air	12	218	2, 6
	H_2-air	0, 3, 6	218	1
	$He-air$	0, 3, 6	218	1
	H_2O-air	0, 3, 6	218	1
	$Air-air$	0, 3, 6	218	1
	CO_2-air	0	218	1
	I_2-air	0	218	1
7	$Xe-air$	0	297, 1666	0.25, 1.1
	$Ar-air$	0	297, 1666	0.25, 1.1
	CO_2-air	0	297, 1666	0.25, 1.1
	$He-air$	0	297, 1666	0.25, 1.1
	H_2-air	0	297, 1666	0.25, 1.1

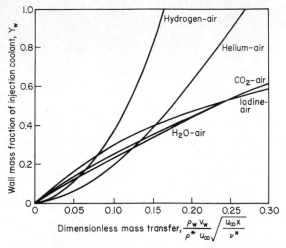

FIG. 12 Wall mass fraction of injected coolant for foreign gas injection into an air boundary layer on a flat plate as a function of the mass transfer rate [5].

d. The Flat Plate—Nonsimilar Solutions

All of the aforementioned solutions assumed a special lengthwise distribution of the injected mass, namely $v_w \sim x^{-1/2}$, and further assumed a constant wall temperature since these assumptions lead to a system of ordinary differential equations. Other distributions of interest, such as constant injection along the length of the plate or constant heat input along the plate, have been studied by Sparrow for a constant-property fluid with a Prandtl number 0.7. These results apply to such systems as air into air, N_2 into air, and N_2 into N_2. Reference 8 gives the dimensionless local heat transfer coefficient $(hx/k)/\sqrt{u_\infty x/\nu}$ as a function of the dimensionless injection rate $(G/\rho_\infty u_\infty)\sqrt{u_\infty x/\nu}$ on Fig. 15 for the following cases

$a.\ T_w = $ constant $\qquad v_w \sim x^{-1/2}$

$b.\ T_w = $ constant $\qquad v_w = $ constant

$c.\ q_w'' = $ constant $\qquad v_w \sim x^{-1/2}$

$d.\ q_w'' = $ constant $\qquad v_w = $ constant

$e.\ q_w'' = Gc_p(T_w - T_c) \qquad v_w = $ constant

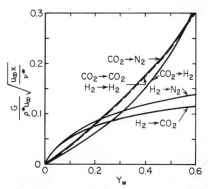

The last boundary condition is of special interest for it considers the realistic case when there is uniform mass injection along the length of the plate. The coolant leaves the reservoir at a constant temperature T_c as in Fig. 16, and the local wall temperature T_w and the local heat rate q_w'' are determined by the external flow conditions.

FIG. 13 Wall mass fraction of injected coolant for mass transfer into a foreign gas-free stream on a flat plate as a function of the mass transfer rate [4].

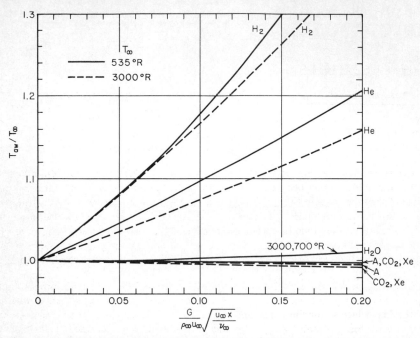

FIG. 14 Adiabatic wall temperatures resulting from diffusion thermo for various gases injected into a laminar air boundary layer on a flat plate [7].

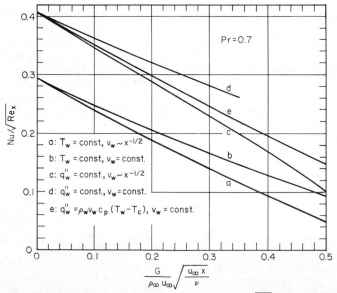

FIG. 15 Dimensionless heat transfer coefficient $\mathrm{Nu}/\sqrt{\mathrm{Re}_x}$ for constant-property transpiration cooling on a flat plate as a function of the blowing rate for various boundary conditions [8].

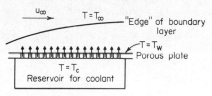

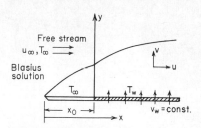

Heat balance: $q_w'' = G c_p (T_w - T_c) = h(T_\infty - T_w)$

FIG. 16 Heat balance on a transpiration-cooled flat plate.

FIG. 17 Transpiration cooling with an unheated solid starting length.

Figure 15 may also be used for variable properties by taking advantage of the Eckert reference procedure described earlier. If foreign gas coolants are used with boundary conditions of the type given in Fig. 15, it is suggested that the heat transfer be first calculated by the procedure developed earlier for the constant wall temperature case with $v_w \sim x^{-1/2}$. This result may then be corrected using Fig. 15 by assuming that the ratio of the heat transfer coefficients $h/h_{T_w=\mathrm{constant},v_w\sim x-1/2}$ is the same for foreign gas injection as for constant-property injection provided it is evaluated at the same value of $(\rho_w v_w / \rho^* u_\infty)\sqrt{u_\infty x / \nu^*}$.

Another case of engineering interest as described on Fig. 17 involves transpiration into a laminar boundary layer on a flat plate under the condition that there is a solid starting length. The solutions for the reduced skin friction coefficients and Stanton numbers are presented on Figs. 18 and 19 [9]. Although the solutions were obtained for a Prandtl number of unity they may be applied to a Prandtl number of 0.7 with engineering accuracy.

e. Plane Stagnation Flow with Constant Properties

The velocity distributions, skin friction coefficients, and dimensionless heat transfer coefficients for constant-property laminar flow in the immediate vicinity of a stagnation line on a two-dimensional body are shown on Figs. 20 to 22 [1]. The blowing (or suction) is constant (i.e., v_w = constant) and the wall is at a constant temperature. It is important to note that in contrast to the flat-plate case the velocity distributions do not reveal inflection points with mass injection. It would appear that blowing is not destabilizing in the presence of a favorable pressure gradient. Further the velocity boundary layer remains

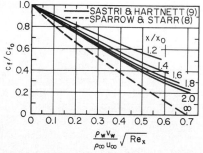

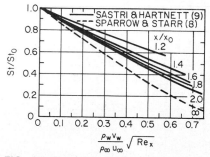

FIG. 18 Normalized skin friction coefficients c_f/c_{f_0} for transpiration into a laminar boundary layer on a flat plate with an unheated starting length [9].

FIG. 19 Normalized Stanton numbers St/St_0 for transpiration into a laminar boundary layer on a flat plate with an unheated starting length [9].

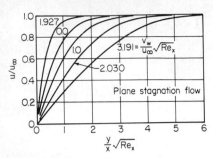

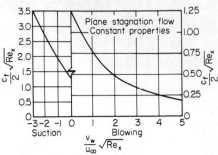

FIG. 20 Velocity distributions for constant-property transpiration cooling in the neighborhood of a plane stagnation point [1].

FIG. 21 Dimensionless skin friction coefficient as a function of the mass transfer rate for constant-property transpiration cooling in the neighborhood of a plane stagnation point [1].

attached even at large blowing rates and the skin friction coefficient remains finite. On the other hand the heat transfer goes to zero, which means that the thermal boundary lifts off the surface. The shearing stress may be calculated from Eq. (8) and Fig. 21. The heat transfer is calculated from Eq. (9) using Fig. 22 for the determination of the heat transfer coefficient. Here it should be noted that T_{aw} is equal to the free-stream temperature T_∞.

f. Plane Stagnation Flow with Variable Properties Including Diffusion-Thermo Effects

Studies have been carried out for the injection of several foreign gases in the region of the plane stagnation point. It has been found for light gases such as hydrogen or helium that the influence of diffusion thermo becomes very important in the case where the wall temperature is close to the free-stream temperature, a situation which may frequently be encountered, particularly in wind-tunnel studies. The heat transfer is again calculated from Eq. (9):

$$q_w'' = h(T_w - T_{aw}) \tag{9}$$

In the presence of diffusion thermo the adiabatic wall temperature (corresponding to the case where there is zero net heat transfer to the wall) is not equal to the free-stream temperature even at low velocities. For light gases the adiabatic wall temperature may be considerably higher than the free-stream value. The predicted adiabatic wall results for helium injected into an air boundary layer are given on Fig. 23 for free-stream temperatures ranging from 520 to 5000°R (289 to 2778 K). Good agreement with experimental results has been reported [11]. The corresponding normalized skin friction coefficient and Nusselt numbers are given on Figs. 24 and 25 for a wide range of free-stream and wall-temperature levels. Although helium transpiration in general reduces the Nusselt number, it may result in an increase in

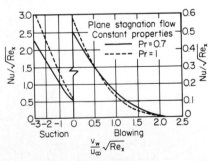

FIG. 22 Dimensionless heat transfer coefficient Nu/Re_x as a function of the mass transfer rate for constant-property transpiration cooling in the neighborhood of a plane stagnation point [1].

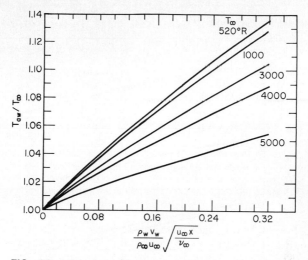

FIG. 23 Adiabatic wall temperature resulting from diffusion thermo (DT) when helium is injected into a plane stagnation flow of air.

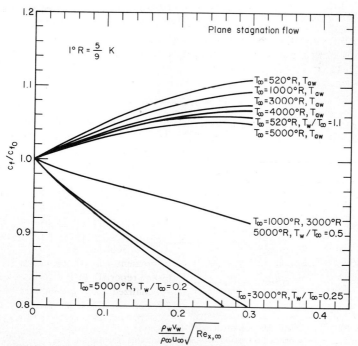

FIG. 24 Normalized skin friction coefficients as a function of the mass transfer rate for helium injected into a plane stagnation flow of air.

Plane stagnation flow

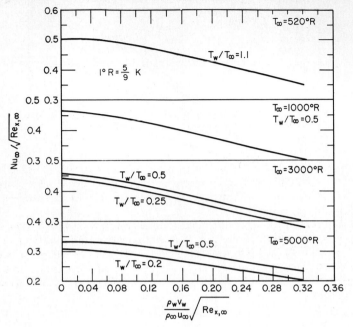

FIG. 25 Dimensionless heat transfer coefficients as a function of the mass transfer rate for helium injected into a plane stagnation flow of air.

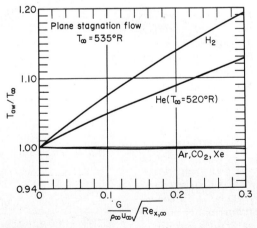

FIG. 26 Adiabatic wall temperature resulting from diffusion thermo for various gases injected into a plane stagnation flow of air [12].

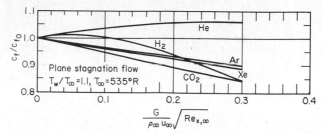

FIG. 27 Normalized skin friction coefficient as a function of mass transfer for foreign gas injection into a plane stagnation flow of air [12].

the skin friction. An increase in skin friction occurs when the wall temperature is greater than or very nearly equal to the adiabatic wall temperature. On the other hand, when the wall is highly cooled, helium transpiration reduces the skin friction coefficient over its solid wall value.

Similar results are available for hydrogen, argon, carbon dioxide, and xenon injected into an air boundary layer, although the range of wall temperatures and free-stream temperatures is limited [12]. These are shown in Figs. 26 to 28. In general it is only the lighter gases which demonstrate an appreciable influence on the adiabatic wall temperatures. Figures 29 and 30 give representative values of the wall concentration of the injected foreign gas as a function of the blowing rate.

g. Axisymmetric Stagnation Flow

For a constant-property flow in the region of a stagnation point on a three-dimensional body, the solid-wall Nusselt value is given by the relation

$$\frac{hx}{k} = 0.767 \sqrt{\frac{u_\omega x}{\nu}} \, Pr^{0.4} \quad (23)$$

The influence of mass transfer on this value is given on Fig. 31 for a Prandtl number of 0.7 [13]. The heat transfer is calculated from Eq. (9) with the adiabatic wall temperature T_{aw} being equal to the free-stream temperature T_∞ for low-speed flow and to the total temperature for high-speed flow.

In a binary system the same difficulties arise as in the two-dimensional stagnation flow, namely the influence of diffusion thermo. The adiabatic wall temperature, skin friction coefficients, and Nusselt numbers for helium injected into an air boundary layer are presented in Figs. 32 through 34 [10]. Other injectant gases are shown in Figs. 35 through 38 [12]. The heat transfer is now calculated from Eq. (9) using the

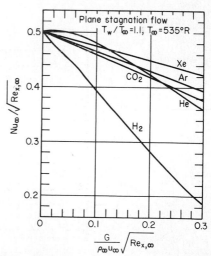

FIG. 28 Dimensionless heat transfer coefficient as a function of the mass transfer rate for foreign gas injected into a plane stagnation flow of air [12].

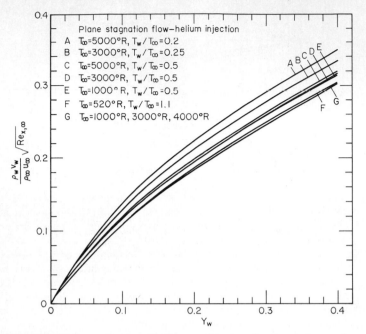

FIG. 29 Wall mass fraction of helium as a function of mass transfer rate for transportation cooling in a plane stagnation flow of air [10].

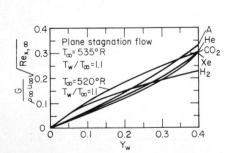

FIG. 30 Wall mass fraction of foreign gas as a function of mass transfer rate for transpiration cooling in a plane stagnation flow of air [10].

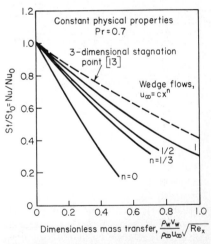

FIG. 31 Normalized Stanton or Nusselt numbers $St/St_0 = Nu/Nu_0$ as a function of the mass transfer rate for constant-property laminar flow in the region of a three-dimensional stagnation point and for laminar wedge flows [5].

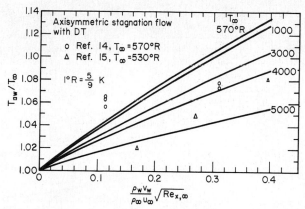

FIG. 32 Adiabatic wall temperature for helium transpiration into an air boundary layer in the region of an axisymmetric stagnation point as a function of the blowing rate [10].

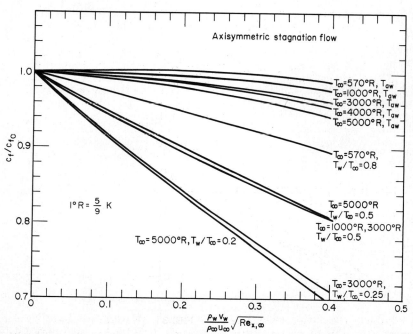

FIG. 33 Normalized skin friction coefficients c_f/c_{f0} for helium transpiration into an air boundary layer in the region of an axisymmetric stagnation point as a function of the blowing rate [10].

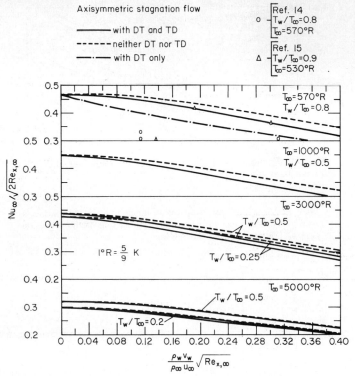

FIG. 34 Dimensionless heat transfer coefficients $\mathrm{Nu}_\infty/\sqrt{2\mathrm{Re}_{x,\infty}}$ for helium transpiration into an air boundary layer in the region of an axisymmetric stagnation point as a function of the blowing rate. Diffusion thermo (DT) and thermal diffusion (TD) are taken into account [10].

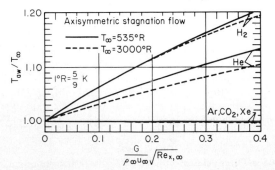

FIG. 35 Adiabatic wall temperature for transpiration of foreign gases into an axisymmetric stagnation flow of air as a function of blowing rate [12].

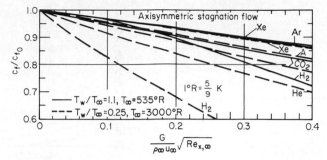

FIG. 36 Reduced skin friction coefficients c_f/c_{f0} for transpiration of foreign gases into an axisymmetric stagnation flow of air as a function of blowing rate [12].

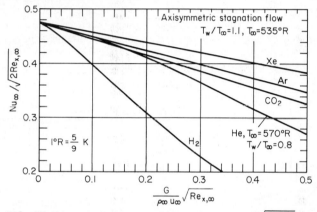

FIG. 37 Dimensionless heat transfer coefficient $Nu_\infty/\sqrt{2Re_{x,\infty}}$ for transpiration of foreign gases into an axisymmetric stagnation flow of air as a function of blowing rate, $T_w/T_\infty = 1.1$, $T_\infty = 535°R$ [12].

appropriate value of T_{aw} determined from Fig. 32 or Fig. 35. Wall mass fraction values for the injected coolant are presented on Figs. 39 and 40.

h. Wedge Flows, $u_\infty = cx^n$, $T_w =$ constant

The general class of flows described as wedge flows or Falkner-Skan flows [16], which treat low-velocity laminar flow over a two-dimensional wedge, has been analyzed for constant physical properties including the influence of mass transfer [17]. The Nusselt number for the solid-wall case for constant wall temperature is shown as a function of the exponent n (or equivalently the wedge angle $\beta\pi$) on Fig. 41. The normalized Stanton or Nusselt value is given on Fig. 31 as a function of the blowing rate [17].

i. Generalized Flows: Two-Dimensional and Axially Symmetric Three-Dimensional Flows with Mass Transfer

Heat transfer and skin friction coefficients are available [18] for the case where the free-stream velocity is given by

$$u_\infty \sim \sqrt{\frac{T_\infty}{T_0}}\, \xi^m \tag{24}$$

and

$$\beta = \frac{2\xi}{u_\infty}\frac{du_\infty}{d\xi}\frac{T_0}{T_\infty} = 2m \tag{25}$$

Here

$$\xi = \int_0^x \rho_w \mu_w u_\infty r_w^{2j}\, dx \tag{26}$$

where

$$j = \begin{cases} 0 \text{ for two-dimensional flow} \\ 1 \text{ for axisymmetric three-dimensional flow} \end{cases}$$

Although rigorously developed for the system of boundary conditions which lead to similar solutions (i.e., the normal velocity $v_w \sim (u_\infty/x)^{1/2}$ and T_w is constant), it is proposed [19] that the following simplified graphical representation be used as an engineering approximation for other cases. It is assumed that the free-stream velocity and temperatures are given and that the wall temperature is a prescribed constant value and further that the blowing distribution is specified. The results may be used for variable-property conditions for air-air, nitrogen-nitrogen, CO_2 into CO_2, etc.

First determine the dimensionless local parameters $\bar{\xi}$ and β:

$$\bar{\xi} = \frac{\displaystyle\int_0^x \rho_w \mu_w u_\infty r_w^{2j}\, dx}{\rho_w \mu_w u_\infty r_w^{2j} x} \tag{27}$$

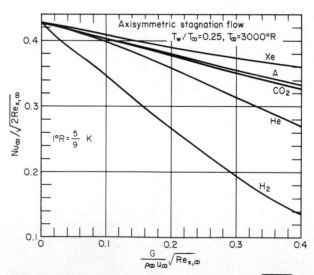

FIG. 38 Dimensionless heat transfer coefficient $\mathrm{Nu}_\infty/\sqrt{2\mathrm{Re}_{x,\infty}}$ for transpiration of foreign gases into an axisymmetric stagnation flow of air as a function of blowing rate, $T_w/T_\infty = 0.25$, $T_\infty = 3000°\mathrm{R}$ [12].

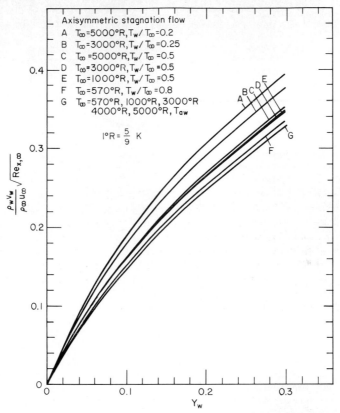

FIG. 39 Wall mass fraction of helium as a function of the blowing rate for transpiration cooling in an axisymmetric stagnation flow of air [10].

and

$$\beta = 2\bar{\xi}\left(\frac{x}{u_\infty}\frac{du_\infty}{dx}\right)\frac{T_0}{T_\infty} \qquad (28)$$

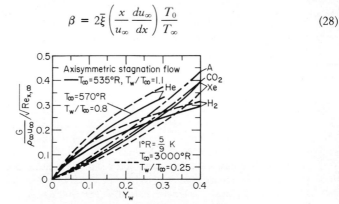

FIG. 40 Wall mass fraction of injected foreign gas as a function of the blowing rate for transpiration cooling in an axisymmetric flow of air [12].

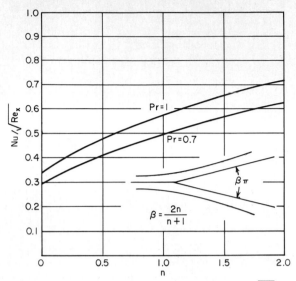

FIG. 41 Dimensionless heat transfer coefficients $\mathrm{Nu}/\sqrt{\mathrm{Re}_x}$ for constant-property wedge flows $u_\infty = cx^n$ as a function of n.

and x is the distance from the leading edge or from the stagnation point. The constant-property solid-wall skin friction coefficients, Nusselt values, and recovery factors may then be determined from Figs. 42 to 45. The starred quantities indicate that all physical properties are to be evaluated at the reference temperature T^* as mentioned above, where

$$T^* = T_w + 0.5(T_w - T_\infty) + 0.22(T_{aw0} - T_\infty) \qquad (11)$$

$$T_{aw0} = T_\infty + r_0^* \frac{u_\infty^2}{2c_p^*} \qquad (12)$$

and where r_0^* is obtained from Fig. 45. The influence of variable physical properties on the solid-wall skin friction and heat transfer is determined from Figs. 46 and 47. Here Ec is the Eckert number:

$$\mathrm{Ec} = \frac{u_\infty^2}{2c_{p\infty}T_0} = \frac{[(\gamma_\infty - 1)/2]\mathrm{Ma}_\infty^2}{1 + [(\gamma_\infty - 1)/2]\mathrm{Ma}_\infty^2} \qquad (29)$$

and $T_w^+ = T_w/T_0$, the ratio of the local absolute wall temperature to the absolute local free-stream total temperature, while ω is the viscosity temperature exponent $\mu \sim T^\omega$. For air the value of ω is approximately 0.7 for typical wind-tunnel conditions and 0.5 for conditions encountered in hypersonic flight.

Finally, the influence of the blowing on the skin friction coefficient, the heat transfer coefficient, and the recovery factors is given in Figs. 48 to 52. The local skin friction coefficient and heat transfer coefficient are then determined. Here the following special subscript notation is used:

I   Corresponds to constant-property solid-wall conditions, $T_w^+ = 1$, Ec $= 0$, f_w $= 0$.

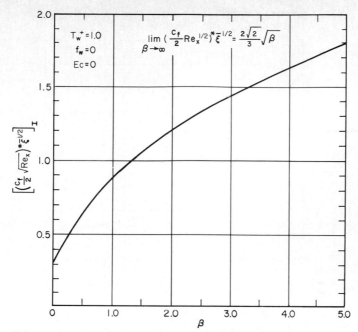

FIG. 42 Skin friction coefficient as a function of the dimensionless pressure gradient β for constant-property flow over a solid wall [19].

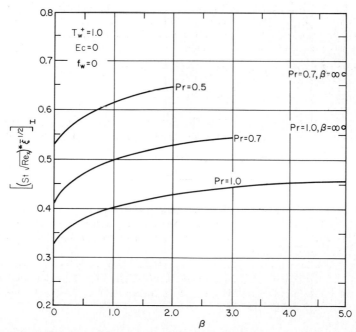

FIG. 43 Stanton number as a function of the dimensionless pressure gradient β for constant-property flow over a solid wall [19].

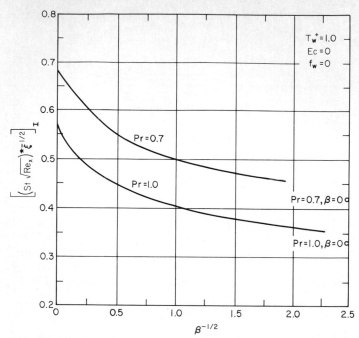

FIG. 44 Solid-wall Stanton number as a function of the asymptotic pressure gradient parameter $\beta^{-1/2}$ [19].

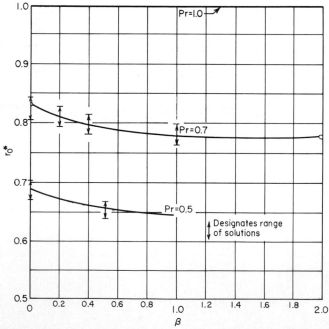

FIG. 45 Dimensionless solid-wall recovery factor as a function of the pressure gradient β [19].

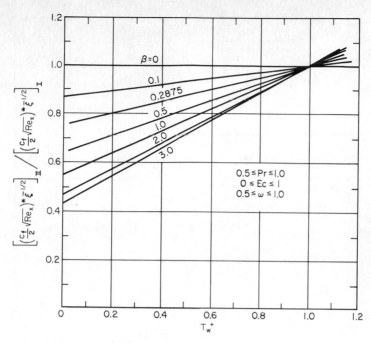

FIG. 46 Normalized solid-wall skin friction coefficient as a function of wall temperature ratio T_w^+ and pressure gradient β [19].

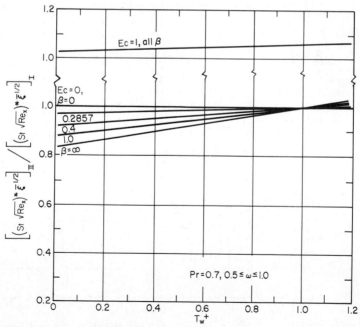

FIG. 47 Normalized solid-wall Stanton number as a function of the wall temperature ratio T_w^+, pressure gradient β, and the Eckert number Ec [19].

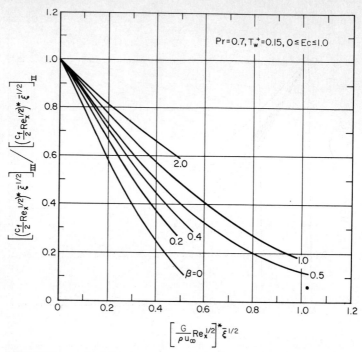

FIG. 48 Normalized skin friction coefficient as a function of the mass transfer rate and pressure gradient for a temperature ratio of 0.15 [19].

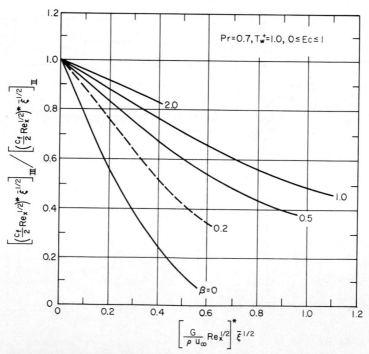

FIG. 49 Normalized skin friction coefficient as a function of the mass transfer rate and pressure gradient for a temperature ratio of 1.0 [19].

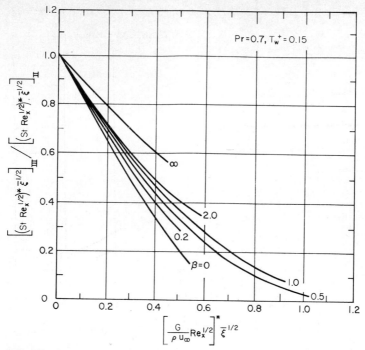

FIG. 50 Normalized Stanton number as a function of the mass transfer rate and pressure gradient for wall temperature ratio of 0.15 [19].

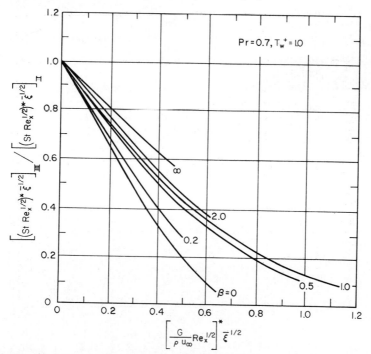

FIG. 51 Normalized Stanton number as a function of the mass transfer rate and pressure gradient for a wall temperature ratio of 1.0 [19].

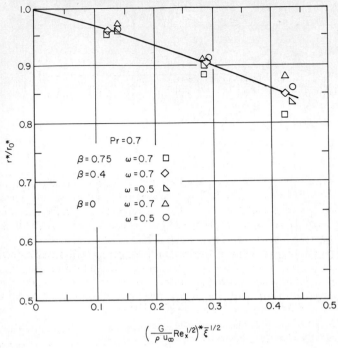

FIG. 52 Normalized recovery factor as a function of the mass transfer rate for a Prandtl number of 0.7 [19].

II   Corresponds to variable-property solid-wall conditions, T_w^+, Ec, $f_w = 0$ where T_w^+ and Ec are specified.

III   Corresponds to variable-property conditions with mass transfer, but with the wall held at the same temperature and with the same free-stream conditions as II, T_w^+, Ec, f_w.

$$\frac{c_f^*}{2} = \left(\frac{c_f^*}{2}\sqrt{\mathrm{Re}_x^*}\sqrt{\bar{\xi}}\right)_{\mathrm{I}} \cdot \underbrace{\frac{[(c_f^*/2)\sqrt{\mathrm{Re}_x^*}\sqrt{\bar{\xi}}]_{\mathrm{II}}}{[(c_f^*/2)\sqrt{\mathrm{Re}_x^*}\sqrt{\bar{\xi}}]_{\mathrm{I}}}}_{} \cdot \underbrace{\frac{[(c_f^*/2)\sqrt{\mathrm{Re}_x^*}\sqrt{\bar{\xi}}]_{\mathrm{III}}}{[(c_f^*/2)\sqrt{\mathrm{Re}_x^*}\sqrt{\bar{\xi}}]_{\mathrm{II}}}}_{} \cdot \frac{1}{\sqrt{\mathrm{Re}_x^*}\sqrt{\bar{\xi}}} \tag{30}$$

Fig. 42   Fig. 46   Figs. 48 & 49

and

$$\mathrm{St}^* = (\mathrm{St}^*\sqrt{\mathrm{Re}_x^*}\sqrt{\bar{\xi}})_{\mathrm{I}} \cdot \underbrace{\frac{(\mathrm{St}^*\sqrt{\mathrm{Re}_x^*}\sqrt{\bar{\xi}})_{\mathrm{II}}}{(\mathrm{St}^*\sqrt{\mathrm{Re}_x^*}\sqrt{\bar{\xi}})_{\mathrm{I}}}}_{} \cdot \underbrace{\frac{(\mathrm{St}^*\sqrt{\mathrm{Re}_x^*}\sqrt{\bar{\xi}})_{\mathrm{III}}}{(\mathrm{St}^*\sqrt{\mathrm{Re}_x^*}\sqrt{\bar{\xi}})_{\mathrm{II}}}}_{} \cdot \frac{1}{\sqrt{\mathrm{Re}_x^*}\sqrt{\bar{\xi}}} \tag{31}$$

Fig. 43 or 44   Fig. 47   Figs. 50 & 51

The local shearing stress and heat transfer are then calculated:

$$\tau_w = \frac{c_f^*}{2}\rho^* u_\infty^2 \tag{8a}$$

$$q_w'' = \mathrm{St}^*\rho^* c_p^* u_\infty(T_w - T_{aw}) \tag{17}$$

and
$$T_{aw} = T_\infty + \underbrace{\left(\frac{r^*}{r_0^*}\right)}_{\text{Fig. 52}} \cdot \underbrace{(r_0^*)}_{\text{Fig. 45}} \cdot \frac{u_\infty^2}{2c_p^*} \qquad (18)$$

It may be noted that these solutions, given on Figs. 42 to 52, include low-velocity two-dimensional flow $u_\infty = cx^n$ where the following relationships apply:

$$\bar{\xi} = \frac{1}{n+1} \qquad (32)$$

$$n = \frac{\beta}{2-\beta} \quad \text{or} \quad \beta = \frac{2n}{1+n} \qquad (33)$$

Thus the flat-plate case ($n = 0$) corresponds to $\beta = 0$ and $\bar{\xi} = 1$ and the plane stagnation case ($n = 1$) to $\beta = 1$ and $\bar{\xi} = \frac{1}{2}$. This particular class of low-velocity wedge flows is confined to values of β less than 2, which corresponds to an infinite–pressure gradient condition. The same restriction does not apply to the more general class of flows given by $u_\infty \sim \sqrt{(T_\infty/T_0)}\xi^m$, where any value of β is possible.

Further it may be noted that the axisymmetric stagnation point is also given on these figures. In this case $u_\infty \sim x$ and $r_w \sim x$ and thus $\bar{\xi} = \frac{1}{4}$ and $\beta = \frac{1}{2}$.

There is little influence of Prandtl number over the range of Pr = 0.5 to 1.0 on the reduced skin friction values shown in Figs. 46, 48, and 49. However, the heat transfer is sensitive to the Prandtl number, as may be seen on Fig. 53.

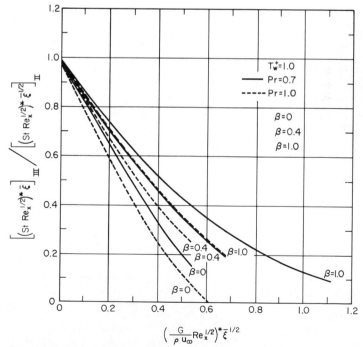

FIG. 53 Influence of Prandtl number on reduced Stanton number as a function of mass transfer rate and pressure gradient [19].

The knowledge of the wall mass fraction as a function of blowing rate is useful for ablation and evaporation studies. This information is given on Figs. 54 to 56 for pressure gradients of $\beta = 0, 0.5,$ and 1.0.

j. Supersonic Laminar Flow over a Cone

Supersonic laminar flow over a cone gives rise to constant pressure along the surface of the cone; the skin friction and heat transfer results are obtainable from Figs. 42 to 53. In this case $u_\infty =$ constant and $r_w \sim x$. Thus $\bar{\xi} = \frac{1}{3}$ and $\beta = 0$.

k. Laminar Flow over a Circular Cylinder

Analytical and experimental results are available for the transpiration of air into the laminar boundary layer on the forward region of a circular cylinder [20–22]. The analysis and experiment both correspond to a constant blowing distribution and a constant coolant temperature within the reservoir. The surface temperature and heat flux are then determined by free-stream conditions. Figure 57 shows the dimensionless temperature distribution along the surface of the cylinder for various blowing rates. There are two different curves for each condition, one corresponding to the velocity distribution proposed by Hiemenz and the other found by Schmidt and Wenner. Comparable results for the Nusselt number are given on Fig. 58. Other methods are available for determining the heat trans-

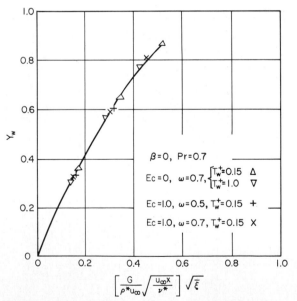

FIG. 54 Wall mass fraction value as a function of the dimensionless mass transfer rate for $\beta = 0$ and Prandtl number $= 0.7$ [19].

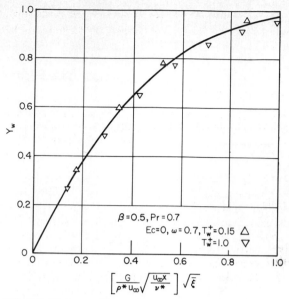

FIG. 55 Wall mass fraction value as a function of the dimensionless mass transfer rate for $\beta = 0.5$ and Prandtl number = 0.7 [19].

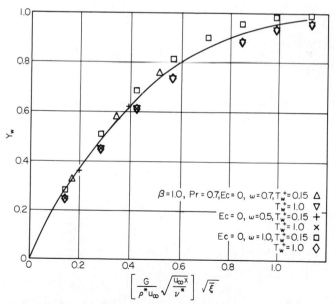

FIG. 56 Wall mass fraction value as a function of the dimensionless mass transfer rate for $\beta = 1.0$ and Prandtl number = 0.7 [19].

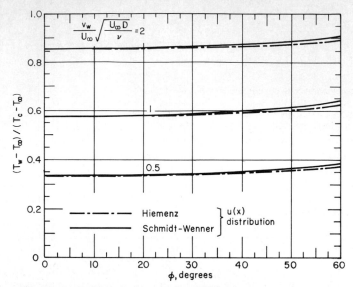

FIG. 57 Dimensionless temperature distribution as a function of angle for several rates of mass transfer through a porous cylinder in cross flow [20].

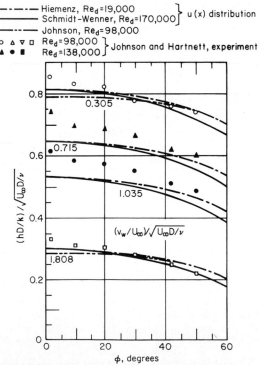

FIG. 58 Analytical and experimental heat transfer results for mass transfer from a porous cylinder in cross flow [21].

fer for laminar flow over cylinders, and other two-dimensional geometries are available [23]. (See Ref. 134 for more details.)

l. Longitudinal Flow along a Circular Cylinder

The influence of curvature on the skin friction and heat transfer coefficient for longitudinal flow along a circular cylinder is given in Figs. 59 [24] and 60 [25] for the case of mass transfer. Two blowing distributions are considered, one corresponding to the similar solution where $v_w \sim x^{-1/2}$ and the other to the realistic but nonsimilar case where v_w is a constant. The solid-wall values of the heat transfer and skin friction for $v_w \sim x^{-1/2}$ may be obtained from Figs. 42 to 45. The heat transfer results, although strictly applicable to a Prandtl number of 0.7, should yield reasonable estimates of curvature effects over the Prandtl number range of 0.5 to 1.0.

m. Rotating Cone

The heat transfer coefficient for a porous cone rotating in an otherwise quiescent fluid with mass transfer to or from the surface is given on Fig. 61 for the constant-property laminar case with a Prandtl number of 0.7 [26, 27]. The normal velocity at the surface $v_{z,w}$ is assumed proportional to $(\Omega \nu)^{1/2}$ and the surface temperature is constant. The special case of a rotating disk is obtained by setting $\sin \alpha = 1$. The local heat transfer rate is determined from the relation $q''_w = h(T_w - T_\infty)$. These same references [26, 27] contain information on the local shearing stress and torque.

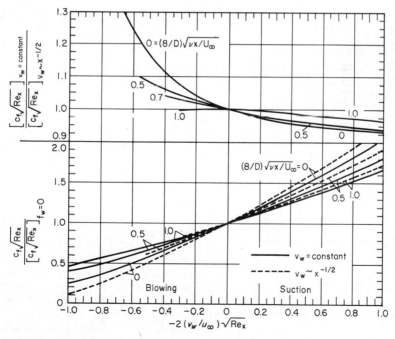

FIG. 59 Influence of curvature and mass transfer on the skin friction coefficient for laminar flow along a porous cylinder [24].

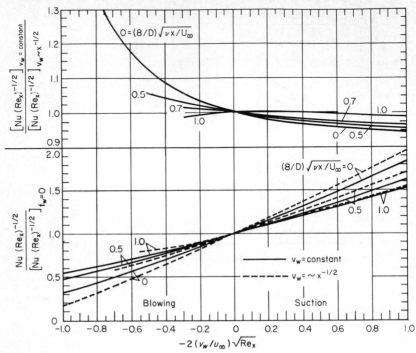

FIG. 60 Influence of curvature and mass transfer on the Nusselt number for flow along a porous cylinder [25].

2. Free-Convection Laminar Flow

a. The Horizontal Circular Cylinder

Experiments and analytical studies have been carried out for helium, hydrogen, carbon dioxide, and Freon-12 injected into a free-convection air boundary layer in the vicinity of

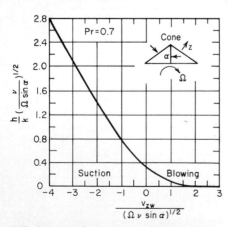

FIG. 61 Nusselt number as a function of mass transfer for a rotating cone.

the plane stagnation point of a horizontal circular cylinder [28]. The stagnation point was at the lowest point on the cylinder circumference for helium and hydrogen injection while for carbon dioxide and Freon-12 it occurred at the highest point on the circumference. This difference occurred since the temperature differences and blowing rates were such that the buoyancy forces created an upflow for the first two gases and conversely a downflow was created for carbon dioxide and Freon-12 injection. The influence of thermal diffusion discussed earlier in this chapter is of critical importance. The adiabatic wall temperatures are given on Fig. 62, while the Nusselt numbers are shown on Figs. 63 and 64. Some data are also presented at positions

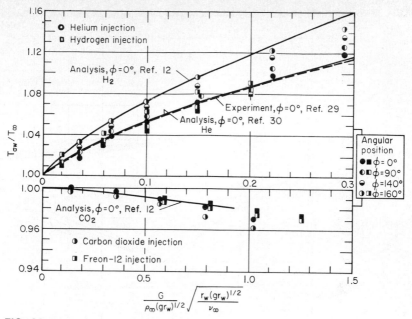

FIG. 62 Dimensionless adiabatic wall temperature resulting from diffusion thermo for a horizontal cylinder in free convection [28].

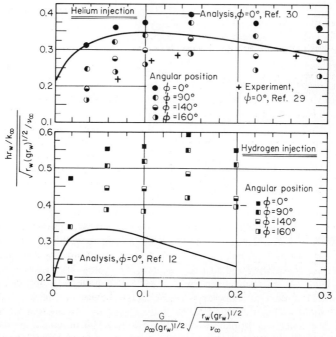

FIG. 63 Dimensionless Nusselt number as a function of mass transfer for helium and hydrogen injected into a free-convection air boundary layer on a horizontal cylinder [28].

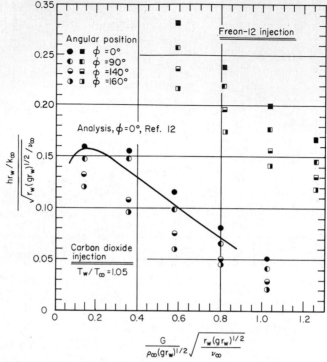

FIG. 64 Dimensionless Nusselt number as a function of mass transfer for Freon-12 and carbon dioxide injected into a free-convection air boundary layer on a horizontal cylinder [28].

around the porous cylinder. The local heat transfer rate is calculated from the relation $q_w'' = h(T_w - T_{aw})$. Results are also available for water vapor and argon injected into the plane stagnation region of an air free-convection boundary layer [12, 31].

3. Forced-Convection Turbulent Flow

a. The Flat Plate with Constant Properties, Air to Air

The experimental studies of W. M. Kays and his colleagues [32, 33] and of Torii et al. [34] provide a basis for predicting heat transfer and skin friction in turbulent low-speed air boundary layers on a flat-plate geometry with blowing or suction. The local Stanton numbers are shown on Figs. 65 and 66 as a function of the length Reynolds number Re_x for a wide range of blowing and suction values F. The values shown are applicable for uniform blowing ($F = $ constant) and for $F \approx x^{-0.2}$. For the constant blowing or suction distribution with a constant temperature of the coolant in the reservoir the wall temperature varies slightly along the length of the plate. The $F \approx x^{-0.2}$ distribution leads to an isothermal condition along the plate surface. Similar results for the skin friction coefficient are given in Fig. 67; they may be used for $F = $ constant or for $F \approx x^{-0.2}$.

The local Stanton number is also given in Fig. 68 as a function of the enthalpy-thick-

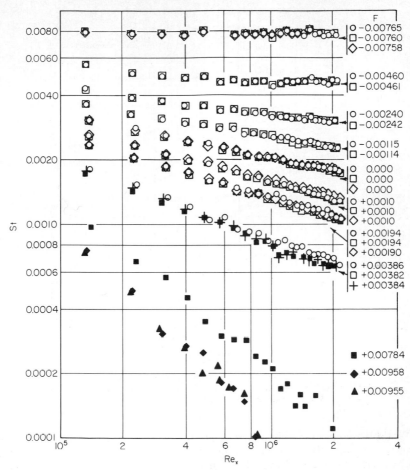

FIG. 65 Stanton number as a function of the length Reynolds number for various rates of mass transfer into a turbulent boundary layer on a flat plate [32].

ness Reynolds number Re_Δ and the blowing parameter F. Here Re_Δ is defined as $u_\infty\Delta/\nu$, where

$$\Delta = \int_0^\infty \frac{u}{u_\infty}\frac{T - T_\infty}{T_w - T_\infty}\,dy \tag{34}$$

This representation is more general than that shown in Figs. 65 and 66 and may be used for modest pressure gradients and for varying blowing or suction. Finally for the flat plate the normalized Stanton number $(St/St_0)_{Re_\Delta}$ is shown on Fig. 69 as a function of the blowing parameter F/St_0. Here the subscript Re_Δ means that the Stanton number with blowing St_{Re_Δ} is evaluated at the same enthalpy-thickness Reynolds numbers as the solid-plate $(St_0)_{Re_\Delta}$. A similar presentation is given for the skin friction coefficient except that the momentum thickness $\theta = \int_0^\infty (u/u_\infty)(1 - u/u_\infty)\,dy$ replaces the enthalpy thickness and the abscissa now reads $F/(c_{f0}/2)$.

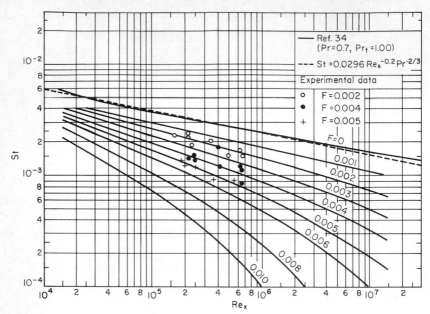

FIG. 66 Local Stanton number as a function of the length Reynolds number for various rates of mass transfer into a turbulent boundary layer on a flat plate [34].

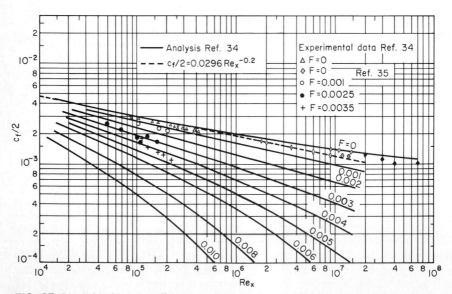

FIG. 67 Local skin friction coefficient as a function of the length Reynolds number for various rates of mass transfer into a turbulent boundary layer on a flat plate [34].

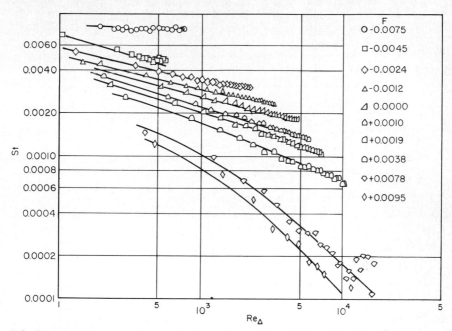

FIG. 68 Local Stanton number as a function of the enthalpy-thickness Reynolds number for various rates of mass transfer into a turbulent boundary layer on a flat plate.

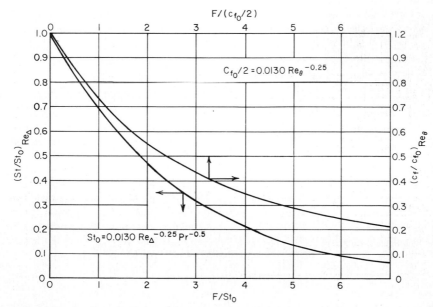

FIG. 69 Normalized Stanton number and normalized skin friction coefficient as a function of the dimensionless mass transfer rate.

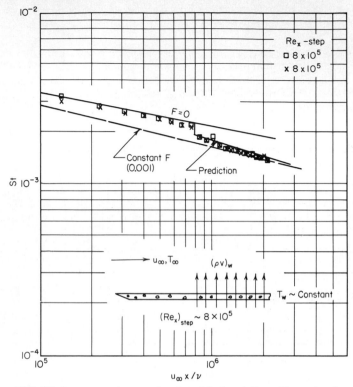

FIG. 70 Stanton number as a function of the length Reynolds number for a step change in mass injection rate for a turbulent boundary layer on a flat plate [33].

b. The Flat Plate with Variable Blowing, Air to Air

Consider a turbulent boundary layer on a flat plate with the plate temperature approximately constant along the length. The initial section of the plate is solid while at some distance downstream the plate is porous and uniform blowing occurs beyond this point. The resulting Stanton numbers St, as a function of the length Reynolds number, are shown on Figs. 70 to 72 for steps in the blowing rate F of 0.001, 0.002, and 0.004. It is seen that there is very rapid adjustment to local conditions.

A variation is shown in Fig. 73 wherein the solid portion of the plate is held at the same temperature as the free stream while at some distance downstream the plate temperature is suddenly changed to a new value and constant blowing simultaneously begins. Again within a short distance downstream the Stanton number adjusts to the same value as for constant blowing from the leading edge. In all these figures (70 to 73) the local Stanton number may be predicted from the integral relationship

$$\frac{d\mathrm{Re}_\Delta}{d\mathrm{Re}_x} = \mathrm{St} + F \qquad (35)$$

Here the Stanton number is taken to be a unique function of Re_Δ and F as given on Fig. 68. If the turbulent boundary layer begins at the leading edge, the equation reads

$$Re_\Delta = \int_0^{Re_x} (St + F)\, dRe_x \qquad (36)$$

and now the Stanton number may be estimated at some incremental distance downstream, the value of Re_Δ then determined, and the Stanton number reevaluated. This process is repeated at this station until the initial guess and final value of the local Stanton number are in agreement. This procedure is continued step by step along the plate. More details may be found in Ref. 33.

c. The Turbulent Boundary Layer on a Flat Plate at High Velocity, Air to Air

The Stanton numbers given above may be used for high-velocity flow, or alternatively the Stanton number results of Ref. 134 may be adopted. The recovery factor is somewhat reduced by blowing, as shown on Fig. 74. The recovery or adiabatic wall temperature is determined by the relation

$$T_{aw} = T_\infty + r_0^* \frac{r}{r_0} \frac{u_\infty^2}{2c_p^*} \qquad (18)$$

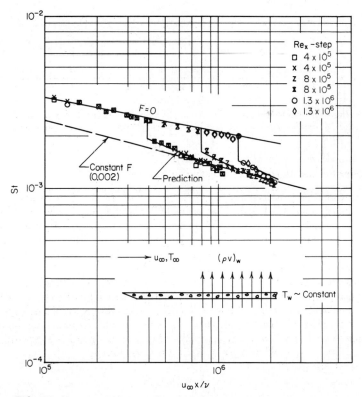

FIG. 71 Stanton number as a function of the length Reynolds number for a step change in mass injection rate for a turbulent boundary layer on a flat plate [33].

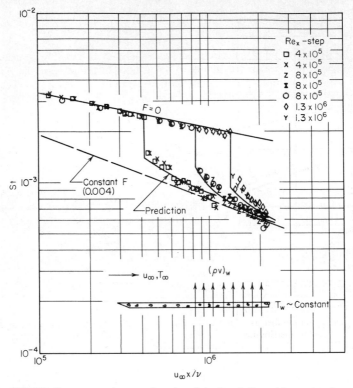

FIG. 72 Stanton number as a function of the length Reynolds number for a step change in the mass injection rate for a turbulent boundary layer on a flat plate [33].

The solid-wall recovery factor r_o^* is given by Eq. (13), and c_p^* is the specific heat evaluated at the reference temperature. The heat transfer is then determined from

$$q_w'' = \text{St } (\rho^* u_\infty c_p^*)(T_w - T_{aw}) \tag{17}$$

d. The Flat Plate with Helium Injected into Air

The reduced Stanton number St/St_0 as a function of the blowing parameter F/St_0 for helium injected into air is given in Fig. 75. The solid-wall Stanton number St_0 may be determined by methods outlined in Ref. 134. At low velocities the influence of diffusion thermo is important for light-gas injection, resulting in an adiabatic wall temperature substantially higher than the free-stream temperature. Experimental values of this adiabatic wall temperature as a function of the dimensionless mass transfer rate are shown on Fig. 76 for helium injection into a turbulent air boundary layer with a free-stream temperature of approximately 75°F (24°C) and a free-stream velocity of 100 ft/s (30.5 m/s). The heat transfer is calculated from the usual relation

$$q_w'' = h(T_w - T_{aw}) \tag{9}$$

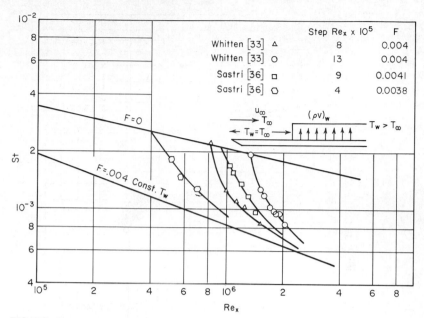

FIG. 73 Stanton number as a function of the length Reynolds number for a simultaneous step in the blowing distribution and in the wall temperature; turbulent boundary layer on a flat plate [36].

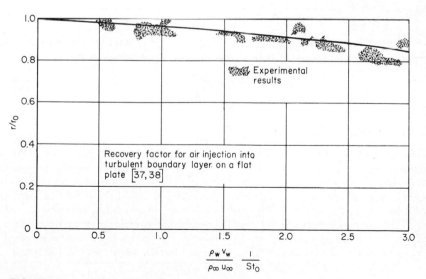

FIG. 74 Normalized recovery factor as a function of the mass transfer rate from air injected into a turbulent air boundary layer on a flat plate.

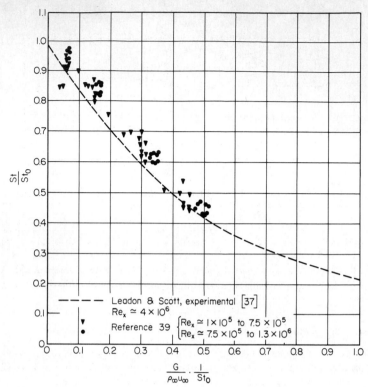

FIG. 75 Reduced Stanton number as a function of the mass transfer rate for helium injected into a turbulent boundary layer on a flat plate [37].

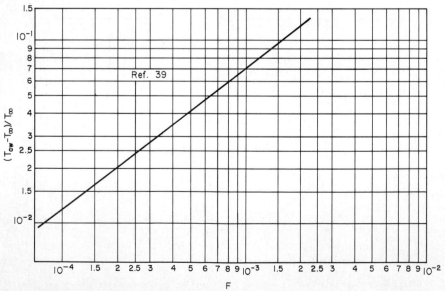

FIG. 76 Dimensionless adiabatic temperature as a function of the mass transfer rate resulting from diffusion thermo for helium injected into a turbulent boundary layer on a flat plate.

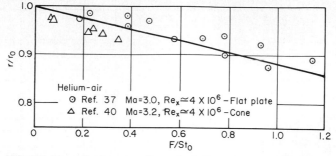

FIG. 77 Normalized recovery factor for helium injected into a turbulent boundary layer on a flat plate and on a cone [40].

At high velocities (Ma $\gg$ 1) the influence of viscous dissipation becomes important, and the measured adiabatic wall temperature, given in terms of the reduced recovery factor, is shown on Fig. 77. The heat transfer is again calculated from Eq. (9).

It is expected that the injection of hydrogen into air boundary layers will also exhibit diffusion-thermo effects at low velocities.

e. Binary Mass Transfer Systems—Flat Plate

The most general presentation of transpiration cooling in binary systems is given by Bartle and Leadon [41]. A wide variety of coolant gases were studied and their results have been verified by later investigators [42, 43]. The so-called effectiveness $(T_w - T_c)/(T_{aw0} - T_c)$ is shown as a function of the modified blowing rate $c_p F/(c_{p\infty} St_0)$ in Fig. 78. The reduced heat transfer q''_w/q''_{w0} (where q''_{w0} is the heat transfer to a flat plate at the

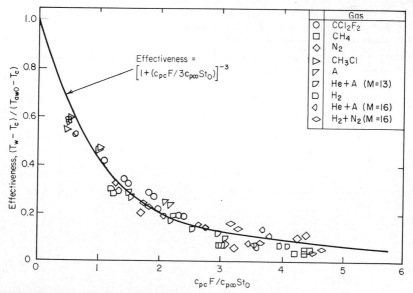

FIG. 78 Effectiveness as a function of the dimensionless blowing rate for the injection of foreign gases into a turbulent air boundary layer on a flat plate [41].

same temperature and exposed to the same free-stream condition, calculated by methods of Ref. 134) is calculated from the relation

$$\frac{T_w - T_c}{T_{aw0} - T_c} = \frac{1}{1 + (c_{pc}/c_{p\infty})(F/St_0)(q''_{w0}/q''_w)}$$ (37)

This method may be applied to the complete Mach number range, provided the wall temperature is not close to the adiabatic wall temperature.

f. Mass Transfer in Turbulent Boundary Layers with Pressure Gradient, Air to Air

Experimental results for air transpiration into a turbulent air boundary layer indicate that it is possible to use the same flat-plate relationship for Stanton number as a function of the enthalpy-thickness Reynolds number if the pressure gradient is moderate [44]. Specifically, results are available for blowing and suction where the initial pressure gradient is zero followed by a pressure gradient described by $K = (\nu/u_\infty^2)(du_\infty/dx)$ = constant, then in some cases followed by a zero pressure gradient. As shown on Fig. 79 for the solid-wall case, the zero pressure gradients shown as a solid line may be used with engineering accuracy for the prediction of the Stanton number for values of K up to 1.45×10^{-6}. When the pressure gradient becomes as large as $K = 2.5 \times 10^{-6}$ there is substantial departure from the zero pressure gradient $St - Re_\Delta$ relation. As shown on Figs. 80 to 83 the influence of blowing reduces the influence of favorable pressure gradient whereas suction increases the effect of the pressure gradient. An alternative approach is to use the analytical method of Baker and Launder [45].

4. Application to Liquid Film Cooling and Ablation

If mass transfer cooling is accomplished by the evaporation of a thin film of liquid distributed along the surface to be cooled, there is a unique relationship between the surface temperature and the wall concentration for the given pressure distribution if thermodynamic equilibrium is assumed. The partial pressure of the diffusing vapor at the surface is approximately equal to the vapor pressure of the liquid at the temperature T_w. Thus when the wall temperature is fixed, the partial pressure is fixed and the wall concentra-

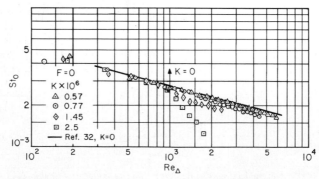

FIG. 79 Stanton number as a function of the enthalpy-thickness Reynolds number for various values of the pressure gradient K [44].

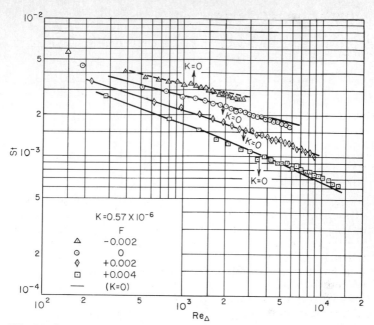

FIG. 80 Dimensionless Stanton number as a function of the enthalpy-thickness Reynolds number for various values of the blowing rate and for a fixed pressure gradient $K = 0.57 \times 10^{-6}$ [44].

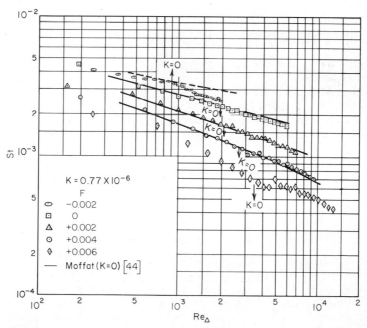

FIG. 81 Dimensionless Stanton number as a function of the enthalpy-thickness Reynolds number for various values of the blowing rate and for a fixed pressure gradient $K = 0.77 \times 10^{-6}$ [44].

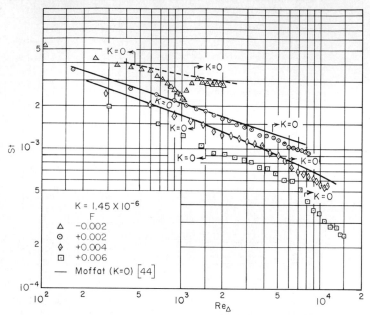

FIG. 82 Dimensionless Stanton number as a function of the enthalpy-thickness Reynolds number for various values of the blowing rate and for a fixed pressure gradient $K = 1.45 \times 10^{-6}$ [44].

tion (i.e., the wall mass fraction) is uniquely determined. Consequently, when the free-stream conditions are known, the wall temperature and the blowing rate are uniquely set by the appropriate combination of the thermodynamics of the system and the conservation equations previously solved. To demonstrate the procedure, if we neglect radiation and conduction along the surface, a local heat balance may be written as

$$Gc_{p_c}(T_w - T_c) + Gi_{lg} = h_x(T_{aw} - T_w) \tag{38}$$

where i_{lg} = latent heat of vaporization. Rearranging,

$$\frac{T_w - T_c + (i_{lg}/c_{p_c})}{T_{aw} - T_w} = \frac{\mathrm{Nu}/\sqrt{\mathrm{Re}_x}}{\mathrm{Pr}(G/\rho_\infty u_\infty)\sqrt{\mathrm{Re}_x}\,(c_{p_c}/c_{p\infty})} \tag{39}$$

or alternatively

$$\frac{T_w - T_c + (i_{lg}/c_{p_c})}{T_{aw} - T_c} = \frac{\mathrm{St}}{(G/\rho_\infty u_\infty)(c_{p_c}/c_{p\infty})} \tag{39a}$$

These relationships may be easily modified to account for variable properties by the use of the Eckert reference method.

If a value for the surface temperature is assumed, the partial pressure at the wall and accordingly the wall mass fraction are fixed. There is a unique relation between this wall mass fraction and the blowing or mass transfer rate as exemplified by Figs. 12, 13, 29, 30, 39, 40, and 54 through 56. The knowledge of the blowing rate allows the determination of the heat transfer coefficient St or $\mathrm{Nu}/\sqrt{\mathrm{Re}_x}$, and these values are substituted

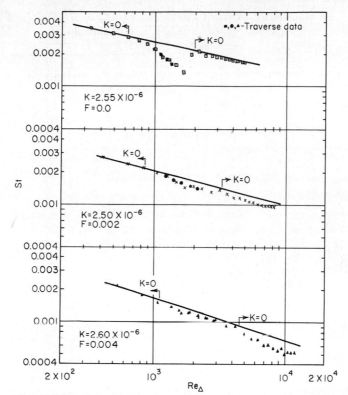

FIG. 83 Dimensionless Stanton number as a function of the enthalpy-thickness Reynolds number for various values of the blowing rate and for a fixed pressure gradient $K = 2.55 \times 10^{-6}$ [44].

into the heat balance (Eq. (39)). If the equation is not satisfied, a new wall temperature is assumed and the procedure continued until the equation is satisfied. The extension to sublimation is straightforward [46].

C. GASEOUS FILM COOLING

1. General Remarks

Film cooling involves the introduction of a cooler secondary fluid through slots or holes for the purpose of protecting the surface immediately downstream of the injection location; alternatively, a hotter gas may be introduced if the purpose is to keep the surface warm (as in prevention of icing on an aircraft wing). The injection geometry may be a slot, a porous section, or a series of holes or louvers. The secondary gas may be the same as the free-stream gas or it may be a foreign gas. The free stream may be subsonic, supersonic, or hypersonic, and the flow over the surface may involve a pressure gradient. With the major exceptions of film cooling near stagnation regions or under very high Mach number conditions, the flow is generally turbulent. In general, the designer is interested primarily in the adiabatic wall temperature downstream of the injection slot. In the event that additional convective cooling or heating is needed, the local heat transfer coefficient along the film-cooled surface is also required. A comprehensive survey of film cooling in turbulent flow including information on the geometries studied up to 1970 is given by Goldstein [47].

2. Forced-Convection Turbulent Flow—Two-Dimensional Film Cooling

a. Film Cooling of a Flat Plate in Subsonic Flow—Analytical Predictions

A number of semiempirical analytical procedures have been developed to predict the adiabatic wall temperature on a film-cooled flat plate. These analyses are two-dimensional and assume that the boundary layer on the plate is turbulent and that the injected coolant enters through a slot. The effect of the mass addition on the boundary-layer growth is taken into account, but from an energy viewpoint the coolant is treated as a heat sink. In general, these analyses have the same final form and yield predictions which do not differ too greatly. Some of the more widely used predictions are:

Librizzi and Cresci [48]

$$\eta = \frac{1}{1 + (c_{p\infty}/c_{p_c})[0.329(4.01 + \zeta)^{0.8} - 1]} \tag{40}$$

Kutateladze and Leont'ev [49]

$$\eta = \frac{1}{1 + 0.329(c_{p\infty}/c_{p_c})\zeta^{0.8}} \tag{41}$$

Goldstein and Haji-Sheikh [50]

$$\eta = \frac{1.9\mathrm{Pr}^{2/3}}{1 + 0.329(c_p\infty/c_pc)\zeta^{0.8}} \tag{42}$$

Here the adiabatic wall temperature in the presence of film cooling, T_{aw}, is given in terms of the film cooling effectiveness η, which in subsonic flow is defined:

$$\eta = \frac{T_{aw} - T_\infty}{T_c - T_\infty} \tag{43}$$

The dimensionless blowing rate parameter ζ is defined:

$$\zeta = \frac{x_c}{Fs}\left(\frac{\mu_c}{\mu_\infty}\,\mathrm{Re}_c\right)^{-1/4} \tag{44}$$

where

$$\mathrm{Re}_c = \frac{\rho_c U_c s}{\mu_c} \tag{45}$$

and

U_c = coolant mean velocity leaving the slot

x_c = distance from injection slot

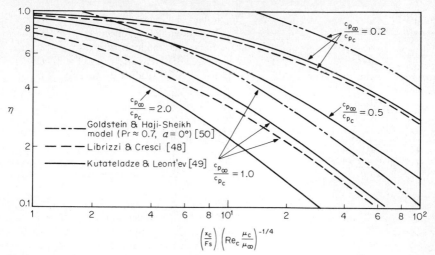

FIG. 84 Predictions of film cooling effectiveness as a function of the generalized distance ζ with the specific heat ratio as a parameter [48, 49, 50].

A comparison of these predictions is presented on Fig. 84 for several values of the specific heat ratio $c_{p\infty}/c_{p_c}$.

Note that the prediction of Goldstein and Haji-Sheikh is generally higher than the other two predictions and has the shortcoming of predicting effectiveness values higher than unity at low values of ζ (since this is impossible, the effectiveness is taken as unity under such circumstances).

b. The Flat Plate with Injection through a Porous Slot— Various Coolants into a Low-Velocity Air Boundary Layer

Figure 85 compares representative experimental data [51] with the available analytical predictions for film cooling through a porous slot into a turbulent air boundary layer. It is seen that the predictions give good estimates of the effectiveness values for the two coolant gases, air and helium. Comparisons with other available data [52–55] lead to the conclusion that a prediction midway between those of Goldstein and Haji-Sheikh, on the one hand, and Kutateladze and Leont'ev, on the other, may be used with some confidence for the porous slot geometry under subsonic flow conditions. Pedersen confirms this conclusion over a wide range of coolant densities [56]. Additional information on these porous slot experimental studies is given in Table 2, part *a*.

It is believed that the local heat transfer coefficient with porous slot cooling will not differ substantially from the solid-wall values. The basis for this contention is that it corresponds to the experimental findings for many other film cooling conditions (see subsequent sections). Therefore, it is recommended that the solid-wall heat transfer coefficient calculated for the same free-stream conditions as the film-cooled wall be used to estimate local heat transfer coefficients along a porous slot-cooled wall.

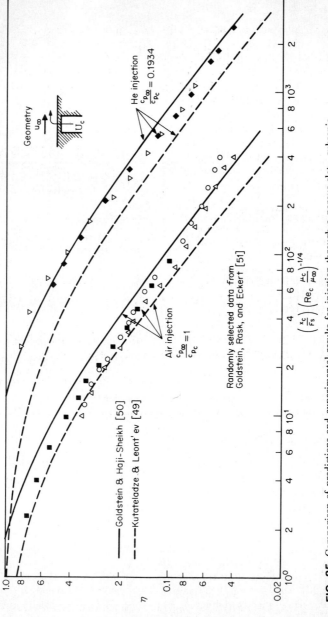

FIG. 85 Comparison of predictions and experimental results for injection through a porous slot—subsonic flow over a flat plate [49–51].

c. The Flat Plate with Tangential Injection through an Open Slot—Air into a Low-Velocity Turbulent Air Boundary Layer

Considerable experimental data are available for air-to-air film cooling for various slot geometries, including the four representative configurations shown on Fig. 86 [57–61]. Although such geometries attempt to approximate the limiting condition of tangential injection, experimental results indicate that the actual adiabatic wall temperature distribution (i.e., the effectiveness) is sensitive to variations in slot geometry, especially to the angle of injection and to the thickness of the lip. Furthermore, the effectiveness is moderately influenced by the turbulence level of the free stream. Increasing turbulence level is accompanied by a slight decrease in effectiveness as a result of the increased mixing of the injected coolant with the main stream.

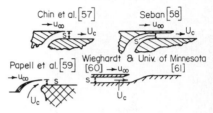

FIG. 86 Four typical slot configurations for gaseous film cooling [57–61].

Notwithstanding all of these concerns the predictions shown on Fig. 84 give a reasonable estimate of the effectiveness if the exit slot angles are in the order of 20° to 30°, the dimensionless blowing rate F is of the order of unity or lower, and the lip thickness is zero to avoid flow separation effects. This is confirmed in Fig. 87, which presents a comparison of selected experimental data [58, 62–64] with the predicted values.

If the slot lip thickness is finite the effectiveness will be reduced from the above-cited predicted values as a result of the flow separation, which in turn causes increased mixing of the injected coolant with the main flow. An estimate of the reduction due to slot lip

(Text continues on page 1–68)

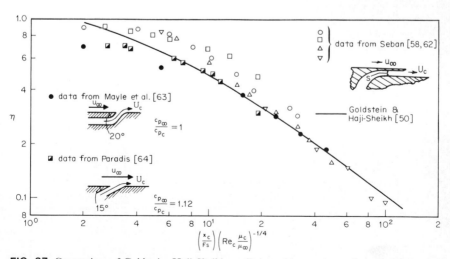

FIG. 87 Comparison of Goldstein–Haji-Sheikh prediction with experimental results [58, 62–64] for nearly tangential injection through a slot on a flat plate in subsonic flow.

TABLE 2 Experimental Studies of Two-Dimensional Film Cooling

Table 2 is an update of a table prepared by R. J. Goldstein for Ref. 47. The author gratefully acknowledges this contribution.

Reference	Geometry	Injection gas[a]	Density ratio ρ_c/ρ_∞	Temp. ratio T_i/T_∞	Blowing rate $F = \dfrac{\rho_c U_c}{\rho_\infty u_\infty}$	Free-stream Mach no.	Velocity ratio U_i/u_∞
						a. Two-dimensional film cooling through	
Nishiwaki, Hirata, Tsuchida [52]		Air	0.825	1.21	0.0 to 0.0975	0.017 to 0.086	0.0 to 0.118
Goldstein, Shavit, Chen [53]		Air	0.83 to 0.95	1.05 to 1.207	0.012 to 0.040	0.0961 to 0.1615	0.013 to 0.042
Mabuchi [54]		Air	0.788 to 0.878	1.14 to 1.27	0.02 to 0.146	0.02 to 0.031	0.026 to 0.185
Goldstein, Rask, Eckert [51]		Air	0.84 to 0.88	1.12 to 1.20	0.013 to 0.052	0.095 to 0.16	0.015 to 0.0605
		He	0.12	1.12 to 1.20	0.0022 to 0.0076	0.095 to 0.16	0.018 to 0.063
Escudier, Whitelaw [55]		Air	1.0	1.0	0.023 to 0.074	0.0702	0.023 to 0.074
Pedersen, Eckert, Goldstein [56]		Air with CO_2 tracer	1.01 to 1.03		0.028	0.08	0.028
		Air with Refrigerant 12 tracer	0.96 to 4.17		0.0125 to 0.0288	0.08	0.003 to 0.029
		Air with He tracer	0.96		0.0282	0.08	0.029
						b. Two-dimensional film cooling through	
Wieghardt [60]		Air	0.78 to 0.91	1.1 to 1.2	0.22 to 1.9 + wall jet	0.046 to 0.092	0.246 to 2.44
Scesa [67]		Air	0.81 to 0.91	1.09 to 1.23	0.2 to 1.14	0.025 to 0.065	0.286 to 1.35

1-56

Velocity u_∞, m/s	Slot size, mm	Starting-length Reynolds number Re_x	δ^*/s	q_w'', T_{aw}, or C_w	Effectiveness	Slot Reynolds number Re_c	Range of x_c/s
a porous slot—subsonic main stream							
6.0 to 30.0	5.0 20.0 50.0			T_{aw}	0.7 to 0.035	0 to 2200	1.4 to 180
30.5 to 55.0	35.6	0.59×10^5 to 1.06×10^5	0.0359 to 0.0403	T_{aw}	0.85 to 0.05	850 to 5800	1.5 to 17.44
7.0 to 10.0	15.1 to 51.0	8.0×10^5 to 22.0 to 10^5	0.011 to 0.092	T_{aw}	0.95 to 0.06	358 to 1320	1 to 35
33.2 to 55.0	25.4	8.0×10^5 to 14.0×10^5	0.037 to 0.046	T_{aw}	0.80 to 0.04	700 to 4000	0.78 to 31.0
33.2 to 55.0	25.4	8.0×10^5 to 14.0×10^5	0.037 to 0.046	T_{aw}	0.80 to 0.03	150 to 300	0.78 to 31.0
24.4	25.4			C_w	1.0 to 0.1	958 to 3000	0.625 to 20.6
27.7	24.34	1.69×10^6		C_w	0.48 to 0.033		
27.7	24.34	1.69×10^6		C_w	0.5 to 0.018		
27.7	24.34	1.69×10^6		C_w	0.50 to 0.043		
a slot—subsonic main stream							
15.8 to 32.0	5.0 10.0	8.0×10^5 to 15.0×10^5	0.16 to 0.36	T_{aw}	1.0 to 0.05	360 to 1200	80 to 800
9.7 to 20.4	3.175	1.2×10^5 to 2.9×10^5	0.22 to 0.33	q_w'' T_{aw}	0.60 to 0.095	550 to 2500	10 to 130

TABLE 2 Experimental Studies of Two-Dimensional Film Cooling (*Continued*

Reference	Geometry	Injection gas[a]	Density ratio ρ_c/ρ_∞	Temp. ratio T_c/T_∞	Blowing rate $F = \dfrac{\rho_c U_c}{\rho_\infty u_\infty}$	Free-stream Mach no.	Velocity ratio U_c/u_∞
Seban, Chan, Scesa [69]	u_∞	Air	0.81 to 0.92	1.09 to 1.23	0.08 to 0.916	0.037 to 0.108	0.097 to 1.06
	u_∞	Air	0.81 to 0.92	1.09 to 1.23	0.19 to 1.14	0.025 to 0.065	0.286 to 1.35
Chin, Shirvin, Hayes, Silver [57]	u_∞	Air	0.83 to 1.17	0.85 to 1.20	0.25 to 2.5	0.056 to 0.145	0.26 to 2.85
Papell, Trout [59]	u_∞	Air	0.52 to 3.54	0.32 to 1.8	0.0 to 12.9	0.15 to 0.80	
Hatch, Papell [70]	u_∞	He	2.45 to 6.5	0.34 to 0.67	0.018 to 1.55	0.53 to 0.57	0.095 to 3.84
Papell [71]	u_∞ At angles 90°, 80°, and 45°	Air	1.2 to 4.8	0.37 to 0.85	0.05 to 12.0	0.20 to 0.70	0.036 to 2.55
Seban [58]	u_∞	Air	0.88	1.14	0.17 to 20.8	0.0045 to 0.13	0.193 to 23.6
Seban [72]	u_∞	Air	0.88 to 1.0	1.0 to 1.14	0.27 to 0.76	0.09	0.31 to 0.865
Hartnett, Birkebak, Eckert [61]	u_∞	Air	0.78 to 0.98	1.02 to 1.27	0.265 to 0.288	0.145	0.294 to 0.333

Velocity u_∞ m/s	Slot size, mm	Starting-length Reynolds number Re_x	δ^*/s	q''_w, T_{aw}, or C_w	Effectiveness	Slot Reynolds number Re_c	Range of x_c/s
13.0 to 37.0	3.175	1.3×10^5 to 3.8×10^5	0.205 to 0.216	q''_w T_{aw}	1.0 to 0.05	580 to 2600	10 to 130
9.7 to 20.4	3.175	1.0×10^5 to 2.09×10^5	0.212 to 0.222	q''_w T_{aw}	0.7 to 0.095	550 to 2500	10 to 130
18.9 to 54.0	2.7	15.0×10^5 to 160.0×10^5	0.94 to 2.36	T_{aw}	1.0 to 0.13	1400 to 8200	9 to 233
168.0 to 425.0	1.59 3.175 6.35 12.7				1.0 to 0.0	0 to 450,000	0 to 540
256.0 to 316.0	3.175 6.35 12.7			T_{aw}	1.0 to 0.16	672 to 46,802	3.18 to 271
147.0 to 395.0	6.35 slot 6.35 holes			T_{aw}	0.95 to 0.22	2,500 to 260,000	4 to 152
1.52 to 45.8	1.59 3.175 6.35	0.4×10^5 to 1.1×10^5	0.03 to 0.12	q''_w T_{aw}	0.95 to 0.04	620 to 7950	5 to 300
30.5	1.59 3.175 6.35	0.76×10^5 to 13.4×10^5	0.03 to 1.2	q''_w T_{aw}	0.75 to 0.05	760 to 4400	2 to 300
50.0	3.12	6.1×10^5	0.244	q''_w T_{aw}	0.85 to 0.125	1510 to 2880	4 to 140

TABLE 2 Experimental Studies of Two-Dimensional Film Cooling (*Continued*)

Reference	Geometry	Injection gas[a]	Density ratio ρ_c/ρ_∞	Temp. ratio T_c/T_∞	Blowing rate $F = \dfrac{\rho_c U_c}{\rho_\infty u_\infty}$	Free-stream Mach no.	Velocity ratio U_c/u_∞
Hartnett, Birkebak, Eckert [73]		Air	0.875 to 0.935	1.07 to 1.14	0.28 to 1.23	0.1185	0.31 to 1.37
Seban, Back [74]		Air	0.90 to 0.97	1.03 to 1.11	4.95 to 12.6	0.00965 to 0.0149	5.1 to 14.0
Seban, Back [75]		Air	0.88	1.13	0.0 to 0.70	0.04 to 0.095	0.0 to 0.795
Seban, Back [62]		Air	0.87 to 1.00	1.0 to 1.15	0.2 to 0.9	0.083 to 0.110	0.20 to 1.03
Eckert, Birkebak [76]		Air	0.87	1.15	0.19 to 0.93	0.14	0.218 to 1.07
Samuel, Joubert [77]		Air	1.11 to 1.28	0.78 to 0.905	0.25 to 3.18	0.040 to 0.085	0.23 to 2.48
Nicoll, Whitelaw [78]		Air with He tracer	1.0	1.0	0.47 to 2.26	0.06	0.47 to 2.26
Metzger, Carper, Swank [79]		Air	0.96	1.05	0.25 to 1.49	0.04 to 0.07	0.26 to 1.55
Kacker, Whitelaw [80]		Air with He tracer	1.0	1.0	0.3 to 2.1	0.06	0.3 to 2.1

Velocity u_∞ m/s	Slot size, mm	Starting-length Reynolds number Re_x	δ^*/s	q_w'', T_{aw}, or C_w	Effectiveness	slot Reynolds number Re_c	Range of x_c/s
42.0	3.11	4.97×10^5	0.2	T_{aw}	0.96 to 0.19	2200 to 10,000	6 to 198
3.4 to 5.2	1.59 3.175 6.35			q_w'' T_{aw}	1.0 to 0.2	3530 to 6960	17 to 300
15.8 to 29.5	1.59 3.175 6.35	0.41×10^5 to 0.76×10^5	0.3 to 1.2	q_w'' T_{aw}	0.95 to 0.06	0 to 6600	10 to 150
29.0 to 38.0	1.59 3.175 6.35	3.3×10^5 to 16.8×10^5	0.24 to 1.36	T_{aw}	0.95 to 0.06	1000 to 7000	5 to 280
50.5	3.2	6.5×10^5	0.2	T_{aw}	1.0 to 0.10	1800 to 8500	2 to 275
15.2 to 30.5	3.175 6.35 9.525			T_{aw}	1.0 to 0.2	1420 to 22,900	3.6 to 275.0
21.4	6.425	2.4×10^5	0.107	C_w	0.95 to 0.20	4035 to 19,500	4 to 218
15.2 to 14.4	0.907 2.54			q_w'' T_{aw}		312 to 2420	35 to 70
21.4	1.87	1.8×10^5 to 11.0×10^5	0.3 to 1.28	C_w	0.7 to 0.1	730 to 5000	50 to 200

Reference	Geometry	Injection gas[a]	Density ratio ρ_c/ρ_∞	Temp. ratio T_c/T_∞	Blowing rate $F = \dfrac{\rho_c U_c}{\rho_\infty u_\infty}$	Free-stream Mach no.	Velocity ratio U_c/u_∞
Whitelaw [81]		Air with He tracer	1.0	1.0	0.47 to 2.24	0.06	0.47 to 2.24
Carlson, Talmor [82]		Nitrogen[b]	2.76	0.363	0.5 to 1.98	0.1 to 0.5	0.17 to 0.67
		Nitrogen[b]	2.76	0.363	0.563 to 2.66	0.115 to 0.242	0.19 to 0.90
Kacker, Whitelaw [83]		Air with He tracer	1.0	1.0	0.288 to 2.66	0.06	0.288 to 2.66
Pai, Whitelaw [84]		Air, hydrogen, air with He tracer, Refrig. 12	0.07 to 4.17	1.0	0.021 to 6.87	>0.13	0.55 to 2.21
Kacker, Whitelaw [66]		Air with He tracer	1.0	1.0	0.2 to 2.4	0.055	0.2 to 2.4
Burns, Stollery [85]		Arcton 12 (Refrig. 12)	4.17	1.0	2.21 to 16.7	0.017 to 0.050	0.53 to 4.00
		He	0.14	1.0	0.071 to 0.236	0.050	0.51 to 1.68
		Mixture air-He or air-Arcton	4.17 to 0.14	1.0	0.14 to 4.17	0.050	1.0
Metzger, Fletcher [86]		Air	0.96	1.05	0.25 to 0.75	0.04 to 0.07	0.26 to 0.78

Velocity u_∞ m/s	Slot size, mm	Starting-length Reynolds number Re_x	δ^*/s	q''_w, T_{aw}, or C_w	Effectiveness	Slot Reynolds number Re_c	Range of x_c/s
21.2	6.425 7.25	2.4 × 10⁵	0.095 to 0.107	C_w	1.0 to 0.26	4035 to 19,500	4 to 218
	1.59			T_{aw}	1.0 to 0.11	10,350 to 565,000	
	1.59			T_{aw}	1.0 to 0.11	13,300 to 368,000	
21.4	1.88 3.35 6.35, 12.7	2.4 × 10⁵	0.0542 to 0 366	C_w	0.95 to 0.15	745 to 44,000	25 to 150
10.0 to 20.8	2.54			C_w	1.0 to 0.005	70.8 to 14,250	2.5 to 212.5
20.8	6.26	2.17 × 10⁵ to 4.35 × 10⁵	0.107 to 0.191	C_w	0.97 to 0.20	1500 to 18,400	12 to 210
6.1 to 17.4	1.59	0.3 × 10⁵ to 0.85 × 10⁵	0.236 to 0.28	C_w	1.0 to 0.3	2220 to 17,420	0 to 512
16.8	1.59	0.85 × 10⁵	0.236	C_w	0.9 to 0.05	113 to 368	0 to 512
17.6	1.59	0.85 × 10⁵ to 3.03 × 10⁵	0.236 to 0.66	C_w	1.0 to 0.05	228 to 11,300	0 to 512
15.2 to 24.4	1.27 2.54			q''_u, T_{aw}		325 to 3500	5 to 70

Reference	Geometry	Injection gas[a]	Density ratio ρ_c/ρ_∞	Temp. ratio T_c/T_∞	Blowing rate $F = \dfrac{\rho_c U_c}{\rho_\infty u_\infty}$	Free-stream Mach no.	Velocity ratio U_c/u_∞
Pai, Whitelaw [87]	u_∞	Hydrogen Arcton 12	0.069 to 4.17	1.0	0.021 to 6.85	0.03 to 0.06	0.55 to 2.21
Williams [88]	u_∞	Nitrogen[c]	2.38 to 3.07	0.33 to 0.42	0.308 to 2.99	0.04 to 2.5	0.127 to 1.09
Matthews, Whitelaw [89]	u_∞	Air with helium and Arcton 12	1.0 to 2.0		0.7 to 2.0	0.05	0.35 to 2.0
Kikkawa, Fujii [90]	u_∞	Air with CO_2	1.0		0.444 to 1.0	0.02 to 0.035	0.444 to 1.0
Foster, Haji-Sheikh [68]	u_∞	Air (no pressure gradient)	0.92	1.09	0.15 to 1.14	0.09	0.16 to 1.24
		Air (with pressure gradient)	0.92	1.09	0.15 to 1.14	0.09	0.16 to 1.24
Haering [91]	u_∞	Air (no pressure gradient)			0.32 to 0.97		
		Air (with pressure gradient)			0.3 to 1.0		
Mayle, Kopper, Blair, Bailey [63]	u_∞	Air	0.9 to 1.1	0.91 to 1.11	0.50 to 0.91	0.06	0.83 to 1.22
Paradis [64]	u_∞	Air	2.0 to 3.0	0.313 to 0.586	0.8 to 4.8	0.33 to 0.55	0.23 to 1.85
Ko Shao Yen et al. [92]	u_∞	Air	0.914 to 1.40	0.71 to 1.09	0.2 to 3.2	0.16 to 0.3	0.15 to 2.65

Velocity u_∞, m/s	Slot size, mm	Starting-length Reynolds number Re_x	δ^*/s	q''_w, T_{aw}, or C_w	Effectiveness	Slot Reynolds number Re_c	Range of x_c/s
10.0 to 20.7	2.54			C_w	1.0 to 0.01	70.8 to 14,400	0 to 212
38.8	0.635 1.522			T_{aw}	1.0 to 0.16	3100 to 24,100	12.7 to 138
18.29	6.35			C_w	0.98 to 0.16		
8.0 to 12.0				C_w	1.0 to 0.16		5.0 to 150
31.7	1.58 3.17 6.35	2.0×10^6		T_{aw}	0.72 to 0.18	1800 to 3400	6.93 to 633
31.7	1.58 3.17 6.35	2.0×10^6		T_{aw}	0.69 to 0.20	1800 to 3400	6.67 to 593
	5.334			T_{aw}	0.95 to 0.15		8.56 to 1000
	5.334		0.327 to 3.40	T_{aw}	0.92 to 0.12		5.2 to 1440
21.0	3.2	1.1×10^6 to 1.52×10^6	0.296	T_{aw}	0.70 to 0.11	2100 to 4100	15.8 to 1232
112.0 to 187.0	0.43 0.76 1.04	9.5×10^5 to 1.8×10^6		T_{aw}	1.0 to 0.2	4.3×10^2 to 1.2×10^4	0 to 1380
54 to 101	2.3	1.9×10^6 to 5.7×10^6		T_{aw}	0.95 to 0.15	1.73×10^3 to 2.0×10^4	0.217 to 8.26

Reference	Geometry	Injection gas[a]	Density ratio ρ_c/ρ_∞	Temp. ratio T_c/T_∞	Blowing rate $F = \dfrac{\rho_c U_c}{\rho_\infty u_\infty}$	Free-stream Mach no.	Velocity ratio U_c/u_∞
Chin, Skirvin, Hayes, Burggraf [93]	Multiple slots	Air	1.15 to 1.13	0.87 to 0.887	0.0512 to 1.026	0.0822 to 0.152	0.057 to 1.13

c. Two-dimensional film cooling through

Reference	Geometry	Injection gas[a]	Density ratio ρ_c/ρ_∞	Temp. ratio T_c/T_∞	Blowing rate F	Free-stream Mach no.	Velocity ratio U_c/u_∞
Goldstein, Eckert, Wilson [96]		Air	0.33 to 0.48	0.77 to 1.14	0.0085 to 0.0223	2.9	
Goldstein, Jabbari [97]		Helium	0.88 to 1.28	0.77 to 1.14	0.0018 to 0.0034	3.0	0.0014 to 0.0039
		Refrigerant 12	0.82 to 0.88	1.11 to 1.20	0.0082 to 0.0115	3.0	0.0093 to 0.014
Wilson, Goldstein [98]		Air	~1.0		0.01 to 0.02	3.0	0.01 to 0.02

d. Two-dimensional film cooling through

Reference	Geometry	Injection gas[a]	Density ratio ρ_c/ρ_∞	Temp. ratio T_c/T_∞	Blowing rate F	Free-stream Mach no.	Velocity ratio U_c/u_∞
Goldstein, Eckert, Tsou, Haji-Sheikh [99]		Air	3.4 to 2.04	0.8 to 1.25	0.0 to 0.408	3.01	
		He	0.3 to 0.4	0.31 to 0.39	0.01 to 0.02	3.01	
Cary, Hefner [100]		Air	1.59 to 2.34	0.43 to 0.63	0.03 to 1.62	6.0	0.010 to 1.039
Hefner [101]		Air	1.59	0.63	0.041 to 0.157	6.0	0.026 to 0.098
		Air	1.59	0.63	0.016 to 0.174	6.0	0.010 to 0.109
Richards, Stollery [102][d]		Air, Refrig. 12 Helium, Argon	3.57 to 4.47	0.22 to 0.28		10.0	

[a]Unless otherwise noted, main stream is air.

[b]Nitrogen used as main-stream gas.

[c]Air-hydrogen combustion products form main-stream gas.

[d]Laminar film cooling in hypersonic flow.

Velocity u_∞, m/s	Slot size, mm	Starting-length Reynolds number Re_x	δ^*/s	q_w'', T_{aw}, or C_w	Effectiveness	Slot Reynolds number Re_c	Range of x_c/s
30.0 to 56.0	2.92	10.3×10^5 to 19.5×10^5		T_{aw}	1.00 to 0.174	413 to 6100	7.9 to 177.8

a porous slot—supersonic main stream

Velocity u_∞, m/s	Slot size, mm	Starting-length Reynolds number Re_x	δ^*/s	q_w'', T_{aw}, or C_w	Effectiveness	Slot Reynolds number Re_c	Range of x_c/s
1025	12.6	6.0×10^5	0.026	T_{aw}	0.9 to 0.12	1200 to 3700	0.25 to 8.0
1030	12.6	8.8×10^6		T_{aw}	0.9 to 0.35	126 to 652	0.07 to 2.96
1030	12.6	8.4×10^8		T_{aw}	0.65 to 0.11	3.6×10^4 to 6.2×10^4	
	14.0			q_w''			

a slot—supersonic main stream

Velocity u_∞, m/s	Slot size, mm	Starting-length Reynolds number Re_x	δ^*/s	q_w'', T_{aw}, or C_w	Effectiveness	Slot Reynolds number Re_c	Range of x_c/s
1050	1.63 3.12 4.62	4.0×10^5	0.045 to 0.127	T_{aw}	1.0 to 0.1	0 to 11,100	0 to 73
1050	1.63	4.0×10^5	0.127	T_{aw}	1.0 to 0.1	0 to 231	0 to 73
2630	1.59 4.76 11.10	2.4×10^7	2.3 to 16.0	T_{aw}	1.0 to 0.48	4.2×10^3 to 3.3×10^6	1.06 to 423
2670	1.60 4.78	2.87×10^7	1.85 to 5.54	T_{aw}	0.93 to 0.60	2.44×10^3 to 7.94×10^4	15.5 to 193
2670	4.78	2.87×10^7	1.85	T_{aw}	0.98 to 0.55	2.44×10^3 to 7.94×10^4	5.43 to 158
	0.84 1.22 1.60	4.1×10^6 to 6.3×10^6		q_w''			

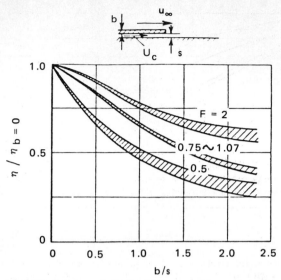

FIG. 88 Influence of slot lip thickness on film cooling effectiveness for nearly tangential injection through a slot on a flat plate in subsonic flow [65].

thickness is given by Mukherjee [65] as shown on Fig. 88, bringing out the importance of this parameter on the film cooling effectiveness. Additional results may be found in Kacker and Whitelaw [66]. It is recommended that Figs. 87 and 88 be used for design estimates of slot cooling for injection angles not exceeding 30°.

d. Effect of Injection Angle on Subsonic Slot Film Cooling

The experimental results of Scesa [67] and of Foster and Haji-Sheikh [68] for air injected at 90° to the flow direction provide some insight into the influence of the injection angle on subsonic film cooling. Figure 89 reveals the flat-plate predictions for tangential injection [49, 50], which have been shown to be in good agreement with experimental results as discussed above. Also shown on the figure is the proposed correlation of Scesa [67], which has been converted to a more generalized presentation. The experimental results of Foster and Haji-Sheikh [68] are in agreement with the proposed curve. It may be noted that the extension of the Goldstein and Haji-Sheikh equation which attempts to account for the prediction angle [50] underestimates the effect while Mukherjee's formula [65] overestimates the influence of injection angle (i.e., it predicts an effectiveness considerably lower than the experimental results). It is recommended that Fig. 89 be used to estimate the influence of injection angle on low-velocity film cooling.

e. Effects of Curvature on Open-Slot Film Cooling

It has been determined that curvature has a considerable influence on film cooling, as shown on Fig. 90 for air injected into an air boundary layer [63]. Since film cooling is often used in applications such as gas turbines or airfoil configurations, this effect may be very important. In general, a convex curvature results in an increase in effectiveness over

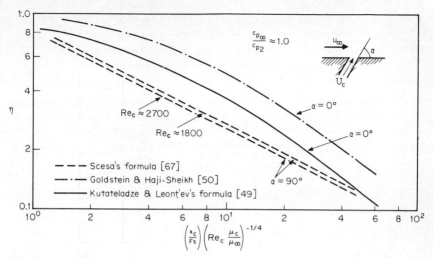

FIG. 89 Recommended correlation for film coolant inejction at 90° to the flow direction—subsonic flow over a flat plate.

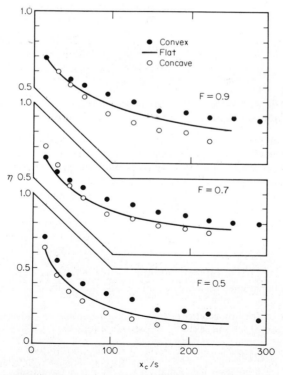

FIG. 90 Influence of curvature on film cooling effectiveness—subsonic flow [63].

the flat-plate value, while a concave curvature causes a decrease in effectiveness relative to the flat-plate value.

f. Effects of Large Property Variations on Open-Slot Film Cooling in Subsonic Flow

It has been demonstrated that large property variations resulting from large temperature differences can be taken into account by the reference temperature method. In this case, the dimensionless blowing rate ζ is replaced by ζ^*, with all properties in ζ^* noted by an asterisk (*) to be evaluated at the reference temperature T^*:

$$T^* = T_\infty + 0.5(T_{aw} - T_\infty) \tag{46}$$

and

$$\zeta^* = \frac{x_c}{Fs}\left(\mathrm{Re}_c \frac{\mu_c}{\mu^*}\right)^{-1/4} \frac{\rho^*}{\rho_\infty} \tag{47}$$

Thus, the formulation of Librizzi and Cresci becomes:

$$\eta = \frac{1}{1 + (c_{p\infty}/c_{p_c})[0.329(4.01 + \zeta^*)^{0.8} - 1]} \tag{40a}$$

Comparable expressions can be written for Eqs. (41) and (42). Thus, Fig. 84 can be used for conditions involving large property variations provided that the ordinate is taken to be ζ^* and properties are evaluated at T^*. In using this procedure, some iteration is necessary since the specification of the adiabatic wall temperature requires a knowledge of the effectiveness η, which in turn depends on ζ^*. However, ζ^* requires a knowledge of the reference temperature, which depends on T_{aw}. The process can be started by assuming T_{aw} to be equal to the coolant temperature, allowing the evaluation of T^* and ζ^*. The value of η can then be obtained from the curve, leading to an estimate of the adiabatic wall temperature, which can be compared with the original estimate. If they agree, the calculated adiabatic wall temperature is the predicted value; if they disagree, the process is repeated using the new estimate of the adiabatic wall temperature until convergence is established.

g. Other Considerations—Open-Slot Film Cooling under Subsonic Main-Stream Conditions

The recommendations given in this section on slot film cooling under subsonic main-stream conditions should yield reasonable estimates for design purposes. However, the actual values of the film cooling effectiveness and local heat transfer depend on the detailed slot geometry, the nature of the coolant, and the flow conditions of the external stream. Table 2, part b, which provides additional information on a number of the available experimental studies of subsonic slot film cooling [57–64, 66–93], may be of value to the designer if the geometry, choice of coolant, and flow conditions correspond to some of the reported studies.

The influence on the film cooling effectiveness of favorable and adverse pressure gradients has been studied [14, 46, 48, 91]. In general, a favorable pressure gradient results in a slight decrease in effectiveness while adverse pressure gradients cause a slight increase in the effectiveness. Studies on the effects of rotation [94] and of swirl [95] on the film-cooled surface have been reported. Rotation was found to have little influence on the effectiveness, while fluid swirl was reported to decrease the effectiveness.

h. Heat Transfer—Open-Slot Film Cooling under Subsonic Main-Stream Conditions

Table 2, part *b*, reveals that a considerable number of studies on local heat transfer in the presence of slot film cooling have been reported. These studies involved different slot geometries with different injection angles. Although most of the heat transfer investigations were carried out for turbulent flow over a film-cooled flat plate, local heat transfer data are also available for film-cooled surfaces in a flow with a pressure gradient [73, 75, 87]. These studies reveal that the local heat transfer coefficient is not appreciably influenced by the presence of film cooling at dimensionless blowing rates (that is, F) of the order of 0.5 or less except in the immediate vicinity of the slot. At higher blowing rates, the influence of film cooling on local heat transfer is more substantial. This is revealed in Fig. 91, which applies to approximately tangential injection into a zero–pressure gradient subsonic flow. Similar results have been reported by Foster and Haji-Sheikh for injection normal to the surface [68]. It is, therefore, recommended that the local heat transfer coefficient along a slot film-cooled surface in subsonic flow be calculated under the assumption that the surface is solid and exposed to the same free-stream conditions as the film-cooled surface (i.e., assume that film cooling does not influence the local heat transfer coefficient). Some correction may be necessary if the blowing rate is high or if the location is close to the injection slot. However, the recommended procedure should yield reasonable design estimates of the local heat transfer.

i. Film Cooling through a Porous Slot—Supersonic Main Stream

Measurements of the effectiveness and local heat transfer have been carried out by the University of Minnesota group [96–98, Table 2, part *c*] for porous slot cooling under supersonic main-stream conditions. Several different coolant gases (air, helium, and Refrigerant 12) were injected through a porous slot into a supersonic boundary layer at

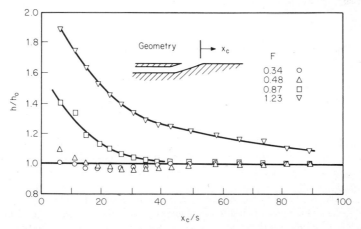

FIG. 91 Influence of film cooling on the local heat transfer coefficient on a flat plate as a function of the blowing rate F and the distance from the slot [73].

a nominal Mach number of 3. In this case, the effectiveness is so defined that the adiabatic wall temperature for the solid wall T_{aw0} replaces the free-stream temperature in Eq. (43):

$$\eta = \frac{T_{aw} - T_{aw0}}{T_c - T_{aw0}} \tag{43a}$$

It has been demonstrated that in such cases Eqs. (40), (41), and (42) may be used for predicting the effectiveness if the properties appearing in the dimensionless distance parameter are evaluated at a suitable reference temperature T^*:

$$T^* = T_\infty + 0.72(T_{aw} - T_\infty)$$

and Eq. (40) becomes

$$\eta = \frac{1}{1 + (c_{p\infty}/c_{pc})[0.329(4.01 + \zeta^*)^{0.8} - 1]} \tag{40a}$$

where

$$\zeta^* = \frac{x_c}{Fs}\left(\mathrm{Re}_c \frac{\mu_c}{\mu^*}\right)^{-1/4} \frac{\rho^*}{\rho_\infty}$$

For supersonic flow, the other two equations, Eqs. (41) and (42), may be written in this same form with ζ^* replacing ζ and the properties evaluated at T^*:

$$\eta = \frac{1}{1 + 0.329(c_{p\infty}/c_{pc})\zeta^{*0.8}} \tag{41a}$$

and

$$\eta = \frac{1.9\mathrm{Pr}^{2/3}}{1 + 0.329(c_{p\infty}/c_{pc})\zeta^{*0.8}} \tag{42a}$$

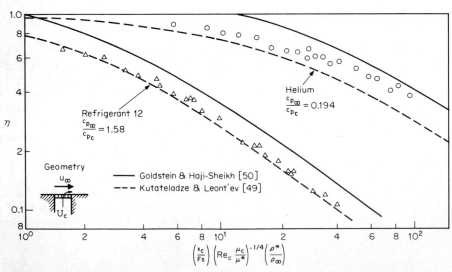

FIG. 92 Comparison of predictions and experimental results for injection through a porous slot—supersonic flow over a flat plate [98]. Data from Goldstein and Jabbari [97].

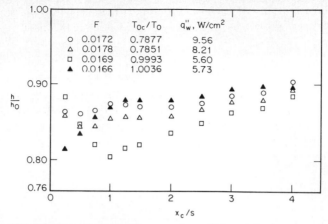

FIG. 93 Normalized local heat transfer coefficient, h/h_0, for injection through a porous slot—supersonic flow over a flat plate [98].

Figure 92 compares the predictions of Goldstein and Haji-Sheikh (Eq. (42a)) and Kutateladze and Leont'ev (Eq. 41a)) with experimental data for Refrigerant 12 and helium injection into a supersonic boundary layer [98]. On the basis of these results, it is recommended that Eqs. (41a) and (42a) be used to estimate film cooling effectiveness for porous-plug film cooling in supersonic flow.

For porous-plug film cooling, the local heat transfer coefficient has been found to be approximately 10 percent lower than that predicted for a solid surface in a supersonic air stream, as shown on Fig. 93 [98]. It is recommended that a value of h/h_0 between 0.9 and 1.0 be used to estimate local heat transfer for porous-plug film cooling in a supersonic flow.

j. Film Cooling through an Open Slot—Supersonic Main Stream

A modest number of experimental studies have been carried out for open-slot film cooling of a surface exposed to a supersonic main stream [99–102, Table 2, part d]. In one case [102] the main stream and boundary-layer flow is reported to be laminar, while the other studies report the boundary layer to be turbulent. These studies reveal that the film cooling effectiveness and local heat transfer are much more sensitive to the details of the slot geometry than in subsonic flow since the boundary layer–shock wave interactions strongly influence these factors. It may be helpful to present available slot cooling effectiveness results for the turbulent supersonic boundary layer. In this case the effectiveness is defined by Eq. (43a). Figure 94 compares the results of Goldstein et al. (99) and Hefner [100, 101] for air injected through a rearward-facing open slot into a supersonic boundary layer. The shaded area indicates the experimental results for a Mach number of 3, while the data points are for a Mach number of 6. It is recommended that this figure be used to estimate the effectiveness for air injected into a supersonic stream through a rearward-facing open slot.

Hefner reported that the effectiveness for two different rearward-facing step geometries could be correlated by the use of the independent parameter $(x_c/s)F^{-0.8}$ as demonstrated on Fig. 95. In general, Fig. 94 reveals that the local effectiveness increases with Mach number for a given geometry and for a fixed coolant flow rate.

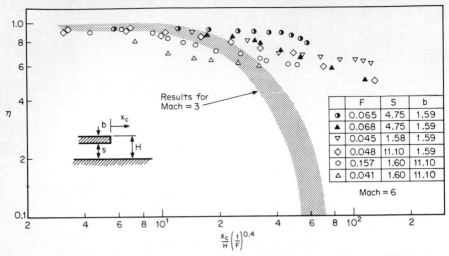

FIG. 94 Comparison of experimental effectiveness for air injection through a rearward-facing open slot—supersonic main-stream Mach numbers of 3 and 6 [99–101]. ◑, ▲, ▽, ◇: Ref. 100. O, △: Ref. 101.

The use of helium as a film coolant leads to effectiveness values which are higher than for the case of air injection. This is not unexpected from the subsonic results already discussed. Figure 96 presents the correlation proposed by Goldstein et al. [99] for helium, while the shaded region represents the data for air injection. This figure can be used as a guide for foreign gas injection through a rearward-facing step into a turbulent supersonic boundary layer.

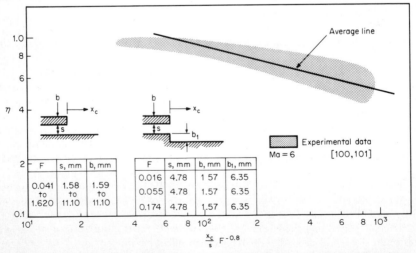

FIG. 95 Comparison of experimental effectiveness values for air injection through two rearward-facing open-slot geometries—supersonic main-stream Mach number of 6 [100, 101].

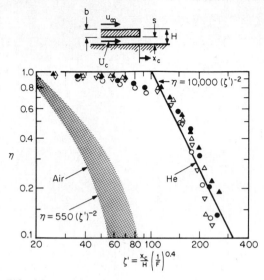

FIG. 96 Comparison of experimental effectiveness for air and helium injected through a rearward-facing open slot—supersonic main-stream Mach number of 3 [99].

The local heat transfer coefficient should be calculated for a solid wall exposed to the same free-stream conditions as the porous wall. This solid-wall heat transfer value should yield engineering estimates for the film-cooled wall.

3. Forced-Convection Turbulent Flow—Three-Dimensional Film Cooling

a. Film Cooling with Injection through a Single Hole

Considerable data exist for film cooling through a single hole, providing insight into the general behavior of three-dimensional film cooling in subsonic flow. For subsonic flow the effectiveness is sensitive to many variables, including the angle of injection of the coolant, the coolant injection rate, the displacement thickness at the location of the hole, the properties of the coolant gas and the free-stream gas, the pressure gradient, the free-stream turbulence level, and the Reynolds number. The range of these parameters for which data are available is given in Table 3 for a number of reported studies [56, 103–114].

Figures 97 and 98 present the effectiveness distribution in the flow direction at a fixed lateral position and in the lateral direction at a fixed value of x for two different values of the coolant injection rate under the condition that the coolant is injected normal (90°) to the flow direction [106]. The other test variables are shown on the figure and in the table. A comparison of the two figures brings out the conclusion that a higher value of the effectiveness occurs with the lower blowing rate of 0.496. This is consistent with other studies which reveal an optimum blowing rate of approximately 0.5 for air injected into an air boundary layer. At higher values of the blowing rate the momentum of the coolant

TABLE 3 Summary of Experimental Conditions[a]—Single Hole, Subsonic Turbulent Mainstream Flow

Reference	Injection angle	δ^*/D	ρ_c/ρ_∞	F	$\rho_\infty u_\infty D/\mu_\infty$	Pressure gradient K	Measured parameter q''_w, T_{aw}, or C_w
Goldstein, Eckert, Ramsey, 1968 [103][b]	35°, 90°	0.029–0.059	0.85	0.1–2.0	4.4–8.8 × 10⁴	0	T_{aw}
Ramsey, Goldstein, 1970 [104][c]	35°, 90°	0.029–0.059	0.85	0.1–2.0	4.4–8.8 × 10⁴	0	T_{aw}
Goldstein, Eckert, Eriksen, Ramsey, 1970 [105]	35°	0.075–0.124	0.85	0.1–2.0	2.2–4.4 × 10⁴	0	T_{aw}
Eriksen, 1971 [106]	90°	0.058	0.86	0.1–2.18	4.4–8.8 × 10⁴	0	q''_w, T_{aw}
Jabbari, Goldstein, 1974 [107][d]	35°	0.058	0.84	0.35–1.50	~8 × 10⁴	0–1.05 × 10⁻⁶	T_{aw}
Goldstein, Eckert, Burggraf, 1974 [108][c]	35°	0.159	0.9–3.5	0.5–2.0	1–2 × 10⁴	0	T_{aw}
Eriksen, Goldstein, 1974 [109]	35°	0.12–0.18	0.85	0.1–1.95	2.2–4.4 × 10⁴	0	q''_w, T_{aw}
Eriksen, Goldstein, 1974 [110]	90°	0.060–0.90	0.85	0.1–2.2	4.4–8.8 × 10⁴	0	q''_w
Goldstein, Ericksen, Ramsey, 1974 [111]	90°	0.058	0.83	0.1–2.0	4.4–8.8 × 10⁴	0	T_{aw}

Bergeles, Gosman, Launder, 1976 [112][f]	90°	0.051	1.0	0.046–0.5	2.8×10^4	0	C_w
Bergeles, Gosman, Launder, 1977 [113]	30°	0.122	1.0	0.1–1.5	3.3×10^4	0	C_w
Pedersen, Eckert, Goldstein, 1977 [56]	35°	0.081–0.163	0.86	0.5–1.0	$1.1–2.2 \times 10^4$	0	C_w
Brown, Saluja, 1979 [114][g]	30°	0.133	1.1	0.3–1.35	$\sim 10^4$	0	T_{aw}

[a] Injected coolant is air except where noted.

[b] No values of η reported for injection angle of 90°. Spreading angle of 90° injection is same as for 35° injection at low values of F (approximately 0.1); for larger values of F, spreading angle is greater for 90° injection than for 35° injection.

[c] Primarily a flow diagnostics study. Also measures detailed temperature, velocity, and turbulent intensity profiles.

[d] Reported effectiveness values lower than Goldstein-Ramsey and Bergeles et al.; main conclusion is that a favorable pressure gradient increases the centerline effectiveness values over the zero pressure gradient. However, if favorable pressure gradient is high the effectiveness is decreased below the zero pressure gradient value. The off-centerline effectiveness values show a decrease with favorable pressure gradient as compared to the zero pressure gradient case.

[e] Air and Refrigerant 12 used as coolant gas. Reports that increasing ρ_c/ρ_∞ results in increases in the centerline effectiveness; in addition, the optimum value of the dimensionless blowing rate F increases. Also reports higher values of effectiveness for shaped holes.

[f] Measured values of off-centerline effectiveness at low values of F are lower than the results of the University of Minnesota, suggests that conduction in the wall could have contributed since Minnesota measures T_{aw} while Bergeles et al. measure C_w.

[g] Report effectiveness results for several different values of the free-stream turbulence.

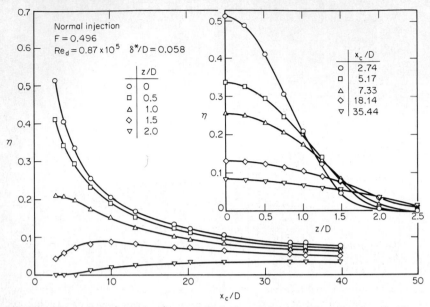

FIG. 97 Film cooling effectiveness for normal injection, air into an air boundary on a flat plate, $F = 0.5$ [106].

carries it farther out into the stream where it can mix with the turbulent free stream, resulting in lower values of the effectiveness.

The effectiveness decreases in the flow direction near the centerline. However, at lateral distances of z/D equal to 1.5 or 2.0 the effect of the injected coolant is not felt for some distance downstream as the injected coolant gradually spreads out; consequently the

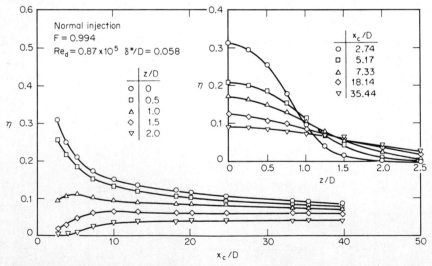

FIG. 98 Film cooling effectiveness for normal injection, air into an air boundary layer on a flat plate, $F = 1.0$ [106].

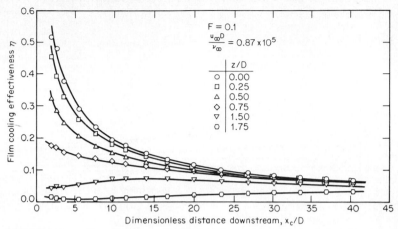

FIG. 99 Film cooling effectiveness for 35° injection angle, air into an air boundary layer on a flat plate, $F = 0.1$ [103].

effectiveness initially increases with distance at these lateral locations, gradually leveling off and ultimately decreasing. The total lateral spread of the coolant injected in a normal direction through a single hole is approximately 5 diameters in width.

Figures 99 to 102 reveal local film coolant effectiveness values as a function of position for values of the injection rate ranging from 0.1 to 2.0 for coolant injection at 35° to the flow direction [103]. Comparing Figs. 100 and 101 with Figs. 97 and 98 brings out the

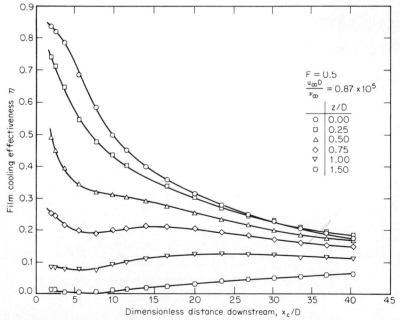

FIG. 100 Film cooling effectiveness for 35° injection angle, air into an air boundary layer on a flat plate, $F = 0.5$ [103].

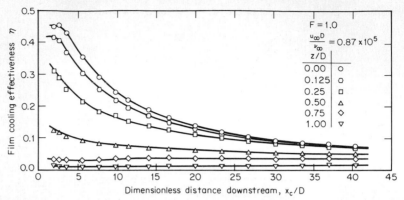

FIG. 101 Film cooling effectiveness for 35° injection angle, air into an air boundary layer on a flat plate, $F = 1.0$ [103].

important conclusion that the effectiveness is considerably higher for the 35° injection than for the 90° injection case. This is understandable since injection at a low angle keeps the coolant closer to the wall where it is not as diluted by the free stream as in the case of injection normal to the flow. In light of this finding the majority of the film cooling studies dealing with injection through single or multiple holes concentrate on injection angles of 25° to 35°, which are in the range of optimum injection angles for film cooling through circular holes in practical applications. Consequently, the remaining sections on film cooling through holes will concentrate on these injection angles.

A comparison of the local effectiveness results for the 35° injection studies brings out the conclusion that the optimum effectiveness under conditions where the density of the injected coolant is approximately the same as the free-stream gas occurs at a blowing rate of approximately 0.5, as revealed in Fig. 103. Injection at an angle of 35° results in a total lateral spreading of the coolant of 3 to 4 diameters (Fig. 104), considerably less than

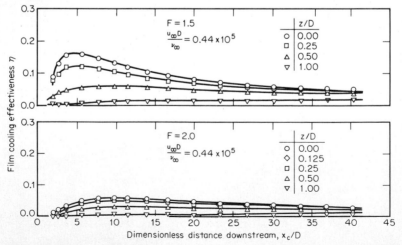

FIG. 102 Film cooling effectiveness for 35° injection angle, air into an air boundary layer on a flat plate, $F = 1.5, 2.0$ [103].

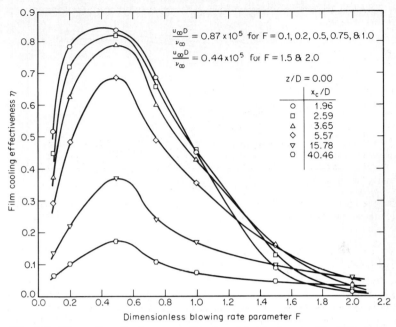

FIG. 103 Film cooling effectiveness as a function of the blowing rate F for 35° injection angle, air into an air boundary layer on a flat plate [103].

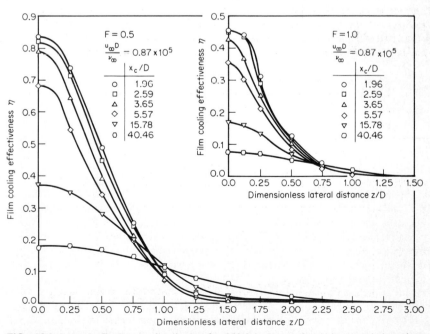

FIG. 104 Lateral effectiveness distribution for 35° injection angle, air into an air boundary layer on a flat plate, $F = 0.5$ and $F = 1.0$ [103].

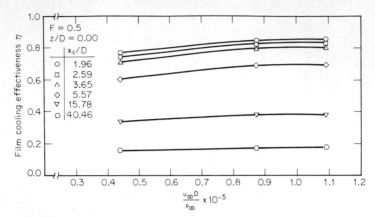

FIG. 105 Effect on centerline effectiveness of Reynolds number based on injection tube diameter, $u_\infty D/\nu_\infty$. Air injected into an air boundary layer on a flat plate, $F = 0.5$ [103].

the 5 diameters found for the normal-injection case (Figs. 97 and 98). The fact that the coolant remains nearer the surface and is restricted to a narrower lateral region explains, in part, why the 35° injection geometry results in higher effectiveness values than the normal-injection case.

The influence on the centerline film cooling effectiveness of the Reynolds number based on the injection tube diameter, $u_\infty D/\nu_\infty$, is shown on Fig. 105 for a 35° injection angle and the dimensionless coolant flow rate at the optimum level [103]. It is seen that an increase in the Reynolds number results in a slight increase in the centerline effectiveness at locations close to the hole, while at distances far downstream there is very little effect of the Reynolds number. Based on these results it appears that the Reynolds number, $u_\infty D/\nu_\infty$, has a relatively small influence on the film cooling effectiveness.

The local dimensionless displacement thickness δ^*/D has a considerable influence on the effectiveness, as revealed in Fig. 106 [108]. As the displacement thickness increases, the centerline effectiveness decreases. This occurs because an increase in displacement thickness results in a decrease of the velocity near the wall, which in turn results in an increase in the average boundary-layer temperature (for coolant injection). This is reflected in decreased centerline effectiveness values. It should be stressed, however, that the effectiveness at other lateral positions away from the centerline would show a different pattern since an increase in the displacement thickness will influence the rate of spreading of the coolant.

In many practical applications the density of the coolant may be higher than that of the free-stream gas. Under such conditions the optimum dimensionless blowing rate is generally higher than 0.5 and the effectiveness values, in general, will increase. An example of this behavior is shown on Fig. 107, which compares centerline effectiveness values for two coolants, air and Refrigerant 12. A major factor in the observed behavior is that the higher-density gas (Refrigerant 12) has a lower momentum at a fixed value of the dimensionless blowing rate F and remains closer to the surface; in contrast the injected air has a higher momentum, which carries it into the main stream, where it mixes with the main-stream gas, reducing its effectiveness.

The limited data available on the effect of free-stream turbulence level on the effectiveness reveal a slight decrease in film cooling effectiveness with increasing turbulence

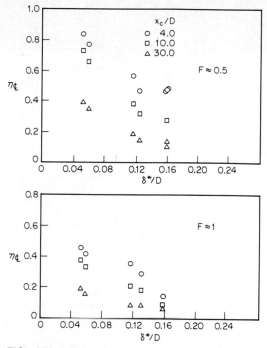

FIG. 106 Effect on centerline effectiveness of displacement thickness, δ^*/D, air injected into an air boundary layer on a flat plate, $F = 0.5$, 1.0 [108].

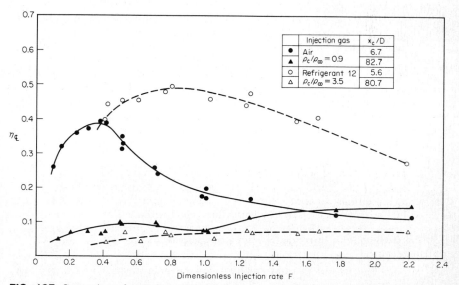

FIG. 107 Comparison of centerline effectiveness for injection of air and injection of Refrigerant-12 into a turbulent air boundary layer on a flat plate, injection angle of 35° [108].

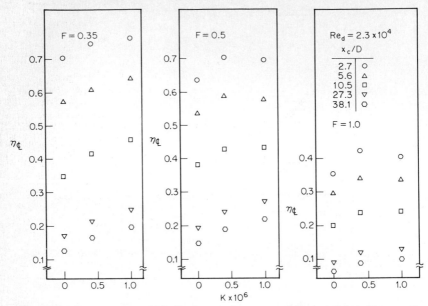

FIG. 108 Influence on centerline effectiveness of pressure gradient, air injected into an air boundary layer, $F = 0.35, 0.5, 1.0$ [107].

level [104, 108]. This results from the greater mixing and the increased dilution of the coolant with the main stream as the turbulence level increases.

With respect to the influence of a pressure gradient on film cooling effectiveness there exists some controversy. Some representative results are shown on Fig. 108, revealing a nonlinear behavior of the centerline effectiveness with the pressure gradient parameter K [107]. For example, at the dimensionless blowing rate of 0.5 the centerline effectiveness at low values of x_c/D at first increases with K, reaches a maximum, and then decreases at large values of the pressure gradient parameter. This behavior may explain why some investigators report an increase in effectiveness with favorable pressure gradient while others report a decrease.

Turning to the heat transfer results, Fig. 109 presents the ratio of the local heat transfer coefficient along the centerline to the heat transfer coefficient on a solid-plate surface evaluated at the same free-stream conditions [110]. Results for coolant injection at 35° and 90° to the flow direction for several different values of the dimensionless coolant flow are shown. The normalized heat transfer coefficients for normal injection are clearly much higher than for the shallower injection angles. For the 35° injection angle with the coolant injection rate at the optimum value of 0.5, the normalized value is close to 0.9 near the coolant entrance and gradually increases toward unity, reaching a value of 0.95 at x_c/D equal to 40.

b. Film Cooling with Injection through a Single Row of Holes

Injection through a single row of holes inclined at 35° to the flow direction and with a lateral spacing of 3 diameters has been studied extensively by the University of Minnesota group led by Professors Goldstein and Eckert.

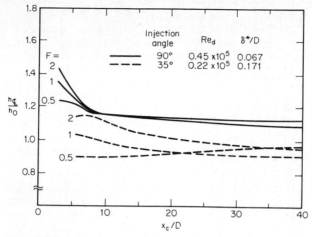

FIG. 109 Normalized heat transfer coefficient, h/h_0, for film cooling, air injected into an air boundary layer on a flat plate, injection angles of 90° and 35° [110].

These studies, along with those of other researchers, are identified in Table 4, which provides information on the range of variables studied by various research groups [56, 86, 105, 106, 108, 109, 115–120]. Figure 110 shows the streamwise distribution of the laterally averaged effectiveness at an injection rate of air equal to 0.5, which has been found by Kadotani and Goldstein [119] to be relatively independent of the Reynolds number

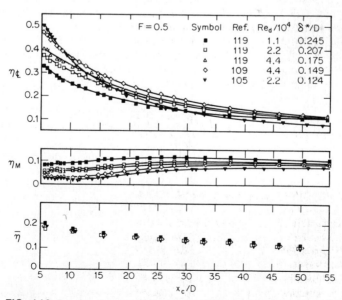

FIG. 110 Local and average effectiveness of film cooling with single row of holes, spacing $S_z/D = 3$, injection angle = 35°, air injected into an air boundary layer on a flat plate, dimensionless coolant flow $F = 0.5$ [119].

TABLE 4 Summary of Experimental Conditions[a]—Single Row of Holes, Subsonic Main-stream Flow

Author	Injection angle	δ^*/D	S_z/D	ρ_c/ρ_∞	F	$\rho_\infty u_\infty D/\mu_\infty$	Pressure gradient K	Measured parameter q''_w, T_{av}, or C_w
Goldstein, Eckert, Eriksen, Ramsey, 1970 [105][b]	35°	0.064–0.124	3	0.85	0.1–2.0	2.2–4.4 × 10⁴	0	T_{av}
Metzger, Fletcher, 1971 [86][c]	20°, 60°		1.55, 1.71	~0.9	0.25–0.75		0	q''_w
Eriksen, 1971 [106]	35°	0.14–0.21	3	0.85	0.1–1.18	2.2–8.8 × 10⁴	0	q''_w, T_{av}
Goldstein, Eckert, Burggraf, 1974 [108][d]	35°	0.160	3	0.9–3.5	0.5–2.0	1–2 × 10⁴	0	T_{av}
Eriksen, Goldstein, 1974 [109]	35°	0.119–0.190	3	0.85	0.1–1.95	2.2–4.4 × 10⁴	0	q''_w, T_{av}
Liess, 1975 [115][e]	35°	0.04–0.62	3	1.0	0.1–2.0	See footnote e.	See footnote e.	q''_w, T_{av}, C_w
Foster, Lampard, 1975 [116]	90°	0.62	3	1.5–4.26	0.38–1.48	0.5 × 10⁴	0	C_w
Pedersen, Eckert, Goldstein, 1977 [56]	35°	0.165	3	0.14–4.17	0.2–2.0	1.1–2.2 × 10⁴	0	C_w

Kadotani, Goldstein, 1977, 1979 [117, 118][f]	0.020–0.190	35°	3	0.85	0.2–1.5	$1.1\text{--}1.4 \times 10^4$	0	T_{aw}
Kadotani, Goldstein, 1979 [119][g]	0.024–0.20	35°	3	0.85	0.2–1.5	$1.1\text{--}4.4 \times 10^4$	0	T_{aw}
Brown, Saluja, 1979 [114][h]	0.132	30°	2.67, 3.2, 4.0, 5.33, 8.0	1.1	0.2–1.6	2×10^4	-0.58×10^{-6}, 0, 1.14×10^{-6}	T_{aw}
Foster, Lampard, 1980 [120]	0.165	35°, 55°, 90°	1.25, 2.50, 3.0, 3.75, 5.0	2.0	0.5–2.5	$\sim\!10^4$	0	C_w

[a]Injected coolant is air except for references 56, 108, 116, and 120.

[b]Results also presented for injection 15° and 35° lateral to the flow direction.

[c]Used transient technique to measure heat transfer, with and without injection. Reported average values of h/h_0.

[d]Shaped injection channels were also studied and were found to yield substantial improvement in effectiveness.

[e]Attempts to obtain laminar flow downstream of injection failed and in general results in this region were turbulent. Free-stream Mach numbers 0.3, 0.6, and 0.9 were used in this study. No effect of Mach number on effectiveness or reduced heat transfer coefficient over the range investigated. Pressure gradients of zero, 75, and 150 mm-Hg/cm were studied. Transient technique used for measurement of heat transfer coefficients.

[f]Main-stream turbulence intensity was varied from 0.3 percent to 20.6 percent, while the turbulence scale was varied from 0.06D to 0.33D. Part A gives details on the flow field, while Part B reports the effectiveness values.

[g]Main-stream turbulence was varied from 0.3 percent to 8.2 percent with scale of turbulence equal to 0.3D.

[h]Main-stream turbulence intensity was varied from 2 percent to 12 percent.

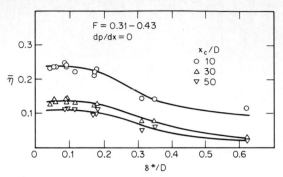

FIG. 111 Film cooling with single row of holes, spacing $S_z/D = 3$, injection angle $= 35°$. Influence of boundary-layer displacement thickness on laterally averaged effectiveness. Air into an air boundary layer on a flat plate [115].

based on the hole diameter, $u_\infty D/\nu_\infty$, over the range of 1.1×10^4 to 4.4×10^4 and independent of the dimensional displacement thickness at the start of injection, δ^*/D from 0.12 to 0.25.

The insensitivity of the laterally averaged effectiveness to initial displacement thickness is explained by the detailed measurements shown on Fig. 110, which reveal that a decrease occurs in the centerline effectiveness as the initial displacement thickness increases; however, this is compensated by a rise in the effectiveness values with increased

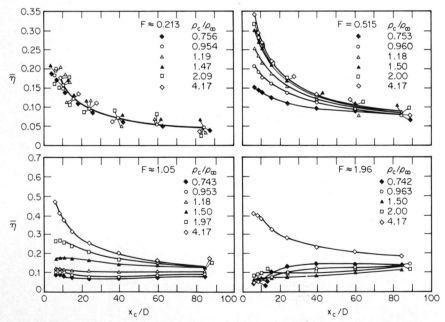

FIG. 112 Effectiveness of film cooling with single row of holes, spacing $S_z/D = 3$, injection angle $= 35°$, various gases injected into an air boundary layer on a flat plate. Dimensionless coolant flows $F = 0.213, 0.515, 1.05, 1.95$ [56].

Scale of turbulence \ Blowing rate	$F = 0.2$	$F = 0.5$	$F = 1.5$
Small scale	○	□	▽
Large scale	●	■	▼

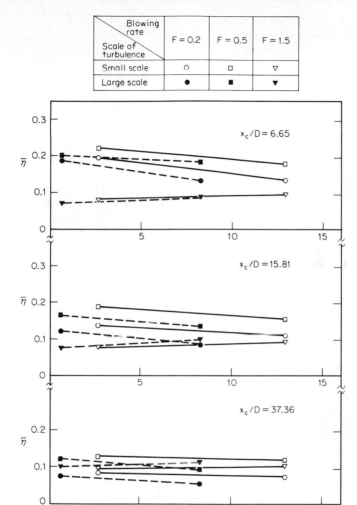

FIG. 113 Film cooling with single row of holes, spacing $S_z/D = 3$, injection angle $= 35°$. Influence of main-stream turbulence intensity on the laterally averaged effectiveness. Air into an air boundary layer on a flat plate. NOTE: Tu $= (\overline{u'^2})^{1/2}/u_\infty$. [117]

displacement thickness at a lateral distance of z/D equal to 1.5. Caution is in order, however, since this insensitivity to displacement thickness apparently does not extend beyond δ^*/D equal to approximately 0.25. As the displacement thickness increases beyond that value the laterally averaged effectiveness decreases, as shown on Fig. 111 [115].

The influence of varying the density ratio ρ_c/ρ_∞ from 0.75 to 4.2 is shown on Fig. 112 for dimensionless coolant rates ranging from 0.2 to 2.0 with injection angle of 35° and lateral spacing of 3.0 diameters [56]. The results are seen to be consistent with those found for the single hole. The optimum value of the blowing rate F increases with increasing density ratio and is accompanied by an increase in the laterally averaged effectiveness values. For example, as the density ratio changes from a value of 0.75 to 4.2 the optimum blowing rate shifts from approximately 0.5 to 2.0 while the corresponding maximum average effectiveness (evaluated at x_c/D equal to 10) increases from 0.2 to approximately 0.5.

The influence of the free-stream turbulence level on the average effectiveness is indicated on Fig. 113, which shows some representative values obtained with various grid geometries having different turbulence scales [117]. In general, an increase in turbulence intensity or in turbulence scale results in a moderate decrease in effectiveness. This is not unexpected since increased turbulence is associated with more effective mixing of the coolant with the free stream, resulting in greater dilution of the coolant.

Similar results have been reported by Brown and Saluja [114], who show a decrease in the average effectiveness with increasing free-stream turbulence intensity over a wide range of injection rates for an injection angle of 30° with a lateral spacing of 5.33 diameters on a flat-plate geometry (Fig. 114). This figure also brings out the effect of favorable and unfavorable pressure gradients. A modest decrease in average effectiveness is associated with a favorable pressure gradient, while an unfavorable pressure gradient results in a substantial increase in the value of the average effectiveness.

Figure 115 also gives some insight into the effect of a favorable pressure gradient on the laterally averaged effectiveness for coolant injection at 35° with a lateral spacing of 3.0 diameters for Mach number ranging from 0.3 to 0.9 [115]. The average effectiveness is seen to decrease with increasing magnitudes of the favorable pressure gradient. As men-

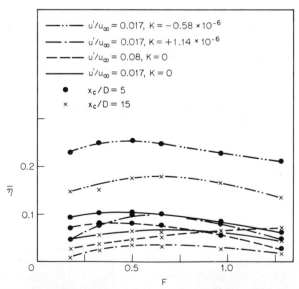

FIG. 114 Film cooling with single row of holes, spacing $S_z/D = 5.33$, injection angle = 30°. Influence of turbulence intensity and pressure gradient on laterally averaged effectiveness. Air into an air boundary layer on a flat plate [114].

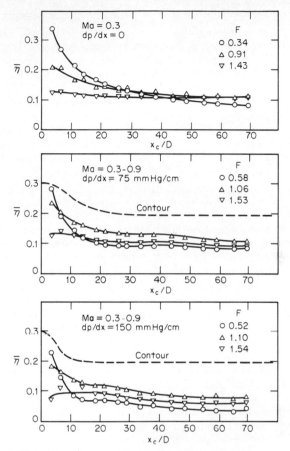

FIG. 115 Film cooling with single row of holes, spacing $S_z/D = 3.0$, injection angle = $35°$. Influence of pressure gradient on laterally averaged effectiveness. Air into an air boundary layer on a flat plate [115].

tioned earlier, some controversy still exists on the detailed behavior of the effectiveness in the presence of a favorable pressure gradient, but the results given here should yield a fair estimate of the behavior.

The laterally averaged heat transfer coefficient in the presence of film cooling through a row of holes is normalized with respect to the solid-wall heat transfer coefficient evaluated for the same free-stream conditions and presented on Fig. 116 [109]. The data are for air injection at $35°$ with a lateral spacing of 3.0 diameters for a flat-plate geometry. In general, the normalized values are close to unity up to a dimensionless injection rate of 1.0 (i.e., the heat transfer coefficient in the presence of film coolant is the same as that on a flat plate under comparable conditions). However, at high injection rates the normalized heat transfer coefficients reach values of 1.10 to 1.25, indicating that the heat transfer coefficients are 10 to 25 percent higher than the solid-wall values.

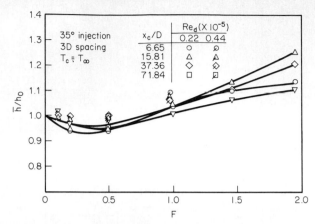

FIG. 116 Normalized heat transfer coefficient h/h_0 for film cooling with single row of holes, spacing $S_z/D = 3.0$, injection angle $= 35°$. Air into an air boundary layer on a flat plate [109].

c. Film Cooling with Injection through a Double Row of Holes

Limited data exist for the film cooling performance where air is injected through a double row of holes. The range of variables in these studies is presented on Table 5 [121–124]. Most of the investigators used the staggered row configuration shown on Fig. 117, although one group of investigators [124] reported data for the in-line configuration, which they compared with data from two staggered rows of holes as shown on Fig. 118. For the two mass injection rates studied, dimensionless coolant rates of either 0.5 or 1.0 at all hole locations, it is clear that the staggered configuration gives higher effectiveness values, especially at the higher injection rates. This accounts for the fact that greater emphasis has been placed on the staggered-row geometry.

The superior performance of two staggered rows of holes over a single row is brought out in Figs. 119 and 120 [123] for conditions where air is injected at an angle of 35° at a dimensionless coolant rate of 0.5, with lateral spacing of 3 diameters and longitudinal spacing of 2.6 diameters for the double-row configuration. Figure 119 reveals that the distribution of effectiveness is much more uniform for the double-row geometry. Figure 120 shows that 10 holes in two staggered rows yield far higher effectiveness values than 5 holes in one row for a fixed total coolant flow rate.

Muska et al. [122] adopted closer lateral spacing (S_z/D equal to 2) and wider streamwise spacing (S_x/D equal to 3.5) than Jabbari and Goldstein [121, 123] and as a result their effectiveness values shown on Fig. 121 do not correspond to those in Fig. 120. A comparison of these two figures gives some indication of the importance of hole spacing on the film cooling effectiveness.

The effect of independently controlling the coolant injectant rate from each row of holes for two staggered rows of holes has been studied for a 35° injection angle and a lateral spacing of 3 diameters. The longitudinal spacing ranged from 10 to 40 diameters [124]. Some representative results are shown on Fig. 122, which brings out the complex behavior of the laterally averaged effectiveness. If it is necessary to use a high blowing rate at one row the authors concluded that it is better to do so at the upstream row with the lower blowing rate at the downstream row; this arrangement gives a better overall

TABLE 5 Summary of Experimental Conditions[a]—Two Rows of Holes, Subsonic Main-Stream Flow

Author(s)	Injection angle	In-line or staggered	δ^*/D	S_z/D	S_x/D	ρ_c/ρ_∞	F	$\rho_\infty u_\infty D/\mu_\infty$	Pressure gradient K	Measured parameter q''_w, T_{aw}, or C_w
Jabbari, Goldstein, 1974 [121][b]	35°	Staggered	0.231	3	2.6	0.84	0.35–1.50	10^4	0, 0.45×10^{-6}, 1.05×10^{-6}	q''_w, T_{aw}
Muska, Fish, Suo, 1976 [122][c]	30°	Staggered	0.267	2	3.46	1.10	0.10–1.30	2×10^4	0	T_{aw}
Jabbari, Goldstein, 1978 [123][d]	35°	Staggered	0.177, 0.231	3	2.6	0.84	0.20–1.50	$1–2.1 \times 10^4$	0	q''_w, T_{aw}
Afejuku, Hay, Lampard, 1980 [124][e]	35° 90°	Staggered; in-line	0.160	3	10, 20, 30, 40	2.0	0.5–3.0	3.4×10^4	0	C_w

[a]Injected coolant was air, except for Ref. 124, which used a mixture of Refrigerant 12 and air.

[b]Blowing rates at all holes were equal. Centerline effectiveness decreased with acceleration, while centerline effectiveness near injection point generally increases with acceleration. Laterally averaged effectiveness generally decreases slightly with modest acceleration. Normalized laterally averaged heat transfer coefficient h/h_0 is reduced by acceleration.

[c]Blowing rates at all holes were equal. Experiments also carried out on airfoils in addition to flat-plate experiments. Measurements also made on two groups of holes with two staggered rows per group, for which it was reported that the predicted effectiveness values using Sellars superposition approach were in excellent agreement with measurements.

[d]Blowing rates at all holes were equal. Study concludes that two staggered rows may be more effective than a single row when compared at the same total coolant flow.

[e]Blowing rates of the two rows were separately controlled, allowing various combination of M values for the two rows. Study concludes that staggered rows yield higher performance than in-line rows, that 35° injection angle is better than 90°, that blowing rate should be approximately unity at both rows for maximum average effectiveness. Concludes Sellars superposition approach is conservative for staggered rows, but for in-line rows the superposition approach may predict too high a value of effectiveness.

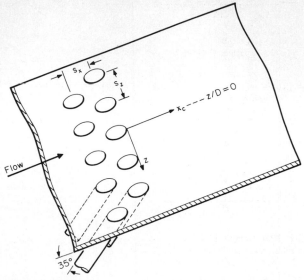

FIG. 117 Sketch of injection geometry for two rows of holes, staggered configuration.

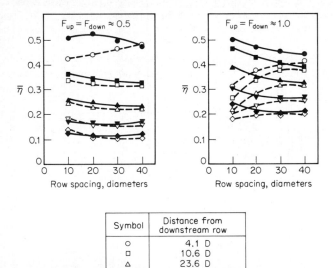

Symbol	Distance from downstream row
○	4.1 D
□	10.6 D
△	23.6 D
▽	55.7 D
◇	109.6 D

Solid symbols = staggered configuration
Open symbols = in-line configuration

FIG. 118 Comparison of laterally averaged effectiveness for two rows of holes, in-line (open symbols) and staggered (solid symbols) configuration [124].

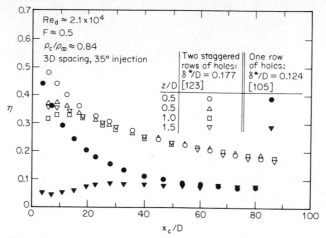

FIG. 119 Comparison of local effectiveness values, two staggered rows of holes versus one row of holes [123].

average effectiveness than if the blowing rates are reversed. Such behavior can be anticipated by applying the superposition principle, which will be discussed in a later section.

The influence of a favorable pressure gradient on the film cooling effectiveness downstream of a double staggered row of holes is shown on Fig. 123, which reveals a relatively small decrease in average effectiveness with increasing values of K for dimensionless blowing rates near the optimum design values of $F \approx 0.5$ under the conditions studied [121].

Finally, the laterally averaged heat transfer coefficients $\bar{h}$ are given as $\bar{h}/h_0$ where h_0 is the local heat transfer coefficient for a solid wall under similar free-stream conditions

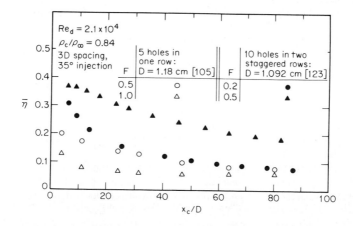

FIG. 120 Comparison of laterally averaged effectiveness values, two staggered rows of holes versus one row of holes [123].

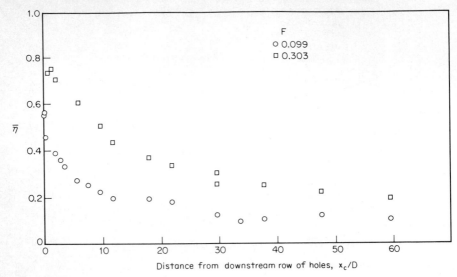

FIG. 121 Laterally averaged effectiveness values for a double row of staggered holes, $F = 0.099$, 0.303 [122].

[121, 123]. Figure 124 presents normalized heat transfer coefficients for the zero–pressure gradient case [123], while Fig. 125 shows the effect of a favorable pressure gradient on the normalized heat transfer coefficients [121]. Again, it is seen that the solid-wall value calculated for the same free-stream conditions as applies to the film-cooled wall may be used with good accuracy if the dimensionless injection rate is below 0.8. At coolant injection rates higher than this value the heat transfer coefficient increases rapidly, particularly near the injection location, reaching a value over 50 percent higher than the solid-wall heat transfer coefficient at F equal to 1.6 and x_c/D equal to 6.

d. Other Considerations—Three-Dimensional Film Cooling

The effects of curvature are as important in three-dimensional film cooling as in the two-dimensional case. This is demonstrated in the experiments of Ito et al. [125], which were carried out for mixtures of helium or Refrigerant 12 and air injected at 35° through a single row of holes, with center-to-center spacing of 3 diameters, into a subsonic air boundary layer. As shown on Fig. 126 these results reconfirm the earlier observation that the average effectiveness is higher for flow over a convex surface than for flow over a flat plate, while a concave curvature results in decreased effectiveness. This is important for the application of film cooling to gas turbine systems.

It has been possible in this section to review only the highlights of available information on three-dimensional film cooling. Other studies have dealt with injection through holes inclined in a direction lateral to the main flow direction [105], through holes located in a step-down rearward-facing surface [125], through shaped holes [108], through rectangular holes [126], and through holes covering the full surface [127, 128]. Flow visualization studies [129] and effectiveness values [130, 131] are available for the film cooling of gas turbine blades. Three-dimensional film cooling is clearly in its infancy; considerably more experimental work is necessary before the design of such systems is on a firm

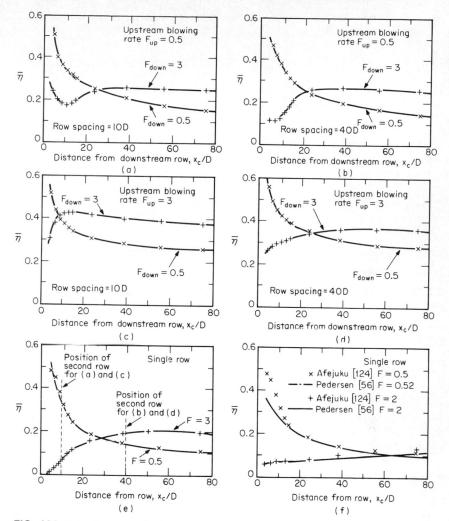

FIG. 122 Laterally averaged effectiveness values for double row of staggered holes for various upstream and downstream blowing rates [124]. (*e*) Single-row data for comparison. (*f*) Single-row data, comparisons with Refs. 56 and 124.

foundation. In the meantime, this section has laid out some guidelines for the practicing engineer.

4. Use of the Superposition Principle for Predicting Film-Cooling Effectiveness

Sellars [132] proposed that results from a single slot may be extended to multiple rows of slots by using the principle of superposition. For example, if there are two rows of slots

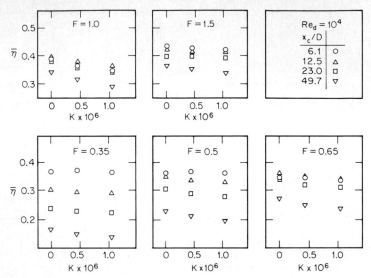

FIG. 123 Laterally averaged effectiveness for double row of staggered holes as a function of pressure gradient for various blowing rates [121].

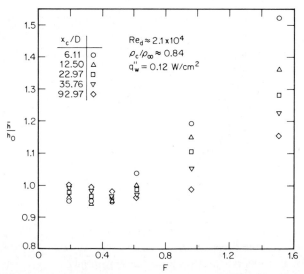

FIG. 124 Normalized heat transfer coefficient $\bar{h}/h_0$ as a function of blowing rate F and distance downstream of last row of holes [123].

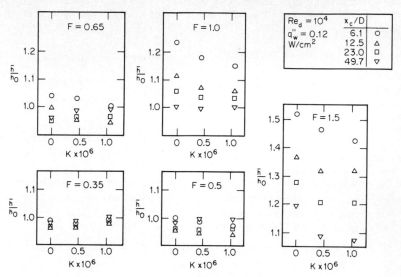

FIG. 125 Normalized heat transfer coefficient $\bar{h}/h_0$ as a function of pressure gradient, blowing rate, and distance downstream of last row of holes [121].

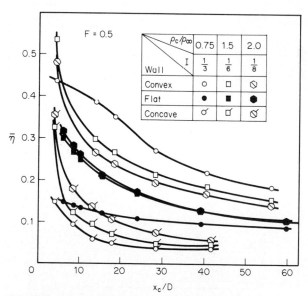

FIG. 126 Influence of curvature on laterally averaged effectiveness as a function of density ratio and distance downstream of injection holes [125].

the effectiveness downstream of the second slot, η_2, may be calculated by the following expression:

$$\eta_2 = \eta_1' + \eta_2'(1 - \eta_1')$$

where

$$\eta_2 = \frac{T_{aw2} - T_\infty}{T_c - T_\infty}$$

Here η_1' is the effectiveness of the first slot flowing alone evaluated at the local position downstream of the second slot, while η_2' is the effectiveness of the second slot flowing alone (this is the same distribution as η_1' but displaced to begin at the exit of the second slot). The assumption underlying the superposition technique for low-velocity flow is that the adiabatic wall temperature (which is assumed to be close to the mean film temperature) resulting from the upstream film cooling slots is the effective free-stream temperature downstream of the last slot. This is shown schematically on Fig. 127.

The extension to three slots is straightforward and yields

$$\eta_3 = \eta_1' + \eta_2'(1 - \eta_1') + \eta_3'(1 - \eta_1')(1 - \eta_2')$$

and for N slots

$$\eta_N = \eta_1' + \eta_2'(1 - \eta_1') + \eta_3'(1 - \eta_1')(1 - \eta_2') + \cdots$$
$$+ \eta_N'(1 - \eta_1')(1 - \eta_2')(1 - \eta_3') \cdots (1 - \eta_{N-1}')$$

This superposition approach will predict too high an effectiveness if the slots are so close together that considerable interaction occurs between the coolant flow from an upstream slot and the next downstream slot; this problem is amplified for high coolant flow rates. The comparisons of experimental data and superposition predictions given by Sellars [132] offer some guidance on the accuracy of the approach for multiple-row slot cooling.

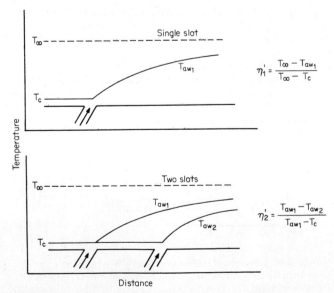

FIG. 127 Scheme of superposition principle.

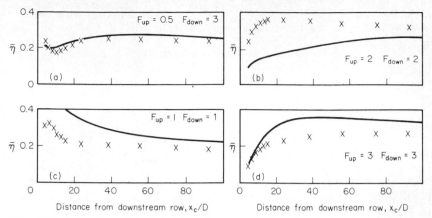

FIG. 128 Comparison of measured laterally averaged effectiveness values for two rows of holes with superposition predictions using single-row measurements [124]. (*a, b*) Staggered configuration. (*c, d*) In-line configuration. Solid line: predicted. Data points: measured.

This same approach can be used to estimate the laterally averaged film cooling effectiveness for multiple rows of holes if data are available for a single row of holes. Afejuku et al. [124] demonstrated this with a staggered hole configuration for a lateral spacing of 3 diameters and a longitudinal spacing of 10 diameters. As shown on Fig. 128 the effectiveness downstream of the second row of holes could be conservatively estimated for moderate rates of blowing. It appears that the superposition prediction will in general yield a low estimate of the effectiveness for staggered rows of holes, which will result in some excess use of injected coolant. Nevertheless, this estimate is on the conservative side, and therefore the superposition approach should provide a useful design estimate for staggered rows of holes.

In the case of two in-line rows of holes the superposition approach yielded effectiveness values which were considerably higher than the experimental values, especially at high blowing rates [124]. Thus, caution is advised in applying the superposition method to predict the performance of in-line rows of holes.

The same general conclusions were reached by Metzger et al. [133] for two rows of holes with lateral and longitudinal spacing of 4.8 diameters with injection normal to the flow direction. As shown on Fig. 129 superposition yielded excellent agreement with the

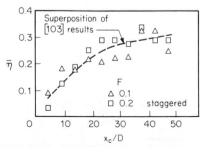

FIG. 129 Comparison of laterally averaged effectiveness measurements for a double row of staggered holes with superposition predictions using single-row measurements [133].

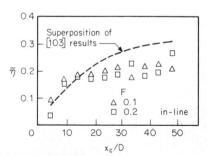

FIG. 130 Comparison of laterally averaged effectiveness measurements for double row of in-line holes with superposition predictions using single-row measurements [133].

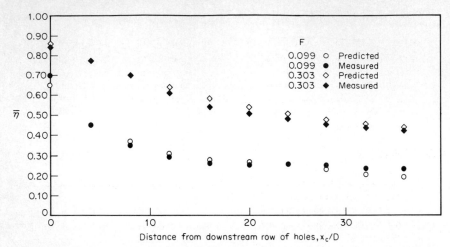

FIG. 131 Comparison of measured laterally averaged effectiveness values for two staggered double rows of holes with superposition prediction using measurements of double row staggered holes [122].

measured effectiveness values for the staggered configuration. In the case of the in-line geometry the predicted effectiveness was considerably higher than the measured values, as shown on Fig. 130.

The superposition approach has also been used to estimate film cooling effectiveness for conditions where the geometry shown on Fig. 117 (i.e., two rows of staggered holes) is repeated some distance downstream [122]. The results are shown for the condition that the second group of holes is 16.7 diameters (Fig. 131) and 25.0 diameters (Fig. 132) downstream of the first group. Again, the agreement is acceptable for design purposes.

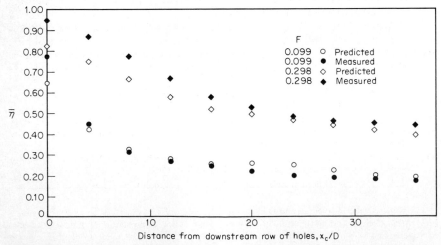

FIG. 132 Comparison of measured laterally averaged effectiveness values for two staggered double rows of holes with superposition predictions using measurements of a double row of holes [122].

ACKNOWLEDGMENTS

The author acknowledges with appreciation the assistance of his colleagues, including Don Bless for help in locating the necessary research publications, Marion Deloney for preparation of the tables, Claudette Eldridge and Renata Szandra for typing the manuscript, Eug Y. Kwack and Won Young Lee for preparing a number of the figures and for checking the final manuscript, and Jim Wiet for general assistance.

REFERENCES

1. J. P. Hartnett and E. R. G. Eckert, Mass Transfer Cooling in a Laminar Boundary Layer with Constant Fluid Properties, *ASME,* **79**:247 (1957).

2. E. R. G. Eckert, Engineering Relations for Heat Transfer and Friction in High-Velocity Laminar and Turbulent Boundary Layer Flow over Surfaces with Constant Pressure and Temperature, *Trans. ASME,* **78**:1273 (1956).

3. H. A. Simon, C. S. Liu, and J. P. Hartnett, The Eckert Reference Formulation Applied to High Speed Boundary Layers of Nitrogen and Carbon Dioxide, *Int. J. Heat Mass Transfer,* **10**:406 (1967).

4. H. A. Simon, J. P. Hartnett, and C. S. Liu, Transpiration Cooling Correlations for Air and Non-air Free Streams, in "Progress in Heat and Mass Transfer" (T. F. Irvine and W. Ibele, eds.), vol. 2 (Special Eckert Volume), Pergamon, New York, 1969.

5. J. F. Gross, J. P. Hartnett, D. J. Masson, and Carl Gazley, Jr., A Review of Binary Laminar Boundary Layer Characteristics, *Int. J. Heat Mass Transfer,* **3**:198 (1961).

6. C. S. Liu, J. P. Hartnett, and H. A. Simon, Mass Transfer Cooling of Laminar Boundary Layers with Hydrogen Injected into Nitrogen and Carbon Dioxide Free Streams, *3d Int. Heat Transfer Conf. Paper 83,* Chicago, August 1966.

7. E. M. Sparrow, W. J. Minkowycz, and E. R. G. Eckert, Mass Transfer Cooling of a Flat Plate with Various Transpiring Gases, *AIAA J.,* **3**:1341 (1965).

8. E. M. Sparrow and J. B. Starr, The Transpiration-Cooled Flat Plate with Various Thermal and Velocity Boundary Conditions, *Int. J. Heat Mass Transfer,* **9**:508 (1966).

9. V. K. M. Sastri and J. P. Hartnett, Effect of an Unheated Solid Starting Length on Skin Friction and Heat Transfer in a Transpired Laminar Boundary Layer, in "Progress in Heat and Mass Transfer" (T. F. Irvine and W. Ibele, eds.), vol. 2 (Special Eckert Volume), Pergamon, New York, 1969.

10. E. M. Sparrow, W. J. Minkowycz, E. R. G. Eckert, and W. E. Ibele, The Effect of Diffusion Thermo and Thermal Diffusion for Helium Injected into Plane and Axisymmetric Stagnation Flow of Air, *J. Heat Transfer, Trans. ASME,* **E86**:311 (1964).

11. O. E. Tewfik, E. R. G. Eckert, and L. S. Jurewicz, Diffusion Thermo Effects on Heat Transfer from a Cylinder in Cross Flow, *AIAA J.,* **1**:1537 (1963).

12. E. M. Sparrow, W. J. Minkowycz, and E. R. G. Eckert, Diffusion-Thermo Effects on Stagnation Point Flow of Air with Injection of Gases of Various Molecular Weights into the Boundary Layer, *AIAA J.* **2**:652 (1964).

13. E. Reshotko and C. B. Cohen, Heat Transfer at the Forward Stagnation Point of Blunt Bodies, National Advisory Committee, *Aeronaut. Tech. Note 3513,* July 1955.

14. A. F. Gollnick, Jr., Thermal Effects on a Transpiration-Cooled Hemisphere, *J. Aerosp. Sci.,* **29**:583 (1962).

15. G. E. Anderson, C. J. Scott, and D. R. Elgin, Mass Transfer Cooling Experiments on a Hemisphere at M = 5, Rosemount Aeronautical Laboratories, University of Minnesota, research report, August 1959.

16. V. M. Falkner and S. W. Skan, Some Approximate Solutions of the Boundary Layer Equations, *Phil. Mag.,* **12**:865 (1931); *Aeronaut. Res. Comm. R & M,* 1314 (1930).

17. W. B. Brown and P. L. Donoughe, Tables of Exact Laminar Boundary Layer Solutions When the Wall Is Porous and Fluid Properties Are Variable, National Advisory Committee, *Aeronaut. Tech. Note 2479,* September 1951.

18. C. F. Dewey, Jr., and J. F. Gross, Exact Similar Solutions of the Laminar Boundary Layer Equations, in "Advances in Heat Transfer" (T. F. Irvine and J. P. Hartnett, eds.), vol. 4, Academic, New York, 1967.

19. J. P. Hartnett, A Simple Graphical Presentation of the Dewey Gross Boundary Layer Solutions, The Rand Corp., *Rep. R-1062-Pr,* September 1972.

20. E. M. Sparrow, K. E. Torrance, and L. Y. Hung, Flow and Heat Transfer on Bodies in Cross Flow with Surface Mass Transfer, *Proc. 3d Int. Heat Transfer Conf.,* vol. 3, p. 23, AIChE, 1966.

21. B. V. Johnson and J. P. Hartnett, Heat Transfer from a Cylinder in Cross Flow with Transpiration Cooling, *J. Heat Transfer, Trans. ASME,* **C85**:173 (1963).

22. E. R. G. Eckert and J. N. B. Livingood, Method for Calculation of Laminar Heat Transfer in Air Flow around Cylinders of Arbitrary Cross Section (Including Large Temperature Differences and Transpiration Cooling), National Advisory Committee, *Aeronaut. Rep. 1118,* Washington, D.C., 1953.

23. D. B. Spalding and W. M. Pun, A Review of Methods for Predicting Heat-Transfer Coefficients for Laminar Uniform-Property Boundary Layer Flows, *Int. J. Heat Mass Transfer,* **5**:239 (1962).

24. D. J. Wanous and E. M. Sparrow, Longitudinal Flow over a Circular Cylinder with Surface Mass Transfer, *AIAA J.,* **3**:147 (1965).

25. D. J. Wanous and E. M. Sparrow, Heat Transfer for Flow Longitudinal to a Cylinder with Surface Mass Transfer, *J. Heat Transfer, Trans. ASME,* **C87**:317 (1965).

26. E. M. Sparrow and J. L. Gregg, Mass Transfer, Flow and Heat Transfer about a Rotating Disk, *J. Heat Transfer, Trans. ASME,* **C82**:294 (1960).

27. C. L. Tien, Mass Transfer in Laminar Flow about a Rotating Cone, *J. Heat Transfer, Trans. ASME,* **C83**:514 (1961).

28. E. M. Sparrow, C. J. Scott, R. J. Forstrom, and W. A. Ebert, Experiments on the Diffusion Thermo Effect in a Binary Boundary Layer with Injection of Various Gases, *J. Heat Transfer, Trans. ASME,* **C87**:321 (1965).

29. O. E. Tewfik and J. W. Yang, The Thermodynamic Coupling between Heat and Mass Transfer in Free Convection, *Int. J. Heat Mass Transfer,* **6**:915 (1963).

30. E. M. Sparrow, W. J. Minkowycz, and E. R. G. Eckert, Transpiration-Induced Buoyancy and Thermal Diffusion/Diffusion Thermo in a Helium-Air Free Convection Boundary Layer, *J. Heat Transfer, Trans. ASME,* **C86**:508 (1964).

31. E. M. Sparrow, J. W. Yang, and C. J. Scott, Free Convection in an Air-Water Vapor Boundary Layer, *Int. J. Heat Mass Transfer,* **9**:53 (1966).

32. R. J. Moffat and W. M. Kays, The Turbulent Boundary Layer on a Ferrous Plate: Experimental Heat Transfer with Uniform Blowing and Suction, *Int. J. Heat Mass Transfer,* **11**:1547 (1968).

33. D. G. Whitten, "The Turbulent Boundary Layer on a Porous Plate: Experimental Heat Transfer with Variable Suction, Blowing and Surface Temperature," doctoral dissertation, Stanford University, 1967.

34. K. Torii, N. Nishiwaki, and M. Hirata, Heat Transfer and Skin Friction in Turbulent Boundary Layer with Mass Injection, *Proc. 3d Int. Heat Transfer Conf.,* 1966.

35. D. Coles, "Measurements in the Boundary Layer on a Smooth Flat Plate in Supersonic Flow," doctoral dissertation, California Institute of Technology, 1953.

36. V. M. K. Sastri, "Analytical and Experimental Study of the Influence of an Unheated Solid Starting Length in a Transpired Boundary Layer," doctoral dissertation, University of Delaware, June 1968.

37. B. M. Leadon and C. J. Scott, Measurement of Recovery Factors and Heat Transfer Coefficients with Transpiration Cooling in a Turbulent Boundary Layer at M = 3 Using Air and

Helium as Coolant, Rosemount Aeronautical Laboratory, University of Minnesota, *Tech. Rep. 126,* February 1956.

38. M. W. Rubesin, The Influence of Surface Injection on Heat Transfer and Skin Friction Associated with the High-Speed Turbulent Boundary Layer, National Advisory Committee, *Aeronaut. Res. Memo, A 55 L 13,* February 1956.

39. O. E. Tewfik, E. R. G. Eckert, and C. J. Shirtliffe, Thermal Diffusion Effects in a Turbulent Boundary Layer with Helium Injection, *Proc. 1962 Heat Transfer and Fluid Mech. Inst.,* p. 42, Stanford University Press, Stanford, Calif., 1962.

40. B. Rockover and A. F. Gollnick, Jr., Mass Transfer Cooling of a Cone, *M.I.T. Naval Supersonic Lab. TR 424,* 1961.

41. E. Roy Bartle and B. M. Leadon, The Effectiveness as a Universal Measure of Mass Transfer Cooling for a Turbulent Boundary Layer, *Proc. 1962 Heat Transfer and Fluid Mech. Inst.,* Stanford University Press, Stanford, Calif., 1962.

42. A. L. Laganelli, "Mass Transfer Cooling on a Porous Flat Plate in Carbon Dioxide and Air Streams," doctoral dissertation, University of Delaware, June 1966.

43. L. W. Woodruff and G. C. Lorenz, Hypersonic Turbulent Transpiration Cooling Including Downstream Effects, *AIAA J.,* **4**:969 (1966).

44. W. H. Thielbahr, W. M. Kays, and R. J. Moffatt, The Turbulent Boundary Layer: Experimental Heat Transfer with Blowing, Suction and Favorable Pressure Gradient, *Rep. No. HMT-5,* Thermosciences Division, Department of Mechanical Engineering, Stanford University, 1969.

45. R. J. Baker and B. E. Launder, The Turbulent Boundary Layer with Foreign Gas Injection— II. Predictions and Measurements in Severe Streamwise Pressure Gradients, *Int. J. Heat Mass Transfer,* **17**:293 (1974).

46. J. F. Gross, J. P. Hartnett, D. J. Masson, and Carl Gazley, Jr., A Review of Binary Boundary Layer Characteristics, The Rand Corp., *R.M. 2516,* Santa Monica, Calif., June 1959.

47. R. J. Goldstein, Film Cooling, in "Advances in Heat Transfer" (T. F. Irvine, Jr., and J. P. Hartnett, eds.), vol. 8, Academic, New York, 1971.

48. J. Librizzi and R. J. Cresci, Transpiration Cooling of a Turbulent Boundary Layer in an Axisymmetric Nozzle, *AIAA J.,* **2**:617 (1964).

49. S. S. Kutateladze and A. I. Leont'ev, Thermal Physics of High Temperatures, **1**(2):281, 1963.

50. R. J. Goldstein and A. Haji-Sheikh, Prediction of Film Cooling Effectiveness, *Semi-int. Symp. Eng.,* Tokyo, 1967.

51. R. J. Goldstein, R. B. Rask, and E. R. G. Eckert, Film Cooling with Helium Injected into an Incompressible Air Flow, *Int. J. Heat Mass Transfer,* **9**:1341 (1966).

52. N. Nishiwaki,, M. Hirata, and A. Tsuchida, in "International Developments in Heat Transfer," pt. IV, p. 675, ASME, New York, 1961.

53. R. J. Goldstein, G. Shavit, and T. S. Chen, Film Cooling with Injection through a Porous Section, *J. Heat Transfer, Trans. ASME,* **C87**:353 (1965).

54. I. Mabuchi, *Japan Soc. Mech. Eng.* **8**:406 (1965).

55. M. P. Escudier and J. H. Whitelaw, *Int. J. Heat Mass Transfer,* **11**:1289 (1968).

56. D. R. Pedersen, E. R. G. Eckert, and R. J. Goldstein, Film Cooling with Large Density Differences between the Mainstream and the Secondary Coolant Measured by the Heat, Mass Transfer Analogy, *J. Heat Transfer, Trans. ASME,* **99**:625 (1977).

57. J. Chin, S. Skirvin, L. Hayes, and A. Silver, Adiabatic Wall Temperature Downstream of a Single Slot, *ASME Paper 58-A-104,* 1958.

58. R. A. Seban, Heat Transfer and Effectiveness for a Turbulent Boundary Layer with Tangential Fluid Injection, *J. Heat Transfer, Trans. ASME,* **82**:303 (1960).

59. S. Papell and A. M. Trout, Experimental Investigation of Air-Film Cooling Applied to an Adiabatic Wall by Means of an Axially Discharging Jet, *NASA Tech. Note D-9,* 1959.

60. K. Wieghardt, Ueber das Ausblasen von Warmluft fur Enteisen, *Z.B.W. Res. Rep. 1900,* 1943; *AAF Transl. No. 1 F-TS-919-RE,* Wright Field, 1946.

61. J. P. Hartnett, R. C. Birkebak, and E. R. G. Eckert, Velocity Distributions, Temperature Distributions, Effectiveness and Heat Transfer for Air Injected through a Tangential Slot into a Turbulent Boundary Layer, *J. Heat Transfer, Trans. ASME*, **83**:293 (1961).

62. R. A. Seban and L. H. Back, Velocity and Temperature Profiles in Turbulent Boundary Layers with Tangential Injection, *J. Heat Transfer, Trans. ASME*, **84**:45 (1962).

63. R. E. Mayle, F. C. Kopper, M. F. Blair, and D. A. Bailey, Effect of Streamline Curvature on Film Cooling, *J. Eng. Power, Trans. ASME*, **99**:77 (1977).

64. M. A. Paradis, Film Cooling of Gas Turbine Blades: A Study of the Effect of Large Temperature Differences on Film Cooling Effectiveness, *J. Eng. Power, Trans. ASME*, **99**:11 (1977).

65. D. K. Mukherjee, Film Cooling with Injection through Slots, *J. Eng. Power, Trans. ASME*, **98**:556 (1976).

66. S. C. Kacker and J. H. Whitelaw, An Experimental Investigation of the Influence of Slot-Lip Thickness on the Impervious Wall Effectiveness of the Uniform Density Two-Dimensional Wall Jet, *Int. J. Heat Mass Transfer*, **12**:1196 (1969).

67. S. Scesa, "Effect of Normal Blowing on Flat Plate Heat Transfer," doctoral dissertation, University of California, Berkeley, 1954.

68. R. C. Foster and A. Haji-Sheikh, An Experimental Investigation of Boundary Layer and Heat Transfer in the Region of Separated Flow Downstream of Normal Injection Slots, *J. Heat Transfer, Trans. ASME*, **97**:260 (1975).

69. R. A. Seban, H. W. Chan, and S. Scesa, Heat Transfer to a Turbulent Boundary Layer Downstream of an Injection Slot, *ASME Paper 57-A-36*, 1957.

70. J. E. Hatch and S. S. Papell, Use of a Theoretical Flow Model to Correlate Data for Film Cooling or Heating an Adiabatic Wall by Tangential Injection of Gases of Different Fluid Properties, *NASA Tech. Note D-130*, 1959.

71. S. S. Papell, Effect on Gaseous Film Cooling of Coolant Injection through Angled Slots and Normal Holes, *NASA Tech. Note D-299*, 1960.

72. R. A. Seban, Effects of Initial Boundary-Layer Thickness on a Tangential Injection System, *J. Heat Transfer*, **82**:392–393, (November 1960).

73. J. P. Hartnett, R. C. Birkebak, and E. R. G. Eckert, Velocity Distributions, Temperature Distributions, Effectiveness and Heat Transfer in Cooling of a Surface with a Pressure Gradient, "International Developments in Heat Transfer," pt. IV, p. 682, ASME, New York, 1961.

74. R. A. Seban and L. H. Back, Velocity and Temperature Profiles in a Wall Jet, *Int. J. Heat Mass Transfer*, **3**:255–265 (1961).

75. R. A. Seban and L. H. Back, Effectiveness and Heat Transfer for a Turbulent Boundary-Layer with Tangential Injection and Variable Free Stream Velocity, *J. Heat Transfer, Trans. ASME*, **84**:235 (1962).

76. E. R. G. Eckert and R. C. Birkebak, "Heat Transfer Thermodynamics and Education: Boelter Anniversary Volume" (H. A. Johnson, ed.), p. 150, McGraw-Hill, New York, 1964.

77. A. E. Samuel and P. N. Joubert, *ASME Paper No. 64WA/HT-48*, 1964.

78. W. B. Nicoll and J. H. Whitelaw, The Effectiveness of the Uniform Density, Two-Dimensional Wall Jet, *Int. J. Heat Mass Transfer*, **10**:623–639 (1967).

79. D. E. Metzger, H. J. Carper, and L. R. Swank, Heat Transfer with Film Cooling near Nontangential Injection Slots, *J. Eng. Power*, **90**:157–163 (April 1968).

80. S. C. Kacker and J. H. Whitelaw, The Dependence of the Impervious Wall Effectiveness of a Two-Dimensional Wall-Jet on the Thickness of the Upper Lip Boundary Layer, *Int. J. Heat Mass Transfer*, **10**:1623–1624 (1967).

81. J. H. Whitelaw, *Aeronaut. Res. Council Paper No. 492*, London, 1967.

82. L. W. Carlson and E. Talmor, "Gaseous Film Cooling at Various Degrees of Hot-Gas Acceleration and Turbulence Levels, *Int. J. Heat Mass Transfer*, **11**:1695–1713 (1968).

83. S. C. Kacker and J. H. Whitelaw, The Effect of Slot Height and Slot Turbulence Intensity on the Effectiveness of the Uniform Density, Two-Dimensional Wall Jet, *J. Heat Transfer, Trans. ASME*, **90**:469 (1968).

84. B. R. Pai and J. H. Whitelaw, *Aeronaut. Res. Council Paper 29929 HMT 182,* London, 1967; also, Imperial College, Department of Mechanical Engineering *Rep. EHT/TN/8,* London, 1967.

85. W. K. Burns and J. L. Stollery, The Influence of Foreign Gas Injection and Slot Geometry on Film Cooling Effectiveness, *Int. J. Heat Mass Transfer,* **12**:935 (1969).

86. D. E. Metzger and D. D. Pletcher, Evaluation of Heat Transfer for Film Cooled Turbine Components, *J. Aircraft* **8**:33–38 (1971).

87. B.R. Pai and J. H. Whitelaw, Imperial College, Department of Mechanical Engineering *Rep. EHT TN/A/15,* London, 1969.

88. J. J. Williams, "The Effect of Gaseous Film Cooling on the Recovery Temperature Distribution in Rocket Nozzles," Ph.D. thesis, University of California, Davis, 1969.

89. L. Matthews and J. H. Whitelaw, Film Cooling Effectiveness in the Presence of a Backward-Facing Step, *J. Heat Transfer, Trans. ASME,* **95**:135 (1973).

90. S. Kikkawa and M. Fujii, Experimental and Theoretical Investigation on Two Dimensional Film Cooling of a Flat Plate, *Heat Transfer Jpn. Res.,* **8**(3):52 (1973).

91. G. W. Haering, Film Cooling in Adverse Pressure Gradients, *Int. J. Heat Mass Transfer,* **19**:117 (1976).

92. Ko Shao Yen et al., Experimental Investigation of the Turbulent Film Effectiveness over an Adiabatic Flat Plate, *Eng. Thermophys. China,* **1**:243 (1980).

93. J. H. Chin, S. C. Skirvin, L. E. Hayes, and F. Burggraf, Film Cooling with Multiple Slots and Louvers, *J. Heat Transfer,* **83**:281 (1961).

94. V. M. Repukhov and K. A. Bogachuk-Kozachuk, Film Cooling and Film Heating of a Rotating Cylindrical Surface with Flat Blades, *Heat Transfer Sov. Res.,* **8**(1):56 (1976).

95. V. M. Repukhov and K. A. Bogachuk-Kozachuk, Effect of Swirl of the Main Air Flow on the Efficiency of Film Cooling in Axisymmetric Air Flow past a Rotating Cylinder, *Heat Transfer Sov. Res.,* **9**(2):100 (1977).

96. R. J. Goldstein, E. R. G. Eckert, and D. J. Wilson, Film Cooling with Normal Injection into a Supersonic Flow, *J. Eng. Ind., Trans. ASME,* **90**:584 (1968).

97. R. J. Goldstein and M. Y. Jabbari, Film Cooling Effectiveness with Helium and Refrigerant-12 Injected into a Supersonic Flow, *AIAA J.,* **8**:2273 (1970).

98. D. J. Wilson and R. J. Goldstein, Effect of Film Cooling Injection on Downstream Heat Transfer Coefficients in High Speed Flow, *J. Heat Transfer, Trans. ASME,* **95**:505 (1973).

99. R. J. Goldstein, E. R. G. Eckert, F. K. Tsou, and A. Haji-Sheikh, Film Cooling with Air and Helium Injection through a Rearward-Facing Slot into a Supersonic Air Flow, *AIAA J.,* **4**:981 (1966).

100. A. M. Cary, Jr., and J. N. Hefner, Film Cooling Effectiveness and Skin Friction in Hypersonic Turbulent Flow, *AIAA J.,* **10**:1188 (1972).

101. J. N. Hefner, "Effect of Geometry Modifications on Effectiveness of Slot Injection in Hypersonic Flow, *AIAA J.,* **14**:817 (1976).

102. B. E. Richards and J. L. Stollery, Laminar Film Cooling Experiments in Hypersonic Flow, *J. Aircraft,* **16**:177 (1979).

103. R. J. Goldstein, E. R. G. Eckert, and J. W. Ramsey, Film Cooling with Injection through Holes: Adiabatic Wall Temperatures Downstream of a Circular Hole, *J. Eng. Power Trans. ASME,* **90**:384–395 (1968).

104. J. W. Ramsey and R. J. Goldstein, Interaction of a Heated Jet with a Deflecting Flow, *J. Heat Transfer, Trans. ASME, ser. C,* **92**:365–372 (1971) (see also *NASA CR-72613* and University of Minnesota Heat Transfer Laboratory *TR No. 92,* April 1970).

105. R. J. Goldstein, E. R. G. Eckert, V. L. Eriksen, and J. W. Ramsey, Film Cooling Following Injection through Inclined Circular Tubes, *Israel J. Technol.,* **8**(1–2):145–154 (1970).

106. V. L. Eriksen, "Film Cooling Effectiveness and Heat Transfer with Injection through Holes," Ph.D. thesis, University of Minnesota, 1971 (also *NASA CR-72991* and University of Minnesota *HTL TR No. 102*).

107. M. Y. Jabbari and R. J. Goldstein, Effect of Mainstream Acceleration on Adiabatic Wall

Temperature and Heat Transfer Downstream of Gas Injection, *Proc. Fifth Int. Heat Transfer Conf.,* Tokyo, 1974.

108. R. J. Goldstein, E. R. G. Eckert, and F. Burggraf, Effects of Hole Geometry and Density on Three-Dimensional Film Cooling, *Int. J. Heat Mass Transfer,* **17**:595–607 (1974).

109. V. L. Eriksen and R. J. Goldstein, Heat Transfer and Film Cooling Following Injection through Inclined Circular Tubes, *J. Heat Transfer, Trans. ASME,* **96**:235 (1974).

110. V. L. Eriksen and R. J. Goldstein, Heat Transfer and Film Cooling Following Normal Injection through a Round Hole, *J. Eng. Power, Trans. ASME,* **9**:329 (1974).

111. R. J. Goldstein, V. L. Eriksen, and J. W. Ramsey, Flow and Temperature Fields Following Injection of a Jet Normal to a Cross Flow, *Proc. Int. Heat Transfer Conf.,* Tokyo, 1974.

112. G. Bergeles, A. D. Gosman, and B. E. Launder, The Near-Field Character of a Jet Discharged Normal to a Main Stream, *J. Heat Transfer, Trans. ASME,* **98**:373 (1976).

113. G. Bergeles, A. D. Gosman, and B. E. Launder, Near Field Character of a Jet Discharged through a Wall at 30° to a Mainstream, *AIAA J.,* **15**:499 (1977).

114. A. Brown and C. L. Saluja, Film Cooling from a Single Hole and a Row of Holes of Variable Pitch to Diameter Ratio, *Int. J. Heat Mass Transfer,* **22**:525 (1979).

115. C. Liess, Experimental Investigation of Film Cooling with Ejection from a Row of Holes for the Application to Gas Turbine Blades, *J. Eng. Power, Trans. ASME,* **97**:21 (1975).

116. N. W. Foster and D. Lampard, Effects of Density and Velocity Ratio on Discrete Hole Film Cooling, *AIAA J.,* **13**:1112 (1975).

117. K. Kadotani and R. J. Goldstein, On the Nature of Jets Entering a Turbulent Flow: Part B, Film Cooling Performance, *Proc. Tokyo Gas Turbine Cong.,* p. 55, 1977.

118. K. Kadotani and R. J. Goldstein, On the Nature of Jets Entering a Turbulent Flow: Part A, Jet-Mainstream Interaction, *J. Eng. Power, Trans. ASME,* **101**:459 (1979).

119. K. Kadotani and R. J. Goldstein, Effect of Mainstream Variables on Jets Issuing from a Row of Inclined Holes, *J. Eng. Power, Trans. ASME,* **101**:298 (1979).

120. N. W. Foster and D. Lampard, The Flow and Film Cooling Effectiveness Following Injection through a Row of Holes, *J. Eng. Power, Trans. ASME,* **102**:584 (1980).

121. M. Y. Jabbari and R. J. Goldstein, Effect of Mainstream Acceleration on Adiabatic Wall Temperature and Heat Transfer Downstream of Gas Injection, *Proc. Fifth Int. Heat Transfer Conf.,* p. 249, Tokyo, 1974.

122. J. F. Muska, R. W. Fish, and M. Suo, The Additive Nature of Film Cooling Rows of Holes, *J. Eng. Power, Trans. ASME,* **98**:457 (1976).

123. M. Y. Jabbari and R. J. Goldstein, Adiabatic Wall Temperatures and Heat Transfer Downstream of Injection through Two Rows of Holes, *J. Eng. Power, Trans. ASME,* **100**:303 (1978).

124. W. O. Afejuku, N. Hay, and D. Lampard, The Film Cooling Effectiveness of Double Rows of Holes, *J. Eng. Power, Trans. ASME,* **102**:601 (1980).

125. S. Ito, R. J. Goldstein, and E. R. G. Eckert, Film Cooling of a Gas Turbine, *J. Eng. Power, Trans. ASME,* **100**:476 (1978).

126. A. K. Rastogi and J. H. Whitelaw, The Effectiveness of Three Dimensional Film Cooling Slots, *Int. J. Heat Mass Transfer,* **16**:1665 (1973).

127. I. T. Schvets et al., Effectiveness of Film Cooling of an Adiabatic Wall Downstream of a Perforated Section, *Heat Transfer Sov. Res.,* **5**(3):57 (1973).

128. D. E. Metzger, D. I. Takeuchi, and P. A. Kuenstler, Effectiveness and Heat Transfer with Full Coverage Film Cooling, *J. Eng. Power, Trans. ASME,* **95**:180 (1973).

129. S. Yavuzkurt, R. J. Moffat, and W. M. Kays, Full Coverage Film Cooling—Part 1, Three Dimensional Measurements of Turbulence Structure, *J. Fluid Mech.* **101**:129 (1980).

130. R. S. Colladay and L. M. Russell, Streamline Flow Visualization of Discrete Hole Film Cooling for Gas Turbine Application, *J. Heat Transfer, Trans. ASME,* **98**:245 (1976).

131. R. P. Dring, M. F. Blair, and H. D. Joslyn, An Experimental Investigation of Film Cooling on a Turbine Rotor Blade, *J. Eng. Power, Trans. ASME,* **102**:81 (1980).

132. J. P. Sellars, Jr., Gaseous Film Cooling with Multiple Injection Stations, *AIAA J.,* **1**:2154 (1963).

133. D. E. Metzger, D. I. Takeuchi, and P. A. Kuenstler, "Effectiveness and Heat Transfer with Full Coverage Film Cooling, *J. Eng. Power, Trans. ASME,* **95**:18 (1973).

134. M. W. Rubesin, M. Inouye, and P. G. Parikh, Forced Convection, External Flows, in "Handbook of Heat Transfer Fundamentals" (W. M. Rohsenow, J. P. Hartnett, and E. N. Ganić, eds.), McGraw-Hill, New York, 1985.

NOMENCLATURE

Symbol, Definition, SI Units, English Units

a speed of sound: m/s, ft/s

b slot lip thickness: m, ft

b_1 slot step dimension shown on Fig. 95: mm, in

C Chapman-Rubesin constant $(\rho_w\mu_w)/(\rho_\infty\mu_\infty)$

C_w concentration of injected gas at the wall: kg/m^3, lb_m/ft^3

c_f local skin friction coefficient $2\tau_w/\rho u_\infty^2$

c_p specific heat at constant pressure: $J/(kg\cdot K)$, $Btu/(lb_m\cdot °F)$

c_v specific heat at constant volume: $J/(kg\cdot K)$, $Btu/(lb_m\cdot °F)$

D diameter: m, ft

D_{12} binary diffusion coefficient: m^2/s, ft^2/s

Ec   Eckert number or dissipation parameter (Eq. (29))

F dimensionless blowing rate $G/\rho_\infty u_\infty$

f_w dimensionless blowing rate $\dfrac{\rho_w v_w \sqrt{2\xi}}{C\rho_\infty u_\infty \mu_\infty r_w^j}$

G mass flow of injected coolant per unit area per unit time $\rho_w v_w$ or $\rho_c U_c$: $kg/(m^2\cdot s)$, $lb_m/(h\cdot ft^2)$

g gravitational acceleration: m/s^2, ft/s^2

H step height for rearward-facing slot: m, ft

h local heat transfer coefficient: $W/(m^2\cdot K)$, $Btu/(h\cdot ft^2\cdot °F)$

$\overline{h}$ laterally averaged heat transfer coefficient $(1/S_z)\displaystyle\int_{Sz/2}^{Sz/2} h\,dz$ $W/(m^2\cdot K)$, $Btu/(h\cdot ft^2\cdot °F)$

I momentum flux ratio $\rho_c U_c^2/\rho_\infty u_\infty^2$

i_{lg} latent heat of vaporization: J/kg, Btu/lb_m

K pressure gradient parameter $(\nu/u_\infty^2)(du_\infty/dx)$

k thermal conductivity: $W/(m\cdot K)$, $Btu/(h\cdot ft\cdot °F)$

k_T thermal diffusion ratio

M molecular weight of gas: kg/mol, lb_m/mol

M_1 molecular weight of injected gas: kg/mol, lb_m/mol

M_2 molecular weight of free-stream gas: kg/mol, lb_m/mol

Ma   Mach number u_∞/a

Nu   local Nusselt number hx/k

P pressure: N/m^2, lb_f/ft^2

Pr   Prandtl number $\mu c_p/k$

Pr_t turbulent Prandtl number ϵ_m/ϵ_h

q_w'' heat transferred per unit area per unit time from the wall to the surrounding stream: W/m^2, $\text{Btu}/(\text{h}\cdot\text{ft}^2)$

R universal gas constant: $\text{J}/(\text{mol}\cdot\text{K})$, $\text{Btu}/(\text{mol}\cdot°\text{F})$

Re_c slot Reynolds number $Gs/\mu_c = \rho_c U_c s/\mu_c$

Re_d Reynolds number $U_\infty D/\nu$

Re_x local Reynolds number $u_\infty x/\nu$

Re_Δ enthalpy-thickness Reynolds number $u_\infty \Delta/\nu$

Re_θ momentum-thickness Reynolds number $u_\infty \theta/\nu$

r recovery factor defined in Eq. (10)

r_w radial distance from axis of symmetry to surface: m, ft

S_x spacing between holes in the x direction: m, ft

S_z spacing between holes in the z direction: m, ft

S_c Schmidt number ν/D_{12}

St local Stanton number $h/\rho c_p u_\infty$

s slot exit dimension: m, ft

T thermodynamic (absolute) temperature: K, °R

T_{aw} adiabatic wall temperature (see footnote pp. 1-4): K, °R

T_0 total thermodynamic temperature of free stream: K, °R

T_{0c} total thermodynamic temperature of injected coolant: K, °R

T_w^+ ratio of absolute wall temperature to absolute total temperature of free stream, T_w/T_0

Tu turbulence intensity of main stream $(\overline{u'^2})^{1/2}/u_\infty$

U_c mean velocity of coolant at exit of slot or hole: m/s, ft/s

U_∞ approach velocity of free stream: m/s, ft/s

u velocity component parallel to surface: m/s, ft/s

u_∞ free-stream velocity: m/s, ft/s

u' velocity fluctuation in free stream: m/s, ft/s

$\overline{u'^2}$ mean value of u'^2: m^2/s^2, ft^2/s^2

v velocity component normal to surface: m/s, ft/s

v_{zw} velocity normal to surface of cone evaluated at the surface (Fig. 61): m/s, ft/s

x distance along the surface: m, ft

x_c distance in the main flow direction measured from the film cooling slot or hole: m, ft

x_0 unheated solid starting length (Fig. 17): m, ft

Y mass fraction of injected coolant

y distance normal to surface: m, ft

z distance normal to cone surface (Fig. 61), distance transverse to flow direction (Fig. 117): m, ft

Greek Symbols

α angle: rad, deg

β relates to dimensionless pressure gradient, Eq. (25) or Eq. (28); for low-speed wedge flows is related to wedge angle (Fig. 41)

γ	ratio of specific heats c_p/c_v
Δ	enthalpy thickness, Eq. (34): m, ft
$\delta*$	boundary-layer displacement thickness at leading edge of slot or hole

$$= \int_0^\infty [1 - (u/u_\infty)] \, dy: \text{m, ft}$$

ϵ_h	eddy diffusivity of heat: m²/s, ft²/s
ϵ_m	eddy diffusivity of momentum: m²/s, ft²/s
ζ	dimensionless blowing rate parameter defined in Eq. (44)
η	film cooling effectiveness, Eqs. (43) and (43a)
η_M	film cooling effectiveness evaluated along the position midway between two injection holes

$$\bar{\eta} \quad \text{laterally averaged effectiveness} = (1/S_z) \int_{-S_z/2}^{S_z/2} \eta \, dz$$

$$\theta \quad \text{momentum thickness} = \int_0^\infty (u/u_\infty)[1 - (u/u_\infty)] \, dy: \text{m, ft}$$

μ	dynamic viscosity: N·s/m², lb$_m$/(h·ft)
ν	kinematic viscosity: m²/s, ft²/s
ξ	generalized distance parameter, Eq. (26): kg²/(m²·s²), lb²$_m$/(h²·ft²) (for j = 0); kg²/s², lb²$_m$/h² (for j = 1)
$\bar{\xi}$	normalized distance parameter, Eq. (27)
ρ	density: kg/m³, lb$_m$/ft³
τ_w	local shearing stress: N/m², lb$_f$/ft²
ϕ	angle measured from forward stagnation point: rad, deg
Ω	rotational velocity: rad/s

Subscripts

aw	adiabatic wall conditions
c	refers to injected coolant gas
down	downstream row of holes
up	upstream row of holes
w	wall conditions
0	solid wall conditions
∞	free-stream conditions
	evaluated along centerline of injection hole

Superscripts

j	equals zero for two-dimensional flow, equals unity for axisymmetric flow
m	exponent in generalized flows, Eq. (24)
n	exponent in wedge-type flows $u_\infty = cx^n$
ω	viscosity temperature exponent $\mu \approx T^\omega$
*	all properties to be evaluated at reference conditions
$-$	laterally averaged values

CHAPTER 2

Nonnewtonian Fluids

By Y. I. Cho

J. P. Hartnett

Energy Resources Center
University of Illinois, Chicago

A. INTRODUCTION

1. Overview

It is well known that the addition of small quantities of a high-molecular-weight polymer
to a solvent results in a viscoelastic fluid possessing both viscous and elastic properties.
Toms [1] and Mysels [2] discovered that the friction drag of such a viscoelastic fluid under
turbulent-flow conditions is lower than the value associated with the pure solvent. This
initiated a great deal of interest in the use of small amounts of polymers in various transport systems of liquids. The possible areas of application of drag reduction include the
transport of liquids or liquid-solid mixtures in pipelines, fire-fighting systems, torpedoes
and ships, and rotating surfaces of hydraulic machines.

An understanding of the heat transfer behavior of these nonnewtonian fluids is important inasmuch as most of the industrial chemicals and many fluids in the food processing
and biochemical industries are viscoelastic in nature and undergo heat exchange processes
either during preparation or in their application.

Early reviews by Metzner [3], Porter [4], Skelland [5], Hoyt [6], Dimant and Poreh
[7], and White and Hemmings [8] present broad surveys of hydrodynamic and heat transfer behavior of nonnewtonian fluids including viscoelastic and purely viscous fluids.
Recent reviews by Shenoy and Mashelkar [9] and Cho and Hartnett [10] reflect the considerable progress in the understanding of the heat transfer characteristics of nonnewtonian fluids in external and internal flows made in the last decade.

2. Classification of Nonnewtonian Fluids

Fluids treated in the classical theory of fluid mechanics and heat transfer are the ideal (or
perfect) fluid and the newtonian fluid. The former is completely frictionless so that shear
stress is absent. The latter simply has a linear relationship between the shear stress and
shear rate.

Since World War II, the study of real fluids used in the mechanical and chemical
industries has become increasingly important, mainly because of severe limitations in the
application of ideal and newtonian theories to real situations. Most real fluids exhibit so-

called nonnewtonian behavior, which means that the shear stress is no longer linearly proportional to the velocity gradient.

Metzner [3] classified fluids into three broad groups:

1. Purely viscous fluids
2. Viscoelastic fluids
3. Time-dependent fluids

This classification of fluids is essentially the same as that of Skelland [5].

Newtonian fluids are a subclass of purely viscous fluids. The purely viscous nonnewtonian fluids can be divided into two categories: (1) shear-thinning fluids and (2) shear-thickening fluids. We can describe purely viscous fluids by a constitutive equation of the general form

$$\tau_{ij} = \eta_{ij}(I, II, III)d_{ij} \tag{1}$$

where η is the viscosity of the fluid. Here η is a decreasing function of the invariants I, II, and III of the strain tensor d_{ij} for shear-thinning fluids and an increasing function of those invariants for shear-thickening fluids. Characteristic flow curves of shear-thinning and shear-thickening fluids are shown in Fig. 1. While most nonnewtonian fluids used in the study of drag and heat transfer are shear-thinning, the study of shear-thickening (or dilatant) fluids is very rare. Typical shear-thickening fluids used in laboratories and industries are summarized in a recent paper [11].

While the stress tensor component τ_{ij} for purely viscous fluids can be determined from the instantaneous values of the rate of deformation tensor d_{ij}, the past history of deformation together with the current value of d_{ij} becomes an important factor in determining τ_{ij} for viscoelastic fluids. We also need a constitutive equation to describe stress relaxation and normal stress phenomena. Two general classes of equations have been introduced in viscoelastic fluids with some success:

1. Rate equations (differential type)
2. Integral equations

The details of these constitutive equations can be found elsewhere [12–15].

The rod-climbing effect (or Weissenberg effect), die swell, recoil, tubeless siphon, normal stress difference, drag and heat transfer reduction in turbulent flow, etc., are typical experimental demonstrations to show elastic effects in real flows.

Time-dependent fluids are those for which the components of the stress tensor are a function of both the magnitude and the duration of the rate of deformation at constant temperature and pressure [5]. These fluids are usually classified into two groups: thixotropic fluids and rheopectic fluids, depending on whether the shear stress decreases or increases with time at a given shear rate. Thixotropic and rheopectic behavior are common to slurries and suspensions of solids or colloidal aggregates in liquids. Figure 2 shows the general behavior of these fluids.

The study of nonnewtonian fluids generally starts with the purely viscous fluid case and then proceeds to viscoelastic fluids, which were found to yield substantial reduction in the drag and heat transfer under turbulent-flow conditions. In addition, various fiber suspensions were also found to produce drag and heat transfer reduction under comparable conditions [16–20]. Additives such as soap, salt, base, or acid also influence the drag and heat transfer performance of polymer solutions under turbulent-flow conditions [21–25].

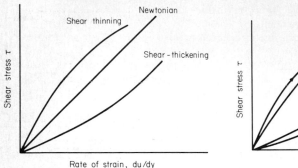

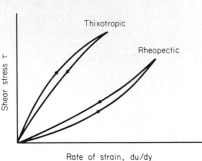

FIG. 1 Flow curves for newtonian fluid and shear-thinning and shear-thickening nonnewtonian fluids [5].

FIG. 2 Flow curves for thixotropic and rheopectic fluids in continuous experiments [5].

3. Empirical Models of Shear Stress

Since the viscosity of many nonnewtonian fluids may change by a factor of 10 to 1000 as the shear rate varies, the shear-rate-dependent viscosity is an important rheological property of nonnewtonian fluids in most engineering design problems dealing with nonnewtonian fluids. Also, in the analysis of internal or external flows with nonnewtonian fluids, the specification of the viscous term in the Navier-Stokes equation is the first step to take into account the nonnewtonian effect.

Based on the experimental observations for simple shear flows, several empirical models to describe the shear-rate-dependent viscosity η have been proposed in the literature [26–31]. Table 1 shows six of the most popular models as well as the characteristic time which can be calculated from each model. For example, for a simple shear flow in which there is only one nonzero velocity component in the flow field [13], we can write:

$$\tau_{yx} = \eta \dot{\gamma}$$
$$\dot{\gamma} = \frac{du}{dy} \tag{2}$$

where η is a function of $\dot{\gamma}$.

These models are often called generalized newtonian fluid models [15, 28] from the fact that they are modifications of the newtonian law of viscosity to allow the viscosity to become a function of shear rate.

4. Rheological Properties of Nonnewtonian Fluids

Since most nonnewtonian fluids used in the study of drag and heat transfer reduction exhibit not only a shear-rate-dependent viscosity but also a normal stress difference, the specification of these viscous and elastic properties is essential for the prediction of the behavior of these fluids.

For moderately and highly concentrated aqueous polymer solutions there are many types of viscometers available to determine the rheological properties, including the rotating viscometer and the familiar capillary-tube viscometer. The rotating viscometer can produce viscosity results over a wide range of shear rate (10^{-2} to 10^3 s^{-1}), while the use of the capillary-tube viscometer is restricted to the high-shear-rate range (greater than

TABLE 1 Generalized Newtonian Models

Model	η	Characteristic time
Power law [26]	$\eta = K(\dot{\gamma})^{n-1}$	None
Bingham [27]	$\eta = \eta_0 + \dfrac{\tau_0}{\dot{\gamma}} \quad \tau \geq \tau_0{}^{*}$ $\dot{\gamma} = 0 \quad \tau \leq \tau_0$	None
Ellis [28]	$\dfrac{1}{\eta} = \dfrac{1}{\eta_0}\left[1 + \left(\dfrac{\tau}{\tau_{1/2}}\right)^{1/n-1} \right]$	$\dfrac{\eta_0}{\tau_{1/2}} \dagger$
Powell-Eyring [29]	$\eta = \eta_\infty + (\eta_0 - \eta_\infty)\left(\dfrac{\sinh^{-1} t_p\dot{\gamma}}{t_p\dot{\gamma}}\right)$	t_p
Sutterby [30]	$\eta = \eta_0 \left(\dfrac{\sinh^{-1} t\dot{\gamma}}{t\dot{\gamma}}\right)^{\alpha}$	t
Carreau A [31]	$\eta = \eta_\infty + (\eta_0 - \eta_\infty)[1 + (t\dot{\gamma})^2]^{n-1/2}$	t

*τ_0 is the yield stress.

†$\tau_{1/2}$ is the value of the shear stress at which $\eta = \eta_0/2$.

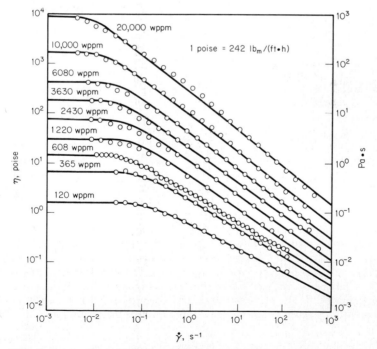

FIG. 3 Steady shear viscosity versus shear rate for polyacrylamide solutions (Separan AP-273), using distilled water as solvent: solid lines are Carreau model B predictions [32].

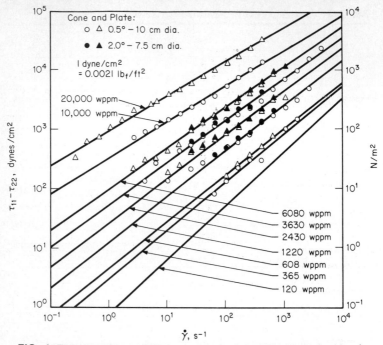

FIG. 4 First normal stress difference versus shear rate for polyacrylamide solutions (Separan AP-273) using distilled water: solid lines are Carreau model B predictions [32].

100 s^{-1}). Although the geometric configuration in each viscometer is quite different, the common goal is to determine the relationship between shear stress and shear rate using available analytical solutions derived under the assumption of laminar-flow conditions. The elastic properties of nonnewtonian fluids—generally the first normal stress differences—are usually measured with a rotating viscometer.

It is worth noting that the accuracy of viscous and elastic property measurement is critically dependent upon the concentration of polymer solutions. Low-concentration polymer solutions (less than 100 wppm) pose a special problem since commonly available rotating viscometers give only limited readings of the viscosity measurements (in the shear rate ranging from 10 to 300 s^{-1}) and relatively poor and inaccurate readings of the first normal stress difference. At present there are no generally accepted procedures for accurately determining the rheological properties of dilute polymer solutions.

Figures 3 and 4 show some typical measurements of rheological properties of aqueous solutions of polyacrylamide (Separan AP-273), including the steady shear viscosity η and the first normal stress difference measured on a model R-18 Weissenberg rheogoniometer [32]. The solid lines are the results of nonlinear curve fitting of these measurements using the Carreau model B [31]. These figures generally demonstrate that the rheological properties are critically dependent on the shear rate as well as the concentration of polymer.

5. Thermophysical Properties of Nonnewtonian Fluids

The physical properties of nonnewtonian fluids necessary for the study of heat transfer are the thermal conductivity, density, specific heat, viscosity, and elasticity. In general,

TABLE 2 Data of Thermal Conductivities k_l, W/(m·K)*

Liquid	c, wppm†	T, °C			
		20°C	30°C	40°C	50°C
Water (current data)	—	0.593	0.612	0.627	0.645
Water [38]	—	0.600	0.615	0.629	0.640
Polyethylene oxide (WSR-301)	100	0.599	0.619	0.630	0.651
	1,000	0.597	0.619	0.638	0.646
	10,000	0.604	0.624	0.634	0.656
Polyacrylamide (Separan AP-273)	100	0.590	0.602	0.611	0.648
	1,000	0.590	0.609	0.616	0.646
	10,000	0.592	0.610	0.632	0.648
Carboxymethyl cellulose (CMC)	1,000	0.576	0.603	0.632	0.648
	10,000	0.583	0.611	0.637	0.665
Carbopol-960	100	0.585	0.614	0.634	0.648
	1,000	0.595	0.606	0.629	0.651
	10,000	0.616	0.644	0.650	0.679
Attagel-40	1,000	0.594	0.605	0.625	0.650
	10,000	0.604	0.614	0.636	0.645
Polyacrylamide (with 4% NaCl)	1,000	0.588	0.604	0.637	0.643

*1 W/(m·K) = 0.5778 Btu/(h·ft·°F).

†wppm = parts per million by weight.

the properties of dilute and concentrated aqueous solutions of drag-reducing materials, other than the viscous and elastic properties, have been taken to be the same as those of water.

Christiansen and Craig [33], Oliver and Jenson [34], and Yoo [35] found experimentally that the thermal conductivities of dilute aqueous solutions of Carbopol-934, carboxymethyl cellulose (CMC), polyethylene oxide, and polyacrylamide are no more than 5 percent lower than those of pure water at corresponding temperature. However, Bellet et al. [36] observed substantial decreases in the thermal conductivity measurements for much higher concentrations of aqueous solutions of Carbopol-960 and CMC (i.e., beyond 10 to 15 percent by weight).

Recently, Lee et al. [37] measured thermal conductivities of various nonnewtonian fluids at four different temperatures using a conventional thermal conductivity cell [35]. These results, shown in Table 2, confirm the common practice of assuming that the thermal conductivity of aqueous polymer solution is equal to that of pure water at corresponding temperature [38] if the concentration of the polymer is less than 10,000 wppm (that is, 1 percent by weight).

6. Use of Reynolds and Prandtl Numbers

In the presentation of experimental results describing the fluid mechanics and heat transfer behavior of nonnewtonian fluids flowing through circular tubes, at least five different definitions of the Reynolds number have been used by various investigators:

1. A generalized Reynolds number Re′, introduced by Metzner and Reed
2. A Reynolds number based on the apparent viscosity at the wall, Re_a
3. A generalized Reynolds number Re_{gen}, derived from the nondimensional momentum equation
4. A Reynolds number based on the solvent viscosity, Re_s
5. A Reynolds number based on the effective viscosity, Re_{eff}

Table 3 summarizes all of these Reynolds numbers as well as the corresponding Prandtl numbers for easy comparison.

As a consequence of the use of different Reynolds numbers from one investigator to another, the comparison of different sets of data becomes quite difficult. The merits and demerits of the five definitions are discussed below. It was pointed out by Skelland [5] that for laminar circular-tube flow of nonnewtonian fluids, $8V/d$ is a unique function of τ_w only. This may be expressed as

$$\tau_w = K'\left(\frac{8V}{d}\right)^n \tag{3}$$

where K' and n vary with $8V/d$ for most polymeric solutions. The generalized Reynolds number of Metzner and Reed [39] is derived from the definition of the Fanning friction factor, which is

$$f = \frac{\tau_w}{\tfrac{1}{2}\rho V^2} \tag{4}$$

Substituting for τ_w from Eq. (3), we obtain for laminar circular-tube flow of nonnewtonian fluids:

$$f = \frac{16}{\rho V^{2-n} d^n / K' 8^{n-1}} \tag{5}$$

If the denominator is defined as the generalized Reynolds number Re′, all the laminar friction data lie on the line $f = 16/\text{Re}'$. Thus, the f versus Re′ graph can be used for checking laminar-flow friction results prior to carrying out turbulent-flow measurements. As a result, the generalized Reynolds number Re′ has had the widest use, particularly in the study of laminar friction factors and heat transfer with nonnewtonian fluids in circular-tube flow.

It should be noted that K' is a function of n. To relate K' to a more fundamental property, the power-law constant K, it is necessary to return to the standard form for the power-law model:

$$\tau_{ij} = K(d_{ij})^n \tag{6}$$

For laminar circular-tube flow, this equation can be evaluated at the wall [5]:

$$\tau_w = K\left(\frac{3n+1}{4n}\frac{8V}{d}\right)^n \tag{7}$$

Therefore, the relation between K' and K becomes

$$K' = K\left(\frac{3n+1}{4n}\right)^n \tag{8}$$

It should be noted that both K and K' are equal to the viscosity η in the newtonian case where $n = 1$.

TABLE 3 Definitions of Reynolds and Prandtl Numbers: Circular-Tube Flow

	Shear stress–shear rate	Reynolds number	Prandtl number	Pe $(= \rho c_p V d / k_l)$
1	$\tau_w = K' \left(\dfrac{8V}{d}\right)^n$	$\mathrm{Re}' = \dfrac{\rho V^{2-n} d^n}{K' 8^{n-1}}$	$\mathrm{Pr}' = \dfrac{c_p K' \left(\dfrac{8V}{d}\right)^{n-1}}{k_l}$	$\mathrm{Re}' \, \mathrm{Pr}'$
2	$\tau_w = \eta_a \dot{\gamma}_w$ $\dot{\gamma}_w = \dfrac{3n+1}{4n}\dfrac{8V}{d}$	$\mathrm{Re}_a = \dfrac{\rho V d}{\eta_a}$	$\mathrm{Pr}_a = \dfrac{\eta_a c_p}{k_l}$	$\mathrm{Re}_a \, \mathrm{Pr}_a$
3	$\tau_{ij} = K(d_{ij})^n$	$\mathrm{Re}_{\mathrm{gen}} = \dfrac{\rho V^{2-n} d^n}{K}$	$\mathrm{Pr}_{\mathrm{gen}} = \dfrac{c_p K \left(\dfrac{V}{d}\right)^{n-1}}{k_l}$	$\mathrm{Re}_{\mathrm{gen}} \, \mathrm{Pr}_{\mathrm{gen}}$
4	$\eta_s = $ solvent viscosity	$\mathrm{Re}_s = \dfrac{\rho V d}{\eta_s}$	$\mathrm{Pr}_s = \dfrac{\eta_s c_p}{k_l}$	$\mathrm{Re}_s \, \mathrm{Pr}_s$
5	$\tau_w = \eta_{\mathrm{eff}} \dfrac{8V}{d}$	$\mathrm{Re}_{\mathrm{eff}} = \dfrac{\rho V d}{\eta_{\mathrm{eff}}}$	$\mathrm{Pr}_{\mathrm{eff}} = \dfrac{\eta_{\mathrm{eff}} c_p}{k_l}$	$\mathrm{Re}_{\mathrm{eff}} \, \mathrm{Pr}_{\mathrm{eff}}$

The second choice of Reynolds number [40] is based on the apparent viscosity at the wall, $\mathrm{Re}_a = \rho V d / \eta_a$, which is a simple modification of the usual definition of Reynolds number for newtonian fluids. The apparent viscosity at the wall is calculated from the following approximating expression of the shear stress at the wall:

$$\tau_w = \eta_a \dot{\gamma}_w \tag{9}$$

where $\dot{\gamma}_w$ becomes $[(3n + 1)/4n]8V/d$ for capillary-tube flow. Accordingly, if one applies the definition of the Fanning friction factor, it can be shown for the laminar circular-tube flow that

$$f = \frac{3n+1}{4n}\frac{16}{\mathrm{Re}_a} \tag{10}$$

demonstrating that f is a function not only of Re_a but also of n. In other words, Re_a does not provide a unique line for experimental friction data in laminar flow.

The third approach, using $\mathrm{Re}_{\mathrm{gen}}$, is rarely used in the study of friction and heat transfer in circular-tube flows since $\mathrm{Re}_{\mathrm{gen}}$ does not have any advantage over Re' or Re_a for this case. However, for nonviscometric flows such as noncircular-tube flows or flow over a sphere or a cylinder, the use of $\mathrm{Re}_{\mathrm{gen}}$ defined as $\rho V^{2-n} d^n / K$ is prevalent [41, 42].

Investigators studying the drag-reducing phenomenon in viscoelastic fluids often used Re_s and $\mathrm{Re}_{\mathrm{eff}}$. The former is generally valid only for dilute polymer solutions, in which case the solution viscosity is quite close to that of the solvent. The use of $\mathrm{Re}_{\mathrm{eff}}$ seems inappropriate in the study of drag and heat transfer because it does not represent any physical property of nonnewtonian fluids, although it produces a unique reference line for experimental friction data in laminar flow:

$$f = \frac{16}{\mathrm{Re}_{\mathrm{eff}}} \tag{11}$$

In all five cases the corresponding Prandtl number is defined to be such that the product of the Reynolds and Prandtl numbers yields the Peclet number, $\rho c_p V d / k_l$.

In summary, for the experimental or analytical studies of drag and heat transfer behavior with nonnewtonian fluids in laminar pipe flow, the use of Re' or Re_a and the corresponding Prandtl number is recommended.

However, in the case of turbulent pipe flow, the presentation of heat transfer results based on Re_a is more practical since it allows comparisons of experimental results with predictions from analytical studies. In part, this results from the fact that most analytical studies of friction and heat transfer for nonnewtonian fluids in turbulent pipe flow have been carried out under the assumption of constant viscosity in the radial direction at a fixed flow rate to avoid mathematical complexity. Therefore, most of the turbulent heat transfer results will be presented as a function of the Reynolds number based on the apparent viscosity at the wall, Re_a.

7. Use of the Weissenberg Number

In dealing with viscoelastic fluids in turbulent flow it is necessary to introduce a dimensionless number to take into account the fluid elasticity [43–47]. Either the Deborah or the Weissenberg number, both of which have been used in fluid mechanical studies, satisfies this requirement. These dimensionless groups are defined as follows:

$$De = \frac{t}{t_F} \tag{12}$$

$$Ws = t\frac{V}{d} \tag{13}$$

where t is a characteristic time of the fluid and a measure of the elasticity of the fluid, t_F is a characteristic time of flow, and V/d is a characteristic shear rate. In this chapter, the Weissenberg number will be used to specify the dimensionless elastic effects.

The evaluation of the Weissenberg number requires the determination of the characteristic time of the fluid. This can be accomplished by combining the use of a generalized newtonian model (see Table 1) with steady shear viscosity data [28, 48]. The characteristic time of a given fluid sample is obtained by determining the value of t which gives the best fit to the measured viscosity data over the complete shear rate range. Among the various models, the Powell-Eyring model [29] and the Carreau model A [31] were found to be the most suitable for aqueous solutions of polyethylene oxide and polyacrylamide [10, 48–50]. It should be noted that the absolute value of the calculated relaxation time differs from one newtonian model to another and consequently it is critical that the procedure for determining t be specified when giving numerical values of the Weissenberg numbers.

B. LAMINAR FLOW IN A CIRCULAR TUBE

1. Velocity Profile and Friction Factor in Fully Developed Laminar Flow

An analysis of laminar heat transfer in nonnewtonian fluids requires an understanding of the hydrodynamic behavior of these fluids. The analytical procedure to obtain the velocity profile for nonnewtonian fluids with the power-law model is exactly the same as for newtonian fluids except for the specification of the shear stress in the momentum equation. The assumption of the power-law fluid particularly in the laminar-flow regime is a good approximation for most nonnewtonian fluids, including viscoelastic fluids, resulting from the fact that the elastic nature does not play any significant role in laminar pipe flow. Under the assumption of a fully developed steady-state laminar flow in a straight circular tube, the momentum equation becomes [51]

$$0 = -\frac{dP}{dx} + \frac{1}{r}\frac{d}{dr}(r\tau_{rx}) \tag{14}$$

With the power-law equation for the shear stress in a circular-tube flow,

$$\tau_{rx} = K\left(\frac{du}{dr}\right)^n \tag{15}$$

and the fully developed velocity profile can be shown as:

$$u = u_{\max}\left[1 - \left(\frac{r}{R}\right)^{(n+1)/n}\right] \tag{16}$$

where

$$u_{\max} = \left(\frac{\tau_w}{K}\right)^{1/n}\frac{R}{1 + 1/n} \tag{17}$$

For n less than 1, this gives a velocity that is flatter than the parabolic profile of newtonian fluids. As n approaches zero, the velocity profile predicted by this equation approaches a plug flow profile. Figure 5 shows the velocity profile generated by Eq. (16) for selected

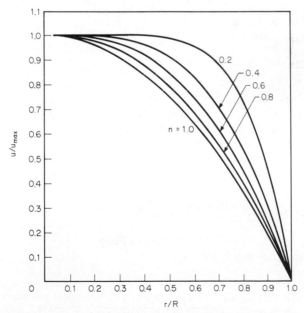

FIG. 5 Velocity profile in fully developed laminar flow for non-newtonian power-law fluids [Eq. (16)].

values of the power-law index n. It should be noted that the velocity profiles given in Fig. 5 are valid in the hydrodynamically fully developed region, where the entrance effect can be neglected.

The Fanning friction factor for fully developed laminar pipe flow can be predicted by the following equation:

$$f = \frac{16}{\text{Re}'} \tag{18}$$

Experimental measurements of pressure drop along a circular tube in the fully developed laminar-flow region confirm this prediction [52, 53]. Equation (18) is recommended for the prediction of pressure drop for nonnewtonian fluids, both purely viscous and visco-elastic, in fully established laminar pipe flow.

2. Hydrodynamic Entrance Length in Laminar Pipe Flow

When a fluid passes into a pipe from a reservoir, it undergoes a development of its velocity profile in the course of its flow through the pipe. The length of pipe required to attain fully developed conditions is called the hydrodynamic entrance length, which is proportional to the Reynolds number for the laminar-flow case. For newtonian fluids, the hydrodynamic entrance length in circular and square-shaped channels is given by the following expression [51, 54, 55]:

$$\frac{L_e}{d} = (0.035 \sim 0.0575) \text{ Re} \tag{19}*$$

For nonnewtonian fluids, Bogue [56] calculated the entrance length theoretically using the von Karman integral method under the assumption of a power-law fluid. Table 4 shows the results for four different n values. Experimental studies generally show that nonnewtonian additives including high-molecular-weight polymers do not affect the entrance length in the laminar region [57–59]. Therefore, it is recommended that Eq. (19) be used to estimate the hydrodynamic entrance length of purely viscous and visco-elastic fluids in laminar channel flow.

3. Fully Developed Laminar Heat Transfer

The fully established laminar heat transfer results from nonnewtonian fluids flowing through a circular tube with a fully developed velocity profile and constant heat flux

TABLE 4 Hydrodynamic Entrance Length in Laminar Pipe Flow [56]

n	$L_e/(d\,\text{Re})$
1.00	0.0575
0.75	0.048
0.50	0.034
0.25	0.017

*For square-shaped channels the hydraulic diameter is taken as the characteristic length.

boundary condition at the wall can be obtained by solving the following energy equation where the heat conduction in the x (axial) direction has been neglected [60]:

$$\rho c_p u \frac{\partial T}{\partial x} = k_l \frac{1}{r} \frac{\partial}{\partial r}\left(r \frac{\partial T}{\partial r}\right) \tag{20}$$

with the boundary conditions:

$$\text{At } r = 0 \qquad T = \text{finite}$$

$$\text{At } r = R \qquad -k_l \left(\frac{\partial T}{\partial r}\right) = q_w'' \tag{21}$$

$$\text{At } x = 0 \qquad T = T_{in}$$

The fully developed velocity profile u, necessary to solve the above energy equation, was calculated for the power-law fluid and presented in Eq. (16).

Applying the separation-of-variables technique to solve the above partial differential equation, the Nusselt number for the constant heat flux case in the thermally fully developed region can be shown to be [15]

$$\text{Nu}_\infty = \frac{8(5n + 1)(3n + 1)}{31n^2 + 12n + 1} \tag{22}$$

This limiting expression for the Nusselt number for nonnewtonian fluids with constant wall heat flux reduces the newtonian value of 4.36 when n is equal to unity. Equation (22) is applicable to the laminar flow of nonnewtonian fluids, both purely viscous and viscoelastic for the constant wall heat flux boundary condition for values of x/d beyond the thermal entrance region. Thus the equation is restricted to the following values of x/d:

$$\frac{x}{d} > 0.04 \text{ Re Pr} \tag{23}$$

The laminar heat transfer results for the constant wall temperature boundary condition were also obtained by the method of separation of variables using the fully developed velocity profile, Eq. (16). The Nusselt number for the constant wall temperature case in the thermally fully developed region was given by the following equation [15]:

$$\text{Nu}_\infty = \beta_1^2 \tag{24}$$

where β_1 is the first (lowest) eigenvalue for the following boundary value problems:

$$\frac{1}{\xi} \frac{1}{d\xi}\left(\xi \frac{dZ_i}{d\xi}\right) + \beta_1^2 \left(\frac{3n + 1}{n + 1}\right)(1 - \xi^{1/n+1})Z_i = 0 \tag{25}$$

with

$$Z_i(1) = 0 \qquad \text{and} \qquad Z_i(0) = 0 \tag{26}$$

where $\xi = r/R$. The values of the Nusselt number for $n = 1.0$, $\frac{1}{2}$, and $\frac{1}{3}$ calculated by Lyche and Bird [61] are 3.657, 3.949, and 4.175, respectively. These values of the Nusselt number are equally valid for purely viscous and viscoelastic fluids for the constant wall temperature case provided that the thermal conditions are fully established. Thus the values are restricted to the following condition:

$$\frac{x}{d} > 0.03 \text{ Re Pr} \tag{27}$$

4. Laminar Heat Transfer in the Thermal Entrance Region

a. Theoretical Studies*

Bird [62] extended the newtonian solution to purely viscous power-law fluids and obtained a series-form solution for the laminar heat transfer in the thermal entrance region for the constant heat flux boundary condition. Bird et al. [51] applied the Lévêque approach [63, 64] to calculate the laminar heat transfer results for the constant wall heat flux as well as the constant wall temperature case for power-law fluids.

These two theoretical methods of predicting the laminar heat transfer results in the thermally developing region for temperature-independent power-law fluids yield almost identical results; they can be expressed very accurately by the following asymptotic relationships [62–69]:

Local Nusselt number–constant wall heat flux:

$$\mathrm{Nu}_x = 1.41 \left(\frac{3n + 1}{4n}\right)^{1/3} \mathrm{Gz}^{1/3} \qquad \text{subject to } \mathrm{Gz} > 25\pi \qquad (28)$$

Local Nusselt number–constant wall temperature:

$$\mathrm{Nu}_x = 1.16 \left(\frac{3n + 1}{4n}\right)^{1/3} \mathrm{Gz}^{1/3} \qquad \text{subject to } \mathrm{Gz} > 33\pi \qquad (29)$$

It is interesting to note that the nonnewtonian effect has been taken into account by a simple multiplication of the corresponding newtonian result by $[(3n + 1)/4n]^{1/3}$.

It should be noted that the mean value of the Nusselt number at any position along the tube is equal to 1.5 times the local values given in Eqs. (28) and (29):

Mean Nusselt number–constant wall heat flux:

$$\mathrm{Nu}_m = 2.11 \left(\frac{3n + 1}{4n}\right)^{1/3} \mathrm{Gz}^{1/3} \qquad \text{subject to } \mathrm{Gz} > 25\pi \qquad (28a)$$

Mean Nusselt number–constant wall temperature:

$$\mathrm{Nu}_m = 1.75 \left(\frac{3n + 1}{4n}\right)^{1/3} \mathrm{Gz}^{1/3} \qquad \text{subject to } \mathrm{Gz} > 33\pi \qquad (29a)$$

b. Experimental Studies

Numerous experimental studies of the heat transfer performance of nonnewtonian fluids in laminar flow with fully etablished velocity profile have been reported in the literature. The empirical correlations of Mizushina et al. [70], Mahalingam et al. [71], and Oliver and Jenson [34] provide a correction for temperature-dependent viscosity [72], while those of Mahalingam et al. [71] and Oliver and Jenson [34] account for natural convection, which can be significant in laminar flow for less viscous nonnewtonian fluids.

Based on these empirical observations, the local heat transfer coefficients for nonnewtonian fluids for large bulk-to-wall temperature conditions (taking into account the presence of significant natural-convection effects) can be predicted by the following equation [34, 70, 71, 73]:

$$\mathrm{Nu}_x = a \left(\frac{3n + 1}{4n}\right)^{1/3} [\mathrm{Gz} + 0.0083(\mathrm{Gr}\,\mathrm{Pr})_w^{0.75}]^{1/3} \left(\frac{K_b}{K_w}\right)^{0.14} \qquad (30)$$

*In this section it is assumed that a fully developed velocity profile exists.

where

$$a = \begin{cases} 1.41 & \text{for constant wall heat flux} \\ 1.16 & \text{for constant wall temperature} \end{cases}$$

The definition of K is given in Eq. (6), the subscript b designates bulk fluid conditions, and the subscript w corresponds to wall conditions for both temperature and shear rate.

The mean value of the Nusselt number for nonnewtonian fluids at any position along the tube is approximately 1.5 times the local value given in Eq. (30).

Mean Nusselt number:

$$\mathrm{Nu}_m = a' \left(\frac{3n + 1}{4n} \right)^{1/3} [\mathrm{Gz} + 0.0083(\mathrm{Gr\ Pr})_w^{0.75}]^{1/3} \left(\frac{K_b}{K_w} \right)^{0.14} \tag{30a}$$

where

$$a' = \begin{cases} 2.11 & \text{for constant wall heat flux} \\ 1.75 & \text{for constant wall temperature} \end{cases}$$

For relatively small bulk-to-wall temperature differences, Eqs. (28) and (29) are recommended for the prediction of local heat transfer. This is supported by a number of investigations [10, 35, 62, 74] as shown in Fig. 6.

It should be emphasized that the laminar heat transfer measurements for viscoelastic fluids such as concentrated aqueous solutions of polyacrylamide and polyethylene oxide show results identical to those for a purely viscous fluid [10], supporting the conclusion that the elastic nature of a viscoelastic fluid does not play any role in laminar pipe flow.

C. TURBULENT FLOW—PURELY VISCOUS FLUIDS IN CIRCULAR-TUBE FLOW

1. Hydrodynamics

A major advance in the hydrodynamic study of nonnewtonian fluids in the turbulent-flow region was made by Dodge and Metzner [57], who proposed the following turbulent pipe flow correlation to predict the friction factor for purely viscous fluids:

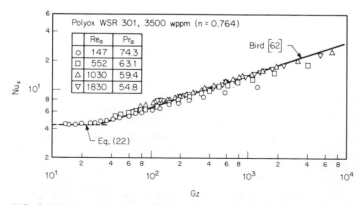

FIG. 6 Experimental results of laminar heat transfer for constant wall heat flux boundary conditions [10].

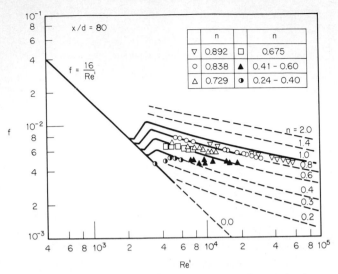

FIG. 7 Experimental pressure-drop measurements for purely viscous nonnewtonian fluids by Yoo [35]: dashed lines extrapolated from Eq. (31).

$$\frac{1}{\sqrt{f}} = \frac{4.0}{n^{0.75}} \log_{10}(\text{Re}' \, f^{(2-n)/2}) - \frac{0.4}{n^{1.2}} \tag{31}$$

where the use of this equation is subject to the restriction that $(\text{Pr} \, \text{Re}^2)f > 5 \times 10^5$.

Experimental measurements [35] for aqueous solutions of Carbopol and slurries of Attagel, both purely viscous fluids, gave good agreement with the predictions of the Dodge-Metzner equation, as shown in Fig. 7.

The hydrodynamic entrance length for purely viscous fluids in turbulent pipe flow is approximately the same as for newtonian fluids, being of the order of 10 to 15 pipe diameters [35].

2. Heat Transfer

Metzner and Friend [75] measured turbulent heat transfer rates with aqueous solutions of Carbopol, corn syrup, and slurries of Attagel in circular-tube flow. They developed a semitheoretical correlation to predict the Stanton number for purely viscous fluids as a function of the friction factor and Prandtl number, applying Reichardt's general formulation for the analogy between heat and momentum transfer in turbulent flow:

$$\text{St} = \frac{f/2}{1.2 + 11.8(f/2)^{1/2}(\text{Pr} - 1)\,\text{Pr}^{-1/3}} \tag{32}$$

where f is as given in Eq. (31). The use of Eq. (32) is limited to $(\text{Pr} \, \text{Re}^2)f > 5 \times 10^5$ and to a Prandtl number range of 0.5 to 600.

A simple correlation has been given by Yoo [35], who compared his results for Carbopol and Attagel solutions with those of previous investigators as shown on Fig. 8. Yoo's empirical equation for predicting turbulent heat transfer for purely viscous fluids is given by

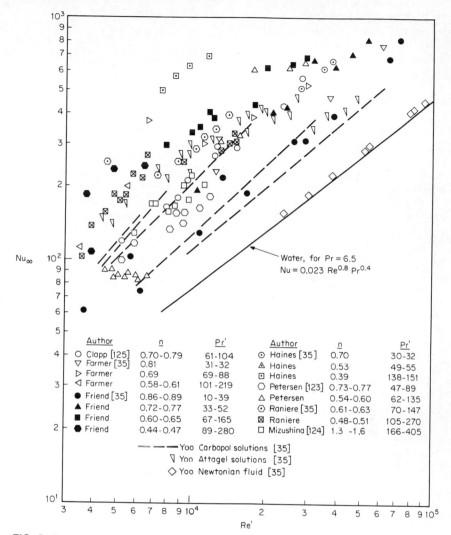

FIG. 8 Experimental heat transfer measurements for purely viscous nonnewtonian fluids [35].

$$St = 0.0152 \, Re_a^{-0.155} \, Pr_a^{-2/3} \qquad (33)$$

This equation describes all of the data with a mean deviation of 2.3 percent. It is recommended that Eq. (33) be used to predict the heat transfer for purely viscous fluids in turbulent pipe flow for values of the power-law exponent n between 0.2 and 0.9 and over the Reynolds number range from 3000 to 90,000.

The thermal entrance lengths for purely viscous nonnewtonian fluids in turbulent pipe flow are of the order of 10 to 15 pipe diameters, the same order of magnitude as for newtonian fluids [76].

D. TURBULENT FLOW—VISCOELASTIC FLUIDS IN CIRCULAR-TUBE FLOW

1. Hydrodynamics

The hydrodynamic behavior of viscoelastic fluids in turbulent pipe flow is quite different from that of the solvent or of a purely viscous nonnewtonian fluid. The friction drag of such a viscoelastic fluid under turbulent-flow conditions is substantially lower than the values associated with the pure solvent or purely viscous nonnewtonian fluids. In general, for turbulent channel flow, this drag reduction increases with higher flow rate, higher polymer molecular weight, and higher polymer concentration. In addition, the diameter of the pipe, the degree of degradation of the polymer, and the chemistry of the solvent are important parameters in the determination of the drag reduction.

It should be noted that the extent of the drag reduction is ultimately limited by a unique asymptote which is independent of the polymer concentration, the solvent chemistry, or the degree of polymer degradation and is solely dependent on the dimensionless axial distance x/d and the Reynolds number [77]. Since polymer concentration, solvent chemistry, and polymer degradation are related to the fluid elasticity it is postulated that these effects can be incorporated in the dimensionless Weissenberg number and that the friction factor is in general a function of the axial location x/d, the Reynolds number, and the Weissenberg number [45, 50, 78]. However, beyond a certain critical value of the Weissenberg number, $(Ws)_f^*$, the friction factor reaches a minimum asymptote value which is dependent solely on the axial distance x/d and the Reynolds number.

In operational terms this can be expressed by the following functional relationships:

$$f = f\left(\frac{x}{d}, \text{Re}_a, \text{Ws}\right) \qquad \text{for Ws} < (\text{Ws})_f^* \tag{34}$$

$$f = f\left(\frac{x}{d}, \text{Re}_a\right) \qquad \text{for Ws} > (\text{Ws})_f^* \tag{35}$$

This behavior can be seen in Fig. 9, which shows the fully established turbulent friction factor as a function of Reynolds number Re_a for concentrations ranging from 10 to 1000 wppm of polyacrylamide in Chicago tap water. This series of measurements, which were taken in a tube 1.30 cm in diameter, revealed that the hydrodynamic entrance length varied with concentration, reaching a maximum of 100 pipe diameters at the higher concentrations. Therefore, the friction factors shown in Fig. 9 were measured at values of x/d greater than 100. The asymptotic friction factor is reached at concentrations of approximately 50 wppm of polyacrylamide in tap water for the tube diameters used in the test program [50, 93].

The steady shear viscosity measurements of the solutions used in the study of the friction factor behavior are given in Fig. 10. For concentrations ranging from 50 to 1000 wppm the viscosity is shear-rate-dependent. The viscosities for 10-wppm polyacrylamide solutions are relatively independent of shear rate.

Relaxation times can be calculated for each of the polyacrylamide solutions used in the measurements shown on Figs. 9 and 10. This may be accomplished by combining the experimentally measured viscosity results with an appropriate generalized newtonian model containing relaxation time as a parameter. The Powell-Eyring model [29] has been used to fit the data, and the resulting values of the relaxation time t_p are shown in the table on Fig. 10. As expected, the relaxation times increase with increasing concentration. The measured fully established friction factors of Fig. 9 are shown in Fig. 11 as a function

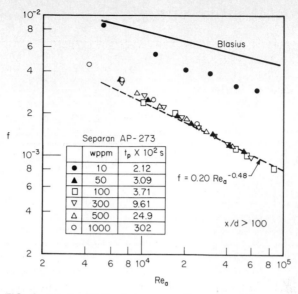

FIG. 9 Fanning friction factor versus Re_a measured in once-through flow system with polyacrylamide (Separan AP-273) solutions. t_p is the characteristic time calculated from the Powell-Eyring model.

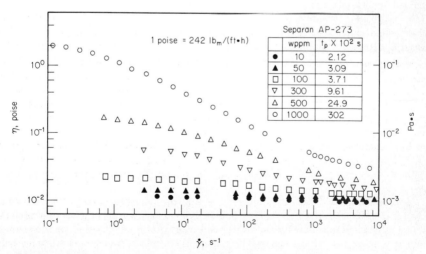

FIG. 10 Steady-shear-viscosity measurements for polyacrylamide (Separan AP-273) solutions from Weissenberg rheogoniometer and capillary-tube viscometer. t_p is the characteristic time calculated from the Powell-Eyring model [50].

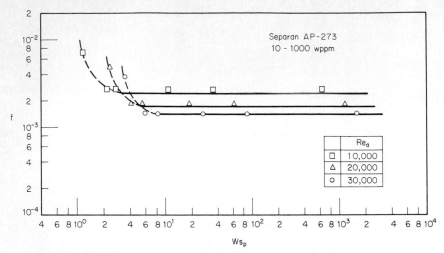

FIG. 11 Friction factor results in once-through flow system as a function of the Weissenberg number: solid lines, predictions from Eq. (36).

of the Weissenberg number based on the Powell-Eyring relaxation time for fixed values of the Reynolds number for the aqueous solutions of polyacrylamide. The critical Weissenberg number for friction, $(Ws_p)_f^*$, is seen to be of the order of 5 to 10. When the Weissenberg number exceeds 10, it is clear that the fully developed friction factor is a function only of the Reynolds number.

Figures 12 and 13 show the lower asymptotic values of the fully developed friction factors for highly concentrated aqueous solutions of polyacrylamide and polyethylene oxide as a function of the Reynolds numbers Re_a and Re', respectively [53, 79]. These measurements, taken at values of x/d greater than 100, were obtained in tubes of 0.98, 1.30, and 2.25 cm inside diameter. These two figures bring out the fact that the laminar-flow region extends to values of the Reynolds number, Re_a or Re', of 5000 to 6000. Beyond a Reynolds number of 6000 the flow may be considered to be turbulent.

The results on Fig. 12 may be correlated by the following simple expression [79]:

$$f = 0.20 \, Re_a^{-0.48} \tag{36}$$

Figure 13, presented in terms of the generalized Reynolds number, validates the measurement program by demonstrating that the laminar-flow measurements are in agreement with the theoretical predictions, $f = 16/Re'$. In the turbulent region above a value of Re' equal to 6000, the experimental data are correlated by the following expression [53, 79]:

$$f = 0.332(Re')^{-0.55} \tag{37}$$

It is recommended that either Eq. (36) or Eq. (37) be used to predict the fully developed friction factor (that is, x/d greater than 100) of viscoelastic aqueous polymer solutions in turbulent pipe flow for Reynolds numbers greater than 6000 and for Weissenberg numbers above the critical value. The critical Weissenberg number for aqueous polyacrylamide solutions based on the Powell-Eyring relaxation time is of the order of 5 to 10 [50]. In the absence of experimental data for other polymers this value should be used for other viscoelastic fluids with the appropriate caution.

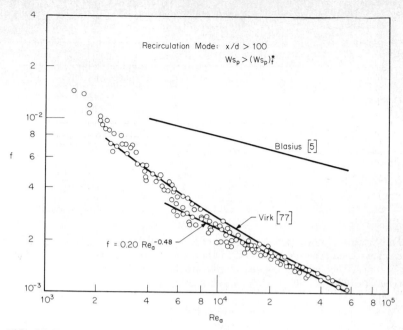

FIG. 12 Fanning friction factor versus Re_a with polyethylene oxide and polyacrylamide solutions.

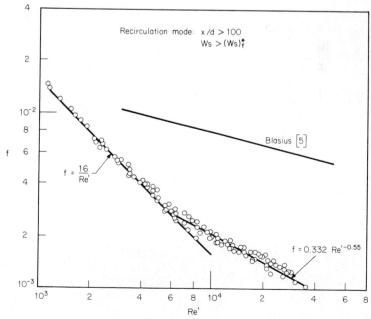

FIG. 13 Fully established friction factor versus Re_a with concentrated polyethylene oxide and polyacrylamide solutions.

TABLE 5 Various Techniques Used in the Local Velocity Measurements with Dilute Polymer Solutions

	Method	Polymer
Seyer [80]	Bubble tracer method	Polyacrylamide AP-30, 1000 wppm
Khabakhpasheva and Perepelitsa [81]	Stroboscopic flow visualization	Polyacrylamide, 120 wppm
Rudd [82]	Laser anemometry	Polyacrylamide AP-30, 100 wppm
Arunachalam et al. [83]	Dye injection	Polyethylene oxide coagulant, 5.5 wppm

Direct measurements of the velocity profile corresponding to asymptotic friction conditions have been reported by several investigators. Table 5 summarizes the techniques and polymers used in obtaining the velocity profiles [80–83]. The velocity measurements reported by Seyer [80], Khabakhpasheva and Perepelitsa [81], and Rudd [82] are shown in Fig. 14.

On the other hand, the velocity profile has been predicted by using friction-factor results in conjunction with modeling procedures. These phenomenological models for describing the velocity profile in channel flows can be classified into three basic categories [10]: Prandtl's mixing length model, Deissler's continuous eddy diffusivity model, and van Driest's damping factor model. Table 6 summarizes typical results for each approach associated with asymptotic friction factor behavior [77, 84–86]. These results are also shown in Fig. 14.

For the asymptotic friction factor case, all of the above-mentioned velocity profiles, either from direct measurements or from modeling, are in fairly good agreement, as shown

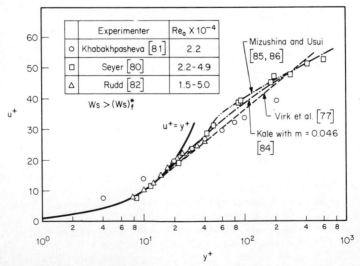

FIG. 14 Experimental measurements of fully established local turbulent velocity profile for the minimum-drag asymptotic case.

TABLE 6 Asymptotic Velocity Profiles Corresponding to the Minimum-Drag Asymptotic Case

Models and investigators	Asymptotic velocity profile
Prandtl's mixing length model, Virk et al. [77]	$u^+ = y^+$, $\qquad\qquad\qquad\qquad\qquad\qquad\qquad\qquad y^+ < 11.6$ $u^+ = 11.7 \ln y^+ - 17.0 \qquad\qquad\qquad\qquad\qquad 11.6 < y^+$
Deissler's continuous eddy diffusivity model, Kale [84]	$u^+ = \displaystyle\int_0^{y^+} \dfrac{dy^+}{1 + m^2 u^+ y^+[1 - \exp(-m^2 u^+ y^+)]} \qquad \begin{array}{l} y^+ < 150 \\[2pt] m = 0.046\dagger \end{array}$ $u^+ = 2.5 \ln y^+ + 30.5 \qquad\qquad\qquad\qquad\qquad\quad 150 < y^+$
Van Driest's damping factor model, Mizushina and Usui [85, 86]	$u^+ = \displaystyle\int_0^{y^+} \dfrac{2(1 - y^+/R^+)\, dy^+}{1 + \{1 + [4 f_n(y^+, R^+) DF_M^2(1 - y^+)/R^+]\}^{1/2}}$ $f_n(y^+, R^+) = 0.4 y^+ - 0.44(y^+)^2/R^+ + 0.24(y^+)^3/R^{+2} - 0.06(y^+)^4/(R^+)^3$ $DF_M = 1 - \exp\{-(y^+/26)[-\alpha + (\alpha^2 + 1)^{1/2}]^{1/2}\}$, where $\alpha = 60$

†Kale's value of $m = 0.06$ has been changed to 0.046 to conform with experimental data with dilute and concentrated polymer solutions reported in Refs. 10, 53, 74, and 93.

in Fig. 14. These investigations show that the laminar sublayer near the wall is thickened and the velocity distribution in the core region is shifted upward from the newtonian mean velocity profile.

It is noteworthy that the use of Pitot tubes and hot-film anemometry, which are applicable to newtonian fluids, is uncertain for drag-reducing viscoelastic fluids. The anomalous behavior of Pitot tubes and hot-film probes in these fluids has been observed by many investigators [87–92].

2. Heat Transfer

Local heat transfer measurements were carried out in the once-through system for the same aqueous polyacrylamide solutions used in the friction factor and viscosity measurements shown in Figs. 9 and 10 [50, 93]. These heat transfer studies involving a constant–heat flux boundary condition required the measurement of the fluid inlet and outlet temperatures and the local wall temperature along the tube. These wall temperatures are presented in terms of a dimensionless wall temperature θ in Fig. 15 for four selected concentrations. Here θ is defined as

$$\theta = \frac{(T_w - T_b)_{x/d}}{(T_w - T_b)_{ex}} \tag{38}$$

For a given concentration, the values of x/d associated with values of θ less than unity are referred to as the thermal entrance region; in this region the thermal boundary layer is not fully developed and the heat transfer coefficient is greater than the value in the thermally developed region.

The figure reveals that the thermal entrance length of the 20-wppm polyacrylamide aqueous solution is almost the same as that of newtonian fluids, which is on the order of 5 to 15 pipe diameters [94–96]. The thermal entrance length increases with increasing concentration (i.e., increasing Weissenberg number), reaching a value of 400 to 500 diameters for the 1000-wppm solutions. It is important to take note of the long entrance lengths of viscoelastic fluids, a fact which has been overlooked in many studies.

The measured dimensionless heat transfer factors j_H (that is, $St\, Pr^{2/3}$) are shown in

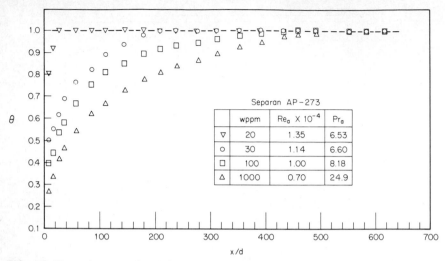

FIG. 15 Thermal entrance length for drag-reducing viscoelastic fluids. Dimensionless wall temperature versus dimensionless axial distance [93].

Fig. 16 as a function of the Reynolds number Re_a for concentrations ranging from 10 to 1000 wppm polyacrylamide [50, 93]. These measurements were made at x/d equal to 430, which corresponds approximately to thermally fully developed conditions as shown in Fig. 15. The asymptotic values of the fully established heat transfer coefficients are reached at a concentration of 500 wppm of polyacrylamide, whereas it required less than 50 wppm to reach the asymptotic friction factor values as shown in Fig. 9.

This is brought out more vividly in Fig. 17, which presents the same data in terms of j_H versus the Weissenberg number based on the Powell-Eyring relaxation time for three different values of the Reynolds number [50]. The critical Weissenberg number for heat transfer, $(Ws_p)_h^*$, is approximately 200 to 250, an order of magnitude higher than the critical Weissenberg number for the friction factor, $(Ws_p)_f^*$. Above a Weissenberg number of 250 the dimensionless heat transfer reaches its minimum asymptotic value, Eq. (40). Note that this critical Weissenberg value has been established for aqueous polyacrylamide solutions, and appropriate care should be used in applying it to other polymers until additional confirmation is forthcoming.

Values of the asymptotic heat transfer factors j_H obtained with highly concentrated aqueous solutions of polyacrylamide and polyethylene oxide are shown on Fig. 18, as a function of the Reynolds number Re_a. These values were measured in tubes of 0.98, 1.30, and 2.25 cm (0.386, 0.512, and 0.886 in) inside diameter in a recirculating-flow loop, with appropriate monitoring to ensure that these values were not changing with time (see Sec. D.4) [53]. The asymptotic turbulent heat transfer data in the thermal entrance region are seen to be a function of the Reynolds number Re_a and of the axial position x/d. The following empirical correlation is derived from the data [10, 93]:

$$j_H = 0.13 \left(\frac{x}{d}\right)^{-0.24} Re_a^{-0.45} \qquad \text{for } \frac{x}{d} < 450 \qquad (39)$$

The fully developed minimum asymptotic heat transfer, which is approximated by the experimental data obtained at x/d equal to 430, is correlated by the following equation [10, 93]:

$$j_H = 0.03 \, Re_a^{-0.45} \qquad \text{for } \frac{x}{d} > 450 \qquad (40)$$

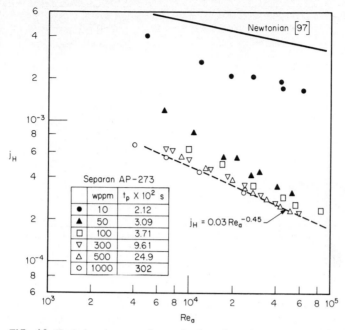

FIG. 16 Turbulent heat transfer results for polyacrylamide solutions measured at $x/d = 430$. t_p is the characteristic time calculated from the Powell-Eyring model.

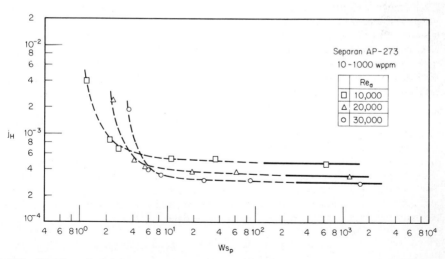

FIG. 17 Dimensionless heat transfer j factor in once-through flow system as a function of the Weissenberg number. Solid lines, predictions from Eq. (40).

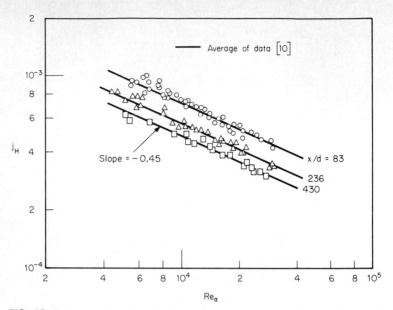

FIG. 18 Experimental results of turbulent heat transfer for concentrated solutions of polyethylene oxide and polyacrylamide in the thermal entrance region.

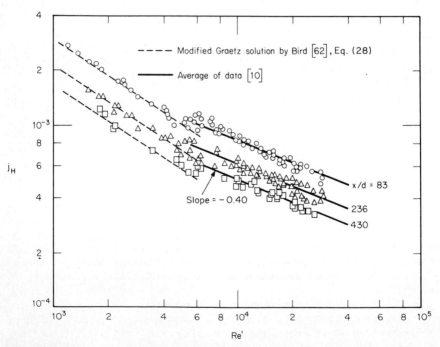

FIG. 19 Experimental results of laminar and turbulent heat transfer for concentrated solutions of polyethylene oxide and polyacrylamide in the thermal entrance region.

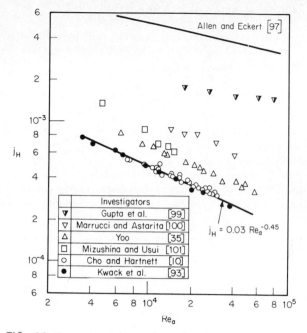

FIG. 20 Comparison of experimental turbulent heat transfer results for drag-reducing viscoelastic fluids.

These same data are shown on Fig. 19 as a function of the generalized Reynolds number Re′. Here it may be noted that the laminar data are in excellent agreement with the theoretical prediction by Bird [62], Eq. (28), lending support to the experimental measurements. Laminar flow extends to a generalized Reynolds number of 5000 to 6000. The empirical correlations resulting from the turbulent-flow data given on Fig. 19 are [10, 98]:

$$j_H = 0.13 \left(\frac{x}{d}\right)^{-0.3} (Re')^{-0.4} \qquad \text{for } x/d < 450 \tag{41}$$

$$j_H = 0.02(Re')^{-0.4} \qquad \text{for } x/d > 450 \tag{42}$$

It is recommended that Eqs. (39) and (40) or equivalently Eqs. (41) and (42) be used to predict the heat transfer performance of viscoelastic aqueous polymer solutions for Reynolds numbers greater than 6000 and for values of the Weissenberg number above the critical value for heat transfer. This critical Weissenberg number for heat transfer based on the Powell-Eyring relaxation time is approximately 250 for aqueous polyacrylamide solutions. Appropriate care should be exercised in using this critical value for other viscoelastic fluids.

Figure 20 supports the conclusion that the experimental results reported by early investigators do not represent minimum asymptotic heat transfer values. This is quite understandable in the cases shown since the test sections of Gupta et al. [99], Marrucci and Astarita [100], Yoo [35], and Mizushina and Usui [101] were 40, 100, 110, and 160 diameters in length, respectively.

3. Failure of the Reynolds-Colburn Analogy

The percentage reduction in friction factor resulting from the addition of a long-chained polymer to water is plotted against the percent reduction in heat transfer coefficient on Fig. 21. Here the percent reduction is defined as follows:

$$\% \text{ heat transfer reduction} = \frac{h_s - h_p}{h_s} \times 100$$

$$\% \text{ friction factor reduction} = \frac{f_s - f_p}{f_s} \times 100$$

where the subscripts s and p designate solvent and polymer, respectively.

The solid line in the figure represents the general trend of the experimental observations of [93] and [101], confirming the fact that the heat transfer reduction always exceeds the friction factor reduction. In particular, the following inequality can be deduced from this figure:

$$j_H < \frac{f}{2} \tag{43}$$

This observation contradicts the common assumption of the validity of the Reynolds or Colburn analogy made in a number of heat transfer studies of viscoelastic fluids [84, 102–103].

4. Degradation

The degradation of the polymer in a viscoelastic polymer solution makes the prediction of the heat transfer and pressure drop extremely difficult, if not impossible, in normal industrial practice. This results from the fact that mechanical degradation, the shearing of the polymer bonds, goes on continuously as the fluid circulates, causing continuous changes in the rheology of the fluid. The elasticity of the fluid is particularly sensitive to

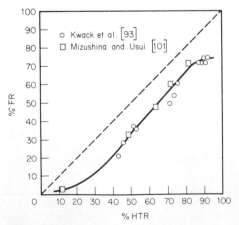

FIG. 21 Comparison of percentage friction reduction and percentage heat transfer reduction.

this mechanical degradation. These changes in the rheology of the fluid ultimately cause changes in the heat transfer and pressure drop.

Notwithstanding the difficulties of accurately predicting the quantitative effects of degradation on the hydrodynamics and heat transfer, it is nevertheless important to qualitatively understand the process if engineering systems are to be designed to handle such fluids.

Systematic studies have been reported on the heat transfer behavior of degrading polymer solutions with highly concentrated polymer solutions: 1000 wppm of polyacrylamide [49, 50] and 1500 wppm of polyethylene oxide [10]. These studies were conducted in test sections with inside diameters of 2.25 cm ($L/d \approx 280$) and 1.30 cm ($L/d \approx 475$). Heat transfer and pressure-drop measurements were carried out at regular time intervals. Although the circulation rate was held approximately constant, periodic flow-rate measurements were carried out using the direct weighing and timing method. Fluid samples were removed at regular time intervals from the flow loop for rheological property measurements in the Weissenberg rheogoniometer (WRG) and in the capillary-tube viscometer.

Figure 22 shows the steady shear viscosity of the polyacrylamide (Separan AP-273) solution as a function of hours of circulation in the flow loop. Chicago tap water was the solvent. This figure brings out very clearly the substantial decrease in the viscosity at low shear rate resulting from the degradation of the polymers, which is accompanied by a decrease in the first normal stress difference and a decrease in the characteristic time [10]. This, in turn, means that a decrease in the Weissenberg number always accompanies degradation. Thus, a circulating aqueous polymer solution experiences a continuing decrease in the Weissenberg number.

The Fanning friction factor f and the dimensionless heat transfer coefficient j_H for the polyacrylamide 1000-wppm solution measured at an x/d of 430 and at the Reynolds number equal to 20,000 [50] are presented in Fig. 23 as a function of hours of circulation. The dimensionless j_H factor is seen to remain relatively constant at its minimum asymp-

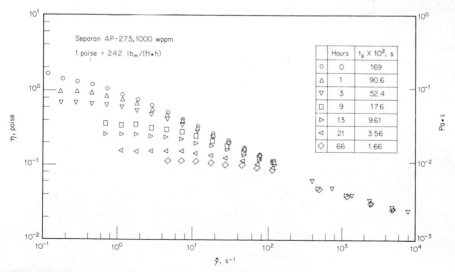

FIG. 22 Degradation effects on steady-shear-viscosity measurements for polyacrylamide 1000-wppm solution at seven different hours of circulation.

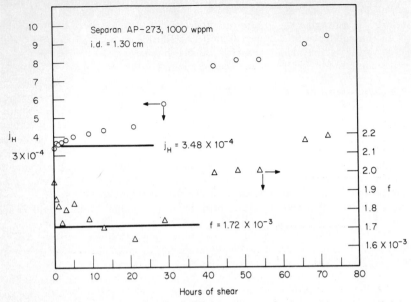

FIG. 23 Fanning friction factor and turbulent heat transfer j factor versus hours of shear for Reynolds number equal to 20,000 and at $x/d = 430$. Separan AP = 273, 1000 wppm. Solid lines are minimum asymptotic values predicted from Eqs. (36) and (40).

totic value until some 3 hours have passed. On the other hand the friction factor does not depart from its asymptotic value until some 30 hours of circulation have occurred. This is consistent with the results of Sec. D.2. Estimates of the critical Weissenberg number based on the Powell-Eyring model yield values that are in good agreement with those given for the once-through system:

Critical Weissenberg number for friction:

$$(\text{Ws}_p)_f^* = 10 \tag{44}$$

Critical Weissenberg number for heat transfer:

$$(\text{Ws}_p)_h^* = 250 \tag{45}$$

Above the corresponding critical Weissenberg number the friction factor and the heat transfer remain at their asymptotic values.

A similar degradation test was conducted with a concentrated polyethylene oxide 1500-wppm solution [10]. The steady shear viscosity showed the same trend as found with the polyacrylamide solution, namely, decreasing zero-shear-rate viscosity resulting in decreasing Weissenberg number as the fluid circulated. The dimensionless friction factor and heat transfer results are shown in Fig. 24 as a function of time. The most interesting feature of this figure is that the departure from asymptotic friction and heat transfer values occurs much more rapidly. For example, the heat transfer coefficient of the polyethylene oxide solution has already increased above its asymptotic value after only 15 minutes of circulation while the polyacrylamide solution required 3 hours of circulation before the heat transfer began to increase. The friction factor began to depart from its asymptotic value after 1 hour for the polyethylene oxide solution while it required 30 hours of circulation for the polyacrylamide solution.

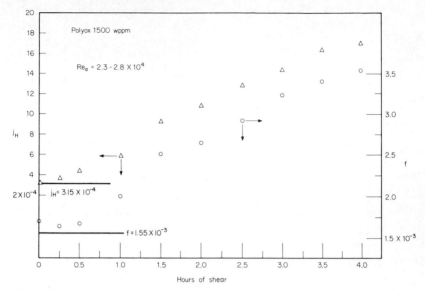

FIG. 24 Degradation effect on friction and heat transfer results for polyethylene oxide 1500-wppm solution at $x/d = 430$. Solid lines, asymptotic predictions from Eqs. (36) and (40).

This behavior may be explained by the fact that the initial elasticity, and consequently the initial Weissenberg number, of the 1500-wppm polyethylene oxide solution is lower than that of the 1000-wppm polyacrylamide solution. Thus, it takes a shorter circulation time to reach the critical Weissenberg number. Preliminary analysis indicates that the critical Weissenberg values for the polyethylene oxide solution are of the same order as for the polyacrylamide solution.

5. Solvent Effects

When an aqueous solution of a high-molecular-weight polymer is used in a practical engineering system, the solvent is generally predetermined by the system. However, the importance of the solvent on the pressure drop and heat transfer behavior with these viscoelastic fluids has often been overlooked. Since the heat transfer performance in turbulent flow is critically dependent upon the viscous and elastic nature of the polymer solution, it is important to understand the solvent effects on the rheological properties of a viscoelastic fluid.

Following the earlier work by Little et al. [23] and Chiou and Gordon [104], Cho et al. [10, 105] measured the rheological properties of the 1000-wppm aqueous solution of polyacrylamide (Separan AP-273) with various kinds of solvents: distilled water, tap water, tap water plus acid or base additives, and tap water plus salt. Figure 25 presents the steady-shear-viscosity data over the shear rate ranging from 10^{-2} to 4×10^{4} s^{-1} using the Weissenberg rheogoniometer and the capillary-tube viscometer. The viscosity in the low shear rate of the 1000-wppm polyacrylamide solution with distilled water is greater than that of the polyacrylamide solution with tap water by a factor of 25. However, when the shear rate is increased the viscosity of the distilled water solution approaches that of the tap water solution. The addition of 100 wppm of NaOH to Chicago tap water results

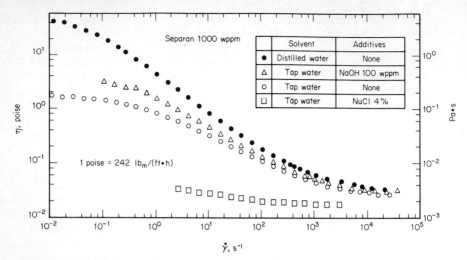

FIG. 25 Steady shear viscosity versus shear rate for polyacrylamide 1000-wppm solutions with four different solvents [105].

in a 100 percent increase of the viscosity in the low-shear-rate range. In contrast, the addition of 4 percent NaCl to the tap water reduced the viscosity of the polyacrylamide solution over the entire range of shear rate by a factor of 4 to 25 depending on the shear rate.

The effect on viscosity of the addition of NaOH, NH_4OH, or H_3PO_4 to Chicago tap

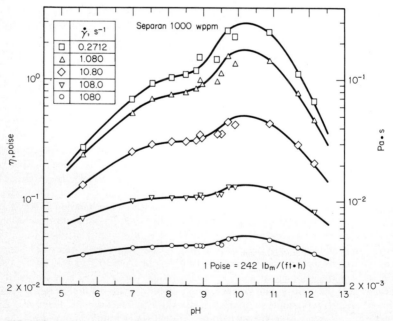

FIG. 26 Steady shear viscosity versus pH for polyacrylamide 1000-wppm solutions with tap water plus acid or base [105].

water also was investigated. The results, as shown in Fig. 26, are presented on viscosity versus pH curves at fixed shear rates [10, 105]. These curves indicate that for base additives there is an optimum pH number (approximately 10) which maximizes the viscosity of the polyacrylamide solution. For acid additives, an increasing concentration of acid is always accompanied by a decrease of viscosity. It is noteworthy that similar observations were made with aqueous solutions of polyethylene oxide.

From the above results together with those of other investigators who used distilled water as a solvent [23, 104], it can be concluded that the rheological properties of polymer solutions may be modified by changing the chemistry of the solvent. It follows that the hydrodynamic and heat transfer performance is sensitive to solvent chemistry.

An example of this is shown on Fig. 27, which presents the dimensionless heat transfer coefficient j_H measured at an x/d of 430 [10, 93]. Three solutions of 20-wppm polyacrylamide (Separan) were used, each having a different solvent: (1) Chicago tap water, (2) Chicago tap water plus 20 wppm NaOH, and (3) Chicago tap water plus 100 wppm NaOH. The figure demonstrates that the addition of 20 and 100 wppm of sodium hydroxide results in substantial reductions in the heat transfer for the dilute polyacrylamide solution, with the 100 wppm showing the larger reduction. It should be noted that there is evidence of some degradation when the Reynolds number exceeds 60,000. Nevertheless, it is clear that the solvent chemistry plays a critical role in the rheological, hydrodynamic, and heat transfer behavior of polymer solutions.

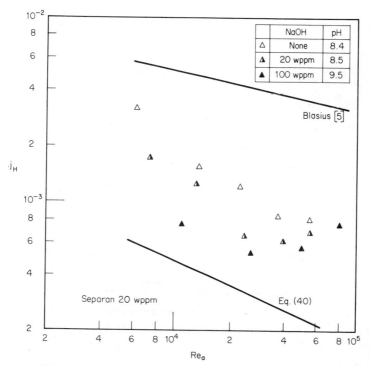

FIG. 27 Solvent effects on turbulent heat transfer performance for a polyacrylamide (Separan AP-273) solution.

E. FAILURE OF THE ANALOGY FOR VISCOELASTIC FLUIDS

1. Introduction

It is well known that for newtonian fluids in turbulent pipe flow, an analogy between momentum, heat, and mass transfer can be drawn and expressed in the following form:

$$j_H = j_D = \frac{f}{2} \tag{46}$$

For drag-reducing polymer solutions, there have been many attempts in the literature to formulate and to apply such an analogy [10]. Most of these works attempted to predict turbulent heat transfer rates for drag-reducing fluids from the use of the friction coefficient measurements. More recently, following the pioneering works of Lin et al. [106] and Hanratty et al. [107, 108] for newtonian fluids, attempts have been made to predict turbulent heat transfer rates from turbulent mass transfer measurements for drag-reducing fluids [109, 110].

To determine the validity of the analogy between heat and mass transfer, it is important to have accurate data for each transport phenomenon representing the fully developed and undegraded polymer solution results. The minimum drag and heat transfer asymptotic lines for drag-reducing viscoelastic fluids have been presented in Eqs. (36) and (40) respectively. Below, the corresponding minimum mass transfer asymptotic results will be examined.

2. Minimum Mass Transfer Asymptote

Compared with the numerous studies relating to momentum and heat transfer phenomena for drag-reducing viscoelastic fluids, relatively few experimental turbulent mass transfer data exist for these fluids. Sidahmed and Griskey [109], Shulman et al. [111], McConaghy and Hanratty [112], and Teng et al. [110] applied the electrochemical method [113] to measure the mass transfer rate at the wall associated with the channel flow of aqueous polymer solutions. In contrast, Virk and Suraiya [114] used both the weight-loss technique [115] and the ultraviolet spectrophotometric technique.

Among these, it is believed that the data of Virk and Suraiya [114], showing the minimum drag asymptote in the pressure-drop measurements, represent the minimum mass transfer asymptote correctly. Their mass transfer measurements were carried out in test sections with L/d equal to 34.5 and 69.0. No difference was found in the mass transfer data obtained in the two different lengths of test section, leading them to conclude that they had fully developed mass transfer at L/d equal to 34.5. Thus, they proposed a minimum mass transfer asymptote for drag-reducing viscoelastic fluids for Reynolds number ranging from 5000 to 35,000:

$$j_D = 0.022 \, Re_a^{-0.29} \tag{47}$$

where j_D is the diffusion (i.e., mass) transfer j factor defined as $St \, Sc^{2/3}$.

3. Comparison of Experimental Results of Momentum, Heat, and Mass Transfer

The empirical results of momentum, heat, and mass transfer for drag-reducing viscoelastic fluids in turbulent pipe flow corresponding to the minimum asymptotic case are shown

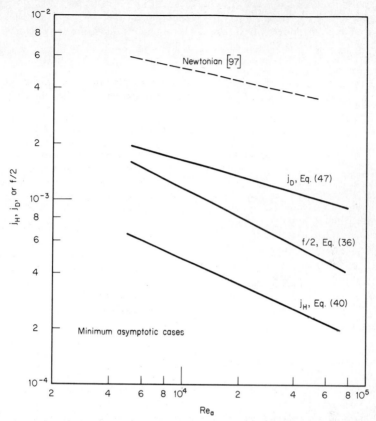

FIG. 28 Comparison of the limiting minimum values of momentum, heat, and mass (or diffusion) transfer for drag-reducing viscoelastic fluids.

in Fig. 28. This figure clearly shows that the mass transfer results are greater than the corresponding heat transfer values by a factor of 3. The experimental values of $f/2$ lie between the heat transfer j factor and mass transfer j factor. The slopes representing the Reynolds number dependency are -0.48, -0.45, and -0.29 for momentum, heat, and mass transfer respectively. These empirical results for the minimum asymptotic cases can be expressed qualitatively with the following inequality:

$$ j_H < \frac{f}{2} < j_D \tag{48} $$

On the basis of these experimental results it can be concluded that the use of the Colburn analogy [60, 116] is not applicable to drag-reducing viscoelastic fluids.

4. Eddy Diffusivities of Momentum, Heat, and Mass Transfer

To further verify the above conclusion on the use of the analogy between momentum, heat, and mass transfer, the approximate values of the eddy diffusivities of momentum, heat, and mass transfer corresponding to the minimum asymptotic cases will be compared.

The eddy diffusivity of momentum corresponding to the minimum asymptotic case was calculated by Kale [84] directly from Deissler's continuous eddy diffusivity model:

$$\frac{\epsilon_M}{\nu} = m^2 u^+ y^+ [1 - \exp(-m^2 u^+ y^+)] \qquad y^+ < 150, \; m = 0.046 \qquad (49)$$

where Kale's value of $m = 0.06$ has been changed to 0.046 to conform with the experimental data [10, 53, 74, 93].

Cho and Hartnett [10] calculated the eddy diffusivity of heat, for drag-reducing viscoelastic fluids, using a successive approximation technique [60], and the result can be expressed in the following polynomial equation with respect to y^+:

$$\frac{\epsilon_H}{\nu} = 2.5 \times 10^{-6} y^{+3} \qquad (50)$$

Shulman and Pokryvailo [117] calculated the eddy diffusivity of mass for viscoelastic fluids corresponding to the minimum asymptotic case, using the data for the developed unsteady-state convective mass transfer behavior of polyethylene oxide (WSR-301) solutions, and the results can be expressed in the following equation:

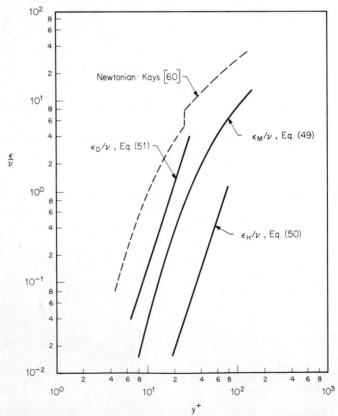

FIG. 29 Comparison of the eddy diffusivities of momentum, heat, and mass (diffusion) transfer for the limiting minimum asymptotic case for drag-reducing viscoelastic fluids.

$$\frac{\epsilon_D}{\nu} = 1.6 \times 10^{-4}y^{+3} \tag{51}$$

Support for this expression is provided in a recent analytical study by Cho and Hartnett [118].

This comparison of eddy diffusivities as shown in Fig. 29 confirms the fact that for drag-reducing viscoelastic fluids, the eddy diffusivity of heat is much smaller than that of momentum. These results for eddy diffusivities of momentum, heat, and mass transfer for viscoelastic fluids can be expressed with the following inequality:

$$\frac{\epsilon_H}{\nu} < \frac{\epsilon_M}{\nu} < \frac{\epsilon_D}{\nu} \tag{52}$$

This is consistent with the inequality, Eq. (48), derived from the empirical results $f/2$, j_H, and j_D for drag-reducing viscoelastic fluids.

In addition, the inequality given in Eq. (52) is consistent with the experimental observations of the hydrodynamic, thermal, and diffusional entrance lengths since the entrance length is inversely proportional to the magnitude of the eddy diffusivity. The entrance lengths of momentum, heat, and mass transfer for viscoelastic fluids corresponding to the minimum asymptotic cases were experimentally observed by Tung et al. [52], by Cho and Hartnett [10], and by Virk and Suraiya [114], respectively. These empirical results of the entrance lengths corresponding to the minimum asymptotes are shown in Table 7 and can be expressed qualitatively with the following inequality:

$$\left(\frac{L_e}{d}\right)_D < \left(\frac{L_e}{d}\right)_M < \left(\frac{L_e}{d}\right)_H \tag{53}$$

Here the subscripts D, M, and H designate diffusion (i.e., mass), momentum, and heat transport, respectively. This is consistent with the inequalities given in Eqs. (48) and (52).

On the basis of these comparisons, it can be concluded that there is no direct analogy between momentum, heat, and mass transfer phenomena for drag-reducing viscoelastic fluids in turbulent pipe flows.

F. SQUARE DUCT

1. Hydrodynamics—Laminar Flow

A variety of noncircular passage geometries, including the square and rectangular duct, etc., have been utilized for internal-flow forced-convection heat transfer applications. Examples include compact heat exchangers and solar collectors, where economy in space and efficiency in heat transfer performance are critically important.

The study of hydrodynamic behavior in a rectangular duct requires a two-dimensional analysis since the axial velocity even in the fully developed region is a function of two independent variables.

TABLE 7 Entrance Lengths: Asymptotic Condition

Transfer mechanism	L_e/d	Re range
Momentum transfer, Tung et al. [52]	100	6000–40,000
Heat transfer, Cho and Hartnett [10]	400–500	6000–40,000
Mass transfer, Virk and Suraiya [114]	35	5000–35,000

TABLE 8 Fanning Friction Factor–Reynolds Number Product for Square ($s = 1$) and Rectangular Ducts

		$f\,\mathrm{Re_{gen}}$‡	
		Wheeler & Wissler [42]	Schechter [119]
n	s†		
1.0	1.0	14.23	14.27
0.9	1.0	11.91	—
0.8	1.0	9.915	—
0.75	1.0	9.055	9.085
0.7	1.0	8.268	—
0.6	1.0	6.883	—
0.5	1.0	5.723	5.755
0.4	1.0	4.743	—
1.0	0.75	—	14.29
1.0	0.5	15.54	15.60
1.0	0.25	—	18.38
0.75	0.75	9.173	9.070
0.75	0.5	9.693	—
0.75	0.25	—	11.11
0.5	0.75	—	5.724
0.5	0.5	—	6.033
0.5	0.25	—	6.641

†When s approaches zero, we have flow between two infinite parallel plates.

‡The original values of $f\,\mathrm{Re}$ from Refs. 42 and 119 have been converted so that they are based on the hydraulic diameter, d_h.

The analytical predictions of the value of $f\,\mathrm{Re_{gen}}$ for nonnewtonian power-law fluids were obtained by Wheeler and Wissler [42] and Schechter [119] using the finite difference technique and the variational principle, respectively. These results obtained by different methods are in excellent agreement.* The analytical results of $f\,\mathrm{Re_{gen}}$ for different values of power-law index n are shown in Table 8, and the numerical results for a square duct can be expressed in the following approximation:

$$f\,\mathrm{Re_{gen}} = 1.8735 \left(\frac{1.7330}{n} + 5.8606 \right)^{n} \qquad 0.4 \leq n \leq 1.0 \qquad (54)$$

Wheeler and Wissler [42] also conducted experimental measurements of laminar friction factor with aqueous solutions of carboxymethyl cellulose (0.28 to 1.0 percent by weight) in a 0.5-in $\times$ 0.5-in (1.27-cm $\times$ 1.27-cm) square duct, which confirmed the validity of the above analytical results.

2. Heat Transfer—Laminar Flow

Chandrupatla and Sastri [120] obtained numerical solutions using the extrapolated Liebmann method for laminar-flow forced-convection heat transfer of nonnewtonian power-

*All results have been converted to $f\,\mathrm{Re_{gen}}$, where f is defined as $\tau_w/(\tfrac{1}{2}\rho V^2)$ and $\mathrm{Re_{gen}}$ is $\rho V^{2-n} d_h^n / K$. Here d_h, the hydraulic diameter, is equal to $4A_c/\mathsf{P}$, where A_c is the cross-sectional area and P is the wetted perimeter.

TABLE 9 Nusselt Numbers for Fully
Developed Velocity and Temperature Profiles
in a Square Duct for Nonnewtonian Power-
Law Fluids [120]

n	Nu_T	Nu_H
1.0	2.975	3.612
0.9	2.997	3.648
0.8	3.030	3.689
0.75	3.050	3.713
0.7	3.070	3.741
0.6	3.120	3.804
0.5	3.184	3.889

law fluids in thermally developed as well as developing regions of a square duct for (1) constant wall temperature and (2) constant axial wall heat flux with uniform peripheral wall temperature.

For hydrodynamically and thermally developed laminar flow of nonnewtonian fluids in a square duct, the limiting Nusselt numbers Nu_T and Nu_H corresponding to the constant wall temperature and constant wall heat flux cases respectively are presented in Table 9. It is observed from Table 9 that Nusselt numbers for nonnewtonian fluids are greater than corresponding values of newtonian fluids by up to 6 to 7 percent.

In the case of a hydrodynamically developed velocity profile at the start of heating, the mean Nusselt numbers in the thermally developing region of a square duct were obtained as functions of Graetz number, $0 \leq \mathrm{Gz} \leq 200$, and the results corresponding to the constant wall temperature and constant wall heat flux cases are presented in Tables 10 and 11 respectively. From a comparison of the numerical solutions, it is concluded that for the same Graetz number and thermal boundary conditions, a nonnewtonian fluid with power-law index less than 1 gives a slightly higher heat transfer coefficient than a newtonian fluid, which is also supported by a recent analytical study by Dunwoody and Hamill [121]. The laminar heat transfer solutions given in Tables 9 and 11 should be used as a lower limit in design since experimental values of the heat transfer coefficients are generally higher due to the secondary flow effects.

TABLE 10 $\mathrm{Nu}_{m,T}$ as a Function of Gz and n for Fully Developed Velocity Profiles in a Square Duct [120]

	$\mathrm{Nu}_{m,T}$						
	Power-law index, n						
Gz	1.0	0.9	0.8	0.75	0.7	0.6	0.5
0	2.975	2.997	3.030	3.050	3.070	3.120	3.184
10	3.514	3.543	3.577	3.597	3.619	3.671	3.739
20	4.024	4.055	4.091	4.112	4.135	4.191	4.263
25	4.253	4.284	4.321	4.343	4.367	4.424	4.499
40	4.841	4.877	4.917	4.941	4.967	5.029	5.110
50	5.173	5.211	5.253	5.278	5.305	5.370	5.455
80	5.989	6.033	6.080	6.107	6.137	6.209	6.304
100	6.435	6.483	6.532	6.561	6.592	6.669	6.768
133.3	7.068	7.123	7.175	7.206	7.240	7.322	7.429
200	8.084	8.150	8.208	8.242	8.280	8.370	8.488

TABLE 11 $Nu_{m,H}$ as a Function of Gz and n for Fully Developed Velocity Profiles in a Square Duct [120]

	$Nu_{m,H}$						
	Power-law index, n						
Gz	1.0	0.9	0.8	0.75	0.7	0.6	0.5
0	3.612	3.648	3.689	3.713	3.741	3.804	3.889
10	4.549	4.586	4.610	4.657	4.666	4.735	4.847
20	5.301	5.340	5.388	5.416	5.447	5.522	5.619
25	5.633	5.674	5.723	5.752	5.784	5.861	5.962
40	6.476	6.521	6.575	6.606	6.641	6.725	6.835
50	6.949	6.996	7.052	7.085	7.122	7.210	7.324
80	8.111	8.163	8.225	8.262	8.302	8.399	8.526
100	8.747	8.801	8.867	8.905	8.948	9.050	9.183
133.3	9.653	9.711	9.780	9.822	9.867	9.975	10.117
160	10.279	10.339	10.412	10.454	10.502	10.614	10.761
200	11.103	11.166	11.241	11.286	11.335	11.453	11.607

3. Hydrodynamics—Turbulent Flow

Kwack et al. [25] measured friction factors for turbulent flows in a square duct of 1 cm $\times$ 1 cm cross section with four different concentrations (20, 70, 200, and 1000 wppm) of polyethylene oxide (Polyox WSR-301), and the results are presented in Fig. 30. It is observed from this figure that the 20-wppm aqueous Polyox solution gives almost no drag reduction over the entire range of Reynolds numbers. For a 200-wppm solution, the minimum drag asymptotic values obtained for the circular-tube flow were reached for Rey-

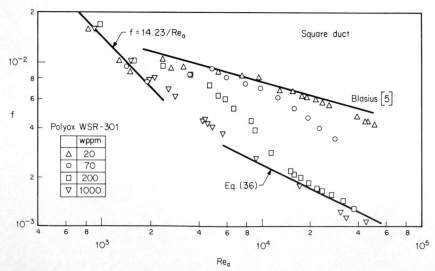

FIG. 30 Fanning friction factor versus Reynolds number for polyethylene oxide (Polyox WSR-301) solutions with Chicago tap water in 1 cm $\times$ 1 cm square duct [25].

nolds number greater than 1.5×10^4. The friction factor results of the 1000-wppm solution are in the minimum-drag asymptote over the entire range of Reynolds number. Therefore, it can be concluded that the minimum-drag asymptote, Eq. (36), derived for the circular-tube flow is applicable for a square-duct flow system for drag-reducing viscoelastic fluids.

No comparable studies for heat transfer reduction have been reported in the literature. However, from the results of drag reduction, it is recommended that the minimum heat transfer asymptote, Eq. (40), developed for the circular-tube flow be used for a square-duct flow system for drag-reducing viscoelastic fluids as a lower limit in a practical engineering design.

G. SUSPENSIONS IN DILUTE POLYMER SOLUTIONS

1. Hydrodynamics—Turbulent Flow

When fibers in suspension are subjected to high shear rates, they agglomerate by mechanical entanglement. Both free fibers and agglomerates are effective in damping turbulence and reducing energy dissipated by viscous shear to below that of the solvent flowing alone under the same conditions [16]. The phenomenon of drag reduction in turbulent fiber suspensions is of significant practical importance to the papermaking industry as well as to coal slurry pipelines.

Lee et al. [16] measured Fanning friction factors as a function of the Reynolds number in circular-tube flows for (1) the asbestos suspension, (2) a polymer solution (Separan AP-30, 150 wppm), and (3) polymer-fiber mixtures. For asbestos fiber solutions, the addition of a surfactant such as Aerosol OT was found to be indispensable for well-dispersed solutions [16]. These results, shown in Fig. 31, showed that 200 to 800 wppm of asbestos suspensions gave almost negligible drag reductions compared to the newtonian

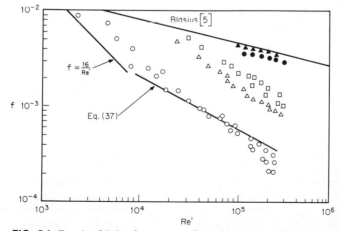

FIG. 31 Fanning friction factor versus Reynolds number. Filled triangles and filled circles, 200- and 800-wppm asbestos suspensions. Open squares, Separan 150-wppm solution. Open triangles and open circles, Separan plus asbestos 200 and 800 wppm, respectively [16].

solvent alone. The 150-wppm Separan solution gave approximately 50 to 64 percent drag reduction depending on the Reynolds number. However, the friction factor results of Separan solution plus asbestos suspensions of 200 to 800 wppm presented surprisingly high levels of drag reduction, about 95 percent drag reduction. Considering that the minimum-drag asymptote developed for dilute and concentrated homogeneous polymer solutions represents approximately 72 percent drag reduction, the polymer plus fiber mixtures exhibit far greater drag reductions than those given by the minimum-drag asymptote, Eq. (37).

2. Heat Transfer—Turbulent Flow

Moyls and Sabersky [18] measured heat transfer coefficients for dilute suspensions of asbestos fibers of 300 wppm flowing in a smooth tube. The results are presented in Fig. 32, which also shows the heat transfer results for 50-wppm Polyox solution obtained in the same tube [122]. Substantial heat transfer reductions for both suspensions of asbestos fibers and Polyox solution were observed at the Reynolds number equal to 2×10^4. However, as Re increased beyond this value, the heat transfer results for suspensions of asbestos fibers began to increase sharply, suggesting that a severe degradation occurred in suspensions of asbestos fibers.

The friction factor results of polymer-fiber mixtures reported by Lee et al. [16] suggest that the heat transfer reduction of these mixtures may be far greater than those predicted by the minimum heat transfer asymptote for aqueous polymer solutions.

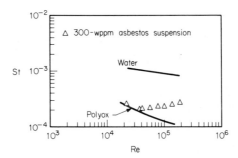

FIG. 32 Stanton number versus Reynolds number for water, Polyox (50 wppm) solution, and 300-wppm asbestos suspension in a smooth circular tube at 29.6°C and Pr = 6.16 [18, 122].

H. FINAL REMARKS

The study of heat transfer to nonnewtonian fluids, particularly viscoelastic fluids, is relatively new. Consequently, there are only relatively few cases in which predictive equations useful for engineering design can be given. Nevertheless, these cases do give some insight into the nonnewtonian behavior to be expected from these fluids. It is in this spirit that this chapter is written. The authors hope that the work will be useful to those working with nonnewtonian fluids.

REFERENCES

1. B. A. Toms, Some Observations on the Flow of Linear Polymer Solutions through Straight Tubes at Large Reynolds Numbers, *Proc. 1st Int. Cong. Rheol.,* Vol. II, p. 135, North Holland, Amsterdam, 1949.

2. K. J. Mysels, Flow of Thickened Fluids, U.S. Patent 2,492,173, Dec. 27, 1949.

3. A. B. Metzner, Heat Transfer in Non-newtonian Fluids, *Advances in Heat Transfer* (ed. J. P. Hartnett and T. F. Irvine, Jr.), Vol. 2, p. 357, Academic, New York, 1965.

4. J. E. Porter, Heat Transfer at Low Reynolds Number (Highly Viscous Liquids in Laminar Flow), Industrial Research Fellow Report, *Trans. Inst. Chem. Eng.,* **49**:1 (1971).

5. A. H. P. Skelland, *Non-newtonian Flow and Heat Transfer,* Wiley, New York, 1967.

6. J. W. Hoyt, The Effect of Additives on Fluid Friction, *Trans. ASME, J. Basic Eng.,* **94**:258 (1972).

7. Y. Dimant and M. Poreh, Heat Transfer in Flows with Drag Reduction, *Advances in Heat Transfer* (ed. T. F. Irvine, Jr., and J. P. Hartnett), Vol. 12, p. 77, Academic, New York, 1976.

8. A. White and J. A. G. Hemmings, *Drag Reduction by Additives—Review and Bibliography, BHRA Fluid Eng.,* Cranfield, U.K., 1976.

9. A. V. Shenoy and R. A. Mashelkar, Thermal Convection in Non-newtonian Fluids, *Advances in Heat Transfer* (ed. T. F. Irvine, Jr., and J. P. Hartnett), Vol. 15, pp. 142–225, Academic, New York, 1981.

10. Y. I. Cho and J. P. Hartnett, Non-newtonian Fluids in Circular Pipe Flow, *Advances in Heat Transfer* (ed. T. F. Irvine, Jr., and J. P. Hartnett), Vol. 15, pp. 59–141, Academic, New York, 1981.

11. W. H. Suckow, P. Hrycak, and R. G. Griskey, Heat Transfer to Non-newtonian Dilatant (Shear-Thickening) Fluids Flowing between Parallel Plates, *AIChE Symp. Ser.,* No. 199, Vol. 76, p. 257 (1980).

12. J. G. Oldroyd, On the Formulation of Rheological Equations of State, *Proc. Roy. Soc. London,* **A200**:523 (1950).

13. S. Middleman, *The Flow of High Polymers,* Interscience, New York, 1968.

14. K. Walters, *Rheometry,* Wiley, New York, 1975.

15. R. B. Bird, R. C. Armstrong, and O. Hassager, *Dynamics of Polymeric Liquids,* Vol. I, p. 207, Wiley, New York, 1977.

16. W. K. Lee, R. C. Vaseleski, and A. B. Metzner, Turbulent Drag Reduction in Polymeric Solutions Containing Suspended Fibers, *AIChE J.,* **20**:128 (1974).

17. I. Radin, J. L. Zakin, and G. K. Patterson, Drag Reduction in Solid-Fluid Systems, *AIChE J.,* **21**:358 (1975).

18. A. L. Moyls and R. H. Sabersky, Heat Transfer to Dilute Asbestos Dispersions in Smooth and Rough Tubes, *Lett. Heat Mass Transfer,* **2**:293 (1975).

19. P. F. W. Lee and G. G. Duffy, Relationships between Velocity Profiles and Drag Reduction in Turbulent Fiber Suspension Flow, *AIChE J.,* **22**:750 (1976).

20. D. J. Jeffrey and A. Acrivos, The Rheological Properties of Suspensions of Rigid Particles, *AIChE J.,* **22**:417 (1976).

21. F. Rodriguez, Drag Reduction by a Polymeric Aluminum Soap, *Nature Phys. Sci.,* **230**:152 (1971).

22. R. Monti, Heat Transfer in Drag-Reducing Solutions, *Progress in Heat and Mass Transfer* (ed. Schowalter et al.), Vol. 5, p. 239, Pergamon, New York, 1972.

23. R. C. Little, R. J. Hansen, D. L. Hunston, O. K. Kim, R. L. Patterson, and R. Y. Ting, The Drag Reduction Phenomenon: Observed Characteristics, Improved Agents and Proposed Mechanisms, *Ind. Eng. Chem. Fund.,* **14**:283 (1975).

24. D. White, Jr., and R. J. Gordon, The Influence of Polymer Conformation on Turbulent Drag Reduction, *AIChE J.,* **21**:1027 (1975).

25. E. Y. Kwack, Y. I. Cho, and J. P. Hartnett, Solvent Effects on Drag Reduction of Polyox Solutions in Square and Capillary Tube Flows, *J. Non-newtonian Fluid Mech.*, **9**:79 (1981).

26. M. Reiner, *Deformation, Strain, and Flow,* Interscience, New York, 1960.

27. E. C. Bingham, *Fluidity and Plasticity,* McGraw-Hill, New York, 1922.

28. R. B. Bird, Experimental Tests of Generalized Newtonian Models Containing a Zero-Shear Viscosity and a Characteristic Time, *Can. J. Chem. Eng.,* **43**:161 (1965).

29. R. E. Powell and H. Eyring, Mechanisms for the Relaxation Theory of Viscosity, *Nature,* **154**:427 (1944).

30. J. L. Sutterby, Laminar Converging Flow of Dilute Polymer Solutions in Conical Sections. II, *Trans. Soc. Rheol.,* **9**:227 (1965).

31. P. J. Carreau, Rheological Equations from Molecular Network Theories, *Trans. Soc. Rheol.,* **16**:99 (1972).

32. A. Argumedo, T. T. Tung, and K. I. Chang, Rheological Property Measurements of Drag-Reducing Polyacrylamide Solutions, *Trans. Soc. Rheol.,* **22**:449 (1978).

33. E. B. Christiansen and S. E. Craig, Heat Transfer to Pseudoplastic Fluids in Laminar Flow, *AIChE J.,* **8**:154 (1962).

34. D. R. Oliver and V. G. Jenson, Heat Transfer to Psuedoplastic Fluids in Laminar Flow in Horizontal Tubes, *Chem. Eng. Sci.,* **19**:115 (1964).

35. S. S. Yoo, Heat Transfer and Friction Factors for Non-newtonian Fluids in Turbulent Pipe Flow, Ph.D. thesis, Univ. of Illinois, Chicago, 1974.

36. D. Bellet, M. Sengelin, and C. Thirriot, Determination of Thermophysical Properties of Non-newtonian Liquids Using a Coaxial Cylindrical Cell, *Int. J. Heat Mass Transfer,* **18**:1177 (1975).

37. W. Y. Lee, Y. I. Cho, and J. P. Hartnett, Thermal Conductivity Measurements of Non-newtonian Fluids, *Lett. Heat Mass Transfer,* **8**:255 (1981).

38. W. Ibele, Thermophysical Properties, *Handbook of Heat Transfer* (ed. W. M. Rohsenow and J. P. Hartnett), McGraw-Hill, New York, 1973.

39. A. B. Metzner and J. C. Reed, Flow of Non-newtonian Fluids—Correlation of the Laminar, Transition, and Turbulent-Flow Regions, *AIChE J.,* **1**:434 (1955).

40. M. F. Edwards and R. Smith, The Turbulent Flow of Non-newtonian Fluids in the Absence of Anomalous Wall Effects, *J. Non-newtonian Fluid Mech.,* **7**:77 (1980).

41. M. L. Wasserman and J. C. Slattery, Upper and Lower Bounds on the Drag Coefficient of a Sphere in a Power-Model Fluid, *AIChE J.,* **10**:383 (1964).

42. J. A. Wheeler and E. H. Wissler, The Friction Factor-Reynolds Number Relation for the Steady Flow of Pseudoplastic Fluids through Rectangular Ducts, *AIChE J.,* **11**:207 (1965).

43. M. Reiner, The Deborah Number, *Physics Today,* **17**:62 (1964).

44. A. B. Metzner, J. L. White, and M. M. Denn, Constitutive Equation for Viscoelastic Fluids for Short Deformation Periods and for Rapidly Changing Flows: Significance of the Deborah Number, *AIChE J.,* **12**:863 (1966).

45. G. Astarita and G. Marrucci, *Principles of Non-newtonian Fluid Mechanics,* McGraw-Hill, New York, 1974.

46. R. R. Huigol, On the Concept of the Deborah Number, *Trans. Soc. Rheol.,* **19**:297 (1975).

47. F. A. Seyer and A. B. Metzner, Turbulent Flow Properties of Viscoelastic Fluids, *Can. J. Chem. Eng.,* **45**:121 (1967).

48. B. Elbirli and M. T. Shaw, Time Constants from Shear Viscosity Data, *J. Rheol.,* **22**:561 (1978).

49. K. S. Ng and J. P. Hartnett, Effects of Mechanical Degradation on Pressure Drop and Heat Transfer Performance of Polyacrylamide Solutions in Turbulent Pipe Flow, *Studies in Heat Transfer* (ed. T. F. Irvine, Jr., et al.), p. 297, McGraw-Hill, New York, 1979.

50. E. Y. Kwack, Y. I. Cho, and J. P. Hartnett, Effect of Weissenberg Number on Turbulent Heat Transfer of Aqueous Polyacrylamide Solutions, *Proc. 7th Int. Heat Transfer Conf.,* Vol. 3, FC11, pp. 63–68, Munich, September 1982.

51. R. B. Bird, W. E. Stewart, and E. N. Lightfoot, *Transport Phenomena,* Wiley, New York, 1960.

52. T. T. Tung, K. S. Ng, and J. P. Hartnett, Pipe Friction Factors for Concentrated Aqueous Solutions of Polyacrylamide, *Lett. Heat Mass Transfer,* **5**:59 (1978).

53. K. S. Ng, Y. I. Cho, and J. P. Hartnett, Heat Transfer Performance of Concentrated Polyethylene Oxide and Polyacrylamide Solutions, *AIChE Symp. Ser.,* No. 199, Vol. 76, p. 250, 1980.

54. J. G. Knudsen and D. L. Katz, *Fluid Dynamics and Heat Transfer,* McGraw-Hill, New York, 1958.

55. J. P. Hartnett, J. C. Y. Koh, and S. T. McComas, A Comparison of Predicted and Measured Friction Factors for Turbulent Flow through Rectangular Ducts, *J. Heat Transfer,* **84**:82 (1962).

56. D. C. Bogue, Entrance Effects and Prediction of Turbulence in Non-newtonian Flow, *Ind. Eng. Chem.,* **51**:874 (1959).

57. D. W. Dodge and A. B. Metzner, Turbulent Flow of Non-newtonian Systems, *AIChE J.,* **5**:189 (1959).

58. S. S. Tung and T. F. Irvine, Jr., Experimental Study of the Flow of a Viscoelastic Fluid in a Narrow Isosceles Triangular Duct, *Studies in Heat Transfer* (ed. T. F. Irvine, Jr., et al.), McGraw-Hill, New York, 1979.

59. H. S. Lee and T. F. Irvine, Jr., Degradation Effects and Entrance Length Studies for a Drag Reducing Fluid in a Square Duct, State Univ. of New York at Stony Brook *TR-382,* 1981.

60. W. M. Kays, *Convective Heat and Mass Transfer,* McGraw-Hill, New York, 1966.

61. B. C. Lyche and R. B. Bird, The Graetz-Nusselt Problem for a Power Law Non-newtonian Fluid, *Chem. Eng. Sci.,* **6**:35 (1956).

62. R. B. Bird, Zur Theorie des Wärmeübergangs an nicht-Newtonsche Flüssigkeiten bei laminarer Rohrströmung, *Chem. Ing. Tech.,* **31**5:69 (1959).

63. M. A. Lévêque, Les lois de la transmission de la chaleur par convection, *Ann. Mines,* **13**:201 (1928).

64. E. R. G. Eckert and R. M. Drake, Jr., *Analysis of Heat and Mass Transfer,* McGraw-Hill, New York, 1972.

65. R. L. Pigford, Nonisothermal Flow and Heat Transfer inside Vertical Tubes, *Chem. Eng. Prog. Symp. Ser.,* No. 17, Vol. 51, p. 79, 1955.

66. A. B. Metzner, R. D. Vaughn, and G. L. Houghton, Heat Transfer to Non-newtonian Fluids, *AIChE J.,* **3**:92 (1957).

67. A. A. McKillop, Heat Transfer for Laminar Flow of Non-newtonian Fluids in Entrance Region of a Tube, *Int. J. Heat Mass Transfer,* **7**:853 (1964).

68. Y. P. Shih and T. D. Tsou, Extended Leveque Solutions for Heat Transfer to Power Law Fluids in Laminar Flow in a Pipe, *Chem. Eng. Sci.,* **15**:55 (1978).

69. S. M. Richardson, Extended Leveque Solutions for Flows of Power Law Fluids in Pipes and Channels, *Int. J. Heat Mass Transfer,* **22**:1417 (1979).

70. T. Mizushina, R. Ito, Y. Kuriwake, and K. Yahikazawa, Boundary Layer Heat Transfer in a Circular Tube to Newtonian and Non-newtonian Fluids, *Kagaku Kogaku,* **31**:250 (1967).

71. R. Mahalingam, L. O. Tilton, and J. M. Coulson, Heat Transfer in Laminar Flow of Non-newtonian Fluids, *Chem. Eng. Sci.,* **30**:921 (1975).

72. E. N. Sieder and G. E. Tate, Heat Transfer and Pressure Drop of Liquids in Tubes, *Ind. Eng. Chem.,* **28**:1429 (1936).

73. S. D. Joshi and A. E. Bergles, Heat Transfer to Laminar In-Tube Flow of Pseudoplastic Fluids, *AIChE J.,* **27**:872 (1981).

74. K. S. Ng, J. P. Hartnett, and T. T. Tung, Heat Transfer of Concentrated Drag Reducing Viscoelastic Polyacrylamide Solutions, *Proc. 17th Nat. Heat Transfer Conf.,* Salt Lake City, 1977.

75. A. B. Metzner and P. S. Friend, Heat Transfer to Turbulent Non-newtonian Fluids, *Ind. Eng. Chem. J.,* **51**:879 (1959).

76. S. S. Yoo and J. P. Hartnett, Thermal Entrance Lengths for Non-newtonian Fluids in Turbulent Pipe Flow, *Lett. Heat Mass Transfer,* **2**:189 (1975).

77. P. S. Virk, H. S. Mickley, and K. A. Smith, The Ultimate Asymptote and Mean Flow Structure in Toms' Phenomenon, *Trans. ASME, J. Appl. Mech.,* **37**:488 (1970).

78. F. A. Seyer and A. B. Metzner, Turbulence Phenomena in Drag Reducing Systems, *AIChE J.,* **15**:426 (1969).

79. Y. I. Cho and J. P. Hartnett, Analogy for Viscoelastic Fluid—Momentum, Heat and Mass Transfer in Turbulent Pipe Flow, *Lett. Heat Mass Transfer,* **7**:339 (1980).

80. F. A. Seyer, Turbulence Phenomena in Drag Reducing Systems, Ph.D. thesis, Univ. of Delaware, Newark, 1968.

81. E. M. Khabakhpasheva and B. V. Perepelitsa, Turbulent Heat Transfer in Weak Polymeric Solutions, *Heat Transfer Sov. Res.,* **5**:117 (1973).

82. M. J. Rudd, Velocity Measurements Made with a Laser Dopplermeter on the Turbulent Pipe Flow of a Dilute Solution, *J. Fluid Mech.,* **51**:673 (1972).

83. Vr. Arunachalam, R. L. Hummel, and J. W. Smith, Flow Visualization Studies of a Turbulent Drag Reducing Solution, *Can. J. Chem. Eng.,* **50**:337 (1972).

84. D. D. Kale, An Analysis of Heat Transfer to Turbulent Flow of Drag Reducing Fluids, *Int. J. Heat Mass Transfer,* **20**:1077 (1977).

85. T. Mizushina, H. Usui, and T. Yoshida, Turbulent Pipe Flow of Dilute Polymer Solutions, *J. Chem. Eng. Jpn.,* **7**:162 (1974).

86. H. Usui, Transport Phenomena in Viscoelastic Fluid Flow, Ph.D. thesis, Kyoto Univ., Japan, 1974.

87. A. B. Metzner and G. Astarita, External Flows of Viscoelastic Materials: Fluid Property Restrictions on the Use of Velocity-Sensitive Probes, *AIChE J.,* **13**:550 (1967).

88. K. A. Smith, E. W. Merrill, H. S. Mickley, and P. S. Virk, Anomalous Pitot Tube and Hot Film Measurements in Dilute Polymer Solutions, *Chem. Eng., Sci.,* **22**:619 (1967).

89. G. Astarita and L. Nicodemo, Behavior of Velocity Probes in Viscoelastic Dilute Polymer Solutions, *Ind. Eng. Chem. Fund.,* **8**:582 (1969).

90. R. W. Serth and K. M. Kiser, The Effect of Turbulence on Hot-Film Anemometer Response in Viscoelastic Fluids, *AIChE J.,* **16**:163 (1970).

91. N. S. Berman, G. B. Gurney, and W. K. George, Pitot Tube Errors in Dilute Polymer Solutions, *Phys. Fluids,* **16**:1526 (1973).

92. N. A. Halliwell and A. K. Lewkowicz, Investigation into the Anomalous Behavior of Pitot Tubes in Dilute Polymer Solutions, *Phys. Fluids,* **18**:1617 (1975).

93. E. Y. Kwack, Y. I. Cho, and J. P. Hartnett, Heat Transfer to Polyacrylamide Solutions in Turbulent Pipe Flow: The Once-Through Mode, *AIChE Symp. Ser.,* No. 208, Vol. 77, p. 123, 1981.

94. V. J. Berry, Non-uniform Heat Transfer to Fluids Flowing in Conduits, *Appl. Sci. Res.,* **A4**:61 (1953).

95. R. G. Deissler, Turbulent Heat Transfer and Friction in the Entrance Regions of Smooth Passages, *Trans. ASME,* **77**:1221 (1955).

96. J. P. Hartnett, Experimental Determination of the Thermal Entrance Length for the Flow of Water and Oil in Circular Pipes, *Trans. ASME,* **77**:1211 (1955).

97. R. W. Allen and E. R. G. Eckert, Friction and Heat Transfer Measurements to Turbulent Pipe Flow of Water (Pr = 7 and 8) at Uniform Wall Heat Flux, *Trans. ASME,* **86**:301 (1964).

98. Y. I. Cho, K. S. Ng, and J. P. Hartnett, Viscoelastic Fluids in Turbulent Pipe Flow—A New Heat Transfer Correlation, *Lett. Heat Mass Transfer,* **7**:347 (1980).

99. M. K. Gupta, A. B. Metzner, and J. P. Hartnett, Turbulent Heat-Transfer Characteristics of Viscoelastic Fluids, *Int. J. Heat Mass Transfer,* **10**:1211 (1967).

100. G. Astarita, Turbulent Heat Transfer in Viscoelastic Liquids, *Ind. Eng. Chem. Fund.,* **6**:470 (1967).

101. T. Mizushina and H. Usui, Reduction of Eddy Diffusion for Momentum and Heat in Viscoelastic Fluid Flow in a Circular Tube, *Phys. Fluids,* **20**:S100 (1977).

102. M. Poreh and U. Paz, Turbulent Heat Transfer to Dilute Polymer Solutions, *Int. J. Heat Mass Transfer,* **11**:805 (1968).

103. R. Smith and M. Edwards, Heat Transfer to Non-newtonian and Drag-Reducing Fluids in Turbulent Pipe Flow, *Int. J. Heat Mass Transfer,* **24**:1059 (1981).

104. C. S. Chiou and R. J. Gordon, Low Shear Viscosity of Dilute Polymer Solutions, *AIChE J.,* **26**:852 (1980).

105. Y. I. Cho, J. P. Hartnett, and Y. S. Park, Solvent Effects on the Rheology of Aqueous Polyacrylamide Solutions, *Chem. Eng. Comm.,* **21**:369 (1983).

106. C. S. Lin, E. B. Denton, H. S. Gaskill, and G. L. Putnam, Diffusion-Controlled Electrode Reactions, *Ind. Eng. Chem.,* **43**:2136 (1951).

107. P. Van Shaw, L. P. Reiss, and T. J. Hanratty, Rates of Turbulent Transfer to a Pipe Wall in the Mass Transfer Entry Region, *AIChE J.,* **9**:362 (1963).

108. D. A. Shaw and T. J. Hanratty, Turbulent Mass Transfer Rates to a Wall for Large Schmidt Numbers, *AIChE J.,* **23**:28 (1977).

109. G. H. Sidahmed and R. G. Griskey, Mass Transfer in Drag Reducing Fluid Systems, *AIChE J.,* **18**:138 (1972).

110. J. T. Teng, R. Greif, I. Cornet, and R. N. Smith, Study of Heat and Mass Transfer in Pipe Flows with Non-newtonian Fluids, *Int. J. Heat Mass Transfer,* **22**:493 (1979).

111. Z. P. Shulman, N. A. Pokryvailo, E. B. Kaberdina, and A. K. Nesterov, Effect of Polymer Additives on Intensity and Spectrum of Pulsations of Velocity Gradient Fluctuations Close to a Solid Surface, A4-63, *Proc. 1st Int. Conf. Drag Reduction, BHRA Fluid Eng.,* Cranfield, U.K., 1974.

112. G. A. McConaghy and T. J. Hanratty, Influence of Drag Reducing Polymers on Turbulent Mass Transfer to a Pipe Wall, *AIChE J.,* **23**:493 (1977).

113. T. Mizushina, The Electrochemical Methods in Transport Phenomena, *Advances in Heat Transfer* (ed. T. F. Irvine, Jr., and J. P. Hartnett), Vol. 7, p. 87, Academic, New York, 1971.

114. P. S. Virk and T. Suraiya, Mass Transfer at Maximum Drag Reduction, G3-41, *Proc. 2d Int. Conf. Drag Reduction, BHRA Fluid Eng.,* Cranfield, U.K., 1977.

115. W. H. Linton, Jr., and T. K. Sherwood, Mass Transfer from Solid Shapes to Water in Streamline and Turbulent Flow, *Chem. Eng. Prog.,* **46**:258 (1950).

116. A. H. P. Skelland, *Diffusional Mass Transfer,* Wiley, New York, 1974.

117. Z. P. Shulman and N. A. Pokryvailo, *Hydrodynamics and Heat and Mass Transfer of Polymer Solutions,* Luikov Heat and Mass Transfer Institute, Byelorussian Academy of Sciences, Minsk, USSR, 1980.

118. Y. I. Cho and J. P. Hartnett, Mass Transfer in Turbulent Pipe Flow of Viscoelastic Fluids, *Int. J. Heat Mass Transfer,* **24**:945 (1981).

119. R. S. Schechter, On the Steady Flow of a Non-newtonian Fluid in Cylinder Ducts, *AIChE J.,* **7**:445 (1961).

120. A. R. Chandrupatla and V. M. K. Sastri, Laminar Forced Convection Heat Transfer of a Non-newtonian Fluid in a Square Duct, *Int. J. Heat Mass Transfer,* **20**:1315 (1977).

121. N. T. Dunwoody and T. A. Hamill, Forced Heat Convection in Lineal Flow of Non-newtonian Fluids through Rectangular Channels, *Int. J. Heat Mass Transfer,* **23**:943 (1980).

122. P. M. Debrule and R. H. Sabersky, Heat Transfer and Friction Coefficients in Smooth and Rough Tubes with Dilute Polymer Solutions, *Int. J. Heat Mass Transfer,* **17**:529 (1974).

123. A. W. Petersen and E. B. Christiansen, Heat Transfer to Non-newtonian Fluids in Transitional and Turbulent Flow, *AIChE J.,* **12**:221 (1966).

124. T. Mizushina and Y. Kuriwaki, Turbulent Heat Transfer in Non-newtonian Fluids, *Mem. Fac. Eng. Kyoto Univ.,* **XXIX**(2):197 (1967).

125. R. M. Clapp, Turbulent Heat Transfer to Pseudoplastic Non-newtonian Fluids, *International Developments in Heat Transfer,* D211, p. 652, ASME, New York, 1963.

NOMENCLATURE

Symbol, Definition, SI Units, English Units

A_c	cross-sectional area: m^2, ft^2
c_p	specific heat at constant pressure: $J/(kg \cdot K)$, $Btu/(lb_m \cdot {}^\circ F)$
D_j	diffusion coefficient of species j: m^2/s, ft^2/s
De	Deborah number $= t/t_F$
d	tube inside diameter: m, ft
d_h	hydraulic diameter: m, ft
d_{ij}	rate of strain tensor: s^{-1}
f	Fanning friction factor $= \tau_w/(\rho V^2/2)$
Gr	Grashof number $= \rho^2 g \beta \, \Delta T \, d^3/\mu^2$
Gz	Graetz number $= W c_p/k_l x$
h	heat transfer coefficient: $W/(m^2 \cdot K)$, $Btu/(h \cdot ft^2 \cdot {}^\circ F)$
j_D	Colburn factor of mass transfer $= St \cdot Sc^{2/3}$
j_H	Colburn factor of heat transfer $= St \cdot Pr^{2/3}$
K, K'	constant in power-law model defined in Eqs. (6) and (3) respectively: $N/(m^2 \cdot s^n)$, $lb_f/(ft^2 \cdot s^n)$
k_l	thermal conductivity of liquid: $W/(m \cdot K)$, $Btu/(h \cdot ft \cdot {}^\circ F)$
L	tube length: m, ft
L_e	entrance length in tube flow: m, ft
m	constant in Deissler's eddy diffusivity model, Eq. (49)
Nu_m	mean Nusselt number $= h_m d/k_l$
Nu_x	local Nusselt number $= h_x d/k_l$
n	power-law index in Eqs. (3) and (6)
P	pressure: N/m^2, lb_f/ft^2
P	perimeter: m, ft
Pe	Peclet number $= Re \cdot Pr$
Pr	Prandtl number $= \mu c_p/k_l$
ΔP	pressure drop along the axial direction: N/m^2, lb_f/ft^2
q_w''	heat flux at the tube wall: W/m^2, $Btu/(h \cdot ft^2)$
R	tube inside radius $= d/2$: m, ft
R^+	normalized tube inside radius $= R u^*/\nu$
Re	Reynolds number $= \rho V d/\mu$
Re_a	Reynolds number based on the apparent viscosity at the wall $= \rho V d/\eta_a$
Re'	Reynolds number defined as $\rho V^{2-n} d^n/K' 8^{n-1}$
r	radial coordinate: m, ft
Sc	Schmidt number $= \mu/\rho D_j$
St	Stanton number $= Nu/(Re\,Pr) = h/\rho V c_p$
s	aspect ratio of a rectangular duct
T	temperature: $^\circ C$, $^\circ F$
t	characteristic time of the viscoelastic fluid, a measure of elasticity, Table 1: s

t_F characteristic time of the flow: s

t_p characteristic time of a viscoelastic fluid calculated using the Powell-Eyring model (see Table 1): s

u axial velocity component: m/s, ft/s

u^* friction velocity $= (\tau_w/\rho)^{1/2}$: m/s, ft/s

u^+ normalized velocity $= u/u^*$

V mean velocity in tube flow: m/s, ft/s

W mass flow rate: kg/s, lb_m/s

Ws   Weissenberg number $= tV/d$

Ws_p Weissenberg number based on Powell-Eyring characteristic time, $t_p V/d$

$(Ws)_f^*, (Ws_p)_f^*$ critical Weissenberg number for friction

$(Ws)_h^*, (Ws_p)_h^*$ critical Weissenberg number for heat transfer

x axial location along the tube: m, ft

y distance normal to the tube wall $= R - r$: m, ft

y^+ normalized distance from the wall $= yu^*/\nu$

Greek Symbols

$\dot{\gamma}$ shear rate: s^{-1}

ϵ_D diffusion eddy diffusivity: m^2/s, ft^2/s

ϵ_H thermal eddy diffusivity: m^2/s, ft^2/s

ϵ_M momentum eddy diffusivity: m^2/s, ft^2/s

η shear rate–dependent viscosity: Pa·s, lb_m/(h·ft)

η_a apparent viscosity: Pa·s, lb_m/(h·ft)

η_0 limiting viscosity at zero shear rate: Pa·s, lb_m/(h·ft)

η_∞ limiting viscosity at infinite shear rate: Pa·s, lb_m/(h·ft)

θ dimensionless temperature defined in Eq. (38)

μ constant viscosity of the fluid: Pa·s, lb_m/(h·ft)

ν kinematic viscosity: m^2/s, ft^2/s

ρ density: kg/m^3, lb_m/ft^3

τ shear stress: N/m^2, lb_f/ft^2

τ_{ij} shear stress tensor: N/m^2, lb_f/ft^2

τ_w shear stress at the wall: N/m^2, lb_f/ft^2

$\tau_{11} - \tau_{22}$ first normal stress difference: N/m^2, lb_f/ft^2

Subscripts

a property based on the apparent viscosity

b bulk fluid condition

ex   condition at the exit of tube

H constant axial heat flux, with peripherally constant wall temperature

in   condition at the inlet of tube

m average value
T constant wall temperature
w wall condition
x local value at axial distance x
∞ fully developed

Superscript

$+$ dimensionless

CHAPTER 3

Techniques to Augment Heat Transfer

By A. E. Bergles

Department of Mechanical Engineering
Iowa State University
Ames, Iowa

A. INTRODUCTION

1. General Background

Most of the burgeoning research effort in heat transfer is devoted to analyzing what might be called the "standard situation." However, the development of high-performance thermal systems has also stimulated interest in methods to improve heat transfer. The study of improved heat transfer performance is referred to as heat transfer augmentation, enhancement, or intensification.

The performance of conventional heat exchangers can be substantially improved by a number of augmentation techniques. On the other hand, certain systems, particularly those in space vehicles, may *require* an augmentation device for successful operation. A great deal of research effort has been devoted to developing apparatus and performing experiments to define the conditions under which an augmentative technique will improve heat (and mass) transfer. Over 3000 technical publications, excluding patents and manufacturers' literature, are listed in a bibliographic report [1]. The recent growth of activity in this area is clearly evident from the yearly distribution of such publications shown in Fig. 1. The most effective and feasible techniques have graduated from the laboratory to full-scale industrial use.

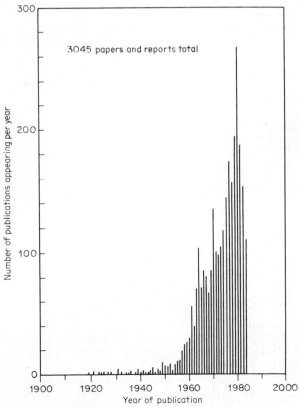

FIG. 1 References on heat transfer augmentation versus year of publication (to late 1983).

The main objective of this chapter is to survey some of the important literature pertinent to each augmentation technique, thus providing guidance for potential users. Wherever possible, correlations for thermal and hydraulic performance will be presented or key sources of design data will be suggested.

2. Classification of Heat Transfer Augmentation Techniques

Augmentation techniques can be classified as *passive* methods, which require no direct application of external power, or as *active* schemes, which require external power. The effectiveness of both types depends strongly on the mode of heat transfer, which might range from single-phase free convection to dispersed-flow film boiling. Brief descriptions of passive techniques follow.

Treated surfaces involve fine-scale alternation of the surface finish or coating (continuous or discontinuous). They are used for boiling and condensing; the roughness height is below that which affects single-phase heat transfer.

Rough surfaces are produced in many configurations ranging from random sand-grain-type roughness to discrete protuberances. The configuration is generally chosen to promote turbulence rather than to increase the heat transfer surface area. The application of rough surfaces is directed primarily toward single-phase flow.

Extended surfaces are routinely employed in many heat exchangers. The development of new types of extended surfaces, such as integral inner-fin tubing, and the improvement of heat transfer coefficients on extended surfaces by shaping or interrupting the surfaces are of particular interest.

Displaced enhancement devices are inserted into the flow channel so as to indirectly improve energy transport at the heated surface. They are used with forced flow.

Swirl-flow devices include a number of geometrical arrangements or tube inserts for forced flow which create rotating and/or secondary flow: coiled tubes, inlet vortex generators, twisted-tape inserts, and axial-core inserts with a screw-type winding.

Surface-tension devices consist of wicking or grooved surfaces to direct the flow of liquid in boiling and condensing.

Additives for liquids include solid particles and gas bubbles in single-phase flows and liquid trace additives for boiling systems.

Additives for gases are liquid droplets or solid particles, either dilute phase (gas-solid suspensions) or dense phase (fluidized beds).

The active techniques are described below.

Mechanical aids stir the fluid by mechanical means or by rotating the surface. Surface "scraping," widely used for viscous liquids in the chemical process industry, can also be applied to duct flow of gases. Equipment with rotating heat exchanger ducts is found in commercial practice.

Surface vibration at either low or high frequency has been used primarily to improve single-phase heat transfer.

Fluid vibration is the most practical type of vibration enhancement, given the mass of most heat exchangers. The vibrations range from pulsations of about 1 Hz to ultrasound. Single-phase fluids are of primary concern.

Electrostatic fields (dc or ac) are applied in many different ways to dielectric fluids. Generally speaking, electrostatic fields can be directed to cause greater bulk mixing of fluid in the vicinity of the heat transfer surface. An electrical field and a magnetic field may be combined to provide a forced convection or electromagnetic pumping.

Injection involves supplying gas to a flowing liquid through a porous heat transfer surface or injecting similar fluid upstream of the heat transfer section. Surface degassing of liquids can produce augmentation similar to gas injection. Only single-phase flow is of interest.

Suction involves either vapor removal through a porous heated surface in nucleate or film boiling, or fluid withdrawal through a porous heated surface in single-phase flow.

Two or more of the above techniques may be utilized simultaneously to produce an enhancement larger than that produced by only one technique. This simultaneous use is termed *compound augmentation*.

It should be emphasized that one reason for studying enhanced heat transfer is to assess the effect of an inherent condition on heat transfer. Some practical examples include roughness produced by standard manufacturing, degassing of liquids with high gas content, surface vibration resulting from rotating machinery or flow oscillations, fluid vibration resulting from pumping pulsation, and electric fields present in electrical equipment.

3. Performance Evaluation Criteria

It seems impossible to establish a generally applicable selection criterion for the use of augmentative techniques since numerous factors influence the ultimate decision. Most of the pertinent considerations are economic: development cost, initial cost, operating cost, maintenance cost, etc.; in addition, other factors such as reliability and safety must be considered. However, to begin the assessment it is useful to consider the relationship between the thermal and hydraulic performance—particularly in the dominant practical case of single-phase forced convection. The typical heat transfer and flow friction data for "turbulence promoters" shown in Fig. 2 illustrate this point. The equivalent plain tube diameter or "envelope diameter" is used in this presentation, as suggested in Ref. 2. The promoters produce a sizable elevation in the heat transfer coefficient at constant velocity; however, there is generally a greater percentage increase in the friction factor.

Common thermal-hydraulic goals include reducing the size of a heat exchanger required for a specified heat duty, increasing the heat duty of an existing heat exchanger, reducing the approach temperature difference for the process streams, or reducing the

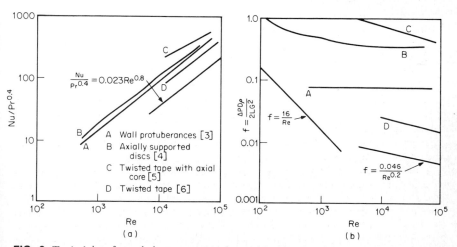

FIG. 2 Typical data for turbulence promoters inserted inside tubes. (*a*) Heat transfer data. (*b*) Friction data.

pumping power. The presence of system and design constraints leads to a number of performance evaluation criteria (PECs). The geometric variables for tube-side flow in a conventional shell-and-tube heat exchanger are tube diameter, tube length, and number of tubes per pass. The primary independent operating variables are the approach temperature difference and the mass flow rate (or velocity). Dependent variables are the heat transfer rate and pumping power (or pressure drop). A PEC is established by selecting for one of the process streams one of the operational variables for the performance objective in accordance with design constraints on the remaining variables.

Table 1 lists PECs for 12 cases of interest concerning enhanced and smooth circular tubes of the same envelope diameter. The table segregates these PECs according to three different geometrical constraints:

1. **FG Criteria.**   The cross-sectional envelope area and tube length are held constant. The FG criteria may be thought of as a retrofit situation in which there is a one-for-one replacement of plain surface with enhanced surfaces of the same basic geometry, e.g., tube envelope diameter, tube length, and number of tubes for in-tube flow. The FG-2 criteria have the same objectives as FG-1, but require the enhanced surface design to operate at the same pumping power as the reference smooth tube design. In most cases this requires the enhanced exchanger to operate at reduced flow rate. The FG-3 criterion seeks reduced pumping power for fixed heat duty.

2. **FN Criteria.**   These criteria maintain fixed flow frontal area and allow the length of the heat exchanger to be a variable. These criteria seek reduced surface area (FN-1, FN-2) or reduced pumping power (FN-3) for constant heat duty.

3. **VG Criteria.**   In many cases a heat exchanger is sized for a required thermal duty with specified flow rate. In these situations the FG and FN criteria are not applicable. Because the tube-side velocity must be reduced to accommodate the higher friction characteristics of the augmented surface, it is necessary to increase the flow area to maintain constant flow rate. This is accomplished by using a greater number of par-

TABLE 1 Performance Evaluation Criteria for $D_{ia}/D_{io} = 1$ [7]

	Fixed						Consequences						
Case	Geom.	**W**	**P**	q	ΔT_i	Objective	$\dfrac{N_a}{N_o}$	$\dfrac{L_a}{L_o}$	$\dfrac{W_a}{W_o}$	$\dfrac{Re_{ia}}{Re_{io}}$	$\dfrac{P_a}{P_o}$	$\dfrac{q_a}{q_o}$	$\dfrac{\Delta T_{ia}}{\Delta T_{io}}$
FG-1a	N, L	X			X	↑q	1	1	1	1[b]	>1	>1	1
FG-1b	N, L	X		X		↓ΔT_i	1	1	1	1[b]	1	1	<1
FG-2a	N, L		X		X	↑q	1	1	<1	<1	1	>1	1
FG-2b	N, L		X	X		↓ΔT_i	1	1	<1	<1	1	1	<1
FG-3	N, L			X	X	↓P	1	1	<1	<1	<1	1	1
FN-1	N		X	X	X	↓L	1	<1	<1	<1	1	1	1
FN-2	N	X		X	X	↓L	1	<1	1	1[b]	<1	1	1
FN-3	N	X		X	X	↓P	1	<1	1	1[b]	<1	1	1
VG-1		X	X	X	X	↓NL	>1[c]	<1	1	<1[c]	1	1	1
VG-2a	NL^a	X	X	X	X	↑q	>1[c]	<1	1	<1[c]	1	>1	1
VG-2b	NL^a	X	X	X		↓ΔT_i	>1[c]	<1	1	<1[c]	1	1	<1
VG-3	NL^a	X		X	X	↓P	<1	<1	1	<1[c]	<1	1	1

[a]The product of N and L is constant in cases VG-2 and 3.

[b]For internal roughness. For internal fins, $Re_{ia}/Re_{io} = D_{ha}A_{fo}/D_{io}A_{fa}$.

[c]Roughness with high-Pr fluids may not result in $N_a/N_o > 1$ (or $Re_{ia}/Re_{io} < 1$).

allel flow circuits. Maintaining a constant exchanger flow rate eliminates the penalty of operating at higher thermal effectiveness encountered in the previous FG and FN cases.

The necessary relations for quantitative formulation of these PECs are summarized in Ref. 7. Evaluation of the objectives is straightforward once the constraints are specified, if the basic heat transfer and flow friction data are available in the range of interest. The calculations can be carried out for any geometry where the data are available; alternatively, using correlations for h and f, the optimum geometry can be determined. Consider the following example.

In proposed ocean thermal energy conversion (OTEC) systems utilizing a closed Rankine cycle, heat exchanger surface is a major consideration, since over half of the plant capital cost is in heat exchangers. The net output is determined primarily by the heat transfer rates in the evaporator and condenser, and by the power consumed by the seawater pumps. Since the seawater flow rate is immaterial, FN-1 is an appropriate PEC. The objective can also be interpreted as reduction of NL (or area), and it is convenient to explore the dependence of NL on N. Figure 3 presents typical results for the surface area ratio for various types of in-tube enhancement (no outside enhancement) for a baseline OTEC shell-and-tube spray film evaporator [8]. The use of rough tubes, inner-fin

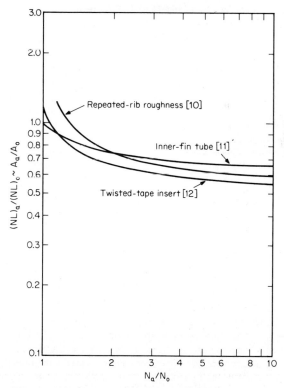

FIG. 3 Composite plots of constant pumping power performance (VG-1, $W \neq$ const) of an OTEC evaporator with various in-tube enhancement techniques [9].

tubes, or tubes with twisted-tape inserts permits a reduction in surface area by almost 30 percent at $N_a/N_o = 2$. This is very significant since heat exchanger surface areas in projected OTEC systems are extremely large—of the order of 10,000 m^2/MW(e) net. Further area reductions can be brought about by enhancing the evaporating side. A recent study by Webb [13] translates the thermal-hydraulic performance of enhanced tubes into cost advantages for OTEC applications.

The extension of these PECs to two-phase heat transfer is complicated by the dependence of the local heat transfer coefficient on the local temperature difference and/or quality. Heat transfer and pressure drop have been considered in the evaluation of internally finned tubes for refrigerant evaporators [14] and internally finned tubes, helically ribbed tubes, and spirally fluted tubes for refrigerant condensers [15]. Pumping power has been incorporated into the evaluation of inserts used to elevate subcooled boiling critical heat flux (CHF) [16, 17].

These PECs will be used occasionally throughout this chapter to demonstrate the advantages of specific enhanced heat transfer techniques. It should be noted, however, that these thermal-hydraulic comparisons apply only to the basic heat exchanger. A full analysis of the augmentation effect may require consideration of the entire system if flow rates or temperature levels change as a result of the augmentation.

B. TREATED AND STRUCTURED SURFACES

1. Boiling

As discussed in Ref. 365, surface material and finish have a strong effect on nucleate and transition pool boiling; thus, optimum surface conditions might be selected for a system operating in these boiling regimes. Certain types of fouling and oxidation, apparently those which improve wettability, produce significant increases in pool-boiling CHF.

A novel technique for promoting nucleate boiling has been proposed by Young and Hummel [18]. Spots of Teflon or other nonwetting material, either on the heated surface or in pits, were found to promote nucleation as shown in Fig. 4. Relatively low wall superheat is required to activate the nonwetting cavities represented by the spots. The spots are placed so that the bubble area of influence includes the whole surface, resulting in a low average superheat. The heat transfer coefficient (at constant heat flux) is increased by factors of 3 to 4. This technique is not effective for refrigerants, as there are no "Freon-phobic" materials (Bergles et al. [19]).

Thin insulating coatings effectively increase rates of heat transfer in pool boiling when heater temperature is the controlling parameter [20]. When surface temperatures are originally in the film-boiling range, a thin coating of Teflon, for example, reduces the fluid-surface interface temperature to the level where transition or nucleate boiling occurs. Bergles and Thompson [21] found large reductions in quench times because of scale or oxide coatings which promote destabilization of film boiling.

With well-wetting fluids (refrigerants, cryogens, organics, alkali liquid metals), doubly reentrant cavities are required to ensure vapor trapping. The probability of having such active nucleation sites present is increased by selective machining, forming, or coating of the surface. Furthermore, large cavities can be created which result in steady-state boiling at low wall superheat. The surfaces may appear either rough or smooth (as if treated), depending on the manufacturing procedure. Hence, the usual "treated" and "rough" classifications are lumped together for purposes of this discussion. Examples of

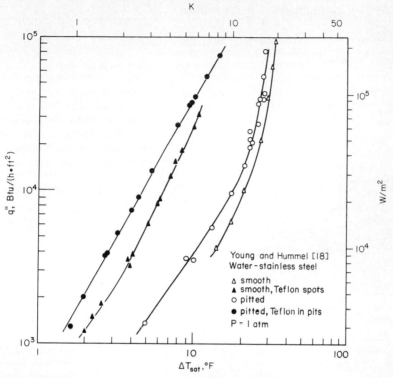

FIG. 4 Influence of surface treatment on saturated pool boiling..

these "structured" boiling surfaces are given in Table 2, which is an extension of a table given in Ref. 22.

Wall superheat reductions up to a factor of 10 have been reported with some of these surfaces. It should be noted that the mechanism of vaporization is different for these surfaces from what it is for normal cavity boiling. Here, the liquid flows via selected paths or channels to the interior, where thin film evaporation occurs over a large surface area; the vapor is then ejected through other paths by "bubbling" (Czikk and O'Neill [38], Nakayama et al. [39]). It should be emphasized that the performance of these special surfaces is quite sensitive to surface geometry and fluid condition. Additionally, very low temperature differences are involved; hence, it is necessary to be especially careful, or at least consistent, in measuring wall temperatures and saturation temperatures (pressures). The first comprehensive comparison of the nucleate-boiling performances of several structured surfaces was reported recently by Yilmaz et al. [40]. As shown in Fig. 5, each of the three surfaces exhibits a boiling curve well to the left of that for a single tube. It is evident that if only low temperature differences are available, high heat fluxes can be realized only with structured surfaces. Note that the heat flux is based on the area of the equivalent smooth tube for a particular outside diameter.

The previously cited studies do not report boiling curves which exhibit hysteresis, i.e., exhibit different characteristics with increasing heat flux from those with decreasing heat flux. It should be noted, however, that with increasing heat flux large temperature over-

TABLE 2 Examples of Structured Boiling Surfaces

Category	Report	Procedure	Result
Machined	Kun and Czikk [23]	Cross-grooved and flattened	Regular matrix of reentrant cavities
	Fujikake [24]*	Low fins knurled and compressed	Reentrant cross grooves
	Hwang and Moran [25]	Laser drilling	Regular matrix of slightly reentrant cavities
Formed	Webb [26]*	Standard low-fin tubing with fins bent to reduce gap	Helical circumferential reentrant cavities
	Zatell [27]	Above with additional variations	Helical circumferential reentrant cavities
	Nakayama et al. [28]*	Rolled, upset, and brushed	Helical circumferential or groove-type reentrant cavities with periodic openings
	Stephan and Mitrovic [29]*	Standard low-fin tubing rolled to form T-shaped fins	Helical circumferential reentrant cavities
Multilayer	Ragi [30]	Stamped sheet with pyramids, open at the top, attached to surface	Regular matrix of reentrant cavities
Coated	Marto and Rohsenow [31]	Poor weld	Irregular matrix of surface and reentrant cavities
	O'Neill et al. [32]*	Sintering or brazing	Irregular matrix of surface and reentrant cavities
	Oktay and Schmeckenbecher [33]	Electrolytic deposition	Irregular matrix which includes reentrant cavities
	Dahl and Erb [34]	Flame spraying	Irregular matrix of surface and reentrant cavities
	Fujii et al. [35]	Particles bonded by plating	Irregular matrix of surface and reentrant cavities
	Janowski et al. [36]*	Metallic coating of a foam substrate	Irregular matrix of surface and reentrant cavities
	Warner et al. [37]	Plasma-deposited polymer	Irregular matrix of surface and reentrant cavities

*Denotes commercial surface.

shoots and boiling-curve hysteresis are common with refrigerants and other highly wetting liquids. Bergles and Chyu [41] provide extensive documentation of such behavior with sintered surfaces; but they note that not all commercial equipment will have start-up problems because of this behavior. Also, in the area of practical application, there is some evidence that tube bundles may exhibit behavior different from single tubes with structured surfaces [42].

Limited evidence indicates that CHF in pool boiling with structured surfaces is usually as high as or higher than that with plain surfaces (O'Neill et al. [32]).

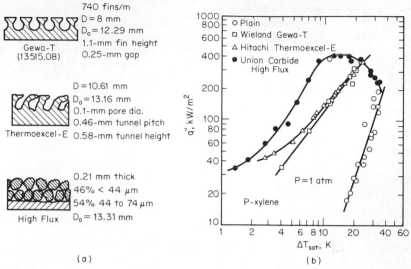

(a)   (b)

FIG. 5 Pool boiling from smooth and structured surfaces on the same apparatus [40]. (a) Sketch of cross sections of three enhanced heat transfer surfaces tested. (b) Boiling curves for three enhanced tubes and smooth tube.

The use of structured surfaces to enhance thin-film evaporation has also been considered recently. Here, in contrast to the flooded-pool experiments noted above, the liquid to be vaporized is sprayed on or dripped on heated horizontal tubes to form a thin film. If the available temperature difference is modest, structured surfaces can be used to promote boiling in the film, thus improving the overall heat transfer coefficient. Chyu et al. [43] found that sintered surfaces yielded nucleate-boiling curves similar to those obtained in pool boiling. T-shaped fins did not exhibit low ΔT boiling; however, a threefold convective enhancement was obtained as a result of the increased surface area.

Boiling heat transfer from structured surfaces is a major growth area in enhanced heat transfer. Most of the processes noted in Table 2 are covered by patents (see also Refs. 44 and 45) and, as noted, many of the surfaces are offered commercially. In general, the behavior of these surfaces is not yet understood to the point where correlations are available which allow custom production of surfaces for a particular fluid and pressure level. In some cases, however, manufacturers have accumulated sufficient experience to provide optimized surfaces for some of the important applications, e.g., flooded-refrigerant evaporators for direct-expansion chillers.

2. Condensing

As noted in Ref. 366, surface treatment for the promotion of dropwise condensation in vapor space environments has been extensively investigated. If dropwise condensation is achieved, the augmentation is 10 to 100 times the filmwise condensation coefficient. Numerous promoters and coatings have been found effective; however, a number of practical problems relate to the method of application, permanence, and compatibility with the rest of the system. Tanasawa's review [47] includes a good discussion of the difficulties that must be overcome if industry is to adopt this condensation process.

It should be noted that the only valid application is for steam condensers, since non-wetting substances are not available for most other working fluids. For example, no drop-wise condensation promoters have been found for refrigerants (i.e., dropwise condensation promoters seem to be "Freon-phobic") (Iltscheff [46]). The augmentation of dropwise condensation, beyond inducing the process by selection of an effective durable promoter, is fruitless since the heat transfer coefficients are already so high.

Glicksman et al. [48] showed that average coefficients for film condensation of steam on horizontal tubes can be improved up to 20 percent by strategically placing horizontal strips of Teflon or other nonwetting material around the tube circumference. The condensate flow is interrupted near the leading edge of a strip and the condensate film is thinner when it re-forms at the downstream edge of the strip.

C. ROUGH SURFACES

1. Single-Phase Flow

Surface roughness is usually not considered for free convection since the velocities are commonly too low to cause flow separation or secondary flow. A recent review [49] of the limited data for free convection from machined or formed roughness, with air, water, and oil, indicates that increases in heat transfer coefficient up to 100 percent have been obtained with air. However, the reported increases with liquids are very small.

Surface roughness was one of the first techniques to be considered seriously as a means of augmenting forced-convection heat transfer. Initially investigators speculated that elevated heat transfer coefficients might accompany the relatively high friction factors characteristic of rough conduits. However, since commercial roughness is not well defined, artificial surface roughness has been employed. Integral roughness may be produced by the traditional manufacturing processes of machining, forming, casting, or welding. Various inserts can also be used to provide surface protuberances. Although the enhanced heat transfer with surface promoters is often due in part to the fin effect, it is difficult to separate the fin contribution from other factors. For the data discussed here, the promoted heat transfer coefficient is referenced to the base or envelope surface area. In view of the infinite number of possible geometric variations, it is not surprising that, even after more than 400 studies [1], no unified treatment is available.

The enhancement of deep laminar flow is of particular interest to the food and chemical process industries. Gluck [50] used spiral wire inserts to improve heat transfer to water and various power-law liquids. Nusselt number increases up to 580 percent were obtained. Blumenkrantz and Taborek [51] found that the heating of Alta-Vis-530 in spirally fluted tubes was improved up to 200 percent; however, there was negligible enhancement for cooling. Rozalowski and Gater [52] tested both Alta-Vis-530 and Zerolene SAE-50 in flexible hoses with helical convolutes. Nusselt numbers were increased up to 200 percent for heating and 100 percent for cooling.

Augmentation is widely utilized in plate-type compact heat exchangers. Pescod [53] reported on a study of the improvements obtained through the use of spikes and ripples to enhance nominally laminar flow of air in parallel plate channels of large aspect ratios. Most plate heat exchangers utilize corrugated surfaces, for structural reasons as well as augmentation. It is generally agreed that the heat transfer and pressure drop characteristics of commerical corrugated surfaces used in plate exchangers are quite similar for both laminar and turbulent flow.

The diversity of results obtained for turbulent-flow heat transfer to water (composite data for other fluids are similar) in all types of roughened tubes is indicated in Figs. 6 and 7. Here, the simplest coordinates are chosen for illustrative purposes. All calculations are based on the base area of the tube with no allowance made for increases in surface area. While the heat transfer coefficients are increased approximately 4 times at the most, friction factors are increased as much as 58 times. Within this matrix of data lie surfaces that are "efficient" as far as both heat transfer and pressure drop are concerned. The PECs noted in Sec. A.3 can be applied to select the best surface for a given application. For example, the transverse-ribbed surface [12] was selected since its heat transfer and pressure drop characteristics combine to yield one of the best VG-1b ratios.

The ideal situation is to have correlations for h and f which can be introduced into appropriate PECs to obtain the optimum geometry for a particular application. At the present time, however, few correlations are available and these are based on analogy. It has been demonstrated that the analogy between heat transfer and friction for rough surfaces in turbulent flow is dependent upon the type of roughness. An analogy solution for a "sand-grain-type roughness" was developed by Dipprey and Sabersky [65]. Recent

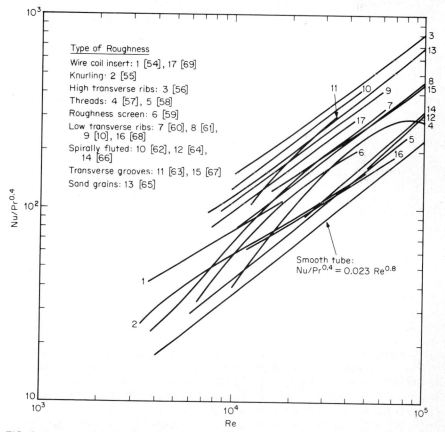

FIG. 6 Summary of heat transfer data for water flowing in internally roughened tubes [9].

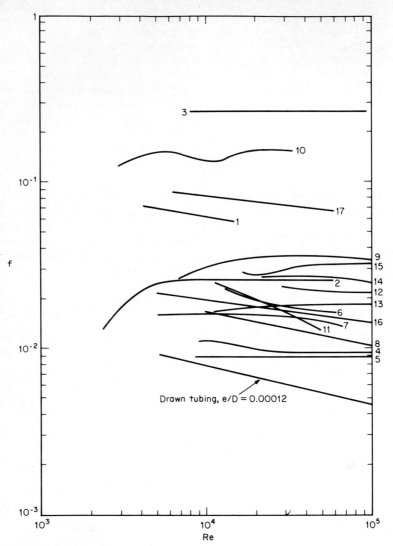

FIG. 7 Summary of friction factor data for water flowing in internally roughened tubes [9].

work has considered surfaces that can be produced commercially. Webb et al. [10] have correlated heat transfer coefficients for tubes with transverse repeated-rib roughness by first interpreting the friction factor data in terms of Nikuradse's similarity function U_e^+, the roughness Reynolds number e^+, and p/e. By application of the heat transfer–momentum transfer analogy, the Stanton number is given by

$$St = \frac{f/2}{1 + (f/2)^{0.5}(\overline{g}F - U_e^+)} \tag{1}$$

where the following correlations and definitions apply:

$$\bar{g} = 4.50(e^+)^{0.28} \quad \text{for} \quad e^+ > 25 \tag{2}$$

$$e^+ = \frac{e}{D} \operatorname{Re}\left(\frac{f}{2}\right)^{0.5} \tag{3}$$

$$F = \operatorname{Pr}^{0.57} \quad \text{for} \quad 0.71 \le \operatorname{Pr} \le 37.6 \tag{4}$$

$$U_e^+ = \left(\frac{2}{f}\right)^{0.5} + 2.5 \ln\frac{2e}{D} + 3.75 \tag{5}$$

$$U_e^+ = 0.95\left(\frac{p}{e}\right)^{0.53} \quad \text{for} \quad e^+ > 35 \tag{6}$$

The correlation for St is actually quite simple for these higher values of e^+, and the calculation is readily carried out for specified roughness, fluid, and velocity. The rather good correlation of data for custom-fabricated repeated-rib roughnesses is demonstrated in Fig. 8.

Equation (3) also successfully predicted air data of other investigations, even for other rib profiles [70]. This similarity correlation method should be valid for any roughness type. It must be recognized, however, that extensive experimental data are required to establish the various functional relations.

Withers [71, 72] applied this technique to commercial single-helix internally-ridged tubes and multiple-helix internally-ridged tubes. The internal profiles are shown in Fig. 9.

Data for single-helix internally-ridged tubes were correlated by

$$St = \frac{(f/2)^{0.5}}{7.22(p/D)^{-0.33} \operatorname{Pr}^{0.5} (e^+)^{0.127} - [U_e^+ - (2/f)^{0.5}]} \tag{7}$$

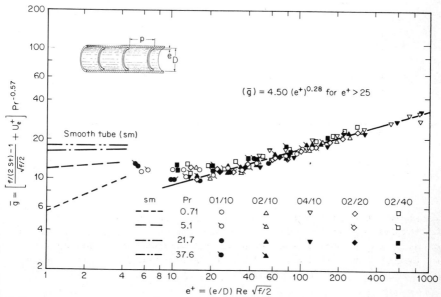

FIG. 8 Correlation of heat transfer data for tubes with internal repeated-rib roughness [10]. Note that 01/10 means $e/D = 0.01$ and $p/e = 10$.

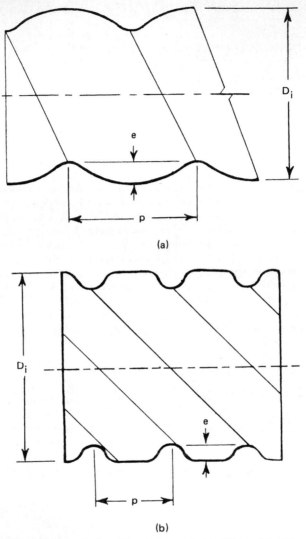

FIG. 9 Internal wall profiles of tubes studied by Withers [71, 72]. (a) Internal wall profile of single-helix internal ridged tube. (b) Internal wall profile of multiple-helix internal ridged tube.

where the friction factor is obtained from

$$\left(\frac{f}{2}\right)^{0.5} = -\frac{1}{2.46 \ln \left[r + (7/\text{Re})^m\right]} \tag{8}$$

and the values of r and m are given for each tube geometry in Ref. 71.

   Similarly, data for the multiple-helix internal-ridged tubes were correlated by

$$\text{St} = \frac{(f/2)^{0.5}}{5.68(e/p)^{-0.125} \, \text{Pr}^{0.5} \, (e^+)^{0.136} - [U_e^+ - (2/f)^{0.5}]} \tag{9}$$

where the friction factor is obtained from Eq. (8) and the tube-specific values of r and m are given in Ref. 72.

A different type of correlation technique has been proposed by Lewis [73]. Basically, the detailed behavior of roughness elements is required: form drag coefficients, heat transfer coefficient distribution, and separation length behind an element. When this information is available, a prediction can be formulated without recourse to data for the actual rough channel. The agreement with experimental data such as those found in Ref. 10 is surprisingly good, considering the "separate effects" character of the model.

The frequently used annular geometry is generally more suitable for the application of surface roughness. Machined surfaces are relatively easy to produce, and increased friction affects only a portion of the wetted surface. The results of Kemeny and Cyphers [74] for a helical groove and a helical protuberance are given in terms of a popular PEC in Fig. 10. The grooved surface is not effective in general, although it does tend to improve with increasing Re. The inferior performance of the coiled wire assembly compared to that of the integral protrusion is probably due to poor thermal contact between the wire and the groove. The results of Bennett and Kearsey [75] for superheated steam flowing in an annulus are included in Fig. 10. The data of Brauer [76] suggest that the optimum L/e for an annular geometry is approximately 3.

Durant et al. [77] summarized an extensive investigation of heat transfer in annuli with the inner, heated tube roughened by means of diamond knurls. At equal pumping power it was found that heat transfer coefficients for the knurled annuli were up to 75 percent higher than those for the smooth annuli.

In commercial gas-cooled nuclear reactors, particularly advanced gas-cooled reactors and gas-cooled fast breeder reactors, increases in core power density have been achieved by artificially roughening the fuel elements. Since bundle experiments are time-consuming and expensive, experiments are usually performed with internally roughened tubes or with annuli having electrically heated roughened inner tubes and smooth outer tubes. In these cases it is necessary to transfer the tube or annulus data to fuel bundle conditions. Dalle Donne and Meyer [78] provide a good general description of this area in their discussion of five popular transformation methods. Considering none of these methods to be entirely satisfactory, they propose another method and apply it to two-dimensional rectangular ribs. A more recent discussion of transformation methods and their applica-

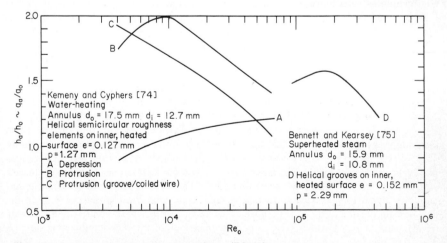

FIG. 10 Performance of annuli with roughness (FG-2a).

tion to rectangular and trapezoidal ribs is given by Hudina [79]. Recent efforts have centered on three-dimensional roughness since it seems to offer more favorable thermal-hydraulic performance.

Dalle Donne [80] notes that at a Reynolds number of 10^5, for two-dimensional roughness, Stanton numbers are typically increased by a factor of 2 while the corresponding increase in friction factor is 4. He also cites data for two 3-dimensional roughnesses which indicate that the Stanton number increased by 3 and 4 with the friction factor ratios 8 and 12, respectively. The first roughness is depicted in Fig. 11. Large-scale computer codes such as SAGAP0 [81] are required to predict accurately the thermal-hydraulic behavior of roughened fuel element bundles.

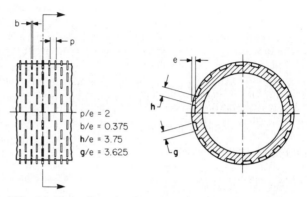

$$p/e = 2$$
$$b/e = 0.375$$
$$h/e = 3.75$$
$$g/e = 3.625$$

FIG. 11 Three-dimensional roughness with alternate studs [80].

It is common practice in gas-cooled reactor technology to interpret data for rough rods in terms of a merit index $(St^3/f)^{1/2}$. The average heat flux for a heat exchanger such as the reactor core is proportional to this parameter for a given ratio of pumping to thermal power and a given average film temperature drop. For the two-dimensional roughness, then, this ratio is 1.4, while for the best three-dimensional roughness the ratio is 2.3.

Turning to cross flow over bundles of tubes with surface roughness, some work has been done in the context of heat exchanger development for gas-cooled reactors and conventional shell-and-tube heat exchangers. This work is backed up by extensive studies of single cylinders such as those with pyramid roughness elements tested in air by Achenbach [82]. Nusselt number increases up to about 150 percent were recorded. Zukauskas et al. [83] obtained similar improvements with a single pyramid-roughened cylinder for cross flow of water.

The preceding discussion indicates that certain types of roughness can improve heat transfer performance considerably. Under nonuniform flow or thermal conditions, however, it may be advantageous to roughen only that portion of the heating surface which has a higher heat flux or lower heat transfer coefficient. In many cases the overall pressure drop will not be greatly affected by roughening the hot spot. Any of the foregoing roughness types are then of interest since they produce large increases in heat transfer coefficient over the smooth-tube value at equal flow rates. The partial roughening technique has been considered for gas-cooled reactors where the thermal limit is reached only in the downstream portion of the core because of the axial heat flux variation [84]. One scheme for achieving selective roughening in large-diameter pipes involves sandblasting through a smaller tube that is transversed inside the pipe [85].

2. Boiling

Pool boiling with rough structured surfaces is discussed in Sec. B.1. Attention here is focused on the effect of surface roughness on forced-convection flow.

Consider first the gravity-driven flows observed in horizontal-tube spray-film evaporators. Longitudinal ribs or grooves may promote turbulence, but they impede film drainage. Knurled surfaces provide turbulence promotion and may also aid liquid spreading over the surface. When nucleate boiling occurs within the film, it appears that bubble motion is favorably affected by the roughness. Knurling increases coefficients by as much as 100 percent [86]. On the basis of experience with single-phase films (trickle coolers), Newson [87] suggests a longitudinal rib profile for horizontal-tube multiple-effect evaporators.

Annular channels with electrically heated inner tubes have been used to study the effects of roughness on forced-convection boiling. Surface conditions do not appear to significantly alter the boiling curve for reasonably high flow velocities; however, certain surface finishes improve flow-boiling CHF. Durant et al. [77, 88] have demonstrated that there is a substantial increase in subcooled CHF with knurls or threads. It was also suggested that the critical fluxes for the rough tubes were up to 80 percent higher than those for smooth tubes at comparable pumping power. Gomelauri and Magrakvelidze [89] found that for two-dimensional roughness, CHF is dependent on subcooling, with decreases observed at low subcooling and increases up to 100 percent observed at high subcooling. Murphy and Truesdale [90] found that subcooled CHF was decreased 15 to 30 percent with large roughness heights.

For bulk boiling of R-12 in commercial helical-corrugated tubing, Withers and Habdas [91] observed up to 100 percent increase in heat transfer coefficient and up to 200 percent increase in CHF. Several investigators have demonstrated that bulk-boiling CHF can be improved by 50 to 100 percent with various other surface modifications: Bernstein et al. [92] (irregular-diameter tubing and slotted helical inserts), Janssen and Kervinen [93] (sandblasting), and Quinn [94] (machined protuberances). Of particular significance for power boilers are the increases in CHF for high-pressure water observed with helical-ribbed tubes. Typical results are given in Fig. 12. Pseudo film boiling is also suppressed with this commercial tubing [96]. Studies of post CHF, or dispersed-flow film boiling, indicate that roughness elements increase the heat transfer coefficient [94, 97].

3. Condensing

Medwell and Nicol [98, 99] were among the first to study the effects of surface roughness on condensate films. They condensed steam on the outside of one smooth and three artificially roughened pipes with pyramid-shaped roughness. All were oriented vertically, and the condensate was drained under gravity alone. The mean heat transfer coefficients were found to increase significantly with roughness height, the values of the roughest tube being almost double those of the smooth tube. Carnavos [100] recently reported that condensing-side coefficients for knurled tubes were 4 to 5 times the smooth-tube values. Part of this, of course, can be attributed to the area increase.

Cox et al. [101] used several kinds of augmented tubes to improve the performance of horizontal-tube multiple-effect plants for saline water conversion. Overall heat transfer coefficients (forced-convection condensation inside and spray-film evaporation outside) were reported for tubes internally augmented with circumferential V grooves (35 percent maximum increase in U) and protuberances produced by spiral indenting from the out-

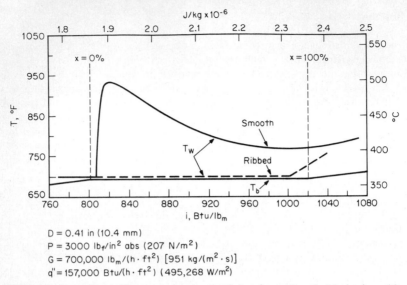

D = 0.41 in (10.4 mm)
P = 3000 lbf/in² abs (207 N/m²)
G = 700,000 lbm/(h·ft²) [951 kg/(m²·s)]
q″ = 157,000 Btu/(h·ft²) (495,268 W/m²)

FIG. 12 Comparison of heat transfer characteristics of smooth and ribbed tubes with once-through boiling of water [95].

side (4 percent increase). No increases were obtained with a knurled surface. Prince [102] obtained a 200 percent increase in U with internal circumferential ribs; however, the outside (spray-film evaporation) was also enhanced. Luu and Bergles [15] recently reported data for enhanced condensation of R-113 in tubes with helical repeated-rib internal roughness. Average coefficients were increased 80 percent above smooth-tube values. Coefficients with deep spirally fluted tubes (envelope diameter basis) were increased by 50 percent.

Random roughness consisting of attached metallic particles (50 percent area density and $e/D = 0.031$) were proposed by Fenner and Ragi [103]. With R-12 the condensing coefficient was increased 300 percent for qualities greater than 0.60, and 140 percent for lower qualities.

D. EXTENDED SURFACES

1. Single-Phase Flow

a. Free Convection

The topic of finned surfaces in free convection is covered in Sec. C.2 of Ref. 367. From the standpoint of augmentation, current interest focuses on interrupted extended surfaces such as the wire-loop fins used for baseboard hot water heaters or "convectors," finned arrays or "heat sinks" used for cooling electronic components, and serrated fins used in process cooler tube banks. While natural circulation is an important normal or off-design condition in these applications, the flow is basically forced convection due to the chimney effect of the ducting. Hence, it is appropriate to include comments on this subject under the topic of compact heat exchangers with forced convection.

b. Compact Heat Exchangers

Compact heat exchangers have large surface-area-to-volume ratios, primarily through the use of finned surfaces. (See Chap. 4, Part 3, Sec. A.) An informative collection of articles related to the development of compact heat exchangers is presented by Shah et al. [104].

Compact heat exchangers of the plate-fin, tube and plate-fin, or tube and center variety use several types of augmented surfaces: offset strip fins, louvered fins, perforated fins, or wavy fins [105]. The flow (usually gases) in these channels is very complex and few generalized correlations or predictive methods are available. Overall heat exchanger information is often available from manufacturers of surfaces for the automotive, air conditioning, and power industries; however, the air-side coefficients cannot be readily deduced from the published information. In general, many proprietary fin configurations have been developed that have heat transfer coefficients 50 to 100 percent above those of flat fins. The improvements are the result of flow separation, secondary flow, or periodic starting of the boundary layer. The latter is illustrated in Fig. 13. It should be emphasized that the tube geometry and arrangement strongly affect the heat transfer and pressure drop. For example, heat transfer coefficients are increased with staggered tubes and pressure drop is reduced with flattened tubes (in the flow direction). Design data for augmented compact heat exchanger surfaces are given by Kays and London [107]. A recent overview of mechanisms was presented by Webb [108].

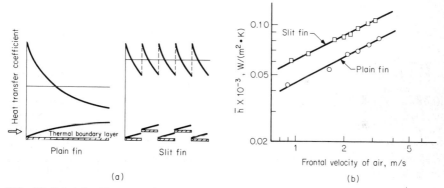

FIG. 13 Principle of interrupted fins and data for auto air conditioner core [106]. (*a*) Local heat transfer coefficients for plain fin and slit fin. (*b*) Average heat transfer coefficients for plain fin and slit fin.

Circular or oval finned-tube banks utilize a variety of enhanced surfaces, as illustrated in Fig. 14. With the exception of material pertaining to smooth helical fins, data are rather limited and no generalized correlations presently exist. Webb [108] provides a guide to the available literature.

Recently, some complex compact heat exchanger surfaces have been studied using mass transfer methods, e.g., naphthalene sublimation [109] and chemical reaction between a surface coating and ammonia added to the air stream [110]. These elegant but tedious methods yield local mass transfer coefficients that can be used to infer heat transfer coefficients by the usual analogy. This detailed information, in turn, should aid in the development of more efficient surfaces. Numerical studies are also yielding useful predictions for laminar flows [111, 112].

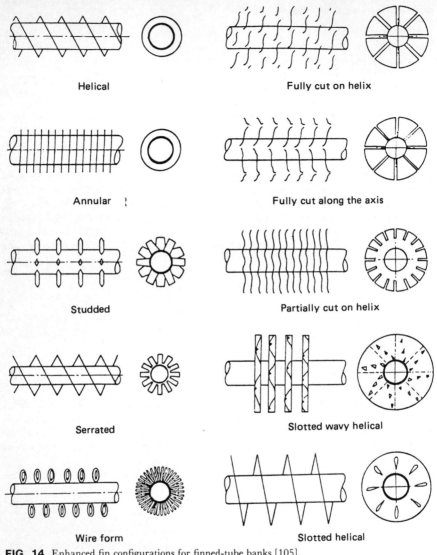

Helical

Fully cut on helix

Annular

Fully cut along the axis

Studded

Partially cut on helix

Serrated

Slotted wavy helical

Wire form

Slotted helical

FIG. 14 Enhanced fin configurations for finned-tube banks [105].

c. Internal Flow

Internally finned circular tubes are commercially available in aluminum and copper (or copper alloys). For laminar flow, the following correlations are available [113]:

Spiral-fin tubes:

$$\frac{\text{Nu}_h}{\text{Pr}^{1/3}} \left(\frac{D_h}{L} \right)^{1/3} \left(\frac{\mu_b}{\mu_w} \right)^{0.14} \phi = 19.2 \left(\frac{b}{p} \right)^{0.5} \text{Re}_h^{0.26} \tag{10}$$

where

$$\phi = 2.25 \frac{1 + 0.01 \text{ Gr}_h^{1/3}}{\log \text{Re}_h} \tag{11}$$

Straight-fin tubes:

$$\frac{\text{Nu}_h}{\text{Pr}^{1/3}} \left(\frac{D_h}{L}\right)^{1/3} \left(\frac{\mu_b}{\mu_w}\right)^{0.14} \phi = 2.43 \left(\frac{1}{n}\right)^{0.5} \text{Re}_h^{0.46} \tag{12}$$

Isothermal friction factors for all tubes:

$$f_h = \frac{16.4(D_h/D)^{1.4}}{\text{Re}_h} \tag{13}$$

These correlations are based on data for oil in horizontal tubes having approximately uniform temperature (steam heating). Other data for both water and ethylene glycol in both steam-heated and electrically heated tubes are in approximate agreement with the correlations [113]. As noted in Ref. 115, the analytical results for uniformly heated tubes are not in good agreement with data.

The following equations are recommended for turbulent flow in straight- and spiral-fin tubes [116]:

$$\text{Nu}_h = 0.023 \text{ Pr}^{0.4} \text{ Re}_h^{0.8} \left(\frac{A_F}{A_{Fi}}\right)^{0.1} \left(\frac{A_i}{A}\right)^{0.5} (\sec \alpha)^3 \tag{14}$$

$$f_h = 0.046 \text{ Re}_h^{-0.2} \left(\frac{A_F}{A_{Fi}}\right)^{0.5} (\sec \alpha)^{0.75} \tag{15}$$

These correlations are based on data for air (cooling), water (heating), and ethylene glycol–water (heating). It is noted that fin inefficiency corrections must be incorporated when applying the equations.

Hilding and Coogan [117] provide data for longitudinal interrupted fins, with air as the working fluid. They conclude that regular interruptions of the fins improve the relative heat transfer–flow friction performance in the laminar and transition regions, but that little advantage is apparent in the turbulent region. Several manufacturers now provide this type of surface.

The first analytical study to predict the performance of tubes with straight inner fins for turbulent airflow was conducted by Patankar et al. [118]. The mixing length in the turbulence model was set up so that just one constant was required from experimental data. Expansion of analytical efforts to fluids of higher Prandtl number, tubes with practical contours, and tubes with spiraling fins is desirable. It would be particularly significant if the analysis could predict with a reasonable expenditure of computer time the optimum fin parameters for a specified fluid, flow rate, etc.

Internally finned tubes can be "stacked" to provide multiple internal passages of small hydraulic diameter. Carnavos [119] demonstrated the large increases in heat transfer coefficient (based on outer tube nominal area) that can be obtained in these tubes with airflow. Of course, pressure drop is also increased greatly in these tubes. Finned annuli represent one case of considerable practical interest.

Finned concentric annuli, with the fins extending about 85 percent of the way across the gap, have been studied by several investigators in connection with widespread process industry application.

Augmented fins (interrupted, cut and twisted, perforated) are frequently used [105,

120]. Gunter and Shaw [121] demonstrate that cut-and-twisted finned tubes (Fig. 15) have substantially higher coefficients than continuously finned tubes.

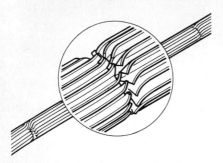

Having assembled the available data, Clarke and Winston [122] recommend the correlation curve shown in Fig. 16. The Reynolds number is based on the equivalent diameter, L is the length of the finned tube, and $\mathcal{P}$ is the wetted perimeter of the channel between two longitudinal fins. In the laminar range, Re < 2000, the curve is valid for both continuous and cut fins if L is interpreted as the distance between cuts, L_c.

FIG. 15 Close-up of finned tube, showing cut-and-twist construction [122].

Multiple longitudinal-fin tubes are often placed in a single pipe. According to Kern and Kraus [120], the correlation given in Fig. 16 is valid, but the entire flow cross section is considered when the hydraulic diameter is evaluated.

Similar longitudinal fins have been used for gas-cooled nuclear reactors. A discussion of this application is given by El-Wakil [123].

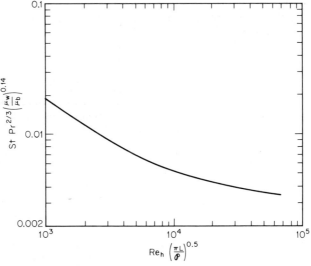

FIG. 16 Correlation of heat transfer data for longitudinally finned tubes in annuli [122].

2. Boiling

Low and medium fin tubes with external, circumferential fins are produced by many manufacturers for pool boiling of refrigerants and organics. Gorenflo [124] reported single-tube boiling of R-11, which indicated that the heat transfer coefficients based on total area were higher for all six tubes tested than for the reference plain tube. As shown in Fig. 17, Hesse [125] found that the heat transfer coefficient for pool boiling of R-114 was

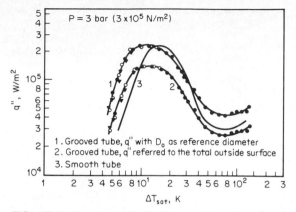

FIG. 17 Boiling curves for R-114, smooth and low-finned tubes [125].

higher for finned tubes than for a plain tube; however, q''_{cr} was lower. Both comparisons were made on a total-area basis. When referenced to the projected area at the maximum outside diameter, the critical heat fluxes were about equal to those of the plain tube. The reduction in local CHF is apparently due to bubble interference between fins. Westwater [126] suggests that the fins may be spaced as close as the departure diameter of a nucleate-boiling bubble ($\sim$1.55 mm for R-113).

Katz et al. [127] and Nakajima and Shiozawa [128], for example, found that coefficients on the upper finned tubes in a bundle are higher than coefficients on the lower tubes, as a result of bubble-enhanced circulation. Similar results are reported by Arai et al. [129] for a bundle of Thermoexcel-E tubes. It is quite probable that with certain types of augmented tubes it is sufficient to use the special tubes only in the lower rows since the augmented circulation in the upper rows is so high that augmentation is not effective there.

Data for a three-dimensional finned (square, straight-sided fins in a square array) tubular test section were reported by Corman and McLaughlin [130]. Significant reductions in temperature differences were observed in low-flux nucleate boiling; however, temperature differences were larger at high fluxes. The critical heat flux was increased.

A thorough survey of large-scale finned surfaces utilized for pool boiling is given by Westwater [126]. Properly shaped or insulated fins promote very large heat transfer rates if the base temperature is in the film-boiling range.

Circumferential fins with various fin profiles have been suggested for enhancement of horizontal-tube spray-film evaporation. Area increases are typically 2:1, and in certain cases the fins promote redistribution of the liquid so that thin films are present at the peaks. V-shaped grooves (threads) produce improvements in evaporating heat transfer coefficients of up to 200 percent [131, 132]. An improvement of 200 percent in the overall coefficient was reported by Prince [102] for tubes with straight circumferential flutes outside and shadow ribs inside (condensing). On the basis of analytical results, Sideman and Levin [133] concluded that square-edged grooves should have the best operational characteristics of flow rate and heat transfer. Cox et al. [134] concluded that spirally fluted tubes offered no particular advantage for this service.

With falling film evaporation inside vertical tubes, Thomas and Young [135] found that heat transfer coefficients could be increased by more than a factor of 10 with loosely

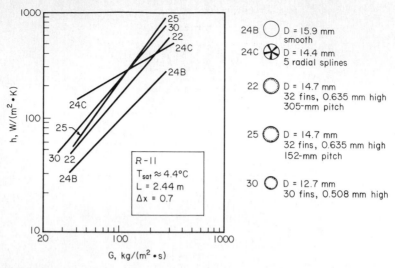

FIG. 18 Heat transfer coefficients for evaporation in internally finned tubes [138].

attached internal fins. While distorted tubes (e.g., doubly fluted and spirally corrugated) have been developed primarily for augmentation of condensation on the outside wall, it is expected that heat transfer coefficients for the evaporating fluid on the inside of the tube should also be increased. This has been confirmed in large-scale tests by Lorenz et al. [136], who obtained a 150 percent increase in vaporization coefficient (external falling film of ammonia) for a doubly fluted tube as compared to the prediction for an equivalent smooth tube. This is greater than the area enhancement of 57 percent.

Tubes with integral or inserted internal fins increase heat transfer rates for horizontal forced-convection vaporization of refrigerants by as much as several hundred percent more than smooth-tube values (Schlünder and Chwala [137], Kubanek and Miletti [138], and Ito and Kimura [139]). Data from the second study are shown in Fig. 18; the heat transfer coefficients are based on the surface area of the smooth tube of the same diameter.

Plate-fin heat exchangers are widely used in process heat exchanges. As described in the review by Robertson [140], the various forms of enhanced fins used for single-phase compact heat exchanger cores are also used for evaporators. These include perforated fins, offset strip fins (serrated), and herringbone fins. Both forced-convection and falling-film evaporation modes in an offset strip fin compact heat exchanger were tested by Panchal et al. [141] under expected OTEC conditions. A composite heat transfer coefficient of 2525 Btu/(h·ft²·°F), or 14,338 W/(m²·K), was obtained. In the subsequent analysis of Yung et al. [142], the enhancement is attributed to splitting of the film. The thinner film results in a higher heat transfer coefficient than would be obtained with plain fins of the same maximum channel width. This situation is illustrated in Fig. 19. This plate-fin heat exchanger also performs well for forced-convection (vertical upflow) vaporization [141]. An analysis for this situation was recently presented by Chen et al. [143]. Recognizing the periodic redevelopment of the flow, the local heat transfer coefficient was assumed to be determined by the local ΔT. The computer solution is in good agreement with test data.

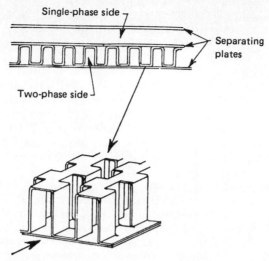

FIG. 19 Sections of an offset strip fin evaporator [142].

3. Condensing

Surface extensions are widely employed for augmentation of condensation on horizontal or vertical tubes. Consider first horizontal arrangements. Integral low-fin tubing is produced by many manufacturers for shell-and-tube air conditioning and process condensers. The increased area and thin condensate film near the fin tips result in heat transfer coefficients several times those of a plain tube with the same base diameter. Coefficients based on total area are also higher. However, condensate may bridge the fins and render the enhancement ineffective if the fin spacing is small or if the liquid has a high surface tension. Fins are used with refrigerants and other organic fluids with low surface tension since the condensing side often represents the dominant thermal resistance. Normally, finning would not be used for steam power plant condensers because of the high surface tension of water and the relatively low thermal resistance of the condensing side.

Beatty and Katz [144] proposed the following equation for the average heat transfer coefficient (total-area basis) for single low-fin tubes on the basis of data for a variety of fluids:

$$h = 0.689 \left(\frac{k_f^3 \rho_f^2 g i_{lg}}{\mu_f \Delta T_{gw}} \right)^{1/4} \left(\frac{A_r}{A_e} \frac{1}{D_r^{0.25}} + 1.3 \frac{\eta A_F}{A_e} \frac{1}{L_f^{0.25}} \right) \tag{16}$$

For the copper-finned tubes normally recommended for commercial condensers, the effective surface area A_e is taken to be the total outside surface area. For similar conditions, the mean effective length of a fin is given by $\pi(D_o^2 - D_r^2)/4D_o$. This semiempirical equation is based on the assumption that condensate is readily drained by gravity. Young and Ward [145] suggest that turbulence and vertical row effects can be included by simply using a different constant for a specific fluid. Rudy and Webb [146] recently suggested that the success of Eq. (16) may be fortuitous in that surface flooding due to condensate retention is compensated for by surface-tension-induced drainage.

There has been recent interest in three-dimensional surfaces for horizontal-tube condensers. A finned and machined surface having three-dimensional character is described by Nakayama et al. [28] and Arai et al. [129]. As shown in Fig. 20, the surface resembles a low-fin tube with notched fins. Coefficients (envelope-area basis) are as much as 7 times the smooth-tube values. The considerable improvement relative to conventional low fins is apparently due to multidirectional drainage at the fin tips.

Circular pin fins have been tested by Chandran and Watson [147]. Their average coefficients (total-area basis) were as much as 200 percent above the smooth-tube values. Square pins have been proposed by Webb and Gee [148]; a 60 percent reduction of fin material as compared to integral-fin tubing is predicted using a gravity drainage model. A new concept of three-dimensional surface was described by Notaro [149]. Small metal particles are bonded randomly to the surface. The upper portions of the particles promote effective thin-film condensation, and the condensate is drained along the uncoated portion of the tube.

The analytical foundations of contoured surfaces are based on the 1954 paper of Gregorig [150]. The shape suggested by Gregorig is shown in Fig. 21. Condensation occurs primarily at the tops of convex ridges. Surface tension forces then pull the condensate into the grooves, which act as drainage channels. The average heat transfer coefficient is substantially greater than that for a uniform film thickness. While manufacturing considerations severely limit the practical realization of such optimum shapes, the surface-tension-driven cross flows are very much in evidence in the finned tubes mentioned above.

Thomas et al. [151] reported a simple resolution of the manufacturing problem. A smooth tube was wrapped with wire so that surface tension pulls the condensate to the

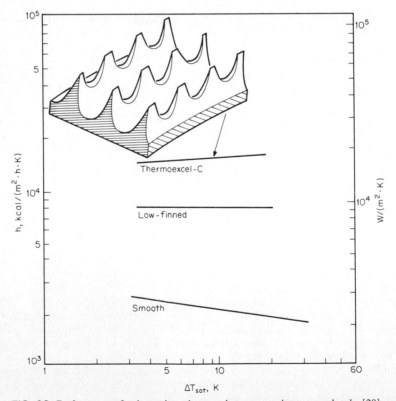

FIG. 20 Performance of enhanced condenser tubes compared to a smooth tube [28].

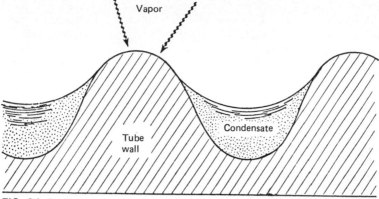

FIG. 21 Profile of the condensing surface developed by Gregorig [150].

base of the wire. The spaces between wires then act as runoff channels. Tests with ammonia indicated a condensing coefficient about 3 times that predicted for the smooth tube.

Several studies considered commercial deep spirally fluted tubes for horizontal single-tube and tube-bundle condensers [152–154]. Commercially available configurations bear some resemblance to the Gregorig profile shown in Fig. 21. The derived condensing coefficients (envelope basis) range from essentially no improvement to over 300 percent above the plain-tube values. More work is clearly needed in this area.

Carnavos [155] reported typical *overall* improvements that can be realized with a variety of commercially available enhanced horizontal condenser tubes. The heat flux for single 130-mm-long tubes, in most cases 19-mm outside diameter, is plotted in Fig. 22 against ΔT_{lm} for 12 tubes qualitatively described in the accompanying table. The overall heat transfer performance gain of the enhanced tubes over the smooth tube is as high as 175 percent. Internal enhancement is a substantial contributor to the overall performance since the more effective external enhancements produce a large decrease in the shell-side thermal resistance. These results for a refrigerant, and the data of Marto et al. [154] and Mehta and Rao [156] for steam, provide good practical guidance for the use of enhanced tubes in surface condensers.

Some detailed results are available for condensing in bundles of enhanced tubes. Withers and Young [157] found, for example, that the vertical row effect for corrugated tubes was different from that for bare tubes; in particular, the augmented tubes were less sensitive to the number of rows.

The enhancement of vertical condensers remains an area of high interest due to potential large-scale power and process industry applications, e.g., desalination, reboilers, and OTEC power plants. Tubes with exterior longitudinal fins or flutes, spiral flutes, and flutes on both the interior and exterior (doubly fluted) have been developed and tested. The common objectives are to use the Gregorig effect to create thin-film condensation at the tips of the flutes and to drain effectively.

Vertical wires, loosely attached and spaced around the tube circumference, provide a simple realization of the desired profile. Thomas reported increases in heat transfer coefficient of up to 800 percent for circular wires [158]. Square wires were found to have a greater condensate-carrying capacity than circular wires of the same dimension [159].

The recent study of Mori et al. [160] represents a good example of the type of sophisticated analysis that can be performed to obtain the optimum geometry. According to their numerical analysis, the optimum geometry is characterized by four factors: sharp leading edge, gradually changing curvature of the fin surface from tip to root, wide

Tube	Outside	Inside
S	smooth	smooth
WT-1	integral helical fins	ribbed inside
WT-2	integral helical fins	ribbed inside
W-1	integral helical fins	plain inside
W-2	integral helical fins	plain inside
HC	interrupted helical fins	plain inside
HP	integral helical fins	plain inside
T	deep spiral flutes	deep spiral flutes
N-1	continuous trapezoidal flutes	helical ribs
N-2	continuous trapezoidal flutes	plain inside
FC-1	trapezoidal pin fins	helical ribs
FC-2	trapezoidal pin fins	plain inside

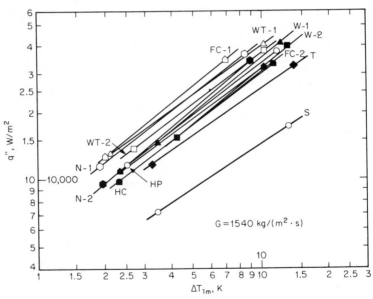

FIG. 22 Overall performance of condenser tubes tested by Carnavos [155].

grooves between fins to collect condensate, and horizontal disks attached to the tube to periodically strip off condensate. The recommended geometry is shown in Fig. 23. Figure 24 presents typical results which illustrate the character of the optimum. The periodic removal of condensate resolves the drainage problem with uniform axial geometry. Ideally, the flute size should be changed axially to allow for condensate buildup; however, this compounds the manufacturing difficulty.

Barnes and Rohsenow [161] present a simplified analytical procedure for determining the performance of vertical fluted-tube condensers. For a sine-shaped flute, the average condensing heat transfer coefficient depends on tube geometry approximately as follows:

$$\overline{h} \approx \frac{a^{0.231}}{pL_s^{0.0774}} \tag{17}$$

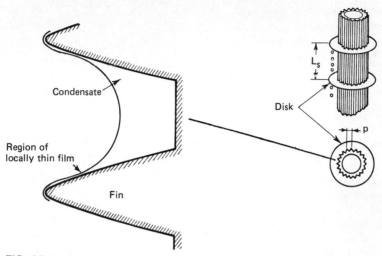

FIG. 23 Recommended flute profile and condensate strippers according to Mori et al. [160].

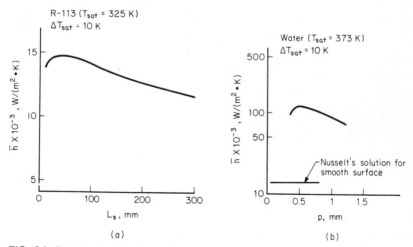

FIG. 24 Optimization of stripper distance and flute pitch for configuration shown in Fig. 23 [160]. (a) Average heat transfer coefficient versus stripper distance. (b) Average heat transfer coefficient versus flute pitch.

for a given fluid and ΔT. The recommended design procedure includes guidelines for placing the strippers. The design can be easily carried out with a hand calculator.

An important large-scale test of a doubly fluted tube for OTEC condenser service was reported by Lewis and Sather [162]. The tube is the same one noted earlier under enhancement of falling-film evaporation [136]. The ammonia-side heat transfer coefficient was enhanced several times to a value of 3730 Btu/(h·ft²·°F), or 21,181 W/(m²·K).

A major study of condensing on the outside of vertical enhanced tubes has been carried out at Oak Ridge National Laboratory in connection with geothermal Rankine cycle condensers. About 12 tubes were tested with ammonia, isobutane, and various fluorocarbons. The most recent report by Domingo [163] on R-11 concluded that the best surface was the axially fluted tube, followed, in order, by the deep spirally fluted tube, spiral tubes, and roped tubes. The composite (vapor and tube wall) heat transfer coefficient was as much as 5.5 times the smooth-tube value. This high performance was further improved to a factor of 7.2 by using skirts to periodically drain off the condensate.

A guide to the *overall* performance of a wide variety of typical vertical evaporator tubes with condensing outside and vaporization inside is given in Fig. 25. This survey by Alexander and Hoffman [164] was specifically directed at vertical-tube evaporators for desalination systems. It is seen that the best surfaces yield increases in overall coefficient up to 200 percent.

The offset strip fin compact heat exchanger configuration has also been suggested for OTEC condensers. The flow situation is similar to that of the evaporator noted in Fig. 19. A composite ammonia-side heat transfer coefficient of 4600 Btu/(h·ft^2·°F), or 26,122 W/(m^2·K), was reported.

Vrable et al. [165] studied horizontal in-tube condensation of R-12 in internally finned tubes. The heat transfer coefficient (envelope basis) was increased by about 200 percent. Reisbig [166] also condensed R-12 (with some oil present) in internally finned tubes having an increased area of up to 175 percent. The nominal heat transfer coefficient was increased by up to 300 percent. Royal and Bergles [167, 168] presented heat transfer and pressure drop data, with correlations, for condensation of steam inside tubes with straight or spiraled fins. Up to 150 percent increases in average coefficients were observed for complete condensation. Luu and Bergles [169, 170] tested similar tubes with complete condensation of R-113 and found that average coefficients were elevated by up to 120 percent. Grooved tubes were also effective, with increases of 250 percent being reported for longitudinal grooves with steam at high velocities [171] and 100 percent for spiral

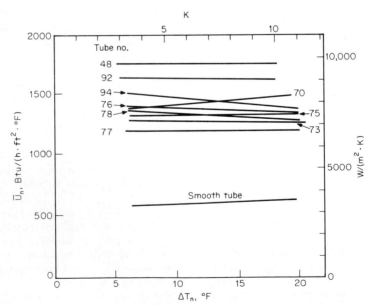

FIG. 25 Heat transfer enhancement in vertical tube evaporator (tubes 48, 70, 94: axial flutes; others; spiral flutes) [163].

grooves with R-113 [172]. The relative heat transfer–pressure drop performance of the grooved tubes is considered superior to that of the finned tubes.

The provisionally recommended design equation for steam is [167]:

$$\bar{h}_i = 0.0265 \frac{k_l}{D_h} \left(\frac{G_e D_h}{\mu_l} \right)^{0.8} Pr_l^{0.33} \left[160 \left(\frac{H^2}{lD_i} \right)^{1.91} + 1 \right] \tag{18}$$

where

$$G_e = G \left[(1 - \bar{x}) + \bar{x} \sqrt{\frac{\rho_l}{\rho_g}} \right]$$

For refrigerants the design equation is [169]:

$$\bar{h}_i = 0.024 \frac{k_l}{D_h} \left(\frac{GD_h}{\mu_l} \right)^{0.8} Pr_l^{0.43} \left(\frac{H^2}{lD_i} \right)^{-0.22} \times \frac{1}{2} \left[\left(\frac{\rho}{\rho_m} \right)_{in}^{0.5} + \left(\frac{\rho}{\rho_m} \right)^{0.5} \right] \tag{19}$$

where

$$\frac{\rho}{\rho_m} = 1 + \frac{\rho_l - \rho_g}{\rho_g} \bar{x}$$

E. DISPLACED ENHANCEMENT DEVICES

1. Single-Phase Flow

A rather large variety of tube inserts fall into the category of displaced enhancement devices. The heated surface is left essentially intact, and the fluid flow near the surface is altered by the insert, which might be metallic mesh, static mixer elements, rings, disks, or balls. Laminar heat transfer data for uniform-wall-temperature tubes and uniformly heated tubes are plotted in Figs. 26 and 27, respectively. The isothermal friction factors are plotted in Fig. 28.

Koch [4] employed suspended rings and disks as inserts, as well as tubes packed with Raschig rings and round balls. The disks give maximum augmentation with moderate increases in friction factors, as indicated in Figs. 26 and 28 (curve d). Augmentation of heat transfer with rings and round balls is quite comparable to that with disks, but rings and balls increase the friction factor more than 1600 percent (curves c and d). For further comments on packed tubes, see Chap. 6, Sec. B.1.c.

Sununu [173] and Genetti and Priebe [174] used Kenics static mixers for heating of viscous oils (curves e and f). These mixers consist of 360° segments of twisted tapes; every second element is inverted and the segments are tack-welded together. The augmentation of heat transfer is about 150 to 200 percent, but the increase in friction factor is almost 900 percent. Sununu proposed a correlation for Nu, but his data exhibit large scatter about the correlation. Genetti and Priebe have also correlated their heat transfer data, with more success.

Van der Meer and Hoogendoorn [175] used Sulzer mixers for the heating of silicon oil (curves g and h). Each mixer element consists of several layers of corrugated sheet. An increase in heat transfer coefficient of about 400 percent is reported.

In the case of uniformly heated tubes, very high heat transfer coefficients have been obtained with the SMV Sulzer mixer (curves i and j of Fig. 27). Comparable heat transfer augmentation is also obtained with Kenics static mixers.

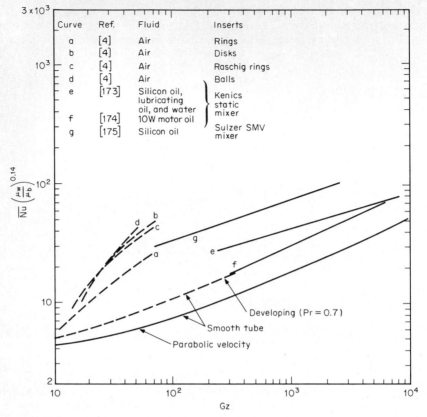

FIG. 26 Representative heat transfer data for displaced promoters in tubes with uniform wall temperature.

Presently, many companies around the world are involved in the manufacture of static mixers for liquids, to promote either heat transfer or mass transfer. The variety of these mixers, their construction, and other characteristics are described in a comprehensive review article by Pahl and Muschelknautz [178]. There are no broad-based correlations available, because of the many geometrical arrangements and the strong influence of fluid properties and heating conditions.

Similar inserts or packings have been used for turbulent flow; however, this application is usually considered only for short sections since the pressure drop is so high. The problem is illustrated by the results of Koch [4], who placed thin rings or disks in a tube. The typical basic data shown in Fig. 2 translate to performance data in Fig. 29, where it is seen that rings are effective only in the lower Reynolds number range. These data, as well as the data of Evans and Churchill [179] (disks and streamlined shapes in tubes), Thomas [180] (rings in annuli), and Maezawa and Lock [181] ("Everter" and disk inserts), indicate that these inserts are not particularly effective for turbulent flow.

Mesh or spiral brush inserts were used by Megerlin et al. [182] to enhance turbulent heat transfer in short channels subjected to high heat flux. The largest recorded improve-

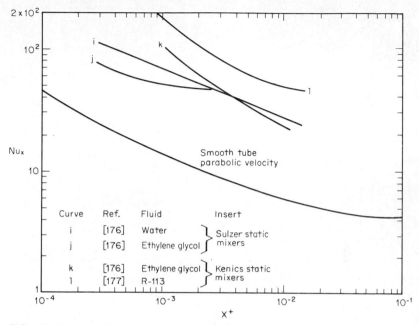

FIG. 27 Representative heat transfer data for displaced promoters with uniform heat flux.

ments in turbulent heat transfer coefficients were obtained: up to 8.5 times; however, the pressure drop was up to 2800 times larger. In general, it appears that these displaced enhancement devices are useful in very few practical turbulent situations for reasons of pressure drop, plugging or fouling, and structural considerations.

2. Flow Boiling

Janssen and Kervinen [93] reported on bulk boiling CHF with displaced turbulence promoters. Flow-disturbing rings were located on the outer tube of an annular test section. Figure 30 shows that critical heat fluxes for quality boiling with the rough liner were as much as 60 percent greater than those for the smooth liner. This is to be expected since the rings strip the liquid from the inactive surface, thereby increasing the film flow rate on the heated surface. Moeck et al. [183] performed an extensive investigation of CHF for annuli with rough outer tubes. Steam-water mixtures were introduced at the test-section inlet so as to obtain high outlet steam qualities. It was found that the critical heat flux increased as the roughness height (1.3 mm maximum) increased and spacing (38 to 114 mm) decreased, with a maximum increase of over 600 percent based on similar inlet conditions. The pressure drop with the most optimum promoter was about 6 times the smooth-annulus value for similar inlet conditions at the critical condition.

Rough liners were also found to produce significant increases in critical power for a simulated boiling-water reactor (BWR) rod bundle [184]. As reported by Quinn [185], rings of stainless steel wire, $e = 1.12$ mm and $p = 25.4$ mm, were spot-welded to the channel wall of a two-rod assembly. Both CHF and film-boiling heat transfer coefficients were improved.

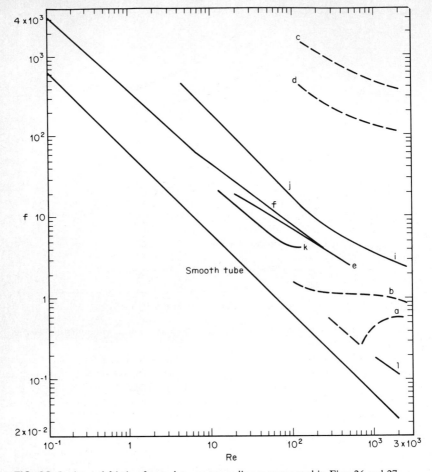

FIG. 28 Isothermal friction factor data corresponding to tests noted in Figs. 26 and 27.

Ryabov et al. [186] summarized a major study of increasing critical power in rod bundles by use of special spacers, inserts, etc. A comprehensive review of the effects of spacing devices on CHF was recently presented by Groeneveld and Yousef [187]. Figure 31 qualitatively illustrates the expected effect. The majority of the studies cited report beneficial effects of spacing devices on CHF; however, several investigations also report detrimental effects.

Megerlin et al. [182] reported subcooled boiling data for tubes with mesh and brush inserts. Critical heat fluxes were increased by about 100 percent; however, wall temperatures were very high on account of the onset of partial film boiling.

3. Condensing

Azer et al. [188] reported data for condensation in tubes with Kenics static mixer inserts. Substantial improvements in heat transfer coefficients were reported; however, the

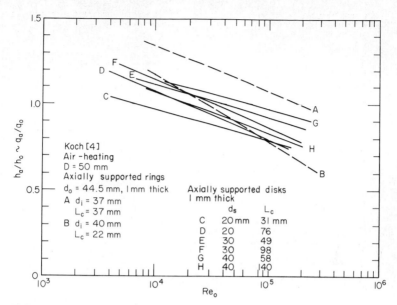

FIG. 29 Performance of tubes with ring or disk inserts (FG-2a).

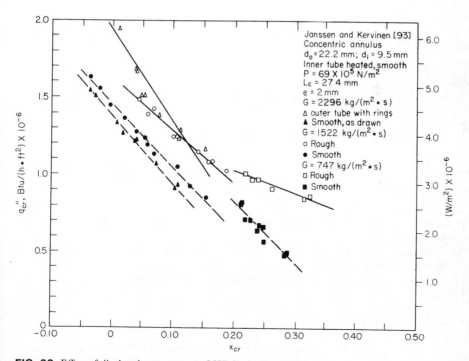

FIG. 30 Effect of displaced promoters on CHF for bulk boiling.

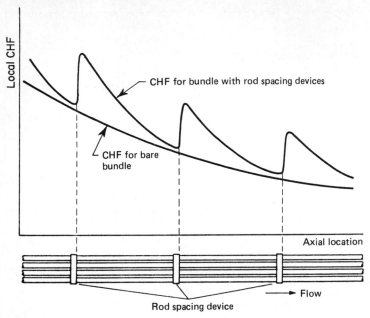

FIG. 31 Effect of rod spacing devices on CHF [187].

increases in pressure drop were very large. A subsequent paper [189] presents a surface renewal model for the condensing heat transfer coefficient. With one experimentally determined constant, the correlation derived from this model is in good agreement with the experimental data.

F. SWIRL-FLOW DEVICES

1. Single-Phase Flow

Swirl-flow devices have been used for more than a century to improve heat transfer in industrial heat exchangers. These devices include inlet vortex generators, twisted-tape inserts, and axial-core inserts with screw-type windings. The augmentation is attributable to several effects: increased path length of flow, secondary flow effects, and, in the case of the tapes, fin effects. Phenomenologically, these devices are part of the general area of confined swirl flows, which also includes curved and rotating systems. The recent survey by Razgaitis and Holman [190] provides a comprehensive discussion of the entire field. See Sec. H of Ref. 368 for a discussion of curved ducts and coils.

Data for uniform-wall-temperature heating are plotted in Fig. 32, and the isothermal friction factors are plotted in Fig. 33. Twisted tapes and propellers were used by Koch [4] to heat air (curves a, b, c, and d). Propellers produce higher heat transfer coefficients than twisted tapes; however, this enhancement is at the expense of a rather large increase in friction factor as seen in Fig. 33. Up to $Re = 200$, the friction factor for the twisted tape is the same as that for the empty half tube ($y = \infty$). The twisted-tape data of Marner and Bergles [114] with ethylene glycol exhibit an augmentation of about 300

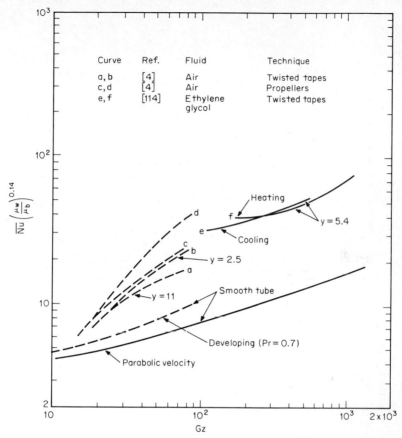

FIG. 32 Representative heat transfer data for swirl flow devices with uniform wall temperature.

percent above the smooth-tube values. Swirl at the pipe inlet does not produce any effective enhancement [192].

The following correlation is recommended for fully developed laminar flow in a uniformly heated tube [191]:

$$\mathrm{Nu}_i = 5.172[1 + 5.484 \times 10^{-3}\,\mathrm{Pr}^{0.7}\,(\mathrm{Re}_i/y)^{1.25}]^{0.5} \qquad (20)$$

Note that the correlation was established for a tape with no heat transfer; considerable increases in heat transfer are predicted with effective fins [193]. The correlation does not seem to be applicable to heating or cooling with a constant wall temperature [114]. At Re < 100, the isothermal friction factors can be approximated by the expression for a semicircular tube:

$$f_i = 42.2\,\mathrm{Re}_i^{-1} \qquad (21)$$

This line lies slightly below the experimental result for a nontwisted tape (curve r in Fig. 33) because the tape thickness is not included in the analysis. There is no substantial increase in friction factor above the empty half-tube results until Re ≅ 300. A lengthy

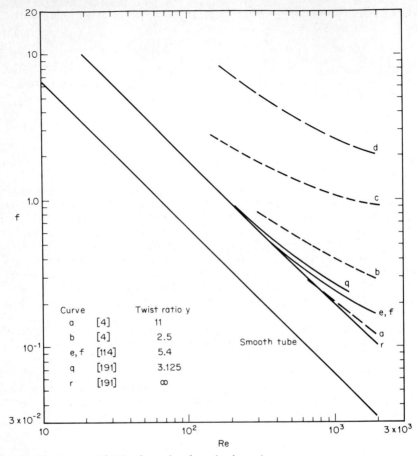

FIG. 33 Isothermal friction factor data for twisted-tape inserts.

set of correlations for the higher–Reynolds number region is presented by Shah and London [194].

Turbulent-flow heat transfer in tubes with twisted-tape inserts has been correlated by [195]

$$\mathrm{Nu}_h = F \left\{ 0.023 \left[1 + \left(\frac{\pi}{2y} \right)^2 \right]^{0.4} \mathrm{Re}_h^{0.8} \, \mathrm{Pr}^{0.4} + 0.193 \left[\left(\frac{\mathrm{Re}_h}{y} \right)^2 \frac{D_h}{D_i} \frac{\Delta \rho}{\rho} \, \mathrm{Pr} \right]^{1/3} \right\} \quad (22)$$

This correlation postulates that the average heat transfer coefficient can be represented essentially as a superposition of heat transfer coefficients for spiral convection and centrifugal convection. The fin factor F, which represents the ratio of total heat transfer to the heat transferred by the walls alone, can be estimated from conduction calculations. The value of F is close to unity for a loose tape fit, and may be as high as 1.25 for a tight tape fit. Equation (22) is accurate for water heating and cooling (with the second term deleted) and for much gas data as well. An equation specifically for gases, which accounts for large radial temperature gradients, is given by Thorsen and Landis [196]. Isothermal friction factors are given by the following expression [195]:

$$f_{h,\text{iso}} = 0.1276y^{-0.406}\text{Re}_h^{-0.2} \tag{23}$$

Diabatic friction factors are obtained in the usual manner by applying a viscosity- or temperature-ratio correction. For heating of water at bulk temperatures below 200°F (93.3°C), the following correction to Eq. (23) has been suggested [195]:

$$f_h = f_{h,\text{iso}} \left(\frac{\mu_w}{\mu_b}\right)^{0.35(D_h/D)} \tag{24}$$

Thorsen and Landis [196] have determined a temperature-ratio correction factor for their correlation of the friction factor data for air.

It is appropriate to conclude this discussion of single-phase data by presenting a constant-pumping-power comparison. Actual friction and heat transfer data from the various investigations are utilized in the computations leading to Fig. 34, for air, and Fig. 35, for water. Because of the diversity in heat transfer and friction data, the performance curves exhibit rather wide scatter; however, the general consensus is that the tapes provide a substantial improvement in performance. Tape twist is, of course, an important parameter; however, even for similar geometries, differences are to be expected due to variations in the fin effect and centrifugal convection effect. The data for axial core assemblies for both air (A, B) and water (K, L, and M) suggest that the tightest twist ratio is not necessarily the best.

Several studies (e.g., Ref. 197) have considered tapes that do not extend the length of the heated section. For uniformly heated tubes, intermittent tapes do not perform as well as full-length tapes and are not used. However, intermittent tapes are particularly useful in cases involving nonuniform heat fluxes. The tapes can be placed at the hot-spot location, thus producing the desired improvement in heat transfer with little effect on the overall pressure drop. This technique has been used to eliminate the burnouts caused by degeneration in heat transfer in certain supercritical boiler systems [203].

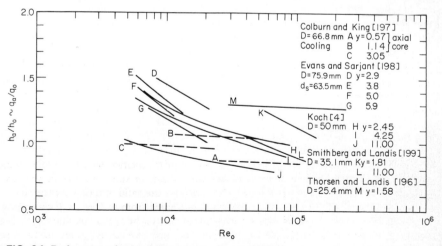

FIG. 34 Performance of twisted-tape inserts with air (FG-2a).

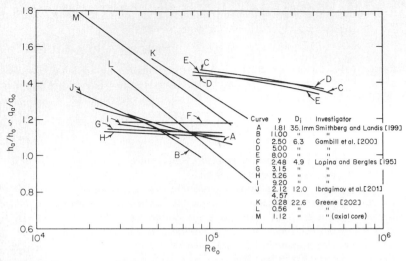

FIG. 35 Performance of twisted-tape inserts with water (FG-2a).

2. Boiling

A variety of devices have been proposed to augment flow boiling by imparting a swirling or secondary motion to the flow. Inlet vortex generators of the spiral-ramp or tangential-slot variety have been used to accommodate very large heat fluxes for subcooled flow boiling of water. One of the highest fluxes on record, $q''_{cr} = 1.73 \times 10^8$ W/m², or 5.48×10^7 Btu/(h·ft²), has been obtained with this technique by Gambill and Greene [204]. Inlet swirl is effective in increasing CHF for subcooled boiling of water in a tube (Mayinger et al. [205]) or in an annulus with a heated inner tube (Ornatskiy et al. [206]).

Twisted tapes are quite popular because of their simplicity and their adaptability to existing heat exchange equipment. They are ideal for hot-spot applications since a short tape can cure the thermal problem with little effect on the overall pressure drop. Boiling curves for subcooled boiling with twisted tapes are similar to those for empty tubes (Lopina and Bergles [207]); however, CHF can be increased by up to 100 percent (Gambill et al. [200]), as shown in Fig. 36. Because of a dramatic reduction in the momentum contribution to the pressure drop in swirl flow [207], CHF for swirl flow is higher than that for straight flow at the same test-section pumping power. This is demonstrated in Fig. 37. Loose-fitting tape inserts have been used by Sephton [208] in tubes which functioned as vertical-tube evaporators for seawater desalination. These inserts are also effective for once-through vaporization of cryogenic fluids (Bergles et al. [209]) or steam (Cumo et al. [210], Hunsbedt and Roberts [211]) since all two-phase regimes are beneficially affected. Twisted-tape inserts have also been considered for rod clusters, with eventual application to nuclear reactor cores [212].

Coiled-tube vapor generators have advantages in terms of packing and generally higher heat transfer performance. As indicated in the literature survey of Jensen [213], the augmentation of boiling is very sensitive to geometrical and flow conditions. Modest improvements in α (circumferential average) for forced-convection vaporization are obtained, with an increase in improvement as coil diameter is decreased. In the subcooled

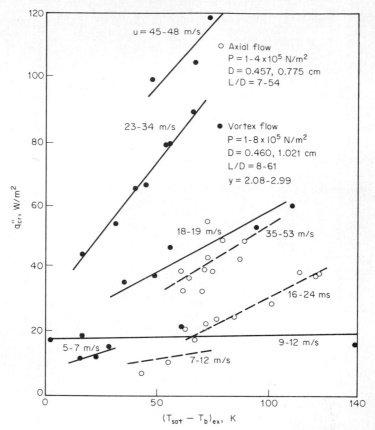

FIG. 36 Influence of twisted-tape inserts on subcooled boiling CHF of water (data of Ref. 200).

region, q''_{cr} is lower than it is for a comparable straight tube; however, q''_{cr} or x_{cr} is usually substantially higher than the straight-tube value at outlet qualities of 0.2 and higher (Fig. 38). The post-dryout heat transfer coefficient is also increased with helical coils.

3. Condensing

Royal and Bergles [167, 168] found that twisted-tape inserts improved heat transfer coefficients for in-tube condensation of water by 30 percent; however, the pressure drop was quite high. Luu and Bergles [169, 170] report similar results for R-113. The following heat transfer correlations are recommended:

For steam:

$$\bar{h}_i = 0.0265 \frac{k_l}{D_h} \left(\frac{G_e D_h}{\mu_l} \right)^{0.8} \mathrm{Pr}_l^{0.33} \left[160 \left(\frac{H^2}{l D_i} \right)^{1.91} + 1 \right] \tag{25}$$

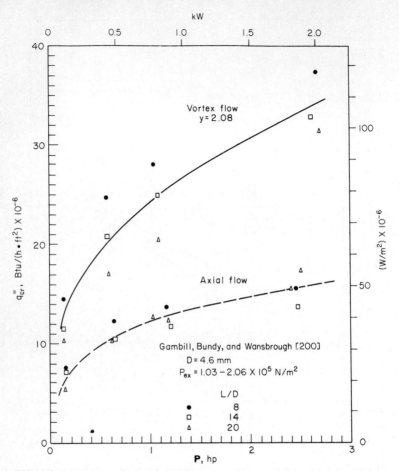

FIG. 37 Dependence of subcooled CHF on pumping power for swirl and axial flow.

where:

$$G_e = G\left(\bar{x}\frac{\rho_l}{\rho_g} + 1 - \bar{x}\right)$$

For R-113:

$$\bar{h}_i = 0.024 \frac{k_e}{D_h} F\left(\frac{F_t G D_h}{\mu_l}\right)^{0.8} \mathrm{Pr}_l^{0.43} \times \left[\frac{(\rho/\rho_m)_{\mathrm{in}}^{0.5} + (\rho/\rho_m)_{\mathrm{ex}}^{0.5}}{2}\right] \qquad (26)$$

where ρ/ρ_m is defined in connection with Eq. (19), F is the same factor used with Eq. (22), and

$$F_t = \frac{8y^2}{3\pi^2}\left\{\left[\left(\frac{\pi}{2y}\right)^2 + 1\right]^{1.5} - 1\right\}$$

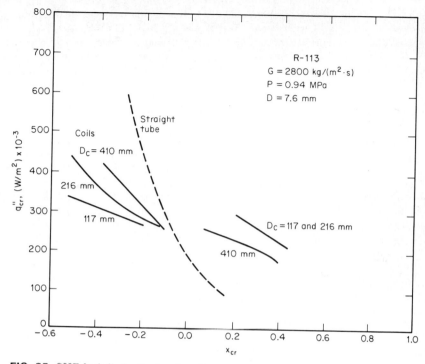

FIG. 38 CHF for helical coils of various diameter compared with straight tube [214].

Shklover and Gerasimov [215] used an interesting baffling technique to create a spiralling motion in the vapor condensing outside a tube bundle. With vapor velocities approaching sonic velocities, the condensing coefficients were high; however, no base data were included for comparison.

Condensation in coiled tubes was studied by Miropolskii and Kurbanmukhamedov [216], and condensation in tube bends was studied by Traviss and Rohsenow [217]. In both cases, modest increases in condensation rates were observed relative to straight tubes.

G. SURFACE-TENSION DEVICES

Within the context of the present classification of augmentation techniques, surface-tension devices are those which involve the application of relatively thick wicking materials to heated surfaces. This technique is distinct from that applied to heat pipes (see Chap. 5). Wicking is usually considered for situations where coolant is unable to reach the heater surface without the wicking material, e.g., the cooling of electronics in aircraft undergoing violent maneuvers or in spacecraft operating in a near-zero-gravity environment. Wicking has also been shown to be effective in augmenting boiling heat transfer from submerged surfaces.

When a heater was completely enclosed with wicking and submerged, Allingham and McEntire [218] found that the heat transfer coefficient for saturated pool boiling was improved at low heat fluxes, but that the reverse was true at moderate fluxes. Costello and Redeker [219] investigated higher heat fluxes and found that the heat flux corresponding to a temperature excursion was only about 10 percent of the normal critical heat flux. It was concluded that proper vapor venting was necessary to avoid blockage of the liquid flow. Subsequent tests [220] indicated that the critical heat flux could be raised by as much as 200 percent when the wicking was not too dense and a narrow channel was maintained at the top for easy escape of vapor. More recently, Corman and McLaughlin [221] presented extensive data for wick-augmented surfaces which qualitatively confirm these observations.

Gill [222] spiraled wicking around cylindrical heaters. In all cases, boiling commenced at a superheat of about 1 K, apparently because the wicking provided large nucleation sites. The boiling curve was generally displaced to lower superheat than the normal curve, to a degree dependent on the diameter and pitch of the wicking. No significant change in the critical heat flux was observed with the wicking. The stable film-boiling region was investigated by quenching a copper calorimeter in liquid nitrogen. It appeared that capillary action effectively transported liquid through the vapor film to the heated surface, since the heat transfer coefficient was increased by about 100 percent.

H. ADDITIVES FOR LIQUIDS

1. Solid Particles in Single-Phase Flow

Watkins et al. [223] studied suspensions of polystyrene spheres in laminar flow of oil. Maximum improvements of 40 percent were observed.

2. Gas Bubbles in Single-Phase Flow

Tamari and Nishikawa [224] observed increases in average heat transfer coefficient of up to 400 percent when air was injected into either water or ethylene glycol. The injection point was at the base of the vertical heated surface, and up to three injection nozzles were used. Other studies are reviewed by Hart [225], who proposed a correlation to fit his own data as well as the results of other investigators for free-convection augmentation.

Kenning and Kao [226] noted heat transfer increases up to 50 percent when nitrogen bubbles were injected into turbulent water flow. A similar level of augmentation was observed by Baker [227], who created slug flow in small rectangular channels with simulated microelectronic chips on one of the wide sides.

Surface degassing, which is initiated when wall temperatures are below the saturation temperature, produces an agitation comparable to that of injected bubbles or even of boiling. In a recent study, Behar et al. [228] found that the wall superheat for saturated pool boiling of nitrogen-pressurized *meta*-terphenyl was reduced by as much as 50°F (27.8°C), while in subcooled flow boiling, the reduction was as much as 30°F (16.7°C). In general, the surface degassing is effective only at lower heat fluxes; once the nucleate boiling becomes well established, there is negligible reduction in the wall superheat.

3. Liquid Additives for Boiling

Trace liquid additives have been extensively investigated in pool boiling and, to a lesser extent, in subcooled flow boiling. A great many additives have been investigated, and some have been found to produce substantial heat transfer improvements. With the proper concentration of certain additives (wetting agents, alcohols), increases of about 20 to 40 percent in the heat transfer coefficient for saturated nucleate pool boiling can be realized [229–233]. This occurs in spite of thermodynamic analyses for boiling of binary mixtures which indicate that boiling performance should be decreased [234].

Most additives increase CHF, but the concentration of the additive and heater geometry have major effects on the augmentation. The typical results of van Stralen et al. [235, 236], as shown in Fig. 39, indicate a sharp increase in CHF at some low concentration of 1-pentanol, and rather rapid decrease as the concentration is increased. The optimum concentration varies with the mixture, and to some extent with the pressure. For a similar water-pentanol system, Carne [237] obtained an increase of only 25 percent in CHF with a 3.2-mm heater, which is small compared to the 240 percent increase that van Stralen obtained with a 0.2-mm heater. Van Stralen's extensive program of testing and modeling of mixtures has been extended to film boiling [238]. In this study, it was found that a 4.1 weight-percent mixture of 2-butanone in water improved coefficients by up to 80 percent.

Of related interest are the results of Gannett and Williams [239], who found that small quantities of certain polymers dissolved in water increase nucleate-boiling coefficients. As reported by Jensen et al. [240], oily contaminants generally decrease boiling coefficients; however, increases are observed under certain conditions.

In subcooled flow boiling, the improvement (if any) in heat transfer is modest. Leppert et al. [241] found that the main advantage of alcohol-water mixtures was an improvement in smoothness of boiling. The influence of a volatile additive on subcooled flow-boiling CHF in tubes has been investigated by Bergles and Scarola [242]. As shown in Fig. 40, there is a distinct reduction in CHF at low subcooling with the addition of 1-pentanol.

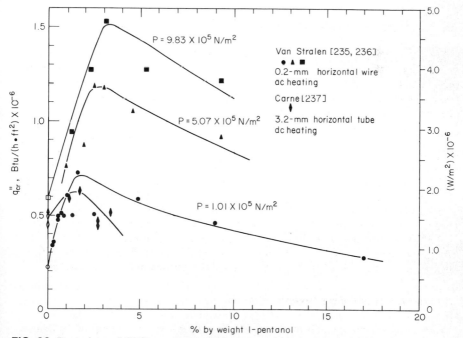

FIG. 39 Dependence of CHF on volatile additive concentration for saturated pool boiling.

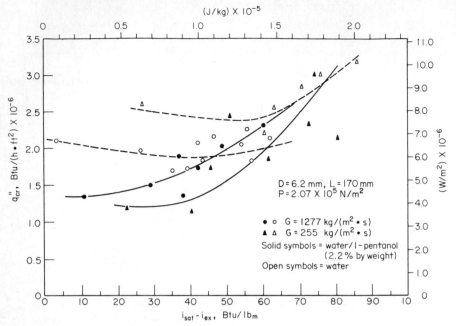

FIG. 40 Influence of volatile additive on subcooled CHF [242].

On the other hand, Sephton [243] found that the overall heat transfer coefficient was doubled when a surfactant was added to seawater evaporating in a vertical (upflow) tube.

In general, the improvements in heat transfer and CHF offered by additives are not sufficient to make them useful for practical systems. There are difficulties involved in maintaining the desired concentration, particularly when the additive is volatile. In many cases, the additives, even in small concentrations, are somewhat corrosive and require special piping or seals.

I. ADDITIVES FOR GASES

1. Solid Particles in Single-Phase Flow

Dilute gas-solid suspensions have been considered as working fluids for gas turbine and nuclear reactor systems. Solid particles in the micron-to-millimeter size range are dispersed in the gas stream at loading ratios W_s/W_g ranging from 1 to 15. The solid particles, in addition to giving the mixture a higher heat capacity, are highly effective in promoting enthalpy transport near the heat exchange surface. Heat transfer is further enhanced at high temperatures by means of the particle-surface radiation. A summary of typical data for air-solid suspensions is given in Fig. 41.

Extensive experimental work was undertaken at Babcock and Wilcox to obtain detailed heat transfer and pressure drop information as well as operating experience. Summary articles by Rhode et al. [245] and Schluderberg et al. [246] elaborate on the conclusions of this work. Heat transfer coefficients for heating were improved by as much as a factor of 10 through the addition of graphite. The suspensions were also shown to

Curve	Re	Particle	Size (microns)	Gas	d_p/D
A	18,000	Glass	60	Air	0.0011
B	18,000	Glass	120	Air	0.0027
C	19,000	Sand	230	Air	0.0060
D	19,000	Sand	80	Air	0.0021
E	15,000	Graphite	65	Air	0.0085
F	53,000	Zinc	40	Air	0.0005
G	53,000	Zinc	40	Air	0.0008
H		Graphite			
I	53,000	Zinc	40	Air	0.016
J	15,000	Al_2O_3	65	Air	0.0085
K	13,500	Glass	30	Air	0.0016
L	13,500	Glass	200	Air	0.0111

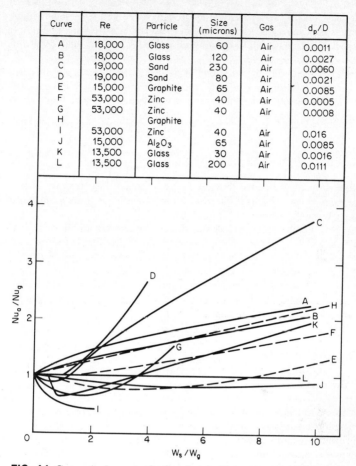

FIG. 41 Composite heat transfer data for air-solid suspensions [244].

be far superior to gas coolants on the basis of pumping power requirements, especially when twisted-tape inserts were used. There was relatively little settling, plugging, or erosion in the system. With helium suspensions, however, there was serious fouling of the loop coolers, which was attributed to brownian particle motion induced by temperature gradient.

Abel and his coworkers [247] demonstrated that the cold-surface deposition is a very serious problem with micronized graphite. This occurred with both helium and nitrogen suspensions and could be alleviated only with very high gas velocities. An economic comparison was presented in terms of a system pumping power to heat transfer rate ratio as a function of gas flow rate. This comparison indicated that the pure gas was generally more effective than the suspension at both low and high gas flow rates. In all probability, the loop heater is very effective; however, this gain is offset by the low performance of the cooler.

A comprehensive analysis of much of the data for dilute gas-solid suspension was reported by Pfeffer et al. [248]. Correlations for both heat transfer coefficient and friction factor were developed. These investigators recently presented a feasibility study of using suspensions as the working fluid in a Brayton space power generation cycle [249]. A more recent presentation of design information and guide to the literature is given by Depew and Kramer [250].

Fluidized beds represent the other end of the spectrum in terms of solids loading (this subject is covered in Chap. 6). The very considerable augmentation in heat transfer coefficients, up to a factor of 20 compared to pure gas flow at the same flow rate, has led to applications in such areas as flue gas heat recovery.

2. Liquid Drops in Single-Phase Flow

When liquid droplets are added to a flowing gas stream, heat transfer is enhanced by sensible heating of the two-phase mixture, evaporation of the liquid, and disturbance of the boundary layer.

Thomas and Sunderland [251] demonstrated that heat transfer coefficients can be increased by as much as a factor of 30 if a continuous liquid film is formed on the heated surface. A more realistic indication of practical enhancement was provided by Yang and Clark [252], who applied spray cooling to a compact heat exchanger core. The maximum improvement of 40 percent was attributed to formation of a partial liquid film and sensible heating of that film. In general, the large volume of liquid required tends to limit practical application of this technique.

J. MECHANICAL AIDS

1. Stirring

Attention is now directed toward active techniques which require direct application of external power to create the augmentation. Heating or cooling of a viscous liquid in batch processing is often enhanced by stirrers or agitators built into the tank. Uhl [253] presents a comprehensive survey of this area, including descriptions of hardware and experimental results.

For forced flow in ducts, a "spiralator" has been proposed. This consists of a loose-fitting twisted tape secured at the downstream end in a bearing. Penney [254] found that with heating of corn syrup, the heat transfer coefficient increased by 95 percent at 100 rev/min.

The augmented convection provided by stirring dramatically improves pool boiling at low superheat. However, once nucleate boiling is fully established, the influence of the improved circulation is small. Pramuk and Westwater [255] found that the boiling curve for methanol was favorably altered for nucleate, transition, and film boiling, with the improvement increasing as agitator speed increased.

2. Surface Scraping

Close-clearance scrapers for viscous liquids are included in the review by Uhl [253]. An application of scraped-surface heat transfer to airflows is reported by Hagge and Junkhan [256]; a tenfold improvement in heat transfer coefficient was reported for laminar flow over a flat plate. Scrapers were also suggested for creating thin evaporating films. Lustenader et al. [257] outline the technique, and Tleimat [258] presents performance data. The heat transfer coefficients are much higher than those observed for pool evaporation (without nucleate boiling).

3. Rotating Surfaces

Rotating heat exchanger surfaces occur naturally in rotating electrical machinery, gas turbine rotor blades, and other industrial systems.

Substantial increases in heat transfer coefficients have been reported for laminar flow in (1) a straight tube rotating about its own axis (McElhiney and Preckshot [259]), (2) a straight tube rotating around a parallel axis (Mori and Nakayama [260]), (3) a rotating circular tube (Vidyanidhi et al. [261]), and (4) the rotating, curved, circular tube (Miyazaki [262]). Reference 260 also presents an analysis of turbulent flow and data for both laminar and turbulent flows. Maximum improvements of 350 percent were recorded for laminar flow, but for turbulent flow the maximum increase was only 25 percent.

Tang and McDonald [263] found that when heated cylinders are rotated at high speeds in saturated pools, convective coefficients are so high that boiling can be suppressed. This constitutes an augmentation of pool boiling. Marto and Gray [264] found that critical heat fluxes were elevated in a rotating-drum boiler where the vaporization occurred at the inside of the centrifuged liquid annulus. With proper liquid feed conditions to the heated surface, exit qualities in excess of 99 percent were obtained.

Studies have been reported on condensation in a rotating horizontal disk [265], a rotating vertical cylinder [266], and a rotating horizontal cylinder [267]. Improvements are several hundred percent above the stationary case for water and organic liquids.

Weiler et al. [268] subjected a tube bundle, with nitrogen condensing inside the tube, to high accelerations (essentially normal to the tube axis). At $325g$, the overall heat transfer coefficient was increased by a factor of over 4.

K. SURFACE VIBRATION

1. Single-Phase Flow

It has long been recognized that transport processes can be significantly affected by inherent or induced oscillations. In general, sufficiently intense oscillations improve heat transfer; however, decreases in heat transfer have been recorded on both a local and an average basis. A wide range of effects is to be expected as a result of the large number of variables necessary to describe the vibrations and the convective conditions.

In discussing the interactions between vibration and heat transfer, it is convenient to distinguish between vibrations that are applied to the heat transfer surface and those that are imparted to the fluid. The most direct approach is to vibrate the surface mechanically, usually by means of an electrodynamic vibrator or a motor-driven eccentric. In order to achieve an adequate amplitude of vibration, frequencies are generally kept well below 1000 Hz.

The predominant geometry employed in vibrational studies has been the horizontal heated cylinder vibrating either horizontally or vertically. When the ratio of the amplitude to the diameter is large, it is reasonable to assume that the convection process which occurs in the vicinity of the cylinder is quasi-steady. The heat transfer may then be described by conventional correlations for steady convection. In order to achieve the quasi-steady convection, characterized by $a/D_o \gg 1$, it is necessary to use small-diameter cylinders. The data presented in Fig. 42 illustrate this situation. The data fall into three rather distinct regions depending on the intensity of vibration: the region of low Re_v, where free convection dominates; a transition region where free convection and the "forced" convection caused by vibration interact; and finally the region of dominant forced convection. The data in this last region are in good agreement with a standard correlation for forced flow normal to a cylinder.

When cylinders of large diameter, typically those found in heat exchange equipment, are used, a different type of behavior is expected. When $a/D_o \lesssim 1$, there is no longer a significant displacement of the cylinder through the fluid to provide enthalpy transport. Natural convection should then dominate. However, where the vibrational intensity reaches a critical value, a secondary flow, commonly called acoustic or thermoacoustic streaming, develops, which is able to effect a net enthalpy flux from the boundary layer. Since the coordinates of Fig. 42 are inappropriate to describe streaming data, a simple heat transfer coefficient ratio is used in Fig. 43 to indicate typical improvements in heat transfer observed under these conditions. The heat transfer coefficient remains at the natural-convection value until a critical intensity is reached and then increases with growing intensity. The rate of improvement in heat transfer appears to decrease as Re_v is increased. If these data were plotted on the coordinates of Fig. 42, they would lie below the quasi-steady prediction, except at very high Re_v, where they are generally higher.

Several studies have been done concerning the effects of transverse or longitudinal vibrations on heat transfer from vertical plates. Analyses indicate that laminar flow is virtually unaffected; however, experimental observations indicate that turbulent flow is induced by sufficiently intense vibrations. The improvement in heat transfer appears to be rather small, with the largest values of $h_a/h_o < 1.6$ [280].

From an efficiency standpoint, it is important to note that the improvements in heat transfer coefficient with vibration may be quite dramatic, but they are only relative to natural convection. The average velocities are actually quite low: for example, $4af \approx 1.8$

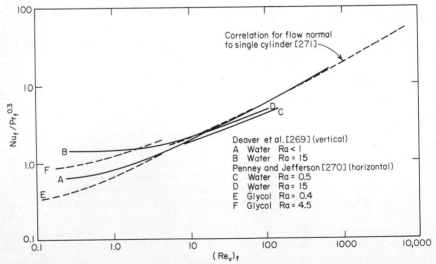

FIG. 42 Influence of mechanical vibration on heat transfer from horizontal cylinders—$a/D_o \gg 1$.

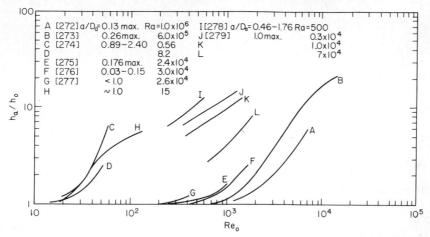

FIG. 43 Influence of mechanical vibration on heat transfer from horizontal cylinders— $a/D_o < 1$.

m/s for the highest-intensity data of Mason and Boelter [273] in Fig. 42. For most systems, it would appear to be more convenient and economical to provide steady forced flow to achieve the desired increase in heat transfer coefficient.

Substantial improvements in heat transfer have also been recorded when vibration of the heated surface is used in forced-flow systems. No general correlation has been obtained; however, this is not surprising in view of the diverse geometrical arrangements. Figure 44 presents representative data for heat transfer to liquids. The effect on heat transfer varies from slight degradation to over 300 percent improvement, depending on the system and the vibrational intensity. One problem is cavitation when the intensity becomes too large. As indicated by A and B, the vapor blanketing causes a sharp degradation of heat transfer.

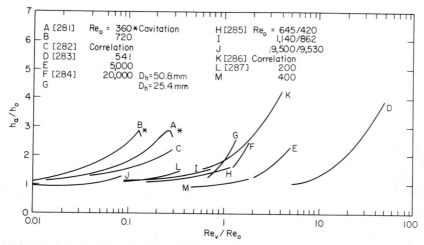

FIG. 44 Effect of surface vibration on heat transfer to liquids with forced flow.

Hsieh and Marsters [288] extended the extensive single-tube experience to a vibrating vertical array of five horizontal cylinders. They found that the average of the heat transfer coefficients increased by 54 percent at the highest vibrational intensity. The bottom cylinder showed the highest increase; the relatively poor performance of the top cylinders is apparently due to wake interaction.

These experiments indicate that vibrations can be effectively applied to practical heat exchanger geometries; however, economic evaluation is difficult since sufficient data are not available. Apparently no comparative pressure drop data have been reported for forced flow. In any case, it appears that the overriding consideration is the cost of the vibrational equipment and the power to run it. Ogle and Engel [285] found for one of their runs that about 20 times as much energy was supplied to the vibrator as was gained in improved heat transfer. Even though the vibrator mechanism was not optimized in this particular investigation, the result suggests that heat-surface vibration will not be practical.

2. Boiling

Experiments by Bergles [279] have established that vibrations have little effect on sub-cooled or saturated pool boiling. It was found that the coefficients characteristic of single-phase vibrational data govern the entry into boiling conditions. Once boiling is fully established, however, vibration has no discernible effect. The maximum increase in critical heat flux was about 10 percent at an average velocity of 0.25 m/s. Experiments by Parker et al. [289, 290], run over the frequency range 50 to 2000 Hz, have further confirmed that fully developed nucleate boiling is essentially unaffected by vibration.

Fuls and Geiger [291] studied the effect of enclosure vibration on pool boiling. A slight increase in the nucleate-boiling heat transfer coefficient was observed.

Raben et al. [284] reported a study of flow surface boiling with heated-surface vibration. A large improvement was noted at low heat flux, but this improvement decreased with increasing heat flux. This is consistent with those pool-boiling results which indicate no improvement in the region of fully established boiling. Pearce [292] found insignificant changes in bulk-boiling CHF when a boiler tube was vibrated transversely.

3. Condensing

The few studies in this area include those of Dent [293] and Brodov et al. [294], who obtained maximum increases of 10 to 15 percent by vibrating a horizontal condenser tube.

L. FLUID VIBRATION

1. Single-Phase Flow

In many applications, it is difficult to apply surface vibration because of the large mass of the heat transfer apparatus. The alternative technique is then utilized, whereby vibrations are applied to the fluid and focused toward the heated surface. The generators which have been employed range from the flow interrupter to the piezoelectric transducer, thus covering the range of pulsations from 1 Hz to ultrasound of 10^6 Hz. The description of the interaction between fluid vibrations and heat transfer is even more complex than it is in the case of surface vibration. In particular, the vibrational variables are more difficult to define because of the remote placement of the generator. Under certain conditions, the flow fields may be similar for both fluid and surface vibration, and analytical results can be applied to both types of data.

A great deal of research effort has been devoted to studying the effects of sound fields on heat transfer from horizontal cylinders to air. Intense plane sound fields of the progressive or stationary type have been generated by loudspeakers or sirens. The sound fields have been oriented axially and transversely, in either the horizontal or vertical plane. With plane transverse fields directed transversely, improvements of 100 to 200 percent over natural-convection heat transfer coefficients were obtained by Sprott et al. [295], Fand and Kaye [296], and Lee and Richardson [297].

It is commonly observed that increases in average heat transfer occur at a sound pressure level of about 134 to 140 dB (well above the normal human tolerance of 120 dB), and that these increases are associated with the formation of an acoustically induced flow (acoustic or thermoacoustic streaming) near the heated surface. Large circumferential variations in heat transfer coefficient are present [298], and it has been observed that local improvements in heat transfer occur at intensities well below those that affect the average heat transfer [299]. Correlations have been proposed for individual experiments; however, an accurate correlation covering the limits of free convection and fully developed vortex motion has not been developed.

In general, it appears that acoustic vibrations yield relatively small improvements in heat transfer to gases in free convection. From a practical standpoint, a relatively simple forced-flow arrangement could be substituted to obtain equivalent improvements.

When acoustic vibrations are applied to liquids, heat transfer may be improved by acoustic streaming as in the case of gases. With liquids, though, it is possible to operate with ultrasonic frequencies given favorable coupling between a solid and a liquid. At frequencies of the order of 1 megahertz, another type of streaming called *crystal wind* may be developed. These effects are frequently encountered; however, intensities are usually high enough to cause cavitation, which may become the dominant mechanism.

Seely [300], Zhukauskas et al. [301], Larson and London [302], Robinson et al. [303], Fand [304], Gibbons and Houghton [305], and Li and Parker [306] have demonstrated that natural-convection heat transfer to liquids can be improved from 30 to 450 percent by the use of sonic and ultrasonic vibrations. In general, cavitation must occur before significant improvements in heat transfer are noted. In spite of these improvements there appears to be some question regarding the practical aspects of acoustic augmentation. When the difficulty of designing a system to transmit acoustic energy to a large heat transfer surface is considered, it appears that forced flow or simple mechanical agitation will be a more attractive means of improving natural-convection heat transfer.

Low-frequency pulsations have been produced in forced-convection systems by partially damped reciprocating pumps and interrupter valves. Quasi-steady analyses suggest that heat transfer will be improved in transitional or turbulent flow with sufficiently intense vibrations. However, heat transfer coefficients are usually higher than predicted, apparently due to cavitation. Figure 45 indicates the improvements that have been reported for pulsating flow in channels. The improvement is most significant in the transitional range of Reynolds numbers, as might be expected since the pulsations force the transition to turbulent flow. Interrupter valves are a particularly simple means of generating the pulsations. The valves must be located directly upstream of the heated section to produce cavitation, which appears to be largely responsible for the improvement in heat transfer [310, 311].

A wide variety of geometrical arrangements and more complex flow fields are encountered when sound fields are superimposed upon forced flow of gases. In general, the improvement is dependent upon the relative strengths of the acoustic streaming and the

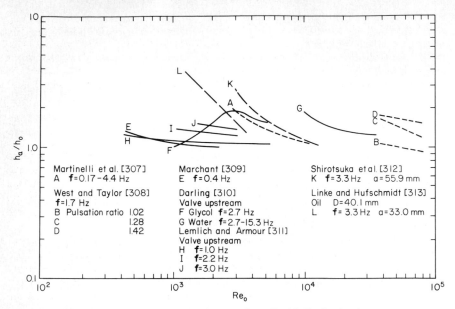

FIG. 45 Effect of upstream pulsations on heat trnsfer to liquids flowing in pipes.

forced flows. The reported improvements in average heat transfer are limited to about 100 percent. Typical data obtained with sirens or loudspeakers placed at the end of a channel are indicated in Fig. 46. The improvements in average heat transfer coefficient are generally significant only in the transition range where the vibrating motion acts as a turbulence trigger. The experiments of Moissis and Maroti [318] are important in that they demonstrate practical limitations. Even with high-intensity acoustic vibrations, the gas-side heat transfer coefficient was improved by only 30 percent for a compact heat exchanger core.

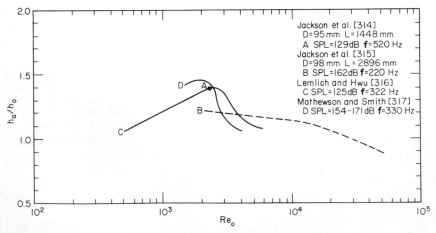

FIG. 46 Influence of acoustic vibrations on heat transfer to air flowing in tubes.

The data of Zhukauskas et al. [301] and Larson and London [302] suggest that ultrasonic vibration has no effect on forced-convection heat transfer once the flow velocity is raised to about 0.3 m/s. However, Bergles [319] demonstrated that lower-frequency vibrations (80 Hz) can produce improvements of up to 50 percent. This experiment was carried out at higher surface temperatures where it was possible to achieve cavitation.

2. Boiling

The available evidence indicates that fully established nucleate pool boiling is unaffected by ultrasonic vibration, apparently because of the dominance of bubble agitation and attenuation of acoustic energy by the vapor [320, 321]. However, enhanced vapor removal can improve CHF by about 50 percent [320, 322]. Transition and film boiling can also be substantially improved, since the vibration has a strong tendency to destabilize film boiling [323].

In channel flow it is usually necessary to locate the transducer upstream or downstream of the test channel, with the result that the sound field is greatly attenuated. Tests with 80-Hz vibrations [319] indicate no improvement of subcooled boiling heat transfer or critical heat flux. Romie and Aronson [324], using ultrasonic vibrations, found that subcooled critical heat flux was unaffected. Even where intense ultrasonic vibrations were applied to the fluid in the immediate vicinity of the heated surface, boiling heat transfer was unaffected [325]. The severe attenuation of the acoustic energy by the two-phase coolant appears to render this technique ineffective for flow-boiling systems.

3. Condensing

Mathewson and Smith [317] investigated the effects of acoustic vibrations on condensation of isopropanol vapor flowing downward in a vertical tube. A siren was used to generate a sound field of up to 176 dB at frequencies ranging from 50 to 330 Hz. The maximum improvement in condensing coefficient was found to be about 60 percent at low vapor flow rates. The condensate film under these conditions was normally laminar; thus, an intense sound field produced sufficient agitation in the vapor to cause turbulent conditions in the film. The effect of the sound field was considerably diminished as the vapor flow rate increased.

M. ELECTRIC AND MAGNETIC FIELDS

A comprehensive discussion of the fundamental effects improved electric and magnetic fields have on heat transfer is given in Ref. 369. A magnetic force field retards fluid motion; hence, heat transfer coefficients decrease. On the other hand, if electromagnetic pumping is established with a combined magnetic and electric field, heat transfer coefficients can be increased far above those expected for gravity-driven flows. For example, an analysis by Singer [326] indicates that electromagnetic pumping can increase laminar film condensation rates of a liquid metal by a factor of 10.

Electric fields are particularly effective in increasing heat transfer coefficients in free convection. The configuration may be a heated wire in a concentric tube maintained at a high voltage relative to the wire, or a fine wire electrode may be utilized with a horizontal plate. Reported increases are as much as a factor of 40; however, several hundred percent is normal. Recent activity has centered on the application of corona discharge cooling to practical problems. The cooling of cutting tools by point electrodes was proposed by Blomgren and Blomgren [327], while Reynolds and Holmes [328] have used parallel wire electrodes to improve the heat dissipation of a standard horizontal finned tube. Heat transfer coefficients can be increased by several hundred percent when sufficient electrical power is supplied. It appears, however, that the equivalent effect could be produced at lower capital cost and without the hazards of 10 to 100 kV by simply providing forced convection with a blower or fan.

Some very impressive enhancements have been recorded with forced laminar flow. The recent studies of Porter and Poulter [329], Savkar [330], and Newton and Allen [331] demonstrated improvements of at least 100 percent when voltages in the 10-kV range were applied to transformer oil. While it is desirable to take advantage of any naturally occurring electric fields in electrical equipment, augmentation appears to be a risky business. Mizushina et al. [332] found that even with intense fields, the enhancement disappeared as turbulent flow was approached in a circular tube with a concentric inner electrode.

The typical effects of electric fields on pool boiling are shown in Fig. 47. These data of Choi [333] were taken with a horizontal electrically heated wire located concentrically within a charged cylinder. Because of the large augmentation of free convection, boiling is not observed until relatively high heat fluxes. Once nucleate boiling is initiated, the electric field has little effect. However, CHF is elevated substantially, and large increases in the film-boiling heat transfer coefficient are obtained.

Durfee and coworkers [334] conducted an extensive series of tests to evaluate the feasibility of applying electrohydrodynamic (EHD) augmentation to boiling-water nuclear reactors. Tests with water in electrically heated annuli indicated that wall temperatures

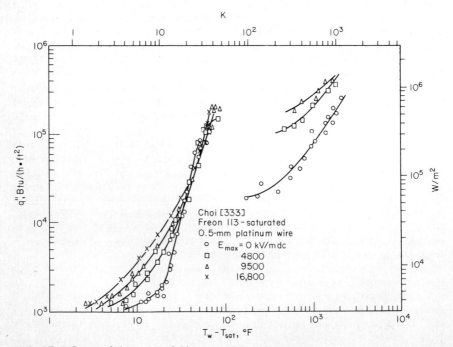

FIG. 47 Influence of electrostatic fields on pool-boiling heat transfer.

for flow bulk boiling were slightly reduced through application of the field. Increases in CHF were observed for all pressures, flow rates, and inlet subcoolings, with the improvement falling generally in the 15 to 40 percent range for applied voltages up to 3 kV. On the basis of limited pressure drop data, it was suggested that greater steam energy flow was obtained with the EHD system than with the conventional system at the same pumping power.

Velkoff and Miller [335] investigated the effect of uniform and nonuniform electric fields on laminar film condensation of Freon-113 on a vertical plate. With screen grid electrodes providing a uniform electric field over the entire plate surface, a 150 percent increase in the heat transfer coefficient was obtained with a power expenditure of a fraction of one watt. Choi and Reynolds [336] and Choi [337] recently reported data for condensation of Freon-113 on the outside wall of an annulus in the presence of a radial electric field. With the maximum applied voltage of 30 kV, the average heat transfer coefficients for a 25.4-mm outside diameter by 12.7-mm inside diameter annulus were increased by 100 percent.

N. INJECTION

Injection and suction have been considered primarily in connection with retarding of heat transfer to bodies subject to aerodynamic heating. On the augmentative side, some thought has been given to intensifying heat transfer by injecting gas through a porous heat transfer surface. The bubbling produces an agitation similar to that of nucleate boiling. Gose et al. [338, 339] bubbled gas through sintered or drilled heated surfaces to stimulate nucleate pool and flow boiling. Sims et al. [340] analyzed the pool data and found that Kutateladze's pool boiling relationship correlated the porous-plate data quite well. For their limited forced-circulation tests with a sintered pipe, Gose et al. found that heat transfer coefficients were increased by as much as 500 percent in laminar flow and about 50 percent in turbulent flow. Kudirka [341] found that heat transfer coefficients for flow of ethylene glycol in porous tubes increased by as much as 130 percent by the injection of air. The practical application of injection appears to be rather limited because of the difficulty of supplying and removing the gas.

Tauscher et al. [347] have demonstrated up to fivefold increases in local heat transfer coefficients by injecting similar fluid into a turbulent tube flow. The effect is comparable to that produced by an orifice plate; in both cases the effect has died out after about 10 L/D.

Bankoff [342] suggested that heat transfer coefficients in film boiling could be substantially improved by continuously removing vapor through a porous heated surface. Subsequent experimental work [343, 344] demonstrated that coefficients could be increased by as much as 150 percent, provided that a porous block was placed on the surface to stabilize the flow of liquid toward the surface. Wayner and Kestin [345] extended this concept to nucleate boiling and found that wall superheats could be maintained at about 3 K (5.4°F) for heat fluxes over 300,000 W/m^2, or 95 × 10^3 Btu/(h·ft^2). The need for a porous heated surface and a flow control element appear to limit the application of suction boiling.

O. SUCTION

Large increases in heat transfer coefficient are predicted for laminar flow (Kinney [348]) and turbulent flow (Kinney and Sparrow [349]) with surface suction. The general characteristics of the latter predictions were confirmed by the experiments of Aggarwal and Hollingsworth [350]. However, suction is difficult to incorporate into practical heat exchange equipment.

The typical studies of laminar film condensation by Antonir and Tamir [351] and Lienhard and Dhir [352] indicate that heat transfer coefficients can be improved by as much as several hundred percent when the film thickness is reduced by suction. This is expected, as the thickness of the condensate layer is the main parameter affecting the heat transfer rate in film condensation.

P. COMPOUND AUGMENTATION

Compound techniques are a slowly emerging area of enhancement which holds promise for practical applications since heat transfer coefficients can usually be increased above any of the several techniques acting alone. Some examples for single-phase flows are

   Rough tube wall with twisted-tape insert, Bergles et al. [353]

   Rough cylinder with acoustic vibrations, Kryukov and Boykov [354]

   Internally finned tube with twisted-tape insert, Van Rooyen and Kroeger [355]

   Finned tubes in fluidized beds, Bartel and Genetti [356]

   Externally finned tubes subjected to vibrations, Zozulya and Khorunzhii [357]

   Gas-solid suspension with an electric field, Min and Chao [358]

   Fluidized bed with pulsations of air, Bhattacharya and Harrison [359]

It is interesting to note that some compound attempts are unsuccessful. Masliyah and Nandakumar [360], for example, found analytically that average Nusselt numbers for internally finned coiled tubes were lower than they were for plain coiled tubes.

Compound augmentation has also been studied to a limited extent with phase-change heat transfer. For instance, the addition of surface roughness to the evaporator side of a rotating evaporator-condenser increased the overall coefficient by 10 percent (Bromley et al. [361]). Sephton [208, 243] found that overall coefficients could be doubled by the addition of a surfactant to seawater evaporating in spirally corrugated or doubly fluted tubes (vertical upflow). However, Van der Mast et al. [362] found only slight improvements with a surfactant additive for falling film evaporation in spirally corrugated tubes.

Compound augmentation as it is used with vapor space condensation includes rotating finned tubes (Chandran and Watson [147]), rotating rough disks (Bromley et al. [361]), and rotating disks with suction (Chary and Sarma [363]). Moderate increases in condensing coefficient are reported. Weiler et al. [268] condensed nitrogen inside rotating tubes treated with a porous coating, which increased coefficients above those for a rotating smooth tube.

Q. PROSPECTS FOR THE FUTURE

As energy and material shortages become more important factors in the overall cost of thermal systems, it is expected that heat transfer augmentation techniques will be called upon with increasing frequency. As noted in Ref. 364, it is appropriate to view augmentation techniques as "second-generation" heat transfer technology. Such new technology normally requires several phases of development for successful commercialization. These include the following steps:

1. Basic performance data for heat transfer and pressure drop, if applicable, must be obtained. General correlations should be developed to predict heat transfer and pressure drop as a function of the geometrical characteristics.

2. Design methods (PECs) must be developed to facilitate selection of the optimum surface geometry for the various techniques and particular applications.

3. Manufacturing technology and cost of manufacture must be available for the desired surface geometry and material.

4. Pilot plant tests of the proposed surface are required to permit a complete economic evaluation and to establish the design, including long-term fouling and corrosion characteristics.

This chapter indicates that many augmentation techniques have gone through all of the steps required for commercialization. The prognosis is for the exponential growth curve of Fig. 1 to level off, not from a lack of interest, but from a broader acceptance of enhancement techniques in industrial practice.

REFERENCES

1. A. E. Bergles, V. Nirmalan, G. H. Junkhan, and R. L. Webb, Bibliography on Augmentation of Convective Heat and Mass Transfer-II, Heat Transfer Lab. Rep. *HTL-31, ISU-ERI-Ames-84221*, DE 84010848, Iowa State University, Ames, Iowa, 1983.

2. W. J. Marner, A. E. Bergles, and J. M. Chenoweth, On the Presentation of Performance Data for Enhanced Tubes Used in Shell-and-Tube Heat Exchangers, *Advances in Enhanced Heat Transfer—1981*, R. L. Webb, T. C. Carnavos, E. L. Park, and K. M. Hostetler, eds., pp. 1–10, ASME, New York, 1981.

3. W. Nunner, Wärmeübergang und Druckabfall in rauhen Rohren, *Forschungsh. Ver. dt. Ing.*, Vol. B22, No. 455, pp. 5–39, 1956. Also, *Atomic Energy Research Establishment (United Kingdom) Lib./Trans.* 786, 1958.

4. R. Koch, Druckverlust und Wärmeübergang bei verwirbelter Strömung, *Forschungsh. Ver. dt. Ing.*, Vol. B24, No. 469, pp. 1–44, 1958.

5. N. D. Greene, Convair Aircraft, private communication to W. R. Gambill, May, 1960. Cited in W. R. Gambill and R. D. Bundy, An Evaluation of the Present Status of Swirl Flow Heat Transfer, *ASME Paper 61-HT-42*, 1961.

6. R. F. Lopina and A. E. Bergles, Heat Transfer and Pressure Drop in Tape Generated Swirl Flow, *J. Heat Transfer*, Vol. 94, pp. 434–442, 1969.

7. R. L. Webb and A. E. Bergles, Performance Evaluation Criteria for Selection of Heat Transfer Surface Geometries Used in Low Reynolds Number Heat Exchangers, *Low Reynolds Number Convection in Channels and Bundles*, S. Kakac, R. H. Shah, and A. E. Bergles, eds., Hemisphere, Washington, D.C., and McGraw-Hill, New York, 1982.

8. L. C. Trimble, B. L. Messinger, H. E. Ulbrich, G. Smith, and T. Y. Lin, Ocean Thermal Energy Conversion System Study Report, *Proc. 3d Workshop Ocean Thermal Energy Conversion (OTEC)*, APL/JHU SR 75-2, pp. 3–21, August 1975.

9. A. E. Bergles and M. K. Jensen, Enhanced Single-Phase Heat Transfer for OTEC Systems, *Proc. 4th Conf. Ocean Thermal Energy Conversion (OTEC)*, University of New Orleans, pp. VI-41 to VI-54, July 1977.

10. R. L. Webb, E. R. G. Eckert, and R. J. Goldstein, Heat Transfer and Friction in Tubes with Repeated Rib Roughness, *Int. J. Heat Mass Transfer*, Vol. 14, pp. 601–618, 1971.

11. A. E. Bergles, G. S. Brown, Jr., and W. D. Snider, Heat Transfer Performance of Internally Finned Tubes, *ASME Paper 71-HT-31,* 1971.

12. E. Smithberg and F. Landis, Friction and Forced Convection Heat Transfer Characteristics in Tubes with Twisted Tape Swirl Generators, *J. Heat Transfer,* Vol. 86, pp. 39–49, 1964.

13. R. L. Webb, Performance, Cost Effectiveness and Water Side Fouling Considerations of Enhanced Tube Heat Exchangers for Boiling Service with Tube-Side Water Flow. *Heat Transfer Engineering,* Vol. 3, Nos. 3–4, pp. 84–98, 1982.

14. G. R. Kubanek and D. L. Miletti, Evaporative Heat Transfer and Pressure Drop Performance of Internally-Finned Tubes with Refrigerant 22, *J. Heat Transfer,* Vol. 101, pp. 447–452, 1979.

15. M. Luu and A. E. Bergles, Augmentation of In-Tube Condensation of R-113 by Means of Surface Roughness, *ASHRAE Trans.,* Vol. 87, Pt. 2, 1981.

16. W. R. Gambill, R. D. Bundy, and R. W. Wansbrough, Heat Transfer, Burnout, and Pressure Drop for Water in Swirl Flow Tubes with Internal Twisted Tapes, *Chem. Eng. Prog. Symp. Ser.,* Vol. 57, No. 32, pp. 127–137, 1961.

17. F. E. Megerlin, R. W. Murphy, and A. E. Bergles, Augmentation of Heat Transfer in Tubes by Means of Mesh and Brush Inserts, *J. Heat Transfer,* Vol. 96, pp. 145–151, 1974.

18. R. K. Young and R. L. Hummel, Improved Nucleate Boiling Heat Transfer, *Chem. Eng. Prog.,* Vol. 60, No. 7, pp. 53–58, 1964.

19. A. E. Bergles, N. Bakhru, and J. W. Shires, Cooling of High-Power-Density Computer Components, *EPL Rep. 70712-60,* Massachusetts Institute of Technology, Cambridge, Mass., 1968.

20. V. M. Zhukov, G. M. Kazakov, S. A. Kovalev, and Y. A. Kuzmakichta, Heat Transfer in Boiling of Liquids on Surfaces Coated with Low Thermal Conductivity Films, *Heat Transfer Sov. Res.,* Vol. 7, No. 3, pp. 16–26, 1975.

21. A. E. Bergles and W. G. Thompson, Jr., The Relationship of Quench Data to Steady-State Pool Boiling Data, *Int. J. Heat Mass Transfer,* Vol. 13, pp. 55–68, 1970.

22. A. E. Bergles, Principles of Heat Transfer Augmentation. II: Two-Phase Heat Transfer, *Heat Exchangers, Thermal-Hydraulic Fundamentals and Design,* S. Kakac, A. E. Bergles, and F. Mayinger, eds., pp. 857–881, Hemisphere, Washington, D.C., and McGraw-Hill, New York, 1981.

23. L. C. Kun and A. M. Czikk, Surface for Boiling Liquids, *U.S. Pat. 3,454,081,* July 8, 1969.

24. J. Fujikake, Heat Transfer Tube for Use in Boiling Type Heat Exchangers and Method of Producing the Same, *U.S. Pat. 4,216,826,* Aug. 12, 1980.

25. U. P. Hwang and K. P. Moran, Boiling Heat Transfer of Silicon Integrated Circuits Chip Mounted on a Substrate, in *Heat Transfer in Electronic Equipment, HTD Vol. 20,* M. D. Kelleher and M. M. Yovanovich, eds., pp. 53–59, ASME, New York, 1981.

26. R. L. Webb, Heat Transfer Surface Having a High Boiling Heat Transfer Coefficient, *U.S. Pat. 3,696,861,* Oct. 10, 1972.

27. V. A. Zatell, Method of Modifying a Finned Tube for Boiling Enhancement, *U.S. Pat. 3,768,290,* Oct. 30, 1973.

28. W. Nakayama, T. Daikoku, H. Kuwahara, and K. Kakizaki, High-Flux Heat Transfer Surface "Thermoexcel," *Hitachi Rev.,* Vol. 24, pp. 329–333, 1975.

29. K. Stephan and J. Mitrovic, Heat Transfer in Natural Convective Boiling of Refrigerants and Refrigerant-Oil-Mixtures in Bundles of T-Shaped Finned Tubes, *Advances in Enhanced Heat Transfer—1981, HTD-Vol. 18,* R. L. Webb, T. C. Carnavos, E. L. Park, Jr., and K. M. Hostetler, eds., pp. 131–146, ASME, New York, 1981.

30. E. Ragi, Composite Structure for Boiling Liquids and Its Formation, *U.S. Pat. 3,684,007,* Aug. 15, 1972.

31. P. J. Marto and W. M. Rohsenow, Effects of Surface Conditions on Nucleate Pool Boiling of Sodium, *J. Heat Transfer,* Vol. 88, pp. 196–204, 1966.

32. P. S. O'Neill, C. F. Gottzmann, and C. F. Terbot, Novel Heat Exchanger Increases Cascade Cycle Efficiency for Natural Gas Liquefaction, *Advances in Cryogenic Engineering,* Vol. 17, pp. 421–437, 1972.

33. S. Oktay and A. F. Schmeckenbecher, Preparation and Performance of Dendritic Heat Sinks, *J. Electrochem. Soc.,* Vol. 21, pp. 912–918, 1974.

34. M. M. Dahl and L. D. Erb, Liquid Heat Exchanger Interface Method, *U.S. Pat. 3,990,862,* Nov. 9, 1976.

35. M. Fujii, E. Nishiyama, and G. Yamanaka, Nucleate Pool Boiling Heat Transfer from Micro-Porous Heating Surfaces, *Advances in Enhanced Heat Transfer,* J. M. Chenoweth, J. Kaellis, J. W. Michel, and S. Shenkman, eds., pp. 45–51, ASME, New York, 1979.

36. K. R. Janowski, M. S. Shum, and S. A. Bradley, Heat Transfer Surface, *U.S. Pat. 4,129,181,* Dec. 12, 1978.

37. D. F. Warner, K. G. Mayhan, and E. L. Park, Jr., Nucleate Boiling Heat Transfer of Liquid Nitrogen from Plasma Coated Surfaces, *Int. J. Heat Mass Transfer,* Vol. 21, pp. 137–144, 1978.

38. A. M. Czikk and P. S. O'Neill, Correlation of Nucleate Boiling from Porous Metal Films, *Advances in Enhanced Heat Transfer,* J. M. Chenoweth, J. Kaellis, J. W. Michel, and S. Shenkman, eds., pp. 53–60, ASME, New York, 1979.

39. W. Nakayama, T. Daikoku, H. Kuwahara, and T. Nakajima, Dynamic Model of Enhanced Boiling Heat Transfer on Porous Surface—Parts I and II, *J. Heat Transfer,* Vol. 102, pp. 445–456, 1980.

40. S. Yilmaz, J. J. Hwalck, and J. N. Westwater, Pool Boiling Heat Transfer Performance for Commercial Enhanced Tube Surfaces, *ASME Paper 80-HT-41,* July 1980.

41. A. E. Bergles and M.-C. Chyu, Characteristics of Nucleate Pool Boiling from Porous Metallic Coatings, *Advances in Enhanced Heat Transfer—1981, HTD Vol. 18,* R. L. Webb, T. C. Carnavos, E. L. Park, Jr., and K. M. Hostetler, eds., pp. 61–71, ASME, New York, 1981.

42. S. Yilmaz, J. W. Palen, and J. Taborek, Enhanced Surfaces as Single Tubes and Tube Bundles, *Advances in Enhanced Heat Transfer—1981, HTD-Vol. 18,* R. L. Webb, T. C. Carnavos, E. L. Park, Jr., and K. M. Hostetler, eds., pp. 123–129, ASME, New York, 1981.

43. M.-C. Chyu, A. E. Bergles, and F. Mayinger, Enhancement of Horizontal Tube Spray Film Evaporators, *Proceedings 7th Int. Heat Trans. Conf.,* Vol. 6, pp. 275–280, Hemisphere Publishing Co., Washington, D.C. 1982.

44. R. L. Webb, G. H. Junkhan, and A. E. Bergles, Bibliography of U.S. Patents on Augmentation of Convective Heat and Mass Transfer-II, Heat Transfer Lab. Rep. *HTL-32, ISU-ERI-Ames-84257, DE 84014865,* Iowa State University, Ames, Iowa, September 1980.

45. R. L. Webb, The Evolution of Enhanced Surface Geometries for Nucleate Boiling, *Heat Transfer Eng.,* Vol. 2, Nos. 3–4, pp. 46–49, 1981.

46. S. Iltscheff, Über einige Versuche zur Erzielung von Tropfkondensation mit fluorierten Kältemitteln, *Kältetech. Klim.,* Vol. 23, pp. 237–241, 1971.

47. I. Tanawasa, Dropwise Condensation: The Way to Practical Applications, *Heat Transfer 1978, Proc. 6th Int. Heat Transfer Conf.,* Vol. 6, pp. 393–405, Hemisphere, Washington, D.C., 1978.

48. L. R. Glicksman, B. B. Mikic, and D. F. Snow, Augmentation of Film Condensation on the Outside of Horizontal Tubes, *AIChE J.,* Vol. 19, pp. 636–637, 1973.

49. A. E. Bergles, G. H. Junkhan, and R. L. Webb, Energy Conservation via Heat Transfer Enhancement, Heat Transfer Lab. Rep. *COO-4649-5,* Iowa State University, Ames, Iowa, 1979.

50. D. F. Gluck, The Effect of Turbulence Promotion on Newtonian and Non-Newtonian Heat Transfer Rates, M.S. thesis, University of Delaware, Newark, Del., 1959.

51. A. R. Blumenkrantz and J. Taborek, Heat Transfer and Pressure Drop Characteristics of Turbotec Spirally Grooved Tubes in the Turbulent Regime, Heat Transfer Research, Inc., *Rep. 2439-300-7,* 1970.

52. G. R. Rozalowski and R. A. Gater, Pressure Loss and Heat Transfer Characteristics for High Viscous Flow in Convoluted Tubing, *ASME Paper 75-HT-40,* 1975.

53. D. Pescod, The Effects of Turbulence Promoters on the Performance of Plate Heat Exchangers, *Heat Exchangers: Design and Theory Sourcebook,* N. H. Afghan and E. U. Schlünder, eds., pp. 601–616, Scripta, Washington, D.C., 1974.

54. Z. Nagaoka and A. Watanabe, Maximum Rate of Heat Transfer with Minimum Loss of Energy, *Proc. 7th Int. Cong. Refrigeration,* Vol. 3, pp. 221–245, 1936.

55. W. F. Cope, The Friction and Heat Transmission Coefficients of Rough Pipes, *Proc. Inst. Mech. Eng.,* Vol. 145, pp. 99–105, 1941.

56. D. W. Savage and J. E. Myers, The Effect of Artificial Surface Roughness on Heat and Momentum Transfer, *AIChE J.,* Vol. 9, pp. 694–702, 1963.

57. V. Kolar, Heat Transfer in Turbulent Flow of Fluids through Smooth and Rough Tubes, *Int. J. Heat Mass Transfer,* Vol. 8, pp. 639–653, 1965.

58. V. Zajic, Some Results on Research of Intensified Water Cooling by Roughened Surfaces and Surface Boiling at High Heat Flux Rates, *Acta Technica CSAV,* No. 5, pp. 602–612, 1965.

59. R. A. Gowen, A Study of Forced Convection Heat Transfer from Smooth and Rough Surfaces, Ph.D. thesis in chemical engineering and applied chemistry, University of Toronto, 1967.

60. E. K. Kalinin, G. A. Dreitser, and S. A. Yarkho, Experimental Study of Heat Transfer Intensification under Condition of Forced Flow in Channels, *Jpn. Soc. Mech. Eng. 1967 Semi-Int. Symp.,* Paper 210, September 1967.

61. D. Eissenberg, Tests of an Enhanced Horizontal Tube Condenser under Conditions of Horizontal Steam Cross Flow, *Heat Transfer 1970,* Vol. I, Paper HE2.1, Elsevier, Amsterdam, 1970.

62. J. M. Kramer and R. A. Gater, Pressure Loss and Heat Transfer for Non-Boiling Fluid Flow in Convoluted Tubing, *ASME Paper 73-HT-23,* 1973.

63. G. Grass, Verbesserung der Wärmeuebertragung an Wasser durch künstliche Aufrauhung der Oberflächen in Reaktoren Wärmetauschern, *Atomkernenergie,* Vol. 3, pp. 328–331, 1958.

64. A. R. Blumenkrantz and J. Taborek, Heat Transfer and Pressure Drop Characteristics of Turbotec Spirally Grooved Tubes in the Turbulent Regime, Heat Transfer Research, Inc., *Rep. 2439-300-7,* 1970.

65. D. G. Dipprey and R. H. Sabersky, Heat and Momentum Transfer in Smooth and Rough Tubes at Various Prandtl Numbers, *Int. J. Heat Mass Transfer,* Vol. 6, pp. 329–353, 1963.

66. A. Blumenkrantz, A. Yarden, and J. Taborek, Performance Prediction and Evaluation of Phelps Dodge Spirally Grooved Tubes. Inside Tube Flow Pressure Drop and Heat Transfer in Turbulent Regime, Heat Transfer Research, Inc., *Rep. 2439-300-4,* 1969.

67. E. C. Brouillette, T. R. Mifflin, and J. E. Myers, Heat Transfer and Pressure Drop Characteristics of Internal Finned Tubes, *ASME Paper 57-A-47,* 1957.

68. J. W. Smith, R. A. Gowan, and M. E. Charles, Turbulent Heat Transfer and Temperature Profiles in a Rifled Pipe, *Chem. Eng. Sci.,* Vol. 23, pp. 751–758, 1968.

69. P. Kumar and R. L. Judd, Heat Transfer with Coiled Wire Turbulence Promoters, *Can. J. Chem. Eng.,* Vol. 8, pp. 378–383, 1970.

70. R. L. Webb, E. R. G. Eckert, and R. J. Goldstein, Generalized Heat Transfer and Friction Correlations for Tubes with Repeated-Rib Roughness, *Int. J. Heat Mass Transfer,* Vol. 15, pp. 180–184, 1972.

71. J. G. Withers, Tube-Side Heat Transfer and Pressure Drop for Tubes Having Helical Internal Ridging with Turbulent/Transitional Flow of Single-Phase Fluid. Pt. 1. Single-Helix Ridging, *Heat Transfer Eng.,* Vol. 2, No. 1, pp. 48–58, 1980.

72. J. G. Withers, Tube-Side Heat Transfer and Pressure Drop for Tubes Having Helical Internal Ridging with Turbulent/Transitional Flow of Single-Phase Fluid. Pt. 2. Multiple-Helix Ridging, *Heat Transfer Eng.,* Vol. 2, No. 2, pp. 43–50, 1980.

73. M. J. Lewis, An Elementary Analysis for Predicting the Momentum and Heat-Transfer Characteristics of a Hydraulically Rough Surface, *J. Heat Transfer,* Vol. 97, pp. 249–254, 1975.

74. G. A. Kemeny and J. A. Cyphers, Heat Transfer and Pressure Drop in an Annular Gap with Surface Spoilers, *J. Heat Transfer,* Vol. 83, pp. 189–198, 1961.

75. A. W. Bennett and H. A. Kearsey, Heat Transfer and Pressure Drop for Superheated Steam Flowing through an Annulus with One Roughened Surface, *Atomic Energy Research Establishment (United Kingdom) 4350,* 1964.

76. H. Brauer, Strömungswiderstand und Wärmeübergang bei Ringspalten mit rauhen Rohren, *Atomkernenergie,* No. 4, pp. 152–159, 1961.

77. W. S. Durant, R. H. Towell, and S. Mirshak, Improvement of Heat Transfer to Water Flowing in an Annulus by Roughening the Heated Wall, *Chem. Eng. Prog. Symp. Ser.,* Vol. 60, No. 61, pp. 106–113, 1965.

78. M. Dalle Donne and L. Meyer, Turbulent Convective Heat Transfer from Rough Surfaces with Two-Dimensional Rectangular Ribs, *Int. J. Heat Mass Transfer,* Vol. 20, pp. 583–620, 1977.

79. M. Hudina, Evaluation of Heat Transfer Performances of Rough Surfaces from Experimental Investigation in Annular Channels, *Int. J. Heat Mass Transfer,* Vol. 22, pp. 1381–1392, 1979.

80. M. Dalle Donne, Heat Transfer in Gas Cooled Fast Reactor Cores, *Ann. Nucl. Energy,* Vol. 5, pp. 439–453, 1978.

81. M. Dalle Donne, A. Martelli, and K. Rehme, Thermo-Fluid-Dynamic Experiments with Gas-Cooled Bundles of Rough Rods and Their Evaluations with the Computer Code SAGAPØ, *Int. J. Heat Mass Transfer,* Vol. 22, pp. 1355–1374, 1979.

82. E. Achenbach, The Effect of Surface Roughness on the Heat Transfer from a Circular Cylinder to the Cross Flow of Air, *Int. J. Heat Mass Transfer,* Vol. 20, pp. 359–369, 1977.

83. A. Zhukauskas, J. Ziugzda, and P. Daujotas, Effects of Turbulence on the Heat Transfer of a Rough Surface Cylinder in Cross-Flow in the Critical Range of Re, *Heat Transfer 1978,* Vol. 4, pp. 231–236, Hemisphere, Washington, D.C., 1978.

84. G. B. Melese, Comparison of Partial Roughening of the Surface of Fuel Elements with Other Ways of Improving Performance of Gas-Cooled Nuclear Reactors, *General Atomics 4624,* 1963.

85. Heat Transfer Capability, *Mech. Eng.,* Vol. 89, p. 55, 1967.

86. R. B. Cox, A. S. Pascale, G. A. Matta, and K. S. Stromberg, Pilot Plant Tests and Design Study of a 2.5 MGD Horizontal-Tube Multiple-Effect Plant, *Off. Saline Water Res. Dev. Rep. No. 492,* October 1969.

87. I. H. Newson, Heat Transfer Characteristics of Horizontal Tube Multiple Effect (HTME) Evaporators—Possible Enhanced Tube Profiles, *Proc. 6th Int. Symp. Fresh Water from the Sea,* Vol. 2, pp. 113–124, 1978.

88. W. S. Durant and S. Mirshak, Roughening of Heat Transfer Surfaces as a Method of Increasing Heat Flux at Burnout, *E. I. Dupont de Nemours and Co. 380,* 1959.

89. V. I. Gomelauri and T. S. Magrakvelidze, Mechanism of Influence of Two Dimensional Artificial Roughness on Critical Heat Flux in Subcooled Water Flow, *Therm. Eng.,* Vol. 25, No. 2, pp. 1–3, 1978.

90. R. W. Murphy and K. L. Trucsdale, The Mechanism and the Magnitude of Flow Boiling Augmentation in Tubes with Discrete Surface Roughness Elements (III), *Raytheon Co. Rep. B12-7294,* November 1972.

91. J. G. Withers and E. P. Habdas, Heat Transfer Characteristics of Helical Corrugated Tubes for Intube Boiling of Refrigerant R-12, *AIChE Symp. Ser.,* Vol. 70, No. 138, pp. 98–106, 1974.

92. E. Bernstein, J. P. Petrek, and J. Meregian, Evaluation and Performance of Once-Through, Zero-Gravity Boiler Tubes with Two-Phase Water, *Pratt and Whitney Aircraft Co. 428,* 1964.

93. E. Janssen and J. A. Kervinen, Burnout Conditions for Single Rod in Annular Geometry, Water at 600 to 1400 psia, *General Electric Atomic Power 3899,* 1963.

94. E. P. Quinn, Transition Boiling Heat Transfer Program, *5th Q. Prog. Rep., General Electric Atomic Power 4608,* 1964.

95. H. S. Swenson, J. R. Carver, and G. Szoeke, The Effects of Nucleate Boiling versus Film Boiling on Heat Transfer in Power Boiler Tubes, *J. Eng. Power,* Vol. 84, pp. 365–371, 1962.

96. J. W. Ackerman, Pseudoboiling Heat Transfer to Supercritical Pressure Water in Smooth and Ribbed Tubes, *J. Heat Transfer,* Vol. 92, pp. 490–498, 1970.

97. A. J. Sellers, G. M. Thur, and M. K. Wong, Recent Developments in Heat Transfer and Development of the Mercury Boiler for the SNAP-8 System, *Proc. Conf. Application of High Temperature Instrumentation to Liquid-Metal Experiments, Argonne National Laboratory 7100,* pp. 573–632, 1965.

98. J. O. Medwell and A. A. Nicol, Surface Roughness Effects on Condensate Films, *ASME Paper 65-HT-43,* 1965.

99. A. A. Nicol and J. O. Medwell, The Effect of Surface Roughness on Condensing Steam, *Can. J. Chem. Eng.,* Vol. 44, No. 6, pp. 170–173, 1966.

100. T. C. Carnavos, An Experimental Study: Condensing R-11 on Augmented Tubes, *ASME Paper 80-HT-54,* 1980.

101. R. B. Cox, G. A. Matta, A. S. Pascale, and K. G. Stromberg, Second Report on Horizontal Tubes Multiple-Effect Process Pilot Plant Tests and Design, *Off. Saline Water Res. Dev. Rep. No. 592,* May 1970.

102. W. J. Prince, Enhanced Tubes for Horizontal Evaporator Desalination Process, M.S. thesis in engineering, University of California, Los Angeles, 1971.

103. G. W. Fenner and E. Ragi, Enhanced Tube Inner Surface Device and Method, *U.S. Pat. 4,154,293,* May 15, 1979.

104. R. K. Shah, C. F. McDonald, and C. P. Howard, eds., *Compact Heat Exchangers—History, Technological Advancement and Mechanical Design Problems, HTD Vol. 10,* ASME, New York, 1980.

105. R. K. Shah, Classification of Heat Exchangers, *Thermal-Hydraulic Fundamentals and Design,* S. Kakac, A. E. Bergles, and F. Mayinger, eds., pp. 9–46, Hemisphere/McGraw-Hill, New York, 1981.

106. M. Ito, H. Kimura, and T. Senshu, Development of High Efficiency Air-Cooled Heat Exchangers, *Hitachi Rev.,* Vol. 20, pp. 323–326, 1977.

107. W. M. Kays and A. L. London, *Compact Heat Exchangers,* 2d ed., McGraw-Hill, New York, 1964.

108. R. L. Webb, Air-Side Heat Transfer in Finned Tube Heat Exchangers, *Heat Transfer Eng.,* Vol. 1, No. 3, pp. 33–49, 1980.

109. L. Goldstein, Jr., and E. M. Sparrow, Experiments on the Transfer Characteristics of a Corrugated Fin and Tube Heat Exchanger Configuration, *J. Heat Transfer,* Vol. 98, pp. 26–34, 1976.

110. S. W. Krückels and V. Kottke, Untersuchung über die Verteilung des Wärmeübergangs an Rippen und Rippen Rohr-Modellen, *Chem. Ing. Tech.,* Vol. 42, pp. 355–362, 1970.

111. E. M. Sparrow, B. R. Baliga, and S. V. Patankar, Heat Transfer and Fluid Flow Analysis of Interrupted-Wall Channels, with Applications to Heat Exchangers, *J. Heat Transfer,* Vol. 99, pp. 4–11, 1977.

112. S. V. Patankar and C. Prakash, An Analysis of the Effect of Plate Thickness on Laminar Flow and Heat Transfer in Interrupted-Plate Passages, *Advances in Enhanced Heat Transfer—1981, HTD Vol. 18,* R. L. Webb, T. C. Carnavos, E. L. Park, Jr., and K. M. Hostetler, eds., pp. 51–59, ASME, New York, 1981.

113. A. P. Watkinson, D. C. Miletti, and G. R. Kubanek, Heat Transfer and Pressure Drop of Internally Finned Tubes in Laminar Oil Flow, *ASME Paper 75-HT-41,* 1975.

114. W. J. Marner and A. E. Bergles, Augmentation of Tubeside Laminar Flow Heat Transfer by Means of Twisted-Tape Inserts, Static-Mixer Inserts, and Internally Finned Tubes, *Heat Transfer 1978,* Vol. 2, pp. 583–588, Hemisphere, Washington, D.C., 1978.

115. A. E. Bergles, Enhancement of Heat Transfer, *Heat Transfer 1978,* Vol. 6, pp. 89–108, Hemisphere, Washington, D.C., 1978.

116.. T. C. Carnavos, Heat Transfer Performance of Internally Finned Tubes in Turbulent Flow, *Advances in Enhanced Heat Transfer,* pp. 61–67, ASME, New York, 1979.

117. W. E. Hilding and C. H. Coogan, Jr., Heat Transfer and Pressure Loss Measurements in Internally Finned Tubes, *Symp. Air-Cooled Heat Exchangers,* pp. 57–85, ASME, New York, 1964.

118. S. V. Patankar, M. Ivanovic, and E. M. Sparrow, Analysis of Turbulent Flow and Heat Transfer in Internally Finned Tube and Annuli, *J. Heat Transfer,* Vol. 101, pp. 29–37, 1979.

119. T. C. Carnavos, Cooling Air in Turbulent Flow with Internally Finned Tubes, *Heat Transfer Eng.,* Vol. 1, No. 2, pp. 41–46, 1979.

120. D. Q. Kern and A. D. Kraus, *Extended Surface Heat Transfer,* McGraw-Hill, New York, 1972.

121. A. Y. Gunter and W. A. Shaw, Heat Transfer, Pressure Drop and Fouling Rates of Liquids for Continuous and Noncontinuous Longitudinal Fins, *Trans. ASME,* Vol. 64, pp. 795–802, 1942.

122. L. Clarke and R. E. Winston, Calculation of Finside Coefficients in Longitudinal Finned-Tube Heat Exchangers, *Chem. Eng. Prog.,* Vol. 51, No. 3, pp. 147–150, 1955.

123. M. M. El-Wakil, *Nuclear Energy Conversion,* American Nuclear Society, La Grange Park, Ill., 1978.

124. D. Gorenflo, Zum Wärmeübergang bei Blasenverdampfung an Rippenrohren, dissertation, Technische Hochschule, Karlsruhe, 1966.

125. G. Hesse, Heat Transfer in Nucleate Boiling, Maximum Heat Flux and Transition Boiling, *Int. J. Heat Mass Transfer,* Vol. 16, pp. 1611–1627, 1973.

126. J. W. Westwater, Development of Extended Surfaces for Use in Boiling Liquids, *AIChE Symp. Ser.,* Vol. 69, No. 131, pp. 1–9, 1973.

127. D. L. Katz, J. E. Meyers, E. H. Young, and G. Balekjian, Boiling outside Finned Tubes, *Petroleum Refiner,* Vol. 34, pp. 113–116, 1955.

128. K. Nakajima and A. Shiozawa, An Experimental Study on the Performance of a Flooded Type Evaporator, *Heat Transfer Jpn. Res.,* Vol. 4, No. 4, pp. 49–66, 1975.

129. N. Arai, T. Fukushima, A. Arai, T. Nakajima, K. Fujie, and Y. Nakayama, Heat Transfer Tubes Enhancing Boiling and Condensation in Heat Exchanger of a Refrigerating Machine, *ASHRAE Trans.,* Vol. 83, Pt. 2, pp. 58–70, 1977.

130. J. C. Corman and M. H. McLaughlin, Boiling Heat Transfer with Structured Surfaces, *ASHRAE Trans.,* Vol. 82, Pt. 1, pp. 906–918, 1976.

131. V. N. Schultz, D. K. Edwards, and I. Catton, Experimental Determination of Evaporative Heat Transfer Coefficients on Horizontal, Threaded Tubes, *AIChE Symp. Ser.,* Vol. 73, No. 164, pp. 223–227, 1977.

132. R. J. Conti, Experimental Investigations of Horizontal Tube Ammonia Film Evaporators with Small Temperature Differentials, *Proc. 5th Ocean Thermal Energy Conversion Conf.,* Miami Beach, Fla., pp. VI-161 to VI-180, 1978.

133. S. Sideman and A. Levin, Effect of the Configuration on Heat Transfer to Gravity Driven Films Evaporating on Grooved Tubes, *Desalination,* Vol. 31, pp. 7–18, 1979.

134. R. B. Cox, G. A. Matta, A. S. Pascale, and K. G. Stromberg, Second Report on Horizontal-Tubes Multiple-Effect Process Pilot Plant Tests and Design, *Off. Saline Water Res. Dev. Prog. Rep. No. 592,* May 1970.

135. D. G. Thomas and G. Young, Thin Film Evaporation Enhancement by Finned Surfaces, *Ind. Eng. Chem. Proc. Des. Dev.,* Vol. 9, pp. 317–323, 1970.

136. J. J. Lorenz, D. T. Yung, D. L. Hillis, and N. F. Sather, OTEC Performance Tests of the Carnegie-Mellon University Vertical Fluted-Tube Evaporator, *ANL/OTEC-PS-5,* Argonne National Laboratory, July 1979.

137. E. U. Schlünder and M. Chwala, Ortlicher Wärmeübergang und Druckabfall bei der Strömung verdampfender Kältemittel in innenberippten, waggerechten Rohren, *Kältetech. Klim.,* Vol. 21, No. 5, pp. 136–139, 1969.

138. G. R. Kubanek and D. L. Miletti, Evaporative Heat Transfer and Pressure Drop Performance of Internally-Finned Tubes with Refrigerant 22, *J. Heat Transfer,* Vol. 101, pp. 447–452, 1979.

139. M. Ito and H. Kimura, Boiling Heat Transfer and Pressure Drop in Internal Spiral-Grooved Tubes, *Bull. JSME,* Vol. 22, No. 171, pp. 1251–1257, 1979.

140. J. M. Robertson, Review of Boiling, Condensing and Other Aspects of Two-Phase Flow in Plate Fin Heat Exchangers, *Compact Heat Exchangers—History, Technological Advances and Mechanical Design Problems, HTD-Vol. 10,* R. K. Shah, C. F. McDonald, and C. P. Howard, eds., pp. 17–27, ASME, New York, 1980.

141. C. B. Panchal, D. L. Hillis, J. J. Lorenz, and D. T. Yung, OTEC Performance Tests of the Trane Plate-Fin Heat Exchanger, *ANL/OTEC-PS-7,* April 1981.

142. D. Yung, J. J. Lorenz, and C. Panchal, Convective Vaporization and Condensation in Ser-

rated-Fin Channels, *Heat Transfer in Ocean Thermal Energy Conversion [OTEC] Systems, HTD Vol. 12,* W. L. Owens, ed., pp. 29–37, ASME, New York, 1980.

143. C. C. Chen, J. V. Loh, and J. W. Westwater, Prediction of Boiling Heat Transfer in a Compact Plate-Fin Heat Exchanger Using the Improved Local Technique, *Int. J. Heat Mass Transfer,* Vol. 24, pp. 1907–1912, 1981.

144. K. O. Beatty, Jr., and D. L. Katz, Condensation of Vapors on Outside of Finned Tubes, *Chem. Eng. Prog.,* Vol. 44, No. 1, pp. 55–70, 1948.

145. E. H. Young and D. J. Ward, How to Design Finned Tube Shell and Tube Heat Exchangers, *The Refining Engineer,* pp. C-32 to C-36, November 1957.

146. T. M. Rudy and R. L. Webb, Condensate Retention of Horizontal Integral-Fin Tubing, *Advance in Enhanced Heat Transfer—1981, HTD Vol. 18,* R. L. Webb, T. C. Carnavos, E. L. Park, Jr., and K. M. Hostetler, eds., pp. 35–41, ASME, New York, 1981.

147. R. Chandran and F. A. Watson, Condensation on Static and Rotating Pinned Tubes, *Trans. Inst. Chem. Eng.,* Vol. 54, pp. 65–72, 1976.

148. R. L. Webb and D. L. Gee, Analytical Predictions for a New Concept Spine-Fin Surface Geometry, *ASHRAE Trans.,* Vol. 85, Pt. 2, pp. 274–283, 1979.

149. F. Notaro, Enhanced Condensation Heat Transfer Device and Method, *U.S. Pat. 4,154,294,* May 15, 1979.

150. R. Gregorig, Hautkondensation an Feingewellten Oberflächen bei Berücksichtigung der Oberflächenspannungen, *Z. Angew. Math. Phys.,* Vol. 5, pp. 36–49, 1954.

151. A. Thomas, J. J. Lorenz, D. A. Hillis, D. T. Young, and N. F. Sather, Performance Tests of 1 Mwt Shell and Tube Heat Exchangers for OTEC, *Proc. 6th OTEC Conf.,* Vol. 2, p. 11.1, Washington, D.C., 1979.

152. A. Blumenkrantz and J. Taborek, Heat Transfer and Pressure Drop Characteristics of Turbotec Spirally Deep Grooved Tubes in the Turbulent Regime, Heat Transfer Research, Inc., *Rep. 2439-300-7,* December 1970.

153. J. Palen, B. Cham, and J. Taborek, Comparison of Condensation of Steam on Plain and Turbotec Spirally Grooved Tubes in a Baffled Shell-and-Tube Condenser, Heat Transfer Research, Inc., *Rep. 2439-300-6,* January 1971.

154. P. J. Marto, R. J. Reilly, and J. H. Fenner, An Experimental Comparison of Enhanced Heat Transfer Condenser Tubing, *Advances in Enhanced Heat Transfer,* J. M. Chenoweth, J. Kaellis, J. W. Michel, and S. Shenkman, eds., pp. 1–9, ASME, New York, 1979.

155. T. C. Carnavos, An Experimental Study: Condensing R-11 on Augmented Tubes, *ASME Paper 80-HT-54,* 1980.

156. M. H. Mehta and M. R. Rao, Heat Transfer and Frictional Characteristics of Spirally Enhanced Tubes for Horizontal Condensers, *Advances in Enhanced Heat Transfer,* J. M. Chenoweth, J. Kaellis, J. W. Michel, and S. Shenkman, eds., pp. 11–21, ASME, New York, 1979.

157. J. G. Withers and E. H. Young, Steam Condensing on Vertical Rows of Horizontal Corrugated and Plain Tubes, *Ind. Eng. Chem. Process Des. Dev.,* Vol. 10, pp. 19–30, 1971.

158. D. G. Thomas, Enhancement of Film Condensation Rate on Vertical Tubes by Longitudinal Fins, *AIChE J.,* Vol. 14, pp. 644–649, 1968.

159. D. G. Thomas, Enhancement of Film Condensation Rate on Vertical Tubes by Vertical Wires, *Ind. Eng. Chem. Fund.,* Vol. 6, pp. 97–103, 1967.

160. Y. Mori, K. Hijikata, S. Hirasawa, and W. Nakayama, Optimized Performance of Condensers with Outside Condensing Surfaces, *J. Heat Transfer,* Vol. 103, pp. 96–102, 1981.

161. C. G. Barnes, Jr., and W. M. Rohsenow, Vertical Fluted Tube Condenser Performance Prediction, *Proc. 7th Int. Heat Trans. Conf.,* Vol. 5, pp. 39–43, Hemisphere Pub. Co., 1982.

162. L. G. Lewis and N. F. Sather, OTEC Performance Tests of the Carnegie-Mellon University Vertical Fluted-Tube Condenser, *ANL/OTEC-PS-4,* Argonne National Laboratory, May 1979.

163. N. Domingo, Condensation of Refrigerant-11 on the Outside of Vertical Enhanced Tubes, *ORNL/TM-7797,* Oak Ridge National Laboratory, August 1981.

164. L. G. Alexander and H. W. Hoffman, Performance Characteristics of Corrugated Tubes for Vertical Tube Evaporators, *ASME Paper 71-HT-30*, 1971.

165. D. L. Vrable, W. J. Yang, and J. A. Clark, Condensation of Refrigerant-12 inside Horizontal Tubes with Internal Axial Fins, *Heat Transfer 1974*, Vol. III, pp. 250–254, Japan Society of Mechanical Engineers, 1974.

166. R. L. Reisbig, Condensing Heat Transfer Augmentation inside Splined Tubes, *ASME Paper 74-HT-7*, July 1974.

167. J. H. Royal and A. E. Bergles, Augmentation of Horizontal In-Tube Condensation by Means of Twisted-Tape Inserts and Internally-Finned Tubes, *J. Heat Transfer*, Vol. 100, pp. 17–24, 1978.

168. J. H. Royal and A. E. Bergles, Pressure Drop and Performance Evaluation of Augmented In-Tube Condensation, *Heat Transfer 1978, Proc. 6th Int. Conf.*, Vol. 2, pp. 459–464, Hemisphere, Washington, D.C., 1978.

169. M. Luu and A. E. Bergles, Experimental Study of the Augmentation of In-Tube Condensation of R-113, *ASHRAE Trans.*, Vol. 85, Pt. 2, pp. 132–145, 1979.

170. M. Luu and A. E. Bergles, Enhancement of Horizontal In-Tube Condensation of R-113, *ASHRAE Trans.*, Vol. 86, Pt. 1, pp. 293–312, 1980.

171. V. G. Rifert and V. Y. Zadiraka, Steam Condensation inside Plain and Profiled Horizontal Tubes, *Therm. Eng.*, Vol. 25, No. 8, pp. 54–57, 1978.

172. Y. Mori and W. Nakayama, High-Performance Mist Cooled Condensers for Geothermal Binary Cycle Plants, in *Heat Transfer in Energy Problems, Proc. Jpn–U.S. Joint Sem.*, Tokyo, pp. 189–196, Sept. 30 to Oct. 2, 1980.

173. J. H. Sununu, Heat Transfer with Static Mixer Systems, Kenics Corp. *Tech. Rep. 1002*, 1970.

174. W. E. Genetti and S. J. Priebe, Heat Transfer with a Static Mixer, AIChE paper presented at the Fourth Joint Chemical Engineering Conference, Vancouver, Canada, 1973.

175. T. H. Van Der Meer and C. J. Hoogenedoorn, Heat Transfer Coefficients for Viscous Fluids in a Static Mixer, *Chem. Eng. Sci.*, Vol. 33, pp. 1277–1282, 1978.

176. W. J. Marner and A. E. Bergles, Augmentation of Tubeside Laminar Flow Heat Transfer by Means of Twisted-Tape Inserts, Static-Mixer Inserts and Internally Finned Tubes, *Heat Transfer 1978, Proc. 6th Int. Heat Transfer Conf.*, Vol. 2, pp. 583–588, Hemisphere, Washington, D.C., 1978.

177. S. T. Lin, L. T. Fan, and N. Z. Azer, Augmentation of Single Phase Convective Heat Transfer with In-Line Static Mixers, *Proc. 1978 Heat Transfer Fluid Mech. Inst.*, pp. 117–130, Stanford University Press, Stanford, Calif., 1978.

178. M. H. Pahl and E. Muschelknautz, Einsatz and Auslegung statischer Mischer, *Chem. Ing. Tech.*, Vol. 51, pp. 347–364, 1979.

179. L. B. Evans and S. W. Churchill, The Effect of Axial Promoters on Heat Transfer and Pressure Drop inside a Tube, *Chem. Eng. Prog. Symp. Ser. 59*, Vol. 41, pp. 36–46, 1963.

180. D. G. Thomas, Enhancement of Forced Convection Heat Transfer Coefficient Using Detached Turbulence Promoters, *Ind. Eng. Chem. Process Des. Dev.*, Vol. 6, pp. 385–390, 1967.

181. S. Maezawa and G. S. H. Lock, Heat Transfer inside a Tube with a Novel Promoter, *Heat Transfer 1978, Proc. 6th Int. Heat Transfer Conf.*, Vol. 2, pp. 596–600, Hemisphere, Washington, D.C., 1978.

182. F. E. Megerlin, R. W. Murphy, and A. E. Bergles, Augmentation of Heat Transfer in Tubes by Means of Mesh and Brush Inserts, *J. Heat Transfer*, Vol. 96, pp. 145–151, 1974.

183. E. O. Moeck, G. A. Wilkhammer, I. P. L. Macdonald, and J. G. Collier, Two Methods of Improving the Dryout Heat-Flux for High Pressure Steam/Water Flow, *Atomic Energy of Canada, Ltd. 2109*, 1964.

184. L. S. Tong, R. W. Steer, A. H. Wenzel, M. Bogaardt, and C. L. Spigt, Critical Heat Flux of a Heater Rod in the Center of Smooth and Rough Square Sleeves, and in Line-Contact with an Unheated Wall, *ASME Paper 67-WA/HT-29*, 1967.

185. E. P. Quinn, Transition Boiling Heat Transfer Program, *6th Q. Prog. Rep., General Electric Atomic Power 4646*, 1964.

186. A. N. Ryabov, F. T. Kamen'shchikov, V. N. Filipov, A. F. Chalykh, T. Yugay, Y. V. Stolyarov, T. I. Blagovestova, V. M. Mandrazhitskiy, and A. I. Yemelyanov, Boiling Crisis and Pressure Drop in Rod Bundles with Heat Transfer Enhancement Devices, *Heat Transfer Sov. Res.,* Vol. 9, No. 1, pp. 112–122, 1977.

187. D. C. Groeneveld and W. W. Yousef, Spacing Devices for Nuclear Fuel Bundles: A Survey of Their Effect on CHF, Post CHF Heat Transfer and Pressure Drop, *Proc. ANS/ASME/NRC Information Topical Meeting on Nuclear Reactor Thermal-Hydraulics, Nuclear Regulatory Commisson/CP-0014,* Vol. 2, pp. 1111–1130, 1980.

188. N. Z. Azer, L. T. Fan, and S. T. Lin, Augmentation of Condensation Heat Transfer with In-Line Static Mixers, *Proc. 1976 Heat Transfer Fluid Mech. Inst.,* pp. 512–526, Stanford University Press, Stanford, Calif., 1976.

189. L. T. Fan, S. T. Lin, and N. Z. Azer, Surface Renewal Model of Condensation Heat Transfer in Tubes with In-Line Static Mixers, *Int. J. Heat Mass Transfer,* Vol. 21, pp. 849–854, 1978.

190. R. Razgaitis and J. P. Holman, A Survey of Heat Transfer in Confined Swirl Flows, *Future Energy Production Systems, Heat and Mass Transfer Processes,* Vol. 2, pp. 831–866, Academic, New York, 1976.

191. S. W. Hong and A. E. Bergles, Augmentation of Laminar Flow Heat Transfer by Means of Twisted-Tape Inserts, *J. Heat Transfer,* Vol. 98, pp. 251–256, 1976.

192. F. Huang and F. K. Tsou, Friction and Heat Transfer in Laminar Free Swirling Flow in Pipes, *Gas Turbine Heat Transfer,* ASME, New York, 1979.

193. A. W. Date, Prediction of Fully-Developed Flow in a Tube Containing a Twisted Tape, *Int. J. Heat Mass Transfer,* Vol. 17, pp. 845–859, 1974.

194. R. K. Shah and A. L. London, *Laminar Flow Forced Convection in Ducts,* p. 380, Academic, New York, 1978.

195. R. F. Lopina and A. E. Bergles, Heat Transfer and Pressure Drop in Tape Generated Swirl Flow of Single-Phase Water, *J. Heat Transfer,* Vol. 91, pp. 434–442, 1969.

196. R. Thorsen and F. Landis, Friction and Heat Transfer Characteristics in Turbulent Swirl Flow Subject to Large Transverse Temperature Gradients, *J. Heat Transfer,* Vol. 90, pp. 87–98, 1968.

197. A. P. Colburn and W. J. King, Heat Transfer and Pressure Drop in Empty, Baffled, and Packed Tubes, III: Relation between Heat Transfer and Pressure Drop, *Ind. Eng. Chem.,* Vol. 23, pp. 919–923, 1931.

198. S. I. Evans and R. J. Sarjant, Heat Transfer and Turbulence in Gases Flowing inside Tubes, *J. Inst. Fuel,* Vol. 24, pp. 216–227, 1951.

199. E. Smithberg and F. Landis, Friction and Forced Convection Heat Transfer Characteristics in Tubes with Twisted Tape Swirl Generators, *J. Heat Transfer,* Vol. 86, pp. 39–49, 1964.

200. W. R. Gambill, R. D. Bundy, and R. W. Wansbrough, Heat Transfer, Burnout, and Pressure Drop for Water in Swirl Flow Tubes with Internal Twisted Tapes, *Chem. Eng. Prog. Symp. Ser.,* Vol. 57, No. 32, pp. 127–137, 1961.

201. M. H. Ibragimov, E. V. Nomofelov, and V. I. Subbotin, Heat Transfer and Hydraulic Resistance with the Swirl-Type Motion of Liquid in Pipes, *Teploenergetika,* Vol. 8, No. 7, pp. 57–60, 1962.

202. N. D. Greene, Convair Aircraft, private communication to W. R. Gambill, May 1969, cited in W. R. Gambill and R. D. Bundy, An Evaluation of the Present Status of Swirl Flow Heat Transfer, *ASME Paper 62-HT-42,* 1962.

203. B. Shiralker and P. Griffith, The Effect of Swirl, Inlet Conditions, Flow Direction, and Tube Diameter on the Heat Transfer to Fluids at Supercritical Pressure, *J. Heat Transfer,* Vol. 42, pp. 465–474, 1970.

204. W. R. Gambill and N. D. Greene, A Preliminary Study of Boiling Burnout Heat Fluxes for Water in Vortex Flow, *Chem. Eng. Prog.,* Vol. 54, No. 10, pp. 68–76, 1958.

205. F. Mayinger, O. Schad, and E. Weiss, Investigations into the Critical Heat Flux in Boiling, *Mannesmann Augsburg Nuernberg (Fed. Republic of Germany) Rep. 09.03.01,* 1966.

206. A. P. Ornatskiy, V. A. Chernobay, A. F. Vasilyev, and S. V. Perkov, A Study of the Heat Transfer Crisis with Swirled Flows Entering an Annular Passage, *Heat Transfer Sov. Res.,* Vol. 5, No. 4, pp. 7–10, 1973.

207. R. F. Lopina and A. E. Bergles, Subcooled Boiling of Water in Tape-Generated Swirl Flow, *J. Heat Transfer,* Vol. 95, pp. 281–283, 1973.

208. H. H. Sephton, Interface Enhancement for Vertical Tube Evaporator: A Novel Way of Substantially Augmenting Heat and Mass Transfer, *ASME Paper 71-HT-38,* 1971.

209. A. E. Bergles, W. D. Fuller, and S. J. Hynek, Dispersed Film Boiling of Nitrogen with Swirl Flow, *Int. J. Heat Mass Transfer,* Vol. 14, pp. 1343–1354, 1971.

210. M. Cumo, G. E. Farello, G. Ferrari, and G. Palazzi, The Influence of Twisted Tapes in Subcritical, Once-Through Vapor Generator in Counter Flow, *J. Heat Transfer,* Vol. 96, pp. 365–370, 1974.

211. A. Hunsbedt and J. M. Roberts, Thermal-Hydraulic Performance of a 2MWT Sodium Heated, Forced Recirculation Steam Generator Model, *J. Eng. Power,* Vol. 96, pp. 66–76, 1974.

212. C. Fouré, C. Moussez, and D. Eidelman, Techniques for Vortex Type Two-Phase Flow in Water Reactors, *Proc. 3d Int. Conf. Peaceful Uses of Atomic Energy,* United Nations, New York, Vol. 8, pp. 255–261, 1965.

213. M. K. Jensen, Boiling Heat Transfer and Critical Heat Flux in Helical Coils, Ph.D. dissertation, Iowa State University, Ames, Iowa, 1980.

214. M. K. Jensen and A. E. Bergles, Critical Heat Flux in Helically Coiled Tubes, *J. Heat Transfer,* Vol. 103, pp. 660–666, 1981.

215. G. G. Shklover and A. V. Gerasimov, Heat Transfer of Moving Steam in Coil-Type Heat Exchangers, *Teploenergetika,* Vol. 10, No. 5, pp. 62–65, 1963.

216. Z. L. Miropolskii and A. Kurbanmukhamedov, Heat Transfer with Condensation of Steam within Coils, *Therm. Eng.,* No. 5, pp. 111–114, 1975.

217. D. P. Traviss and W. M. Rohsenow, The Influence of Return Bends on the Downstream Pressure Drop and Condensation Heat Transfer in Tubes, *ASHRAE Trans.,* Vol. 79, Pt. 1, pp. 129–137, 1973.

218. W. D. Allingham and J. A. McEntire, Determination of Boiling Film Coefficient for a Heated Horizontal Tube in Water Saturated with Material, *J. Heat Transfer,* Vol. 83, pp. 71–76, 1961.

219. C. P. Costello and E. R. Redeker, Boiling Heat Transfer and Maximum Heat Flux for a Surface with Coolant Supplied by Capillary Wicking, *Chem. Eng. Prog. Symp. Ser.,* Vol. 59, No. 41, pp. 104–113, 1963.

220. C. P. Costello and W. J. Frea, The Role of Capillary Wicking and Surface Deposits in the Attainment of High Pool Boiling Burnout Heat Fluxes, *AIChE J.,* Vol. 10, pp. 393–398, 1964.

221. J. C. Corman and M. H. McLaughlin, Boiling Augmentation with Structured Surfaces, *ASHRAE Trans.,* Vol. 82, Pt. 1, pp. 906–918, 1976.

222. R. S. Gill, Pool Boiling in the Presence of Capillary Wicking Materials, S.M. thesis in mechanical engineering, Massachusetts Institute of Technology, Cambridge, Mass., 1967.

223. R. W. Watkins, C. R. Robertson, and A. Acrivos, Entrance Region Heat Transfer in Flowing Suspensions, *Int. J. Heat Mass Transfer,* Vol. 19, pp. 693–695, 1976.

224. M. Tamari and K. Nishikawa, The Stirring Effect of Bubbles upon the Heat Transfer to Liquids, *Heat Transfer Jpn. Res.,* Vol. 5, No. 2, pp. 31–44, 1976.

225. W. F. Hart, Heat Transfer in Bubble-Agitated Systems. A General Correlation, *I&EC Process Des. Dev.,* Vol. 15, pp. 109–111, 1976.

226. D. B. R. Kenning and Y. S. Kao, Convective Heat Transfer to Water Containing Bubbles: Enhancement Not Dependent on Thermocapillarity, *Int. J. Heat Mass Transfer,* Vol. 15, pp. 1709–1718, 1972.

227. E. Baker, Liquid Immersion Cooling of Small Electronic Devices, *Microelectronics and Reliability,* Vol. 12, pp. 163–173, 1973.

228. M. Behar, M. Courtaud, R. Ricque, and R. Semeria, Fundamental Aspects of Subcooled Boiling with and without Dissolved Gases, *Proc. 3d Int. Heat Transfer Conf.,* AIChE, New York, Vol. 4, pp. 1–11, 1966.

229. M. Jakob and W. Linke, Der Wärmeübergang beim Verdampfen von Flüssigkeiten an senkrechten und waagerechten Flächen, *Phys. Z.,* Vol. 36, pp. 267–280, 1935.

230. T. H. Insinger, Jr., and H. Bliss, Transmission of Heat to Boiling Liquids, *Trans. AIChE,* Vol. 36, pp. 491–516, 1940.

231. A. I. Morgan, L. A. Bromley, and C. R. Wilke, Effect of Surface Tension on Heat Transfer in Boiling, *Ind. Eng. Chem.,* Vol. 41, pp. 2767–2769, 1949.

232. E. K. Averin and G. N. Kruzhilin, The Influence of Surface Tension and Viscosity on the Conditions of Heat Exchange in the Boiling of Water, *Izv. Akad. Nauk SSSR Otdel. Tekh. Nauk,* Vol. 10, pp. 131–137, 1955.

233. A. J. Lowery, Jr. and J. W. Westwater, Heat Transfer to Boiling Methanol—Effect of Added Agents, *Ind. Eng. Chem.,* Vol. 49, pp. 1445–1448, 1957.

234. J. G. Collier, Multicomponent Boiling and Condensation, *Two-Phase Flow and Heat Transfer in the Power and Process Industries,* pp. 520–557, Hemisphere, Washington, D.C., and McGraw-Hill, New York, 1981.

235. W. R. van Wijk, A. S. Vos, and S. J. D. van Stralen, Heat Transfer to Boiling Binary Liquid Mixtures, *Chem. Eng. Sci.,* Vol. 5, pp. 68–80, 1956.

236. S. J. D. van Stralen, Heat Transfer to Boiling Binary Liquid Mixtures, *Brit. Chem. Eng.,* Pt. I, Vol. 4, pp. 8–17; Pt. II, Vol. 4, pp. 78–82, 1959.

237. M. Carne, Some Effects of Test Section Geometry, in Saturated Pool Boiling, on the Critical Heat Flux for Some Organic Liquids and Liquid Mixtures, *AIChE Preprint 6 for 7th Nat. Heat Transfer Conf.,* August 1964.

238. S. J. D. van Stralen, Nucleate Boiling in Binary Systems, *Augmentation of Convective Heat and Mass Transfer,* A. E. Bergles and R. L. Webb, eds., pp. 133–147, ASME, New York, 1970.

239. H. J. Gannett, Jr., and M. C. Williams, Pool Boiling in Dilute Nonaqueous Polymer Solutions, *Int. J. Heat Mass Transfer,* Vol. 11, pp. 1001–1005, 1971.

240. M. K. Jensen, A. E. Bergles, and F. A. Jeglic, Effects of Oily Contaminants on Nucleate Boiling of Water, *AIChE Symp. Ser.,* Vol. 75, No. 189, pp. 194–203, 1979.

241. G. Leppert, C. P. Costello, and B. M. Hoglund, Boiling Heat Transfer to Water Containing a Volatile Additive, *Trans. ASME,* Vol. 80, pp. 1395–1404, 1958.

242. A. E. Bergles and L. S. Scarola, Effect of a Volatile Additive on the Critical Heat Flux for Surface Boiling of Water in Tubes, *Chem. Eng. Sci.,* Vol. 21, pp. 721–723, 1966.

243. H. H. Sephton, Upflow Vertical Tube Evaporation of Sea Water with Interface Enhancement; Process Development by Pilot Plant Testing, *Desalination,* Vol. 16, pp. 1–13, 1975.

244. A. E. Bergles, G. H. Junkhan, and J. K. Hagge, Advanced Cooling Systems for Agricultural and Industrial Machines, *SAE Paper 751183,* 1976.

245. G. K. Rhode, D. M. Roberts, D. C. Schluderberg, and E. E. Walsh, Gas-Suspension Coolants for Power Reactors, *Proc. Am. Power Conf.,* Vol. 22, pp. 130–137, 1960.

246. D. C. Schluderberg, R. L. Whitelaw, and R. W. Carlson, Gaseous Suspensions—A New Reactor Coolant, *Nucleonics,* Vol. 19, pp. 67–76, 1961.

247. W. T. Abel, D. E. Bluman, and J. P. O'Leary, Gas-Solids Suspensions as Heat-Carrying Mediums, *ASME Paper 63-WA-210,* 1963.

248. R. Pfeffer, S. Rossetti, and S. Lieblein, Analysis and Correlation of Heat Transfer Coefficient and Friction Factor Data for Dilute Gas-Solid Suspensions, *NASA TN D-3603,* 1966.

249. R. Pfeffer, S. Rossetti, and S. Lieblein, The Use of a Dilute Gas-Solid Suspension as the Working Fluid in a Single Loop Brayton Space Power Generation Cycle, *AIChE Paper 49c,* presented at 1967 national meeting.

250. C. A. Depew and T. J. Kramer, Heat Transfer to Flowing Gas-Solid Mixtures, *Advances in Heat Transfer,* Vol. 9, pp. 113–180, Academic Press, New York, 1973.

251. W. C. Thomas and J. E. Sunderland, Heat Transfer between a Plane Surface and Air Containing Water Droplets, *Ind. Eng. Chem. Fund.,* Vol. 9, pp. 368–374, 1970.

252. W.-J. Yang and D. W. Clark, Spray Cooling of Air-Cooled Compact Heat Exchangers, *Int. J. Heat Mass Transfer,* Vol. 18, pp. 311–317, 1975.

253. V. W. Uhl, Mechanically Aided Heat Transfer to Viscous Materials, *Augmentation of Convective Heat and Mass Transfer,* pp. 109–117, ASME, New York, 1970.

254. W. R. Penney, The Spiralator—Initial Tests and Correlations, *AIChE Preprint 16 for 8th Nat. Heat Transfer Conf.,* Los Angeles, 1965.

255. F. S. Pramuk and J. W. Westwater, Effect of Agitation on the Critical Temperature Difference for Boiling Liquid, *Chem. Eng. Prog. Symp. Ser.,* Vol. 52, No. 18, pp. 79–83, 1956.

256. J. K. Hagge and G. H. Junkhan, Experimental Study of a Method of Mechanical Augmentation of Convective Heat Transfer Coefficients in Air, *HTL-3, ISU-ERI-Ames-74158,* Iowa State University, Ames, Iowa, November 1974.

257. E. L. Lustenader, R. Richter, and F. J. Neugebauer, The Use of Thin Films for Increasing Evaporation and Condensation Rates in Process Equipment, *J. Heat Transfer,* Vol. 81, pp. 297–307, 1959.

258. B. W. Tleimat, Performance of a Rotating Flat-Disk Wiped-Film Evaporator, *ASME Paper 71-HT-37,* 1971.

259. J. E. McElhiney and G. W. Preckshot, Heat Transfer in the Entrance Length of a Horizontal Rotating Tube, *Int. J. Heat Mass Transfer,* Vol. 20, pp. 847–854, 1977.

260. Y. Mori and W. Nakayama, Forced Convection Heat Transfer in a Straight Pipe Rotating around a Parallel Axis, *Int. J. Heat Mass Transfer,* Vol. 10, pp. 1179–1194, 1967.

261. V. Vidyanidhi, V. V. S. Suryanarayana, and V. C. Chenchu Raju, An Analysis of Steady Freely Developed Heat Transfer in a Rotating Straight Pipe, *J. Heat Transfer,* Vol. 99, pp. 148–150, 1977.

262. H. Miyazaki, Combined Free and Forced Convective Heat Transfer and Fluid Flow in a Rotating Curved Circular Tube, *Int. J. Heat Mass Transfer,* Vol. 14, pp. 1295–1309, 1971.

263. S. I. Tang and T. W. McDonald, A Study of Heat Transfer from a Rotating Horizontal Cylinder, *Int. J. Heat Mass Transfer,* Vol. 14, pp. 1643–1658, 1971.

264. P. J. Marto and V. H. Gray, Effects of High Accelerations and Heat Fluxes on Nucleate Boiling of Water in an Axisymmetric Rotating Boiler, *NASA TN D-6307,* 1971.

265. V. B. Astafev and A. M. Baklastov, Condensation of Steam on a Horizontal Rotating Disk, *Therm. Eng.,* Vol. 17, No. 9, pp. 82–85, 1970.

266. A. A. Nicol and M. Gacesa, Condensation of Steam on a Rotating Vertical Cylinder, *J. Heat Transfer,* Vol. 97, pp. 144–152, 1970.

267. R. M. Singer and G. W. Preckshot, The Condensation of Vapor on a Horizontal Rotating Cylinder, *Proc. 1963 Heat Transfer Fluid Mech. Inst.,* pp. 205–221, Stanford University Press, Stanford, Calif., 1963.

268. D. K. Weiler, A. M. Czikk, and R. S. Paul, Condensation in Smooth and Porous Coated Tubes under Multi-g Accelerations, *Chem. Eng. Prog. Symp. Ser.,* Vol. 62, No. 64, pp. 143–149, 1966.

269. F. K. Deaver, W. R. Penney, and T. B. Jefferson, Heat Transfer from an Oscillating Horizontal Wire to Water, *J. Heat Transfer,* Vol. 84, pp. 251–256, 1962.

270. W. R. Penney and T. B. Jefferson, Heat Transfer from an Oscillating Horizontal Wire to Water and Ethylene Glycol, *J. Heat Transfer,* Vol. 88, pp. 359–366, 1966.

271. W. H. McAdams, *Heat Transmission,* 3d ed., p. 267, McGraw-Hill, New York, 1954.

272. R. C. Martinelli and L. M. K. Boelter, The Effect of Vibration on Heat Transfer by Free Convection from a Horizontal Cylinder, *Heat. Pip. Air Cond.,* Vol. 11, pp. 525–527, 1939.

273. W. E. Mason and L. M. K. Boelter, Vibration—Its Effect on Heat Transfer, *Pwr. Pl. Eng.,* Vol. 44, pp. 43–46, 1940.

274. R. Lemlich, Effect of Vibration on Natural Convective Heat Transfer, *Ind. Eng. Chem.,* Vol. 47, pp. 1173–1180, 1955; Errata Vol. 53, p. 314, 1961.

275. R. M. Fand and J. Kaye, The Influence of Vertical Vibrations on Heat Transfer by Free Convection from a Horizontal Cylinder, *International Developments in Heat Transfer,* pp. 490–498, ASME, New York, 1961.

276. R. M. Fand and E. M. Peebles, A Comparison of the Influence of Mechanical and Acoustical Vibrations on Free Convection from a Horizontal Cylinder, *J. Heat Transfer,* Vol. 84, pp. 268–270, 1962.

277. A. J. Shine, Comments on a paper by Deaver et al., *J. Heat Transfer*, Vol. 84, p. 226, 1962.

278. R. Lemlich and M. A. Rao, The Effect of Transverse Vibration on Free Convection from a Horizontal Cylinder, *Int. J. Heat Mass Transfer*, Vol. 8, pp. 27–33, 1965.

279. A. E. Bergles, The Influence of Heated-Surface Vibration on Pool Boiling, *J. Heat Transfer*, Vol. 91, pp. 152–154, 1969.

280. V. D. Blankenship and J. A. Clark, Experimental Effects of Transverse Oscillations on Free Convection of a Vertical, Finite Plate, *J. Heat Transfer*, Vol. 86, pp. 159–165, 1964.

281. J. A. Scanlan, Effects of Normal Surface Vibration on Laminar Forced Convection Heat Transfer, *Ind. Eng. Chem.*, Vol. 50, pp. 1565–1568, 1958.

282. R. Anantanarayanan and A. Ramachandran, Effect of Vibration on Heat Transfer from a Wire to Air in Parallel Flow, *Trans. ASME*, Vol. 80, pp. 1426–1432, 1958.

283. I. A. Raben, The Use of Acoustic Vibrations to Improve Heat Transfer, *Proc. 1961 Heat Transfer Fluid Mech. Inst.*, pp. 90–97, Stanford University Press, Stanford, Calif., 1961.

284. I. A. Raben, G. E. Cummerford, and G. E. Neville, An Investigation of the Use of Acoustic Vibrations to Improve Heat Transfer Rates and Reduce Scaling in Distillation Units Used for Saline Water Conversion, *Off. Saline Water Res. Dev. Prog. Rep. No. 65*, 1962.

285. J. W. Ogle and A. J. Engel, The Effect of Vibration on a Double-Pipe Heat Exchanger, *AIChE Preprint 59 for 6th Nat. Heat Transfer Conf.*, 1963.

286. I. I. Palyeyev, B. D. Kachnelson, and A. A. Tarakanovskii, Study of Process of Heat and Mass Exchange in a Pulsating Stream, *Teploenergetika*, Vol. 10, No. 4, p. 71, 1963.

287. E. D. Jordan and J. Steffans, An Investigation of the Effect of Mechanically Induced Vibrations on Heat Transfer Rates in a Pressurized Water System, *New York Operations Office, Atomic Energy Comm.-2655-1*, 1965.

288. R. Hsieh and G. F. Marsters, Heat Transfer from a Vibrating Vertical Array of Horizontal Cylinders, *Can. J. Chem. Eng.*, Vol. 51, pp. 302–306, 1973.

289. F. C. McQuiston and J. D. Parker, Effect of Vibration on Pool Boiling, *ASME Paper 67-HT-49*, 1967.

290. D. C. Price and J. D. Parker, Nucleate Boiling on a Vibrating Surface, *ASME Paper 67-HT-58*, 1967.

291. G. M. Fuls and G. E. Geiger, Effect of Bubble Stabilization on Pool Boiling Heat Transfer, *J. Heat Transfer*, Vol. 97, pp. 635–640, 1970.

292. H. R. Pearce, The Effect of Vibration on Burnout in Vertical, Two-Phase Flow, *Atomic Energy Research Establishment (United Kingdom) 6375*, 1970.

293. J. C. Dent, Effect of Vibration on Condensation Heat Transfer to a Horizontal Tube, *Proc. Inst. Mech. Eng.*, Vol. 184, Pt. 1, pp. 99–105, 1969–1970.

294. Y. M. Brodov, R. Z. Salev'yev, V. A. Permayakov, V. K. Kuptsov, and A. G. Gal'perin, The Effect of Vibration on Heat Transfer and Flow of Condensing Steam on a Single Tube, *Heat Trans. Sov. Res.*, Vol. 9, No. 1, pp. 153–155, 1977.

295. A. L. Sprott, J. P. Holman, and F. L. Durand, An Experimental Study of the Effects of Strong Progressive Sound Fields on Free-Convection Heat Transfer from a Horizontal Cylinder, *ASME Paper 60-HT-19*, 1960.

296. R. M. Fand and J. Kaye, The Influence of Sound on Free Convection from a Horizontal Cylinder, *J. Heat Transfer*, Vol. 83, p. 133, 1961.

297. B. H. Lee and P. D. Richardson, Effect of Sound on Heat Transfer from a Horizontal Circular Cylinder at Large Wavelength, *J. Mech. Eng. Sci.*, Vol. 7, pp. 127–130, 1965.

298. R. M. Fand, J. Roos, P. Cheng, and J. Kaye, The Local Heat-Transfer Coefficient around a Heated Horizontal Cylinder in an Intense Sound Field, *J. Heat Transfer*, Vol. 84, pp. 245–250, 1962.

299. P. D. Richardson, Local Details of the Influence of a Vertical Sound Field on Heat Transfer from a Circular Cylinder, *Proc. 3d Int. Heat Transfer Conf.*, AIChE, New York, Vol. 3, pp. 71–77, 1966.

300. J. H. Seely, Effect of Ultrasonics on Several Natural Convection Cooling Systems, master's thesis, Syracuse University, Syracuse, N.Y., 1960.

301. A. A. Zhukauskas, A. A. Shlanchyauskas, and Z. P. Yaronees, Investigation of the Influence

of Ultrasonics on Heat Exchange between Bodies in Liquids, *J. Eng. Phys.,* Vol. 4, pp. 58–61, 1961.

302. M. B. Larson and A. L. London, A Study of the Effects of Ultrasonic Vibrations on Convection Heat Transfer to Liquids, *ASME Paper 62-HT-44,* 1962.

303. G. C. Robinson, C. M. McClude III, and R. Hendricks, Jr., The Effects of Ultrasonics on Heat Transfer by Convection, *Am. Ceram. Soc. Bull.,* Vol. 37, pp. 399–404, 1958.

304. R. M. Fand, The Influence of Acoustic Vibrations on Heat Transfer by Natural Convection from a Horizontal Cylinder to Water, *J. Heat Transfer,* Vol. 87, pp. 309–310, 1965.

305. J. H. Gibbons and G. Houghton, Effects of Sonic Vibrations on Boiling, *Chem. Eng. Sci.,* Vol. 15, p. 146, 1961.

306. K. W. Li and J. D. Parker, Acoustical Effects on Free Convective Heat Transfer from a Horizontal Wire, *J. Heat Transfer,* Vol. 89, pp. 277–278, 1967.

307. R. C. Martinelli, L. M. Boelter, E. B. Weinberg, and S. Takahi, Heat Transfer to a Fluid Flowing Periodically at Low Frequencies in a Vertical Tube, *Trans. ASME,* Vol. 65, pp. 789–798, 1943.

308. F. B. West and A. T. Taylor, The Effect of Pulsations on Heat Transfer, *Chem. Eng. Prog.,* Vol. 48, pp. 34–43, 1952.

309. J. M. Marchant, Discussion of a paper by R. C. Martinelli et al., *Trans. ASME,* Vol. 65, pp. 796–797, 1943.

310. G. B. Darling, Heat Transfer to Liquids in Intermittent Flow, *Petroleum,* Vol. 180, pp. 177–178, 1959.

311. R. Lemlich and J. C. Armour, Forced Convection Heat Transfer to a Pulsed Liquid, *AIChE Preprint 2 for 6th Nat. Heat Transfer Conf.,* 1963.

312. T. Shirotsuka, N. Honda, and Y. Shima, Analogy of Mass, Heat and Momentum Transfer to Pulsation Flow from Inside Tube Wall, *Kagaku-Kikai,* Vol. 21, pp. 638–644, 1957.

313. W. Linke and W. Hufschmidt, Wärmeübergang bei pulsierender Strömung, *Chem. Ing. Tech.,* Vol. 30, pp. 159–165, 1958.

314. T. W. Jackson, W. B. Harrison, and W. C. Boteler, Free Convection, Forced Convection, and Acoustic Vibrations in a Constant Temperature Vertical Tube, *J. Heat Transfer,* Vol. 81, pp. 68–71, 1959.

315. T. W. Jackson, K. R. Purdy, and C. C. Oliver, The Effects of Resonant Acoustic Vibrations on the Nusselt Number for a Constant Temperature Horizontal Tube, *International Developments in Heat Transfer,* pp. 483–489, ASME, New York, 1961.

316. R. Lemlich and C. K. Hwu, The Effect of Acoustic Vibration on Forced Convective Heat Transfer, *AIChE J.,* Vol. 7, pp. 102–106, 1961.

317. W. F. Mathewson and J. C. Smith, Effect of Sonic Pulsation on Forced Convective Heat Transfer to Air and on Film Condensation of Isopropanol, *Chem. Eng. Prog. Symp. Ser.,* Vol. 41, No. 59, pp. 173–179, 1963.

318. R. Moissis and L. A. Maroti, The Effect of Sonic Vibrations on Convective Heat Transfer in an Automotive Type Radiator Section, Dynatech Corp. Rep. No. 322, July 1962.

319. A. E. Bergles, The Influence of Flow Vibrations on Forced-Convection Heat Transfer, *J. Heat Transfer,* Vol. 86, pp. 559–560, 1964.

320. S. E. Isakoff, Effect of an Ultrasonic Field on Boiling Heat Transfer—Exploratory Investigation, *Heat Transfer and Fluid Mechanics Institute Preprints,* pp. 16–28, Stanford University, Stanford, Calif., 1956.

321. S. W. Wong and W. Y. Chon, Effects of Ultrasonic Vibrations on Heat Transfer to Liquids by Natural Convection and by Boiling, *AIChE J.,* Vol. 15, pp. 281–288, 1969.

322. A. P. Ornatskii and V. K. Shcherbakov, Intensification of Heat Transfer in the Critical Region with the Aid of Ultrasonics, *Teploenergetika,* Vol. 6, no. 1, pp. 84–85, 1959.

323. D. A. DiCicco and R. J. Schoenhals, Heat Transfer in Film Boiling with Pulsating Pressures, *J. Heat Transfer,* Vol. 86, pp. 457–461, 1964.

324. F. E. Romie and C. A. Aronson, Experimental Investigation of the Effects of Ultrasonic Vibrations on Burnout Heat Flux to Boiling Water, *Advanced Technology Laboratories A-123,* July 1961.

325. A. E. Bergles and P. H. Newell, Jr., The Influence of Ultrasonic Vibrations on Heat Transfer to Water Flowing in Annuli, *Int. J. Heat Mass Transfer,* Vol. 8, pp. 1273–1280, 1965.

326. R. M. Singer, Laminar Film Condensation in the Presence of an Electromagnetic Field, *ASME Paper 64-WA/HT-47,* 1964.

327. O. C. Blomgren, Sr., and O. C. Blomgren, Jr., Method and Apparatus for Cooling the Workpiece and/or the Cutting Tools of a Machining Apparatus, *U.S. Pat. 3,670,606,* 1972.

328. B. L. Reynolds and R. E. Holmes, Heat Transfer in a Corona Discharge, *Mech. Eng.,* pp. 44–49, October 1976.

329. J. E. Porter and R. Poulter, Electro-Thermal Convection Effects with Laminar Flow Heat Transfer in an Annulus, *Heat Transfer 1970,* Vol. 2, Paper FC3.7, Elsevier, Amsterdam, 1970.

330. S. D. Savkar, Dielectrophoretic Effects in Laminar Forced Convection between Two Parallel Plates, *Phys. Fluids,* Vol. 14, pp. 2670–2679, 1971.

331. D. C. Newton and P. H. G. Allen, Senftleben Effect in Insulating Oil under Uniform Electric Stress, *Letters in Heat and Mass Transfer,* Vol. 4, No. 1, pp. 9–16, 1977.

332. T. Mizushina, H. Ueda, and T. Matsumoto, Effect of Electrically Induced Convection on Heat Transfer of Air Flow in an Annulus, *J. Chem. Eng. Jpn.,* Vol. 9, No. 2, pp. 97–102, 1976.

333. H. Y. Choi, Electrohydrodynamic Boiling Heat Transfer, *Mech. Eng. Rep. 63-12-1,* Tufts University, Meford, Mass., December 1961.

334. R. L. Durfee, Boiling Heat Transfer of Electric Field (EHD), *At. Energy Comm. Rep. NYO-24-04-76,* 1966.

335. H. R. Velkoff and J. H. Miller, Condensation of Vapor on a Vertical Plate with a Transverse Electrostatic Field, *J. Heat Transfer,* Vol. 87, pp. 197–201, 1965.

336. H. Y. Choi and J. M. Reynolds, Study of Electrostatic Effects on Condensing Heat Transfer, *Air Force Flight Dynamics Laboratory TR-65-51,* 1966.

337. H. Y. Choi, Electrohydrodynamic Condensation Heat Transfer, *ASME Paper 67-HT-39,* 1967.

338. E. E. Gose, E. E. Peterson, and A. Acrivos, On the Rate of Heat Transfer in Liquids with Gas Injection through the Boundary Layer, *J. Appl. Phys.,* Vol. 28, p. 1509, 1957.

339. E. E. Gose, A. Acrivos, and E. E. Peterson, Heat Transfer to Liquids with Gas Evolution at the Interface, paper presented at AIChE annual meeting, 1960.

340. G. E. Sims, U. Aktürk, and K. O. Evans-Lutterodt, Simulation of Pool Boiling Heat Transfer by Gas Injection at the Interface, *Int. J. Heat Mass Transfer,* Vol. 6, pp. 531–535, 1963.

341. A. A. Kudirka, Two-Phase Heat Transfer with Gas Injection through a Porous Boundary Surface, *ASME Paper 65-HT-47,* 1965.

342. S. G. Bankoff, Taylor Instability of an Evaporating Plane Interface, *AIChE J.,* Vol. 7, pp. 485–487, 1961.

343. P. C. Wayner, Jr., and S. G. Bankoff, Film Boiling of Nitrogen with Suction on an Electrically Heated Porous Plate, *AIChE J.,* Vol. 11, pp. 59–64, 1965.

344. V. K. Pai and S. G. Bankoff, Film Boiling of Nitrogen with Suction on an Electrically Heated Horizontal Porous Plate: Effect of Flow Control Element Porosity and Thickness, *AIChE J.,* Vol. 11, pp. 65–69, 1965.

345. P. C. Wayner, Jr., and A. S. Kestin, Suction Nucleate Boiling of Water, *AIChE J.,* Vol. 11, pp. 858–865, 1965.

346. R. J. Raiff and P. C. Wayner, Jr., Evaporation from a Porous Flow Control Element on a Porous Heat Source, *Int. J. Heat Mass Transfer,* Vol. 16, pp. 1919–1930, 1973.

347. W. A. Tauscher, E. M. Sparrow, and J. R. Lloyd, Amplification of Heat Transfer by Local Injection of Fluid into a Turbulent Tube Flow, *Int. J. Heat Mass Transfer,* Vol. 13, pp. 681–688, 1970.

348. R. B. Kinney, Fully Developed Frictional and Heat Transfer Characteristics of Laminar Flow in Porous Tubes, *Int. J. Heat Mass Transfer,* Vol. 11, pp. 1393–1401, 1968.

349. R. B. Kinney and E. M. Sparrow, Tubulent Flow, Heat Transfer, and Mass Transfer in a Tube with Surface Suction, *J. Heat Transfer,* Vol. 92, pp. 117–125, 1970.

350. J. K. Aggarwal and M. A. Hollingsworth, Heat Transfer for Turbulent Flow with Suction in a Porous Tube, *Int. J. Heat Mass Transfer,* Vol. 16, pp. 591–609, 1973.

351. I. Antonir and A. Tamir, The Effect of Surface Suction on Condensation in the Presence of a Noncondensible Gas, *J. Heat Transfer,* Vol. 99, pp. 496–499, 1977.

352. J. Lienhard and V. Dhir, A Simple Analysis of Laminar Film Condensation with Suction, *J. Heat Transfer,* Vol. 94, pp. 334–336, 1972.

353. A. E. Bergles, R. A. Lee, and B. B. Mikic, Heat Transfer in Rough Tubes with Tape-Generated Swirl Flow, *J. Heat Transfer,* Vol. 91, pp. 443–445, 1969.

354. Y. V. Kryukov and G. P. Boykov, Augmentation of Heat Transfer in an Acoustic Field, *Heat Trans. Sov. Res.,* Vol. 5, No. 1, pp. 26–28, 1973.

355. R. S. Van Rooyen and D. G. Kroeger, Laminar Flow Heat Transfer in Internally Finned Tubes with Twisted-Tape Inserts, *Heat Transfer 1978,* Vol. 2, pp. 577–581, Hemisphere, Washington, D.C., 1978.

356. W. J. Bartel and W. E. Genetti, Heat Transfer from a Horizontal Bundle of Bare and Finned Tubes in an Air Fluidized Bed, *AIChE Symp. Ser. No. 128,* Vol. 69, pp. 85–93, 1973.

357. N. V. Zozulya and Y. Khorunzhii, Heat Transfer from Finned Tubes Moving Back and Forth in Liquid, *Chem. Petroleum Eng.,* Nos. 9–10, pp. 830–832, 1968.

358. K. Min and B. T. Chao, Particle Transport and Heat Transfer in Gas-Solid Suspension Flow under the Influence of an Electric Field, *Nucl. Sci. Eng.,* Vol. 26, pp. 534–546, 1966.

359. S. C. Bhattacharya and D. Harrison, Heat Transfer in a Pulsed Fluidized Bed, *Trans. Inst. Chem. Eng.,* Vol. 54, pp. 281–286, 1976.

360. J. H. Masliyah and K. Nandakumar, Fluid Flow and Heat Transfer in Internally Finned Helical Coils, *Can. J. Chem. Eng.,* Vol. 55, pp. 27–36, 1977.

361. C. A. Bromley, R. F. Humphreys, and W. Murray, Condensation on and Evaporation from Radially Grooved Rotating Disks, *J. Heat Transfer,* Vol. 88, pp. 80–93, 1966.

362. V. C. Van der Mast, S. M. Read, and L. A. Bromley, Boiling of Natural Sea Water in Falling Film Evaporators, *Desalination,* Vol. 18, pp. 71–94, 1976.

363. S. P. Chary and P. K. Sarma, Condensation on a Rotating Disk with Constant Axial Suction, *J. Heat Transfer,* Vol. 98, pp. 682–684, 1976.

364. A. E. Bergles, R. L. Webb, and G. H. Junkhan, Energy Conservation via Heat Transfer Enhancement, *Energy,* Vol. 4, pp. 193–200, 1979.

365. W. M. Rohsenow, Boiling, *Handbook of Heat Transfer Fundamentals,* W. M. Rohsenow, J. P. Hartnett, and E. N. Ganić, eds., Chap. 12, McGraw-Hill, New York, 1985.

366. P. Griffith, Dropwise Condensation, *Handbook of Heat Transfer Fundamentals,* W. M. Rohsenow, J. P. Hartnett, and E. N. Ganić, eds., Chap. 11, Pt. 2, McGraw-Hill, New York, 1985.

367. G. D. Raithby and K. G. T. Hollands, Natural Convection, *Handbook of Heat Transfer Fundamentals,* W. M. Rohsenow, J. P. Hartnett, and E. N. Ganić, eds., Chap. 6, McGraw-Hill, New York, 1985.

368. W. M. Kays and H. C. Perkins, Forced Convection, Internal Flow in Ducts, *Handbook of Heat Transfer Fundamentals,* W. M. Rohsenow, J. P. Hartnett, and E. N. Ganić, eds., Chap. 7, McGraw-Hill, New York, 1985.

369. R. Viskanta, Electric and Magnetic Fields, *Handbook of Heat Transfer Fundamentals,* W. M. Rohsenow, J. P. Hartnett, and E. N. Ganić, eds., Chap. 10, McGraw-Hill, New York, 1985.

NOMENCLATURE

Symbol, Definition, SI Units, English Units

Properties are evaluated at the bulk fluid condition unless otherwise noted.

A heat transfer surface area: m^2, ft^2

A_e effective heat transfer surface area, Eq. (16): m^2, ft^2

A_F heat transfer surface area of fins: m^2, ft^2

A_f cross-sectional flow area: m^2, ft^2

A_r area of unfinned portion of tube: m^2, ft^2

a vibrational amplitude; amplitude of sinusoidally shaped flute: m, ft

b stud or fin thickness: m, ft

c_p specific heat at constant pressure: $J/(kg \cdot K)$, $Btu/(lb_m \cdot {}^\circ F)$

D tube inside diameter: m, ft

D_c diameter of coil: m, ft

D_h hydraulic diameter: m, ft

D_o outside diameter of circular finned tube or cylinder: m, ft

D_r root diameter of finned tube: m, ft

d_i inside diameter of annulus or ring insert: m, ft

d_o outside diameter of annulus or ring insert: m, ft

d_p diameter of particles in air-solid suspensions: m, ft

d_s diameter of spherical packing, disk insert, or loose-fitting twisted tape: m, ft

E electric field strength: V/m, V/ft

e protrusion height: m, ft

e^+ roughness Reynolds number, Eq. (3)

F fin factor, Eq. (22); Prandtl number function, Eq. (4)

F_t convective factor in Eq. (26)

f Fanning friction factor $= \Delta P \, D\rho / 2LG^2$

$\mathbf{f}$ vibrational frequency: s^{-1}

G mass velocity $= W/A_f$: $kg/(m^2 \cdot s)$, $lb_m/(h \cdot ft^2)$

G_e effective mass velocity, Eq. (18): $kg/(m^2 \cdot s)$, $lb/_m/(h \cdot ft^2)$

Gr Grashof number $= g\beta \, \Delta T \, D^3 / \nu^2$

Gz Graetz number $= Wc_p/kL$

g gravitational acceleration: m/s^2, ft/s^2

$\mathbf{g}$ spacing between protrusions: m, ft

$\overline{g}$ heat transfer roughness function, Eq. (2)

H fin height: m, ft

h heat transfer coefficient: $W/(m^2 \cdot K)$, $Btu/(h \cdot ft^2 \cdot {}^\circ F)$

$\overline{h}$ mean value of the heat transfer coefficient: $W/(m^2 \cdot K)$, $Btu/h \cdot ft^2 \cdot {}^\circ F$

$\mathbf{h}$ spacing defined in Fig. 11: m, ft

i_{lg} enthalpy of vaporization: J/kg, Btu/lb_m

k thermal conductivity: $W/(m \cdot K)$, $Btu/(h \cdot ft \cdot {}^\circ F)$

L channel heated length: m, ft

L_c finned length between cuts for interrupted fins or between inserts: m, ft

L_f mean effective length of a fin, Eq. (16): m, ft

L_s distance between condensate strippers: m, ft

l average space between adjacent fins: m, ft

N number of tubes

Nu Nusselt number $= hD/k$

$\overline{\mathrm{Nu}}$	mean value of the Nusselt number $= \overline{h}D/k$
n	number of fins
P	pressure: $\mathrm{N/m^2}$, $\mathrm{lb_f/ft^2}$
P	pumping power: W, Btu/h
$\mathcal{P}$	wetted perimeter of channel between two longitudinal fins (Fig. 16): m, ft
ΔP	pressure drop: $\mathrm{N/m^2}$, $\mathrm{lb_f/ft^2}$
Pr	Prandtl number $= \mu c_p/k$
p	roughness or flute pitch (Fig. 9): m, ft
q	rate of heat transfer: W, Btu/h
q''	heat flux: $\mathrm{W/m^2}$, $\mathrm{Btu/(h \cdot ft^2)}$
q''_{cr}	critical heat flux: $\mathrm{W/m^2}$, $\mathrm{Btu/(h \cdot ft^2)}$
Ra	Rayleigh number $=$ Gr Pr
Re	Reynolds number $= GD/\mu$
Re_v	vibrational Reynolds number $= 2\pi afD_o/\nu$
St	Stanton number $= h/Gc_p$
T	temperature: $°\mathrm{C}$, $°\mathrm{F}$
T_{sat}	saturation temperature: $°\mathrm{C}$, $°\mathrm{F}$
ΔT	temperature difference: $°\mathrm{K}$, $°\mathrm{F}$
ΔT_{gw}	temperature difference from saturated vapor to wall: K, $°\mathrm{F}$
ΔT_i	heat exchanger inlet temperature difference: K, $°\mathrm{F}$
ΔT_{lm}	log mean temperature difference: K, $°\mathrm{F}$
ΔT_n	shell-side to tube-side exit temperature difference: K, $°\mathrm{F}$
ΔT_{sat}	wall-minus-saturation temperature difference: K, $°\mathrm{F}$
U	overall heat transfer coefficient: $\mathrm{W/(m^2 \cdot K)}$, $\mathrm{Btu/(h \cdot ft^2 \cdot °F)}$
U_e^+	Nikuradse similarity function, Eq. (5)
$\overline{U}_n$	average overall heat transfer coefficient based on nominal tube outside diameter: $\mathrm{W/(m^2 \cdot K)}$, $\mathrm{Btu/(h \cdot ft^2 \cdot °F)}$
u	average axial velocity: m/s, ft/s
W	mass flow rate: kg/s, $\mathrm{lb_m/s}$
X^+	dimensionless position $= x/(D \ \mathrm{Re} \ \mathrm{Pr})$
x	axial position: m, ft
$\overline{x}$	average flowing mass quality
x_{cr}	quality at critical heat flux
y	twist ratio, tube diameter per 180° tape twist

Greek Symbols

α	spiral fin helix angle: rad, deg
β	volumetric coefficient of expansion: $\mathrm{K^{-1}}$, $°\mathrm{R^{-1}}$
η	fin efficiency
μ	dynamic viscosity: $\mathrm{N/(m^2 \cdot s)}$, $\mathrm{lb_m/(h \cdot ft)}$
ν	kinematic viscosity: $\mathrm{m^2/s}$, $\mathrm{ft^2/s}$
ρ	density: $\mathrm{kg/m^3}$, $\mathrm{lb_m/ft^3}$

Subscripts

a	augmented heat transfer condition
b	evaluated at bulk or mixed-mean fluid condition
ex	condition at outlet of channel
f	evaluated at film temperature, $(T_w + T_b)/2$
g	based on vapor or gas
h	based on hydraulic diameter
i	based on maximum inside (envelope) diameter
in	condition at inlet of channel
iso	isothermal
l	based on liquid
o	nonaugmentation data
p	particles
s	standard condition; refers to solids
x	local value
w	evaluated at wall temperature

CHAPTER 4

Heat Exchangers

By R. K. Shah   (Part 1, Part 3)

Harrison Radiator Division
General Motors Corporation
Lockport, New York

A. C. Mueller   (Part 1, Part 2)

Consultant
Church Hill, Maryland

Heat Exchanger Basic Thermal Design Methods

A. INTRODUCTION

A heat exchanger is a device used to transfer thermal energy between two or more fluids at different temperatures. Heat exchangers are classified according to their transfer processes as *direct contact* type and *indirect contact* type. In the direct contact type, two different phase streams come into direct contact, exchange heat, and then are separated. In the indirect contact type, the streams remain separate and the heat transfers through a dividing wall or into and out of a wall in a transient manner. Those heat exchangers in which there is a continuous flow of heat from the hot to the cold fluid through a dividing wall are referred to as *direct transfer*–type exchangers or simply as *recuperators*. Those exchangers in which there is an intermittent flow of heat from the hot to the cold fluid (via heat storage and heat rejection through the exchanger surface) are referred to as *indirect transfer*–type exchangers or storage-type exchangers, or simply as *regenerators*. Part 1 of this chapter is limited to the indirect contact, direct transfer–type exchangers, or recuperators. Regenerators will be covered in Part 3 of this chapter.

    Another arbitrary classification can be based on the surface area to volume ratio, into compact (a limit of 700 m^2/m^3 or 213 ft^2/ft^3 is chosen) and noncompact exchangers. This surface area density classification is made because the physical design of the exchangers, fields of applications, and design techniques differ. Further heat exchanger classification can also be made according to construction, flow arrangement, number of fluids, and heat transfer mechanisms, as shown in Fig. 1. Further details on classification and basic terminology used for a variety of heat exchangers are summarized in Ref. 1.

    The problem of heat exchanger design is complex because of the many compromises of conflicting needs made in selecting the proper type of exchanger, in selecting the allocation of the fluid passages, and finally in selecting the appropriate design equations with due regard to their limitations. There is no unique solution to designing an exchanger, since many qualitative judgments, in addition to the quantitative calculations, must be made, and each designer makes different qualitative judgments. An overall exchanger design methodology is summarized in Ref. 2.

    In this part of Chap. 4, we will describe the exchanger thermal design methods (heat transfer analysis) applicable to direct transfer–type exchangers, namely, tubular, plate, and extended-surface exchangers. Essentially four steady-state design methods will be summarized for two-fluid heat exchangers: the ϵ-NTU, P-NTU$_t$, LMTD, and ψ-P methods. Also, transient solutions will be summarized for batch heating (cooling) and step changes in inlet temperatures or flow ratio. In Part 2, detailed design considerations are presented for shell-and-tube, plate, and other noncompact heat exchangers operating under single-phase or two-phase flow conditions. In Part 3, detailed design considerations are presented for plate-fin, tube-fin, and regenerative compact heat exchangers.

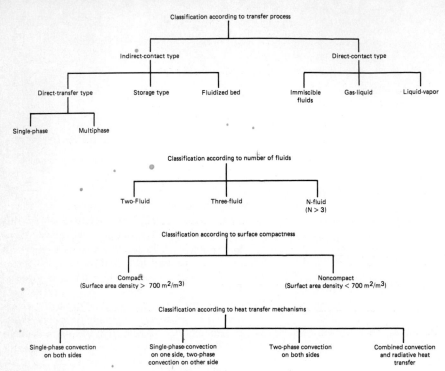

FIG. 1 Classification of heat exchangers.

B. HEAT EXCHANGER VARIABLES AND THERMAL CIRCUIT

In order to develop relationships between the heat transfer rate q, surface area A, fluid terminal temperatures, and flow rates in a heat exchanger, the basic equations used for analysis are the energy conservation and rate equations. Consider the counterflow exchanger of Fig. 2 just as an example to introduce the variables associated with an exchanger. Two energy conservation differential equations for an exchanger having an

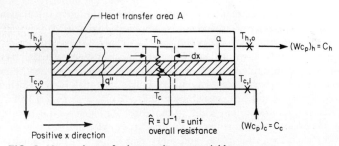

FIG. 2 Nomenclature for heat exchanger variables.

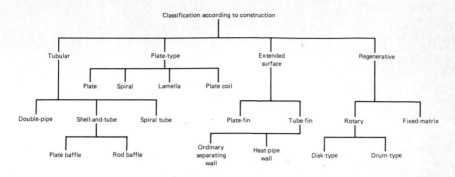

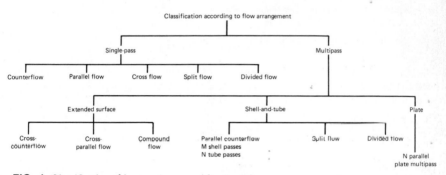

FIG. 1 Classification of heat exchangers. (*Continued*)

arbitrary flow arrangement are†

$$dq = q'' \, dA = -C_h \, dT_h = \pm C_c \, dT_c \tag{1}$$

where the $\pm$ depends upon whether dT_c is increasing or decreasing with increasing dA. The overall rate equation on a local basis is

$$dq = q'' \, dA = U(T_h - T_c)_{\text{local}} \, dA = U \, \Delta T \, dA \tag{2}$$

Integration of Eqs. (1) and (2) across the exchanger surface area results in

$$q = C_h(T_{h,i} - T_{h,o}) = C_c(T_{c,o} - T_{c,i}) \tag{3}$$

and

$$q = UA \, \Delta T_m = \frac{\Delta T_m}{R_o} \tag{4}$$

The true mean temperature difference ΔT_m is dependent upon the exchanger flow arrangement and the degree of fluid mixing within each fluid stream. The inverse of the overall thermal conductance UA is referred to as the overall thermal resistance R_o, and is made up of component resistance in series as shown in Fig. 3:

$$R_o = R_h + R_{s,h} + R_w + R_{s,c} + R_c \tag{5}$$

where R_h = hot-side film-convection resistance = $\dfrac{1}{(\eta_o hA)_h}$

†The idealizations built into this equation are discussed later following Eq. (13).

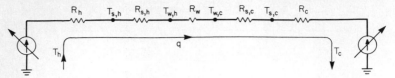

FIG. 3 Thermal circuit for heat transfer in an exchanger.

$$R_{s,h} = \text{hot-side scale (fouling) resistance} = \frac{1}{(\eta_o h_s A)_h}$$

$$R_w = \text{wall thermal resistance expressed by Eq. (7)}$$

$$R_{s,c} = \text{cold-side scale (fouling) resistance} = \frac{1}{(\eta_o h_s A)_c}$$

$$R_c = \text{cold-side film-convection resistance} = \frac{1}{(\eta_o h A)_c}$$

In the foregoing definitions, h is the heat transfer coefficient, h_s is the scale (fouling) coefficient, A represents a total of the primary and the secondary (finned) surface area, η_o is the total surface temperature effectiveness of an extended (fin) surface, and the subscripts h and c are for the hot and cold fluid sides, respectively. η_o is related to the fin efficiency η_f and the ratio of fin surface area A_f to total surface area A as follows:

$$\eta_o = 1 - \frac{A_f}{A}(1 - \eta_f) \tag{6}$$

Note that $\eta_{o,h}$ and $\eta_{o,c}$ are unity for an all prime surface exchanger (no fins). If there is any contact or bond resistance present between the fin and the tube or plate on the hot or the cold side, it is included as an added thermal resistance on the right-hand side of Eq. (5).

The wall thermal resistance R_w is

$$R_w = \begin{cases} \dfrac{\delta}{A_w k_w} & \text{for a flat wall} \\[2ex] \dfrac{\ln(d_o/d_i)}{2\pi k_w L N_t} & \text{for a circular tube with single layer wall} \\[2ex] \dfrac{1}{2\pi L N_t}\left[\sum_j \dfrac{\ln(d_{j+1}/d_j)}{k_{w,j}}\right] & \text{for a circular tube with multiple-layered wall} \end{cases} \tag{7}$$

where δ is the plate thickness, A_w is the total wall area for heat conduction, k_w is thermal conductivity of the wall material, d_o is the tube outside diameter, d_i is the tube inside diameter, L is the tube length, and N_t is the number of tubes. A flat (or plain) wall is generally associated with a plate-fin or an all prime surface plate exchanger. In this case,

$$A_w = L_1 L_2 N_P \tag{8}$$

Here L_1, L_2, and N_P are the length, width, and total number of separating plates.

Equation (5) can be alternately expressed

$$\frac{1}{UA} = \frac{1}{(\eta_o h A)_h} + \frac{1}{(\eta_o h_s A)_h} + R_w + \frac{1}{(\eta_o h_s A)_c} + \frac{1}{(\eta_o h A)_c} \tag{9}$$

In Eq. (9), the overall heat transfer coefficient U may be defined optionally in terms of the surface area of either hot fluid surface, cold fluid surface, or wall conduction area. Thus

$$UA = U_h A_h = U_c A_c = U_w A_w \tag{10}$$

Thus the option of A_h, A_c, or A_w *must* be specified in evaluating U from the product UA. For plain tubular exchangers, U_o based on tube outside surface area, from Eq. (9), reduces to

$$\frac{1}{U_o} = \frac{1}{h_o} + \frac{1}{h_{o,s}} + \frac{d_o \ln(d_o/d_i)}{2k_w} + \frac{d_o}{h_{i,s} d_i} + \frac{d_o}{h_i d_i} \tag{11}$$

The knowledge of wall temperature in a heat exchanger is essential to determine the localized hot spots, freeze points, thermal stresses, local fouling characteristics, or boiling and condensing coefficients. Based on the thermal circuit of Fig. 3, when R_w is negligible, $T_{w,h} = T_{w,c} = T_w$ is computed from

$$T_w = \frac{T_h + [(R_h + R_{s,h})/(R_c + R_{s,c})]T_c}{1 + (R_h + R_{s,h})/(R_c + R_{s,c})} \tag{12}$$

When $R_{s,h} = R_{s,c} = 0$, this further simplifies to

$$T_w = \frac{T_h/R_h + T_c/R_c}{1/R_h + 1/R_c} = \frac{(\eta_o hA)_h T_h + (\eta_o hA)_c T_c}{(\eta_o hA)_h + (\eta_o hA)_c} \tag{13}$$

Note that T_h, T_c, and T_w are *local* temperatures in this equation.

Integration of Eqs. (1) and (2) across the surface area for a specified exchanger flow arrangement yields the expressions for the ϵ-NTU, P-NTU$_t$, LMTD, and ψ-P methods to be described in the following sections. The following idealizations are built into Eqs. (1) and (2), and their subsequent integration:

1. The heat exchanger operates under steady-state conditions (i.e., constant flow rate, and thermal history of fluid particles independent of time).

2. Heat losses to the surroundings are negligible.

3. There are no thermal energy sources in the exchanger.

4. In counterflow and parallel-flow exchangers, the temperature of each fluid is uniform over every flow cross section. From the temperature distribution point of view, in cross-flow exchangers each fluid is considered mixed or unmixed at every cross section depending upon the specifications. For a multipass exchanger, the foregoing statements apply to each pass depending upon the basic flow arrangement of the passes; the fluid is considered mixed or unmixed between passes.

5. Either there are no phase changes (condensation or boiling) in the fluid streams flowing through the exchanger or the phase changes occur under one of the following conditions: (*a*) Phase change occurs at a constant temperature as for a single component fluid at constant pressure; the effective specific heat for the phase-changing fluid is infinity in this case, and hence $C_{\max} \to \infty$. (*b*) The temperature of the phase-changing fluid varies linearly with heat transfer during the condensation or boiling. In this case, the effective specific heat is constant and finite for the phase-changing fluid.

6. The specific heat of each fluid is constant throughout the exchanger so that the heat capacity rate on each side is treated as constant.

7. The velocity and temperature at the entrance of the heat exchanger on each fluid side are *uniform*.

8. For an extended-surface exchanger, the overall extended-surface temperature effectiveness η_o is considered uniform and constant.

9. The overall heat transfer coefficient between the fluids is *constant* throughout the exchanger, including the case of phase-changing fluid in idealization 5.

10. The heat transfer area is distributed uniformly on each fluid side. In a multipass unit, heat transfer surface area is equal in each pass.

11. For a plate-baffled shell-and-tube exchanger, the temperature rise per baffle pass is small compared to the overall temperature rise along the exchanger; i.e., the number of baffles is large.

12. The fluid flow rate is uniformly distributed through the exchanger on each fluid side in each pass. No stratification, flow bypassing, or flow leakages occur in any stream. The flow condition is characterized by the bulk (or mean) velocity at any cross section.

13. Longitudinal heat conduction in the fluid and in the wall is negligible.

Idealizations 1 to 4 are necessary in a theoretical analysis of steady-state heat exchangers. Idealization 5 essentially restricts the analysis to single-phase flow on both sides or on one side with a dominating thermal resistance. For two-phase flows on both sides, many of the foregoing idealizations are not valid, since mass transfer in phase change results in variable properties and variable flow rates of each phase, and the heat transfer coefficients vary significantly. As a result, the heat exchanger cannot be analyzed using the theory of Part 1. The design theory of two-phase exchangers is presented in Part 2 of this chapter.

If idealization 6 is not valid, divide the exchanger into small segments until the specific heats can be treated as constant. Idealizations 7 and 8 are primarily important for compact heat exchangers and will be discussed in Part 3 of this chapter. The influences of idealizations 9 to 13 are discussed in Sec. F of Part 1.

If any of these idealizations are not valid for a particular exchanger application, the best solution is to work directly with either Eqs. (1) and (2) or their modified form by including a particular effect, and to integrate them over a small exchanger segment in which all of the idealizations are valid.

C. EXCHANGER HEAT TRANSFER ANALYSIS METHODS

Four alternate methods for the exchanger heat transfer analysis are the ϵ-NTU, **P-NTU**$_t$, LMTD, and ψ-**P** methods. In Table 1, the exchanger total heat transfer rate, the general functional relationships among the dimensionless groups, and the definition or working equation for each dimensionless group are presented for each of these methods. The relationships between the dimensionless groups of these methods are summarized in Table 2. Clearly, there are three dimensionless groups associated with each method, and there is a direct one-to-one correspondence between each pair of methods, and only the algebraic forms of the resulting equations are different. It must be emphasized that regardless of which method is used for exchanger heat transfer analysis, the solution to the exchanger rating or sizing problem will be identical if it is converged to a desired degree.

TABLE 1 General Functional Relationships and Dimensionless Groups for ϵ-NTU, P-NTU$_t$, LMTD, and ψ-P Methods

ϵ-NTU method	P-NTU$_t$ method†
$q = \epsilon C_{\min}(T_{h,i} - T_{c,i})$	$q = PC_t\lvert T_{s,i} - T_{t,i}\rvert$
$\epsilon = \phi(\text{NTU}, C^*, \text{flow arrangement})$	$\mathbf{P} = \phi(\text{NTU}_t, \mathbf{R}, \text{flow arrangement})$
$\epsilon = \dfrac{C_h(T_{h,i} - T_{h,o})}{C_{\min}(T_{h,i} - T_{c,i})} = \dfrac{C_c(T_{c,o} - T_{c,i})}{C_{\min}(T_{h,i} - T_{c,i})}$	$\mathbf{P} = \dfrac{T_{t,o} - T_{t,i}}{T_{s,i} - T_{t,i}}$
$\text{NTU} = \dfrac{UA}{C_{\min}} = \dfrac{1}{C_{\min}} \displaystyle\int_A U\, dA$	$\text{NTU}_t = \dfrac{UA}{C_t} = \dfrac{\lvert T_{t,o} - T_{t,i}\rvert}{\Delta T_m}$
$C^* = \dfrac{C_{\min}}{C_{\max}} = \dfrac{(Wc_p)_{\min}}{(Wc_p)_{\max}}$	$\mathbf{R} = \dfrac{C_t}{C_s} = \dfrac{T_{s,i} - T_{s,o}}{T_{t,o} - T_{t,i}}$
LMTD method†	ψ-P method†
$q = UAF\,\Delta T_{lm}$	$q = UA\psi(T_{h,i} - T_{c,i})$
$\text{LMTD} = \Delta T_{lm} = \dfrac{\Delta T_1 - \Delta T_2}{\ln(\Delta T_1/\Delta T_2)}$	$\psi = \phi(\mathbf{P}, \mathbf{R}, \text{flow arrangement})$
$\Delta T_1 = T_{h,i} - T_{c,o} \qquad \Delta T_2 = T_{h,o} - T_{c,i}$	$\psi = \dfrac{\Delta T_m}{T_{h,i} - T_{c,i}}$
$F = \phi(\mathbf{P}, \mathbf{R}, \text{flow arrangement})$	
$F = \dfrac{\Delta T_m}{\Delta T_{lm}}$	**P** and **R** are defined in the **P-NTU**$_t$ method.
P and **R** are defined in the **P-NTU**$_t$ method.	

†Although **P**, **R**, and NTU$_t$ are defined on the basis of C_t, it must be emphasized that all the results of the P-NTU$_t$, LMTD, or ψ-P method are valid if the definitions of **P**, NTU$_t$, and **R** are consistently based on C_s, C_h, or C_c; the only exception is the results of Fig. 34, where the definitions of **P** and **R** are based on C_t.

1. The ϵ-NTU Method

The dimensionless groups of the ϵ-NTU method have a thermodynamic significance. The heat exchanger effectiveness ϵ is an efficiency factor. It is a ratio of the actual heat transfer rate from the hot fluid to the cold fluid in a given heat exchanger of any flow arrangement to the thermodynamically limited maximum possible heat transfer rate. This $q_{\max}$ is obtained in a *counterflow* heat exchanger (recuperator) of *infinite surface area* operating with the fluid flow rates and fluid inlet temperatures the same as those of an actual exchanger. The number of transfer units NTU is a ratio of the overall conductance UA to the smaller heat capacity rate $C_{\min}$. It designates the dimensionless "heat transfer size" or "thermal size" of the exchanger. It may also be interpreted as the $C_{\min}$ fluid dimensionless residence time, a temperature ratio, or a modified Stanton number [3]. The heat capacity rate ratio C^* is simply a ratio of the smaller to the larger heat capacity rate for

TABLE 2 Relationships between Dimensionless Groups of the P-NTU$_t$, LMTD, and ψ-P Methods and Those of the ϵ-NTU Method

$$\mathbf{P} = \frac{C_{\min}}{C_t}\,\epsilon = \begin{cases} \epsilon & \text{for} & C_t = C_{\min} \\ \epsilon C^* & \text{for} & C_t = C_{\max} \end{cases}$$

$$\mathbf{R} = \frac{C_t}{C_s} = \begin{cases} C^* & \text{for} & C_t = C_{\min} \\ 1/C^* & \text{for} & C_t = C_{\max} \end{cases}$$

$$\mathbf{NTU}_t = \mathbf{NTU}\,\frac{C_{\min}}{C_t} = \begin{cases} \mathbf{NTU} & \text{for} & C_t = C_{\min} \\ \mathbf{NTU}\,C^* & \text{for} & C_t = C_{\max} \end{cases}$$

$$F = \frac{\mathbf{NTU}_{cf}}{\mathbf{NTU}} = \frac{1}{\mathbf{NTU}(1 - C^*)}\ln\frac{1 - C^*\epsilon}{1 - \epsilon}\;\xrightarrow[C^* = 1]{}\;\frac{\epsilon}{\mathbf{NTU}(1 - \epsilon)}$$

$$F = \frac{1}{\mathbf{NTU}_t(1 - \mathbf{R})}\ln\left[\frac{1 - \mathbf{RP}}{1 - \mathbf{P}}\right]\;\xrightarrow[\mathbf{R} = 1]{}\;\frac{\mathbf{P}}{\mathbf{NTU}_t(1 - \mathbf{P})}$$

$$\psi = \frac{\epsilon}{\mathbf{NTU}} = \frac{\mathbf{P}}{\mathbf{NTU}_t} = \frac{F\mathbf{P}(1 - \mathbf{R})}{\ln[(1 - \mathbf{RP})/(1 - \mathbf{P})]}\;\xrightarrow[\mathbf{R} = 1]{}\;F(1 - \mathbf{P})$$

the two fluid streams so that $C^* \leq 1$. The definitions or working equations for ϵ, NTU, and C^* are summarized in Table 1.

As outlined in Table 1, ϵ is a function of NTU, C^*, and the flow arrangements of the two fluids in the exchanger. Closed-form expression for 14 single-pass and multipass flow arrangements are summarized in Table 3A and B [3–11]. Graphical results are presented in Figs. 4 to 6 for three flow arrangements most commonly used in compact heat

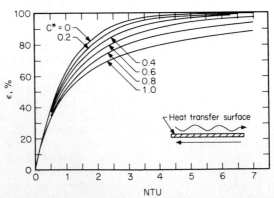

FIG. 4 Heat transfer effectiveness ϵ as a function of NTU and C^* for a counterflow exchanger.

Ref.	Flow arrangement	ϵ-NTU formulas	ϵ-NTU formulas for $C^* = 1$	Asymptotic value of ϵ when NTU $\to \infty$
[4]	Counterflow	$\epsilon = \dfrac{1 - \exp[-\text{NTU}(1 - C^*)]}{1 - C^* \exp[-\text{NTU}(1 - C^*)]}$	$\epsilon = \dfrac{\text{NTU}}{1 + \text{NTU}}$	$\epsilon = 1$ for all C^*
[4]	Parallel flow	$\epsilon = \dfrac{1 - \exp[-\text{NTU}(1 + C^*)]}{1 + C^*}$	$\epsilon = \frac{1}{2}[1 - \exp(-2\,\text{NTU})]$	$\epsilon = \dfrac{1}{1 + C^*}$
[5]	Crossflow, both fluids unmixed	$\epsilon = 1 - \exp[-(1 + C^*)\,\text{NTU}]\left[I_0(2\,\text{NTU}\sqrt{C^*}) + \sqrt{C^*} I_1(2\,\text{NTU}\sqrt{C^*}) - \dfrac{1 - C^*}{C^*} \sum_{n=2}^{\infty} C^{*n/2} I_n(2\,\text{NTU}\sqrt{C^*}) \right]$	$\epsilon = 1 - [I_0(2\,\text{NTU}) + I_1(2\,\text{NTU})]\,e^{-2\text{NTU}}$	$\epsilon = 1$ for all C^*
[4]	Crossflow, one fluid mixed, the other unmixed	For $C_{\min}$ mixed, $C_{\max}$ unmixed, $\epsilon = 1 - \exp\left[-\dfrac{1 - \exp(-\text{NTU}\,C^*)}{C^*} \right]$ For $C_{\max}$ mixed, $C_{\min}$ unmixed, $\epsilon = \dfrac{1}{C^*}(1 - \exp\{-C^*[1 - \exp(-\text{NTU})]\})$	For $C_{\min}$ mixed, $\epsilon = 1 - \exp\{-[1 - \exp(-\text{NTU})]\}$ For $C_{\max}$ mixed, $\epsilon = \dfrac{1 - \exp(-C^*)}{C^*}$	For $C_{\min}$ mixed, $\epsilon = 1 - \exp\left(-\dfrac{1}{C^*} \right)$ For $C_{\max}$ mixed, $\epsilon = \dfrac{1 - \exp(-C^*)}{C^*}$
[4]	Crossflow, both fluids mixed	$\epsilon = \dfrac{1}{\dfrac{1}{1 - \exp(-\text{NTU})} + \dfrac{C^*}{1 - \exp(-\text{NTU}\,C^*)} - \dfrac{1}{\text{NTU}}}$	$\epsilon = \dfrac{1}{\dfrac{2}{1 - \exp(-\text{NTU})} - \dfrac{1}{\text{NTU}}}$	$\epsilon = \dfrac{1}{1 + C^*}$
[4, 6]	1-2 shell-and-tube exchanger; shell fluid mixed; TEMA E shell	$\epsilon = \dfrac{2}{(1 + C^*) + (1 + C^{*2})^{1/2} \coth(\Gamma/2)}$ where $\Gamma = \text{NTU}(1 + C^{*2})^{1/2}$ $\coth \Gamma/2 = \dfrac{1 + e^{-\Gamma}}{1 - e^{-\Gamma}}$	$\epsilon = \dfrac{2}{2 + \sqrt{2}\,\coth(\Gamma/2)}$ where $\Gamma = \sqrt{2}\,\text{NTU}$	$\epsilon = \dfrac{2}{(1 + C^*) + (1 + C^{*2})^{1/2}}$

TABLE 3A ϵ-NTU Formulas and Limiting Values of ϵ for $C^* = 1$ and NTU $\to \infty$ for Various Exchanger Flow Arrangements [3] *(Continued)*

Ref.	Flow arrangement	ϵ-NTU formulas	ϵ-NTU formulas for $C^* = 1$	Asymptotic value of ϵ when NTU $\to \infty$
[7]	Shell fluid / Tube fluid — 1-2 shell-and-tube exchanger; shell fluid unmixed; TEMA E shell	If $C_{min} = C_{tube}$ and $C_{max} = C_{shell}$, $$\epsilon = 1 - \frac{2C^* - 1}{2C^* + 1}\left[\frac{2C^* + \exp[-NTU(C^* + \frac{1}{2})]}{2C^* - \exp[-NTU(C^* - \frac{1}{2})]}\right]$$ If $C_{min} = C_{shell}$ and $C_{max} = C_{tube}$, $$\epsilon = \frac{1}{C^*}\left[\ C_{tube} - \frac{2 - C^*}{C^*(2 + C^*)} \times \frac{2 + C^*\exp[-NTU(1 + C^*/2)]}{2 - C^*\exp[-NTU(1 - C^*/2)]}\right]$$	If $C_{min} = C_{tube}$ and $C^* = \frac{1}{2}$, $$\epsilon = 1 - \frac{1 + e^{-NTU}}{2 + NTU}$$ If $C^* = 1$, $$\epsilon = 1 - \frac{1}{3}\left[\frac{2 + \exp(-\frac{2}{3}\,NTU)}{2 - \exp(-\frac{1}{2}\,NTU)}\right]$$	If $C_{min} = C_{tube}$, $$\epsilon = \begin{cases} \dfrac{2}{1 + 2C^*} & \text{for } C^* \geq 0.5 \\ 1 & \text{for } C^* < 0.5 \end{cases}$$ If $C_{min} = C_{shell}$, $$\epsilon = \frac{2}{2 + C^*}$$
[6,8]	Shell fluid / Tube fluid — 1-4 shell-and-tube exchanger; shell fluid mixed; TEMA E shell	If $C_{min} = C_{tube}$ and $C_{max} = C_{shell}$, $$\epsilon = 4/[2(1 + C^*) + (1 + 4C^{*2})^{1/2}\coth(\Gamma/4) + \tanh(NTU/4)]$$ where $\Gamma = NTU(1 + 4C^{*2})^{1/2}$ If $C_{min} = C_{shell}$ and $C_{max} = C_{tube}$, $$\epsilon = 4/[2(1 + C^*) + (4 + C^{*2})^{1/2}\coth(\Gamma^v/4) + C^*\tanh(NTU\,C^*/4)]$$ where $\Gamma^v = NTU(4 + C^{*2})^{1/2}$	$$\epsilon = \frac{4}{4 + \sqrt{5}\coth(\Gamma/4) + \tanh(NTU/4)}$$ where $\Gamma = \sqrt{5}\,NTU$ $$\epsilon = \frac{4}{4 + \sqrt{5}\coth(\Gamma^v/4) + \tanh(NTU/4)}$$ where $\Gamma^v = \sqrt{5}\,NTU$	If $C_{min} = C_{tube}$, $$\epsilon = \frac{4}{2(1 + C^*) + (1 + 4C^{*2})^{1/2} + 1}$$ If $C_{min} = C_{shell}$, $$\epsilon = \frac{4}{2(1 + C^*) + (4 + C^{*2})^{1/2} + C^*}$$
[9]	Shell fluid / Tube fluid — 1-2 split-flow exchanger; shell fluid mixed	If $C_{min} = C_{tube}$ and $C_{max} = C_{shell}$, $$\epsilon = \frac{(1 + G + 2C^*G) + (2C^* + 1)De^{-\alpha} - e^{-\alpha}}{(1 + G + 2C^*G) + 2C^*(1 - D) + 2C^*De^{-\alpha}}$$ where $D = \dfrac{1 - e^{-\alpha}}{2C^* + 1}$ $G = \dfrac{1 - e^{-\beta}}{2C^* - 1}$ $\;\longrightarrow\; \dfrac{NTU}{2}$	$$\epsilon = \frac{4 - e^{-NTU/2} - e^{-3NTU/2}}{4 - 3e^{-NTU/2} + \frac{2}{3}(2 + 2e^{-3NTU/4}) - e^{-3NTU/2}}$$	For $C^* > \frac{1}{2}$, $$\epsilon = \frac{2C^* + 1}{2C^{*2} + C^* + 1}$$ For $C^* \leq \frac{1}{2}$, $$\epsilon = 1$$

$$\beta = \tfrac{1}{2}\,\mathrm{NTU}(2C^* - 1)$$

Flow arrangement	ϵ	ϵ ($C^*=1$)	ϵ ($\mathrm{NTU}\to\infty$)
 [10] 1-2 divided-flow exchanger, shell fluid mixed; TEMA J shell	If $C_{min} = C_{shell}$ and $C_{max} = C_{tube}$, use the above formula with C^* replaced by $1/C^*$, NTU replaced by NTU C^*, and ϵ replaced by ϵC^*.	(Same as the above formula)	For $C^* < 2$, $$\epsilon = \frac{C^* + 2}{C^{*2} + C^* + 2}$$ For $C^* \geq 2$, $$\epsilon = \frac{1}{C^*}$$
	If $C_{min} = C_{tube}$ and $C_{max} = C_{shell}$, $$\epsilon = \frac{2}{1 + 2C^*\Phi'}$$ where $\Phi' = 1 + \gamma\left(\dfrac{1+\Phi}{1-\Phi}\right)$ $\quad - 2\gamma\left[\dfrac{\gamma\Phi + (1-\Phi)e^{-\mathrm{NTU}\,C^*(\gamma-1)/2}}{(1-\Phi)^2 + \gamma(1-\Phi^2)}\right]$ $\Phi = \exp(-\mathrm{NTU}\,C^*\gamma)$ $\gamma = \dfrac{(1 + 4C^{*2})^{1/2}}{2C^*}$	If $C_{min} = C_{tube}$ and $C_{max} = C_{shell}$, $$\epsilon = \frac{2}{1 + 2\Phi'}$$ where $\Phi' = 1 + \gamma\left(\dfrac{1+\Phi}{1-\Phi}\right)$ $\quad - 2\gamma\left[\dfrac{\gamma\Phi + (1-\Phi)e^{-\mathrm{NTU}(\gamma-1)/2}}{(1-\Phi)^2 + \gamma(1-\Phi^2)}\right]$ $\Phi = \exp(-\mathrm{NTU}\,\gamma)$ $\gamma = \dfrac{\sqrt{5}}{2}$	If $C_{min} = C_{tube}$, $$\epsilon = \frac{2}{1 + 2C^* + (1 + 4C^{*2})^{1/2}}$$
	If $C_{min} = C_{shell}$ and $C_{max} = C_{tube}$, use the above formula with C^* replaced by $1/C^*$, NTU replaced by NTU C^*, and ϵ replaced by ϵC^*.	(Same as the above formula)	If $C_{min} = C_{shell}$, $$\epsilon = \frac{2}{C^* + 2 + (C^{*2} + 4)^{1/2}}$$

TABLE 3B ϵ-NTU Formulas and Limiting Values of ϵ for Various Multipass Exchanger Flow Arrangements

Ref.	Flow arrangement	ϵ-ϵ_p formulas	Special cases of $C^* = 1$ and 0	Number of passes $n \to \infty$
[4]	Fluid 1 / Fluid 2 n-pass exchanger ($n = 3$ as shown), both fluid streams parallel, overall counterflow arrangement, fluids mixed between passes, each pass having the same ϵ_p	$$\epsilon = \frac{[(1 - \epsilon_p C^*)/(1 - \epsilon_p)]^n - 1}{[(1 - \epsilon_p C^*)/(1 - \epsilon_p)]^n - C^*}$$ Note: $C^* = C^*_p$, and NTU $= n\,\mathrm{NTU}_p$. $$\epsilon_p = \frac{[(1 - \epsilon C^*)/(1 - \epsilon)]^{1/n} - 1}{[(1 - \epsilon C^*)/(1 - \epsilon)]^{1/n} - C^*}$$	$$\epsilon = \begin{cases} \dfrac{n\epsilon_p}{1 + (n - 1)\epsilon_p} & \text{for } C^* = 1 \\ 1 - (1 - \epsilon_p)^n & \text{for } C^* = 0 \end{cases}$$ $$\epsilon_p = \begin{cases} \dfrac{\epsilon}{n - (n - 1)\epsilon} & \text{for } C^* = 1 \\ 1 - (1 - \epsilon)^{1/n} & \text{for } C^* = 0 \end{cases}$$	$\epsilon = \epsilon_{cf}$ $\epsilon_p \to 0$
[3]	Fluid 1 / Fluid 2 n-pass exchanger ($n = 3$ as shown), both fluid streams parallel, overall parallel flow arrangement, fluids mixed between passes, each pass having the same ϵ_p	$$\epsilon = \frac{1}{1 + C^*}\{1 - [1 - (1 + C^*)\epsilon_p]^n\}$$ Note: $C^* = C^*_p$, and NTU $= n\,\mathrm{NTU}_p$. $$\epsilon_p = \frac{1}{1 + C^*}\{1 - [1 - (1 + C^*)\epsilon]^{1/n}\}$$	$$\epsilon = \begin{cases} \tfrac{1}{2}[1 - (1 - 2\epsilon_p)^n] & \text{for } C^* = 1 \\ 1 - (1 - \epsilon_p)^n & \text{for } C^* = 0 \end{cases}$$ $$\epsilon_p = \begin{cases} \tfrac{1}{2}[1 - (1 - 2\epsilon)^{1/n}] & \text{for } C^* = 1 \\ 1 - (1 - \epsilon)^{1/n} & \text{for } C^* = 0 \end{cases}$$	$\epsilon = \epsilon_{parallel\ flow}$ $\epsilon_p \to 0$
[11]	C_{max} fluid / C_{min} fluid n-pass exchanger ($n = 3$ as shown), C_{min} stream in parallel, C_{max} stream in series, series fluid mixed between passes, each pass having the same ϵ_p	Parallel stream as C_{min} stream, $$\epsilon = \frac{1}{C^*}\left[1 - \left(1 - \frac{C^*\epsilon_p}{n}\right)^n\right]$$ Note: $C^* = nC^*_p$, and NTU $= \mathrm{NTU}_p$. $$\epsilon_p = \frac{1}{C^*_p}[1 - (1 - nC^*_p\epsilon)^{1/n}]$$	$$\epsilon = \begin{cases} 1 - \left(1 - \dfrac{\epsilon_p}{n}\right)^n & \text{for } C^* = 1 \\ \epsilon_p & \text{for } C^* = 0 \end{cases}$$ $$\epsilon_p = \begin{cases} n[1 - (1 - \epsilon)^{1/n}] & \text{for } C^* = 1 \\ \epsilon & \text{for } C^* = 0 \end{cases}$$	$$\epsilon = \frac{1}{C^*}[1 - \exp(-C^*\epsilon_p)]$$ $$= \frac{1}{C^*}\Big[1 - \exp\big(-C^* \times [1 - \exp(-\mathrm{NTU})]\big)\Big]$$ $$\epsilon_p = 1 - \exp(-\mathrm{NTU}_p)$$ $$U = \mathrm{NTU}_p.$$

[11]

n-pass exchanger (n = 3 as shown), C_{max} stream in parallel, C_{min} stream in series, series fluid mixed between passes, each pass having the same ϵ_p

Diagram labels: C_{min} fluid, t_2, T_2, t_1, T_1, C_{max} fluid

---

Parallel stream as C_{max} stream, and $C_{max}/n < C_{min}$:

$$\epsilon = 1 - \left(1 - \frac{\epsilon_p}{nC^*}\right)^n$$

Note: $C^* = 1/(nC_p^*)$, $NTU = NTU_p/C^*$.

$$\epsilon_p = \frac{(t_1 - t_2)_p}{(t_1 - T_1)_p} = \frac{1}{C_p^*}[1 - (1 - \epsilon)^{1/n}]$$

$$\epsilon = \begin{cases} 1 - (1 - \epsilon_p)^n & \text{for } C_p^* = 1 = \dfrac{1}{nC^*} \\[2mm] 1 - e^{-NTU} & \text{for } n \to \infty,\ C^* = 0 \end{cases}$$

$$\epsilon_p = \begin{cases} 1 - (1 - \epsilon)^{1/n} & \text{for } C_p^* = 1 = \dfrac{1}{nC^*} \\[2mm] 1 & \text{for } n \to \infty,\ C_p^* = 0 \end{cases}$$

$$\epsilon = 1 - \exp\left(-\frac{\epsilon_p}{C^*}\right)$$
$$= 1 - \exp\left\{-\frac{1}{C^*}\right.$$
$$\left. \times\, [1 - \exp(-NTU\, C^*)]\right\}$$

$$\epsilon_p = 1 - \exp(-NTU_p)$$

Parallel stream as C_{max} stream, and $C_{max}/n > C_{min}$:

$$\epsilon = 1 - (1 - \epsilon_p)^n$$

Note: $C^* = C_p^*/n$, and $NTU = n\, NTU_p$.

$$\epsilon_p = 1 - (1 - \epsilon)^{1/n}$$

$C^* = 1$ only when $n = 1$.

$$\epsilon = \begin{cases} 1 - (1 - \epsilon_p)^n & \text{for } C_p^* = 1 = \dfrac{C^*}{n} \\[2mm] 1 - \exp(-NTU) & \text{for } C^* = 0 \end{cases}$$

$$\epsilon_p = \begin{cases} 1 - (1 - \epsilon)^{1/n} & \text{for } C_p^* = 1 \\[2mm] 1 - \exp(-NTU_p) & \text{for } C_p^* = 0 \end{cases}$$

$n \to \infty$ only when $C^* = 0$.

$$\epsilon = 1 - \exp(-NTU)$$

$$\epsilon_p = 1 - \exp(-NTU_p)$$

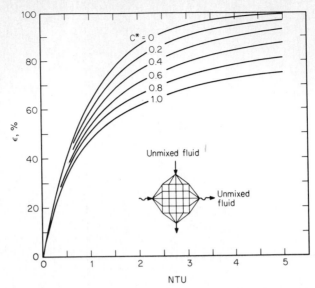

FIG. 5 Heat transfer effectiveness ϵ as a function of NTU and C^* for a cross-flow exchanger with both fluids unmixed.

exchanger applications. ϵ-NTU results for a number of these flow arrangements are presented in tabular and graphical forms by Kays and London [4]. Some of the multipass flow arrangements not considered in Table 3 are summarized in Sec. D.

2. The P-NTU$_t$ Method

One of the drawbacks of the ϵ-NTU method is the need to keep track of the $C_{\min}$ side, particularly for shell-and-tube exchanger flow arrangements, for which the ϵ-NTU expressions are different for $C_{\min}$ on the shell side ($C_{\min} = C_s$) from what they are on the tube side ($C_{\min} = C_t$). Here the subscripts s and t denote shell side and tube side, respectively. In order to avoid possible errors and confusion, ϵ, NTU, and C^* are redefined based on C_t instead of $C_{\min}$ and are designated **P, NTU$_t$,** and **R,** respectively.† Explicit definitions of **P, NTU$_t$,** and **R** are provided in Table 1. **P** is referred to as the temperature effectiveness or thermal effectiveness, and **R** as the heat capacity rate ratio. **P-NTU$_t$** results for one of the most common 1-2 shell-and-tube exchangers‡ are shown in Fig. 7. Note that such results are generally presented on a semilog paper in order to stretch the NTU$_t$ scale in the most useful NTU$_t$ design range of 0.2 to 3.0 for shell-and-tube exchangers. Using the explicit relationships between **P** and ϵ, NTU$_t$ and NTU, and **R** and C^* provided in Table 2, **P-NTU$_t$** expressions and results can be derived from the ϵ-NTU formulas of Table 3. As a result, no further results in terms of **P-NTU$_t$** are provided.

---

†Although **P, NTU$_t$,** and **R,** are defined on the basis of C_t, it must be emphasized that the **P-NTU$_t$** charts are valid if **P, NTU$_t$,** and **R** are consistently defined on the basis of C_s, C_h, or C_c; the only exception is the results of Fig. 34.

‡A one-shell pass two-tube pass shell-and-tube exchanger with TEMA E shell and shell fluid mixed will be referred to simply as a 1-2 exchanger throughout this chapter unless clearly specified otherwise.

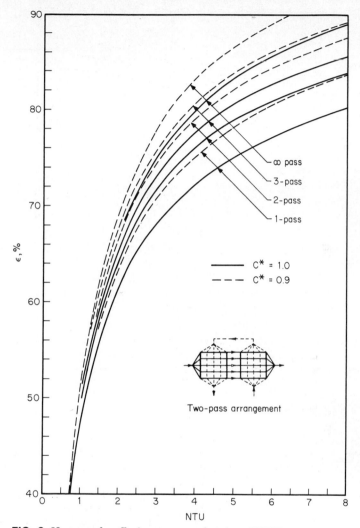

FIG. 6 Heat transfer effectiveness ϵ as a function of NTU and number of passes for a multipass cross-counterflow exchanger with fluid 2 unmixed throughout and fluid 1 unmixed within passes and mixed between passes.

3. The LMTD Method‡

In this method, the log-mean temperature difference correction factor F is a ratio of actual mean temperature difference in an exchanger to the log-mean temperature difference

‡This method is also sometimes referred to as the F-LMTD method. However, in light of the ϵ-NTU, **P**-NTU$_t$, and ψ-**P** methods, this method should be referred to as the F-**P** method since F is plotted against **P**. Since this terminology is not commonly used, we have preferred to call it simply the LMTD method, which is the term commonly used.

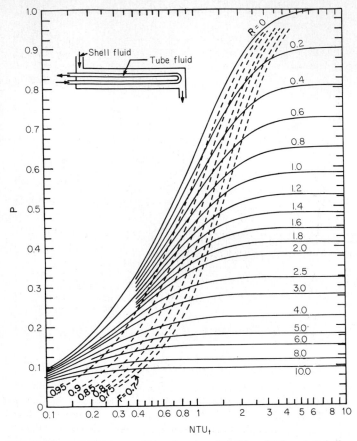

FIG. 7 Thermal effectiveness **P** as a function of NTU_t and **R** for a 1-2 shell-and-tube exchanger with shell fluid mixed.

(LMTD). The LMTD is *defined*

$$\text{LMTD} = \Delta T_{lm} = \frac{\Delta T_1 - \Delta T_2}{\ln(\Delta T_1/\Delta T_2)} \tag{14}$$

where

$$\Delta T_1 = T_{h,i} - T_{c,o} \qquad \Delta T_2 = T_{h,o} - T_{c,i} \text{ for all flow arrangements} \tag{15}$$
$$\text{except parallel flow}$$

and $\Delta T_1 = T_{h,i} - T_{c,i} \qquad \Delta T_2 = T_{h,o} - T_{c,o}$ for parallel flow   (16)

The LMTD represents a true mean temperature difference for a counterflow exchanger under the idealizations listed following Eq. (13). Thus the log-mean temperature difference correction factor F represents a degree of departure for the true mean temperature difference from the counterflow log-mean temperature difference; it does not represent the effectiveness of a heat exchanger. It depends upon two dimensionless groups: the thermal effectiveness **P** and the heat capacity rate ratio **R**.

The definitions or working equations for F, **P**, and **R** are summarized in Table 1. Closed-form relationships of F with ϵ, with NTU and C^*, and with **P**, NTU_t, and **R** are presented in Table 2. The F factors are generally presented as a function of **P** and **R** as shown in Fig. 8 for the 1-2 exchanger. However, since such charts do not provide a straightforward solution to the rating problem, ψ-**P** charts, to be discussed next, are recommended. The F factors are superimposed on the ψ-**P** charts in Figs. 9 to 15 and 17 to 23 for compact representation.

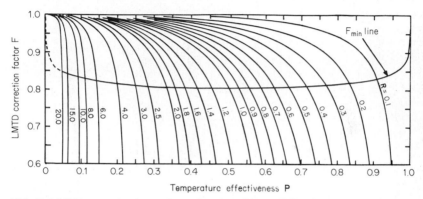

FIG. 8 LMTD correction factor F as a function of **P** and **R** for the 1-2 exchanger with shell fluid mixed.

The F_{min} line in Fig. 8 corresponds to the temperature cross just at the exit of the tube outlet pass [3]. The temperature cross refers to the case when $T_{c,o} > T_{h,o}$. Shell-and-tube exchangers are designed for F factors greater than F_{min} in order to utilize the surface area efficiently. It should be pointed out that although the curves at high values of **R** appear to be too steep compared to the curves for low values **R,** that appearance is misleading. If we consider **P** and **R** to have been based on the tube side as 0.2899 and 2.5, the F factor is 0.7897. If **P** and **R** had been based on the shell side, they would be 0.7247($=$ 0.2899 $\times$ 2.5) and 0.4($= 1/2.5$), and the corresponding F factor would again be 0.7897. A careful review of Fig. 8 indeed indicates this fact (although accuracy is only within two digits in Fig. 8).

4. The ψ-P Method

Both the LMTD and ϵ-NTU methods have some limitations, as discussed in some depth in Ref. 3. In particular, a trial-and-error approach is needed for the solution of the rating problem by the LMTD method and for the solution of the sizing problem (for multipass shell-and-tube exchangers) by the ϵ-NTU method. A method that combines all the variables of the LMTD and ϵ-NTU methods and eliminates their limitations for a hand solution has been proposed by Mueller [12, 13]. In this method, a new grouping ψ is introduced. It is a ratio of the true mean temperature difference to the inlet temperature difference of the two fluids. The other necessary equations for this method are summarized in Table 1. The relationship of ψ with dimensionless groups of other methods is

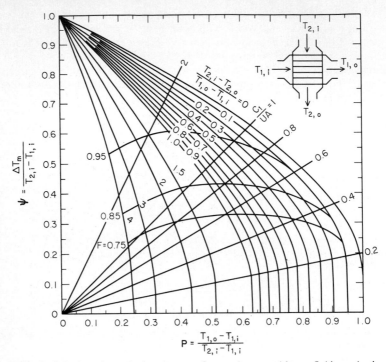

FIG. 9 ψ-**P** chart for a single-pass cross-flow exchanger with one fluid unmixed and the other mixed. **P**, NTU_i, and **R** based on the unmixed side.

presented in Table 2. The graphical ψ-**P**-**R** results† for single-pass cross-flow exchangers with either one or both fluids unmixed are shown in Figs. 9 and 10. The ψ-**P** charts for some multipass shell-and-tube and cross-flow exchangers are shown in Figs. 11 to 15 and 17 to 23.

D. MULTIPASS HEAT EXCHANGERS

Results in terms of one of the four aforementioned basic methods for most of the basic and commonly used exchanger flow arrangements are presented in Table 3, and for some of them in Figs. 4 to 10. However, there are a number of additional multipass flow arrangements that are used in applications. The ϵ-NTU, LMTD, or ψ-**P** results for these are presented next for shell-and-tube exchangers, extended-surface exchangers, and plate exchangers.

1. Shell-and-Tube Exchangers

Various shell types for shell-and-tube exchangers have been standardized by TEMA [14] as shown in Fig. 1 of Part 2. Since the shell-side fluid flow arrangement is unique with

---

†A comment by Shah: The ψ-**P** charts (such as Figs. 9 to 15) are referred to as Mueller's charts in the heat exchanger literature.

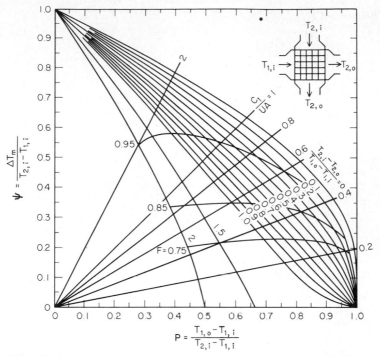

FIG. 10 ψ-P chart for a single-pass cross-flow exchanger with both fluids unmixed.

each shell, the LMTD, ψ-P, or ϵ-NTU relationship is different for each shell even though the tube pass arrangement may be the same. These relationships are presented next, categorized by the type of shell used.

a. Parallel Counterflow Exchanger, TEMA E Shell

The most general exchanger has M shell passes and N tube passes ($N \geq M$), and is simply referred to as an M-N exchanger. The ϵ-NTU expressions for 1-2 and 1-4 exchangers having shell fluid mixed are provided in Table 3. The graphical results for the 1-2 exchanger are provided in Figs. 7, 8, and 11.

The effectiveness ϵ or **P** of the 1-N exchanger ($N > 2$ and even) is lower than that of the 1-2 exchanger at specified values of NTU_t and **R**. The maximum reduction in ϵ or **P** of the 1-N exchanger compared to the 1-2 exchanger occurs at **R** = 1, and is 4.4 percent and 6.8 percent for 1-4 and 1-12 exchangers, respectively, at NTU_t = 6. These differences will be significantly lower for low NTU_t ($NTU_t < 3$) and **R** $\neq$ 1. Hence, ϵ or **P** for the 1-N exchanger is generally calculated using the expression for the 1-2 exchanger. However, if one is really concerned, one may employ the 1-4 exchanger ϵ-NTU expression of Table 3 for the 1-N exchanger for $N > 4$ and even. However, it must be emphasized that the error in ϵ or **P** for a 1-N exchanger ($N > 2$ and even) resulting from the use of this approximation is negligibly small compared to the errors associated with flow bypass and leakage (to be discussed later), and hence one can safely neglect this error in computing ϵ or **P**.

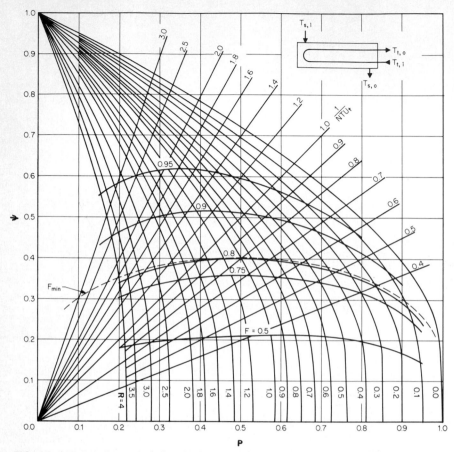

FIG. 11 ψ-P chart for a 1-2 shell-and-tube exchanger, TEMA E shell, shell fluid mixed.

One shell pass and N tube passes with N odd are uncommon in practice. Fisher [15] shows the best performance for the 1-3 exchanger is obtained when the shell fluid flows in countercurrent to the two tube passes (Fig. 12); however, the major benefit is at low F values, where it is generally not desirable to operate. The improvement is not as great as the addition of one more pass to the exchanger. Further recent work for odd tube pass exchangers has been reported by Crozier and Samuels [16].

The overall effectiveness of N-$2N$ exchangers (several identical 1-2 exchangers in series) can be evaluated from

$$\mathbf{P} = \frac{[(1 - \mathbf{P}_1\mathbf{R})/(1 - \mathbf{P}_1)]^N - 1}{[(1 - \mathbf{P}_1\mathbf{R})/(1 - \mathbf{P}_1)]^N - \mathbf{R}} \xrightarrow[\mathbf{R} = 1]{} \frac{N\mathbf{P}_1}{1 + (N - 1)\mathbf{P}_1} \tag{17}$$

where $\mathbf{P}_1$ represents the thermal effectiveness of one 1-2 exchanger and N is the number of 1-2 exchangers in series. The fluids are considered mixed in the headers between exchangers. The total NTU_t of N-$2N$ exchangers is then N times NTU_t for each 1-2 exchanger. Once overall $\mathbf{P}$ and NTU_t are evaluated, the overall F can be computed from the relationship provided in Table 2.

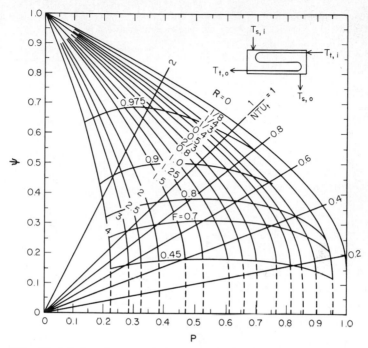

FIG. 12 ψ-P chart for a 1-3 shell-and-tube exchanger, TEMA E shell, shell fluid mixed. Two tube passes in counterflow and one in parallel flow.

In all of the foregoing 1-N or N-2N exchangers, the shell fluid is considered mixed in each shell pass. This represents a good idealization in segmental-baffled (or plate-baffled) exchangers. The analysis of Gardner [7] for the 1-2 exchanger with shell fluid mixed versus unmixed indicates a slight increase in **P** or F when the shell fluid is unmixed. Consequently, relatively higher **P** or F is obtained for rod-baffled heat exchangers in which the shell fluid is unmixed provided no fluid bypasses the bundle.

b. Examples of Use of ψ-P Charts

The use of ψ-**P** charts will be illustrated here for the solution of an exchanger rating or sizing (design) problem. The use of a **P**-NTU$_t$ or ϵ-NTU chart is straightforward and will not be illustrated here. In the exchanger rating problem, the detailed exchanger geometry and dimensions, the heat transfer rates of both fluids, and their inlet temperatures are known. In the design problem, the temperature and flows are specified but the exchanger size and heat transfer coefficients are unknown.

Rating Example†

Two 1-2 exchangers are connected in series. Each exchanger has 20 m² of surface area, the overall coefficient U is 585 W/(m²·K), the shell flow is 2.5 kg/s entering at

---

†This example can also be solved straightforwardly by using the **P**-NTU$_t$ method. If F factors are already superimposed on the **P**-NTU$_t$ chart as in Fig. 7, then the use of the ψ-**P** charts does not provide any operational advantage over the **P**-NTU$_t$ charts, except for a concise presentation of all variables involved.

150°C, and the tube flow is 1.875 kg/s entering at 25°C. The specific heat of both fluids is 2500 J/(kg·K). What are the outlet temperatures?

$$R = \frac{1.875 \times 2500}{2.5 \times 2500} = 0.75$$

$$(NTU_t)_1 = \frac{585 \times 20}{1.875 \times 2500} = 2.50 \text{ (per pass value)}$$

From Fig. 11, for $(NTU_t)_1 = 2.50$ and $R = 0.75$, we obtain the abscissa as $P_1 = 0.64$.† Substituting these values in Eq. (17),

$$P = \frac{[(1 - 0.64 \times 0.75)/(1 - 0.64)]^2 - 1}{[(1 - 0.64 \times 0.75)/(1 - 0.64)]^2 - 0.75} = 0.813 = \frac{T_{t,o} - T_{t,i}}{T_{s,i} - T_{t,i}}$$

Thus

$$T_{t,o} = T_{t,i} + P(T_{s,i} - T_{t,i}) = 25 + 0.813(150 - 25) = 126.6°C$$

$$T_{s,o} = T_{s,i} - R(T_{t,o} - T_{t,i}) = 150 - 0.75(126.6 - 25) = 73.8°C$$

Sizing Example

Design a multipass exchanger to cool one process fluid from 150°C to 100°C with another fluid rising from 80°C to 130°C.

$$R = \frac{150 - 100}{130 - 80} = 1$$

$$P = \frac{130 - 80}{150 - 80} = 0.714$$

Figure 11 indicates that the maximum P for one shell is 0.586 for $R = 1$. Therefore, several shells in series are required. The number of shells N from Eq. (17) for $R = 1$ is

$$N = \frac{1/P_1 - 1}{1/P - 1}$$

An examination of the curve for $R = 1$ in Fig. 11 indicates a rapid drop of ψ or a rapid increase of $(NTU_t)_1$ when P_1 exceeds 0.55. Using $P_1 = 0.55$ as a first approximation for the design value,

$$N = \frac{1/0.55 - 1}{1/0.714 - 1} = 2.04$$

This value should be rounded to the next larger integer, or $N = 3$.

Now calculate the corresponding P_1 from Eq. (17) as

$$P_1 = \frac{P}{N - (N - 1)P} = \frac{0.714}{3 - 2 \times 0.714} = 0.454$$

---

†If the value of $(NTU_t)_1$ were different from the lines in Fig. 11, you could generate a line for any $(NTU_t)_1$ by simply drawing a line from the origin with a slope of $1/(NTU_t)_1$. The subscript 1 in this example designates quantities per pass (in contrast to the overall).

From Fig. 11, $\psi_1 = 0.473$ and $F_1 = 0.83$ for $\mathbf{P}_1 = 0.454$ and $\mathbf{R} = 1$. Hence,

$$(\text{NTU}_t)_1 = \frac{\mathbf{P}_1}{\psi_1} = \frac{0.454}{0.473} = 0.96$$

Hence, the total NTU_t for the three-shell exchanger is $3 \times 0.96 = 2.88$.

The calculated number of shells of 2.04 was based on an assumed maximum $\mathbf{P}_1$ of 0.55. However, the theoretical maximum value is 0.586. Hence two shells could also be a solution. For two shells, $\mathbf{P}_1 = 0.555$. Then from Fig. 11, $\psi_1 = 0.29$ and $F_1 \approx 0.63$. Thus $(\text{NTU}_t)_1 = 0.555/0.29 = 1.91$ and total NTU_t for two shells is 3.82, as opposed to 2.88 for three shells. The final choice depends on costs and other factors.

c. Counterflow Exchanger, TEMA F Shell

This exchanger has two tube passes and two shell passes with shell fluid mixed in each pass. If there are no flow leakages around the longitudinal baffle and heat conduction from one shell pass to the other is neglected, the arrangement represents a true counterflow exchanger with $F = 1$ or the ϵ-NTU results of Fig. 4. However, the design of the actual exchanger can be far from the ideal. Finite heat conduction across the longitudinal baffle reduces the F and $\mathbf{P}$ of the exchanger, as analyzed by Whistler [17]. If the tube bundle is fixed, the longitudinal baffle will probably be welded and the flow leakage across it will be negligible. However, for a removable bundle, there must be a clearance between the longitudinal baffle and the shell through which flow will leak from the first to the second pass. Rosenman and Taborek [18] studied the effect of this leakage and heat conduction through the longitudinal baffle. The results are very complicated and thus will not be presented here. An important conclusion is that a small amount of leakage can significantly reduce the effectiveness of the second shell pass. Hence, care must be exercised in design and fabrication of the longitudinal baffle seal.

d. Split-Flow Exchanger, TEMA G Shell

This exchanger is often used as a horizontal thermosiphon reboiler. In this case, the longitudinal baffle serves to prevent flashing out of lighter components of the shell fluid and provides increased mixing. For single-phase applications ($C^* \neq 0$) of this exchanger, the ϵ-NTU expression with two tube passes is presented in Table 3. The corresponding ψ-$\mathbf{P}$ results are presented in Fig. 13; aligned with the ψ-$\mathbf{P}$ chart in this figure is an F-$\mathbf{P}$ chart presented as an example [18a]. The thermal effectiveness $\mathbf{P}$ of the 1-2 split-flow exchanger is higher than that of the conventional 1-2 exchanger (TEMA E shell) at a given NTU_t and $\mathbf{R}$. For $\text{NTU}_t < 1.5$, this increase is less than 10 percent. The difference increases as NTU_t increases. The pressure drop for the G shell is also generally higher than for the E shell at the same flow conditions because of an additional turn at the midlocation of its flow passage.

e. Double Split-Flow Exchanger, TEMA H Shell

An exchanger with an H shell is often used as a horizontal thermosiphon vaporizer in which flows are large and pressure drop must be kept low. In this service, C^* is effectively zero (idealization 5) and the effectiveness is given by

$$\epsilon = 1 - e^{-\text{NTU}} \tag{18}$$

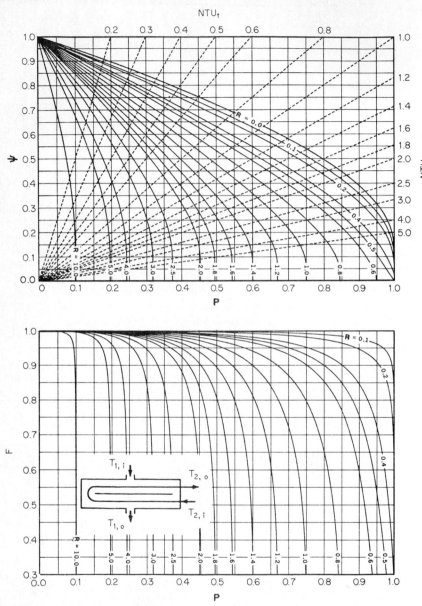

FIG. 13 ψ-**P** chart and F-**P** chart for a 1-2 split-flow exchanger, TEMA G shell, shell fluid mixed [18a].

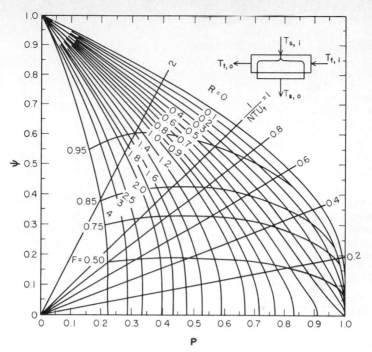

FIG. 14 ψ-P chart for a 1-1 divided-flow exchanger, TEMA J shell.

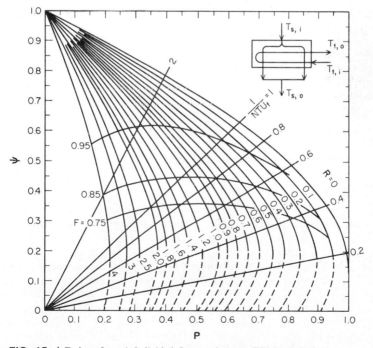

FIG. 15 ψ-P chart for a 1-2 divided-flow exchanger, TEMA J shell.

provided that the overall heat transfer coefficient U can be treated as constant. Note that the F factor will be unity for this case. No ϵ-NTU, LMTD, or ψ-P results are available for this exchanger when $C^* \neq 0$ (single-phase applications).

f. Divided-Flow Exchanger, TEMA J Shell

This exchanger is used for low shell-side pressure drop applications, such as for a condenser in a vacuum, or where the shell flow is too large for convenient baffle spacing. The divided-flow exchanger with one tube pass is the same as the conventional 1-2 exchanger (TEMA E shell) with shell fluid unmixed [7] for which the ϵ-NTU expression is given in Table 3 and the ψ-P chart in Fig. 14. The 1-2 and 1-4 divided-flow exchangers have been analyzed by Jaw [10]. His results for the 1-2 divided-flow exchanger are presented in Table 3 and Fig. 15. For LMTD correction factor $F > 0.8$, Jaw concluded that there is no significant difference between the F-P curves of 1-2 and 1-4 divided-flow exchangers and that of the 1-2 conventional exchanger.

g. Kettle-Type Reboiler, TEMA K Shell

Since evaporation or boiling takes place on the shell side, C^* for the kettle-type reboiler is effectively zero, and the effectiveness is given by Eq. (18) if U can be treated as constant. The LMTD correction factor F for this exchanger is then unity.

h. Cross-Flow Exchanger, TEMA X Shell

This exchanger is used for very low shell-side pressure drop applications, such as gas cooling with finned tubes, or condensation. No baffles are used in the X shell; however, support plates are used to suppress flow-induced vibrations. For a 1-1 cross-flow exchanger, the ϵ-NTU expressions are presented in Table 3 and Figs. 5 and 10 for the single-pass cross-flow exchanger with both fluids unmixed. For a 1-2 cross-flow exchanger, the ϵ-NTU formulas are presented in Table 4 for both fluids unmixed throughout with overall parallel flow (shell fluid entering at the tube inlet pass end) and overall counterflow (shell fluid entering at the tube exit pass end). Figure 20 represents the ψ-P chart for the latter arrangement.

2. Extended Surface Exchangers

a. Multipass Cross-Flow Exchangers

In a single-pass cross-flow exchanger, fluids flow perpendicular to each other and fluids can be mixed or unmixed on each side. As a result, four different formulas exist for single-pass cross-flow exchangers as summarized in Table 3.

In a multipass cross-flow exchanger, the results depend upon how the *unmixed* fluid in the pass return (or header or area between passes) is distributed, whether in identical or inverted order. In Fig. 16, the tube fluid is in two rows per pass and the air flows across the tubes of rows 1 through 4 in sequence. In Fig. 16a, the tube fluid in row 1 is first in contact with the air in pass 1, and the *same* fluid in row 3 is first in contact with the air in pass 2. The tube fluid between passes for this arrangement is then said to be in *identical order*. In Fig. 16b, the tube fluid in row 1 is again first in contact with the

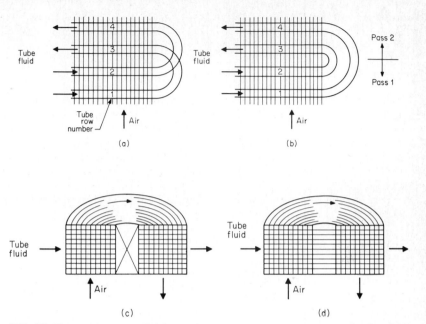

FIG. 16 Two-pass cross-parallel flow exchangers with both fluids unmixed throughout. For tube fluid in the pass return: (*a*) identical order; (*b*) inverted order. Cases *c* and *d* are symbolic representations of cases *a* and *b*, respectively.

air in pass 1; however, the *different* fluid (of row 2) in row 3 is first in contact with the air in pass 2. In this arrangement, the tube fluid between passes is said to be in *inverted order*. In either Fig. 16*a* or Fig. 16*b*, the air-side fluid is in inverted order between passes. This is because the air stream that crossed the tube inlet end in the first pass then crosses the tube exit end in the second pass. Figures 16*a* and *b* are symbolically represented as Figs. 16*c* and *d*, respectively, since one does not always have tubes in the exchanger (e.g., in a plate-fin multipass exchanger).

Now in a multipass cross-flow exchanger, fluids can be mixed or unmixed within each pass as well as between passes. When the fluid is unmixed between passes, it can have either identical or inverted order as just discussed. In addition, when one considers the additional variables (such as fluid 1 as C_{min} or C_{max} fluid, and number of passes), the number of combinations for overall counterflow or parallel-flow arrangements rapidly escalates. The ϵ-NTU formulas for two-pass overall cross-counterflow and cross-parallel flow arrangements with inverted order are summarized in Table 4. The expressions for the functions used in Table 4 are summarized in Table 5. The corresponding graphical ϵ-NTU results are presented in detail by Bačlić and Gvozdenac [19]. The ϵ-NTU formulas and graphical results for two- and three-pass cross-flow exchangers with so-called identical order are presented by Bačlić et al. [20, 21]. However, Bačlić et al. [20, 21] define identical order differently from the Fig. 16*b* and hence their results are not applicable to the cases of Fig. 16*b* and *d*. The ψ-P charts for some important two- and three-pass cross-flow exchangers used in air-cooled and other compact exchangers are summarized in Figs. 17 to 23 [13].

TABLE 4 ε-NTU Formulas for Two-Pass Cross-Parallel Flow and Cross-Counterflow Exchangers [19]†

Overall cross-parallel flow two-pass exchangers			Overall cross-counterflow two-pass exchangers		
	Fluid 1 unmixed throughout Fluid 2 unmixed throughout	$\epsilon = \mu_{1/2} + \dfrac{1}{C^*}\,\bar{\mu}_{1/2}$		Fluid 1 unmixed throughout Fluid 2 mixed throughout	$\epsilon C^* = 1 - 2/\{1 - \exp(-NTU/2) + [1 + \exp(-NTU/2)]/\bar{\nu}_{1/2}^*\}$
	Fluid 2 unmixed throughout Fluid 1 unmixed in each pass and mixed between passes	$\epsilon = \mu_{1/2} + \nu_{1/2} - \nu_{1/2}^2$		Fluid 2 unmixed throughout Fluid 1 mixed throughout	$\epsilon = 1 - 2/\{1 - \exp(-C^* NTU/2) + [1 + \exp(-C^* NTU/2)]\nu_{1/2}^*\}$
	Fluid 2 unmixed throughout Fluid 1 unmixed only in one pass	$\epsilon = \nu_{1/2}^*(2\bar{g} - \nu_{1/2})$		Fluid 1 unmixed throughout Fluid 2 unmixed only in one pass	$\epsilon C^* = 1 - \dfrac{\bar{\nu}_{1/2}}{g}$
	Fluid 2 unmixed throughout Fluid 1 mixed throughout	$\epsilon = 1 - \dfrac{1}{2}\left\{1 - \exp\left(-\dfrac{C^* NTU}{2}\right)\right\} + \nu_{1/2}^{*2}\left[1 + \exp\left(-\dfrac{C^* NTU}{2}\right)\right]$		Fluid 2 unmixed throughout Fluid 1 unmixed only in one pass	$\epsilon = 1 - \dfrac{\nu_{1/2}}{g}$

 Fluid 1 unmixed throughout Fluid 2 unmixed in each pass and mixed between passes $\epsilon C^* = \bar{\mu}_{1/2} + \bar{\nu}_{1/2} - \bar{\nu}_{1/2}^2$	 Fluid 1 unmixed throughout Fluid 2 unmixed throughout Formula too complex to present here. See Ref. 19.
 Fluid 1 unmixed throughout Fluid 2 unmixed only in one pass $\epsilon C^* = \bar{\nu}_{1/2}^*(g - \bar{\nu}_{1/2})$	 Fluid 1 unmixed throughout Fluid 2 unmixed in each pass and mixed between passes $\epsilon C^* = 1 - \dfrac{\bar{\nu}_{1/2}^2}{\bar{\nu}_{1/2} + \bar{\mu}_{1/2}}$
 Fluid 1 unmixed throughout Fluid 2 mixed throughout $\epsilon C^* = 1 - \dfrac{1}{2}\left\{1 - \exp\left(-\dfrac{\text{NTU}}{2}\right)\right.$ $\left. + \bar{\nu}_{1/2}^{*2}\left[1 + \exp\left(-\dfrac{\text{NTU}}{2}\right)\right]\right\}$	 Fluid 2 unmixed throughout Fluid 1 unmixed in each pass and mixed between passes $\epsilon = 1 - \dfrac{\nu_{1/2}^2}{\nu_{1/2} + \mu_{1/2}}$

†The expressions for $\mu_{1/2}$, $\bar{\mu}_{1/2}$, g, $\bar{g}$, $\nu_{1/2}$, $\bar{\nu}_{1/2}$, $\nu_{1/2}^*$, and $\bar{\nu}_{1/2}^*$ in terms of NTU and C^* are presented in Table 5.

TABLE 5 Expressions for the Functions Used in Table 4 [19]

$$\nu_{1/2}^* = \exp\left(-\frac{1 - \exp(-C^* \,\text{NTU}/2)}{C^*}\right)$$

$$\bar{\nu}_{1/2}^* = \exp\left\{-C^*\left[1 - \exp\left(-\frac{\text{NTU}}{2}\right)\right]\right\}$$

$$\nu_{1/2} = \exp\left[-(1 + C^*)\frac{\text{NTU}}{2}\right]\left[I_0(\text{NTU }\sqrt{C^*}) + \sqrt{C^*}I_1(\text{NTU }\sqrt{C^*})\right.$$
$$\left. - \frac{1 - C^*}{C^*}\sum_{n=2}^{\infty}(C^*)^{-n/2}I_n(\text{NTU }\sqrt{C^*})\right]$$

$$\bar{\nu}_{1/2} = \exp\left(-\frac{1 + C^*}{C^*}\frac{C^*\,\text{NTU}}{2}\right)\left[I_0(\text{NTU }\sqrt{C^*}) + \sqrt{\frac{1}{C^*}}\,I_1(\text{NTU }\sqrt{C^*})\right.$$
$$\left. + (1 - C^*)\sum_{n=2}^{\infty}(C^*)^{-n/2}I_n(\text{NTU }\sqrt{C^*})\right]$$

$$g = 1 + \sum_{j=2}^{\infty}\left(\frac{1 - e^{-\text{NTU}/2}}{\text{NTU}/2}\right)^{j-1}V_j\left(\frac{\text{NTU}}{2}, \frac{C^*\,\text{NTU}}{2}\right)$$

$$\bar{g} = 1 + \sum_{j=2}^{\infty}\left(\frac{1 - e^{-C^*\text{NTU}/2}}{C^*\,\text{NTU}/2}\right)^{j-1}V_j\left(\frac{C^*\,\text{NTU}}{2}, \frac{\text{NTU}}{2}\right)$$

$$\mu_{1/2} = \frac{2}{C^*\,\text{NTU}}\sum_{m=0}^{\infty}\sum_{n=0}^{\infty}\frac{(-1)^m(n + m)!}{n!m!}V_{m+2}\left(\frac{C^*\,\text{NTU}}{2}, \frac{\text{NTU}}{2}\right)V_{n+2}\left(\frac{\text{NTU}}{2}, \frac{C^*\,\text{NTU}}{2}\right)$$

$$\bar{\mu}_{1/2} = \frac{2}{\text{NTU}}\sum_{m=0}^{\infty}\sum_{n=0}^{\infty}\frac{(-1)^m(n + m)!}{n!m!}V_{m+2}\left(\frac{\text{NTU}}{2}, \frac{C^*\,\text{NTU}}{2}\right)V_{n+2}\left(\frac{C^*\,\text{NTU}}{2}, \frac{\text{NTU}}{2}\right)$$

where $V_0(\xi, \eta) = e^{-(\xi+\eta)}I_0(2\sqrt{\xi\eta})$

$$V_m(\xi, \eta) = e^{-(\xi+\eta)}\sum_{n=m-1}^{\infty}\binom{n}{m-1}\left(\frac{\eta}{\xi}\right)^{n/2}I_n(2\sqrt{\xi\eta}) \qquad m \geq 1$$

Compound multipass cross-flow arrangements have been analyzed by Worsøe-Schmidt and Høggard Knudsen [22]. They presented the complicated ϵ-NTU expressions by polynomial approximations as follows:

$$\epsilon = \mathbf{Z} + B_2\mathbf{Z}^2 + \cdots + B_m\mathbf{Z}^m \tag{19}$$

where

$$B_2 = b_{20} + b_{21}C^* + b_{22}C^{*2} + b_{23}C^{*3} + b_{24}C^{*4}$$
$$B_3 = b_{30} + b_{31}C^* + b_{32}C^{*2} + b_{33}C^{*3} + b_{34}C^{*4}$$
$$B_4 = b_{40} + b_{41}C^* + b_{42}C^{*2} + b_{43}C^{*3} + b_{44}C^{*4} \tag{20}$$
$$B_5 = b_{50} + b_{51}C^* + b_{52}C^{*2} + b_{53}C^{*3} + b_{54}C^{*4}$$

and

$$\mathbf{Z} = \frac{\text{NTU}}{1 + \text{NTU}} \tag{21}$$

Hence each pass is considered to have the same NTU_p, and total NTU is the sum of the NTU_p's of all passes. The coefficients b_{ij} ($i = 2, 3, 4, 5$, and $j = 0, 1, 2, 3$, and 4) for

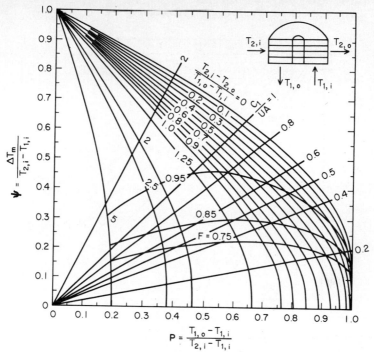

FIG. 17 ψ-P chart for a two-pass overall cross-counterflow exchanger. Fluid 1 mixed throughout, fluid 2 unmixed throughout, inverted order between passes.

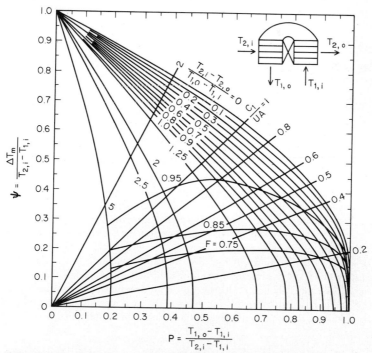

FIG. 18 ψ-P chart for a two-pass overall cross-counterflow exchanger. Fluid 1 mixed throughout, fluid 2 unmixed throughout, identical order between passes.

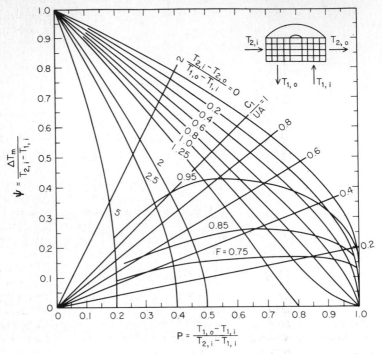

FIG. 19 ψ-P chart for a two-pass overall cross-counterflow exchanger. Both fluids unmixed within passes. Fluid 1 mixed between passes, fluid 2 unmixed between passes, with inverted order.

seven compound cross-flow arrangements are presented in Table 6. Each compound flow arrangement is designated by u, m-C or u, m-P, where u denotes number of unmixed passes, m stands for number of mixed passes, and C and P indicate overall counterflow and parallel-flow arrangements, respectively.

b. Counterflow Exchanger with Cross-Flow Headers

Since it is difficult to separate two fluids at each end in a pure counterflow plate-fin recuperator, special inlet and outlet headers are sometimes designed such that the fins in this section turn the flow and separate the fluids as in a cross-flow heat exchanger. One such example is shown in Fig. 24a. The ϵ-NTU results for such a counterflow core with two cross-flow headers, one on each end, are tabulated by Kays et al. [23]. They drew one important conclusion: if the composite heat exchanger is treated as a counterflow heat exchanger in series with two cross-flow heat exchangers (in the header area) with fluids mixed between passes as shown in Fig. 24b, the approximate result is very close to the exact solution over a wide range of operating conditions. This means that the composite exchanger can be analyzed as a multipass unit with different NTUs for each pass by the method of Domingos [24]. Since this arrangement is an important one for compact counterflow exchangers, the overall effectiveness as a function of individual pass effectiveness ϵ_j and C^* is

$$\epsilon = \frac{X - 1}{X - C^*} \qquad (22)$$

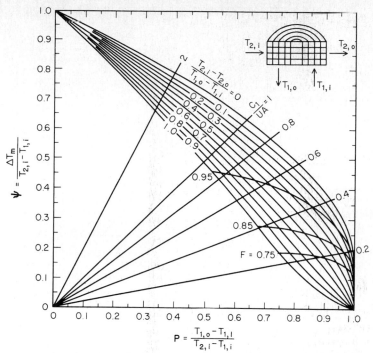

FIG. 20 ψ-P chart for a two-pass overall cross-counterflow exchanger. Both fluids unmixed throughout, both in inverted order between passes.

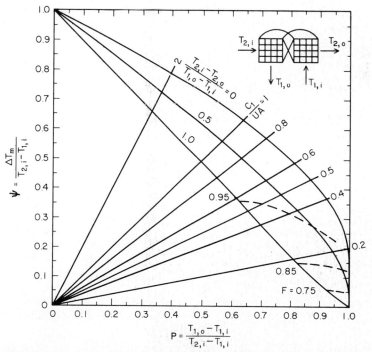

FIG. 21 ψ-P chart for a two-pass overall cross-counterflow exchanger. Both fluids unmixed throughout, both in identical order between passes.

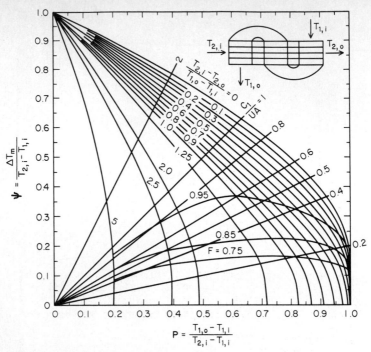

FIG. 22 ψ-P chart for a three-pass overall cross-counterflow exchanger. Fluid 1 mixed throughout, fluid 2 unmixed throughout, with inverted order between passes.

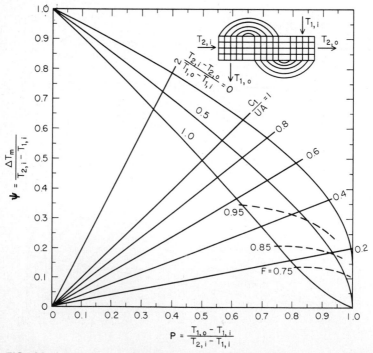

FIG. 23 ψ-P chart for a three-pass overall cross-counterflow exchanger. Both fluids unmixed throughout and both in inverted order between passes.

TABLE 6 Coefficients b_{ij} of Eq. (20) for ϵ-NTU Expressions for Some Compound Multipass Cross-Flow Arrangements [22]

Flow arrangement†		$C_{mixed} = C_{min}$				$C_{mixed} = C_{max}$			
		$i=2$	3	4	5	$i=2$	3	4	5
1,2-C	$j=0$	0.19794	1.74722	−2.24308	0.24147	0.19912	1.74102	−2.23340	0.23709
	1	1.26186	−10.74641	18.17005	−8.66721	0.83538	−8.18038	13.46378	−6.03456
	2	−3.46457	21.80493	−41.23901	23.68910	−1.49958	9.29936	−16.60739	8.41024
	3	3.48005	−22.52697	44.99923	−28.02999	0.54599	−3.58490	7.02318	−3.90223
	4	−1.41921	9.28865	−18.85441	12.06158	−0.02083	0.26731	−0.76512	0.55486
1,2-P	$j=0$	0.55389	−0.36261	1.62067	−1.98460	0.19935	1.73987	−2.23184	0.23659
	1	0.39379	−5.63183	11.92904	−8.55585	−1.36947	5.29665	−13.24089	8.37732
	2	−3.41744	23.65947	−50.05650	35.17258	1.40359	−8.62311	17.22740	−9.35529
	3	3.58623	−24.51218	52.10017	−35.49589	−0.82200	4.92214	−8.92216	4.51567
	4	−1.24898	8.48664	−17.90895	12.00492	0.19623	−1.14200	1.97953	−0.96287
2,1-C	$j=0$	0.19622	1.75814	−2.26444	0.25463	0.19910	1.74117	−2.23376	0.23732
	1	1.28998	−10.99955	18.82052	−9.13452	0.83387	−8.17810	13.46870	−6.04132
	2	−3.35198	21.35322	−40.89075	23.82678	−1.24416	7.89864	−14.32282	7.39032
	3	3.24839	−21.30406	43.09425	−27.19047	0.19261	−1.68746	4.02622	−2.61667
	4	−1.29292	8.55714	−17.55109	11.34538	0.11083	−0.42509	0.30024	0.11384
2,1-P	$j=0$	0.18772	1.80958	−2.36144	0.31270	0.19929	1.74014	−2.23222	0.23672
	1	−0.71259	1.23735	−5.50913	3.82742	−1.36743	5.28616	−13.22586	8.37205
	2	0.88756	−5.70134	12.29798	−6.96007	1.62916	−9.91114	19.42106	−10.39301
	3	−1.74014	11.06687	−21.81180	13.21668	−1.00751	5.94617	−10.58126	5.25525
	4	1.08665	−6.81873	13.27086	−8.12750	0.24707	−1.41561	2.40755	−1.14432
2,2-C	$j=0$	0.20297	1.71836	−2.19254	0.21411	0.19963	1.73729	−2.22630	0.23322
	1	−0.35897	−0.44241	−2.62602	3.75510	−0.22798	−1.20335	−1.26544	2.99206
	2	−1.68148	9.53506	−15.22451	6.99944	−0.45475	2.20387	−1.63964	−0.94730
	3	3.46362	−20.37194	36.60118	−20.20814	0.68497	−3.86902	6.24228	−2.59389
	4	−1.71731	10.14770	−18.42533	10.39007	−0.28917	1.69543	−2.93725	1.44318

†Flow arrangements are designated by u, m-C or u, m-P, where u denotes number of unmixed passes, m stands for number of mixed passes, and C and P indicate overall counterflow or parallel flow.

TABLE 6 Coefficients b_{ij} of Eq. (20) for ϵ-NTU Expressions for Some Compound Multipass Cross-Flow Arrangements [22] (Continued)

Flow arrangement†		$C_{mixed} = C_{min}$					$C_{mixed} = C_{max}$			
		$i = 2$	3	4	5		$i = 2$	3	4	5
2, 2-P	$j = 0$	0.38617	0.66067	−0.30749	−0.84199	$j = 0$	0.38429	0.67231	−0.32951	−0.82894
	1	1.38804	−12.15623	21.57513	−12.82830	1	−0.21759	−2.17779	2.36006	−1.14057
	2	−4.24750	26.47152	−50.06994	30.35643	2	−0.46896	3.15290	−5.53104	3.52742
	3	3.36415	−20.66071	39.11527	−23.31902	3	0.24889	−1.57648	2.97677	−1.77165
	4	−0.98824	6.02894	−11.34283	6.69126	4	−0.04255	0.26434	−0.48940	0.26223
1, 4-C	$j = 0$	0.19416	1.77187	−2.29297	0.27337	$j = 0$	0.19741	1.75234	−2.25672	0.25219
	1	1.09005	−10.05298	17.85796	−8.98111	1	0.96472	−9.29464	16.44208	−8.15328
	2	−1.26798	9.31597	−19.86277	12.56781	2	−0.88150	6.88757	−15.13055	9.65598
	3	−0.71901	2.55974	−1.00766	−1.78078	3	−0.85148	3.53845	−3.20980	−0.23589
	4	0.83562	−4.43098	6.90487	−3.10445	4	0.69934	−3.69428	5.70701	−2.51679
1, 4-P	$j = 0$	0.23691	1.51979	−1.83317	0.00913	$j = 0$	0.19907	1.74072	−2.23203	0.23579
	1	1.78387	−12.98121	19.23133	−9.93626	1	−1.43890	6.01468	−15.41725	9.99133
	2	−8.78111	51.13204	−91.48753	53.58550	2	2.58274	−15.73705	30.23803	−16.24355
	3	11.79221	−69.51774	127.37140	−74.59910	3	−1.91830	11.09587	−19.22955	9.38980
	4	−5.11137	30.22377	−55.49195	32.35310	4	0.51388	−2.84317	4.62247	−2.07043
4, 1-C	$j = 0$	0.19551	1.76425	−2.27962	0.26600	$j = 0$	0.19770	1.75098	−2.25474	0.25130
	1	1.02662	−9.70653	17.27275	−8.66542	1	0.94461	−9.18898	16.26710	−8.06181
	2	−0.71871	6.15233	−14.21910	9.40485	2	−0.66652	5.66148	−12.94770	8.44063
	3	−1.74159	8.56018	−11.94207	4.49907	3	−1.32643	6.34403	−8.39888	2.80115
	4	1.36245	−7.54606	12.63196	−6.42847	4	0.97069	−5.31833	8.75233	−4.32881

4, 1-P

j								
0	0.19995	1.73589	-2.22363	0.23119	0.19910	1.74081	-2.23238	0.23603
1	-1.45847	6.10892	-15.54943	10.05317	-1.43316	5.98348	-15.36542	9.96511
2	2.77239	-16.80387	32.09742	-17.28278	2.57413	-15.72578	30.28889	-16.30764
3	-2.29618	13.27627	-23.12814	11.60578	-2.00140	11.65023	-20.34860	10.08532
4	0.68649	-3.83879	6.39961	-3.07568	0.56566	-3.17130	5.25515	-2.44818

2, 4-C

j								
0	0.20060	1.73271	-2.21933	0.22983	0.19994	1.73656	-2.22620	0.23370
1	-1.38623	5.37358	-12.76337	8.96690	-1.36052	5.22709	-12.50539	8.82274
2	0.23642	-1.56458	4.51629	-3.74358	0.67045	-4.09321	9.06587	-6.31595
3	1.07711	-6.20501	10.58069	-5.25026	0.22533	-1.25543	1.69810	-0.23777
4	-0.58295	3.37942	-5.90859	3.10780	-0.18956	1.09718	-1.81963	0.80404

2, 4-P

j								
0	0.19959	1.73861	-2.22980	0.23561	0.19993	1.73657	-2.22611	0.23352
1	0.76953	-7.81419	12.25504	-6.25435	0.75174	-7.70808	12.05983	-6.14274
2	-0.94342	5.94377	-10.36635	5.77767	-1.45078	8.88419	-15.63241	8.74778
3	-0.19082	0.80522	-1.21322	0.77926	0.75289	-4.66620	8.58807	-4.74640
4	0.25641	-1.38868	2.41335	-1.40824	-0.16293	1.04074	-1.93483	1.04031

4, 2-C

j								
0	0.20061	1.73251	-2.21880	0.22948	0.19892	1.74063	-2.23156	0.23602
1	-1.39400	5.41255	-12.81294	8.97753	-1.36492	5.24741	-12.53613	8.83780
2	0.15123	-1.06831	3.58853	-3.18397	0.69403	-4.18666	9.18280	-6.36146
3	1.28135	-7.36148	12.65137	-6.43316	0.21271	-1.24235	1.74353	-0.29121
4	-0.69183	3.98876	-6.98368	3.71129	-0.19239	1.13963	-1.92485	0.87490

4, 2-P

j								
0	0.20088	1.73091	-2.21583	0.22771	0.19886	1.74116	-2.23256	0.23649
1	2.16360	-15.92700	26.85743	-14.53371	0.75737	-7.73843	12.11148	-6.17076
2	-4.55521	26.85864	-47.82877	26.90752	-1.48874	9.10876	-16.04068	8.97981
3	3.21353	-18.81817	33.79562	-18.88966	0.88586	-5.45188	10.02729	-5.57830
4	-0.88920	5.18799	-9.28126	5.14255	-0.21876	1.36721	-2.52875	1.38191

†Flow arrangements are designated by u, m-C or u, m-P, where u denotes number of unmixed passes, m stands for number of mixed passes, and C and P indicate overall counterflow or parallel flow.

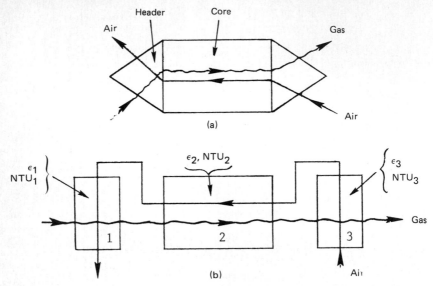

FIG. 24 Counterflow exchanger with crossflow headers: (a) one arrangement; (b) model representation for analysis.

where

$$X = \left(\frac{1 - C^*\epsilon_1}{1 - \epsilon_1}\right)\left(\frac{1 - C^*\epsilon_2}{1 - \epsilon_2}\right)\left(\frac{1 - C^*\epsilon_3}{1 - \epsilon_3}\right) \tag{23}$$

Here ϵ_1, ϵ_2, and ϵ_3 are individual pass effectivenesses and are dependent upon C^* and individual pass NTU_1, NTU_2, and NTU_3, respectively. Equation (22) can be derived straightforwardly by the operating-line–equilibrium-line technique of Kays and London [4].

3. Plate Heat Exchangers

In a plate exchanger, there exist a large number of feasible multipass flow arrangements. Some of these are shown in Fig. 25. A plate exchanger is designated by the number of passes each stream makes in the exchanger. For example, 2 pass–1 pass plate exchanger means fluid 1 makes two passes and fluid 2 makes one pass in the exchanger. In each pass, there can be any equal or unequal number of thermal plates.† Essentially these passes are combinations of parallel-flow and counterflow arrangements with heat transfer taking place in adjacent channels. These arrangements can be obtained simply by properly gasketing around the ports in the plates. The results for single and multipass plate exchangers are available in terms of ϵ-NTU or F factors. The time permitted to complete this chapter is not sufficient to recompute and present all the results in terms of one method only. We will describe them as they appear in the literature.

Jackson and Troupe [25] presented the ϵ-NTU results for a 1 pass–1 pass exchanger with counterflow arrangement (Fig. 26) and parallel-flow arrangement (Fig. 27). Since

---

†In the plate exchanger, the two outer plates serve as end plates and ideally do not transfer heat, while the remaining plates, known as thermal plates, transfer heat.

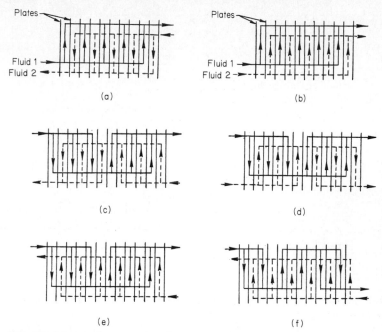

FIG. 25 Some single-pass and multipass arrangements of plate exchangers: (*a*) 1 pass–1 pass counterflow; (*b*) 1 pass–1 pass parallel flow; (*c*) 2 pass–2 pass parallel flow; (*d*) 2 pass–2 pass counter flow; (*e*) 2 pass–1 pass; (*f*) 3 pass–1 pass.

the fluid streams near the end plates transfer heat only on one side, the number of plates influences the ϵ-NTU results. In Fig. 26, the ϵ-NTU results are presented by 1 pass–1 pass counterflow arrangements with the number of thermal plates given as 2, 3, 4, 5, 6, and ∞. In Fig. 27, the ϵ-NTU results are presented for 1 pass–1 pass parallel-flow arrangements with the number of thermal plates as 2, 3, 4, 5, 6, 7, and ∞.

Jackson and Troupe [25] also analyzed the 2 pass–2 pass arrangements with overall counterflow and overall parallel flow, and the 4 pass–4 pass arrangement with overall counterflow. In each pass, only one thermal plate was considered. The reader should refer to Ref. 25 for the ϵ-NTU results.

Foote [26] analyzed the 2 pass–1 pass, 3 pass–1 pass, and 4 pass–1 pass arrangements for a finite and an infinite number of thermal plates (no end effects) in each pass of the fluid having multiple passes. She presented the log-mean temperature difference correction factor F in terms of $\mathbf{P}/(1 - \mathbf{P})$ and $\mathbf{R}$ as shown in Figs. 28 to 31. Note that $\mathbf{P}$ is the temperature effectiveness on n-pass side while $\mathbf{R}$ is defined specifically as $\mathbf{R} = (C$ on one-pass side$)/(C$ on n-pass side$)$. It can be clearly seen that the influence of the number of plates in each pass does not depend only on the number of plates but also depends on $\mathbf{R}$.

E. HEAT EXCHANGER ARRAYS

The overall effectiveness or performance of identical or nonidentical heat exchangers arranged in overall counterflow, parallel flow, and mixed flows (that incorporate combinations of counterflow, parallel-flow, and cross-flow heat exchangers) can be analyzed by the powerful method proposed by Domingos [24]. In this method, the analysis is made

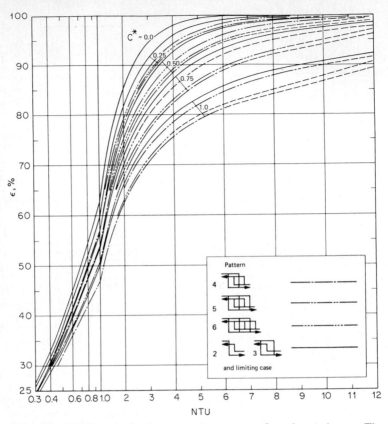

FIG. 26 ϵ-NTU results for the 1 pass–1 pass counterflow plate exchanger. The influence of the number of thermal plates is shown.

by the ϵ-NTU method by matrix algebra after introducing the concepts of the static transfer matrix and the thermal transfer factor of a heat exchanger. The idealizations made in the analysis are: (1) each fluid is completely mixed before the inlet and after the outlet of each exchanger, and (2) the fluid capacity rate ratio C^* or its specified fraction is constant. As far as an individual heat exchanger in an array is concerned, it can have any flow arrangement (single-pass or multipass) described in Sec. C and D, but its effectiveness and heat capacity rate ratios need to be known for an array analysis.

For an array of N exchangers having an overall counterflow arrangement as shown in Fig. 32a, the overall effectiveness ϵ_o is related to the individual effectiveness ϵ_j of each exchanger as†

$$\epsilon_o = \frac{1 - \prod_{j=1}^{N} \dfrac{1 - \epsilon_j C^*}{1 - \epsilon_j}}{C^* - \prod_{j=1}^{N} \dfrac{1 - \epsilon_j C^*}{1 - \epsilon_j}} \quad \xrightarrow{C^* = 1} \quad \frac{1 - \sum_{j=1}^{N} \dfrac{\epsilon_j}{1 - \epsilon_j}}{1 + \sum_{j=1}^{N} \dfrac{\epsilon_j}{1 - \epsilon_j}} \qquad (24)$$

†Π refers to the product of terms with index j. $\prod_{j=1}^{N} a_j$ means $a_1 a_2 \ldots a_N$.

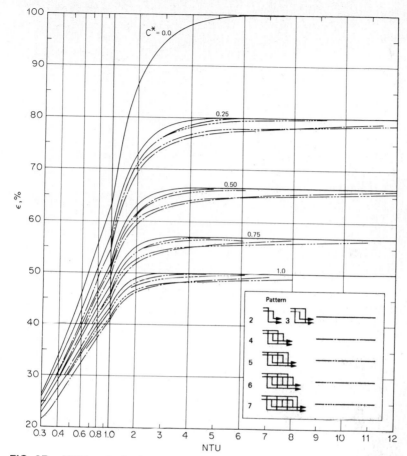

FIG. 27 ϵ-NTU results for the 1 pass–1 pass parallel-flow plate exchanger. The influence of the number of thermal plates is shown.

Here it must be emphasized that individual exchangers can have any flow arrangements and the proper ϵ_j should be used in Eq. (24). If all the exchangers in the array are identical, then Eq. (24) reduces to that presented in Table 3B for the N-pass exchanger with overall counterflow arrangement.

For an overall parallel-flow arrangement of N exchangers (Fig. 32b), the overall effectiveness ϵ_o is related to the individual effectiveness ϵ_j of each exchanger as

$$\epsilon_o = \frac{1 - \prod_{j=1}^{N} [1 - (1 + C^*)\epsilon_j]}{1 + C^*} \tag{25}$$

For all identical exchangers in the array, Eq. (25) reduces to that presented in Table 3B for the N-pass exchanger with overall parallel-flow arrangement.

If one stream is in series and the other in parallel and all exchangers are identical, the results presented in Table 3B for N-pass exchangers are applicable with each pass treated as an individual exchanger. For mixed-flow arrangements of an array of exchangers, refer to the method and results presented in Ref. 24.

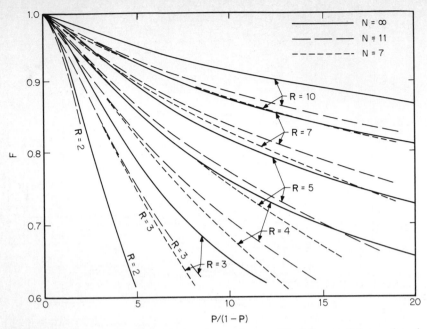

FIG. 28 F factors for the 2 pass–1 pass plate exchanger. The influence of the number of thermal plates on F is shown.

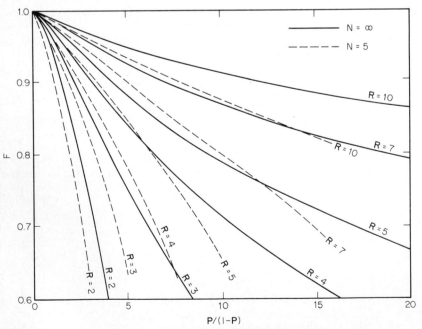

FIG. 29 F factors for the 3 pass–1 pass plate exchanger with parallel flow at inlet (Fig. 25f with the direction of fluid 1 reversed). The influence of the number of thermal plates on F is shown.

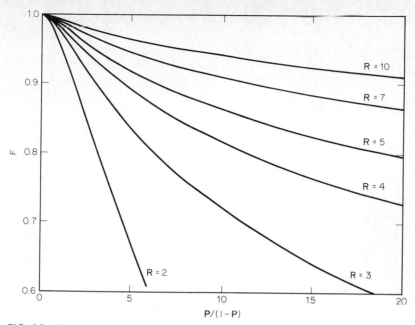

FIG. 30 *F* factors for the 3 pass–1 pass plate exchanger with counterflow at inlet (Fig. 25*f*) and infinite number of plates.

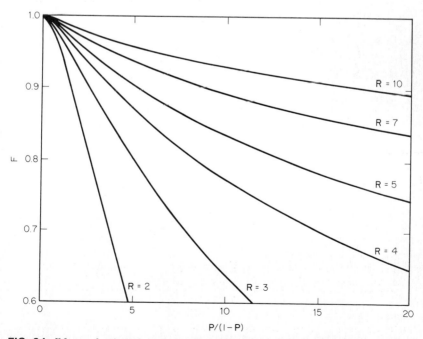

FIG. 31 *F* factors for the 4 pass–1 pass plate exchanger with infinite number of plates.

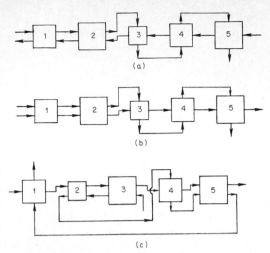

FIG. 32 Heat exchanger arrays: (*a*) overall counterflow arrangement; (*b*) overall parallel-flow exchanger; and (*c*) mixed flow arrangement.

F. EXTENSIONS OF THE BASIC THEORY

Any of the four methods described in Sec. C can be used to rate or design exchangers, and will provide identical results for the third dimensionless group when the other two are specified. From operational and practical viewpoints, the LMTD and ψ-P methods have some advantages in the design of power and process industry exchangers, while the ϵ-NTU method has some advantages and is traditionally used by the automotive, aircraft, industrial, air conditioning, and cryogenics industries. The results presented in Sec. C, Tables 3 to 6, and Figs. 4 to 31 are based on the idealizations presented in Sec. B. Since a number of these idealizations are violated in reality, the influence of relaxing idealizations 9 to 13 (p. 4-8) will now be discussed.

1. Variable Overall Coefficients

In all of the four methods of Table 1, it is idealized that the overall heat transfer coefficient U is constant throughout the exchanger. In reality, U varies with location and is strongly dependent on the flow Reynolds number, heat transfer surface geometries, fluid physical properties, and temperature difference. Particularly for viscous liquid exchangers, a tenfold variation in h or U is possible when the flow pattern encompasses laminar, transition, and turbulent regions on one side. U also varies significantly for evaporators and condensers. Methods to account for specific variations in U are summarized next for counterflow, cross-flow, and multipass shell-and-tube exchangers.

a. Counterflow Heat Exchangers

Consider a counterflow 1-1 exchanger having (1) U varying linearly with a temperature of either fluid, or (2) U varying linearly with local T_x. Colburn [27] has shown that for either case,

$$\frac{q}{A} = U_m \, \Delta T_{lm} = \frac{U_2 \, \Delta T_1 - U_1 \, \Delta T_2}{\ln[(U_2 \, \Delta T_1)/(U_1 \, \Delta T_2)]} \tag{26}$$

where U_1 and U_2 are the values on the ends of the exchanger having ΔT_1 and ΔT_2 temperature differences, respectively.

When $1/U$ varies linearly with T_x, Butterworth [28] has shown that

$$\frac{q}{A} = U_m \, \Delta T_{lm} \tag{27}$$

where

$$\frac{1}{U_m} = \frac{1}{U_1} \frac{\Delta T_{lm} - \Delta T_2}{\Delta T_1 - \Delta T_2} + \frac{1}{U_2} \frac{\Delta T_1 - \Delta T_{lm}}{\Delta T_1 - \Delta T_2} \tag{28}$$

For very viscous liquids and partial condensation, the Colburn equation (Eq. (26)) is a better representation. For some condenser applications, the Butterworth equation (Eq. (28)) may be appropriate. If a question of which of Eqs. (26) and (28) should be used is not resolved, divide the exchanger into a number of segments, and treat U as either a constant or an arithmetic mean for each segment.

b. Cross-Flow Exchangers

When U varies linearly with T_c, T_h, or T_x, Sieder and Tate [29] proposed a simple approximate procedure by modifying Colburn's solution, Eq. (26), by the log-mean temperature correction factor:

$$q = U_m A F \, \Delta T_{lm} \tag{29}$$

where U_m is obtained from Eq. (26) treating the exchanger as counterflow. According to Gardner and Taborek [30], this approximation provides results that are within 10 percent accuracy for all but extreme temperature approach cases. The latter are avoided since they result in low values of F.

Roetzel [31] considered several cases of variable coefficients (laminar flow and temperature dependency) and developed simplified methods for calculating the mean overall coefficients. The paper should be consulted for the details since the method is beyond the scope of this chapter.

For arbitrary variations in U, the heat exchanger is divided into many segments. A constant and same or different value of U is then assigned to each segment, and the solution is carried out by a finite difference analysis.

c. Multipass Shell-and-Tube Exchangers

The analytical solutions to the variable U problem are very complicated for multipass exchangers. Gardner and Taborek [30] recommend the approximate Sieder-Tate method for linear variations in U. In this method, U_m is given by Eq. (26) and q by Eq. (29) with the appropriate value of F for multipass exchangers. This approximate method provides results accurate within about ± 10 percent for all but extreme temperature approach cases.

If a greater accuracy is desired for the 1-2 exchanger, Gardner [32] provides the results for U linearly varying with the shell fluid temperature T_s and the shell thermal resistance controlling (low h_{shell}). Results in terms of ψ-**P** charts for several ratios of U_1/U_2 are presented in Fig. 33. Here U_1 and U_2 are evaluated at the shell inlet and outlet, respectively. The values read from these charts are used with the arithmetic average U to compute q. The results of Fig. 33 can be used for any even tube passes with only a 1 to 2 percent error in the design ranges. Note that ψ in Fig. 33 does not approach unity as it does for constant U. Ramalho and Tiller [33] present **P**-NTU$_t$ results for $U_2'/U_1' = 1$, 1.2, 1.4, 1.6, 1.8, and 2 for the case in which the tube-side coefficient varies linearly with T_t and is controlling (low h_t). Here U_1' and U_2' are evaluated at the tube inlet and outlet, respectively.

For a nonlinear variation of U, Gardner [32] suggests an analytical expression, but no specific results are provided. However, Kao [34] has recently obtained numerical results for several cases of 1-2N exchangers with tube or shell fluid controlling and tube inlet pass either counter or parallel to the shell fluid flow.

2. Unequal Pass Areas

The case of unequal pass area for multipass exchangers has been analyzed only for the 1-2 exchanger. Gardner's results [35] are presented in Fig. 34 for the case of different counterflow and parallel-flow tube pass areas. Actually, the parameter in Fig. 34 is $U_c A_c / U_p A_p$, where the subscripts c and p denote the counterflow and parallel-flow tube passes. As the value of $U_c A_c / U_p A_p$ increases (from Fig. 34a to Fig. 34d), the value of **P** increases for specified ψ and **R**. Note that **P** and **R** in Fig. 34 are strictly based on C_t (see Table 1), and their definitions cannot be interchanged from C_t to C_s as noted in the footnote of Table 1 since the results of Fig. 34 are then no more applicable.

Since $U_c A_c / U_p A_p$ is a parameter of Fig. 34, it means not only that the influence of unequal tube pass areas can be accounted for, but also the influence of variable tube-side heat transfer coefficient can be evaluated. An interesting result is that for equal tube pass areas, one would like to have a higher value of U_c/U_p. This can be achieved by having the shell fluid enter at the fixed end when heating the tube fluid and at the floating end when cooling the tube fluid. This is because higher temperatures generally mean higher coefficients. Thus it should be clear that F or **P** would be somewhat different depending upon whether the shell fluid entered at the fixed or the floating end.

Results similar to Fig. 34 but for the 1-4 exchanger indicate that the **P** values for a specified ψ and **R** are higher than those for the 1-2 exchanger with $U_c A_c / U_p A_p = 1$ [35]. For the 1-4 exchanger having $U_c A_c / U_p A_p > 1$, the use of the 1-2 exchanger will be conservative.

3. Shell-Side Baffle Passes

Idealization 11 following Eq. (13) indicates that the number of baffles is large so that the shell fluid temperature over any cross section normal to the shell axis may be considered uniform. This means, for example, the F-**P** and ψ-**P** charts of Figs. 8 and 11 are derived for $N_b \to \infty$, where N_b is the number of baffles. Gardner and Taborek [30] investigated the validity of this idealization. They concluded for the 1-2 exchanger that four or more baffles are needed before the F-**P** or ψ-**P** curves approach asymptotically the case of $N_b \to \infty$ (Figs. 8 and 11). For $N_b < 4$, the curves bracket the standard curve ($N_b \to \infty$), but oscillate from one side to the other as the curves for increasing N_b approach the standard curve. Also, there is a significant difference depending upon whether the shell fluid first encounters the tube inlet or outlet pass. For a 1-1 exchanger (plate-baffled counterflow), even more baffles ($N_b > 10$) are needed before an F of unity is reached.

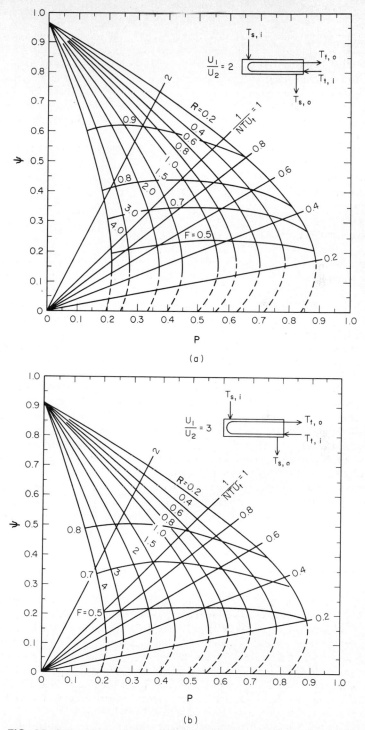

FIG. 33 ψ-P charts for 1-2 (or 1-2N) exchangers having a linearly varying overall coefficient U dependent upon the shell fluid.

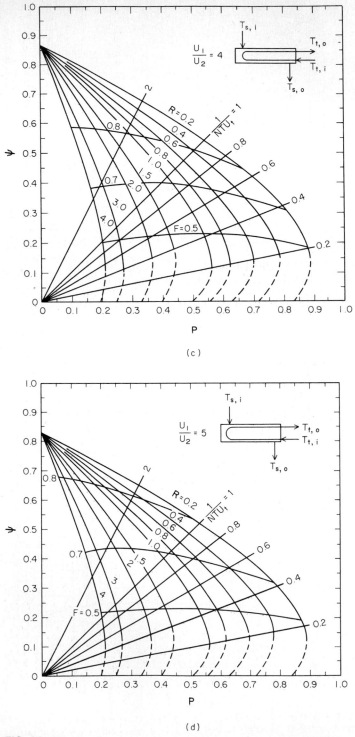

FIG. 33 (*Continued*)

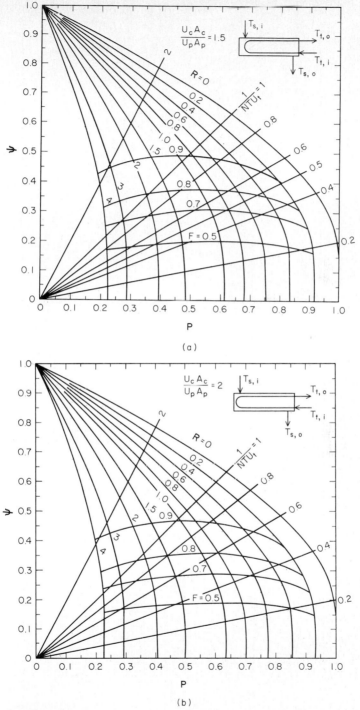

FIG. 34 ψ-P charts for 1-2 (or 1-2N) exchangers having a linearly varying overall coefficient dependent upon the tube fluid or unequal areas in the tube passes.

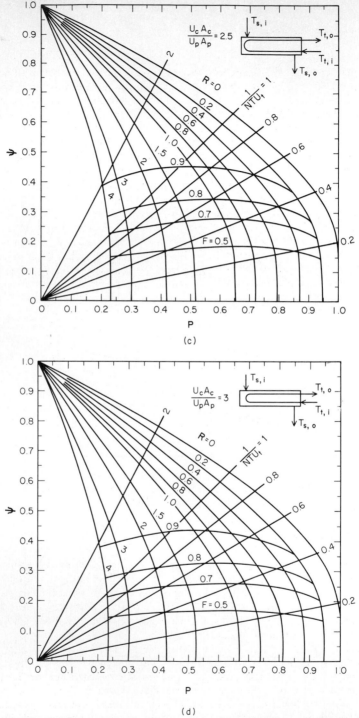

FIG. 34 (*Continued*)

4. Shell Fluid Bypassing

Various clearances are required for the construction of the plate-baffled shell-and-tube exchanger. The shell fluid leaks or bypasses through these clearances with or without flowing past the tubes. Three clearances associated with a plate baffle are tube-to-baffle hole clearance, bundle-to-shell clearance, and baffle-to-shell clearance. Various leakage streams associated with these clearances are identified in Part 2 of this chapter.

Gardner and Taborek [30] have summarized the effect of various bypass and leakage streams on the mean temperature difference. As shown in Fig. 35, the baffle-to-shell leakage stream E experiences practically no heat transfer; the bundle-to-shell bypass stream C indicates some heat transfer, and the cross-flow stream B shows a large temperature change and a possible pinch or temperature cross ($T_{B,o} > T_{t,o}$). The mixed mean outlet temperature $T_{s,o}$ is much lower than the B stream outlet temperature $T_{B,o}$, thus resulting in an indicated temperature difference larger than is actually present; the overall exchanger performance will be lower than the design value. Since the bypass and leakage streams can exceed 30 percent of the total flow, the effect on the mean temperature difference can be very large especially for close temperature approaches. The Bell-Delaware method of designing shell-and-tube exchangers that includes the effect of leakage and bypass streams is described in Part 2 of this chapter.

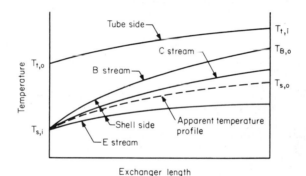

FIG. 35 Effect of bypass and leakage streams on the temperature profile of a shell-and-tube exchanger.

5. Longitudinal Wall Heat Conduction Effects

All four methods discussed in the preceding sections are based on the idealizations of zero longitudinal heat conduction both in the wall and in the fluid in the flow direction. Longitudinal heat conduction in the fluid is negligible for Pe $>$ 10 and $x^* \geq 0.005$ [36], where Pe $=$ Re Pr and $x^* = x/(D_h \text{ Pe})$. For most heat exchangers, except for liquid metal exchangers, Pe and x^* are higher than the above indicated values, and hence longitudinal heat conduction in the fluid is negligible.

Longitudinal heat conduction in the wall reduces the exchanger effectiveness and thus reduces the overall heat transfer performance. The reduction in the exchanger performance could be important and thus significant for heat exchangers designed for effectivenesses greater than about 75 percent. This would be the case for counterflow and single-pass cross-flow exchangers. For high-effectiveness multipass exchangers, the exchanger effectiveness per pass is generally low, and thus longitudinal conduction effects for each pass are generally negligible. The influence of longitudinal wall heat conduction on the exchanger effectiveness is dependent mainly upon the longitudinal conduction parameter $\lambda = k_w A_k / L C_{min}$ (where k_w is the wall material thermal conductivity, A_k is the conduction cross-sectional area, and L is the exchanger length for longitudinal conduction). It would also depend upon the convection-conductance ratio $(\eta_o h A)^*$, a ratio of $\eta_o h A$ on the C_{min} to that on the C_{max} side, if it varies significantly from unity. The influence of longitudinal conduction on ϵ is summarized next for counterflow and single-pass cross-flow exchangers.

a. Counterflow Exchanger

Kroger [37] analyzed extensively the influence of longitudinal conduction on counterflow exchanger effectiveness. He found the influence of longitudinal conduction is the largest for $C^* = 1$. For a given C^*, increasing λ decreases ϵ. Longitudinal heat conduction has a significant influence on the counterflow exchanger size (i.e., NTU) for a given ϵ when NTU > 10 and λ > 0.005. Kroger's solution in terms of exchanger ineffectiveness (1 − ϵ) for $C^* = 1$, $0.1 \leq (\eta_o h A)^* \leq 10$, and NTU ≥ 3 is

$$1 - \epsilon = \cfrac{1}{1 + \text{NTU} \cfrac{1 + \lambda[\lambda \, \text{NTU}/(1 + \lambda \, \text{NTU})]^{1/2}}{1 + \lambda \, \text{NTU}}} \tag{30}$$

The results for 1 − ϵ from this equation are presented in Fig. 36.

Kroger [37] also obtained the detailed results for 1 − ϵ for $0.8 \leq C^* \leq 0.98$ for the counterflow exchanger. Illustrative results for $C^* = 0.95$ are presented in Fig. 37. Refer

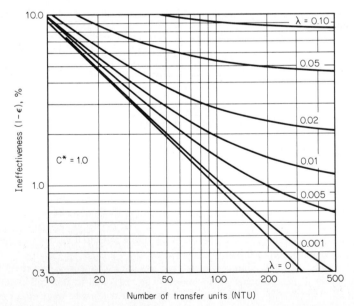

FIG. 36 Counterflow exchanger ineffectiveness as a function of NTU and λ for $C^* = 1.0$.

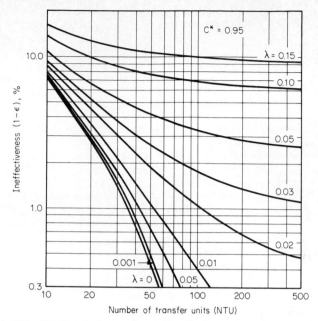

FIG. 37 Counterflow exchanger ineffectiveness as a function of NTU and λ for $C^* = 0.95$.

to Ref. 37 for other values of C^*. Kroger correlated his results for $1 - \epsilon$ for $0.8 \le C^* \le 1$ as follows:

$$1- \epsilon = \frac{1 - C^*}{\Psi \exp(r_1) - C^*} \tag{31}$$

where

$$r_1 = \frac{(1 - C^*)\,\text{NTU}}{1 + \lambda\,\text{NTU}\,C^*} \tag{32}$$

$$\Psi = \frac{1 + \gamma\Psi^*}{1 - \gamma\Psi^*} \qquad \Psi^* = \left(\frac{\alpha}{1 + \alpha}\right)^{1/2} \frac{1 + \gamma}{1/\alpha - \gamma - \gamma^2} \tag{33}$$

$$\gamma = \frac{1 - C^*}{1 + C^*}\frac{1}{1 + \alpha} \qquad \alpha = \lambda\,\text{NTU}\,C^* \tag{34}$$

Note that here γ and α are local dimensionless variables as defined in Eq. (34). The approximate Eq. (31), although derived for $(\eta_o h A)^*/C^* = 1$, could be used for values of this parameter other than unity. For $0.5 < (\eta_o h A)^*/C^* \le 2$, the error introduced in the ineffectiveness is within 0.8 percent and 4.7 percent, for $C^* = 0.95$ and 0.8, respectively.

b. Cross-Flow Exchanger with Both Fluids Unmixed

Chiou [38] analyzed extensively the influence of longitudinal conduction on the effectiveness of a cross-flow exchanger with both fluids unmixed. In this case, the exchanger effectiveness is dependent upon five dimensionless groups: NTU, C^*, λ_x, λ_y, and $(\eta_o hA)^*$, where

$$\lambda_x = \left(\frac{k_w A_k}{LC}\right)_x \qquad \lambda_y = \left(\frac{k_w A_k}{LC}\right)_y \qquad (35)$$

Here $A_{k,c}$ and $A_{k,h}$ for a plate-fin exchanger are†

$$A_{k,c} = (2N_p + 2)\, L_h \delta \qquad A_{k,h} = 2N_p L_c \delta \qquad (35a)$$

His tabular results are presented in Table 7 for $C_x/C_y = 1.0$. In this table, x can be hot or cold side, with y thus cold or hot side. From this table and tables for $C_x/C_y = 0.25$ and 0.5 in Ref. 38, the following qualitative conclusions could be drawn: The loss in ϵ due to longitudinal conduction increases with increasing NTU up to a certain magnitude, and then the loss decreases with further increase in NTU. The loss in ϵ increases with increasing C^* (maximum at $C^* = 1$), decreasing λ_x/λ_y, and decreasing λ_x. The influence of $(\eta_o hA)_x/(\eta_o hA)_y$ on ϵ is relatively small and is of a nonmonotonic nature.

G. TRANSIENT SOLUTIONS

All the results presented in preceding sections are for the steady-state operation of heat exchangers. The transient performance of heat exchangers is also important for system startup and shutdown, for the maintenance-free desired life of the exchanger, and/or for proper control in a system. Solutions for basically two types of transient problems are summarized in this section: (1) batch heating or cooling with a fluid through a coil (heat exchanger) at constant inlet temperature; the batch fluid temperature is continuously rising or lowering with time; and (2) a step change in the inlet temperature or flow rate of one of the two fluids. Frequency response and impulse response are not considered; however, they can be determined from the step response.

1. Batch Heating or Cooling

Some processes are operated in a batch or transient manner. For example, a tank of fluid at an initial temperature $T_{b,i}$ is to be heated (cooled) to a final temperature $T_{b,f}$ in a period of τ hours. The heating (cooling) can be accomplished by means of a coil immersed in the tank or circulating the tank fluid through an external exchanger. The heating or cooling fluid flows continuously through the coil or the exchanger. The problem may be one of sizing the coil or exchanger or determining the heating (cooling) time for a specified exchanger. This is a transient heat transfer problem; the batch temperature is changing with time and is a function of the amount of heating (cooling); hence, the mean temperature difference in the exchanger is also a function of time. Further complications arise when additional heat sources (such as power from an agitator or a chemical reaction) are present, or an additional process fluid flows through the tank system. Solutions to several of these cases are presented in Table 8. For additional solutions, refer to Kern [39].

In deriving the equations in Table 8, all of the idealizations (except 1 and 3) made following Eq. (13) are invoked; and those idealizations for both fluids are now considered only for the exchanger (coil) fluid. In addition, it is idealized for all cases that the batch

---

†Here it is idealized that the number of passages for the hot fluid is N_p and for the cold fluid, $N_p + 1$ to minimize heat losses to ambient.

TABLE 7 Reduction in Cross-Flow Exchanger Effectiveness ($\Delta\epsilon/\epsilon$) Due to Longitudinal Wall Heat Conduction for $C^* = 1$

$\dfrac{\lambda_x}{\lambda_y}$	$\dfrac{(\eta_o hA)_x}{(\eta_o hA)_y}$	NTU	ϵ $\lambda_x = 0$	$\Delta\epsilon/\epsilon$ $\lambda_x =$ 0.005	0.010	0.015	0.020	0.025	0.030	0.0400	0.0600	0.0800	0.1000	0.2000	0.4000
0.5	0.5	1.00	0.4764	0.0032	0.0062	0.0089	0.0115	0.0139	0.0162	0.0204	0.0276	0.0336	0.0387	0.0558	0.0720
		2.00	0.6147	0.0053	0.0103	0.0150	0.0194	0.0236	0.0276	0.0350	0.0481	0.0592	0.0689	0.1026	0.1367
		4.00	0.7231	0.0080	0.0156	0.0227	0.0294	0.0357	0.0418	0.0530	0.0726	0.0892	0.1036	0.1535	0.2036
		6.00	0.7729	0.0099	0.0192	0.0279	0.0360	0.0437	0.0510	0.0644	0.0877	0.1072	0.1239	0.1810	0.2372
		8.00	0.8031	0.0114	0.0220	0.0319	0.0411	0.0497	0.0578	0.0728	0.0984	0.1197	0.1377	0.1988	0.2580
		10.00	0.8238	0.0127	0.0243	0.0351	0.0451	0.0545	0.0633	0.0793	0.1066	0.1290	0.1480	0.2115	0.2724
		50.00	0.9229	0.0246	0.0446	0.0616	0.0764	0.0897	0.1017	0.1229	0.1569	0.1838	0.2057	0.2765	0.3427
		100.00	0.9476	0.0311	0.0543	0.0732	0.0893	0.1034	0.1160	0.1379	0.1729	0.2001	0.2223	0.2942	0.3665
	1.0	1.00	0.4764	0.0029	0.0055	0.0079	0.0102	0.0123	0.0143	0.0180	0.0244	0.0296	0.0341	0.0490	0.0634
		2.00	0.6147	0.0055	0.0097	0.0141	0.0182	0.0221	0.0258	0.0327	0.0449	0.0553	0.0643	0.0959	0.1286
		4.00	0.7231	0.0078	0.0151	0.0220	0.0284	0.0346	0.0404	0.0512	0.0702	0.0863	0.1002	0.1491	0.1991
		6.00	0.7729	0.0097	0.0188	0.0273	0.0353	0.0428	0.0499	0.0630	0.0857	0.1049	0.1213	0.1779	0.2344
		8.00	0.8031	0.0113	0.0217	0.0314	0.0404	0.0489	0.0569	0.0716	0.0968	0.1178	0.1356	0.1965	0.2560
		10.00	0.8238	0.0125	0.0240	0.0347	0.0446	0.0538	0.0624	0.0782	0.1052	0.1274	0.1462	0.2096	0.2708
		50.00	0.9229	0.0245	0.0445	0.0614	0.0763	0.0895	0.1015	0.1226	0.1566	0.1834	0.2053	0.2758	0.3405
		100.00	0.9476	0.0310	0.0543	0.0731	0.0892	0.1033	0.1159	0.1378	0.1727	0.1999	0.2221	0.2933	0.3619
	2.0	1.00	0.4764	0.0027	0.0051	0.0074	0.0095	0.0116	0.0135	0.0170	0.0232	0.0285	0.0330	0.0489	0.0652
		2.00	0.6147	0.0048	0.0092	0.0134	0.0173	0.0211	0.0247	0.0313	0.0432	0.0533	0.0621	0.0938	0.1274
		4.00	0.7231	0.0076	0.0147	0.0213	0.0277	0.0336	0.0393	0.0499	0.0685	0.0844	0.0982	0.1468	0.1971
		6.00	0.7729	0.0095	0.0184	0.0268	0.0346	0.0420	0.0490	0.0619	0.0844	0.1033	0.1196	0.1760	0.2328
		8.00	0.8031	0.0111	0.0214	0.0309	0.0398	0.0482	0.0561	0.0706	0.0956	0.1164	0.1342	0.1949	0.2548
		10.00	0.8238	0.0124	0.0238	0.0343	0.0440	0.0532	0.0617	0.0774	0.1041	0.1262	0.1450	0.2083	0.2698
		50.00	0.9229	0.0245	0.0444	0.0613	0.0761	0.0893	0.1013	0.1223	0.1563	0.1831	0.2050	0.2754	0.3401
		100.00	0.9476	0.0310	0.0542	0.0731	0.0891	0.1032	0.1158	0.1377	0.1725	0.1997	0.2219	0.2928	0.3600

TABLE 7 Reduction in Cross-Flow Exchanger Effectiveness ($\Delta\epsilon/\epsilon$) Due to Longitudinal Wall Heat Conduction for $C^* = 1$ (Continued)

$\frac{\lambda_x}{\lambda_y}$	$\frac{(\eta_o hA)_x}{(\eta_o hA)_y}$	NTU	ϵ $\lambda_x = 0$	$\Delta\epsilon/\epsilon$ $\lambda_x = 0.005$	0.010	0.015	0.020	0.025	0.030	0.0400	0.0600	0.0800	0.1000	0.2000	0.4000
1.0	0.5	1.00	0.4764	0.0020	0.0039	0.0057	0.0074	0.0090	0.0106	0.0136	0.0190	0.0237	0.0280	0.0436	0.0609
		2.00	0.6147	0.0034	0.0067	0.0098	0.0128	0.0157	0.0185	0.0238	0.0336	0.0423	0.0501	0.0803	0.1154
		4.00	0.7231	0.0053	0.0103	0.0152	0.0198	0.0243	0.0287	0.0369	0.0520	0.0653	0.0773	0.1229	0.1753
		6.00	0.7729	0.0066	0.0129	0.0189	0.0246	0.0302	0.0355	0.0456	0.0637	0.0797	0.0940	0.1472	0.2070
		8.00	0.8031	0.0076	0.0148	0.0217	0.0283	0.0346	0.0406	0.0520	0.0723	0.0900	0.1057	0.1634	0.2270
		10.00	0.8238	0.0085	0.0165	0.0241	0.0313	0.0382	0.0448	0.0571	0.0790	0.0979	0.1145	0.1751	0.2410
		50.00	0.9229	0.0170	0.0316	0.0445	0.0562	0.0667	0.0765	0.0940	0.1233	0.1474	0.1677	0.2378	0.3092
		100.00	0.9476	0.0218	0.0395	0.0543	0.0673	0.0789	0.0894	0.1080	0.1385	0.1632	0.1840	0.2546	0.3273
	1.0	1.00	0.4764	0.0020	0.0038	0.0055	0.0072	0.0088	0.0103	0.0132	0.0183	0.0228	0.0268	0.0412	0.0567
		2.00	0.6147	0.0034	0.0066	0.0097	0.0127	0.0156	0.0183	0.0236	0.0331	0.0417	0.0493	0.0786	0.1125
		4.00	0.7231	0.0053	0.0103	0.0152	0.0198	0.0243	0.0286	0.0368	0.0517	0.0650	0.0769	0.1220	0.1742
		6.00	0.7729	0.0066	0.0129	0.0189	0.0246	0.0301	0.0354	0.0455	0.0636	0.0795	0.0936	0.1466	0.2064
		8.00	0.8031	0.0076	0.0149	0.0217	0.0283	0.0346	0.0406	0.0519	0.0722	0.0898	0.1054	0.1630	0.2266
		10.00	0.8238	0.0085	0.0165	0.0241	0.0313	0.0382	0.0448	0.0571	0.0789	0.0978	0.1143	0.1749	0.2407
		50.00	0.9229	0.0170	0.0316	0.0445	0.0562	0.0667	0.0765	0.0940	0.1233	0.1473	0.1677	0.2377	0.3090
		100.00	0.9476	0.0218	0.0395	0.0543	0.0673	0.0789	0.0894	0.1080	0.1385	0.1632	0.1840	0.2546	0.3270
	2.0	1.00	0.4764	0.0020	0.0039	0.0057	0.0074	0.0090	0.0106	0.0136	0.0190	0.0237	0.0280	0.0436	0.0609
		2.00	0.6147	0.0034	0.0067	0.0098	0.0128	0.0157	0.0185	0.0238	0.0336	0.0423	0.0501	0.0803	0.1154
		4.00	0.7231	0.0053	0.0103	0.0152	0.0198	0.0243	0.0287	0.0369	0.0520	0.0653	0.0773	0.1229	0.1753
		6.00	0.7729	0.0066	0.0129	0.0189	0.0246	0.0302	0.0355	0.0456	0.0637	0.0797	0.0940	0.1472	0.2070
		8.00	0.8031	0.0076	0.0148	0.0217	0.0283	0.0346	0.0406	0.0520	0.0723	0.0900	0.1057	0.1634	0.2270
		10.00	0.8238	0.0085	0.0165	0.0241	0.0313	0.0382	0.0448	0.0571	0.0790	0.0979	0.1145	0.1751	0.2410
		50.00	0.9229	0.0170	0.0316	0.0445	0.0562	0.0667	0.0765	0.0940	0.1233	0.1474	0.1677	0.2378	0.3092
		100.00	0.9476	0.0218	0.0395	0.0543	0.0673	0.0789	0.0894	0.1080	0.1385	0.1632	0.1840	0.2546	0.3273

2.0	0.5	1.00	0.4764	0.0014	0.0027	0.0039	0.0051	0.0063	0.0074	0.0095	0.0135	0.0170	0.0203	0.0330	0.0489
		2.00	0.6147	0.0024	0.0048	0.0070	0.0092	0.0113	0.0134	0.0173	0.0247	0.0313	0.0375	0.0621	0.0938
		4.00	0.7231	0.0039	0.0076	0.0112	0.0147	0.0181	0.0213	0.0277	0.0393	0.0499	0.0596	0.0982	0.1468
		6.00	0.7729	0.0049	0.0095	0.0141	0.0184	0.0227	0.0268	0.0346	0.0490	0.0619	0.0736	0.1196	0.1760
		8.00	0.8031	0.0057	0.0111	0.0163	0.0214	0.0262	0.0309	0.0398	0.0561	0.0706	0.0837	0.1342	0.1949
		10.00	0.8238	0.0063	0.0124	0.0182	0.0238	0.0291	0.0343	0.0440	0.0617	0.0774	0.0914	0.1448	0.2083
		50.00	0.9229	0.0130	0.0245	0.0349	0.0444	0.0531	0.0613	0.0761	0.1013	0.1223	0.1404	0.2050	0.2754
		100.00	0.9476	0.0168	0.0310	0.0433	0.0542	0.0641	0.0731	0.0891	0.1158	0.1377	0.1563	0.2219	0.2928
	1.0	1.00	0.4764	0.0015	0.0029	0.0042	0.0055	0.0067	0.0079	0.0102	0.0143	0.0180	0.0213	0.0341	0.0490
		2.00	0.6147	0.0026	0.0050	0.0074	0.0097	0.0119	0.0141	0.0182	0.0258	0.0327	0.0391	0.0643	0.0559
		4.00	0.7231	0.0040	0.0078	0.0115	0.0151	0.0186	0.0220	0.0284	0.0404	0.0512	0.0611	0.1002	0.1491
		6.00	0.7729	0.0050	0.0097	0.0144	0.0188	0.0231	0.0273	0.0353	0.0499	0.0630	0.0749	0.1213	0.1779
		8.00	0.8031	0.0057	0.0113	0.0166	0.0217	0.0266	0.0314	0.0404	0.0569	0.0716	0.0848	0.1356	0.1965
		10.00	0.8238	0.0064	0.0125	0.0184	0.0240	0.0295	0.0347	0.0446	0.0624	0.0782	0.0924	0.1462	0.2096
		50.00	0.9229	0.0130	0.0245	0.0350	0.0445	0.0533	0.0614	0.0763	0.1015	0.1226	0.1407	0.2053	0.2758
		100.00	0.9476	0.0169	0.0310	0.0434	0.0543	0.0641	0.0731	0.0892	0.1159	0.1378	0.1564	0.2221	0.2933
	2.0	1.00	0.4764	0.0016	0.0032	0.0047	0.0062	0.0076	0.0089	0.0115	0.0162	0.0204	0.0241	0.0387	0.0558
		2.00	0.6147	0.0027	0.0053	0.0078	0.0103	0.0127	0.0150	0.0194	0.0276	0.0350	0.0418	0.0689	0.1026
		4.00	0.7231	0.0041	0.0080	0.0119	0.0156	0.0192	0.0227	0.0294	0.0418	0.0530	0.0632	0.1036	0.1535
		6.00	0.7729	0.0051	0.0099	0.0146	0.0192	0.0256	0.0279	0.0360	0.0510	0.0644	0.0766	0.1239	0.1810
		8.00	0.8031	0.0058	0.0114	0.0168	0.0220	0.0270	0.0319	0.0411	0.0578	0.0728	0.0862	0.1377	0.1988
		10.00	0.8238	0.0065	0.0127	0.0186	0.0243	0.0298	0.0351	0.0451	0.0633	0.0793	0.0936	0.1480	0.2115
		50.00	0.9229	0.0130	0.0246	0.0350	0.0446	0.0534	0.0616	0.0764	0.1017	0.1229	0.1410	0.2057	0.2765
		100.00	0.9476	0.0169	0.0311	0.0434	0.0543	0.0642	0.0732	0.0893	0.1160	0.1379	0.1566	0.2223	0.2942

TABLE 8 Transient Solutions for Batch Heating or Cooling

Case	Transient solutions
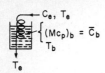	A batch tank with a coil. A constant-temperature fluid (condensing or evaporating) flows through the coil. $$\frac{UA\tau}{\overline{C}_b} = \frac{UA\tau}{(Mc_p)_b} = \ln\frac{T_e - T_{b,i}}{T_e - T_b}$$
	A batch tank with a coil. A single-phase fluid flows through the coil. $$\frac{UA}{C_e} = -\ln\left(1 - \frac{\overline{C}_b}{\tau C_e}\ln\frac{T_{b,i} - T_{e,i}}{T_b - T_{e,i}}\right)$$
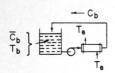	A batch tank with a heat source and a coil. A single-phase fluid flows through the coil. $$\frac{UA}{C_e} = -\ln\left[1 - \frac{\overline{C}_b}{\tau C_e}\ln\frac{b(T_{b,i} - T_{e,i}) - q_a}{b(T_b - T_{e,i}) - q_a}\right]$$ where $b = C_e\left[1 - \exp\left(-\frac{UA}{C_e}\right)\right]$
	A batch tank with an external exchanger. A constant-temperature fluid flows through the external exchanger. $$\frac{UA}{C_b} = -\ln\left(1 - \frac{\overline{C}_b}{\tau C_b}\ln\frac{T_{b,i} - T_e}{T_b - T_e}\right)$$
	A batch tank with an external exchanger having a single-phase fluid flowing through it. $$\frac{UA}{C_e} = -\frac{C_b/C_e}{1 - C_b/C_e}\ln\left[\frac{C_e}{C_b}\left(1 - \frac{1 - C_b/C_e}{1 + \frac{\overline{C}_b}{\tau C_e}\ln\frac{T_b - T_{e,i}}{T_{b,i} - T_{e,i}}}\right)\right]$$ When $C_b = C_e$, this equation reduces to $$\frac{UA}{C_b} = \frac{(\overline{C}_b/\tau C_b)\ln[(T_b - T_{e,i})/(T_{b,i} - T_{e,i})]}{1 + (\overline{C}_b/\tau C_b)\ln[(T_b - T_{e,i})/(T_{b,i} - T_{e,i})]}$$
	A batch tank with a heat source and an external exchanger having a single-phase fluid flowing through it. $$B\tau = \ln\frac{B(T_{b,i} - T_{e,i}) - q_a/\overline{C}_b}{B(T_b - T_{e,i}) - q_a/\overline{C}_b}$$ where $B = \frac{C_b}{\overline{C}_b}\left[\frac{1 - \exp[(-UA/C_b)(1 - C_b/C_e)]}{1 - (C_b/C_e)\exp[(-UA/C_b)(1 - C_b/C_e)]}\right]$

TABLE 8 Transient Solutions for Batch Heating or Cooling *(Continued)*

Case	Transient solutions
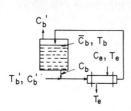	A batch tank with an external heat exchanger. A constant-temperature fluid flows through the exchanger. The batch is continuously drained with W_b flow rate. The same fluid is added at W_b flow rate and temperature T_b' as shown in the sketch and is well mixed. $$\frac{UA}{C_b + C_b'} = -\ln\left[1 - \frac{\overline{C_b}}{\tau C_b}\ln\frac{T_e - T_b' + (C_b/C_b')(T_e - T_{b,i})}{T_e - T_b' + (C_b/C_b')(T_e - T_b)}\right]$$
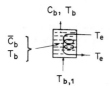	A batch tank with a coil. A constant-temperature fluid flows through the coil. The batch fluid continuously flows at rate W_b. The batch temperature is uniform and constant at T_b at any instant of time. $$\frac{(UA + C_b)\tau}{\overline{C_b}} = \ln\frac{UA(T_e - T_{b,i}) - C_b(T_{b,i} - T_{b,1})}{UA(T_e - T_b) - C_b(T_b - T_{b,1})}$$

fluid in the tank is well mixed and is at a uniform temperature at any instant of time, that the wall heat capacity is negligible, and that the heating or cooling fluid enters the exchanger or coil at a constant temperature $T_{e,i}$. For the cases of an external exchanger, the fluid holdup in the exchanger circuits is considered negligible and the exchanger is considered counterflow when the external fluid is single-phase.

From the solutions of Table 8, the surface area A is determined for the exchanger or coil sizing problem. Alternatively, for a specified exchanger (coil), the heating or cooling time τ is evaluated to heat or cool the batch by a predetermined ΔT. Note that the overall thermal conductance UA is computed from Eq. (9) or (11). In Table 8, $C_b = (Mc_p)_b$ is the batch fluid heat capacity and M_b is the mass of the batch fluid; $T_{b,i}$ is initial batch temperature; and T_b is the batch temperature at any time τ. The fluid in the external exchanger or coil flows with the heat capacity rate C_e and at a constant inlet temperature $T_{e,i}$ or T_e.

2. Step Response

Three types of transient responses are of interest: (1) step response, (2) frequency response, and (3) impulse response. The step response characterizes the behavior of the heat exchanger subjected to a sudden change in operating conditions (inlet temperatures or flow rates). The step response asymptotically approaches the steady-state value corresponding to new operating conditions. The frequency response describes the behavior of the exchanger subjected to a change that varies periodically with time. This response also varies periodically, but its maximum amplitude and phase angle relative to the input change depend upon the input frequency. An impulse response characterizes the behavior of the device subjected to a disturbance having infinite amplitude but infinitesimal duration. The disturbances that occur in most heat exchangers are approximated more closely by step functions and hence those responses are summarized in this section. Refer to Ref. 40 for a more comprehensive review on the subject. The frequency or impulse response can be determined from the step response as described by Raven [41] among others. We will present the transient response solutions only for single-phase two-fluid heat exchangers, first starting with the physical meaning of the dimensionless groups involved.

For steady state, the dimensionless fluid outlet temperatures (directly related to the exchanger effectiveness) are functions of NTU and C^*. During the transient conditions, these temperatures are functions of time and are influenced by the following parameters: the heat capacity of the solid wall, the thermal resistances between the fluids and wall, and the dwell time or residence time of fluid particles on both hot and cold sides. These latter parameters can be presented in dimensionless form as the wall heat capacity ratio $\overline{C}_w^*$, the thermal resistance ratio R^*, and the dwell time ratio τ_d^*, defined as follows:

$$\overline{C}_w^* = \frac{\overline{C}_w}{\overline{C}_{min}} \tag{36}$$

where $\overline{C}_w = (Mc)_w$
$\quad\quad \overline{C}_{min} = (Wc_p\tau_d) \quad$ on C_{min} side

$$R^* = \frac{1}{(\eta_o hA)^*} = \frac{\eta_o hA \text{ on the } C_{max} \text{ side}}{\eta_o hA \text{ on the } C_{min} \text{ side}} \tag{37}$$

$$\tau_d^* = \frac{\tau_d \text{ for the } C_{min} \text{ fluid}}{\tau_d \text{ for the } C_{max} \text{ fluid}} \tag{38}$$

The generalized (dimensionless) time variable is defined

$$\tau^* = \frac{\tau}{\tau_d \text{ on the } C_{min} \text{ side}} \tag{39}$$

Thus, the *dependent* fluid outlet temperature responses $\epsilon_{f,1}^*$ and $\epsilon_{f,2}^*$ to a step input in one of the fluid inlet temperatures are functions of the following groups for a specified exchanger flow arrangement:†

$$\epsilon_{f,1}^*, \epsilon_{f,2}^* = \phi(\tau^*, \text{NTU}, C^*, \overline{C}_w^*, R^*, \tau_d^*) \tag{40}$$

In forming $\epsilon_{f,1}^*$ and $\epsilon_{f,2}^*$, the fluid temperature *change* from the initial temperature is normalized with respect to the ultimate change at time equal to infinity:

$$\epsilon_{f,1}^* = \frac{T_{h,o}(\tau) - T_{h,o}(0)}{T_{h,o}(\infty) - T_{h,o}(0)} \quad\quad \epsilon_{f,2}^* = \frac{T_{c,o}(\tau) - T_{c,o}(0)}{T_{c,o}(\infty) - T_{c,o}(0)} \tag{41}$$

Thus, $\epsilon_{f,1}^*$ and $\epsilon_{f,2}^*$ are initially zero and tend toward unity with increasing time. The subscripts 1 and 2 have a special meaning. The subscript 1 denotes the outlet temperature response of that fluid which has the step input in its inlet temperature. The subscript 2 denotes the response of the other fluid at its outlet section.

Since there are six independent groups in Eq. (40) for the transient problem as opposed to two (NTU and C^*) for the steady-state problem, there are no general solutions to the transient problem available, on account of its complexity. Hence, the available specific solutions are presented next in terms of the fluid temperature responses at the exchanger outlet sections. The fluid temperature responses within the exchanger and the wall temperature response are not presented, because of space limitations. The reader may refer to the original references.

---

†In the batch heating or cooling problem, C^* is replaced by a group having C_e and C_b or $\overline{C}_b\tau$, $\overline{C}_w^*$ drops out because the wall heat capacity is neglected, R^* does not come into the picture since the wall heat capacity is neglected, and τ_d^* disappears because τ_d for the batch fluid is zero.

a. Heat Exchangers with $C^* = 0$

In many heat exchangers, such as condensers, evaporators, intercoolers, precoolers, and liquid-to-gas heat exchangers, the heat capacity rate of one fluid is much larger than that of the other fluid. For these exchangers, the temperature of the C_{max} fluid can be approximated as constant throughout the exchanger, and hence $C_{max} \to \infty$ and $\tau_{d,max} \to 0$. We will consider the temperature responses to a step change in inlet temperature of (1) constant-temperature (C_{max}) fluid, Fig. 38, and (2) variable-temperature (C_{min}) fluid, as shown later in Fig. 42. The results are applicable to counterflow, parallel flow, crossflow, or any other heat exchanger flow arrangement.

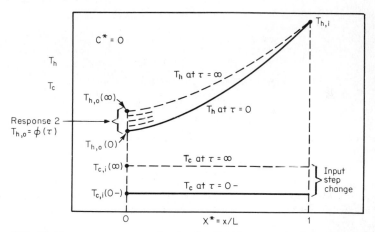

FIG. 38 Temperature response for the C_{min} fluid for a step change in temperature of the C_{max} fluid.

Step Change in the Inlet Temperature of the C_{max} Fluid.   In this case, the sudden changes in the inlet temperature of the condensing or evaporating fluid occur due to sudden changes in the system pressure. Or a step change is imposed on the inlet temperature of the liquid coolant. A typical temperature response for the C_{min} fluid in an evaporator is shown in Fig. 38. In this case, since $C^* = 0$ and $\tau_d^* = \infty$, Eq. (40) for $\epsilon_{f,2}^*$ reduces to

$$\epsilon_{f,2}^* = \phi(\tau^*, NTU, \overline{C}_w^*, R^*) \tag{42}$$

and by problem specification

$$\epsilon_{f,1}^* = 1 \tag{43}$$

The solution for $\epsilon_{f,2}^*$ has been obtained as a function of NTU by Rizika [42], London et al. [43], and Myers et al. [44] for different ranges of τ^*, $\overline{C}_w^*$, and R^*. These solutions are summarized in Table 9, and some specific results are presented in Figs. 39, 40, and 41.

Step Change in the Inlet Temperature of the C_{min} Fluid.   In this case, the response of the outlet temperature of the C_{min} fluid is required for a step change in its inlet temperature as illustrated in Fig. 42. This type of transient problem is more common in practice compared to the one in the preceding section. For the present case, the indepen-

TABLE 9 Transient Solutions for a Step Change in the Inlet Temperature of the C_{max} Fluid for a Heat Exchanger with $C^* = 0$

I. $0 \leq \tau^* < 1$, all $\overline{C}_w^*$ and R^* [42]

$$\epsilon_{f,2}^* = \frac{1 - e^{-X}[Y \sinh(X/Y) + \cosh(X/Y)]}{1 - \exp(-NTU)}$$

where $X = \dfrac{NTU(1 + R^*)(1 + R^* + \overline{C}_w^*)\tau^*}{2R^*\overline{C}_w^*}$

$$Y = \left[1 - \frac{4R^*\overline{C}_w^*}{(1 + R^* + \overline{C}_w^*)^2}\right]^{-1/2}$$

II. $\tau^* > 1$, $1 < \overline{C}_w^* < 2000$, and $R^* \geq 1$ ($R^* < 1$ is not practical) [44]

$$\epsilon_{f,2}^* = 1 - \tilde{A} \exp\left[-\frac{\tilde{B}(\tau^* - 1)}{\overline{C}_w^*}\right]$$

where $\tilde{A} = 1 - \dfrac{1 - \exp(-Z)[Y \sinh(Z/Y) + \cosh(Z/Y)]}{1 - \exp(-NTU)}$

$$\tilde{B} = \frac{1}{\tilde{A}}\left[\frac{2\,NTU(1 + R^*)\overline{C}_w^*}{(1 + R^* + \overline{C}_w^*)}\right]\left[\frac{Ye^{-Z} \sinh(Z/Y)}{1 - \exp(-NTU)}\right]$$

$$Z = \frac{NTU(1 + R^*)(1 + R^* + \overline{C}_w^*)}{2R^*\overline{C}_w^*}$$

III. Any τ^*, $\overline{C}_w^* > 2000$, and any R^* [44]

$$\epsilon_{f,2}^* = \frac{1}{1 - \exp(-NTU)}\left\{1 - e^{-\zeta_1 NTU} - e^{-NTU}\phi_0\left[\frac{NTU(1 + R^*)\zeta_1}{R^*}, \frac{NTU}{R^*}\right]\right.$$

$$\left. + e^{-\zeta_1 NTU}\phi_0\left[\frac{\zeta_1 NTU}{R^*}, \frac{NTU(1 + R^*)}{R^*}\right]\right\}$$

where $\zeta_1 = \dfrac{(1 + R^*)(\tau^* - 1)}{\overline{C}_w^*}$

$\phi_0(x, y) = e^{-y}\displaystyle\int_{\tau=0}^{x} e^{-\tau}I_0(2\sqrt{y}\sqrt{\tau})\,d\tau$, tabulated in Ref. 45

IV. τ^* as indicated, any $\overline{C}_w^*$, and $R^* = \infty$ [43]

$$\epsilon_{f,2}^* = \begin{cases} \dfrac{1 - \exp(-NTU\,\tau^*)}{1 - \exp(-NTU)} & \text{for } \tau^* \leq 1 \\ 1 & \text{for } \tau^* \geq 1 \end{cases}$$

V. $\tau^*/(1 + \overline{C}_w^*)$ grouping as indicated, any $\overline{C}_w^*$, and $R^* = 0$ [43]

$$\epsilon_{f,2}^* = \begin{cases} \dfrac{1 - \exp\{-NTU[\tau^*/(1 + \overline{C}_w^*)]\}}{1 - \exp(-NTU)} & \text{for } \dfrac{\tau^*}{1 + \overline{C}_w^*} \leq 1 \\ 1 & \text{for } \dfrac{\tau^*}{1 + \overline{C}_w^*} \geq 1 \end{cases}$$

VI. Any $(\tau^* - 1)/\overline{C}_w^*$, $\overline{C}_w^*$, NTU, and R^*: See Figs. 39, 40, and 41.

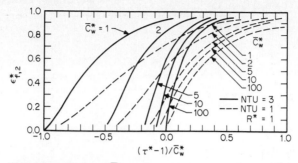

FIG. 39 Influence of $\overline{C}_w^*$ on $\epsilon_{f,2}^*$ for $C^* = 0$ with a step change in the temperature of the $C_{\max}$ fluid [40].

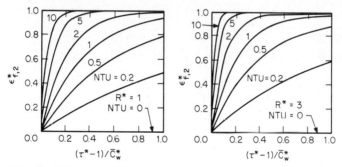

FIG. 40 Influence of R^* on $\epsilon_{f,2}^*$ for $C^* = 0$ with a step change in the temperature of the $C_{\max}$ fluid [40].

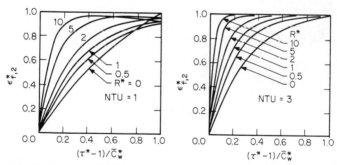

FIG. 41 Influence of NTU on $\epsilon_{f,2}^*$ for $C^* = 0$ with a step change in the temperature of the $C_{\max}$ fluid [40].

dent dimensionless groups of the solution $\epsilon_{f,1}^*$ are the same as those of the preceding section, Eq. (42).

$$\epsilon_{f,1}^* = \phi(\tau^*, \text{NTU}, \overline{C}_w^*, R^*) \tag{44}$$

The $C_{\max}$ fluid has infinite heat capacitance and hence its temperature does not change at all in this case; that is, $\epsilon_{f,2}^* = 1$. Alternatively, if the $C_{\max}$ fluid is the cold fluid as shown in Fig. 42, then $T_{c,o}(\infty) = T_{c,o}(0)$.

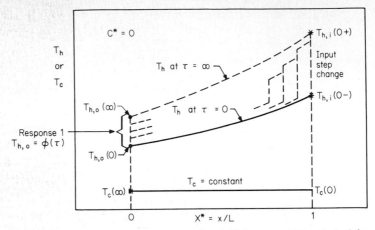

FIG. 42 Temperature response for the C_{min} fluid for a step change in its inlet temperature.

Kays and London [4] outline the explicit solutions of Eq. (44). However, refinements in these solutions have been made by Myers et al. [46], who analyzed the problem analytically and showed that the four independent variables of Eq. [44] can be reduced to two:[†]

$$\epsilon_{f,1}^* = \phi\left(\zeta_2, \frac{NTU}{R^*}\right) \tag{45}$$

where

$$\zeta_2 = (1 + R^*)^2 \left(\frac{\tau^* - 1}{\overline{C}_w^*}\right) \tag{46}$$

Myers et al. [44] provided the following exact solution for $\tau^* > 1$:[‡]

$$\epsilon_{f,1}^* = \phi_0\left(\frac{\zeta_2\,NTU}{R^*}, \frac{NTU}{R^*}\right) + \exp\left(\frac{-NTU}{R^*}\right)\exp\left(\frac{-\zeta_2\,NTU}{R^*}\right) I_0\left(\frac{2\,NTU}{R^*}\,\zeta_2^{1/2}\right) \tag{47}$$

Here $I_0^*(\ )$ represents the modified Bessel function of the first kind and zero order and the function $\phi_0(x, y)$ is defined in case III of Table 9. The temperature response expressed by Eq. (47) is shown in Fig. 43.

Step Change in the Flow Rate of the C_{min} Fluid for $C^* = 0$.   London et al. [43] conducted analog tests and found that a 100 percent temperature response to a flow rate change in the C_{min} fluid is virtually attained in one dwell time. A step change in the flow rate of the C_{min} fluid will change NTU. The new steady-state effectiveness is then calculated from

$$\epsilon = 1 - \exp(-NTU) \tag{48}$$

[†]The variable R^* of Myers et al. [44, 46] in the present terminology is $1/R^*$. Throughout this chapter, R^* is consistently defined as in Eq. (37) unless specified otherwise.
[‡]Note that for $\tau^* < 1$, $\epsilon_{f,1}^* = 0$, and $\tau^* = 0$ represents a point of singularity.

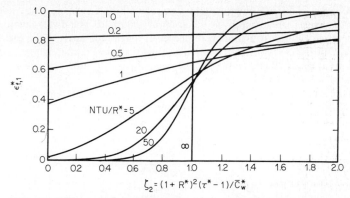

FIG. 43 Outlet temperature response for a step change in inlet temperature of the $C_{\min}$ fluid for the $C^* = 0$ case [44].

(see Ref. 4). It is seen from this relationship that a sizable flow rate change has only a small influence on the steady-state effectiveness, and, as a result, only a small direct influence on the outlet fluid temperature.

b. Counterflow Heat Exchangers with $C^* = 1$

In gas turbine regenerators and some heat exchangers in process industries, the heat capacity rate of both fluids is approximately the same, and $C^* \approx 1$ is a good approximation. A typical transient response of outlet fluid temperatures is shown in Fig. 44 for the case when a step change is imposed on the hot fluid inlet temperature. The results presented next are valid regardless of whether the step change is imposed on the hot or cold fluid inlet temperature. Although the temperature responses are dependent upon five dimensionless groups as shown in Eq. (40)(since $C^* = 1$), the number of these groups is significantly reduced for some specific cases of technical interest presented next.

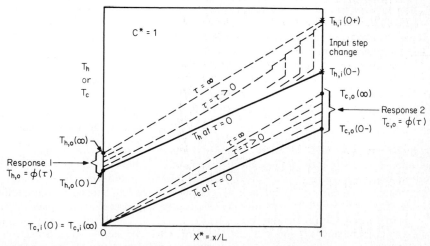

FIG. 44 Identification of the temperature responses for a counterflow exchanger.

Step Change in Inlet Temperature: The $\overline{C}_w^* > 100$ Case.   In a gas-to-gas heat exchanger, $\overline{C}_w \gg \overline{C}_h$ or $\overline{C}_c$ and is on the order of 1000-fold. The dwell-time ratio τ_d^* is generally between ¼ and 4. Based on the order of magnitude analysis for each term of appropriate differential equations, London et al. [43] showed that τ_d^* is not a significant parameter for large $\overline{C}_w^*$. Based on the electromechanical analog tests, London et al. further demonstrated that $\epsilon_{f,1}^*$ can be presented as a function of $\tau^*/\overline{C}_w^* - 0.4[(R^* - 1)/(R^* + 1)]$ and NTU, and $\epsilon_{f,2}^*$ as a function of $\tau^*/(1.5 + \overline{C}_w^*)$ and NTU for $0.25 \leq R^* \leq 4$ and $\overline{C}_w^* > 100$. Their temperature responses are presented in Fig. 45.

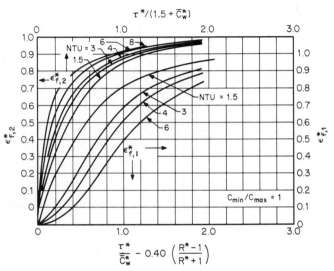

FIG. 45   Temperature responses for a step change in the temperature of one fluid for a counterflow exchanger with $C^* = 1$, $\overline{C}_w^* > 100$, and $0.25 \leq R^* \leq 1$ [4].

Step Change in Inlet Temperature: The $\overline{C}_w^* = 0$ Case.   For a water-cooled steam condenser, $\overline{C}_w^* \approx 0.3$. Hence the $\overline{C}_w^* = 0$ case is useful as a limiting case for some applications. London et al. [43] showed that $\epsilon_{f,2}^*$ of Fig. 45 is also applicable for this case with the restriction as shown on the following parameters: $C^* = 1$, $\tau_d^* = 1$, and no limitation on R^*. No solution is available for $\epsilon_{f,1}^*$.

Step Change in Inlet Temperature: The $0 \leq \overline{C}_w^* \leq 100$ Case.   The temperature response $\epsilon_{f,2}^*$ of Fig. 45 is also applicable for this case provided that $R^* = 1$, $\tau_d^* = 1$. No solution is available for $\epsilon_{f,1}^*$.

Step Change in the Flow Rate.   When a step change in the flow rate is imposed at the inlet, its response at the outlet is instantaneous for incompressible fluids and very fast for compressible fluids so that it is also considered instantaneous. Of course, the changes in the flow rate on one side do not have any influence on the flow rate on the other side. Hence, of prime interest is to determine the temperature responses $\epsilon_{f,1}^*$ and $\epsilon_{f,2}^*$ due to the change in the flow rate on one side.

Cima and London [47] obtained analog solutions for two step-function changes in NTU from 1.5 to 1 and from 1 to 1.5. A step change in NTU is achieved by a step change in the flow rate of the $C_{\min}$ fluid. The temperature response is presented in Fig. 46. The results were obtained for constant $\overline{C}_w^*$ magnitudes. It is interesting to note from Fig. 46 that a 90 percent response is achieved within one dwell time ($\tau^* \approx 1$). For a gas turbine regenerator application, one dwell time is about 0.5 to 1 s.

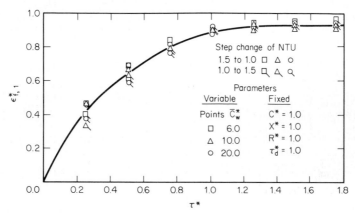

FIG. 46 $\epsilon_{f,1}^*$ response to a step change in NTU [47].

Example of a Transient Problem

A counterflow gas turbine recuperator has inlet gas and air temperatures of 540°C and 200°C, respectively. The outlet air temperature is 489°C and the flow rate is 30 kg/s. Consider $C^* = 1$, $R^* = 1$, $\tau_{d,c} = 0.40$ s, and $\tau_{d,h} = 0.08$ s. The core mass is 31,000 kg and the specific heat of the core material is 0.46 kJ/(kg·°C). The specific heat of air is 1.017 kJ/(kg·°C). If the gas temperature is suddenly increased by 50°C, determine (1) the new steady-state gas outlet temperature, (2) the new steady-state air outlet temperature, (3) how long it will take to attain 90 percent response on the gas outlet temperature, (4) what the gas temperature will be at this time (for $\epsilon_{f,1}^*$ = 0.90), (5) how long it will take to attain 90 percent response on the air outlet temperature, and (6) what the air outlet temperature will be at this time (for $\epsilon_{f,2}^*$ = 0.90).

Solution

If we neglect the changes in c_p of air and gas due to the changes in their mean temperatures after a step change, NTU and C^* after the step change in $T_{h,i}$ will be the same as those before the step change in $T_{h,i}$. Since $\epsilon = \phi(\text{NTU}, C^*)$ for a specified heat exchanger, ϵ will not change as a result of the step change in $T_{h,i}$. Let us calculate ϵ_{old} and equate it to ϵ_{new} and subsequently obtain the new steady-state outlet temperatures. Since $C^* = 1$, we will treat $C_{\min} = C_a$.

$$\epsilon = \frac{T_{a,o} - T_{a,i}}{T_{g,i} - T_{a,i}} = \frac{489 - 200}{540 - 200} = 0.85$$

and $T_{g,o} = T_{g,i} - \epsilon(T_{g,i} - T_{a,i}) = 540 - 0.85(540 - 200) = 251°C$. Here the subscripts g and a denote the gas and air sides, respectively. Since the new $T_{g,i} = 540$

$+ 50 = 590°C$, the new steady-state $T_{g,o}$ and $T_{a,o}$ are

$$T_{g,o} = T_{g,i} - \epsilon(T_{g,i} - T_{a,i}) = 590 - 0.85(590 - 200) = 258.5°C$$
$$T_{a,o} = T_{a,i} + \epsilon(T_{g,i} - T_{a,i}) = 200 + 0.85(590 - 200) = 531.5°C$$

To determine τ for $\epsilon_{f,1}^* = 0.90$, we need to evaluate the dimensionless groups required for the use of Fig. 45.

$$\text{NTU} = \frac{\epsilon}{1 - \epsilon} = \frac{0.85}{1 - 0.85} = 5.67$$

$$\overline{C}_w^* = \frac{\overline{C}_w}{C_{min}} = \frac{M_w c_w}{C_{min} \tau_{d,c}} = \frac{31{,}000 \times 0.46}{30 \times 1.017 \times 0.40} = 1168.5$$

Since $\overline{C}_w^* > 100$, the results of Fig. 45 are applicable. For $\epsilon_{f,1}^* = 0.9$ and NTU = 5.67, we get the abscissa from Fig. 45 as 2.25:

$$\frac{\tau^*}{\overline{C}_w^*} - 0.40 \frac{R^* - 1}{R^* + 1} = 2.25$$

In this equation, substituting $\overline{C}_w^* = 1168.5$ and $R^* = 1$, we get

$$\tau^* = \frac{\tau}{\tau_{d,c}} = 2629.1$$

Hence

$$\tau = 1052 \text{ s} = 17.5 \text{ min}$$

This is the time required to attain $\epsilon_{f,1}^* = 0.90$.

Using the definition of $\epsilon_{f,1}^*$ from Eq. (41), the corresponding gas outlet temperature is found from

$$0.90 = \frac{T_{g,o}(\tau) - 251.0}{258.5 - 251.0}$$

Thus

$$T_{g,o} = 257.75°C$$

For $\epsilon_{f,2}^* = 0.90$ and NTU = 5.67, we get the abscissa from Fig. 45 as

$$\frac{\tau^*}{1.5 + \overline{C}_w^*} = 1.05$$

Hence

$$\tau = 0.40 \times 1.05(1.5 + 1168.5) = 491.4 \text{ s} = 8.2 \text{ min}$$

This is the time required to attain $\epsilon_{f,2}^* = 0.90$.

Using the definition of $\epsilon_{f,2}^*$ from Eq. (41), the corresponding air outlet temperature is

$$0.90 = \frac{T_{a,o}(\tau) - 489}{531.5 - 489}$$

Thus

$$T_{a,o} = 527.25°C$$

Notice that the time required to attain a 90 percent response in the gas outlet temperature is about twice that for a 90 percent response in the air outlet temperature.

Alternatively, the air outlet temperature response is faster. Since one is more interested in $\epsilon^*_{f,2}$ for the fuel control in a gas turbine cycle, it is fortuitous that the response in $\epsilon^*_{f,2}$ is fast compared to $\epsilon^*_{f,1}$ for a step change in the gas inlet temperature.

c. Cross-Flow Heat Exchangers

The analysis for the transient response of a cross-flow heat exchanger is much more difficult than that of a counterflow heat exchanger. This is because the temperature is a function of time and two position variables (x and z). All of the dimensionless groups for a counterflow exchanger are still applicable to the cross-flow exchanger.

Myers et al. [46] analyzed by an integral method the transient response of a cross-flow heat exchanger with one fluid mixed and the other unmixed. They considered the fluid with a "stepped" inlet temperature as mixed, and the "unstepped" fluid as unmixed. For the large wall capacitance ($\overline{C}^*_w$ large), they presented the following approximate solution:

$$\epsilon^*_{f,1} = \phi_0[B'(\tau^* - 1), A'] + e^{-A'}e^{-B'(\tau^*-1)}I_0\{2[B'(\tau^* - 1)A']^{1/2}\} \text{ for } \tau^* > 1 \quad (49)$$

$$\epsilon^*_{f,2} = \frac{P'}{1 - e^{-P}} \int_{X^*=0}^{\tau^*} e^{-PX^*}\phi_0[B'(\tau^* - X^*), A'X^*] \, dX^* \qquad \text{for } \tau^* \leq 1 \quad (50)$$

$$\epsilon^*_{f,2} = \frac{P'}{1 - e^{-P}} \int_{X^*=0}^{1} e^{-PX^*}\phi_0[B'(\tau^* - X^*), A'X^*] \, dX^* \qquad \text{for } \tau^* \geq 1 \quad (51)$$

where

$$A' = \frac{NTU'_1(a_1 + R^*)}{1 + R^*} \qquad B' = \frac{NTU'_1(1 + R^*)}{\overline{C}^*_w(a_1 + R^*)} \quad (52)$$

$$P' = \frac{NTU'_1(1 - a_1)}{1 + R^*} \qquad a_1 = 1 - \frac{1 - e^{-NTU_2}}{NTU_2} \quad (53)$$

$$R^* = \frac{R_2}{R_1} \qquad \overline{C}^*_w = \frac{\overline{C}_w}{\overline{C}_1} \qquad NTU'_1 = \left(\frac{\eta_o hA}{C}\right)_1 \qquad NTU_2 = \frac{UA}{C_2} \quad (54)$$

Here the subscript 1 denotes values for the stepped fluid and the subscript 2 denotes values for the unstepped fluid. The function $\phi_0(\ )$ is defined in Case III of Table 9. Myers et al. [46] presented graphically the criterion for the magnitude of how large $\overline{C}^*_w$ should be in terms of functions of $\overline{C}^*_w$, NTU, R^*, and C^*.

Since Eqs. (49) to (51) incorporate a large number of parameters, a compact graphical presentation of the results is not feasible. A summary of the results is presented in terms of the time required to attain 90 percent response for $\tau^* > 10$ and large $\overline{C}^*_w$ as shown in Fig. 47. The dimensionless groups A', B', and P' of this figure are defined by Eqs. (52) and (53).

Yamashita et al. [48] analyzed the transient response of a cross-flow exchanger with both fluids unmixed. They imposed a step change in the inlet temperature of the hot fluid. Using finite difference methods,

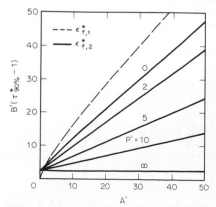

FIG. 47 90 percent response times for $\tau^* > 10$ for a cross-flow heat exchanger with "stepped" fluid mixed, "unstepped" fluid unmixed [46].

they determined and graphically presented the dimensionless outlet temperatures as functions of τ^* or $\tau^*/\overline{C}_w^*$. They considered one of five groups (τ_d^*, $1/\overline{C}_w^*$, R^*, NTU, and C^*) as a parameter, the value of the other four groups being unity in their graphical presentation.

d. Shell-and-Tube Exchangers

One of the most comprehensive solutions for the transient response of shell-and-tube heat exchangers was obtained by Tan and Spinner [49] by employing the Laplace transformation technique. They considered the shell-side fluid to have an infinite heat capacity rate so that its temperature does not change. Thus, they effectively considered the exchanger having $C^* = 0$ instead of the commonly used one shell pass–two tube pass shell-and-tube exchanger. However, they considered finite shell-side thermal resistance and finite tube-wall heat capacity. They presented closed-form solutions for the transient response of the tube-side outlet fluid temperature for the following cases: (1) a step change in the inlet temperature of the tube fluid, (2) a step change in the inlet temperature of the shell fluid, and (3) a step change in the tube fluid velocity (flow rate) with the tube-side heat transfer coefficient either constant or variable dependent upon the velocity. Since the results are complicated, the reader is referred to Ref. 49 in order to avoid any confusion.

ACKNOWLEDGMENTS

The authors are grateful to Dr. E. N. Ganić of the University of Illinois at Chicago and Dr. M. K. Jensen of the University of Wisconsin-Milwaukee for providing constructive suggestions on this article. The chapter is based on literature up to August 1981.

REFERENCES

1. R. K. Shah, Classification of Heat exchangers, in *Heat Exchangers: Thermal-Hydraulic Fundamentals and Design,* ed. S. Kakaç, A. E. Bergles, and F. Mayinger, pp. 9–46, Hemisphere/McGraw-Hill, Washington, D.C., 1981.

2. R. K. Shah, Heat Exchanger Design Methodology—An Overview, in *Heat Exchangers: Thermal-Hydraulic Fundamentals and Design,* ed. S. Kakaç, A. E. Bergles, and F. Mayinger, pp. 455–459, Hemisphere/McGraw-Hill, Washington, D.C., 1981.

3. R. K. Shah, Heat Exchanger Basic Design Methods, in *Low Reynolds Number Flow Heat Exchangers,* ed. S. Kakaç, R. K. Shah, and A. E. Bergles, Hemisphere/McGraw-Hill, Washington, D.C., 1982.

4. W. M. Kays and A. L. London, *Compact Heat Exchangers,* 2d ed., McGraw-Hill, New York, 1964.

5. B. S. Bačlić, A Simplified Formula for Cross-Flow Heat Exchanger Effectiveness, *J. Heat Transfer,* Vol. 100C, pp. 746–747, 1978.

6. W. M. Nagle, Mean Temperature Differences in Multipass Heat Exchangers, *Ind. Eng. Chem.,* Vol. 25, pp. 604–609, 1933.

7. K. A. Gardner, Mean Temperature Difference in Multipass Exchangers—Correction Factors with Shell Fluid Unmixed, *Ind. Eng. Chem.,* Vol. 33, pp. 1495–1500, 1941.

8. A. D. Kraus and D. Q. Kern, The Effectiveness of Heat Exchangers with One Shell Pass and Even Numbers of Tube Passes, *ASME Paper No. 65-HT-18,* 1965.

9. D. L. Schindler and H. T. Bates, True Temperature Difference in a 1-2 Divided Flow Heat Exchanger, *Chem. Eng. Prog. Symp. Ser.* No. 30, Vol. 56, pp. 203–206, 1960.

10. L. Jaw, Temperature Relations in Shell and Tube Exchangers Having One-Pass Split-Flow Shells, *J. Heat Transfer,* Vol. 86C, pp. 408–416, 1964.

11. K. A. Gardner, Mean Temperature Difference in an Array of Identical Exchangers, *Ind. Eng. Chem.,* Vol. 34, pp. 1083–1087, 1942.

12. A. C. Mueller, New Charts for True Mean Temperature Difference in Heat Exchangers, *AIChE Paper No. 10, 9th Nat. Heat Transfer Conf.,* Seattle, 1967.

13. A. C. Mueller, Heat Exchangers, in *Handbook of Heat Transfer,* ed. W. M. Rohsenow and J. P. Hartnett, 1st ed., Chap. 18, McGraw-Hill, New York, 1973.

14. Tubular Exchanger Manufacturers Association, *Standards of TEMA,* 6th ed., New York, 1978.

15. F. K. Fisher, Mean Temperature Difference Correction in Multipass Exchangers, *Ind. Eng. Chem.,* Vol. 30, pp. 377–383, 1938.

16. R. Crozier, Jr., and M. Samuels, Mean Temperature Difference in Odd-Tube-Pass Heat Exchangers, *J. Heat Transfer,* Vol. 99C, pp. 487–489, 1977.

17. A. M. Whistler, Correction for Heat Conduction through Longitudinal Baffle of Heat Exchanger, *Trans. ASME,* Vol. 69, pp. 683–685, 1947.

18. T. Rosenman and J. Taborek, The Effect of Leakage through the Longitudinal Baffle on the Performance of Two-Pass Shell Exchangers, *AIChE Symp. Ser.* No. 118, Vol. 68, pp. 12–20, 1971.

18a. J. Taborek, Charts for Mean Temperature Difference in Industrial Heat Exchanger Configurations, in *Heat Exchanger Design Handbook,* ed. E. U. Schlünder, Vol. 1, Sec. 1.5, pp. 1.5.2–1 to 1.5.2–15, Hemisphere, Washington, D.C., 1982.

19. B. S. Bačlić and D. D. Gvozdenac, ε-NTU-ω Relationships for Inverted Order Flow Arrangements of Two-Pass Crossflow Heat Exchangers, in *Regenerative and Recuperative Heat Exchangers,* ed. R. K. Shah and D. E. Metzger, *Book No. H00207, HTD-Vol. 21,* pp. 27–41, ASME, New York, 1983.

20. B. S. Bačlić and D. D. Gvozdenac, Exact Explicit Equations for Some Two- and Three-Pass Cross-Flow Heat Exchangers Effectiveness, in *Heat Exchangers: Thermal-Hydraulic Fundamentals and Design,* ed. S. Kakaç, A. E. Bergles, and F. Mayinger, pp. 481–494, Hemisphere/McGraw-Hill, Washington, D.C., 1981.

21. B. S. Bačlić, D. P. Sekulić, and D. D. Gvozdenac, Exact Explicit Equations for Some Two- and Three-Pass Cross-Flow Exchangers Effectiveness—Part II, in *Low Reynolds Number Forced Convection in Channels and Bundles, Proc. 4th NATO Advanced Study Inst.,* Ankara, pp. 863–876, 1981.

22. P. Worsøe-Schmidt and H. J. Høggard Knudsen, Thermal Modeling of Heat Exchangers for Simulation Purposes, *Proc. 1976 Heat Transfer Fluid Mech. Inst.,* ed. A. A. McKillop, J. W. Baughn, and H. A. Dwyer, pp. 495–511, Stanford University Press, Stanford, Calif., 1976.

23. W. M. Kays, R. K. Jain, and S. Sabberwal, The Effectiveness of a Counterflow Heat Exchanger with Cross-Flow Headers, *Int. J. Heat Mass Transfer,* Vol. 11, pp. 772–774, 1968.

24. J. D. Domingos, Analysis of Complex Assemblies of Heat Exchangers, *Int. J. Heat Mass Transfer,* Vol. 12, pp. 537–548, 1969.

25. B. W. Jackson and R. A. Troupe, Plate Exchanger Design by ε-NTU Method, *Chem. Eng. Prog. Symp. Ser.* No. 64, Vol. 62, pp. 185–190, 1966.

26. M. R. Foote, Effective Mean Temperature Differences in Multi-Pass Plate Heat Exchangers, *NEL Rep. No. 303,* National Engineering Laboratory, East Kilbride, Glasgow, 1967.

27. A. P. Colburn, Mean Temperature Difference and Heat Transfer Coefficient in Liquid Heat Exchangers, *Ind. Eng. Chem.,* Vol. 25, pp. 873–877, 1933.

28. D. Butterworth, Condensers: Thermohydraulic Design, in *Heat Exchangers: Thermal-Hydraulic Fundamentals and Design,* ed. S. Kakaç, A. E. Bergles, and F. Mayinger, pp. 647–679, Hemisphere/McGraw-Hill, Washington, D.C., 1981.

29. E. N. Sieder and G. E. Tate, Heat Transfer and Pressure Drop of Liquids in Tubes, *Ind. Eng. Chem.,* Vol. 28, pp. 1429–1435, 1936.

30. K. Gardner and J. Taborek, Mean Temperature Difference: A Reappraisal, *AIChE J.*, vol. 23, pp. 777–786, 1977.

31. W. Roetzel, Heat Exchanger Design with Variable Transfer Coefficients for Crossflow and Mixed Flow Arrangements, *Int. J. Heat Mass Transfer,* Vol. 17, pp. 1037–1049, 1974.

32. K. A. Gardner, Variable Heat-Transfer Rate Correction in Multipass Exchangers: Shell-Side Film Controlling, *Trans. ASME,* Vol. 67, pp. 31–38, 1945.

33. R. S. Ramalho and F. M. Tiller, Improved Design Method for Multipass Exchangers, *Chem. Eng.,* Vol. 72, pp. 87–92, Mar. 29, 1965.

34. S. Kao, Analysis of Multipass Heat Exchangers with Variable Properties and Transfer Rate, *J. Heat Transfer,* Vol. 97C, pp. 509–515, 1975.

35. K. A. Gardner, Mean Temperature Difference in Unbalanced-Pass Exchangers, *Ind. Eng. Chem.,* Vol. 33, pp. 1215–1223, 1941.

36. R. K. Shah and A. L. London, *Laminar Flow Forced Convection in Ducts,* Supplement 1 to *Advances in Heat Transfer,* pp. 111, 132, Academic, New York, 1978.

37. P. G. Kroger, Performance Deterioration in High Effectiveness Heat Exchangers Due to Axial Heat Conduction Effects, *Advances in Cryogenics Engineering,* Vol. 12, pp. 363–372, Plenum, New York, 1967; condensed from a paper presented at the 1966 Cryogenic Engineering Conference, Boulder, Colo.

38. J. P. Chiou, The Advancement of Compact Heat Exchanger Theory Considering the Effects of Longitudinal Heat Conduction and Flow Nonuniformity, in *Symposium on Compact Heat Exchangers—History, Technological Advancement and Mechanical Design Problems,* ed. R. K. Shah, C. F. McDonald, and C. P. Howard, *Book No. G00183, HTD-Vol. 10,* pp. 101–121, ASME, New York, 1980.

39. D. Q. Kern, *Process Heat Transfer,* Chap. 18, McGraw-Hill, New York, 1950.

40. R. K. Shah, The Transient Response of Heat Exchangers, in *Heat Exchangers: Thermal-Hydraulic Fundamentals and Design,* ed. S. Kakaç, A. E. Bergles, and F. Mayinger, pp. 915–953, Hemisphere/McGraw-Hill, Washington, D.C., 1981.

41. F. H. Raven, *Automatic Control Engineering,* pp. 345–347, McGraw-Hill, New York, 1961.

42. J. W. Rizika, Thermal Lags in Flowing Incompressible Fluid Systems Containing Heat Capacitors, *Trans. ASME,* Vol. 78, pp. 1407–1413, 1956.

43. A. L. London, F. R. Biancardi, and J. W. Mitchell, The Transient Response of Gas-Turbine Plant Heat Exchangers—Regenerators, Intercoolers, Precoolers, and Ducting, *J. Eng. Power,* Vol. 81A, pp. 433–448, 1959.

44. G. E. Myers, J. W. Mitchell, and C. F. Lindeman, Jr., The Transient Response of Heat Exchangers Having an Infinite Capacity Rate Fluid, *J. Heat Transfer,* Vol. 92C, pp. 269–275, 1970.

45. S. R. Binkley, H. E. Edwards, and R. W. Smith, Table of the Temperature Distribution Function for Heat Exchange between a Fluid and a Porous Solid, U.S. Bureau of Mines, Pittsburgh, 1952.

46. G. E. Myers, J. W. Mitchell, and R. F. Norman, The Transient Response of Crossflow Heat Exchangers, Evaporators and Condensers, *J. Heat Transfer,* Vol. 89C, pp. 75–80, 1967.

47. R. M. Cima and A. L. London, The Transient Response of a Two-Fluid Counterflow Heat Exchanger—The Gas-Turbine Regenerator, *Trans. ASME,* Vol. 80, pp. 1169–1179, 1958.

48. H. Yamashita, R. Izumi, and S. Yamaguchi, Analysis of the Dynamic Characteristic of Cross-Flow Heat Exchangers with Both Fluids Unmixed, *Bull. JSME,* Vol. 21, No. 153, pp. 479–485, 1978.

49. K. S. Tan and I. H. Spinner, Dynamics of a Shell-and-Tube Heat Exchanger with Finite Tube-Wall Heat Capacity and Finite Shell-Side Resistance, *Ind. Eng. Chem. Fund.,* Vol. 17, pp. 353–358, 1978.

50. S. V. Patankar and D. B. Spalding, A Calculation Procedure for the Transient and Steady-State Behavior of Shell-and-Tube Heat Exchangers, in *Heat Exchangers: Design and Theory Sourcebook,* ed. N. Afgan and E. U. Schlünder, pp. 155–176, McGraw-Hill, New York, 1974.

NOMENCLATURE
Symbol, Definition, SI Units, English Units

A exchanger total heat transfer area (primary plus secondary, if any) on one side: m^2, ft^2

A_f area of fins: m^2, ft^2

A_k total wall cross-sectional area for longitudinal conduction: m^2, ft^2

A_w total area for heat conduction from hot to cold fluid: m^2, ft^2

C flow-stream heat capacity rate $= Wc_p$: W/K, Btu/(h·°F)

C^* heat capacity rate ratio $= C_{min}/C_{max}$, dimensionless

$\overline{C}$ flow-stream heat capacity $= Mc_p = C\tau_d$: W·s/°C, Btu/°F

C_b heat capacity rate of batch fluid $= (Wc_p)_b$: W/K, Btu/(h·°F)

$\overline{C}_b$ heat capacity of batch fluid $= (Mc_p)_b$: W·s/°C, Btu/°F

C_c flow-stream heat capacity rate of cold fluid $= (Wc_p)_c$: W/K, Btu/(h·°F)

C_e heat capacity rate of coil or exchange fluid: W/K, Btu (h·°F)

C_h flow-stream heat capacity rate of hot fluid $= (Wc_p)_h$: W/K, Btu/(h·°F)

C_{max} maximum of C_c and $C_h = (Wc_p)_{max}$: W/K, Btu/(h·°F)

C_{min} minimum of C_c and $C_h = (Wc_p)_{min}$: W/K, Btu/(h·°F)

$\overline{C}_{min}$ heat capacity of the C_{min} side: W·s/°C, Btu/°F

$\overline{C}_w$ wall heat capacity $= M_w c_w$: W·s/°C, Btu/°F

$\overline{C}_w^*$ a ratio of $\overline{C}_w$ to $\overline{C}_{min}$, dimensionless

c_p specific heat of fluid at constant pressure: J/(kg·°C), Btu/(lb$_m$·°F)

c_w specific heat of wall material: J/(kg·°C), Btu/(lb$_m$·°F)

D_h hydraulic diameter: m, ft

d_i inside diameter of a circular tube: m, ft

d_o outside diameter of a circular tube: m, ft

F log-mean temperature difference correction factor, dimensionless

h heat transfer coefficient on one side of the exchanger: W/(m^2·K), Btu/(h·ft^2·°F)

I_0, I_1 modified Bessel functions

k thermal conductivity of the fluid if no subscript: W/(m·K), Btu/(h·ft·°F)

k_w thermal conductivity of the wall material: W/(m·K), Btu/(h·ft·°F)

L exchanger length: m, ft

M mass of the fluid in the heat exchanger at any instant of time: kg, lb$_m$

M_w mass of the heat exchanger core: kg, lb$_m$

N number of passes

N_b number of baffles

N_p number of flow passages on one side in a plate-fin exchanger

N_t number of tubes

NTU   number of heat transfer units $= UA/C_{min}$, dimensionless

NTU_t number of heat transfer units based on the tube-side heat capacity rate $= UA/C_t$, dimensionless

P temperature effectiveness of the tube-side stream (see Table 1 for definition), dimensionless

Pe	Péclet number $=$ Re Pr, dimensionless
Pr	Prandtl number $= \mu c_p/k$, dimensionless
q	heat transfer rate, heat duty: W, Btu/h
q_a	heat input by an agitator: W, Btu/h
q''	heat flux, heat transfer rate per unit area: W/m^2, Btu/(h·ft^2)
R	thermal resistance based on area A, $R_o = 1/UA$ (see text after Eq. (5) for the definitions of other thermal resistances), K/W, h·°F/Btu
R	heat capacity rate ratio (see Table 1 for definition), dimensionless
R^*	ratio of thermal resistances, defined by Eq. (37), dimensionless
Re	Reynolds number $= \rho V D_h/\mu$, dimensionless
T	temperature relative to an arbitrary datum: °C, °F
ΔT_{lm}	log-mean temperature difference: °C, °F
ΔT_m	true mean temperature difference $= F \Delta T_{lm}$: °C, °F
U	overall heat transfer coefficient: W/(m^2·K), Btu/(h·ft^2·°F)
V	mean fluid velocity through the exchanger on one side: m/s, ft/h
W	fluid mass flow rate: kg/s, lb$_m$/h
X^*	axial distance $= x/L$, dimensionless
x	cartesian coordinate along the flow direction: m, ft
x^*	axial coordinate for the thermal entrance region $= x/(D_h\text{RePr})$, dimensionless
Z	NTU/(1 + NTU), dimensionless

Greek Symbols

δ	wall thickness: m, ft
ϵ	exchanger effectiveness, dimensionless
$\epsilon^*_{f,1}$	temperature self-response, defined by Eq. (41), dimensionless
$\epsilon^*_{f,2}$	temperature cross response, defined by Eq. (41), dimensionless
η_f	fin efficiency, see p. 4-183 for definition, dimensionless
η_o	total surface temperature effectiveness, defined by Eq. (6), dimensionless
$(\eta_o hA)^*$	convection-conductance ratio, defined by Eq. (37), dimensionless
λ	longitudinal wall heat conduction parameter $= k_w A_k/LC_{min}$, dimensionless
μ	viscosity: Pa·s, lb$_m$/h·ft
ρ	density: kg/m^3, lb$_m$/ft^3
τ	time variable: s
τ_d	dwell time, residence time, or transit time of fluid particle: s
$\tau_{d,max}$	dwell time of the C_{max} fluid; s
$\tau_{d,min}$	dwell time of the C_{min} fluid; s
τ^*	time variable, $\tau/\tau_{d,min}$, dimensionless
τ^*_d	dwell time ratio, $\tau_{d,min}/\tau_{d,max}$, dimensionless
ϕ	function of
ψ	ratio of true mean temperature difference to inlet temperature difference $= \Delta T_m/(T_{h,i} - T_{c,i})$, dimensionless

Subscripts

b	batch fluid
c	cold fluid side
cf	counterflow
e	external exchanger or coil
f	fin
h	hot fluid side
i	initial (at time $\tau = 0$) for the transient problem
i	inlet to the exchanger
m	mean value
o	outlet to the exchanger; overall when used with R or ϵ
s	scale or fouling when used with R or h; otherwise, shell side
t	tube side
w	wall
x	local value
x, y	denotes two sides of the exchanger regardless of hot and cold side identification
1	one section (inlet or outlet) of the exchanger; one pass in a multipass exchanger; fluid 1
2	other section (outlet or inlet) of the exchanger; fluid 2

PART 2

Process Heat Exchangers

A. SHELL-AND-TUBE HEAT EXCHANGERS

The tubular-type heat exchangers are extensively used in industry because of the wide range of designs possible, the materials of construction available, the ease of fabrication and maintenance, and low cost. The selection of the type of shell and head depends upon whether single-phase or multiphase fluids are transferring heat and on process conditions discussed in more detail below.

1. Types of Exchangers

Shell-and-tube heat exchangers are classified and constructed in accordance with the widely used TEMA standards [1] (in Europe the DIN standards are used). The sizes and type are designated by numbers and letters as described below. Sizes of shells (and tube bundles) are designated by numbers describing the shell (and tube bundle) diameters and tube lengths. The nominal diameter is the inside diameter of the shell in inches rounded off to the nearest integer. For kettle reboilers the nominal diameter is the port diameter followed by the shell diameter, each rounded off to the nearest integer. The nominal length is the actual overall straight tube length; for U tubes the straight length from the end of the tube to the bend tangent is used. Types are designated by letters describing the stationary head, shell, and rear head in that order and as designated in Fig. 1. Some examples are:

1. Split-ring floating-head exchanger with removable channel and cover, single-pass shell, 23¼-in inside diameter with 16-ft tubes. Size 23-192 type AES.
2. U-tube exchanger with bonnet stationary head, split-flow shell, 19-in inside diameter with 7-ft straight-length tubes. Size 19-84 type BGU.
3. Pull-through floating-head kettle-type reboiler, having stationary head integral with tube sheet, 23-in port diameter and 37-in inside shell diameter with 16-ft tubes. Size 23/37-192 type CKT.

The nomenclature for the heat exchanger components is illustrated in Fig. 2.

The exchangers are built in accordance with three mechanical standards which specify design, fabrication, and materials of unfired shell-and-tube heat exchangers. Class R is

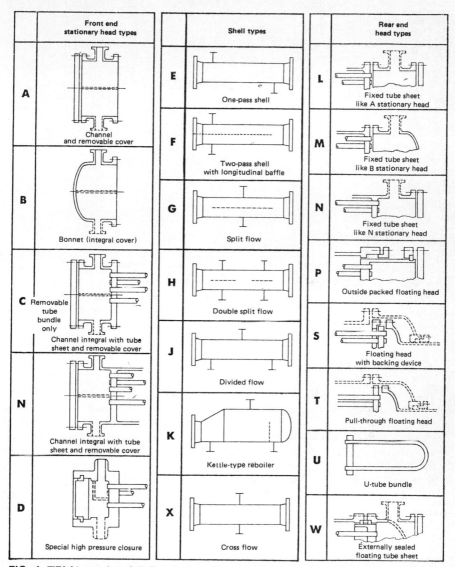

FIG. 1 TEMA notation of shell-and-tube exchangers [1] (courtesy of Tubular Exchangers Manufacturers Association).

for generally severe requirements of petroleum and related processing applications. Class C is for generally moderate requirements for commercial and general process applications. Class B is for chemical process service. The exchangers are built to comply with the applicable ASME (American Society of Mechanical Engineers) Boiler and Pressure Vessel Code, Section VIII. The TEMA standards supplement and define the ASME code for heat exchanger applications. In addition, the state and local codes applicable to the plant location must also be met. In this chapter we use the TEMA standards, but there are other standards in the metric system of units such as DIN 28 008.

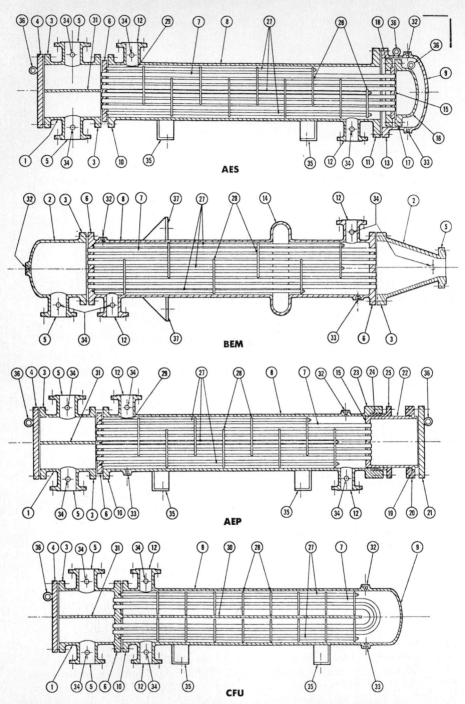

FIG. 2 Nomenclature of exchanger components and illustrations of several types of exchangers [1]. Typical parts and connections, for illustrative purposes only, are numbered for identification.

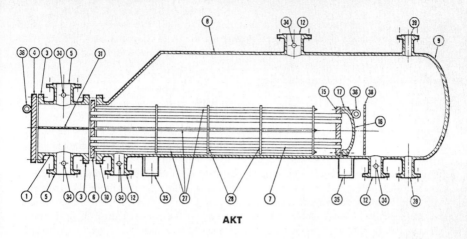

AKT

AJW

FIG. 2 (*Continued*)

1	Stationary head--channel
2	Stationary head--bonnet
3	Stationary head flange-- channel or bonnet
4	Channel cover
5	Stationary head nozzle
6	Stationary tubesheet
7	Tubes
8	Shell
9	Shell cover
10	Shell flange--stationary head end
11	Shell flange--rear head end
12	Shell nozzle
13	Shell cover flange
14	Expansion joint
15	Floating tubesheet
16	Floating head cover
17	Floating head flange
18	Floating head backing device
19	Split shear ring
20	Slip-on backing flange
21	Floating head cover--external
22	Floating tubesheet skirt
23	Packing box
24	Packing
25	Packing gland
26	Lantern ring
27	Tierods and spacers
28	Transverse baffles or support plates
29	Impingement plate
30	Longitudinal baffle
31	Pass partition
32	Vent connection
33	Drain connection
34	Instrument connection
35	Support saddle
36	Lifting lug
37	Support bracket
38	Weir
39	Liquid level connection

The TEMA standards specify the manufacturing tolerances for the various mechanical classes, the range of tube sizes and pitches, baffling and support plates, pressure classification, tube sheet thickness formulas, etc., and must be consulted for all these details.

a. Criteria for Mechanical Selection

Shells.   The E shell is a single-pass shell which is economical and usually the most efficient thermal arrangement; i.e., it has the highest mean temperature difference correction factor (MTD-F factor). However, for multipass tube-side exchangers, if the MTD-F factor is low enough to require two E shells in series then the F shell (two-pass shell) can be used as an equivalent but more economical unit. However, the F shell baffle is subject to fluid and thermal leakage and must be carefully designed and constructed. It also provides more problems in removing or replacing the tube bundle. If shell-side pressure drop becomes limiting then the divided-flow J shell is used; however, there is some loss in thermal efficiency (lower MTD-F factor). If the pressure drop in an F shell is limiting then a split flow as in the G or H shell can be used with some sacrifice in F factor. The X shell is used for large shell flows. In the X shells full-size support plates spaced to prevent tube vibration are used.

Stationary Heads.   There are two basic types of stationary heads: the bonnet and the channel. The bonnet (B) is used for generally clean tube-side fluids; it has fewer joints but does require breaking the piping joints in order to clean or inspect the tubes. The channel head can be removable (A) or integral with the tube sheet (C). It has a removable cover plate allowing easy access to the tubes without disturbing the piping.

Rear End Heads.   The fixed tube sheet (L, M, or N) is a rigid design but allows the least clearance between the shell and the tube bundle. It is limited by differential thermal expansion to moderate temperature differences ($<100°$F or 56 K) between the tubes and shell. Use of a shell expansion joint can raise this limit to $150°$F (83 K). Any number of tube passes can be used. However, the shell side can only be chemically cleaned. Individual tubes can be replaced. This is a low-cost exchanger but slightly higher in cost than the U-tube exchanger.

The U-tube head (U) is a very simple design requiring only one tube sheet but no expansion joints and allowing easy removal; however, individual tube replacement is not possible (an even number of tube passes is required), and tube-side cleaning of the bends is difficult. It is the lowest-cost design.

The outside packed floating head (P) provides for expansion and can be designed for any number of passes. Shell and tube fluids cannot mix if gaskets or packing develop leaks, but the head requires a larger bundle-to-shell clearance as well as side strip baffles. This is a high-cost design.

The split-ring floating head (S) has the tube sheet sandwiched between a removable split ring and the cover, which has a larger diameter than the shell. This permits a smaller clearance between the shell and bundle, and side strip sealing is required for only selected applications. On account of the floating head location the minimum outlet baffle spacing is the largest of any design. Gasket failure is not visible and allows mixing of tube and shell fluids. To remove the bundle or clean the tubes both ends of the exchanger must be disassembled. Cleaning costs somewhat more than for the pull-through type (T), and the exchanger cost is relatively high.

The pull-through floating head (T) can be removed from the shell by disassembling the stationary head. Because of the floating-head flange bolting this design has the largest bundle-to-shell clearance and thus sealing strips are necessary. Even-numbered multipassing is imposed. Again gasket leakage allows mixing of the shell and tube fluids and is not externally visible. Cost is relatively high.

The packed floating head with lantern ring (W) has the lantern ring packing compressed by the rear head bolts. Bundle-to-shell clearance is relatively small. A single- or two-pass arrangement is possible. Potential leakage of either shell or tube fluid is to the atmosphere; however, mixing of these two is possible in the leakage area. The bundle is easily removed, but this design is not recommended, on account of severe thermal fluctuations which can loosen the packing. This is the lowest cost of all floating head designs.

Baffles. Longitudinal baffles in the shell are used to form a two-pass shell flow and are used in place of two E shells in series. Transverse baffling is used to support the tubes, to prevent vibration, and to increase the fluid velocity for improved heat transfer. The transverse baffling can be segmental with or without tubes in the window, multisegmental, or disk-and-doughnut. The single-segmental baffle is most common and is formed by cutting a segment from a disk. As shown in Fig. 3 the cuts are alternately 180° apart and cause the shell fluid to flow back and forth across the tubes more or less perpendicularly.

One disadvantage of the segmental types is the bypassing that occurs between the outer edges of the bundle and the shell. If the pressure drop is too high or more tube supports

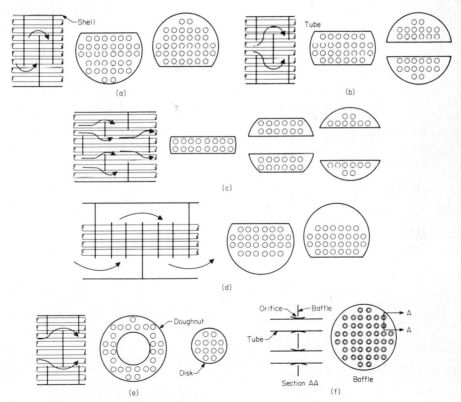

FIG. 3 Plate baffle types. (*a*) Single-segmental baffle. (*b*) Double-segmental baffle. (*c*) Triple-segmental baffle. (*d*) No-tubes-in-window segmental baffle. (*e*) Disk-and-doughnut baffle. (*f*) Orifice baffle.

are needed to prevent vibration the segmental baffle is further subdivided into a double-segmental or strip-baffle arrangement. An alternate method of improving tube support for vibration prevention is to eliminate the window tubes, in which case intermediate support baffles can be used. This requires a larger shell to contain the same number of tubes, but lower pressure drops and improved heat transfer can help to reduce this diameter increase.

Another form of tube support is the rod baffle support (Fig. 4), which is a patented arrangement [2]; however, it is an excellent tube support design. In this case the shell fluid flows more or less parallel to the tubes and shell.

Tube Layouts.   The types of tube layouts used in exchangers are illustrated in Fig. 5. Of these the 30° layout is most common as it gives a good ratio of heat transfer per unit pressure drop and is a compact layout. The 60° layout (not shown) is poorer than the 30°. The 45° and 90° layouts are used when mechanical cleaning is required for the shell side. The 90° layout gives the best performance for a given pressure drop in turbulent flow, but in laminar flow it is the worst. The 45° layout is intermediate, being slightly better than the 30° but not enough to overcome its lower area per unit volume.

Tube Pitch.   The pitch ratio (pitch divided by tube diameter) for TEMA exchangers ranges from 1.25 to 1.5; an additional requirement is a minimum ligament of ⅛ in (3.2 mm) for clean services, but where mechanical cleaning is required the minimum ligament

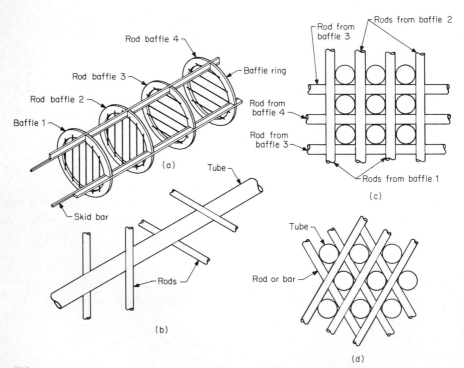

FIG. 4 Rod baffle supports. (*a*) Four rod baffles supported by skid bars (no tubes shown). (*b*) A tube supported by four rods at 90° angle around the periphery. (*c*) A square layout of tubes with rods. (*d*) A triangular layout of tubes with rods.

Cross flow ⟶	θ_{tp}	L_{pn}	L_{pp}
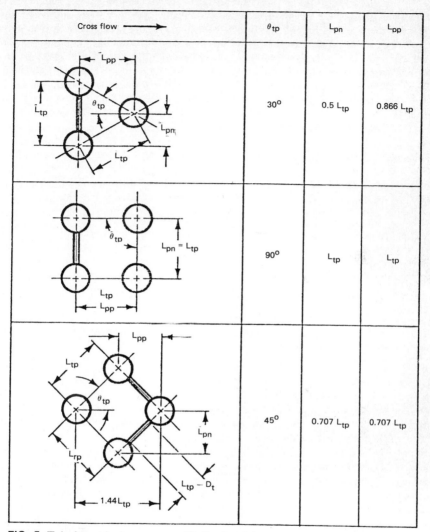	30°	0.5 L_{tp}	0.866 L_{tp}
	90°	L_{tp}	L_{tp}
	45°	0.707 L_{tp}	0.707 L_{tp}

FIG. 5 Tube layout basic parameters [8]. Reprinted, by permission, from *Heat Exchanger Design Handbook,* E. U. Schlünder, editor-in-chief, Hemisphere, Washington, D.C., 1982. Copyright © 1982 by Hemisphere Publishing Corporation.

is ¼ in (6.4 mm). Some companies insist on a ¼-in (6.4-mm) minimum ligament in order to have sufficient strength to hold the tubes and not to distort the ligament under tube rolling pressures.

Tubes. The tubes used are either plain or finned with low fins (about 0.05 in or 1.3 mm) or high fins (generally 0.63 to 0.75 in, or 15.88 to 19.05 mm) with 16 to 19 fins to the inch (630 to 750 fins to the meter). Consult manufacturers' catalogs for dimensions. The plain tubes range from 0.5 to 2 inches in outside diameter. For small exchangers, of less than 8-in (203-mm) shell diameter, smaller tubes and pitches are used, but these exchangers fall outside the range of TEMA standards. Tube diameter and length are based on the type of cleaning to be used. If a drilling operation is required then ¾- or 1-in (19- or 25.4-mm) tubes about 16 ft (4.9 m) long are considered the minimum diameter and maximum length. Longer exchangers are made with tubes up to 40 ft (12.2 m), the length limited by the ability to handle such long exchangers in the shop and in the field.

Tubes are fastened to tube sheets by welding, mechanical rolling, or both. However, these joints are susceptible to thermal and pressure stresses and may develop leaks. In those instances where mixing of the shell and the tube fluids would result in corrosive or other hazardous conditions, then special designs such as double tube sheets are used, with the space between the tube sheets vented. A double tube sheet can be used only in the following rear head designs: fixed tube sheets (L, M, or N) and the outside packed head (P).

Bimetal tubes are used when corrosive conditions of the shell and tube fluids require the use of different metals.

Pass Arrangements.   The number of tube passes per exchanger can range from 1 to 16. If more than one pass is used some loss in efficiency results because of the effect of flow pattern on the mean temperature difference. A design for large numbers of passes results from the need to compensate for low flow rates or the need to maintain high velocities to reduce fouling and get good heat transfer. However, large temperature changes in the tube fluid can, by thermal expansion, cause the floating tube sheets to cock and bind. The passes should be so arranged as to minimize the number of lanes between the passes that are in the same direction as the shell fluid flow. Also the passes should be arranged so that the tube side can be drained and vented.

Shell Nozzles and Impingement Methods.   Whenever a high-velocity two-phase flow is entering the shell some type of impingement protection is required to avoid tube erosion and vibration. The forms such devices can take are shown in Fig. 6: annular distributors (*d*), impingement plates (*a*,*c*), and impingement rods (*b*). The nozzles must also be sized with the understanding that the tube bundle will partially block the opening. In order to provide escape area with impingement plates some tubes may have to be removed. The annular distributor is an excellent design which allows any orientation of the nozzles, provides impingement protection, and allows baffling closer to the tube sheets and thus higher velocities; however, it is a very expensive design. Dummy rods or extra-heavy walled tubes near the nozzles are also good impingement devices.

Drains and Vents.   All exchangers need to be drained and vented; therefore, care should be taken to properly locate and size drains and vents. The proper location depends upon the exchanger design and orientation. Additional openings may be required for instruments such as pressure gauges and thermocouples.

2. Selection Procedure

The selection procedure for a specific design of exchanger involves the consideration of many and often conflicting requirements of process conditions, operation and maintenance. Depending upon the relative importance of these factors as determined by the designer one or several designs may be selected for evaluation.

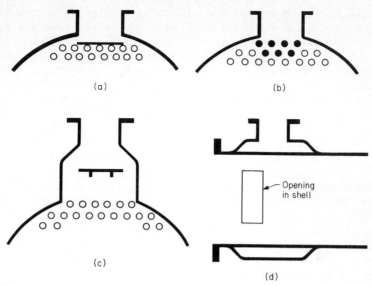

FIG. 6 Impingement protection designs: (*a*) impingement plate; (*b*) impingement rods; (*c*) nozzle impingement baffle; (*d*) vapor belt.

a. Selecting Tube-Side Fluid

The choice of the fluid to be on the tube side will influence the selection of the type of exchanger and requires evaluation of the following factors to arrive at a satisfactory compromise:

1. **Cleanability.** The shell is difficult to clean and requires the cleaner fluid.

2. **Corrosion.** Corrosion or process cleanliness may dictate the use of expensive alloys; therefore, the more corrosive fluids are placed inside the tubes in order to save the cost of an alloy shell.

3. **Pressure.** High-pressure shells, because of their diameters, are thick-walled and expensive; therefore, high-pressure fluids are placed in the tubes.

4. **Temperatures.** The high-temperature fluid should be inside the tubes. High temperatures reduce the allowable stresses in materials and the effect is similar to high pressure in determining shell thickness. Furthermore, safety of personnel may require the additional cost of insulation if the high-temperature fluid is in the shell.

5. **Hazardous or Expensive Fluids.** The more hazardous or expensive fluid should be placed on the tighter side of the exchanger, which is the tube side of some types of exchangers.

6. **Quantity.** A better overall design may be obtained when the smaller quantity of fluid is placed in the shell. This effect may be due to avoiding multipass construction with consequent loss of temperature drop efficiency or to obtaining turbulent flow in the shell at low Reynolds number.

7. **Viscosity.** The critical Reynolds number for turbulent flow in the shell is about 200; hence, when the flow of a fluid in the tubes is laminar, it may be turbulent if that

same fluid is placed in the shell. However, if the flow is still laminar when in the shell, then it is best to place the fluid back inside the tubes as it will be somewhat easier to predict both heat transfer and flow distribution.

8. Pressure Drop.   If the pressure drop of one fluid is critical and must be accurately predicted, then place that fluid inside the tubes. Pressure drop inside tubes can be calculated with less error as the pressure drop in the shell will deviate widely from theoretical values depending upon the shell leakage clearances in the particular exchanger.

b. Selecting Shell and Head

The E shell is the best arrangement; however, if shell-side pressure drop is too high a divided-flow J or G shell may be used. The F shell is a possible alternative when a temperature cross occurs and more than one shell pass is required. Accessibility to the tubes governs the selection of the stationary head, while thermal stress, need for cleaning, possible gasket problems, leakage, plant maintenance experience, and cost are factors influencing the rear head selection. See comments on specific heads above.

c. Selecting Tube Size and Layout

The best ratio of heat transfer to pressure drop is obtained with the smallest tubes; however, the minimum size is determined by the ability to clean the tube. Pressure drop, tube vibration, tube-sheet joints, and cost are several factors limiting the minimum size. Also a reasonable balance between the tube-side and shell-side heat transfer is desired. The ligament between tubes is governed by the pitch ratio and tube size selected; however, for tube-sheet strength, drilling tolerances, and ability to roll a tight tube joint a minimum ligament of ⅛ in (3.2 mm) to ¼ in (6.4 mm) is recommended, the more conservative design using the larger ligament. The pitch ratio and ligament thickness also affect the shell-side fluid velocity and hence the heat transfer and pressure drop.

Tube layouts are either triangular or square, the choice usually depending on the need for shell-side cleaning. The square pitch is particularly suitable for cleaning; however, a larger triangular pitch can also be used. For example, a 1-in (25.4-mm) tube on a 1⅜-in (34.9-mm) triangular pitch will have essentially the same tube count, shell velocities, and heat transfer coefficients and almost the same clearances for cleaning as a 1-in (25.4-mm) tube on a 1¼-in (31.8-mm) square pitch, but the ⅜-in (9.5-mm) ligament will be 50 percent stronger. Other factors including number of tubes and heat transfer for different flow angles (30°, 45°, etc.) are discussed above.

d. Selecting Baffles

The segmental baffle is commonly used unless problems of pressure drop, tube vibration, or tube support dictate the use of double, strip, or rod baffling or a no-tube-in-window configuration. Note that these alternate choices also seriously affect the reliability of the correlations for heat transfer and pressure drop. The segmental baffles are spaced at a minimum distance of 2 in (50.8 mm) or $0.2D_s$, whichever is larger, and a maximum spacing of $1D_s$. The baffle cut also depends linearly on baffle spacing and should be 20 percent of D_s at $0.2D_s$ spacing and 33 percent at $1D_s$. The maximum spacing is also determined by the need for tube support. The TEMA maximum unsupported length depends on tube size and material but ranges from 50 to 80 tube diameters (see standards for specific values). This maximum length usually occurs at the ends of the exchanger in the window area of the first or last baffle since the end baffle spacing generally is greater (due to nozzle location) than the central baffle spacing. Baffle spacing is selected to obtain a high velocity within a pressure drop limit.

e. Selecting Nozzles

Nozzle sizes and impingement devices are related by the TEMA rule-of-thumb value of ρV^2. If $\rho V^2 > 1500 \, \text{lb}_m/(\text{ft} \cdot \text{s}^2)$ or $2250 \, \text{kg}/(\text{m} \cdot \text{s}^2)$, for noncorrosive or nonabrasive fluids, or 500 (750) for other liquids or for any vapor-liquid mixture or saturated vapor, then an impingement device is needed. There are several possible configurations indicated in Fig. 6. Also the entrance into the tube bundle should have a ρV^2 less than $4000 \, \text{lb}_m/(\text{ft} \cdot \text{s}^2)$, or $6000 \, \text{kg}/(\text{m} \cdot \text{s}^2)$. The entrance area is the total free area between a nozzle and the projected area on the tube bundle. Meeting these requirements may require removal of some tubes. Usually such dimensions and area are not available until the mechanical drawings have been made. In the design stage an estimate of these effects is made or a final check calculation is made based on final drawings if the shell pressure drops are marginal. Nozzle locations with respect to the shell flange are governed by pressure vessel codes.

f. Selecting Tube Passes

The number of tube passes is kept as low as possible in order to get simple head and tube-sheet designs. The number of passes if other than a single pass is in even multiples, as then no off-center nozzle on the floating head is required. The flow quantity and the desired minimum tube-side velocity determine the number of tubes per pass, and the total area and tube length then fix the number of passes. However, the number of tube passes must be an even integer; hence, the tube length is the variable.

3. Design Procedures

Large, complex, proprietary computer programs are now used to design heat exchangers. The main usefulness of the design methods given below is to give the computer-aided designer a simple means of checking the computer output and to provide the designer with a "feel" for the effect of design variables. The modified Bell-Delaware method given below is the best available for this purpose; however, it cannot compete in accuracy with computer programs, especially for unusual design conditions. For most standard designs the Bell-Delaware method is quite good and vastly superior to the methods used in the precomputer era.

Two types of calculations are made on exchangers: (1) a rating calculation, which means determining the heat transfer capacity of a specified exchanger; and (2) a design calculation, which determines the size of the exchanger for a specified duty. The design calculations are essentially a series of trial-and-error rating calculations made on an assumed design and modified as a result of these calculations until a satisfactory design is achieved. The overall procedure is outlined below.

1. Calculate the required duty.
2. If not already specified then select the unknown temperatures or fluid flow rates.
3. Guess an overall heat transfer coefficient based on experience or with the help of Table 1.

TABLE 1 Overall Heat Transfer Coefficients for Estimating Exchangers
Values include some allowances for fouling and limiting pressure drop on the
controlling stream to moderate values.

Hot fluid	Cold fluid	Overall heat transfer coefficient	
		Btu/(h·ft²·°F)	W/(m²·K)
Coolers			
Water	Water	250–500	1220–2440
Ammonia	Water	250–500	1220–2440
Aqueous solutions	Water	250–500	1220–2440
Light organics*	Water	75–150	370– 730
Medium organics†	Water	50–125	240– 610
Heavy organics‡	Water	5– 75§	25– 370
Gases	Water	2– 50¶	10– 240
Water	Brine	100–200	490– 980
Light organics	Brine	40–100	200– 490
Heaters			
Steam	Water	200–700	980–3400
Steam	Ammonia	200–700	980–3400
Steam	Aqueous solutions		
	Less than 2 cP	200–700	980–3400
	More than 2 cP	100–500	490–2940
Steam	Light organics	100–200	490– 980
Steam	Medium organics	50–100	240– 490
Steam	Heavy organics	6– 60	30– 300
Steam	Gases	5– 50¶	25– 240
Exchangers			
Water	Water	250–500	1220–2440
Aqueous solutions	Aqueous solutions	250–500	1220–2440
Light organics	Light organics	40– 75	190– 370
Medium organics	Medium organics	20– 60	100– 290
Heavy organics	Heavy organics	10– 40	50– 200
Heavy organics	Light organics	30– 60	150– 300
Light organics	Heavy organics	10– 40	50– 200

*Fluids with viscosities less than 0.5 centipoise.

†Viscosities between 0.5 to 1.0 centipoise.

‡Viscosities greater than 1.0 centipoise = 0.001 N·s/m².

§Pressure drop 20 to 30 lb$_f$/in².

¶These rates are influenced by operating pressures.

Source: D. Q. Kern, *Process Heat Transfer,* McGraw-Hill, New York, 1950.

4. Calculate the area.

5. Based on this area, the flow rates, and the process conditions select suitable types of exchangers for analysis—i.e., shell-and-tube, plate, etc.

6. For shell-and-tube exchangers determine whether a multipass exchanger is required and, if so, how many shells.

7. Select tube size, pitch, and layout.

8. Select tube-side fluid. Select number of passes, shell size, and baffle spacing.

9. Calculate heat transfer and pressure drop.

10. Inspect the results and judge whether the requirements have been met. If not, repeat the steps to step 8 with an estimated change in design until a suitable design is obtained.

11. Get a cost estimate.

12. Based on total costs and sizing of alternate designs determine the best economic choice acceptable to the plant.

13. Submit the design to a selected fabricator for final thermal and mechanical design and fabrication.

4. Design Methods

The basis for current design methods was developed by Tinker [3], but because of the complexity of his equations the method was never widely used in its original form. Later refinements by Palen and Taborek [4] led to the stream analysis method, which is the basis of computer programs. However, many proprietary improvements have been included in these programs. Since later discussions refer to the various flow streams as designated by Tinker a very brief explanation of the Tinker method follows.

a. Tinker Method

As shown in Fig. 7, Tinker divided the shell flow into four basic streams: A is the leakage stream through the tube-baffle clearance in one baffle, B is the cross-flow stream through the tube bundle, C is the bypass stream between the shell and the outside of the tube bundle, and E is the leakage stream through the baffle-shell clearance. An F stream (not shown) occurs only in multipass exchangers and is otherwise considered an equivalent C stream. No D stream was defined. The total flow Q_{TOT} is then

$$Q_{TOT} = 2Q_A + Q_B + 2Q_C + 2Q_F \tag{1}$$

Tinker considered the C and E streams ineffective and only one-fourth of the A streams, or 0.5A as effective in the B stream (note that two A streams are indicated). Thus the ratio of the effective cross-flow stream Q_{eff} to the total stream is

$$\frac{Q_{eff}}{Q_{TOT}} = \frac{Q_B + 0.5Q_A}{Q_{TOT}} \tag{2}$$

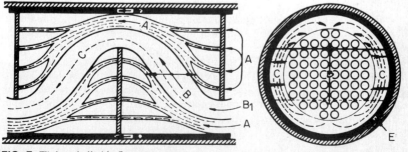

FIG. 7 Tinker shell-side flow streams.

The direct determination of these streams is impossible, but by defining a "relative" stream flow the effective stream ratio can be determined. A relative flow quantity can be calculated as

$$Q_{RA} = A_a \left(\frac{1 + x'}{K_A} \right)^{0.5} \tag{3}$$

where Q_{RA} = relative flow of the A stream
A_a = total leakage area through one baffle
K_A = flow resistance through the clearance expressed in terms of velocity head
x' = ratio of window pressure drop to pressure drop across one baffle pass

Likewise the E stream is

$$Q_{RE} = A_e \left(\frac{1 + x'}{K_E} \right)^{0.5} \tag{4}$$

while the B stream is

$$Q_{RB} = A_b \left(\frac{1}{K_B} \right)^{0.5} \tag{5}$$

and the C stream

$$Q_{RC} = A_c \left(\frac{1}{K_C} \right)^{0.5} \tag{6}$$

Then

$$\frac{Q_{\text{eff}}}{Q_{\text{TOT}}} = \frac{Q_{RB} + 0.5Q_{RA}}{2Q_{RA} + Q_{RB} + 2Q_{RC} + 2Q_{RE}} \tag{7}$$

The value of x', the window-to-bundle pressure drop ratio, must be initially assumed but will quickly converge after a few iterations. The K_A and K_E values can be obtained from sources such as Bell [5]. The value of K_B is obtained from friction factor charts and K_C from data on bypass flows as in Ref. 6. These flow resistance values, however, are functions of the leakage and flow-path Reynolds number and thus require further iteration procedures. Thus a complex computer program is required to utilize the stream analysis method, and many of the details have not been published. Once the B stream is determined then the ideal tube-bundle j and f factors are used to calculate heat transfer and pressure drop.

b. Bell-Delaware Method and Further Improvements by Taborek

A simpler method suitable for hand calculation and possible programming on the newer calculators was developed by Bell [6] and is known as the Bell-Delaware method. The original method has been further refined by Bell [7], and other modifications and improvements have been made by Taborek [8], principally in the R_b and R_l curves and some lesser changes in J_b and J_l. The approach used by Bell differs significantly from the Tinker method in that the ideal heat transfer and pressure drop are calculated on the basis of total flow across the bundle and then five correction factors are applied to approximate the effect of leakages and bypassing. A detailed step-by-step sequence is given below.

The basic equation is

$$h_s = h_i \times J_c J_l J_b J_s J_r \tag{8}$$

where h_i = coefficient for pure cross flow in an ideal tube bank assuming the entire shell-side stream flows across the tube array at the centerline of the exchanger.

J_c = correction factor for baffle cut. This correction includes the effect of window and bundle heat transfer. This value is 1.0 for an exchanger with no tubes in the window and increases to 1.15 for small baffle cuts and decreases to 0.65 for large baffle cuts. For a typical well-designed heat exchanger the value is near 1.0.

J_l = correction factor for baffle leakage effects including both tube-baffle and baffle-shell leakage (A and E stream) with a heavy weight given to the latter. It is a function of the ratio of leakage to cross-flow area and a function of the clearances. A typical value of J_l is in the range of 0.7 to 0.8.

J_b = correction factor for bundle bypass (C and F stream). For a fixed tube-sheet construction $J_b \approx 0.9$ and for a pull-through $J_b \approx 0.7$, which can be increased to approximately 0.9 by the use of side strips.

J_s = correction factor for variable baffle spacing at the inlet and outlet sections. The nozzle locations result in larger baffle spacing and lower velocities and thus lower heat transfer coefficients. J_s usually ranges from 0.85 to 1.0.

J_r = correction factor for any adverse temperature gradient buildup in laminar flows. This correction applies only for shell-side Reynolds numbers of less than 100.

The combined effect of all these corrections is typically around 0.6 for well-designed exchangers but can be as low as 0.4 in other designs.

Shell-side pressure drop is the summation of pressure drops in the inlet and outlet sections and the baffle central section, consisting of the cross-flow and window pressure drop. The following correction factors are used:

R_l Correction for leakage (A and E) streams. This correction is based on the same factor as J_l but is of different magnitude. Usually $R_l \approx 0.4$ to 0.5 although lower values are possible with small baffle spacing.

R_b Correction factor for bypass flow (C and F stream). It is different in magnitude from J_b and ranges from 0.5 to 0.8.

R_s Correction for inlet and outlet sections having a baffle spacing different from the central section.

Now let ΔP_{bi} be the pressure drop for cross flow in an ideal tube bank and ΔP_{wi} for the window section. These are based on total flow as in the case of the heat transfer calculation.

The three components of the total pressure drop (excluding nozzles) are

$$\Delta P_{\text{tot}} = \Delta P_c + \Delta P_w + \Delta P_e \tag{9}$$

1. Pressure drop for cross flow in the central section between baffle tips is

$$\Delta P = \Delta P_{bi} (N_b - 1) R_l R_b \tag{10}$$

where N_b is number of baffles.

2. Pressure drop in the window is

$$\Delta P_w = \Delta P_{wi} N_b R_l \tag{11}$$

3. Pressure drop in inlet and outlet sections is

$$\Delta P_e = \Delta P_{bi} \frac{N_c + N_{tcw}}{N_{tcc}} R_s R_b \tag{12}$$

where N_{tcc} and N_{tcw} are the number of tube rows between baffle tips and in the baffle window, respectively.

While each of the corrections varies over a wide range, the total shell-side pressure drop of a typical exchanger is about 20 to 30 percent of the pressure drop for a similar exchanger but without leakage or bypass.

5. Auxilary Calculations

Before the correction factors can be determined it is necessary to calculate various leakage and flow areas.

a. Segmental Baffle Window Calculations

The segmental baffle geometry in relation to the tube field is shown in Fig. 8. The central angles of the baffle cut with the inside of the shell, θ_{ds}, and the angle intersecting the tube diameter $D_{ctl} = D_{otl} - D_t$ are

$$\theta_{ds} = 2 \cos^{-1} \left(1 - 2 \frac{B_c}{100} \right) \tag{13}$$

$$\theta_{ctl} = 2 \cos^{-1} \left[\frac{D_s}{D_{ctl}} \left(1 - 2 \frac{B_c}{100} \right) \right] \tag{14}$$

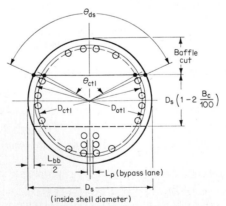

FIG. 8 Basic baffle geometry relations [8]. Reprinted, by permission, from *Heat Exchanger Design Handbook*, E. U. Schlünder, editor-in-chief, Hemisphere, Washington, D.C., 1982. Copyright © 1982 by Hemisphere Publishing Corporation.

If the baffles are of the "window type," where the area between D_s and D_{otl} is blocked, then the angle is referenced to D_{otl}, or

$$\theta_{otl} = 2 \cos^{-1} \left[\frac{D_s}{D_{otl}} \left(1 - 2 \frac{B_c}{100} \right) \right]$$ (15)

b. Baffle Window Flow Areas

The gross window flow area—i.e., without tubes—A_{wg}, is

$$A_{wg} = \frac{\pi}{4} D_s^2 \left(\frac{\theta_{ds}}{360°} - \frac{\sin \theta_{ds}}{2\pi} \right)$$ (16)

If "window baffles" are used then replace θ_{ds} with θ_{otl}. Assuming a uniform tube field the fraction of tubes in one window, F_w, and the fraction in cross flow between baffle tips is

$$F_w = \frac{\theta_{ctl}}{360°} - \frac{\sin \theta_{ctl}}{2\pi}$$ (17)

$$F_c = 1 - 2F_w$$ (18)

If the tube field is not uniform because of pass lanes or tubes removed in the nozzle entry zone then this correction has to be separately calculated or neglected and a minor error accepted.

The baffle window area occupied by the tubes, A_{wt}, is

$$A_{wt} = N_t F_w \frac{\pi D_t^2}{4}$$ (19)

and the net flow area through one window, A_w, is

$$A_w = A_{wg} - A_{wt}$$ (20)

c. Equivalent Hydraulic Diameter for the Window, D_w

This value is required only for pressure drop calculation in the laminar region ($Re_s <$ 100). It is the classical hydraulic diameter except that the baffle edge is omitted.

$$D_w = \frac{4 A_w}{\pi D_t F_w N_t + \pi D_s (\theta_{ds}/360°)}$$ (21)

d. Number of Effective Tube Rows in Cross Flow, N_{tcc} and N_{tcw}

The number of tube rows is a function of tube layout and pitch parameters. The value of L_{pp} is defined in Fig. 5. For one cross-flow section between baffle tips,

$$N_{tcc} = \frac{D_s}{L_{pp}} \left(1 - 2 \frac{B_c}{100} \right)$$ (22)

The effective number of tube rows in the baffle window, N_{tcw}, is based on interpretation of the window flow pattern and some data. Although elaborate proprietary methods are used, the effective penetration based on Delaware and other data can be approximated as 0.4 times the tube field in the window. This distance is crossed twice in each window; hence

$$N_{tcw} = \frac{0.8}{L_{pp}} D_s \left(\frac{B_c}{100} - \frac{D_s - D_{ctl}}{2} \right)$$ (23)

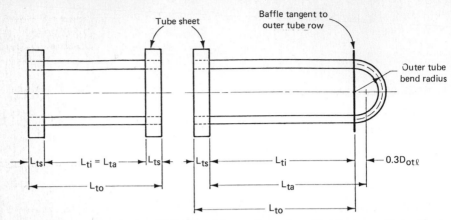

FIG. 9 Tube length definitions [8]. Reprinted, by permission, from *Heat Exchanger Design Handbook*, E. U. Schlünder, editor-in-chief, Hemisphere, Washington, D.C., 1982. Copyright © 1982 by Hemisphere Publishing Corporation.

e. Number of Baffles, N_b

The number of baffles, N_b, is determined from drawings; however, in design-type calculations the number is determined from tube length and the central baffle spacing even if larger end spacings are used:

$$N_b = \frac{L_{ti}}{L_{bc}} - 1 \tag{24}$$

The tube length L_{ti} is defined in Fig. 9; note the different interpretation for U tubes. The value of N_b should be an integer, and if larger end spacings are used it should be rounded off downwards.

f. Bundle-Shell Bypass

The flow between the shell and the tube bundle can attain high velocities due to the lower resistance and amount to 20 to 30 percent of the flow. Bypass sealing strips are used to restrict this flow. There can be an additional bypass stream through the bundle in multipass exchangers if the pass partition lanes are parallel to the cross flow. Usually these lanes are made perpendicular to the cross flow, or tie rods or baffles are placed in the lane to restrict this bypass. Accordingly these F-stream bypasses are usually neglected. The bypass area within one baffle, A_{ba}, is

$$A_{ba} = L_{bc}\left[(D_s - D_{otl}) + L_{pl}\right] \tag{25}$$

where L_{pl} is the tube lane partition bypass width as follows:

$$L_{pl} = \begin{cases} 0 \text{ for all standard calculations.} \\ \text{½ of real dimension of tube lane partition, } L_p. \text{ For estimation purposes} \\ \text{assume } L_p = D_t. \end{cases}$$

g. Shell-Baffle Leakage Area A_{sb}

The diametral clearance between the shell diameter D_s and the baffle diameter D_b is designated L_{sb}. See the TEMA standards [1] for allowable clearances. The shell-baffle leakage area within the circle segment of the baffle is

$$A_{sb} = \pi D_s \frac{L_{sb}}{2} \frac{360° - \theta_{ds}}{360°} \tag{26}$$

h. Tube-Baffle Leakage Area A_{tb}

The diametral clearance L_{tb} can be estimated from Ref. 1 or can be determined from specified manufacturing tolerances. The total leakage area for a baffle is

$$A_{tb} = \frac{\pi}{4} [(D_t + L_{tb})^2 - D_t^2] N_t (1 - F_w) \tag{27}$$

i. Bundle Cross-Flow Area A_{mb}

This is the minimum cross-flow area for one baffle at the shell centerline determined as follows:

$$A_{mb} = L_{bc} \left[L_{bb} + \frac{D_{ctl}}{(L_{tb})_{eff}} (L_{tp} - D_t) \right] \tag{28}$$

where L_{bc} is the central baffle spacing; $(L_{tp})_{eff}$ is L_{tp} for 30° and 90° layouts and $0.707 L_{tp}$ for the 45° layout (Fig. 5); and D_{ctl} is defined in Fig. 8.

6. Shell-Side Heat Transfer Coefficient

The shell-side heat transfer coefficient is calculated by the following sequence:

1. Calculate maximum mass velocity, G_s:

$$G_s = \frac{W_T}{A_{mb}^*} \qquad \text{kg/(m}^2 \cdot \text{s) or lb}_\text{m}/(\text{h} \cdot \text{ft}^2) \tag{29}$$

where A_{mb}^* is A_{mb} corrected to m² or ft².

2. Calculate shell-side Reynolds number:

$$\text{Re}_s = \frac{D_t G_s}{\mu} \tag{30}$$

3. From the appropriate curve for tube layout and pitch in Fig. 10 read the j and f factors.

4. Calculate the ideal bundle heat transfer coefficient, h_i:

$$h_i = j c_p G_s \, \text{Pr}_s^{-2/3} \, \phi_s \tag{31}$$

where Pr_s = Prandtl number based on property values at average shell temperature
    ϕ_s = correction factor for viscosity gradient

For liquids, $\phi_s = (\mu_s/\mu_w)^{0.14}$, where μ_w is the viscosity at wall temperature. For gases being cooled $\phi_s = 1$. For gases being heated, $\phi_s = [(T_s + 273)/(T_w + 273)]^{0.25}$ where T_s is average shell temperature and T_w is the wall temperature in °C.

The ϕ_s correction is a weak function and even an approximate wall temperature will suffice.

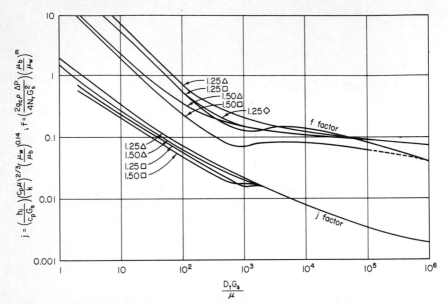

FIG. 10 Heat transfer and friction factors for cross flow in tube bundles. Here $m = 0.57\ Re_s^{-0.25}$

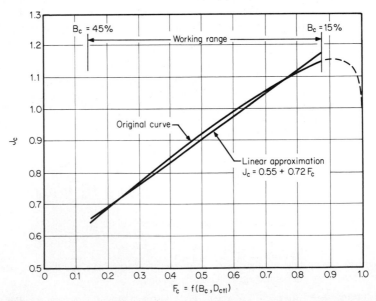

FIG. 11 Segmental baffle window correction factor J_c as function of fraction of tubes in cross flow F_c [8]. Reprinted, by permission, from *Heat Exchanger Design Handbook*, E. U. Schlünder, editor-in-chief, Hemisphere, Washington, D.C., 1982. Copyright © 1982 by Hemisphere Publishing Corporation.

5. Determine the segmental baffle window correction J_c from Fig. 11 or the equation $J_c = 0.55 + 0.72F_c$, where F_c is calculated in Eq. (18).

6. Determine the baffle leakage correction for heat transfer, J_l, and pressure drop R_l from Figs. 12 and 13. The ratios r_{lm} and r_s are determined from area values calculated in Secs. A.5.g, h, and i, where

$$r_{\ell m} = \frac{A_{sb} + A_{tb}}{A_{mb}} \tag{32}$$

and.

$$r_s = \frac{A_{sb}}{A_{sb} + A_{tb}} \tag{33}$$

7. Determine the bundle bypass correction factors J_b and R_b from Figs. 14 and 15. The ratios r_b and N_{ss}^+, below, are determined from area values calculated in Secs. A.5.f and i and the number of effective rows in Sec. A.5.d and the specified number (pairs) of sealing strips, N_{ss}. The ratios are

$$r_b = \frac{A_{ba}}{A_{mb}} \tag{34}$$

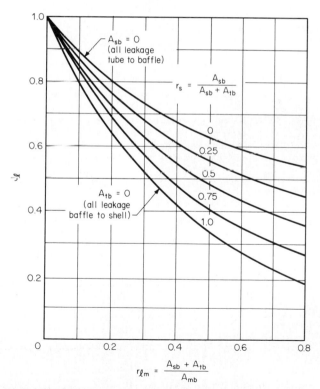

FIG. 12 Baffle leakage heat transfer correction factor J_l [8]. Reprinted, by permission, from *Heat Exchanger Design Handbook*, E. U. Schlünder, editor-in-chief, Hemisphere, Washington, D.C., 1982. Copyright © 1982 by Hemisphere Publishing Corporation.

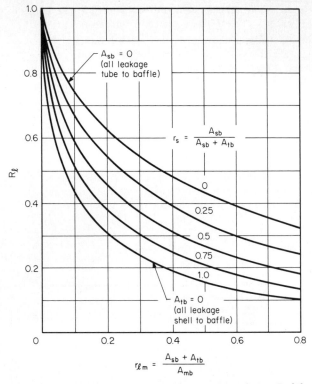

FIG. 13 Baffle leakage pressure drop correction factor R_l [8]. Reprinted, by permission, from *Heat Exchanger Design Handbook*, E. U. Schlünder, editor-in-chief, Hemisphere, Washington, D.C., 1982. Copyright © 1982 by Hemisphere Publishing Corporation.

and

$$N_{ss}^+ = \frac{N_{ss}}{N_{tcc}} \tag{35}$$

8. Determine for laminar-flow cases the adverse temperature correction J_r. This is a function of the number of tube rows crossed as calculated below:

$$N_c = N_{tcc} + N_{tcw} \tag{36}$$

with limits

$$J_r = \begin{cases} 1 & \text{for } \mathrm{Re}_s > 100 \\ \left(\dfrac{10}{N_c}\right)^{0.18} & \text{for } \mathrm{Re}_s \le 20 \end{cases} \tag{37}$$

Use a linear interpolation for $20 < \mathrm{Re} < 100$.

9. Determine the heat transfer correction for unequal end spacing J_s from Fig. 16. Here N_b is from Sec. A.5.e and $L^+ = L_{bo}/L_{bc} = L_{bi}/L_{bc}$. The figure is based on turbulent flow and $L^+ = L_i^+ = L_o^+$ where $L_i^+ = L_{bi}/L_{bc}$ and $L_o^+ = L_{bo}/L_{bc}$. L_{bi} is the inlet

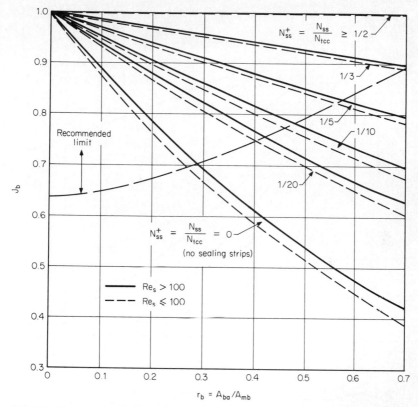

FIG. 14 Heat transfer correction factor J_b for shell-to-bundle bypass as function of r_b, with effect of sealing strips per tube rows crossed, N_{ss}^+ as parameter [8]. Reprinted, by permission, from *Heat Exchanger Design Handbook*, E. U. Schlünder, editor-in-chief, Hemisphere, Washington, D.C., 1982. Copyright © 1982 by Hemisphere Publishing Corporation.

baffle spacing, L_{bo} is the outlet spacing, and L_{bc} is the central spacing (see Fig. 17). Usually the inlet and outlet spacings are equal. However,

$$J_s = \frac{(N_b - 1) + (L_i^+)^{1-n} + (L_o^+)^{1-n}}{(N_b - 1) + L_i^+ + L_o^+} \tag{38}$$

where $n = 0.6$ for turbulent flow.

For laminar flow the correction is about halfway between 1 and J_s computed for turbulent conditions.

10. Calculate the shell-side heat transfer coefficient:

$$h_s = h_i \times J_c J_l J_b J_r J_s \tag{8}$$

11. Calculate the overall heat transfer coefficient, adding in the effects of fouling, tube-wall, and tube-side coefficients.

12. Based on the exchanger and flow geometry and the fluid temperatures determine the true mean temperature difference.

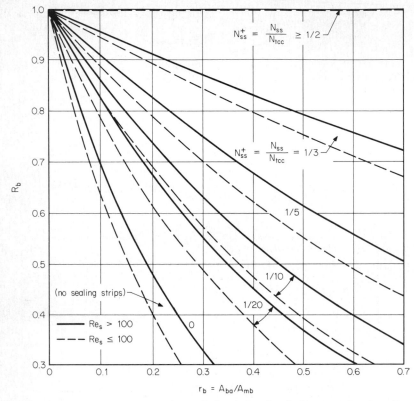

FIG. 15 Pressure drop correction factor R_b for shell-to-bundle bypass as function of r_b, with effect of sealing strips per tube rows crossed, N_{ss}^+ as parameter [8]. Reprinted, by permission, from *Heat Exchanger Design Handbook*, E. U. Schlünder, editor-in-chief, Hemisphere, Washington, D.C., 1982. Copyright © 1982 by Hemisphere Publishing Corporation.

13. For the specified heat load q, calculate the theoretical exchanger area $A = q/(U \, \Delta T)$. If this area differs from the actual area in the rated exchanger beyond a desired safety factor then modify the trial design and repeat the calculations from step 1.

7. Shell-Side Pressure Drop

The shell-side pressure drop of this method is the sum of the cross-flow pressure drop ΔP_c, the baffle window pressure drop ΔP_w, and the pressure drop associated with the inlet and outlet end zones, ΔP_e. The procedure continues as follows:

14. Calculate the ideal bundle pressure drop:

$$\Delta P_{bi} = 2fN_{tcc}\frac{G_s^2}{\rho_s}\phi_s \tag{39}$$

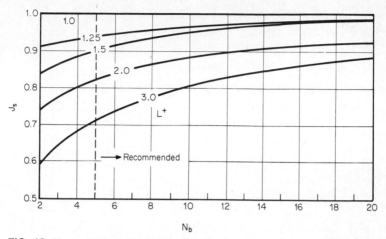

FIG. 16 Unequal inlet/outlet baffle spacing correction factor J_s for heat transfer as function of number of baffles N_b and baffle spacing ratio L^+ [8]. Reprinted, by permission, from *Heat Exchanger Design Handbook*, E. U. Schlünder, editor-in-chief, Hemisphere, Washington, D.C., 1982. Copyright © 1982 by Hemisphere Publishing Corporation.

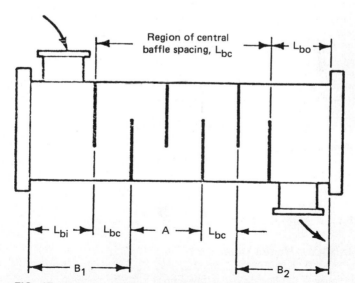

FIG. 17 Schematic sketch of baffle distribution. Maximum unsupported tube span $(L_b)_{max}$ in L_{bc} region (A) and in the inlet/outlet region (B_1 and B_2) [8]. Reprinted, by permission, from *Heat Exchanger Design Handbook*, E. U. Schlünder, editor-in-chief, Hemisphere, Washington, D.C., 1982. Copyright © 1982 by Hemisphere Publishing Corporation.

where f is from step 3, G_s is from Eq. (29), N_{tcc} is from Eq. (22), and ϕ_s is as used in step 4.

15. Calculate the cross-flow pressure drop:

$$\Delta P_c = \Delta P_{bi}(N_b - 1)R_b R_l \qquad (10)$$

where N_b is from Eq. (24) and R_b and R_l are from Figs. 15 and 13, respectively.

16. Calculate the window pressure drop. For turbulent flow ($\text{Re}_s > 100$),

$$\Delta P_w = N_b(2 + 0.6N_{tcw})\frac{G_w^2}{2\rho_s} R_l \qquad (40)$$

Here

$$G_w = \frac{W_T}{\sqrt{A_{mb}A_w}} \qquad (41)$$

N_b and N_{tcw} are from Eqs. (24) and (23) and R_l is from Fig. 13. For laminar flow the equation is more complicated:

$$\Delta P_w = N_b \left\{ 26 \frac{G_w \mu_s}{\rho_s} \left[\frac{N_{tcw}}{L_{tp} - D_t} + \frac{L_{bc}}{D_w^2} \right] + 2(10^{-3})\frac{G_w^2}{2\rho_s} \right\} R_l \qquad (42)$$

The turbulent and laminar equations will not quite agree at the breakpoint ($\text{Re}_s = 100$), but the larger value is the conservative approach.

17. Calculate the end-zone pressure drops:

$$\Delta P_e = \Delta P_{bi}\left(1 + \frac{N_{tcw}}{N_{tcc}}\right)R_b R_s$$

where R_b is from Fig. 15, ΔP_{bi} is from Eq. (39), and N_{tcw} and N_{tcc} are from Eqs. (23) and (22). Here

$$R_s = \left(\frac{L_{bc}}{L_{bo}}\right)^{2-n} + \left(\frac{L_{bc}}{L_{bi}}\right)^{2-n}$$

where $n = 0.2$ for turbulent flow and 1.0 for laminar flow.

18. Calculate shell-side pressure drop:

$$\Delta P_s = \Delta P_c + \Delta P_w + \Delta P_e \qquad (9)$$

If this pressure drop is outside the desired range then modify the trial exchanger and repeat from step 1.

8. External Low-Finned Tubes

The above method can be extended to low-finned tubes by modifying some of the dimensions and areas as described below. After making these modifications the heat transfer coefficient and pressure drop are calculated the same as for plain tubes. The low-finned tubes have 630 to 1000 fins per meter (16 to 26 fins per inch) and the usual fin height is 1 to 2 mm (1/16 in). The surface area of the finned tube is 2 to 3 times the surface of a

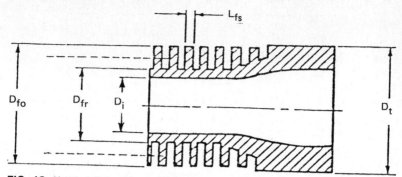

FIG. 18 Sketch defining finned-tube nomenclature [8]. Reprinted, by permission, from *Heat Exchanger Design Handbook,* E. U. Schlünder, editor-in-chief, Hemisphere, Washington, D.C., 1982. Copyright © 1982 by Hemisphere Publishing Corporation.

plain tube. Figure 18 gives the basic geometry. An "equivalent projected diameter" is used for some modifications and is

$$D_{feq} = D_{fr} + 2L_{fh}N_fL_{fs}$$

where L_{fh} is the fin height $(D_{fo} - D_{fr})/2$

Based on the above D_{feq} and dimensions shown in Fig. 18 the following changes are made in the above equations:

Eq. (16)   Flow area A_{wt}. In place of D_t use D_{fo}.

Eq. (21)   Diameter D_w. In place of D_t use D_{fo}.

Eq. (27)   Tube-baffle leakage A_{tb}. Use D_{fo} in place of D_t to determine L_{tb} as well as in the equation.

Eq. (28)   Bundle cross-flow area A_{mb}. Use D_{feq} in place of D_t.

Eq. (30)   Shell Reynolds number. Use D_{feq} for D_t.

Step 3. Use the plain-tube curves to get j_{plain} and f_{plain}. If $Re_s \leq 1000$ apply the J_f correction from Fig. 19 so that

$$j = J_f j_{plain} \tag{43}$$

This correction is necessary because a boundary-layer overlap occurs, making the fins less effective.

Step 4. With the above j the ideal bundle heat transfer coefficient is calculated. However, this coefficient must be corrected for a fin efficiency as described elsewhere in the handbook. For most services the fin efficiency will be 0.9 or higher.

Step 14. The ideal bundle pressure drop is calculated using 1.4 times f_{plain} as recommended by Taborek [8], which is different from Bell's recommendation [7].

9. Extensions to Other Shell Geometries

The above method is easily adapted to the divided-flow J shell and can be used for the two-pass shell (F shell) provided the geometrical parameters are modified to recognize

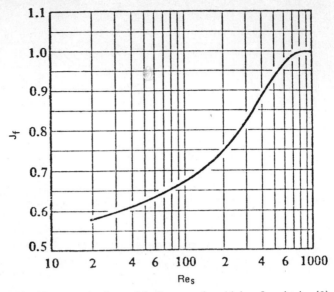

FIG. 19 Correction factor J_f for heat transfer with low-finned tubes [8]. Reprinted, by permission, from *Heat Exchanger Design Handbook,* E. U. Schlünder, editor-in-chief, Hemisphere, Washington, D.C., 1982. Copyright © 1982 by Hemisphere Publishing Corporation.

that the segmental baffle has been split in two halves. The no-tube-in-window, double-segmental, and disk-and-doughnut baffling are geometries not easily adapted to this method and have not been tested.

B. SPIRAL HEAT EXCHANGERS

The spiral plate exchanger is assembled by taking two long strips of plate which are parallel but spaced a fixed distance apart by means of studs, and wrapping them together to form two concentric spiral passages as shown in Fig. 20. Normal design calls for closing

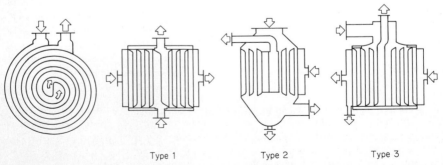

Type 1   Type 2   Type 3

FIG. 20 Spiral heat exchanger (courtesy Alfa-Laval AB, Sweden).

one passage at one side and the other passage on the opposite side. Then by use of covers on each side the unit is sealed. Headers and nozzles are attached to the ends of the spiral passages. Passage closures and covers can be altered to form condensers and vaporizers. These exchangers can be fabricated from any material that can be cold-worked and welded.

The strips for the spiral are from 4 to 72 in (100 to 1800 mm) wide. Spacing of either passage can vary from ³⁄₁₆ to 1 in (4.8 to 25.4 mm). The maximum diameter is 56 in (1422 mm) so that the maximum area per unit is 1800 ft² (167 m²). The maximum design pressure is 150 psi (1 MN/m²).

The advantages of the spiral design are that only a single passage for each fluid is used, and that the length of passage can be up to 200 ft (61 m) and thus close approaches and close control can be obtained. Furthermore, because of the single passage it can handle slurries and sludges, and is very good for cooling or heating viscous fluids. The exchangers are very compact. Major limitations are the pressure limit, available cross section and surface areas, and the difficulty of repairing the system. However, repairs to the sides of the spirals (the welding) can be made.

There is little published information concerning heat transfer and pressure drop, but see Refs. 9, 10, and 11. For preliminary rating the conventional equations for tubular flow can be used with the hydraulic diameter or twice the plate clearance. However, in Ref. 10 the following equation for the transition and turbulent regions is recommended:

$$\text{Nu} = \text{Pr}^{0.25} \left(\frac{\mu}{\mu_w} \right)^{0.17} \left[0.0315 \ \text{Re}_{av}^{0.8} - 6.65(10)^{-7} \left(\frac{L}{l} \right)^{1.8} \right] \qquad (44)$$

where L is the length of the plate and l is the space between plates.

The presence of the stud spacers and the effect of the curvature result in higher heat transfer coefficients than are normally obtained for smooth surfaces. In Ref. 11 the predicted heat transfer coefficients are approximately twice those of in-tube. It appears that the transition region extends from $\text{Re}_{av} = 100$ to $\text{Re}_{av} = 10,000$ and that the curve is nearly linear in this region and is substantially higher than the in-tube curves. Above $\text{Re}_{av} = 10,000$ the curves appear about 50 to 60 percent greater than the in-tube curve. In the spiral plate the temperature difference is different for each face; the outer face difference is different from the inner face by the effect of the temperature rise in one spiral turn. Baird et al. [11] discuss this effect and propose a means of predicting it. Normally these differences are ignored [10] and the conventional LMTD is used; also, for condensation or boiling the usual equations for tubes are used, but modified for plates.

There is another type of spiral exchanger which consists of tubes coiled into a flat spiral and stacked one on top of another so that another spiral passage is formed on the outside of the tubes. These are smaller exchangers that are mostly used as oil coolers on engines, tractors, etc. They are available in a range of standard sizes in a number of materials and are suitable for higher-pressure operation. They can be approximately rated by using equations for flow inside coiled tubes.

C. PLATE HEAT EXCHANGERS

A plate heat exchanger (Fig. 21) resembles a plate-and-frame filter press. It has both a fixed and a movable end plate which are not heat transfer surfaces. Pressed between these end plates are corrugated or embossed plates having ports in the corners and gasketed, as shown in Fig. 22. The fluids flow in alternate spaces between the plates. The embossing patterns are so arranged that the plates are supported every few inches.

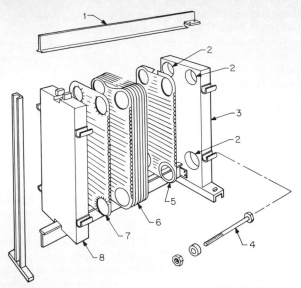

FIG. 21 Exploded view of plate exchanger [13]. (1) Carrying bar. (2) Connections. (3) Fixed frame. (4) Tightening bolt. (5) End plate with four holes. (6) Channel plate with four holes. (7) End plate with all four ports blind. (8) Pressure plate.

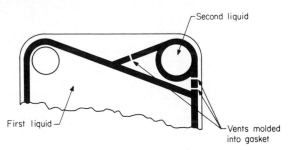

FIG. 22 Plate exchanger gasketing detail [13]. Double gasketing prevents fluid mixing.

Plates are available in a wide variety of metals and alloys; gaskets are available in nitrile, butyl, silicone, fluorocarbon rubber, and in certain cases compressed asbestos. Exchangers have been made with 1500 m² (16,000 ft²) of surface, with up to 700 plates, and with ports up to 40 cm (15.7 in). Plates range from 0.03 to 2.5 m² (0.3 to 26.9 ft²), 0.5 to 1.2 mm (0.02 to 0.05 in) thick, and 1.5 to 5 mm (0.06 to 0.2 in) in spacing. The operating pressure ranges from 0.1 to 1.5 MPa (14.5 to 217 lb$_f$/in²). Temperatures are limited by the gaskets and for rubbers range from -25 to 150°C (-13 to 300°F) and up to 260°C (500°F) for asbestos. Port velocities range up to 5 m/s (16.4 ft/s), with maximum flows of 2500 m³/h (1471 ft³/min). The number of transfer units (NTU) ranges from 0.3 to 4 per pass, and optimum pressure drops are 30 kPa/NTU (4.4 psi).

The advantages of plate exchangers are as follows: they have higher heat transfer rates and they produce less fouling than shell-and-tube exchangers, they are easy to clean and it is easy to change their plates if process conditions are changed, they can handle slurries provided the particles are less than 0.5 mm (0.02 in), they require less space, and they are generally less expensive than shell-and-tube exchangers. The disadvantages are that the choice of fluids is limited by the chemical resistance and temperature limits of the gaskets, that the large amount of gasketing leakage can be serious, that pressure is limited to 1.5 MPa (217 psi), and that pressure drop in the ports limits flow rates.

The principal use of the plate exchanger is in liquid-liquid heat exchange. A wide range of viscosities are handled but the cooling of very viscous fluids can result in some maldistribution. Some condensation can be done, depending upon allowable pressure drops. In general, pressure drop in the ports makes gas-to-gas heat exchange undesirable.

Equations for approximating heat transfer and pressure drop [14] are:

For turbulent flow:

$$Nu = 0.25 \, Re_{av}^{0.65} \, Pr^{0.4} \tag{45}$$

For laminar flow:

$$h = 0.74 c_p G \, Re_{av}^{-0.62} \, Pr^{-2/3} \left(\frac{\mu_{av}}{\mu_w} \right)^{0.14} \tag{46}$$

where D_h is the hydraulic equivalent diameter or approximately 2 times the clearances and $Re_{av} = D_h G / \mu$.

For pressure drop:

$$\Delta P = \frac{2 f G^2 L}{g_c D_e \rho} \tag{47}$$

where $f = 2.5 / Re_{av}^{0.3}$.

An extensive discussion of design equations and methods is given in the second of a series of papers by Raju and Bansal [14].

Many different flow patterns are possible such as single-pass, multipass with equal passes, and multipass with unequal passes. If the flow volumes of the streams differ widely, then the unequal-pass arrangement is used. These various patterns also affect the LMTD corrections as shown in Figs. 25 to 31 in Part 1 of this chapter.

Distribution of fluid between the various plates is affected by the end connections. The U loop with all connections at the fixed head tends to give unequal flow distribution among the plates arranged for parallel flows. The Z connection (one set of connections on the movable head) provides for more uniform distribution. The LMTD correction charts are based on an assumed uniform distribution.

The plate design, whether an intermated corrugated design or the chevron design, affects both heat transfer and pressure drop. The angle of the chevrons is also important. Intermediate results are obtained by mixing the plates, e.g., by using a combination of wide- and narrow-angle chevron plates.

Plate dimensions are dependent upon port size, which governs the plate width because of the flange dimensions on the nozzles, and the plate length, which is determined by the desired aspect ratio (heat transfer length to flow width), which has a lower limit of 1.8.

Plate designs vary from manufacturer to manufacturer and so will their performance. All this information is proprietary; hence, available equations in the literature [12, 13, 14, 15] can only approximate the performance; therefore, one must go to the manufacturer to obtain a specific design.

Recommendation: If your process conditions fall within the limitations of the plate exchanger then try to get access to some of the computer programs made available by the manufacturer. Most likely the plate exchanger will have less area and be less expensive than the shell-and-tube exchanger for the same duty.

D. CONDENSERS

Condensation occurs in heaters, vaporizers, or condensers but this section is concerned with shell-and-tube exchangers in which the major purpose of the heat transfer is to condense a vapor. Air-cooled condensers are discussed later, and compact exchangers in Part 3 of this chapter. Condensing heat transfer coefficients in heaters or vaporizers are calculated in the same manner as for condensers for the equivalent physical conditions but the exchanger physical design is governed by the requirements of the other fluid, e.g., boiling liquids in vaporizers.

1. Selection of Condenser Types

The principal factors involved in selecting suitable types of condensers for evaluation and design depend on whether condensation is total or partial, whether the vapors are single-component or multicomponent, and whether some components are noncondensable. A vaporizing coolant or the need for subcooling the condensate also affects the selection. The various types of condensers suitable for the various classes of condensation are indicated in Table 2.

Table 2 is a guide to the selection of suitable types of condensers for specific condensation cases. Here a letter G, F, P, or X gives an indication whether the physical arrangement is suitable, while a keyed symbol indicates the probable accuracy of the predictive methods.

In addition to the factors discussed previously the selection of the tube-side fluid involves the consideration of the need for condensate-vapor contacting, freezing of the condensate, and venting. Condensate control is necessary when condensing multicomponent mixtures having a substantial boiling- or dew-point range or when soluble gases are present. To condense low-temperature boilers the condensate and vapor must be in good contact with each other and with the cooling surface, or when stripping low-temperature boilers contact between the vapor and cool condensate must be minimized; thus tube-side condensation is preferred. Condensing a high–melting point compound is best done on the shell side, as a solid layer can build up to an equilibrium thickness if the tube pitch is large enough, while tube-side condensation would plug. Venting noncondensables is more positive for tube-side condensation.

The decision of tube-side, shell-side, or direct condensation depends upon pressure, allowable pressure drop, corrosion, fouling, and temperature as discussed previously in Sec. A. In addition, for condensers that are susceptible to freezing, condensate control and venting are important.

2. Discussion of Condenser Types

a. Inside Tubes—Vertical Downflow

A vertical in-tube condenser is shown in Fig. 23 with an outside packed head and a separating head. It is possible to entrap air between the upper shell nozzle and the upper tube sheet and this requires special vents usually drilled into the tube sheets. The lower,

TABLE 2 Guide to Selection of Suitable Types of Condensers

	Shell-and-tube condenser							Direct contact
	In-tube			Shell-side				
		Vertical		Horizontal		Vertical		
	Horizontal	Upflow	Downflow	Cross	Baffled	Upflow	Downflow	
Total condensation								
Single component	G ○	F ◐	G ○	G ◐	G ○	G ●	G ○	G ◐
Multicomponent	F ◐	F ●	G ◐	G ◐	G ○	F ●	G ●	P ●
Subcooled condensate	P ⊗	X ⊗	G ○	P ●	P ●	X ⊗	F ●	Has external cooler
ΔP: high	G ◐	X	G ●	G ○	G ◐	X ●	G ●	X ⊗
low	P	G ●	F	G	F	F	G	G
Partial condensation								
Single component	G ○	X ⊗	G ○	P ○	P ○	P ◐	F ○	X ⊗
Multicomponent	P ◐	X ⊗	G ◐	F ◐	P ●	X ⊗	P ○	F ●
Noncondensable	G ○	X ⊗	G ○	G ○	G ◐	X ⊗	F ○	F ◐
ΔP: high	G ◐	P ●	G ●	G ○	G ◐	X ●	G ○	X ⊗
low	P	F	G	G	P	F	G	G

Physical arrangement:

G = good
F = fair
P = poor
X = not applicable

Probable accuracy of predictive methods:

O = average, ≈25 percent
◐ = fair, <50 percent
● = poor, 50+ percent
⊗ = no method or not recommended

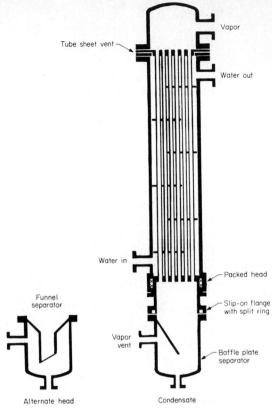

FIG. 23 Vertical in-tube downflow condenser.

separating head has either a baffle or a funnel to reduce entrainment of condensate into the vent system. The condensate level is maintained below the baffle.

Condensate forms on the tube wall and drains to the bottom in a thin film. At the end of the condensing zone there is a vapor–noncondensable gas interface, and below this interface the condensate is cooled as a falling film. As the vapor load varies this interface moves up or down and changes the condensing and subcooling length. Pressure control is obtained by control of the vent gas pressure.

The advantages of this design are that the condensate is always in contact with a cold wall and with the vapor, thus promoting condensation of low-temperature boilers from a wide-range boiling mixture; thus the tubes are completely washed by condensate and sprays can be added in the upper head when needed; and that subcooling is good and predictable.

The disadvantages include fouling by the coolant, venting through the upper tube sheet, and at low absolute pressures (below 25 mmHg) pressure drops that become high and thus require the use of large-diameter tubes.

b. Inside Tubes—Vertical Upflow

A vertical reflux condenser is shown in Fig. 24; here, the vapor flows up and the condensate returns by gravity. The lower ends of the tubes extend below the tube sheet and are

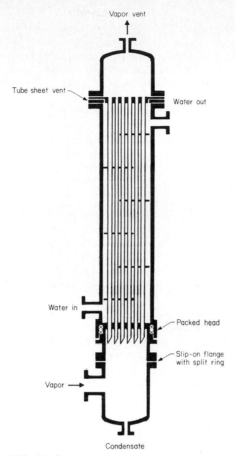

Vapor vent

Tube sheet vent

Water out

Water in

Packed head

Slip-on flange
with split ring

Vapor →

Condensate

FIG. 24 Reflux condenser.

cut at an angle of 60 to 75° to provide drip points for the condensate and to increase the vapor capacity of the tube before the tubes begin to load or flood.

Reflux condensers are used when a hot condensate return is necessary, or to strip out small amounts of low–boiling temperature components. The effectiveness of separation is not good, and as pointed out in Ref. 16 it is not possible to specify fractionation, heat requirements, and coolant temperatures independently.

The major disadvantage is the capacity limitation due to flooding which occurs when the upward vapor flow prevents the free return of condensate. In this situation the tubes will periodically load with condensate and then dump. At very high vapor velocities all the condensate can be blown out of the top of the condenser, but then only a fraction is condensed.

A number of studies have been made on flooding velocities in tubes, resulting in different equations or graphs. Among the earliest were the graphs of Holmes [17]. The English et al. [18] equation

$$G_f = \frac{80 D_i^{0.3} \rho_l^{0.46} \sigma^{0.09} \rho_g^{0.5}}{\mu^{0.14} (\cos \theta)^{0.32} (\mathbf{LG})^{0.07}} \qquad \text{kg}/(\text{m}^2 \cdot \text{s}) \qquad (48)$$

gives results close to Holmes' curves for beveled tubes. Here, D_i = tube inside diameter (mm), ρ_l and ρ_g are density (kg/m³), σ is surface tension (dyn/cm), μ is viscosity in centipoise, θ is the taper of the tube end from the horizontal, and **LG** is the weight ratio of condensate to vapor at the bottom of the tube. This equation is based on condensing data in a single-diameter tube with bevel cuts of 0, 30, 60, and 75°. The diameter term was apparently obtained from Holmes' data on 60° bevels; however, Holmes shows a much greater diameter effect for the square end tube (0° bevel).

Large bevel cuts should be used; the above equation is then satisfactory. For square end tubes use the curves of Holmes [17]. In all cases the design velocities should not exceed 75 percent of the flooding velocity. A further increase in flooding velocity occurs if the tubes are inclined about 30° from the vertical.

c. Inside Horizontal Tubes

Horizontal in-tube condensers are commonly an air-cooled design. In shell-and-tube exchangers condensation is of secondary importance, as the principal uses are as heaters or vaporizers. Single-pass or multipass is used, but more than two passes is unusual; hence, U-tube bundles are frequently used. However, Boyko and Kruzhilin [19] have shown that tubes of unequal lengths with common terminal pressures (such as in a U-tube bundle) have a lower condensation rate than if the same total length were used in a single-pass straight tube bundle.

Although condensing heat transfer coefficients are high the subcooling coefficients are poor. At high vapor loads a condensate wave or slug can form and then an oscillating flow may be encountered. In the design of these condensers it is important to know the two-phase flow patterns and avoid those patterns that can harm the condenser.

d. Outside Horizontal Tubes

Horizontal condensers are of a baffled or a cross-flow design. Figure 25 is a conventional design with an impingement plate opposite the inlet nozzle and two outlet nozzles, one for condensate and one for vent discharge. Baffles or support plates have vertical cuts to cause the vapor to flow from side to side. The bottoms of the baffles are notched to allow condensate drainage. Baffles may be given a variable spacing in order to maintain a vapor velocity. If the pressure drop is too high then double-segmental baffles are used or other shells such as split flow (J shells), or cross flow (X shells) are substituted.

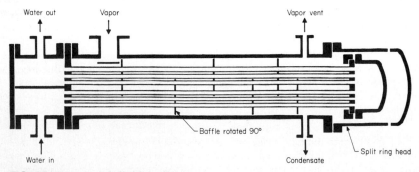

FIG. 25 Horizontal shell-side condenser.

This design is acceptable for total condensation of a single component or a narrow–boiling range mixture, and for partial condensation with a noncondensable gas. For partial condensation of wide–boiling range mixtures it is important to maintain good contact between the vapor and condensate; therefore the flow pattern must be maintained in the spray region. A horizontal baffle cut, so that the vapor-condensate mixture flows up and down through the exchanger, gives a good contact between the phases but requires a high pressure drop to maintain a spray flow pattern.

Low-finned tubes (1 to 2 mm) are used when condensing coefficients are low and if the condensate surface tension is below 40 dyn/cm.

Subcooling is sometimes attempted by partially flooding the shell; however, the pressure gradient through the condenser affects the liquid level and consequently the available surface area is very uncertain. Therefore, if subcooling is important a separate cooler should be considered.

The advantages of these condensers are low pressure drop, a fouling coolant that is on the tube side, a tube side that can be multipassed to get high velocities, reduced fouling, and greater temperature rises. Baffle spacing can be varied to get high velocities. Finned and enhanced tubes can be used.

The disadvantages are high shell cost if alloys are required; poor and unpredictable subcooling; a tendency for vapor and condensate to stratify, thus creating problems in condensing mixtures; and the possibility that tube vibration may occur.

e. Outside Vertical Tubes

Condensation outside vertical tubes is most often encountered in vaporizers or heaters; however, a falling-film condenser as shown in Fig. 26 is sometimes used. Here the water flows down the inside of the tube as a falling film.

The advantages of this condenser are low pressure (static head only) for the water, low water consumption, high heat transfer coefficients and large temperature rise, and the ability to clean the water side while in operation. However, there are problems in getting good water distribution; small distributors such as shown in Fig. 26 will help but can plug with dirt. Also a sump and possibly a pump may be required for the discharged water.

3. Heat Transfer

The effect of vapor velocity on the condensate film cannot be ignored in the design of heat exchangers. Since the velocity in most exchangers is continuously changing during condensation it is necessary to evaluate the condensing heat transfer coefficients and temperature differences as a function of the vapor fraction and then numerically or graphically integrate to determine the surface area. The equations for calculating the condensing heat transfer coefficients are available in Ref. 71; a few additional factors in their application to condenser design are given below.

a. Inside Tubes

At high vapor shear the condensing heat transfer coefficient is independent of tube orientation, while at low or zero shear the coefficient depends on tube inclination. Palen and Breber et al. [20, 21] recommend that the Wallis j_g^* parameter be used to judge the flow

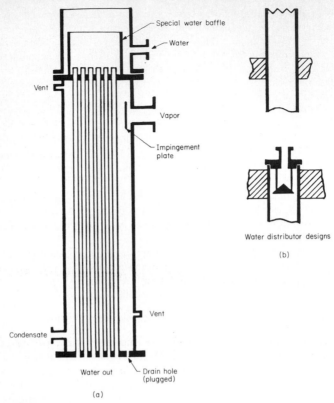

FIG. 26 Vertical falling-film condenser (*a*) and water distributors (*b*).

pattern as annular when $j_g^* \geq 1.5$ and stratified when $j_g^* \leq 0.5$ and to use a linear interpolation for intermediate values of j_g^*. Here

$$j_g^* = \frac{x G_t}{\sqrt{g D_i \rho_g (\rho_l - \rho_g)}} \qquad (49)$$

The high shear equations are then used for either horizontal or vertical tubes. At low shear for downflow in vertical tubes the vertical wall equations for condensing films are used, but in horizontal tubes the usual Nusselt equation for condensation on the outside of horizontal tubes is modified to allow for the buildup of the condensate layer on the bottom of the tube. The suggested multiplying correction factor for this coefficient is [33]

$$\left[1 + \frac{1-x}{x} \left(\frac{\rho_g}{\rho_l} \right)^{2/3} \right]^{-3/4}$$

For reflux condensers where the upward vapor flow affects condensate drainage some reduction in coefficient is expected; however, the vapor velocities must be kept low to avoid flooding. Kutateladze and Borishanski [22] show a small reduction of 0.7 when $V_g^2 \rho_g h_c / g \rho_l k_l = 200$ and a curve that is nearly linear between 0 and 200. Here V_g is the vapor velocity and the group is dimensionless. Most often the correction is small (approximately 0.9) and is neglected.

b. Outside Tubes—Horizontal Bundles

Condensation on tube bundles involves both vapor velocity and condensate dripping from row to row; these effects have been only partially investigated. In the absence of vapor velocity, experiments have shown the average value of the condensing heat transfer coefficients $\bar{h}_c$ to be higher than the theoretical prediction of Nusselt and the empirical correction to the single-tube heat transfer coefficient h_1 to be $\bar{h}_c = h_1 N_{tv}^{-1/6}$ where N_{tv} is the number of tubes in a vertical path. In a circular bundle the number of tubes N_{tv} can vary from the edge to the center and some theoretical studies have been made to derive a theoretical average, but in view of the complex interaction of vapor flow and condensate drippage the simple estimate of $N_{tv} = 0.78 \, D_B/Y_v$ should be adequate. Here D_B is the bundle diameter and Y_v is the vertical tube pitch.

The effect of vapor velocity is currently being researched; Ref. 71 will contain some recommendations. A simple equation has been recommended by Taborek [23] and for an ideal tube bundle is

$$\mathrm{Nu} = 0.3 \, \mathrm{Re}^{0.6} \, \mathrm{Pr}^{0.4} \, \sqrt{\frac{\rho_l}{\rho_g} + 1} \tag{50}$$

In a baffled exchanger the effect of leakage and bypass streams on the ideal flow must be included.

c. Finned Tubes

Low-finned tubes are frequently used in condensers; equations for heat transfer are available in Rohsenow [71]. Other tube enhancements are by grooving or corrugating the tubes but these are not standardized and the manufacturer must be consulted for heat transfer data.

d. Outside Vertical Bundles

If the vapor flows parallel to the tubes, treat the condensation the same as if the flow were inside the tubes, using a hydraulic equivalent diameter. For baffled exchangers the cross-flow equations for horizontal bundles are used.

4. Heat Transfer—Mixtures

The condensation of a mixture involves a vapor film across which heat and mass transfer occurs, as well as a condensate film. Thus interfacial temperatures, compositions, and equilibrium data are required to calculate the local heat flux. The local fluxes are then numerically integrated to obtain the required area. A complete theory for mixtures exists only for one condensable component from one noncondensable gas. The theory for two condensable components is incomplete since it must be combined with mass transfer in the condensate film. See Rohsenow [71] for more discussion.

Most industrial condensers involve many components for which both theory and available physical data are inadequate; therefore, approximation methods are used for design. The Bell-Ghaly [24] method or some modification thereof is commonly used for all types of mixtures. The following procedure assumes a single-pass arrangement. The condensation curve T_v versus heat removed Q_t is calculated by the usual thermodynamic procedures; also, by heat balance there is a unique relationship between T_v and T_c.

1. Calculate the condensing curves:
   a. Total condensing stream enthalpy versus local vapor stream temperature T_v, assuming thermodynamic equilibrium between phases
   b. Vapor mass flow rate W_v versus T_v, where $W_v = W_T x$
   c. Total heat removed from vapor from inlet to given point, Q_t versus T_v, where Q_t is defined as follows:

   $$Q_t = Q_{sv} + W_T (1 - x)[i_{lg} + (i_l)_{in} - (i_l)_T]$$

   and

   $$Q_{sv} = W_T x[(i_v)_{in} - (i_v)_T]$$

   d. Coolant temperature T_c versus T_v
2. Calculate as a function of total heat removed Q_t, $dQ_{sv} = -W_v\, di_v$, or in finite increments $\Delta Q_{sv} = \overline{W_v}\, \Delta i_v$, where ΔQ_{sv} is the vapor sensible heat removed in the increment and Δi_v is the decrease in vapor enthalpy per unit weight in the increment.
3. Calculate $Z = dQ_{sv}/dQ_t$ or $\Delta Q_{sv}/\Delta Q_t$ as a function of Q_t.
4. Calculate the condensing coefficient h_c and the vapor (sensible) coefficient h_{sv} as functions of G_l and G_v, respectively, and then as functions of Q_t.
5. Calculate U_i as a function of Q_t, where $1/U_i = 1/h_w + \mathbf{R}_t + 1/h_c$, $\mathbf{R}_t$ is the combined wall and fouling resistance, and h_w is the coolant coefficient.
6. Calculate $(1 + ZU_i/h_{sv})/U_i(T_v - T)$ as a function of Q_t.
7. Calculate the total area A_o by integrating the function of step 6 numerically from $Q_t = 0$ to $Q_t = Q_t^*$, where Q_t^* is the total heat load in Btu/h (W).

5. Subcooling

Subcooling is predictable only for vertical tubes using the falling-film equations in Rohsenow [71, 72]. Inside horizontal tubes the condensate layer thickness, area, and heat transfer coefficient cannot be predicted with any accuracy and therefore this arrangement is not recommended for subcooling.

Subcooling on the outside of horizontal bundles is frequently attempted by creating a pool by means of a dam. However, the pressure gradient in the condenser also affects the liquid level and, hence, the area. Heat transfer coefficients and area can usually only be approximated by assuming a static liquid level. Separate subcoolers should be used if subcooling is critical.

6. Pressure Drop†

The calculation of pressure drop in a condenser is at best only an approximation because of changing velocities and flow pattern. The overall pressure drop is

$$\Delta P = \Delta P_e + \Delta P_s + \Delta P_m + \Delta P_{tp} \tag{51}$$

where ΔP_e is the various inlet and exit losses due to nozzles and headers, ΔP_s is the static head, ΔP_m is the momentum change, and ΔP_{tp} is the two-phase friction loss.

†See also Griffith [73].

The expansion and contraction losses are calculated using the homogenous flow model, i.e., assuming equal liquid and vapor velocities.

The static head is usually insignificant in condensers, but would be calculated as

$$\Delta P_s = \rho_{tp} \frac{g}{g_c} (\sin \theta) \Delta L \tag{52}$$

where θ is the angle of inclination and ρ_{tp} is the two-phase density. The true density is unknown but can be approximated by assuming homogeneous flow, or with increasing accuracy by using appropriate void fraction correlations.

The momentum change ΔP_m during condensation results in a pressure regain. The magnitude of this regain is usually small except in vacuum operations, where the pressure regain can approach or exceed the friction loss. It is calculated from

$$\Delta P_m = \frac{G_t^2}{g_c} \{[(1 - x)v_l + xv_g]_2 - [(1 - x)v_l + xv_g]_1\} \tag{53}$$

In this form both liquid and gas velocities need to be known. Again homogeneous or separated flow models can be assumed to obtain a relationship between liquid and gas velocities. The momentum pressure drop is usually small. Furthermore, it is a pressure regain and ignoring it results in a conservative overall pressure drop calculation.

The two-phase friction loss ΔP_{tpf} is calculated by stepwise calculations along the condensing path. Basically the separated-flow model of Martinelli is used for in-tube horizontal flow although it underpredicts for large vapor fractions and overpredicts at small vapor fractions. In vertical tubes with either upward or downward vapor flow the Martinelli-type equations result in larger errors. For downward vapor flow a number of researchers have developed different correlations which apply to their specific apparatus [25, 26, 27, 28] but do not correlate with other data.

Pressure drop for two-phase flow across horizontal tube banks has been studied by a number of authors [29, 30, 31] and all present different correlations, but the equations from Grant and Chisholm [32] seem best:

$$\frac{\Delta P_{tp}}{\Delta P_{LO}} = 1 + (\Gamma - 1)[Bx^{(2-n)/2}(1 - x)^{(2-n)/2} + x^{2-n}] \tag{54}$$

where $\Gamma = \Delta P_{GO}/\Delta P_{LO}$ and $n = $ slope of the friction factor curve in Fig. 10 [that is, for $(D_t G_s/\mu)^n$]. The values of B are listed below and the flow patterns are identified in Fig. 27. ΔP_{GO} and ΔP_{LO} are pressure drops for each phase if it alone were flowing.

	B values	
	Horizontal	Vertical
Cross flow		
Spray or bubble	0.75	1.0
Stratified or stratified spray	0.25	
Window, $n = 0$	$\dfrac{2}{\Gamma + 1}$	$\left(\dfrac{\rho_h}{\rho_l}\right)^{1/4}$

where $\rho_h = $ homogeneous flow density $= 1/[x/\rho_g + (1 - x)/\rho_l]$.

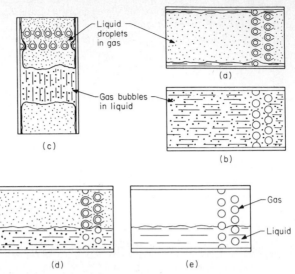

FIG. 27 Shell-side flow patterns (*a* to *e*) and map (*f*) [32]. (*a*) Spray flow. (*b*) Bubbly flow, vertical and horizontal flow. (*c*) Chugging flow, vertical flow. (*d*) Stratified-spray flow and (*e*) stratified flow, horizontal flow. (*f*) Shell-side flow pattern maps obtained by Grant.

7. Mean Temperature Difference

For the simple case of condensation at constant temperature and constant overall heat transfer coefficient in a single-pass arrangement the logarithmic mean temperature difference is used. When stepwise calculations are made, the local temperature difference is used with the local overall heat transfer coefficients and numerically integrated. In a multipass arrangement it would be necessary to use the local overall heat transfer coefficient and local temperature difference for each pass. Iterative calculations are required in order to have the stream temperatures match at the turnaround. Butterworth [33] developed a method for shell-side condensation assuming the overall heat transfer coefficient is constant for each pass at a given location along the shell.

For shell-side condensation and condensate subcooling Bell [34] points out the need to calculate separately the temperature rise for each zone and the resulting temperature in the turnaround head for two-pass condensers.

8. Design Procedure

All calculations start with an assumed design which is then rated to determine its capacity. Suitable modifications are then made to the assumed design in order to approach the required capacity and the final design. To obtain the assumed design,

1. Determine suitable types of condensers; see Table 2.
2. Determine the heat load.

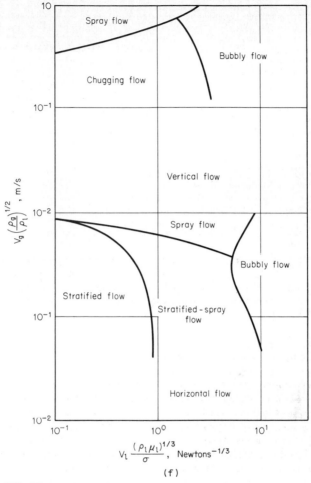

FIG. 27 (*Continued*)

3. Select the coolant temperatures and calculate an overall logarithmic mean temperature difference.

4. Estimate an overall heat transfer coefficient from Table 3 or by prior experience using estimated individual coefficients.

5. Calculate the area.

6. Select a tube size, pitch, and length and determine the number of tubes, shell size, and baffling, if required.

7. The above results in a guessed design. This design is now subjected to detailed rating calculations either by hand calculations or by computer programs and the design is modified iteratively until an acceptable condenser design is obtained.

Note that condensers can be designed to perform the required duty while still having unacceptable physical dimensions—e.g., very short but large-diameter shells. Thus con-

TABLE 3 Overall Heat Transfer Coefficients for Estimating Condensers

Vapor	Coolant	Overall heat transfer coefficient	
		$W/(m^2 \cdot K)$	$Btu/(h \cdot ft^2 \cdot °F)$
Alcohol	Water	550–1100	95– 200
Dowtherm	Tall oil	340– 450	60– 80
Dowtherm	Dowtherm	450– 680	80– 120
High–boiling temperature hydrocarbons under vacuum	Water	100– 280	20– 50
Low–boiling temperature hydrocarbons	Water	450–1140	80– 200
Hydrocarbons	Oil	140– 230	25– 40
Organic solvents	Water	550–1140	100– 200
Kerosene	Water	170– 370	30– 65
Naphtha	Water	280– 430	50– 80
Naphtha	Oil	110– 170	20– 30
Steam	Feed water	2200–5700	390–1000
Vegetable oils	Water	110– 280	20– 50
Organic-steam azeotrope	Water	220– 450	40– 80

denser design involves not only duty performance, but also economic considerations and suitable physical arrangement for plant installations.

9. Operational Problems

The diversity of condenser applications makes impossible a complete listing of operational problems; however, most problems arise when condensers are operated at conditions other than the design point. Variation in vapor or coolant flows or temperatures will cause pressure changes or affect the condensing area, and attempts to control these changes can create other operational problems. Overdesign or less fouling will also result in excess area and add to the control problem. Calculations should be made to determine process control when operating at other than the design conditions.

a. Venting

The accumulation of a few percent of noncondensable gases in condensers can significantly reduce the condensing heat transfer coefficients. Venting for in-tube condensation is simple, as the flow path is fixed; however, for shell-side condensation, the gases can segregate in pockets and are difficult to remove unless sufficient pressure drop (vapor velocity) is used to force these gases to the vent outlet. The venting problem is most severe in the typical cross-flow condenser with either horizontal or vertical tubes.

The literature is sparse on this subject and very few rules have been proposed. Grant [35] has studied the effect of vent location on the performance of a small (19-m^2) condenser, as have Hampson and Thain [36] on a single horizontal-tube laboratory condenser.

Khan [37] studied steam condensation at two temperature levels (54 and 104°C) for various amounts of noncondensable gases added to the steam and at various venting rates. The basic conclusion is that the minimum vapor velocity leaving the last tube row in a bundle should be 1.5 m/s.

Eissenberg and Bogue [38] state that the minimum mass velocity should be that rate at which the Reynolds flux just equals the condensing rate, and for their tube bundle with steam this corresponded to 600 to 1000 $kg/(m^2 \cdot h)$.

b. Fogging

In condensing mixtures with noncondensable gases, the Colburn and Edison [39] equation closely predicts the condensation path. When this path crosses the vapor pressure equilibrium curve, supersaturation occurs and the possibility of fogging exists. Plant experience has shown that under identical conditions of temperature and vapor composition a fog may or may not be present. According to theory [40], above certain supersaturation ratios spontaneous fogging will occur but this ratio is on the order of 2 to 10. Nuclei, if present in substantial numbers (1000 per cm^3) and if soluble or wetted by the condensate, can cause fogging at lower supersaturations. Apparently the erratic plant experience is associated with the presence of nuclei. The fog is very dense, being opaque in thickness of a few decimeters.

The tendency to fog is an inherent property of the mixture and determined by the ratio of Prandtl number to Schmidt number. While some methods of prevention and elimination are known, none is a universally acceptable answer to the problem.

E. VAPORIZERS

The term *vaporizer* applies to any apparatus in which a vapor is generated from its liquid. This includes boilers, callandrias, falling-film vaporizers, evaporators, or reboilers, all of which are, with varying degrees of specificity, different types of vaporizers. In this section only those vaporizers heated by a liquid or a condensing vapor and of a basic shell-and-tube design are discussed. Boiling on the inside of horizontal tubes and the outside of vertical tubes will not be considered.

The phase change occurring in vaporizers has a strong influence on the heat transfer coefficients, the two-phase flow characteristics, and the fluid dynamics of the system. Further, two modes of boiling can occur: the nucleate or the film. In addition there is a transition region between nucleate and film boiling and occasionally a preheating zone before nucleation begins. The nucleate region is defined as where bubbles are formed at discrete spots (nuclei) on the surface, and the film region is where a thin continuous vapor film covers the surface and thus the liquid does not contact the surface. The theory of film boiling is not completely explained (see Ref. 72). Nevertheless, the heat transfer and operating control in the transition and film-boiling regions are such that for design these boiling regions are avoided. In the preheating zone (required to superheat the fluid in order to activate the nuclei) heat transfer is by natural convection and is a weak function of temperature difference (approximately the 0.25 power). In the nucleate region the heat transfer coefficient is a very strong function of ΔT (typically a power of 2 to 4). In the transition region there is a reversal of the effect of temperature difference in that the heat transfer coefficient decreases with increasing ΔT. In the film region the coefficient decreases with increasing ΔT as a result of increasing vapor film thickness; then after the film reaches a maximum thickness the coefficient begins to increase as convective and radiative effects begin to dominate. Another cause for decreased heat transfer occurs when the liquid content of the two-phase stream is too low to adequately wet the heating surfaces.

1. Selection Criteria for Types of Vaporizers

The selection of a type of vaporizer is complicated by the number of factors which can influence the heat transfer, such as surface characteristics (i.e., nucleation sites), type of fluid, fouling characteristics, heating medium, temperatures, and pressure. Other non–heat transfer factors such as space limitations, materials of construction, and process control must also be evaluated. However, the major criteria used to make the preliminary selection are operating pressure, design temperature difference, mixture boiling range, and fouling.

a. Operating Pressure

The heat transfer coefficients, maximum heat fluxes, and critical ΔT_{cr} (surface temperature at maximum flux—boiling point) are all functions of the reduced pressure P_r^+ (operating pressure divided by critical pressure). The maximum flux and critical ΔT are nearly zero for $P_r^+ = 0$ or 1.0 and reach a maximum at $P_r^+ \cong \frac{1}{3}$. At very low absolute pressures (less than 10 mmHg) the effects of system pressure drop and hydrostatic head are important and can be high enough to suppress boiling; therefore, a falling-film evaporator or mechanical equipment such as a forced-circulation or thin-film-wiped evaporator is used. At very high pressures ($P_r^+ > 0.8$) small density differences reduce the circulation effect. For supercritical pressure local density differences can occur and can cause violent pressure surges; therefore, this range of operation is beyond the scope of this section. In the intermediate pressure range the relative merits of kettle, thermosiphon, or forced-circulation types are somewhat dependent upon the pressure level, but the other criteria begin to dominate.

b. Design Temperature Difference

The design ΔT should be less than the critical temperature difference ΔT_{cr} and, if not, should be high enough to insure film boiling. The transition boiling region should be avoided because of control difficulty. However, the film-boiling region should also be avoided unless the fluid has a small boiling range and is very clean and nonfouling. The minimum design ΔT should be greater than the superheat required to initiate boiling (less than 5°C); otherwise a falling-film or mechanically aided device would be required. However, the superheat required for nucleation is strongly affected by the type of surface—porous, sintered, fouling, etc. If this design value of ΔT in a vapor-heated system is too high, it may be easily reduced by means of a pressure controller on the heating vapor; but if the resulting pressure is too low then condensate removal may be a problem. The boiling range of the fluid is also an important consideration in the selection of a design ΔT.

c. Mixture Boiling Range

A single-component fluid has a zero boiling range, while a mixture containing a nonvolatile substance has a very high or almost infinite boiling range. If the highest boiling point of the mixture at the absolute pressure existing on the lowest heating surface is less than the heating medium temperature then nearly all types of vaporizers can be used. However, if the heating medium temperature is not sufficient to cause boiling in the concentrated mixture then circulation could cease when the low-temperature boilers are stripped out. In this case the design of the vaporizer and means of feed addition become critical.

d. Fouling

In severe fouling service the forced-circulation evaporator is used because of its high veloc-ities and ease of cleaning. A vertical tube-side thermosiphon is also good. The choice between these two depends upon the mixture boiling range and the available ΔT. For clean or low fouling rates the kettle types or shell-side thermosiphons can also be used. The final choice depends upon the evaluation of the other criteria.

2. Types of Vaporizers

a. Kettle or Internal Reboilers

A kettle reboiler consists of a horizontal bundle of U tubes placed in a shell. A short nozzle section is of small diameter sufficient to contain the tube bundle. The remaining shell is much larger so as to act as a vapor-disengaging space. A baffle just beyond the U-bend end serves as a liquid level dam. Beyond the dam is a small holdup space for controlling the removal of the liquid overflow (see Fig. 28). Depending upon the length of the kettle one or more vapor nozzles are used. Multiple nozzles are needed to reduce entrainment from the kettle. For large bundles and large vapor volume the tube pitch ratio may be increased to 1.5 or 2.0 and in some cases additional vapor release lanes are added.

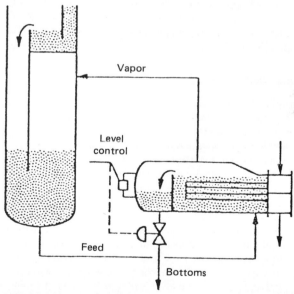

FIG. 28 Kettle reboiler [42].

The tube bundles are fabricated with plain tubes, low-finned tubes, or other types of enhanced surfaces.

An internal reboiler is the same as a kettle reboiler except that the shell is replaced by another vessel (see Fig. 29).

Kettle reboilers are relatively insensitive to hydrodynamics and, therefore, are sized on the basis of pool-boiling data. These reboilers are used to obtain a high concentration

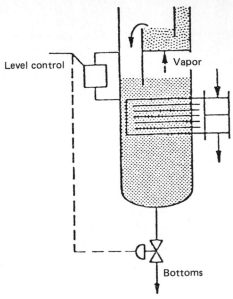

FIG. 29 Internal reboiler [42].

of the bottom waste stream. This may result in lower circulation in the bundle causing increased fouling, which is difficult to remove from the outside of the tubes.

b. Thermosiphon Vaporizers

Vertical Tubes. These vaporizers are basically a TEMA E shell with single-pass construction using tubes 1.0 to 2.0 in (25.4 to 50.8 mm) in diameter and 8 to 12 ft (2.4 to 3.7 m) long (see Fig. 30). The exit vapor pipe is usually a short horizontal takeoff from a channel head and has a cross-section area equal to the sum of the tube cross-sectional areas. It is important not to choke down on the exit line as its size strongly influences the stability of flow. If the exit pressure drop is too high (more than 30 percent of driving head) then an unsteady surging action can occur at high fluxes. The cross section of the inlet liquid line can be smaller, but it is best to have at least 50 percent of the tube sectional area. This area may seem somewhat large, but it will still allow for some throttling action of orifices or valves (in case of instability one cure is to throttle the liquid feed). An external vessel, usually the base of a distillation column, is utilized as a vapor separator and a liquid holdup tank. Normally the liquid level is controlled in the external tank (see Fig. 30) to keep the effective liquid level at the top tube sheet. For vacuum service, large-diameter tubes are used and the liquid level is dropped to about 0.4 to 0.6 times the tube length.

The difference between the external driving head of liquid and the tube-side vapor-liquid mixture is available to overcome the friction and acceleration losses. The circulation is quite high, with exit vapor fractions of 0.02 to 0.1 for water mixtures and 0.1 to 0.25 for hydrocarbon mixtures. The high circulation rates help to reduce fouling.

The vertical construction requires a higher skirt for the columns. Also the high static head has a large effect on the boiling point at low column pressures.

The design of these units requires the simultaneous solution of heat transfer and pressure drop. The heat transfer can consist of a preheat section, a subcooled boiling zone, a

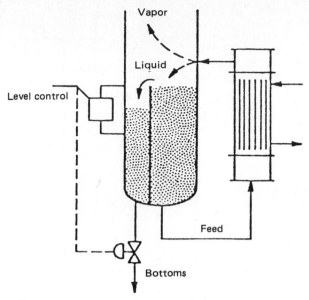

FIG. 30 Vertical in-tube thermosiphon [42].

nucleate boiling zone, and a two-phase and mist-flow zone. Each of these zones has a different heat transfer coefficient. The complete hydraulic circuit needs to be defined and its pressure drop calculated. Such calculations require a good computer program. The liquid temperature also goes through a maximum and thus the computer must also calculate the temperature profile and the resulting temperature difference.

In spite of these problems the vertical-tube thermosiphon is widely used because of its compact arrangement, cleanability, and cost.

Sometimes these vaporizers are made with long tubes, 1 in × 16 ft (25.4 mm × 4.9 m) or 2 in × 35 ft (50.8 mm × 10.7 m), and are operated as a once-through vaporizer. The effective liquid level is low, usually 2 to 4 ft (0.6 to 1.2 m), and the vaporization per pass is high. In this arrangement a higher effective ΔT is obtained and because of the vaporization the flow is in the climbing-film or mist-flow range. Holdup is very low, which is an advantage for heat-sensitive liquids. Heat transfer is high because of the high vapor velocity. These units require a lot of headroom and are not well suited to fouling-type fluids.

Vertical Shell-Side Thermosiphon.   Although this type of flow is occasionally encountered (Fig. 31), the flow arrangement is mostly found in reactors, where the primary purpose is cooling. Vaporizers of this type would be designed much as the in-tube unit but would have additional problems of tube support and control of the vapor-liquid level under the top tube sheet.

c. Horizontal Thermosiphon Reboilers

Here the boiling occurs on the shell side. TEMA X, G, H, or E shells are used. The tube bundle fills the shell; hence the vapor-liquid separation occurs externally much as in the vertical tube unit; see Fig. 32. The circulation is developed by means of the piping arrangement, and high circulation rates can be developed. Heat transfer and pressure

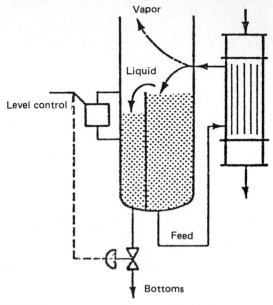

FIG. 31 Vertical shell-side thermosiphon [42].

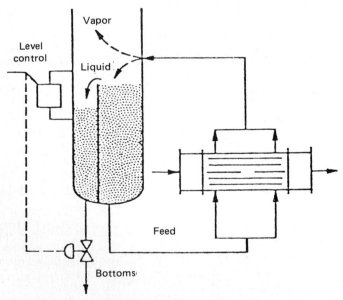

FIG. 32 Horizontal shell-side thermosiphon [42].

drops would be calculated much the same as for vertical in-tube units; however, much less is known about these relationships for this arrangement. The principal advantages are a lower headroom requirement, a more effective ΔT due to higher circulation, and a lower fouling potential.

Cleaning is again a problem. On large units multiple nozzles and expensive manifolding may be required. With a mixture having a wide boiling point range horizontal baffles (G or H shells) are needed to prevent flashing of light components with accumulation of heavy components at the ends. Very little information is available regarding this design.

d. Forced-Circulation Vaporizers

Forced-circulation vaporizers are essentially heaters (Fig. 33), as very little boiling occurs within the tubes. These are usually horizontal multipass exchangers operating with velocities greater than 6 ft/s (1.8 m/s). They are of major use with high-viscosity, wide-range boiling mixtures operating at very low pressures (less than 10 mmHg abs). They are designed on the basis of sensible heat transfer equations. The only important requirement is that boiling is suppressed until the liquid leaves the exchanger or until after it enters the last pass. If vaporization occurs before the last pass then the liquid distribution in this pass is nonuniform and more fouling can take place. Pumping costs are the major disadvantage.

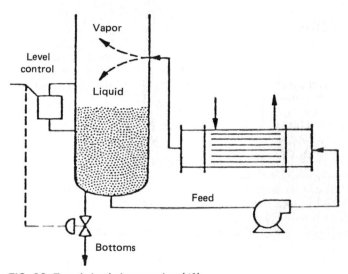

FIG. 33 Forced-circulation vaporizer [42].

e. Falling-Film Vaporizers

In a vertical falling-film vaporizer, liquid is fed to the upper tube sheet and flows down the inside walls of the tube in a thin film. The distribution and flow rate per tube are controlled by the design of the distributors, which are similar to those indicated in Fig. 26. Evaporation occurs at the exposed film surface and the vapors can flow either cocurrent with or countercurrent to the falling film. Cocurrent flow is usually used unless it is necessary to remove all the low-temperature boiler from the bottom product, in which case a countercurrent vapor flow is used. In general the vertical falling-film vaporizer is similar to the falling-film condenser (Fig. 26) except for the direction of heat flow and the addition of heads. In a horizontal falling-film vaporizer the feed is sprayed or distributed to the top rows of the tubes and then drips down across all the lower tubes, much as in a horizontal shell-side condenser.

Falling-film condensers have high heat transfer coefficients, no hydrostatic head to affect the boiling point, low pressure drop, and low holdup. They are used in high-vacuum evaporation processes, for operation at low temperature differences, for stripping trace amounts of low-temperature boilers from bottom products, and for handling heat-sensitive products.

The principal problem is the uniform distribution of feed to all the tubes. In addition, for upward vapor flow a flooding condition can occur, usually at the top of the tube, and is a function of distributor design. Distributors are designed and fabricated for each job, as there are no commercially available distributors. It is important to keep the tubes wetted, as the formation of dry patches due to insufficient flow or too high a temperature difference will reduce the capacity and may increase fouling. A recirculation of the bottom product is done to improve the flow rate and reduce the possibility of dry patches, but this has the effect of reducing the average temperature difference.

Falling-film evaporators operating at small temperature differences are used in sea-water desalinization plants and in low-temperature energy conversion projects.

3. Thermal Design

The more data one has on physical properties and experimentally determined coefficients, the better the design. The simplified procedures given below are used when there are inadequate data to calculate the more fundamental equations in Rohsenow [72].

a. Kettle and Internal Reboilers

A kettle reboiler is a bundle of tubes immersed in a pool; hence, the basis of the design correlations is the single-tube pool-boiling equations, to which correction factors are applied to take care of the bundle geometry. For heat transfer two sets of calculations are required, the first set to determine the maximum heat flux, and the second to determine the operating coefficients or fluxes. Finally the required disengaging space and number of nozzles are determined.

The maximum flux for bundles is calculated by the Palen and Small [41] correlation

$$q''_{b,\max} = q''_{1,\max}\phi_b \tag{55}$$

The maximum flux for a single tube, $q''_{1,\max}$, can be obtained from correlations in Ref. 72 or can be estimated with sufficient accuracy for design [42] from

$$q''_{1,\max} = KP_{cr}\left(\frac{P}{P_{cr}}\right)^{0.35}\left(\frac{1-P}{P_{cr}}\right)^{0.9} \tag{56}$$

where $q''_{1,\max}$ = maximum heat flux for a single tube
K = 367 for flux in W/m² and pressures in kPa
K = 803 for flux in Btu/(h·ft²) and pressures in lb$_f$/in²

The bundle correction factor ϕ_b as obtained from Ref. 41 can be simplified to

$$\phi_b = K\frac{L_{tp}}{D_t(D_B/L_{tp})^{1.1}} \tag{57}$$

where K = 4.12 for square pitch
K = 3.56 for triangular pitch

with a maximum limit of $\phi_b = 1$. The basic relation was developed from field data [41] and later reported [44] to be conservative. In lieu of other methods this relation can be used but the circulation in reboilers is not well understood and further research is needed.

The bundle heat transfer coefficient h_b is found from

$$h_b = h_{nb1}F_bF_m + h_{nc} \tag{58}$$

where h_{nc} = natural-convection heat transfer coefficient
h_{nb1} = single-tube nucleate-boiling heat transfer coefficient
F_b = factor for bundle geometry
F_m = factor for mixtures

The minimum heat transfer coefficient is the liquid-phase natural-convection heat transfer coefficient h_{nc} and ranges around 250 W/(m²·K), or 44 Btu/(h·ft²·°F) for hydrocarbons and around 1000 W/(m²·K), or 176 Btu/(h·ft²·°F), for water. It is significant only at low ΔT (less than 4 K or 7°F). For more accuracy the natural-convection heat transfer coefficient can be determined from methods given in Ref. 74.

The single-tube heat transfer coefficient h_{nb1} can be determined in accord with the methods given in Ref. 72 or by a simple [42, 43] Mostinski equation for pure liquids as given below:

$$h_{nb1} = KP_c^{0.69}(q'')^{0.7}F_p \tag{59}$$

where h_{nb1} = single-tube nucleate-boiling heat transfer coefficient.

K	q''	P_{cr}	h_{nb1}
0.00417	W/m²	kPa	W/(m²·K)
0.00658	Btu/(h·ft²)	lb$_f$/in²	Btu/(h·ft²·°F)

The pressure correction factor F_p given below can be simplified for reboiler design by omitting the last two terms [42, 43]:

$$F_p = 1.8\left(\frac{P}{P_{cr}}\right)^{0.17} + 4\left(\frac{P}{P_{cr}}\right)^{1.2} + 10\left(\frac{P}{P_{cr}}\right)^{10} \tag{60}$$

The bundle factor F_b corrects for the improved heat transfer coefficient in bundles due to the high circulation rates. F_b is a complicated function of bundle effect [44, 45] and proprietary methods are used to determine it. However, F_b can range as high as 2 to 3 for large bundles, but a conservative value of 1.5 is recommended. It can be low at high fluxes and can exceed 3 for special evaporators.

The apparent heat transfer coefficients for boiling mixtures can be lower than for single-component fluids. For further discussion of this effect see Rohsenow [72]. While some theory has been developed for binary mixtures the correction for process fluids containing more than two components is usually made by an approximation developed by Palen and Small [41]. The following equation gave results that were reasonable when compared to data in Ref. 44 but were not as good as results given by other proprietary equations:

$$F_m = \exp(-0.027 \, \Delta T_{BR}) \tag{61}$$

where F_m = mixture correction factor (minimum $F_m = 0.1$)
ΔT_{BR} = boiling range (difference between dew point and bubble point), K

Vapor-Liquid Disengagement. A disengagement space is provided to allow the vapor to separate from the liquid; however, a certain amount of entrainment will always occur. If a really dry vapor is required then additional mist eliminators are needed. However, for most cases of low entrainment several empirical methods have been used. One is to make the distance from the centerline of the top tube to the top of the shell not less than 40 percent of the shell diameter. Another empirical equation [41] is

$$VL = 2290\rho_g \left(\frac{\sigma}{\rho_l - \rho_g}\right)^{0.5}$$

(62)

where VL = vapor load, $lb_m/(h \cdot ft^2)$(vapor rate in lb_m/h divided by volume of vapor space, ft^3)

σ = surface tension, dyn/cm

ρ_g, ρ_l = density of vapor and of liquid, lb_m/ft^3

Also, to aid distribution and reduce entrainment the number of liquid and vapor nozzles should be increased for long shell lengths (L). A rule of thumb [42] is

$$N_N \approx \frac{L}{5D_B}$$

(63)

The sizes of vapor and liquid lines should be large enough that the liquid level in the kettle is not depressed by flow resistance in the piping [41].

Enhanced or finned tubes can be effectively used in kettle-type reboilers. However, while the effectiveness of these enhancements is high at low ΔT, they lose their effectiveness at high fluxes because of vapor blanketing and circulation.

b. Thermosiphon Vaporizers

Both the vertical and the horizontal thermosiphon evaporators require a hydrodynamic and heat transfer analysis in order to make the two-phase flow-type calculations. In thermosiphons several different types of boiling zones may occur within a tube, ranging from a preheating zone of subcooled boiling to a boiling or two-phase flow zone, a mist-flow zone, or at high temperature differences a film-boiling zone. All these different zones have different heat transfer coefficients and must be separately calculated. At the same time a pressure drop calculation is made in order to determine the local temperature difference for each zone. Because of the trial-and-error nature of the calculations and the need to determine the circulation of the entire system, these calculations are done in complex computer programs which are proprietary. However, hand calculations can be made to obtain a good approximation to the computer results by dividing the tube into the several boiling zones and calculating the heat transfer coefficients on the basis of the average conditions for each zone.

Vertical-Tube Vaporizers. The best present method is the Chen equation [46], which combines a convective and a nucleate boiling mechanism:

$$h_b = s h_{nb} + h_{cb}$$

(64)

where h_b = boiling heat transfer coefficient

h_{nb} = nucleate-boiling heat transfer coefficient

h_{cb} = convective-boiling heat transfer coefficient

s = suppression factor

The convective-boiling heat transfer coefficient is a function of the Martinelli two-phase flow parameter X_{tt}. The Chen correlation below is one of the best in the open literature:

$$\frac{h_{cb}}{h_l} = f(X_{tt}) = F_{ch} \tag{65}$$

where $F_{ch} = 2.35 \left(\frac{1}{X_{tt}} + 0.213 \right)^{0.736}$ $\tag{66}$

$$X_{tt} = \left(\frac{1-x}{x} \right)^{0.9} \left(\frac{\rho_g}{\rho_l} \right)^{0.5} \left(\frac{\mu_l}{\mu_g} \right)^{0.1} \tag{67}$$

h_l = liquid-phase heat transfer coefficient based on the amount of liquid present

The nucleate-boiling heat transfer coefficient h_{nb} is determined as

$$h_{nb} = h_{nb1} F_m \tag{68}$$

where h_{nb1} is obtained from Eq. (59) and F_m from Eq. (61). The suppression factor s is determined as follows:

1. Calculate liquid-phase Reynolds number $Re_l = 4W/\pi\mu_l N_t D_i$.
2. Calculate two-phase Reynolds number $Re_{tp} = Re_l F_{ch}^{1.25}$
3. Calculate $s = 1/[1 + 2.53(10^{-6})\, Re_{tp}^{1.17}]$.

Now h_b, Eq. (64), can be calculated and the overall heat transfer coefficient for the nucleate-boiling zone determined. The LMTD can be used for this zone.

At the tube entrance the liquid is usually subcooled because of the static head of liquid and so a subcooled boiling zone of a varying length exists before the two-phase region starts. An estimate of the subcooled boiling heat transfer coefficient can be obtained by again utilizing Eq. (64) but defining s as equal to $(\Delta T_b/\Delta T_o)$, h_{cb} as equal to h_l, and h_{nb} as based on the temperature difference ΔT_b between the tube wall and the saturation temperature of the liquid at the given local pressure. The difference between the tube wall temperature and the subcooled bulk temperature is ΔT_o. Hence, for this calculation Eq. (59) is transformed to

$$h_{nb1} = 1.4(10^{-8}) P_{cr}^{2.3} (\Delta T_b)^{2.33} F_p^{3.33} \tag{69}$$

and Eq. (68) to

$$h_{nb} = h_{nb1} F_m^{3.33} \tag{70}$$

where F_m is from Eq. (61).

Heat Flux Limitation.   In vertical-tube thermosiphons there are several limits of operations such as surging, heat flux, and mist flow. The surging instability occurs as ΔT increases but is controlled by the physical arrangement. Blumenkrantz and Taborek [47] discuss this phenomenon. The surging can be controlled by increasing the frictional resistance in the inlet piping and decreasing the exit piping losses. At high reduced pressures (P/P_{cr}) the critical wall temperature for film boiling is reached before instability occurs.

As the temperature difference increases, the evaporation rate of a given tube will increase, pass through a maximum, and then decrease. This was investigated by Lee [48] and later confirmed by Palen et al. [49]. The Lee correlation involves a number of physical properties (see Refs. 48 and 49 for details) but Palen et al. [49] present a simpler correlation, as follows, which gives equally good results:

$$q''_{max} = 23,660 \left(\frac{D_i^2}{L}\right)^{0.35} P_{cr}^{0.61} \left(\frac{P}{P_{cr}}\right)^{0.25} \left(1 - \frac{P}{P_{cr}}\right) \tag{71}$$

where P_{cr} = critical pressure, kPa
$\quad D_i$ = tube inside diameter, m
$\quad L$ = tube length, m

However, this Eq. (71) does not allow for lower maximum fluxes caused by small exit piping. The exit piping should have a cross-sectional area at least equal to the total cross-section area of the tubes. At low fluxes a smaller exit pipe could be used, but the frictional pressure drop must be less than 30 percent of the total reboiler pressure drop [42].

Another limit is the mist-flow zone; however, this is an arbitrary limit in that operation in the mist zone is poor on account of the low heat transfer coefficient. The mist-flow criterion is determined from the Fair flow map (Fig. 34), or from the equation

$$G_{mm} = 500X_{tt} \tag{72}$$

where G_{mm} is the maximum mass velocity before mist flow, in $lb_m/(ft^2 \cdot s)$. As evaluation of X_{tt} requires the vapor fraction x, the additional recommendations [42] are that for hydrocarbons the weight fraction vapor not exceed 0.35 for pressurized reboilers or 0.5 for vacuum operation, while for water the vapor fraction should be less than 0.1.

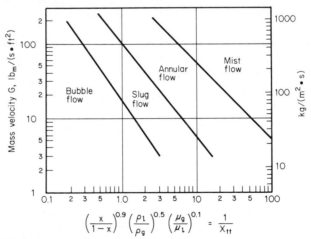

FIG. 34 Modified Fair flow regime map for vertical in-tube flow [34].

Unusual Design Conditions.   The above criteria will keep the thermosiphon design in its best operating range; however, there may be occasions where operation in the mist or film regimes is desired or unavoidable. Mist-flow heat transfer coefficients can be estimated by using the single-phase gas convection coefficient based on total mass velocity and on vapor properties (also see Refs. 72 and 73).

Film boiling should be avoided if possible. If the fluid is very clean and the entire tube is in the film region then film boiling can be considered. Heat transfer can be calculated by the Glickstein and Whiteside correlation [50]

$$\text{Nu} = 0.106 \ \text{Re}^{0.641} \ \text{Pr}^{0.4} \left(\frac{\rho_b}{\rho_g}\right)^{0.5} \tag{73}$$

where ρ_b is the bulk average density on a no-slip basis:

$$\rho_b = \frac{\rho_l}{x(\rho_l/\rho_g - 1) + 1} \tag{74}$$

The major problem with the correlation is determining the liquid velocity, since a thin vapor film (see Ref. 72, Sec. O.1) with low viscosity surrounds the liquid core. Two-phase pressure drop correlations for this flow pattern are not available in the open literature.

Horizontal Thermosiphons.   The general approach to horizontal thermosiphons is like that for the vertical thermosiphons, but design techniques and performance data for horizontal units are lacking. A flow and heat transfer analysis must be combined. The pressure drop balance between the available static head and the pressure drop in the reboiler and piping will determine the circulation rate. The convective-boiling heat transfer coefficient h_{cb} is then estimated [51] by

$$h_{cb} = \frac{\Delta P_{tp}}{\Delta P_{LO}} \ h_l \tag{75}$$

where h_l = heat transfer coefficient based on liquid alone in flow across an ideal tube bank

$\dfrac{\Delta P_{tp}}{\Delta P_{LO}}$ = ratio of two-phase pressure drop in the reboiler to pressure drop for liquid alone

c. Forced-Circulation Reboilers

Forced-circulation reboilers are designed as though they were a liquid sensible heat exchanger.

d. Falling-Film Vaporizers

The heat transfer equations for falling-film vaporizers are similar to the condensation equations except for the direction of heat flow; therefore, follow the procedures given in Ref. 71. There is an extensive literature on the hydrodynamics, the heat and mass transfer, and the "dry-patch" phenomena, but Refs. 52, 53, and 54 will provide a good summary.

In the evaporation of wide–boiling range mixtures, especially for viscous liquids, the mass transfer within the liquid film must be allowed for, as it may significantly affect the vapor-liquid interfacial composition and temperatures and, hence, evaporation rate. The study of mass transfer within the film involves research on gas absorption and stripping, and references dealing with those subjects should be searched. Some progress has been made in sponsored research in this field but has not been published at this time.

4. Temperature Differences for Reboilers

The temperature differences used to calculate the heat transfer rate depend upon the type of reboiler and heat transfer correlations. For the kettle or internal reboiler with a narrow–boiling range mixture the log mean temperature difference is used; for a wide–boiling range mixture the log mean difference is based on the exit vapor temperature. For horizontal thermosiphon reboilers Palen [42] suggests the cocurrent flow log mean ΔT be used or, as a less conservative alternative, the corrected log mean ΔT for an X or G TEMA shell. For very large bundles or horizontal thermosiphons with a large static head the boiling temperatures used for the MTD calculation should be based on the mean pressure within the bundle. Forced-circulation reboilers are calculated on the applicable LMTD and, if necessary, correction factor F for the particular flow pattern. In vertical-tube thermosiphons, however, the fluid temperatures change throughout the tube and a temperature profile has to be determined. This profile together with the varying coefficients needs to be calculated stepwise and numerically integrated.

5. Pressure Drop

Two-phase pressure drop calculations are the basis for calculating flows in thermosiphon reboilers, and other sections should be consulted for correlation details. The system pressure drop ΔP is calculated for short increments over which the vapor fraction is assumed constant, and these incremental values are summed:

$$\Delta P = \Delta P_f + \Delta P_m + \Delta P_s \tag{76}$$

where the friction-loss pressure drop ΔP_f is found using the Martinelli relations. The static head loss ΔP_s is

$$\Delta P_s = \rho_{tp} \, \Delta H \sin \theta \tag{77}$$

but for vertical exchangers $\sin \theta = 1$. The two-phase density ρ_{tp} must be calculated from void-volume correlations, which allow for vapor slip. The momentum loss ΔP_m must also be based on vapor volume (or liquid holdup) correlations allowing for vapor slip. However, in horizontal thermosiphons the mixing in the shell side is good and a homogeneous theory is better.

F. FOULING

Fouling is the undesired accumulation of solids on the heat transfer surfaces which results in additional resistances to heat transfer, thus reducing exchanger performance. These deposits are formed by one or more of the following mechanisms:

1. Precipitation fouling—the precipitation of dissolved substances on the heat transfer surface
2. Particulate fouling—the accumulation of finely divided solids on the heat transfer surface
3. Chemical reaction fouling—deposits formed on the surface by a chemical reaction not involving the surface material

4. Corrosion fouling—corrosion of the heat transfer surface that produces products fouling the surface and/or roughens the surface, promoting attachment of other foulants

5. Biological fouling—the attachment of macro- and/or microorganisms to the heat transfer surfaces

6. Freezing fouling—the solidification of a liquid or some of its higher–melting point constituents on the heat transfer surface

Fouling in heat exchangers usually involves several of the above processes, often with synergistic results.

The heat exchanger designer must make an allowance for exchanger fouling such that a reasonable period of operation can be maintained between shutdowns for cleaning. Two methods are used: (1) Derate the clean overall heat transfer coefficient by a single specified cleanliness factor; this procedure is successful for power plant condensers where the range of operation is narrow. (2) Add a separate fouling heat transfer coefficient specific for each fluid and its operation parameters. The ratio U_{fouled}/U_{clean} is equivalent to the cleanliness factor but involves consideration of both fouling coefficients. Except for power plant condensers, heat exchanger fouling is calculated by the second method. Here the separate fouling heat transfer coefficients often mask the real overall penalty, which is obtained by examination of the cleanliness factor. The above current practice is not necessarily correct. A proper fouling heat transfer coefficient is determined by considering the additional cost (direct and indirect) of the surface versus the cost of cleaning, including allowances for lost production, and the company accounting practices in evaluating investment return. To do a proper economic analysis [55, 56] requires data on the time rate curve of fouling heat transfer coefficients.

While fouling has been acknowledged as the major unknown factor in heat transfer [57], very little research had been done until the late 1970s; however, recent research is beginning to give some insight into the complex phenomenon of fouling. The proceedings of an international conference [58] are an excellent source of current knowledge. Careful experiments of specific fouling processes are explaining some of the observed plant anomalies of fouling, e.g., that increased velocity reduces some fouling and increases other fouling. Much more research is required before any significant change in fouling predictions can be made and used in design.

Several theories [57, 59] have proposed a deposition and removal mechanism which eventually comes to equilibrium and results in an asymptotic fouling coefficient. However, not all fluids exhibit such behavior or follow the expected effect of velocity. In the absence of any reliable theories the designer is forced to rely on experience, data from plant tests, and the TEMA tables in order to arrive at a suitable value for the fouling heat transfer coefficient. Plant tests are often unreliable because of uncertainty of measurements, uncertainty regarding the initial assumed state of cleanliness, and the variability of process operation due to process changes or upsets. Therefore, better test data may be obtained from a test unit mounted in a controlled side stream; however, further development of such test chambers and understanding of the relationship of their results with plant exchanger performance are needed.

The TEMA tables of fouling heat transfer coefficients in Tables 4A and 4B are widely used even though the basis of these tables is undocumented, the sources of the values are unknown, and many important factors, such as tube material, are ignored. The tables are used because they have not been challenged by other sources. The principal advantage of the TEMA tables is their age. These tables have been in use for over 30 years and cannot be underpredicting the fouling; otherwise complaints would have forced revisions. The

TABLE 4A Water Fouling Heat Transfer Coefficients [1]
In Btu/(h·ft²·°F). Quantity in parentheses in W/(m²·K).

Temperature of heating medium	Up to 240°F (115°C)		240–400°F (115–205°C)	
Temperature of water	125°F (52°C) or less		Over 125°F (52°C)	
	Water velocity, ft/s (m/s)		Water velocity, ft/s (m/s)	
Types of water	3 (1) and less	Over 3 (1)	3 (1) and less	Over 3 (1)
Seawater	2000 (9760)	2000 (9760)	1000 (4880)	1000 (4880)
Brackish water	500 (2440)	1000 (4880)	330 (1630)	500 (2440)
Cooling tower and artificial spray pond:				
Treated makeup	1000 (4880)	1000 (2440)	500 (2440)	500 (2440)
Untreated	330 (1630)	330 (1630)	200 (980)	250 (1220)
City or well water (such as Great				
Lakes)	1000 (4880)	1000 (4880)	500 (2440)	500 (2440)
River water:				
Minimum	500 (2440)	1000 (4880)	330 (1630)	500 (2440)
Average	330 (1630)	500 (2440)	250 (1220)	330 (1630)
Muddy or silty	330 (1630)	500 (2440)	250 (1220)	330 (1630)
Hard (over 15 grains/gal)	330 (1630)	330 (1630)	200 (980)	200 (980)
Engine jacket	1000 (4880)	1000 (4880)	1000 (4880)	1000 (4880)
Distilled or closed cycle condensate	2000 (9760)	2000 (9760)	2000 (9760)	2000 (9760)
Treated boiler feedwater	1000 (4880)	2000 (9760)	1000 (4880)	1000 (4880)
Boiler blowdown	500 (2440)	500 (2440)	500 (2440)	500 (2440)

fouling coefficients could be overly conservative without generating any complaints from plant personnel, who generally favor excess equipment capacity. It is important to recognize that these fouling coefficients are for shell-and-tube exchangers. Plate exchangers foul less than shell-and-tube types, having approximately twice the shell-and-tube fouling heat transfer coefficients. It is important not to be overly conservative in fouling allowances since in start-up or clean condition a grossly oversized exchanger may result in low velocities and then foul even more rapidly.

Reduction or prevention of water fouling is aided by use of water treatment chemicals in cooling tower systems; by filtration; and by chlorination to kill biological growths. Process fouling reduction depends upon the process fluid and the allowable treatments; e.g., removal of salt and water from crude oil will decrease the rate of fouling.

Cleaning methods for the exchanger depend upon the type and characteristics of the deposit. Cleaning can range from something as simple as washing the salt off an evaporator or melting ice from a coil or dissolving the fouling with chemicals, on the one hand, to the more complex methods of disassembling the exchanger and physically removing the solids by means of drills, high-pressure hydraulic systems, or sandblasting, on the other. There are many companies that offer cleaning services in the several methods. The type of cleaning must be carefully chosen to suit the type of fouling, the materials of construction, and the design of the exchangers. The tube side is usually considered the easier to clean; therefore, the fluid that is the more fouling is placed on the tube side unless other factors such as corrosion, pressure, or temperatures are more important considerations.

G. FLOW-INDUCED VIBRATIONS

Two types of vibration can occur in shell-and-tube exchangers due to shell-side flows: (1) tube vibrations and (2) acoustic vibration for vapor flows. These vibrations can cause

TABLE 4B Industrial-Fluid Fouling Heat Transfer Coefficients [1]

	Btu/(h·ft²·°F)	W/(m²·K)
Oils		
Fuel oil	200	980
Transformer oil	1000	4880
Engine lube oil	1000	4880
Quench oil	250	1220
Gases and vapors		
Manufactured gas	100	490
Engine exhaust gas	100	490
Steam (non–oil-bearing)	2000	9760
Exhaust steam (oil-bearing)	1000	4880
Refrigerant vapors (oil-bearing)	500	2440
Compressed air	500	2440
Industrial organic heat transfer media	1000	4880
Liquids		
Refrigerant liquids	1000	4880
Hydraulic fluid	1000	4880
Industrial organic heat transfer media	1000	4880
Molten heat transfer salts	2000	9760
Chemical processing fouling		
Gases and vapors		
Acid gas	1000	4880
Solvent vapors	1000	4880
Stable overhead products	1000	4880
Liquids		
MEA and DEA solutions	500	2440
DEG and TEG solutions	500	2440
Stable side draw and bottom product	1000	4880
Caustic solutions	500	1220
Vegetable oils	330	1630
Natural gas–gasoline processing fouling		
Gases and vapors		
Natural gas	1000	4880
Overhead products	1000	4880
Liquids		
Lean oil	500	2440
Rich oil	1000	4880
Natural gasoline and liquefied petroleum gases	1000	4880
Oil refinery fouling		
Crude and vaccuum unit gases and vapors		
Atmospheric tower overhead vapors	1000	4880
Light naphthas	1000	4880
Vacuum overhead vapors	500	2440
Crude and vacuum liquids		
Crude oil (depends on temperatures, velocity, dry, salt; see TEMA standards for details)	140–500	680–2440
Gasoline, naphtha, and light distillates	1000	4880

TABLE 4B Industrial-Fluid Fouling Heat Transfer Coefficients [1] *(Continued)*

	Btu/(h·ft²·°F)	W/(m²·K)
Oil refinery fouling *(Continued)*		
Kerosene	1000	4880
Light gas oil	500	2440
Heavy gas oil	330	1630
Heavy fuel oils	200	980
Asphalt and residuum	100	490
Cracking and coking unit streams		
Overhead vapors	500	2440
Light cycle oil	500	2440
Heavy cycle oil	330	1630
Light coker gas oil	330	1630
Heavy coker gas oil	250	1220
Bottom slurry oil (4.5 ft/s minimum)	330	1630
Light liquid products	500	2440
Catalytic reforming, hydrocracking, and Hydrodesulfurization		
Reformer charge	500	2440
Reformer effluent	1000	4880
Hydrocracker charge and effluent	500	2440
Recycle gas	1000	4880
Hydrodesulfurization charge and effluent	500	2440
Overhead vapors	1000	4880
Liquid product over 50° A.P.I.	1000	4880
Liquid product 30–50° A.P.I.	500	2440
Light ends processing stream		
Overhead vapors and gases	1000	4880
Liquid products	1000	4880
Absorption oils	500	2440
Alkylation trace acid streams	500	2440
Reboiler streams	330	1630
Lube oil processing streams		
Feed stock	500	2440
Solvent feed mix	500	2440
Solvent	1000	4880
Extract, wax slurries (no wax deposited on cold tube)	330	1630
Raffinate	1000	4880
Asphalt	200	980
Refined lube oil	1000	4880

tubes to leak or break, cause very high noise levels (greater than 150 decibels), and result in as much as a twofold increase in shell-side pressure drop. High fluid velocities and improper design are the major causative factors. As a result of improved pressure drop predictability allowing the use of higher fluid velocities the number of exchangers experiencing vibration has been increasing but the total number of documented cases of failure is small (about 100); however, the actual number of failures, unreported and undocumented, may be in the thousands.

Several different theories are used to evaluate the potential for a vibration problem and appear to be reasonably successful. Tests are now being made at Argonne National Laboratory on a specially designed and instrumented large exchanger (24 inches in diameter and 12 ft long, or 610 by 3660 mm, with 499 tubes ¾ in, or 19 mm, in diameter) to provide data to check the theories. A report summarizing the current status of tube vibration work was prepared by Chenoweth [60].

1. Tube Vibration

Tubes are usually the most flexible parts of an exchanger and can vibrate. If the vibration is severe enough, damage will occur. Collision damage occurs if the vibration amplitude is large enough for the tubes to collide with each other or with the shell. The tube walls are worn thin and eventually split. Baffle holes are slightly larger than the tube and thus vibrating tubes can be cut by the edges of baffle holes, especially if these are harder than the tubes. Vibration can result in fatigue failure. Failure also can occur at the tube sheet due to a cutting action of a sharp edge of the tube-sheet hole.

Although tube failures have been reported in many locations within the exchanger, most of the failures occur in the long-span tubes in the windows and are concentrated at the bundle edges, where the high-velocity bypass stream C flows, and in a few rows at or near the baffle tip. Preliminary tests at Argonne Laboratory show very little vibration of tubes directly below the nozzles, and early results indicate current prediction methods are conservative. The flow pattern is very complex in the shell; therefore, it is no surprise that no single criterion can reliably predict the velocity limits for vibration. The important forces affecting vibration are vortex shedding, turbulent buffeting, and fluid-elastic whirling referenced to the tube natural frequency.

a. Natural Frequency

For straight tubes the frequency depends upon the type of supports and on span lengths. For exchangers an adequate approach is to assume clamped ends and simple intermediate supports, and for unequal span lengths to use the maximum span. If there are fewer than four spans, then the type of end support must be considered; however, most exchangers will have more than four spans. The natural frequency is then

$$f_n = 0.04944 C_n \sqrt{\frac{EIg}{W_{ef}L^4}} \tag{78}$$

where f_n = natural frequency, Hz

$\quad C_n$ = frequency constant, dimensionless (see Table 5)

$\quad I$ = sectional moment of inertia, in^4 (m^4)

$\quad\quad = \pi \dfrac{D_o^4 - D_i^4}{64}$

$\quad g$ = gravitational constant, 386.4 in/s^2 (9.81 m/s^2)

$\quad E$ = modulus of elasticity of tube material, lb$_f$/in^2 (N/m^2)

$\quad L$ = length of each span, in (m)

$\quad W_{ef}$ = effective weight per unit length of tube itself, lb$_f$/in (N/m)

$\quad\quad = W_m + W_t + W_v$

$\quad W_m$ = weight per unit length of tube itself, lb$_f$/in (N/m)

$\quad W_t$ = weight per unit length for tube-side fluid, lb$_f$/in (N/m)

TABLE 5 Values of the Frequency Constant C_n for Uniform Beams of Equal Span Length Simply Supported with Extreme Ends Clamped [60]

Number of spans	Frequency constant C_n in Eq. (78), for mode numbers 1 through 5				
	1	2	3	4	5
1	72.36	198.34	388.75	642.63	959.98
2	49.59	72.36	160.66	198.34	335.20
3	40.52	59.56	72.36	143.98	178.25
4	37.02	49.59	63.99	72.36	137.30
5	34.99	44.19	55.29	66.72	72.36
6	34.32	40.52	49.59	59.56	67.65
7	33.67	38.40	45.70	53.63	62.20
8	33.02	37.02	42.70	49.59	56.98
9	33.02	35.66	40.52	46.46	52.81
10	33.02	34.99	39.10	44.19	49.59
11	32.37	34.32	37.70	41.97	47.23
12	32.37	34.32	37.02	40.52	44.94

W_v = virtual weight per unit length for the shell-side fluid displaced by the tube, lb_f/in (N/m)

$$= k\rho_s \frac{\pi}{4} D_o^2$$

k = experimentally determined hydrodynamic inertia coefficient (see Fig. 35)

ρ_s = shell-side fluid density, lb_m/in^3 (kg/m^3)

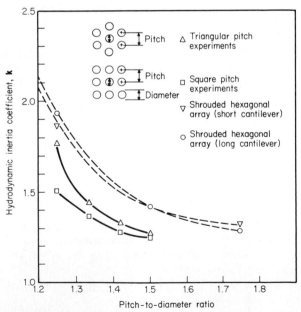

FIG. 35 Experimental measurement of hydrodynamic inertia [60].

D_i = inside tube diameter, in (m)

D_o = outside tube diameter, in (m)

An axial stress will change the natural frequency; the computation can then be adjusted as:

$$(f_n)_s = f_n \sqrt{1 + \frac{P_a L^2}{EI\pi^2}} \tag{79}$$

where $(f_n)_s$ = stressed frequency, Hz

f_n = unstressed frequency, Hz

P_a = axial load, negative if compressive, positive if tensile, lb_f (N)

$= S_n A_m$

S_n = axial stress, psi (N/m^2)

A_m = cross-section metal area, in^2 (m^2)

However, the axial stress is not well known, as it depends on how the exchanger was assembled (e.g., how the tubes were rolled) and how it is installed and operated. In the design stage much of the required information is unavailable and the effect of axial stress is ignored.

For U bends the natural frequency can be estimated by using the longest bend length in the straight-tube formula and then adjusting for in-plane and out-of-plane vibrations as

$$f_{ni} = 1.985 f_n \tag{80}$$

$$f_{no} = 0.829 f_n \tag{81}$$

This assumes no intermediate supports.

For finned tubes the natural frequency can be calculated from Eq. (78) by the following changes:

$$I = \frac{\pi}{64}(D_e^4 - D_{if}^4) \tag{82}$$

$$D_e = D_o + 1.08(D_{fr} - D_{if}) \tag{83}$$

where D_{fr} = tube diameter at root of fin, in (m)

D_{if} = tube inside diameter under finned section, in (m)

D_o = tube outside diameter, in (m)

The weight of tube per unit length, W_m, should be the actual weight for the finned section. The multiplying factor k is taken as 1.0 and the over-fin diameter is used for calculating W_s.

b. Amplitude of Tube Vibration

Usually the amplitude in heat exchangers is small so that vibration is not a problem. For the amplitude to become appreciable requires an appreciable static deflection and a large magnification factor, which is a function of the ratio of forcing frequency to natural frequency and of the system damping factor. The design rules used for exchangers can simplify the effects of high vibration modes and damping values. The mid-span static deflection force is

$$F_x = C_D \frac{\rho_s V_c^2}{2g_c} D_o \tag{84}$$

where F_x = force per unit length due to flow across tube, lb_f/ft (N/m)
$\quad C_D$ = lift or drag coefficient, dimensionless
$\quad \rho_s$ = shell-side fluid density, lb_m/ft^3 (kg/m^3)
$\quad V_c$ = shell-side cross-flow velocity, ft/s (m/s)
$\quad g_c$ = conversion constant, 32.17 $lb_m \cdot ft/(lb_f \cdot s^2)$ (1 in SI units)
$\quad D_o$ = tube outside diameter, ft (m)

Here C_d can range from 0.5 to 2.0 for lift and 0.2 to 1.35 for drag; however, a conservative value of 2.0 is suggested. The static deflection is then

$$X_s = \frac{5}{384} \frac{F_x L^4}{EI} \qquad (85)$$

where X_s = mid-span static deflection, in (m).

Finally, the vibration amplitude is

$$X_v = X_s C_{MF} \qquad (86)$$

where C_{MF} is the magnification factor. This factor is strongly dependent on the frequency ratio (forcing frequency divided by natural frequency) and damping; see Fig. 36. In exchangers C_{MF} will range from 1.2 to 3 for a frequency ratio of 0.5 to 0.8, but as the frequency ratio approaches 1 the C_{MF} rises steeply to infinity. Amplitudes of half the minimum gap between tubes can lead to impact collisions, and smaller amplitudes can still produce problems, but amplitudes of less than 0.05 in (1.3 mm) are acceptable for long life in some exchangers.

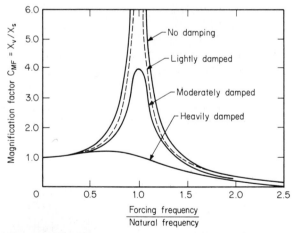

FIG. 36 Magnification factor for calculating forced-vibration amplitude from static deflection [60].

c. Vortex Shedding Frequencies

The vortex shedding frequency is calculated from

$$f_{vs} = \frac{Sl_c \, V_c}{D_o} \qquad (87)$$

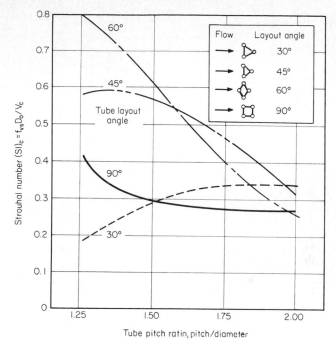

FIG. 37 Strouhal numbers for equilateral tube layouts from curves of Chen [60].

where Sl_c = Strouhal number (see Fig. 37)

V_c = velocity past the tube row, in/s (m/s)

D_o = tube outside diameter, in (m)

Note that V_c is different from the cross-flow velocities used in heat transfer and pressure drop, where the velocity is based on flow through the minimum cross section. Here the velocity is based on flow past the tube row. For the 30 and 90° layouts the velocities are the same, but they are different for the 45 and 60° tube layouts.

d. Turbulent Buffeting Frequencies

Turbulent buffeting is characterized by a spectrum of frequencies distributed around a dominant frequency. An empirical equation which is based on gas flow only is

$$f_{tb} = \frac{V_c D_o}{L_{pp}L_{tp}} \left[3.05 \left(1 - \frac{D_o}{L_{tp}} \right)^2 + 0.28 \right] \tag{88}$$

where f_{tb} = dominant turbulent buffeting frequency, Hz

V_c = cross-flow velocity through tube row, in/s (m/s)

L_{pp} = longitudinal pitch, in (m)

L_{tp} = transverse pitch, in (m)

D_o = tube outside diameter, in (m)

No data or equations are available for turbulent buffeting frequencies for liquids.

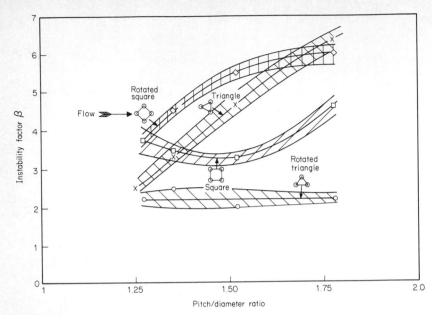

FIG. 38 Effect of tube layout on β [63].

e. Fluid-Elastic Whirling

The Conners equation [61] for the critical cross-flow velocity to initiate fluid-elastic whirling is as follows:

$$V_{crit} = \beta f_n D_o \sqrt{\frac{W_{ef}\delta_0 g_c}{\rho_s D_o^2 g}} \tag{89}$$

where V_{crit} = critical cross-flow velocity, in/s (m/s)

$\quad\quad\ \beta$ = instability threshold constant (see Fig. 38)

$\quad\quad\quad$ = 9.9 for Conners experiments with a single tube row

$\quad\quad\ \delta_0$ = log decrement of the tube bundle in the shell-side fluid under static (no-flow) conditions

$\quad\quad\ D_o$ = tube outside diameter, in (m)

$\quad\quad\ \rho_s$ = shell fluid density, lb_m/in^3 (kg/m^3)

The log decrement value δ_0 is difficult to predict, although a value of 0.033 is suggested; nevertheless, some tests and experiments indicate values ranging from 0.001 to 0.2 are possible.

f. Tube Vibration Damage Numbers

Thorngren [62] proposed two dimensionless numbers be used to predict damage due to cutting action at the baffle or to collision between tubes.

$$N_{BD} = \frac{D_o \rho_s V_c^2 L^2 g}{F_B S_f g_c A_m B_t} \tag{90}$$

$$N_{CD} = \frac{0.625 D_o \rho_s V_c^2 L^4}{F_B^4 g_c A_m (D_o^2 + D_i^2) C_t E} \tag{91}$$

where N_{BD} is the baffle damage number and N_{CD} is the collision damage number. Here

F_B = tube-to-baffle clearance factor, dimensionless
$\quad$ = 1.00 for $\frac{1}{32}$-in clearance (0.079 cm)
$\quad$ = 1.25 for $\frac{1}{64}$-in clearance (0.040 cm)
S_f = fatigue stress, lb_f/in^2 (N/m^2)
C_t = minimum gap between adjacent tubes, in (m)
B_t = baffle thickness, in (m)
g_c = conversion constant, 32.17 $lb_m \cdot ft/(lb_f \cdot s^2)$ = 1 in SI units
A_m = metal cross-section area, in^2 (m^2)
D_i = tube inside diameter, in (m)
D_o = tube outside diameter, in (m)
V_c = cross-flow velocity, in/s (m/s)
ρ_s = shell fluid density, lb_m/in^3 (kg/m^3)
L = tube span length, in (m)

Based on the ERDA report [60] these numbers are plotted in Fig. 39 to find out if they lie in a vibration or in a no-vibration region. The original Thorngren value was 1.

2. Acoustic Vibration

For vapors or gases only an acoustic vibration can occur, in an effect similiar to an organ pipe. The acoustic frequency is

$$f_a = \frac{ma}{2Y} \tag{92}$$

where f_a = acoustic frequency, Hz
$\quad m$ = mode number, dimensionless integer
$\quad a$ = velocity of sound in shell-side fluid, ft/s (m/s)
$\quad Y$ = characteristic length, usually shell diameter, ft (m)

The velocity of sound can be calculated from

$$a = \sqrt{\frac{Z\gamma g_c R T}{M}} \tag{93}$$

where Z is the compressibility factor, γ the specific heat ratio, g_c the conversion constant, R the gas constant, T the absolute gas temperature, and M the molecular weight. However, for most gas mixtures encountered in practice the specific heat ratio is unknown and by assuming a value of 1.4 and making other simplifications the following equation can be used to predict the velocity of sound:

$$a = b \sqrt{\frac{P_s g_c}{\rho_s}} \tag{93a}$$

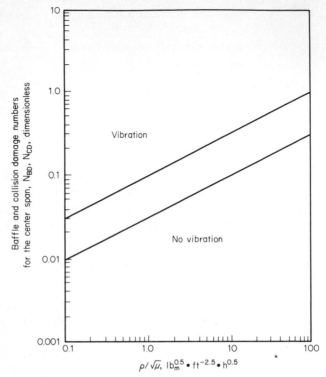

FIG. 39 Thorngren collision and baffle damage numbers for center spans [60].

where b = constant = $\left(\dfrac{c_p}{c_v}\right)^{0.5}$

P_s = system pressure, psia (N/m^2)

ρ_s = density of gas, $\text{lb}_\text{m}/\text{ft}^3$ (kg/m^3)

The acoustic frequencies can be excited by either vortex shedding or turbulent buffeting, and if their frequencies are within 20 percent of the acoustic frequency a loud noise is produced. The principal waves are the fundamental (m = 1) and the first overtone (m = 2), or the inscribed square can also be a characteristic length which leads to an acoustic frequency 1.41 times that based on shell diameter.

3. Procedure for Vibration Prediction

a. Primary Checks

No single method can reliably predict vibration; therefore, the recommended procedure is a four-step primary check followed by some secondary checks. In any procedure it is as important to predict nonvibration as it is to predict vibration; the balance between these two is affected by the choice of limits of the various predictive equations. The following procedure has successfully predicted 80 percent of the existing data bank. The steps are:

1. Acoustic vibration for shell-side vapor flows is possible within the following limits of ratios with vortex shedding and turbulent buffeting:

$$0.8 < \frac{f_{vs}}{f_a} < 1.2$$

$$0.8 < \frac{f_{tb}}{f_a} < 1.2$$

2. For either gas or liquid shell-side fluid, vibration is possible when the cross-flow velocity exceeds the critical fluid-elastic velocity ($V_c > V_{crit}$).

3. For either gas or liquid shell-side fluid, compare the vortex shedding frequency based on the velocity at the baffle edge. Vibration is probable when $f_{vs}/f_n > 0.5$. Check the cross-flow amplitude to see if it is large enough to cause tube damage.

4. For gas shell-side fluid compare the turbulent buffeting frequency based on the window velocity V_{cw} to the lowest natural tube frequency. Vibration is probable if $f_{tb}/f_n > 0.5$. Check the cross-flow amplitude for tube damage.

Notice that the primary check procedure predicts the probability of vibration and not damage. Also these checks should be made for the inlet and outlet regions of the exchanger as well as the central region.

b. Secondary Checks

The primary procedure checks for only the most obvious causes of vibration. Other high velocities in the region of the nozzles and bypass streams should be checked for local vibration problems. The collision and baffle damage numbers should be compared to Fig. 39. A check should be made to insure that acoustic vibration for the inscribed square and first overtone will not occur. And the acoustic frequency should be compared to the tube natural frequency to insure there is no resonance between them.

The above procedures are not perfect but have a high degree of success. Current Argonne Laboratory research should supply reliable data to improve these methods but additional theoretical analysis is also required.

H. AIR-COOLED EXCHANGERS

1. Types of Air Coolers

There are many designs of air-cooled heat exchangers, but they are broadly classified according to the flow of air. In forced-draft exchangers the air is blown through the coils by a fan. In induced-draft exchangers the air first passes through the coils, then through the fan, and finally a natural draft exchanger, where the air is drawn through the coils by means of a chimney on the hot exhaust air. Characteristics of air coolers are shallow tube bundles and large face areas to compensate for the low heat capacity of air and the pressure drop limitations of the fans. Forced draft results in better heat transfer due to the turbulence imparted by the fans, but offsetting this advantage are possibly poor distribution through the coil and greater chances of air recirculation. With induced designs the air leaves the fan with sufficient velocity to carry the discharge stream high into the atmosphere, thus preventing local air recirculation. Forced-draft fans handle the cool, higher-density air and are more efficient.

There are numerous designs of tubes and bundles. The tube bundles may be horizontal, vertical, or in V- or A-frame designs, as well as in cylindrical shapes. The tubes have extended surfaces of many types: spiked, slit, extruded, and wrapped radial fins; and plate fins. In addition to the heat transfer and pressure drop characteristics of the finned tubes, the other important factors in their selection are cost and material temperature limitation, as well as possible loosening of the bond between tube and fin, or corrosion at this point.

In the process industries the air coolers have almost reached a standard form. A 1-in (25.4-mm) outside diameter tube having ⅝-in (15.9-mm) high aluminum fins with 8 to 11 fins per inch (315 to 435 per meter) is very common. The fins are extruded on the tube, or wound on with a tension contact, or pinched in a groove, an L-foot contact, or an overlapped L foot. Severity of service (temperature, corrosive atmospheres, etc.) affects the selection of fin-to-tube bond. These fins are usually 0.017 to 0.020 in (0.4 to 0.5 mm) thick. If tube-side pressure drops become excessive, 1.5- to 2.0-in (38- to 50.8-mm) tubes are used. The 2-in (50.8-mm) tube will use ⅞- or 1-in (22- or 25-mm) high fins. An equilateral staggered pitch is used with a fin tip clearance of about ¼ in (6.4 mm). Individual bundles are 8, 10, or 12 ft (2.4 to 3.7 m) wide, up to 40 ft (12 m) long, and generally 4 to 8 rows deep. A number of bundles are connected in parallel to provide the necessary surface. The tubes are horizontal but the bundles may be horizontal or mounted in a V- or A-type exchanger. Elliptical tubes with steel fins are also used, but mostly in Europe.

There is no simple means of selecting the best design for a specific application because of the interplay of the numerous factors entering into the selection, such as heat transfer, pressure drop, air recirculation, cost, space and location, means of control, corrosion, range of operation and climatic conditions, and mechanical problems with fan drives and controls.

2. Heat Transfer and Pressure Drop

The air-side heat transfer and pressure drop for final design are based on experimental curves for the specific tube, fin, and layout pattern. The most useful curves are the air heat transfer coefficient based on the tube diameter and the pressure drop in inches of water versus the face velocity at standard temperature and pressure. The heat transfer coefficient then includes the effects of fin area, fin efficiency, and fin contact resistance. For the usual range of air cooler tubes as given above, the heat transfer and pressure drop curves would lie within the broadband curves given in Fig. 40. Here the heat transfer coefficient is based on the fin area after allowing for fin efficiencies and is plotted against the maximum mass velocity (based on free area between tubes and fins). To get the air heat transfer coefficient on the bare-tube basis requires calculation of the fin efficiency η_f (usually Gardner's equations [64] are used, or see Ref. 75). Then

$$h_o = h_f \left(1 + \eta_f \frac{A_f}{A_{bt}} \right) \tag{94}$$

Having gotten h_o the overall heat transfer coefficient is obtained in the usual manner.

The pressure drop per row in Fig. 40 is only the static drop due to friction across the tubes. To this pressure drop the acceleration loss, including the effect of heating, must be

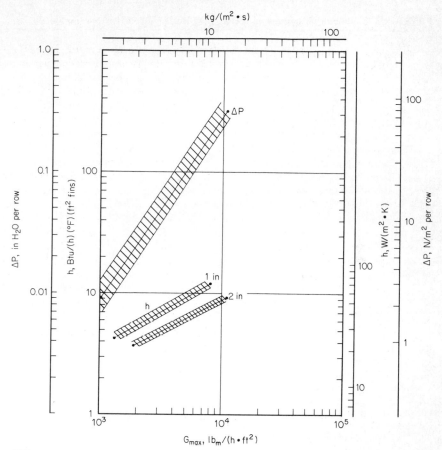

FIG. 40 Heat transfer coefficients and pressure drop for radial finned-tube bundles (1-in tube outside diameter, 2.25-in fin outside diameter, 2.5-in triangular pitch and 2-in tube outside diameter, 4-in fin outside diameter, 4.5-in triangular pitch, 9 to 11 fins per inch).

added, as well as any other losses due to the screens or guards at the fan. Also the design of the fan ring and plenum affect fan performance.

Table 6 [65] lists typical ranges of overall heat transfer coefficients which can be used for preliminary estimating. Actual overall coefficients can be calculated through use of Fig. 40 and suitable equations for inside coefficients.

The principal problem in calculating heat transfer is that the size of the cooler and hence airflow are not known, which normally leads to trial-and-error solutions. However, by specifying the face velocity and number of tube rows and tube type and fin tube dimensions, a direct solution is available by means of Fig. 41. In this figure a heat capacity ratio of the two streams is plotted against a ratio of temperature differences with curves of constant number of transfer units per square foot of face area, or UA^+/Wc_p. This avoids a trial-and-error solution and does not require the exit air temperature. Figure 41 is for cross flow with unmixed fluids, Fig. 42 is based on counterflow pattern, and Fig. 43 is a slightly different chart for condensers. While Figs. 41 and 42 have in them a table of A^+

TABLE 6 Typical Overall Heat Transfer Coefficients U for Air-Cooled Exchangers* [65]

Service	U, Btu/(h·ft²·°F) (bare tube surface)	W/(m²·K)
Liquid service		
Jacket water	120–130	680–740
50% glycol-water	95–105	540–600
Engine lube oil with retarders	20–30	114–170
Engine lube oil without retarders	15–20	85–114
Light hydrocarbons	75–95	425–540
Light naphtha	70–80	397–455
Hydroformers and platformers liquids	70	397
Light gas oil—viscosity, less than 1 cP	60–75	340–397
Heavy gas oil—viscosity, 2 to 30 cP	20–25	114–142
Heavy lube distillate—viscosity, 10 to 300 cP	8–20	45–114
Residuum—viscosity 50 to 1000 cP	10–20	57–114
Tar	5–10	28–57
Process water	105–120	600–680
Fuel oil	20–30	114–170
Gas cooling		
Flue gas at 10 psi, $\Delta P = 1$ psi	10	57
Flue gas at 100 psi, $\Delta P = 5$ psi	30	170
Air at 30 to 40 psi	20	114
Air at 50 to 100 psi	20–30	114–170
Air at 100 to 300 psi	30–35	170–200
Air at 300 to 600 psi	35–40	200–227
Air at 600 to 1000 psi	40–50	227–284
Air at 1000 to 3000 psi	50–65	284–370
Ammonia reactor stream	80–90	455–511
Hydrocarbon gases at 15–50 psi ($\Delta P = 1$ psi)	30–40	170–227
Hydrocarbon gases at 50–250 psi ($\Delta P = 3$ psi)	50–60	284–340
Hydrocarbon gases at 250–1500 psi ($\Delta P = 5$ psi)	70–90	397–511
Hydrocarbon gases at 1500–2500 psi ($\Delta P = 7$ psi)	80–100	455–570
Condensing		
Steam (0–20 psig)	130–140	740–795
Ammonia	100–120	570–680
Amine reactivator	90–100	511–570
Light hydrocarbons	80–95	455–540
Light gasoline	80	455
Light naphtha	70–80	377–455
Freon-12	60–80	340–455
Heavy naphtha	60–70	340–397
Reactor effluent—platformers, rexformers, hydroformers	60–80	340–455
Still overhead—light naphtha, steam, and noncondensable gases	60–70	340–397

*Based on Wolverine-type finned tybes, 2⅜-in triangular pitch, 1-in outside diameter, 8 fins per inch (2.25-in outside diameter).

values (square feet of tube surface area divided by square feet of face area), these charts are not restricted to those listed tubes: their A^+ values can be determined and used for other tubes and pitches. Table 7 gives some typical air velocities for various rows of tubes and values of A^+ and M for a typical tube. Use of these charts is illustrated in the problem below.

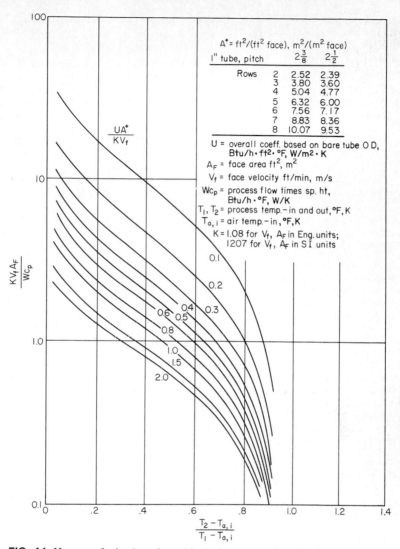

FIG. 41 Heat transfer in air coolers with single-pass cross-flow—both sides unmixed.

3. Design Considerations

a. Field of Use and Economics

Air coolers can be used in all climates and are not limited to those areas where water supply is low or poor. The economics of using air versus water is very complicated, since the relevant costs include space, electric power, piping, sewers, water costs, maintenance, fouling, etc. In general, air is usually overwhelmingly economical for high–temperature level cooling and can still be economical (depending on local conditions) when temperature levels are lower and the process discharge approaches within 30 to 50°F (16 to 28°C) of the ambient air temperature. The surface area is usually larger than with water-cooled

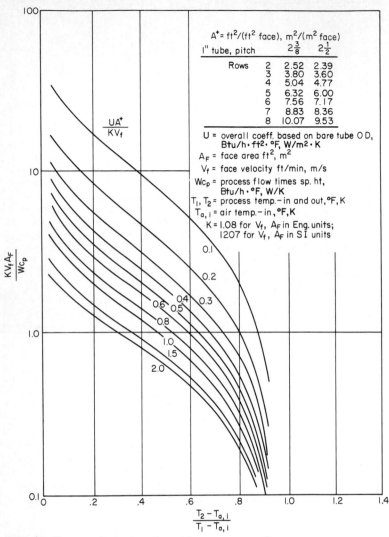

FIG. 42 Heat transfer in air coolers with countercurrent flow.

TABLE 7 Typical Values of Air Velocities and Surface Area Ratios for Air-Cooled Exchangers

Rows deep	3	4	5	6	7
Area ratio, bare tube to unit face, A^+†	3.80	5.04	6.32	7.60	8.84
Face velocity, standard ft/min	700	650	600	550	530
$M = A^+/KV_f$	0.0050	0.0072	0.0097	0.0128	0.0154
Face velocity, standard m/s	3.6	3.3	3.0	2.8	2.7
$M = A^+/KV_f$	8.7×10^{-4}	1.27×10^{-3}	1.75×10^{-3}	2.25×10^{-3}	2.7×10^{-3}

†Based on 1-in finned tube on 2⅜-in triangular pitch, 2.25 in over fins, 8 fins per inch.

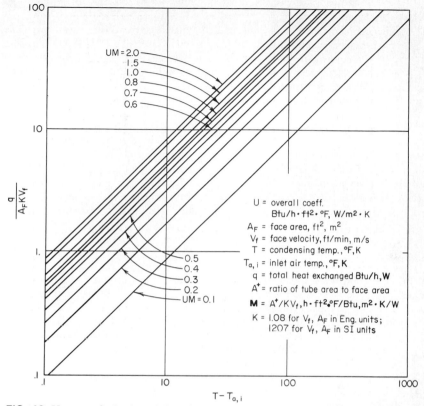

FIG. 43 Heat transfer in air-cooled condensers.

units, but the ratio is not large. Air coolers can have air-side heat transfer coefficients of 150 to 200 Btu/[h·(ft² tube area)·°F], or 850 to 1140 W/(m²·K), including fouling; while water heat transfer coefficients may be 400 to 1200 Btu/(h·ft²·°F), or 2270 to 6814 W/(m²·K), but these must include allowance for fouling, making the range of the water plus fouling coefficient from 170 to 550 Btu/(h·ft²·°F), or 965 to 3120 W/(m²·K). By the time the process side is included, the net effect may be only an area increase of 10 to 30 percent.

8	9	10	11	12
10.08	11.36	12.64	13.92	15.20
510	490	470	450	425
0.0183	0.0215	0.025	0.0286	0.0332
2.6	2.5	2.4	2.3	2.2
3.21×10^{-3}	3.76×10^{-3}	4.36×10^{-3}	5.01×10^{-3}	5.72×10^{-3}

b. Sizes

The exchangers come in all sizes from small automobile or refrigerator radiators to large power plant radiators approximately 40 ft high (excluding chimney) and 300 ft in diameter. However, most industrial exchangers are built in modules, usually in an 8-ft (2.4-m) wide bundle and 24 to 40 ft (7.3 to 12.2 m) long. When a number of small exchangers are needed, then several small bundles are mounted together over the fans. The tube bundles usually range from 3 to 6 rows deep but on occasion may extend to 12 rows.

c. Design Temperature

The ambient air temperature has a large range throughout the year. Usually the maximum air temperature selected for design is that temperature that is exceeded about 2 to 3 percent of the annual hours. Lower design temperatures can be used, especially if for peak temperatures (only a few hours per day and few days per year) some good grade of water is sprayed into the inlet air stream, cooling it so as to approach the wet-bulb temperature.

d. Fans

Fans are usually 4 to 12 ft (1.2 to 3.7 m) in diameter, and one or more may be used with a bundle. Larger and smaller sizes have been used. Multiple blades are used, and blade pitch is adjustable either manually or automatically. Fan tip speed is usually limited to less than 12,000 ft/min (60 m/s) in order to keep the noise level down. These fans handle large volumes of air at relatively low pressure drops of 1 to 2 inches of water (248 N/ m^2). The distance between the fan and the bundle, for structural and economic reasons, is far from the ideal and thus has an important effect on fan performance. For best performance the ratio of bundle face area to the fan cylinder area should be kept below 2 and the bundle aspect ratio per fan should approach 1; e.g., a 10-ft by 30-ft bundle should have three fans approximately 8.5 ft in diameter.

e. Control

Due to rapid ambient temperature changes the control system has to be engineered for the process conditions. Many means of air control are used, e.g., multiple-speed fans, turning fans on or off when several are used, pitch control, use of louvers, recirculation, and process bypass control. Control can be affected by the design of the cooler. For instance, in a forced-draft exchanger, air recirculation can occur, particularly if the unit is downwind from and next to a larger building. Sun and rain can make a marked effect. Sometimes many of these problems can be avoided by using induced-draft units. Winter conditions and possible freezing must be considered. Here auxiliary heating coils or reversing the pitch on some fans to induce recirculation can prevent freezing, or air recirculation systems can be designed. In the event of power failure, air coolers are still partially effective because of the natural draft created through the bundle. In the medium temperature ranges, a cooling capacity of one-third normal is obtainable with the fans turned off. In winter this effect may become a disadvantage.

f. Drives

Fans are mostly driven by electric motors, either directly or through V belts or gear boxes. The choice depends on operating conditions, although gear drives are common for motors of 10 hp or larger.

g. Induced versus Forced Draft

Some of the advantages and disadvantages of these two types are:

	Induced	Forced
Heat transfer coefficient	Lower: Effect of number of rows	High because of fan turbulence
Power	Higher: Fan in high-temperature stream	Lower
Drive and lubrication	In hot air: Limit to 170°F discharge above bundle	In ambient: Below bundle
Recirculation due to wind	Very low due to high upward velocity from fan	High possibility due to low bundle discharge velocity
Air distribution	Good	Poor: Depends on fan location and design
Rain and sun effects	None: Cover and high air velocity prevent rain from entering	Large unless otherwise protected by louvers
Power failure	Some natural convection cooling due to chimney effect of plenum and fan stack	Small amount of cooling

h. Space Problems

Physically air coolers take up more space than an equivalent water-cooled exchanger. Thus more ingenuity is required by the design engineer in laying out the plant. Ground space can be reduced by going to A-frame or V designs. Space is saved by placing the coolers over pipe bridges or on top of buildings, or placing them high enough that space under the coolers can be utilized by other equipment.

i. Condenser Venting

A single-pass condenser requires excess area in order to completely condense the vapor. For approximately equal flows in each tube row the vapors will condense in a shorter length in the first row, where the temperature difference is high, than in the last row, where the temperature difference is much lower. Because of the accumulation of noncondensable gases at the end of the condensing zone a back flow of excess vapor from the last row into the first rows will not occur. On the assumption of equal vapor flow into each row, a simple analysis will give the equation for minimum excess area for complete condensation in the last row as

$$\frac{\text{Total required tube length}}{\text{Total condensing length}} = \frac{1 + (N_r - 1)K}{1 + K(N_r - 1)/2} \tag{95}$$

where N_r = number of tube rows

$$K = 1 - \exp(-UA^+/Wc_p)$$

and UA^+/Wc_p is based on 1 ft^2 of face area of bundle (as in Figs. 41 and 42). Techniques to overcome this problem are described in and a more complicated analysis was made by Berg and Berg [66].

4. Air Cooler Problem

a. Use of Figure 42

Cool 65,000 lb_m/h of a fluid with specific heat equal to 0.8 from 200°F to 120°F in a 4-row–4-pass countercurrent bundle. An overall heat transfer coefficient of 90 Btu/(h·ft²·°F) is calculated for this flow and with air at 650 ft²/min face velocity. Air temperature used for design is 90°F. Using the 1-inch outside diameter finned tube (Table 7), **M** = 0.0072,

$$UM = 90 \times 0.0072 = 0.647$$

$$\frac{T_2 - T_{a,i}}{T_1 - T_{a,i}} = \frac{120 - 90}{200 - 90} = 0.273$$

From the chart read

$$\frac{1.08 V_f A_F}{W c_p} = 2.45$$

$$A_F = \frac{2.45 \times 0.8 \times 650,000}{1.08 \times 650} = 181 \text{ ft}^2$$

A bundle (more or less standard size) of 8 ft by 24 ft or 192 ft² face area will be satisfactory and have a 6 percent safety factor. To check,

$$\text{Heat load } q = 65,000 \times 0.8(200 - 120) = 4,150,000 \text{ Btu/h}$$

$$\text{Air rise for 181 ft}^2 \text{ face} = \frac{4,150,000}{1.08 \times 181 \times 650} = 32.6°F$$

$$\Delta T_{lm} = \frac{(200 - 122.6) - (120 - 90)}{\ln 2.58} = 49.2°F$$

$$\text{Required surface area} = \frac{4,150,000}{49.2 \times 90} = 936 \text{ ft}^2$$

Surface area of 181 ft² face = 181 × 5.04 = 914 ft² (close enough check for a graph).

b. Use of Figure 43

Suppose for a given type of surface and air conditions we wished to condense at 150°F a vapor with a total condensing load q = 2,500,000 Btu/h. How big a bundle is required? Suppose the calculated U = 100 Btu/(h·ft²·°F); then UM = 0.72 and $T - T_{a,i}$ = 150 − 90 = 60. From Fig. 43,

$$\frac{q}{A_F 1.08 V_f} = 31$$

$$A_F = \frac{2,500,000}{31 \times 1.08 \times 650} = 115 \text{ ft}^2$$

An 8-ft x 16-ft bundle has 128 ft² face area and an 11 percent safety factor. To check,

$$\text{Air temperature rise} = \frac{2,500,000}{1.08 \times 650 \times 128} = 27.8°F$$

$$\Delta T_{lm} = \frac{(150 - 90) - (150 - 117.8)}{\ln 1.865} = 44.6°F$$

$$\text{Surface area required} = \frac{2,500,000}{44.6 \times 100} = 560 \text{ ft}^2$$

Surface area in 115 ft^2 FA $= 115 \times 5.04 = 580$ ft^2.

Example

Air-side coefficient: Calculate air-side coefficient for 1-in outside diameter tube with 8 fins per inch 0.017 in thick and ⅝ in high. Tube pitch is 2⅜ in and air-face velocity is 500 ft/min at 70°F. ($\rho = 0.075$ lb$_m$/ft^3.)

Calculate minimum area for flow. Per foot of tube,

$$\text{Gross free area } 12 (2.375 - 1) = 16.5 \ \text{in}^2$$
$$\text{Fin blockage } 12 \times 8 \times 2 \times 0.017 \times 0.625 = \underline{2.04 \text{ in}^2}$$
$$\text{Net free area} = 14.46 \text{ in}^2$$

$$\frac{\text{Free area}}{\text{Face area}} = \frac{14.46}{2.375 \times 12} = 0.508$$

Hence

$$G_m = \frac{500 \times 0.075 \times 60}{0.508} = 4430 \text{ lb}_m/(\text{ft}^2 \cdot \text{h})$$

From Fig. 40, $h = 9$ Btu/(h·ft^2·°F). *Fin efficiency:* For aluminum fins,

$$\sqrt{\frac{2h}{kt}} \times l = \sqrt{\frac{2 \times 9 \times 12}{120 \times 0.017}} \times \frac{0.625}{12} = 0.535$$

$$\frac{D_{fo}}{D_t} = \frac{2.25}{1} = 2.25$$

From fin efficiency curves (see Sec. C.2 of Ref.75), efficiency = 86 percent. *Area ratio (approximately):* Per foot of tubes,

$$\text{Fin area} = \frac{12 \times 8 \times 2}{144} \times \frac{\pi}{4} (2.375^2 - 1^2) = 4.85 \text{ ft}^2$$

$$\text{Tube area/ft} = 0.2618 \text{ ft}^2$$

$$\text{Ratio fin area to tube (OD) area} = \frac{4.85}{0.2618} = 18.55$$

Air-side heat transfer coefficient based on tube outside diameter =

$$9(18.55 \times 0.86 + 1) = 153 \text{ Btu}/(\text{h} \cdot \text{ft}^2 \cdot °\text{F})$$

I. DESIGN FACTORS FOR HEAT EXCHANGERS

The usual impulse is to specify for design the worst heat transfer conditions under which the heat exchanger is to operate, e.g., a cooler operating under maximum load at maximum fouling with maximum water temperatures. However, the minimum operating conditions should also be evaluated since it is just as bad to lose control of processes due to excess surface as not to make production with insufficient surface. Both maximum and minimum conditions should be examined to be certain that process control is always possible and there is no possibility of freezing process streams.

The so-called safety factor is really a factor allowing for uncertainty in the heat transfer equations. The safety factor should be applied to the length of the exchanger if possible and not to the number of tubes since increasing the number reduces the velocities and heat transfer coefficients and hence the safety factor would not be as large as anticipated. There is no standard in defining the safety or ignorance factor. The factor may be applied to the overall heat transfer coefficient or to the individual fluid-film coefficients. In the latter case the factor usually is not applied to the dirt coefficients, but this procedure then implies that the dirt coefficients are known with greater accuracy than the fluid-film coefficients, which is incorrect, or that the dirt coefficient already has a very conservative factor in it, which may be true. However, one should determine the overall coefficient on the best available information and then apply a single factor if the uncertainty on each coefficient is equal or weight the result by applying the variable factor to each film. The normal heat transfer equations are accurate to ± 20 to 25 percent. Of course, if each film has a 25 percent error and if all are of equal magnitude, the probability of all errors occurring in the same direction is small and a lower overall safety factor can be applied. However, if one film is controlling, then the safety factor is the full value. Assuming all films of equal magnitude, then the average safety factor can be determined from the following equation. Average safety factor is

$$\left(\sqrt{e_1^2 + e_2^2 + \cdots e_n^2}\right)\left(\frac{1}{n}\right)$$

where e_1 is the $\pm$ error of one film, e_n of the nth film.

When water is used for cooling, it may be that only the inlet temperature is specified and the designer must select the quantity or temperature rise. In these days of rising costs and with water resources becoming limited, an economic balance must be made. The optimum exit water temperature given in Fig. 44 is the same as given by Colburn [67] but with additional cost factors in the abscissa term to allow for earning a profit on the

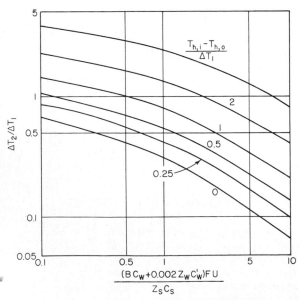

FIG. 44 Optimum exit water temperature for exchangers.

additional investments involved. Here B is the fraction of a year unit to be used; C_s is investment for exchanger surface ($/ft^2); C_w is incremented cost of water ($/1000 gal); C_w^1 is investment for water facilities ($/gpm); F is correction factor for log mean temperature difference; U is overall heat transfer coefficient in Btu/(h·ft^2·°F); Z is gross return required including allowances for taxes, repair, and maintenance; subscript s is for exchanger surface, w for water facilities; $T_{h,o}$ is outlet process temperature; $T_{h,i}$ is inlet process temperature; ΔT_1 is temperature difference between process and water at water inlet end; ΔT_2 is temperature difference between process and water at water outlet end. There is, however, a practical upper limit of about 70°C (or about 160°F) for water discharge temperatures. At high temperatures, air begins to come out of the water, corrosion rates increase, and scale formation becomes more rapid and serious.

In the design of heat exchangers for laminar flow the peculiarities of the system should be kept in mind. Heating of viscous solutions results in a stable system, but in cooling, the system is unstable when parallel paths of flow are present. If the flow or fluid temperature in one tube should become less than that in the other tubes, then the flow will slow up further because the flow rate is inversely proportional to viscosity. Also, an increase of flow in one tube will result in less cooling; thus the viscosity of that flow is less, which, in turn, results in even faster flow rates. Mueller [68] shows that a minimum pressure drop is required to avoid this maldistribution and provides a method for predicting it.

Tube sizes, thicknesses, lengths, and pitches all enter into the calculations of heat transfer and must be chosen with care. Sizes used in exchangers range from ¼-in outside diameter to 2-in outside diameter, but the most popular sizes are ⅝-in, ¾-in, and 1-in outside diameter. Some new plastic tube exchangers have a tube of 0.1-in diameter. These sizes are a compromise between the knowledge that the highest heat transfer occurs in the smallest tubes and results in the smallest exchanger, on the one hand, and the practical viewpoint, on the other, that large tubes are easier to clean and are rugged. In dirty service where tubes must be cleaned, especially by mechanical cleaners, the smallest practical size is ¾ in, but 1 in would be preferred. If chemical cleaning can be done, then the smaller tubes can be used, providing that the tubes will never plug. If the tubes plug, then chemical solutions cannot reach these points, and mechanical rodding is necessary to open the tubes for circulation. Tube lengths are based on an even number of feet; and although the TEMA standards are 8, 12, and 16 ft, many exchangers are built with tube lengths ranging from 6 to 30 ft in 2-ft increments. Choice of tube size and length may be further limited by plant standardization. In the new plants emphasis is on standardization to reduce the stock of repair tubes. In order to follow such standards, the safety factor may be greater or less, and the designer must examine the other uncertainties involved in the design to determine the extent of the real gamble in reducing the safety factor.

Tube thickness is governed by many factors such as pressure, corrosion, cost, and the need for rolling the tubes into the tube sheet. The usual thickness used is sufficiently strong to hold pressures of at least 200 lb$_f$/in^2. Steel tubes are usually several gauges heavier than alloy or nonferrous tubes, principally to get more corrosion protection. The ⅝-in tubes are usually 16 or 18 BWG (Birmingham Wire Gauge), ¾-in tubes average from 14 for steel to 16 or 18 for other materials, and 1-in tubes range from 10 to 16 BWG with 12 BWG preferred for steel and 16 BWG for other materials. Now the wall thickness is a nominal dimension, and two types of walls are available. The minimum wall tube for ferrous materials has a minimum thickness of the gauge and may be 22 percent thicker while the average wall tube has an average thickness of the gauge with a ±10 percent tolerance. The tolerances of nonferrous tubes are closer and nominal wall tubes will be used. The choice of type of wall will influence the cost and will influence the pressure drop through the tubes. As an approximation the minimum wall tube is equivalent to an average wall tube of one gauge heavier. Wall thickness also affects the rolling operation, as too heavy or too thin walls will give difficulty in getting tight joints. The gauges listed above with the tube size will give satisfactory joints for most metals.

Tube pitch is also a compromise. The best heat transfer is obtained with close pitches but such bundles plug easily and are difficult to clean. The tube-sheet ligament for close pitches may become too weak for proper rolling of the tubes and result in leaky joints. The current minimum pitch is 1.25 times the tube diameter or, if a cleaning lane is to be provided, ¼ in on square pitch. This is the nominal pitch and the actual ligament may be smaller due to drill drift, drill centering, and the need for oversizing the hole for tube insertion. For instance, with a 1-in tube on a 1¼-in pitch the TEMA minimum permissible ligament is 0.12 in, which for a 10 BWG tube is too thin and is not recommended. When the exchanger must be tight, it would be best to increase the tube pitch. The two standard types of pitches are the square and the equilateral triangle. The square pitch is mostly used when cleaning is necessary on the bundle. The triangular pitch gives a smaller bundle and shell but cannot be mechanically cleaned.

A number of special designs are available to overcome inherent weaknesses in exchangers. One weak point is leakage through the tube joints, and double tube sheets are used where intermixing of tube and shell fluid must be prevented. The tubes are rolled into each tube sheet and the space between tube sheets is vented or open. The problems in this design are to satisfactorily roll the tube joint on the inner tube sheet, to be able to locate a leak for rerolling operations, and to take care of additional stresses. These stresses are due to each tube sheet being at a different temperature, thus causing a deflection or shearing action on the tubes, and to the expansion of the tubes causing compression stresses in the tubes if both tube sheets are fastened or held together at their edges. Additional protection to prevent leakage caused by pinholes or corrosion through the tube wall can be secured either by using a double tube with a liquid metal filling the annulus and vented to a space where a change in the volume of this intermediate fluid can be detected [69] or by having small longitudinal grooves in one of the tubes so that when the tubes are drawn together to improve contact and heat transfer the small grooves can still carry any leaking fluid to the space between the double tube sheets for detection before the leak penetrates both tubes [70]. When the tube material is not satisfactory for both the tube and shell fluids then a bimetallic tube may be used. Various combinations of metals are available and tests indicate no contact resistance to heat flow if the tubes have been drawn together to effect a good bond. Use of bimetallic tubes usually also requires double tube sheets or a clad tube sheet. Special techniques are used to strip back the outer material for a portion of the tube sheet thickness, and a ferrule may be used as a filler so that the tube fluid is sealed from the jacket metal of the bimetal tube.

ACKNOWLEDGMENTS

The author is indebted to Jerry Taborek of Heat Transfer Research, Inc., for his help in supplying material for Part 2 of Chap. 4 and for his review comments, and to J. W. Palen and J. M. Chenoweth of HTRI for their reviews.

REFERENCES

1. *Tubular Exchanger Manufacturers Association Standards,* 6th ed., New York, 1978.
2. U.S. Patents 3,708,142, 4,127,165, and 4,136,736.

3. T. Tinker, Shell Side Characteristics of Shell and Tube Heat Exchangers pts. I, II, and III, *General Discussion of Heat Transfer, Proc. Inst. Mech. Eng.,* London, 1951.

4. J. W. Palen and J. Taborek, Solution of Shell Side Flow Pressure Drop and Heat Transfer by Stream Analysis Method, *Chem. Eng. Prog. Symp. Ser.* 65, no. 92, 1969.

5. K. J. Bell and O. P. Bergelin, Flow through Annular Orifices, *Trans. ASME,* vol. 79, pp. 595–603, 1957.

6. K. J. Bell, Final Report of the Cooperative Research Program on Shell and Tube Heat Exchangers, *U. Del. Eng. Exp. Sta. Bull.,* no. 5, 1963; also see Bulletin no. 4.

7. K. J. Bell, Delaware Method for Shell Side Design, *Heat Exchangers—Thermal Hydraulic Fundamentals and Design,* ed. S. Kakaç, A. E. Bergles, and F. Mayinger, pp. 581–618, Hemisphere/McGraw-Hill, Washington, D.C., 1981.

8. J. Taborek, Shell and Tube Exchanger Design—Sensible Heat, *Heat Exchanger Design Handbook,* sec. 3.3, Hemisphere, Washington, D.C., 1982.

9. K. W. Coons, A. M. Hargis, P. Q. Hewes, and F. T. Weems, *Chem. Eng. Prog.,* vol. 43, pp. 405–414, 1947.

10. A. M. Hargis, A. T. Beckman, and J. J. Loiancono, Application of Spiral Plate Heat Exchangers, AIChE Houston meeting, February 1967.

11. M. H. I. Baird, W. McCrae, F. Rumford, and C. G. M. Slesser, Some Considerations on Heat Tranfer in Spiral Plate Heat Exchangers, *Chem. Eng. Sci.,* vol. 7, pp. 112–115 and 196, 1957.

12. A. Cooper and J. D. Usher, Thermal and Hydraulic Design of Plate Exchangers, *Heat Exchanger Design Handbook,* sec. 3.6, Hemisphere, Washington, D.C., 1982.

13. K. S. N. Raju and J. C. Bansal, Consider the Plate Exchanger, *Chem. Eng.,* pp. 133–144, Aug. 11, 1980.

14. K. S. N. Raju and J. C. Bansal, (1) Plate Heat Exchangers and their Performance, (2) Design of Plate Heat Exchangers, *Low Reynolds Number Forced Convection in Channels and Bundles,* ed. S. Kakaç, R. K. Shah, and A. E. Bergles, Hemisphere, Washington, D.C., 1982.

15. M. F. Edwards, Heat Transfer in Plate Heat Exchangers at Low Reynolds Numbers, *Low Reynolds Number Forced Convection in Channels and Bundles,* ed. S. Kakaç, R. K. Shah, and A. E. Bergles, Hemisphere, Washington, D.C., 1982.

16. E. R. Kent and R. L. Pigford, Fractionation during Condensation of Vapor Mixtures, *AIChE J.,* vol. 2, p. 363, 1956.

17. R. C. Holmes, in *Chemical Engineers' Handbook,* ed. R. H. Perry, 3d ed., p. 686, McGraw-Hill, New York, 1950.

18. K. G. English, W. T. Jones, R. C. Spillers, and V Orr, Criteria of Flooding and Flooding Correlation Studies with a Vertical Updraft Partial Condenser, *AIChE preprint 9, 6th Nat. Heat Transfer Conf.,* 1963.

19. L. D. Boyko and G. N. Kruzhilin, Heat Transfer and Hydraulic Resistance during Condensation of Steam in a Horizontal Tube and in a Bundle of Tubes, *Int. J. Heat Mass Transfer,* vol. 10, pp. 361–373, 1967.

20. J. W. Palen, G. Breber, and J. Taborek, Predictions of Flow Regimes in Horizontal Tubeside Condensation, *AIChE preprints, 17th Nat. Heat Transfer Conf.,* p. 38, Salt Lake City, 1977.

21. G. Breber, J. W. Palen, and J. Taborek, Prediction of Horizontal Tubeside Condensation of Pure Components Using Flow Regime Criteria, *Condensation Heat Transfer, 18th Nat. Heat Transfer Conf.,* 1979; ASME, San Diego, also in *Heat Transfer Eng.,* vol. 1, no. 2, pp. 47–57, 1979.

22. S. S. Kutateladze and V. M. Borishanski, *Concise Encyclopedia of Heat Transfer,* p. 181, Pergamon, New York, 1966.

23. J. Taborek, Design Method for Heat Transfer Equipment, *Heat Exchanger Design and Theory Source Book,* ed. N. Afgan and E. U. Schlünder, Scripta Book, Washington, D.C., 1974.

24. K. J. Bell and A. M. Ghaly, An Approximate Generalized Design Method for Multicomponent/Partial Condensers, *AIChE Symp. Ser.* 69, no. 131, pp. 72–79, 1972.

25. O. P. Bergelin, P. A. Kegal, F. G. Carpenter, and C. Gazley, *Heat Transfer and Fluid Mech. Inst.,* Berkeley, Calif., 1964.

26. D. A. Charvonia, Experimental Investigation of the Mean Liquid Film Thicknesses and the Characteristics of the Interfacial Surface in Annular, Two-Phase Flow, *ASME paper 61-WA-243,* 1961.

27. S. F. Chien and W. Ibele, Pressure Drop and Liquid Film Thickness of Two Phase Annular and Annular-Mist Flow, *ASME paper 62-WA-170,* 1962.

28. L. Y. Zhivaikin and B. P. Volgin, Hydraulic Resistance in Descending Two-Phase Flow in Film-Type Equipment, *Int. Chem. Eng.,* vol. 4, pp. 80–84, 1964.

29. T. Fujii, H. Uehara, K. Hiratu, and K. Oda, Heat Transfer and Flow Resistance in Condensation of Low Pressure Steam Flowing through Tube Banks, *Int. J. Heat Mass Transfer,* vol. 15, pp. 247–260, 1972.

30. I. D. R. Grant, Flow and Pressure Drop with Single Phase and Two Phase Flow in Shell Side of Segmentally Baffled Shell and Tube Heat Exchangers, *NEL Rep. No. 590,* pp. 1–22, National Engineering Laboratory, Glasgow, 1975.

31. J. E. Diehl, Calculate Condenser Pressure Drop, *Pet. Refiner,* vol. 36, pp. 147–153, 1957.

32. I. D. R. Grant and D. Chisholm, Two Phase Flow on the Shell Side of a Segmentally Baffled Shell-and-Tube Heat Exchanger, *ASME paper 77-WA/HT-22,* presented in Atlanta, 1977.

33. D. Butterworth, Developments in the Design of Shell and Tube Condensers, *ASME paper 77-WA/HT-24,* 1977.

34. K. J. Bell and A. C. Mueller, Condensation Heat Transfer and Condenser Design, *AIChE Today Series,* American Institute of Chemical Engineers, New York, 1971.

35. I. D. R. Grant, Condenser Performance—The Effect of Different Arrangements for Venting Noncondensable Gases, *Brit. Chem. Eng.,* vol. 14, pp. 651–653, 1969.

36. H. Hampson and W. Thain, The Effect of Venting Arrangements during the Condensation on a Horizontal Tube from Static Vapor Gas Mixture, *Proc. 3d Int. Heat Transfer Conf.,* vol. I, pp. 118–129, 1966.

37. R. A. Khan, Effect of Vent Rate, Noncondensable Loading and Vent Location on Heat Transfer in a Large Vertical Tube Bundle of a Sea Water Evaporator, *Symp. Special Problems in Process Heat Transfer, paper 14G,* AIChE national meeting, Dallas, 1972.

38. D. Eissenberg and D. Bogue, Tests of an Enhanced Horizontal Tube Condenser under Conditions of Horizontal Steam Cross Flow, *Heat Transfer,* I, HE 2.1, *Proc. 4th Int. Heat Transfer Conference,* 1970.

39. A. P. Colburn and A. G. Edison, Prevention of Fog in Cooler Condensers, *Ind. Eng. Chem.,* vol. 33, p. 457, 1941.

40. J. L. Katz and B. J. Ostermier, *J. Chem. Phys.,* vol. 47, p. 482, 1967.

41. J. W. Palen and W. M. Small, A New Way to Design Kettle and Internal Reboilers, *Hydrocarbon Processing,* vol. 43, no. 11, p. 199, 1964.

42. J. W. Palen, Shell and Tube Reboilers, *Heat Exchanger Design Handbook,* sec. 3.6, Hemisphere, Washington, D.C., 1981.

43. J. G. Collier, Boiling and Evaporation, *Heat Exchangers—Thermal-Hydraulic Fundamentals and Design,* ed. S. Kakaç, A. E. Bergles, and F. Mayinger, pp. 235–288, Hemisphere/McGraw-Hill, Washington, D.C., 1981.

44. J. W. Palen, A. Yarden, and J. Taborek, Characteristics of Boiling outside Large Scale Multitube Bundles, *Chem. Eng. Prog. Symp. Ser.* 68, no. 118, pp. 50–61, 1972.

45. R. Wallner, Heat Transfer in Flooded Shell and Tube Evaporators, *5th Int. Heat Transfer Conf., paper HE 2.4,* Tokyo, 1974.

46. J. C. Chen, A Correlation for Boiling Heat Transfer to Saturated Fluids in Convective Flow, *ASME paper 63-HT-34,* 1963; also, *Ind. Eng. Chem. Process Des. Dev.,* vol. 5, no. 3, pp. 322–329, 1966.

47. A. Blumenkrantz and J. Taborek, Applications of Stability Analysis for Design of Natural Circulation Boiling Systems and Comparison with Experimental Data, *AIChE Symp. Ser.* 68, no. 118, 1971.

48. D. C. Lee, J. W. Dorsey, G. Z. Moore, and F. D. Mayfield, Design Data for Thermosyphon Reboilers, *Chem. Eng. Prog.,* vol. 52, no. 4, p. 160, 1956.

49. J. W. Palen, C. C. Shih, A. Yarden, and J. Taborek, Performance Limitations in a Large Scale Thermosyphon Reboiler, *Proc. 5th Int. Heat Transfer Conf., paper HE 2.2,* vol. 5, pp. 204–208, Tokyo, 1974.

50. M. R. Glickstein and R. H. Whitesides, Forced Convection Nucleate and Film Boiling of Several Aliphatic Hydrocarbons, *ASME paper 67-HT-7, 9th Nat. Heat Transfer Conf.,* Seattle, 1967.

51. J. Taborek, Design Methods for Heat Transfer Equipment—A Critical Survey of the State-of-the-Art, *Heat Exchangers: Design and Theory Sourcebook,* ed. N. Afgan and E. U. Schlünder, McGraw-Hill, New York, 1974.

52. G. D. Fulford, The Flow of Liquids in Thin Films, *Advances in Chemical Engineering,* vol. 5, pp. 151–236, Academic, New York, 1964.

53. R. A. Seban, Transport to Falling Films, *Heat Transfer 1978,* vol. 6, pp. 417–428.

54. E. N. Ganic and K. Mastanaish, Hydrodynamics and Heat Transfer in Falling Film Flow, *Low Reynolds Number Heat Exchangers,* ed. S. Kakaç, R. K. Shah, and A. E. Bergles, Hemisphere, Washington, D.C., 1982.

55. A. C. Mueller, Thermal Design of Heat Exchangers, *Purdue Univ. Exp. Sta. Bull.* no. 121, September 1954; also *Handbook of Heat Transfer,* ed. W. M. Rohsenow and J. P. Hartnett, 1st ed., McGraw-Hill, New York, 1973.

56. N. Epstein, Optimum Evaporator Cycles with Scale Formation, *Can. J. Chem. Eng.,* vol. 57, pp. 659–661, 1979.

57. J. Taborek, J. Knudsen, T. Aoki, R. B. Ritter, and J. W. Palen, Fouling—The Major Unresolved Problem in Heat Transfer, pts. I and II, *Chem. Eng. Prog.,* vol. 68, no. 2, pp. 59–67, and vol. 68, no. 7, pp. 69–78, 1972.

58. E. F. C. Somerscales and J. G. Knudsen, eds., *Fouling of Heat Transfer Equipment,* Hemisphere, Washington, D.C., 1972.

59. D. Q. Kern and R. E. Seaton, A Theoretical Analysis of Thermal Surface Fouling, *Brit. Chem. Eng.,* vol. 4, no. 5, pp. 258–262, 1959.

60. J. M. Chenoweth, Flow Induced Tube Vibrations in Shell-and-Tube Heat Exchangers, final report on contract no. EY-76-C-03-1273 for Division of Conservation and Technology of ERDA, *SAN/1273-1, UC-93,* February 1977.

61. H. J. Conners, Fluidelastic Vibration of Tube Arrays Excited by Cross Flow, presented at the ASME Winter Annual Meeting, New York, Dec. 1, 1970.

62. J. T. Thorngren, Predict Exchanger Tube Damage, *Hydrocarbon Processing,* vol. X, pp. 129–131, April 1970.

63. B. M. H. Sopper, The Effect of Tube Layout on the Fluidelastic Instability of Tube Bundles, *Flow-Induced Heat Exchanger Tube Vibration—1980,* ed. J. M. Chenoweth and J. R. Stenner, *ASME HTD vol. 9, G00182,* pp. 1–10, November 1980.

64. K. A. Gardner, Efficiency of Extended Surfaces, *Trans. ASME,* vol. 67, pp. 621–631, 1945.

65. E. C. Smith, Air-Cooled Heat Exchangers, *Chem. Eng.* vol. 63, no. 23, p. 145, 1958.

66. W. F. Berg and J. L. Berg, Flow Patterns for Isothermal Condensation in One Pass Air Cooled Heat Exchangers, *Heat Transfer Eng.,* vol. 1, no. 4, pp. 21–31, 1980.

67. A. P. Colburn, Heat Transfer by Natural and Forced Convection, *Purdue Univ. Eng. Exp. Sta. Bull.* 84, 1942.

68. A. C. Mueller, Criteria for Maldistribution in Viscous Flow Coolers, *Proc. 5th Int. Heat Transfer Conf., paper HE 1.4,* vol. 5, pp. 170–174, Tokyo, 1974.

69. T. Trocki and D. B. Nelson, Report on a Liquid Metal Heat Transfer and Steam Generation System for Nuclear Power Plant, *J. Mech. Eng., Trans. ASME,* vol. A52, pp. 140, 472, and 927, 1953.

70. J. T. Cullen, A Leak-proof Heat Exchanger, *ASME paper A50,* p. 125, 1950; *Mech. Eng.,* vol. 73, p. 425, 1951 (abstract).

71. W. M. Rohsenow, Film Condensation, *Handbook of Heat Transfer Fundamentals,* ed. W. M. Rohsenow, J. P. Hartnett, and E. N. Ganić, chap. 11, pt. 1, McGraw-Hill, New York, 1985.

72. W. M. Rohsenow, Boiling, *Handbook of Heat Transfer Fundamentals,* ed. W. M. Rohsenow, J. P. Hartnett, and E. N. Ganić, chap. 12, McGraw-Hill, New York, 1985.

73. P. Griffith, Two-Phase Flow, *Handbook of Heat Transfer Fundamentals,* ed. W. M. Rohsenow, J. P. Hartnett and E. N. Ganić, chap. 13, McGraw-Hill, New York, 1985.

74. G. D. Raithby and K. G. T. Hollands, Natural Convection, *Handbook of Heat Transfer Fundamentals,* ed. W. M. Rohsenow, J. P. Hartnett, and E. N. Ganić, chap. 6, McGraw-Hill, New York, 1985.

75. P. J. Schneider, Conduction, *Handbook of Heat Transfer Fundamentals,* ed. W. M. Rohsenow, J. P. Hartnett, and E. N. Ganić, chap. 4, McGraw-Hill, New York, 1985.

NOMENCLATURE

Symbol, Definition, SI Units, English Units

A	surface area: m^2, ft^2
A_a	total leakage area of one baffle: mm^2, in^2
A_b, A_c, A_e	cross-sectional areas associated with B, C, and E flows in Fig. 7: mm^2, in^2
A_{ba}	bypass area of one baffle: mm^2, in^2
A_{bt}	area of bare tube: m^2, ft^2
A_{ef}	effective area: m^2, ft^2
A_F	face area of bundle: m^2, ft^2
A_f	actual fin area: m^2, ft^2
A_m	cross-section metal area of tube: m^2, ft^2
A_{mb}	minimum flow area at centerline of one baffle: mm^2, in^2
A_o	total heat transfer area: m^2, ft^2
A_{sb}	shell-baffle leakage area: mm^2, in^2
A_{tb}	tube-baffle leakage area: mm^2, in^2
A_w	net flow area in window: mm^2, in^2
A_{wg}	gross window area: mm^2, in^2
A_{wt}	window area occupied by tubes: mm^2, in^2
A_b'	tube area per unit length: m^2/m, ft^2/ft
A_{mb}^*	A_{mb} converted to m^2 or ft^2
A^+	tube surface per unit face area
a	velocity of sound: m/s, ft/s
B	constant defined at Eq. (54)
B_c	baffle cut, percent of diameter
B_t	baffle thickness: m, ft
C_D	drag coefficient
C_{max}	larger heat capacity flow rate $(Wc_p)_{max}$: $J/(s \cdot K)$, $Btu/(h \cdot °F)$
C_{MF}	magnification factor
C_{min}	smaller heat capacity flow rate $(Wc_p)_{min}$: $J/(s \cdot K)$, $Btu/(h \cdot °F)$
C_n	frequency constant
C_t	minimum gap between adjacent tubes: m, in
c_p	specific heat at constant pressure: $J/(kg \cdot K)$, $Btu/(lb_m \cdot °F)$
c_v	specific heat at constant volume: $J/(kg \cdot K)$, $Btu/(lb_m \cdot °F)$
D	diameter: m, ft

D_B	bundle diameter: m, ft
D_b	baffle diameter: mm, in
D_{ctl}	defined in Fig. 8: mm, in
D_e	effective diameter of finned tube, Eq. (83): mm, in
D_{eq}	equivalent diameter of finned tube: m, ft
D_{feq}	equivalent projected diameter of finned tube, defined in Sec. A.8: mm, in
D_{fo}	diameter over fins: mm, in
D_{fr}	diameter at fin root: mm, in
D_h	hydraulic diameter: m, ft
D_i	inside tube diameter: m, ft
D_o	outside tube diameter: m, ft
D_{otl}	defined in Fig. 8: mm, in
D_s	shell diameter: mm, in
D_t	tube diameter: mm, in
D_w	equivalent diameter in window: mm, in
E	modulus of elasticity: N/m^2, lb_f/in^2
F	LMTD correction factor
F_B	tube-baffle clearance factor
F_b	tube bundle boiling correction factor
F_c	fraction of tubes in cross flow
F_{ch}	h_{cb}/h_l
F_m	mixture correction factor
F_p	pressure correction factor for nucleate boiling
F_w	fraction of tubes in window
$\mathbf{F}_x$	force per unit length due to flow, Eq. (84): N/m, lb_f/ft
f	friction factor, $(2g_c \rho\ \Delta P/4N_r G_s^2)(\mu_b/\mu_w)^m$
f_a	acoustic frequency: Hz
f_n	straight tube natural frequency: Hz
$(f_n)_s$	natural frequency under axial stress: Hz
f_{tb}	turbulent buffeting frequency: Hz
f_{vs}	vortex shedding frequency: Hz
G	mass velocity: $kg/(m^2 \cdot s)$, $lb_m/(ft^2 \cdot h)$
G_f	mass velocity at flooding, Eq. (48): $kg/(m^2 \cdot s)$, $lb_m/(ft^2 \cdot h)$
G_{mm}	maximum mass velocity above which mist flow occurs: $kg/(m^2 \cdot s)$, $lb_m/(ft^2 \cdot h)$
G_s	maximum mass velocity in shell: $kg/(m^2 \cdot s)$ $lb_m/(ft^2 \cdot h)$
G_t	actual total mass velocity: $kg/(m^2 \cdot s)$, $lb_m/(ft^2 \cdot h)$
G_v	vapor mass flow rate: $kg/(m^2 \cdot s)$, $lb_m/(ft^2 \cdot h)$
G_w	window mass velocity: $kg/(m^2 \cdot s)$, $lb_m/(ft^2 \cdot h)$
g	gravitational acceleration: $9.81\ m/s^2$, $32.2\ ft/s^2$
g_c	gravitational conversion constant: $4.18 \times 10^8\ lb_m \cdot ft/(lb_f \cdot h^2)$; 1 in SI units
H	half distance, baffle tip to shell: mm, in
h_b	corrected boiling heat transfer coefficient: $W/(m^2 \cdot K)$, $Btu/(h \cdot ft^2 \cdot °F)$
h_c	condensing heat transfer coefficient: $W/(m^2 \cdot K)$, $Btu/(h \cdot ft^2 \cdot °F)$
h_{cb}	convective boiling heat transfer coefficient: $W/(m^2 \cdot K)$, $Btu/(h \cdot ft^2 \cdot °F)$

h_f heat transfer coefficient based on fin surface: $W/(m^2 \cdot K)$, $Btu/(h \cdot ft^2 \cdot {}^\circ F)$

h_i ideal bundle heat transfer coefficient: $W/(m^2 \cdot K)$, $Btu/(h \cdot ft^2 \cdot {}^\circ F)$

h_l convective heat transfer coefficient for liquid phase: $W/(m^2 \cdot K)$, $Btu/(h \cdot ft^2 \cdot {}^\circ F)$

h_{nb} nucleate boiling heat transfer coefficient: $W/(m^2 \cdot K)$, $Btu/(h \cdot ft^2 \cdot {}^\circ F)$

h_{nb1} single-tube nucleate boiling heat transfer coefficient: $W/(m^2 \cdot K)$, $Btu/(h \cdot ft^2 \cdot {}^\circ F)$

h_{nc} natural-convection heat transfer coefficient: $W/(m^2 \cdot K)$, $Btu/(h \cdot ft^2 \cdot {}^\circ F)$

h_o heat transfer coefficient based on tube outside diameter: $W/(m^2 \cdot K)$, $Btu/(h \cdot ft^2 \cdot {}^\circ F)$

h_s corrected shell-side heat transfer coefficient: $W/(m^2 \cdot K)$, $Btu/(h \cdot ft^2 \cdot {}^\circ F)$

h_w coolant heat transfer coefficient: $W/(m^2 \cdot K)$, $Btu/(h \cdot ft^2 \cdot {}^\circ F)$

h_1 single-tube heat transfer coefficient: $W/(m^2 \cdot K)$, $Btu/(h \cdot ft^2 \cdot {}^\circ F)$

I moment of inertia: m^4, in^4

i_l enthalpy of liquid: J/kg, Btu/lb_m

i_{lg} latent heat of vaporization: J/kg, Btu/lb_m

i_v enthalpy of vapor: J/kg, Btu/lb_m

Δi enthalpy change: J/kg, Btu/lb_m

J correction factor: subscript b = bypass, c = baffle cut, l = leakage, r = adverse temperature gradient, s = end spacing

j heat transfer factor, $(h_i/c_p G_s)\, Pr^{2/3}\, (\mu_w/\mu_b)^{0.14}$

j_g^* Wallis factor, defined in Eq. (49)

K constant in equations

$K_A, K_B,$
K_C, K_E flow resistance in velocity heads for flow streams

k thermal conductivity of the fluid: $W/(m \cdot K)$, $Btu/(h \cdot ft \cdot {}^\circ F)$

k_l thermal conductivity, liquid: $W/(m \cdot K)$, $Btu/(h \cdot ft \cdot {}^\circ F)$

k_w thermal conductivity, wall: $W/(m \cdot K)$, $Btu/(h \cdot ft \cdot {}^\circ F)$

$\mathbf{k}$ constant in vibration equation

L length: m, ft

L_b baffle spacing (L_{bi} = inlet, L_{bo} = outlet, L_{bc} = central): mm, in

L_{bb} bundle bypass diametral gap: mm, in

L_{fh} fin height: mm, in

L_p bypass lane, Fig. 8: mm, in

L_{pl} tube lane partition bypass width, Eq. (25): mm, in

$L_{pn}, L_{pp},$
L_{tp} tube pitch dimensions in Fig. 5: mm, in

L_{sb} diametral clearance, shell to baffle: mm, in

L_{tb} diametral clearance, tube to baffle: mm, in

$L_{ta}, L_{ti},$
L_{to}, L_{ts} tube lengths defined in Fig. 9: mm, in

$L^+, L_i^+,$
L_o^+ dimensionless spacings defined in the paragraph preceding Eq. (38)

$\mathbf{LG}$ weight ratio of condensate to vapor, Eq. (48)

l	spiral plate spacing: m, ft
M	A^+/KV_f: $\text{h} \cdot \text{ft}^2 \cdot {}^\circ\text{F/Btu}$, $\text{m}^2 \cdot \text{K/W}$
m	mode number
N_{BD}, N_{CD}	baffle, collision damage number, Eqs. (90), (91)
N_b	number of baffles
N_N	number of nozzle pairs
N_r	number of tube rows
N_{ss}	number of pairs of sealing strips
N_t	number of tubes
N_{tcc}	effective number of tube rows between baffle tips
N_{tcw}	effective number of tube rows in window
N_{tr}	number of condenser tubes in a vertical path, Sec. D.3
N_{tt}	total tubes in shell
N_f'	number of fins per unit length: mm^{-1}, in^{-1}
N_{ss}^+	ratio of N_{ss} to N_{tcc}
Nu	Nusselt number, hD/k
P	operating pressure: N/m^2, $\text{lb}_\text{f}/\text{in}^2$
P_{cr}	critical pressure for substance: N/m^2, $\text{lb}_\text{f}/\text{in}^2$
P_s	shell-side pressure: N/m^2, $\text{lb}_\text{f}/\text{in}^2$
ΔP	pressure drop (subscript bi = ideal bundle, c = cross-flow, e = end zone, f = friction, GO = gas only, LO = liquid only, m = momentum, s = static, tot = total, tp = two-phase, w = window, wi = ideal window): N/m^2, $\text{lb}_\text{f}/\text{in}^2$
P_r^+	ratio of operating pressure to critical pressure
P$_a$	axial load: N, lb_f
Pr	Prandtl number, $\mu c_p/k$
p	exponent
Q	with subscript TOT, A, B, C, E: flow-stream quantity; with subscript RA, RB, RC, RE: relative flow quantity
Q_{sv}	vapor sensible heat: W, Btu/h
Q_t	total heat flow: W, Btu/h
$q''_{b,\text{max}}$	maximum heat flux, tube bundle: W/m^2, $\text{Btu/(h} \cdot \text{ft}^2)$
q''_{max}	maximum heat flux for thermosiphon: W/m^2, $\text{Btu/(h} \cdot \text{ft}^2)$
$q''_{1,\text{max}}$	maximum heat flux for single tube: W/m^2, $\text{Btu/(h} \cdot \text{ft}^2)$
R_b, R_l,	pressure drop correction factors for bypass, leakage flows,
R_s	and end spacing
R$_t$	thermal resistance of tube wall and fouling: $\text{m}^2 \cdot \text{K/W}$, $\text{h} \cdot \text{ft}^2 \cdot {}^\circ\text{F/Btu}$
Re$_{\text{av}}$	Reynolds number based on hydraulic diameter, $D_h G/\mu$
Re$_s$	shell-side Reynolds number $D_t G_s/\mu$, Eq. (30)
r_b	A_{ba}/A_{mb}
$r_{\ell m}$	area ratio $(A_{sb} + A_{tb})/A_{mb}$
r_s	area ratio $A_{sb}/(A_{sb} + A_{tb})$
S_f	fatigue stress: N/m^2, $\text{lb}_\text{f}/\text{in}^2$
Sl$_c$	Strouhal number, $f_{vs}/D_o V_c$

s	suppression factor, Eq. (64)
T_a	air temperature: K, °F
T_c	coolant temperature: K, °F
T_s	average shell temperature: °C, °F
T_v	vapor temperature: K, °F
T_w	wall temperature: K, °F
ΔT_{BR}	boiling temperature range, Eq. (61): K, °F
ΔT_b	temperature difference between wall and saturation temperature: K, °F
ΔT_{cr}	critical temperature difference in boiling corresponding to maximum heat flux: K, °F
ΔT_o	temperature difference between wall and bulk boiling liquid: K, °F
U	overall heat transfer coefficient: W/(m²·K), Btu/(h·ft²·°F)
V_c	cross velocity: m/s, in/s
V_{crit}	critical cross-flow velocity: m/s, in/s
V_{cw}	cross-flow window velocity: m/s, in/s
V_f	face velocity: m/s, ft/s
V_g	vapor flow velocity: m/s, ft/s
v_g, v_l	specific volume of vapor or liquid: m³/kg, ft³/lb$_m$
W	flow rate: kg/s, lb$_m$/h
W_s	total shell flow: kg/s, lb$_m$/h
W_T	total flow rate: kg/s, lb$_m$/h
$\mathbf{W}_{ef}$	effective weight of tube per unit length: N/m, lb$_f$/in
$\mathbf{W}_v$	weight of displaced fluid, Eq. (78): N/m, lb$_f$/in
X_s	mid-span deflection: m, in
X_{tt}	Martinelli parameter
X_v	vibration amplitude: m, in
X_0	static deflection: m, in
x	weight fraction of vapor
x'	ratio of window pressure drop to pressure drop across one baffle pass
Y	characteristic length for acoustic vibration: m, ft
Y_v	vertical tube pitch: mm, in
Z	dQ_{sv}/dQ_t

Greek Symbols

β	constant in Conners equation, Eq. (89)
Γ	ratio of pressure drops, $\Delta P_{GO}/\Delta P_{LO}$
δ	
$_0$	damping log decrement
η_f	fin effectiveness
θ	various angles as defined locally

μ viscosity (subscript b = bulk, g = gas, l = liquid, s = shell fluid, w = wall): $N \cdot s/m^2$,† $lb_m/(h \cdot ft)$

ρ density (subscript b = bulk, g = gas or vapor, h = homogeneous, l = liquid, s = shell fluid, tp = two-phase): kg/m^3, lb_m/ft^3

σ surface tension: N/m,‡ lb_f/ft

ϕ_b tube bundle critical flux factor, Eq. (57)

ϕ_s viscosity ratio correction factor, defined in Eq. (31)

ψ tube bundle geometry factor

†These are the official units. If centipoise are used, this is indicated at the equation.

‡If dynes per centimeter are used, this is noted at the equation.

Compact Heat Exchangers

A. INTRODUCTION

A compact heat exchanger incorporates a heat transfer surface having a high area density, somewhat arbitrarily 700 m^2/m^3 (213 ft^2/ft^3) and higher. The area density β is a ratio of heat transfer surface area A to volume **V**. A compact heat exchanger is not necessarily of small bulk and mass. However, if it did not incorporate a surface of high area density, it would be much more bulky and massive.

The convective heat transfer coefficient h for gases is generally one or two orders of magnitude lower than that for water, oil, and other liquids. Hence, for equivalent hA products on both sides of the exchanger, the heat transfer surface on the gas side needs to have a much larger area than can be practically realized with circular tubes. Thus, the major applications of compact heat exchangers are the gas side of gas-to-gas, gas-to-liquid, and gas-to-condensing or evaporating fluid heat exchangers. Compact surfaces have also been used for two-phase applications or on the liquid side for some applications.

The unique characteristics of compact exchangers, as compared to the conventional shell-and-tube exchangers, are: (1) many surfaces available having different orders of magnitude of surface area density; (2) flexibility in distributing surface area on the hot and cold sides as warranted by design considerations; and (3) generally substantial cost, weight, or volume savings.

The important design and operating considerations for compact exchangers are: (1) usually at least one of the fluids is a gas; (2) fluids must be clean and relatively noncorrosive; (3) the fluid pumping power (i.e., pressure drop) design constraint is often as important as the heat transfer rate; (4) operating pressures and temperatures are somewhat limited compared to shell-and-tube exchangers due to construction features such as brazing, mechanical expansion, etc.; (5) with the use of highly compact surfaces, the resultant shape of the exchanger is one having a large frontal area and a short flow length; the header design of a compact heat exchanger is thus important for a uniform flow distribution; and (6) the market potential must be large enough to warrant the sizable manufacturing research and tooling costs.

We will now summarize the surface geometries, flow arrangements, and overall design methodology for compact heat exchangers.

1. Surface Geometries

The basic construction types employed in the design of compact heat exchangers are: extended-surface recuperators employing fins on one or more sides, regenerators employing small–hydraulic diameter surface geometries, and tubular exchangers employing small-diameter tubes. The two most common types of extended-surface exchangers are the plate-fin and tube-fin exchangers. The description of surface geometries for these exchangers is provided next.

a. Plate-Fin Exchangers

In a plate-fin exchanger, fins are sandwiched between parallel plates (referred to as plates or parting sheets) as shown in Fig. 1. Sometimes fins are incorporated in a flat tube with rounded corners (referred to as a formed tube) thus eliminating the need for the side bars. Fins are attached to the plates by brazing, soldering, gluing, welding, mechanical fit, or extrusion. The plate fins are categorized as (1) plain (i.e., uncut) and straight fins such as plain triangular and rectangular fins, (2) plain but wavy fins, and (3) interrupted fins such as offset strip, louver, perforated, and pin fins. Examples of these fins are shown in Fig. 2 and further description is provided in Ref. 1. Typical fin densities are 120 to 700 fins per meter (3 to 18 fins per inch). However, in some applications, 2100 fins per meter

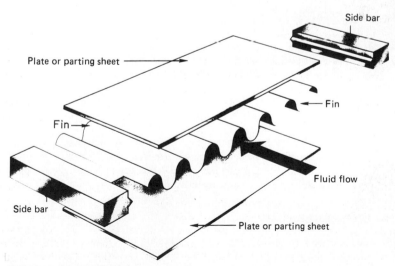

FIG. 1 A plate-fin assembly.

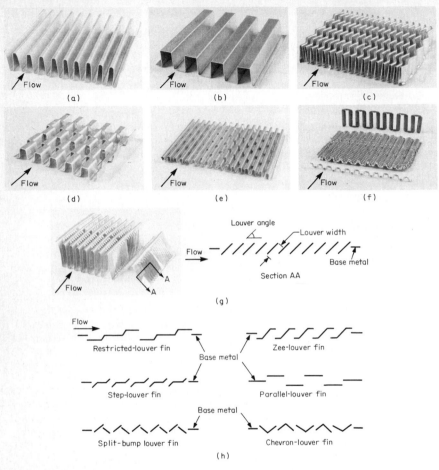

FIG. 2 Fin geometries for plate-fin heat exchangers. (*a*) Plain triangular fin. (*b*) Plain rectangular fin. (*c*) Wavy fin. (*d*) Offset strip fin. (*e*) Round perforated fin. (*f*) Pin fins. (*g*) Multilouver fin. (*h*) Various louver fins.

(53 fins per inch) fin density has been employed. Fin thicknesses of 0.05 to 0.25 mm (0.002 to 0.01 in) are common. Fin heights may range from 2 to 25 mm (0.08 to 1.0 in). A plate-fin exchanger with 600 fins per meter (15.2 fins per inch) provides about 1300 m^2 of heat transfer surface area per cubic meter volume (400 ft^2/ft^3) occupied by the fins.

b. Tube-Fin Exchangers

In a tube-fin exchanger, round and rectangular tubes are most commonly used, although elliptical tubes are also being used. When fins are used, they are employed either on the outside, or on the inside, or on both outside and inside of the tubes, depending upon the application. The fins are attached to the tubes by a tight mechanical fit, tension winding, soldering, brazing, welding, gluing, or extrusion. Fins on the outside tubes may be cate-

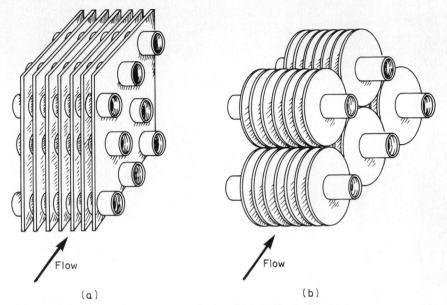

FIG. 3 (*a*) Continuous fins on an array of tubes. (*b*) Individually finned tubes.

gorized as (1) continuous (plain, wavy, or interrupted) external fins on an array of tubes (Figs. 3*a* and 4);† (2) normal fins on individual tubes, referred to as individually finned tubes or simply as finned tubes (Figs. 3*b* and 5); and (3) longitudinal fins on individual tubes. Fins inside the tubes are either integral or attached fins. Since an exchanger with longitudinally finned tubes or internally finned tubes is generally not a compact exchanger, we will not consider such finned tube exchangers here. The typical fin densities for continuous fins vary from 250 to 630 fins per meter (6 to 16 fins per inch), fin thickness from 0.09 to 0.25 mm (0.0035 to 0.010 in), and fin flow lengths from 25 to 250 mm (1 to 10 in). A tube-fin exchanger with 400 fins per meter (10 fins per inch) has about 720 m^2/m^3 (220 ft^2/ft^3) surface area density.

c. Compact Regenerators

In a compact regenerator, the matrix consists of cylindrical passages in parallel, usually of noncircular cross section as shown in Fig. 6*a* and *b*. Interrupted surfaces are not used in a rotary regenerator, in order to prevent transverse flow leakage. Either the metal matrix surface is stacked in baskets, as in a Ljungstrom air preheater, or it is brazed, as in a highly compact vehicular regenerator. The ceramic matrices are made by the wrapped coated paper process, extrusion, or the wrapped calendaring or embossing process. Some regenerators for process heating and air conditioning applications are made

---

†An exchanger having continuous fins on tubes is also referred to as a *plate-fin and tube* exchanger. In order to avoid confusion with plate-fin surfaces, we will refer to it as a tube-fin exchanger having continuous fins.

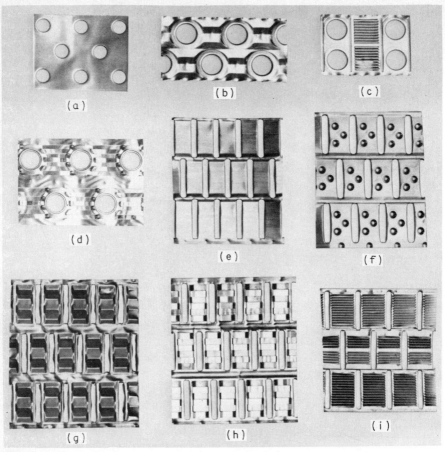

FIG. 4 Continuous fins on an array of round or flat tubes. In all photos, the fluid direction is vertical, from bottom to top or vice versa. (*a*) Plain fin for staggered round tubes. (*b*) Wavy fin for staggered round tubes. (*c*) Multilouver fin for in-line round tubes. (*d*) Parallel-louver fin for staggered round tubes. (*e*) Plain fin for staggered round tubes. (*f*) Dimple fin for staggered flat tubes. (*g*) Wavy fin for staggered flat tubes. (*h*) Parallel-louver fin for flat tubes. (*i*) Multilouver fin for staggered flat tubes.

from polyester film wound spirally over spacers held in spokes. Typically, the surface area density is about 5200 m^2/m^3 (1600 ft^2/ft^3) for a vehicular rotary regenerator, and is 2100 m^2/m^3 (640 ft^2/ft^3) for an air conditioning regenerator of the wound film type.

2. Flow Arrangements

For an extended-surface compact heat exchanger, crossflow is the most common flow arrangement. This is because it greatly simplifies the header design at the entrance and exit of each fluid. If the desired heat exchanger effectiveness is high (say greater than 75 to 80 percent), the size of a crossflow unit may become excessive, and an overall cross-counterflow multipass unit or a pure counterflow unit may be preferred. However, there

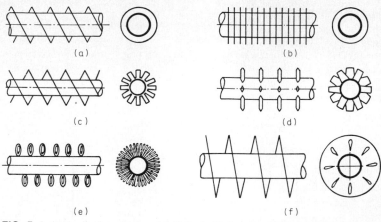

FIG. 5 Individually finned tubes. (*a*) Helical. (*b*) Annular (disk). (*c*) Segmented. (*d*) Studded. (*e*) Wire loop. (*f*) Slotted helical.

are manufacturing difficulties associated with a true counterflow arrangement in a compact exchanger as it is necessary to separate the fluids at each end. Thus, the header design is more complex for a counterflow heat exchanger. Multipassing retains the header and ducting advantages of the simple crossflow heat exchanger, while it is possible to approach the thermal performance of a counterflow unit. A parallel-flow arrangement having the lowest exchanger effectiveness for a given NTU is seldom used as a compact heat exchanger for single-phase fluids.

For regenerators, the most common flow arrangement is counterflow; parallel flow is rarely used. There are no counterparts of single-pass or multipass crossflow arrangements so common in compact recuperators.

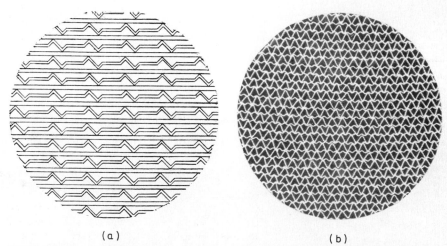

FIG. 6 Rotary regenerator matrix passage geometries: (*a*) notched plate, and (*b*) triangular. Gas flow is perpendicular to the plane of the paper.

3. Overall Design Methodology

A design methodology for an optimum compact heat exchanger is illustrated in Fig. 7 and is discussed in some detail in Ref. 2. It may be best characterized as a case study method based on a particular *surface selection*. Information on the key components of this

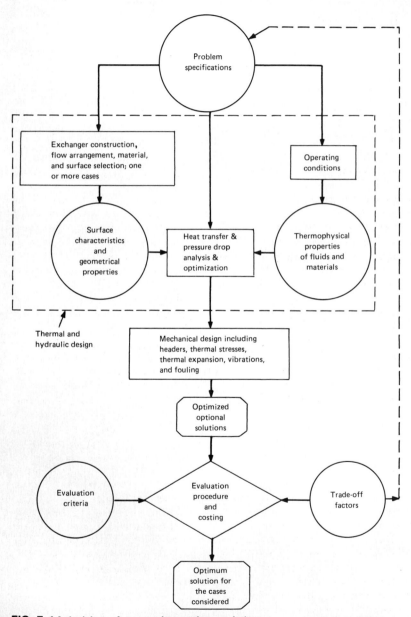

FIG. 7 Methodology of compact heat exchanger design.

methodology for thermal-hydraulic design and some mechanical design will be presented in Secs. B to H. This design methodology is also applicable to shell-and-tube heat exchangers, except that the mechanical design is done after pricing the unit, the optimization of the exchanger is done only when required, and performance calculations are seldom redone to investigate the effect of changing initial specifications.

B. HEAT TRANSFER ANALYSIS

Basic heat transfer analyses are presented separately for direct transfer–type exchangers (recuperators) and storage-type exchangers (regenerators) in the following sections.

1. Direct Transfer–Type Exchangers (Recuperators)

a. ϵ-NTU Results

The basic heat transfer analysis for compact recuperators is generally done by the ϵ-NTU method. The definitions and working equations for ϵ, NTU, and C^* are presented in Table 1 on p. 4-9. The ϵ-NTU expressions and numerical results have been obtained for most of the practically important flow arrangements, and they are presented in Tables 3, 4, and 6 on pp. 4-12, 4-30, and 4-32. Results for three practically important flow arrangements are also presented graphically in Figs. 4 to 6 of Part 1. The influence of longitudinal wall heat conduction may be important for high-effectiveness compact heat exchangers; it is presented on p. 4-53. The only consideration for compact extended-surface exchangers not covered in Part 1 is the fin efficiency and surface efficiency, which are discussed next.

b. Fin Efficiency and Extended-Surface Efficiency

Extended-surface compact heat exchangers employ fins to increase the surface area, and consequently to increase the total rate of heat transfer. Because of conduction along the fin, its temperature differs from the base (prime surface) temperature T_0, as shown in Fig. 8. This in turn reduces the temperature potential between the fin and the fluid for convection heat transfer as shown in Fig. 8a for the fin cooling situation and in Fig. 8b for the fin heating situation. The associated reduction in heat transfer is accounted for by the fin temperature effectiveness or fin efficiency η_f, which is defined as a ratio of the actual heat transfer rate q_0 through the base divided by the maximum possible heat transfer rate q_{max}. The latter is obtained by a "perfect fin," having the same fin geometry and operating conditions, but with infinite thermal conductivity which maximizes the temperature potential for convection.

In order to derive a one-dimensional expression for the fin efficiency, the following idealizations are made: (1) Heat transfer through the fin is steady-state and there are no thermal energy sources in the fin. (2) The thermal conductivity of the fin material is uniform and constant in the x direction; it is infinity in the y and z directions. Hence, heat flow is one-dimensional. (3) There is negligible thermal resistance between the fin and the fin base. (4) Radiation heat transfer to and from the fin is neglected. (5) The heat transfer coefficient is uniform and constant over the entire fin surface. (6) The temperatures of the primary surface and the fin base are the same. (7) The temperature of the surrounding fluid between fins is uniform at any one section. Idealizations 1 to 3 are common for the derivation of η_f for any fin geometry. Idealization 4 restricts the appli-

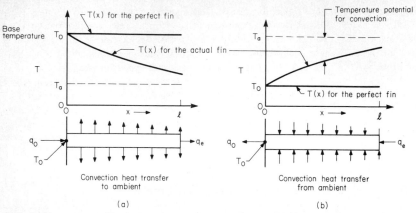

FIG. 8 Temperature distributions for actual and perfect fins: (a) fin is being cooled; (b) fin is being heated. The L_f or ξ dimension is in the direction perpendicular to the plane of the paper. q_e is the heat transfer rate at the fin tip.

cation of η_f to close fin spacings. Idealizations 5 to 7 have questionable validity and will be discussed in some detail.

The fin efficiencies for common plate-fin geometries and individually finned surfaces, based on the aforementioned idealizations, are summarized in Table 1. For other fin geometries, refer to Refs. 3, 4, and 5. The fin efficiencies for straight (first and third from the top in Table 1) and circular (seventh from the top in Table 1) fins of uniform thickness δ are presented in Fig. 9.

The fin efficiency for continuous plain fins on an array of circular tubes is obtained

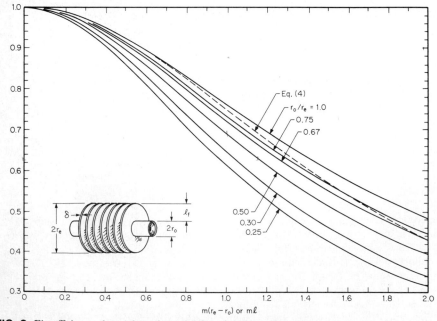

FIG. 9 Fin efficiency of straight and circular fins of uniform thickness.

by an equivalent annulus method [3] or by a sector method [6]. In the equivalent annulus method, the rectangular or hexagonal fin around the tube (see Fig. 10a and b) is hypothetically represented as a radial or circular fin (seventh from the top in Table 1) having the same fin surface area. The fin efficiency is then computed for this circular fin using the formula of Table 1. In the sector method, the rectangular or hexagonal fin (or its smallest symmetrical section) is divided into N sectors. Each sector is then considered as a circular fin with the radius $r_{e,i}$ equal to the length of the centerline of the sector. The fin efficiency of each sector is subsequently computed using the circular fin formula of Table 1. The fin efficiency η_f for the whole fin is then the surface area weighted average of $\eta_{f,i}$ of each sector.

$$\eta_f = \frac{\displaystyle\sum_{i=1}^{N} \eta_{f,i} A_{f,i}}{\displaystyle\sum_{i=1}^{N} A_{f,i}} \tag{1}$$

This approximation improves as N becomes large. Generally, only a few segments N will suffice to provide η_f within the desired accuracy, such as 0.1 percent. An implicit idealization made in the sector method is that the heat flow is only in the radial direction, and not in the path of the least thermal resistance. Hence, η_f calculated by the sector method will be somewhat more conservative (lower) than the true value. However, the equivalent annulus method yields an η_f that may be considerably higher than that by the sector method, particularly when the fin geometry around the circular tube becomes more and more rectangular (departing from a square fin); as a result, the computed heat transfer rate will be too high.

In an extended-surface exchanger, heat transfer takes place from both the fins ($\eta_f <$ 1) and the primary surface ($\eta_f = 1$). This total heat transfer is evaluated through a concept of total surface temperature effectiveness or surface efficiency η_o defined as

$$\eta_o = \frac{A_p}{A} + \eta_f \frac{A_f}{A} = 1 - \frac{A_f}{A}(1 - \eta_f) \tag{2}$$

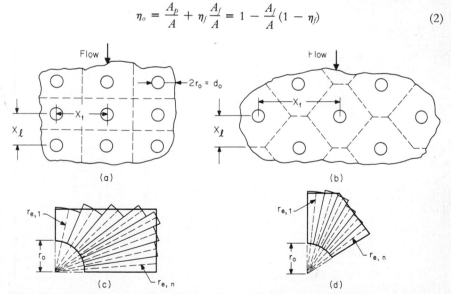

FIG. 10 Continuous fin over (a) an inline and (b) a staggered tube arrangement; the smallest representative segment of the fin for (c) an inline and (d) a staggered tube arrangement.

TABLE 1 Fin Efficiency for Plate-Fin and Tube-Fin Geometries of Uniform Fin Thickness

Geometry	Fin efficiency formula
	$$m_i = \left[\frac{2h}{k_f \delta_i}\left(1 + \frac{\delta_i}{\xi}\right)\right]^{1/2} \qquad E_i = \frac{\tanh(m_i l_i)}{m_i l_i} \qquad i = 1, 2, 3, 4$$
Plain, wavy, or offset strip fin of rectangular cross section	$$\eta_f = E_1$$ $$l_1 = \frac{b}{2} - \delta_1 \qquad \delta_1 = \delta$$
Triangular fin heated from one side	$$\eta_f = \frac{hA_1(T_0 - T_a)\dfrac{\sinh(m_1 l_1)}{m_1 l_1} + q_e}{\cosh(m_1 l_1)\left[hA_1(T_0 - T_a) + q_e \dfrac{T_0 - T_a}{T_i - T_a}\right]}$$
Plain, wavy, or louver fin of triangular cross section	$$\eta_i = E_1$$ $$l_1 = l, \qquad \delta_1 = \delta$$
Double sandwich fin	$$\eta_f = \frac{E_1 l_1 + E_2 l_2}{l_1 + l_2}\frac{1}{1 + m_1^2 E_1 E_2 l_1 l_2}$$ $$\delta_1 = \delta \qquad \delta_2 = \delta_3 = \delta + \delta_s$$ $$l_1 = b - \delta + \frac{\delta_s}{2} \qquad l_2 = l_3 = \frac{p_f}{2}$$

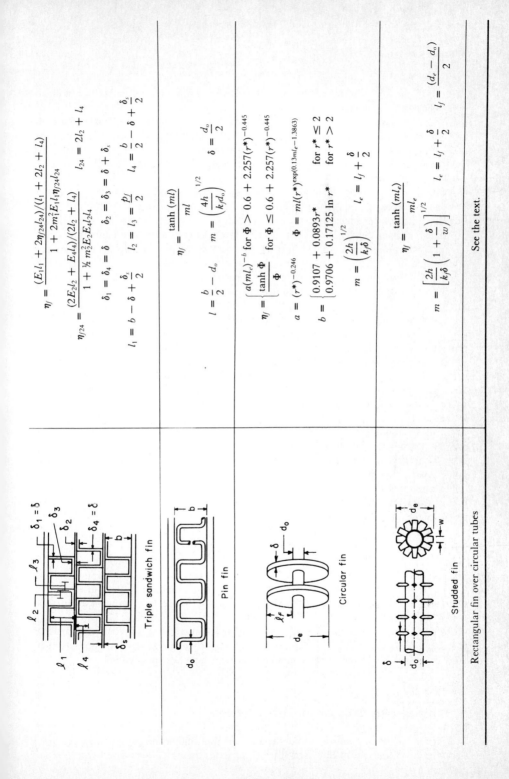

Triple sandwich fin

$$\eta_f = \frac{(E_1 l_1 + 2\eta_{f24}l_{24})/(l_1 + 2l_2 + l_4)}{1 + 2m_1^2 E_1 l_1 \eta_{f24}l_{24}}$$

$$\eta_{f24} = \frac{(2E_2 l_2 + E_4 l_4)/(2l_2 + l_4)}{1 + \tfrac{1}{2}m_2^2 E_2 E_4 l_2^2 l_4} \qquad l_{24} = 2l_2 + l_4$$

$$\delta_1 = \delta_4 = \delta \qquad \delta_2 = \delta_3 = \delta + \delta_s$$

$$l_1 = b - \delta + \frac{\delta_s}{2} \qquad l_2 = l_3 = \frac{l_f}{2} \qquad l_4 = \frac{b}{2} - \delta + \frac{\delta_s}{2}$$

Pin fin

$$\eta_f = \frac{\tanh(ml)}{ml}$$

$$l = \frac{b}{2} - d_o \qquad m = \left(\frac{4h}{k_f d_o}\right)^{1/2} \qquad \delta = \frac{d_o}{2}$$

Circular fin

$$\eta_f = \begin{cases} a(ml_e)^{-b} & \text{for } \Phi > 0.6 + 2.257(r^*)^{-0.445} \\[4pt] \dfrac{\tanh\Phi}{\Phi} & \text{for } \Phi \le 0.6 + 2.257(r^*)^{-0.445} \end{cases}$$

$$a = (r^*)^{-0.246} \qquad \Phi = ml(r^*)^{\exp(0.13 ml_e - 1.3863)}$$

$$b = \begin{cases} 0.9107 + 0.0893r^* & \text{for } r^* \le 2 \\ 0.9706 + 0.17125 \ln r^* & \text{for } r^* > 2 \end{cases}$$

$$m = \left(\frac{2h}{k_f \delta}\right)^{1/2} \qquad l_e = l_f + \frac{\delta}{2}$$

Studded fin

$$\eta_f = \frac{\tanh(ml_e)}{ml_e}$$

$$m = \left[\frac{2h}{k_f \delta}\left(1 + \frac{\delta}{w}\right)\right]^{1/2} \qquad l_e = l_f + \frac{\delta}{2} \qquad l_f = \frac{(d_e - d_o)}{2}$$

Rectangular fin over circular tubes

See the text.

with $A = A_f + A_p$, so that

$$q = \eta_o h A (T_0 - T_a) \tag{3}$$

In Eq. (3), the heat transfer coefficients over the finned and unfinned surfaces are idealized to be equal. Note that η_o is always greater than η_f for an extended-surface exchanger.

We will now discuss the violation of several idealizations for η_f listed earlier on p. 4-183. In idealization 5, the heat transfer coefficient is considered uniform over the entire fin surface. However, recent experiments [7, 8] show a significant variation in h over the fin surface. This nonuniformity in h results in a significantly lower η_f for η_f below 80 percent based on constant h.

Now let us turn to idealization 6. Because of the substantial amount of heat transfer through the fin compared to the prime surface, the heat transfer rate at the fin-base interface is much higher than that through a comparable surface area on the primary (unfinned) surface. Such a large heat transfer rate results in a fin base temperature that is depressed or elevated compared to the prime surface temperature T_0 depending upon whether the fin is being cooled or heated. This effect is similar to the "fin effect" (Fig. 8), and will yield actual heat transfer rates through the fin which are lower than when the fin base temperature is considered the same as the prime (unfinned) surface temperature [9, 10, 11]. This reduction in the heat transfer rate can be substantial depending upon the fin geometry. In contrast, the heat transfer rate through the prime (unfinned) surface is higher than the case when there is no fin base temperature depression [9]. For straight fins, the total heat transfer rate through finned and unfinned surfaces by a two-dimensional analysis is always higher than that by one-dimensional analysis [11]; hence, the use of one-dimensional fin efficiency (of Table 1) will predict conservative heat transfer rates. For circular fins on round tubes, the two-dimensional analysis predicts either higher or lower heat transfer rates depending upon the fin geometry and operating conditions [11].

In idealization 7, the temperature T_a of the fluid surrounding the fin is considered constant (as shown by dashed lines in Fig. 8) along the fin length l. However, for laminar flow, the transverse mixing along the fin length l may be negligible after a short distance along the fin flow length L_f. In such a case, the surrounding fluid temperature profile tends to parallel the fin temperature profile; the difference between the fin and ambient temperature, $T - T_a$, at any x will be constant, independent of x, and the fin efficiency of a straight fin of rectangular cross section can be derived as

$$\eta_f = \frac{1}{1 + (m^2 l^2/3)} \tag{4}$$

in contrast to $\eta_f = (\tanh ml)/ml$ (Table 1). The fin efficiency of Eq. (4) is shown in Fig. 9 as a dashed line. The reduction in η_f due to the above variations in T_a (vertical difference between the $r_o/r_e = 1$ line and the dashed line in Fig. 9) can be significant of $\eta_f < 80$ percent.

In summary, the violations of idealizations 5 to 7 can reduce η_f significantly from that calculated from Table 1, if $\eta_f < 80$ percent. In a good design, η_f is usually greater than 80 percent, and hence the aforementioned effects are generally negligible.

2. Storage-Type Exchangers (Regenerators)

In order to have continuous operation in a regenerator, either the matrix must be moved periodically in and out of the fixed streams of gases as in a *rotary regenerator*, or gas flows must be diverted to and from the fixed matrices as in a *fixed-matrix regenerator*. Thus, for a continuous operation, a fixed-matrix regenerator has at least two matrices, but usually three or four [12]. Operation of valves provides the desired periodicity of the hot and cold flows over each matrix in turn.

Heat transfer analysis for recuperators needs to be modified for regenerators in order to take into account the additional effects of the periodic thermal energy storage characteristics of matrix wall, and the establishment of wall temperature distribution dependent upon $(hA)_h$ and $(hA)_c$. This adds two additional dimensionless groups to the analysis to be discussed in the following subsection. All idealizations, except for numbers 8 and 11 listed on p. 4-8 for recuperators, are also invoked for the regenerator heat transfer analysis. In addition, it is idealized that regular periodic (steady-state periodic) conditions are established; wall thermal resistance in the wall thickness direction is zero, and it is infinity in the flow direction; no mixing of the fluids occurs during the switch from hot to cold flows or vice versa; and the fluid carryover and bypass rates are negligible relative to the flow rates of the hot and cold fluid. The influence of all these additional effects will be summarized later in this section. Note that negligible carryover means the dwell (residence) times of the fluids are negligible compared to the hot and cold gas flow periods.

a. ϵ-NTU$_0$ and Λ-Π Methods

Two methods for the regenerator heat transfer analysis are the ϵ-NTU$_0$ and Λ-Π methods [12]. The dimensionless groups associated with these methods are defined in Table 2, the relationship between the two sets of dimensionless groups is presented in Table 3A, and these dimensionless groups are defined in Table 3B for rotary and fixed-matrix regenerators. Notice that the regenerator effectiveness is dependent upon four dimensionless groups, in contrast to the two parameters NTU and C^* for recuperators (see Table 1 on p. 4-9). The additional parameters C_r^* and $(hA)^*$ for regenerators denote the dimensionless heat storage capacity rate of the matrix and the convection-conductance ratio of the cold and hot sides, respectively.

Extensive theory and results in terms of the Λ-Π method have been provided by Hausen [13] and Schmidt and Willmott [14]. The ϵ-NTU$_0$ method has been mainly used for rotary regenerators and the Λ-Π method for fixed-matrix regenerators. In a rotary regenerator, the outlet fluid temperatures vary across the flow area and are independent of time. In a fixed-matrix regenerator, the outlet fluid temperatures vary with time but are uniform across the flow area at any instant of time.† In spite of these subtle differences, if the elements of a regenerator (either rotary or fixed-matrix) are fixed relative to the observer by the selection of the appropriate coordinate systems, the heat transfer analysis is identical for both types of regenerators for arriving at the regenerator effectiveness.

In the Λ-Π method, several different designations are used to classify regenerators depending upon the values of Λ and Π. Such designations and their equivalent dimensionless groups of the ϵ-NTU$_0$ method are summarized in Table 4.

Because of the complexity of the problem, no closed-form generalized solutions are available in terms of the ϵ-NTU$_0$ or Λ-Π method for counterflow or parallel-flow regenerators. Extensive numerical results have been obtained by Lambertson [15, 16] for a counterflow regenerator and by Theoclitus and Eckrich [17] for a parallel-flow regenerator for a wide range of NTU$_0$, C^*, and C_r^*. Their results for $C^* = 1$ are presented in Figs. 11 and 12. Note that longitudinal heat conduction in the wall is neglected in these results since infinite thermal resistance is specified for the matrix in the flow direction.

---

†The difference between the outlet temperatures of the heated air (cold fluid) at the beginning and end of a given period is referred to as the *temperature swing* δT.

TABLE 2 General Functional Relationships and Basic Definitions of Dimensionless Groups for ϵ-NTU$_0$ and Λ-Π Methods for Counterflow Regenerators

ϵ-NTU$_0$ method	Λ-Π method
$q = \epsilon C_{\min}(T_{h,i} - T_{c,i})$	$Q = \epsilon_h C_h \mathcal{P}_h (T_{h,i} - T_{c,i}) = \epsilon_c C_c \mathcal{P}_c (T_{h,i} - T_{c,i})$
$\epsilon = \phi\{\text{NTU}_0, C^*, C_r^*, (hA)^*\}$	$\epsilon_r, \epsilon_h, \epsilon_c = \phi(\Lambda_m, \Pi_m, \gamma, R^*)$
$\epsilon = \dfrac{C_h(T_{h,i} - T_{h,o})}{C_{\min}(T_{h,i} - T_{c,i})} = \dfrac{C_c(T_{c,o} - T_{c,i})}{C_{\min}(T_{h,i} - T_{c,i})}$	$\epsilon_h = \dfrac{Q_h}{Q_{\max,h}} = \dfrac{C_h \mathcal{P}_h (T_{h,i} - \overline{T}_{h,o})}{C_h \mathcal{P}_h (T_{h,i} - T_{c,i})} = \dfrac{T_{h,i} - \overline{T}_{h,o}}{T_{h,i} - T_{c,i}}$
$\text{NTU}_0 = \dfrac{1}{C_{\min}}\left[\dfrac{1}{1/(hA)_h + 1/(hA)_c}\right]$	$\epsilon_c = \dfrac{Q_c}{Q_{\max,c}} = \dfrac{C_c \mathcal{P}_c (\overline{T}_{c,o} - T_{c,i})}{C_c \mathcal{P}_c (T_{h,i} - T_{c,i})} = \dfrac{\overline{T}_{c,o} - T_{c,i}}{T_{h,i} - T_{c,i}}$
$C^* = \dfrac{C_{\min}}{C_{\max}}$	$\epsilon_r = \dfrac{Q_h + Q_c}{Q_{\max,h} + Q_{\max,c}} = \dfrac{2Q}{Q_{\max,h} + Q_{\max,c}}$
$C_r^* = \dfrac{C_r}{C_{\min}}$	$\dfrac{1}{\epsilon_r} = \dfrac{1}{2}\left(\dfrac{1}{\epsilon_h} + \dfrac{1}{\epsilon_c}\right)$
$(hA)^* = \dfrac{hA \text{ on the } C_{\min} \text{ side}}{hA \text{ on the } C_{\max} \text{ side}}$	$\dfrac{1}{\Pi_m} = \dfrac{1}{2}\left(\dfrac{1}{\Pi_h} + \dfrac{1}{\Pi_c}\right) \qquad \dfrac{1}{\Lambda_m} = \dfrac{1}{2\Pi_m}\left(\dfrac{\Pi_h}{\Lambda_h} + \dfrac{\Pi_c}{\Lambda_c}\right)$
	$\gamma = \dfrac{\Pi_c/\Lambda_c}{\Pi_h/\Lambda_h} \qquad R^* = \dfrac{\Pi_h}{\Pi_c} \qquad \Lambda_h = \dfrac{(hA)_h}{C_h}$
	$\Lambda_c = \dfrac{(hA)_c}{C_c} \qquad \Pi_h \approx \left(\dfrac{hA}{C_r}\right)_h \qquad \Pi_c \approx \left(\dfrac{hA}{C_r}\right)_c$

For counterflow regenerators, the results can also be presented by the following empirical formula [16]:

$$\epsilon = \epsilon_{cf}\left[1 - \frac{1}{9(C_r^*)^{1.93}}\right] \tag{5}$$

TABLE 3A Relationship between Dimensionless Groups of ϵ-NTU$_0$ and Λ-Π Methods for $C_c = C_{\min}$†

$$\epsilon = \epsilon_c = \frac{\epsilon_h}{\gamma} = (\gamma + 1)\frac{\epsilon_r}{2\gamma} \text{ for } C_c = C_{\min}$$

ϵ-NTU$_0$	Λ-Π
$\text{NTU}_0 = \dfrac{\Lambda_m(1 + \gamma)}{4\gamma} = \dfrac{\Lambda_c/\Pi_c}{1/\Pi_h + 1/\Pi_c}$	$\Lambda_h = C^*\left[1 + \dfrac{1}{(hA)^*}\right]\text{NTU}_0$
$C^* = \gamma = \dfrac{\Pi_c/\Lambda_c}{\Pi_h/\Lambda_h}$	$\Lambda_c = \left[1 + (hA)^*\right]\text{NTU}_0$
$C_r^* = \dfrac{\Lambda_m(1 + \gamma)}{2\gamma\Pi_m} = \dfrac{\Lambda_c}{\Pi_c}$	$\Pi_h = \dfrac{1}{C_r^*}\left[1 + \dfrac{1}{(hA)^*}\right]\text{NTU}_0$
$(hA)^* = \dfrac{1}{R^*} = \dfrac{\Pi_c}{\Pi_h}$	$\Pi_c = \dfrac{1}{C_r^*}\left[1 + (hA)^*\right]\text{NTU}_0$

†If $C_h = C_{\min}$, the subscripts c and h in this table should be changed to h and c, respectively.

TABLE 3B Working Definitions of Dimensionless Groups for Regenerators in Terms of Dimensional Variables of Rotary and Fixed-Matrix Regenerators for $C_c = C_{min}$†

Dimensionless group	Rotary regenerator	Fixed-matrix regenerator
NTU_0	$\dfrac{h_c A_c}{C_c}\dfrac{h_h A_h}{h_h A_h + h_c A_c}$	$\dfrac{h_c A}{C_c}\dfrac{h_h P_h}{h_h P_h + h_c P_c}$
C^*	$\dfrac{C_c}{C_h}$	$\dfrac{C_c P_c}{C_h P_h}$
C_r^*	$\dfrac{M_w c_w \omega}{C_c}$	$\dfrac{M_w c_w}{C_c P_c}$
$(hA)^*$	$\dfrac{h_c A_c}{h_h A_h}$	$\dfrac{h_c P_c}{h_h P_h}$
$\dfrac{1}{\Lambda_m}$	$\dfrac{C_c + C_h}{4}\left(\dfrac{1}{h_h A_h} + \dfrac{1}{h_c A_c}\right)$	$\dfrac{C_c P_c + C_h P_h}{4A}\left(\dfrac{1}{h_h P_h} + \dfrac{1}{h_c P_c}\right)$
$\dfrac{1}{\Pi_m}$	$\dfrac{M_w c_w \omega}{2}\left(\dfrac{1}{h_h A_h} + \dfrac{1}{h_c A_c}\right)$	$\dfrac{M_w c_w}{2A}\left(\dfrac{1}{h_h P_h} + \dfrac{1}{h_c P_c}\right)$
γ	$\dfrac{C_c}{C_h}$	$\dfrac{C_c P_c}{C_h P_h}$
R^*	$\dfrac{h_h A_h}{h_c A_c}$	$\dfrac{h_h P_h}{h_c P_c}$

†If $C_h = C_{min}$, the subscripts c and h in this table should be changed to h and c, respectively. The definitions are given for one rotor (disk) of a rotary regenerator or for one matrix of a fixed-matrix regenerator.

where

$$\epsilon_{cf} = \frac{1 - \exp[-NTU_0(1 - C^*)]}{1 - C^* \exp[-NTU_0(1 - C^*)]} \tag{6}$$

is the closed-form solution for the direct transfer type counterflow exchanger. The applicable ranges of dimensionless groups for ± 1 percent accuracy in ϵ are: (1) $3 \leq NTU_0 \leq 9$, $0.90 \leq C^* \leq 1$, $1.25 \leq C_r^* \leq 5$; (2) $2 < NTU_0 < 14$, $C^* = 1$, $C_r^* \geq 1.5$; (3) $NTU_0 \leq 20$, $C^* = 1$, $C_r^* = 2$; and (4) the complete range of NTU_0, $C^* = 1$, and $C_r^* \geq 5$. In all cases, the range of $(hA)^*$ is within 0.25 to 4.0. Note that the range of C^* is limited to between 0.9 and 1.0.

TABLE 4 Designation of Various Types of Regenerators Depending upon the Values of Dimensionless Groups

Terminology	Λ-Π method	ϵ-NTU_0 method
Balanced regenerators	$\Lambda_h/\Pi_h = \Lambda_c/\Pi_c$ or $\gamma = 1$	$C^* = 1$
Unbalanced regenerators	$\Lambda_h/\Pi_h \neq \Lambda_c/\Pi_c$	$C^* \neq 1$
Symmetric regenerators	$\Pi_h = \Pi_c$ or $R^* = 1$	$(hA)^* = 1$
Unsymmetric regenerators	$\Pi_h \neq \Pi_c$	$(hA)^* \neq 1$
Symmetric and balanced regenerators	$\Lambda_h = \Lambda_c$, $\Pi_h = \Pi_c$	$(hA)^* = 1$, $C^* = 1$
Unsymmetric but balanced regenerators	$\Lambda_h/\Pi_h = \Lambda_c/\Pi_c$	$(hA)^* \neq 1$, $C^* = 1$
Long regenerators	$\Lambda/\Pi > 5$	$C_r^* > 5$

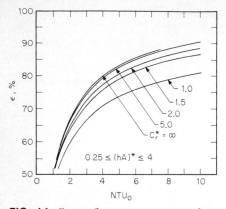

FIG. 11 *Counterflow regenerator:* ϵ *as a function of* NTU_0 *and* C_r^* *for* $C^* = 1$ [16].

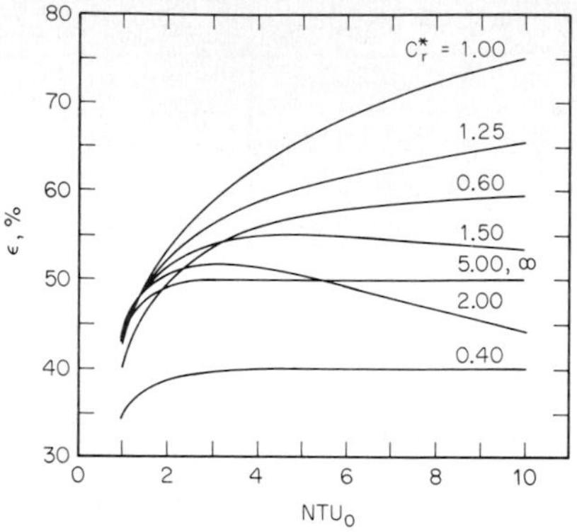

FIG. 12 *Parallel-flow regenerator:* ϵ *as a function of* NTU_0 *and* C_r^* *for* $C^* = 1$ *and* $(hA)^* = 1$ [17].

Razelos [12, 18] proposed the following approximate procedure to calculate the counterflow regenerator effectiveness ϵ for $C_r^* \geq 1$, $0.25 \leq (hA)^* \leq 4$, and the complete range of C^* and NTU_0, thus having no restrictions on C^*. For the known values of NTU_0, C^*, and C_r^*, calculate "appropriate values" of NTU_0 and C_r^* for an "equivalent" balanced regenerator ($C^* = 1$), designated with a subscript m, as follows:

$$NTU_{0,m} = \frac{2\,NTU_0\,C^*}{1 + C^*} \tag{7}$$

$$C_{r,m}^* = \frac{2C_r^*C^*}{1 + C^*} \tag{8}$$

With these values of $NTU_{0,m}$ and $C_{r,m}^*$, obtain ϵ_r from

$$\epsilon_r = \frac{NTU_{0,m}}{1 + NTU_{0,m}}\left[1 - \frac{1}{9(C_{r,m}^*)^{1.93}}\right] \tag{9}$$

For $C_{r,m}^* < 1$, the regenerator effectiveness ϵ_r can be obtained from Hausen's effectiveness chart, Fig. 6 of Ref. 12 or Fig. 5.4 of Ref. 14, using $\Lambda = 2\,NTU_{0,m}$ and $\Pi = 2\,NTU_{0,m}/C_{r,m}^*$.

Finally, calculate the desired regenerative effectiveness from

$$\epsilon = \frac{1 - \exp\{\epsilon_r(C^{*2} - 1)/[2C^*(1 - \epsilon_r)]\}}{1 - C^*\exp\{\epsilon_r(C^{*2} - 1)/[2C^*(1 - \epsilon_r)]\}} \tag{10}$$

A comparison of ϵ from Eqs. (7) to (10) with the tabular values of ϵ of Kays and London [16] indicates that the Razelos approximation yields more accurate values of ϵ compared to that from Eq. (5) for $C^* < 1$.

b. Longitudinal Heat Conduction in Wall

Longitudinal heat conduction in the wall was neglected in deriving the results of the preceding section. However, it may not be negligible, particularly for a high-effectiveness regenerator having a short flow length L producing a large temperature gradient in the axial direction. It reduces the regenerator effectiveness and the overall heat transfer rate.

For example, for regenerators designed for $\epsilon > 85$ percent, a 1 percent reduction in ϵ would reduce gas turbine power plant efficiency by about 1 to 5 percent depending upon the load conditions, which could translate into a significant economic penalty. The reduction in ϵ due to longitudinal conduction in the wall can be 1 percent or higher and hence must be properly accounted for in the design. Based on extensive numerical results by Bahnke and Howard [19], this effect can be taken into account by an additional parameter λ, referred to as the longitudinal conduction parameter:

$$\lambda = \frac{k_w A_{k,t}}{L C_{\min}} \qquad (11)$$

where k_w is the thermal conductivity of the matrix wall, and $A_{k,t}$ is the total solid area for longitudinal conduction:

$$A_{k,t} = A_{k,h} + A_{k,c} = A_{\text{fr}} - A_o = A_{\text{fr}}(1 - \sigma) \qquad (12)$$

The ineffectiveness $(1 - \epsilon)$ as a function of NTU_0 and λ is shown in Figs. 36 and 37 on pp. 4-54 and 4-55 for $C^* = 1$ and 0.95 and $C_r^* > 5$. Note that the counterflow regenerator ϵ for $C_r^* > 5$ is almost identical to the counterflow recuperator ϵ. Kays and London [16] also present similar results in their Figs. 2-36 and 2-37 for values of λ that are somewhat different from Figs. 36 and 37 of Part 1. Bahnke and Howard's results have been correlated by Shah [12] as

$$\epsilon = \epsilon_i = \epsilon_{cf} \left(1 - \frac{1}{9 C_r^{*1.93}} \right) \times$$

$$\left\{ 1 - \frac{1}{2 - C^*} \left[\frac{1}{1 + \text{NTU}_0 (1 + \lambda\Phi)/(1 + \lambda\,\text{NTU}_0)} - \frac{1}{1 + \text{NTU}_0} \right] \right\} \qquad (13)$$

where

$$\Phi = \left(\frac{\lambda\,\text{NTU}_0}{1 + \lambda\,\text{NTU}_0} \right)^{1/2} \tanh \left\{ \frac{\text{NTU}_0}{[\lambda\,\text{NTU}_0/(1 + \lambda\,\text{NTU}_0)]^{1/2}} \right\} \qquad (14)$$

$$\approx \left(\frac{\lambda\,\text{NTU}_0}{1 + \lambda\,\text{NTU}_0} \right)^{1/2} \text{ for NTU}_0 \geq 3$$

The regenerator effectiveness ϵ of Eq. (13) agrees within ± 0.5 percent with the results of Bahnke and Howard for the following range of parameters: $3 \leq \text{NTU}_0 \leq 12, 0.9 \leq C^* \leq 1, 2 \leq C_r^* \leq \infty, 0.5 \leq (hA)^* \leq 2$, and $0 \leq \lambda \leq 0.04$. It agrees within ± 1 percent for the following range of parameters: $1 \leq \text{NTU}_0 \leq 20, 0.9 \leq C^* \leq 1, 2 \leq C_r^* \leq \infty, 0.25 \leq (hA)^* \leq 4$, and $0 \leq \lambda \leq 0.08$.

c. Transverse Heat Conduction in Wall

The thermal resistance for transverse heat conduction† in the wall (in the wall thickness direction) was treated as zero for the results of Sec. B.2.a. This is a good idealization for metal matrices having high thermal conductivity and thin walls. However, for ceramic matrices (with thick walls) of fixed-matrix regenerators, the wall thermal resistance may not be negligible. This transverse heat conduction reduces the regenerator effectiveness; the reduction is more for a planar surface as opposed to a cylindrical or spherical geometry of the matrix surface (checkerwork) [20].

---

†This is also sometimes referred to as intraconduction.

The wall thermal resistance needs to be determined separately during the hot- and cold-gas flow periods in a regenerator since there is no continuous flow of heat from the hot gas to the cold gas in a given matrix (or a fraction) of a regenerator. This wall thermal resistance on a unit area basis has been determined by Hausen [13] as

$$\hat{R}_w = R_w A = \frac{a}{6k_w} \Phi^* \tag{15}$$

so that the effective heat transfer coefficients during the hot- and cold-gas flow periods (designated by a superscript bar) are

$$\frac{1}{\bar{h}_h} = \frac{1}{h_h} + \frac{a}{6k_w} \Phi^* \qquad \frac{1}{\bar{h}_c} = \frac{1}{h_c} + \frac{a}{6k_w} \Phi^* \tag{16}$$

where the value of Φ^* for a plain wall is given by

$$\Phi^* = \begin{cases} 1 - \dfrac{1}{15}\left(\dfrac{\text{Bi}_h}{\Pi_h} + \dfrac{\text{Bi}_c}{\Pi_c}\right) & \text{for } \dfrac{\text{Bi}_h}{\Pi_h} + \dfrac{\text{Bi}_c}{\Pi_c} \le 5 \quad (17a) \\[3mm] 2.142\left[0.3 + 2\left(\dfrac{\text{Bi}_h}{\Pi_h} + \dfrac{\text{Bi}_c}{\Pi_c}\right)\right]^{-1/2} & \text{for } \dfrac{\text{Bi}_h}{\Pi_h} + \dfrac{\text{Bi}_c}{\Pi_c} > 5 \quad (17b) \end{cases}$$

where $\text{Bi}_h = h_h(a/2)/k_w$, $\Pi_h = h_h A_h / C_{r,h}$, and Bi_c and Π_c are defined in a similar manner. The ratio Bi/Π is the reciprocal of the Fourier number Fo. The range of Φ^* for Eq. (17a) is $\frac{2}{3}$ to 1, and for Eq. (17b) from 0 to $\frac{2}{3}$. When $\text{Bi} \to 0$, the transverse thermal resistance approaches zero as expected.† In order to maintain reasonable wall thermal resistance, Bi_h and Bi_c should be kept below 2, which is also the limit of applicability of Eqs. (17a) and (17b). For $\text{Bi} > 2$, use the numerical results of Heggs et al. [20]. The accuracy of Eq. (17) decreases with increasing Bi/Π and decreasing C_r^*.

Formulas for Φ^* for cylindrical and spherical wall geometries are also summarized by Hausen [13]. Φ^* for hollow cylinders may be obtained from Razelos and Lazaridis [21], who showed that Φ^* not only is a function of the harmonic mean Fourier number $\text{Fo}_m = 2/(\text{Bi}_h/\Pi_h + \text{Bi}_c/\Pi_c)$, but also depends upon the period ratio $\mathcal{P}_h/\mathcal{P}_c$. Solid cylinders and plain walls are the limiting cases of hollow cylinders.

In order to account for the wall thermal resistance in a regenerator, the heat transfer coefficients h_h and h_c used in the definitions of NTU_0, the Λ's, the Π's, $(hA)^*$, and R^* should be replaced by $\bar{h}_h$ and $\bar{h}_c$ computed from Eq. (16) before doing any rating or sizing calculations. Note that for a "thin" wall, $\Phi^* \to 1$ and $\hat{R}_w = a/6k_w$ during each period. For a rotary regenerator with a 1:1 flow area split or for a fixed-matrix regenerator with $\mathcal{P}_h = \mathcal{P}_c$, the total unit thermal resistance during a cycle will be $a/3k_w$, which is one-third the value for a recuperator.

The prediction of the temperature swing δT in a fixed-matrix regenerator will not be accurate by the foregoing approximate method. The numerical analysis of the type made by Heggs et al. [20] is essential for accurate δT determination. It may be noted that the δT values of Table 1 of Ref. 20 had a typing error and all should be multiplied by a factor of 10; also all the charts in Fig. 1 of Ref. 20 are poorly drawn, as a result of which the δT values shown are approximate [20a].

---

†Note that when $\text{Bi} \to 0$, $\Phi^* \to 1$ but $R_w \to 0$. This is because $\text{Bi} \to 0$ only when $k_w/a \to \infty$ for finite h, and as a result, $R_w \propto a/k_w$ approaches zero.

d. Fluid Bypass

In both rotary and fixed-matrix regenerators, flow mixing from the cold to hot fluid stream and vice versa occurs due to fluid carryover and bypass. The influence of fluid carryover will be presented in the next subsection. A fluid stream can wholly or partly bypass the matrix in a number of ways and mix with the other stream, thus affecting the performance of a regenerator. Cross bypass flows are due to the pressure-driven leakage through seals and to the fluid trapped in header volumes between seals and matrix, and may occur in both rotary and fixed-matrix regenerators. The "seals" are at the matrix faces for a rotary regenerator and at the fluid stream switching valves for a fixed-matrix regenerator. Side bypass flows occur in a rotary regenerator, and are caused by pressure leakage through the circumferential seals between the disk and the housing.

The bypass flow analysis has been made by Banks and Ellul [22] and Klopfer [23, 12] among others. Banks and Ellul considered the cross bypass and side bypass flows separately in the analysis; the cross bypass flows occur due to the pressure leakage in either direction (cold to hot side, or hot to cold side) as in an air conditioning regenerator at approximately ambient pressures. Hence, they treated the same fluid density and specific heat in each fluid stream, and $C^* = 1$. Their results indicate that (1) the cross bypass flows into the inlet and into the outlet of a regenerator side cause an increase in the temperature effectiveness on that side, (2) the cross bypass flow leaving the outlet of a side decreases its temperature effectiveness, (3) the side bypass flow always reduces the temperature effectiveness, and (4) the cross bypass flows due to header volumes, which occur in both directions at each matrix face, increase the temperature effectiveness for equal header volumes at each face.

Klopfer lumped the cross bypass and side bypass flows into a single pressure leakage term, and considered the pressure leakage from the cold side (high pressure) to the hot side (low pressure), as in a gas turbine regenerator, occurring just outside (inlet and outlet side) of the regenerator as shown in the idealized model of Fig. 13b. Klopfer considered a different fluid density for each fluid stream in the regenerator, the same specific heat in each stream, and any C^*. Since Klopfer's analysis was incomplete, it has been extended here as suggested by Banks [23a]. The following procedure may be used to obtain an effective reduction in the regenerator effectiveness due to the pressure leakage. Refer to Fig. 13a and b for the idealized models having zero and finite pressure leakages. The cold fluid side is considered to be the $C_{\min}$ side in these models.

1. For the original zero–pressure leakage NTU_0, C_r^*, and C^*, compute the ideal regenerator effectiveness ϵ_i from Fig. 11 or Eqs. (7) to (10).

2. For the specified pressure leakage factor χ_p, determine the new values of C^*, NTU_0, and C_r^*, designated with a subscript p, as

$$C_p^* = \frac{1 - \chi_p/2}{1 + C^*\chi_p/2}\, C^*$$

$$NTU_{0,p} = \frac{1}{1 - \chi_p/2}\, NTU_0 \qquad C_{r,p}^* = \frac{1}{1 - \chi_p/2}\, C_r^* \tag{18}$$

Here $\chi_p = \Delta W_p/W_c$ and ΔW_p is the total mass flow leakage rate from the high-pressure cold side to the low-pressure hot side.

3. Now compute the ideal regenerator effectiveness in the presence of the pressure leakage, designated as $\epsilon_{i,p}$, for the values of C_p^*, $NTU_{0,p}$ and $C_{r,p}^*$, using Fig. 11 or Eqs.

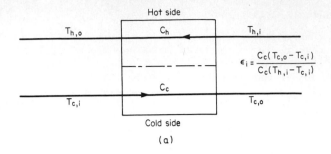

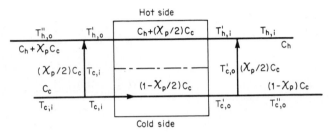

FIG. 13 Regenerator idealized flow diagram and associated temperatures: (a) zero pressure leakage or zero carryover, (b) finite pressure leakage and zero carryover.

(7) to (10). The increase in the ideal regenerator effectiveness at C_p^* ($C_p^* < C^*$), $\Delta\epsilon_1$, is then

$$\Delta\epsilon_1 = \epsilon_{i,p} - \epsilon_i$$

4. The reduction in the actual regenerator effectiveness ϵ_p over $\epsilon_{i,p}$ due to the loss of cold fluid and the reduced hot fluid inlet temperature, $T'_{h,i}$ in Fig. 13b, is given by [23, 12]†

$$\Delta\epsilon_2 = \epsilon_{i,p} - \epsilon_p \qquad (19)$$
$$= \frac{(\chi_p/2)[(1 - \chi_p)T^* - 1](1 - \epsilon_{i,p})\epsilon_{i,p} + (\chi_p/2 + 1/C^*)\chi_p(1 - \epsilon_{i,p})}{(\chi_p/2)[(1 - \chi_p)T^* - 1](1 - \epsilon_{i,p}) + [(1 - \chi_p)T^* - 1]/C^*}$$

Here $T^* = T_{h,i}/T_{c,i}$ and $T_{h,i}$ and $T_{c,i}$ are the hot and cold gas inlet temperatures on the *absolute* scale.

5. The net *increase* in the regenerator effectiveness due to the finite pressure leakage is then

$$\Delta\epsilon = \epsilon_p - \epsilon_i = \Delta\epsilon_1 - \Delta\epsilon_2 \qquad (19a)$$

†This is the correct equation. The same equation presented in Ref. 12 had a typographical error.

6. Even though, there may be a net increase in the regenerator effectiveness due to the pressure leakage, the actual heat transfer rate in the regenerator decreases because the reduction in C_{min} is greater than the increase in ϵ. The actual heat transfer rate in the regenerator is expressed as (see the definition of ϵ_p in Fig. 13)

$$q = \epsilon_p C_c [(1 - \chi_p) T_{h,i} - T_{c,i}] = \epsilon_{eff} C_c (T_{h,i} - T_{c,i}) \qquad (19b)$$

where the expression on the right-hand side of the second equality sign defines the "effective" regenerator effectiveness based on the original C_{min}. The effective decrease in the regenerator effectiveness is then expressed as

$$\Delta \epsilon_{eff} = \epsilon_i - \epsilon_{eff} = \epsilon_i - \left(1 - \frac{T^*}{T^* - 1}\chi_p\right)\epsilon_p \qquad (19c)$$

Based on this procedure, Fig. 14 shows $\Delta\epsilon_{eff}$ as a function of χ_p and ϵ_i for typical values of $C_r^* = 4$, $C^* = 1$, and $T^* = 2$. $\Delta\epsilon_{eff}$ seems to be similar in magnitude to χ_p and is little affected by ϵ_i.

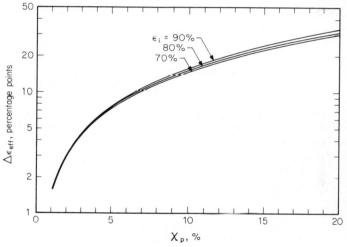

FIG. 14 Reduction in the "effective" counterflow regenerator effectiveness $\Delta\epsilon_{eff}$ as a function of the pressure leakage rate χ_p and ϵ_i for $T^* = 2$ and $C^* = 1$. $\Delta\epsilon_{eff}$ (percentage points) $= \epsilon(\%) - \epsilon_{eff}(\%)$.

If the influence of longitudinal heat conduction needs to be included, the results of Fig. 11 or Eqs. (7) to (10) in the above procedure should be replaced by the appropriate solution such as that of Bahnke and Howard [19].

e. Fluid Carryover

The fluid remaining in the matrix at the end of a period will leave with the other fluid stream during the following period and is referred to as carryover. The fluid carryover occurs continually in a rotary regenerator and just after the valve switching in a fixed-matrix regenerator. Carryover differs from bypass flows in that it generally changes in state in passing from one fluid stream to the other [23a].

The carryover analysis has been performed by Banks [24], Maclaine-cross [24a], and Klopfer [23, 12] among others. The influence is presented in terms of a carryover ratio defined as a fraction of each fluid stream carried over to other stream, and can be expressed as

$$\kappa_h = \frac{\Delta W_h}{W_h} = \frac{L}{V_h \mathscr{P}_h} \qquad \kappa_c = \frac{\Delta W_c}{W_c} = \frac{L}{V_c \mathscr{P}_c} \qquad (19d)$$

In addition, if the fluid densities of the hot and cold fluids are different, the temperature ratio $T^* = T_{h,i}/T_{c,i}$ also appears as an additional independent parameter.

Maclaine-cross [24a] has provided an approximate method to account for the influence of fluid carryover by correcting the matrix wall specific heat c_w and the heat transfer coefficients (for both hot and cold sides) as follows for the case of hot and cold fluid densities about the same:

$$c_{w,\text{corr}} = \frac{1 + \overline{C}_r^*}{\overline{C}_r^*} c_w \qquad h_{\text{corr}} = \left(\frac{1 + \overline{C}_r^*}{\overline{C}_r^*}\right)^2 h \qquad (20)$$

where $\overline{C}_r^* = \overline{C}_r/\overline{C}_{\min} = M_w c_w / M c_p$ and M is the mass of the fluid contained in the void volume of the matrix with M and c_p for the $C_{\min}$ fluid stream. After comparing with the exact numerical solution for a symmetric regenerator with $\overline{C}_r^* > 10$, Maclaine-cross indicated that the above approximation will introduce less than 0.1 percent error in ϵ for $NTU_0 > 5$, $\kappa_{\max} < 0.03$, and $C^* > 0.5$; it will introduce less than 1 percent error in ϵ for $NTU_0 > 2.5$, $\kappa_{\max} < 0.1$, and $C^* > 0.5$. Note that $\kappa = C_r^*/\overline{C}_r^*$.

Banks [23a] provided a design chart, based on a numerical solution, to account for the influence of fluid carryover over a wider range of operating conditions than Maclaine-cross' method, but it is limited to the $C^* = 1$ case. His design chart clearly shows the regenerator effectiveness increasing with increasing fluid carryover, and is accurate within 0.1 percent in ϵ for $\kappa < 0.1$ and $\overline{C}_r^* > 50$ for the case when hot and cold fluid densities and specific heats are considered the same.

Klopfer [23] considered the hot and cold fluid densities different and provided an approximate procedure to evaluate the influence of the fluid carryover on the regenerator effectiveness. He represents the fluid carryover by cross bypass flows into matrix outlets as shown in Fig. 15; in contrast, a zero–fluid carryover model is shown in Fig. 13a. The cold fluid side is considered as the $C_{\min}$ side. In the absence of results from numerical solutions for different hot and cold fluid densities, the accuracy of Klopfer's procedure cannot be checked. However, Banks [23a] shows that the procedure may be applicable

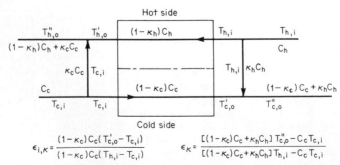

FIG. 15 Regenerator idealized flow diagram and associated temperatures with fluid carryover.

for practical operating conditions, but that Klopfer's analysis is incomplete. The complete analysis is given here, and the following procedure may be used to obtain an effective reduction in the regenerator effectiveness:

1. For the original zero–fluid carryover NTU_0, C_r^*, and C^*, compute the ideal regenerator effectiveness ϵ_i from Fig. 11 or Eqs. (7) to (10).

2. For finite fluid carryover, compute new values of C^*, NTU_0, and C_r^*, designated with a subscript κ, as

$$ C_\kappa^* = \frac{1 - \kappa_c}{1 - \kappa_h} C^* \qquad NTU_{0,\kappa} = \frac{1}{1 - \kappa_c} NTU_0 \qquad C_{r,\kappa}^* = \frac{1}{1 - \kappa_c} C_r^* \quad (20a) $$

Note that κ_c and κ_h are related by the equation $\kappa_h = \kappa_c(C^*/P^*T^*)$ for the case of hot and cold fluid specific heats about the same. Here $P^* = P_{c,i}/P_{h,i}$. For a gas turbine regenerator, $P^* > 1$, $T^* > 1$, so that $\kappa_c > \kappa_h C^*$.

3. Now calculate the ideal regenerator effectiveness with finite fluid carryover, designated as $\epsilon_{i,\kappa}$, for the values of C_κ^*, $NTU_{0,\kappa}$, and $C_{r,\kappa}^*$ using Fig. 11 or Eqs. (7) to (10). The increase in the ideal regenerator effectiveness $\Delta\epsilon_1$ is then

$$ \Delta\epsilon_1 = \epsilon_{i,\kappa} - \epsilon_i \qquad (20b) $$

4. The reduction in the actual regenerator effectiveness ϵ_κ over $\epsilon_{i,\kappa}$ due to the net loss of cold fluid stream is given by [23, 12]

$$ \Delta\epsilon_2 = \epsilon_{i,\kappa} - \epsilon_\kappa = \frac{(1 - \epsilon_{i,\kappa})(P^* - 1)}{P^*(T^* - 1)/\kappa_c - P^*T^* + 1} \qquad (21) $$

5. The net *increase* in the regenerator effectiveness due to the finite fluid carryover is then

$$ \Delta\epsilon = \epsilon_\kappa - \epsilon_i = \Delta\epsilon_1 - \Delta\epsilon_2 $$

6. Since the regenerator heat transfer rate will reduce with the decrease of $C_{min} (= C_c)$ to $(1 - \kappa_c)C_c + \kappa_h C_h$ as shown in Fig. 15, the "effective" regenerator effectiveness based on C_c will decrease as mentioned earlier, and it is given by

$$ \Delta\epsilon_{eff} = \epsilon_i - \epsilon_{eff} = \epsilon_i - \left[1 - \frac{T^*}{T^* - 1}\left(1 - \frac{1}{P^*T^*}\right)\kappa_c \right]\epsilon_\kappa \qquad (21a) $$

Based on this procedure, Fig. 16 shows $\Delta\epsilon_{eff}$ values as a function of κ_c (and C_r^*) and ϵ_i for typical values of $C^* = 1$, $\overline{C}_r^* = 200$, $P^* = 4$, and $T^* = 2$. It should be noted that the influence of fluid carryover may not be negligible for high-effectiveness gas turbine rotary regenerators.

The foregoing analysis is not applicable to the Stirling engine regenerator, for which the period of switching the gases is greater than the residence time for gas flow, since κ_c becomes greater than unity. In this case, the results of Heggs and Carpenter [24b] should be referred to for fluid density the same in each fluid stream.

In the foregoing analyses, the influences of fluid bypass flows and carryover are presented separately to show clearly what are the individual effects. However, they would occur simultaneously in a regenerator. In this case, it is suggested that a model combining Figs. 13b and 15 be developed for the analysis. If the influences of longitudinal and transverse wall heat conductions are important, they can be incorporated through the appro-

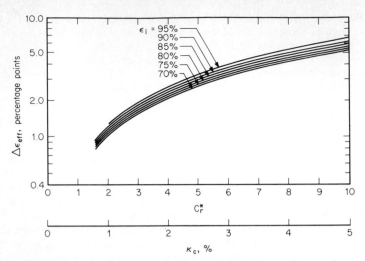

FIG. 16 Reduction in the "effective" regenerator effectiveness $\Delta\epsilon_{eff}$ as a function of C_r^* or κ_c and ϵ_i for $C^* = 1$, $\overline{C}_r^* = 200$, $P^* = 4$, and $T^* = 2$. Temperatures $T'_{c,o}$, $T'_{h,o}$, $T''_{c,o}$, and $T''_{h,o}$ in this figure are conceptually different variables from those in Fig. 14. $\Delta\epsilon_{eff}$ (percentage points) $= \epsilon(\%) - \epsilon_{eff}(\%)$.

priate solutions replacing the solutions of Fig. 11 or Eqs. (7) to (10) in the foregoing procedures.

C. PRESSURE DROP ANALYSIS

Fluid pumping power is a design constraint in many applications. This pumping power is proportional to the pressure drop in the exchanger in addition to the pressure drops associated with inlet or outlet headers, manifolds, nozzles, or ducting. The fluid pumping power P is related to the pressure drop in the exchanger as

$$P = \frac{W \Delta P}{\rho} \approx \begin{cases} \dfrac{1}{2g_c} \dfrac{\mu}{\rho^2} \dfrac{4L}{D_h} \dfrac{W^2}{D_h A_o} f\,\mathrm{Re} & \text{for laminar flow} \qquad (22a) \\[3mm] \dfrac{0.046}{2g_c} \dfrac{\mu^{0.2}}{\rho^2} \dfrac{4L}{D_h} \dfrac{W^{2.8}}{A_o^{1.8} D_h^{0.2}} & \text{for turbulent flow} \qquad (22b) \end{cases}$$

Only the core friction term is considered in the right-hand side approximation for discussion purposes. Now consider the case of specified flow rate and geometry (i.e., specified W, L, D_h, and A_o). As a first approximation, f Re in Eq. (22a) is constant for fully developed laminar flow, while $f = 0.046\,\mathrm{Re}^{-0.2}$ is used in deriving Eq. (22b) for fully developed turbulent flow. It is evident that P is strongly dependent on ρ in laminar and turbulent flow and on μ in laminar flow, and weakly dependent on μ in turbulent flow. For high-density moderate-viscosity liquids, the pumping power is generally so small that it has only a minor influence on the design. For laminar flow of highly viscous liquids in large L/D_h exchangers, pumping power is an important constraint; and this is also the case for gases, both in turbulent and laminar flow, because of the great impact of $1/\rho^2$.

In addition, when blowers and pumps are used for the fluid flow, they are generally head-limited, and the pressure drop itself can be a major consideration. As shown in Eq.

(147), the pressure drop is proportional to D_h^{-3} and hence it is strongly influenced by the passage hydraulic diameter.

Now we will summarize the core pressure drop formulas for plate-fin, tube-fin, and regenerative exchangers. Pressure drop evaluation associated with bends, valves, and fittings has been summarized in Ref. 168.

1. Plate-Fin Exchangers

The flow through one passage in a heat exchanger is shown in Fig. 17. The flow upstream of the passage is idealized as uniform. As it enters the passage, it contracts due to an area change. Flow separation takes place at the entrance, followed by an irreversible free expansion. In the core, the fluid experiences skin friction; it may also experience form drag at the leading and trailing edges of an interrupted fin surface; it may also experience internal contractions and expansions within the core as in a perforated fin core. If heating or cooling takes place in the core, as in any heat exchanger, the fluid density and mean velocity change along the flow length. The flow accelerates or decelerates depending upon the heating or cooling. At the core exit, flow separation takes place, followed by an expansion due to the area change. The total pressure drop on one side of the exchanger, based on the model of Fig. 17, is [16, 25]

$$\frac{\Delta P}{P_i} = \frac{G^2}{2g_c}\frac{1}{P_i\rho_i}\left[\underbrace{(1-\sigma^2+K_c)}_{\text{entrance effect}} + \underbrace{f\frac{L}{r_h}\rho_i\left(\frac{1}{\rho}\right)_m}_{\text{core friction}}\right.$$

$$\left.+ \underbrace{2\left(\frac{\rho_i}{\rho_o}-1\right)}_{\text{flow acceleration}} - \underbrace{(1-\sigma^2-K_e)\frac{\rho_i}{\rho_o}}_{\text{exit effect}}\right] \quad (23)$$

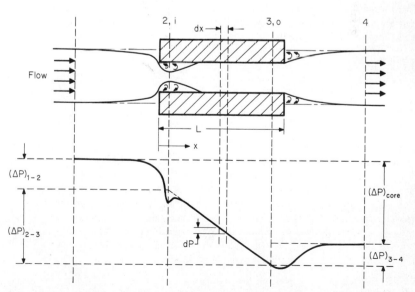

FIG. 17 Pressure drop components associated with one passage of a plate-fin exchanger.

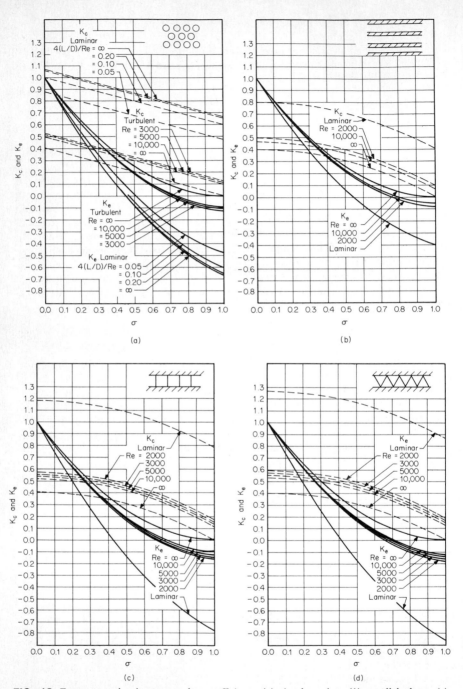

FIG. 18 Entrance and exit pressure loss coefficients: (*a*) circular tubes, (*b*) parallel plates, (*c*) square passages, and (*d*) triangular passages [16]. For each of these flow passages, the fluid flows perpendicular to the plane of the paper into the flow passages.

Here σ is a ratio of minimum free flow area to frontal area, f is the Fanning friction factor for the surface (see Sec. D, p. 4-205), and K_c and K_e are the flow contraction and expansion loss coefficients. These loss coefficients are presented in Fig. 18 for four different entrance flow passage geometries [16]. The mean specific volume v_m or $(1/\rho)_m$ is expressed as follows. For a liquid with any flow arrangement, or for a perfect gas with $C^* = 1$ and any flow arrangement (except for parallel flow),

$$\left(\frac{1}{\rho}\right)_m = v_m = \frac{v_i + v_o}{2} = \frac{1}{2}\left(\frac{1}{\rho_i} + \frac{1}{\rho_o}\right) \tag{24}$$

Here v represents specific volume in m^3/kg or ft^3/lb_m. For a perfect gas with $C^* = 0$ and any flow arrangement,

$$\left(\frac{1}{\rho}\right)_m = \frac{\tilde{R}}{P_{ave}} T_{lm} \tag{25}$$

where $\tilde{R}$ is the gas constant in $J/(kg \cdot K)$ or $lb_f \cdot ft/(lb_m \cdot R)$, $P_{ave} = (P_i + P_o)/2$, and $T_{lm} = T_{const} \pm \Delta T_{lm}$. Here T_{const} is the arithmetic average temperature of the fluid on the other side of the exchanger, and $\Delta T_{\ell m}$ is the log-mean temperature difference defined in Eq. (14) on p. 4-18.

Generally, the core frictional pressure drop in Eq. (23) is a dominating term accounting for about 90 percent or more of ΔP. Hence, Eq. (23) may be approximated as

$$\Delta P \approx \frac{4fLG^2}{2g_c D_h}\left(\frac{1}{\rho}\right)_m \tag{26}$$

The entrance and exit losses in Eq. (23) are important at low values of σ and L (short cores), at high values of Re, and for gases; they are negligible for liquids. The values of K_c and K_e presented in Fig. 18 apply to long tubes for which flow is fully developed at the exit. For partially developed flows, K_c is lower and K_e is higher than for fully developed flows. For interrrupted surfaces, the flow is generally not fully developed. For highly interrupted fin geometries, the entrance and exit losses are generally small compared to the core pressure drop, and the flow is mixed very well; hence, K_c and K_e for $Re \rightarrow \infty$ should represent a good approximation.

2. Tube-Fin Exchangers

a. Flow Inside Tubes

The pressure drop inside the tubes is determined in the same manner as for plate-fin surfaces using Eq. (23). Appropriate values of the f factor and K_c and K_e are used in this expression for flow inside the tubes with or without fins.

b. Continuous Fins on an Array of Tubes

For these fins (Fig. 4), the components of the total core pressure drop on the fin side are all the same as those for plate-fin surfaces. The only difference is that the evaluation of entrance and exit losses is based on the flow area at the leading (or trailing) edges of the fins, and not on the minimum free flow area A_o. Note that A_o is dependent upon the tube pitches and tube arrangement. If the ratio of free flow area to frontal area at the fin leading edges is designated as σ', and the mass velocity at the leading edges as G', then by continuity, $G'\sigma' = G\sigma$, with the latter quantities based on minimum free flow area between tubes. Now K_c and K_e are evaluated for σ' using Fig. 18. The total pressure drop for the core is then given by

$$\frac{\Delta P}{P_i} = \frac{G^2}{2g_c} \frac{1}{P_i \rho_i} \left[f \frac{L}{r_h} \rho_i \left(\frac{1}{\rho}\right)_m + 2 \left(\frac{\rho_i}{\rho_o} - 1\right) \right]$$

$$+ \frac{G'^2}{2g_c} \frac{1}{P_i \rho_i} \left[(1 - \sigma'^2 + K_c) - (1 - \sigma'^2 - K_e) \frac{\rho_i}{\rho_o} \right] \qquad (27)$$

c. Individually Finned Tubes

Equation (23) needs to be modified as regards entrance and exit losses for this case (flow normal to the tubes, such as those of Fig. 5). Each tube row consists of a contraction and an expansion due to the flow area change; hence, the pressure losses associated with a tube row within the core are of the same order of magnitude as those at the entrance with the first tube row and those at the exit with the last tube row. Consequently, the entrance and exit pressure drops are generally lumped into the friction factor for individually finned tubes, and Eq. (23) reduces to

$$\frac{\Delta P}{P_i} = \frac{G^2}{2g_c} \frac{1}{P_i \rho_i} \left[f \frac{L}{r_h} \rho_i \left(\frac{1}{\rho}\right)_m + 2 \left(\frac{\rho_i}{\rho_o} - 1\right) \right] \qquad (28)$$

3. Regenerators

For a matrix having continuous cylindrical passages (Fig. 6), the pressure drop consists of the same components as that for the plate-fin exchanger, Eq. (23). For a matrix made up of any porous material (such as randomly packed screens, cross rods, bricks, tiles, spheres, copper wool, etc.), the entrance and exit pressure losses are accounted for in the friction factor. The pressure drop for these matrices is computed from Eq. (28).

4. Additional Pressure Drop for Liquid Flows

In the case of vertical liquid flows through heat exchangers, the pressure drop or rise due to an elevation change may not be negligible. This pressure drop or rise, sometimes referred to as "static head," is given by

$$\Delta P = \pm \frac{\rho_m g L}{g_c} \qquad (29)$$

where the $+$ case stands for vertical upflow (equivalent pressure drop due to the elevation change), the $-$ case denotes vertical downflow (equivalent pressure rise), g is the gravitational acceleration, and L is the exchanger length.

D. SURFACE BASIC HEAT TRANSFER AND FLOW FRICTION CHARACTERISTICS

The dimensionless heat transfer and fluid flow friction (pressure drop) characteristics of a heat transfer surface are simply referred to as the surface basic characteristics, or surface basic data.† Generally, the dimensionless experimental heat transfer characteristics are presented in terms of the Colburn factor $j = \text{St Pr}^{2/3}$ versus Reynolds number Re, and the theoretical characteristics in terms of Nusselt number Nu versus Re or $x^* = x/(D_h \text{ Re Pr})$. The dimensionless pressure drop characteristics are presented in terms of the Fanning friction factor f versus Re, or modified friction factor per tube row f' versus Re_d.

Since the majority of basic data for compact surfaces are obtained experimentally, the dimensionless heat transfer and pressure drop characteristics of these surfaces are presented in terms of j and f versus Re. Here the Reynolds number Re is based on the hydraulic diameter D_h. This approach is somewhat arbitrary since several variations of one basic types of surface geometry will not generally correlate on the j and f versus Re basis. This is because geometric variables, other than the hydraulic diameter, may have a significant effect on surface performance. Because the values of j, f, and Re are dimensionless, the test data are applicable to surfaces of any hydraulic diameter, providing complete geometric similarity is maintained.

The limitations of the j versus Re plot, commonly used in presenting compact heat exchanger surface basic data, should be understood. In fully developed laminar flow, as will be discussed, the Nusselt number is theoretically constant, independent of Pr (and also Re). Since $j = \text{St Pr}^{2/3} = \text{Nu Pr}^{-1/3}/\text{Re}$, then j will be dependent upon Pr in the fully developed laminar region, and hence the j factors presented in Chap. 7 of Kays and London [16] for gas flows in the fully developed laminar region should be first converted to a Nusselt number (using $\text{Pr} = 0.70$), which can then be used directly for liquid flows as constant-property results. Based on theoretical solutions for thermally developing laminar flow (to be discussed), $\text{Nu} \propto (x^*)^{-1/3}$. This means $\text{Nu Pr}^{-1/3}$ is independent of Pr and hence j is independent of Pr for thermally developing laminar flows.‡ For fully developed turbulent flow, $\text{Nu} \propto \text{Pr}^{0.4}$ [26], and hence, $j \propto \text{Pr}^{0.07}$. Thus j is again dependent upon Pr in the fully developed turbulent region.§ All of the foregoing comments apply to either constant-property theoretical solutions, or almost constant-property (low–temperature difference) experimental data. The influence of property variations, discussed in Sec. D.4 on p. 4-240, must be accounted for by correcting the aforementioned constant-property j or Nu when designing a heat exchanger.

1. Experimental Methods

Primarily, three different test techniques are used to determine the surface heat transfer characteristics. These techniques are based on the steady-state, transient, and periodic nature of heat transfer modes through the test sections. Generally, the isothermal steady-

---

†We will not use the terminology "surface performance data" since performance in industry means a dimensional plot of heat transfer rate and pressure drop as a function of the fluid flow rate for an exchanger. Note that we need to distinguish between the performance of a surface geometry and the performance of a heat exchanger.

‡If a slope of -1 for the log-log j-Re characteristic is used as a criterion for *fully developed laminar flow*, none of the surfaces reported in Chap. 10 of Kays and London [16] would qualify as being in a fully developed laminar condition. Data for most of these surfaces indicate thermally developing flow conditions for which j is independent of Pr as indicated, and hence the j-Re characteristic should not be converted to the Nu-Re characteristics for the data of Chap. 10 of Ref. 16.

§Colburn in 1933 [27] proposed $j = \text{St Pr}^{2/3}$ as a correlating parameter to include the effect of Prandtl number based on the then available data for turbulent flow. Based on presently available experimental data, however, the j factor is clearly dependent upon Pr for fully developed turbulent flow and for fully developed laminar flow, but not for developing laminar flows.

state technique is used for the determination of f factors. These test techniques are now briefly described.

a. Steady-State Test Techniques

This is one of the most common test techniques used to establish the j versus Re characteristics of a recuperator surface. Different data acquisition and reduction methods are used depending upon whether the test fluid is a gas (air) or a liquid. The method for liquids is generally referred to as the Wilson plot technique. These methods are summarized separately next.

Steady-State Test Technique for Gases.   Generally, a crossflow heat exchanger is employed as a test section. On one side, a surface for which the j versus Re characteristic is known is employed; a fluid with high heat capacity rate flows on this side. On the other side of the exchanger, a surface for which the j versus Re characteristic is to be determined is employed; the fluid which flows over this "unknown" surface is the one which is used in a particular application of the unknown-side surface. Generally, air is used on the unknown side, and steam, hot water, chilled water, or oils are used on the known side. A typical test setup used by Kays and London [16] is shown in Fig. 19 to provide some ideas on the air-side (unknown-side) components of the test rig. For further details, refer to Ref. 28.

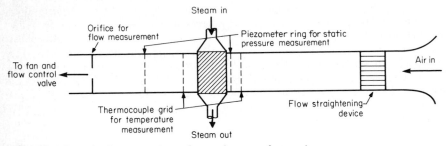

FIG. 19 Schematic of steam-to-air steady-state heat transfer test rig.

In the experiments, the fluid flow rates on both sides of the exchanger are set constant at predetermined values. Once the steady-state conditions are achieved, fluid temperatures upstream and downstream of the test section on both sides are measured, as well as all pertinent measurements for the determination of the fluid flow rates. The upstream pressure and pressure drop across the core on the unknown side are also recorded to determine the "hot" friction factors.† The tests are repeated with different flow rates on the unknown side to cover the desired range of the Reynolds number.

In order to determine the j factor on the unknown side, the exchanger effectiveness is determined from the temperature measurements, and the heat capacity rate ratio is determined from the flow measurements and specific heats. NTU is subsequently computed from the appropriate ϵ-NTU relationship for the test core flow arrangement. Generally, the test section is a new exchanger core and fouling resistances are negligible; $\eta_o hA$ on the unknown side is determined from the following thermal resistance equation where

---

†The friction factor determined from the ΔP measurement taken during the heat transfer testing is referred to as the "hot" friction factor.

UA is found from NTU:

$$\frac{1}{UA} = \frac{1}{(\eta_o h A)_{\text{unknown side}}} + R_w + \frac{1}{(\eta_o h A)_{\text{known side}}} \qquad (30)$$

Once the surface area and the geometry are known for the extended surface (if any) h and η_o are computed iteratively. Then the j factor is calculated from its definition. The Reynolds number on the unknown side for the test point is determined from its definition for the known flow rate.

The test core is designed with two basic considerations in mind to reduce the experimental uncertainty in the j factors: (1) the magnitudes of thermal resistances on each side as well as the wall; and (2) the range of NTU.

The thermal resistances in a heat exchanger are related by Eq. (30). To reduce the uncertainty in the determination of the thermal resistance of the unknown side (with known overall thermal resistance, $1/UA$), the thermal resistances of the exchanger wall and the known side should be kept small by design. The wall thermal resistance is usually negligible and may further be minimized through the use of a thin material with high thermal conductivity. On the known side, the thermal resistance is minimized by the use of a liquid (hot or cold water) at high flow rates, or a condensing vapor, to achieve a high h, and also by extending the surface area. The thermal boundary condition achieved during steady-state testing is generally a close approach to a uniform wall temperature condition.

The NTU range for testing is generally restricted between 0.5 and 3 or between 40 percent and 90 percent in terms of the exchanger effectiveness. In order to understand this and point out precisely the problem areas, consider the test fluid on the unknown side to be cold air being heated in the test section and the fluid on the known side to be hot water (steam replaced by hot water and its flow direction reversed in Fig. 19). The high NTU occurs at low test fluid flows for a given test core. Both temperature and airflow measurements become critical at low flows, and the resultant heat unbalances ($q_w - q_a)/q_a$ increase sharply at low airflows with decreasing airflows. In this subsection, the subscripts w and a denote water and air sides, respectively. Now, the exchanger effectiveness can be computed in two different ways:

$$\epsilon = \frac{q_a}{C_a(T_{w,i} - T_{a,i})} \quad \text{or} \quad \epsilon = \frac{q_w}{C_a(T_{w,i} - T_{a,i})} \qquad (31)$$

Thus a large variation in ϵ will result at low airflows depending upon whether it is based on q_a or q_w. Since ϵ-NTU curves are very flat at high ϵ (high NTU), there is a very large error in the resultant NTU, h, and j. The j versus Re curve drops off consistently with decreasing Re as shown by a dashed line in Fig. 20. This phenomenon is referred to as *rollover* or *drop-off* in j. Some of the problems causing the rollover in j are the errors in temperature and airflow measurements as follows:

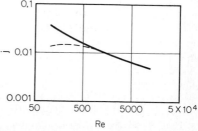

FIG. 20 The rollover phenomenon for j versus Re characteristic of a heat exchanger surface at low airflows. The dashed curve indicates the rollover phenomenon; the solid curve represents the accurate characteristics.

1. Basically, the heat transfer coefficients associated with the thermocouple junction or resistance thermometer are quite low at low airflows. Hence, what we measure is the junction temperature and not the ambient temperature.

Thus the "measured" air temperature downstream of the test core, $T_{a,o}$, may be too low due to heat conduction along the thermocouple wire. This error is not so pronounced for the upstream temperature measurement since air is at a lower temperature. The measured air temperature upstream of the test core, $T_{a,i}$, may be too high due to the radiation effect from the hot core and from the hot walls of the wind tunnel because of heat conduction in the duct wall from the hot test core. This error is negligible for core downstream since the duct walls are at about the same temperature as outlet air temperature. Both the aforementioned errors in $T_{a,i}$ and $T_{a,o}$ will decrease the calculated q_a.

2. At low airflows, temperature stratification in the vertical direction would be a problem both upstream and downstream of the test core. Thus it becomes difficult to obtain true bulk mean temperatures $T_{a,i}$ and $T_{a,o}$.

3. On the water side, the temperature drop is generally very small, and hence it will require very accurate instrumentation for ΔT_w measurements. Also, care must be exercised to ensure good mixing of water at the core outlet before ΔT_w is measured.

4. There are generally some small leaks in the wind tunnel between the test core and the point of airflow measurement. These leaks, although small, are approximately independent of the airflow rate. These leaks will represent an increasing fraction of the measured flow rate W_a at low airflows. A primary leak test is essential at the lowest encountered test airflow before any testing is conducted.

5. Heat losses to the ambient are generally small for a well-insulated test section. However, they could represent a good fraction of the heat transfer rate in the test section at low airflows. A proper calibration is essential to determine these heat losses.

6. For some test core surfaces, longitudinal heat conduction in the test core surface wall may be important and should be accounted for in the data reduction.

The first five factors cause heat unbalances $(q_w - q_a)/q_a$ to increase sharply at low airflows with decreasing airflow rates. In order to minimize or eliminate the rollover in j factors, the data should be reduced based on $q_{ave} = (q_w + q_a)/2$, and whenever possible, by reducing the core flow length by half and then retesting the core.

The uncertainty in the j factors obtained from the steady-state tests ($C^* \approx 0$ case) for a given uncertainty in $\Delta_2 (= T_{w,o} - T_{a,o})$ is given by [28]

$$\frac{d(j)}{j} = \frac{d(\Delta_2)}{\Delta_0} \frac{\text{NTU}_c}{\text{NTU}} \frac{e^{\text{NTU}}}{\text{NTU}} \qquad (32)$$

Here $\Delta_0 = T_{w,i} - T_{a,i}$ and $\text{NTU}_c/\text{NTU} \approx 1.1$. Thus, a measurement error in the outlet temperature difference [i.e., $d(\Delta_2)$] magnifies the error in j by the foregoing relationship both at high NTU (NTU > 3) and low NTU (NTU < 0.5). The error at high NTU due to the error in Δ_2 and the error in other factors was discussed above. The error at low NTU due to the error in Δ_2 can also be significant and hence a careful design of the test core is essential for obtaining accurate j data.

In addition to the foregoing measurement errors, erroneous j data are obtained for a surface if the test core is not constructed properly. The problem areas are: poor thermal bond between the fins and the primary surface, gross blockage (gross flow maldistribution) on the air side or water (steam) side, and passage-to-passage nonuniformity (or maldistribution) on the air side. These factors influence the measured j and f factors differently in different Reynolds number ranges. Qualitative effects of these factors are presented in Fig. 21 to show the trends. The solid lines in these figures represent the j data of an ideal core having a perfect thermal bond, no gross blockage, and perfect uni-

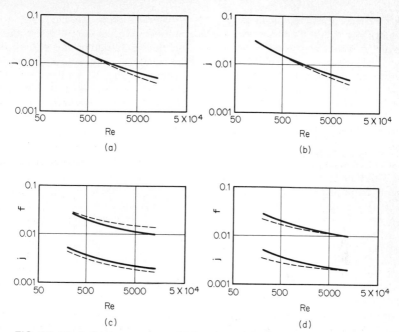

FIG. 21 The influence on measured j data due to (a) poor thermal bond between fins and primary surface, (b) water- (steam-) side gross blockage, (c) air-side gross blockage, and (d) air-side passage-to-passage nonuniformity. The solid lines are for the perfect core, the dashed lines for the specified imperfect core.

formity. The dashed lines represent what happens to j factors when the specified imperfections exist. It is imperative that a detailed air temperature distribution be measured at the core outlet to ensure none of the foregoing problems are associated with the core.

The experimental uncertainty in the j factor for the foregoing steady-state method is usually within ± 5 percent when the temperatures are measured accurately to within $\pm 0.1°C$ and none of the aforementioned problems exist in the test core. The uncertainty in the Reynolds number is usually within ± 2 percent when the flow is measured accurately within ± 0.7 percent.

Wilson Plot Technique for Liquids. In order to obtain highly accurate j factors, one of the considerations for the design of a test core in the preceding method was to have the thermal resistance on the test fluid (gas) side dominant, i.e., the test fluid side $\eta_o hA$ significantly lower compared to that on the other (known) side. This is achieved by using either steam or hot or cold water at high flows on the known side. However, if the test fluid is water or another liquid and it has a high heat transfer coefficient, it may not represent a dominant thermal resistance even if condensing steam is used on the other side. This is because the test fluid resistance may be of the same order of magnitude as the wall resistance. Hence for liquids, the Wilson plot test technique is employed to obtain j factors. This method is restricted to turbulent flow of liquids through circular or noncircular flow passages. A modification of the Wilson plot technique is also presented later that could be used for transition and laminar-flow regimes.

In this method, liquid (test fluid) flows on one side for which j versus Re characteristics are being determined; condensing steam, liquid, or air flows on the other side, for which we may or may not know the j versus Re characteristics. The fluid flow rate on this other side and the log-mean average temperature are kept constant so that its thermal resistance remains approximately constant.† The fluid flow rate on the unknown side (liquid side) is varied systematically. The fluid flow rates and temperatures upstream and downstream of the test core on each side are measured for each point. Thus when ϵ and C^* are known, NTU and UA are computed. For discussion purposes, consider the test fluid side to be cold and the other fluid side to be hot. Then U_c is given by

$$\frac{1}{U_c} = \frac{A_c}{(\eta_o h A)_h} + R_{s,h} A_c + R_w A_c + R_{s,c} A_c + \frac{1}{h_c} \tag{33}$$

Note that η_o is not incorporated on the liquid (test fluid) side since fins are generally not used with liquid flows. However, if there are fins, the equation can be modified easily by using $\eta_{o,c} h_c$ in place of h_c, but the solution will then be obtained iteratively. The test conditions are maintained such that the fouling (scale) resistances $R_{s,h}$ and $R_{s,c}$ remain approximately constant though not necessarily zero. Since h is maintained constant on the hot side, the first four terms on the right-hand side of the equality sign of Eq. (33) are constant, let us say equal to C_2. For fully developed turbulent flow through constant cross-sectional ducts, the Nusselt number correlation is of the form‡

$$\text{Nu} = C_0 \, \text{Re}^{0.8} \, \text{Pr}^{0.4} \tag{34}$$

where C_0 is a constant. By substituting the definitions of Nu, Re, and Pr in Eq. (34),

$$h_c = (C_0 k^{0.6} \rho^{0.8} c_p^{0.4} \mu^{-0.4} D_h^{-0.2}) V^{0.8} = C_1 V^{0.8} \tag{35}$$

Here V represents the mean fluid velocity through the minimum free flow area and C_1 is a constant independent of V. Substitution of Eq. (35) for h_c in Eq. (33) results in

$$\frac{1}{U_c} = C_2 + \frac{1}{C_1 V^{0.8}} \tag{36}$$

Wilson [29] proposed a plot of $1/U_c$ versus $V^{-0.8}$ on a linear scale, as shown in Fig. 22a, by varying the test fluid velocity. The intercept on the y axis then represents the constant C_2, and the slope of the curve is $1/C_1$. As long as Eq. (34) or the equivalent applies, the "Wilson plot" of Fig. 22a will be a straight line. The Wilson plot technique has been modified by Briggs and Young [110], and their technique with some further modifications is described next.

Since the fluid temperatures on both sides could vary from one test point to another, the foregoing plot of $1/U_c$ versus $V^{-0.8}$ will not take into account variations in the fluid properties from one test point to another. An alternative is to work directly with the Nu expression as follows:

$$\text{Nu}_c = C_3 \, \text{Re}_c^{0.8} \, \text{Pr}_c^{0.4} \left(\frac{\mu_w}{\mu_m}\right)^n \tag{37}$$

Here, C_3 is a constant to be determined and n is presented in Table 12. Evaluating h from Eq. (37), substituting it in Eq. (33), and rearranging in $y = m'x + C_4$ form, we

---

†The condensing coefficients may vary widely within the test exchanger, and hence care must be exercised that the thermal resistance on that side is kept very small.

‡Here the exponent n' on the Reynolds number is considered to be 0.8 for illustration. However, later it is shown that a more appropriate value of n' may be obtained by a linear regression. For the Wilson plot technique, the Nusselt number correlation has to be in the explicit form; an implicit form, such as the Gnielinski correlation of Eq. (50), cannot be utilized.

get

$$
\frac{1}{U_c} = \frac{1}{C_3} \left\{ \frac{1}{\left(\dfrac{k}{D_h} \right)_c \mathrm{Re}_c^{0.8}\, \mathrm{Pr}_c^{0.4} \left(\dfrac{\mu_w}{\mu_m} \right)^n} \right\} + \left[\left(\frac{1}{(\eta_o h A)_h} + R_{s,h} + R_w + R_{s,c} \right) A_c \right]
$$

$$
= \frac{1}{C_3} x + C_4
$$

(38)

so that $y = 1/U_c$, $m' = 1/C_3$, and the quantities in the brackets { } and [ ] are x and C_4 respectively. A plot of y versus x will then have a slope $m' = 1/C_3$ and the intercept on the y axis as C_4. Once C_3 is known, the desired Nusselt number on the liquid side is known, from which the j factor can be computed directly. For most experimental data, the correction $(\mu_w/\mu_m)^n$ will be close to unity and can be treated as unity in the first computation to obtain h_c. Once h_c and the other resistances are found by processing all test points, the wall temperature can be calculated for each point to incorporate the viscosity correction, if necessary. If a linear regression computer program is available, the slope can be determined through a linear regression. However, proper care and judgment should be exercised for not carrying out linear regression when the curve representing Eq. (38) is not a straight line on a linear grid graph paper.

Since the foregoing technique is restricted to the type of correlation of Eq. (37), the exponent n' on the Reynolds number is taken as 0.8. Actually, this exponent is a function of the Prandtl number and Reynolds number; it varies from 0.78 at $\mathrm{Pr} = 0.7$ to 0.90 at $\mathrm{Pr} = 100$ for $\mathrm{Re} = 5 \times 10^4$, based on the best available correlation of Gnielinski or Petukhov-Kirillov for fully developed turbulent flow through a circular tube [26]. In order to determine an accurate exponent for Re, a linear regression analysis needs to be carried out on the modified Wilson plot of Fig. 22b as indicated in the preceding paragraph with different values of n'.† The value of n' which provides the least error in the linear regression is then the most appropriate one for the surface geometry under consideration. Here, it is idealized that the modified Wilson plot is linear before investigating the appropriate value of n'.

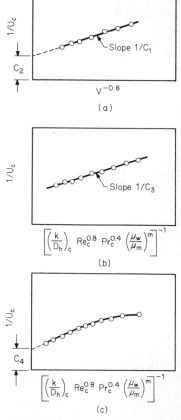

FIG. 22 (a) A Wilson plot, (b) a modified Wilson plot for variable thermophysical fluid properties, and (c) a modified Wilson plot for transition and laminar regimes.

†In Fig. 22b and c, $n' = 0.8$ is used for illustration.

If the data points extend into transition and laminar-flow regimes, a Wilson plot will not be a straight line, but it will be as shown in Fig. 22c. In this case, it is recommended that at least a few test data points be obtained in turbulent flow that can be extrapolated by a straight line as shown in Fig. 22c to get the intercept C_4. This value of C_4 is then used in Eq. (38) to obtain h on the liquid side for each test point.

b. Transient Test Techniques

This is one of the most common test techniques used to establish the j versus Re characteristic of a matrix-type or a high-NTU surface. One variant is also referred to as the "single-blow" test technique. The test section is a single-fluid exchanger (matrix) built up from the heat transfer surface for which the j versus Re characteristic is to be determined. Generally, air is used as a working fluid. In the test facilities, an electric heater (resistance heating screen) is employed upstream of the test section to obtain a step change in the air temperature at the inlet of the test section.

In the experiments, the airflow rate is set constant at a predetermined value. The air is heated with the resistance heating screen to about 10°C (20°F) above the ambient temperature and in turn heats the matrix. The heating is continued until the core reaches a uniform temperature exhibited by a negligible difference between the air temperatures at the inlet and exit of the matrix. Once the stable condition is reached, the power to the heating screen is turned off. The temperature-time history of the air leaving the matrix is continuously recorded during the matrix cooling period.

This temperature-time history of air depends upon the heat transfer rate from the matrix and is thus a function of the matrix NTU. Several theoretical methods are available to determine NTU, such as: (1) maximum slope data reduction method; (2) zero intercept method; (3) direct curve matching method; and (4) first moment of area method. Each of these methods has some restrictions on the NTU range to obtain an accurate j factor. For the applicable range of NTU and further details, refer to Refs. 30 to 32. Since the total heat transfer in the core is dependent upon the heat capacity of the matrix, care must be exercised to minimize the heat capacity of the test section walls. Materials having low volumetric specific heats (ρc_p) and low thermal conductivity k_w are preferable for the test section walls. If the single-blow technique is used, the heater design should be such that the time constant of the heater is negligible compared to the time constant for the core.

During the experiments, sometimes both the cooling and heating temperature-time histories of the exit air are recorded and the average of the pertinent information is then used in data reduction. The experiment is repeated at different flow rates to cover the desired range of the Reynolds number. Experience shows that the thermal boundary condition achieved during the single-blow testing is between the constant wall temperature and constant wall heat flux conditions.

The maximum slope data reduction method is most commonly used for high-NTU (≥ 3) surfaces. With this method, the experimental uncertainty in the j factors is usually within ± 13 percent for NTU ≥ 3.

It should be emphasized that the determination of core geometrical properties (porosity σ and compactness β) must be accurate since the experimentally determined j, f, and Re are sensitive to σ and β as follows:

$$j \propto \frac{\sigma}{\beta} \qquad f \propto \frac{\sigma^3}{\beta} \qquad \text{Re} \propto \frac{1}{\beta} \tag{39}$$

Thus a 2 percent error in the core porosity or flow area will result in a 2 percent error in j and a 6 percent error in f. A 2 percent error in surface compactness will result in a 2 percent error in each of $j, f,$ and Re.

c. Periodic Test Technique

This method is another variant of the transient technique. In the single-blow method, a step change in the fluid temperature is achieved at the inlet of the matrix. In the periodic method, the temperature of the inlet air is continuously changed by a periodic (approximately sinusoidal) power input to the electric heater. The phase shift and/or the amplitude change between inlet and outlet air temperatures is used to determine h and hence j factors. This method is applicable to a wide range of NTU. The inlet air temperature variation needs only to be approximately sinusoidal, and a Fourier analysis is performed to extract the behavior of the first harmonic for analysis. This method is described in detail by Stang and Bush [33].

d. Test Technique for Friction Factors

The experimental determination of flow friction characteristics of compact heat exchanger surfaces is relatively straightforward. Regardless of the core construction and the method of heat transfer testing, the determination of f is made under steady fluid flow rates with or without heat transfer. At a given fluid flow rate on the unknown side, the following measurements are made: core pressure drop, core inlet pressure and temperature, core outlet temperature for "hot" friction data, fluid flow rate, and the core geometrical properties. The Fanning friction factor f is then determined from the following equation:

$$f = \frac{r_h}{L} \frac{1}{(1/\rho)_m} \left[\frac{2g_c \, \Delta P}{G^2} - \frac{1}{\rho_i} (1 - \sigma^2 + K_c) \right.$$
$$\left. - 2\left(\frac{1}{\rho_o} - \frac{1}{\rho_i}\right) + \frac{1}{\rho_o}(1 - \sigma^2 - K_e) \right] \quad (40)$$

This equation is an inverted form of the core pressure drop, Eq. (23). For the isothermal pressure drop data, $\rho_i = \rho_o = 1/(1/\rho)_m$. The friction factor thus determined includes the effects of skin friction, form drag, and local flow contraction and expansion losses, if any, within the core. Tests are repeated with different flow rates on the unknown side to cover the desired range of the Reynolds number. The experimental uncertainty in the f factors is usually within ± 5 percent when ΔP is measured accurately within ± 1 percent.

Generally, the Fanning friction factor f is determined from isothermal pressure drop data (no heat transfer across the core). The hot friction factor f versus Re curve should be close to the isothermal f versus Re curve, particularly when the variations in the fluid properties are small, i.e., the average fluid temperature for the hot f data is not significantly different from the wall temperature. Otherwise, the hot f data must be corrected to account for the temperature-dependent fluid properties (see Sec. D.4 on p. 4-240).

2. Analytical Solutions and Correlations for Simple Geometries†

Compact exchangers employ surfaces having either continuous-flow passages or flow passages with frequent boundary-layer interruptions. The velocity and temperature profiles

†All the results presented in this section are for constant fluid properties. The influence of temperature-dependent properties is taken into account by the property ratio method presented in Sec. D.4, on p. 4-240.

across the flow cross section are generally fully developed in the continuous-flow passages, while they are generally developing at each boundary-layer interruption in an interrupted surface. The heat transfer and flow friction characteristics of flow passages are in general substantially different for fully developed flows and developing flows. The analytical results are discussed separately next for developed and developing flows for simple flow passage geometries. For complex surface geometries, the surface basic characteristics are obtained by experiments. They are discussed in Sec. D.3 starting on p. 4-221.

a. Fully Developed Laminar Flow

The constant-property fully developed laminar flow Nusselt numbers are constant and independent of Re and Pr, but dependent upon the flow passage cross-section geometry and thermal boundary conditions. The constant-property product of Fanning friction factor and Reynolds number is also constant and independent of Re and Pr, but dependent upon the flow passage geometry, for fully developed laminar flow. The laminar-flow analytical results will be summarized for the three important thermal boundary conditions $\textcircled{T}$, $\textcircled{H1}$, and $\textcircled{H2}$, defined in Fig. 23. The $\textcircled{T}$ boundary condition refers to constant wall temperature both axially and peripherally throughout the duct (or passage) length. For the $\textcircled{H1}$ boundary condition, the wall heat transfer rate per unit length q' is constant in the axial direction while the wall temperature at any cross section is constant in the peripheral direction. For the $\textcircled{H2}$ boundary condition, the wall heat flux q'' is constant in the peripheral direction *as well as* in the axial direction. Shah and London [34] describe these and other thermal boundary conditions in detail. The Nusselt numbers for these boundary conditions have subscripts T, $H1$, and $H2$, respectively. Nu_T is lower than Nu_{H1} for all passage geometries, and Nu_{H2} is lower than Nu_{H1} for noncircular flow passages [34].

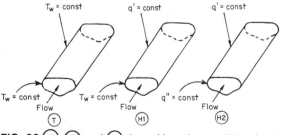

FIG. 23 $\textcircled{T}$, $\textcircled{H1}$, and $\textcircled{H2}$ thermal boundary conditions for the duct flow.

Shah and London [34] present a compilation of analytical solutions for laminar heat transfer and friction of a total of 40 different cylindrical duct cross-section shapes, out of which about 20 shapes could be applicable for compact heat exchanger design. The results include information on solutions for developed and developing velocity and temperature profiles, and the entry length required for fully developed flow. Solutions are given for a wide range of aspect ratios for each basic channel shape, e.g., triangular and rectangular channels. Some of these results are summarized in Ref. 168.

The Nusselt numbers and f Re for some duct shapes of interest in compact heat exchanger design are summarized in Table 5. In addition to Nu_{H1}, Nu_{H2}, Nu_T, f Re, and the flow area goodness factor j_{H1}/f,† included in Table 5 are the analytical (not experi-

---

†The significance of the flow-area goodness factor is discussed in Sec. E.2.a on p. 4-246.

TABLE 5 Solutions for Heat Transfer and Friction for Fully Developed Laminar Flow through Specified Ducts [34]

Geometry ($L/D_h > 100$)	Nu_{H1}	Nu_{H2}	Nu_T	$f\,Re$	$\dfrac{j_{H1}\dagger}{f}$	$K(\infty)\ddagger$	$L_{hy}^+\S$
Isosceles triangle, $\dfrac{2b}{2a} = \dfrac{\sqrt{3}}{2}$	3.014	1.474	2.39	12.630	0.269	1.739	0.04
$60°$ triangle, $\dfrac{2b}{2a} = \dfrac{\sqrt{3}}{2}$	3.111	1.892	2.47	13.333	0.263	1.818	0.04
Square, $\dfrac{2b}{2a} = 1$	3.608	3.091	2.976	14.227	0.286	1.433	0.090
Hexagon	4.002	3.862	3.34	15.054	0.299	1.335	0.086
Rectangle, $\dfrac{2b}{2a} = \dfrac{1}{2}$	4.123	3.017	3.391	15.548	0.299	1.281	0.085
Circle	4.364	4.364	3.657	16.000	0.307	1.25	0.056
Rectangle, $\dfrac{2b}{2a} = \dfrac{1}{4}$	5.331	2.94	4.439	18.233	0.329	1.001	0.078
Rectangle, $\dfrac{2b}{2a} = \dfrac{1}{6}$	6.049	2.93	5.137	19.702	0.346	0.885	0.070
Rectangle, $\dfrac{2b}{2a} = \dfrac{1}{8}$	6.490	2.94	5.597	20.585	0.355	0.825	0.063
Parallel plates, $\dfrac{2b}{2a} = 0$	8.235	8.235	7.541	24.000	0.386	0.674	0.011

†$j_{H1}/f = Nu_{H1}\,Pr^{-1/3}/(f\,Re)$ with $Pr = 0.7$. Similarly, values of j_{H2}/f and j_T/f may be computed.

‡$K(\infty)$ for sine and equilateral triangular channels may be too high [34]; $K(\infty)$ for some rectangular and hexagonal channels is interpolated based on the recommended values in Ref. 34.

§L_{hy}^+ for sine and equilateral triangular channels is too low [34], so use with caution. L_{hy}^+ for rectangular channels is based on the faired curve drawn through the recommended value in Ref. 34. L_{hy}^+ for a hexagonal channel is an interpolated value.

mental) values of L_{hy}^+ and $K(\infty)$. The hydrodynamic entrance length L_{hy} [dimensionless form is $L_{hy}^+ = L_{hy}/(D_h\,Re)$] is the duct length required to achieve a maximum channel section velocity of 99 percent of that for fully developed flow when the entering fluid velocity profile is uniform. Since the flow development region precedes the fully developed region, the entrance region effects could be substantial even for channels having fully developed flow along a major portion of the channel. This increased friction in the entrance region and the change of momentum rate is taken into account by the incremental pressure drop number $K(\infty)$ defined by

$$\Delta P = \left[\frac{4f_{fd}L}{D_h} + K(\infty) \right] \frac{G^2}{2g_c\rho} \qquad (41)$$

where the subscript fd denotes the fully developed value.

For most channel shapes, the mean Nusselt number and friction factor will be within 10 percent of the fully developed value if $L/D_h > 0.2$ Re Pr (see Fig. 24). If $L/D_h < 0.2$ Re Pr, the fully developed analytical solutions may not be adequate, since Nu and f are higher in the developing flow region compared to their fully developed values. In contrast, Nu can be substantially lower than the theoretical value in a heat exchanger that consists of many flow passages in parallel when these flow passages are not uniform. Passage-to-passage nonuniformity, discussed on p. 4-277, generally reduces Nu and heat transfer by more than the gain by the thermal entrance effect, and hence the latter effect is generally neglected if $L/D_h \geq 100$ for gas flows. This passage-to-passage nonuniformity also reduces the friction factor (and ΔP), although generally by a negligible amount. The entrance length effect for pressure drop may not be neglected even for $L/D_h \approx 100$ for gas flows, since it could be substantial.

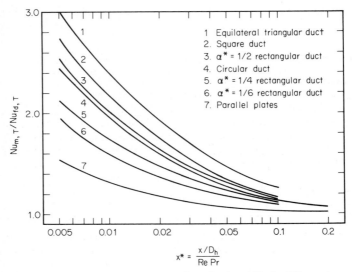

FIG. 24   The ratio of laminar developing to developed Nu for different ducts; the velocity profile developed for both Nu's.

The following observations may be made from the solutions of Table 5 for laminar-flow surfaces having fully developed flows: (1) There is a strong influence of thermal boundary conditions (T), (H1), and (H2) on the convective behavior. Depending on the flow cross-sectional geometry, the j factor for the (H1) boundary condition may be on the order of 50 percent greater than that for the (H2) boundary condition and about 20 percent greater than that for the (T) boundary condition. (2) As Nu $= hD_h/k$, a constant Nu implies a convective coefficient h that is independent of the flow velocity. (3) An increase in h is best achieved by reducing D_h, or by a change in the type of geometry, e.g., a change from a triangular to an 8:1 rectangular geometry. (4) The friction factor varies inversely with the velocity so that the fluid pressure drop tends to vary with the first power of the velocity.

The theoretical and analytical solutions presented in Ref. 34 have been verified by many researchers for flow and heat transfer in a single channel. Thus these results provide a valuable guideline for compact exchangers that employ many such channels in parallel. However, passage-to-passage nonuniformity could invalidate the application of the analytical solutions. Also the thermal boundary condition for heat transfer in an actual exchanger may not correspond to any of the previously described boundary conditions. Hence, accurate j and f versus Re data are generally determined experimentally even for the simple geometries used in highly compact exchangers. For gas flow through triangular passages ($40 < \text{Re} < 800$), the following are the experimental correlations [35]:

$$j = \frac{3.0}{\text{Re}} \qquad f = \frac{14}{\text{Re}} \tag{42}$$

in contrast to the theoretical values from Table 5 as $j_{H1} = 3.50/\text{Re}$, $j_T = 2.78/\text{Re}$, and $f = 13.33/\text{Re}$. For gas flow through hexagonal passages ($80 < \text{Re} < 800$), the following are the experimental correlations [36]:

$$j = \frac{4.0}{\text{Re}} \qquad f = \frac{17}{\text{Re}} \tag{43}$$

The corresponding theoretical values from Table 5 are $j_{H1} = 4.51/\text{Re}$, $j_T = 3.76/\text{Re}$, and $f = 15.05/\text{Re}$.

b. Developing Laminar Flow

If the channel length is sufficiently short (such as in interrupted surfaces), it may be feasible to use the entrance length solutions to predict the mean Nu and f over the flow length. Shah and London [34] have summarized the thermal and hydrodynamic entry length solutions for a large number of practically important flow passage geometries.

The ratio $\text{Nu}_m/\text{Nu}_{fd}$ is shown in Fig. 24 for several channel geometries having a constant wall temperature boundary condition. Here the abscissa $x^* = x/(D_h \, \text{Re} \, \text{Pr})$ is a dimensionless length for the entrance region. Several important observations may be made from this figure: (1) The entrance-region Nusselt numbers and hence the heat transfer coefficients can be 2 to 3 times higher than the fully developed values depending upon the value of dimensionless interrupted length $x^* = l_{ef}^* = l_{ef}/(D_h \, \text{Re} \, \text{Pr})$.† (2) At $x^* \approx 0.1$, although the *local* Nusselt number approaches the fully developed value, the value of the *mean* Nusselt number can be significantly higher for a channel of length $x^* = 0.1$. For example, for a triangular tube of length $l_{ef}^* = 0.1$, while the *local* Nusselt number at $x^* = 0.1$ attains a value within 4 percent of the fully developed value [34], the *mean* Nusselt number Nu_m over the developing region is about 25 percent higher than Nu_{fd} as found in Fig. 24. Thus, observations 1 and 2 imply that the entrance length effect could be substantial and advantageous in obtaining high heat transfer rates. (3) The order of increasing $\text{Nu}_m/\text{Nu}_{fd}$ as a function of the channel shape at a given x^* in Fig. 24 is just the opposite of Nu_{fd} in Table 5. For a highly interrupted surface, this means a basic inferior duct shape for fully developed flow (such as triangular) will not be penalized in terms of low Nu or low h in developing flow. (4) A higher value of $\text{Nu}_m/\text{Nu}_{fd}$ at $x^* = 0.1$ means that the flow channel has a longer entrance region.

Shah and London [34] proposed the following correlations for thermal entrance solutions for both *circular and noncircular* channels having laminar developed velocity profiles and developing temperature profiles:

$$\text{Nu}_{x,T} = 0.427(f \, \text{Re})^{1/3}(x^*)^{-1/3} \tag{44}$$

$$\text{Nu}_{m,T} = 0.641(f \, \text{Re})^{1/3}(x^*)^{-1/3} \tag{45}$$

$$\text{Nu}_{x,H1} = 0.517(f \, \text{Re})^{1/3}(x^*)^{-1/3} \tag{46}$$

$$\text{Nu}_{m,H1} = 0.775(f \, \text{Re})^{1/3}(x^*)^{-1/3} \tag{47}$$

---

†For continuous uninterrupted flow passages, $l_{ef} = L_f$.

where f is the Fanning friction factor for fully developed flow for a given channel, Re is the Reynolds number, and $x^* = x/(D_h \, \text{Re} \, \text{Pr})$. For interrupted surfaces, $x = l_{ef}$. The above equations are recommended for $x^* < 0.001$, which represents the x^* range below that of Fig. 24. The functional relationship $\text{Nu} \propto (x^*)^{-1/3}$ of Eqs. (44) to (47) is quite useful for predicting the j data for a new surface if they are known for another surface of the same family and l_{ef}/D_h is known for both surfaces.

The following observations may be made from Eqs. (44) to (47) and the solutions in Ref. 34 for laminar flow surfaces having developing laminar flows: (1) The influence of thermal boundary conditions on the convective behavior appears to be of the same order as that for the fully developed flow. (2) Since Nu is proportional to $(x^*)^{-1/3} = [x/(D_h \, \text{Re} \, \text{Pr})]^{-1/3}$, Nu is proportional to $\text{Re}^{1/3}$, and hence h varies as $V^{1/3}$. (3) The influence of the duct shape on thermally developing Nu is not as great as that for the fully developed Nu.

For simultaneously developing laminar velocity and temperature profiles, the entrance region Nusselt numbers are theoretically even higher than those of Fig. 24. These solutions are based on uniform velocity and temperature profiles at the entrance of a channel, and no wake effects or secondary flow effects (which are present in interrupted surfaces) are included in the analysis. However, based on the experimental data, the compact interrupted surfaces *do not* achieve the higher heat transfer coefficients predicted for the simultaneously developing flows; the results of Fig. 24 or Eqs. (44) to (47) with developed velocity profiles are in better agreement with the experimental data for interrupted surfaces and hence those are recommended for design guidelines.

The analytical solutions for the friction factor in the developing laminar flow region have been summarized by Shah and London [34] for a circular tube, parallel plates, rectangular channels, isosceles triangular channels, and concentric and eccentric annular channels. A correlation for f_{app} Re factors, by combining the entrance region and fully developed region ΔP^*, is given by the following equation [34]:

$$f_{app} \, \text{Re} = \frac{3.44}{(x^+)^{1/2}} + \frac{K(\infty)/(4x^+) + f\,\text{Re} - 3.44(x^+)^{-0.5}}{1 + C'(x^+)^{-2}} \qquad (48)$$

Here f_{app} is the apparent Fanning friction factor that takes into account both the skin friction and the change in momentum rate in the hydrodynamic entrance region. It is based on the static pressure drop from $x = 0$ to L. The definitions for f_{app} and ΔP^* are

$$f_{app} \frac{4L}{D_h} = \frac{\Delta P}{G^2/2g_c\rho} = \Delta P^* \qquad (49)$$

$K(\infty)$, f Re, and C' of Eq. (48) for rectangular, triangular, and concentric annular channels are presented in Table 6. For the entrance region $x^+ < 0.001$, the first term on the right-hand side of Eq. (48) is dominant. Considering this term alone, and using the definitions of f_{app} from Eq. (49) and x^+, it can be shown that $\Delta P \propto V^{1.5}$ for the developing laminar boundary layer region.

Although Eq. (48) may provide some guidelines for friction factors for interrupted surfaces, it includes only the effect of skin friction. The form drag associated with the blunt (smooth or burred) edges of the surface also contributes to the pressure drop. Hence, analytical values of apparent friction factors are generally *not* used in designing interrupted surface compact exchangers. But, as a rule of thumb, $f \approx 4j$ may be used to predict f factors for interrupted surfaces for which the j factors are already known either from the theory or from experiments.

TABLE 6 $K(\infty)$, $f\,Re$, and C' for use in Eq. (48), from Shah and London [34]

	$K(\infty)$	$f\,Re$	C'
α^*	Rectangular channels		
1.00	1.43	14.227	0.00029
0.50	1.28	15.548	0.00021
0.20	0.931	19.071	0.000076
0.00	0.674	24.000	0.000029
2ϕ	Equilateral triangular channels		
60°	1.69	13.333	0.00053
r_i/r_o	Concentric annular channels		
0	1.25	16.000	0.000212
0.05	0.830	21.567	0.000050
0.10	0.784	22.343	0.000043
0.50	0.688	23.813	0.000032
0.75	0.678	22.967	0.000030
1.00	0.674	24.000	0.000029

c. Fully Developed Turbulent Flow

Turbulent flow is realized in only a limited number of compact heat exchanger applications. In such flow, the constant-property Nusselt number is independent of thermal boundary conditions for $Pr > 0.5$, but it is dependent upon both Re and Pr. For liquid metals ($Pr < 0.1$), the turbulent flow Nusselt number is also dependent upon the thermal boundary condition. The ratio of Nu_H/Nu_T is provided as a function of Re and Pr by Kays and Perkins [168].

Circular Tube. The constant-property experimental data for $Pr > 0.7$ are well correlated for a smooth circular tube by the following Gnielinski correlation [37, 26]:

$$Nu = \frac{(f/2)(Re - 1000)\,Pr}{1 + 12.7(f/2)^{1/2}(Pr^{2/3} - 1))} \tag{50}$$

where

$$f = (1.58 \ln Re - 3.28)^{-2} \tag{51}$$

or f is given by Eq. (56), (57), or (58). The Nusselt numbers predicted by Eq. (50) are within 10 percent of the experimental results for $10^4 < Re < 5 \times 10^6$ and $0.5 < Pr < 2000$. Equation (50) also predicts Nu satisfactorily for $2300 < Re < 10^4$. The Gnielinski correlation is a modification of the well-known Petukhov-Kirillov correlation [38] to cover a lower Reynolds number range down to about 2300. A simpler power-law form correlation by Sleicher and Rouse [39] that agrees within ± 10 percent with the Gnielinski correlation for $Re > 10^4$ is

$$Nu = 5 + 0.015\,Re^a\,Pr^b \tag{52}$$

where $a = 0.88 - \dfrac{0.24}{4 + \text{Pr}}$ (53a)

$\quad\; b = \frac{1}{3} + 0.5 \exp\,(-0.6\,\text{Pr})$ (53b)

Kays and Crawford [40] proposed a correlation for *gases* as

$$\text{Nu} = 0.022\,\text{Re}^{0.8}\,\text{Pr}^{0.5} \tag{54}$$

It agrees with the Gnielinski correlation within 0 to 4 percent for $\text{Re} \geq 5000$. The afore-mentioned correlations agree with accurate experimental data better than the classical Dittus-Boelter, Colburn, and other correlations [26].

The turbulent flow thermal entry length L_{th} for gases and liquids is almost independent of Re and thermal boundary conditions. L_{th}/D_h varies from 8 to 15 for air and is less than 3 for liquids [26]! Hence, if the entry length effect is small, it may be ignored for conservatism. Otherwise, the following Gnielinski correlation for the thermal entry length Nusselt number Nu_m is recommended:

$$\text{Nu}_m = \text{Nu}_{fd}\left[1 + \left(\frac{L}{D_h}\right)^{-2/3}\right] \tag{55}$$

Here Nu_{fd} represents the fully developed Nusselt number, such as that of Eq. (50) or (52). Since this equation does not incorporate any effect of Pr, it should be used with caution.

The classical and standard correlation for Fanning friction factor for fully developed turbulent flow through a *smooth* circular tube is the Kármán-Nikuradse correlation as follows:

$$\frac{1}{\sqrt{f}} = 4.0 \log_{10}\,(\text{Re}\,\sqrt{f}) - 0.4 \tag{56}$$

or

$$\frac{1}{\sqrt{4f}} = 2.0 \log_{10}\,(\text{Re}\,\sqrt{4f}) - 0.8 \tag{57}$$

This semitheoretical correlation agrees with the experimental data within ± 2 percent for $4 \times 10^3 \leq \text{Re} \leq 3 \times 10^6$. Since it is theoretically based, it may be extrapolated to arbitrarily large Reynolds numbers. Because Eq. (56) or (57) is implicit in f, several explicit correlations have been proposed in the literature and are summarized in Ref. 26. The Filonenko correlation, Eq. (51), agrees with Eq. (56) within ± 0.5 percent for $3 \times 10^4 \leq \text{Re} \leq 10^7$ and within 1.8 percent at $\text{Re} = 10^4$. Other simple equations are

$$f = \begin{cases} 0.079\,\text{Re}^{-0.25} & \text{for } 4 \times 10^3 \leq \text{Re} \leq 10^5 \\ 0.046\,\text{Re}^{-0.2} & \text{for } 3 \times 10^4 \leq \text{Re} \leq 10^6 \end{cases}$$

$\qquad\qquad\qquad\qquad\qquad\qquad\qquad\qquad\qquad\qquad$ (58a)
$\qquad\qquad\qquad\qquad\qquad\qquad\qquad\qquad\qquad\qquad$ (58b)

The f factor from Eq. (58a) agrees with those from Eq. (56) within ± 2.5 percent, and f factors from Eq. (58b) agree with those from Eq. (56) within $+2.2$ percent and -0.4 percent for the indicated range of Re.

For fully developed laminar flow, $f = 16/\text{Re}$ for $\text{Re} < 2300$. The foregoing Eq. (56) or (58a) is valid for fully developed turbulent flow, for $\text{Re} \geq 4000$. The flow is transitional for $2300 < \text{Re} < 4000$. Since the experimental friction factors for this Re range fall roughly on a straight line on a log-log scale [41], they may be approximated by

$$f = 0.000133\,\text{Re}^{0.52} \qquad \text{for} \qquad 2300 < \text{Re} < 4000 \tag{59}$$

Noncircular Ducts.   A careful observation of accurate experimental friction factors for all noncircular smooth ducts reveals that ducts with laminar f Re < 16 have turbulent f factors lower than those for the circular tube; whereas ducts with laminar f Re > 16 also have turbulent f factors higher than those for the circular tube. For most noncircular smooth ducts, the accurate constant-property experimental friction factors (pressure drop) are within ± 10 percent of those predicted using Eq. (56) with the hydraulic diameter as a characteristic dimension. The constant-property experimental Nusselt numbers are also within ± 10 to 15 percent except for some rod bundle geometries and sharp-cornered and narrow ducts. This order of accuracy is adequate for most engineering calculations for overall heat transfer and pressure drop, although it may not be adequate for detailed flow distribution and local temperature distribution as required, for example, in a nuclear reactor.

Several generalized correlations for Nu and f for noncircular ducts and specific correlations for f factors for rectangular, triangular, elliptical, and concentric annular ducts are available in the open literature [26]. Some of these are summarized in Ref. 168. These correlations provide an indication of the magnitude of the secondary influence of the flow passage geometry on Nu and f relative to the circular tube correlations, using D_h as the characteristic dimension.

Some information on the critical Reynolds number, flow friction, and heat transfer in transition flow for noncircular ducts is provided in Ref. 168.

3. Analytical Solutions and Correlations for Extended-Surface Geometries

As mentioned on p. 4-205, compact exchanger surface basic data are usually presented in terms of j and f versus Re. In these groups, directly or indirectly the hydraulic diameter D_h is used arbitrarily but consistently as a characteristic dimension. Although the j and f data will be applicable to any geometrically similar surface, as soon as the basic geometry is changed (such as an offset strip-fin geometry changed from 400 to 600 fins per meter), the original j and f data are no longer applicable. This is because D_h is not a universal characteristic dimension. A great variety of compact surface geometries are available, and generally these different geometries will have different j and f data. Such surface basic data have been provided by Kays and London [16] for over 90 different surface geometries, summarized in Table 7. These surface basic data will not be presented here because of their ready availability and the present space limitations. Some later published data are summarized in the following sections on plate-fin and tube-fin geometries [42]:

a. Plate-Fin Surfaces

A partial list of plate-fin surface basic data or correlations since 1964 [16] is as follows: plain fins [43], wavy fins [44], offset strip fins [45–54], louver fins [55, 56], perforated fins [30, 32, 57–61], and pin fins [62, 63]. These fin geometries are shown in Fig. 2. A brief description and discussion based on Ref. 42 follows.

Plain Fins.   These are straight fins that are uninterrupted (uncut) in the fluid flow direction. Although triangular and rectangular passages are more common, any desired complex shape can be formed depending upon how the fin material is folded. While the

TABLE 7 A Summary of the Number of Cores for Which Surface Basic Data Have Been Reported by Kays and London [16]

Surface geometry	Number of cores
Flow inside circular/rectangular tubes	7
Flow normal to circular tube banks	13
Flow normal to flat tube banks	2
Plate-fin surfaces:	
Plain fins	19
Louver fins	14
Offset strip fins	13
Wavy fins	3
Perforated fins	1
Finned circular tube banks	10
Finned flat tube banks	5
Regenerative surfaces	5

triangular (corrugated) fin is less expensive, can be manufactured at high speeds, and has an added flexibility of adjusting the fin pitch, it is generally not structurally as strong as the rectangular fin for the same passage size and fin thickness. Also, the triangular fins have lower j, f, and j/f factors compared to those for rectangular fins, particularly in laminar flow.

Surface basic data for rectangular and triangular plain plate-fin geometries are available in Ref. 16. Analytical solutions exist for most plate-fin geometries in the laminar flow region as discussed in the preceding section. These solutions provide valuable guidelines in laminar flow when no experimental results are available. In the turbulent flow region, the hydraulic diameter is the significant characteristic dimension. One may predict the turbulent region j and f data for plain fin geometries using the circular tube correlations with the hydraulic diameter as a characteristic dimension [26]. This procedure will generally yield good results, except for isosceles triangular channels having a very small apex angle. The transition region characteristics are unpredictable. No theories or correlations exist that would predict the transition region j and f data. In most cases, the experimental results are the only reliable sources.

Plain fins are used in those applications where the pressure drop is very critical and the augmented interrupted surfaces cannot meet the design requirements of a fixed frontal area. Also, plain fins are preferred for very low Reynolds number applications. This is because in interrupted fins, when the flow approaches the fully developed state at such low Re, the advantage of the high h of the interrupted fins is diminished. Plain fins are also preferred for high Reynolds number applications where the ΔP for interrupted fins become excessively high.

Wavy Fins. These are also uncut surfaces in the flow direction, and have cross-sectional shapes similar to plain surfaces. However, they are wavy in the flow direction, while the plain fins are straight in the flow direction (see Fig. 2). The wave form in the flow direction provides effective interruptions to flow and induces very complex flows.

No specific correlations or prediction methods exist for the j and f data for wavy or herringbone fins. The only data available in the literature are for three wavy fin geometries [16]. Goldstein and Sparrow [44] used a mass transfer technique to measure the local Sherwood number Sh† for a wavy fin having two complete waves, a wave angle of 21°, a spacing between fins of 1.65 mm, and a total horizontal (projected) fin length in the flow direction of 18.5 mm. They measured local and average distributions of the Sherwood number and identified complex flow phenomena. They found that the enhancement in the transfer coefficients due to the waviness of the wall was small at low Re (about 25 percent at Re = 1000), but was significant in the low Re turbulent regime (about 200 percent at Re = 6000 to 8000). The enhancement is due to Goertler vortices, which form as the fluid passes over the concave wave surfaces. These are counterrotating vortices, which produce a corkscrewlike flow pattern.

No performance or fouling comparisons are available between wavy fins and interrupted fins such as offset strip or louver fins. It appears that offset strip or louver fins are superior since these fins have replaced wavy fins, which were extensively used in industry for a long period.

Offset Strip Fins.‡ This is one of the most widely used enhanced fin geometries in compact heat exchangers. The fin has a rectangular cross section, it is cut into small strips of length l_{ef}, and every alternate strip is displaced (offset) by about 50 percent of the fin pitch in the transverse direction (see Fig. 2). In addition to the fin spacing and fin height, the major variables are the fin thickness and fin strip length in the flow direction.

In addition to the surface basic data [45–48], considerable work has been reported in the open literature on the generalized correlation and analysis of local performance. The information will be summarized now. Note that a typical strip length is 3.2 mm (1/8 in), and typical design Reynolds numbers (based on the strip length) are well within the laminar region.

Wieting [49] correlated available experimental heat transfer and flow friction data for 22 offset strip fin surfaces as follows:

For Re ≤ 1000:

$$f = 7.661\alpha^{*-0.092}\left(\frac{l_{ef}}{D_h}\right)^{-0.384} \text{Re}^{-0.712} \tag{60}$$

$$j = 0.483\alpha^{*-0.184}\left(\frac{l_{ef}}{D_h}\right)^{-0.162} \text{Re}^{-0.536} \tag{61}$$

For Re ≥ 2000:

$$f = 1.136\left(\frac{l_{ef}}{D_h}\right)^{-0.781}\left(\frac{\delta}{D_h}\right)^{0.534} \text{Re}^{-0.198} \tag{62}$$

$$j = 0.242\left(\frac{l_{ef}}{D_h}\right)^{-0.322}\left(\frac{\delta}{D_h}\right)^{0.089} \text{Re}^{-0.368} \tag{63}$$

Here l_{ef} is the strip length or interrupted fin flow length, δ is the fin thickness, D_h is the hydraulic diameter of the passages, and α^* is the ratio of width to height of the passage. The following ranges were covered: $0.7 \leq l_{ef}/D_h \leq 5.6$, $0.030 \leq \delta/D_h \leq 0.166$, $0.162 \leq \alpha^* \leq 1.196$, and $0.65 \leq D_h \leq 3.41$ mm. Although 85 percent of all data are correlated within ±15 percent for f and ±10 percent for j, a few data have a maximum

---

†The Sherwood number for the convective mass transfer problem is analogous to the Nusselt number for the convective heat transfer problem.

‡This fin is also sometimes referred to as a serrated, lance-offset, segmented, multientry, or step fin.

discrepancy as high as 40 percent. The deviation is probably influenced by burrs on the leading and trailing edges of the fin, whose effect is not considered in the correlation.

To obtain the f and j factors for a transitional Reynolds number, Wieting suggested the following procedure. Determine the reference Reynolds number for f and j from the following equations:

$$\text{Re}_f^* = 41\alpha^{*-0.179}\left(\frac{l_{ef}}{D_h}\right)^{0.772}\left(\frac{\delta}{D_h}\right)^{-1.04} \tag{64}$$

$$\text{Re}_j^* = 61.9\alpha^{*-1.1}\left(\frac{l_{ef}}{D_h}\right)^{0.952}\left(\frac{\delta}{D_h}\right)^{-0.53} \tag{65}$$

Here Re_f^* is the Reynolds number at the intersection point of the two f versus Re curves, one for $\text{Re} \leq 1000$, and the other for $\text{Re} \geq 2000$. Similarly, Re_j^* is the Reynolds number at the intersection point of the two j versus Re curves, one for $\text{Re} \leq 1000$, and the other for $\text{Re} \geq 2000$. If the Reynolds number of interest Re is lower than Re_f^*, use Eq. (60) for f; otherwise use Eq. (62). If $\text{Re} < \text{Re}_j^*$, use Eq. (61) for j; otherwise use Eq. (63). It should be emphasized that the Wieting correlation is strictly based on a limited amount of reported test data. It does not have a "burr" parameter for burrs on the leading and trailing edges. Care must be exercised in extrapolating data for fin geometries that have geometrical parameters outside the range of those for the correlations.

Recently, Webb and Joshi [50] have provided an improved correlation for friction factors for the offset strip fins based on a power-law linear multiple regression analysis of the test data from their eight scaled-up offset strip-fin cores. Since there is no theoretical basis for such a correlation, Webb and Joshi [51] obtained another refined correlation based on the method of Churchill and Usagi [52]. In this approach, two asymptotic values of f are needed; they correspond to only the skin friction contribution at $\text{Re} \to 0$ (from the fully developed laminar flow solutions for rectangular passages) and to only the form drag contribution at $\text{Re} \to \infty$ with the drag coefficient as 0.8. The appropriate coefficients were computed based on their test data from eight scaled-up offset strip fin cores. The correlation is

$$f = \frac{(1 - \delta/s')(1 + \alpha^*)}{\text{Re}\,(1 + \delta/s')(1 + \alpha^* + \delta/l_{ef})^2}\left\{[K(\alpha^*)]^n\right.$$
$$\left. + \left[\frac{0.4\,\text{Re}\,(1 + \alpha^* + \delta/l_{ef})}{1 + \alpha^*}\frac{\delta}{l_{ef}}\right]^n\right\}^{1/n} \tag{66}$$

where

$$n = 0.787\left(\frac{l_{ef}}{D_h}\right)^{0.153}\left(\frac{\delta}{l_{ef}}\right)^{0.178}(\alpha^*)^{-0.046} \tag{67}$$

and

$$K(\alpha^*) = 24(1 - 1.3553\alpha^* + 1.9467\alpha^{*2}$$
$$- 1.7012\alpha^{*3} + 0.9564\alpha^{*4} - 0.2537\alpha^{*5}) \tag{67a}$$

This correlation is based on the following ranges of geometrical parameters: $0.94 \leq l_{ef}/D_h \leq 4.17$, $0.054 \leq \delta/D_h \leq 0.148$, $0.11 \leq \alpha^* \leq 0.25$, and $6.1 \leq D_h \leq 13.5$ mm. In this correlation, the minimum free flow area A_o occurs within the core and the heat transfer surface area includes the surface area of the edges of the strip fins. A method to compute A_o and A is outlined on p. 4-254. The friction factors from this correlation agree with 99 percent of their data within ± 15 percent. The data for $\alpha^* < 0.25$ from Kays and London [16] and London and Shah [45] are predicted within ± 17 percent for 500 $<$ Re $<$ 3000. Predictions for $\alpha^* > 0.25$ show increasing error as α^* increases. Hence, the correlation is not recommended for use for $\alpha^* > 0.25$. Until a better correlation is available, the Wieting correlation is recommended for j factors for offset strip fins.

Several flow visualization and heat and mass transfer basic studies have been conducted on a few scaled-up strip fins [50, 53]. A simple model of heat transfer enhancement is the periodic growth and destruction of laminar boundary layers over each strip. However, the boundary layers are not completely destroyed within the wake region. Some details on the modeling and results are provided in Ref. 42.

Some advances in the numerical prediction of j and f data for offset strip-fin geometries have been made and a partial success in terms of trends has been achieved [54].

A careful examination of all good data that are published has revealed the ratio j/f ≤ 0.25 for strip fin, louver fin, and other, similar interrupted surfaces. This can be approximately justified as follows. The flow is developing along each interruption in such a surface. Based on the Reynolds analogy for flow over a flat plate, in the absence of form drag, j/f should be 0.5 for Pr ≈ 1. Since the contribution of form drag is of the same order of magnitude as the skin friction for such an interrupted surface, j/f will be about 0.25. Published data for strip and louver fins are questionable if $j/f > 0.3$, and such is the case for the results of Mochizuki and Yagi [48], who measured j/f up to 0.6 (their f factors appeared to be too low). All pressure and temperature measurements and possible sources of flow leaks and heat losses must be checked thoroughly for all those basic data having $j/f > 0.3$ for strip and louver fins.

The heat transfer coefficients for the offset strip fins are 1.5 to 4 times higher than those of plain fin geometries. The corresponding friction factors are also high. The ratio of j/f for an offset strip fin to j/f for a plain fin is about 80 percent. If properly designed, the offset strip fin would require substantially lower heat transfer surface area than that of plain fins at the same ΔP, but about 10 percent larger flow area [refer to Eq. (77)]. The heat transfer enhancement for an offset strip fin is caused mainly by the developing laminar boundary layers for Re $< 10,000$ (or Re$_l$ = Re $l_{ef}/D_h < 100,000$ and $l_{ef}/D_h \leq$ 10). The thick strip fins provide further enhancement in h due to the fin blockage effect that results in a high effective fluid velocity between the fins. The friction factors are also high because of the developing boundary layers over the strips and increased form drag due to the finite thickness of the fins. The shorter the strip length or the larger the fin thickness, the higher the heat transfer enhancement over plain fins.

Offset strip fins are used in the approximate Re range of 500 to 10,000, where the enhancement over plain fins is significant. For specified heat transfer and pressure drop requirements, the offset strip fin requires a somewhat higher frontal area than a plain fin, but a shorter flow length and lower overall volume. Offset strip fins are extensively used by industry in many compact heat exchanger applications.

Louver Fins.   Louvers are formed by cutting the metal and either turning, bending, or pushing out the cut elements from the plane of the base metal. Louvers can be made in many different forms and shapes, some of which are shown in Fig. 2. Note that the parallel louver fin and offset strip fin both have small strips aligned parallel to the flow. The strips of parallel louvers are not pushed out as far as those of the offset strip fin, and they do not extend to the full height of the fin; the strips of offset strip fins do extend to the full height of the fin. The louver fin gauge is generally thinner compared to the offset strip fin. The louver width l_{ef} in the flow direction for modern high-performing louver fin geometries is significantly shorter than the offset strip fin length l_{ef}. The louver fin base metal has a triangular (or corrugated) shape. It is generally not as strong structurally as the offset strip fin for the equivalent fin geometry, since the latter has a "large" flat area for brazing, thus providing strength. Louver fins may have a slightly higher potential for fouling than offset strip fins.

Louver fins are amenable to high speed mass production manufacturing technology, and as a result are less expensive compared to offset strip fins and other interrupted fins when produced in very large quantities. Desired fin spacing can be achieved by squeezing or stretching the fin; hence, the louver fin allows some flexibility in fin spacing, without changes in the tools and dies. This flexibility is not possible with the offset strip fin. A wide range in performance can be achieved by changing the louver angle and louver width l_{ef}. These fins are extensively used in automotive heat exchangers.

Various louver fins have been tested extensively, and a considerable amount of proprietary data exists that is unavailable in the open literature. The surface basic data on early forms of 14 louver fin geometries are presented by Kays and London [16]. The multilouver fin (see Fig. 2) has the highest heat transfer enhancement relative to the pressure drop in comparison with most other louver forms, and hence it is now extensively used in industry. Some results for multilouver fins have been reported by Mori and Nakayama [55]. However, unfortunately, they are in dimensional form with insufficient details. Recently, Davenport [56] reported data for eight multilouver fin geometries that are presented in Figs. 25 and 26 with the geometrical information in Table 8.

Modern high-performing multilouver fins have heat transfer coefficients 1.3 to 1.8 times higher than those of high-performing offset strip fins. Also, the friction factors are

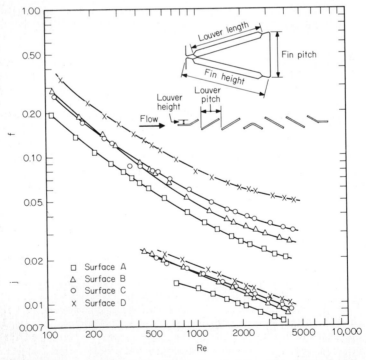

FIG. 25 Heat transfer and flow friction characteristics of four louver fin surfaces having fin height of 12.7 mm [56].

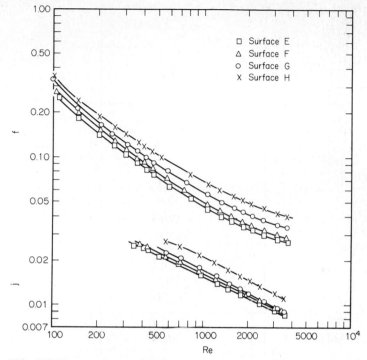

FIG. 26 Heat transfer and flow friction characteristics of four louver fin surfaces having fin height of 7.8 mm [56].

2 to 2.5 times higher than those of the offset strip fins. If properly designed, the multilouver fins require lower surface area than offset strip fins at the same pressure drop.

The operating Reynolds number range for louver fins is 100 to 5000, depending upon the type of louver geometry employed. Although more form drag may be associated with the louver fin compared to an offset strip fin, the performance of a well-designed multilouver fin exchanger can approach that of an offset strip exchanger with possibly increased surface compactness and reduced manufacturing cost.

Perforated Fins.   A perforated fin has either round or rectangular perforations with their size, shape, and longitudinal and transverse spacings as major perforation variables. The perforated fin has either triangular or rectangular flow passages. When used as a plate-fin surface, generally it is brazed.

After the pioneering study on a perforated fin, Kays [64, 16] claimed that perforated heat exchanger surfaces have a substantial increase in heat transfer performance without introducing a pronounced form drag due to frequent boundary layer interruptions. During the 15 years following that study, seven groups of investigators tested 68 perforated cores for heat transfer, pressure drop, flow phenomena, noise, and vibration. All these results are summarized by Shah [58, 59]. The j and f data for seven perforated surfaces are presented in Figs. 27 and 28 with the geometrical properties in Table 9. Subsequently, some additional results have been reported [60, 61].

TABLE 8 Geometrical Properties of Louver Fin Surfaces of Figs. 25 and 26 [56]

Surface desig-nation	Fin height, mm†	γ, fins per meter	Fin thickness δ, mm‡	Louver pitch l_{ef}, mm†	Louver length, mm†	Louver height, mm†	$\dfrac{A_n}{A_{fr}}$	$\dfrac{A_f}{A}$	α, m²/m³‡	D_h, mm	Plate spacing b, mm	$\dfrac{l_{ef}}{D_h}$	$\dfrac{L^\ddagger}{D_h}$
A	12.7	641.0	0.075	3.0	9.5	0.29	0.816	0.891	1248	2.80	12.6	1.1	14.3
B	12.7	641.0	0.075	2.25	9.5	0.31	0.816	0.891	1248	2.80	12.6	0.80	14.3
C	12.7	645.2	0.075	1.8	9.5	0.29	0.816	0.892	1255	2.78	12.6	0.65	14.4
D	12.7	630.9	0.075	1.5	9.5	0.29	0.817	0.890	1229	2.84	12.6	0.53	14.1
E	7.8	651.5	0.075	3.0	7.1	0.36	0.749	0.837	1239	2.61	7.65	1.1	15.3
F	7.8	662.3	0.075	2.25	7.1	0.34	0.748	0.839	1257	2.58	7.65	0.87	15.5
G	7.8	651.5	0.075	1.8	7.1	0.31	0.749	0.837	1239	2.61	7.65	0.69	15.3
H	7.8	645.2	0.075	1.5	7.1	0.33	0.749	0.836	1229	2.64	7.64	0.57	15.1

†See insert in Fig. 25 for the definitions of louver height, pitch, length, and fin height. Fins/m = 2/fin pitch, m.

‡All cores were of $L = 40$ mm long in the airflow direction. The frontal area A_{fr} was for the whole core including that due to water tubes.

The conversion units are: 1 mm = 0.03937 in and 1 m²/m³ = 0.3048 ft²/ft³.

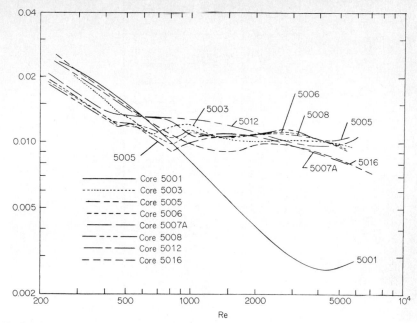

FIG. 27 Heat transfer characteristics of round-hole perforated surfaces [59].

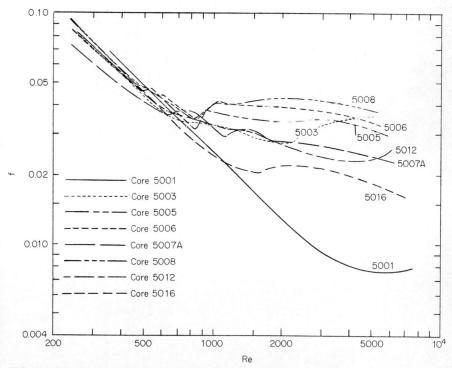

FIG. 28 Flow friction characteristics of round-hole perforated surfaces [59].

TABLE 9 Geometrical Properties of Perforated Plate Surfaces of Figs. 27 and 28 [59]†

Surface desig-nation	Plate spacing b, mm	Plate or fin thickness δ, mm	Porosity σ	α, m²/m³	D_h, mm	Hole diameter d, mm	$\frac{d}{\delta}$	% perforated plate open area	$\frac{X_t}{D_h}$	$\frac{b}{\delta}$	$\frac{b}{d}$
5001	1.57	0.38	0.762	958	3.10	0	0	0	-	4.1	—
5003	1.57	0.62	0.681	858	3.10	0.79	1.3	16.1	0.3	2.6	2.0
5005	1.59	0.38	0.764	959	3.12	1.59	4.2	22.7	0.9	4.2	1.0
5006	1.59	0.38	0.767	967	3.14	1.59	4.2	30.7	0.8	4.2	1.0
5007A	1.61	0.36	0.773	962	3.14	3.18	8.8	40.3	1.3	4.5	0.5
5008	1.59	0.38	0.765	951	3.15	1.59	4.2	29.6	0.4	4.2	1.0
5012	1.57	0.38	0.764	967	3.09	0.51	1.3	16.0	0.2	4.1	3.1
5016	1.58	0.38	0.763	958	3.12	1.59	4.2	5.7	1.0	4.1	1.0

†In all of the test cores, the perforated plates were stacked parallel to each other and were separated by three phenolic plastic spacers between plates to achieve the desired plate spacing. The j and f data were obtained by the transient single-blow test technique. All cores had $L/D_h \approx 98$, and the aspect ratio of rectangular flow passages was 46. The conversion units are: 1 mm = 0.03937 in, 1 m²/m³ = 0.3048 ft²/ft³.

Conclusions of Shah are: Microscopic perforations ($d_{\text{hole}} \leq 0.8$ mm) may provide enhanced heat transfer in the laminar region, if the plate open area due to perforations is greater than 20 percent. Macroscopic perforations ($d_{\text{hole}} > 1$ mm) do not enhance performance in the laminar region. Perforations induce early transition. However, the transition region j and f versus Re characteristics are quite complex and unpredictable. Contrary to expectations, an increase in f is often higher than an increase in j. The turbulent flow j and f data are usually much higher than those of unperforated surfaces, and are usually associated with noise and vibration in transition and turbulent flow regimes if the fins are not rigidly attached (brazed) to the plates or parting sheets. Rectangular slot perforations yield slightly better performance compared to round perforations. Since the punched-out material is wasted and eliminates surface area, the performance of the perforated surfaces is not superior as was originally claimed. Offset strip fins and multilouver fins are higher-performing surfaces, and are preferred.

Perforated fins are now used in only a limited number of applications, such as "turbulators" in oil coolers. Perforated fins were once used in two-phase cryogenic air separation exchangers, but now they have been replaced by the offset strip fins.

Pin Fins.   From the foregoing discussion, we know that the shorter the strip length is, the higher the enhancement characteristics will be. From these viewpoints, a limiting strip fin geometry with good mechanical strength is a pin fin geometry (see Fig. 2). Hence, in a pin fin exchanger, a large number of small pins are sandwiched between plates in either an inline or a staggered arrangement. Pins may have a round, an elliptical, or a rectangular cross section. Pin fins can be manufactured at a very high speed continuously from a wire of proper diameter. After the wire is formed into rectangular passages, the top and bottom horizontal wire portions are flattened for brazing or soldering to the plates.

Because of the potential for high enhancement characteristics of pin fins in plate-fin exchangers, they have been investigated extensively by industry. However, only a limited amount of data is published. The j and f data for five round pin fin geometries and one elliptical pin fin geometry (all but one inline) are presented by Kays and London [16]. The range of the aspect ratios covered for round pins was $6 \leq b/d_o \leq 24$, where b is the pin length (plate spacing) and d_o is the pin diameter. Theoclitus [62] presented j and f data for nine pin fin geometries with inline square pin spacing and $4 < b/d_o < 12$.

Pin fins may be considered to be tube banks having very small diameter tubes. As a result, low Reynolds number correlations may provide a first-order prediction of the j and f characteristics of the pin fin geometry. Indeed, the friction factor results of Theoclitus [62] "bracketed" the tube bank correlations. However, the heat transfer results were 20 to 35 percent lower for pin fins than predicted by the tube bank correlations that have "long" tubes.

In recent years, pin fins (short circular cylinders) have been used in internally cooled gas turbine engine blades of airfoil shape. The pin length-to-diameter ratio b/d_o is typically of the order of unity in this application. Extensive work has been reported on these short fins, $b/d_o \leq 2$, and has been summarized by Metzger et al. [63]. The overall heat transfer performance for these short pin fins is considerably lower than that for the long cylinders used in tube banks. However, the friction factors are in reasonably good agreement with those for long cylinder tube banks.

Pin fins have not become widely used in compact exchangers for the following reasons: (1) The compactness β achieved by the pin fin geometry is much less than that by multilouver fins or offset strip fins. (2) The cost of wire is considerably higher than that of a thin strip per unit surface area. (3) From manufacturing considerations, only the inline arrangement of the pins is desirable. This arrangement may not yield high heat transfer coefficients. Also, parasitic form drag is associated with the flow normal to the pins. (4) Due to vortex shedding behind the pins, noise and flow-induced vibration are possible [62], which are generally not acceptable in most heat exchangers.

The potential application of pin fins is at very low flows (Re < 500), for which the pressure drop is generally negligible. Pin fins are used as electronic cooling devices generally with free-convection flow on the pin fin side.

b. Tube-Fin Surfaces

Heat transfer and flow friction data are summarized separately for two types of tube-fin surfaces: continuous fins on a tube array, and individually finned tubes. A more detailed review on the performance of tube-fin surfaces has been provided by Webb [65] and Shah and Webb [42].

Continuous Fins on a Tube Array. This type of tube-fin geometry (shown in Figs. 3a and 4) is most commonly used in air conditioning and refrigeration exchangers† in which high pressure needs to be contained on the refrigerant side. As mentioned earlier, even though this type of tube-fin geometry is not as compact (having high α) as the plate-fin geometries, its use is becoming widespread in the current energy conservation era. This is because the bond between the fin and the tube is readily made by mechanically or hydraulically expanding the tube against the fin. Formation of this mechanical bond thus requires a very small energy consumption compared to the energy required to solder, braze, or weld the fin to the tube. Because of the mechanical bond, the applications are restricted to those cases in which the differential thermal expansion between the tube and fin material is small, and preferably the tube expansion is greater than the fin expansion. Otherwise, the loosened bond may have a significant thermal resistance.

The flow structure within a finned tube bank is more complex than that in the previously discussed plate-fin channels. The presence of a circular tube causes flow acceleration over the fin surface and flow separation on the back side of the tube which may yield low velocity wake regions. Secondary flows may be associated with wavy fins, while flow separation and reattachment may be associated with the louver fins. The performance of plain, wavy, and louver fins is discussed separately next.

Plain Fins on a Round Tube Array. Many geometric variables (d_o, X_t, $X_{l,s}$, L) are associated with even plain fins, and the flow structure is too complex to permit analytical predictions of the heat transfer and friction characteristics. Rich [66] obtained heat transfer and friction data for four row deep plain fin coils with the fin density varying from 115 to 811 fins per meter (2.92 to 20.6 fins per inch). The j and f data for eight plain fin surfaces are presented in Fig. 29 with the geometrical properties in Table 10. The friction factors for surfaces 7 and 8 of this figure may be questionable, since these surfaces show smaller j/f values than for the other fin spacings at low Re. Normally, the j/f ratio will increase as the fin spacing is decreased because the fractional parasitic drag associated with the tube is reduced.

In a later study, Rich [67] used the same exchanger geometry with 571 fins per meter (14.5 fins per inch) to determine the effect of the number of tube rows on the j factor. The geometrical properties of these cores are the same as those of the 571 fins per meter core in Table 10. Figure 30 shows the j factors (faired data fit) for each coil as a function

---

†These exchangers are simply referred to as "coils."

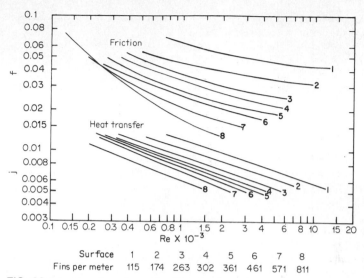

Surface	1	2	3	4	5	6	7	8
Fins per meter	115	174	263	302	361	461	571	811

FIG. 29 Heat transfer and pressure drop characteristics of continuous plain fin surfaces (of the type of Fig. 3a) [66].

of Re_l. The numbers on the figure indicate the number of tube rows in each coil. Since the original data for coils with three or more rows had a rollover in j for $Re_l \lesssim 6000$, those data were reprocessed and straightened out, and these corrected data are plotted in Fig. 30 [42].

Saboya and Sparrow [68, 69, 7] have used a mass transfer technique to establish the local mass transfer coefficient distribution on one-, two-, and three-row coils having a staggered tube arrangement. The results were interpreted in terms of heat transfer through the heat and mass transfer analogy. They employed the following geometry: dis-

TABLE 10 Geometrical Properties of Continuous Plain Fin Surfaces of Fig. 29 [66]†

Surface designation	Fins per meter (fins per inch)	Hydraulic diameter, mm (ft)	Ratio of free flow to frontal area, σ	Ratio of heat transfer area to total volume, α, m^2/m^3 (ft^2/ft^3)	Ratio of fin area to total area
1	115 (2.92)	9.63 (0.0316)	0.577	240 (73.3)	0.806
2	174 (4.42)	6.77 (0.0222)	0.566	335 (102)	0.862
3	263 (6.67)	4.63 (0.0152)	0.558	482 (147)	0.905
4	302 (7.67)	4.00 (0.01312)	0.555	554 (169)	0.919
5	361 (9.17)	3.39 (0.01112)	0.550	650 (198)	0.932
6	461 (11.7)	2.78 (0.00912)	0.543	781 (238)	0.943
7	571 (14.5)	2.13 (0.00700)	0.536	1004 (306)	0.957
8	811 (20.6)	1.46 (0.00478)	0.510	1401 (427)	0.972

†All the cores had electrolytic tough pitch copper fins of 0.15-mm (0.006-in) thickness and 110.5-mm (4.35-in) flow depth. All cores had four rows of 12.7 mm (0.50 in) outside-diameter tubes (40 tubes in total) in staggered arrangement with $X_t = 31.8$ mm (1.25 in) and $X_l = 27.5$ mm (1.083 in). In the data reduction, a constant value of the extended surface efficiency η_o was used; this was obtained by averaging η_o for the range of air flow rates for each coil.

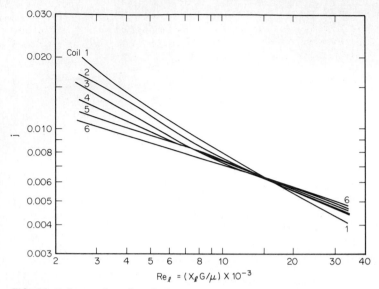

FIG. 30 Influence of number of tube rows on heat transfer characteristics of continuous plain fins on staggered round tubes.

tance between fins 1.65 mm (0.065 in),† tube outside diameter 8.53 mm (0.336 in), transverse tube spacing 21.3 mm (0.839 in), and longitudinal tube spacing 18.5 mm (0.728 in). They covered the Reynolds number range of 160 to 1270. They found different transfer mechanisms operating in different portions of the fin. For the fin associated with the first tube row, two factors provide high heat transfer rates: the boundary layer on the forward part (leading edge) of the fin, and a vortex system on the fin portion in front and at the sides of the tube. Low heat transfer coefficients are found on the fins downstream of the minimum flow cross section at the first tube row. The boundary layer development is the most important factor for the first-row fin, with the vortex-induced transfer mechanism becoming an important factor at higher Reynolds number ($\sim$1000 to 1200). For the second- and third-row fins, there is no region of boundary layer development; the vortex system alone is responsible for high heat transfer. At low Reynolds numbers (Re $\approx$ 200), the fin associated with the first row transfers about 50 percent of the total heat, the fin with the second row 28 percent, and the fin with the third row 22 percent. As the Reynolds number increases, stronger vortices are activated, and the relative contributions of the second- and third-row fins increase. At Re $\approx$ 1100, the respective heat transfers by individual fins of the rows of a three-row coil are nearly equal.

McQuiston [70] developed the following simple empirical correlation for four-row staggered banks of plain fins based on his data [71, 72], Rich's data [66], and the Kays-London data [16], a total of 17 cores [42]:

$$j = 0.0014 + 0.2618\ \text{Re}_d^{-0.4} \left(\frac{A}{A_t}\right)^{-0.15} \tag{68}$$

†For an assumed fin thickness of 0.15 mm (0.006 in), this translates into 556 fins per meter (14.12 fins per inch).

where $\text{Re}_d = Gd_o/\mu$ and d_o is the tube outside diameter, A is the total (fin + tube) air-side surface area, and A_t is the outside area of the bare tubes (without the fins). Ninety percent of data were correlated within ± 10 percent. McQuiston also provided the following correlation for Fanning friction factor f based on D_h:

$$f = 0.004904 + 1.382\ \text{Re}_d^{-0.50} \left(\frac{r_o}{r^*}\right)^{0.50} \left[\frac{(X_t - d_o)\gamma}{4(1 - \gamma\delta)}\right]^{-0.8} \left(\frac{X_t}{2r^*} - 1\right)^{-1} \quad (69)$$

where r^* is defined by

$$\frac{r^*}{r_o} = \frac{A/A_t}{(X_t - d_o)\gamma + 1} \quad (70)$$

and γ is fin density and δ is the fin thickness. Equation (69) correlated the data of the same 17 coil geometries within ± 35 percent.

Krückels and Kottke [73] used a mass transfer technique to establish the local heat transfer coefficient distribution on plain fins on two-row inline tubes. They found a low-velocity wake region between the tube rows and low heat transfer coefficients in this region as expected.

There appear to be no surface basic data for multirow plain fin coils having an inline tube arrangement. This is because the flow is bypassed between the tube rows in the flow direction (the tube bypass effect), which substantially degrades the performance of an inline tube arrangement. This flow bypassing has been shown by Fukui and Sakamoto [74] through flow visualization. The degree of performance degradation has been established for only circular fins, as will be seen on p. 4-240.

Plain continuous fins are used in those applications in which the pressure drop is critical, although a larger amount of surface area is required on the tube outside for specified heat transfer compared to the wavy or interrupted fins. Plain continuous fins have the lowest pressure drop compared to any other tube-fin surfaces for the same surface area.

Wavy Fins on a Round Tube Array.   This continuous fin geometry, also sometimes referred to as corrugated fin geometry, is a most commonly used design for air conditioning condensers and commercial heat exchangers because of its superior performance over plain fins and its ruggedness. This fin is probably the most investigated by industry. However, actual performance data have been published for only one geometry in dimensional form [75], and the performance comparison is made only with the parallel louver fin to be discussed in the next subsection.

Goldstein and Sparrow [76] used a mass transfer technique to measure the local mass transfer coefficients. Their test geometry simulated a one-row wavy fin design having 556 fins per meter (for an assumed fin thickness of 0.15 mm) on an 8.53-mm-diameter tube.†
The wave configuration is the same as that tested by Goldstein and Sparrow [44] without the tube present; those test results were discussed on p. 4-222. Goldstein and Sparrow presented detailed local Sherwood number distribution on each upper and lower facet of the wavy fin. They identified the presence of several vortex systems and concluded that the windward facets of the wavy fins are primarily responsible for the enhancement of the transfer coefficients; the leeward facets have lower transfer coefficients and are strongly affected by the flow separation. A comparison of average transfer coefficients of

---

†The dimensions s, X_t, X_b and d_o are the same as those of plain fins tested by Saboya and Sparrow [68] and are listed earlier on p. 4-234.

this wavy fin with those of a similar plain fin showed that the enhancement due to the wavy fin surface increased with the Reynolds number. At Re = 1000, this one-row design provided a 45 percent higher average transfer coefficient than a plain fin exchanger of the same basic geometry.

Louver Fins on a Round Tube Array.   As discussed earlier, both strip fin and louver fin geometries have high performance and are most commonly used for plate-fin exchangers. One would expect that the concepts of strips and louvers could be extended to continuous fins in order to obtain high performance. Many different forms of louvers on circular tubes have been investigated by the heat exchanger industry both with staggered and inline tube arrangements. Some of them are shown in Fig. 4. Unfortunately, only a limited amount of data is published and it is in dimensional form. Hosoda et al. [75] showed that parallel louver fins on staggered tubes increased the heat transfer coefficient by about 50 percent over that by wavy fins on the same staggered tube array. Ito et al. [77] showed that another parallel louver fin geometry on staggered tubes increased the heat transfer coefficient by 35 percent over a plain fin geometry. Fukui and Sakamoto [74] showed that single-cut louvers in continuous fins on inline tubes increased the heat transfer coefficient by about 50 percent over flat fins.

Continuous Fins on a Flat Tube Array.   Evaluation of Rich's data [66] for plain fins on circular tubes indicates that form drag on the tubes accounts for approximately 40 to 60 percent of the total pressure loss depending upon the fin density. In comparison, the use of flattened tubes (rectangular tubes with rounded or sharp corners) yields a lower pressure loss due to lower form drag, and avoids the low performance wake region behind the tubes. Also, the heat transfer coefficient is higher for flow inside flat tubes than inside circular tubes, particularly at low Re. The use of flat tubes is limited to low-pressure applications, such as vehicular "radiators," unless the tubes are extruded with integral fins outside.

Kays and London [16] provide j and f data for one core having plain fins and inline flat tubes, and one core having plain fins and echelon† flat tubes. They also provide j and f data for one core having wavy fins and inline flat tubes, and two cores having wavy fins and echelon flat tubes. The wavy fins provided 30 percent higher heat transfer coefficients than the plain fins. Vlădea et al. [78] tested 0.1 mm (0.004 in) thick plain fins on two-, three-, and four-row inline and staggered flat tubes of 18.7- × 2.5-mm size. They showed that the heat transfer and friction factors were higher for the staggered tube arrangement than for the inline tube arrangement, as expected. Some data on parallel louver fins and perforated fins on flat tubes are summarized in Refs. 79 to 81.

Individually Circular Finned Tubes.   This tube-fin geometry (shown in Figs. 3b and 5) is generally much more rugged than the continuous fin geometry, but has lower compactness (surface area density). Correlations and information on basic data are summarized separately for circular plain fins, segmented and spine fins, studded fins, slotted fins, and wire-loop fins.

Circular Fins.   These fins are the simplest and most common. They are manufactured by tension-wrapping the fin material around a tube to form a continuous helical

---

†An echelon arrangement is a staggered arrangement with the tube pattern repeating every fourth (or higher) row; in the commonly used "staggered" arrangement, the tube pattern repeats every third row.

fin, or by mounting circular disks on the tube. Several different methods have been used to attach the circular fin to the tube.

Substantial data have been published on the circular (helical or disk) fin of Fig. 5. Most of the reported data have been taken with a staggered tube arrangement, six or more tube rows deep. Webb [65] presents a survey of the published data and correlations. Heat transfer and friction correlations are usually empirically based upon a multiple regression analysis of the basic dimensionless groups. Such correlations must account for five geometric parameters, which include the tube diameter d_o, the fin parameters δ, l_f, and s (or s'), and the tube bundle geometry X_t and/or X_l for inline or staggered arrangements. A number of correlations have been proposed, which differ in the choice of dimensionless groups and of the characteristic dimension used in the Reynolds number.

For heat transfer, the recommended equation is that of Briggs and Young [82] as follows:

$$j = 0.134 \ \mathrm{Re}_d^{-0.319} \left(\frac{l_f}{s'}\right)^{-0.2} \left(\frac{\delta}{s'}\right)^{-0.11} \tag{71}$$

This equation is based on the test data for airflow over 14 equilateral triangular tube banks. The following ranges were covered: $\mathrm{Re}_d \sim 1100$ to 18,000; $s'/l_f \sim 0.13$ to 0.63; $s'/\delta \sim 1.01$ to 6.62; $l_f/d_o \sim 0.09$ to 0.69; $\delta/d_o \sim 0.011$ to 0.15; $X_t/d_o \sim 1.54$ to 8.23; fin root diameter $d_o \sim 11.1$ to 40.9 mm; and fin density 246 to 768 fins per meter. The standard deviation for Eq. (71) was 5.1 percent. Although Briggs and Young tried to correlate the friction factor data, they were not successful. The standard deviation for friction factor correlation was ±40 percent, and hence, a subsequent study [83] was conducted for the ΔP correlation with a limited range of geometrical variables.

For isothermal pressure drops, the recommended equation is that of Robinson and Briggs [83], as follows:

$$f' = \frac{\Delta P}{N_r} \frac{2g_c \rho}{4G^2} = 9.465 \ \mathrm{Re}_d^{-0.316} \left(\frac{X_t}{d_o}\right)^{-0.937} \tag{72}$$

Here f' is a modified Fanning friction factor per tube row. This equation is based on the test data for airflow over 15 equilateral triangular and 2 isosceles triangular tube banks. The following ranges were covered: $\mathrm{Re} \sim 2000$ to 50,000; $s'/l_f \sim 0.15$ to 0.19; $s'/\delta \sim 3.75$ to 6.03; $l_f/d_o \sim 0.35$ to 0.56; $\delta/d_o \sim 0.011$ to 0.025; $X_t/d_o \sim 1.86$ to 4.60; fin root diameter $d_o \sim 18.6$ to 40.9 mm; and fin density 311 to 431 fins per meter. The standard deviation for Eq. (72) was 7.8 percent.

A review of Eqs. (71) and (72) reveals that the tube pitch X_t (or $X_l = 0.866X_t$) has no effect on the heat transfer coefficient, but the pressure drop is strongly affected by it. The pressure drop decreases with an increase in the tube pitch.

The Briggs and Young data included tubes having low fins, e.g., $l_f/d_o \approx 0.10$, and high fin density (750 fins per meter). Recently, Rabas et al. [84] have established an improved empirical correlation as follows for these high fin density low-finned tubes using the data of five investigators including those of Briggs and Young and their own new data:

$$j = 0.292 \ \mathrm{Re}_d^{-0.415 + 0.0346 \ \ln (d_e/s)} \left(\frac{s}{d_o}\right)^{1.115} \left(\frac{s}{l_f}\right)^{0.257} \left(\frac{\delta}{s}\right)^{0.666} \left(\frac{d_e}{d_o}\right)^{0.473} \left(\frac{d_e}{\delta}\right)^{0.772} \tag{73}$$

$$f' = 3.805 \ \mathrm{Re}^{-0.234} \left(\frac{s}{d_e}\right)^{0.251} \left(\frac{l_f}{s}\right)^{0.759} \left(\frac{d_o}{d_e}\right)^{0.729} \left(\frac{d_o}{X_t}\right)^{0.709} \left(\frac{X_t}{X_l}\right)^{0.379} \tag{74}$$

These equations are valid for the following ranges: $l_f \leq 6.35$ mm, $1000 \leq \mathrm{Re}_d \leq 25,000$, $X_l \leq X_t$, $N_r \geq 6$, fin root diameter $d_o \sim 4.76$ to 31.75 mm, fin density $\gamma \sim 246$ to 1181 fins per meter, $X_t \sim 15.08$ to 111.0 mm, and $X_l \sim 10.32$ to 96.11 mm. These correlations predicted 94 percent of j data and 90 percent of f data within ± 15 percent.

There is considerably less choice of correlations for inline tube banks. This is because a strong bypass stream exists in the open and finned zones between tube rows; this bypass stream is dependent on the fin tip clearance. A bypass fluid stream is one which does not at all come in direct contact with the heat transfer surface. Schmidt [85] made an attempt to develop a heat transfer correlation for such tube banks based on data from 11 sources.

A staggered tube arrangment is preferred for multirow finned tube banks. The heat transfer coefficient of an inline finned tube bank is substantially lower than that for a staggered tube arrangement. The difference between staggered and inline tube arrangements is mentioned on p. 4-240.

Figure 5 shows some of the enhanced surface geometries used on circular tubes. Webb [65] in his Table 2 provides references to information on the performance of these enhanced surface geometries for individually finned circular tubes. All of the concepts provide enhancement by the periodic development of thin boundary layers on small-diameter wires or flat strips, followed by dissipation in the wake region between elements.

Segmented and Spine Fins.   These fins are the counterpart of the strip fins used in plate-fin exchangers. Segmented and spine fins have essentially the same basic geometry as shown in Fig. 5. They are used in a wide range of applications from boiler economizers to air conditioning. A segmented fin is generally rugged, has "heavy-gauge" metal, and is usually less compact compared to a spine fin. A segmented fin is also referred to as a serrated fin. A continuous strip of metal, after being partially cut into narrow sections, is wound helically around the tubes. Upon winding, the narrow sections separate and form the narrow strips that are connected at the base. The steel segmented fin is attached to a tube by continuous welding of the base of the strip to the tube. The aluminum spine fin may be attached to the tube by epoxy.

A steel segmented fin geometry is used for boiler economizers and waste heat recovery boilers. The j and f' versus Re_d characteristics for a four-row staggered tube geometry are shown in Fig. 31 [86]. In this figure, also shown are the predicted j and f' factors for an equivalent plain fin tube bank using Eqs. (71) and (72). Weierman [87] presents design correlations for steel segmented fins and plain circular fins. The correlations, presented in graphical form, are for both staggered and inline banks. Based on the heat transfer correlation, the segmented fin heat transfer coefficient is 40 percent greater than that of the plain fin when the fin height is 10 times its spacing. The enhancement ratio decreases with reduced fin height-to-spacing ratio. Rabas and Eckels [88] present additional data on steel segmented fins; they found that the inline arrangement yields heat transfer coefficients that are substantially lower than those for the staggered arrangement.

The spine fin is used extensively by General Electric in air conditioning condensers. The details of geometry, manufacturing, and history are provided by Abbott et al. [89]. Moore [90] provides St and f' versus Re_d for one-, two-, and three-row staggered spine-finned tube layouts.

Studded Fins.   This fin geometry is made by welding individual "studs" around the base of the tube (see Fig. 5). The shape of the studs and the number of studs around the tube and along the tube are the variables. Because of ruggedness, this finned tube is used in steam generator economizers. Ackerman and Brunsvold [91] presented $\mathrm{Nu}_d \times \mathrm{Pr}^{-1/3}$ versus Re_d and $4f'$ versus Re_d for five staggered and one inline tube bank arrangement,

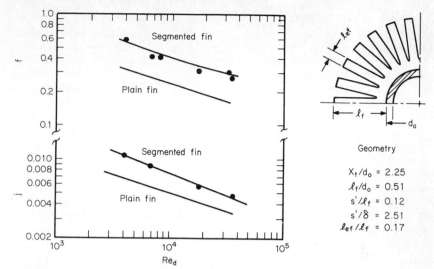

FIG. 31 Heat transfer and flow friction characteristics of staggered four-row segmented fin [86].

and also provided correlations in closed-form equations separately for each geometry. They observed that the heat transfer performance was dependent upon the transverse pitch of the tube bank; the highest performance was given by the bank having the largest X_t. The inline tube arrangement was the worst performer.

Slotted Fins.   These fins have slots in the radial direction (see Fig. 5). When radially slitted fin material is wound on the tube, the slits open, forming slots whose width increases in the radial direction. This fin geometry offers an enhancement over tension-wound plain fins. Precce et al. [92] found the heat transfer coefficient as much as 40 percent greater than that of a smooth fin.

Wire-Loop Fins.   This fin surface is formed by spirally wrapping a flattened helix of wire around the tube. The wire loops are held to the tube by a tensioned wire within the helix, or by soldering. The enhancement characteristics of small-diameter wires is important at low flows when the enhancement of other interrupted fins diminishes. Wall [93] describes the practical aspects of a commercially available wire-loop design. He claims that the heat transfer coefficient is 2.5 times that of a plain fin, and that $h \propto d^{-0.4}$, where d is the wire diameter.

Individually Finned Flat Tubes.   Oval and flat cross-sectional tube shapes are also applied to individually finned tubes. Figure 32 compares the performance of staggered banks of oval and circular finned tubes tested by Brauer [94]. Both banks have 313 fins per meter, and 10-mm-high fins on approximately the same transverse and longitudinal pitches. The oval tubes gave a 15 percent higher heat transfer coefficient and 25 percent less pressure drop than the circular tubes. The performance advantage of the oval tubes results from lower form drag on the tubes and the smaller wake region on the fin behind the tube. The use of oval tubes may not be practical unless the tube-side design pressure is low enough. Higher design pressures are possible using flattened aluminum tubes made

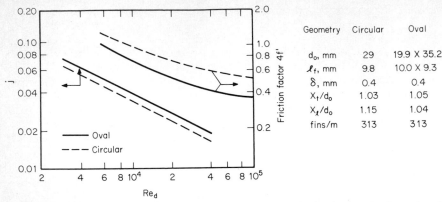

Geometry	Circular	Oval
d_o, mm	29	19.9 X 35.2
ℓ_f, mm	9.8	10.0 X 9.3
δ, mm	0.4	0.4
X_t/d_o	1.03	1.05
X_ℓ/d_o	1.15	1.04
fins/m	313	313

FIG. 32 Heat transfer and flow friction characteristics of finned tubes having circular and oval tubes in a staggered arrangement [94].

by an extrusion process [65]. Such tubes can be made with internal full-height ribs which strengthen the tube and allow for a high tube-side design pressure. A variety of geometrical fin-and-tube shapes may be made by aluminum extrusions of different shapes.

Staggered Versus Inline Tube Arrangements.   There is a basic difference in the flow phenomena over staggered versus in-line finned tube banks. In the inline arrangement, the flow region upstream of the circular tubes of the second and subsequent rows is the low-velocity wake region. This results in low heat transfer through these tubes. Also there are straight through-flow channels (of width $X_t - d_o$) that offer a path of the least flow resistance; and as a result, a significant portion of the flow can bypass without touching the tubes (the heat transfer surfaces). Both these phenomena, which yield low heat transfer and pressure drop, are essentially not present in the staggered tube arrangement. Further details on the comparisons are provided in Refs. 42 and 65.

Row Effect in Finned Tube Banks.   The published correlations are generally for "deep" tube banks ($N_r \geq 4$), and do not account for row effects. The heat transfer coefficient will decrease with increasing number of tube rows in an inline bank due to the flow bypass effect. But the heat transfer coefficient will increase with rows in a staggered bank. This is because the turbulent eddies shed from the tube cause good mixing in the fin region of the downstream row. The tube row effect is a complex phenomenon and is dependent upon at least the Reynolds number, tube pitches, and basic fin geometry (circular or continuous fin). Only a very limited amount of information is available; it is summarized in [42, 65].

4. Influence of Temperature-Dependent Fluid Properties

One of the basic idealizations made in the theoretical solutions for Nu and f of Sec. D.2 is that the fluid properties remain constant throughout the flow field. Most of the experimental j and f data obtained in the preceding section involve small temperature differences so that the fluid properties generally do not vary significantly. In certain heat exchanger applications, fluid temperatures vary significantly. At least two questions arise: (1) Can we use the j and f data obtained for air at 50 to 100°C (100 to 200°F) for air at 500 to 600°C (900 to 1100°F)? (2) Can we use the j and f data obtained with air (such as all data in Ref. 16) for water, oil, or viscous liquids? The answer is yes, by modifying the constant-property j and f data to account for variations in the fluid properties within a heat exchanger. The influence of fluid property variations and the correction schemes for j or Nu and f factors are presented in detail in Ref. 168. The "property ratio" method is most commonly used to account for the fluid property variations in the heat exchanger. In this method, the Nusselt number and friction factors for the variable-fluid property case are related to the constant-property values for gases and liquids as follows:†

For gases:

$$\frac{\mathrm{Nu}}{\mathrm{Nu}_{cp}} = \left(\frac{T_w}{T_m}\right)^n \qquad \frac{f}{f_{cp}} = \left(\frac{T_w}{T_m}\right)^m \tag{75}$$

For liquids:

$$\frac{\mathrm{Nu}}{\mathrm{Nu}_{cp}} = \left(\frac{\mu_w}{\mu_m}\right)^n \qquad \frac{f}{f_{cp}} = \left(\frac{\mu_w}{\mu_m}\right)^m \tag{76}$$

Here the subscript cp refers to the constant-property variable, and all temperatures in Eq. (75) are *absolute*. All of the properties in the dimensionless groups of Eqs. (75) and (76) are evaluated at the *bulk mean* temperature. The values of the exponents n and m for fully developed laminar and turbulent flows in a circular tube are summarized in Tables 11 and 12 for heating and cooling situations. These correlations, Eqs. (75) and (76), with exponents from Tables 11 and 12, are derived for the constant heat flux boundary condition. The variable-property effects are generally not important for fully developed flow having constant wall temperature boundary condition, since T_m approaches T_w for fully developed flow. Therefore, in order to take into account minor influence of property variations for the constant wall temperature boundary condition, the correlations of Eqs. (75) and (76) are adequate.

The Nu and f factors are also dependent upon the duct cross-sectional shape in laminar flow, and are practically independent of the duct shape in turbulent flow. The influence of variable fluid properties on Nu and f for fully developed laminar flow through rectangular ducts has been investigated by Nakamura et al. [95]. They concluded that the velocity profile is strongly affected by the μ_w/μ_m ratio, and the temperature profile is weakly affected by the μ_w/μ_m ratio. They found that the influence of the aspect ratio on

---

†The j factor for variable properties is then computed from the variable property Nu as $j = \mathrm{Nu}\ \mathrm{Pr}^{-1/3}/\mathrm{Re}$.

TABLE 11 Property Ratio Method Exponents of Eqs. (75) and (76) for Laminar Flow [106]

Fluid	Heating	Cooling
Gases	$n = 0.0, m = 1.00$	$n = 0.0, m = 0.81$
	for $1 < T_w/T_m < 3$	for $0.5 < T_w/T_m < 1$
Liquids	$n = -0.14, m = 0.58$	$n = -0.14, m = 0.54$
	for $\mu_w/\mu_m < 1$	for $\mu_w/\mu_m > 1$

TABLE 12 Property Ratio Method Correlations or Exponents of Eqs. (75) and (76) for Turbulent Flow [26]

Fluid	Heating	Cooling
Gases	$n = -[\log_{10} (T_w/T_m)]^{1/4} + 0.3$ for $1 < T_w/T_m < 5, 0.6 < \text{Pr} < 0.9$, $10^4 < \text{Re} < 10^6$, and $L/D_h > 40$.	$n = 0.$
	$m = -0.1$ for $1 < T_w/T_m < 2.4$.	$m = -0.1$ (tentative).
Liquids	$n = -0.11†$ for $0.08 < \mu_w/\mu_m < 1$.	$n = -0.25†$ for $1 < \mu_w/\mu_m < 40$.
	$f/f_{cp} = (7 - \mu_m/\mu_w)/6‡$ or $m \approx 0.25$ for $0.35 < \mu_w/\mu_m < 1$.	$m = 0.24‡$ for $1 < \mu_w/\mu_m < 2$.

†Valid for $2 \leq \text{Pr} \leq 140, 10^4 \leq \text{Re} \leq 1.25 \times 10^5$.

‡Valid for $1.3 \leq \text{Pr} \leq 10, 10^4 \leq \text{Re} \leq 2.3 \times 10^5$.

the correction factor $(\mu_w/\mu_m)^m$ for the friction factor is negligible for $\mu_w/\mu_m < 10$. For the heat transfer problem, the Sieder-Tate correlation ($n = -0.14$) is valid only in the narrow range of $0.4 < \mu_w/\mu_m < 4$.

5. Influence of Superimposed Free Convection

The influence of superimposed free convection over pure forced convection flow is important when either the flow velocity is low or a high temperature difference ($T_w - T_m$) is employed, or the passage geometry has a large hydraulic diameter D_h. The effect of the superimposed free convection is generally important in the laminar flow of a noncompact heat exchanger; it is quite negligible for compact heat exchangers [34] and hence it will not be covered here. The reader may refer to Ref. 169 for further details.

It should be emphasized that for laminar flow of liquids in tubes, the influence of viscosity variations and density variations (bouyancy or free-convection effects) must be accounted for simultaneously for heat exchanger applications. Some correlations and work in this area have been summarized by Bergles [96].

E. SURFACE SELECTION METHODS AND OPTIMIZATION

As mentioned in the introduction, compact exchanger design offers flexibility in distributing surface area on the hot or cold side as warranted by design considerations. Many compact surfaces are available having different orders of magnitude of surface area density, and each surface has a large number of geometrical variables. Some guidelines are provided in this section for the selection of particular surfaces and numerical optimization techniques for compact heat exchanger designs.

A proper selection of a surface is one of the most important considerations in compact heat exchanger design. There is no such thing as a surface that is "best" for all applications. The particular application strongly influences the selection of the surface to be used. The objectives of the system in which the heat exchanger is to be used also influence the surface selection. Both qualitative and quantitative considerations are now presented for surface selection.

1. General Qualitative Considerations for Surface Selection

Let us first discuss general guidelines for the selection of a plate-fin, tube-fin, or regen-erative-type surface. Some qualitative design aspects of specific surfaces have already been presented in Sec. D.3 starting on p. 4-221.

a. Plate-Fin Surfaces

Plate-fin construction is commonly used in gas-to-gas exchanger applications. It offers high area densities (up to about 6000 m^2/m^3 or 1800 ft^2/ft^3) and a considerable amount of flexibility. The passage height on each side can be easily varied and different fins can be used between plates for different applications. On each fluid side, the fin thickness and the number of fins can be varied independently. If a corrugated fin (such as the plain triangular, louver, perforated, or wavy fin) is used, the fin can be squeezed or stretched to vary the fin pitch, thus providing an added flexibility. The fins on each side can be easily arranged such that the overall flow arrangement of the two fluids can result in cross-flow, counterflow, or parallel flow. Even the construction of a multifluid plate-fin exchanger is relatively straightforward except for the inlet and outlet headers for each fluid.

Plate-fin exchangers are generally designed for low-pressure applications, with oper-ating pressures limited to about 1000 kPa gauge (150 psig).† The maximum operating temperatures are limited by the type of fin-to-plate bonding and the materials employed. Plate-fin exchangers have been designed from low cryogenic operating temperatures up to about 800°C (1500°F).

Fouling is generally not a severe problem with gases as it is with liquids. A plate-fin exchanger is generally not designed for applications involving heavy fouling since there is no easy method of cleaning the exchanger. If an exchanger is made of small modules (stacked in height, width, and length directions), and if it can be cleaned with a detergent, with a high-pressure air jet, or by baking it in an oven, it could be designed for those applications having considerable fouling.

Fluid contamination (mixing) is generally not a problem in plate-fin exchangers since there is practically zero fluid leakage from one to the other side of the exchanger.

Cost is a very important factor in the selection of exchanger construction type and surface. The plate-fin surface in general is less expensive per unit of heat transfer area than a tube-fin surface.

b. Tube-Fin Surfaces

When an extended surface is needed on only one side (such as in a gas-to-liquid exchanger) or when the operating pressure needs to be contained on one side, a tube-fin

---

†Some cryogenic plate-fin exchangers have been designed for operating pressures up to about 8300 kPa gauge (1200 psig).

exchanger is selected with tubes that are round, flat, or elliptical in shape. Flat or ellip-tical tubes, instead of round tubes, are used for increased heat transfer in the tube and reduced pressure drop outside the tubes; however, the operating pressure is limited com-pared to that of round tubes. Tube-fin exchangers usually have lower compactness than a plate-fin unit, with a maximum of about 3300 m^2/m^3 (1000 ft^2/ft^3) surface area density.

A tube-fin exchanger may be designed for a wide range of tube fluid operating pres-sures (up to about 3100 kPa gauge or 450 psig) with the other fluid at low pressure (up to about 100 kPa or 15 psig). The highest operating temperature is again limited by the type of bonding and the materials employed. Tube-fin exchangers are designed to cover the operating temperature range from the low cryogenic temperatures to about 870°C (1600°F).

Reasonable fouling can be tolerated on the tube side if the tubes can be cleaned. Foul-ing is generally not a problem on the gas side (fin side); plain uninterrupted fins are used when "moderate" fouling is expected. Fluid contamination (mixing) is generally not a problem since there is essentially no fluid leakage between the two fluid streams.

Since tubes are generally more expensive than extended surfaces, the tube-fin exchanger is in general more expensive. In addition, the heat transfer surface area density of a tube-fin core is generally lower than that of a plate-fin exchanger, as mentioned earlier.

c. Regenerative Surfaces

Regenerators, exclusively used in gas-to-gas heat exchanger applications, could have the most compact surface area density compared to plate-fin or tube-fin surfaces. While rotary regenerators have been designed for surface area density α of up to about 8800 m^2/m^3 (2700 ft^2/ft^3), compact fixed-matrix regenerators have been designed for α of up to about 16,400 m^2/m^3 (5000 ft^2/ft^3).

Regenerators are usually designed for low-pressure applications, with operating pres-sures limited to about 620 kPa gauge (90 psig) for rotary regenerators, and even lower pressures for fixed-matrix regenerators. The regenerators are designed to cover an oper-ating temperature range from low cryogenic to very high temperatures. Metal regenera-tors are used for operating temperatures up to about 870°C (1600°F); ceramic regener-ators are used for higher temperatures up to about 2000°C (3600°F).

Regenerators have self-cleaning characteristics because the hot and cold gases flow in the opposite directions periodically through the same passage. As a result, compact regen-erators have minimal fouling problems with relatively clean gases, and usually have very small hydraulic diameter passages. If severe fouling is anticipated, regenerators are not used.

Carryover and bypass leakages from the hot fluid to the cold fluid (or vice versa) occur in the regenerator. While this fluid contamination can be minimized, if it is absolutely not permissible, the regenerator is not used.

The cost of the regenerator surface per unit of transfer area is generally substantially lower than that of a plate-fin or tube-fin surface.

d. Manufacturing Considerations

In addition to the foregoing guidelines for selecting a compact surface, there are other manufacturing considerations that must be evaluated before selecting a surface.

Selection of a surface depends upon the operating temperature, with reference to bond-ing of fins to plates or tubes and the choice of material. For low-temperature applications, a mechanical joint, soldering, or brazing may be adequate. Fins can be made from copper, brass, or aluminum, and thus maintain high fin efficiency. For high-temperature applications, only special brazing techniques and welding may be used; stainless steel and other expensive alloys may be needed for fins, with possibly a resultant reduction in the fin efficiency. Consequently, suitable high-performance surfaces may be selected to offset the reduction in fin efficiency.

There are two types of manufacturing limitations for forming and fabricating a surface: industry-oriented and technology-oriented. Most of the heat exchanger manufacturing industries make only a limited number of surfaces due to limited availability of tools, machines, and plant facilities or due to company policy, competition, and market potential. Hence, the selection of a surface for a given application will be from the available stock, even though there may be some superior surfaces available elsewhere. Theory and analysis can be used to arrive at some high-performance surfaces; however, it may not be feasible to manufacture them due to technological limitations. The highly interrupted louver and offset strip fins are the result of manufacturing advances of the past decade. Further advances in manufacturing technology are needed to manufacture high-performance new surfaces.

Cost is one of the single most important factors in the selection of surfaces such that the overall heat exchanger is least expensive either in initial cost or in both initial and operating costs. If a plain fin surface can do the job for an application, the more "efficient" louver or offset strip-fin surface is not used because it is more expensive to manufacture. In most applications, one does not choose a high-performance surface, but chooses the least expensive surface, if it can meet the performance criteria within specified constraints.

2. Quantitative Considerations for Surface Selection

Quantitative considerations for surface selection are based on the performance comparison of various heat exchanger surfaces, and choosing the best under some specified criteria for a given heat exchanger application.

A variety of methods have been proposed in the literature for surface performance comparisons. Over 30 dimensional or nondimensional comparison methods have been critically reviewed by Shah [97]. Most of these comparisons are for the surfaces only as if they were on *one side* of a heat exchanger.

When a complete exchanger design is considered that lends itself to having both sides with approximately equal $\eta_o hA$, the best surface selected by the foregoing methods may not be an optimum surface for a given application. This is because the selection of the other side surface and its thermal resistance, flow arrangement, overall exchanger envelope, and other criteria (not necessarily related to the surface characteristics) influence the overall performance of a heat exchanger. Hence, there is no need of fine-tuning the surface selection quantitatively by any one of the comparison methods, but instead surfaces should be screened by some simple and meaningful methods. The final selection should be based on optimization procedures (that will consider surfaces on both sides and other constraints), mechanical design, and other qualitative criteria as summarized in Fig. 7.

Basically, the quantitative surface selection methods may be categorized as screening methods and performance comparison methods. When one needs to consider the heat transfer and non–heat transfer related constraints imposed by the application along with numerical optimization for the design, the screening methods are used. In these methods, a few "best surfaces" are selected on each fluid side considering only one side of the exchanger at a time, generally followed by numerical optimization of the exchanger design. However, when one needs to compare the performance of actual surfaces under specified criteria considering both sides of the exchanger, the performance comparison methods are used; generally, no numerical optimization is followed, but one is not precluded.

a. Screening Methods

The selection of a surface depends upon the exchanger design criteria. For a specified heat transfer rate and pressure drop on one side, two important design criteria for compact exchangers are the minimum frontal area requirement and the minimum surface area (or volume) requirement. Let us first discuss the significance of these criteria.

With more compact surfaces (reduced D_h), higher heat transfer coefficients are achieved. At the same time, the friction factors are also high. In fully developed laminar flow, a decrease in G linearly reduces ΔP without reducing h. For developing laminar flows, a reduction in G reduces ΔP with only a slight decrease in h [1]. In very compact heat exchangers, generally developed or developing laminar flows are realized. Thus with increased friction factors in a compact heat exchanger, the specified pressure drop requirement would be satisfied by reducing G without a significant reduction in h. For a specified constant flow rate, a reduction in G means an increase in the flow area A_o and hence the frontal area. Thus one of the characteristics of highly compact surfaces is that the resultant shape of the exchanger becomes awkward, having a large frontal area and a short flow length. Hence, it is important to determine which of the compact surfaces will have a minimum frontal area requirement. This is achieved by the flow area goodness factor comparison to be discussed.

The surface having the highest heat transfer coefficient at a specified flow rate will require the minimum heat transfer surface area. However, the allowed pressure drop is not unlimited. Therefore, one chooses the surface having the highest heat transfer coefficient for a specified fluid pumping power. The exchanger with the minimum surface area will tend to have the minimum overall volume requirement. For this purpose, surface area (or volume) goodness factor comparisons are made.

Flow Area Goodness Factor Comparison.   The flow area goodness factor is defined as j/f. In this method, a plot of j/f versus Re is made for different surfaces. It can be shown that

$$\frac{j}{f} = \frac{\text{Nu Pr}^{-1/3}}{f\,\text{Re}} = \frac{1}{A_o^2}\left[\frac{\text{Pr}^{2/3}}{2g_c c_p \rho^2}\,\frac{hA}{\text{P}}\,W^2\right] \tag{77}$$

The second equality in Eq. (77) provides the significance of j/f as being inversely proportional to A_o^2 (A_o = core minimum free flow area) with the bracketed quantities (heat transfer hA, flow rate W, pumping power P, and fluid properties) constant. A surface having a high j/f factor is "good," because it will require a lower free flow area and hence a lower frontal area for the exchanger. This comparison method is independent of the *scale* of the geometry (i.e., the hydraulic diameter), because of the dimensionless j, f, and Re. j_{H1}/f is presented in Fig. 33 for fully developed laminar flow through some idealized passage geometries [34]. It varies from 0.265 for the equilateral triangular duct to 0.390 for the parallel-plates duct. Thus, parallel plates, relative to the triangular duct, have a 47 percent $(0.390/0.265 - 1)$ higher j/f. Based on Eq. (77), then, the parallel-plates duct geometry would require an 18 percent $(1 - 1/\sqrt{1.47})$ lower free flow area. The porosity of the exchanger must be considered in order to translate this free flow area advantage into a frontal area improvement. Note that in the free flow area goodness factor comparison, no estimate of total heat transfer area or volume can be inferred. Such esti-

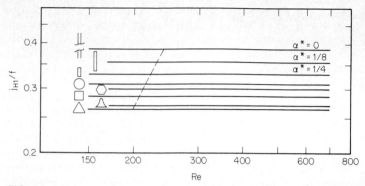

FIG. 33 Theoretical laminar flow area goodness factors for some simple duct geometries.

mates may be derived from the surface and volume "goodness" factors to be described shortly.

Since we have kept hA, P, and W constant in the flow area goodness factor comparison, the operating Reynolds numbers will be different for different surfaces having the *same* hydraulic diameter D_h. Then from Eq. (77),

$$\frac{j}{f} \propto \frac{W^2}{A_o^2} \propto Re^2 \tag{78}$$

The dashed line having the slope of 2 on a log–log scale in Fig. 33 indicates the required Reynolds numbers for different ducts when $Re = 200$ for the triangular duct for the same hA, P, W, and D_h. Thus the parallel-plates duct will operate at $Re = 243$ ($= 200\sqrt{1.472}$), with an 18 percent lower free flow (and hence frontal) area.

Surface Area Goodness Factor Comparison.   In this method, the heat transfer power per unit surface area and per unit temperature difference† is plotted against the fluid pumping power per unit surface area. These dimensional quantities h and P''_{std} are usually evaluated at some standard conditions and expressed on *one* side of the exchanger as

$$h_{std} = \frac{1}{A}\left[\frac{1}{\eta_o}\frac{q}{T_w - T_m}\right] = \frac{1}{\beta V}\left[\frac{1}{\eta_o}\frac{q}{T_w - T_m}\right] \tag{79}$$

$$P''_{std} = \frac{P}{A} = \frac{P}{\beta V} = \frac{W\,\Delta P/\rho}{A} \tag{80}$$

The first equality in Eq. (79) indicates

$$h \propto \frac{1}{A} \tag{81}$$

with the bracketed quantities and P''_{std} constant. Thus the minimum surface area requirement is investigated under the following constraints: (1) same heat transfer q, (2) same temperature difference between the wall and fluid, and (3) same fluid flow rate. The

---

†Note that the heat transfer power per unit temperature difference and per unit surface area is the same as the heat transfer coefficient h.

surface having the higher h_{std} at a specified P''_{std} is "good" because it will require a lower amount of surface area for the required heat transfer. Because of the dimensional plot, the comparisons are made for surfaces having the same scale (i.e., the same D_h) in order to eliminate the size variable of the flow passages and focus on the influence of the geometry alone. Also, the extended surface efficiency η_o is not incorporated with h_{std} in order to eliminate the influence of the fin variables (material and thickness).

Considering the gas turbine regenerator application, the h_{std} versus P''_{std} plot for fully developed constant-property laminar flow through some constant cross-sectional ducts is presented in Fig. 34 for $D_h = 0.5$ mm (0.0016 ft). From this figure, it is found that h_{std} varies from 256.4 to 700.6, a factor of 2.7, with $h_{std} = 264.7$ W/(m$^2\cdot°$C) for the equilateral triangular duct. Thus from Eq. (81), the idealized parallel-plate heat exchanger would require 62 percent $(1 - 264.7/700.6)$ less heat transfer area than the equilateral triangular passage exchanger. For the foregoing comparisons, we have kept hA, **P**, and W constant. Hence, the design points for various geometries will lie on a 45° slope line (shown as a dashed line in Fig. 34), which represents common hA and $P''A$ magnitudes.

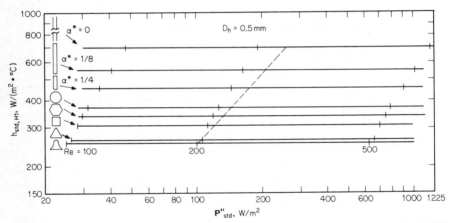

FIG. 34 Theoretical laminar surface area goodness factors for some simple duct geometries. 1 W/(m$^2\cdot°$C) = 0.1761 Btu/(h·ft$^2\cdot°$F).   1 W/m^2 = 124.6 × 10^{-6} hp/ft^2.

A plot such as Fig. 34, when made for a fixed D_h, clearly shows the influence of the passage shape. The parallel-plate heat exchanger may prove impractical, but it is clear that there are several other configurations that possess significant advantages over the triangular and sine duct geometries. Based on this plot, the development of a rectangular passage geometry is being continued for the vehicular gas turbine regenerator, instead of the first generation triangular passage geometry.

For a specified surface, h and P'' are determined from the j and f versus Re characteristics from the following expressions:†

$$h = \frac{c_p\mu}{Pr^{2/3}}\frac{1}{D_h}\,j\,Re \tag{82}$$

$$P'' = \frac{P}{A} = \frac{1}{2g_c}\frac{\mu^3}{\rho^2}\frac{1}{D_h^3}\,f\,Re^3 \tag{83}$$

---

†It is idealized that all other components, except for the core pressure drop, are negligibly small. Also the fluid is treated as having constant physical properties.

From these equations, it is evident that the dimensional h_{std} versus P''_{std} performance is strongly dependent upon the scale of the surface geometry, i.e., D_h. Thus, this comparison method reveals the incentive of increased performance (reduced surface area requirement) by going to a smaller D_h surface. This will also result in a much more compact surface.

Volume Goodness Factor Comparison. This method is just a variant of the preceding method in which the h_{std} versus P''_{std} plot is modified to allow a comparison of actual surfaces (having different D_h) on a volume basis.† No scale-up or scale-down of the actual surface geometries is performed. In this method, a plot of $\eta_o h \beta$ versus P''_{std} is made for comparison. For a specified $P''_{std}\beta$ ($= P/V$), one compares $\eta_o hA$ per unit volume (since $\beta = A/V$). And the higher this quantity, the lower will be the overall volume requirement for specified q, $T_w - T_m$, and W. By incorporating η_o, the actual fin geometry, fin material, etc., are taken into account for comparison.

b. Performance Comparison Methods

In these methods, the surface on one of the two fluid sides is generally chosen and the surface on the other side is required to be selected. For example, in a tube-fin exchanger, plain fins are chosen on one side, and a "best" fin geometry is to be selected on the other side. Comparisons between different surfaces and a reference surface are then made, with one of operational variables as the performance objective subject to design constraints on the remaining variables.‡ Operational variables considered are: geometry (number of tubes N_t, and tube length L), flow rate W, fluid pumping power P, heat transfer rate q, and fluid inlet temperature T_i (or inlet temperature difference ΔT_i). A total of 11 surface performance evaluation criteria (PEC), as summarized in Table 13, have been developed by Webb [98] to define performance merits of tubular surfaces developed to augment heat transfer in single-phase flow.

For a given PEC, a ratio of the design objective for a surface of interest to that for a reference surface is then calculated as a function of a similar ratio of a design variable. For example, consider that one is interested in minimizing the exchanger volume (or surface area) for fixed values of flow rate, fluid pumping power, heat transfer rate, and fluid inlet temperatures (PEC case VG-1 in Table 13). One selects an augmented surface and a reference surface, and calculates the exchanger volume ratio V/V_r as a function of the Reynolds number ratio Re/Re_r or G/G_r. Here the subscript r denotes the reference surface. Such performance ratio plots are made for all surfaces of interest. The surface having the highest position on such a plot is considered the best for the selected PEC.

Note that Table 13 also summarizes the order of magnitude for the ratios of other variables of interest. Webb provides all necessary equations for all PECs. This work has been extended by Webb and Bergles [99] for tube inserts and gas-side extended surfaces, and by Webb [100] for air-cooled exchanger tube-fin surfaces.

3. Heat Exchanger Optimization

Surfaces on each side of an exchanger may be selected by one of the methods in the preceding section. When such surfaces are incorporated into a heat exchanger, the resultant exchanger may not be optimum, since criteria other than those related to j and f may play

---

†The unvarying quantities mentioned just after Eq. (81) are also kept constant in this method.

‡In the performance comparison methods, the thermal resistance on both sides of the exchanger may be accounted for in the comparison.

TABLE 13 Performance Evaluation Criteria for Augmented Tubular Surfaces, $d/d_r = 1$ [98]

PEC Case	Fixed Geom.	W	P	q	T_i	Objective $N_{i,r}$	Consequences $\dfrac{N_i}{N_{i,r}}$	$\dfrac{L}{L_r}$	$\dfrac{W}{W_r}$	$\dfrac{Re}{Re_r}$	$\dfrac{P}{P_r}$	$\dfrac{q}{q_r}$	$\dfrac{\Delta T_i}{\Delta T_{i,r}}$
FG-1a	N_i, L	X			X	$\uparrow q$	1	1	1	1	>1	>1	1
FG-1b	N_i, L	X		X		$\downarrow \Delta T_i$	1	1	1	1	1	1	<1
FG-2A	N_i, L		X	X		$\uparrow q$	1	1	<1	<1	1	>1	1
FG-2b	N_i, L		X	X	X	$\downarrow \Delta T_i$	1	1	<1	<1	<1	1	<1
FG-3	N_i, L		X	X		$\downarrow P$	1	1	<1	<1	1	1	1
FN-1	N_i		X	X		$\downarrow L$	1	<1	<1	<1	<1	1	1
FN-2	N_i		X	X		$\downarrow P$	1	<1	1	1	1	1	1
VG-1	N_i, L	X	X	X	X	$\downarrow N_i, L$	>1	<1	1	<1	1	1	1
VG-2a	N_i, L	X	X	X		$\uparrow q$	>1	1	1	<1	1	>1	1
VG-2b	N_i, L	X	X		X	$\downarrow \Delta T_i$	>1	1	1	<1	1	1	<1
VG-3	N_i, L	X		X		$\downarrow P$	>1	1	1	<1	<1	1	1

an important role. For example, in addition to prescribed heat transfer and pressure drop, the core frontal area may have been restricted on one side of the exchanger. If the surface selected by one of the preceding methods requires larger than specified frontal area, it will not meet the design specifications and hence it will not be considered in the final selection.

In general, heat exchangers are designed for many varied applications, and hence may involve many different performance criteria such as minimum initial cost, minimum initial and operating costs, minimum weight or material, minimum volume or heat transfer surface area, minimum mean temperature difference, and so on. When a single performance measure has been defined quantitatively and is to be minimized or maximized, it is called an "objective function" in a design optimization. A particular design may also be subjected to certain requirements such as required heat transfer, allowable pressure drop, limitations on height, width and/or length of the exchanger, and so on. These requirements are called "constraints" in a design optimization. A number of different surfaces could be incorporated in a specific design problem, and there are many geometrical parameters that could be varied for each surface geometry. In addition, operating flows and temperatures could also be changed. Thus a large number of "design variables" are associated with a heat exchanger design. The question arises as to how one can effectively adjust these design variables within imposed constraints and come up with a design having optimum objective function. This may be achieved by a mathematical optimization.

Once the general configuration and surfaces are selected, an optimized heat exchanger design may be arrived at if the objective function and constraints can be expressed mathematically, and if all of the variables are automatically and systematically changed on some statistical or mathematical basis. A complete mathematical optimization of heat exchanger design is presently neither practical nor possible. Many engineering judgments based on experience are involved in different stages of the thermal and mechanical design of a heat exchanger. Even during the thermal design, if many constraints are imposed, it may not be possible to satisfy all constraints. It is the designer's job to make judgments on what constraints can be relaxed to obtain a good design.

A large number of optimization (search) techniques are available. These are "canned" computer programs that can be used either with heat exchanger optimization or any other engineering optimization problem. Some of these canned computer programs have been used by different investigators [101, 102, 103] for heat exchanger optimization. Cited references in these papers provide a thorough history on heat exchanger optimization.

In order to provide some insight into the heat exchanger optimization procedure, a flowchart [104] is shown in Fig. 35. It consists of three major blocks: (1) the formulation of the problem that must be made before any exchanger design and optimization are carried out, (2) a set of single-point design (rating) calculations through heat exchanger design programs, and (3) the checking of objective function and constraints, and the determination of a new set of design variables if the objective function is not optimized and the constraints are not satisfied. This third step is done through an optimization package. It should be emphasized that the heat exchanger programs must be working properly before any attempt is made to implement the optimization package. Even with the computer programs, heat exchanger optimization is an art since the designer will need to make decisions on the exchanger design based on experience and judgment when all the constraints are not met or when the sensitivity of the optimum solution to the associated variables is revealed.

The procedure of Fig. 35 is referred to as the *case study method*. In this method, each possible surface geometry and construction type is considered as an alternative design. In

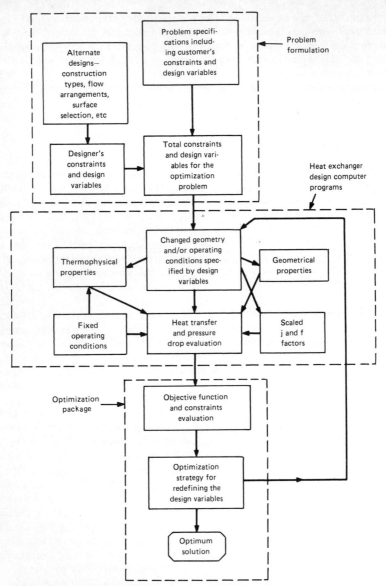

FIG. 35 Methodology for heat exchanger optimization [104].

order to make a legitimate comparison of these alternatives, each design must be optimized for the specified application. Thus there may be several independent optimized solutions satisfying the problem requirements. Engineering judgment, comparison of objective function values, and other evaluation criteria are then applied to select a final optimum solution for implementation.

During heat exchanger optimization, one of the most important but least known inputs for the heat transfer and pressure drop evaluation is the magnitude for "scaled" j and f factors. As soon as one of the surface geometrical dimensions is changed (such as fin pitch, height, or thickness), the surface is no longer geometrically similar to the original surface for which experimental j and f data are available. In such cases, either theoretical or experimental correlations should be incorporated in the computer program to arrive at scaled j and f factors for the new geometry. Some of these correlations, available in the literature, were presented earlier in Sec. D. The designer must use experience and judgment regarding which correlations should be used to obtain the scaled j and f factors. In addition, care must be exercised to avoid excessive extrapolations.

F. SURFACE GEOMETRICAL PROPERTIES

For the solution of the exchanger rating and sizing problems to be discussed, the following surface geometrical properties are needed on each side of a two-fluid exchanger: minimum free flow area, fin surface area, total heat transfer surface area, surface area density α based on the overall volume of the exchanger, hydraulic diameter, fin flow length, exchanger flow length, and the ratio of minimum free flow area to frontal area (or porosity). In addition, the surface area density β between plates may be needed for plate-fin exchangers. The surface geometrical properties in terms of basic core dimensions are now summarized for plate-fin, tube-fin, and regenerative surfaces.

1. Plate-Fin Surfaces

General relationships for geometrical properties are first summarized and are followed by specific detailed geometrical relationships for an offset strip-fin surface as an example.

a. General Geometrical Relationships

For any plate-fin geometry, the basic geometrical quantities required on each fluid side are: the minimum free flow area A_o, frontal area A_{fr}, heat transfer surface area A, core length L_1, width L_2 or L_3, core plate spacing b, and total number of fluid passage N_p. The hydraulic diameter D_h, surface area densities β and α, and the ratio of free flow area to frontal area σ on fluid side 1 (with subscript 1) are computed from the following relationships:

$$D_{h,1} = \frac{4A_{o,1}L_1}{A_1} = \frac{4\sigma_1}{\alpha_1} \tag{84}$$

$$\beta_1 = \frac{A_1}{L_1L_2b_1N_{p,1}} \tag{85}$$

$$\alpha_1 = \frac{A_1}{V} = \frac{A_1}{L_1L_2L_3} \tag{86}$$

and $$\sigma_1 = \frac{A_{o,1}}{A_{fr,1}} = \frac{A_{o,1}}{L_2L_3} \approx \frac{A_{o,1}}{L_2(b_1 + b_2 + 2a)N_{p,1}} = \frac{b_1\beta_1D_{h,1}/4}{b_1 + b_2 + 2a} \tag{87}$$

From the foregoing equations, it is also found that

$$\alpha_1 = \frac{\sigma_1}{D_{h,1}/4} = \frac{b_1\beta_1}{b_1 + b_2 + 2a} \tag{88}$$

$$\sigma_1 = \frac{\alpha_1 D_{h,1}}{4} \tag{89}$$

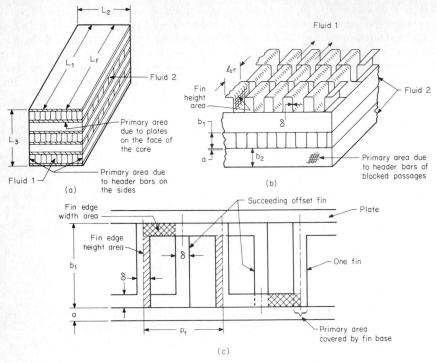

FIG. 36 (*a*) A crossflow plate-fin exchanger, (*b*) offset strip fins, and (*c*) idealized fin geometry.

b. Offset Strip-Fin Surface

We need to determine A_o, A_{fr}, and A for computing D_h, β, α, and σ as mentioned in the preceding section. As an illustration, the detailed geometrical properties are now presented for the idealized offset strip-fin surface of Fig. 36. The minimum free flow area is evaluated at the front face of the core (to avoid ambiguity at any row of strips within the core). The blocked area in one fin pitch p_f can be computed from Fig. 36c so that the minimum free flow area on the fluid 1 side is given by

$$A_{o,1} = b_1 L_2 N_p - [(b_1 - \delta) + p_f]\delta\, n_f \qquad (90)$$

where b_1 is the plate spacing, L_2 is the width of the fluid 1 passage, N_p is the total number of fluid 1 passages, δ is the fin thickness, p_f is the fin pitch, and n_f is a total number of fins on the fluid 1 side at any flow cross section.

The total heat transfer area consists of that area which is swept by fluid 1. That area consists of primary and secondary (fin) surface area. Primary area $A_{p,1}$ is a sum of: (1) plate area minus area covered by the fin base, (2) exposed header bar area on the sides of each fin passage, and (3) surface area of header bars and plates of fluid 2 that is exposed at core inlet and outlet faces.

$$A_{p,1} = \underbrace{2L_1 L_2 N_p - 2\delta L_f n_f}_{\substack{\text{plate area minus fin}\\ \text{base area}}} + \underbrace{2b_1 L_1 N_p}_{\substack{\text{header bar area}\\ \text{on sides}}} + \underbrace{2(b_2 + 2a)(N_p + 1)L_2}_{\substack{\text{header bar and plate}\\ \text{area at core faces}}} \qquad (91)$$

Here L_1 is the fluid 1 flow length, L_f is the fin flow length, and b_2 is the plate spacing on the fluid 2 side. It is idealized that the total number of passages on the fluid 2 side is one more than that on the fluid 1 side. In an actual core, sometimes the fin flow length L_f is slightly shorter than the core flow length L_1, and hence that distinction is made here.

The secondary (fin) area consists of the fin surface area parallel to the flow direction (to be referred to as fin height area) plus the fin edge area (two edges with each strip fin). The fin edge area has two components: the area associated with the edge height and the area associated with the edge width.

$$A_{f,1} = \underbrace{2(b_1 - \delta)L_f n_f}_{\text{fin height area}} + \underbrace{2(b_1 - \delta)\delta n_{\text{off}} n_f}_{\text{fin edge height area}} + \underbrace{(p_f - \delta)\delta(n_{\text{off}} - 1)n_f + 2p_f \delta n_f}_{\text{fin edge width area}} \quad (92)$$

where n_{off} is the total number of strips (offsets) in the flow direction.

The total heat transfer area on the fluid 1 side is

$$A_1 = A_{p,1} + A_{f,1} \quad (93)$$

Other geometrical properties of interest for the fluid 1 side are then computed from Eqs. (84) to (89).

2. Tube-Fin Surfaces

For heat exchanger calculations, the geometrical properties needed are the same as those for the plate-fin exchanger except for β, which loses its significance. The detailed geometrical properties are presented for *plain* individually finned tubes and continuous fins on a circular tube array.

a. Individually Finned Tubes

For the individually finned tubes of Fig. 3b, the fin root has an effective diameter d_o and the fin tip has a diameter d_e. Depending upon the manufacturing techniques, d_o may be the tube outside diameter or tube outside diameter plus two collar thicknesses. Consider the idealized inline and staggered arrangements of Fig. 37. The total number of tubes N_t is given by

$$N_t = \begin{cases} \dfrac{L_1 L_3}{X_t X_l} & \text{inline and staggered arrangements} \\[2ex] \dfrac{L_3}{X_t}\dfrac{L_1/X_l + 1}{2} + \left(\dfrac{L_3}{X_t} - 1\right)\dfrac{L_1/X_l - 1}{2} & \begin{array}{l}\text{staggered arrangement if} \\ \text{half tubes in alternate rows removed}\end{array} \end{cases} \quad (94)$$

The primary surface area is a sum of: (1) tube surface area minus area covered by the fins, and (2) header plate surface area on each side:

$$A_p = \pi d_o(L_2 - \delta N_f L_2)N_t + 2\left(L_1 L_3 - \frac{\pi d_o^2}{4} N_t\right) \quad (95)$$

Here L_1, L_2, and L_3 are core dimensions as shown in Fig. 37, δ is the fin thickness, and N_f is the number of fins per unit length. The secondary surface area is the fin surface area plus the fin edge area:

$$A_f = \left[\frac{2\pi(d_e^2 - d_o^2)}{4} + \pi d_e \delta\right] N_f L_2 N_t \quad (96)$$

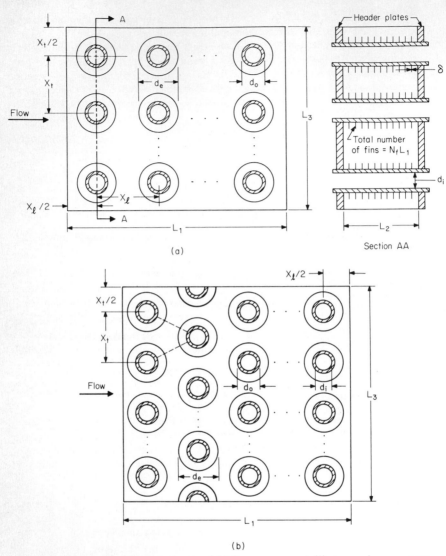

FIG. 37 A circular finned tube exchanger: (a) inline arrangement, (b) staggered arrangement.

Subsequently, the total surface area is

$$A = A_p + A_f \tag{97}$$

The expressions for A_p and A_f are valid for both inline and staggered arrangements as long as the proper value of N_t is used.

For the inline arrangement, the minimum free flow area is that area for a tube bank minus the area blocked by the fins:

$$A_o = [(X_t - d_o)L_2 - (d_e - d_o)\delta N_f L_2]\frac{L_3}{X_t} \tag{98}$$

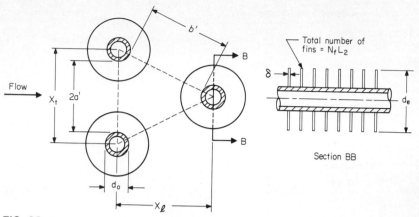

FIG. 38 A unit cell of staggered finned tube arrangement.

For the staggered arrangement, the minimum free flow area could occur either through the front row ($2a'$ in Fig. 38) or through the diagonals ($2b'$ in Fig. 38). Including the influence of the circular fins, the flow area per unit cell and unit tube length is associated with the $2a'$ and $2b'$ dimensions. Approximate magnitudes are shown in Fig. 38 and precise values are

$$2a' = (X_t - d_o) - (d_e - d_o)\delta N_f \tag{99}$$

$$b' = \left[\left(\frac{X_t}{2}\right)^2 + X_l^2 \right]^{1/2} - d_o - (d_e - d_o)\delta N_f \tag{100}$$

Now define c' as the minimum of $2a'$ and $2b'$. Then

$$A_o = \underbrace{\left(\frac{L_3}{X_t} - 1\right) c'L_2}_{\substack{\text{open area within} \\ \text{tubes}}} + \underbrace{[(X_t - d_o) - (d_e - d_o)\delta N_f]L_2}_{\substack{\text{open area beyond} \\ \text{the end tubes of the} \\ \text{first row}}} \tag{101}$$

Other geometrical properties of interest for the fin side are

$$D_h = \frac{4A_oL_1}{A} = \frac{4\sigma}{\alpha} \tag{102}$$

$$\alpha = \frac{A}{L_1L_2L_3} \tag{103}$$

$$\sigma = \frac{A_o}{A_{\mathrm{fr}}} = \frac{A_o}{L_2L_3} \tag{104}$$

b. Continuous Plain Fins on a Circular Tube Array

For this case, the only difference from the foregoing formulas is the different influence of the fins on geometrical properties. Continuous plain fins on staggered tubes are shown in Fig. 3a, for example. The expressions without explanation for A_p and A_f for both the inline and the staggered arrangement are

$$A_p = \pi d_o (L_2 - \delta N_f L_2) N_t + 2 \left(L_1 L_3 - \frac{\pi d_o^2}{4} N_t \right) \tag{105}$$

$$A_f = 2 \left(L_1 L_3 - \frac{\pi d_o^2}{4} N_t + L_3 \delta \right) N_f L_2 \tag{106}$$

The minimum free flow area for the inline arrangement is

$$A_o = [(X_t - d_o)L_2 - (X_t - d_o)\delta N_f L_2] \frac{L_3}{X_t} \tag{107}$$

and for the staggered arrangement, it is

$$A_o = \left(\frac{L_3}{X_t} - 1 \right) c'' L_2 + [(X_t - d_o) - (X_t - d_o)\delta N_f)] L_2 \tag{108}$$

where c'' is minimum of $2a''$ and $2b''$ and

$$2a'' = (X_t - d_o) - (X_t - d_o)\delta N_f \tag{109}$$

$$b'' = \left[\left(\frac{X_t}{2} \right)^2 + X_l^2 \right]^{1/2} - d_o - (X_t - d_o)\delta N_f \tag{110}$$

Subsequently, D_h, α, and σ can be calculated from Eqs. (102) to (104).

3. Compact Regenerators

In a compact regenerator, the matrix may consist of cylindrical passages in parallel as shown in Fig. 6 with passage cross section either circular or noncircular. Surface geometrical properties for some idealized geometries are provided in Table 14 [105]. Some general relationships are

$$D_h = \frac{4 A_o L}{A} = \frac{4\sigma}{\alpha} \qquad \sigma = \frac{A_o}{A_{fr}} \qquad \alpha = \frac{A}{V} = \frac{4\sigma}{D_h} \tag{111}$$

G. EXCHANGER RATING AND SIZING PROBLEMS

Two most common heat exchanger design problems are rating and sizing problems. For an existing exchanger, the performance evaluation problem is referred to as the *rating problem*. To arrive at a design of a new exchanger to meet the specified performance is referred to as a *sizing problem*. A detailed procedure is outlined to obtain solutions to these problems for a direct transfer–type two-fluid compact heat exchanger. For further details and a numerical example, refer to Ref. 106. Customarily, the ϵ-NTU method of heat exchanger analysis is employed by the designers and manufacturers of compact heat exchangers. Hence, the solution procedure is presented using the ϵ-NTU method.†

---

†Since the dimensionless groups of the ϵ-NTU method and the log-mean temperature difference (LMTD) method are uniquely related (see Table 2 on p. 4-10), the solutions to the rating and sizing problems by the LMTD method will be identical (within the specified convergence criteria) to those by the ϵ-NTU method, although the detailed steps of calculations for the two methods will be different.

TABLE 14 Surface Geometrical Properties for Some Idealized Passages Used in Rotary Regenerators [105]

Geometry	Cell density N_c cells/m²	Porosity σ	Surface area density β, m²/m³	Hydraulic radius r_h, m
	—	0.37–0.39	$\dfrac{6(1-\sigma)}{b}$	$\dfrac{b\sigma}{6(1-\sigma)}$
	$\dfrac{1}{(b+\delta)^2}$	$\dfrac{b^2}{(b+\delta)^2}$	$\dfrac{4b}{(b+\delta)^2}$	$\dfrac{b}{4}$
	$\dfrac{2}{\sqrt{3}(b+\delta)^2}$	$\dfrac{b^2}{(b+\delta)^2}$	$\dfrac{4b}{(b+\delta)^2}$	$\dfrac{b}{4}$
	$\dfrac{2}{\sqrt{3}(b+\delta)^2}$	$\dfrac{\pi b^2}{2\sqrt{3}(b+\delta)^2}$	$\dfrac{2\pi b}{\sqrt{3}(b+\delta)^2}$	$\dfrac{b}{4}$
	$\dfrac{1}{(b/6+\delta)(b+\delta)}$	$\dfrac{b^2/6}{(b/6+\delta)(b+\delta)}$	$\dfrac{7b/3}{(b/6+\delta)(b+\delta)}$	$\dfrac{b}{14}$
	$\dfrac{4\sqrt{3}}{(2b+3\delta)^2}$	$\dfrac{4b^2}{(2b+3\delta)^2}$	$\dfrac{24b}{(2b+3\delta)^2}$	$\dfrac{b}{6}$

1. The Rating Problem

The rating problem, also referred to as a *performance problem,* is analyzed for predicting the performance of either an existing exchanger or an already sized exchanger. The objective here is either to verify the vendor's specifications or to determine the performance under off-design conditions. In this problem, the following quantities are specified: the exchanger construction type, flow arrangement, overall core dimensions, complete details on the materials and surface geometries on both sides including their heat transfer and pressure drop characteristics, fluid flow rates, inlet temperatures, and fouling factors. The designer's task is to predict the fluid outlet temperatures, total heat transfer rate, and pressure drops on each side.

In the following, basic steps for heat transfer and pressure drop calculations are summarized for the rating problem for a recuperator.

a. Heat Transfer Calculations

1. Compute Surface Geometrical Properties. The following geometrical properties on each side are needed for the heat transfer and pressure drop analysis: the primary and secondary (if any) surface area A, minimum free flow area A_o, frontal area A_{fr}, hydraulic diameter D_h, flow length L for the ΔP calculations, ratio of free flow area to frontal area σ for exchanger entrance and exit losses, the fin length l, and other pertinent dimensions for the fin. Whatever information is not specified directly is computed based on the specified surface geometries and core dimensions as outlined as an example for the offset strip-fin surface on p. 4-254.

2. Determine Bulk Mean Temperature and Fluid Properties. Exchanger transfer rate and fluid pumping power are strongly dependent upon the type of fluid flowing through the exchanger. It is essential that fluid properties be established accurately for the appropriate temperature and pressure conditions. The fluid properties needed are density, specific heat, viscosity, thermal conductivity, and Prandtl number. These properties, available for a large number of fluids in the literature (see, for example, Ref. 170) are determined at an appropriate temperature on each side of the exchanger.

For the heat transfer analysis, we need to obtain a "single" temperature value to represent the flow of fluid through each side of the exchanger. For counterflow and parallel-flow exchangers, the bulk fluid temperature varies in the flow direction as well as over the cross section of the individual flow passages. In a crossflow heat exchanger, it also varies in the flow direction of the fluid on the other side. In more complex arrangements, the fluid temperature variation is generally three-dimensional. The temperature changes in the flow direction affect the fluid bulk properties for the exchanger analyses. The temperature changes across the individual flow passages influence the velocity and temperature profiles, and thereby influence the friction factor and the convective heat transfer coefficient. If the fluid properties vary substantially within an exchanger, an "average" temperature may not be adequate to accurately determine the heat transfer rate and pressure drop. In that case, the exchanger is divided into several segments, and the conventional ϵ-NTU analysis is carried out for each segment with an average temperature representative of each segment.

The influence of temperature variations across the individual flow passages is accounted for by the property ratio method as discussed in Sec. D.4, p. 4-240. Here the temperature is characterized by both the wall and bulk mean temperatures, and corrections to Nu and f are applied by appropriate property ratio fractions.

TABLE 15 Approximate Bulk Mean Temperatures on the Hot and Cold Fluid Sides of a Two-Fluid Compact Exchanger

C_{max} = hot fluid, C_{min} = cold fluid	C_{max} = cold fluid, C_{min} = hot fluid
$C^* < 0.5$	
$T_{h,m} = \dfrac{T_{h,i} + T_{h,o}}{2}$	$T_{c,m} = \dfrac{T_{c,i} + T_{c,o}}{2}$
$T_{c,m} = T_{h,m} - \Delta T_{lm}$	$T_{h,m} = T_{c,m} + \Delta T_{lm}$
$\Delta T_{lm} = \dfrac{(T_{h,m} - T_{c,i}) - (T_{h,m} - T_{c,o})}{\ln\,[(T_{h,m} - T_{c,i})/(T_{h,m} - T_{c,o})]}$	$\Delta T_{lm} = \dfrac{(T_{h,i} - T_{c,m}) - (T_{h,o} - T_{c,m})}{\ln\,[(T_{h,i} - T_{c,m})/(T_{h,o} - T_{c,m})]}$
$C^* \geq 0.5$	
$T_{h,m} = \dfrac{T_{h,i} + T_{h,o}}{2}$	$T_{c,m} = \dfrac{(T_{c,i} + T_{c,o})}{2}$

For the flow length "average" (or bulk mean) temperature on each side of a two-fluid heat exchanger, the suggested method is presented in Table 15 for two cases: $C^* < 0.5$ and $C^* \geq 0.5$ [106]. It should be emphasized that this procedure is somewhat arbitrary. Since ΔT_{lm} is used directly or indirectly for the exchanger heat transfer analysis, it is used on the C_{min} side to compute the bulk mean temperature as summarized in Table 15.

Since the outlet temperatures are not known for this problem, an approximate method is used first to best estimate them. For example, assume 60 percent effectiveness for shell-and-tube or plate heat exchangers, 75 percent effectiveness for most extended-surface crossflow heat exchangers, and 85 percent effectiveness for regenerators and extended-surface counterflow exchangers. If the designer knows a more accurate value for the application, that value of the effectiveness is used. Based on the assumed effectiveness, determine the outlet temperatures. Subsequently arrive at the mean temperature on both sides by the method of Table 15. Evaluate the physical properties of the fluids (μ, ρ, c_p, k, and Pr) on each side at these mean temperatures for the first trial.

3. Calculate Reynolds Numbers and Surface Basic Characteristics. Once the surface geometrical and fluid physical properties are known, determine G and Re from their definitions:

$$G = \frac{W}{A_o} \qquad \text{Re} = \frac{GD_h}{\mu} \tag{112}$$

The exchanger surface basic characteristics are generally presented in terms of j and f versus Re or Nu versus Re or x^*, and f_{app} Re versus x^+ as mentioned in Sec. D. Thus depending upon the correlations, other independent parameters such as $L^* = L/(D_h\,\text{Re}\,\text{Pr})$, $L^+ = L/(D_h\,\text{Re})$, etc., may need to be evaluated to arrive at j or Nu and f. Some correlations may involve more complicated functions or geometrical parameters.

The j or Nu factor thus determined is generally for constant fluid properties. The j or Nu correction factor for variable fluid properties is computed from Eqs. (75) and (76) with the property ratio indexes n and m chosen from Table 11 or 12. However, for these corrections, we need to calculate T_w, which is unknown and is dependent upon the thermal resistances on the hot and cold sides [see Eqs. (12) and (13) on p. 4-7]. Fortunately, for laminar flow, $n = 0$ from Table 11, indicating that the correction factor for the variable-property j factor is unity. For turbulent flow, unless an approximate value of T_w is known, assume the property ratio correction factor to be unity for this first trial.

After the following heat transfer calculations are completed, the thermal resistances on the hot and cold fluid sides will be known, and we will be able to compute T_w to correct for the constant-property f factors.

4. Evaluate Heat Transfer Coefficients and Fin Efficiencies. Once the Colburn factor j or the Nusselt number Nu is known for each side and is corrected for the temperature-dependent fluid properties, determine the heat transfer coefficient from one of the definitions

$$h = jGc_p \, \mathrm{Pr}^{-2/3} \quad \text{or} \quad h = \frac{\mathrm{Nu} \, k}{D_h} \tag{113}$$

If fins are used on any side, the temperature effectiveness of the fin, η_f, is evaluated using h from Eq. (113) and the geometry and material of the fin. η_f formulas for some important fin geometries are presented in Table 1 on p. 4-186. The extended surface efficiency η_o is then calculated from Eq. (2).

The overall thermal conductance UA is then computed from

$$\frac{1}{UA} = \frac{1}{(\eta_o hA)_h} + \frac{1}{(\eta_o h_s A)_h} + R_w + \frac{1}{(\eta_o h_s A)_c} + \frac{1}{(\eta_o hA)_c} \tag{114}$$

Here h_s is the known scale or fouling coefficient, and R_w is the wall thermal resistance to be determined from Eq. (7) of Part 1 of this chapter, p. 4-6.

5. Compute C^* and NTU. First calculate the fluid heat capacity rate $C = Wc_p$ on each side, and then C^* and NTU from their definitions:

$$C^* = \frac{C_{\min}}{C_{\max}} \qquad \mathrm{NTU} = \frac{UA}{C_{\min}} \tag{115}$$

6. Exchanger Effectiveness and Outlet Temperatures. Once NTU and C^* are determined and the flow arrangement is known, the exchanger effectiveness is determined from the ϵ-NTU relationships of Table 3 on p. 4-11, or from the tabular data presented by Kays and London [16]. If the calculated effectiveness is greater than about 80 percent, the influence of longitudinal heat conduction must be evaluated as summarized on pp. 4-53 to 4-56. The pertinent parameters such as λ, $(\eta_o hA)^*$, etc., then are first calculated to determine ϵ.

The outlet temperatures are subsequently calculated from the definition of ϵ:

$$T_{h,o} = T_{h,i} - \epsilon \frac{C_{\min}}{C_h} (T_{h,i} - T_{c,i}) \tag{116}$$

$$T_{c,o} = T_{c,i} + \epsilon \frac{C_{\min}}{C_c} (T_{h,i} - T_{c,i}) \tag{117}$$

If these outlet temperatures are significantly different from those assumed in the foregoing step 2, then steps 3 through 6 are repeated with the latest values of the outlet temperatures of Eqs. (116) and (117). Generally, one iteration would be sufficient. The heat transfer rate is then computed from

$$q = \epsilon C_{\min}(T_{h,i} - T_{c,i}) \tag{118}$$

b. Pressure Drop Calculations

Once the heat transfer calculations are complete, the evaluation of pressure drop on each fluid side is straightforward. The pressure drop associated with the core on each side is computed from Eq. (23) or the appropriate equation from Sec. C. The pressure drops associated with inlet or outlet tanks, manifolds, headers, or nozzles are determined from empirical relationships such as those in Ref. 168.

For the core pressure drop of Eq. (23), K_c and K_e are evaluated from Fig. 18 at the Re and σ computed earlier. The constant-property friction factor f first must be corrected for variable properties using Eq. (75) or (76) and Table 11 or 12 for the exponent m. Here the required T_m and T_w are evaluated as follows: The bulk mean temperature T_m for the fluid on each side is computed from the relationships of Table 15. Subsequently, the mean wall temperature is calculated from Eq. (12) or (13) on p. 4-7.

2. The Sizing Problem

The sizing problem, also referred to as a *design problem*, involves a design of a new heat exchanger for specified performance within known constraints and minimum objective function. In this problem, the fluid inlet and outlet temperatures and flow rates are generally specified as well as the pressure drop on each side. The designer's task is to select construction type, flow arrangement, materials, and surfaces on each side, and to determine the core dimensions to meet the specified heat transfer and pressure drop. We have already briefly discussed the selection of compact exchanger construction type (Sec. E.1), flow arrangement (Sec. A.2), and surface selection (Sec. E.2). The sizing problem then reduces to the determination of the core dimensions for the specified heat transfer and pressure drop performance. Of course, one could reduce this problem to the rating problem by tentatively specifying the dimensions, and then predicting the performance for comparison with that specified. With successive approximations, one could finally arrive at the required size. However, this process is accelerated if a coupling equation between the specified heat transfer and pressure drop is utilized to determine the first guess of the core mass velocity G on one side of a two-fluid exchanger.

This equation, referred to as the core mass velocity equation, is [16, 106]

$$G_1 = \left\{ \left[\frac{2g_c P_i \eta_o}{(1/\rho)_m \, \mathrm{Pr}^{2/3}} \right]_1 \left[\frac{\Delta P/P_i}{\mathrm{NTU}} \right]_1 \left(\frac{j}{f} \right)_1 \right\}^{1/2} \tag{119}$$

Here, the subscript 1 refers to one side of interest in a two-fluid heat exchanger. This equation is based on the idealization that most of the pressure drop is associated with the core friction term of Eq. (23). Since $\Delta P/P_i$ is specified on one side of the exchanger, the unknown quantities in Eq. (119) are η_o, NTU on that side, and the j/f ratio for the surface selected on that side.

The feature that makes this equation so useful is that the ratio j/f is a relatively flat function of the Reynolds number (see Fig. 2-38 of Ref. 16). Thus one can readily estimate an accurate magnitude of j/f based on a "ball park" estimate of Re.

If there are no fins, $\eta_o = 1$ in Eq. (119). For a "good design," the fin geometry is chosen such that η_o is in the range 70 to 90 percent. Therefore, $\eta_o = 0.80$ is suggested as a first approximation to determine G from Eq. (119).

The overall NTU is determined for the chosen flow arrangement from the required exchanger effectiveness ϵ (a known quantity) and the known heat capacity rate ratio C^*. Now NTU_1 (on one side) can be estimated from the overall NTU for two specific cases as follows [106]:

$$\text{NTU}_1 = \begin{cases} 2\ \text{NTU} & \text{for a gas-to-gas exchanger} \\ \\ 1.1\ \text{NTU} & \text{for a gas-to-liquid exchanger} \end{cases} \tag{120}$$

Thus any value of NTU_1 on the gas side from 1.1 NTU to 2 NTU will be adequate. A better estimate of NTU_1 will provide the first estimate of G closer to the actual value. Our purpose is to get a good estimate on G for the first iteration, and the foregoing guidelines on η_o, NTU, and j/f will provide the estimated G within 10 to 20 percent of the actual value and sometimes even much closer. Once the core mass velocity is determined, the solution to the sizing problem is carried out in a manner similar to the rating problem. We will now outline a step-by-step procedure for sizing two specific exchangers: (1) a counterflow or $C^* \approx 0$ exchanger, and (2) a crossflow gas-to-gas exchanger.

a. Solution for a Counterflow or $C^* \approx 0$ Exchanger

In a single-pass compact counterflow heat exchanger, if the core dimensions on one side are fixed, the core dimensions for the other side (except for the passage height) are also fixed. Therefore, the design problem for this case is solved for the side that has the more stringent $\Delta P/P$ specification. For the $C^* \approx 0$ case, i.e., gas-to-liquid or phase-changing fluid exchanger, the thermal resistance is primarily on the gas side and the pressure drop is more stringent (critical) on the gas side. As a result, the core dimensions arrived at are based on the gas side ΔP and NTU. The dimensions on the other side are then chosen so that they meet the specified ΔP. Thus either for counterflow or for a $C^* = 0$ exchanger, the core dimensions are arrived at for the side having the more stringent ΔP. This side will be referred to as side 1 in the following. The following is a step-by-step procedure for the solution.

1. Determine the fluid outlet temperatures from the specified exchanger effectiveness. Obtain the fluid mean temperature on each side by the method of Table 15 and evaluate the fluid physical properties ρ_i, ρ_o, $(1/\rho)_m$, c_p, μ, k, and Pr.

2. Calculate C^* and ϵ, and determine NTU for the exchanger for the selected flow arrangement. The influence of longitudinal heat conduction, if any, is ignored in the first approximate calculation.

3. Determine NTU_1 by the approximations discussed with Eq. (120).

4. For the selected surface on the stringent ΔP side, plot the j/f versus Re curve and obtain a "ball park" value of j/f. If fins are employed on this side, assume $\eta_o = 0.80$.

5. Evaluate G from Eq. (119) using the information from steps 1 to 4 and the input value of $\Delta P/P$.

6. Calculate Reynolds number Re, and determine j and f for this Re.

7. Compute h, η_f, and η_o. In order to determine U_1, either lump the fouling resistances with the convective film resistances on the respective sides or ignore them. Similarly, either divide and lump the wall thermal resistance with the respective hot- and cold-side resistances or ignore it. U_1 from Eq. (114), after replacing the subscripts h and c with 1 and 2, is

$$\frac{1}{U_1} = \frac{1}{(\eta_o h)_1} + \frac{(A_1/A_2)}{(\eta_o h)_2} \tag{121}$$

where

$$\frac{A_1}{A_2} = \frac{\alpha_1}{\alpha_2} \qquad (122)$$

Here α_1 and α_2 (the ratios of heat transfer area on the respective side divided by the total exchanger volume) are known from the input geometry specifications.

8. Now calculate the core dimensions. Determine A_1 from NTU for this U_1 and known C_{min}

$$A_1 = \text{NTU} \frac{C_{min}}{U_1} \qquad (123)$$

and A_o from known W and G

$$A_{o,1} = \left(\frac{W}{G}\right)_1 \qquad (124)$$

so that

$$A_{fr,1} = \frac{A_{o,1}}{\sigma_1} \qquad (125)$$

where σ_1 is known as computed from Eq. (87). Finally evaluate the core length L_1 from known $A_{o,1}$, A_1, and $D_{h,1}$ (known for input geometry) as

$$L_1 = \left(\frac{D_h A}{4A_o}\right)_1 \qquad (126)$$

Since $A_{fr,1} = A_{fr,2}$ for the counterflow exchanger, once $A_{fr,1}$ is calculated, the minimum free flow area on side 2 is computed from $A_{o,2} = \sigma_2 A_{fr,1}$, with σ_2 known for side 2. Then the mass velocity $G_2 = W_2/A_{o,2}$ should be less than G_2 calculated by applying the core velocity equation on side 2, in order to assure satisfying the ΔP requirement on side 2. For the known frontal area $A_{fr,1}$, the reasonable core dimensions are chosen considering the header design and available space.

9. Now correct the f factor for the variable-property effects as mentioned in Sec. G.1.b, p. 4-263, and calculate $\Delta P/P$ from Eq. (23) or an appropriate equation.

10. If the calculated value of $\Delta P/P$ is within the input specification, the approximate solution to the sizing problem is finished. Further refinements in the core dimensions may be carried out at this time.

   If the calculated value of $\Delta P/P$ is larger than the input value, compute an improved value of G from Eq. (23) using the specified $\Delta P/P$ and the values of f and L from the preceding step. With this new value of G, repeat steps 6 to 10 until both heat transfer and pressure drops are met as specified.

   If the influence of longitudinal heat conduction is important, the longitudinal conduction parameter and other appropriate dimensionless groups are calculated based on the core geometry from the preceding iteration and input operating conditions. Subsequently, a new value of NTU, accounting for longitudinal conduction, is calculated. This new value of NTU is then used in step 8.

b. Solution for a Crossflow Exchanger

For a crossflow exchanger, determining the core dimensions on one side (A_{fr} and L) does not fix the dimensions on the other side. In such a case, the design problem is solved simultaneously on both sides. The solution procedure follows closely with that of the preceding subsection, and is outlined next through detailed steps.

1. Determine G on each side by following Steps 1 to 5 of the preceding subsection G.2.a.

2. Follow Steps 6 and 7 of the Sec. G.2.a, and compute for each side the following: Re, j, f, h, η_f, and η_o. Subsequently, compute U_1 using Eq. (121).

3. The core dimensions are then calculated as follows: A_1, $A_{o,1}$, $A_{fr,1}$, and L_1 using Eqs. (123) to (126). Note that here $A_{fr,1} = L_2L_3$. Similarly calculate A_2, $A_{o,2}$, $A_{fr,2}$, and L_2 using Eqs. (123) to (126) by replacing the subscript 1 with 2. Note that here $A_{fr,2} = L_1L_3$. Thus the no-flow (or stack) height L_3 can be determined from either $A_{fr,1}$ or $A_{fr,2}$ and known L_2 or L_1. It can be rigorously shown that L_3 calculated either from $A_{fr,1}$ or from $A_{fr,2}$ will be identical.

4. Now correct the f factor for the variable-property effects as mentioned in the preceding Sec. G.1.b, and calculate $\Delta P/P$ from Eq. (23) or an appropriate equation on each side.

5. If the calculated value of $\Delta P/P$ is within the input specification on each side, the approximate solution to the sizing problem is complete. Further refinements in the core dimensions may be carried out at this time.

   If the calculated value of $\Delta P/P$ is larger on one or both sides, compute an improved value of G on each side from Eq. (23) using the specified $\Delta P/P$ and the values of f and L from the preceding step. With these new values of G, repeat the foregoing steps 2 to 5 until both heat transfer and pressure drops are met as specified.

   If the influence of longitudinal heat conduction is important, the longitudinal conduction parameters and other appropriate dimensionless groups are calculated based on the core geometry from the preceding iteration and input operating conditions. Subsequently, a new value of NTU, accounting for longitudinal conduction, is calculated. This new value of NTU is then used in Eq. (123).

Since we have not imposed any restrictions on the core dimensions, a unique set of core dimensions for a crossflow exchanger will be obtained by the aforementioned procedure that will meet the specified heat transfer and pressure drops on each side.

3. Computer-Aided Thermal Design

At the present state of the art, a complete thermal design can be carried out by hand calculations. However, the process is time-consuming, tedious, and prone to human errors and neglects many second-order influences to simplify calculations. In the present computer era, thermal design is exclusively performed using computers by most industries. Chenoweth and Kistler [107] and Bell [108] discuss computer-aided thermal design methods for shell-and-tube exchangers, and Shah [1] for compact exchangers. A sophisticated computer program not only has the options of solving straightforward rating and sizing problems, but also has optimization procedures incorporated. Such a computer program is complicated and rather large depending upon the options made available. Some details on the major subroutines of such a computer program are presented in Ref. 1.

H. HEADER DESIGN

Headers or manifolds connect the supply and return pipes to the heat exchanger core. An inlet header is a transition "duct" joining the inlet face of the exchanger core or matrix to the inlet (supply) pipe. The outlet header joins the outlet face of the exchanger core to the outlet (return) pipe. The header is variously referred to as a tank, box, or distributor. Headers may be classified as normal, turning, and oblique flow headers, as shown in Figs. 39 to 42.

A manifold (or header) is a flow channel in which the fluid enters at one end (or in the middle) and leaves through side walls, or vice versa, due to the action of a differential pressure. Some examples of manifolds will be shown in Figs. 47 and 48. In the literature and in practical use, generally there is no sharp distinction between the terms header and manifold; they are used interchangeably.

In all heat transfer and pressure drop analyses so far presented, it is presumed that the fluid is uniformly distributed through the core. A serious deterioration in heat exchanger performance may result when the flow distribution through the core is non-uniform, particularly when an area increase from the inlet pipe to the core face is 5 to 50 times. Uniformity of flow distribution over the core face is the primary function of the headers and manifolds.

The design of the outlet header should match that of the inlet header (or vice versa) so that the pressure drop across the core is uniform, resulting in a uniform flow distribution. To minimize pressure losses due to flow separation, area contraction in the outlet header should be smooth. Also sharp turns in the outlet header should be avoided. Turns and bends create centrifugal forces, resulting in nonuniform pressure at the core face, which may lessen flow uniformity in the core.

From the foregoing viewpoints, important considerations for header and manifold design are: (1) Flow should be distributed as uniformly as possible at the core face, (2) pressure drop in the headers and manifolds should be as low as possible by minimizing flow separation and flow impact effects, and (3) for liquid flows, flow nonuniformity in the headers and manifolds should be minimized to avoid high velocity regions that can produce localized erosion. Design considerations for oblique-flow headers and normal-flow headers are summarized next.

1. Oblique-Flow Headers

Unlike normal-flow headers, the inflow is at an angle to the heat transfer core face in an oblique-flow header. However, design theory is available only for those oblique-flow headers for which inflow is parallel to the core face as shown in Figs. 40, 41, and 42. For

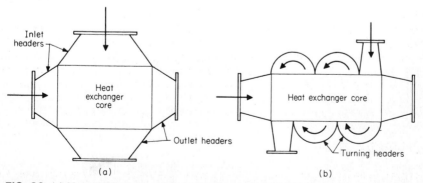

FIG. 39 (*a*) Normal-flow inlet and outlet headers and (*b*) turning headers between passes.

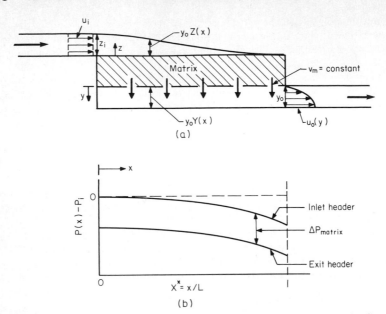

FIG. 40 (*a*) Parallel-flow headers and core and (*b*) pressure distribution in the inlet and outlet headers.

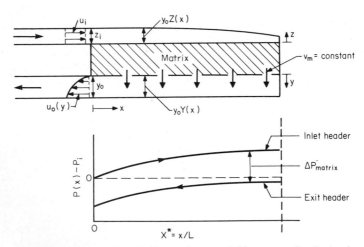

FIG. 41 (*a*) Counterflow headers and core and (*b*) pressure distribution in the inlet and outlet headers.

simplicity of design, either the outlet header is considered as a box type (Figs. 40 and 41) or there is a free discharge into an infinite reservoir (no outlet header).† For uniform flow

---

†The free-discharge header configuration is commonly encountered in air conditioning heat exchanger installations.

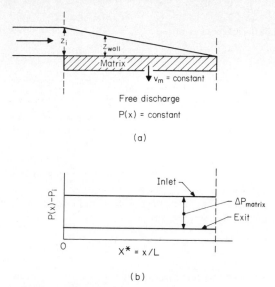

FIG. 42 (*a*) Free discharge header and core and (*b*) pressure distribution at the core inlet and outlet.

through the core with a free discharge, the inlet header has to be triangular. Other shapes are required when a box-type outlet header is used. For the overall parallel-flow arrangement (in inlet and outlet headers) of Fig. 40, flow continuously accelerates in the inlet header as in a nozzle; the flow continuously decelerates in the inlet header as in a diffuser for the overall counterflow arrangement of Fig. 41. For parallel-flow headers, the inlet header dimension z_i can be either larger or smaller than the outlet dimension y_o. The theoretical shape of the inlet header (see Fig. 40) is given by [109]

$$\frac{z}{y_o} = \frac{1 - X^*}{[\,(\rho_i/\rho_o)\,(\pi^2/4)X^{*2} + (y_o/z_i)^2\,]^{1/2}} \tag{127}$$

where ρ_i and ρ_o are the fluid densities at the entrance of the inlet header and the exit of the outlet header, respectively, and $X^* = x/L$. The pressure drop in the inlet and outlet headers is

$$\frac{\Delta P_i}{H_i} = 1 + 0.822 \frac{H_o}{H_i} \tag{128}$$

$$\frac{\Delta P_o}{H_i} = 0.645 \frac{H_o}{H_i} \tag{129}$$

$$\frac{\Delta P_t}{H_i} = 1 + 1.467 \frac{H_o}{H_i} \tag{130}$$

where H_i and H_o are the velocity heads at the entrance of the inlet header and at the exit of the outlet header, respectively,

$$H_i = \left(\frac{\rho V^2}{2g_c}\right)_i, \qquad H_o = \left(\frac{\rho V^2}{2g_c}\right)_o \tag{131}$$

and $\Delta P_t = \Delta P_i + \Delta P_o$. For $z_i = y_o$, $H_i = H_o$; the pressure drop in inlet, outlet, and in both inlet and outlet headers is thus 1.822, 0.645, and 2.467 times the inlet velocity head, respectively.

For the counterflow headers of Fig. 41, the shape of the inlet header is given by [109]

$$\frac{z}{y_o} = 0.636 \left(\frac{\rho_o}{\rho_i}\right)^{1/2} \tag{132}$$

in order to assure matching pressure distribution (constant ΔP across the core) and a minimum header loss. The pressure drops associated with the inlet, outlet, and inlet plus outlet headers are

$$\frac{\Delta P_i}{H_i} = 0.333 \qquad \frac{\Delta P_o}{H_i} = 0.261 \qquad \frac{\Delta P_t}{H_i} = 0.594 \tag{133}$$

For the free discharge case (Fig. 42), the profile of the inlet header is linear to ensure uniform flow distribution through the core.

$$\frac{z}{z_i} = 1 - \frac{x}{L} \tag{134}$$

The pressure loss associated with the inlet header is one velocity head.

$$\frac{\Delta P_i}{H_i} = \frac{\Delta P_t}{H_i} = 1 \tag{135}$$

In summary, the parallel-flow configuration has a total header loss, exclusive of core ΔP, equal to 2.47 times the inlet velocity head based on the velocity u_i as shown in Fig. 40. In contrast, the free-discharge configuration has only one inlet velocity head loss and the counterflow configuration has 0.59 times the inlet velocity head loss, one-fourth that of the parallel-flow configuration! Clearly, if the option is available, a counterflow header arrangement is the preferred geometry.

In practice, a variety of oblique-flow headers are used in heat exchangers. As a practical solution, guide vanes are sometimes used to turn the flow and ensure reasonable uniform distribution (such as shown in Fig. 43). However, no design theory is available at present for vaned headers.

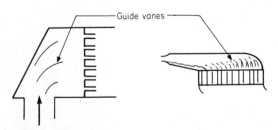

FIG. 43 Oblique-flow headers with turning vanes.

A highly compact exchanger generally requires a large frontal area and a short flow length due to the pressure drop constraint. One way to accommodate a large frontal area and still reduce the header volume is the folded core concept as shown in Fig. 44. The arrows indicate the desired fluid flow directions. However, the fluid does not turn and flow perpendicular to the core. This configuration results in very nonuniform flow distribution through the core and a serious reduction in heat exchanger performance.

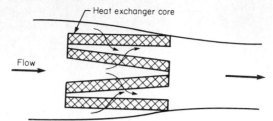

FIG. 44 A folded exchanger core illustrating an oblique-flow header configuration.

2. Normal-Flow Headers

In normal-flow headers, the inflow is perpendicular to the heat transfer core as shown in Fig. 39. The inlet header acts as a diffuser because of a large increase in the flow area from the inlet pipe to the exchanger core face. An extensive study on the flow regimes and design correlations for planar, annular, and conical diffusers with a free or duct discharge has been conducted by Kline [111]. Also some guidelines on the diffuser design are presented by Miller [112]. The diffuser performance data presented in Refs. 111 and 112 may not be applicable to the design of heat exchanger headers since a relatively high flow resistance is imposed at the diffuser outlet (i.e., inlet header exit) in an exchanger; this resistance in turn affects the header performance.

However, for reference, the major flow regimes that occur in a two-dimensional diffuser (Fig. 45a) with a ducted discharge (no appreciable flow resistance downstream) are shown in Fig. 45b. The following four regimes are obtained by increasing the total diffuser angle 2θ for a constant N/W_1: (1) *No appreciable stall* occurs in Fig. 45b in the region below line a-a. The main flow is steady, well behaved, and unseparated. (2) *Large transitory stall* occurs in the region between lines a-a and b-b. In this regime, the boundary layer separates from one wall but remains attached on the other. The separation location varies with time. The flow is rapidly varying with distorted and unsteady exit conditions. As the angle increases, the average position of detachment moves along the wall toward the throat. (3) *Fully developed stall* occurs between lines b-b and d-d. In this regime, there is a large recirculating zone on one wall which extends from the exit to close to the throat. The main flow follows smoothly along the other wall, and flow is relatively steady. (4) *Jet flow* occurs for those diffusers that lie above line c-c. The main flow is separated from both walls and detachment occurs slightly downstream of the diffuser throat. This regime is also relatively steady. The regime between lines c-c and d-d is referred to as the hysteresis zone. Experimental correlations for the pressure recovery in each flow regime are summarized in Ref. 111.

For gas flows, the header volume is considerably reduced and a fairly uniform flow distribution is attained at the core inlet with the use of a baffle or a perforated plate as shown in Fig. 46. For a perforated plate, the desired perforation size is small-diameter holes in the central region with increasing-diameter holes away from the center. No systematic study on the hole pattern, baffle size, inlet pipe diameter, and header size has been reported.

3. Manifolds

Manifolds can be classified as two basic types: simple dividing-flow and combining-flow as shown in Fig. 47. The parallel and reverse flow systems are then the combinations of

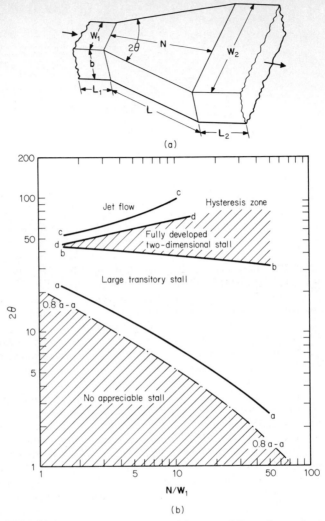

FIG. 45 (a) A planar diffuser and (b) planar diffuser flow regime chart.

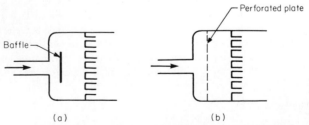

FIG. 46 (a) Baffle and (b) perforated plate in the plenum chamber of the inlet header.

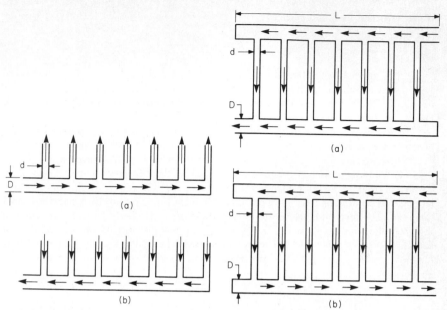

FIG. 47 (*a*) Dividing flow manifold and (*b*) combining flow manifold.

FIG. 48 (*a*) Parallel-flow manifold system and (*b*) reverse-flow manifold system.

the basic dividing and combining flow manifolds interconnected by lateral branches or a heat exchanger core as shown in Fig. 48.

For simple open-ended dividing- and combining-flow manifolds, Majumdar [113] obtained a numerical solution to predict the flow distribution in the lateral circular tube branches and the pressure distribution in the circular tube header. He found an excellent agreement between the numerical and experimental results. He identified two dimensionless parameters on which the aforementioned flow and pressure distributions depend. After a parametric study, he presented design charts for dividing- and combining-flow manifolds.

Bajura and Jones [114] investigated parallel-flow and reverse-flow manifold systems both analytically and experimentally. They found an excellent agreement between the experimental results and analytical predictions for pressure distributions in the headers and flow distribution in the lateral branches. The analysis of Bajura and Jones has been extended by Datta and Majumdar [115] by a finite difference method. They found that the results are dependent upon three dimensionless groups:

Branch-to-header flow area ratio

$$A_o^* = \frac{Nd^2}{D^2} \tag{136}$$

Header friction parameter

$$F = \frac{4fLD}{Nd^2} \tag{137}$$

Branch pressure loss coefficient

$$K = \frac{\Delta P}{(\rho V^2 / 2 g_c)} \tag{138}$$

Here N is the number of lateral branches; d, D, and L are defined in Fig. 48; f is the Fanning friction factor for flow in the header pipe (diameter D); and ΔP is the pressure drop in the lateral branches. For a heat exchanger, the branch pressure loss coefficient K is equal to the bracketed term [ ] of Eq. (23).

The results of a parametric study of A_o^* and F by Datta and Majumdar [115] for parallel- and reverse-flow systems are presented in Figs. 49 and 50 for the case of 10 lateral branches and the branch pressure loss coefficient $K = 1.72$. These results may be used for N branches with $N > 10$ if the abscissas of these figures are interpreted as x/L ranging from 0.1 to 1.0 instead of the 1 to 10 shown. The following observations may be made from these figures: (1) The ratio of branch flow area to the header flow area (area of the header pipe before the lateral branches), A_o^*, should be less than unity to minimize flow maldistribution. (2) More uniform flow distribution through the lateral branches (core) is achieved by a reverse-flow manifold system than by a parallel-flow manifold system. (3) In parallel- and reverse-flow manifold systems, the maximum flow

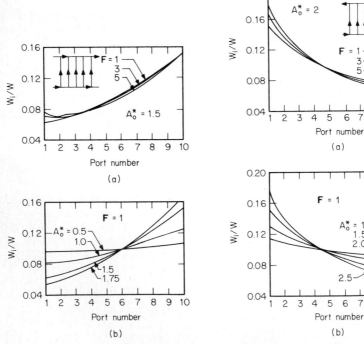

FIG. 49 Influence of (a) header friction parameter F and (b) branch-to-header flow area ratio A_o^* on the flow distribution through the lateral branches of the parallel-flow manifold system ($K = 1.72$).

FIG. 50 Influence of (a) header friction parameter F and (b) branch-to-header flow area ratio A_o^* on the flow distribution through the lateral branches of the reverse-flow manifold system ($K = 1.72$).

occurs through the last port and first port, respectively. (4) The influence of the friction parameter F is less significant than that of the flow area ratio A_o^*.

The following additional observations may be made from Bajura and Jones' work [114]. (5) The pressure drop within the core should be significantly greater than that in the headers for uniform flow distribution through the core. (6) The flow area of a combining-flow header should be larger than that for the dividing-flow header for a more uniform flow distribution through the core in the absence of heat transfer within the core. If there is heat transfer in the lateral branches (core), the fluid densities will be different in the inlet and outlet headers, and the flow areas should be adjusted first for this density change, and then the flow area of the combining header should be made larger than that calculated previously. (7) Flow reversal is more likely to occur in parallel-flow manifold systems that are subject to poor flow distribution.

I. FLOW MALDISTRIBUTION

In most heat transfer and pressure drop analyses, it is presumed that the fluid is uniformly distributed through the core. A significant reduction in heat exchanger performance may result when the flow distribution through the core is nonuniform. Two geometrically related maldistributions are gross and passage-to-passage. In gross maldistribution, the fluid flow distribution at the core inlet is nonuniform on a macroscopic level. This occurs because of either poor header design or gross blockage in the core during manufacturing. Passage-to-passage maldistribution occurs within a core on a microscopic level in a highly compact exchanger (such as a rotary regenerator) when the small flow passages are not identical as a result of manufacturing tolerances. Differently sized and shaped passages exhibit different flow resistances resulting in a nonuniform flow distribution even when the flow approaching the core face is uniform. Gross maldistribution generally reduces heat transfer and increases the pressure drop significantly. Passage-to-passage maldistribution reduces heat transfer significantly, and also *reduces* the pressure drop slightly.

1. Gross Flow Maldistribution

An analysis for one-dimensional nonuniformity on one side of a two-fluid exchanger is straightforward for counterflow, parallel-flow, and single-pass unmixed–mixed crossflow exchangers (flow nonuniformity on unmixed side only). The one-dimensional inlet velocity distribution can be divided into N step functions, and correspondingly the exchanger is divided into N parallel small exchangers each representing uniform velocity distribution at the entrance. The analysis of N exchangers will then provide the overall exchanger effectiveness, which can be compared with the effectiveness if the flow were totally uniform at the exchanger inlet.

As an illustration, consider a counterflow exchanger with a two-step function flow distribution on the hot side and uniform flow distribution on the cold side as shown in Fig. 51. In this figure, two heat exchangers, 1 and 2, are in parallel and are analyzed separately [116]. Knowing the temperature effectiveness $\epsilon_{h,1}$ and $\epsilon_{h,2}$ on the hot side for exchangers 1 and 2 allows the heat transfer rates q_1 and q_2 to be calculated. The overall effectiveness ϵ_h, based on the total heat transfer rate $q(= q_1 + q_2)$, is then

$$\epsilon_h = \frac{C_{h,1}\epsilon_{h,1} + C_{h,2}\epsilon_{h,2}}{C_h} \tag{139}$$

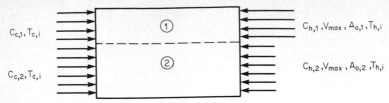

FIG. 51 Idealized flow nonuniformity on the hot side of a counterflow exchanger.

Subsequently, the reduction in exchanger effectiveness $\Delta\epsilon^*$ is computed once the exchanger effectiveness $\epsilon_{h,o}$ is known for a perfectly uniform flow:

$$\Delta\epsilon^* = \frac{\epsilon_{h,o} - \epsilon_h}{\epsilon_{h,o}} \qquad (140)$$

$\Delta\epsilon^*$ has been computed for a counterflow exchanger having $C^* = 1$, $V_{max}/V_m = 1.5$ and 2, and $A_{o,1}/A_o = 0.50$, 0.35, and 0.25 as shown in Fig. 52. In the computation, it was idealized that the flows were fully developed laminar and hence there was no change in h and hence U. It is found from Fig. 52 that a greater maldistribution ($V_{max}/V_m = 2.00$) yields a greater reduction in the effectiveness ($\Delta\epsilon^*$), as expected. Similar results can be obtained for a flow maldistribution consisting of an N-step velocity function ($N > 2$).

The theoretical results for the influence of flow maldistribution on the unmixed side of an unmixed–mixed single-pass crossflow exchanger are available in Ref. 116. For all other flow arrangements, no closed-form solution for $\Delta\epsilon^*$ is possible. The decrease in exchanger effectiveness can only be determined numerically.

No analysis is available for an increase in the pressure drop due to gross flow maldis-

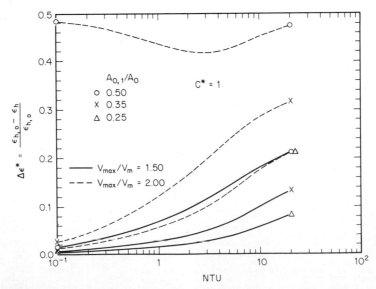

FIG. 52 Influence of gross flow maldistribution on the counterflow exchanger effectiveness.

tribution. Since such nonuniform flow is associated with poor header design or gross core blockage, the static pressure distributions at the core inlet and outlet faces may not be uniform. Hence, no simple modeling for the pressure drop evaluation is possible. As a conservative approach, it is recommended that the core pressure drop be evaluated based on the highest-velocity component of the flow maldistribution. For example, compute ΔP_1 and ΔP_2 in exchangers 1 and 2 of Fig. 51, and consider the higher of two ΔP's as the core pressure drop as a result of flow maldistribution.

Flow maldistribution in general may be present over the entire core face; hence the foregoing simplified analysis may not be valid. In that case, the heat transfer problem can only be analyzed numerically. Chiou has investigated the influence of flow nonuniformity on the thermal performance of a single-pass crossflow exchanger for two cases: flow non-uniformity on only one side [117], and flow nonuniformity on both sides [118]. In each case, four different two-dimensional velocity profiles were considered at the core entrance. Chiou employed the successive substitution technique for numerical analysis and presented the deterioration in the exchanger effectiveness, $\Delta \epsilon^*$, as a function of NTU, C^*, and $(\eta_o hA)^*$. Depending upon the type of flow nonuniformity, the reduction in the exchanger effectiveness $\Delta \epsilon^*$ was found to vary from 0.1 percent to over 20 percent.

Recently, Chiou [119] has also applied the same numerical technique to determine the influence of nonuniform inlet temperature with uniform flow on one side of a single-pass crossflow exchanger. He considered 20 specific nonuniform temperature distributions and found that the exchanger effectiveness could increase or decrease depending upon the location of the nonuniformity. Of all the cases he considered, a maximum increase in ϵ was about 12 percent for one case while a maximum decrease in ϵ was about 6 percent for other cases. For most of the cases, for which ϵ decreased, the reduction in ϵ was 2 to 3 percent. It may be concluded that the reduction in ϵ is less significant for nonuniform inlet temperature than for nonuniform inlet flow.

2. Passage-to-Passage Flow Maldistribution

Neighboring passages in a compact heat exchanger are geometrically never identical, because of manufacturing tolerances. It is especially difficult to control precisely the passage size when small dimensions are involved (e.g., a rotary regenerator with $D_h = 0.5$ mm or 0.020 in). Since differently sized and shaped passages exhibit different flow resistances and the flow seeks a path of least resistance, a nonuniform flow through the matrix results. This passage-to-passage flow nonuniformity can result in a significant penalty on heat transfer performance with only a small compensating effect of reduced pressure drop. This effect is especially important for laminar flows in continuous cylindrical passages, but is of lesser importance for interrupted surfaces where transverse flow mixing can occur.

The theoretical analysis for this flow maldistribution for low Re laminar flow surfaces has been carried out by London [120] for a two-passage model, and extended by Shah and London [121] for an N-passage model. In the latter model, there are N different-size passages of the same basic shape, either rectangular or triangular. In Fig. 53, a reduction in NTU for rectangular passages is shown when 50 percent of the flow passages are large $(c_2 > c_r)$ and 50 percent of the passages are small $(c_1 < c_r)$ compared to the reference or nominal passages. The results are presented for the passages having a nominal aspect ratio α_r^* of 1, 0.5, 0.25, and 0.125, for the (H1) and (T) boundary conditions and for a reference NTU of 5.0. Here, a percentage loss in NTU and the channel deviation param-

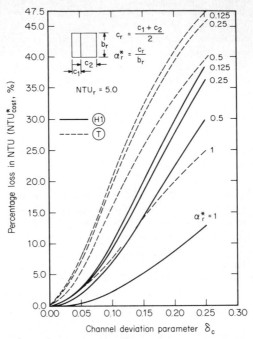

FIG. 53 Percentage loss in NTU as a function of δ_c and α_r^* for two-passage nonuniformities in rectangular passages.

eter δ_c are defined as

$$\text{NTU}^*_{\text{cost}} = \left(1 - \frac{\text{NTU}_e}{\text{NTU}_r}\right) \times 100 \qquad (141)$$

$$\delta_c = 1 - \frac{c_1}{c_r} \qquad (142)$$

where NTU_e is the effective NTU when two-passage model passage-to-passage nonuniformity is present, and NTU_r is the reference or nominal NTU. It can be seen from Fig. 53 that a 10 percent channel deviation (which is common for a highly compact surface) results in a 10 and a 21 percent reduction in NTU_{H1} and NTU_T, respectively, for $\alpha_r^* = 0.125$ and $\text{NTU}_r = 5.0$. In contrast, a gain in the pressure drop due to passage-to-passage nonuniformity is only 2.5 percent for $\delta_c = 0.10$ and $\alpha_r^* = 0.125$, as found from Fig. 54. Here ΔP^*_{gain} is defined as

$$\Delta P^*_{\text{gain}} = \left(1 - \frac{\Delta P_{\text{actual}}}{\Delta P_{\text{nominal}}}\right) \times 100 \qquad (143)$$

The results of Figs. 53 and 54 are also applicable to an N-passage model in which there are N different-size passages in a normal distribution about the nominal passage size. The channel deviation parameter needs to be modified for this case to

$$\delta_c = \left[\sum_{i=1}^{N} \chi_i \left(1 - \frac{c_i}{c_r}\right)^2\right]^{1/2} \qquad (144)$$

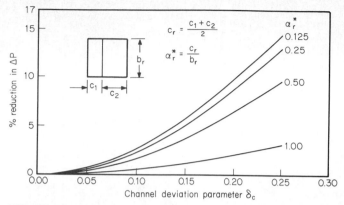

FIG. 54 Percentage reduction in ΔP as a function of δ_c and α_r^* for two-passage nonuniformities in rectangular passages.

Here χ_i is the fractional distribution of the ith shaped passage. For $N = 2$, Eq. (144) reduces to Eq. (142).

The following observations may be made from Fig. 53 and additional results presented in Ref. 121: (1) The loss in NTU is more significant for the $\textcircled{T}$ boundary condition than for the $\textcircled{H1}$ boundary condition. (2) The loss in NTU increases with higher nominal NTU. (3) The loss in NTU is much more significant compared to the gain in ΔP at a given δ_c. (4) The deterioration in performance is the highest for the two-passage model compared to the N-passage model with $N > 2$ for the same value of c_{max}/c_r.

Results similar to those in Figs. 53 and 54 are summarized in Fig. 55 for the N-passage model of nonuniformity associated with equilateral triangular passages. In this case, the definition of the channel deviation parameter δ_c is modified to

$$\delta_c = \left[\sum_{i=1}^{N} \chi_i \left(1 - \frac{r_{h,i}}{r_{h,r}} \right)^2 \right]^{1/2} \tag{145}$$

J. FOULING AND CORROSION

1. General

Fouling refers to the undesired accumulation (deposits) of solid material on heat exchanger surfaces which increases thermal resistance to heat transfer. Liquid-side fouling is covered in Part 2 of this chapter and in Ref. 122. In this section, we will discuss only gas-side fouling. In addition to Refs. 123 and 124, Marner and Webb [125] cover many aspects of gas-side fouling and provide an extensive list of references on the subject.

Fouling on heat transfer surfaces may cause a significant loss in heat transfer performance. Fouling may also increase the pressure drop. An increase in the pressure drop may reduce gas flows. For some applications, such as waste heat recovery from paint oven exhausts, this could cause solvent concentrations to increase above acceptable levels. In a fossil-fired exhaust environment, fouling produces a potential fire hazard in heat exchangers [126]. Such fires have resulted in catastrophic lost production and repair costs [127]. Fouling increases the maintenance cost of cleaning heat exchangers. Fouling may promote corrosion, severe plugging, and eventual failure of uncleaned exchangers.

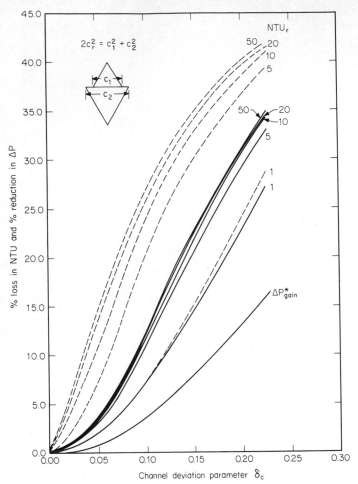

FIG. 55 Percentage loss in NTU and percentage reduction in ΔP as functions of δ_c for N-passage nonuniformities in equilateral triangular passages.

Like liquid-side fouling [122], gas-side fouling is time-dependent (zero at time equals zero) and increases along some pseudo-asymptotic or linear relationship a long time after initiation. In Eq. (9) on p. 4-6 and generally in design, a constant value of the fouling resistance or fouling factor $1/h_s$ is used. The heat exchanger will be "oversized" for the initial clean condition and "undersized" just before cleaning, if the design allowance for the fouling resistance represents an average for the period of operation.

In order to understand the influence of fouling on compact heat exchanger performance, the following equations for h and ΔP are derived from the equations presented earlier for fully developed gas flow in a circular or noncircular tube:

$$
h = \begin{cases} \dfrac{\text{Nu } k}{D_h} \text{ with Nu } = \text{constant} & \text{for laminar flow} \\[2ex] \dfrac{k}{D_h}\left[0.022\left(\dfrac{4LW}{A\mu}\right)^{0.8}\text{Pr}^{0.5}\right] & \text{for turbulent flow} \end{cases} \tag{146}
$$

$$\Delta P = \begin{cases} \dfrac{1}{D_h^3} \left[\dfrac{1}{2g_c} \dfrac{\mu}{\rho} \dfrac{16L^2}{A} W(f\,\text{Re}) \right] & \text{for laminar flow} \\[2em] \dfrac{1}{D_h^3} \left[\dfrac{0.046}{2g_c} \dfrac{\mu^{0.2}}{\rho} \dfrac{(4L)^{2.8}}{A^{1.8}} W^{1.8} \right] & \text{for turbulent flow} \end{cases} \qquad (147)$$

For constant W, L, A, and fluid properties, from Eqs. (146) and (147),

$$h \propto \frac{1}{D_h} \qquad \Delta P \propto \frac{1}{D_h^3} \qquad (148)$$

As fouling will reduce the flow area A_o and hence the passage D_h, it will increase h to some extent, but the pressure drop is increased more strongly. The thermal resistance of the fouling film will generally result in an overall reduction in heat transfer in spite of a slight increase in h.

Generally, gas-side fouling does not represent a significant fraction of the total thermal resistance in a gas-to-gas compact exchanger, but it can affect the pressure drop significantly.† For example, consider $U = 250$ Btu/(h·ft²·°F),‡ as in a process plant liquid-to-liquid heat exchanger. $R = 1/U = 0.004$. If the fouling resistances $(R_{s,h} + R_{s,c})$ together amount to 0.002 h·ft²·°F/Btu, 50 percent of the transfer area requirement, A, for a given heat transfer rate q is chargeable to fouling. However, a compact heat exchanger generally involves gas flow on either or both the hot and cold sides. Then U is generally less than 50 Btu/(h·ft²·°F), and the same fouling resistance 0.002 would be only of the order of 10 percent of the total. From the heat transfer performance point of view, gas-side fouling may call for a compensating 5 to 10 percent increase in the design surface area for a compact exchanger. The design pressure drop increment may be of the order 15 to 30 percent and higher depending upon the fouling induced surface roughness. Thus the major cost involved is in the increased fan or blower power requirement and in the necessary cleaning provisions rather than the extra surface requirement.

Highly compact exchangers are rarely used in moderate to severe gas-side fouling cases since they plug rapidly and they cannot be cleaned inexpensively. Both rotary regenerators and fluidized-bed exchangers, however, have self-cleaning characteristics and can tolerate moderate fouling.

Gas-side fouling has received research attention only recently. There are hardly any systematic studies conducted to understand the basic phenomena and to obtain experimental design data. Hence, an overview of the subject has been made in the following sections from the designer's point of view with a summary of available design data.

2. Types of Gas-Side Fouling

Gas-side fouling may be classified according to its immediate cause, the type of heat transfer service, or the mechanisms which control the deposition rate. The following classification is according to its immediate cause: particulate, chemical reaction, corrosion, scaling, and freezing fouling. Strong interactions may exist between some of these mechanisms.

†For noncompact individually finned tube exchangers such as those used in waste heat recovery (plain helical fins or segmented fins having up to 236 fins per meter or 6 fins per inch), the gas-side fouling may represent up to about 35 percent of the total thermal resistance, but with a negligible effect on ΔP due to the large D_h [135a].

‡1 Btu/(h·ft²·°F) = 5.678 W/(m²·K).

a. Particulate Fouling

The accumulation of particulates (which are suspended in the gas flow) onto a heat transfer surface is termed particulate fouling. Some measure of adhesion is required for accumulation (except in the stagnant areas); otherwise aerodynamic forces would tend to clean the surface. In particulate fouling, condensation of a sticky substance first covers the surface; then soot or fly ash in the exhaust (process) stream sticks to the surface and accumulates, which eventually may plug the flow passages. If the fouling is due to plant residues, flowers, weeds, or insects in the airflows, it may occur only at the front face of the core, as compact surfaces function as a screen. Once particulate plugging begins, it increases at a rapidly accelerating rate often leading to an emergency shutdown or a fire.

b. Corrosion Fouling

Corrosion fouling is due to corrosion products produced by a chemical reaction between the heat transfer surface and the fluid. Condensation of acid vapors, such as sulfuric or phosphoric acid, promotes corrosion fouling, which is a frequent problem in waste heat recovery exchangers that operate on exhaust gases from fossil fuel combustion. Oxidation of the heat transfer surface and fluxing of the protective oxide film by molten ash, which results in glassy, corrosive deposits, are examples of high-temperature corrosion. Corrosion due to salty air is an example of low-temperature corrosion.

c. Chemical Reaction Fouling

Fouling deposits in this mechanism are formed by chemical reactions, either in the gas stream or at the heat transfer surface, in which the surface material is not a reactant. As an example, the excess SO_3 in the exhaust of a glass furnace combines with Na_2SO_4 and water vapor to form $Na_2S_2O_7$ and $NaHSO_4$. All three sodium compounds combine together to produce a sticky eutectic that fouls the heat transfer surface.

d. Scaling

The formation of a hard scale may result when condensed salts or compounds from the gas stream solidify as a result of a sufficiently low temperature of the heat transfer surface. As an example, the sticky eutectic mentioned in the preceding paragraph forms a hard scale on the heat transfer surface at sufficiently low temperatures.

e. Freezing Fouling

Freezing fouling is due to the solidification of vapor condensing from the gas stream onto a below–freezing temperature heat transfer surface. A common example is the frost layer in the evaporator of an automotive air conditioner when the surface temperature falls below $0°C$ ($32°F$). The evaporator then loses its performance rapidly since a portion of the core becomes completely blocked.

Solid precipitation (i.e. crystallization from solution) fouling is not too common from gases because most gases contain few dissolved salts (mainly in mists) and even fewer inverse-solubility salts. Biofouling is not too common for gases, because there are generally no nutrients (e.g., hydrocarbons, organic matter) present in air except possibly with moist air in a marine environment. No biofouling problem with gases is reported in the literature.

3. Mechanisms and Models for Particulate and Corrosion Fouling

a. Particulate Fouling

There are a number of mechanisms that drive suspended particles from the main flow to the surface on which they are deposited. The mechanisms that do not depend upon the temperature gradient are: brownian diffusion for submicron-size particles [128, 129], gravitational settling for micron-size particles in a laminar stream [130], eddy diffusion for a turbulent stream [131], and inertial and diffusional impaction. The mechanism that depends upon the temperature gradient is referred to as *thermophoresis,* in which aerosol particles move toward the lower-temperature region, fouling the cold surface. The theory of thermophoresis was pioneered by Epstein [132] in 1929, and recently Nishio et al. [133] have advanced the theory with some experimental verification. The mechanism that depends upon electrical forces is referred to as *electrophoresis,* in which aerosol particles deposit onto a solid surface as a result of electrostatic charge. The particles collect electric charges by contact with a surface, from atmospheric electricity, from an ionized gas, or by emission. Electrical forces become increasingly important on charged surfaces for particle sizes below about 0.1 μm. Soo [134] presents a criterion for when the electrostatic deposition will always occur.

b. Corrosion Fouling

The mechanism of corrosion fouling depends upon the constituents in the gas stream, gas temperature, surface temperature, and surface material. The elements, originating in the fossil fuel and present in the flue gas, that cause corrosion [135] will be discussed first. Then a summary of the corrosion fouling mechanisms will be presented.

Sulfur, chlorine, and hydrogen† are the most common elements that cause corrosion at high temperatures (above about 500°F or 260°C), while sulfur and chlorine cause corrosion at low temperatures (below about 300°F or 150°C). High-temperature sulfur corrosion is due to an attack of hydrogen sulfide on carbon steel at metal temperatures above 500°F. Chlorine and chlorides attack steel at metal temperatures above 400°F (204°C) and 450°F (232°C) respectively. In contrast to high-temperature sulfur corrosion, chlorine and chloride attacks are more violent with increasing gas temperatures, even if the metal temperature remains the same. However, the addition of sulfur decreases the severity of this corrosion. Chrome alloys are resistant to high-temperature sulfur and chlorine corrosion, while nickel alloys are susceptible. Alonizing‡ protects against high-temperature sulfur corrosion. Hydrogen present in the gases decarburizes steel above 450°F (232°C) and 200 psi (1380 kPa), causing blistering and eventual failure. Molybdenum and chrome alloys protect against this form of corrosion.

---

†Sulfur can be found in practically all fossil fuels; even certain types of natural gas have some sulfur. Some chlorine could be found in coals and especially in refuse (garbage). Hydrogen is one of the main components of process gas from the catalyst pipe of reformer furnaces in the steam-methane reforming process (hydrogen, ammonia, and methanol production).

‡Alonizing is a process of vapor deposition and diffusion of aluminum on carbon steel surfaces, in contrast to an aluminizing process, in which the part is submerged into a molten aluminum bath.

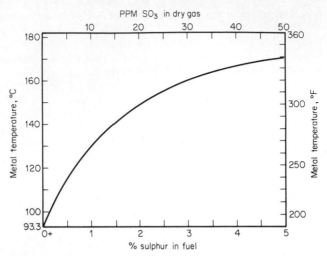

FIG. 56 Minimum heat transfer surface temperature to avoid cold-end sulfuric acid corrosion [135].

At low temperature, below the dew point, SO_3 forms sulfuric acid, which is very corrosive. The dew point is a function of fuel sulfur content as shown in Fig. 56.† The best way to protect against this "cold-end" corrosion is to maintain the metal temperature above the dew point of the gas. Dry chlorine and hydrogen chloride are harmless between their dew points and below 400°F (204°C). The chlorine content of the gas does not affect the dew point. From the foregoing discussion, there is a narrow range between 300 and 400°F (150 and 204°C) where neither high- nor low-temperature corrosion is expected. Maintaining the metal temperatures in this range is the best protection against corrosion. The corrosion fouling mechanisms for some specific gas streams will now be summarized.

In coal-fired boilers [136], the exhaust gas stream contains Na_2O, K_2O, SO_2, and SO_3 in addition to H_2O and other constituents. These sodium and potassium oxides react with SO_3 either in the gas stream or on the exchanger surface to form sodium and potassium sulfates, which melt at relatively "low" temperatures. They deposit on the tube and attract ash particles, eventually building up a moderately thick deposit on the tube. In addition, SO_2 in the flue gas is catalytically oxidized to SO_3 on iron oxide surfaces, present either in the ash deposits or on the heat transfer surface as fouling layers. The alkali sulfates, iron oxide, and SO_3 then react to form complex alkali iron sulfates (melting point about 600°C) in the presence of a high SO_3 atmosphere. These molten alkali iron sulfates are corrosive and act as a bonding medium between the exchanger surfaces and the slag deposits. Condensation of SO_3 in the gas stream as sulfuric acid (at about 130 to 160°C, depending upon its concentration) causes corrosion of metal surfaces, forming primarily

†This figure represents a practical guideline for engineers rather than a scientific diagram. The top and bottom abscissa scales are independent. A 1 percent sulfur in the fuel does not mean that you will have 10 ppm SO_3 in the dry combustion products of the same fuel. SO_3 formation depends upon many things. The two curves (metal temperature versus the sulfur content of different fuels or ppm SO_3 content of different dry gases) reported are surprisingly similar. Hence, both such curves are represented as a single curve in this figure [135a].

ferrous sulfate, which in turn causes plugging deposits. Also the absorption of SO_3 by particles in the flue gas stream, especially unburned carbon, causes the particles to become adherent, which further increases deposition and plugging.

With diesel exhaust formed from using no. 6 fuel oil, the following fouling mechanism is active [137]. Initially a thin layer of sulfuric acid is condensed on the cold bare metal of the exchanger. After the surface becomes wetted, soot particles attach to the surface, and a fouling layer is built up. Increase in the layer thickness produces higher thermal resistance to heat flow, causing the surface temperature to be above the acid dew point and thus preventing additional acid condensation. A further buildup in the fouling layer thickness is attributed to the condensation of unburned hydrocarbons in the low-temperature gas boundary layer. The resulting fouled surface (soot plus hydrocarbons) is tacky, and the fouling layer continues to grow until plugging occurs or the soot surface temperature exceeds the hydrocarbon condensation temperature.

With glass furnace exhaust gas [138], the following corrosion fouling mechanisms are active in the fixed-matrix regenerator, depending upon the temperature range.† The sodium sulfate formed by Na_2O and SO_x gases condenses as a liquid around 1000°C (1800°F). Although liquid Na_2SO_4 is not corrosive, it dissolves SO_3, which in turn attacks MgO in the bricks and weakens them. The liquid Na_2SO_4 forms a eutectic with $MgSO_4$ and $CaSO_4$ which drips down the bricks and becomes a white slag after cooling (after several flow reversals in the regenerator). Below 884°C (1623°F), Na_2SO_4 condenses in a fluffy powder form which plugs heat exchanger passages downstream. At low temperatures, Na_2SO_4, SO_3, and water vapor combine to form a hard scale (eutectic) on the brick surface upon which the fluffy powder continues to build up the fouling layer.

4. Influence of Some Operating Variables

The influence of some operating variables (associated with the gas turbine exhaust [139–141], diesel engine exhaust [137], and the different types of exchanger surfaces) on gas-side fouling is presented here. Some quantitative results are given in the foregoing references. However, no generalized results are available at present.

a. Gas Velocity

This variable has a significant effect on particulate fouling in the exchanger, particularly for flow parallel to the heat transfer surface. The fouling deposits generally increase with decreasing gas velocity.‡ Decreasing the velocity from 60 to 20 ft/s (18.3 to 6.1 m/s) increased ΔP by 250 percent and decreased gas-side hA by 40 percent in 25 hours of test time on a plain triangular plate-fin unit with gas turbine exhaust [139]. The highest fouling rates observed were for gas velocity less than 30 ft/s (9.1 m/s).

In some applications, erosion can be a problem when there is a heavy concentration of particulates and the gas velocity is relatively high.

---

†The temperature of the exhaust gases is typically 1400 to 1600°C (2550 to 2900°F). The average compositions of the particulates and gaseous constituents are provided in Ref. 138.

‡The fouling deposits can increase with increasing gas velocities in some applications, if the gas impinges on the surface. The particles impact on the surface and build up as a result of high velocity. If these particles are heavy and sticky, they will not be removed by the high-velocity gas.

b. Gas Temperature

This variable also has a significant effect on particulate and corrosion fouling. A lower gas inlet temperature produces accelerated fouling and a very rapid degradation of heat transfer and pressure drop characteristics. In general, the fouling rate decreases with increased exhaust gas temperatures to the heat exchanger.

c. Local Temperature of Heat Transfer Surface

Generally, there is more condensation, more particulate buildup, and increased film sulfur content in the colder regions of the exchanger compared to the hotter regions. Hence, the "cold-end" corrosion is one of the major gas-side fouling problems. If the wall temperature in the exchanger does not drop below the dew point of any vapor in the gas stream, no measurable amount of fouling is detected.

d. Fuel Type

Combat gasoline (leaded fuel) produces greater fouling than JP-4. The lead deposits adhere more firmly to the metal surfaces and do not blow off or burn off as easily as deposits from JP-4 [139]. The fouling associated with no. 2 diesel fuel is higher than with JP-5 fuel because there are more unburned hydrocarbons [141]. The fouling rate increases with an increase in the fuel-air ratio for a given combustion efficiency.

e. Surface Roughness

Exchanger surface roughness is important only for a clean surface. After a fouling layer is deposited on a rough surface, the original roughness no longer exists. Thus the original surface roughness influences the fouling only during the induction period. Two tubular units, with surfaces mechanically roughened, were tested in a 25-hour fouling test and compared to a standard unit. The resultant differences were insignificant [139]. Plain tubes and finned tubes were coated with a low-temperature vitreous coating, and no difference in the soot layer thickness was observed [141].

f. Exchanger Surface Geometry

A staggered tube bank (no fins) has a greater tendency to particulate fouling than an inline tube bank, based on extensive testing and operational experience with steam generators [142]. Hence, the advantage of a staggered tube bank in terms of higher heat transfer over an inline tube bank is nullified to a large extent by higher pressure drop due to fouling.† In addition, the inline arrangement is easier to clean and would suffer less abrasion. It is preferable to arrange bare tube bundles vertically (and not horizontally) from the fouling point of view, because there is less solid buildup and any buildup which does occur is easier to remove. From the fouling deposit rate point of view, conflicting information exists for the plain helical and segmented circular fins [143]. However, De Anda [137] reports the same fouling rates for plain helical and segmented circular fins for iden-

---

†In the waste heat boiler industry, where 50.8-mm (2-in) outside-diameter tubes on 101.4-mm (4-in) centers are typical, a 2.5-mm (0.1-in) deposit on the tubes may drastically reduce heat transfer but has a negligible effect on ΔP.

tical fin spacing, fin height, and period of operation. Fins of greater density fouled worse in terms of flow area reduction than more widely spaced fins. For the plain helical fin, complete bridging between fins occurred for a fin density of 197 fins per meter (5 fins per inch) when used with a dirty exhaust gas using no. 6 fuel oil [137]. For segmented fins, localized bridging occurred both in fin gaps and across the segments. The plugging occurs sooner on shorter fins than on longer plain or segmented fins, with equal tube spacing.

Mort [139] tested the following four types of heat exchanger modules for fouling with gas turbine exhaust using JP-4 and combat fuels: plain triangular plate-fin, wavy plate-fin, plain helical finned tubes, and bare staggered tube banks. He found no significant reduction in heat transfer (hA) for these different exchangers. However, the pressure drop increase due to fouling was different. The highest pressure drop increase was for the plain triangular plate-fin followed by wavy plate-fin units, bare tube bundles, and plain helical finned tubes.

Miller [129] tested the following six plate-fin surface geometries for fouling with a marine diesel-fueled gas turbine exhaust: plain triangular, straight and wavy (in the flow direction) surfaces, plain rectangular, and three offset strip fins. He found that the heat transfer performance degradation was not substantially influenced by the differences in the surface geometries, but the pressure drop increase was sensitive to the geometry variation. The offset strip fin surface in general had the highest reduction in heat transfer and the lowest increase in pressure drop.

Cowell and Cross [144] tested for particulate fouling the multilouver plate-fin (Fig. 2) and tube-fin geometries used in automotive radiators. A precipitated calcium carbonate powder was used for the study with the fin surface initially sprayed with the diesel fuel to simulate the field conditions. The following results were found: (1) Fouling has a negligible effect on reducing the air-side heat transfer but a significant effect on increasing the air-side ΔP. (2) The particulate fouling is confined only to the front face of the core. (3) The influence of fouling is strongly dependent upon the hydraulic diameter of the flow passages.

Eastwood [145] tested an offset strip fin and a rectangular plain fin surface for particulate fouling using glass furnace exhaust dust. The pressure drop increased with time for the offset strip-fin core while ΔP reached an asymptotic value for the plain fin core, when both were cleaned by an air lancing technique. Eastwood reports that water or steam cleaning has shown favorable results for the offset fin surface.

5. Fouling Factors

As mentioned earlier, there are at least five types of gas-side fouling, there is a wide range of potential foulants, and there are a large number of operating variables and geometrical parameters for each extended surface that affect the fouling mechanisms. Therefore, it is not possible to find unique methods to predict the fouling processes. Some of the modeling and experimental results are summarized in the preceding sections. However, no thorough understanding of quantitative fouling behavior (rates of deposition and removal) exists.

Empirical fouling factors ($1/h_s$), to be used in Eq. (9) on p. 4-6, are known for only limited fouling applications; also conflicting information exists for some fouling factors. As a design guideline, some approximate fouling factors are summarized in Table 16 [143] for three types of fossil fuel flue gases characterized as clean gas, average gas, and dirty gas that are defined in a footnote for clarity. Which type of fuel will produce such a gas is also summarized under each category. Although burning no. 2 fuel oil usually

TABLE 16 Fouling Factors and Related Design Parameters for Extended Surfaces in Fossil Fuel Flue Gases [143]

Type of flue gas	Fouling factor $1/h_s$, h·ft²·°F/Btu‡	Minimum spacing between fins or plates, in‡	Maximum gas velocity to avoid erosion, ft/s‡
Clean gas†			
Natural			
gas	0.0005–0.003	0.050–0.118	100–120
Propane	0.001–0.003	0.070	
Butane	0.001–0.003	0.070	
Gas turbine	0.001		
Average gas†			
No. 2 Oil	0.002–0.004	0.120–0.151	85–100
Gas turbine	0.0015		
Diesel	0.003		
engine			
Dirty gas†			
No. 6 oil	0.003–0.007	0.180–0.228	60–80§
Crude oil	0.004–0.015	0.200	
Residual oil	0.005–0.02	0.200	
Coal	0.005–0.05	0.231–0.340	50–70§

†Clean, average, and dirty gases may be defined in terms of total ppm (parts per million) as less than 100 ppm, between 100 and 500 ppm, and greater than 500 ppm, respectively. Some industrial processes for these exhaust gases are as follows. Clean gas: steel heat treat and aluminum heat treat; average gas: aluminum remelt and vitreous melting; and dirty gas: cupola, glass, and Cowper.

‡1 h·ft²·°F/Btu = 0.176 m²·K/W, 1 in = 25.4 mm, 1 ft/s = 0.305 m/s.

§Minimum recommended gas velocity to minimize fouling is 30 ft/s (9.1 m/s).

produces an average flue gas, it is reported in the literature that the fouling factor is 0.015 h·ft²·°F/Btu in a gas turbine exhaust while it is double that in the diesel engine exhaust [143]. So it appears that the type of fuel, the air-fuel ratio, and how the fuel is burned are important and can affect the equipment design.

Since the pressure drop is critical for gas flows, information on the effect of fouling film thickness buildup on pressure drop is necessary. Unfortunately, only scattered information is available in the literature, and the user should be prepared to evaluate on-site conditions.

A systematic quantitative study of particulate fouling on plain helical circular fins has been performed by Bott and Bemrose [146]. They use 15.9-mm (0.625-in) high fins at 410 fins per meter (10.5 fins per inch) fin density. Fins were coated with a thin layer of liquid (0.017-mm thickness), and precipitated calcium carbonate particles (3 to 30 μm, or 1.2×10^{-4} to 1.2×10^{-3} inch size) were then injected into the air stream. They obtained quantitative j and f factors (including the fouling effects) as functions of Re and time. They found that the asymptotic f factor increased to 1.4 to 2.5 times the initial value, while the asymptotic j factor decreased by only about 15 percent. For an air-cooler application, they showed that the increase in the fan power requirement due to fouling is of the same order as the expected return on the capital investment.

6. Cleaning Methods

Methods used in the industry to clean gas-fouled heat exchangers are mechanical, sonic, chemical, and thermal cleaning. It is very important to match the cleaning technique with the type of fouling layer which is to be removed. For example, sonic horns may remove a light deposit of ash but not a highly sticky deposit.

a. Mechanical Cleaning

Mechanical cleaning is used for particulate fouling. Soot blowers are the most effective. Either high-pressure air, high-pressure steam, or water is directed through a nozzle at the exchanger surface. High-pressure steam or water is used only when it will not corrode the exchanger surface. Rotary soot blowers are used for nonsticky particulate deposits and for "low-temperature" applications, and retractable soot blowers for stubbornly sticking particles and internal bundle cleaning. Motor-driven air lances are used to clean heavily fouled surfaces *without* interrupting or shutting down the process. Hand lances, air cannon, steel or ceramic shots, or hanging chains have also been used for cleaning the surfaces. Mechanical shock or vibration of tubes to free deposits is another alternative used in industry. Use of a detergent solution mixed with high-pressure steam or compressed air sprayed over compact plate-fin units is the common method for cleaning particulate fouling. Back blowing is used when the core is fouled at the front face due to particulates, such as in an agricultural service.

b. Sonic Cleaning

Sonic cleaning is another method for cleaning the particulate fouling. Medium- and low-frequency (20 to 250 Hz and 100 to 140 dB) sonic horns are used to knock the deposits loose from the surface. The loosened deposit is then carried away by the gas stream.

c. Chemical Cleaning

The chemical treatment or cleaning method is used for scaling, chemical reaction, and corrosion fouling. Dilute acids, detergents, or some solvents are used as cleaning agents. Small plate-fin recuperator modules may be immersed in a tank filled with a strong chemical mix and left to soak for a period of time. The selection of cleaning agent and method depends upon the nature of the scale or condensate film, the type of heat exchanger, and the application.

d. Thermal Cleaning

In the thermal cleaning method, the temperature of the heat exchanger core is raised to the extent that most of the deposits oxidize and decompose or vaporize without damaging the heat exchanger surfaces. This is achieved by either blowing hot air (or gas), shutting the cold gas supply in a gas-to-gas exchanger (thus raising the surface temperature), reversing the gases, or baking the exchanger (if small) in a temperature-controlled oven. Sometimes only selected portions of the exchanger are heated.

7. Prevention of Corrosion Fouling

Heat exchanger surfaces may disintegrate and/or plug due to corrosion fouling. In order to prevent failures, steps should be taken before the corrosion fouling is initiated. There is no inexpensive method available for prevention of corrosion due to sulfuric acid condensation on the exchanger surfaces at temperatures below the acid vapor dew point. Here are some of the methods used by industry: Maintaining the minimum metal temperature above the dew point is the best way to avoid acid corrosion. If practical, use a parallel-flow arrangement, which has the highest minimum metal temperatures when compared to all other flow arrangements at the same exchanger effectiveness if that effectiveness is achievable by a parallel-flow exchanger. If it is necessary to operate below the acid dew point, the most effective way to reduce corrosion fouling is a chemical treatment (injection of neutralizing agent) directly into the gas stream. Magnesium oxide in fine powder form has been proved to be an effective way in neutralizing SO_3 and to reduce subsequent corrosion [136], if injected into the gas stream (but less effective if the same amount is injected into the fuel). However, the magnesium oxide may be abrasive and also contribute to fouling of the exchanger surfaces. A Teflon coating over low-temperature exchanger surfaces is used to prevent corrosion due to phosphoric acid vapor condensation. Paints and special alloy coatings are used to prevent salt corrosion. Ceramic surfaces are used to prevent high-temperature corrosion. Sometimes, already preheated air is used to dilute the exhaust gas stream to minimize corrosion. Some other specific corrosion prevention methods are presented earlier on p. 4-284.

Johnson et al. [147] have surveyed and summarized in depth aqueous and air-side corrosion aspects of candidate materials in dry cooling tower applications. They report that dry coolers having aluminum fins on aluminum tubes have functioned over 30 years in clean atmospheres with regular cleaning with air jets, steams jets, or water spray with detergents. In adverse environments, design factors such as crevices, horizontal fins, imbedded fins, and wrap-on fins with dissimilar metals between the fins and the tubes tend to promote corrosion under the caked particulate layer. However, in clean environments, corrosion does not appear to be a major problem.

K. BOILING AND CONDENSATION†

Compact extended surfaces can be designed to yield very high heat transfer rates per unit volume with boiling and condensation, and may be used for many applications, except for those having heavy fouling. Typical examples of current use are plate-fin exchangers for separating oxygen and nitrogen from air at cryogenic temperatures, separating olefin products from the cracking of hydrocarbons in ethylene plants, and condensing water vapor from the mixture of gases in fuel-cell power plants.

The design theory for vaporizers or condensers that use compact surfaces is essentially the same as that used for shell-and-tube exchangers. A heat exchanger is divided into a number of segments, and energy balance and rate equations are applied to each segment to compute outlet temperatures, qualities, and pressures. Such stepwise calculations are performed either by a marching procedure (crossflow exchanger) or by an iterative procedure (counterflow exchanger) until outlet conditions are known. Variations in local h, ΔT, and velocity effects can be taken into account in such calculations. Some details on such procedures are given in Part 2 of this chapter as well as in Refs. 171 to 173. One of the important inputs to this design procedure is to provide proper correlations for heat transfer coefficients and friction factors for two-phase flows. For compact heat exchanger surfaces, only a very limited amount of heat transfer coefficient data is available and this is summarized here. A procedure for the two-phase pressure drop evaluation is presented on p. 4-295.

For the design of plate-fin partial condensers (which have a mixture of condensables and noncondensables), the following method has been used in industry. First, experimental data are obtained on a model core during partial condensation covering a wide range

---

†The author is grateful to Mr. J. M. Robertson of Heat Transfer and Fluid Flow Service for a critical review of this section.

of Reynolds number, quality, and mixture component proportions. An attempt is then made to predict the experimental results using the method of partial condenser design for the circular tube core (for example, see Ref. 171), with the circular tube single-phase j and f factors replaced by the corresponding factors for the plate-fin surface. If necessary, the predictions are made to agree with the measured results by adjusting the single-phase j and f data for the plate-fin surface with correction factors. These adjusted j and f factors are then used with the circular tube theory for the design of a partial condenser with plate-fin surfaces.

The determination of the "fin efficiency" during boiling or condensation is important for fins in compact heat exchangers because of the high heat transfer coefficients. It is current practice to evaluate the fin efficiency during boiling or condensation by using the single-phase formula; e.g., see Table 1. In addition to the variation in the boiling or condensing heat transfer coefficient along a fin, it will vary in the direction of fluid flow. Particularly, during condensation, the condensate film thickness increases along the flow length. This yields markedly low heat transfer rates and low heat transfer coefficients, and consequently high fin efficiencies some distance along the flow length, as found from a numerical analysis by Patankar and Sparrow [148]. Collier et al. [149] point out that the fin efficiency evaluated assuming constant heat transfer coefficient may not be accurate when the heat transfer coefficient in reality may vary over a fin by a large factor. However, based on their example, if the fin efficiency is maintained over 80 percent, the error in computing η_f based on constant h may be small. Evaluation of the fin efficiency is not possible at present if the fin is not completely wet. Until further analysis is available, it is recommended that the single-phase concept of fin efficiency also be cautiously applied to the two-phase flow situation. The exchanger in the flow direction can be divided into several segments, and separate fin efficiency magnitudes can be determined locally for each segment.

A very limited amount of published performance data is available for boiling and condensation in plate-fin exchangers. As a result, it is not possible to provide either generalized correlations for design purposes, or correction schemes to apply to the circular tube correlations. This section is therefore limited to a survey of the available information.

A survey on boiling and condensation in plate-fin exchangers has been provided by Robertson [150]. A review on boiling and condensation in narrow passages and plate-fin surfaces has been recently made by Westwater [151]. Recent correlations and data on plate-fin surfaces are summarized in the following subsections. Additional material is presented in Chap. 3, where offset strip fins are regarded as augmented surfaces. Boiling and evaporation and condensation in low-finned tubing and other augmented-tube geometries have also been summarized by Bergles in Chap. 3.

1. Boiling

a. Heat Transfer

The main features of boiling of cryogenic fluids in very small passages of plate-fin heat exchangers, based on local temperature measurements in evaporating nitrogen and Freon-11, may be summarized as follows [150, 152–155].[†]

[†]The following ranges were covered in the tests: for nitrogen, mass velocities from 11.5 to 110 kg/(m²·s) or 8480 to 81,100 $lb_m/(h \cdot ft^2)$, heat fluxes up to 4 kW/m² or 1268 Btu/(h·ft²), wall-to-bulk temperature difference up to 1.5°C or 2.7°F, and operating pressures from 2 to 7 bars; and for Freon-11, mass velocities from 34.4 to 159 kg/(m²·s) or 25,360 to 1.17×10^5 $lb_m/(h \cdot ft^2)$, operating pressures from 3 to 7 bars, and other ranges the same as that for nitrogen.

1. At the transition between the single-phase (heating of subcooled liquid) and boiling regions, a large liquid superheating occurs before boiling commences. This delay in boiling is mainly a result of the good ability of the cryogenic fluids to wet the aluminum surface thoroughly and to penetrate the potential nucleation sites (estimated to be about 1 μm in size) on the metal. The peak wall temperatures and hence large superheating are dependent on the heat flux, that is, to the ΔT required to activate nucleation sites, and to a lesser extent on pressure [154].

2. Once vapor production has commenced in a channel, convective boiling (film evaporation) is dominant with little nucleate boiling present at the low heat flux levels (small temperature difference between streams, $\Delta T < 3°C$) typical of cryogenic process applications [152].

3. The dependence of boiling heat transfer coefficients on the mass velocity at low and high levels in offset strip fin passages is different from that of the corresponding single-phase heat transfer coefficients. At low mass velocities, the convective boiling coefficients for nitrogen [152], Freon [153], and ammonia [156] flowing through an offset strip-fin test section appear to be independent of the mass velocity. At high mass velocities, the boiling coefficients are dependent on the mass velocity to the power of 0.8, as shown in Fig. 57. These two ranges of mass velocities appear to correspond to the fully developed laminar and turbulent flow single-phase forced convection regimes in cylindrical flow passages. However, for offset strip fins with single-phase fluids, the flow regimes are periodically developing laminar flow at low Reynolds numbers and fully developed turbulent flow at very high Reynolds numbers.

From Fig. 57, it can be seen that the effect of quality on the convective-boiling coefficient is even greater than that of the mass velocity. The smaller effect of quality at low mass velocities is also apparent. The high boiling coefficients (based on A) observed with offset strip fins indicate the maintaining of a thin liquid film on each strip. With evaporation as the main boiling mechanism, the thickness (and turbulence level) of this film will directly affect the boiling coefficient. Particularly, the liquid droplet entrainment in the vapor core from the trailing edge of the strip fin will yield a thinner film and hence the high observed boiling heat transfer coefficients.

Robertson [152] in 1979 proposed a dimensional correlation for the forced convective boiling coefficient h for nitrogen and Freon-11 in an offset strip-fin exchanger.† However, he has recently attempted to correlate the same data based on an annular two-phase film-flow model [155]. He computed the film thickness δ_f of the annular flow by the same theory as that for the annular flow in a circular tube [157]. However, the experimental pressure gradient for the offset strip fin was used for the calculation of δ_f. The experimental data of Fig. 57 are replotted in terms of the film Nusselt number Nu_f versus W^+ ($= \text{Re}_f/4$) in Fig. 58. Nu_f and Re_f are defined as

$$\text{Nu}_f = \frac{h_{TP}\delta_f}{k_l} \qquad \text{Re}_f = \frac{G(1-x)D_h}{\mu_l} \qquad (149)$$

where h_{TP} is the boiling (two-phase) heat transfer coefficient, x is the vapor quality, and the subscript l denotes the liquid phase value.

The following observations may be made from Fig. 58: (1) The results for low qualities tend to fall on the line AA. Also at low mass velocities, the results from all qualities

---

†The geometry of the offset strip fin was: 591 fins per meter (15 fins per inch), 6.35 mm (0.25 in) in fin height, 0.2 mm (0.008 in) in fin thickness, and 3.18 mm (⅛ in) in strip length.

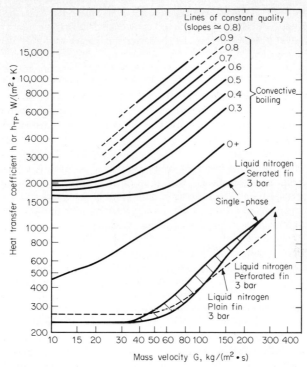

FIG. 57 Film evaporation heat transfer coefficients for boiling nitrogen at 300 kPa in an offset strip-fin exchanger [152].

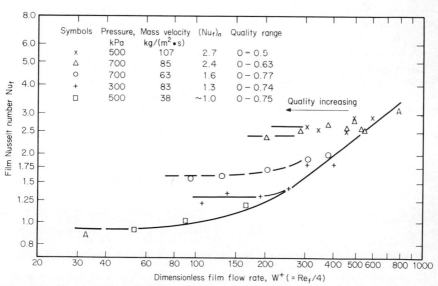

FIG. 58 Film Nusselt number as a function of W^+ for boiling nitrogen in an offset strip-fin exchanger [155].

fall on line AA. This line represents results similar to but higher than the single-phase values in a circular tube. Here, again, forced convective boiling is dominant over nucleate boiling and hence the results display the characteristic features of forced convection heat transfer. (2) At high qualities, the Nusselt number approaches an asymptotic value $(Nu_f)_a$ which increases for higher values of mass velocities. $(Nu_f)_a > 1$ provides an estimate of the amount of liquid entrained in the vapor core. (3) The line AA indicates a laminar film for $W^+ \leq 75$, a turbulent film for $W^+ \geq 250$ (the slope is 0.8 on a log–log scale), and a transition film in between. Robertson [155] also reports similar data for Freon-11 and obtains a limiting curve similar to the line AA but somewhat displaced, indicating a Prandtl number effect.

Yung et al. [156] in 1980 developed an analytical model for condensation or evaporation in an offset strip-fin exchanger. They considered annular two-phase flow with laminar liquid film near the vertical wall, but either laminar or turbulent vapor core. Their model predicted well the Robertson data [152] of low mass velocities as well as their own experimental data (overall measurements only) with ammonia evaporating in an offset strip-fin exchanger.† In their model, no liquid entrainment is idealized and hence the model will apply only for $Re_l\ (=\ GD_h/\mu_l) < 1000$.

Panitsidis et al. [158] in 1975 measured thermosiphon-type boiling‡ overall heat transfer rates from an offset strip-fin exchanger§ with Freon-113 and isopropanol as test fluids at atmospheric pressure. The measured heat transfer rates varied from 300 to 98,600 W/m² or 95 to 31,280 Btu/(h·ft²), and $\Delta T\ (=\ T_{wall} - T_{sat})$ from 2 to 80°C. Since no local measurements were made, no estimate of the heat transfer coefficient is possible for their test geometry and fluids as a function of the quality, ΔT, and mass velocity. Hence, they used the pool boiling curve, obtained for Freon-113 and isopropanol over a 6.35-mm (0.25-in) diameter horizontal tube, for prediction of the overall heat transfer rate based on a single-fin analysis. The fin was divided into 26 segments and the temperature of each segment was considered from the point of view of a one-dimensional fin analysis. The total heat transfer rate in the exchanger, based on the one-fin analysis, was in good agreement with the experimental data for low ΔT. Chen et al. in 1981 [159] refined the prediction method of Panitsidis et al. by modifying the aforementioned pool-boiling curves for the velocity effect. Basically, Chen et al. considered the boiling two-phase mixture as homogeneous, and evaluated the boiling heat transfer coefficient from the boiling curve for the small circular cylinder, taking into account the effect of local ΔT and local velocity. The predicted overall heat transfer rates were in good agreement with the measured value. Thus, it appears that the performance of a thermosiphon reboiler with the compact plate-fin exchanger may be computed fairly accurately if the appropriate flow boiling curves are available for a small diameter horizontal cylinder. However, further experimental verification is necessary.

It is interesting to compare the results of Robertson [152–155] and Westwater and coworkers [158, 159] for boiling in offset strip-fin exchangers. While Robertson used forced flow of Freon-11 and nitrogen during flow boiling, Westwater had induced flow

---

†The geometry of the offset strip fin was: 775 fins per meter (19.7 fins per inch), 4.68 mm (0.184 in) in fin height, 0.41 mm (0.016 in) in fin thickness, 3.18 mm (⅛ in) in strip length, and 1.98 mm (0.078 in) in hydraulic diameter.

‡This type of boiling is referred to as either pool boiling with finite velocity or "low" velocity forced convective boiling.

§The geometry of the offset strip fin was: 641 fins per meter (16.3 fins per inch), 3.88 mm (0.153 in) in fin height, 0.101 mm (0.004 in) in fin thickness, and 4.67 mm (0.184 in) in strip length.

(due to thermosiphon) of Freon-113 and isopropanol during pool boiling. The induced velocities were high, and hence effectively, the mechanism was flow boiling. Apart from the different test fluids and different geometrical dimensions of the offset strip fins, Robertson employed $\Delta T = T_{\text{wall}} - T_{\text{sat}}$ of less than 2°C, while Westwater had ΔT varying from 2 to 80°C. Forced convective boiling (thin film evaporation) was dominant in the low ΔT tests of Robertson, while nucleate boiling was dominant in the high ΔT tests of Westwater. When Robertson's tentative correlation [152] is plotted on Westwater's data, the agreement is good for $\Delta T < 10$°C and for $\Delta T > 60$°C. In between, the prediction by Robertson's correlation is as much as 50 percent too low. It appears that ΔT and flow mass velocity may be two important parameters for forced convective (film and nucleate) boiling. The influence of flow passage geometry, although it does not seem to be significant, should be investigated with a systematic test program.

Galezha et al. [160] performed pool boiling tests with Freon-12 and Freon-22 in five offset strip-fin geometries. Since the flow rates were very low [less than 16 kg/(m²·s) or 11,800 lb$_m$/(h·ft²)], the boiling heat transfer coefficients did not depend upon the flow rates, but were functions of the geometry, pressure, and fluid type. They provided correlations for their test data.

b. Two-Phase Pressure Drop

Yung et al. [156] measured two-phase overall pressure drop across the test section during evaporation of ammonia. They predicted the pressure drop during evaporation based on the annular film-flow model mentioned earlier. Their measured ΔP's were up to 100 percent higher than the prediction. This difference was attributed to the form drag associated with the strip fins which was not accounted for in the modeling.

The current practice for calculating two-phase pressure drop on one side of an exchanger is to idealize the pressure drop as consisting of three components: two-phase friction ΔP_{TP}, momentum change (flow acceleration or deceleration), and gravity.† The pressure drop due to momentum change and gravity is computed using a method identical to that for the circular tube (see Ref. 174). For circular tubes, the two-phase friction component is calculated by Lockhart-Martinelli, Grant-Chisholm, or similar correlations (see Part 2 of this chapter or Refs. 171 to 173) in which form drag is generally ignored. However, based on the results of Yung et al. [156], form drag in interrupted surfaces under evaporation or condensation is not negligible. The following procedure is recommended to compute the two-phase friction component for compact surfaces: Compute first the Lockhart-Martinelli parameter ϕ_{lo}^2 by the Grant-Chisholm method (see Part 2 of this chapter). Then calculate the single-phase pressure drop ΔP_{lo} (considering total fluid flow as liquid flow) using the experimental single phase f versus Re curve for the surface. This f does include the form drag contribution. Now $\Delta P_{TP} = \phi_{lo}^2 \, \Delta P_{lo}$. In addition to this ΔP_{TP}, evaluate the single-phase ΔP's due to the fluid friction, considering total fluid flow as liquid and as gas. Reviewing these three values of ΔP, select the most appropriate value of ΔP for the friction component depending upon the application. Then add the contributions due to the momentum change and gravity to arrive at the two-phase ΔP for the particular compact exchanger application.

---

†The entrance and exit losses, usually included in single-phase flow ΔP's [e.g., see Eq. (23)], are generally small for two-phase flows (smaller than two-phase ΔP uncertainties), and no information is available for their evaluation. Hence, they are generally neglected in the two-phase ΔP evaluation.

2. Condensation

No systematic study has been reported to obtain heat transfer coefficients during condensation over plate-fin exchangers, and hence no generalized design correlations can be deduced. For noncompact surfaces (large passage hydraulic diameter), the condensation process is generally gravity-controlled with a finite interfacial shear stress. However, for compact surfaces, the surface-tension effects become important in the corner regions, particularly at low Reynolds numbers when the passages are not "flooded." The surface-tension forces pull the condensate to the corners (which act locally as drainage channels) and maintain thin films on the rest of the surface. The resultant heat transfer coefficients can be significantly higher than those for the gravity-controlled condensation with finite interfacial shear stress [161]. Therefore, the Nusselt analysis [162] with a proper modeling of the fin geometry should provide a conservative value for the heat transfer coefficient for compact surfaces. The results of Refs. 156, 163, and 164 for condensation in offset strip-fin exchangers are now summarized.

Gopin et al. [163] measured local condensing coefficients for a vertical downflow of Freon-22 in four plate-fin geometries at low vapor velocities ($\lesssim$ 0.3 m/s): one plain rectangular and three offset strip fins. The test section was 220 mm (8.66 in) long. They reported length average Nusselt numbers for condensation. The vapor quality did not appear as a variable in the results and the fin efficiency was treated as unity. For the plain rectangular plate-fin surface, they reported experimental condensing coefficients that were 20 to 25 percent higher than those obtained by the Nusselt equation for laminar film condensation on a plain vertical surface [162]. This increase was attributed to the surface tension effects by Gopin et al. For offset strip-fin surfaces, the measured condensing coefficients were 1.8 to 2 times higher than those obtained by the Nusselt equation. They also tested the best of four fin geometries with Freon-12 and Freon-142 as working fluids. They presented a closed-form Nusselt number correlation for their four fin geometries and three test fluids. Westwater [151] showed that their correlation is a modification to the Nusselt equation for condensation on a vertical wall with an added group L/D_h to take into account different geometries.

The results of Gopin et al. [163] for an offset strip fin with a longest strip length (but not the most compact surface) yielded the highest condensing coefficients, compared with those for short strip fins, and hence appear to be questionable.

Yung et al. [156] measured overall condensing coefficients (three data points) for vertical downflow of ammonia in an offset strip-fin geometry (see the first footnote on p. 4-294). The measured heat transfer coefficients on the ammonia side were within about 30 percent of the predicted results using an annular film-flow analysis. No surface tension effects were incorporated in the analysis and that may be one of the reasons for the differences in the slopes of experimental versus measured data lines.

Haseler [164] measured local condensing coefficients for vertical downflow of nitrogen in a plain rectangular plate-fin geometry.† He covered the range of mass velocity and heat flux conditions commonly encountered in the cryogenic process industry. In evaluating the heat transfer coefficient during condensation, a single-phase definition of fin efficiency was used. He presented the dimensional heat transfer coefficient during condensation as a function of vapor quality for two values of the mass velocity G. Typical results for G = 58.5 kg/(m²·s) or 43,100 lb$_m$/(h·ft²), are shown in Fig. 59. The corre-

---

†The geometry of the plain rectangular fin was: 709 fins per meter (18 fins per inch), 6.4 mm (0.252 in) in fin height, 0.2 mm (0.008 in) in fin thickness, and 3.2 m (10.5 ft) in fin length.

lations of Shah (S) [165], Boyko and Kruzhilin (B) [166], and Nusselt (N) [162], developed for condensation in a circular tube, are also compared in this figure. The Shah and Boyko–Kruzhilin correlations predict the experimental data reasonably well for $0.2 < x \leq 0.9$, while the Nusselt correlation is consistently low. However, at low mass velocities [$G \leq 30.6$ kg/(m$^2 \cdot$s) or 22,600 lb$_m$/(h·ft^2)], all three correlations [164, 165, 166] underpredict the data substantially (not shown in Fig. 59). More test data with systematic variations of the associated variables are needed before any firm design guidelines are recommended.

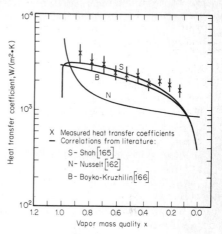

FIG. 59 Heat transfer coefficients for condensing nitrogen in an offset strip-fin exchanger [164].

L. HEAT EXCHANGER IRREVERSIBILITIES†

In the current energy conservation era, there is a great need to design heat exchangers most efficiently and to optimize them from the thermodynamics viewpoint. It is essential that the designer be aware of all important irreversibilities associated with the exchanger so that proper trade-offs can be made in the design. Four important irreversibilities associated with a heat exchanger are [167]: (1) heat transfer across a finite temperature difference, (2) fluid friction in flow over solid surfaces, (3) mixing of similar fluids at different temperatures, and (4) heat leakage to the surroundings. The last two irreversibilities may occur outside the exchanger, but are chargeable to the exchanger.

The entropy measure of the irreversibilities for a given system at an instant of time is defined as

$$\sum \dot{\mathrm{Irrev}}_s \equiv \sum \dot{\mathrm{OUTFLOW}} \text{ of entropy} - \sum \dot{\mathrm{INFLOW}} \text{ of entropy}$$
$$+ \sum \dot{\mathrm{INCREASE\ OF\ STORAGE}} \text{ of entropy} \qquad (150)$$
$$- \left[\sum \left(\frac{q}{T} \right)_i - \sum \left(\frac{q}{T} \right)_o \right]$$

By the second law of thermodynamics,

$$\dot{\mathrm{Irrev}}_s \geq 0 \qquad (151)$$

The energy measure of irreversibilities, useful for economic considerations to develop trade-off factors in the design process of Fig. 7, is defined as

$$\dot{\mathrm{Irrev}}_E \equiv T_{WF} \dot{\mathrm{Irrev}}_s \qquad (152)$$

Here T_{WF} is a temperature-weighting factor to be specified by the analyst based on his or her judgment of its relevance to the system being analyzed. It can be demonstrated that if T_{WF} equals T_o (the temperature of the "natural thermal sink" located in the surroundings of the system), $\dot{\mathrm{Irrev}}_E$ becomes identical to an "availability" or an "exergy" measure

†All temperatures in this section are based on the absolute scale.

of the irreversibility in question. In many cases, the analyst will elect to take $T_{WF} = T_o$, either as a judgment demand or because an alternative temperature is not apparent. The expressions for the energy measure of irreversibilities associated with an exchanger, normalized with respect to the exchanger heat transfer rate q, are [167]:

- *Heat Transfer Irreversibility:*

$$\left.\frac{\overset{\bullet}{\text{Irrev}}_E}{q}\right|_q = \frac{T_{WF}}{T_{c,lm}}\left(1 - \frac{T_{c,lm}}{T_{h,lm}}\right) \tag{153}$$

where T_{lm} is defined as

$$T_{lm} = \frac{T_i - T_o}{\ln(T_i/T_o)} \tag{154}$$

- *Flow Friction Irreversibility.*   For a perfect gas, it is

$$\left.\frac{\overset{\bullet}{\text{Irrev}}_E}{q}\right|_{\Delta P} = \frac{T_{WF}}{T_i}\frac{1}{q}\frac{W\,\Delta P}{\rho_i} \tag{155}$$

For an incompressible liquid, T_i in Eq. (155) should be replaced by T_{lm} as defined by Eq. (154).
For condensation or evaporation, it is

$$\left.\frac{\overset{\bullet}{\text{Irrev}}_E}{q}\right|_{\Delta P} = \frac{T_{WF}}{T_{ave}}\frac{\Delta P/\rho_g}{2i_{lg}} \tag{156}$$

where ρ_g is the density of saturated vapor evaluated at P_i for the condenser and P_o for the evaporator. Also, the latent heat of evaporation i_{lg} is evaluated at these pressures. T_{ave} is the average temperature during condensation or evaporation.

- *Fluid Mixing Irreversibility:*

$$\left.\frac{\overset{\bullet}{\text{Irrev}}_E}{q}\right|_{mix} = \frac{W_{exh}c_P T_{WF}}{q}\left(\frac{T_{exh}}{T_o} - 1 - \ln\frac{T_{exh}}{T_o}\right) \tag{157}$$

where W_{exh} and T_{exh} are the mass flow rate and temperature of the exhaust fluid stream into the surroundings at temperature T_o.

- *Heat Leak Irreversibility:*

$$\left.\frac{\overset{\bullet}{\text{Irrev}}_E}{q}\right|_{q,leak} = T_{WF}\left(\frac{1}{T_c} - \frac{1}{T_h}\right)\frac{q_{leak}}{q} \tag{158}$$

The total normalized energy measure of irreversibility in a heat exchanger is

$$\left.\frac{\overset{\bullet}{\text{Irrev}}_E}{q}\right|_{total} = \left.\frac{\overset{\bullet}{\text{Irrev}}_E}{q}\right|_q + \left.\frac{\overset{\bullet}{\text{Irrev}}_E}{q}\right|_{\Delta P_h} + \left.\frac{\overset{\bullet}{\text{Irrev}}_E}{q}\right|_{\Delta P_c}$$

$$+ \left.\frac{\overset{\bullet}{\text{Irrev}}_E}{q}\right|_{mix} + \left.\frac{\overset{\bullet}{\text{Irrev}}_E}{q}\right|_{q,leak} \tag{159}$$

If the temperature-weighting factor T_{WF} were selected as T_o for all the component irreversibilities, the total irreversibility would agree with the results obtained by the conventional available energy ("exergy") analysis. However, when the individual irreversibility components are recognized as in Eq. (159), their relative importance can be assessed, and a monetary value in terms of energy input requirements and capacity costs can be assigned. Then trade-off factors can be developed to provide the designer with the insight needed in considering options for reduction of the major irreversibility components.

ACKNOWLEDGMENTS

The author is grateful to Professor A. L. London of Stanford University for providing a critical review of this article, and for his constant guidance, encouragement, and motivation. The author also appreciates the constructive suggestions and critical review of this article by Dr. M. K. Jensen of the University of Wisconsin-Milwaukee, Dr. E. N. Ganić of the University of Illinois at Chicago, and Dr. P. J. Banks of CSIRO Division of Energy Technology, Australia. This article is based on the literature up to December 1981.

REFERENCES

1. R. K. Shah, Compact Heat Exchanger Surface Selection, Optimization, and Computer-Aided Thermal Design, in *Low Reynolds Number Flow Heat Exchangers*, ed. S. Kakaç, R. K. Shah, and A. E. Bergles, pp. 845–874, Hemisphere/McGraw-Hill, Washington, D.C., 1983.
2. R. K. Shah, Heat Exchanger Design Methodology—An Overview, in *Low Reynolds Number Flow Heat Exchangers*, ed. S. Kakaç, R. K. Shah, and A. E. Bergles, pp. 15–19, Hemisphere/McGraw-Hill, Washington, D.C., 1983.
3. D. Q. Kern and A. D. Kraus, *Extended Surface Heat Transfer*, McGraw-Hill, New York, 1972.
4. R. K. Shah, Temperature Effectiveness of Multiple Sandwich Rectangular Plate-Fin Surfaces, *J. Heat Transfer*, Vol. 93C, pp. 471–473, 1971.
5. T. C. Scott and I. Goldschmidt, Accurate, Simple Expressions for the Efficiency of Single and Composite Extended Surfaces, in *Advances in Enhanced Heat Transfer, ASME Symp. Vol. No. 100122*, pp. 79–85, ASME, New York, 1979.
6. D. G. Rich, The Efficiency and Thermal Resistance of Annular and Rectangular Fins, *Proc. 3d Int. Heat Transfer Conf.*, Vol. III, pp. 281–289, 1966.
7. F. E. M. Saboya and E. M. Sparrow, Experiments on a Three-Row Fin and Tube Heat Exchanger, *J. Heat Transfer*, Vol. 98C, pp. 520–522, 1976.
8. T. V. Jones and C. M. B. Russell, Local Heat Transfer Coefficients on Finned Tubes, in *Regenerative and Recuperative Heat Exchangers*, ed. R. K. Shah and D. E. Metzger, *ASME Symp. Book No. H00207, HTD-Vol. 21*, pp. 17–25, ASME, New York, 1981.
9. E. M. Sparrow and L. Lee, Effects of Fin Base-Temperature Depression in a Multifin Array, *J. Heat Transfer*, Vol. 97C, pp. 463–465, 1975.
10. N. V. Suryanarayana, Two-Dimensional Effects on Heat Transfer Rates from an Array of Straight Fins, *J. Heat Transfer*, Vol. 99C, pp. 129–132, 1977.
11. P. J. Heggs and P. R. Stones, The Effects of Dimensions on Heat Flowrate through Extended Surfaces, *J. Heat Transfer*, Vol. 102, pp. 180–182, 1980.
12. R. K. Shah, Thermal Design Theory for Regenerators, in *Heat Exchangers: Thermal-Hydraulic Fundamentals and Design*, ed. S. Kakaç, A. E. Bergles, and F. Mayinger, pp. 721–763. Hemisphere/McGraw-Hill, Washington, D.C., 1981.
13. H. Hausen, *Heat Transfer in Counterflow, Parallel Flow and Cross Flow*, 2d ed., McGraw-Hill, New York, 1983.
14. F. W. Schmidt and A. J. Willmott, *Thermal Energy Storage and Regeneration*, chaps. 5–9, Hemisphere/McGraw-Hill, Washington, D.C., 1981.
15. T. J. Lambertson, Performance Factors of a Periodic-Flow Heat Exchanger, *Trans. ASME*, Vol. 80, pp. 586–592, 1958.
16. W. M. Kays and A. L. London, *Compact Heat Exchangers*, 2d ed., McGraw-Hill, New York, 1964.
17. G. Theoclitus and T. L. Eckrich, Parallel Flow through the Rotary Heat Exchanger, *Proc. 3d Int. Heat Transfer Conf.*, Vol. I, pp. 130–138, 1966.

18. P. Razelos, An Analytical Solution to the Electric Analog Simulation of the Regenerative Heat Exchanger with Time-Varying Fluid Inlet Temperatures, *Wärme Stoffübertrag.*, Vol. 12, pp. 59–71, 1979.

19. G. D. Bahnke and C. P. Howard, The Effect of Longitudinal Heat Conduction on Periodic-Flow Heat Exchanger Performance, *J. Eng. Power*, Vol. 86A, pp. 105–120, 1964.

20. P. J. Heggs, L. S. Bansal, R. S. Bond, and V. Vazakas, Thermal Regenerator Design Charts Including Intraconduction Effects, *Trans. Inst. Chem. Eng.*, Vol. 58, pp. 265–270, 1980.

20 *a*. P. J. Heggs, personal communication, Department of Chemical Engineering, University of Leeds, Leeds, U.K., June 1982.

21. P. Razelos and A. Lazaridis, A Lumped Heat-Transfer Coefficient for Periodically Heated Hollow Cylinders, *Int. J. Heat Mass Transfer*, Vol. 10, pp. 1373–1387, 1967.

22. P. J. Banks and W. M. J. Ellul, Predicted Effects of By-pass Flows on Regenerator Performance, *Mech. Chem. Eng. Trans. Inst. Eng. Australia*, Vol. MC9, Nos. 1 and 2, pp. 10–14, 1973.

23. G. H. Klopfer, The Design of Periodic-Flow Heat Exchangers for Gas Turbine Engines, *TR HE-1*, Department of Mechanical Engineering, Stanford University, Stanford, Calif. 94305, August 1969.

23 *a*. P. J. Banks, The Representation of Regenerator Fluid Carryover by Bypass Flows, *J. Heat Transfer*, Vol. 106, pp. 216–220, 1984.

24. P. J. Banks, Effect of Fluid Carryover on Regenerator Performance, *J. Heat Transfer*, Vol. 104, pp. 215–217, 1982.

24 *a*. I. L. Maclaine-cross, Effect of Interstitial Fluid Heat Capacity on Regenerator Performance, *J. Heat Transfer*, Vol. 102, pp. 572–574, 1980.

24 *b*. P. J. Heggs and K. J. Carpenter, The Effect of Fluid Hold-up on the Effectiveness of Contraflow Regenerators, *Trans. Inst. Chem. Eng.*, Vol. 54, pp. 232–238, 1976.

25. R. K. Shah, Heat Exchanger Basic Design Methods, in *Low Reynolds Number Flow Heat Exchangers*, ed. S. Kakaç, R. K. Shah, and A. E. Bergles, pp. 21–72, Hemisphere/McGraw-Hill, Washington, D.C., 1983.

26. R. K. Shah and R. S. Johnson, Correlations for Fully Developed Turbulent Flow through Circular and Noncircular Channels, *Proc. 6th Nat. Heat Mass Transfer Conf.*, pp. D-75 to D-95, Indian Inst. Technol., Madras, 1981.

27. A. P. Colburn, A Method of Correlating Forced Convection Heat Transfer Data and a Comparison with Fluid Friction, *Trans. AIChE*, Vol. 29, pp. 174–210, 1933; reprinted in *Int. J. Heat Mass Trans.*, Vol. 7, pp. 1359–1384, 1964.

28. W. M. Kays and A. L. London, Heat Transfer and Flow Friction Characteristics of Some Compact Heat Exchanger Surfaces—Part I: Test System and Procedure, *Trans. ASME*, Vol. 72, pp. 1075–1085, 1950; also Description of Test Equipment and Method of Analysis for Basic Heat Transfer and Flow Friction Tests of High Rating Heat Exchanger Surfaces, *TR No. 2*, Department of Mechanical Engineering, Stanford University, Stanford, Calif., 1948.

29. E. E. Wilson, A Basis for Rational Design of Heat Transfer Apparatus, *Trans. ASME*, Vol. 37, pp. 47–82, 1915.

30. P. F. Pucci, C. P. Howard, and C. H. Piersall, Jr., The Single Blow Transient Testing Technique for Compact Heat Exchanger Surfaces, *J. Eng. Power*, Vol. 89A, pp. 29–40, 1967.

31. A. J. Wheeler, Single-Blow Transient Testing of Matrix-Type Heat Exchanger Surfaces at Low Values of N_{tu}, *TR. No. 68*, Department of Mechanical Engineering, Stanford University, Stanford, Calif., 1968.

32. J. R. Mondt and D. C. Siegla, Performance of Perforated Heat Exchanger Surfaces, *J. Eng. Power*, Vol. 96A, pp. 81–86, 1974.

33. J. H. Stang and J. E. Bush, The Periodic Method of Testing Compact Heat Exchanger Surfaces, *J. Eng. Power*, Vol. 96A, pp. 87–94, 1974.

34. R. K. Shah and A. L. London, *Laminar Flow Forced Convection in Ducts*, Supplement 1 to *Advances in Heat Transfer*, Academic, New York, 1978.

35. A. L. London, M. B. O. Young, and J. H. Stang, Glass-Ceramic Surfaces, Straight Triangular

Passages—Heat Transfer and Flow Friction Characteristics, *J. Eng. Power,* Vol. 92A, pp. 381–389, 1970.

36. A. L. London and R. K. Shah, Glass-Ceramic Hexagonal and Circular Passage Surfaces— Heat Transfer and Flow Friction Design Characteristics, *SAE Trans.,* Vol. 82, Sec. 1, pp. 425–434, 1973.

37. V. Gnielinski, New Equations for Heat and Mass Transfer in Turbulent Pipe and Channel Flow, *Int. Chem. Eng.,* Vol. 16, pp. 359–368, 1976.

38. B. S. Petukhov, Heat Transfer and Friction in Turbulent Pipe Flow with Variable Physical Properties, in *Advances in Heat Transfer,* Vol. 6, pp. 503–564, Academic, New York, 1970.

39. C. A. Sleicher and M. W. Rouse, A Convenient Correlation for Heat Transfer to Constant and Variable Property Fluids in Turbulent Pipe Flow, *Int. J. Heat Mass Transfer,* Vol. 18, pp. 677–683, 1975.

40. W. M. Kays and M. Crawford, *Convective Heat and Mass Transfer,* 2d ed., McGraw-Hill, New York, 1980.

41. H. Schlichting, *Boundary Layer Theory,* 4th ed., pp. 504, 521, McGraw-Hill, New York, 1960.

42. R. K. Shah and R. L. Webb, Compact and Enhanced Heat Exchangers, in *Heat Exchangers: Theory and Practice,* ed. J. Taborek, G. F. Hewitt, and N. Afgan, pp. 425–468, Hemisphere/ McGraw-Hill, Washington, D.C., 1983.

43. Z. V. Tishchenko, V. N. Bondarenko, and L. I. Golechek, Heat Transfer and Pressure Drop in Gas-Carrying Ducts Formed by Smooth-Finned Plate-Type Heat-Exchange Surfaces, *Heat Transfer Sov. Res.,* Vol. 11, No. 5, pp. 117–124, 1979.

44. L. Goldstein, Jr., and E. M. Sparrow, Heat/Mass Transfer Characteristics for Flow in a Corrugated Wall Channel, *J. Heat Transfer,* Vol. 99C, pp. 187–195, 1977.

45. A. L. London and R. K. Shah, Offset Rectangular Plate-Fin Surfaces—Heat Transfer and Flow Friction Characteristics, *J. Eng. Power,* Vol. 90A, pp. 218–228, 1968.

46. R. K. Shah and A. L. London, Influence of Brazing on Very Compact Heat Exchanger Surfaces, *ASME Paper No. 71-HT-29,* 1971.

47. E. V. Dubrovskii and A. I. Fedotova, Investigation of Heat-Exchanger Surfaces with Plate Fins, *Heat Transfer Sov. Res.,* Vol. 4, No. 6, pp. 75–79, 1972.

48. S. Mochizuki and Y. Yagi, Heat Transfer and Friction Characteristics of Strip Fins, *Heat Transfer Jpn. Res.,* Vol. 6, No. 3, pp. 36–59, 1977.

49. A. R. Wieting, Empirical Correlations for Heat Transfer and Flow Friction Characteristics of Rectangular Offset-Fin Plate-Fin Heat Exchangers, *J. Heat Transfer,* Vol. 97C, pp. 488–490, 1975.

50. R. L. Webb and H. M. Joshi, A Friction Factor Correlation for the Offset Strip-Fin Matrix, *Heat Transfer 1982,* Vol. 6, pp. 257–262, Hemisphere, Washington, D.C., 1982.

51. R. L. Webb and H. M. Joshi, Prediction of the Friction Factor for the Offset Strip Fin Matrix, *Proc. ASME/JSME Thermal Eng. Joint Conf.,* Vol. 1, pp. 461–469, ASME, New York, 1983.

52. S. W. Churchill and R. Usagi, A General Expression for the Correlation of Rates of Transfer and Other Phenomena, *AIChE J.,* Vol. 18, pp. 1121–1128, 1972.

53. E. M. Sparrow and A. Hajiloo, Measurements of Heat Transfer and Pressure Drop for an Array of Staggered Plates Aligned Parallel to Air Flow, *J. Heat Transfer,* Vol. 102, pp. 426–432, 1980.

54. S. V. Patankar and C. Prakash, An Analysis of the Effect of Plate Thickness on Laminar Flow and Heat Transfer in Interrupted-Plate Passages, *Int. J. Heat Mass Transfer,* Vol. 24, pp. 1801–1810, 1981.

55. Y. Mori and W. Nakayama, Recent Advances in Compact Heat Exchangers in Japan, in *Compact Heat Exchangers—History, Technological Advancement and Mechanical Design Problems,* ed. R. K. Shah, C. F. McDonald, and C. P. Howard, *ASME Symp. Book No. G00183, HTD-Vol. 10,* pp. 5–16, ASME, New York, 1980.

56. C. J. Davenport, Heat Transfer and Flow Friction Characteristics of Louvered Heat Exchanger Surfaces, in *Heat Exchangers: Theory and Practice,* ed. J. Taborek, G. F. Hewitt, and N. Afgan, pp. 397–412, Hemisphere/McGraw-Hill, Washington, D.C., 1983.

57. H. L. Miller and C. A. Leeman, Heat Transfer and Pressure Drop Characteristics of Several Compact Plate Surfaces, in *Heat Transfer: Fundamentals and Industrial Applications, AIChE Symp. Ser.* 131, Vol. 39, pp. 63–71, 1973.

58. R. K. Shah, Perforated Heat Exchanger Surfaces: Part I—Flow Phenomena, Noise and Vibration Characteristics, *ASME Paper No. 75-WA/HT-8,* 1975.

59. R. K. Shah, Perforated Heat Exchanger Surfaces: Part 2—Heat Transfer and Flow Friction Characteristics, *ASME Paper No. 75-WA/HT-9,* 1975.

60. C. Y. Liang and W. J. Yang, Heat Transfer and Friction Loss Performance of Perforated Heat Exchanger Surfaces, *J. Heat Transfer,* Vol. 97C, pp. 9–15, 1975.

61. C. P. Lee and W. J. Yang, Augmentation of Convective Heat Transfer from High-Porosity Perforated Surfaces, *Heat Transfer 1978,* Vol. 2, pp. 589–594, Hemisphere, Washington, D.C., 1978.

62. G. Theoclitus, Heat-Transfer and Flow-Friction Characteristics of Nine Pin-Fin Surfaces, *J. Heat Transfer,* Vol. 88C, pp. 383–390, 1966.

63. D. E. Metzger, Z. X. Fan, and W. B. Shepard, Pressure Loss and Heat Transfer through Multiple Rows of Short Pin Fins, *Heat Transfer 1982,* Vol. 3, pp. 137–142, Hemisphere, Washington, D.C., 1982.

64. W. M. Kays, The Heat Transfer and Flow Friction Characteristics of Six High Performance Heat Transfer Surfaces, *J. Eng. Power,* Vol. 82A, pp. 27–34, 1960.

65. R. L. Webb, Air-Side Heat Transfer in Finned Tube Heat Exchangers, *Heat Transfer Eng.,* Vol. 1, No. 3, pp. 33–49, 1980.

66. D. G. Rich, The Effect of Fin Spacing on the Heat Transfer and Friction Performance of Multi-row, Smooth Plate Fin-and-Tube Heat Exchangers, *ASHRAE Trans.,* Vol. 79, Pt. 2, pp. 137–145, 1973.

67. D. G. Rich, The Effect of the Number of Tube Rows on Heat Transfer Performance of Smooth Plate Fin-and-Tube Heat Exchangers, *ASHRAE Trans.,* Vol. 81, Pt. 1, pp. 307–319, 1975.

68. F. E. M. Saboya and E. M. Sparrow, Local and Average Transfer Coefficients for One-Row Plate Fin and Tube Heat Exchanger Configurations, *J. Heat Transfer,* Vol. 96C, pp. 265–272, 1974.

69. F. E. M. Saboya and E. M. Sparrow, Transfer Characteristics of Two-Row Plate Fin and Tube Heat Exchanger Configurations, *Int. J. Heat Mass Transfer,* Vol. 19, pp. 41–49, 1976.

70. F. C. McQuiston, Correlation of Heat, Mass and Momentum Transport Coefficients for Plate-Fin-Tube Heat Transfer Surfaces with Staggered Tubes, *ASHRAE Trans.,* Vol. 84, Pt. 1, pp. 294–309, 1978.

71. F. C. McQuiston and D. R. Tree, Heat-Transfer and Flow-Friction Data for Two Fin-Tube Surfaces, *J. Heat Transfer,* Vol. 93C, pp. 249–250, 1971.

72. F. C. McQuiston, Heat, Mass and Momentum Transfer Data for Five Plate-Fin-Tube Heat Transfer Surfaces, *ASHRAE Trans.,* Vol. 84, Pt. 1, pp. 266–293, 1978.

73. S. W. Krückels and V. Kottke, Untersuchung über die Verteilung des Wärmeübergangs an Rippen und Rippenrohr-Modellen (Investigation of the Distribution of Heat Transfer on Fins and Finned Tube Models), *Chem. Ing. Tech.,* Vol. 42, pp. 355–362, 1970.

74. S. Fukui and M. Sakamoto, Some Experimental Results on Heat Transfer Characteristic of Air Cooled Heat Exchangers for Air Conditioning Devices, *Bull. JSME,* Vol. 11, No. 44, pp. 303–311, 1968.

75. T. Hosoda, H. Uzuhashi, and N. Kobayashi, Louver Fin Type Heat Exchangers, *Heat Transfer Jpn. Res.,* Vol. 6, No. 2, pp. 69–77, 1977.

76. L. Goldstein, Jr., and E. M. Sparrow, Experiments on the Transfer Characteristics of a Corrugated Fin and Tube Heat Exchanger Configuration, *J. Heat Transfer,* Vol. 98C, pp. 26–34, 1976.

77. M. Ito, H. Kimura, and T. Senshu, Development of High Efficiency Air-Cooled Heat Exchangers, *Hitachi Rev.,* Vol. 26, No. 10, pp. 323–326, 1977.

78. I. Vlădea, H. Theil, and F. Neiss, Wärmeaustauscher mit Flachrohren und durchlaufenden Rippen, *Heat Transfer 1970,* Vol. 1, Paper HE 1.5, Elsevier, Amsterdam, 1970.

79. G. Kovacs, Application of Short-Finned Heat Exchanger as Air-Cooled Condenser (in French), *Rev. Gén. Froid*, pp. 159–168, February 1963.

80. N. V. Zozulya, A. A. Khavin, and B. L. Kalinin, Effect of Perforated Transverse Plate Fins on Oval Pipes on the Rate of Heat Transfer, *Heat Transfer Sov. Res.*, Vol. 2, No. 1, pp. 77–79, 1970.

81. N. V. Zozulya, A. A. Khavin, and B. L. Kalinin, Effect of Fin Deformation on Heat Transfer and Drag in Bundles of Oval Tubes with Transverse Fins, *Heat Transfer Sov. Res.*, Vol. 7, No. 2, pp. 95–98, 1975.

82. D. E. Briggs and E. H. Young, Convection Heat Transfer and Pressure Drop of Air Flowing across Triangular Pitch Banks of Finned Tubes, *Chem. Eng. Prog. Symp. Ser.* No. 41, Vol. 59, pp. 1–10, 1963.

83. K. K. Robinson and D. E. Briggs, Pressure Drop of Air Flowing across Triangular Pitch Banks of Finned Tubes, *Chem. Eng. Prog. Symp. Ser.* No. 64, Vol. 62, pp. 177–184, 1966.

84. T. J. Rabas, P. W. Eckles, and R. A. Sabatino, The Effect of Fin Density on The Heat Transfer and Pressure Drop Performance of Low-Finned Tube Banks, *ASME Paper No. 80-HT-97*, 1980.

85. E.Schmidt, Heat Transfer at Finned Tubes and Computations of Tube Bank Heat Exchangers (in German), *Kältetech.*, Vol. 15, No. 4, pp. 98–102, 1963; *Kältetech.*, Vol. 15, No. 12, pp. 370–378, 1963.

86. C. Weierman, J. Taborek, and W. J. Marner, Comparison of the Performance of Inline and Staggered Banks of Tubes with Segmented Fins, *AIChE Paper No. 6*, presented at the 15th National Heat Transfer Conference, San Francisco, 1975.

87. C. Weierman, Correlations Ease the Selection of Finned Tubes, *Oil Gas J.*, pp. 94–100, Sept. 6, 1976.

88. T. J. Rabas and P. W. Eckels, Heat Transfer and Pressure Drop Performance of Segmented Surface Tube Bundles, *ASME Paper No. 75-HT-45*, 1975.

89. R. W. Abbott, R. H. Norris, and W. A. Spofford, Compact Heat Exchangers for General Electric Products—Sixty Years of Advances in Design and in Manufacturing Technologies, in *Compact Heat Exchangers—History, Technological Advancement and Mechanical Design Problems*, ed. R. K. Shah, C. F. McDonald, and C. P. Howard, *ASME Symp. Book No. G00183, HTD-Vol. 10*, pp. 37–55, ASME, New York, 1980.

90. F. K. Moore, Analysis of Large Dry Cooling Towers with Spine-Fin Heat Exchanger Elements, *ASME Paper NO. 75-WA/HT-46*, 1975.

91. J. W. Ackerman and A. R. Brunsvold, Heat Transfer and Draft Loss Performance of Extended Surface Tube Banks, *J. Heat Transfer*, Vol. 92C, pp. 215–220, 1970.

92. R. J. Preece, J. Lis, and J. A. Hitchcock, Comparative Performance Characteristics of Some Extended Surfaces for Air-Cooler Applications, *Chem. Eng. (London)*, pp. 238–244, June 1972.

93. A. J. Wall, Extended Surface Tubes, *Chem. Process Eng. (London)*, Vol. 48, No. 8, pp. 136–138, 1967.

94. H. Brauer, Compact Heat Exchangers, *Chem. Process Eng. (London)*, Vol. 45, No. 8, pp. 451–460, 1964.

95. H. Nakamura, A. Matsuura, J. Kiwaki, N. Matsuda, S. Hiraoka, and I. Yamada, The Effect of Variable Viscosity on Laminar Flow and Heat Transfer in Rectangular Ducts, *J. Chem. Eng. Jpn.*, Vol. 12, No. 1, pp. 14–18, 1979.

96. A. E. Bergles, Experimental Verification of Analyses and Correlation of the Effects of Temperature-Dependent Fluid Properties, in *Low Reynolds Number Flow Heat Exchangers*, ed. S. Kakaç, R. K. Shah, and A. E. Bergles, pp. 473–486, Hemisphere/McGraw-Hill, Washington, D.C., 1983.

97. R. K. Shah, Compact Heat Exchanger Surface Selection Methods, *Heat Transfer 1978*, Vol. 4, pp. 193–199, Hemisphere, Washington, D.C., 1978.

98. R. L. Webb, Performance Evaluation Criteria for Use of Enhanced Heat Transfer Surfaces in Heat Exchanger Design, *Int. J. Heat Mass Transfer*, Vol. 24, pp. 715–726, 1981.

99. R. L. Webb and A. E. Bergles, Performance Evaluation Criteria for Selection of Heat Transfer Surface Geometries Used in Low Reynolds Number Heat Exchangers, in *Low Reynolds Number Flow Heat Exchangers,* ed. S. Kakaç, R. K. Shah, and A. E. Bergles, pp. 735–752, Hemisphere/McGraw-Hill, Washington, D.C., 1983.

100. R. L. Webb, Performance Evaluation Criteria for Air-Cooled Finned-Tube Heat Exchanger Surface Geometries, *Int. J. Heat Mass Transfer,* Vol. 25, p. 1770, 1982.

101. H. J. Fontein and J. G. Wassink, The Economically Optimal Design of Heat Exchangers, *Eng. Process Econ.,* Vol. 3, pp. 141–149, 1978.

102. C. M. Johnson, G. N. Vanderplaats, and P. J. Marto, Marine Condenser Design Using Numerical Optimization, *J. Mech. Design,* Vol. 102, pp. 469–475, 1980.

103. C. P. Hedderich, M. D. Kelleher, and G. N. Vanderplaats, Design and Optimization of Air-Cooled Heat Exchangers, *J. Heat Transfer,* Vol. 104, pp. 683–690, 1982.

104. R. K. Shah, K. A. Afimiwala, and R. W. Mayne, Heat Exchanger Optimization, *Heat Transfer 1978,* Vol. 4, pp. 185–191, Hemisphere/McGraw-Hill, Washington, D.C., 1978.

105. J. R. Mondt, Regenerative Heat Exchangers: The Elements of Their Design, Selection and Use, *Rep. No. GMR-3396,* Research Laboratories, GM Technical Center, Warren, Mich., 1980 (report available on request).

106. R. K. Shah, Compact Heat Exchanger Design Procedures, in *Heat Exchangers: Thermal-Hydraulic Fundamentals and Design,* ed. S. Kakaç, A. E. Bergles, and F. Mayinger, pp. 495–536, Hemisphere/McGraw-Hill, Washington, D.C., 1981.

107. J. M. Chenoweth and R. S. Kistler, Computer Program as a Tool for Heat Exchanger Rating and Design, *ASME Paper NO. 76-WA/HT-4,* 1976.

108. K. J. Bell, Preliminary Design of Shell and Tube Heat Exchangers, in *Heat Exchangers: Thermal-Hydraulic Fundamentals and Design,* ed. S. Kakaç, A. E. Bergles, and F. Mayinger, pp. 559–579, Hemisphere/McGraw-Hill, Washington, D.C., 1981.

109. A. L. London, G. Klopfer, and S. Wolf, Oblique Flow Headers for Heat Exchangers, *J. Eng. Power,* Vol. 90A, pp. 271–286, 1968.

110. D. E. Briggs and E. H. Young, Modified Wilson Plot Techniques for Obtaining Heat Transfer Correlations for Shell and Tube Heat Exchangers, *Chem. Eng. Prog. Symp. Ser.* 92, Vol. 65, pp. 35–45, 1969.

111. S. J. Kline, Diffusers—Flow Phenomena and Design, Department of Mechanical Engineering, Stanford University, Stanford, Calif., 1978; see also *Tech. Reps. PD-22, PD-23, PD-24,* and *PD-25* available through Prof. Kline, 1981.

112. D. S. Miller, *Internal Flow Systems,* Vol. 5 in the BHRA Fluid Engineering Series, published by The British Hydromechanics Research Association, 1978; available also through Air Science Co., P. O. Box 143, Corning, NY 14830.

113. A. K. Majumdar, Mathematical Modelling of Flows in Dividing and Combining Flow Manifolds, *Appl. Math. Modelling,* Vol. 4, pp. 424–432, 1980.

114. R. A. Bajura and E. H. Jones, Jr., Flow Distribution Manifolds, *J. Fluids Eng.,* Vol. 98, pp. 654–666, 1976.

115. A. B. Datta and A. K. Majumdar, Flow Distribution in Parallel and Reverse Flow Manifolds, *Int. J. Heat Fluid Flow,* Vol. 2, pp. 253–262, 1980.

116. R. K. Shah, Compact Heat Exchangers, in *Heat Exchangers: Thermal-Hydraulic Fundamentals and Design,* ed. S. Kakaç, A. E. Bergles, and F. Mayinger, pp. 111–151, Hemisphere/McGraw-Hill, Washington, D.C., 1981.

117. J. P. Chiou, Thermal Performance Deterioration in Crossflow Heat Exchangers Due to Flow Nonuniformity, *J. Heat Transfer,* Vol. 100, pp. 580–587, 1978.

118. J. P. Chiou, Thermal Performance Deterioration in Crossflow Heat Exchanger Due to Flow Nonuniformity on Both Hot and Cold Sides, *Heat Transfer 1978,* Vol. 4, pp. 279–284, Hemisphere/McGraw-Hill, Washington, D.C., 1978.

119. J. P. Chiou, The Effect of Nonuniformity of Inlet Fluid Temperature on the Thermal Performance of Crossflow Heat Exchanger, *Heat Transfer 1982,* Vol. 6, pp. 179–184, Hemisphere, Washington, D.C., 1982.

120. A. L. London, Laminar Flow Gas Turbine Regenerators—The Influence of Manufacturing Tolerances, *J. Eng. Power,* Vol. 92A, pp. 46–56, 1970.

121. R. K. Shah and A. L. London, Effects of Nonuniform Passages on Compact Heat Exchanger Performance, *J. Eng. Power,* Vol. 102A, pp. 653–659, 1980.

122. N. Epstein, Fouling in Heat Exchangers, *Heat Transfer 1978,* Vol. 6, pp. 235–253, Hemisphere/McGraw-Hill, Washington, D.C., 1978.

123. W. T. Reid, *External Corrosion and Deposits—Boilers and Gas Turbines,* Elsevier, New York, 1971.

124. R. W. Bryers, ed., *Ash Deposits and Corrosion Due to Impurities in Combustion Gases,* Hemisphere/McGraw-Hill, Washington, D.C., 1978.

125. W. J. Marner and R. L. Webb, eds., Workshop on an Assessment of Gas-Side Fouling in Fossil Fuel Exhaust Environments, *JPL Publ. 82/67,* Jet Propulsion Laboratory, California Institute of Technology, Pasadena, July 1982.

126. G. Theoclitus, Heat Exchanger Fires and the Ignition of Solid Metals, in *Regenerative and Recuperative Heat Exchangers,* ed. R. K. Shah and D. E. Metzger, *HTD-Vol. 21,* pp. 51–57, ASME, New York, 1981.

127. C. F. McDonald, The Potential Danger of Fire in Gas Turbine Heat Exchangers, *ASME Paper No. 69-GT-38,* 1969.

128. N. A. Fuchs, *The Mechanics of Aerosols,* pp. 204–212, Pergamon, New York, 1964.

129. J. A. Miller, Mechanisms of Gas Turbine Regenerator Fouling, *ASME Paper No. 67-GT-26,* 1967.

130. J. W. Thomas, Gravity Settling of Particles in a Horizontal Tube, *J. Air Pollution Control Assn.,* Vol. 8, pp. 32–34, 1958.

131. S. K. Beal, Deposition of Particles in Turbulent Flow on Channel or Pipe Walls, *Nucl. Sci. Eng.,* Vol. 40, pp. 1–11, 1970.

132. P. S. Epstein, Zur Theorie des Radiometers, *Z. Phys.,* Vol. 54, pp. 537–563, 1929.

133. G. Nishio, S. Kitani, and K. Takahashi, Thermophoretic Deposition of Aerosol Particles in a Heat-Exchanger Pipe, *Ind. Eng. Chem., Process Des. Dev.,* Vol. 13, pp. 408–415, 1974.

134. S. L. Soo, Pneumatic Conveying, in *Handbook of Multiphase Systems,* ed. G. Hetsroni, pp. 7-3 to 7-28, Hemisphere/McGraw-Hill, Washington, D.C., 1980.

135. D. Csathy, Latest Practice in Industrial Heat Recovery, presented at the Energy-Source Technology Conference and Exhibition, New Orleans, Feb. 4, 1980.

135 *a.* D. Csathy, personal communication, Deltak Corp., P. O. Box 9496, Minneapolis, MN 55440, 1982.

136. J. E. Radway and T. Boyce, Reduction of Coal Ash Deposits with Magnesia Treatment, *Combustion,* Vol. 49, No. 10, pp. 24–30, 1978.

137. E. De Anda, Heat Exchanger Fouling and Corrosion Evaluation, *Rep. No. 81-18003,* prepared for DOE Contract No. DE-AC22-77ET11296, AiResearch Manufacturing Co., 2525 W. 190th St., Torrance, CA 90509, October 1981.

138. A. K. Kulkarni, Y.-J. Wang, and R. L. Webb, Fouling and Corrosion by Glass Furnace Exhausts, *Proc. Int. Conf. Fouling of Heat Exchange Surface,* White Haven, Pa., Oct. 31 to Nov. 5, 1982.

139. C. B. Mort, Research on Exhaust Gas Effects on Heat Exchangers, *Wright-Patterson AFB Tech. Rep. SEG-TR-66-30,* prepared by Hamilton Standard Division of United Aircraft Corp. under Contract AF33(615)2471, July 1966.

140. R. D. Rogalski, Fouling Effects of Turbine Exhaust Gases on Heat Exchanger Tubes for Heat Recovery Systems, *Rep. No. DTNSRDC/PAS-79/5,* David W. Taylor Naval Ship Research and Development Center, Bethesda, Md., February 1979.

141. P. B. Roberts and H. D. Marron, Soot and the Combined Cycle Boiler, *ASME Paper No. 79-GT-67,* 1979.

142. P. Profos and H. N. Sharan, Effect of Tube Spacing and Arrangement upon the Fouling Characteristics of Banks of Tubes, *Sulzer Tech. Rev.,* Vol. 42, No. 4, pp. 31–43, 1960.

143. C. Weierman, Design of Heat Transfer Equipment for Gas Side Fouling Service, a paper published in Ref. 125, 1982.

144. T. Cowell and D. A. Cross, Airside Fouling of Internal Combustion Engine Radiators, *SAE Trans.,* Vol. 89, Paper No. 801012, 1980.

145. J. C. Eastwood, High-Performance Metallic Recuperator for Industrial Flue Gas Heat Recovery, *Proc. Aluminum Ind. Energy Conservation Workshop VI,* pp. 111–128, The Aluminum Association, Washington, D.C., November 1981.

146. T. R. Bott and C. R. Bemrose, Particulate Fouling on the Gas-Side of Finned Tube Heat Exchangers, in *Fouling in Heat Exchange Equipment,* ed. J. M. Chenoweth and M. Impagliazzo, *HTD-Vol. 17,* pp. 83–88, ASME, New York, 1981; also published in *J. Heat Transfer,* Vol. 105, pp. 178–183, 1983.

147. A. B. Johnson, Jr., D. R. Pratt, and G. E. Zima, A Survey of Materials and Corrosion Performance in Dry Cooling Applications, *Rep. No. BNWL-1958, UC-12,* Battelle, Pacific Northwest Laboratories, Richland, WA 99352, March 1976.

148. S. V. Patankar and E. M. Sparrow, Condensation on an Extended Surface, *J. Heat Transfer,* Vol. 101, pp. 434–440, 1979.

149. J. G. Collier, T. D. A. Kennedy, and J. A. Ward, The Thermal Design of Plate Fin Reboilers, *Cryotech. 73 Prod. Use Ind. Gases Proc. Conf.,* Vol. 18, pp. 95–100, IPC Technological Press, Guilford, U.K., 1974.

150. J. M. Robertson, Review of Boiling, Condensing and Other Design Aspects of Two-Phase Flow in Plate-Fin Heat Exchangers, in *Compact Heat Exchangers,* ed. R. K. Shah, C. F. McDonald, and C. P. Howard, *ASME Symp. Book NO. G00183, HTD-Vol. 10,* pp. 17–27, ASME, New York, 1980.

151. J. W. Westwater, Boiling Heat Transfer in Compact and Finned Heat Exchangers, in *Advances in Two-Phase Flow and Heat Transfer,* ed. S. Kakaç and M. Ishii, Vol. II, pp. 827–857, Martinus Nijhoff, Boston, 1983.

152. J. M. Robertson, Boiling Heat Transfer with Liquid Nitrogen in Brazed-Aluminium Plate-Fin Heat Exchangers, *AIChE Symp. Ser.* 189, Vol. 75, pp. 151–164, 1979.

153. J. M. Robertson and P. C. Lovegrove, Boiling Heat Transfer with Freon 11 in Brazed-Aluminium Plate-Fin Heat Exchangers, *ASME Paper No. 80-HT-58,* 1980.

154. J. M. Robertson and R. H. Clarke, The Onset of Boiling of Liquid Nitrogen in Plate-Fin Heat Exchangers, *AIChE Symp. Ser.* 208, Vol. 77, pp. 86–95, 1981.

155. J. M. Robertson, The Correlation of Boiling Coefficients in Plate-Fin Heat Exchanger Passages with a Film-Flow Model, *Heat Transfer 1982,* Vol. 6, pp. 341–345, Hemisphere, Washington, D.C., 1982.

156. D. Yung, J. J. Lorenz, and C. Panchal, Convective Vaporization and Condensation in Serrated-Fin Channels, in *Heat Transfer in Ocean Thermal Energy Conversion [OTEC] Systems,* ed. W. L. Owens, *ASME Symp. HTD-Vol. 12,* pp. 29–37, ASME, New York, 1980.

157. G. F. Hewitt and N. S. Hall-Taylor, *Annular Two-Phase Flow,* pp. 192–194, Pergamon, Oxford, 1970.

158. H. Panitsidis, R.D. Gresham, and J. W. Westwater, Boiling of Liquids in a Compact Plate-Fin Heat Exchanger, *Int. J. Heat Mass Transfer,* Vol. 18, pp. 37–42, 1975.

159. C. C. Chen, J. V. Loh, and J. W. Westwater, Prediction of Boiling Heat Transfer Duty in a Compact Plate-Fin Heat Exchanger Using the Improved Local Assumption, *Int. J. Heat Mass Transfer,* Vol. 24, pp. 1907–1912, 1981.

160. V. B. Galezha, I. P. Usyukin, and K. D. Kan, Boiling Heat Transfer with Freons in Finned-Plate Heat Exchangers, *Heat Transfer Sov. Res.,* Vol. 8, No. 3, pp. 103–110, 1976.

161. R. L. Webb, S. T. Keswani, and T. M. Rudy, Investigation of Surface Tension and Gravity Effects in Film Condensation, *Heat Transfer 1982,* Vol. 5, pp. 175–180, Hemisphere, Washington, D.C., 1982.

162. W. Nusselt, Die Oberflächenkondensation des Wasserdampfes, *VDI Z.,* Vol. 60, pp. 541 and 569, 1916.

163. S. R. Gopin, I. P. Usyukin, and I. G. Aver'yanov, Heat Transfer in Condensation of Freons on Finned Surfaces, *Heat Transfer Sov. Res.,* Vol. 8, No. 6, pp. 114–119, 1976.

164. L. Haseler, Condensation of Nitrogen in Brazed Aluminium Plate-Fin Heat Exchangers, *ASME Paper No. 80-HT-57,* 1980.

165. M. M. Shah, A General Correlation for Heat Transfer during Film Condensation inside Pipes, *Int. J. Heat Mass Transfer,* Vol. 22, pp. 547–556, 1979.

166. E. P. Ananiev, L. D. Boyko, and G. N. Kruzhilin, Heat Transfer in the Presence of Steam Condensation in a Horizontal Tube, *Proc. 1961 Int. Heat Transfer Conf.,* Boulder, Colo., Pt. II, pp. 290–295, 1961.

167. A. L. London and R. K. Shah, Costs of Irreversibilities in Heat Exchanger Design, *Heat Transfer Eng.,* Vol. 4, No. 2, pp. 59–73, 1983.

168. W. M. Kays and H. C. Perkins, Forced Convection, Internal Flow in Ducts, in *Handbook of Heat Transfer Fundamentals,* ed. W. M. Rohsenow, J. P. Hartnett, and E. N. Ganić, chap. 7, McGraw-Hill, New York, 1985.

169. G. D. Raithby and K. G. T. Hollands, National Convection, in *Handbook of Heat Transfer Fundamentals,* ed. W. M. Rohsenow, J. P. Hartnett, and E. N. Ganić, chap. 6, McGraw-Hill, New York, 1985.

170. P. E. Liley, Thermophysical Properties, in *Handbook of Heat Transfer Fundamentals,* ed. W. M. Rohsenow, J. P. Hartnett, and E. N. Ganić, chap. 3, McGraw-Hill, New York, 1985.

171. W. M. Rohsenow, Film Condensation, in *Handbook of Heat Transfer Fundamentals,* ed. W. M. Rohsenow, J. P. Hartnett, and E. N. Ganić, chap. 11, pt. 1, McGraw-Hill, New York, 1985.

172. P. Griffith, Dropwise Condensation, in *Handbook of Heat Transfer Fundamentals,* ed. W. M. Rohsenow, J. P. Hartnett, and E. N. Ganić, chap. 11, pt. 2, McGraw-Hill, New York, 1985.

173. W. M. Rohsenow, Boiling, in *Handbook of Heat Transfer Fundamentals,* ed. W. M. Rohsenow, J. P. Hartnett, and E. N. Ganić, chap. 12, McGraw-Hill, New York, 1985.

174. P. Griffith, Two-Phase Flow, in *Handbook of Heat Transfer Fundamentals,* ed. W. M. Rohsenow, J. P. Hartnett, and E. N. Ganić, chap. 13, McGraw-Hill, New York, 1985.

NOMENCLATURE

Symbol, Definition, SI Units, English Units

English Symbols

A — total heat transfer surface area (both primary and secondary, if any) on one side of a direct transfer–type exchanger; also total heat transfer surface area of a rotary regenerator; heat transfer surface area of one matrix of a fixed-matrix regenerator: m^2, ft^2

A_c — heat transfer area on the cold side of the exchanger: m^2, ft^2

A_f — fin or extended surface area on one side of the exchanger: m^2, ft^2

A_{fr} — frontal or face area of the exchanger: m^2, ft^2

A_h — heat transfer area on the hot side of the exchanger: m^2, ft^2

A_k — total wall cross-sectional area for longitudinal conduction (subscripts c, h, and t denote cold side, hot side, and total for the regenerator): m^2, ft^2

A_k^* — ratio of $A_{k,c}$ to $A_{k,h}$, dimensionless

A_o — minimum free flow (or open) area: m^2, ft^2

A_o^* — branch to header flow area ratio, dimensionless

A_p — primary surface area on one side of the exchanger: m^2, ft^2

A_t — outside surface area of bare tubes (without any fins): m^2, ft^2

A_w — total wall area for heat conduction from hot to cold fluid, or total wall area for transverse heat conduction (in the matrix wall thickness direction): m^2, ft^2

a — wall thickness: m, ft

Bi — Biot number $= h(a/2)/k_w$, dimensionless

b	plate spacing; distance between two plates (fin height) in a plate-fin exchanger: m, ft
C	flow stream heat capacity rate with a subscript c or h, $C = Wc_p$: W/°C, Btu/(h·°F)
C^*	heat capacity rate ratio C_{min}/C_{max}, dimensionless
C_{max}	maximum of C_c and C_h: W/°C, Btu/(h·°F)
C_{min}	minimum of C_c and C_h: W/°C, Btu(h·°F)
C_r	total heat capacity rate of a regenerator: $C_r = M_w c_w \omega$ for a rotary regenerator, $C_r = M_w c_w / P_t$, for a fixed matrix regenerator, $C_r = C_{r,h} = C_{r,c}$: W/°C, Btu(h·°F)
C_r^*	total matrix heat capacity rate ratio $= C_r / C_{min}$, $C_{r,h}^* = C_{r,h}/C_h$, $C_{r,c}^* = C_{r,c}/C_c$, dimensionless
$C_{r,h}$	matrix heat capacity rate during the hot gas flow period $= \overline{C}_{r,h}/P_h = \overline{C}_r/P_t = \overline{C}_r/\omega$, where $\overline{C}_{r,h} = M_{w,h} c_w$ and $\overline{C}_{r,c} = M_{w,c} c_w$: W/°C, Btu/(h·°F)
c_p	specific heat of fluid at constant pressure: J/(kg·°C), Btu/(lb$_m$·°F)
c_w	specific heat of wall or matrix material: J/(kg·°C), Btu/(lb$_m$·°F)
D	circular header inside diameter: m, ft
D_h	hydraulic diameter of flow passages $= 4A_o L/A = 4\sigma/\alpha$: m, ft
d	branch inside diameter (see Fig. 47): m, ft
d_e	fin tip diameter of an individually finned tube: m, ft
d_o	tube (or pin) outside diameter: m, ft
F	header friction parameter, defined by Eq. (137), dimensionless
f	Fanning friction factor $= \tau_w/(\rho V^2/2g_c)$, dimensionless
f'	modified friction factor per tube row $= \Delta P/(4N_r G^2/2g_c\rho)$, dimensionless
f_{app}	apparent Fanning friction factor, defined by Eq. (49), dimensionless
f_D	Darcy friction factor $= 4f$, dimensionless
G	mass velocity based on minimum free flow area $= W/A_o$: kg/(m²·s), lb$_m$/(h·ft²)
g	gravitational acceleration: m/s², ft/s²
g_c	proportionality constant in Newton's second law of motion: $g_c = 1$ and is dimensionless in SI units; $g_c = 32.174$ lb$_m$·ft/(lb$_f$·s²)
H	velocity pressure $= \rho V^2/2g_c$: Pa, lb$_f$/ft²
Ⓗ	thermal boundary condition referring to constant axial as well as peripheral wall heat flux, also constant peripheral wall temperature; boundary condition valid only for the circular tube, parallel plates, and concentric annular ducts when symmetrically heated
Ⓗ1	thermal boundary condition referring to constant axial wall heat flux with constant peripheral wall temperature
Ⓗ2	thermal boundary condition referring to constant axial wall heat flux with constant peripheral wall heat flux
h	heat transfer coefficient: W/(m²·°C), Btu(h·ft²·°F)
$(hA)^*$	convection-conductance ratio, defined in Table 2, dimensionless
i_{lg}	latent heat of evaporation: J/kg, Btu/lb$_m$
j	Colburn factor $= $ St Pr$^{2/3}$, dimensionless

K	pressure loss coefficient $= \Delta P/(\rho V^2/2g_c)$, dimensionless
$K(\infty)$	incremental pressure drop number for fully developed flow, defined by Eq. (41), dimensionless
K_c	contraction loss coefficient for flow at exchanger entrance, dimensionless
K_e	expansion loss coefficient for flow at exchanger exit, dimensionless
k	fluid thermal conductivity: W/(m·°C), Btu/(h·ft·°F)
k_f	thermal conductivity of the fin material: W/(m·°C), Btu/(h·ft·°F)
k_w	thermal conductivity of the wall material: W/(m·°C), Btu/(h·ft·°F)
L	fluid flow (core) length on one side of the exchanger: m, ft
L_f	fin flow length ($L_f \leq L$): m, ft
L_{hy}	hydrodynamic entrance length: m, ft
L_{hy}^+	dimensionless hydrodynamic entrance length $= L_{hy}/(D_h \, \text{Re})$
L_1	flow (core) length for fluid 1 of a two-fluid heat exchanger: m, ft
L_2	flow (core) length for fluid 2 of a two-fluid heat exchanger: m, ft
L_3	no-flow height (stack height) of a two-fluid heat exchanger: m, ft
l	fin length for heat conduction from primary surface to the midpoint between plates for symmetric heating (l with this meaning used only in the fin analysis): m, ft
l_{ef}	effective flow length between major boundary layer disturbances; distance between interruptions: m, ft
l_f	fin height for individually finned tubes $= (d_e - d_o)/2$: m, ft
M_w	mass of the heat exchanger core, mass of a complete regenerator (all matrices) if no other subscript—$M_{w,h}$ ($= \rho_w A_h a/2$) means mass of the regenerator associated during the hot gas flow period: kg, lb$_\text{m}$
m	fin parameter, defined in Table 1 for each fin geometry (m with this meaning used only in the fin analysis): m^{-1}, ft^{-1}
m	exponent for property ratio method for f (refer to Eqs. (75) and (76)), dimensionless
N	number of lateral branches
N_f	number of fins per unit length in a tube-fin exchanger: m^{-1}, ft^{-1}
N_p	number of flow passages on one side in a plate-fin exchanger
N_r	number of tube rows in the flow direction
N_t	total number of tubes in the exchanger
NTU	number of heat transfer units $= UA/C_{\min}$, dimensionless
NTU$_c$	number of heat transfer units on cold side $= (\eta_o hA)_c/C_c$, dimensionless
NTU$_h$	number of heat transfer units on hot side $= (\eta_o hA)_h/C_h$, dimensionless
NTU$_0$	modified Nusselt number, defined in Table 2, dimensionless
Nu	Nusselt number $= hD_h/k$, dimensionless
n	exponent for property ratio method for Nu (refer to Eqs. (75) and (76)), dimensionless
n_f	total number of fins on one side of a plate-fin exchanger
P	fluid static pressure: Pa, lb$_f$/ft^2
P^*	$P_{c,i}/P_{h,i}$, dimensionless
ΔP	fluid static pressure drop on one side of a heat exchanger core: Pa, lb$_f$/ft^2
P	fluid pumping power $= W\,\Delta P/\rho$: W, hp

P'' fluid pumping power per unit surface area, $W\,\Delta P/\rho A$: W/m^2, hp/ft^2

$\mathcal{P}_c$ cold-gas flow period, duration of the cold gas stream in the matrix or duration of matrix in the cold gas stream: s

$\mathcal{P}_h$ hot-gas flow period, duration of the hot gas stream in the matrix or duration of matrix in the hot gas stream: s

$\mathcal{P}_t$ total period $= \mathcal{P}_c + \mathcal{P}_h$: s

Pr   Prandtl number $= \mu c_p/k$, dimensionless

p_f fin pitch: m, ft

Q heat transfer in a specified period of time: J, Btu

q heat transfer rate in the exchanger: W, Btu/h

q_e heat transfer rate at fin tip: W, Btu/h

q_0 heat transfer rate through the fin base: W, Btu/h

q'' heat transfer rate per unit surface area $= q/A$: W/m^2, Btu/(h·ft^2)

R^* a ratio of the hot to cold reduced periods $= \Pi_h/\Pi_c$, dimensionless

R_w wall thermal resistance, expressed by Eq. (7) on p. 4-6: °C/W, h·°F/Btu

Re   Reynolds number based on the hydraulic diameter $= GD_h/\mu = \rho VD_h/\mu$, dimensionless

Re$_d$ Reynolds number based on the tube outside diameter $= Gd_o/\mu = \rho Vd_o/\mu$, dimensionless

Re$_l$ Reynolds number based on the interrupted length $= Gl_{ef}/\mu$, $\rho Vl_{ef}/\mu$, GX_l/μ, dimensionless

r_h hydraulic radius $= A_oL/A = D_h/4$: m, ft

r_o tube outside radius: m, ft

St   Stanton number $= h/Gc_p$, dimensionless

s fin spacing or fin pitch; also transverse spacing for offset-strip fins: m, ft

s' distance between two fins $= s - \delta$: m, ft

T temperature: °C or K, °F or R

T^* $= T_{h,i}/T_{c,i}$, both temperatures in K or R, dimensionless

T_a ambient fluid temperature: °C, °F

T_0 fin base temperature: °C, °F

T_{WF} temperature-weighting factor (see discussion after Eq. (152)): K, R

ⓉＴ   thermal boundary condition referring to constant wall temperature, both axially and peripherally

U overall heat transfer coefficient; U_o = modified overall heat transfer coefficient for regenerators: W/(m^2·°C), Btu/(h·ft^2·°F)

V fluid mean axial velocity occurring at the minimum free flow area: m/s, ft/s

V heat exchanger total volume: m^3, ft^3

V_p heat exchanger volume between plates on one side: m^3, ft^3

v specific volume: m^3/kg, ft^3/lb$_m$

W fluid mass flow rate $= \rho VA_o$: kg/s, lb$_m$/s

W^+ total liquid flow rate through the film, dimensionless

X^* axial distance $= x/L$, dimensionless

X_l longitudinal tube pitch: m, ft

X_t transverse tube pitch: m, ft

x cartesian coordinate along the flow direction: m, ft

x	vapor mass quality, vapor weight fraction, used in Sec. K, dimensionless
x^*	axial distance $= x/(D_h \, \text{Re} \, \text{Pr})$, dimensionless
x^+	axial distance $= x/(D_h \, \text{Re})$, dimensionless
y, z	cartesian coordinates: m, ft

Greek Symbols

α	ratio of total heat transfer area on one side of the exchanger to the total volume of the exchanger $= A/\mathbf{V}$: m^2/m^3, ft^2/ft^3
α^*	ratio of width to height of a rectangular duct $= (s - \delta)/(b - \delta)$, dimensionless
α_w	thermal diffusivity of the matrix material $= k_w/\rho_w c_w$: m^2/s, ft^2/s
β	ratio of total heat transfer area on one side of a plate-fin exchanger to the volume between the plates on that side $= A/\mathbf{V}_p$: m^2/m^3, ft^2/ft^3
γ	ratio of heat capacity rates, defined in Table 2, dimensionless
γ	fin density: fins per meter, fins per inch
δ	fin thickness: m, ft
δ_c	channel deviation parameter, defined by Eqs. (145) and (146), dimensionless
ϵ	heat exchanger effectiveness (ϵ, ϵ_c, ϵ_h, and ϵ_r are defined in Table 2), dimensionless
ϵ_i	ideal regenerator effectiveness when pressure and carryover leakages are zero, dimensionless
η_f	temperature effectiveness of fins or an extended surface, dimensionless
η_o	total surface temperature effectiveness, defined by Eq. (2), dimensionless
κ	a parameter defined by Eq. (19d), dimensionless
Λ	reduced length for a regenerator, defined in Table 2, dimensionless
λ	longitudinal wall heat conduction parameter based on the total conduction area, $\lambda = k_w A_{k,t}/LC_{\min}$, $\lambda_c = k_w A_{k,c}/L_c C_c$, $\lambda_h = k_w A_{k,h}/L_h C_h$, dimensionless
μ	fluid dynamic viscosity coefficient: Pa·s, $\text{lb}_\text{m}/(\text{h}\cdot\text{ft})$
ξ	interrupted fin length in the fluid flow direction, $\xi = L_f$ for an uninterrupted fin: m, ft
Π	reduced period for a regenerator, defined in Table 2, dimensionless
ρ	density, for fluid if no subscript w: kg/m^3, $\text{lb}_\text{m}/\text{ft}^3$
σ	ratio of free flow area to frontal area $= A_o/A_{\text{fr}}$; or the volumetric porosity for regenerators $= r_h\beta$: dimensionless
τ_w	equivalent fluid shear stress at wall: Pa, $\text{lb}_\text{f}/\text{ft}^2$
χ_p	pressure leakage factor in a regenerator, defined in the text after Eq. (18), dimensionless
ω	rotational speed for a rotary regenerator: rev/s

Subscripts

c	cold fluid side
cp	constant property
fd	fully developed flow value
g	gas phase in Secs. K and L
$H, H1, H2$	constant axial wall heat flux boundary conditions

h	hot fluid side
i	inlet to the exchanger
l	liquid phase in Sec. K, fin length l at all other places
lm	logarithmic mean
m	mean or bulk mean
o	outlet to the exchanger
r	reference or nominal passage
s	scale or fouling
T	constant wall temperature boundary condition
x	a local value along the flow length
w	wall or property at the wall temperature
1, 2	fluids 1 and 2 in a two-fluid exchanger

CHAPTER 5

Heat Pipes

By C. L. Tien

Department of Mechanical Engineering
University of California, Berkeley

A. INTRODUCTION

The heat pipe is a thermal device for the efficient transport of thermal energy. It is a closed structure (Fig. 1) containing a working fluid that transports thermal energy from one part, called the evaporator, where heat is supplied to the device, to another part, called the condenser, where heat is extracted from the device. This energy transport is accomplished by means of liquid vaporization in the evaporator, vapor flow in the core region, vapor condensation in the condenser, and condensate return to the evaporator by capillary action in the wick. The basic idea of the heat pipe was first suggested by Gaugler [1] in 1942, but it was not until 1963 that Grover [2] independently built an actual device based on this idea and named the device the heat pipe.

Since pressure variations inside the vapor core are normally small, the heat pipe is usually very nearly isothermal and close to the saturated vapor temperature corresponding to the vapor pressure. The capability of transporting large amounts of thermal energy between two terminals (evaporator and condenser) with a small temperature difference is equivalent to having an extra-high thermal conductivity according to Fourier's law. It is like a superconductor in the thermal sense. In addition to its superior heat transfer characteristics, the heat pipe is structurally simple, relatively inexpensive, insensitive to the gravitational field, and silent and reliable in its operation. It can be made into different shapes and can operate at temperatures from the cryogenic regions up to structurally limited high-temperature levels by using various working fluids, ranging from cryogens to liquid metals.

The heat pipe in many aspects is similar to the thermal siphon which has been used for many years. If no wick is used for capillary pumping and the condensate is returned to the evaporator by gravity, the heat pipe becomes indeed a two-phase closed thermal siphon, which is sometimes called the gravity-assisted wickless heat pipe.

The immense potential of heat pipes for engineering applications was quickly realized,

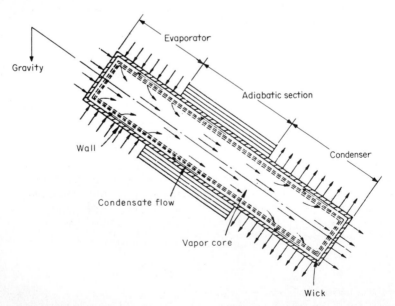

FIG. 1 The heat pipe.

especially in industry. Heat pipes have already been employed successfully in many applications, including spacecraft thermal control, electronic-systems cooling, and ordinary commercial thermal devices ranging from isothermal furnaces to meat-cooking pins. Intensive heat pipe research and development have resulted in rapid advancement in a variety of directions: from a screen mesh to multicomponent capillary-wick structures; from a passive variable-temperature conduction link to self-controlled, variable-conductance, constant-temperature devices; from the high-conductance function to a multitude of thermal-control functions (e.g., switching transformer, diode, damping, etc); and from simple analytical considerations to calculations of complex multidimensional momentum, energy, and mass transfer. Considerable progress has also been made in materials compatibility, fabrication technology, and operational handling.

This short chapter is not intended as an exhaustive review of the general subject of heat pipes, nor is it an account of recent research advances in this area. For more comprehensive reviews and compilations of research literature, readers should consult other books [3–6] and perhaps the proceedings of international heat pipe conferences (Stuttgart, 1973; Bologna, 1976; Palo Alto, 1978; London, 1981; Tsukuba, 1984). The purpose of this chapter is twofold: first, to provide a general orientation for those interested in this technology but not familiar with the subject, and second, to present background information in a concise but usable form for engineering applications.

B. OPERATING PRINCIPLES

Crucial to heat pipe operation is the circulation of the working fluid in two phases, which can only be analyzed by the combined laws of thermodynamics, fluid mechanics, and heat transfer. The first heat pipe analysis was put forward by Cotter [7]; it was followed subsequently by numerous other studies [3]. Generally speaking, adequate theoretical understanding has been achieved over the years, but due to the complex transport phenomenon involved in heat pipes, there exists considerable room for improvement in the quantitative description of heat pipe performance [8].

1. Thermodynamics

As a device involving liquid-vapor phase transition, the heat pipe can operate only in the temperature range between the thermodynamic triple point and the critical point of the working fluid. These two endpoints represent the theoretical lower and upper bounds of the operating temperature, which are seldom approached in practice. Operating near the low end is often hampered by adverse vapor dynamics (such as sonic limit, entrainment limit, or excessive vapor-pressure drop, which are discussed later) due to low vapor densities and correspondingly high vapor velocities. Near the critical point, the high vapor pressure can be a major concern in the mechanical design of the pipe structure.

At the liquid-vapor interface inside the heat pipe, the phase-equilibrium condition must be satisfied. From the Clapeyron equation it can be shown under the assumptions of constant latent heat and ideal gas for vapor that

$$\frac{P}{P_0} = \exp\left(-\frac{i_{lg}}{R_0 T}\right) \tag{1}$$

where P is pressure, P_0 pressure at some reference point, i_{lg} latent heat, R_0 ideal gas constant, and T temperature. The equation, which approximates well the vapor-liquid equilibrium line, can be used for approximate calculation when the phase-equilibrium $P(T)$ data are not available. The fact that thermodynamic equilibrium exists at the liquid-vapor interface is central to the understanding of the thermodynamic state of the working fluid in the heat pipe. The liquid within the wick in the evaporator is superheated because of the radial temperature gradient and the meniscus curvature. When the superheat becomes sufficiently large, boiling will occur in the wick. By the same reasoning, the vapor in the evaporator is generally subcooled, while the vapor in the condenser is superheated. Thus, depending on flow conditions, there is likely to be fogging in the evaporator as well as in the condenser. In any quantitative description of the thermodynamic state of the working fluid, the phase-equilibrium condition must be incorporated into the theoretical framework, particularly when multicomponent mixtures are involved such as in the gas-controlled [9] and two-component heat pipes [10].

2. Flow and Pressure Variations

The circulation fo the working fluid inside the heat pipe involves pressure drops in both the liquid- and vapor-flow regions. For steady-state operation, the capillary pressure head is balanced by the sum of the pressure drops in the liquid and vapor regions and the body-force head, if any:

$$\Delta P_k = \Delta P_l + \Delta P_g + \Delta P_b \tag{2}$$

These pressure variations as shown in Fig. 2c are generally small as compared with the absolute pressure level.

The capillary head originates from the pressure difference across a curved liquid-vapor interface, given by

$$\Delta P_k = \sigma \left(\frac{1}{r_1} + \frac{1}{r_2} \right) \tag{3}$$

where σ is the surface tension and r_1 and r_2 are the two orthogonal radii of curvature of the interface. This relation, sometimes called the Laplace-Young equation, is valid for any liquid-vapor interface regardless of the wetting conditions of the solid. The meniscus radii are smallest at the evaporator end, as illustrated in Fig. 2a. For circular pores, $r_1 = r_2 = r_K/(\cos \theta)$, where r_K is the pore radius and θ the wetting angle. For rectangular grooves, a common wickless capillary structure used in heat pipes, $r_1 = \infty$ and $r_2 = w/(\cos \theta)$, where w is the groove width. In general neither r nor the capillary head is constant. Vaporization results in a decrease in r, while condensation has an opposite effect. The capillary pumping head self-adjusts to meet the heat transfer and flow conditions until it reaches the minimum of r or the maximum of ΔP_k. Any further increase in heat transfer will cause drying and overheating in the wick at the evaporator end.

The pressure drop in liquid can be expressed in accordance with Darcy's law, since liquid flow in heat pipe wicks is generally characterized by low Reynolds numbers:

$$\frac{dP_l}{dx} = \frac{\mu_l W_l(x)}{K A \rho_l} \tag{4}$$

where μ is viscosity, W mass flow rate, K permeability, A wick-flow cross-sectional area, and ρ density. This relation can be regarded as the definition of K, which is usually empirically determined. It should be noted, however, that for thin wicks the free interface as well as the wall boundary could affect K appreciably.

Calculations of the vapor-pressure drop are generally much more complicated since

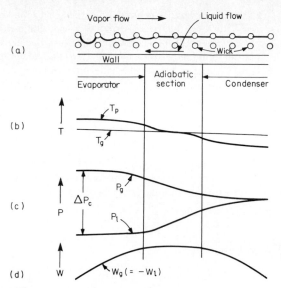

FIG. 2 (*a*) Fluid flow; (*b*) wall and vapor temperature distributions; (*c*) vapor and liquid pressure distributions; (*d*) vapor and liquid flow rates.

the analysis must consider not only the viscous effect, but also the inertia effect, which may cause substantial pressure recovery in the condenser region such as in the liquid-metal heat pipes. In addition, evaporation and condensation as well as the phase-equilibrium relation at the vapor-liquid interface may exert a significant influence on the vapor-pressure drop. A rigorous treatment must also account for the elliptic nature of the vapor flow field since the heat pipe is a closed structure.

3. Heat Transfer and Temperature Drops

Heat can enter or leave the heat pipe by many different modes: conduction from a heat source or sink, environmental convection and/or radiation, or electrical means such as Joule heating, electron-bombardment heating, and electron-emission cooling. The total temperature drop across the heat pipe, ΔT_t, can be approximately subdivided into several components along the main heat path:

$$\Delta T_t = \Delta T_{pe} + \Delta T_{we} + \Delta T_{ie} + \Delta T_g + \Delta T_{ic} + \Delta T_{wc} + \Delta T_{pc} \qquad (5)$$

where subscript p refers to pipe wall, e evaporator, w wick, i liquid-vapor interface, g vapor core between evaporator and condenser, and c condenser. There also exists a parallel heat path due to axial conduction along the pipe wall. The associated heat flows q and temperature drops ΔT can be conveniently expressed in an analogous electrical network of equivalent thermal resistances R.

Among the component ΔT's, ΔT_g can only be evaluated from complex flow-field computations and is relatively small in most cases, and ΔT_i is another quantity small but difficult to estimate accurately. In general, the most important ΔT's are ΔT_p, which can be calculated easily from the simple conduction relation with the given heat flow information ($q = \Delta T/R$), and ΔT_w, whose evaluation, however, requires a knowledge of the effective thermal conductivity k_E of the liquid-saturated wick. There exist several heat transfer models for various heterogeneous wick materials [3, 4], but the simplest configurations consist of wick and liquid either in series or in parallel. The solutions for these cases are, respectively:

$$k_E = \frac{k_l k_w}{\alpha k_w + k_l(1 - \alpha)} \tag{6}$$

and

$$k_E = \alpha k_l + (1 - \alpha)k_w \tag{7}$$

where α is the volume fraction of liquid in the liquid-saturated wick. The actual k_E of conduction should lie somewhere between the two limits given above. Convection currents in the wick, however, will tend to increase k_E. Moreover, in the evaporator, boiling in the wick may increase or decrease k_E substantially, depending on whether bubbles can escape from or be trapped in the wick.

4. Enclosure Conditions

The enclosure conditions are consequences of the fact that the heat pipe is a closed vessel and at steady state there are no energy and mass accumulations. For example, the mass of the vapor generated and the energy transferred into the system in the evaporator section must be equal to the mass of vapor condensed and the energy rejected out of the condenser section, respectively. The closed ends also require, at any cross section x, $W_g(x) = -W_l(x)$, as indicated in Fig. 2d.

C. OPERATING LIMITS

Although the heat pipe is a device of very high thermal conductance, there are limits to the maximum heat transfer rate it can achieve. These limits have their origins mostly in fluid mechanics and result from a breakdown or a rate limit in the circulation of the working fluid. Figure 3 shows a schematic diagram of five known limits for a liquid-metal heat pipe. These limits show limitations due to vapor flow (the viscous and sonic limits), to liquid flow (the wicking limit), to interactions of liquid and vapor flows (the entrainment limit), and to phase change in the evaporator (the boiling limit). The possible, although unlikely and experimentally unconfirmed, condensation limit due to phase change in the condenser has also been suggested [8]. For room-temperature and low-temperature heat pipes the working fluids have a narrower operating temperature range, and only two or three limits, usually the wicking and boiling limits, will be in effect.

For vapor-flow limitations, either the viscous limit or the sonic limit may occur depending on the relative magnitude of inertia and viscous forces in the vapor flow. In the viscous-flow regime, the vapor-pressure drop increases with the axial heat flux. According to Busse's definition [11], the viscous limit of heat transfer is reached when the vapor pressure at the condenser end drops to zero. The viscous limit lies below the sonic limit as shown in Fig. 3 and thus warrants particular attention in the analysis of the heat-pipe start-up process [3]. Some questions have been raised [12], however, that the viscous limit is rather unattainable in actual operation since, before the vapor pressure at the condenser end reaches zero, the large pressure drop accompanied by a large temperature drop incapacitates the heat pipe as an efficient heat transfer device. Accordingly, instead

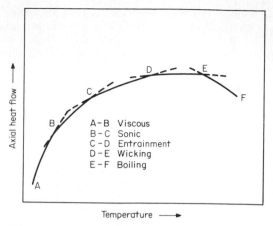

FIG. 3 Heat pipe operating limits.

ot the viscous limit, the minimum-heat-flux limit, below which the heat pipe ceases its designed function, was suggested.

In the inertia-flow regime where viscous forces are negligible, the vapor flow is limited by the well-known gas-dynamic choking phenomenon. It can easily be recognized that the vapor flow in a constant-cross-section heat pipe with mass addition and removal is equivalent to the constant-mass flow in a convergent-divergent nozzle, and the evaporator exit end corresponds to the nozzle throat. The maximum mass flow rate that relates directly to the maximum heat transfer rate is attained when the evaporator exit velocity reaches the local sonic velocity. The sonic limit is often encountered in liquid-metal heat pipes operating near the low-vapor-pressure end, where low vapor densities and high vapor velocities exist. There have been many different vapor models proposed to analyze the gas-dynamic choking phenomenon [8]. They differ basically in the assumptions concerning the thermodynamic and chemical state of the vapor that is initially in a nearly saturated, chemical-equilibrium condition and is subcooled in its expanding flow toward the condenser. The first model is the perfect-gas model [13], in which the vapor is assumed to be a perfect gas and all effects of supersaturation as well as chemical reaction (i.e., dissociation and recombination) are neglected. A simpler but still accurate physical model is the isothermal perfect-gas model [11], in which vapor-temperature variations are neglected in addition to the assumptions involved in the perfect-gas model. Based on simple continuity and momentum considerations, it gives the axial heat flux

$$q'' = 0.5 i_{lg} (\rho_0 P_0)^{1/2} \qquad (8)$$

where subscript 0 refers to the end condition of the evaporator. This implies that the sonic heat transfer limit is predominantly momentum-controlled.

The entrainment limit has its origin in the interfacial shear due to the flows of vapor and liquid in opposite directions. When the relative velocity betwen the vapor and liquid is sufficiently large, the interface becomes unstable, and liquid droplets will be entrained in the vapor. This loss of liquid in the wick, which thus never reaches the evaporator, amounts to a certain reduction in the maximum evaporation heat transfer and this lowered heat transfer rate defines the entrainment limit. The entrainment phenomenon is governed by the Weber number that compares the vapor inertial forces to the liquid surface-tension forces. Formulas have been established for entrainment limits for various heat pipes [14], but because of the complexity introduced by the presence of the wick and the curved interface, they can at best be regarded as semiempirical and must be used with caution.

The wicking limit has already been touched on in the discussion of pressure variations in heat pipes. The maximum-heat-transfer limit takes place when the capillary head reaches its maximum value at the evaporator end [see Eq. (2)]. In the early development of heat pipe technology, the wicking limit was a major design concern. This situation, however, has been greatly altered by the improved design of new composite wick structures (to be discussed later), all of which have a greatly enhanced capillary pumping capacity.

The boiling limit was not regarded as a serious limitation in the early development, but is now becoming an increasingly important concern. As stated previously, because of phase equilibrium at the interface, liquid within the wick in the evaporator is always in a superheated state. Becasue of the complex nature of wick structures and the thin liquid film, it is hard to characterize the level of liquid superheat at which nucleation of bubbles will occur, nor is it certain that nucleation will take place at the pipe wall, where the greatest superheat exists. Most important of all probably is the fact that bubbling action, when held at a low level, enhances heat transfer in the evaporator. This is also in agreement with the finding that the liquid-saturated wick often has a much higher thermal conductivity at the evaporator than at the condenser. Only when bubbles start to get trapped in the wick and local vapor blanketing occurs will the boiling limit be in effect. The entire phenomenon is also very much influenced by whether the wick is in close contact with the pipe wall.

D. DESIGN CONSIDERATION

The three basic elements in a heat pipe structure are the working fluid, the wick, and the pipe (a container of any shape, not restricted to tubular form). The optimal design strives for a best combination of the three, although conflicting requirements often occur from design criteria for each individual element.

The first requirement for a suitable working fluid is the proper operating temperature range (Table 1). Next, the thermophysical properties of a candidate fluid msut be considered with respect to the various operating limits. In general, the fluid should have a comparably high latent heat, high liquid thermal conductivity, low liquid and vapor viscosities, and high surface tension. The possibility of thermal degradation that may occur particularly in certain organic fluids must not be overlooked. The compatibility of the working fluid with wick and pipe materials will be discussed later.

Selection of the wick hinges primarily on the wick's capability of providing sufficient capillary pumping pressure to transport the working liquid from the condenser to the evaporator. From a fundamental viewpoint, the liquid capillary flow in heat pipe wicks is an extremely complicated problem that involves, in addition to the complexities normally encountered in flows through a porous medium, surface-tension effects at the liquid-vapor interface, pronounced boundary effects (because heat pipe wicks are usually thin), pressure interactions between liquid and vapor phases, and evaporation-condensation effects. Very little is known about these various effects. It is well known, however, that the maximum capillary head of a wick increases with decreasing pore size, suggesting the use of wicks with small pores, but smaller pores result in larger frictional resistance to flow or a small permeability, an undesirable effect. Table 2 shows some typical values of experimentally determined wick properties. It should be noted that different test meth-

TABLE 1 Heat Pipe Working Fluids

Fluid	T_0, K	T_{op}, K	i_{lg}, 10^3 J/kg	ρ_g, kg/m^3	ρ_l, kg/m^3	k_l, W/(m·K)	μ_l, 10^{-4} Pa·s	σ, 10^{-3} N/m
Helium	4	2–4	21	10.0	128	2.77	0.03	0.09
Nitrogen	77	70–113	199	4.6	810	0.14	1.6	8.9
Ammonia	240	213–373	1370	0.9	683	0.55	2.8	33
Freon-113	321	263–373	146	8.2	1508	0.094	5.1	16
Methanol	337	273–403	1100	1.1	751	0.201	3.4	19
Water	373	303–473	2258	0.6	958	0.680	2.8	59
Thermex	530	423–668	297	5.4	850	0.112	2.6	15
Mercury	634	523–923	298	5.3	12,740	12.2	8.8	380
Cesium	943	673–1173	491	17	1681	17.5	1.7	42
Potassium	1047	773–1273	1935	0.51	672	35.8	1.1	50
Sodium	1165	873–1473	3920	0.28	747	53.8	1.7	110
Lithium	1613	1273–2073	19,700	0.06	420	69	2.3	270

T_0—atmospheric boiling temperature; T_{op}—practical operating temperature range; all properties at T_0. 1 J/kg = 4.31 × 10^{-4} Btu/lb$_m$; 1 kg/m^3 = 0.0642 lb$_m$/ft^3; 1 W/(m·K) = 0.578 Btu/(h·ft·°F); 1 Pa·s = 2419.3 lb$_m$/(ft·h); 1 N/m = 0.0685 lb$_f$/ft.

ods and sample preparation often result in different values. The composite wicks that have found increasing usage are designed to decouple capillary pumping from flow resistance in the wick.

Some representative composite wicks are shown in Fig. 4. Artery wicks are particularly attractive in applications as they further decouple radial heat transfer from axial liquid flow (i.e., they give rise to smaller temperature drops across the heat pipe wall). It is crucial, however, to maintain the artery primed (filled). Should the artery be partially or fully depleted of liquid, it must have the capability to self-refill. In other words, the artery should be self-priming under the surface-tension forces. This self-priming requirement often restricts the size of the artery. In many situations it is advantageous to use wickless capillary channels for liquid pumping, such as rectangular and triangular grooves.

TABLE 2 Experimentally Determined Wick Properties

Wick type	Porosity	Pore radius, 10^{-6} m	Permeability, 10^{-10} m^2
Screen, stainless steel, 200 mesh	0.733	58	0.52
Screen, nickel, 200 mesh, sintered	0.676	64	0.77
Screen, nickel, 50 mesh, sintered	0.625	305	6.63
Felt, stainless steel, sintered	0.822	110	11.61
Felt, stainless steel, sintered	0.808	65	1.96
Felt, nickel, sintered	0.868	<37	0.40
Felt, copper, sintered	0.895	229	12.4
Foam, copper, Am Por Cop, 220-5	0.912	241	23.2
Foam, nickel, Am Por Cop, 220-5	0.960	229	37.2
Powder, nickel, sintered	0.658	61	2.73
Beads, Monel, 30–40 mesh	0.40	252	4.12
Beads, Monel, 70–80 mesh	0.40	97	0.78
Beads, Monel, 140–200 mesh	0.40	45	0.11

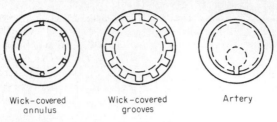

| Wick–covered annulus | Wick–covered grooves | Artery |

FIG. 4 Representative composite wicks.

Selection of the pipe material depends generally on the following criteria: large strength-to-weight ratio, high thermal conductivity, ease of fabrication, impermeability, good wettability, and perhaps most important of all, compatibility with the working fluid, the wick, and the external environment. Two major concerns of incompatibility are corrosion and the generation of noncondensable gas. Some compatibility data are available in the literature, but it has become common practice to conduct life tests on representative heat pipes for compatibility studies. Generally speaking, stainless steel pipe and wick are found compatible with acetone, ammonia, and liquid metals, but they have a relatively low thermal conductivity, for instance, as compared to copper, which is particularly attractive for mass-produced units using water as the working fluid. For common working fluids, water is recommended for copper and Monel but not for stainless steel, aluminum, nickel, and Inconel; ammonia for aluminum, stainless steel, and nickel but not for copper; methanol for copper and stainless steel but not for aluminum; and liquid metals for stainless steel and Inconel but not for titanium.

In the manufacture, assembly, and testing of heat pipes, care must be exercised for a number of necessary but relatively simple operations. Foremost is the cleanliness of all parts, since contaminants can affect significantly the performance and life of heat pipes in many detrimental ways. The normal cleaning procedure consists of: (1) degreasing—vapor degreasing, solvent treatment, or emulsifiable solvent treatment; (2) solid particle removal—alkaline bath treatment in spray, solvent circulation and brushing, mechanical agitation, or chemical passivation; (3) deoxidizing—acid bath treatment, chemical reduction, or hydrogen reduction; and (4) degassing—evacuation, vacuum bake-out, or cyclic baking and freezing. To ensure that there is no intrusion of external contaminants such as foreign gases, the heat pipe after assembly must be leaktight to a very high degree. During the assembly of parts, the most important operation is wick forming and insertion, which affect directly the capillary pumping capacity. Also, it should be certain that the wick and the pipe wall are in good contact and will be wetted by the working fluid. Other preparatory operations involve working-fluid purification, fluid filling, adequate fluid inventory inside the pipe, heat pipe sealing, performance evaluation, and life tests.

E. SPECIAL TYPES OF HEAT PIPES

There exist numerous different types of heat pipes. In terms of geometric configuration, most common is the straight tubular form, but heat pipes are relatively easy to build, according to applications, into other shapes or forms, such as flat plate, curved, or flexible. Also, many heat pipes are built for special thermal-control functions; these include variable-conductance heat pipes, two-component heat pipes, and thermal diodes and switches. Another broad category of heat pipes employs special body force (other than capillary) to return the condensate to the evaporator. The rotating heat pipe and various types of the electrohydrodynamic heat pipe belong to this category. In the following, a brief description of the special thermal-control heat pipes and the special body-force heat pipes is given.

Adequate thermal control of heat pipes is important in many applications because the conventional heat pipe is passively determined by heat source and sink conditions. In particular, for instance, the operating temperature of heat pipe rises with an increase in heat load. In practice, it is often desirable or required (1) to maintain the pipe at a very narrow temperature range in spite of source or sink conditions, (2) to have two temperature zones along the pipe, (3) to allow heat flow in one direction but not the other, or (4) to cut off heat flow when the source temperature becomes too high. These modes of operation are called variable-conductance, two-temperature (or two-component), thermal diode, and thermal switch, respectively.

The variable-conductance heat pipe, which can be regarded as a special case of the two-component heat pipe with noncondensable gas as one component, has received considerable attention because of its attractive feature as a self-controlled, constant-temperature (thus variable-conductance) device. Great progress has been made in the theory and design of the gas-loaded heat pipe [9], and it has been widely used in applications. Basically it involves a noncondensable gas and a gas reservoir added to the conventional heat pipe at the end of the condenser. Presence of the noncondensable gas in a portion of the condenser prevents the vapor from condensing in that section, thus adjusting the effective area of heat removal in accordance with the heat load and maintaining the operating temperature constant at a designed level. To have both components actively participating in the heat pipe processes requires two working fluids of overlapping operating temperature ranges. For instance, a water-ethanol heat pipe could have both components functioning in the temperature range 516 to 273 K, or 469 to 32°F, in which two operating temperature ranges (642 to 273 K, or 696 to 32°F, for water and 516 to 158 K, or 469 to −176°F, for ethanol, defined theoretically by their critical and triple points) overlap. The two-component heat pipe provides two uniform-temperature zones (with the more volatile one on the condenser side) and a transition zone in between [10]. The function of thermal diode or thermal switch can be achieved by an adequate design of wick or groove configurations such that sufficient liquid capillary pumping exists only in one direction or at a preset level, respectively.

Among the special body-force heat pipes, the rotating heat pipe [15] is a wickless, closed, hollow shaft with a slight internal taper containing a certain mass of working fluid. The liquid condensate returns to the evaporator by virtue of the centrifugal force generated by rotation. The rotating heat pipe eliminates many limitations associated with liquid capillary flows and is especially attractive for applications in systems where rotating shafts are present. The electrohydrodynamic heat pipe, like the rotating heat pipe, utilizes another form of body force to drive the liquid return flow. Two types of electrohydrodynamic driving forces have been investigated for heat pipe applications. One is the electroosmotic force that arises from the presence of a naturally occurring potential or a charge accumulation at the interface of the dielectric liquid and the capillary surface (wick or grooves). By applying a potential across two porous electrodes, the electroosmotic force is generated to further enhance capillary pumping [16]. The continued reliance on a capillary structure as well as the added complexity in installing electrodes and associated accessories greatly restricts its application potential. Another suggestion [17] to use the polarization force (sometimes called dielectrophoresis) appears to have the additional merit of removing completely the conventional capillary structure. The dielectric liquid tends to collect in regions of higher electric field intensity with vapor expelled to regions of lower electric field intensity. With proper design, the electrode structure will hold the liquid, and the receding meniscus in the evaporator will cause the liquid to flow continuously from the condenser, in close analogy to the wick functions in conventional heat pipes. This heat pipe has the advantages of greatly reduced liquid-flow friction, electrohydrodynamically enhanced evaporation and condensation heat transfer, and a voltage on-off control, but also many negative aspects, including the requirement of a high-voltage, low-current electrode and power supply, the need for high-voltage insulation, and the long-term degradation of dielectric fluids.

REFERENCES

1. R. S. Gaugler, Heat Transfer Device, U.S. Patent 2,350,348, appl. Dec. 21, 1942, publ. June 6, 1944.
2. G. M. Grover, Evaporation-Condensation Heat Transfer Device, U.S. Patent 3,229,759, appl. Dec. 2, 1963, publ. Jan. 18, 1966.
3. P. Dunn and D. A. Reay, *Heat Pipes,* 2d ed., Pergamon, New York, 1978.
4. S. W. Chi, *Heat Pipe Theory and Practice,* McGraw-Hill, New York, 1976.
5. D. Chisholm, *The Heat Pipe,* Mills & Boom, London, 1971.
6. C. L. Tien (ed.), *Heat Pipes, AIAA Sel. Reprint Ser.,* Vol. 16, AIAA, New York, 1973.
7. T. P. Cotter, Theory of Heat Pipes, *Los Alamos Sci. Lab. Rep. LA-3246-MS,* 1965; reprinted in Ref. 6.
8. C. L. Tien, Fluid Mechanics of Heat Pipes, *Ann. Rev. Fluid Mech.* Vol. 7, pp. 167–185, 1975.
9. B. D. Marcus, Theory and Design of Variable Conductance Heat Pipes, *NASA CR-2018,* 1972.
10. C. L. Tien and A. R. Rohani, Theory of Two-Component Heat Pipes, *J. Heat Transfer,* Vol. 94, pp. 479–484, 1972.
11. C. A. Busse, Theory of Ultimate Heat Transfer Limit of Cylindrical Heat Pipes, *Int. J. Heat Mass Transfer,* Vol. 16, pp. 169–186, 1973.
12. A. R. Rohani and C. L. Tien, Minimum Heat Transfer Limit in Simple and Gas-Loaded Heat Pipes, *AIAA J.,* Vol. 13, pp. 530–532, 1975.
13. E. K. Levy, Theoretical Investigation of Heat Pipes Operating at Low Vapor Pressures, *J. Eng. Ind.,* Vol. 90, pp. 547–552, 1968.
14. C. L. Tien and K. S. Chung, Entrainment Limits in Heat Pipes, *AIAA J.,* Vol. 17, pp. 643–646, 1979.
15. P. J. Marto and L. L. Wagenseil, Augmenting the Condenser Heat Transfer Performance of Rotating Heat Pipes, *AIAA J.,* Vol. 17, pp. 647–652, 1979.
16. M. M. Abu-Romia, Possible Application of Electro-osmotic Flow Pumping in Heat Pipes, *AIAA Paper No. 71-423,* 1971.
17. T. B. Jones, An Electrohydrodynamic Heat Pipe, *Mech. Eng.,* Vol. 96, No. 1, pp. 27–32, 1974.

NOMENCLATURE

Symbol, Definition, SI Units, English Units

A wick-flow cross-sectional area: m^2, ft^2

i_{lg} latent heat of evaporation: J/kg, Btu/lb_m

K permeability: m^2, ft^2

k thermal conductivity: W/(m·K), Btu/(h·ft·°F)

P pressure: Pa (N/m^2), lb_f/ft^2

q	heat transfer rate: W, Btu/h
q''	heat flux: W/m^2, $Btu/(h \cdot ft^2)$
R	thermal resistance: K/W, $h \cdot °F/Btu$
R_0	gas constant: $J/(kg \cdot K)$, $Btu/(lb_m \cdot °R)$
r	radius: m, ft
r_K	pore radius: m, ft
T	temperature: K, °R
W	mass flow rate: kg/s, lb_m/h
w	groove width: m, ft
x	axial distance: m, ft

Greek Symbols

α	liquid volume fraction in liquid-saturated wick
μ	dynamic viscosity: $Pa \cdot s$, $lb_m/(h \cdot ft)$
ρ	density: kg/m^3, lb_m/ft^3
σ	surface tension: N/m, lb_f/ft

Subscripts

b	body force
c	condenser
e	evaporator
E	effective
g	vapor
i	liquid-vapor interface
k	capillary
l	liquid
p	pipe
t	total
w	wick

CHAPTER 6

Heat Transfer in Fluidized and Packed Beds

By J. D. Gabor

Reactor Analysis and Safety Division
Argonne National Laboratory
Argonne, Illinois

J. S. M. Botterill

Department of Chemical Engineering
University of Birmingham
Birmingham, England

INTRODUCTION

The inclusion of a chapter on heat transfer in packed and fluidized beds in the *Handbook of Heat Transfer Applications* is particularly a recognition of their importance to engineering practice in the conversion and combustion of fossil fuels and of their potential in energy conversion schemes. In application, the solid material in the packed or fluidized bed may enter directly into a chemical reaction, serve as a catalyst, be the inert medium which enhances heat and/or mass transfer, or be the medium in which to store energy. Thus, packed beds have been used as reactors and regenerative heat exchangers and have a potential in large-scale and domestic energy storage schemes [1]. Where temperature control presents problems, fluidized or packed-fluidized beds are used and hot solids constitute a very convenient high-temperature heat transfer medium by virtue of their low effective vapor pressure [2].

Particulate solids have a very high surface area, which is advantageous for fluid-solid heat and mass transfer operations. Thus, for example, with the finer powders used in fluidized catalytic reactors, a cubic meter of 100-μm diameter particles has a surface area greater than 30,000 m^2, which is of similar magnitude to that of the surface area of the Great Pyramid of Cheops. If the velocity of flow upwards through an unrestrained packed bed is increased, the drag force exerted on the particles will increase until it is sufficient to support the weight of the material in the bed. The bed will then begin to show fluidlike properties and, beyond the velocity called the *minimum fluidizing velocity* V_{mf}, it will become fluidized (Sec. A.2). The pressure drop across the bed then stays practically constant and equal to its weight per unit area with further increase in velocity. The range of operating velocities can be extended until the elutriation of solids becomes a limitation. This depends upon the mean size and size range of the material within the bed. Higher fluid flow rates, however, can be sustained through a restrained packed bed but at the cost of the pumping power required to overcome the increased pressure drop through the bed (Sec. A.1). In the fluidized condition it is comparatively easy to transport material from one vessel to another. This and the good heat transfer properties of fluidized beds [3] are two other important advantages of gas fluidized systems (Sec. B.2). It is, however, often difficult to design and scale up fluidized bed systems because of the very wide range of behavior that can be encountered.

At the onset of fluidization it makes very little difference to the general bed behavior whether the fluid is a gas or a liquid. With further increase in the fluid superficial velocity, a liquid fluidized bed continues to expand uniformly whereas a gas fluidized bed becomes unstable and cavities containing few solids are formed. These cavities or bubbles are responsible for generating solids mixing, which is such an important feature of gas fluidized beds. The classification suggested by Geldart [4] is helpful in categorizing the range of behavior that can be encountered. Geldart's group A powders generally have densities less than 1400 kg/m^3 (87.4 lb$_m$/ft^3) and fall within the size range of 20 to 100 μm (6.6 $\times$ 10^{-5} to 3.3 $\times$ 10^{-4} ft). These powders exhibit a degree of stable bed expansion until the minimum fluidization velocity is exceeded by a factor of perhaps 3 or even more. With further increase in gas velocity, however, the bed will eventually collapse back to a less expanded state and begin to bubble; this condition is termed *aggregative* fluidization, as distinct from *particulate* fluidization when the bed is uniformly expanded as with liquid fluidized systems. Geldart's group B powders tend to have a mean size within the range 40 to 500 μm (1.3 $\times$ 10^{-4} to 1.6 $\times$ 10^{-3} ft) and a density between 1400 and 4000 kg/m^3 (87.4 and 250 lb$_m$/ft^3). Bubbling occurs in beds of these powders when the gas velocity increases beyond that for minimum fluidization. Group C materials are of small mean size and of low density so that interparticle forces have greater effect than that of gravity. They cannot be fluidized in the ordinary sense. Group D materials are composed of larger and/or denser particles than type B. D-type materials fluidize less stably than types A and B and their bubbling character is different so that there is less effective solids mixing within the bulk of the bed. Bubbles in fluidized beds grow primarily by coalescence as they rise through the bed, and there seems to be a smaller maximum stable bubble size with group A materials than B. This has a bearing on scale-up problems because, in small-scale tests, maximum bubble size is restricted by the diameter of the bed, whereas with larger equipment, very much larger bubbles may be obtainable and this will affect the overall fluidization behavior and the gas residence time distribution in particular [5].

Where it is not possible to fluidize a solid directly because of its large size and/or shape but it is desirable to have the good thermal properties of a fluidized bed, or where axial gas and solids mixing is too great to obtain adequate reactant conversion in a freely fluidized bed, recourse can be made to fluidized-packed beds (Sec. C). In these, a finer powder is fluidized within the interstices of the larger solid packing.

Apart from the very wide range of behavior that can be encountered with fluidized beds and the difficulties these present for scale-up, it must also be appreciated that there are other problems inherent in any fluid-solid operation. Particularly, there is the problem of characterizing the particulate material itself (see, for example, Chap. 4 in the book by Allen [6]). Thus, according to the physical situation, different definitions of mean particle diameter will be relevant. There is no satisfactory way by which particle shape can be categorized numerically, and it is even difficult to measure the voidage of a loosely packed bed (and especially so if the particles are porous). Accordingly, it is very important to ascertain the way in which particle properties have been defined when using different correlations. It must also be constantly stressed that correlations should only be used within the ranges of variables from which they were developed. Often these are too restricted to be of relevance in industrial design.

Some variations from common fluid-particle system usage are being used in this chapter so that the nomenclature throughout this volume shall be uniform. In particular, V_{mf} is used instead of U_{mf} for minimum fluidization velocity and α instead of ϵ for bed voidage. Consistent SI or English units are used as indicated in the dimensional equations and listed in the nomenclature. The form adopted has often been chosen according to historic usage.

A. HYDRODYNAMICS

1. Pressure Drop—Packed Beds

The pressure drop of fluids flowing through the bed of particles is a function of bed properties as well as the fluid properties. An important parameter is particle size. In most operations the particles are not uniform in size. Sieving is a common technique for determining particle size distribution. There are many bases for determining a mean particle diameter from sieve fractions [6, 7]. A diameter based on the surface area to volume ratio a is recommended for estimating friction losses [8, 9]. For spherical particles,

$$a = \frac{\pi d_p^2}{(\pi/6)d_p^3} = \frac{6}{d_p} \tag{1}$$

The mean value of a from sieve weight fractions is:

$$\bar{a} = \sum w_i a_i = \sum \frac{6w_i}{d_{p_i}} = \frac{6}{\bar{d}_p} \tag{2}$$

The mean particle diameter on the basis of mean surface area to volume ratio is then

$$\bar{d}_p = \frac{1}{\Sigma(w_i/d_{p_i})} \tag{3}$$

Bed porosity is another controlling parameter of bed pressure drop. The void fraction for randomly packed beds of uniformly sized spheres in containers of diameter about 50 times the particle diameter is in the range of 0.36 to 0.43. For spheres in regular array the voidage ranges from 0.2595 (rhombohedral) to 0.476 (cubic) as shown in Fig. 1. Representative void fractions for various materials are given in Table 1.

Particle shape is important in that it affects the surface area per unit volume. The

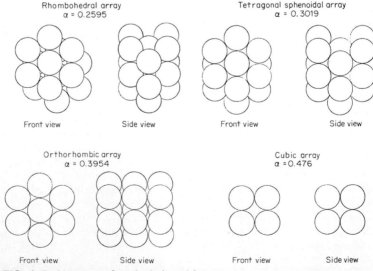

Rhombohedral array
α = 0.2595

Front view   Side view

Tetragonal sphenoidal array
α = 0.3019

Front view   Side view

Orthorhombic array
α = 0.3954

Front view   Side view

Cubic array
α = 0.476

Front view   Side view

FIG. 1 Packing arrays for spherical particles.

TABLE 1 Void Fractions for Packed Beds

Material	Voidage α	Reference
Spheres—rhombohedral	0.2595	
Spheres—tetragonal sphenoidal	0.3019	
Spheres—orthorhombic	0.3954	
Spheres—cubic	0.476	
Spheres—random	0.36–0.43	Bernard and Wilhelm [10]
Sand	0.37–0.50	Carman [11]
Granular crushed rock	0.44–0.45	Bernard and Wilhelm [10]

particle shape factor ϕ or sphericity is usually defined in terms of a spherical particle having the minimum surface area per unit volume in this context. Thus

$$\phi = \frac{\text{surface area of sphere per unit volume}}{\text{surface area of particle per unit volume}} \tag{4}$$

and some typical data are given in Table 2. For nonspherical particles the real equivalent diameter will tend to be less than the value of $\overline{d}_p$ given by sieve analysis (Abrahamsen and Geldart [16]) but the effect of this is usually allowed for in the constants of correlations. However, as stressed in the introduction, it is important to ascertain the way in which powder properties have been measured before using any correlations.

Pressure drop through packed beds of uniformly sized particles has been correlated by Ergun [17] by combining the Blake-Kozeny correlation for the laminar-flow region and the Burke-Plummer correlation for the turbulent-flow region:

$$\frac{\Delta P}{L} g_c = \underbrace{150 \frac{(1-\alpha)^2}{\alpha^3} \frac{\mu_f V_0}{(\phi d_p)^2}}_{\text{laminar}} + \underbrace{1.75 \frac{(1-\alpha)}{\alpha^3} \frac{\rho_f V_0^2}{\phi d_p}}_{\text{turbulent}} \tag{5}$$

The above equation combines the properties of the fluid (ρ_f and μ_f) with the bed parameters (d_p, α, and ϕ).

Additional reference can be made to Brownell and Katz [18, 19], Coulson [20], Ranz [21], and Scheidegger [22].

TABLE 2 Particle Shape Factors

Material	Shape factor ϕ	Reference
Spheres	1.000	
Cubes	0.806	
Equilateral pyramids	0.671	
Cylinder ($D = H$)	0.874	
Sand	0.534–0.861	Leva et al. [12, 13], Shirai [15]
Silica	0.554–0.628	Shirai [15]
Coal	0.625–0.696	Leva et al. [12, 13], Shirai [15]
Iron catalyst	0.578	Leva et al. [12, 13]
Celite cylinders	0.861–0.877	Leva et al. [12, 13]
Broken solids	0.63	Uchida and Fujita [14]

2. Minimum Fluidization

The minimum fluidization velocity V_{mf} can be estimated as the flow velocity through a packed bed of voidage equal to that of a bed at incipient fluidization, α_{mf}, which gives a pressure drop equal to the weight of the bed per unit area. The following relationship can then be obtained from the Ergun equation (Eq. (5)):

$$\text{Ar} = 150 \frac{1 - \alpha_{mf}}{\phi^2 \alpha_{mf}^3} \text{Re}_{mf} + \frac{1.75}{\phi \, \alpha_{mf}^3} \text{Re}_{mf}^2 \tag{6}$$

This reflects the very strong influence of voidage on pressure drop. An error of 10 percent in the estimate of voidage for laminar-flow conditions produces an error of 40 percent in the estimate of V_{mf}. In the turbulent regime, the corresponding error is 15 percent [23].

The form of Eq. (6) suggests that V_{mf} should be independent of pressure for gas-fluidized systems while the flow through the bed is laminar, but inversely proportional to the square root of the static pressure for turbulent gas flow conditions. This is approximately so. However, there is the complication when operating temperature is changed that α_{mf} is temperature-dependent for beds of Geldart's A and B group materials [4] and its variation cannot be predicted with confidence (Fig. 2) [23, 24]. Goroshko et al. [25] derived an interpolation form of the Ergun equation. In this they have incorporated the effect of particle shape into the estimate of particle diameter. Through omitting the product term to simplify the solution of the quadratic, the form derived makes some fortuitous compensation for the change in voidage for Ar $< 10^4$ if the effect of change in gas physical properties with temperature is incorporated and a constant value of α_{mf} "chosen" to fit the more readily measurable value of V_{mf} at ambient conditions to the correlation. Their equation is:

$$\text{Re}_{mf} = \frac{\text{Ar}}{150(1 - \alpha_{mf})/\alpha_{mf}^3 + \sqrt{1.75 \, \text{Ar}/\alpha_{mf}^3}} \tag{7}$$

It is notable that many fluid-bed operations with B and D group materials are conducted over a range of conditions which fall within the transitional region between laminar and turbulent flow conditions where the Ergun correlation, in any case, is least reliable. The transition between B- and D-type behavior occurs at $\text{Re}_{mf} \approx 12.5$ and $\text{Ar} \approx 26{,}000$ [23, 26].

Workers have attempted to avoid the problem of particle characterization by "averaging out" the effect of shape and voidage and so deriving a modified form of the Ergun-type correlation. Typically, Baeyens and Geldart [27] suggest the equation:

$$\text{Ar} = 1823 \, \text{Re}_{mf}^{1.07} + 21.7 \, \text{Re}_{mf}^2 \tag{8}$$

While this can be recommended for prediction of V_{mf} under ambient conditions for group A powders when the flow will be consistently within the laminar regime and only the first term of the correlation is significant, it should only be used with caution beyond ambient conditions and for materials of groups B and D. Generally, however, when only the ambient-conditions value is required, it is better to determine it experimentally. Non-cohesive, closely sized materials display a sharp transition from the fixed to the fluidized state when the pressure drop across the bed is plotted as a function of the superficial gas velocity upward through the bed. With a wide size distribution the transition is less distinct and some hysteresis may be apparent between increasing and decreasing gas velocities. Extrapolation back of the pressure drop equivalent to the buoyant weight per unit area of bed to its intersection with the straight fixed-bed pressure-drop line gives a reliable

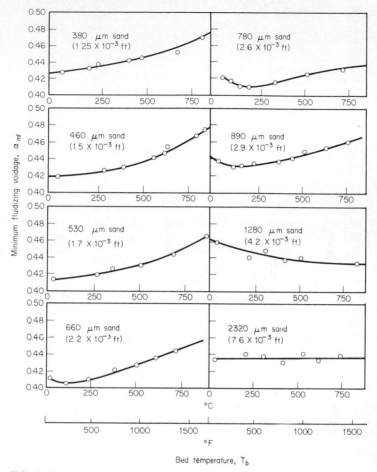

FIG. 2 Variation in minimum fluidization voidage with increasing bed temperature for beds with various mean particle diameters [23].

measurement when operating in the laminar-flow regime. (For group D materials, where the flow is turbulent, the initial packed-bed pressure-drop line will be parabolic.)

3. Bed Expansion

a. Liquid Fluidized Beds

Liquid-solid fluidized systems are generally characterized by their even expansion as the liquid flow rate is progressively increased beyond that necessary to fluidize the bed. However, if there is a wide particle size or density range there will be a marked tendency for segregation to occur within the bed [28]. Romero and Johanson [29] presented the following criteria for predicting whether the bed will undergo particulate (nonbubbling) fluidization or aggregative (bubbling) fluidization:

Particulate fluidization

$$\text{Fr}_{mf}\,\text{Re}_{mf}\frac{\rho_s - \rho_f}{\rho_f}\frac{L_{mf}}{D_T} < 100 \tag{9}$$

Aggregate fluidization

$$\text{Fr}_{mf}\,\text{Re}_{mf}\frac{\rho_s - \rho_f}{\rho_f}\frac{L_{mf}}{D_T} > 100 \tag{10}$$

Richardson and Zaki [30], following other workers, correlated their experimental results for particulate fluidization by plotting the fluidizing velocity V against voidage α on a semilog plot and obtained a linear relationship of the form:

$$\frac{V}{V_i} = \alpha^n \tag{11}$$

where V_i, the velocity at a voidage of unity, is approximately equal to the free-falling velocity of the particles for sedimentation. The index n is a function of the ratio of particle to bed diameter d_p/D_T. For fluidization,

$$\log V_i = \log V_t - \frac{d_p}{D_T} \tag{12}$$

and for spherical particles,

$$n = 4.65 + 20\,\frac{d_p}{D_T} \qquad\qquad \text{Re}_t < 0.2 \tag{13}$$

$$n = \left(4.4 + 18\,\frac{d_p}{D_T}\right)\text{Re}_t^{-0.03} \qquad 0.2 < \text{Re}_t < 1 \tag{14}$$

$$n = \left(4.4 + 18\,\frac{d_p}{D_T}\right)\text{Re}_t^{-0.1} \qquad 1 < \text{Re}_t < 200 \tag{15}$$

$$n = 4.4\,\text{Re}_t^{-0.1} \qquad\qquad 200 < \text{Re}_t < 500 \tag{16}$$

$$n = 2.4 \qquad\qquad\qquad \text{Re}_t > 500 \tag{17}$$

For nonspherical particles n has higher values than for spheres [30].

The particle terminal velocity may be estimated from the following relationships [9]:

$$V_t = \frac{g(\rho_s - \rho_f)d_p^2}{18\mu} \qquad\qquad \text{Re}_p < 0.4 \tag{18}$$

$$V_t = \left[\frac{4}{225}\frac{(\rho_s - \rho_f)^2 g^2}{\rho_f \mu}\right]^{1/3} d_p \qquad 0.4 < \text{Re}_p < 500 \tag{19}$$

$$V_t = \left[\frac{3.1g(\rho_s - \rho_f)d_p}{\rho_f}\right]^{1/2} \qquad 500 < \text{Re}_p < 200{,}000 \tag{20}$$

b. Gas-Fluidized Beds

The situation with gas fluidized beds is more complicated [31]. As noted in the introduction, powders of Geldart's group A have the capacity to sustain continuous phase expansion beyond the onset of fluidization, whereas with B- and D-type materials, bubbling begins to occur when the minimum fluidizing velocity is exceeded, and the continuous phase voidage remains approximately equal to that at minimum fluidization. Bed expansion, therefore, with A group materials is predominantly a property of the powder and gas physical properties at lower fluidizing velocities. At higher velocities or with B- and D-type powders, the expansion is primarily caused by the bubble volume holdup within the bed. This latter depends in turn on the bubble size distribution within the bed and the size to which bubbles can grow in rising through the bed (bigger bubbles rise faster). In small-scale beds maximum bubble size is limited by the bed diameter, but with B- and D-type materials bubbles may grow to very large sizes in beds of high cross-sectional area. This behavior is the source of the scale-up problem so well illustrated by the work of de Groot and his colleagues [5].

Abrahamsen and Geldart [16] found that the maximum nonbubbling bed expansion ratio for group A powders at atmospheric pressure, H_{mb}/H_{mf}, was related to the ratio V_{mb}/V_{mf} by the simple relationship

$$\frac{H_{mb}}{H_{mf}} = \left(\frac{V_{mb}}{V_{mf}}\right)^{0.22} \tag{21}$$

and H_{mb}/H_{mf} was correlated with the physical properties of the gas and powder in SI units by:

$$\frac{H_{mb}}{H_{mf}} = \frac{5.5 \exp (0.158F)\rho_g^{0.028}\mu_g^{0.115}}{d_p^{0.176}g^{0.205}(\rho_p - \rho_g)^{0.205}} \tag{22}$$

where F is the fines fraction defined as the proportion of the powder less than 45 μm (1.5 $\times$ 10^{-3} ft) mean diameter. The minimum bubbling velocity V_{mb}, again using SI units, is correlated by

$$V_{mb} = 2.07 \exp (0.716F) \frac{d_p\rho_g^{0.06}}{\mu_g^{0.347}} \tag{23}$$

They also give an equation for the estimation of bed voidage at velocities between incipient fluidization and bubbling. There is evidence that the dense phase voidage is close to α_{mf} at the top of a deep, vigorously bubbling bed and increases toward the distributor [32]. Increasing the static operating pressure increases the density of the gas, and the observed increase in dense phase expansion [33] is commensurate with the predictions of Eq. (23) above.

TABLE 3 Variation of Y, the Proportion of Gas in Excess of V_{mf} Flowing in the Bubble Phase, with Mean Particle Sieve Sizes d_p

d_p, μm	d_p, ft	Y
<60	<2 $\times$ 10^{-4}	1
100	3.3 $\times$ 10^{-4}	0.85
150	5 $\times$ 10^{-4}	0.73
200	6.6 $\times$ 10^{-4}	0.61
250	8.2 $\times$ 10^{-4}	0.52
300	9.8 $\times$ 10^{-4}	0.43
350	1.1 $\times$ 10^{-3}	0.37
400	1.3 $\times$ 10^{-3}	0.30
500	1.6 $\times$ 10^{-3}	0.25
>500	>1.6 $\times$ 10^{-3}	0.25

With freely bubbling beds, overall expansion can be calculated if it is known how much gas flows through the bed in the bubble phase together with the average bubble velocity [31]. To a first approximation, it has often been assumed that the bubble volume flow rate is equal to the gas velocity flow rate in excess of V_{mf}. However, the proportion Y of gas flowing as the bubble phase reduces as the mean particle size increases (Table 3).

The degree of bed expansion over that at minimum fluidization, H_{mf}, is then given by

$$H - H_{mf} = \frac{Y(V - V_{mf})H}{\overline{V}_b} \tag{24}$$

An additional problem is that the average bubble rise velocity $\overline{V}_b$ is not generally known and the interactions between bubbles in a swarm is particularly complex.

B. HEAT TRANSFER

1. Packed Beds

Although the paths of heat flow and fluid flow are quite complex in packed beds, the system can be well described because the particles are fixed in place. Therefore the various modes of heat transfer for packed beds are more readily treated than for fluidized beds. An effort is therefore made in this section to use correlations based on first principles that satisfactorily fit the data and which provide designers with a higher degree of confidence when extrapolating to conditions beyond the range of existing data than would strictly empirical correlations.

Conductance, convection, and radiation will be considered. These modes of heat transfer will be assumed to be additive in their contribution to the overall rate of heat transfer. It is recognized that there are interactive effects, but these are considered negligible here. A reference on the treatment of combining contributing mechanisms in transfer processes is the paper by Churchill and Usagi [34].

a. Particle-to-Fluid Heat Transfer

A single particle in an infinite fluid medium is a limiting condition for packed-bed heat transfer. The rate of heat transfer q from a single spherical particle in an infinite stagnant fluid is determined by the conductivity of the fluid, k_f:

$$q = -k_f 4\pi r^2 \frac{dT}{dr} \tag{25}$$

Integrating from the surface of the spherical particle of radius r_p to infinity yields the rate of heat transfer in terms of the difference between the surface temperature T_s and the fluid temperature T_∞:

$$\int_{r_p}^{\infty} \frac{q}{r^2} = -k_f 4\pi \int_{T_s}^{T\infty} dT \tag{26}$$

$$q = k_f 4\pi r_p (T_s - T_\infty) \tag{27}$$

Alternatively, if the rate of heat transfer is defined in terms of a heat transfer coefficient h, then:

$$q = h 4\pi r_p^2 (T_s - T_\infty) \tag{28}$$

Combining Eqs. (27) and (28) gives:

$$\frac{hr_p}{k_f} = 1 \qquad (29)$$

or, in terms of the particle diameter d_p,

$$\frac{hd_p}{k_f} = \mathrm{Nu}_p = 2 \qquad (30)$$

The Nusselt number of 2 based on conductive heat transfer from a sphere is considered to be the minimum since any convection in the fluid and/or radiation from the sphere would add to the rate of heat transfer.

If the fluid is flowing past the sphere, the Nusselt number can be correlated by adding a convection component to the conduction term (Ranz and Marshall [35]):

$$\mathrm{Nu}_p = \frac{hd_p}{k_f} = \underset{\text{conduction}}{2} + \underset{\text{convection}}{0.6\ \mathrm{Pr}^{1/3}\ \mathrm{Re}_p^{1/2}} \qquad (31)$$

In the particle-bed case the heat transfer from a particle and the fluid flow about the particle are affected by the adjacent particles. A significant effect Ranz [21] suggested is that the local fluid velocity V_{loc} in the particle vicinity is much higher than the readily measured superficial velocity V_0, which is based on the total cross-sectional area of the bed (in other words, on an empty container). Ranz [21] estimated the relationship between V and V_0 for a bed of spherical particles in rhombohedral array to be:

$$\frac{V_{\mathrm{loc}}}{V_0} = \frac{1}{1 - \pi/2\sqrt{3}} = 10.74 \qquad (32)$$

Applying this to the Ranz and Marshall [35] correlation for a single sphere, Eq. (31), gives:

$$\mathrm{Nu}_p = 2 + 2.0\ \mathrm{Pr}^{1/3}\ \mathrm{Re}_p^{1/2} \qquad (33)$$

A better fit to the experimental data for randomly packed beds of spheres, which have a higher void fraction ($\alpha \approx 0.4$) than spheres in rhombohedral array ($\alpha = 0.26$), is:

$$\mathrm{Nu}_p = \frac{hd_p}{k_f} = 2 + 1.8\ \mathrm{Pr}^{1/3}\ \mathrm{Re}_p^{1/2} \qquad \mathrm{Re}_p > 50 \qquad (34)$$

Equation (34) fits the data for large particles. However, data for small particles and low Reynolds numbers fall well below this correlation as well as below the minimum (conduction) Nusselt number of 2 for a single particle. Several explanations have been offered of why the measured Nusselt numbers for particles in a packed bed are below the minimum Nusselt number for a single sphere. Cornish [36] explained this in terms of an altered temperature gradient, Kunii and Suzuki [37] theorized that this difference was because of fluid flow channeling in the bed, and Glicksman and Joos [38] demonstrated that axial diffusion would reduce the particle Nusselt number below 2. On the basis of the experimental evidence and the theoretical rationalization it is recommended that a correlation based on experimental data for low Reynolds number situations be used. Nu_p is correlated in terms of $\mathrm{Pr}^{1/3}\ \mathrm{Re}_p^{1/2}$ in Fig. 3. Equation (34) can be used for Reynolds numbers greater than 50 (Fig. 3).

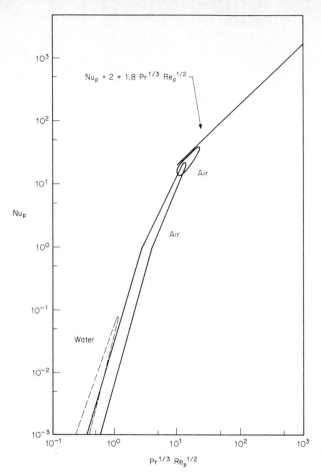

FIG. 3 Particle-to-bed heat transfer (data taken from Kunii and Suzuki [37]).

b. Effective Thermal Conductivity

Heat transport through packed beds can readily be calculated by using an effective bed thermal conductivity. This combines the transport properties of the separate solid and fluid phases into one term convenient for mathematical manipulation. The effective thermal conductivity can be used accurately for steady-state heat transfer and for unsteady-state heat transfer if t/d_p^2 (time divided by the square of particle diameter) is greater than 500 h/ft^2 (1.94×10^7 s/m^2). Its use for low values of t/d_p^2 will lead to overpredicting the rate of heat transfer (Gabor [39]). The thermal response of the separate phases must then be individually treated as the thermal wave penetrates into the bed at low values of t/d_p^2. Only after a longer time period can the properties of the separate phases be combined in terms of a single constant (such as k_e) to satisfactorily determine unsteady-state heat transfer.

An analytical solution for effective thermal conductivities of packed beds that has

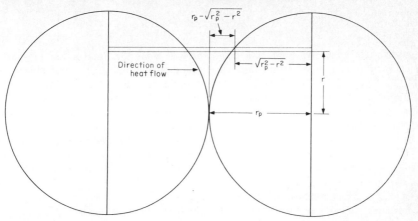

FIG. 4 Interparticle conduction model for packed-bed effective thermal conductivity.

found wide application (Kunii and Smith [40], Swift [41], Kobayashi [42], Baskakov [43], and Shimokawa [44]) is based on the following assumption that all heat transfer is in the axial direction (Fig. 4), i.e., that no temperature gradients exist in the radial direction. The heat flux can then be related to the solid thermal conductivity k_s and the fluid thermal conductivity k_f combined in a single term:

$$\int dq = \int_0^{r_p} \frac{2\pi r(T_1 - T_2)}{\dfrac{r_p - \sqrt{r_p^2 - r^2}}{k_f} + \dfrac{\sqrt{r_p^2 - r^2}}{k_s}} \, dr$$

$$= 2\pi r_p (T_1 - T_2) \left(\frac{k_s}{k_s - k_f}\right)^2 \left(\ln \frac{k_s}{k_f} - \frac{k_s - k_f}{k_s}\right) k_f \qquad (35)$$

For spheres in a cubic array the fraction of the cross-sectional area, as viewed from the heat transfer surface, occupied by the solid and fluid phases is $(\pi/4)d_p^2/d_p^2 = 0.7854$, and the area fraction occupied by only the fluid phase is 0.2146. This leads to the following expression for the effective thermal conductivity for a bed in a cubic array ($\alpha = 0.4764$) with a stagnant fluid:

$$\frac{k_e^\circ}{k_f} = (0.7854)(2) \left(\frac{k_s}{k_s - k_f}\right)^2 \left(\ln \frac{k_s}{k_f} - \frac{k_s - k_f}{k_s}\right) + 0.2146 \qquad (36)$$

For an orthorhombic array ($\alpha = 0.3954$) the effective thermal conductivity is:

$$\frac{k_e^\circ}{k_f} = (0.9069)(2) \left(\frac{k_s}{k_s - k_f}\right)^2 \left(\ln \frac{k_s}{k_f} - \frac{k_s - k_f}{k_s}\right) + 0.0931 \qquad (37)$$

For both the cubic and orthorhombic array the direction of heat transfer is modeled to flow through particles directly in line with one another. This of course is a highly idealized orientation. For different orientations and arrays such as rhombohedral or random packing the direction of heat transfer is generally not parallel to the axis of the contacting spheres. Kunii and Smith [40] considered this in their modeling of effective thermal conductivity. They summed the resistance to various heat transfer paths to give the following equation:

$$\frac{k_e^\circ}{k_f} = \frac{1 - \alpha}{1/\delta_f + 2/3(k_s/k_f)} + \alpha \tag{38}$$

where δ_f is an effective fluid path length derived along the lines of Eq. (35), which accounts for multiple contact points at various angles to the overall direction of heat flow. Figure 5 gives a plot of k_e°/k_f versus k_s/k_f for a cubic array, Eq. (36); for an orthorhombic array, Eq. (37); and for beds with void fractions of 0.4 (random packing), 0.3, and 0.26 for the rhombohedral array (the most dense packing arrangement) using the Kunii and Smith [40] correlation. The model for the orthorhombic array has been extended by Swift [41] to include coatings on the particle surface such as oxides, and by Kobayashi [42] to particles separated by a small finite distance as might be envisioned for a bed at incipient fluidization. Reference can be made to Kuzay [47] and Godbee and Ziegler [48] for a survey of various alternate correlations for effective bed thermal conductivities and additional data. Bauer and Schlünder [49, 50] have considered the effects of pressure, particle contacting, particle size distribution, and packing shape in their modeling.

Figure 5 describes k_e° for $k_s/k_f \geq 1$. The correlations could be extended to the region of $k_s/k_f < 1$, where the fluid phase has a higher thermal conductivity than the solid phase. Heat is then primarily conducted through the continuous (fluid) phase. The system then is similar to a porous body, and it is suggested that the thermal conductivity be determined using a porosity model such as that by Kampf and Karsten [51] for $k_s/k_f < 1$:

$$\frac{k_e^\circ}{k_f} = 1 - \frac{(1 - \alpha)(k_f/k_s - 1)}{1 + (1 - \alpha)^{1/3}(k_f/k_s - 1)} \tag{39}$$

Equation (39) has been extensively used in liquid-metal nuclear reactor work [47].

If the fluid is in motion, the heat transfer is enhanced by convective heat removal. Yagi and Kunii [52] have correlated this effect by the addition of a convective term to the stagnant effective thermal conductivity:

$$\frac{k_e}{k_f} = \frac{k_e^\circ}{k_f} + 0.11 \text{ Pr Re}_p \qquad \frac{d_p}{D_T} < 0.04 \tag{40}$$

where D_T is the diameter of the vessel containing the bed.

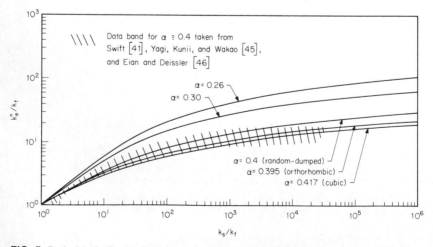

FIG. 5 Packed-bed effective thermal conductivity with motionless gas.

Equation (40) applies to heat transfer in the radial direction (normal to the direction of fluid flow). If heat transfer is in the axial direction (the same direction as the fluid flow), the following correlation by Yagi et al. [53] is recommended:

$$\frac{k_e}{k_f} = \frac{k_e^\circ}{k_f} + 0.75 \text{ Pr Re}_p \tag{41}$$

c. Wall-to-Bed Heat Transfer

Wall-to-bed heat transfer under steady-state conditions is essentially that of heat absorption by the fluid flowing through the packed bed. For large particles (>5 mm or 1.6×10^{-2} ft) the mechanism is primarily that of fluid eddy diffusion from within the bed to the wall, heat absorption (or desorption, depending on the process) at the wall, and then eddy diffusion back into the bed again (Hanratty [54]). For smaller particles Sec. B.2.b on interphase gas convective heat transfer for particle beds at or below minimum fluidization should be referred to. The mechanism of eddy diffusion in a packed bed near the wall is basically the same as within the bed (Ranz [21]) and is therefore directly dependent on packing size.

Yagi and Kunii [55] obtained the following correlation for the heat transfer coefficient at the wall h_w for $\text{Re}_p < 2000$:

$$\frac{h_w d_p}{k_f} = \frac{h_w^\circ d_p}{k_f} + 0.054 \text{ Pr Re}_p \qquad \text{Re}_p < 2000 \tag{42}$$

h_w°, the wall coefficient at zero flow, can be determined from:

$$\frac{1}{h_w^\circ d_p} = \frac{1}{k_w^\circ} - \frac{0.5}{k_e^\circ} \tag{43}$$

where k_e° is the effective bed thermal conductivity at stagnant conditions and can be obtained from the techniques indicated in Sec. B.1.b. k_w°, the bed conductivity near the wall, is obtained from:

$$\frac{k_w^\circ}{k_f} = 2\alpha_w + \frac{1 - \alpha_w}{\delta_w + k_f/3k_s} \tag{44}$$

α_w can be assumed to be 0.7, and δ_w is found from the conduction model, Eq. (35), to be:

$$\delta_w = \frac{1}{4} \frac{\left[\dfrac{k_s - k_f}{k_s}\right]^2}{\ln \dfrac{k_s}{k_f} - \dfrac{k_s - k_f}{k_s}} - \frac{k_f}{3k_s} \tag{45}$$

At high Reynolds numbers (>2000) Yagi and Kunii [56] recommend the following relationship:

$$\frac{h_w d_p}{k_f} = \frac{h_w^\circ d_p}{k_f} + \frac{1}{k_f/h_w^* d_p + 1/(0.054 \text{ Re}_p \text{ Pr})} \tag{46}$$

h_w^* is a fluid film coefficient, which accounts for turbulent destruction of the laminar boundary layer at the wall and is correlated by

$$\frac{h_w^* d_p}{k_p} = 2.6 \text{ Pr}^{1/3} \text{ Re}_p^{1/2} \tag{47}$$

d. Radiant Heat Transfer

At elevated temperatures ($>600°C$ or $1100°F$) radiation becomes a contributor to heat transfer in packed beds. The radiant component is most readily treated by addition to the conduction and convection contributions to the heat transfer process. While it is recognized that there are interaction effects, this procedure has been quite successful in determining heat transfer rates (see Churchill and Usagi [34]).

Single-Particle Heat Transfer.   The total heat transfer from a single sphere which includes the contributions of conduction, convection, and radiation is correlated by:

$$\text{Nu}_p = \frac{h_p d_p}{k_f} = 2.0 + 0.6\ \text{Pr}^{1/3}\ \text{Re}_p^{1/2} + d_p^2 \epsilon(T_s^4 - T_\infty^4) \tag{48}$$

$$\underbrace{}_{\text{conduction}} \qquad \underbrace{\phantom{0.6\ \text{Pr}^{1/3}\ \text{Re}_p^{1/2}}}_{\text{convection}} \qquad \underbrace{}_{\text{radiation}}$$

Particle-to-Bed Heat Transfer.   The heat transmitted by radiation from a particle in a bed of particles is primarily absorbed by the surrounding particles. The rate of heat transfer to the surrounding particles is dependent on the temperature differences. Therefore particle-to-bed heat transfer requires a temperature gradient in the bed. Techniques for determining radiant heat transfer in which temperature gradients in the bed exist can be obtained by referring to Wakao and Kato [57] and Whitaker [58].

Effective Thermal Conductivity.   The contribution of radiant heat transport to the effective thermal conductivity can be derived from the following equations (Schotte [59]) and Bauddour and Yoon [60]:

$$k_r = \frac{1 - \alpha}{\dfrac{1}{k_s} + \dfrac{1}{k_r^\circ}} + \alpha k_r^\circ \tag{49}$$

where

$$k_r^\circ = 0.692 \epsilon \alpha d_p \frac{T^3}{10^8} \qquad \text{Btu/(h·ft·°F)}\ (T \text{ in °R and } d_p \text{ in ft}) \tag{50}$$

or

$$k_r^\circ = 0.229 \epsilon \alpha d_p \frac{T^3}{10^6} \qquad \text{W/(m·K)}\ (T \text{ in K and } d_p \text{ in m})$$

The contribution of radiation, k_r, can be combined with the stagnant bed conductivity and the convective contribution to determine the overall bed conductivity:

$$k_e = k_e^\circ + k_{cv} + k_r \tag{51}$$

It is significant that, although the basic packed-bed models differ markedly in detail, they all predict effective thermal conductivities agreeing within a factor of 2 if the radiant component is omitted (Fig. 6). The shape of the dependence upon temperature depends primarily on the geometry of the cell considered and the method of calculation. Thus, while the cell for the Godbee and Ziegler [48] model is identical to that considered by Imura and Takegoshi [116], the former workers assumed parallel isotherms and the latter a constant linear heat flux in deriving the effective thermal conductivity. Bauer and Schlünder [49, 50] consider a quite different unit cell and assume a constant linear heat flux. However, even with the inclusion of the provision for radiant heat transfer (made in different ways in the models), the models consistently underpredict the observed variation in effective bed thermal conductivity with temperature; e.g., see the results of Bot-

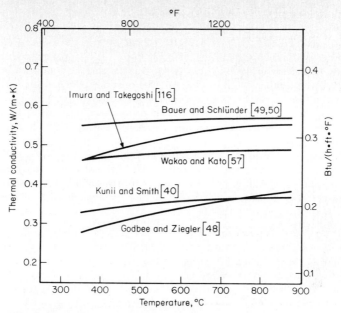

FIG. 6 Predicted conduction component of effective thermal conductivity for 410-μm silica sand bed.

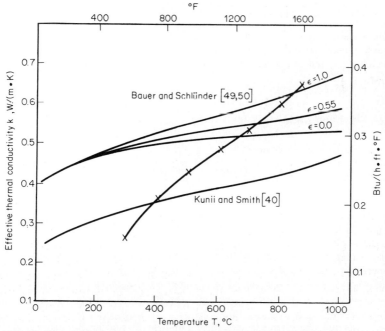

FIG. 7 Comparison between measured effective thermal conductivity of 410-μm sand bed [117] and predictions of Kunii and Smith [40] and Bauer and Schlünder [49, 50] models.

terill, Salway, and Teoman [117] for 410 μm (1.25×10^{-3} ft) silica sand over the temperature range 400 to 900°C (750 to 1650°F), Fig. 7. It is apparent that the strong temperature dependence of the conductivity for beds of sands and alumina is a consequence of radiative transfer, but it can be seen that even a maximum emissivity of unity is insufficient to predict the temperature dependence observed. Acccordingly, it can be concluded that the transmission of radiant energy by these materials at temperatures within this range, a factor omitted from the models, needs also to be taken into account.

Wall-to-Bed Heat Transfer.   Heat radiation from the wall is absorbed by the particles next to the wall and within the nearby voids. Yagi and Kunii [56] correlated the radiation contribution of wall-to-bed heat transfer in terms of a heat transfer coefficient for solid surface to solid surface radiation h_{rs}, and for radiation through void spaces h_{rv}:

$$h_{rs} = 0.686 \frac{\epsilon}{2 - \epsilon} \frac{T^3}{10^{10}} \quad \text{Btu/(h·ft}^2\text{·°F)} \; (T \text{ in °R})$$

$$ \tag{52}$$

or
$$h_{rs} = 0.227 \frac{\epsilon}{2 - \epsilon} \frac{T^3}{10^6} \quad \text{W/(m}^2\text{·K)} \; (T \text{ in K})$$

and
$$h_{rv} = \frac{0.686}{1 + \dfrac{\alpha}{2(1 - \alpha)} \dfrac{1 - \epsilon}{\epsilon}} \frac{T^3}{10^{10}} \quad \text{Btu/(h·ft}^2\text{·°F)}(T \text{ in °R})$$

$$ \tag{53}$$

or
$$h_{rv} = \frac{0.227}{1 + \dfrac{\alpha}{2(1 - \alpha)} \dfrac{1 - \epsilon}{\epsilon}} \frac{T^3}{10^6} \quad \text{W/(m}^2\text{·K)} \; (T \text{ in K})$$

Substitution of k_w^o into Eq. (43) gives h_w^o, which includes the contribution of radiation. The overall wall heat transfer coefficient h_w, which includes the contributions of radiation, conduction, and convection, is then obtained from Eq. (42) ($Re_p < 2000$) or Eq. (46) ($Re_p > 2000$).

2. Gas Fluidized Beds

a. Gas-to-Particle Heat Transfer

Values for the gas-to-particle heat transfer coefficient based on the total particle surface area are generally only small, typically of the order of 6 to 23 W/(m²·K), or 34 to 130 Btu/(h·ft²·°F). However, as noted in the introduction, the total particle surface area within the bed can be very large. It is thus to be concluded that by no means all that surface area is directly accessible to the gas, and this would account for the very low limiting Nusselt numbers obtainable. Kunii and Levenspiel [9] have correlated published results of experiments analyzed on the basis of a gas plug flow model (the situation most nearly corresponding to the actual situation despite the high degree of solids mixing that can occur within a bubbling fluidized bed) and give the equation:

$$\text{Nu}_p = 0.03 \, \text{Re}_p^{1.3} \tag{54}$$

The experimental measurements are difficult to make [3] and there remains doubt as to what an immersed temperature-sensing element may measure. Singh and Ferron [61] have reported differences between gas and particle temperatures and suggested that the reading indicated by a thermocouple corresponds to a weighted mean reflecting 80 percent of the particle temperature and 20 percent of that of the gas for measurements in a bed of catalyst of 230 μm (7.5 $\times$ 10^{-4} ft) being cooled from 200°C (392°F) by air at ambient temperature.

Using SI units, the height $l_{1/2}$ in meters over which the gas-particle temperature difference will fall by half can be shown to be

$$l_{1/2} \approx \frac{5.55\mu_g^{1.3}d_p^{0.7}c_{pg}}{\rho_g^{0.3}V^{0.3}k_g(1-\alpha)} \tag{55}$$

in the simplified situation of steady-state operation with a well-mixed bed of particles of low Biot number, i.e., particles of negligible internal thermal resistance.

b. Wall-to-Bed Heat Transfer

To a first approximation, heat transfer between a bubbling fluidized bed and its containing surface or other immersed heat transfer surface can be thought to consist of three additive components [3]:

1. The particle convective component h_{pc} is dependent upon heat transferred by particle circulation between the bulk of the bed and the region directly adjacent to the heat transfer surface. This is predominant for beds of smaller particles (40 to 800 μm or 1.3 $\times$ 10^{-4} to 2.6 $\times$ 10^{-3} ft), within Geldart's group A and B powders.

2. The interphase gas convective component h_{gc} is that by which particle-to-surface heat transfer is augmented by interphase gas convective heat transfer. This is significant for denser and larger particles and also, at higher static pressures, group D materials.

3. The radiant component of heat transfer h_r contributes significantly at bed temperatures above 600°C or 1100°F and with larger bed-to-surface temperature differences. Thus:

$$h \quad = \quad h_{pc} \quad + \quad h_{gc} \quad + \quad h_r \tag{56}$$

approx.	40 μm → 1 mm	>800 μm and	higher
range of		at higher	temperatures
significance		static	(>600°C)
		pressures	

Although there are mechanistic models which give insight into the limiting factors and particularly so with the particle convective component [3, 62], the relevant parameters are rarely known a priori [26], so the application of these models for predictive purposes is severely limited. Indeed, it is also true that most of the empirical correlations have the limitation of having been developed to give a close representation of results within a specific apparatus and then only for tests close to ambient conditions. They should only be used with considerable care, as was earlier demonstrated by van Heerden [63].

Small Particles. The particle convective component h_{pc} is responsible for the characteristic rapid increase in bed-to-surface heat transfer when a gas fluidized bed passes from the quiescent to the bubbling or aggregatively fluidized state (Fig. 8). With finer powders, those coming within Geldart's group A and expanding stably until the minimum fluidizing velocity is exceeded by a sizable factor, there is an additional minor peak in the rising portion of the curve as the bed becomes fluidized, and diffusive particle mixing can occur before the stable bed collapses back and effective bubble-induced circulation is

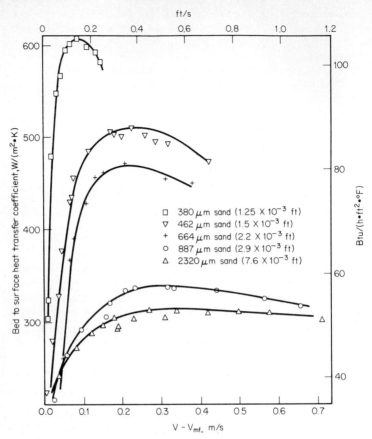

FIG. 8 Variation in bed-to-surface heat transfer coefficient with excess gas velocity $V - V_{mf}$ for beds of various mean particle size operating at about 620°C (1150°F) [26].

developed [64]. With continuing increase in the fluidizing gas flow rate, the coefficient passes through a maximum; then, with further increase in superficial gas velocity, the blanketing effect of the increasing bubble flow across the transfer surface causes the heat transfer coefficient to drop [3, 65]. Because contact areas between particles and surface and between the particles themselves are too small for significant heat transfer through points of solid contact, heat must flow by conduction through the gas (with group A and B behavior, the fluidizing gas velocity is too low for significant heat transfer by convective transfer through the gas). There is also evidence of additional resistance to heat transfer in the vicinity of the heat transfer surface [3]. It is the low thermal conductivity of the fluidizing gas which limits the maximum achievable bed-to-surface heat transfer coefficients. The strong inverse dependence of the coefficient upon particle size over the range 40 to 800 μm or 1.3×10^{-4} to 2.6×10^{-3} ft (Fig. 9) is a consequence of the increase in the effective proportion of heat transfer surface area through which heat can flow to or from the particles by short gas conduction paths. As would be expected, bed-to-surface

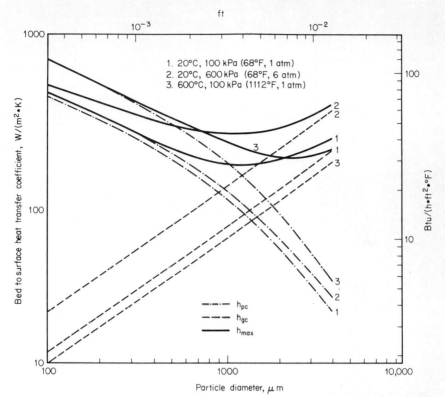

FIG. 9 Effect of particle size on maximum bed-to-surface heat transfer coefficient h_{max} for operation at (1) 20°C and 100 kPa, (2) 20°C and 600 kPa, and (3) 600°C and 100 kPa. Interphase gas convective component of heat transfer h_{gc} taken as that for quiescent bed; particle convective component h_{pc} estimated by difference [26].

transfer coefficients increase with increase in the gas thermal conductivity resulting from operating beds at higher temperatures. However, the actual bed-to-surface heat transfer coefficient is generally dependent upon the particle residence time adjacent to the transfer surface, as pointed out by Mickley and his coworkers [66]. Solids circulation generated by the rising bubbles is the key factor.

Most of the published correlations for bed-to-surface heat transfer have been derived from tests in small-scale equipment and reflect the simple circulation patterns obtained in them. Work reported by de Groot [5] illustrates the pronounced change in bed behavior consequent upon changing bubble development patterns which can result from a change in the equipment scale and particularly so with beds of B-type materials. The design of the distributor and of any internals will also affect the particle circulation. As stressed in the introduction to this section, correlations should only be used with great caution. Their predictions scatter widely. For example, Table 4 compares predictions of some of the simpler correlations with experimentally measured maximum bed-to-surface heat transfer coefficients for B- and D-type materials [26]. Nevertheless, the simple empirical correlation proposed by Zabrodsky and his colleagues from tests with calorimetric probes of

TABLE 4 Comparison between Measured Bed-to-Surface Heat Transfer Coefficients for Operation at ~ 640°C (1184°F) and Prediction of Correlations [26]

	Material				
	462 μm (1.5 × 10⁻³ ft) sand	887 μm (2.9 × 10⁻³ ft) sand	820 μm (2.7 × 10⁻³ ft) ash	1148 μm (3.8 × 10⁻³ ft) alumina	2320 μm (7.6 × 10⁻³ ft) sand
	Experimental conditions				
Archimedes number for test conditions	635	4500	2150	14,700	80,500
Measured minimum fluidizing velocity V_{mf}, m/s (ft/s)	0.10 (0.33)	0.47 (1.54)	0.34 (1.12)	0.98 (3.21)	1.10 (3.61)
Fluidizing velocity for h_{max}, V_{opt}, m/s (ft/s)	0.29 (0.95)	0.67 (2.20)	0.60 (1.97)	1.57 (5.15)	1.48 (4.86)
Measured bed-to-surface heat transfer coefficient, W/(m²·K)[Btu/(h·ft²·°F)]	510 (90)	325 (57)	310 (55)	290 (51)	300 (53)
	h_{max} predictions				
Zabrodsky [67]	526 (93)	416* (73)	389 (69)	—	—
Botterill and Denloye [68]	—	—	—	266 (47)	220 (39)
Grewal and Saxena [69]	507 (89)	402 (71)	325 (57)	273 (48)	252 (44)
Zabrodsky et al. [70]	482 (85)	385 (68)	352 (62)	378 (67)	269 (47)
Varygin and Martyushin [71]	433 (76)	338 (60)	311 (55)	327 (58)	227 (40)
Maskaev and Baskakov [72]	377 (66)	226 (40)	191 (34)	234 (41)	215 (38)
Sarkits [73]	502 (88)	291 (51)	200 (35)	342 (60)	361 (64)

h as a function of V for value of V_{opt}

Vreedenberg [74]	198 (35)	—	—	—	—
Andeen and Glicksman [75]	211 (37)	152 (27)	183 (32)	—	—
Petrie et al. [76]	421 (74)	214 (38)	242 (43)	—	—
Einstein [77]	512 (90)	461 (81)	438 (77)	412 (73)	308 (54)
Gel'perin et al. [78]	480 (85)	424 (75)	397 (70)	340 (60)	271 (48)
Ternovskaya and Korenberg [79]	407 (72)	393 (69)	379 (67)	406 (72)	297 (52)

Predictions for tube array

Gel'perin [80] in line	385 (68)	—	284 (50)	—	—
Gel'perin [80] "staggered"	331 (58)	—	244 (43)	—	—
Chekansky [81] "staggered"	517 (91)	359 (63)	370 (65)	—	—

*Outside recommended range for prediction.

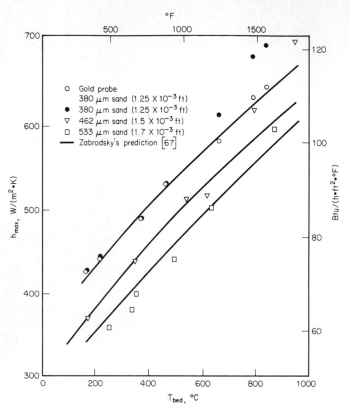

FIG. 10 Variation in maximum bed-to-surface heat transfer coefficient h_{max} with operating temperature compared with the prediction of the Zabrodsky correlation, Eq. (57), for beds of sand of different mean particle size [26]. Note that $h_{max} \approx h_{pcmax} + h_r$

low heat capacity [67] is surprisingly good and has the double merit of both giving reasonable predictions for the maximum bed-to-surface heat transfer coefficient and being very simple to use. This takes the form:

$$h_{pcmax} = 35.8 \rho_g^{0.2} k_g^{0.6} d_p^{-0.36} \qquad W/(m^2 \cdot K) \qquad (57)$$

where all parameters are in SI units and k_g is the gas thermal conductivity at bed temperature. Figure 10 compares predictions of this correlation with experimental results over a range of operating temperatures. It is to be recommended for use over the particle size range ~100 to 800 μm (3.3 × 10⁻⁴ to 2.6 × 10⁻³ ft). (Above 600°C, or 1100°F, the measured values lie increasingly above the predictions because of the increasing contribution of the radiant component h_r; with a gold transfer surface of low emissivity (denoted by the open circles), the agreement of the correlation continues closely; see results for 380-μm sand in Fig. 10.)

The effective bed-to-surface heat transfer coefficient will be higher to an object of low heat capacity (within an order of magnitude of that of the bed particles). This has implications for heat transfer to coal fed to a combuster or in the heat treatment of small components. Further, particularly with group B materials, there is an effect of relative bed-to-surface temperature (Fig. 11). Thus, if a horizontal cooling tube is immersed within a bed, the bed-to-tube coefficient will be lower because of two factors; the limiting gas thermal conductivity adjacent to the tube will be lower than that generally obtaining in the bed because of the lower surface temperature and because of the disturbance caused to the local fluidization through insertion of the tube (see Sec. B.2.c).

For the case of smaller, less dense powders, Khan et al. [64] give the correlation:

$$Nu_{max} = 0.157\ Ar^{0.475} \tag{58}$$

from experiments with group A powders of mean diameters between 40 and 96 μm (1.3 $\times$ 10^{-4} and 3.1 $\times$ 10^{-4} ft). This has not been tested over a range of operating temperatures and it is to be expected that change in gas physical properties resulting from change in the temperature of operation will affect the fluidization behavior and hence the bed-to-surface heat transfer coefficient.

As indicated in Sec. A.2 above, the level of operating temperature affects the bed voidage at minimum fluidization with group B and probably A-type materials. It can also be expected that it will affect the conditions under which maximum bed-to-surface heat transfer will be obtained. In any case, although correlations for predicting optimum conditions have been proposed, even under ambient conditions their precision is very limited. A typical correlation is that by Todes [82]:

$$Re_{opt}' = \frac{Ar}{18 + 5.22\sqrt{Ar}} \tag{59}$$

This does not allow for the effect of voidage variation and is not to be recommended for use with powders less than 400 μm (1.3 $\times$ 10^{-3} ft) in mean diameter. (As the mean particle size increases, Re_{opt} approaches more closely Re_{mf}.) It is a matter of experience that the behavior of industrial catalytic reactors using A-type particles can change and

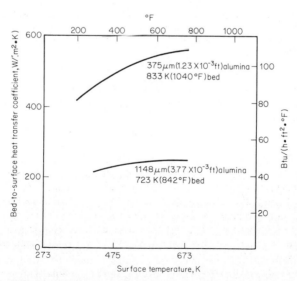

FIG. 11 Effect of heat transfer surface temperature on bed-to-surface heat transfer coefficient.

bed-to-surface heat transfer coefficients may vary by as much as a factor of 2 over a period of operation with "difficult" materials.

Because of the overall complexities in behavior, correlations which attempt to include allowance for variation in the optimum fluidizing velocity, bed expansion, and other such obviously important factors should be treated with particular caution. They are more likely to introduce additional error than compensate for changes in behavior when applied over a range of operating temperatures and pressures.

Heat Transfer to Large Particles and at Minimum Fluidization.   With larger particles, Geldart's group D, and at minimum fluidization, the time constant associated with heat transfer for the particles close to the transfer surface will exceed their residence time there. The process then becomes quasi steady-state. With large, dense particles and at higher static pressures, the interphase gas convective component h_{gc} becomes significant. The interstitial gas flow conditions are then within the transitional to turbulent regime. A transition in behavior between the characteristics of a bed of group D to a group B material has been observed as the operating temperature increased and Re_{mf} reduced to ~ 12.5, the corresponding Archimedes number being $\sim 26,000$ [26]. With D-type behavior, crosswise bubble coalescence tends to occur and particle mixing is less effective. The bed-to-surface heat transfer coefficient is then less sensitive to operating velocity (Fig. 8) and this is a feature of the correlations given below for h_{gc}.

Baskakov and Suprun [83] first estimated the interphase gas convective component by analogy from mass transfer measurements and recommended the following correlations:

$$\text{Nu}_{gc} = 0.0175 \, \text{Ar}^{0.46} \, \text{Pr}^{0.33} \qquad \text{for } V > V_m \tag{60}$$

and
$$\text{Nu}_{gc} = 0.0175 \, \text{Ar}^{0.46} \, \text{Pr}^{0.33} \left(\frac{V}{V_m} \right)^{0.3} \qquad \text{for } V_{mf} < V < V_m \tag{61}$$

An alternative empirical correlation was derived by Denloye and Botterill [68] from tests over a range of conditions including operating pressures up to 10 atmospheres (1 MPa) on the assumption that the bed-to-surface coefficient at the onset of fluidization, h_{mf}, is a measure of h_{gc}. (This is not strictly true, because there will be some conductive heat transfer between the particle and surface to augment the interphase gas convective component.) The correlation, using SI units, took the form:

$$\frac{h_{gc}d_p^{1/2}}{k_g} = 0.86 \, \text{Ar}^{0.39} \qquad \text{for } 10^3 < \text{Ar} < 2 \times 10^6 \tag{62}$$

having dimensions of $\text{m}^{-1/2}$. This fitted the experimental results at higher operating pressures better than the Baskakov-Suprun correlations. The corresponding maximum particle convective component was correlated against the Archimedes number by

$$\frac{h_{pc\text{max}}d_p}{k_g} = 0.843 \, \text{Ar}^{0.15} \tag{63}$$

assuming that the two components are additive. Their relative magnitude as a function of particle diameter is illustrated in Fig. 9.

Theoretical models for the prediction of h_{gc} are generally complicated and of doubtful value. The simple model by Gabor [84] for heat transfer to the packed or quiescent bed is nearly as precise although it ignores the effect of the increased voidage adjacent to the transfer surface and assumes constant physical properties for the system.

It is reported [85] that measurements have shown little influence of particle size on the heat transfer coefficient in tests with four beds of mean particle diameters between

800 and 2700 μm (2.6 $\times$ 10^{-3} and 8.9 $\times$ 10^{-3} ft) with top fractions ranging from 1410 to 5600 μm (4.6 $\times$ 10^{-3} to 1.8 $\times$ 10^{-2} ft). It is significant with these tests that the fines distribution (1410 μm or 4.6 $\times$ 10^{-3} ft downward) was very similar in all four cases, which suggests that the fines present limited the contribution of the interphase gas convective component, possibly by damping down the interstitial turbulence. This is an important possibility which needs to be examined further.

Radiative Transfer.   The estimation of the radiative component h_r presents the most difficulty. For conditions where this is significant (above ~600°C or 1100°F), particle packing and gas conductivity tend to become of lessening importance and there is evidence [3, 26] that the radiant mode tends to transfer some heat at the expense of the particle convective component so that these two components are not strictly additive. The effect is more pronounced with larger particles because the heat transfer surface receives and radiates energy from and to the whole of the particle's surface visible to it. Thus, as particle size increases, relatively less heat is left to flow by conduction through the particle to pass to the exchange surface by conduction through the shortest gas transfer paths close to the "point of contact" between particle and surface. The most serious problem in estimating the radiant component, though, is that the temperature of the bed particles immediately adjacent to the surface will have been affected through exchange of heat with the transfer surface and so be different from the temperature of the bulk of the bed, which is the temperature more generally known. Baskakov and his colleagues [65] have attempted to deal with these difficulties using a modified bed emissivity. This (Fig. 12) reduces as the temperature difference between the surface and the bulk of the bed increases in order to compensate for the increasingly strong interaction between the bed and surface. Particle

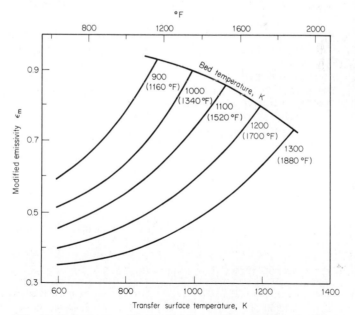

FIG. 12 Variation in modified emissivity ϵ_m of a nonisothermal bed with both the temperature of the transfer surface and bed, after Baskakov et al. [65].

emissivity also varies with temperature, and there is need to take into account the different emissive properties of the bed and surface.

For rule-of-thumb estimates, the radiative component can be estimated using absolute temperatures and an adaptation of the Stefan-Boltzmann equation in the form:

$$h_r = \frac{\sigma \epsilon_r (T_{\text{bed}}^4 - T_{\text{surface}}^4)}{T_{\text{bed}} - T_{\text{surface}}} \tag{64}$$

where σ is the Stefan-Boltzmann constant and has the value 5.67×10^{-8} in SI units and 1.71×10^{-9} in English units and where ϵ_r is the reduced emissivity to take into account the different emissive properties of the surface, ϵ_s, and the effective emissivity of the bed, ϵ_b, as follows:

$$\epsilon_r = \frac{1}{1/\epsilon_s + 1/\epsilon_b - 1} \tag{65}$$

(the effective emissivity of the bed, ϵ_b, is higher than that for the individual particles because of the increase in the self-view factor within the bed).

As indicated above, because of the exchange between the transfer surface and bed, the bed temperature adjacent to the surface will differ from that of the bulk of the bed. Allowance can be made for this when using the bulk bed temperature in Eq. (64) by replacing ϵ_b by a modified emissivity ϵ_m following Baskakov et al. [65]. A reasonable value for ϵ_m would be 0.6 to 0.7, although ϵ_r may be as low as 0.1 with a shiny metal transfer surface. Equation 64 suggests that h_r should vary as T^3.

It may be expected that the total bed-to-surface heat transfer coefficient will be of the order of 600 W/(m²·K) or ~100 Btu/(h·ft²·°F), for an immersed tube in a bed operating at a temperature of 800°C (1470°F), which is of similar magnitude to the likely steam-side coefficient for a unit raising steam at 300°C and 2 MPa (20 atmospheres pressure). Under such circumstances the tube surface temperature may approach 550 to 600°C (1020 to 1110°F), leading to an effective emissivity of ~0.6, which seems a reasonable value.

The heat transfer surface will obtain a sight deeper into the bed when the bed is bubbling vigorously. This favors the radiant component because the bed interior so exposed by bubbles passing up the transfer surface has not been influenced by the temperature of the transfer surface and the cavity will behave like a blackbody radiator. The consequent bursts of radiant energy have been measured directly by Makhorin and his colleagues [118] using a fast response probe. In this work they also made a series of measurements of apparent bed emissivity taking into account the influence of an immersed cooling surface set by the operating temperature of the cooled window of their probe. Similar measurements of apparent bed emissivity have also been made by Ozkaynak, Chen, and Frankenfield [119]. They found negligible effect of window temperature over the range 30 to 220°C (85 to 430°F) and gas fluidizing velocity on the radiative component in experiments with a bed of sand of mean particle diameter 1.03 mm operated at temperatures between 250 to 760°C (480 to 895°F). The radiant component of the coefficient, defined on the basis of the bed temperature minus the cold wall temperature, varied between 10 and 80 W/(m²·K) [2 and 4 Btu/(h·ft²·°F)] over this temperature range, being between 7 and 35 percent of the total overall bed-to-surface coefficient.

c. Heat Transfer to Plain Tubes

The insertion of tubes in a bed to increase the available heat transfer surface area affects the local fluidization behavior. This, in turn, affects the particle convective component of heat transfer, h_{pc}. When horizontal tubes are immersed within a bed, although much heat transfer surface is exposed to the cross flow of solids, which will lead to low local particle residence times and higher particle convective components of heat transfer, a stagnant layer of defluidized particles will form on top of the tube and bubbles may tend to shroud the downward-facing surface, reducing heat transfer in those regions and the average over all of the heat transfer surface. It is generally found that the region of maximum bed-to-surface coefficient moves to higher positions around the tube as the fluidizing velocity is increased and the stagnant cap of particles is increasingly disturbed.

There will be some influence of the geometrical arrangement of arrays of tubes on the overall bed behavior, on particle circulation, and hence on bed-to-surface heat transfer. Near-neighbors within the array will also affect the average heat transfer coefficient to an individual tube by already having exchanged some heat with the incident particle during its approach to the tube. However, there is evidence that tests on single tubes can be relevant to the behavior of a tube within a bank. Certainly, bed-to-tube heat transfer coefficients are comparatively insensitive to tube spacing. Thus, although very close spacing seriously affects the particle circulation and, in the extreme, the bed may rise above the tubes so that they become the effective gas distributor, McLaren and Williams [86] found no drastic reduction in heat transfer coefficients as the tube spacing was progressively reduced in low-temperature tests. Results for both in-line and staggered arrays fell on a single curve when plotted in terms of the narrowest gap between the tubes, and coefficients fell by about 25 percent as the gap reduced from 280 to 15 mm (0.92 to 0.05 ft) in tests with fluidized ash of wide size distribution and mean particle diameters between 400 and 500 μm (1.31 $\times$ 10^{-3} and 1.64 $\times$ 10^{-3} ft) but including small fractions of greater than 1.7 mm (5.6 $\times$ 10^{-3} ft) diameter (Fig. 13). McLaren and Williams commented on obtaining coefficients some 20 percent lower when the tube bundle was located

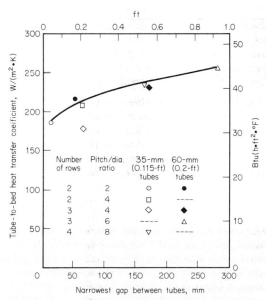

FIG. 13 Variation in heat transfer coefficient with narrowest gap between tubes [86].

close to the distributor, a region strongly influenced by distributor design and its influence on bubble development.

It can be expected that beds with inserts will be less sensitive to the overall effects of changing scale briefly mentioned above (and see, for example, Ref. 5). Indeed, the use of vertical tubular inserts has been advocated as a means of reducing the effect of scale on bed behavior, and the beneficial effect of horizontal tubes in breaking up slugs has also been reported. The final choice between horizontal or vertical tube arrangements primarily depends upon constructional convenience: how to introduce the tubes into the bed, support them, and allow for thermal expansion. There is the danger that tubes exposed within the freeboard may be particularly prone to erosion damage, although one means of obtaining turndown within a heat-generating bed is to arrange to vary the operating airflow rate in order to control the bed expansion and so vary the number of horizontal tubes immersed within the bed.

The very full review by Saxena et al. [87] presents the wide range of results that have been reported, but it should again be stressed that most of the tests have been carried out in comparatively small-scale beds, where the containing walls exert a strong influence on bed behavior. Usually the test range will be outside the relevant range of Archimedes numbers for higher-temperature operation (the Archimedes number decreases some twelvefold as the operating temperature increases from ambient to 600°C or 1110°F, and 29-fold from ambient to 1000°C or 1830°F).

The most extensive series of measurements on heat transfer to a single immersed tube was carried out by Vreedenberg [74, 88], although he worked predominantly over the rising range of the heat transfer versus gas flow rate curve, where behavior is very sensitive to variation in the fluidization condition of the bed. He found it more difficult to correlate the results to a vertical tube than to a horizontal one, which is to be expected. In the latter instance, complicated though it be, the prevailing conditions of strong vertical solids circulation past horizontal tubes is a simpler process than that obtaining at a vertical surface when local particle residence time, under the prevailing bed circulation conditions, can be expected to be a more complex function of bed operating conditions.

Typical of the general approach, Gel'perin and Einstein [80] studied heat transfer from bundles of horizontal tubes of 20 mm (6.6×10^{-2} ft) diameter in in-line and staggered arrays. The horizontal-tube pitch-to-diameter ratio was varied from 2 to 4 for staggered tubes and 2 to 9 for in-line. For vertical tubes it was varied from 1 to 2 and 1 to 10 respectively. Tests were carried out with five- and nine-row bundles. Data for particles <500 μm (1.6×10^{-3} ft) were correlated in the form

$$\mathrm{Nu}_{max} = 0.79 \, \mathrm{Ar}^{0.22} \left(1 - \frac{d_T}{X'_t} \right)^{0.25} \tag{66}$$

for in-line tube arrangements, where X'_t is the horizontal pitch, and

$$\mathrm{Nu}_{max} = 0.74 \, \mathrm{Ar}^{0.22} \left[1 - \frac{d_T}{X'_t} \left(1 + \frac{d_T}{X'_l + D_T} \right) \right]^{0.25} \tag{67}$$

for staggered tube arrays with vertical spacing X'_l. For an in-line tube arrangement, the heat transfer coefficient decreased with decreasing horizontal pitch but did not change with vertical pitch until the tubes were nearly touching. For staggered tube arrangements, however, the heat transfer coefficients were affected both by horizontal and vertical spacing by up to 10 percent. Their overall observations on the effect of intertube spacing were not dissimilar to those of McLaren and Williams [86] referred to above and shown in Fig. 13.

Chekansky et al. [81] have correlated data for staggered arrays of horizontal tubes using the ratio of the minimum spacing between the tubes to the particle diameter (X'_m/d_p) over the range of values from 16 to 63. Their correlation took the form of a modification of the Zabrodsky equation [Eq. (57) above]:

$$h_{max} = 28.2\rho_p^{0.2}k_g^{0.6}d_p^{-0.36}\left(\frac{X'_m}{d_p}\right)^{0.04}\left(\frac{d_T}{d_{20}}\right)^{-0.12} \qquad W/(m^2\cdot K) \qquad (68)$$

where d_T is the tube diameter, all parameters are in SI units, and comparison is made with a reference tube, d_{20}, of diameter 2×10^{-2} m (6.6×10^{-2} ft). This shows only small sensitivity to tube spacing and generally gives values of approximately 70 to 85 percent of those to be obtained from using the Zabrodsky equation directly. As with the latter correlation, it would be imprudent to use Eq. (68) for a mean particle size greater than 800 μm (2.6×10^{-3} ft).

With vertical tubes, overall average heat transfer coefficients generally increase with decrease in tube length and diameter and with decrease in the mean particle size [62]. With particles $<$600 μm (2×10^{-3} ft), this is consistent with the influence of particle circulation on the residence time distribution at the transfer surface. Increase in tube spacing increased the heat transfer coefficients, and higher coefficients were observed to tubes located nearer the center of the experimental beds, again reflecting the bed circulation patterns within their equipment. Gel'perin and Einstein [80] gave the correlation

$$Nu_{max} = 0.75\, Ar^{0.22}\left(1 - \frac{d_T}{X'_t}\right)^{0.14} \qquad (69)$$

for bundles of vertical tubes, where X'_t is the tube spacing. This was based on experiments with particles in the size range 160 to 350 μm (5.2×10^{-4} to 1.1×10^{-3} ft) using 20 to 40 mm (6.6×10^{-2} to 1.3×10^{-1} ft) diameter tubes and 1.25 to 5 pitch to tube diameter ratio. It can be expected that the strong dependence on particle residence time will reduce when working with larger mean particle diameters so that the average coefficient will become less dependent on tube length and diameter.

Predictions, within their recommended range of usefulness, based on some of the quoted correlations are given in Table 4 and comparison is made with experimental measurements for heat transfer to a small spherical heat transfer element [26]. Some of the tested correlations attempted to allow for the variation of heat transfer coefficient with fluidizing velocity, which, in turn, is a complex function of operating temperature with group B materials (see Sec. A.2.b above). The predictions from these were made at the fluidizing velocities to give maximum heat transfer coefficients. The last three correlations were tested for the hypothetical situation of in-line or staggered arrays of 9.5 mm (3×10^{-2} ft) diameter tubes on a 20-mm (6.6×10^{-2} ft) pitch. Generally, few correlations give reasonable predictions over all their recommended range and none adequately predicts the effect of changing the relative bed-to-surface temperature over a range of up to 950°C (1750°F). The use of the simple Eq. (57) or Eqs. (62) and (63) is therefore recommended because they will not introduce larger errors than are incurred with the more complicated correlations attempting to make allowance for the complex factors involved in the operating bed, and the user will not have exaggerated expectations of their power. Thus, a conservative estimate of heat transfer to tubes immersed within beds of group B materials would be $0.7h_{pc_{max}}$ as estimated by the Zabrodsky equation, Eq. (57), while for group D powders (because the maximum is much flatter, since interphase gas convective heat transfer is a significant factor), using between 70 and 100 percent of the sum of the simple correlation Eqs. (62) and (63) would be applicable. These factors arise because, as noted above in this section, insertion of tubes affects the local fluidization behavior;

stagnant particles on the upper surface of horizontal tubes and bubbles tending to collect below them or to track up vertical tubes reduce the effective heat transfer surface while, particularly with B-type materials where the gas thermal conductivity is the limiting heat transfer resistance, the effect of the insertion of a cooled surface on this is significant (see the preceding section and Fig. 11). Again, at temperatures above 600°C (1100°F) allowance must also be made for the contribution of the radiant component [Eq. (64)].

An understanding of the heat transfer component mechanisms is suggestive of what is required in order to obtain good bed-to-surface heat transfer, but the challenge remains to engineer the system so that it can be obtained. "Unit-cell" tests are to be advocated when the full scale incorporates a series of units—i.e., to test, at full scale, equipment which reproduces all the features of one unit of the proposed design; the area of bed fed from one fuel distribution point, for example. Lateral mixing may be very important in a system with "large" local heat sources. Because of the change in bed behavior that can occur with change in temperature, cold tests need to be interpreted with care. Thus, it is possible for a change in bed behavior to alter the predominant heat transfer mechanisms as well as directly to affect the particle convective component of heat transfer through its influence on gas thermal conductivity. However, it is to be expected that heat transfer coefficient estimates from ambient-condition tests will underestimate those obtainable at higher operating temperatures and pressures.

The necessary bed volume in order to be able to immerse a given transfer surface area will depend upon the chosen tube diameter and pitch. Higher fluidizing gas pumping costs are the price to be paid for unnecessarily deep beds.

A different overall form of bed behavior has been recognized when fluidizing large particles at high velocities and static pressures. In the so-called turbulent condition, bubbles in the bed are smaller, much more extensive lateral solids mixing occurs, and the whole behavior is much smoother than when fluidizing D-type materials close to V_{mf}. Staub and Canada [89] reported tests in such beds with 650 and 2600 μm (2.1×10^{-3} and 8.5×10^{-3} ft) particles and at pressures up to 10 atmospheres (1 MPa). Maximum heat transfer coefficients were obtained at gas velocities of ¼ to ½ the particle terminal velocity, typical coefficients being about 230 W/(m²·K), or 40 Btu/(h·ft²·°F), for 650 μm (2.1×10^{-3} ft) particles fluidized under atmospheric pressure conditions, and the coefficient varied as the 0.2 to 0.3 power of the gas density, depending on the particle size and tube geometry. Staub [90] developed a credible heat transfer model based on correlations established for flowing gas-solid suspension. It uses gas and solid superficial velocities pertinent to the upflow and downflow situations as predicted within his flow model. The simplified form predicts the ratio of Nusselt numbers (based on the tube diameter) for the conditions with bed present, Nu, and without particles, Nu_g, in the form

$$\frac{\text{Nu}}{\text{Nu}_g} = \left\{ 1 + \left(\frac{150}{d_p \times 10^{-6}} \right)^{0.73} \left[\frac{0.42 \rho_p (1 - \alpha_{av}) l_m^{0.4}}{\rho_g V} \right] \right\}^{0.45} \tag{70}$$

for 20 μm $< d_p <$ 1000 μm (6.6×10^{-5} ft $< d_p < 3.3 \times 10^{-3}$ ft); for 1000 μm $< d_p$ < 3000 μm (3.3×10^{-3} ft $< d_p < 9.8 \times 10^{-3}$ ft), the value $d_p = 10^{-3}$ m should be used in Eq. (70) (all units being SI). l_m in the expression is a mixing length in meters, which is the average distance between tube centers in this model. The average bed holdup $(1 - \alpha_{av})$ is taken from bed expansion data or the correlation data given by Staub and Canada [89]. The model has been extended to heat transfer tubes in the splash zone of a fluidized bed of large particles [91].

d. Finned Tubes

When the principal resistance to heat flow is on the fluidized bed side the available transfer surface may be increased by using finned tubes [76]. Coefficients have been measured for both single horizontally and vertically positioned finned tubes and for ones within arrays. Although the heat transfer is controlled basically by the same parameters as with bare-tube heat transfer in deep bed applications, coefficients based on the total exposed surface area are reduced over those obtainable per unit area of bare tube. Thus, about a 30 percent reduction was observed after allowance had been made for fin efficiency [89]. Overall effectiveness will depend on fin configuration, the relative size of fin spacing, and the tube arrangement itself. Total heat transfer to a single horizontal discontinuous finned tube was observed to increase with fin height until, in addition to the expected fall in the fin efficiency due to the limited thermal conductivity of the tube material, reduction in the degree of particle penetration between the fins caused significant changes in the local heat transfer coefficients from the base of the fin to the tip. This was especially important with steel tubes. At low superficial velocities, highest heat transfer rates were obtained for triangular transverse fin profiles, while at the higher flow rates, for parabolic fins (Gel'perin et al. [92]). Rectangular fins gave the poorest performance. The fin effectiveness generally increases with superficial velocity but decreases with increasing particle size. With serrated transverse finned tubes, fin spacing had little effect on the heat transfer coefficient provided that it was greater than 10 particle diameters [93]. Continuous fins hinder overall particle movement. Generally, however, although decrease in fin spacing reduces the fin effectiveness, it is more than compensated by the increased exposed area to give actual transfer rates of 3 to 5 times those obtainable with bare tubes [76, 89]. Bartel and Genetti [94] observed that the heat transfer rate was sensitive to tube spacing for a bundle of seven serrated finned tubes with short fins (height < 22 mm or 7.2×10^{-2} ft) at up to about 50 mm (0.16 ft) between the tube centers. For tubes with long fins the heat transfer rate became independent of center-to-center tube distance as the resistance to particle penetration toward the center became the principal resistance.

Vertically positioned finned tubes behave similarly. Chen and Withers [95] obtained heat transfer coefficients on the basis of the bare tube in the range of 0.5 to 1.2 times the coefficients for plain tubes operating under similar fluidization conditions. Higher heat transfer rates were obtained with smaller particles (130 μm $< d_p < 610$ μm or 4.3×10^{-4} ft $< d_p < 2.0 \times 10^{-3}$ ft) and the selection of optimum fin geometry was found to be dependent on particle size, fin height, fin gap, and gas flow rate. Longitudinal fin geometry could be used for vertically positioned tubes in deep beds, but this may further encourage slug formation and reduce radial particle mixing.

In investigating the effect of transfer surface roughness on bed-to-surface heat transfer coefficient, it was found that the heat transfer rates, compared to those obtained with smooth surfaces, showed first a decrease with increasing average roughness height to particle diameter ratio because the effective gas gap between the surface and the particle and the retarding effect of the irregularities on particle residence times both increased (Korolev and Syromyatnikov [96] and Grewal and Saxena [97]). However, when the roughness height to particle diameter ratio increased beyond about 1, higher heat transfer coefficients based on the smooth tube area were observed.

Very efficient shallow-bed waste heat recovery units have been developed [98] using horizontal finned-tube heat transfer surfaces close to the distributor. In this application, the regions between the fins behave virtually as separate, dilute fluidized beds. Bubbles remain restricted in size by the closeness of fin spacing, which may be as small as 15 to

20 times the particle diameter; below this the particle circulation may be impaired. In operation, bed expansion as high as 400 percent is usual while still obtaining high bed to surface heat transfer coefficients, as good as those for a bare tube, because of the very short particle residence times at the heat transfer surface; particle velocities are possibly an order of magnitude higher than in a deep bed and this more than compensates for the greatly reduced bed density. Based on the outside surface area of the tube, obtainable coefficients are in the range of 1 to 4 kW/(m²·K), about 175 to 700 Btu/(h·ft²·°F). Because the particles in the bed present a very high surface area to the gas, very good heat recovery can be achieved from a hot fluidizing bed.

Properly designed finned tubes would also seem to offer advantages in turbulent fluidized beds [89]. However, data are needed to decide whether or not the erosion and corrosion effects on fins would be a limiting factor in such applications.

3. Liquid Fluidized Beds

The heat transfer mechanisms of liquid fluidized beds differ from gas fluidized beds in two important respects: (1) liquid fluidized beds are generally particulate (nonbubbling), and (2) the liquid can transfer about a thousand times more heat than gas at atmospheric pressure for the same temperature difference. As a result liquid convection predominates in heat removal rather than particle convection, which dominates with all but Geldart's group D gas fluidized beds. The presence of solids in the liquid stream can enhance surface heat transfer coefficients up to a factor of 8 in that they erode the laminar boundary layer. The solid particles have a significant effect although not as dramatic as the 50- to 100-fold increase for gas systems. As a result heat transfer data for gas fluidized beds have been of more interest, and heat transfer studies on liquid fluidized beds are relatively limited compared to those for gas fluidized beds. However, developing technology is increasing the applications of liquid fluidized beds.

a. Particle-to-Fluid Heat Transfer

Holman et al. [99] measured particle-to-fluid heat transfer coefficients for stainless steel and lead spherical particles fluidized in water and heated by radio-frequency induction. They correlated their data and data for air fluidized beds by:

$$\mathrm{Nu}_p = 0.33 \times 10^{-6}\, \mathrm{Re}_p^{1.75}\, \mathrm{Pr}^{0.67}\, \frac{D_T}{d_p}\, \frac{\rho_p}{\rho_f} \tag{71}$$

b. Wall-to-Bed Heat Transfer

The limiting film thickness restricting heat transfer reduces as the liquid velocity increases, so wall-to-bed coefficients increase with increase in liquid flow rate and hence bed voidage. They also increase with increase in the size of the particles at a given voidage because of the consequential increase in liquid velocity.

Wasmund and Smith [100] obtained data for heat transfer from the wall of the container to the fluidized bed. Brea and Hamilton [101] and Richardson et al. [102] obtained heat transfer data for an internal heater and correlated the data of Wasmund and Smith in the same form as their data. These are the resulting correlations for heat transfer from the column wall [102]:

$$\text{Nu} = 3.38 \ \text{Re}^{0.565} \ \text{Pr}^{0.33} \ \alpha^{0.565n}(1 - \alpha)^{0.435} \left(\frac{d_p}{D_T}\right)^{0.57} \tag{72}$$

and from an internal heater:

$$\text{Nu} = 0.67 \ \text{Re}^{0.62} \ \text{Pr}^{0.33} \ \alpha^{0.62n-1}(1 - \alpha)^{0.38} \frac{d_p}{D_T} \tag{73}$$

where n is the exponent of α in the bed expansion correlation [Eq. (11), Sec. A.3.a].

c. Effective Thermal Conductivity

Karpenko et al. [103] measured effective radial thermal conductivities for glass and Alundum particles fluidized by water with concentrations of glycerol varying from 0 to 70 percent by weight. They obtained the following correlation:

$$k_e = k_{max}5.05\left(\frac{\text{Re}}{\text{Re}_{opt}} - 0.25\right)e^{-1.33\text{Re}/\text{Re}_{opt}} \tag{74}$$

where $k_{max} = k_f 89.4 \ \text{Ar}^{0.2}$ (maximum effective radial thermal conductivity) and $\text{Re}_{opt} = 0.1 \ \text{Ar}^{0.66}$ (optimum Reynolds number).

C. FLUIDIZED-PACKED BEDS

Fluidization of small particles in the voids of larger fixed packing can provide certain advantages over operation with either a packed bed or a fluidized bed. Small inert fluidized particles improve the heat transfer from the surface of large particles either undergoing reaction or acting as catalysts which cannot, themselves, be fluidized. The packing, either solid or open mesh, can also be used as a baffle to reduce slugging and disperse bubbles [104]. The transport properties of this system not only depend on the fluidized particle and fluidizing gas properties but are also controlled by the size and shape of the fixed packing. Lateral mixing is directly dependent on the size of the fixed packing [105]. The minimum fluidization velocity in a fluidized-packed bed, V_{mfp}, is altered by the presence of the packing.* Sutherland et al. [104] compared V_{mfp} to the minimum fluidization velocity without packing, V_{mf}. For smooth particles fluidized in the voids of a spherical packing they obtained the following empirical correlation in English units:

$$V_{mfp} = \left(1 + \frac{0.0141}{D_P}\right)V_{mf} \qquad \text{ft/s} \tag{75}$$

For angular particles fluidized in a spherical packing

$$V_{mfp} = \left(1.15 + \frac{0.0333}{D_P}\right)V_{mf} \qquad \text{ft/s} \tag{76}$$

There are relatively few references to combined fluidized-packed bed systems compared to the extensive literature on either fluidized or packed beds by themselves. As a

---

*In this section on fluidized-packed beds the velocity V_{mfp} is based on the voidage of the fixed packing, that is, $V_{mfp} = V_0/\alpha$.

result, heat transfer studies on fluidized-packed beds have been primarily experimental, and the data for this system will be presented in an empirical manner.

1. Packing-to-Bed Heat Transfer

Baskakov and Vershinina [106] and Gabor [107] reported data for heat transfer from the surfaces of heated particles which make up the fixed packing of fluidized-packed beds. These heated particles were identical in size and shape to the unheated particles making up the rest of the packing in the beds and were distributed at various locations. Baskskov and Vershinina [106] also obtained data for heat transfer from fixed particles surrounded by other heated particles. These data were lower than for heated particles surrounded by unheated particles. The reasons for this difference are similar to those discussed in Sec. B.1.a of why particle-to-bed Nusselt numbers are less than 2.0 (the conduction limit) in packed beds. Since most chemical reaction processes would result in heat generation by all of the fixed particles, the data of Baskakov and Vershinina [106] for particles surrounded by heated particles are most significant and are presented on Fig. 14.

2. Wall-to-Bed Heat Transfer

Heat transfer coefficients for wall-to-bed heat transfer are shown on Figs. 15 and 16 for 3.9×10^{-4} ft (119 μm) glass beads fluidized in various types of packing. The data are

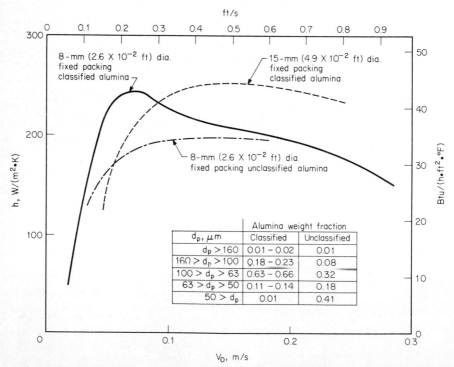

d_p, μm	Alumina weight fraction	
	Classified	Unclassified
$d_p > 160$	0.01 – 0.02	0.01
$160 > d_p > 100$	0.18 – 0.23	0.08
$100 > d_p > 63$	0.63 – 0.66	0.32
$63 > d_p > 50$	0.11 – 0.14	0.18
$50 > d_p$	0.01	0.41

FIG. 14 Packing-to-bed heat transfer for fluidized-packed beds (Baskakov and Vershinina [106]).

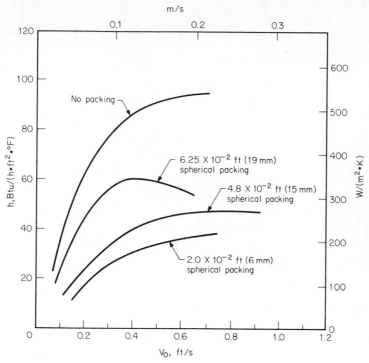

FIG. 15 Wall-to-bed heat transfer for fluidized-packed beds with spherical packing and 3.9×10^{-4} ft (119 μm) glass beads (Sutherland et al. [104]).

those reported by Sutherland et al. [104] for a 0.333-ft (0.1-m) diameter column and by Gabor [107] for a 0.24-ft (7.3×10^{-2} m) diameter column. Subsequent experiments by Capes et al. [108] with a 1-ft (0.3-m) diameter column indicated that the heat transfer is little affected by column diameter. Particle movement and bubble size are controlled by the size and shape of the fixed packing. The packing prevents the formation of large bubbles and reduces the rate of particle circulation. Wall-to-bed coefficients for fluidized-packed beds are therefore lower than for unobstructed fluidized beds.

3. Effective Thermal Conductivity

a. Lateral Direction

In fluidized-packed beds, the major contributor to heat transfer through the bed is the mixing of the fluidized particles. As with unobstructed beds, it is the rising bubbles which carry streams of particles upward in their wakes (Rowe [109]).

In fluidized-packed beds the streams of particles are laterally deflected as they impinge upon the fixed packing. This is similar to the mechanism of lateral gas mixing in a packed bed (Ranz [21] and Baron [110]). The rate of particle mixing depends both on the distance of lateral deflection and on the particle velocity. It is assumed that, to a first approximation, gas flow in excess of that required for minimum fluidization, $V - V_{mfp}$, forms

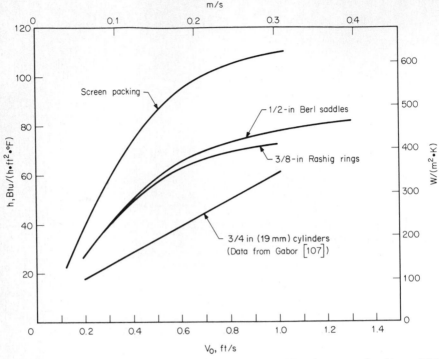

FIG. 16 Wall-to-bed heat transfer for fluidized-packed beds with various types of packing and 3.9 $\times 10^{-4}$ ft (119 μm) glass beads (Sutherland et al. [104]).

bubbles and the amount of particles moved in the bubble wake is a function of the drag force on each particle divided by the weight of the particle (Gabor [105]). Rowe [109] correlated the minimum fluidization velocity in the same manner as the terminal velocity for a single sphere by correcting the drag coefficient for a single sphere in an infinite fluid medium by a factor of 68.5. It follows that the amount of particles carried in the bubble wake is inversely proportional to the minimum fluidization velocity resulting in the following correlation for the diffusivity for lateral particle movement in a fluidized-packed bed (Gabor [105]):

$$D = (3.35 \times 10^{-3})D_p \, \frac{V - V_{mfp}}{V_{mfp}} \qquad \text{ft}^2/\text{s} \; (D_p \text{ in ft})$$

$$\text{or} \qquad D = (1.02 \times 10^{-3})D_p \, \frac{V - V_{mfp}}{V_{mfp}} \qquad \text{m}^2/\text{s} \; (D_p \text{ in m})$$

(77)

However, the thermal diffusivities $\alpha_t \equiv (k_e/\rho_{BC_p})$ are about ⅔ that of the particle mixing diffusivities (Gabor et al. [111]) because the transfer of sensible heat from the fluidized particles is not instantaneous as they move from point to point within the bed. Gabor [111] obtained a reasonable correlation based on a modified thermal diffusivity $\alpha_t' \equiv (k_e/\rho_p c_p)$.

The modified thermal diffusivity is based on the fluidized particle density ρ_p, which

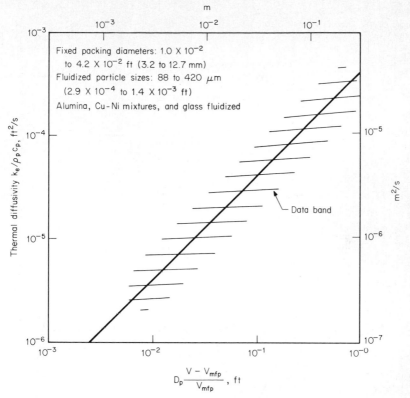

FIG. 17 Thermal diffusivities for lateral heat transfer in fluidized-packed beds.

is more readily obtained than the fluidized bed density ρ_B, the latter being difficult to estimate without direct measurement (although it is not easy to measure ρ_p for porous particles). This correlation is:

$$\alpha_t' \equiv \frac{k_e}{\rho_p c_p} = (4.14 \times 10^{-4})D_p \frac{V - V_{mfp}}{V_{mfp}} \qquad \text{ft}^2/\text{s} \ (D_p \text{ in ft})$$

$$\text{or} \qquad \alpha_t' = (1.17 \times 10^{-5})D_p \frac{V - V_{mfp}}{V_{mfp}} \qquad \text{m}^2/\text{s} \ (D_p \text{ in m})$$

(78)

The data band for various materials is shown on Fig. 17. If bed expansion (or density) data are available, a better estimate of k_e would be obtained from a correlation of α based on ρ_B:

$$\alpha_t \equiv \frac{k_e}{\rho_B c_p} = (2.23 \times 10^{-3})D_p \frac{V - V_{mfp}}{V_{mfp}} \qquad \text{ft}^2/\text{s} \ (D_p \text{ in ft})$$

$$\text{or} \qquad \alpha_t = (6.31 \times 10^{-5})D_p \frac{V - V_{mfp}}{V_{mfp}} \qquad \text{m}^2/\text{s} \ (D_p \text{ in m})$$

(79)

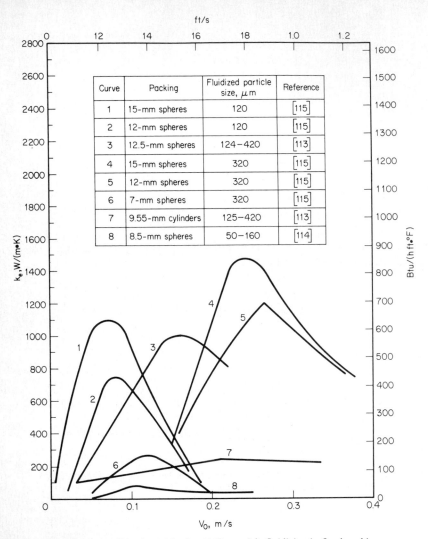

FIG. 18 Axial thermal conductivities for Al_2O_3 particle fluidizing in fixed packing.

b. Axial Direction

Particle movement in the axial (vertical) direction is an order of magnitude greater than in the lateral direction (Gabor [112]). The bubble movement is upward and therefore the particle movement is primarily in the vertical direction. The rate of axial particle mixing depends on the complicated interaction of bubbles, packing geometry, and fluidized particle properties. Basically it depends on the duration of the particle in the bubble wake. As a result, axial solids mixing and therefore thermal conductivity cannot be correlated as simply as those in the lateral direction. Data on axial thermal conductivities for heat transport by the fluidized particulate phase only by Gabor and Mecham [113], Baskakov

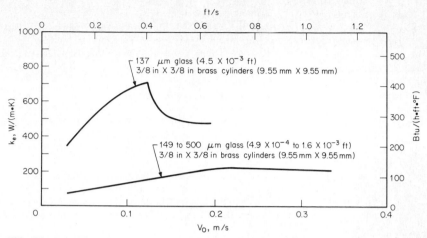

FIG. 19 Axial thermal conductivities for glass particles fluidizing in fixed packing (Gabor and Mecham [113]).

et al. [114], and Pakhaluyev [115] for various systems are shown on Figs. 18 and 19. For heat flow in the direction of the fluidizing gas flow the heat transported by the fluidizing gas should be added to that transported by the particulate phase.

REFERENCES

1. M. A. Bergougnou, J. S. M. Botterill, J. R. Howard, D. C. Newey, A. G. Salway, and Y. Teoman, High Temperature Heat Storage, in *Future Energy Concepts,* IEEE Conference Publication No. 192, pp. 61–64, London, 1981.

2. J. R. Howard, ed., *Thermal Application of Fluidized Bed Technology,* Applied Science Publishers, London, 1983.

3. J. S. M. Botterill, *Fluid-Bed Heat Transfer,* Academic Press, London, 1975.

4. D. Geldart, Types of Gas Fluidization, *Powder Technol.,* Vol. 7, pp. 285–292, 1973.

5. J. H. de Groot, Scaling-Up of Gas-Fluidized Bed Reactors, in A. A. H. Drinkenburg, ed., *Proceedings of the International Symposium on Fluidization,* pp. 348–358, Netherlands University Press, Amsterdam, 1967.

6. T. Allen, *Particle Size Measurement,* Chapman and Hall, London, 1975.

7. J. M. Dallavalle, *Micromeritics,* 2d ed., Pitman, New York, 1948.

8. S. S. Zabrodsky, *Hydrodynamics and Heat Transfer in Fluidized Beds,* MIT Press, Cambridge, Mass., 1966.

9. D. Kunii and O. Levenspiel, *Fluidization Engineering,* Wiley, New York, 1969.

10. R. A. Bernard and R. H. Wilhelm, Turbulent Diffusion in Fixed Beds of Packed Solids, *Chem. Eng. Prog.,* Vol. 46, pp. 233–244, 1950.

11. P. C. Carman, *J. Soc. Chem. Ind.,* Vol. 57, p. 225, 1938.

12. M. Leva, M. Grummer, M. Weintraub, and M. Pollchik, Introduction to Fluidization, *Chem. Eng. Prog.,* Vol. 44, pp. 511–520, 1948.

13. M. Leva, M. Grummer, M. Weintraub, and M. Pollchik, Fluidization of Anthracite Coal, *Ind. Eng. Chem.,* Vol. 41, pp. 1206–1213, 1949.

14. S. Uchida and S. Fujita, *J. Chem. Soc. Jpn. (Ind. Eng. Sec.),* Vol. 37, pp. 1578, 1583, 1589, and 1707, 1934.

15. T. Shirai, Ph.D thesis, Tokyo Institute of Technology, 1954.

16. A. R. Abrahamsen and D. Geldart, Behavior of Gas-Fluidized Beds of Fine Powders, Part I: Homogeneous Expansion, *Powder Technol.,* Vol. 26, pp. 35–46, 1980.

17. S. Ergun, Fluid Flow through Packed Columns, *Chem. Eng. Prog.,* Vol. 48, pp. 89–94, 1952.

18. L. E. Brownell and D. L. Katz, Flow of Fluids through Porous Media, 1: Single Homogeneous Fluids, *Chem. Eng. Prog.,* Vol. 43, pp. 537–548, 1947.

19. L. W. Brownell and D. L. Katz, Flow of Fluids through Porous Media, 2: Simultaneous Flow of Two Homogeneous Phases, *Chem. Eng. Prog.,* Vol. 43, pp. 601–612, 1947.

20. J. M. Coulson, The Flow of Fluids through Granular Beds: Effect of Particle Shape and Voids in Streamline Flow, *Trans. Inst. Chem. Eng.,* Vol. 27, pp. 237–257, 1949.

21. W. E. Ranz, Friction and Transfer Coefficients for Single Particles and Packed Beds, *Chem. Eng. Prog.,* Vol. 48, pp. 247–253, 1952.

22. A. E. Scheidegger, *The Physics of Flow through Porous Media,* 3d ed., University of Toronto Press, Toronto, 1974.

23. J. S. M. Botterill, Y. Teoman, and K. R. Yüregir, The Effect of Operating Temperature on the Velocity of Minimum Fluidization, Bed Voidage and General Behaviour, *Powder Technol.,* Vol. 31, pp. 101–116, 1982.

24. G. H. Hong, R. Yamazaki, T. Takahashi, and G. Jimbo, Minimum Fluidization Velocity of a Fluidized Bed at High Temperatures, *Kagaku Kogaku Ronbunshu,* Vol. 6, pp. 557–562, 1980.

25. V. D. Goroshko, R. B. Rozenbaum, and O. H. Todes, Approximate Hydraulic Relationships for Suspended Beds and Hindered Fall, *Izv. Vuzov Neft' i Gaz,* Vol. 1, pp. 125–131, 1958.

26. J. S. M. Botterill, Y. Teoman, and K. R. Yüregir, Temperature Effects on the Heat Transfer Behavior of Fluidized Beds, *AIChE Symp. Ser.,* Vol. 77, No. 208, pp. 330–340, 1981.

27. J. Baeyens and D. Geldart, Predictive Calculations of Flow Parameters in Gas Fluidized Beds and Fluidization Behaviour of Various Powders, *Fluidization et ses Applications,* pp. 263–273, Société chimie industrielle, Toulouse, 1973.

28. C. Y. Wen and Y. H. Yu, Mechanics of Fluidization, *Chem. Eng. Prog. Symp. Ser.,* Vol. 62, No. 62, pp. 100–111, 1966.

29. J. B. Romero and L. N. Johanson, Factors Affecting Fluidized Bed Quality, *Chem. Eng. Prog. Symp Ser.,* Vol. 58, No. 38, pp. 28–37, 1962.

30. J. F. Richardson and W. N. Zaki, Sedimentation and Fluidization, Part I, *Trans. Inst. Chem. Engrs.,* Vol. 32, pp. 35–52, 1954.

31. D. Geldart, Predicting the Expansion of Gas Fluidized Beds, in D. L. Keairns, ed., *Fluidization Technology,* Vol. I, pp. 237–244, Hemisphere, Washington, D.C., 1976.

32. A. R. Abrahamsen and D. Geldart, Behavior of Gas-Fluidized Beds of Fine Powders, Part 2: Voidage of the Dense Phase in Bubbling Beds, *Powder Technol.,* Vol. 26, pp. 47–55, 1980.

33. J. R. F. Guedes de Carvalho, Dense Phase Expansion in Fluidized Beds of Fine Particles: The Effect of Pressure on Bubble Stability, *Chem. Eng. Sci.,* Vol. 36, pp. 413–416, 1981.

34. S. W. Churchill and R. Usagi, A General Expression for the Correlation of Rates of Transfer and Other Phenomena, *AIChE J.,* Vol. 18, No. 6, pp. 1121–1128, 1972.

35. W. E. Ranz and W. R. Marshall, Jr., Evaporation from Drops, Part II, *Chem. Eng. Prog.,* Vol. 48, No. 4, pp. 173–180, 1952.

36. A. R. H. Cornish, Note on Minimum Possible Rate of Heat Transfer from a Sphere when Other Spheres Are Adjacent to It, *Trans. Inst. Chem. Eng.,* Vol 43, pp. T332–T333, 1965.

37. D. Kunii and M. Suzuki, Particle-to-Fluid Heat and Mass Transfer in Packed Beds of Fine Particles, *Int. J. Heat Mass Transfer,* Vol. 10, pp. 845–852, 1967.

38. L. R. Glicksman and F. M. Joos, Heat and Mass Transfer in Fixed Beds at Low Reynolds Numbers, *J. Heat Transfer,* Vol. 102, pp. 736–751, 1980.

39. J. D. Gabor, Wall-to-Bed Heat Transfer in Fluidized and Packed Beds, *Chem. Eng. Prog. Symp. Ser.,* Vol. 66, No. 105, pp. 76–85, 1970.

40. D. Kunii and J. M. Smith, Heat Transfer Characteristics of Porous Rocks, *AIChE J.,* Vol. 6, No. 1, pp. 71–78, 1960.

41. D. L. Swift, The Thermal Conductivity of Spherical Metal Powders Including the Effect of an Oxide Coating, *Int. J. Heat Mass Transfer,* Vol. 9, pp. 1061–1074, 1966.

42. M. Kobayashi, Pulsed-Bed Approach to Fluidization, Argonne National Laboratory, *ANL-7592,* Argonne, Ill., October 1969.

43. A. P. Baskakov, The Mechanism of Heat Transfer between a Fluidized Bed and a Surface, *Int. Chem. Eng.,* Vol. 4, No. 2, pp. 320–324, 1964.

44. J. Shimokawa, Deionization Processes with Ion Exchange and with Electromigration, Japan Atomic Energy Research Institute, *JAERI 1038,* June 1962.

45. S. Yagi, D. Kunii, and N. Wakao, Radially Effective Thermal Conductivities in Packed Beds, 1961 International Heat Transfer Conference, *International Developments in Heat Transfer,* Part IV, pp. 742–749, 1961.

46. C. S. Eian and R. G. Deissler, Effective Thermal Conductivities of MgO, Stainless Steel and Uranium Oxide Powders in Various Gases, National Advisory Committee for Aeronautics, *NACA RM E53 G03,* 1953.

47. T. M. Kuzay, Effective Thermal Conductivity of Porous Solid-Gas Mixtures, *ASME Winter Ann. Mtg., Paper 80-WA/HT-63,* Chicago, 1980.

48. H. W. Godbee and W. T. Ziegler, Thermal Conductivities of MgO, Al_2O_3, and ZrO_2 Powders to 850°C, II: Theoretical, *J. Appl. Phys.,* Vol. 37, pp. 56–65, 1966.

49. R. Bauer and E. V. Schlünder, Effective Radial Thermal Conductivity of Packings in Gas Flow, Part I: Convective Transport Coefficient, *Int. Chem. Eng.,* Vol. 18, pp. 181–188, 1978.

50. R. Bauer and E. V. Schlünder, Effective Radial Thermal Conductivity of Packings in Gas Flow, Part II: Thermal Conductivity of the Packing Fraction without Gas Flow, *Int. Chem. Eng.,* Vol. 18, pp. 189–204, 1978.

51. H. Kampf and G. Karsten, Effects of Different Types of Void Volumes on the Radial Temperature Distribution of Fuel Pins, *Nucl. Appl. Tech.,* Vol. 9, pp. 288–300, 1970.

52. S. Yagi and D. Kunii, Studies on Effective Thermal Conductivities in Packed Beds, *AIChE J.,* Vol. 3, pp. 373–381, 1957.

53. S. Yagi, D. Kunii, and N. Wakao, Studies on Axial Effective Thermal Conductivities in Packed Beds, *AIChE J.,* Vol. 6, pp. 543–546, 1960.

54. T. J. Hanratty, Nature of Wall Heat Transfer in Packed Beds, *Chem. Eng. Sci.,* Vol. 8, pp. 204–214, 1954.

55. S. Yagi and D. Kunii, Studies on Heat Transfer near Wall Surface in Packed Beds, *AIChE J.,* Vol. 6, pp. 97–104, 1960.

56. S. Yagi and D. Kunii, Studies on Heat Transfer in Packed Bed, 1961 International Heat Transfer Conference, *International Developments in Heat Transfer,* Part IV, pp. 750–759, 1961.

57. N. Wakao and K. Kato, Effective Thermal Conductivity of Packed Beds, *J. Chem. Eng. Jpn.,* Vol. 2, pp. 24–33, 1969.

58. S. Whitaker, Radiant Energy Transport in Porous Media, *18th Nat. Heat Transfer Conf., ASME/AIChE Paper 79-HT-1,* San Diego, 1979.

59. W. Schotte, Thermal Conductivity of Packed Beds, *AIChE J.,* Vol. 6, pp. 63–67, 1960.

60. R. F. Baddour and C. Y. Yoon, Local Radial Effective Conductivity and the Wall Effect in Packed Beds, *Chem. Eng. Prog. Symp. Ser.,* Vol. 57, No. 32, pp. 35–50, 1961.

61. A. N. Singh and J. R. Ferron, Particle Temperature Measurement in a Gas-Solid Fluidized Bed, *Chem. Eng. J.,* Vol. 15, pp. 169–177, 1978.

62. S. C. Saxena and J. D. Gabor, Mechanisms of Heat Transfer between a Surface and a Gas-Fluidized Bed for Combustor Applications, *Prog. Energy Combust. Sci.,* Vol. 7, pp. 73–102, 1981.

63. C. van Heerden, Some Fundamental Characteristics of the Fluidized State, *J. Appl. Chem.,* Vol. 2, Suppl. Issue No. 1, pp. S7–S17, 1952.

64. A. R. Khan, J. F. Richardson, and K. J. Shakiri, Heat Transfer between a Fluidized Bed and Small Immersed Surface, in J. F. Davidson and D. L. Keairns, eds., *Fluidization,* pp. 351–356, Cambridge, 1978.

65. A. P. Baskakov, B. V. Berg, O. K. Vitt, N. F. Filippovsky, V. A. Kirakosyan, J. M. Goldobin, and V. K. Maskaev, Heat Transfer to Objects Immersed in Fluidized Beds, *Powder Technol.*, Vol. 8, pp. 273–282, 1973.

66. H. S. Mickley and D. F. Fairbanks, Mechanisms of Heat Transfer to Fluidized Beds, *AIChE J.*, Vol. 1, pp. 374–384, 1955.

67. S. S. Zabrodsky, N. V. Antonishin, and A. L. Parnas, On Fluidized Bed-to-Surface Heat Transfer, *Can. J. Chem. Eng.*, Vol. 54, pp. 52–58, 1976.

68. A. O. O. Denloye and J. S. M. Botterill, Bed to Surface Heat Transfer in a Fluidized Bed of Large Particles, *Powder Technol.*, Vol. 19, pp. 197–203, 1978.

69. N. S. Grewall and S. C. Saxena, Heat Transfer between a Horizontal Tube and a Gas-Solid Fluidized Bed, *Int. J. Heat Mass Transfer,* Vol. 23, pp. 1505–1519, 1980.

70. S. S. Zabrodsky, N. V. Antonishin, G. M. Vasiliev, and A. L. Parnas, *Vesti Akad. Nauk. BSSR, Ser. Fiz. Energy Nauk,* No. 4, p. 103, 1974.

71. N. N. Varygin and I. G. Martyushin, *Khim. Mash.,* No. 5, p. 6, 1959.

72. V. K. Maskaev and A. P. Baskakov, External Heat Transfer Peculiarities in a Coarse-Particle Fluidized Bed, *J. Eng. Phys.,* Vol. 29, pp. 589–593, 1973.

73. V. B. Sarkits, candidate dissertation, Leningrad, 1959.

74. H. A. Vreedenberg, Heat Transfer between a Fluidized Bed and a Horizontal Tube, *Chem. Eng. Sci.,* Vol. 9, pp. 52–60, 1958.

75. B. R. Andeen and L. R. Glicksman, Heat Transfer to Horizontal Tubes in Shallow Fluidized Beds, *ASME/AIChE Heat Transfer Conf.,* Paper No. *76-HT-67,* St. Louis, August 1976.

76. J. C. Petrie, W. A. Freeby, and J. A. Buckman, In-Bed Heat Exchangers, *Chem. Eng. Prog.,* Vol. 64, No. 7, pp. 45–51, July 1968.

77. V. G. Einstein, An Investigation of Heat Transfer Process between Fluidized Beds and Single Tubes Submerged in the Bed, in S. S. Zabrodsky, *Hydrodynamics and Heat Transfer in Fluidized Beds,* pp. 270–272, MIT Press, Cambridge, Mass., 1966.

78. N. I. Gel'perin, V. Ya. Kruglikov, and V. G. Einstein, in Heat Transfer between Fluidized Bed and a Surface by V. G. Einstein and N. I. Gel'perin, *Int. Chem. Eng.,* Vol. 6, pp. 67–73, 1966.

79. A. N. Ternovskaya and Yu. G. Korenberg, *Pyrite Kilning in a Fluidized Bed,* Izd. Khimiya, Moscow, 1971.

80. N. I. Gel'perin and V. G. Einstein, Heat Transfer in Fluidized Beds, Chap. 10 in J. F. Davidson and D. Harrison, eds., *Fluidization,* pp. 471–540, Academic Press, London, 1971.

81. V. V. Chekansky, B. S. Sheinadlin, D. M. Galershtein, and K. S. Antonyuk, *Tr. Spec. Kostr., Byro Automat. Neftepeierab Neftekhim,* Vol. 3, p. 143, 1970.

82. O. M. Todes, *Applications of Fluidized Beds in the Chemical Industry,* Part II, pp. 4–27, Izd. Znanie, Leningrad, 1965.

83. A. P. Baskakov and V. M. Suprun, Determination of the Convective Component of the Heat Transfer Coefficient to a Gas in a Fluidized Bed, *Int. Chem. Eng.,* Vol. 12, pp. 324–326, 1972.

84. J. D. Gabor, Heat Transfer to Particle Beds with Gas Flows Less than or Equal to That Required for Incipient Fluidization, *Chem. Eng. Sci.,* Vol. 25, pp. 979–984, 1970.

85. L. P. Golan, D. C. Cherrington, R. Diener, G. E. Scarborough, and S. C. Wiener, Particle Size Effects in Fluidized Bed Combustion, *Chem. Eng. Prog.,* Vol. 75, pp. 63–72, July 1979.

86. J. McLaren and D. F. Williams, Combustion Efficiency, Sulphur Retention and Heat Transfer in Pilot-Plant Fluidized-Bed Combustors, *J. Inst. Fuel,* Vol. 42, pp. 303–308, 1969.

87. S. C. Saxena, N. S. Grewal, J. D. Gabor, S. S. Zabrodsky, and D. M. Gallershtein, Heat Transfer between a Gas Fluidized Bed and Immersed Tubes, T. F. Irvine and J. P. Hartnett, eds., *Advances in Heat Transfer,* Vol. 14, pp. 149–247, Academic, New York 1978.

88. H. A. Vreedenberg, Heat Transfer between a Fluidized Bed and a Vertical Tube, *Chem. Eng. Sci.,* Vol. 11, pp. 274–285, 1960.

89. F. W. Staub and G. S. Canada, Effect of Tube Bank and Gas Density on Flow Behavior and Heat Transfer in Fluidized Beds, in J. F. Davidson and D. Keairns, eds., *Fluidization—Proceedings of the Second Engineering Foundation Conference,* pp. 339–344, Cambridge, 1978.

90. F. W. Staub, Solids Circulation in Turbulent Fluidized Beds and Heat Transfer to Immersed Tube Banks, *J. Heat Transfer,* Vol. 101, pp. 391–396, 1979.

91. R. T. Wood, M. Kuwata, and F. W. Staub, Heat Transfer to Horizontal Tube Banks in the Splash Zone of a Fluidized Bed of Large Particles, in J. R. Grace and J. M. Matsen, eds., *Proceedings of the 1980 Fluidization Conference,* pp. 235–242, Plenum, New York, 1980.

92. N. I. Gel'perin, V. G. Einstein, I. N. Toskabaev, S. K. Vasiliev, and N. S. Ryspaeu, *Tr. TITCh. Vyp.,* Vol. 1, p. 165, 1974.

93. S. J. Priebe and W. E. Genetti, Heat Transfer from a Horizontal Bundle of Extended Surface Tubes to an Air Fluidized Bed, *AIChE Symp. Ser.,* Vol. 73, No. 161, pp. 38–43, 1977.

94. W. J. Bartel and W. E. Genetti, Heat Transfer from a Horizontal Discontinuous Finned Tube in a Fluidized Bed, *Chem. Eng. Prog. Symp. Ser.,* Vol. 69, No. 128, pp. 85–92, 1973.

95. J. C. Chen and J. G. Withers, An Experimental Study of Heat Transfer from Plain and Finned Tubes in Fluidized Beds, *AIChE Symp. Ser.,* Vol. 74, No. 174, pp. 327–333, 1978.

96. V. N. Korolev and N. I. Syromyatnikov, Heat Transfer from a Surface with Artificial Roughness to a Fluidized Bed, *J. Eng. Phys.,* Vol. 28, pp. 977–980, 1975.

97. N. S. Grewal and S. C. Saxena, Effect of Surface Roughness on Heat Transfer from Horizontal Immersed Tubes in a Fluidized Bed, *J. Heat Transfer,* Vol. 101, pp. 397–403, 1979.

98. M. J. Virr, Commercialisation of Small Scale Fluidized Combustion Techniques, *Proc. 4th Int. Conf. Fluidization,* pp. 631–647, Mitre Corp., 1976.

99. J. P. Holman, T. W. Moore, and V. M. Wong, Particle-to-Fluid Heat Transfer in Water-Fluidized Systems, *Ind. Eng. Chem. Fund.,* Vol. 4, pp. 21–31, 1965.

100. B. Wasmund and J. W. Smith, Wall to Fluid Heat Transfer in Liquid Fluidized Beds, Part 2, *Can. J. Chem. Eng.,* Vol. 45, pp. 156–165, 1967.

101. F. M. Brea and W. Hamilton, Heat Transfer in Liquid Fluidized Beds with a Concentric Heater, *Trans. Inst. Chem. Eng.,* Vol. 49, pp. 196–203, 1971.

102. J. F. Richardson, M. N. Romani, and K. J. Shakiri, Heat Transfer from Immersed Surfaces in Liquid Fluidized Beds, *Chem. Eng. Sci.,* Vol. 31, pp. 619–624, 1976.

103. A. I. Karpenko, N. I. Syromyatnikov, L. K. Vasanova, and N. N. Galimulin, Radial Heat Conduction in a Liquid-Fluidized Bed, *Heat Transfer Sov. Res.,* Vol. 8, pp. 110–116, 1976.

104. J. P. Sutherland, G. Vassilatos, H. Kubota, and G. L. Osberg, The Effect of Packing on a Fluidized Bed, *AIChE J.,* Vol. 9, pp. 437–441, 1963.

105. J. D. Gabor, Lateral Transport in a Fluidized-Packed Bed, Part 1: Solids Mixing, *AIChE J.,* Vol. II, pp. 127–129, 1965.

106. A. P. Baskakov and V. S. Vershinina, An Investigation of Heat Transfer between a Packing and a Fluidized Bed in the Interstices, *Int. Chem. Eng.,* Vol. 4, pp. 119–123, 1964.

107. J. D. Gabor, Fluidized-Packed Beds, *Chem. Eng. Prog. Symp. Ser.,* Vol. 62, No. 62, pp. 32–41, 1966.

108. C. E. Capes, J. P. Sutherland, and A. E. McIlhinney, An Experimental Study of the Effect of Dump Packing on Wall to Fluidized Bed Heat Transfer, *Can. J. Chem. Eng.,* Vol. 46, pp. 473–475, 1968.

109. P. N. Rowe, Drag Forces in a Hydraulic Model of a Fluidized Bed—Part II, *Trans. Inst. Chem. Eng.,* Vol. 39, pp. 175–180, 1961.

110. T. Baron, Generalized Graphical Method for the Design of Fixed Bed Catalytic Reactors, *Chem. Eng. Prog.,* Vol. 48, pp. 118–124, 1952.

111. J. D. Gabor, B. E. Stangeland, and W. J. Mecham, Lateral Transport in a Fluidized-Packed Bed, Part II: Heat Transfer, *AIChE J.,* Vol. 11, pp. 130–132, 1963.

112. J. D. Gabor, Axial Solids Mixing in Fluidized-Packed Beds, *Chem. Eng. Prog. Symp. Ser.,* Vol. 62, No. 67, pp. 35–41, 1966.

113. J. D. Gabor and W. J. Mecham, Fluidized-Packed Beds: Studies of Heat Transfer, Solids-Gas Mixing, and Elutriation, Argonne National Laboratory, *ANL-6859,* Argonne, Ill., March 1965.

114. A. P. Baskakov, V. S. Vershinina, A. P. Lummi, and V. M. Pakhaluyev, Heat Transfer and Vertical Heat Conductivity of a Bed of Fine-Grained Material Fluidized in Packing, *All-Union Conf. Heat Mass Transfer, II-E,* Minsk, May 5–9, 1964.

115. V. M. Pakhaluyev, Investigation of the Thermal Conductivity of Boiling Bed with Spheres as Packing Material, *Heat Transfer Sov. Res.,* Vol. 6, pp. 88–91, 1974.

116. S. Imura and E. Takegoshi, Effect of Gas Pressure on the Effective Thermal Conductivity of Packed Beds, *Heat Transfer Jpn. Res.,* Vol. 34, pp. 13–26, 1974.

117. J. S. M. Botterill, A. G. Salway, and Y. Teoman, Heat Conduction through Slumped Fluidised Beds, in D. Kunii and R. Toei, eds., *Proceedings of the Fourth International Conference on Fluidization,* pp. 315–322, Engineering Foundation, 1984.

118. K. E. Makhorin V. S. Pikashov, and G. P. Kuchin, Heat Transfer in a High Temperature Fluidized Bed, *Nauk. Du,a Kiev,* 1981.

119. T. F. Ozkaynak, J. C. Chen, and T. R. Frankenfield, An Experimental Investigation of Radiation Heat Transfer in a High Temperature Fluidized Bed, in D. Kunii and R. Toei, eds., *Proceedings of the Fourth International Conference on Fluidization,* pp. 371–378, Engineering Foundation, 1984.

NOMENCLATURE

Symbol, Definition, SI Units, English Units

Ar	Archimedes number $= [g d_p^3 \rho_g (\rho_p - \rho_g)]/\mu_g^2$
a	particle surface to volume ratio: m^{-1}, ft^{-1}
c_p	specific heat of particle: $J/(kg \cdot K)$, $Btu/(lb_m \cdot {}^\circ F)$
c_{pg}	specific heat of gas: $J/(kg \cdot K)$, $Btu/(lb_m \cdot {}^\circ F)$
D_p	diameter of fixed packing: m, ft
D_T	diameter of the fluidizing vessel: m, ft
D	particle mixing diffusivity: m^2/s, ft^2/s
d_p	particle diameter: m, ft
d_T	tube diameter: m, ft
d_{20}	tube diameter equivalent to 0.02 m: m, ft
F	fraction of fines (defined as material of $d_p < 45 \ \mu m$)
Fr_{mf}	Froude number $= V_{mf}/d_p g$
g	gravitational acceleration: m/s^2, ft/s^2
g_c	proportionality constant in Newton's second law of motion: $lb_m \cdot ft/(lb_f \cdot s^2)$
H	bed height: m, ft
H_{mb}	bed height at minimum bubbling velocity: m, ft
H_{mf}	bed height at minimum fluidization velocity: m, ft
h	heat transfer coefficient: $W/(m^2 \cdot K)$, $Btu/(h \cdot ft^2 \cdot {}^\circ F)$
h_{gc}	interphase gas convective component of heat transfer coefficient: $W/(m^2 \cdot K)$, $Btu/(h \cdot ft^2 \cdot {}^\circ F)$
h_{max}	maximum bed-to-surface heat transfer coefficient: $W/(m^2 \cdot K)$, $Btu/(h \cdot ft^2 \cdot {}^\circ F)$
h_{mf}	bed-to-surface heat transfer at onset of fluidization: $W/m^2 \cdot K$, $Btu/(h \cdot ft^2 \cdot {}^\circ F)$
h_p	particle heat transfer coefficient: $W/(m^2 \cdot K)$, $Btu/(h \cdot ft^2 \cdot {}^\circ F)$
h_{pc}	particle convective component of heat transfer coefficient: $W/m^2 \cdot K$, $Btu/(h \cdot ft^2 \cdot {}^\circ F)$
$h_{pc\,max}$	maximum particle convective component of heat transfer coefficient: $W/(m^2 \cdot K)$, $Btu/(h \cdot ft^2 \cdot {}^\circ F)$

h_r radiative component of heat transfer coefficient: $W/(m^2 \cdot K)$, $Btu/(h \cdot ft^2 \cdot {}^\circ F)$

h_{rs} heat transfer coefficient for solid-to-solid radiation: $W/(m^2 \cdot K)$, $Btu/(h \cdot ft^2 \cdot {}^\circ F)$

h_{rv} heat transfer coefficient for radiation to voids: $W/(m^2 \cdot K)$, $Btu/(h \cdot ft^2 \cdot {}^\circ F)$

h_w wall heat transfer coefficient: $W/(m^2 \cdot K)$, $Btu/(h \cdot ft^2 \cdot {}^\circ F)$

h_w^* wall heat transfer coefficient for boundary-layer disruption: $W/(m^2 \cdot K)$, $Btu/(h \cdot ft^2 \cdot {}^\circ K)$

h_w° wall heat transfer coefficient with motionless fluid: $W/(m^2 \cdot K)$, $Btu/(h \cdot ft^2 \cdot {}^\circ F)$

k_{cv} fluid convection contribution to effective thermal conductivity: $W/(m \cdot K)$, $Btu/(h \cdot ft \cdot {}^\circ F)$

k_e effective thermal conductivity: $W/(m \cdot K)$, $Btu/(h \cdot ft \cdot {}^\circ F)$

k_e° effective thermal conductivity with stagnant fluid: $W/(m \cdot K)$, $Btu/(h \cdot ft \cdot {}^\circ F)$

k_f fluid thermal conductivity: $W/(m \cdot K)$, $Btu/(h \cdot ft \cdot {}^\circ F)$

k_g gas thermal conductivity: $W/(m \cdot K)$, $Btu/(h \cdot ft \cdot {}^\circ F)$

k_{max} maximum effective radial thermal conductivity: $W/(m \cdot K)$, $Btu/(h \cdot ft \cdot {}^\circ F)$

k_r radiation contribution to effective thermal conductivity: $W/(m \cdot K)$, $Btu/(h \cdot ft \cdot {}^\circ F)$

k_r° contribution by radiation across void space to effective thermal conductivity: $W/(m \cdot K)$, $Btu/(h \cdot ft \cdot {}^\circ F)$

k_s solid-phase thermal conductivity: $W/(m \cdot K)$, $Btu/(h \cdot ft \cdot {}^\circ F)$

k_w° effective bed thermal conductivity near wall: $W/(m \cdot K)$, $Btu/(h \cdot ft \cdot {}^\circ F)$

L length: m, ft

L_{mf} bed depth at minimum fluidization: m, ft

l_m mixing length: m, ft

$l_{1/2}$ height over which gas to particle temperature difference will fall to half: m, ft

Nu_g Nusselt number without particles in bed $= hD_T/k_g$

Nu_{gc} Nusselt number for interphase gas convective heat transfer $= h_{gc}d_p/k_g$

Nu_{max} Nusselt number for maximum heat transfer coefficient $= h_{max}d_p/k_g$

Nu_p Nusselt number for heat transfer from particle $= h_p d_p/k_f$

n value of power in Eq. (11)

P pressure: Pa (N/m^2), lb_f/ft^2

Pr Prandtl number $= c_{pg}\mu_g/k_g$

q heat transfer rate: W, Btu/h

Re_{mf} Reynolds number at minimum fluidization $= d_p\rho_f V_{mf}/\mu_f$

Re_{opt} optimum Reynolds number $= d_p\rho_f V_{opt}/\mu_f$

Re_p particle Reynolds number $= d_p\rho_f V/\mu_f$

Re_t particle Reynolds number at terminal velocity $d_p\rho_f V_t/\mu_f$

r radial distance: m, ft

r_p particle radius: m, ft

T temperature: $^\circ C$, K, $^\circ F$

T_b bed temperature: $^\circ C$, K, $^\circ F$

T_s surface temperature: $^\circ C$, K, $^\circ F$

T_∞ temperature at infinity: $^\circ C$, K, $^\circ F$

t time: s

V	fluidized-bed superficial velocity: m/s, ft/s
V_b	bubble rise velocity: m/s, ft/s
V_i	velocity at a voidage of unity: m/s, ft/s
V_{loc}	local velocity: m/s, ft/s
V_m	superficial velocity for maximum heat transfer: m/s, ft/s
V_{mb}	minimum bubbling velocity: m/s, ft/s
V_{mf}	minimum fluidization velocity: m/s, ft/s
V_{mfp}	minimum fluidization velocity in a fluidized-packed bed: m/s, ft/s
V_{opt}	optimum velocity: m/s, ft/s
V_t	terminal velocity: m/s, ft/s
V_0	packed-bed superficial velocity: m/s, ft/s
w	weight fraction
X'_l	vertical spacing tubes: m, ft
X'_m	minimum spacing between tubes: m, ft
X'_t	horizontal spacing between tubes: m, ft
Y	proportion of gas flowing in bubble phase

Greek Symbols

α	bed voidage
α_{av}	average bed voidage
α_{mf}	bed voidage at minimum fluidization
α_w	bed voidage near wall
α_t	fluidized-packed bed lateral thermal diffusivity based on ρ_B ($\alpha_t \equiv k_e/\rho_B c_p$) m²/s, ft²/s
α'_t	fluidized-packed bed lateral thermal diffusivity based on ρ_p ($\alpha'_t \equiv k_e/\rho_p c_p$) m²/s, ft²/s
δ_f	length of fluid conduction path: m, ft
δ_w	fluid conduction path near wall: m, ft
ϵ	emissivity
ϵ_b	bed emissivity
ϵ_m	modified emissivity
ϵ_r	reduced emissivity
ϵ_s	surface emissivity
μ	dynamic viscosity: Pa·s, lb$_m$/(h·ft)
μ_f	dynamic viscosity of fluid: Pa·s, lb$_m$/(h·ft)
μ_g	dynamic viscosity of gas: Pa·s, lb$_m$/(h·ft)
ρ	density: kg/m³, lb$_m$/ft³
ρ_B	fluidized-bed density: kg/m³, lb$_m$/ft³
ρ_f	fluid density: kg/m³, lb$_m$/ft³
ρ_g	gas density: kg/m³, lb$_m$/ft³
ρ_p	particle density: kg/m³, lb$_m$/ft³
ρ_s	solid density: kg/m³, lb$_m$/ft³
σ	Stefan-Boltzmann constant: W/(m²·K⁴), Btu/(h·ft²·°R⁴)
ϕ	particle shape factor

CHAPTER 7

Solar Energy

By **Frank Kreith**

Thermal Research Branch
Solar Energy Research Institute
Golden, Colorado

Ari Rabl

Center for Energy and Environmental Studies
Princeton University
Princeton, New Jersey

A. CONVERSION PROCESSES

Solar radiation can be converted into useful energy by the following processes:

- Direct conversion to electricity, e.g., photovoltaics
- Photochemical conversion to chemically bound energy, e.g., photosynthesis
- Thermal conversion to heat, e.g., various types of solar collectors

This chapter presents an introduction to the availability of solar energy on earth and to methods of collecting this energy for conversion into heat or electric power. The presentation here emphasizes thermal conversion; photovoltaic and photochemical conversion processes are discussed in Refs. 1 and 2.

Thermal conversion is a simple process whose efficiency depends primarily on the design of the conversion equipment, the solar radiation flux, and the temperature level at which the energy is delivered. The heat from thermal conversion devices can be used to increase the temperature of a working fluid and, by secondary conversion, to produce electricity or chemical energy. For low-temperature applications such as space conditioning and domestic hot water heating, fixed flat-plate collectors can be used with relatively good efficiency, whereas for higher temperatures that are necessary to produce electricity or drive chemical processes concentration of solar radiation is required.

Evaluation of the technical and economic potential of a solar conversion device requires predicting the average yearly useful energy that the device can deliver at a specified temperature and in a given location. The last part of this chapter therefore presents information to evaluate the average yearly output of commercially important types of solar collectors for comparison with a fossil-fuel system for the same end use.

B. SOLAR RADIATION

This section examines the nature of solar radiation, the geometric relation between the sun and the earth, and the characteristics of the solar energy that reaches the surface of the earth.

The sun is a star, of intensely hot gases, with a diameter of 1.4×10^9 m (8.6×10^5 mi), at an average distance of 1.5×10^{11} m (9.3×10^7 mi) from the earth. The temperatures in the interior of the sun are on the order of 8×10^6 to 40×10^6 K, but the surface temperature is much lower, approximately 5800 K [3]. The rate of energy emission from the sun is 3.8×10^{23} kW, of which 1.7×10^{14} kW is intercepted by the earth. Of this amount 30 percent is reflected, 47 percent is converted into low-temperature heat and reradiated into space, and about 23 percent powers the evaporation and precipitation cycle of the biosphere; less than half a percent is transformed into the kinetic energy of wind and waves and photosynthetic storage in green plants.

1. Solar Radiation Characteristics

Figure 1 shows the sun-to-earth relationship. The earth's orbit around the sun is elliptical, the sun-earth distance varying ±1.7 percent over the course of the year. At the mean distance between earth and sun, the sun subtends an angle of about 32 minutes. The extraterrestrial solar irradiance at normal incidence, the so-called solar constant I_0, has been measured at 1373 W/m^2, or 435.5 Btu/(h·ft^2) [4].

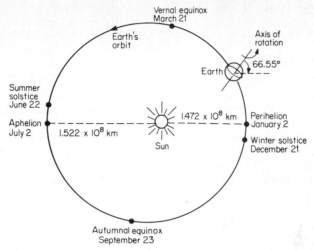

FIG. 1 Geometry of sun-earth system.

Due to atmospheric effects, the solar radiation at the earth's surface consists of two components:

- Beam or direct radiation, received from the sun without change of direction
- Diffuse radiation, the portion received from throughout the sky after its direction has been changed by reflection and scattering in the atmosphere

Total solar radiation refers to the sum of these two. The distinction between these two types of solar radiation is important for application and selection of the most appropriate solar conversion device and for matching the output to its end use. For example, focusing collectors can use only direct radiation, whereas flat-plate collectors can capture the total radiation. Also the instruments to measure total and direct radiation are different. Total radiation is measured with a pyranometer that measures the intensity of radiant energy arriving from a hemisphere of directions, whereas the direct radiation component is measured by a normal-incidence pyrheliometer [5]. The latter uses a collimating tube which is intended to limit the view of the sensor to the direct radiation component. In practice most normal-incidence pyrheliometers have a field of view with a half-angle of 2.8°, even though the half angle of the solar disk is only 0.25°. Consequently, pyrheliometer data include circumsolar radiation, i.e., radiation from the immediate vicinity of the solar disk [6, 7].

As a result of scattering and absorption, solar radiation changes during its passage through the atmosphere. The atmospheric constituents primarily responsible for scattering are gas molecules (Rayleigh scattering), particulates, and water droplets. Scattering is strongly wavelength-dependent. The amount of cloud cover is of particular importance because clouds can reduce the irradiation on earth by up to 80 to 90 percent. Absorption is largely due to ozone in the ultraviolet, and to water vapor and carbon dioxide in the infrared (see Ref. 34). Lesser absorption effects are due to oxygen, other gases, and particulates.

The amount of atmospheric absorption depends on the length of the radiation path through the atmosphere. The path length can be characterized by the so-called air mass, defined as ratio of path length and thickness of atmosphere; it is given by

$$m = \frac{1}{\cos \theta_z} \tag{1}$$

where θ_z is the zenith angle, i.e., the angle between the sun's rays and the normal to the earth's surface. Figure 2 shows the atmospheric transmittance for ($CO_2 + O_2$), O_3, H_2O,

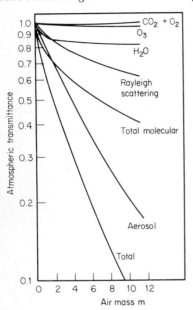

Rayleigh scattering, and aerosols as a function of the air mass for direct radiation, while Fig. 3a shows a comparison of the radiation spectrum from a 5762-K blackbody with the measured solar irradiation outside the atmosphere, i.e., air mass of zero. Figure 3b shows the solar spectrum at the earth's surface on a clear day with air mass of 1, 4, 7, and 10. (Note that Fig. 3 is based on an older value of the solar constant, but the effect on this figure is negligible.) For most solar energy calculations a single standard spectrum can be assumed. The spectral distribution in Table 1 has been recommended as a standard by Lind et al. [8]; it corresponds to typical atmospheric conditions in the United States at an air mass of 1.5.

FIG. 2 Atmospheric transmittance versus air mass [9].

2. Sun-Earth Geometry

The earth revolves in an elliptical orbit around the sun, with the sun at one focus of the ellipse. The period of revolution is defined as one year. The plane containing the earth's elliptical orbit is called the *ecliptic* plane, while the plane containing the earth's equator is called the *equatorial* plane. The angle between these two planes is 23.45°. The seasons are due to the fact that the earth's axis is inclined with respect to the ecliptic plane. This causes radiation to strike the northern hemisphere more directly near aphelion, causing summer there during that portion of the year, but winter in the southern hemisphere.

The earth-sun vector moves in the ecliptic plane, and the angle between this vector and the equatorial plane is called the solar declination δ_s. It varies from $-23.45°$ on December 21 to $+23.45°$ on June 22. By convention, it is positive when the earth-sun vector points northward relative to the equatorial plane. The declination is given by the relation

$$\sin \delta_s = -\sin 23.45° \cos \frac{360°(N + 10)}{365.25} \tag{2}$$

where N is the number of the day of the year, counted from January 1.

The rotation of the earth about its axis causes the day-night cycle. To an observer on earth the sun appears to move 15 degrees per hour. It is convenient to define a solar hour angle ω as 15°/h times the number of hours from solar noon, the latter being the time of day when the sun is highest in the sky. Morning values, i.e., east of south, are negative,

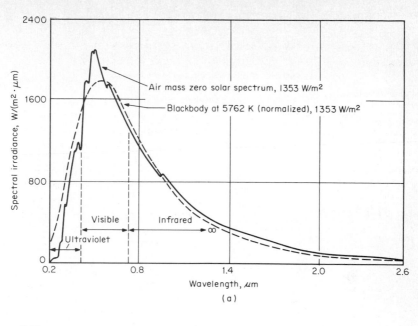

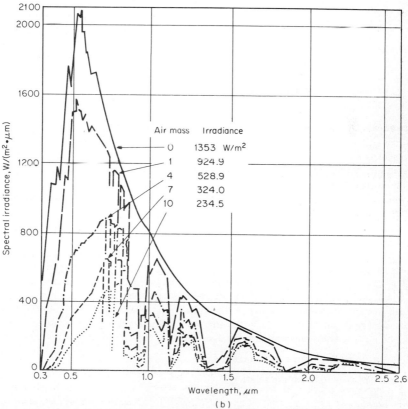

FIG. 3 Spectral distribution of solar radiation: (*a*) air-mass zero solar spectrum and blackbody spectrum; (*b*) solar spectrum for air mass 0, 1, 4, 7, and 10.

TABLE 1 Standard Spectrum Recommended for Solar Energy Calculations by Lind et al. [8]

The wavelengths correspond to the mean of the energy in each of the 20 equal energy intervals. (This reference also gives the spectra with 50 and with 100 intervals.)

Percent	Wavelength, nm
2.50	386.29
7.50	443.39
12.50	480.34
17.50	515.61
22.50	551.38
27.50	587.73
32.50	623.94
37.50	660.57
42.50	697.93
47.50	739.27
52.50	784.78
57.50	832.00
62.50	881.15
67.50	956.35
72.50	1023.73
77.50	1091.65
82.50	1208.54
87.50	1315.22
92.50	1624.27
97.50	2131.58

while afternoon values are positive. For example, at 8 a.m. solar time the solar hour angle ω is $-60°$, and at 6 p.m. $\omega = +90°$.

For engineering calculations it is more convenient to adopt the ptolemaic view and analyze the motion of the sun in a coordinate system fixed to the earth with its origin at the point of interest. In this system the sun is constrained to move with two degrees of freedom on the celestial sphere and its location can be specified by two angles as shown in Fig. 4: the solar altitude ψ, measured from the local horizontal plane and a line to the center of the sun, and the azimuth angle γ_s, measured in the horizontal plane between a line due south and the projection of the site-to-sun line. The azimuth is negative east of south and positive west of south. The zenith angle θ_z is the complement of the altitude angle; that is, $\theta_z = 90° - \psi$.

The solar altitude and azimuth angles are related to the hour angle ω, the solar declination δ_s, and the latitude L, at which the coordinate system is fixed (see Fig. 5). L can be read from an atlas and is positive north of the equator and negative south of it. Using vector algebra it can be shown [12] that the solar altitude angle ψ can be calculated for any time of day, date, and location from the relation

$$\sin \psi = \sin L \sin \delta_s + \cos L \cos \delta_s \cos \omega \qquad (3)$$

and the solar azimuth from the relation

$$\sin \gamma_s = \frac{\cos \delta_s \sin \omega}{\cos \psi} \qquad (4)$$

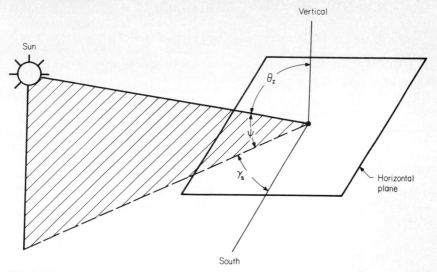

FIG. 4 Diagram showing solar-altitude angle ψ, solar zenith angle θ_z, and solar azimuth angle γ_s.

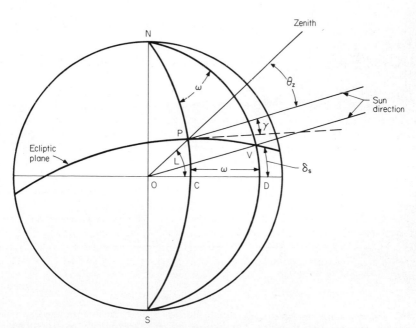

FIG. 5 Definitions of solar hour angle ω (CND), solar declination δ_s (VOD), and latitude L (POC). P is site of interest. Modified from J. F. Kreider and F. Kreith, *Solar Heating and Cooling*, rev. 1st ed., Hemisphere, Washington, D.C., 1977.

The solar altitude angle is given unambiguously by the inverse of the sine function in Eq. (3). But to find the solar azimuth angle from Eq. (4) it is necessary to distinguish the cases where the sun is in the northern half of the sky from those where the sun is situated in the southern half. The complete specification of γ_s is

$$
\gamma_s = \begin{cases} \sin^{-1} \dfrac{\cos \delta_s \sin \omega}{\cos \psi} & \text{if } \cos \omega > \dfrac{\tan \delta_s}{\tan L} \\[4mm] 180° - \sin^{-1} \left(\dfrac{\cos \delta_s \sin \omega}{\cos \psi} \right) & \text{if } \cos \omega < \dfrac{\tan \delta_s}{\tan L} \end{cases}
\tag{5}
$$

When $\cos \omega = \tan \delta_s / \tan L$, $\gamma_s = -90°$ when ω is negative and $\gamma_s = +90°$ when ω is positive. Sunrise and sunset occur when $\psi = 0$. From Eq. (3), the corresponding hour angles are

$$
\omega_{\text{sunset}} = -\omega_{\text{sunrise}} = \omega_s = \cos^{-1}(-\tan L \tan \delta_s)
\tag{6}
$$

3. Incidence Angles for Solar Radiation

The orientation of an inclined surface such as a wall or a window can be specified in terms of its azimuth angle γ and its tilt angle β. As shown in Fig. 6, the vector $\vec{n}$ is perpendicular to the surface; the surface azimuth angle γ is measured from the south, positive toward the west; and the tilt angle β at which the surface is inclined from the horizontal is taken positive for south-facing surfaces.

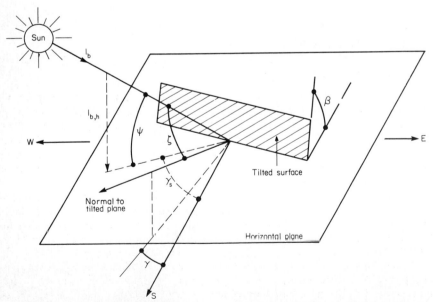

FIG. 6 Definition of incidence angle ζ, surface tilt angle β, solar altitude angle ψ, wall-azimuth angle γ, and solar azimuth angle γ_s for a non-south-facing, tilted surface. Also shown are the beam component of solar radiation I_b and the component of beam radiation $I_{b,h}$ on a horizontal plane.

The solar incidence angle, as shown in Fig. 6, is defined as the angle between the surface normal $\bar{n}$ and a line collinear with the sun's rays. The component of the beam or direct normal radiation from the sun, I_b, that is intercepted by the surface, $I_{b,s}$, is

$$I_{b,s} = I_b \cos \zeta \tag{7}$$

For a non-south-facing, tilted surface the cosine of the incidence angle is

$$\cos \zeta = \cos (\gamma_s - \gamma) \cos \psi \sin \beta + \sin \psi \cos \beta \tag{8}$$

If the right-hand side of Eq. 8 is negative, the sun's rays will not strike the surface. Some useful equations for special cases of practical interest are listed below:

Horizontal surface (tilt $\beta = 0$):

$$\cos \zeta = \cos \theta_z = \sin \psi \tag{9}$$

South-facing surface (azimuth $\gamma = 0$):

$$\cos \zeta = \sin \delta_s \sin (L - \beta) + \cos \delta_s \cos (L - \beta) \cos \omega \tag{10}$$

Vertical south-facing surface (tilt $\beta = 90°$):

$$\cos \zeta = -\sin \delta_s \cos L + \cos \delta_s \sin L \cos \omega \tag{11}$$

Many concentrating collectors move during the day to track the sun. Tracking can be about one or two axes. Table 2 lists the principal tracking schemes and the corresponding equations for $\cos \zeta$, the cosine of the incidence angle. When this expression is negative for a given surface, no direct solar radiation is intercepted by the surface.

4. Solar Radiation on Inclined Surfaces

a. Instantaneous Radiation

Most weather stations report only data for the total insolation on the horizontal surface (H_h). However, in order to calculate the performance of collectors whose aperture is not horizontal, one needs to know how much of the total is direct and how much is diffuse. This is necessary even for a flat-plate collector, because the conversion of radiation from one plane to another is different for direct and for diffuse insolation. For diffuse insolation one may assume isotropy because the error of this assumption is at most 3 percent in calculating total insolation on tilted surfaces [13].

Several procedures have been developed for predicting the diffuse solar radiation component when only the total hemispherical insolation is known [14]. The model developed in Ref. 15 is currently accepted as the most accurate and has been used to calculate the beam insolation "data" on the weather tapes distributed by the National Climatic Center [16]. However, this model is so complicated that it requires a computer.

A much simpler approximate method is based on a correlation first introduced by Liu and Jordan [17] which relates for each day the ratio H_d/H_h, of daily total diffuse insolation H_d and daily total hemispherical insolation H_h (both measured on the horizontal surface) to K_T, defined by

$$K_T = \frac{H_h}{H_0} \tag{12}$$

TABLE 2 Cosine of Incidence Angle ζ for Principal Collector Orientations

Orientation of collector aperture	$\cos \zeta$
Fixed aperture azimuth, $\gamma = 0$ Tilt β = latitude L	$\cos \delta_s \cos \omega$
Tilt $\beta \neq$ latitude L	$\cos \delta_s \cos (\beta - L) (\cos \omega - \cos \omega'_s)$ with $\cos \omega'_s = +\tan (\beta - L) \tan \delta_s$
Fixed aperture azimuth, $\gamma \neq 0$	$\cos \delta_s \cos L \cos \beta \cos \omega + \sin \delta_s \sin L \cos \beta + \cos \delta_s \sin \gamma \sin \beta \sin \omega +$ $\cos \delta_s \cos \gamma \sin L \sin \beta \cos \omega - \sin \delta_s \cos \gamma \sin \beta \cos L$
Aperture tracking about horizontal east-west axis	$\cos \delta_s (\cos^2 \omega + \tan^2 \delta_s)^{1/2}$
Aperture tracking about polar axis	$\cos \delta_s$
Aperture tracking about horizontal north-south axis	$\cos \delta_s [\sin^2 \omega + (\cos L \cos \omega + \tan \delta_s \sin L)^2]^{1/2}$
Aperture tracking about vertical axis	$\cos \delta_s [\sin^2 \omega + (\sin L \cos \omega - \tan \delta_s \cos L)^2]^{1/2}$
Aperture tracking about north-south axis inclined at angle β from horizontal	$\cos \delta_s \{\sin^2 \omega + [\cos (\beta - L) \cos \omega - \tan \delta_s \sin (\beta - L)]^2\}^{1/2}$
Aperture with full two-axis tracking	1

where H_h is the terrestrial insolation and H_0 is the extraterrestrial insolation given by

$$H_0 = \frac{t}{\pi} I_0 \left(1 + 0.033 \cos \frac{360°N}{365.25} \right) \cos L \cos \delta_s \left(\sin \omega_s - \frac{\omega_s \pi}{180°} \cos \omega_s \right)$$

with $t = 24$ h $=$ length of day and $N =$ day of year, starting January 1.

The ratio K_T is called the clearness index and serves as a useful dimensionless measure of the transparency of the atmosphere. On heavily overcast days K_T may be as low as 0.05 to 0.1, while on perfectly clear days it is in the range of 0.75 to 0.8. Monthly averages of K_T, designated by $\overline{K}_T$, range from 0.4 for very cloudy climates (e.g., upstate New York) to 0.7 for very sunny climates (e.g., Albuquerque, New Mexico). An updated version of the Liu and Jordan correlation for H_d/H_h is shown in Fig. 7a. It can also be expressed by the equations

$$\frac{H_d}{H_h} = \begin{cases} 0.99 \\ 1.188 - 2.272K_T + 9.473K_T^2 \\ \quad - 21.856K_T^3 + 14.648K_T^4 \end{cases}$$

$$\begin{matrix} \text{for } K_T \le 0.17 \\ \text{for } 0.17 < K_T < 0.8 \\ \text{for } K_T > 0.8 \end{matrix} \qquad (13)$$

Correlations of this type are reliable only in an average sense. For any particular day or hour the data may differ significantly from the average, as indicated by the error bars in Fig. 7a, which show the rms variation about the average.

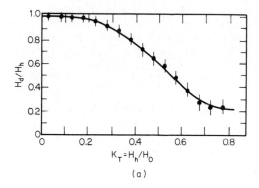

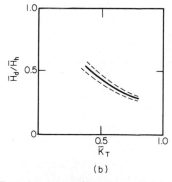

FIG. 7 Correlation of ratio of daily diffuse and daily hemispherical insolation with clearness index K_T: (a) correlation of H_d/H_h with K_T for individual days, Eq. (13); (b) correlation of $\overline{H}_d/\overline{H}_h$ versus $\overline{K}_T$ for monthy average, Eq. (14) (solid line $\omega_s = 90°$, spring and fall, dashed lines $\omega_s = 90° + 11.5°$, top, or summer, and $\omega_s = 90° - 11.5°$, bottom, or winter).

b. Long-Term Average Radiation

For many applications one needs to know the monthly average insolation on tilted or tracking surfaces. This can be calculated with good accuracy by the following algorithm which requires as input only the monthly average clearness index $\overline{K}_T$ (or equivalently, the monthly average hemispherical insolation $\overline{H}_h$ on the horizontal surface, since H_0 during a month can be approximated by a constant). The structure of this algorithm is outlined in Fig. 8.

As the first step the monthly average diffuse insolation $\overline{H}_d$ is calculated from Fig. 7b or Eq. (14):

$$\frac{\overline{H}_d}{\overline{H}_h} = 0.775 + 0.347(\omega_s - 90°)\frac{\pi}{180°}$$

$$- \left[0.505 + 0.261\ (\omega_s - 90°)\frac{\pi}{180°} \right] \cos\frac{360°(\overline{K}_T - 0.9)}{\pi} \tag{14}$$

This equation is the monthly analog of Eq. (13); it differs from Eq. (13) because it depends on the frequency distribution of clear and cloudy days during the month.

As the second step the daily radiation total $\overline{H}_d$ and $\overline{H}_h$ are converted into the instantaneous values $\bar{I}_d$ and $\bar{I}_h$ by means of the correlations

$$\bar{I}_d = r_d(\omega, \omega_s)\overline{H}_d \tag{15}$$

and
$$\bar{I}_h = r_h(\omega, \omega_s)\overline{H}_h \tag{16}$$

where

$$r_d(\omega, \omega_s) = \frac{\pi}{t}\ \frac{\cos\omega - \cos\omega_s}{\sin\omega_s - (\pi\omega_s/180°)\cos\omega_s} \tag{17}$$

and
$$r_h(\omega, \omega_s) = (a + b\cos\omega)r_d(\omega, \omega_s) \tag{18}$$

with $t = 24\ \text{h} = $ length of day
$\omega = $ hour angle
$\omega_s = $ sunset hour angle, Eq. (6)
$a = 0.409 + 0.5015\sin(\omega_s - 60°)$ (19)
$b = 0.6609 - 0.4767\sin(\omega_s - 60°)$ (20)

The correlation functions r_d and r_h are plotted in Fig. 9. Like the other correlations in this section, they are reliable only in an average sense; the ratios I_d/H_d and I_h/H_h for any particular day can be very different.

Of the three radiation types, beam, diffuse, and hemispherical, only two are independent. Given the average diffuse and hemispherical insolation for a particular time of the day and the year, the beam insolation is uniquely determined; at normal incidence it is

$$I_b = \frac{I_h - I_d}{\cos\theta_z} \tag{21}$$

where θ_z is the solar zenith angle. Thus, the average direct and diffuse insolation on any surface and for any time of day can be calculated if one knows the monthly average clearness index $\overline{K}_T$. Explicit expressions for the integrated daily total insolation incident on the most important fixed or tracking surfaces are listed in Ref. 18.

For collectors with heat loss the monthly average insolation on the aperture does not determine the energy delivery completely. To explain this fact, consider a collector with a heat loss equal to the monthly average insolation absorbed by the collector. If all days

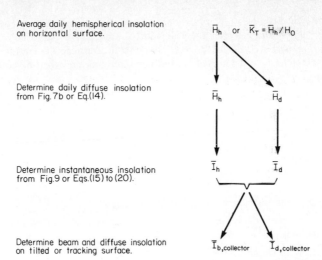

Average daily hemispherical insolation on horizontal surface.

$\overline{H}_h$ or $\overline{K}_T = \overline{H}_h / H_O$

Determine daily diffuse insolation from Fig. 7b or Eq.(14).

$\overline{H}_h$ $\overline{H}_d$

Determine instantaneous insolation from Fig.9 or Eqs.(15) to (20).

$\overline{I}_h$ $\overline{I}_d$

Determine beam and diffuse insolation on tilted or tracking surface.

$\overline{I}_{b,collector}$ $\overline{I}_{d,collector}$

FIG. 8 Structure of algorithm for calculating long-term average insolation on any surface.

were identical, this collector would not deliver any useful energy. If, on the other hand, the month consisted of completely clear days and of completely cloudy days (with a distribution that results in the same monthly average as before), then this collector would deliver useful energy on the clear days. Therefore the performance of solar thermal collectors depends not only on the average insolation but also on the fluctuations about the

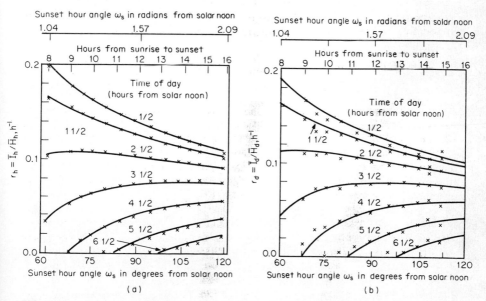

FIG. 9 Correlation between daily total solar irradiation and instantaneous solar irradiance: (a) for hemispherical insolation $r_h = \overline{I}_h / \overline{H}_h$; (b) for diffuse insolation $r_d = \overline{I}_d / \overline{H}_d$.

average. This effect can be analyzed by means of generalized frequency distributions [19, 20], from which one can calculate the utilizability, defined as the fraction of the average insolation which is above a specified level. This so-called utilizability method is described in detail for all collector types in Ref. 18. For flat-plate collectors the utilizability has also been calculated in Refs. 19 and 21.

C. SOLAR THERMAL COLLECTORS

1. Collector Morphology

There are essentially three types of thermal collectors:

1. Nonconcentrating stationary: flat-plate and solar ponds
2. Slightly concentrating, with or without periodic adjustment: compound parabolic concentrator (CPC) [22] and V troughs
3. Concentrating tracking collectors,
   a. One-axis tracking
   b. Two-axis tracking

Figure 10 displays the morphology of solar thermal collector schemes. The most common of the nonconcentrating types of collectors are the flat-plate collector and the solar pond (see Chap. 8). The key elements of a flat-plate collector (see Fig. 11) are a frame, a transparent cover, a collector plate, flow tubes with inlet and outlet headers through

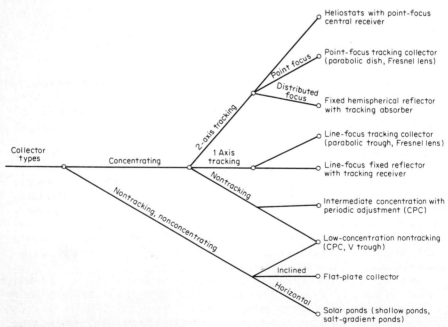

FIG. 10 Morphology of collector types.

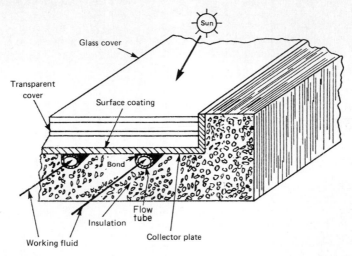

FIG. 11 Schematic diagram of a flat-plate collector with two covers.

which a working fluid passes, and back-side insulation. Table 3 shows the thermal and radiative properties of typical collector cover materials and Table 4 the radiative properties of collector absorber surfaces. The cover is usually made of glass or a plastic that is transparent to radiation in the solar spectral range, but impervious to infrared radiation emanating from the collector plate (see Ref. 34). Thus, solar radiation is absorbed by the plate, which then heats up and transfers heat to the working fluid.

The concentration ratio C is defined in terms of the ratio of aperture area A to receiver area A_r:

$$C = \frac{A}{A_r} \tag{22}$$

Collectors with $C > 1$ are called concentrating collectors. Concentrating collectors can be subdivided into tracking, stationary, and periodically adjustable types. Stationary concentrating collectors are limited to $C \lesssim 1.7$, but by seasonal tilt adjustments this limit can be extended to about 5 [22]. The most common types are V troughs and compound parabolic concentrators, called CPC (see Fig. 12a).

Tracking collectors are capable of much higher concentrations. They can be subdivided into reflecting and refracting concentrators. The most common reflector shapes are parabolic (see Fig. 12b and d). Most refracting concentrators use Fresnel lenses, either in line-focus or in point-focus systems. The optics of a refractive element or lens differ from those of a reflective element or mirror in the off-axis aberrations characteristic of refractive elements and in the effects of surface and tracking errors. Aberrations caused when the incident radiation is not parallel to the optical axis shorten the effective focal length. This produces an increase in the size of the focal spot at the absorber, since the absorber is at a fixed distance from the lens.

Concentrating collectors can have a line focus or a point focus. Most of the former require only one-axis tracking and can achieve concentration ratios up to about 50 in practice. Point-focus devices, on the other hand, require two-axis tracking, but can attain concentration ratios up to several thousand and operate with reasonable efficiency up to temperatures around 1700 K (2500°F).

TABLE 3 Thermal and Radiative Properties of Transparent Collector Cover Materials[a]

Material name	Index of refraction n	τ (solar), %	τ (infrared), %	Expansion coefficient, in/(in·°F)	Temperature limits, °F	Weatherability (comment)	Chemical resistance (comment)
Lexan (polycarbonate)	1.586 (D 542)[c]	125 mil 72.6 (±0.1)	125 mil 2.0 (est)[b]	$3.75 (10^{-5})$ (H 696)[c]	250–270 service temperature	Good: 2-yr exposure in Florida caused yellowing; 5-yr caused 5% loss in τ.	Good: comparable to acrylic
Plexiglas (acrylic)	1.49 (D 542)[c]	125 mil 79.6 (±0.8)	125 mil 2.0 (est)[b]	$3.9 (10^{-9})$ at 60°F; 4.6 (10^{-6}) at 100°F	180–200 service temperature	Average to good: based on 20-yr testing in Arizona, Florida, and Pennsylvania.	Good to excellent: resists most acids and alkalis
Teflon F.E.P. (fluorocarbon)	1.343 (D 542)[c]	5 mil 89.8 (±0.4)	5 mil 25.6 (±0.5)	$5.9 (10^{-5})$ at 160°F; 9.0 (10^{-5}) at 212°F	400 continuous use; 475 short-term use	Good to excellent: based on 15-yr exposure in Florida environment.	Excellent: chemically inert
Tedlar P.V.F. (fluorocarbon)	1.46 (D 542)	4 mil 88.3 (±0.9)	4 mil 20.7 (±0.2)	$2.8 (10^{-5})$ (D 696)[c]	225 continuous use; 350 short-term use	Good to excellent: 10-yr exposure in Florida with slight yellowing.	Excellent: chemically inert
Mylar (polyester)	1.64–1.67 (D 542)[c]	5 mil 80.1 (±0.1)	5 mil 17.8 (±0.5)	$0.94 (10^{-5})$ (D 696-44)[c]	300 continous use; 400 short-term use	Poor: ultraviolet degradation great.	Good to excellent: comparable to Tedlar
Sunlite[d] (fiberglass)	1.54 (D 542)[c]	25 mil 75.4 (±0.1) (P) 25 mil 77.1 (±0.7) (R)	25 mil 7.6 (±0.1) (P) 25 mil 3.3 (±0.3) (R)	$1.4 (10^{-5})$ (D 696)[c]	200 continuous use causes 5% loss in τ (solar)	Fair to good: regular, 7-yr solar life; premium, 20-yr solar life.	Good: inert to chemical atmospheres

Material	Refractive index	Transmission (%)	Thickness (mil)	Thermal expansion coefficient	Temperature limits (°F)	Resistance to UV	Overall rating
Float glass (glass)	1.518 (D 542)[c]	125 mil 78.6 (±0.2)	125 mil 2.0 (est)[b]	4.8 (10^{-5}) (D 696)[c]	1350 softening point; 100 thermal shock	Excellent: time-proved.	Good to excellent: time-proved
Temper glass (glass)	1.518 (D 542)[c]	125 mil 78.6 (±0.2)	125 mil 2.0 (est)[b]	4.8 (10^{-6}) (D 696)[c]	450–500 continuous use; 500–550 short-term use	Excellent: time-proved.	Good to excellent: time-proved
Clear lime sheet glass (low–iron oxide glass)	1.51 (D 542)[c]	125 mil 87.5 (±0.5)	125 mil 2.0 (est)[b]	5.0 (10^{-6}) (D 696)[c]	400 for continuous operation	Excellent: time-proved	Good to excellent: time-proved
Clear lime temper glass (low–iron oxide glass)	1.51 (D 542)[c]	125 mil 87.5 (±0.5)	125 mil 2.0 (est)[b]	5.0 (10^{-6}) (D 696)[c]	400 for continuous operation	Excellent: time-proved.	Good to excellent: time-proved
Sunadex white crystal glass (0.01% iron oxide glass)	1.50 (D 542)[c]	125 mil 91.5 (±0.2)	125 mil 2.0 (est)[b]	4.7 (10^{-6}) (D 696)[c]	400 for continuous operation	Excellent: time-proved.	Good to excellent: time-approved

Conversion factors: 1 °F = 5/9°C; 1 mil = 2.54 × 10^{-5} m.

[a] Abstracted from Ratzel, A. C., and R. B. Bannerot, Optimal Material Selection for Flat-Plate Solar Energy Collectors Utilizing Commercially Available Materials, presented at ASME-AIChE Nat. Heat Transfer Conf., 1976.

[b] Data not provided; estimate of 2% to be used for 125-mil samples.

[c] All parenthesized numbers refer to ASTM test codes.

[d] Sunlite premium data denoted by (P); Sunlite regular data denoted by (R).

TABLE 4 Absorber Test Materials

Code*	Absorber material		Optical properties†	
	Coating	Substrate	Absorptance‡	Emittance‡
A	Black nickel	Steel	0.87	0.13
C	Flat black paint	Copper	0.98	0.92
D	Black chrome	Steel (nickel-flashed)	0.97	0.07
E	Flat black paint	Copper	0.95	0.87
F	Copper oxide	Copper	0.96	0.75
G	Black porcelain enamel	Steel	0.93	0.86
H	Flat black paint	Aluminum	0.95	0.89
I	Black chrome	Stainless steel	0.88	0.19
J	Black chrome	Aluminum	0.98	0.14
L	Lead oxide	Copper	0.99	0.29
M	Oxide anodized	Aluminum	0.94	0.10
N	Oxide conversion coating	Aluminum	0.93	0.51
P	Black chrome	Copper	0.96	0.08

*Code letters A through H indicate materials coupon specimens cut from solar collectors A through H. Codes I through P tested at the materials level only.

†These properties are dependent on the formulations and manufacturing processes used. Other products within a generic class of materials may have significantly different properties.

‡Average values based on a minimum of 10 test specimens.

From *NBS TN 1136*, 1981. NBS Solar Collector Durability and Reliability Test Program Plan by D. Wakman, E. Street, and T. Seiler.

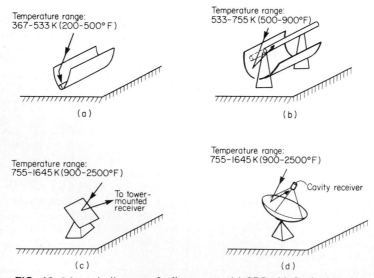

FIG. 12 Schematic diagrams of collector types: (*a*) CPC with fixed orientation or periodic adjustment (ICPA—intermediate concentration with periodic adjustment); (*b*) parabolic trough with one-axis tracking (LFTC—line-focus tracking collector); (*c*) heliostat with two-axis tracking (PFCR—point-focus central receiver); (*d*) paraboloidal dish with two-axis tracking (PFTP—point-focus two-axis parabolic).

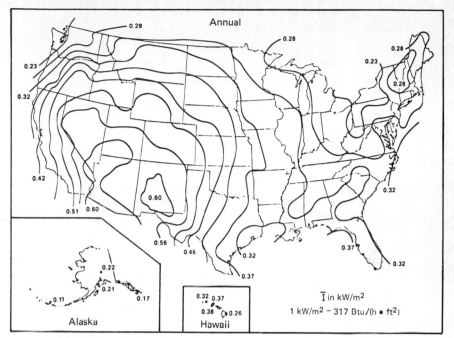

FIG. 13 Yearly average direct normal irradiance $\bar{I}$ during daytime for the United States [31].

2. Instantaneous Collector Efficiency

The output of a thermal collector $\dot{Q}$ can be specified by the rate of heat transfer to the working fluid, or

$$\dot{Q} = Wc_p(T_{\text{out}} - T_{\text{in}}) \tag{23}$$

where W is the mass flow rate of the working fluid passing through the collector, c_p is the specific heat at constant pressure, T_{in} is the inlet temperature of the fluid, and T_{out} is the outlet temperature of the fluid. Table 5 gives the physical properties of working fluids used in solar applications.

It should be noted, however, that the rate of heat transfer to the working fluid in the collector is not equal to the useful energy delivered by a solar system. The system performance is considerably less than the collector performance because of heat losses in connecting piping and valves (see Refs. 35 and 36) and because of transient warm-up effects such as in the morning and after interruptions by clouds. Evaluation of the useful energy delivered by a complete solar system to perform a given task, e.g., heating a home, is a very complex process which must take into account the dynamics of the system as well as variations in the weather. Due to losses and transients in such systems, the useful output of solar systems can be severely degraded if they are not carefully designed. This is especially true at high temperatures.

The instantaneous thermal efficiency η for any type of thermal collectors can be evaluated from the equation

$$\eta = \frac{\dot{Q}}{AI_c} = \eta_0 - U\frac{T_r - T_a}{I_c} \tag{24}$$

TABLE 5 Physical Properties of Some Solar Collector Working Fluids

Commercial name (manufacturer)	Chemical composition	Freezing point, °C	Boiling point, °C	Thermal conductivity, W/(m·°C)	Specific heat, kJ/(kg·°C)	Viscosity, m²·s × 10⁶ at °C	Volume expansion, °C⁻¹ × 10⁴	Specific gravity at °C	Comments for solar applications
Propylene glycol									Glycol fluids have generally good physical properties, but may have problems due to toxicity and oxidation products. Annual check of pH and specific gravity is recommended. Fluid life is 2–4 years. Aqueous propylene glycol solutions are preferred for medium-temperature applications. Corrosion likely at high temperatures.
Dowfrost (Dow)[1] 59 weight percent aqueous solution	Propylene glycol K$_2$HPO$_4$ one percent	−31°	102°	0.39	3.6 at 27°C	3 at 38°C 70 at −18°C	7.3	1.033 at 27°C	
Sunsol (Sunworks)[2] 60 weight percent aqueous solution	Propylene glycol inhibitor	−48°	109°	0.35	3.4 at 25°C	4 at 38°C 140 at −20°C	7.5	1.055 at 25°C	
Ethyelene glycol									
Dowtherm SR-1 (Dow)[3] 59 weight percent aqueous solution	Ethylene glycol K$_2$HPO$_4$ two percent	−37°	110°	0.42	3.4 at 27°C	2.6 at 38°C 22 at −18°C	6.3	1.074 at 200°C	

Polyglycols									
UCON 500 (Union Carbide)	Polyglycol 90 percent inhibitor	−37°	289° rec max temp	0.15	2.2 at 100°C	11.5 at 100°C 61 at 38°C	7.9	0.98 at 100°C	
Paraffinic hydrocarbon oils									Hydrocarbon oils have low toxicity and good high temperature stability, if air supply is limited. Some rubber gaskets and hose materials are attacked. Antioxidant additives should be added for stabilization. Roof spills on asphalt shingles can cause problems.
Caloria HT 43 (Exxon)[4]	Petroleum distillate, moderate molecular weight, low aromatic content	−9.5°	311° rec max temp	0.13	2.1 at 100°C	5 at 100°C 31 at 38°C	10.0	0.80 at 100°C	
Synthetic hydrocarbon oils									
Brayco 888 HF (Bray Oil Co.)[5]	Polymerized 1-decene	−85°	227° flash point	0.13	2.3 at 25°C	4.5 st 100°C 22.5 st at 38°C	4.8	0.83 at 15°C	

TABLE 5 Physical Properties of Some Solar Collector Working Fluids (*Continued*.)

Commercial name (manufacturer)	Chemical composition	Freezing point, °C	Boiling point, °C	Thermal conductivity, W/(m·°C)	Specific heat, kJ/(kg·°C)	Viscosity, $m^2 \cdot s \times 10^6$ at °C	Volume expansion, $°C^{-1} \times 10^4$	Specific gravity at °C	Comments for solar applications
Silicone oils									
Syltherm 444 (Dow Corning)[6]	Polydimethysiloxane	−46°	> 315°	0.14	1.6 at 100°C	7.0 at 100°C 20 at 25°C	10.7	0.95 at 25°C	Silicone fluids are noncorrosive, but incompatible with water. Low surface tension can produce seepage through cracks. Expensive.
Water		0°	100°	0.60	4.2 at 93°C	0.3 at 93°C	2.1	1.00 at 4°C	Good all around, but requires pressurization above 100°C.

Notes: Similar physical properties exist for (1) UCAR 35.50 percent solution (Union Carbide); Sunsafe 230.50 percent solution (NPD Energy Systems); (2) Solar Winter Ban (CAMCO); Corona Solar Fluid (A. O. Smith); (3) UCAR 17.50 percent solution (Union Carbide); (4) Mobitherm 603 (Mobil); Thermia C (Shell); Dowtherm HP (Dow); (5) PAO-20E (Uniroyal); H-30 (Mark Enterprises); (6) SF-96 (General Electric). Information for the following products are taken from manufacturers' data sheets: Sunsol 60, Brayco 888, Syltherm 444, Sunsafe 230, UCAR 35, Solar Winter Ban, Corona Solar Fluid, PAO-20E, and H-30.

Adapted from "Solar Heat Transfer Fluids," by T. Sullivan, *Solar Age*, December 1980.

where η_0 = optical efficiency

$\quad T_r$ = average receiver surface temperature

$\quad T_a$ = ambient temperature

$\quad U$ = total heat transfer (or thermal loss) coefficient between the collector surface and ambient, based on aperture area A

$\quad I_c$ = irradiance (or insolation) on the collector aperture area

The most appropriate specification for I_c, corresponding to the above three types of collectors, is

- Flat-plate collectors: I_c = total hemispherical irradiation I_h.
- Tracking collectors with high concentration ($C > 10$): I_c = beam or direct irradiation I_b.
- Collectors with low concentration ($C < 10$): $I_c = (I_b + I_d/C)$ = irradiation within the acceptance angle [12], where I_d is the diffuse irradiation.

The thermal loss coefficient U depends only weakly on the temperatures T_r and T_a. In many cases it can be treated as a constant. A simple correction procedure for cases when U is temperature-dependent is given in Ref. 23.

In practice it is easier to measure the collector fluid temperature than the receiver temperature. The mean fluid temperature T_m is the average of the inlet and outlet fluid temperature T_{in} and T_{out}:

$$T_m = \frac{T_{in} + T_{out}}{2} \tag{25}$$

If the temperature rise along the collector is sufficiently small that the collector can be treated as if all the fluid were at the mean temperature T_m, then the efficiency equation takes the simple form

$$\eta = F_m\left(\eta_0 - U\frac{T_m - T_a}{I_c}\right) \tag{26}$$

where F_m is a collector parameter which accounts for the heat transfer from the absorber surface to the fluid. F_m is called the collector efficiency factor or heat transfer factor and is also known by the symbol F'. It depends on the construction of the collector but is nearly independent of operating conditions. For flat-plate collectors typical values are in the range of 0.8 to 0.9 for nonevacuated air collectors, 0.9 to 0.95 for nonevacuated liquid collectors, and 0.95 to 1 for evacuated tube collectors.

For some applications it is more convenient to specify the fluid inlet temperature T_{in} than T_m. In terms of T_{in}, the efficiency equation reads

$$\eta = F_{in}\left(\eta_0 - U\frac{T_{in} - T_a}{I_c}\right) \tag{27}$$

where F_{in} is called the heat removal factor, defined by the equation

$$F_{in} = \frac{Wc_p}{UA}\left(1 - \exp\frac{-UAF_m}{Wc_p}\right) \tag{28}$$

When efficiency data are plotted as η versus $(T_m - T_a)/I_c$, they can be approximated by a straight line of intercept $F_m\eta_0$ and slope $-F_mU$. If plotted versus $(T_{in} - T_a)/I_c$, the efficiency curve has intercept $F_{in}\eta_0$ and slope $-F_{in}U$. For practical applications there is no need to determine F_m separately.

Standard thermal test procedures, such as ASHRAE no. 93-77 [24], determine only the instantaneous or average thermal efficiency of a single collector module. To evaluate the performance of a collector field or a collector–heat exchanger system, losses from transfer lines between individual collectors and between the field and the point of end use can be accounted for by modifying η_0 and U, as shown by Beckman [25]. The effect of a heat exchanger can be treated by a similar method, as shown by de Winter [26].

For systems with a heat exchanger, the right-hand side of Eq. (27) is multiplied by the heat exchanger factor F_x:

$$F_x = \frac{1}{1 + [F_{in}UA/(Wc_p)_c]\{(Wc_p)_c/[\epsilon(Wc_p)_{min} - 1]\}} \tag{29}$$

where $(Wc_p)_c$ = fluid capacitance rate in collector field
$(Wc_p)_p$ = fluid capacitance rate in process loop
$(Wc_p)_{min}$ = minimum $[(Wc_p)_c, (Wc_p)_p]$
$(Wc_p)_{max}$ = maximum $[(Wc_p)_c, (Wc_p)_p]$

and ϵ is the heat exchanger effectiveness (see Chap. 4), which for single-pass counterflow is given by

$$\epsilon = \frac{1 - \exp B}{1 - C^* \exp B} \tag{30}$$

where $C^* = \dfrac{(Wc_p)_{min}}{(Wc_p)_{max}}$

$$B = \frac{(UA)_{HX}(C^* - 1)}{(Wc_p)_{min}}$$

where $(UA)_{HX}$ is the product of area and heat transfer coefficient of the heat exchanger. Use of F_x allows one to treat the collector plus heat exchanger combination as if it were a simple collector with efficiency

$$\eta = F_x F_{in}\left(\eta_0 - U\frac{T_{in} - T_a}{I_c}\right) \tag{31}$$

without having to consider the inlet and outlet temperatures of the collector itself.

If there are losses from the ducts or pipes between collector and point of use, one can treat the collector-plus-pipe combination as if it were a simple collector with modified optical efficiency η_0' and modified U value U':

$$\eta_0' = \frac{1}{1 + U_{pipe}A_{out}/Wc_p}\,\eta_0 \tag{32}$$

and

$$U' = \frac{1 - \dfrac{U_{pipe}A_{in}}{Wc_p} + \dfrac{U_{pipe}(A_{in} + A_{out})}{AF_{in}U}}{1 + \dfrac{U_{pipe}A_{out}}{Wc_p}}\,U \tag{33}$$

where U_{pipe} is the heat loss coefficient of the pipe insulation and A_{in} and A_{out} are the surface areas of the pipe on the inlet and outlet side of the collector, respectively. If a system contains both pipes and heat exchanger, one first modifies the collector parameters as described in this section and then one uses these modified parameters in the equation with the heat exchanger factor (Eq. 29).

D. LONG-TERM AVERAGE PERFORMANCE OF SOLAR ENERGY SYSTEMS

Standard collector test procedures [24] provide for measurements of the instantaneous efficiency under specified clear sky conditions. For practical applications, however, one needs to know the long-term average energy delivered by a real solar energy system, taking into account such effects as variable cloudiness and varying incidence angles. Prediction of system performance can be quite complicated. Frequently a computer simulation requiring considerable expense and expertise is necessary [27]. The computing time is often quite long for a meaningful system optimization [28].

Alternatively, shorthand calculational procedures, called design tools, can be used. They are suitable for certain standarized solar energy systems and are simple enough for programmable hand calculators. The best known design tools are the f chart [29] for active space heating systems and the utilizability method [19] for systems without thermal storage. Recently, these methods have been generalized to all collector types [18] and to certain systems with storage [30].

For an economic analysis of solar energy systems only a single performance index is needed, namely the energy saved by the system. This is the useful solar energy delivered by the system minus parasitic energy consumption for running pumps, controls, etc. For many applications the energy delivered by the solar energy system can be calculated by the procedure below [31]. It requires only the reading of a single graph and multiplication of the coordinates by the optical and thermal parameters of the collector field. This simple procedure reproduces the results of hourly computer calculations with an accuracy (rms error) on the order of 2 percent for flat plates and 2 to 4 percent for concentrators. Only three variables are needed: the operating threshold (for thermal collectors this is the average heat loss divided by the optical efficiency), the geographical latitude, and the yearly average direct normal insolation. This remarkable simplification is possible because to a solar collector all climates appear very much alike and differ only in their average direct normal insolation.

To achieve such simplification it is necessary to make certain standardizing assumptions:

- The collector uses all solar radiation above a specified threshold X, and the portion of the solar radiation that is above X is used with constant efficiency.

- Transient effects are neglected. This is of little consequence if the time constant of the collector field is significantly less than one hour.

- Flat plate and CPC are deployed at a tilt equal to the latitude.

- Standard spacing between concentrator modules is assumed to account for shading (no shading for flat plate and CPC; ground cover ratio equal to 0.5 for one-axis and 0.25 for two-axis tracking).

- Standard incidence angle modifiers are assumed [32].*

These assumptions are realistic for a wide range of applications.

*Note on correlations for central receiver: The results for the central receiver vary strongly with optical design. The optical design in turn depends on many factors, in particular the geographic latitude, the relative cost of heliostats and the remainder of the plant, and whether the heliostat field surrounds the tower (typical design for large installations) or whether it lies to the north of the tower (typical design for small installations). Investigation of these points is beyond the scope of a handbook chapter. We have considered only the incidence angle modifier in Table 2-1 of Ref. 31; it corresponds to a heliostat field which surrounds the tower and which is optimized for 35° latitude.

This method can be applied when the collector system operates year-round in such a way that no collected energy is discarded, and, in the case of thermal collectors, where the average threshold is known. This includes

- Photovoltaic systems
- Solar-augmented industrial process heat systems
- Solar thermal power systems

In addition, the method is useful for rating collectors of different types of different manufacturers on the basis of yearly average performance. Such a rating provides a convenient criterion for selecting the most cost-effective collector. The method is also useful for evaluating the effects of collector degradation, the benefits of collector cleaning, and the gains from collector improvements due to measures such as enhanced optical efficiency or decreased heat loss per absorber surface. For most of these applications, the method is accurate enough to replace a detailed system simulation.

Since only the yearly average direct normal irradiance is needed as input, one can use this method for locations where hourly insolation data are not available. The contour map of direct normal irradiance values shown in Fig. 13 permits determination of deliverable energy for any place in the United States. Direct insolation has been monitored only at very few stations in the world, while information for daily total hemispherical irradiation on a horizontal surface is available for a much larger number of locations. To make the present method usable for locations without direct normal insolation data, a correlation between yearly average clearness index $\overline{K}_T$ and direct normal irradiation $\overline{I}$ is provided in Fig. 14. The method is illustrated below by a few examples.

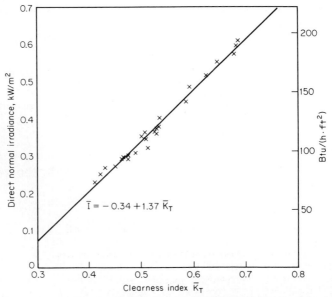

FIG. 14 Correlation between yearly average clearness index $\overline{K}_T$ and yearly average direct normal solar irradiance $\overline{I}$ during daytime [31].

1. Photovoltaic Systems

The conversion efficiency of a photovoltaic cell is nearly independent of insolation. Once the yearly average temperature of the cell has been determined for a given cell design and cell cooling mode, one can approximate the corresponding cell efficiency by a constant η_c. The yearly electricity production is then obtained by multiplying the yearly irradiation on the aperture, Q_0, for the collector type in question by η_c. Figure 15 is convenient for this purpose because it shows Q_0 directly, as a function of yearly average beam irradiance during daylight hours $\bar{I}$, for flat-plate, east-west, and two-axis collectors. Alternatively, Q_0 can be obtained from Figs. 16 to 2^3 by setting $X = 0$ and $F\eta_0 = 1$.

As an example, consider a flat-plate photovoltaic panel at a tilt angle equal to the latitude and with cell efficiency $\eta_c = 10$ percent. From the contour map in Fig. 13, one sees that the sunniest region of the United States has insolation values around $\bar{I} = 0.6$ kW/m^2. Reading the solid line in Fig. 15 corresponding to L = 35° (Albuquerque), one finds an available solar irradiation of 8.0 GJ/m^2. The corresponding electricity production is

$$Q_e = \eta_c Q_0 = 0.80 \text{ GJ/m}^2 \text{ per year} \tag{34}$$

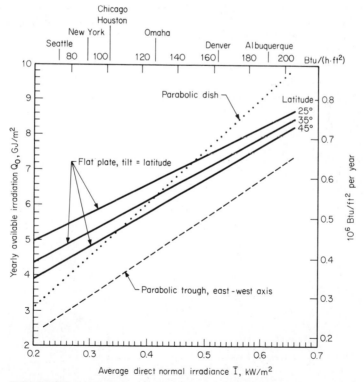

FIG. 15 Yearly total solar irradiation incident on aperture, Q_0, for flat plate, concentrator with east-west tracking axis, concentrator with two-axis tracking, and central receiver, as a function of average direct normal irradiance [31].

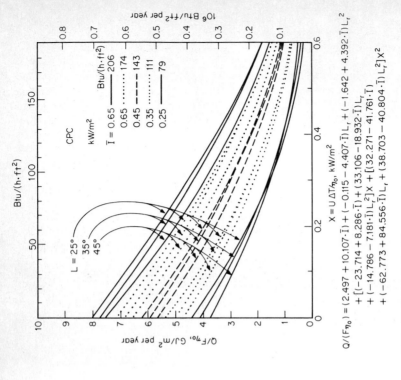

$$Q/(F_{\eta_0}) = (2.497 + 10.107 \cdot \bar{I}) + (-0.115 - 4.407 \cdot \bar{I})L_r + (-1.642 + 4.392 \cdot \bar{I})L_r^2$$
$$+ [(-23.714 + 8.286 \cdot \bar{I}) + (33.106 - 18.932 \cdot \bar{I})L_r$$
$$+ (-14.786 - 7.181 \cdot \bar{I})L_r^2]X + [(32.271 - 41.761 \cdot \bar{I})$$
$$+ (-62.773 + 84.556 \cdot \bar{I})L_r + (38.703 - 40.804 \cdot \bar{I})L_r^2]X^2$$

FIG. 17 Yearly collectible energy for CPC. For polynomial fit the correct units must be used: L_r = latitude in radians, $\bar{I}$ = insolation in kW/m², X = threshold in kW/m², and Q = collectible energy in GJ/m² [31].

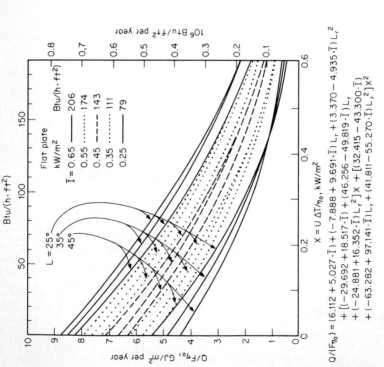

$$Q/(F_{\eta_0}) = (6.112 + 5.027 \cdot \bar{I}) + (-7.888 + 9.691 \cdot \bar{I})L_r + (3.370 - 4.935 \cdot \bar{I})L_r^2$$
$$+ [(-29.692 + 18.517 \cdot \bar{I}) + (46.256 - 49.819 \cdot \bar{I})L_r$$
$$+ (-24.881 + 16.352 \cdot \bar{I})L_r^2]X + [(32.415 - 43.300 \cdot \bar{I})$$
$$+ (-63.282 + 97.141 \cdot \bar{I})L_r + (41.811 - 55.270 \cdot \bar{I})L_r^2]X^2$$

FIG. 16 Yearly collectible energy for flat plate. For polynomial fit the correct units must be used: L_r = latitude in radians, $\bar{I}$ = insolation in kW/m², X = threshold in kW/m², and Q = collectible energy in GJ/m² [31].

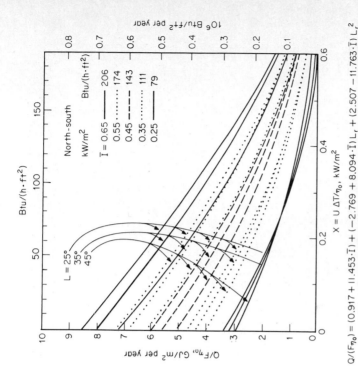

FIG. 19 Yearly collectible energy for concentrator tracking about horizontal north-south axis. For polynomial fit the correct units must be used: L_r = latitude in radians, $\bar{I}$ = insolation in kW/m^2, X = threshold in kW/m^2, and Q = collectible energy in GJ/m^2 [31].

$$Q/(F_{\eta_0}) = (0.917 + 11.453 \cdot \bar{I}) + (-2.769 + 8.094 \cdot \bar{I})L_r + (2.507 - 11.763 \cdot \bar{I})L_r^2$$
$$+ [(-9.480 - 12.976 \cdot \bar{I}) + (0.194 + 21.747 \cdot \bar{I})L_r$$
$$+ (9.298 - 34.109 \cdot \bar{I})L_r^2]X + [(11.954 - 3.755 \cdot \bar{I})$$
$$+ (-0.225 - 37.385 \cdot \bar{I})L_r + (-12.584 + 62.705 \cdot \bar{I})L_r^2]X^2$$

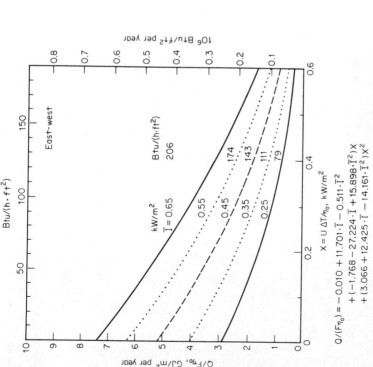

FIG. 18 Yearly collectible energy for concentrator tracking about horizontal east-west axis. For polynomial fit the correct units must be used: $\bar{I}$ = insolation in kW/m^2, X = threshold in kW/m^2, and Q = collectible energy in GJ/m^2 [31].

$$Q/(F_{\eta_0}) = -0.010 + 11.701 \cdot \bar{I} - 0.511 \cdot \bar{I}^2$$
$$+ (-1.768 - 27.224 \cdot \bar{I} + 15.898 \cdot \bar{I}^2)X$$
$$+ (3.066 + 12.425 \cdot \bar{I} - 14.161 \cdot \bar{I}^2)X^2$$

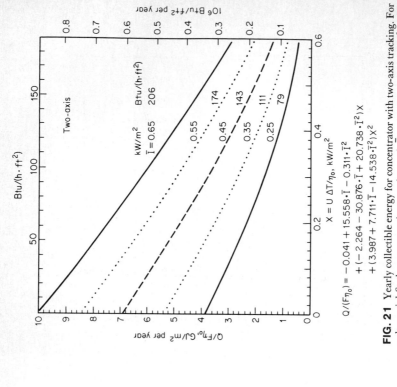

FIG. 20 Yearly collectible energy for concentrator tracking about polar axis. For polynomial fit the correct units must be used: $\bar{I}$ = insolation in kW/m^2, X = threshold in kW/m^2, and Q = collectible energy in GJ/m^2 [31].

FIG. 21 Yearly collectible energy for concentrator with two-axis tracking. For polynomial fit the correct units must be used: $\bar{I}$ = insolation in kW/m^2, X = threshold in kW/m^2, and Q = collectible energy in GJ/m^2 [31].

Figure 20 (Polar):

$$Q/(F\eta_0) = 0.048 + 13.834 \cdot \bar{I} - 0.049 \cdot \bar{I}^2$$
$$+ (-2.362 - 27.986 \cdot \bar{I} + 17.501 \cdot \bar{I}^2)X$$
$$+ (3.976 + 7.004 \cdot \bar{I} - 12.393 \cdot \bar{I}^2)X^2$$

Figure 21 (Two-axis):

$$Q/(F\eta_0) = -0.041 + 15.558 \cdot \bar{I} - 0.311 \cdot \bar{I}^2$$
$$+ (-2.264 - 30.876 \cdot \bar{I} + 20.738 \cdot \bar{I}^2)X$$
$$+ (3.987 + 7.711 \cdot \bar{I} - 14.538 \cdot \bar{I}^2)X^2$$

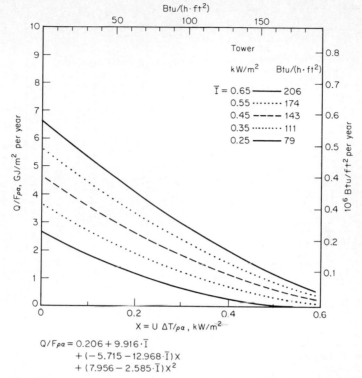

$$Q/F_{\rho a} = 0.206 + 9.916 \cdot \overline{I}$$
$$+ (-5.715 - 12.968 \cdot \overline{I}) x$$
$$+ (7.956 - 2.585 \cdot \overline{I}) x^2$$

FIG. 22 Yearly collectible energy for central receiver at 35° latitude; see p. 7-25, footnote. For polynomial fit the correct units must be used: $\overline{I}$ = insolation in kW/m², X = threshold in kW/m², and Q = collectible energy in GJ/m² [31].

The lowest beam irradiance occurs in the northwest and northeast. Reading the solid line corresponding to L = 45° and $\overline{I}$ = 0.23 kW/m² (northern part of New York State), one finds an electricity production of 0.42 GJ/m² for the same cell efficiency.

2. Thermal Systems

The efficiency equations of a thermal collector can be summarized in the form

$$\eta = F\left(\eta_0 - U\frac{T_c - T_a}{I_c}\right) \tag{35}$$

where $F = 1$ for $T_c = T_r$
$F = F_m$ for $T_c = T_m$
$F = F_{in}$ for $T_c = T_{in}$

The collector will be turned on whenever the insolation I_c exceeds the threshold X defined by

$$X = \frac{U(T_c - T_a)}{\eta_0} \tag{36}$$

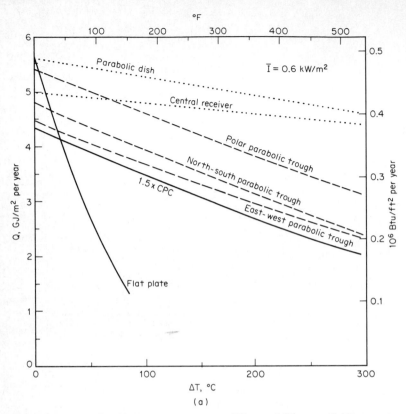

FIG. 23 Yearly collectible energy versus average difference ΔT between fluid temperature and ambient for typical collectors. (a) $\bar{I}$ = 0.6 kW/m². (b) $\bar{I}$ = 0.3 kW/m² [31].

X is the abscissa of Figs. 16 to 23, while the ordinate is the yearly collectible energy Q in the form $Q/F\eta_0$ in units of GJ/m² per year or millions of Btu per square foot per year.

To illustrate how the correlations can be used for thermal collectors, consider a parabolic trough with the following parameters:

$$F_m\eta_0 = 0.65 \quad \text{and} \quad F_m U = 0.67 \text{ W}/(\text{m}^2 \cdot {}^\circ\text{C})$$

If this collector is to operate in Albuquerque, New Mexico, at a mean fluid temperature $T_m = 300°C$, the threshold is

$$X = 0.67 \times \frac{300 - 13}{0.65} \text{ W}/\text{m}^2 = 0.294 \text{ kW}/\text{m}^2$$

since the yearly average ambient temperature is $T_a = 13°C$. The yearly average beam insolation is $\bar{I} = 0.6$ kW/m², from Fig. 13. Interpolation between the $\bar{I} = 0.55$ and 0.65 curves in Fig. 18 yields $Q/F\eta_0 = 3.65$ GJ/m². Hence, this collector can deliver $Q = 2.37$ GJ/m² per year if it is installed with an east-west tracking axis. To illustrate the effect of dirt and degradation, let us assume that the optical efficiency is reduced by 10

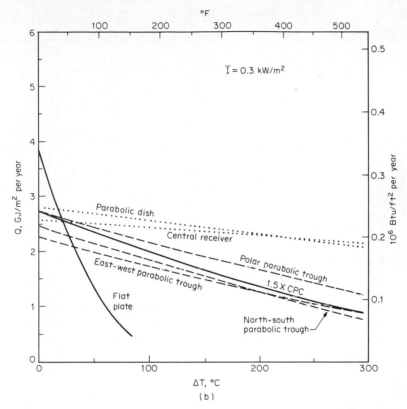

FIG. 23 (*Continued.*)

percent to $F_m\eta_0 = 0.585$, averaged over the lifetime of the collector, while the U value remains constant. Then the threshold increases to 0.327 kW/m^2 and the yearly output is only 1.96 GJ/m^2.

A standard collector test [24] determines only the parameters of a single collector module, but performance calculations for an actual installation must include losses from heat transfer lines between collector and point of use, as discussed in Eqs. (31) to (33) above.

For a direct comparison between several typical collectors, Fig. 23 shows the yearly average energy delivery Q versus operating temperature (as $\Delta T = T_m - T_a = $ difference between fluid and ambient temperatures) at a latitude of $35°$ for two climates: (1) a sunny climate with average beam irradiant $\bar{I} = 0.6 \text{ kW/m}^2$, and (2) a fairly cloudy one with $\bar{I} = 0.3 \text{ kW/m}^2$. The collector parameters, listed in Table 6, are typical values based on standard outdoor tests. One striking conclusion from this comparison concerns the crossover between flat-plate and concentrating collectors. At temperatures more than approximately $25°C$ above ambient, all of the concentrating collectors in Fig. 23 surpass the flat-plate in performance. This conclusion holds even for relatively cloudy climates where concentrating collectors are at a disadvantage because they cannot collect much diffuse insolation. But the high heat loss of flat-plate collectors turns out to be more

TABLE 6 Typical Collector Parameters Assumed for Comparison in Fig. 23

Collector type	$F_m\eta_0$	F_mU W/(m²·°C)	F_mU Btu/(h·ft²·°F)
Flat-plate, double glazed[a]	0.70[b]	5.0	0.88
CPC 1.5X with evacuated tubes	0.60[c]	0.75[c]	0.13
Parabolic trough[d]	0.65[e]	0.67	0.12
Parabolic dish[d]	0.61[e]	0.27	0.048
Power tower[f]	$\rho\alpha = 0.81$[g]	0.16	0.028

[a]Average of four best double-glazed collectors, Florida Solar Energy Center, Summary Test Package, issued Feb. 15, 1978, and Supplements 1 (April 1978) and 2 (June 1978).

[b]With respect to pyranometer.

[c]Typical values based on test results reported in Ref. 33; optical efficiency with respect to radiation within acceptance angle.

[d]Typical test results from Ref. 33.

[e]With respect to pyrheliometer.

[f]Projected performance (see footnote on p. 7-25).

[g]ρ = heliostat reflectance; α = receiver absorptance.

important; at times when the insolation is above the threshold for a flat-plat collector (with $\Delta T > 25°C$) there is enough direct radiation to run a concentrating collector. Of course, energy delivery is only the denominator of the cost per energy ratio. The best choice of a collector depends as much on its capital and operating costs as on its thermal performance.

REFERENCES

1. F. Kreith and J. F. Kreider, *Principles of Solar Engineering,* 1st ed., McGraw-Hill, New York, 1978.

2. J. F. Kreider and F. Kreith, eds., *Solar Energy Handbook,* 1st ed., McGraw-Hill, New York, 1981.

3. M. P. Thekaekara, "The Solar Constant and Spectral Distribution of Solar Radiant Flux," *Sol. Energy* **9**:7 (1965).

4. J. R. Hickey et al., "Extraterrestrial Solar Radiation Variability: Two and One-Half Years of Measurements from Nimbus 7," *Sol. Energy* **28**:443 (1982).

5. K. L. Coulson and Y. Howell, "Solar Radiation Instruments," *Sunworld* **4**:87 (1980).

6. D. F. Grether, A Hunt, and M. Wahlig, "Techniques for Measuring Circumsolar Radiation," *Lawrence Berkely Lab. Rep. LBL-8345* (1977).

7. A. Rabl and P. Bendt, "Effect of Circumsolar Radiation on Performance of Focusing Collectors," *ASME J. Sol. Energy Eng.* **104**:237 (1982).

8. A. M. Lind et al., "The Sensitivity of Solar Transmittance, Reflectance and Absorptance to Selected Averaging Procedures and Solar Irradiance Distributions," *ASME J. Sol. Energy Eng.* **102**:34 (1980).

9. R. E. Bird and R. L. Hulstrom, "A Simplified Clear Sky Model for Direct and Diffuse Insolation on Surfaces," *Rep. SERI/TR-642-761* (1981), Solar Energy Research Institute, Golden, CO 80401. (Also available from NTIS.)

10. A. P. Thomas and M. P. Thekaekara, "Experimental and Theoretical Studies on Solar Energy for Energy Conversion," *Proc. 1976 Ann. Meeting Am. Sec. Int. Sol. Energy Soc.,* Winnipeg, Canada, 1976.

11. C. Frohlich, "Contemporary Measure of the Solar Constant," in O. R. White, ed., *The Solar Output and Its Variations,* pp. 93–109, Colorado Association of University Presses, Boulder, 1977.

12. A. Rabl, "Concentrating Collectors," Chap. 10 of W. C. Dickinson and P. H. Cherminoff, eds., *Solar Energy Technology Handbook,* Marcel Dekker, Inc., New York, 1980.

13. J. E. Hay, "Measurement and Modeling of Shortwave Radiation on Inclined Surfaces," *3d Conf. Atm. Rad.,* pp. 28–30, Davis, Calif. June 1978; published by American Meteorological Society, Boston.

14. M. Collares-Pereira and A. Rabl, "The Average Distribution of Solar Radiation—Correlations between Diffuse and Hemispherical and between Daily and Hourly Insolation Values," *Sol. Energy* **22**:155 (1979).

15. C. M. Randall and M. E. Whitson, "Final Report, Hourly Insolation and Meteorological Data Bases Including Improved Direct Insolation Estimates," *Aerospace Corp. Rep. No. ATR-78 (7592)-1* (December 1977).

16. "SOLMET," Volume 1, User's Manual TD-9724, Hourly Solar Radiation—Surface Meteorological Observations," National Climatic Center, Asheville, NC 18801 (1978).

17. B. Y. H. Liu and R. C. Jordan, "The Interrelationship and Characteristic Distribution of Direct, Diffuse and Total Solar Radiation," *Solar Energy* **4**:1 (1960).

18. M. Collares-Pereira and A. Rabl, "Derivation of Method for Predicting Long-Term Average Energy Delivery of Solar Collectors," *Solar Energy* **23**:223 (1979); and "Simple Procedure for Predicting Long-Term Average Performance of Non-concentrating and of Concentrating Solar Collectors," *Solar Energy* **23**:235 (1979).

19. B. Y. Liu and R. C. Jordan, "A Rational Procedure for Predicting the Long-Term Average Performance of Flat-Plate Solar Energy Collectors," *Solar Energy* **7**:53 (1963).

20. P. Bendt, M. Collares-Pereira, and A. Rabl, "The Frequency Distribution of Daily Insolation Values," *Solar Energy* **27**:1 (1981).

21. S. A. Klein, "Calculation of Flat-Plate Collector Utilizability," *Solar Energy* **21**:393 (1978).

22. A. Rabl, "Comparison of Solar Concentrators," *Solar Energy* **18**:93 (1976).

23. P. I. Cooper and R. V. Dunkle, "A Non-linear Flat-Plate Collector Model," *Solar Energy* **26**:133 (1981).

24. "Collector Test Procedure of the American Society of Heating, Refrigeration and Air Conditioning Engineers," *ASHRAE Standard 93-77,* ASHRAE, 343 W. 43d St., New York, NY 10036.

25. W. A. Beckman, "Duct and Pipe Losses in Solar Energy Systems," *Solar Energy* **21**:53 (1978).

26. F. de Winter, "Heat Exchanger Penalties in Double Loop Solar Water Heating Systems," *Solar Energy* **17**:335 (1975).

27. S. A. Klein et al., "TRNSYS—A Transient Simulation Program: Users Manual," *Engineering Experiment Sta. Rep. 38,* Solar Energy Laboratory, University of Wisconsin, Madison.

28. M. E. Fewell and N. R. Grandjean, "User's Manual for Computer Code SOLTES-1 (Simulator of Large Thermal Energy Systems)," *Sandia Labs. Rep. SAND-78-1315* (June 1979).

29. W. A. Beckman, S. A. Klein, and J. A. Duffie, *Solar Heating Design by the f-chart Method,* Wiley, New York, 1977.

30. J. A. Duffie and W. A. Beckman, *Solar Engineering of Thermal Processes,* Wiley, New York, 1980.

31. A. Rabl, "Yearly Average Performance of the Principal Solar Collector Types," *Solar Energy* **27**:215 (1981).

32. H. W. Gaul and A. Rabl, "Incidence Angle Modifier and Average Optical Efficiency of Parabolic Trough Collectors," *Trans. ASME J. Sol. Energy Eng.* **102**:16 (1980).

33. V. E. Dudley and R. M. Workhoven, "Summary Report: Concentrating Solar Collector Test Results Collector Module Test Facility," *Sandia Labs, Ref. SAND-78-0977,* Albuquerque, N.M. (1979).

34. E. R. G. Eckert, C. L. Tien, and D. K. Edwards, "Radiation," Chap. 14 of W. M. Rohsenow, J. P. Hartnett, and E. N. Ganić, eds., *Handbook of Heat Transfer Fundamentals,* McGraw-Hill, New York, 1985.

35. E. N. Ganić, J. P. Hartnett, and W. M. Rohsenow, "Basic Concepts of Heat Transfer," Chap. 1 of W. M. Rohsenow, J. P. Hartnett, and E. N. Ganić, eds., *Handbook of Heat Transfer Fundamentals,* McGraw-Hill, New York, 1985.
36. P. J. Schneider, "Conduction," Chap. 4 of W. M. Rohsenow, J. P. Hartnett, and E. N. Ganić, eds., *Handbook of Heat Transfer Fundamentals,* McGraw-Hill, New York, 1985.

NOMENCLATURE

Symbol, Definition, SI Units, English Units

A collector aperture area: m^2, ft^2

A_r receiver area: m^2, ft^2

C geometric concentration ratio $= A/A_r$

c_p specific heat at constant pressure: $J/(kg \cdot K)$, $Btu/(lb_m \cdot °F)$

D_λ percentage of solar spectrum at wavelengths shorter than λ: %

E_λ solar spectral irradiance at wavelength λ: $W/(m^2 \cdot \mu m)$, $Btu/(h \cdot ft^2 \cdot \mu m)$

F heat transfer factor for solar collectors, equal to 1 if $T_c = T_r$, F_m if $T_c = T_m$, F_{in} if $T_c = T_{in}$

F_x heat exchanger penalty factor

H_d daily diffuse irradiation on horizontal surface: J/m^2, Btu/ft^2

$\overline{H}_d$ monthly average of daily diffuse irradiation on horizontal surface: J/m^2, Btu/ft^2

H_h daily hemispherical irradiation on horizontal surface: J/m^2, Btu/ft^2

$\overline{H}_h$ monthly average of daily hemispherical irradiation on horizontal surface: J/m^2, Btu/ft^2

H_0 daily extraterrestrial irradiation on horizontal surface: J/m^2, Btu/ft^2

$\overline{I}$ annual average direct normal solar irradiance during daylight hours: kW/m^2, $Btu/(h \cdot ft^2)$

I_b beam (or direct) solar irradiance: W/m^2, $Btu/(h \cdot ft^2)$

I_c irradiance on collector aperture: W/m^2, $Btu/(h \cdot ft^2)$

I_d diffuse solar irradiance (hourly average): W/m^2, $Btu/(h \cdot ft^2)$

I_h hemispherical solar irradiance (hourly average): W/m^2, $Btu/(h \cdot ft^2)$

I_0 solar constant, equal to extraterrestrial solar irradiance at normal incidence: W/m^2, $Btu/(h \cdot ft^2)$

K_T clearness index

$\overline{K}_T$ monthly average clearness index

$\mathbf{L}$ geographic latitude: deg

$\mathbf{L}_r$ geographic latitude: rad

N number of day of year, starting January 1

Q yearly collectible energy: $GJ/(m^2 \cdot yr)$ $Btu/(ft^2 \cdot yr)$

Q_0 yearly irradiation on aperture: $GJ/(m^2 \cdot yr)$ $Btu/(ft^2 \cdot yr)$

$\dot{Q}$ instantaneous rate of heat collection: W, Btu/h

T_a ambient temperature: K, °F

T_c collector temperature: K, °F

T_{in} fluid inlet temperature: K, °F

T_m $(T_{in} + T_{out})/2$: K, °F

T_{out} fluid outlet temperature: K, °F

T_r receiver temperature: K, °F

t length of day $= 24$ h

U collector heat loss coefficient; heat transfer coefficient or U value: $W/(m^2 \cdot K)$, $Btu/(h \cdot ft^2 \cdot °F)$

W flow rate through collector: kg/s, lb_m/h

X threshold: kW/m^2, $Btu/(h \cdot ft^2)$

Greek Symbols

α absorptance for solar radiation

β collector tilt angle: deg

γ azimuth angle of collector: deg

γ_s solar azimuth angle: deg

δ_s solar declination: deg

ϵ heat exchanger effectiveness

η collector efficiency

η_0 collector optical efficiency

θ_z zenith angle: deg

λ wavelength: μm

ζ solar incidence angle: deg

ρ reflectance for solar radiation

ψ solar altitude angle: deg

ω hour angle: deg

ω_s sunset hour angle: deg

CHAPTER 8

Thermal Energy Storage

By **John A. Clark**

Department of Mechanical Engineering and
Applied Mechanics
University of Michigan, Ann Arbor

A. INTRODUCTION

In most energy conversion and/or utilization systems there is seldom a perfect coincidence in time between the availability of an energy source and the demand for its utilization. In the absence of such a coincidence some form of energy storage is both necessary and convenient if the process for which the energy is used is to proceed uninterrupted. This circumstance is so commonplace that it often is overlooked. The gasoline in the fuel tanks of vehicles, a coal pile, natural gas in reservoirs, the kinetic energy in a flywheel, and water behind a hydroelectric dam are common examples of energy storage. In the field of thermal energy utilization, also, storage is a fundamental process, long used. A steam accumulator, a brick-lined regenerator, the thick walls of an adobe house are each an example of the deposition of thermal energy at one point in time for use at a later time.

Since the early 1970s energy conservation and the search for alternative energy sources have become prominent. Two important components of this activity which are dealt with in this chapter from the standpoint of thermal storage are waste heat recovery and solar thermal energy utilization. The emphasis will be on solar thermal conversion systems because of the wide application of this technology. However, the characteristics of thermal storage are quite similar for each of these technical applications. Thermal energy storage includes energy transfer processes in systems having temperatures both above the ambient (heating) and below the ambient (cooling).

A comprehensive discussion of thermal storage systems that includes much useful material of a practical nature has been published by Cole et al. [1]. A thorough technical discussion of thermal storage including its engineering fundamentals is found in a book by Duffie and Beckman [2] devoted to general solar thermal conversion processes. A comprehensive and practically oriented book on thermal storage has been published by Schmidt and Willmott [43].

Energy storage is normally accomplished by one of three methods: in the form of sensible heat of either a solid or a fluid medium, by a process of phase change in a substance, or as chemical energy in a chemical reaction. The choice of the particular substance or process depends largely on the application and the economics of the process. In liquid-based systems water or oil is commonly employed. Air-based systems generally use a packed bed of pebbles or other solid material, usually in a coarse granular form, for thermal storage. Phase-change materials (PCMs), also known as latent-heat materials, can be used with either fluid. Each of these storage systems will be discussed in this chapter from the standpoint of engineering design.

Duffie and Beckman [2] identify the following major characteristics of a thermal storage system: (1) its thermal storage capacity, by volume or by mass, (2) the temperatures at which the system receives thermal energy and the temperature at which the energy is recovered, (3) the means by which energy is stored or recovered by the system and the associated temperature differences, (4) the stratification of temperature within the storage mass, (5) the power consumption during charging and recovery of energy, (6) the containers, tanks, or other structural elements of the storage system, (7) the means for controlling thermal loss (or gain) from the storage system, and (8) its cost.

Two typical solar conversion systems including storage are shown, in Fig. 1 for an air-based system and in Fig. 2 for a liquid-based system [1]. Important factors affecting the performance of such systems may be understood from these figures. The purpose of these solar systems is to convert energy falling on the collectors from the sun into useful thermal energy at approximately the collector temperature T_C, which is then delivered ultimately to the load at temperature T_H through a system of ducts, heat exchangers, and

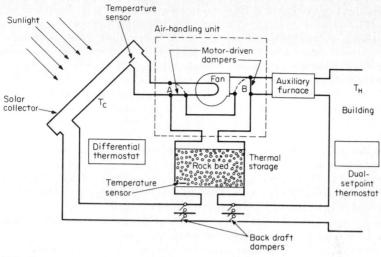

FIG. 1 Typical air-based system [1].

the storage unit. The relationship between the highest collected temperature T_C and the delivery temperature T_H may be expressed generally as [2]

$$T_C - T_H = \Delta T_1 \quad \text{(transport from collector to storage)}$$

$$+ \ \Delta T_2 \quad \text{(into storage, including heat exchanger temperature drop)}$$

$$+ \ \Delta T_3 \quad \text{(storage loss)} \tag{1}$$

$$+ \Delta T_4 \quad \text{(out of storage)}$$

$$+ \Delta T_5 \quad \text{(transport from storage to application)}$$

$$+ \Delta T_6 \quad \text{(into application)}$$

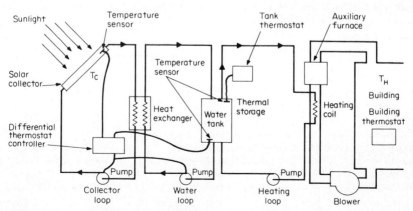

FIG. 2 Typical liquid-based system [1].

The useful energy obtained by thermal conversion in the collector is reduced as T_C is increased for a fixed T_H. Accordingly, the temperature of the collector, which controls the useful gain, is increased above the temperature of delivery (utilization) T_H by a series of temperature differences generated by thermal losses in ductwork, thermal transport in heat exchangers, thermal mixing and transport in storage, and finally thermal delivery to the load. One objective of system design is to reduce this string of temperature differences to their minimum values, thus bringing T_C to a temperature as close to that of the load, T_H, as economically and practically as possible. Temperature differences within the storage system, $\Delta T_2 + \Delta T_3 + \Delta T_4 + \Delta T_5$, though inherent in any energy transport process, represent important contributions to this string.

The thermal storage capacity of sensible heat materials is described by one of two factors: the heat capacity per unit mass, or the specific heat, c_p, in Btu/(lb$_m$·°F) [J/(kg·°C)]; and the heat capacity per unit volume, $\rho c_p (1 - \epsilon)$, in Btu/(ft^3·°F) [MJ/(m^3·°C)], where ρ is the mass density of the homogeneous material and ϵ is the void fraction of an ensemble of particles (in liquid-based systems ϵ is zero, and for packed-bed storage units ϵ is usually in the range 0.30 to 0.40, depending on particle size and packing arrangement). Typical values of these properties are given in Table 1. The large value of the volumetric heat capacity for liquid water in comparison with the other common materials is to be noted. Liquid water has the largest value of ρc_p of all the ordinary materials that would be considered for thermal storage applications. For this reason liquid water is widely used for this purpose and will doubtless always hold its prominent place as a material for sensible heat storage. As can be seen from Table 1 the volume of rock with 30 percent voids required to store a given quantity of thermal energy is 2.57 times that required of liquid water to store the same quantity of energy. This is one of the reasons water is often the preferred storage medium.

A listing of commonly considered phase-change materials (PCM) is given in Table 2. A more complete table is given in Sec. G, on phase-change storage. Those given in Table 2 are the salt hydrates, the most widely used PCMs, and are listed for comparison with the sensible heat materials in Table 1. Sodium sulfate decahydrate (Na$_2$So$_4$·10H$_2$O) has a volumetric heat capacity ρi_{sl} of 9941 Btu/ft^3 (369,380 kJ/m^3). Considering only the process of phase change, the volume of this substance needed to store one million Btu of thermal energy is 100.59 ft^3 (2.85 m^3). The corresponding volumes of liquid water and rock to store the same quantity of energy for a 20°F (11.1°C) change in temperature are 801.28 ft^3 (22.69 m^3) and 2057.61 ft^3 (58.27 m^3), respectively. In other words, the rock

TABLE 1 Sensible Heat Storage Materials [1]

Material	Density ρ_s, lb$_m$/ft^3 (kg/m^3)	Heat capacity c_{p_s}, Btu/(lb$_m$·°F) [J/(kg·°C)]	Volumetric heat capacity, $\rho_s c_{p_s}(1 - \epsilon)$, Btu/(ft^3·°F) [MJ/(m^3·°C)]	
			No voids ($\epsilon = 0$)	30% voids ($\epsilon = 0.30$)
Water	62.4 (1000)	1.00 (4180)	62.4 (4.18)	
Scrap iron	489 (7830)	0.11 (460)	53.8 (3.61)	37.7 (2.53)
Scrap aluminum	168 (2690)	0.22 (920)	36.96 (2.48)	25.9 (1.74)
Scrap concrete	140 (2240)	0.27 (1130)	27.8 (1.86)	26.5 (1.78)
Rock	167 (2680)	0.21 (879)	34.7 (2.33)	24.3 (1.63)
Brick	140 (2240)	0.21 (879)	29.4 (1.97)	20.6 (1.38)

NOTE: SI units in parentheses follow English units.

TABLE 2 Latent Heat Storage Materials [1]

Material	Melting point T_m, °F (°C)	Heat of fusion, i_{sl}, Btu/lb$_m$ (kJ/kg)
Calcium chloride hexahydrate	84.9 (29.4)	73.1 (170)
Sodium carbonate decahydrate	91.0 (33)	108.0 (251)
Disodium phosphate dodecahydrate	97.0 (36)	114.0 (280)
Sodium sulfate decahydrate	90.3 (32.4)	109.0 (253)
Sodium thiosulfate pentahydrate	120.0 (49)	86.0 (200)
N-octodecane	82.4 (28.0)	105.0 (243)
N-eicosane	98.1 (36.7)	106.0 (247)
Polyethylene glycol 600	68–77 (20–25)	63.0 (146)
Stearic acid	156.9 (69.4)	85.5 (199)
Water	32.0 (0.0)	143.1 (333.4)
Tristearin	133.0 (56)	82.1 (190.8)
Sunoco 116 paraffin wax	116.0 (47)	90.0 (209.2)
Sodium sulfate decahydrate/sodium chloride/ammonium chloride eutectic	55.0 (13)	78.0 (181.3)

storage volume is 20 times that of the PCM, and the water storage volume is 8 times that of the PCM. This vastly smaller volume for equivalent thermal storage for the PCM is the principal reason it is considered as a storage material. Although in practice some of the total storage volume in a PCM system must be taken up with the heat transfer fluid (finite voids), the comparisons indicated are valid as a guide.

B. GENERAL SIZING CONSIDERATIONS

The sizing of a thermal storage unit considers the amount and type of storage material to provide a given quantity of thermal energy, the selection of insulation to meet performance criteria, the proportioning of storage mass to collector area for solar applications, and economic factors. The simplest of these design procedures is to employ approximate methods and certain rules of thumb. These approximate methods give generally satisfactory results and may be easily and rapidly used for both final design calculations and initial estimates. More complex calculation procedures are available and usually employ some kind of computer application, especially for economic optimization determination. Such methods and the sources of calculational procedures are given by Cole et al. [1] and Duffie and Beckman [2]. The presentation here will focus on the approximate methods because of their practical simplicity, while still providing acceptable results.

For solar applications in residential space heating and domestic water heating systems (Figs. 1 and 2), detailed computer simulations using hourly calculations have shown that above a certain size further increases in the mass of storage per unit area of collector have virtually no influence on improving the annual thermal performance of the system [1, 2, 3, 4]. A typical result of such calculations is shown in Fig. 3. These results have provided a practical rule of thumb for sizing thermal storage units appropriate to these applications. Like all such approximations, this rule of thumb must be viewed as one that provides a general estimate for sizing that will suffer in precision as it is applied to situations differing from those for which it was developed. It will become less useful for nonresidential heating applications and in climates significantly different from those in the continental United States. As may be seen in Fig. 3, the annual solar fraction becomes insen-

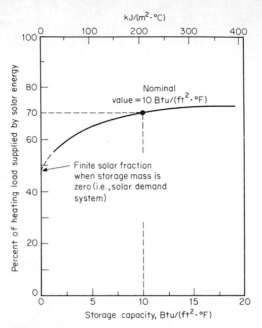

FIG. 3 Effect of varying storage capacity upon percentage of heating load supplied by solar energy (heating load and collector area remain constant) [1].

sitive to increases in the storage mass per unit of collector area when the storage capacity exceeds the value of 10 to 15 Btu/(ft$^2 \cdot °$F) or 200 to 300 kJ/(m$^2 \cdot °$C). Accordingly, the design criterion for a sensible heat storage unit is

$$\frac{m_s c_{p_s}}{A_c} = 10 \text{ to } 15 \text{ Btu/(ft}^2 \cdot °\text{F)} \tag{2}$$

$$= 200 \text{ to } 300 \text{ kJ/(m}^2 \cdot °\text{C)} \tag{3}$$

where m_s is the mass of storage, lb$_m$ (kg); c_{p_s} is the heat capacity of the storage material, Btu/(lb$_m \cdot °$F) [J/(kg$\cdot °$C)]; and A_c is the collector area, ft^2 (m^2). For a daily temperature change of 40°F (22°C), a storage unit can store from 400 to 600 Btu per square foot of collector (4400 to 6600 kJ/m^2), or approximately the energy that a normal solar collection system can convert during a sunny day in a North American winter. For storage masses that exceed this guideline the solar system cannot provide a useful amount of energy; also, both its thermal losses and cost will be increased. However, the penalty for slightly oversizing is not as severe as undersizing, so it is better to select a design that is on the high side of the criterion. Undersizing the storage unit can result in significant derating of the annual system performance. This primarily results from the increase in temperature of the collector and the corresponding loss in its efficiency from inadequate storage size. A small storage unit becomes more fully charged during high sun–low utilization periods and returns fluid to the collector system at a temperature greater than that which would be produced from a storage unit of larger capacity.

The storage size selection for phase-change materials (PCM) can be done using the criterion derived from Eqs. (2) and (3) for sensible heat systems for a selected daily temperature change. For the case indicated above, the corresponding storage capacity for a 40°F (22°C) daily temperature change is 400 to 600 Btu/ft^2 (4400 to 6600 kJ/m^2). Thus, for PCM units the criterion would be

$$\frac{m_s \, \Delta i}{A_c} = 400 \text{ to } 600 \text{ Btu/ft}^2 \tag{4}$$

$$= 4400 \text{ to } 6600 \text{ kJ/m}^2 \tag{5}$$

where Δi is the unit enthalpy change, in Btu/lb$_m$ (kJ/kg), of the PCM from liquid to solid phase, including the melting or freezing phase change, or

$$\Delta i = c_{pl}(T_0 - T_m) + i_{sl} + c_{ps}(T_m - T_{min}) \tag{6}$$

and c_{pl} = liquid-state heat capacity, Btu/lb$_m$·°F [J/(kg·°C)]
  i_{sl} = heat of fusion (latent heat) of the the PCM, Btu/lb$_m$ (kJ/kg)
  c_{ps} = solid-state heat capacity, Btu/(lb$_m$·°F) [J/(kg·°C)]
  T_0 = initial (high) temperature of the PCM, °F (°C)
  T_m = melting or freezing temperature of the PCM, °F (°C)
  T_{min} = minimum (or lower) temperature to which the PCM is cooled, °F (°C)

The thermodynamic properties for several PCMs are given in Table 2 and in Sec. G.

1. Estimate of Storage Volume

The volume of storage needed to supply the total domestic hot water requirements as well as supplying a specified quantity of energy for a given period of time for solar space heating, including energy to offset the effect of storage losses, may be estimated using the following equations for both sensible and latent heat storage units:

Sensible Heat Storage Units

$$V_s = \frac{[(UA + W_i c_{p_i}) + (UA)_s]F\theta + Q_w}{(\rho c_p)_s(1 - \epsilon)(T_0 - T_{min})} \tag{7}$$

Latent Heat (PCM) Storage Units

$$V_s = \frac{[(UA + W_i c_{p_i}) + (UA)_s]F\theta + Q_w}{\rho_s(1 - \epsilon)[c_{pl}(T_0 - T_m) + i_{sl} + c_{ps}(T_m - T_{min})]} \tag{8}$$

where UA = structure thermal load factor, Btu/(°F·day) [kJ/(°C·day)]
  $W_i c_{p_i}$ = daily air infiltration capacity rate* for the structure (corresponding to ½ to 1 air volume change per hour), Btu/(°F·day) [kJ/(°C·day)]
  $(UA)_s$ = storage unit (loss) thermal load factor, Btu/(°F·day) [kJ/(°C·day)]
  F = fraction of the month during which the storage unit is to supply energy (1 or 2 days usually, or ⅟₃₀ or ⅖₀)
  θ = monthly mean number of heating degree-days, °F·day (°C·day), for the location and month
  Q_w = total hot water load to be supplied from storage during this period, Btu (kJ)
  ϵ = void fraction in the storage unit (0 for liquid systems, 0.3 to 0.4 for packed beds, 0.2 to 0.4 for PCM units)

*See Chap. 9 of this handbook for more information on infiltration.

Example 1:

Determine the storage volume required to supply 1 day of energy for a building having a thermal load factor UA of 12,000 Btu/(°F·day) [22,790 kJ/(°C·day)], an air infiltration rate $W_i c_{p_i}$ of 2000 Btu/(°F·day) [3676 kJ/(°C·day)], a storage unit thermal loss factor $(UA)_s$ of 500 Btu/(°F·day) [950 kJ/(°C·day)], and a hot water load Q_w of 80 gal for this 1-day period or 66,733 Btu/day (70,410 kJ/day), in a location having a value of θ of 1400°F·day/month (778°C·day/month) for a maximum temperature change in the storage $(T_0 - T_{min})$ of 40°F (22.2°C), with T_0 of 150°F (65.6°C) and T_{min} of 110°F (43.3°C) for (1) liquid water ($\epsilon = 0$), (2) a rock bed ($\epsilon = 0.40$), and (3) latent heat material (PCM) of $Na_2S_2O_3 \cdot 5H_2O$ ($\epsilon = 0.20$), sodium thiosulfate pentahydrate ("hypo").

Solution

1. $V_s = \dfrac{(12{,}000 + 2000 + 500)(1/30)(1400) + 66{,}733}{(62.4)(1)(1)(40)}$ (9)

$= 298 \text{ ft}^3 \ (8.43 \text{ m}^3) \ (2230 \text{ gal})$

2. $V_s = \dfrac{(12{,}000 + 2000 + 500)(1/30)(1400) + 66{,}733}{(167)(0.21)(1 - 0.4)(40)}$ (10)

$= 883 \text{ ft}^3 \ (25.0 \text{ m}^3) \ (6605 \text{ gal})$

3. $V_s = \dfrac{(12{,}000 + 2000 + 500)(1/30)(1400) + 66{,}733}{(106.0)(1 - 0.2)[0.57(150 - 120) + 86 + 0.346(120 - 110)]}$ (11)

$= 82 \text{ ft}^3 \ (2.33 \text{ m}^3) \ (613 \text{ gal})$

2. Estimate of Thermal Loss

As important consideration for the design of thermal storage units is the insulation of the storage container to reduce the loss of heat. The HUD Intermediate Minimum Property Standards [5] specify that the storage unit should be insulated so that losses during a 24-hour period do not exceed 10 percent of the storage capacity. The specifications of the Sheet Metal and Air Conditioning Contractors' National Association (SMACNA) are somewhat more restrictive in requiring that losses be limited to no more than 2 percent of the storage capacity in 12 h. For sensible heat storage systems, the maximum value for $(UA)_s$ to meet either of these requirements may be determined from

$$\exp\left[-\frac{(UA)_s}{(mc_p)_s} t_0 \right] = 1 - f_0 \frac{T_0 - T_{min}}{T_0 - T_a} \qquad (12)$$

where $(UA)_s$ = loss factor for the storage unit, Btu/(h·°F) [kJ/(h·°C)]
$(mc_p)_s$ = product of the storage mass and its heat capacity, Btu/°F (kJ/°C)
t_0 = period considered for cooling, h
f_0 = fraction of specified heat loss
T_0 = initial temperature of storage, °F (°C)
T_{min} = minimum temperature for storage, °F (°C)
T_a = ambient temperature of the storage unit, °F (°C)

For small values of f (0.02 to 0.05) the result in Eq. (12) may be written to an acceptable approximation as

$$\frac{(UA)_s}{(mc_p)_s} t_0 = f_0 \frac{T_0 - T_{min}}{T_0 - T_a} \tag{13}$$

or

$$U_s = \frac{f_0 (mc_p)_s (T_0 - T_{min})}{A_s t_0 (T_0 - T_a)} \tag{14}$$

In Eqs. (12), (13), and (14) the values for $(UA)_s$ and U_s are the values that these quantities should not exceed in order to satisfy the requirement that in the period t_0 the losses from the storage not exceed the fraction f_0 of the storage capacity. Since the storage capacity Q is written as $(mc_p)_s (T_0 - T_{min})$, Eq. (14) becomes

$$U_s = \frac{f_0 Q / A_s t_0}{T_0 - T_a} \tag{15}$$

Equation (15) may be used to determine the maximum value of the overall heat transfer coefficient U_s, in Btu/(h·ft²·°F) [W/(m²·°C)], and thus the nature and thickness of insulation for the storage unit such that its losses will not exceed the fraction f of its storage capacity Q. Equation (15) may be used as an approximation for latent heat storage units by substituting $(T_m - T_a)$ for $(T_0 - T_a)$ and using $m_s i_{sl}$ for Q. Cole et al. [1] provide a very useful set of tables for the quantity $(f_0 Q / A_s t_0)$ for rectangular and cylindrical (horizontal and vertical) water tanks and rectangular containers for rock bed storage units. Some of the results of these authors for common tank shapes are given in Tables 3, 4, and 5.

TABLE 3 Insulation Factor $f_0 Q / A_s t_0$ for Rectangular Water Tanks [1]

Size, gal (liters)	Shape			
250 (946)	2.87 (9.05)	2.66 (8.39)	2.53 (7.98)	2.51 (7.92)
500 (1893)	3.62 (11.4)	3.35 (10.6)	3.19 (10.1)	3.17 (10.0)
750 (2839)	4.14 (13.1)	3.83 (12.1)	3.65 (11.5)	3.62 (11.4)
1000 (3785)	4.56 (14.4)	4.22 (13.3)	4.02 (12.7)	3.99 (12.6)
1500 (5678)	5.22 (16.5)	4.83 (15.2)	4.60 (14.5)	4.57 (14.4)
2000 (7571)	5.74 (18.1)	5.32 (16.8)	5.06 (16.0)	5.02 (15.8)
3000 (11356)	6.57 (20.7)	6.09 (19.2)	5.80 (18.3)	5.75 (18.1)
4000 (15142)	7.24 (22.8)	6.70 (21.2)	6.38 (20.1)	6.33 (20.0)
5000 (18927)	7.79 (24.6)	7.22 (22.8)	6.87 (21.7)	6.82 (21.5)

NOTE: Table values are for a 2-percent loss in 12 h with an assumed daily temperature range of 60°F (33°C). Table units are Btu/(h·ft²) (W/m²). To obtain the maximum allowable thermal transmittance for the side and top insulation, divide the insulation factor by the difference between the average storage temperature and the ambient temperature. The maximum allowable thermal transmittance for the bottom insulation is assumed to be twice that on the top and sides.

TABLE 4 Insulation Factor $f_0 Q / A_s t_0$ for Vertical Cylindrical Water Tanks [1]

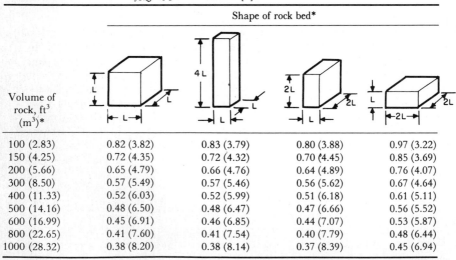

Size, gal (liters)	Shape D to 3D	(1/2)D	(1/3)D	(1/4)D
80 (303)	2.10 (6.62)*†	1.88* (5.93)	2.15 (6.78)	1.97 (6.21)
120 (454)	2.39 (7.54)	2.15 (6.78)	2.46 (7.76)	2.26 (7.13)
250 (946)	3.07 (9.68)	2.74 (8.64)	2.46 (9.91)	2.88 (9.09)
500 (1893)	3.87 (12.2)	3.46 (10.9)	3.96 (12.5)	3.63 (11.5)
750 (2839)	4.43 (14.0)	3.96 (12.5)	4.53 (14.3)	4.16 (13.1)
1000 (3785)	4.87 (15.4)	4.36 (13.8)	4.99 (15.7)	4.57 (14.4)
1500 (5678)	5.58 (17.6)	4.99 (15.7)	5.71 (18.0)	5.24 (16.5)
2000 (7571)	6.13 (19.3)	5.49 (17.3)	6.28 (19.8)	5.76 (18.2)
3000 (11356)	7.03 (22.2)	6.28 (19.8)	7.19 (22.7)	6.60 (20.8)
4000 (15142)	7.73 (24.4)	6.92 (21.8)	7.92 (25.0)	7.26 (22.9)
5000 (18927)	8.33 (26.3)	7.45 (23.5)	8.53 (26.9)	7.82 (24.7)

NOTE: Table values are for a 2 percent loss in 12 h with an assumed daily temperature range of 60°F (33°C). Table units are Btu/(h·ft²) (W/m²). To obtain the maximum allowable thermal transmittance of insulation, divide the insulation factor by the difference between the average storage temperature and the ambient temperature.

*The maximum allowable thermal transmittance of the bottom insulation is assumed to be twice that on the top and sides for tanks specified by the first two columns.

†The first column is applicable to all tanks with height of 1 to 3 times the diameter. Insulation factors for most domestic hot water tanks can be found in the first column.

TABLE 5 Insulation Factor $f_0 Q / A_s t_0$ for Rock Beds [1]

Volume of rock, ft³ (m³)*	Shape of rock bed*			
100 (2.83)	0.82 (3.82)	0.83 (3.79)	0.80 (3.88)	0.97 (3.22)
150 (4.25)	0.72 (4.35)	0.72 (4.32)	0.70 (4.45)	0.85 (3.69)
200 (5.66)	0.65 (4.79)	0.66 (4.76)	0.64 (4.89)	0.76 (4.07)
300 (8.50)	0.57 (5.49)	0.57 (5.46)	0.56 (5.62)	0.67 (4.64)
400 (11.33)	0.52 (6.03)	0.52 (5.99)	0.51 (6.18)	0.61 (5.11)
500 (14.16)	0.48 (6.50)	0.48 (6.47)	0.47 (6.66)	0.56 (5.52)
600 (16.99)	0.45 (6.91)	0.46 (6.85)	0.44 (7.07)	0.53 (5.87)
800 (22.65)	0.41 (7.60)	0.41 (7.54)	0.40 (7.79)	0.48 (6.44)
1000 (28.32)	0.38 (8.20)	0.38 (8.14)	0.37 (8.39)	0.45 (6.94)

NOTE: Table values are for a 2-percent loss in 12 h with an assumed daily temperature range of 50°F (28°C). Table units are in Btu/(ft²·h) (W/m²). To obtain the maximum allowable thermal transmittance for the side and top insulation, divide the insulation factor by the difference between the average storage temperature and the ambient temperature. The maximum allowable thermal transmittance for the bottom insulation is assumed to be twice that on the top and sides.

*The insulation is assumed to cover both plena, but the volume and shapes given are for the rocks only.

Example 2

Determine the value of $(UA)_s$ for the water storage tank in Example 1, part 1, that will limit the thermal loss to 2 percent ($f_0 = 0.02$) of the storage capacity in a period of 12 h for an ambient at 70°F (21.1°C).

Solution

Using Eq. (12),

$$\exp \left[-\frac{(UA)_s t_0}{(mc_p)_s} \right] = 1 - 0.02 \frac{150 - 110}{150 - 70}$$

$$= 0.99000$$

Since for the water storage $(mc_p)_s$ is 18,595 Btu/°F (35,316 kJ/°C),

$$(UA)_s = \frac{18,595}{12} \ln \frac{1}{0.99000}$$

$$= 15.57 \text{ Btu}/(\text{h} \cdot \text{°F})$$

$$= 374 \text{ Btu}/(\text{°F} \cdot \text{day}) \, [710 \text{ kJ}/(\text{°C} \cdot \text{day})]$$

The storage unit specified in Example 1, part 1, would therefore fail to satisfy the requirement of a 2 percent maximum heat loss in 12 h, since its $(UA)_s$ loss factor is given as 500 Btu/(°F·day)]. This loss factor would produce a 2.7 percent loss in 12 h, as may be found from Eq. (12).

C. THERMAL STRATIFICATION

The general sizing calculations outlined above have assumed that the temperatures in the storage unit were uniform throughout the volume at any time. That is, they represent the mean bulk temperatures of the storage mass. Actually, owing to the heat transfer and fluid flow processes within most thermal storage containers the temperature will be more or less distributed; i.e., a stratification of temperatures will exist in the container. Such a stratification is desirable for solar thermal applications, as it allows for the lowest system temperature to be introduced into the collector, thus improving its collection efficiency, while storing the higher-temperature energy for more useful thermal recovery. The effect of thermal stratification is believed to produce a 5 to 10 percent increase in thermal performance in comparison with an unstratified (completely mixed) storage [1]. The establishment of stratification depends on the kind of system.

For rock beds thermal stratification is relatively easy to establish, particularly for vertical, downward flow during charging. In a properly designed bed with heated air entering at the top, most of the energy in the air will be transferred to the rocks in the upper region of the bed. The flow is very complex through the matrix of rock and void, which provides a large surface area for heat transfer. As a consequence, even after several hours of charging, the temperature of the air flowing from the bottom of the bed will be essentially that of the initial bed temperature. When charging of the storage is complete a fairly steep temperature profile will be established in the bed with the higher temperatures in the upper regions. Because the rocks themselves have only point contact, heat transfer within the bed is small, so once charging is finished and the flow stopped, the thermal stratification tends to be maintained.

Thermal exchange by natural convection can cause a diffusion of the energy and destruction of the stratification. Little quantitive information on this is available at present, however. With the stratification so established, the air coming out of the bed at the top during thermal recovery will be close to the rock temperatures in the upper portion of the bed, thus giving a fairly rapid response from the bed.

Stratification in liquid tanks is more difficult to establish. The reason for this is that portions of the liquid at different temperatures have a tendency to mix. Further, the flow of heated liquid into the tank also produces a tendency for mixing. By maintaining the inlet velocities as low as possible and spreading the inlet fluid horizontally in the upper regions of the tank, stratification is enhanced. Flow diffusers and perforated plates installed in the tanks also have been used. Internal heat exchanger will disturb the thermal stratification. Because of its complexity stratification in liquid storage is not well understood at present. However, even with a mixed storage, liquid-based solar systems will perform satisfactorily.

D. OTHER STORAGE APPLICATIONS

Architectural designs that employ passive solar applications can take advantage of thermal storage in masonry walls, which allow both for storage and thermal distribution. Thermal energy collected by an outer surface during sunlight periods is transferred by thermal diffusion through the wall to the interior of a building, thus providing for solar collection, storage, and thermal transfer with appropriate time lag in a single structure. When the wall is separated from the outside by glass panels it is known as a Trombe wall [7]. For small, relatively simple buildings Judkoff and Sokol [8] describe an application of a Trombe wall employing a selective surface on the sun side. This installation, in Boulder, Colorado, required no auxiliary heating or cooling during its first year of operation, yet comfortable interior temperatures were maintained. Thermal storage in passive solar applications is also discussed in Ref. 9. Tamblyn [10] describes an active solar storage system (liquid) for a 60,000-ft^2 building in Toronto, Ontario, that is utilized to reclaim building waste heat, provide suitable capacity for off-peak electric resistance heating, and permit off-peak cooling in summer. Thermal storage associated with solar-hybrid electric power plants is discussed by Wen and Steele [11]. They show the primary controlling factors to be economic and conclude that as the cost of fuel increases at a rate greater than general inflation (taken to be 6 percent per year), thermal storage systems in the $40 to $70 per kWh (thermal) range can be expected to become cost effective.

The storage of thermal energy over a long period, known as annual-cycle thermal energy storage, is reported by Sillman [12] for residences and by Baylin et al. [13] and Baylin and Sillman [14] for community-size systems. For residential application it is shown that system output increases linearly with storage volume up to the point where the storage (liquid) tank is large enough to store all the heat collected during the summer. The ratio of storage volume to collector area under this circumstance ranges from 2 to 3 m^3/m^2. This may be compared with the value of 0.04 m^3/m^2 for the conditions of diurnal storage indicated in Fig. 3, indicating a tank size for annual storage to be approximately 50 times that for diurnal storage.

E. ROCK BED STORAGE DESIGN

The storage of thermal energy in air-based solar sytems and in certain other high-temperature applications using oil is usually accomplished using a bed of selected rocks. This device, sometimes called a pebble bed, is one form of a fixed (as opposed to fluidized) packed bed widely used in the chemical industry for various heat and mass transfer operations. Accordingly, its general performance is well understood and quite thoroughly documented. A rock bed arrangement recommended by the Solaron Corporation is shown in Fig. 4. For most applications in solar thermal storage a general understanding of performance and approximate design methods is satisfactory at present. However, as more refined designs are required for system optimization, greater detail regarding the flow distribution, thermal transfer, thermal stability, and temperature stratification will be required. The flow and transfer processes within a rock bed are so complex, however, that at present bed performance is most practically determined by approximate methods with strong dependence on empirical information. These approximate methods will be cited here primarily. In general, the principal governing parameters for design of a rock bed are rock type, mean size and shape, void fraction within the bed, rock and air (fluid) thermal properties, overall size of bed (cross-sectional area and length), rate of flow, and inlet temperature of the air (fluid).

Air (fluid) flow through the bed should be in the vertical direction, if possible. During the charging mode the flow should be in the downward direction so the rock and air at the top of the bed will at the higher temperatures. This arrangement enables a reasonably stable condition to be maintained during stagnation periods between charging and recovery modes and provides an immediate response of higher-temperature air during recovery, which should occur with upward flow. Although not always possible, it is better to avoid

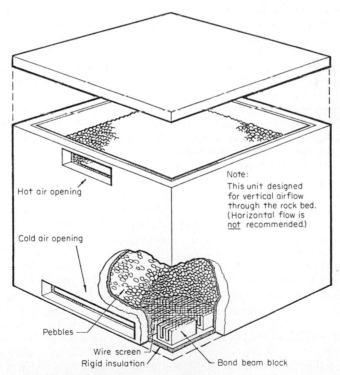

Note:
This unit designed
for vertical airflow
through the rock bed.
(Horizontal flow is
not recommended.)

Hot air opening

Cold air opening

Pebbles

Wire screen
Rigid insulation
Bond beam block

FIG. 4 Pebble-bed heat sotrage unit [15].

horizontal flow and flow in any kind of serpentine or baffled path. In such an arrangement the danger is great that the flow will bypass through clearance voidages caused by settling of the rock. The pressure drop and pumping power are apt to be excessive if the flow path is too complex or long.

1. Rock Type, Size, and Shape

Rocks suitable for thermal storage are natural rocks or stones usually obtained from gravel pits, river beds, or certain beaches. The ideal shape is a rounded rock with a smooth, hard surface. Rocks that conform to ASTM C 33, "Specifications for Concrete Aggregates," will be generally acceptable [1]. Soft materials such as limestone or sandstone should not be used, as they tend to crumble. Marble, limestone (hard), and dolomite, which are acceptable for dry conditions, react with water and carbon dioxide and could cause problems in moist cooling applications [1]. Materials that have large, flat faces or irregular shapes should be avoided to prevent either excessive, nonuniform voidage or consolidation and packing within the bed. During cyclic heating and cooling the bed will expand and contract and this introduces stresses in the rocks and settling of the bed. Certain manufactured rocks, such as hard Al_2O_3 in spherical form, would be acceptable. To eliminate fines, dust, and dirt, the supply of rocks, especially from natural sources, should be carefully screened and thoroughly washed. It should be remembered that in washing a pile of rocks the finer material will accumulate in the lower regions of the pile. The fine material if introduced into the bed will cause flow blockage and flow bypassing. Once a rock bed is loaded, it is very costly and difficult (if not impractical) to unload it. As indicated in Fig. 3 the bed will contain 50 to 75 lb_m of rock per square foot of collector (228 to 341 kg/m^2). Hence, a system having 300 ft^2 (28 m^2) of collector will have a rock bed containing from 15,000 to 22,500 lb_m of rock (6380 to 9550 kg). Fungus or other biological growths are a potential problem, but the author has yet to know of one case that is documented. Rock beds in heating service tend to be self-drying. This problem is discussed by Guyer et al. [44] and by Hogg in Ref. 54.

The size of rock for the most efficient and cost-effective storage unit is between 0.75 and 1.50 inches in diameter (1.91 to 3.81 cm) [15]. This dimension is the mean screened size of the natural stone or the equivalent diameter of the mean sphere from a representative sample from an ensemble of nonuniform rocks to be used in the bed, or

$$D_e = \left(\frac{6V_R}{n\pi}\right)^{1/3} \tag{16}$$

where V_R is the total volume of the sample of n rocks. To relate the deviation of a particle shape from the spherical, a shape factor ϕ may be defined [16] as

$$D_s = D_e\phi \tag{17}$$

where D_s is the effective particle diameter. Values of ϕ are given by Gabor and Botterill in Table 2 of Chap. 6 of this handbook as: spheres, $\phi = 1.00$; cubes, $\phi = 0.806$; equilateral pyramids, $\phi = 0.671$; cylinders (length equals diameter), $\phi = 0.874$; broken solids, $\phi = 0.63$. Haughey and Beveridge [17] indicate the difficulty in obtaining a generalized definition for ϕ, but show that in particular experiments it can be used to correlate data. For rock bed design its value will have to be selected using judgment in each instance but will probably be in the range 0.70 to 1.00 for most carefully worked-out designs.

2. Void Fraction ϵ

Practical rock beds will involve a random packing of nonuniform-size particles. This causes uncertainty in the determination of the void fraction within the bed. The void fraction for most rock beds of suitable design will probably be in the range of 0.35 to 0.45. Where no other information is available a value of 0.40 may be assumed. An estimate of the void fraction can be obtained empirically by measuring the void volume (by water displacement) of a sample of representative *in situ* rock. A summary of information on the void fraction is given in Table 6. Further information is found in Chap. 6 and in Ref. 17.

TABLE 6 Representative Void Fraction ϵ for Packed Beds

Particle type	Packing	Void fraction ϵ
Sphere	Rhombohedral	0.26
Sphere	Tetragonal sphenoidal	0.30
Sphere	Orthorhombic	0.40
Sphere	Cubic	0.48
Sphere	Random	0.36–0.43
Crushed rock	Granular	0.44–0.45
Sphere	Very loose, random	0.46–0.47
Sphere	Loose, random	0.40–0.41
Sphere	Poured, random	0.37–0.39
Sphere	Close, random	0.36–0.38
Sphere	Close, random (author's measurement)	0.386

At the wall and in the regions near a wall the void fractions will be greater than in the bulk of the bed. The effect of this is to increase the flow near walls and create the effect of bypass. Accordingly, careful attention during loading should be given to placement of rocks near walls to avoid unnecessary bridging. Rock containers should have widths or diameters that exceed 20 to 50 times the rock size to minimize this effect.

3. Flow Conditions in the Bed, G_0, V_o, Re

The airflow through a rock bed can be estimated for a collector airflow rate of 2 ft^3/min per square foot of collector area [0.60 m^3/(min·m^2)], a rule of thumb for design [3, 4]. Using this guideline and the storage mass estimate from Eq. (2) as 10 Btu/(ft^2·°F) [200 kJ/(m^2·°C)], the mass flow rate of air W_0 per unit area of bed cross section A_0 (Fig. 5) is written

$$\frac{W_0}{A_0} \equiv G_0 = 0.85 \rho_s c_{p_s}(1 - \epsilon)L \qquad \text{lb}_\text{m}/(\text{h·ft}^2) \tag{18}$$

This quantity is known as the *superficial mass velocity*. For a bed with a void fraction of 0.40 and a length of 5 ft composed of ordinary rock (Table 1), G_0 becomes 85.7 lb$_\text{m}$/(h· ft^2) [418.3 kg/(h·m^2)]. This corresponds to a volumetric flow rate of air at 100°F (37.8°C) of 1207 ft^3/(h·ft^2) [367.8 m^3/(h·m^2)]. An important parameter in bed design is the Reynolds number based on mean rock size and G_0, or

$$\text{Re} = \frac{D_s G_0}{\mu} \tag{19}$$

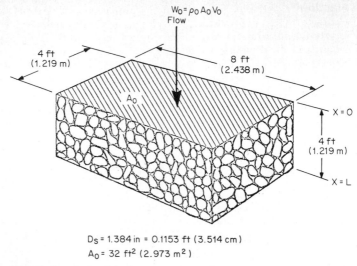

$$D_S = 1.384 \text{ in} = 0.1153 \text{ ft} (3.514 \text{ cm})$$
$$A_0 = 32 \text{ ft}^2 (2.973 \text{ m}^2)$$

FIG. 5 A typical rock bed energy storage unit.

Thus, for the conditions above using $\mu = 0.0463 \text{ lb}_m/(\text{ft}\cdot\text{h})$ (0.1914 Pa·s) the corresponding value of Reynolds number for a mean rock size of 1.375 inches in diameter ($\phi = 1.00$) is

$$\text{Re} = \frac{(1.375/12)(85.7)}{0.0463} = 212 \tag{20}$$

Accordingly, for the range of usual bed sizes the flow conditions that may be expected would fall in the following ranges: $G_0 = 30$ to $150 \text{ lb}_m/(\text{h}\cdot\text{ft}^2)$ [146 to 732 kg/(h·m²)] and $\text{Re} = 75$ to 370. Another quantity of importance is the velocity of the air V_0 entering the bed, known as the *face velocity*. For air at 100°F (37.8°C) the density is 0.071 lb_m/ft3 (1.137 kg/m³), so the face velocity corresponding to a G_0 of 85.7 $\text{lb}_m/(\text{h}\cdot\text{ft}^2)$ is

$$V_0 = \frac{G_0}{\rho_0} = \frac{85.7}{0.071} = 1207 \text{ ft/h} (367.8 \text{ m/h})$$
$$= 20.1 \text{ ft/min} (6.1 \text{ m/min}) \tag{21}$$

Hence, the corresponding range of face velocities would be $V_0 = 7.0$ to 35.2 ft/min (2.1 to 10.7 m/min). Considerations of bed pressure drop, discussed below, will influence the selection of V_0. The range of face velocities of 20 to 30 ft/min (6.0 to 7.6 m/min) will be representative of most rock bed designs.

4. Length of Bed, L

The length L of a rock bed should be established so that for the conditions of flow and bed properties the bed is capable of storing the thermal energy from the air flowing through it. For a bed of fixed volume (mass), its length can be varied by changing its cross-sectional area A_0. Generally, it is better to have a relatively short, flat bed than a long, narrow one [3]. A guide to selecting a bed length is given by Balcomb et al. [3] in terms of what is called the *relaxation length* λ, where

$$\lambda \equiv \frac{D_s/6(1 - \epsilon)}{(h/c_{p_i}G_0)} \tag{22}$$

where h is the heat transfer coefficient in Btu/(h·ft²·°F) [W/(m²·°C)] between the air and the rock (discussed below) and c_{p_i} is the specific heat of the air. For effective diurnal storage Balcomb et al. [3] recommend that the ratio L/λ be greater than 3.0. Some estimates of λ may be made using heat transfer data that are given later. From Eq. (28), for air, with Pr = 0.70.

$$\frac{h}{c_{p_i}G_0} = 1.97 \left(\frac{D_s G_0}{\mu}\right)^{-0.43} \tag{23}$$

Hence,

$$\frac{\lambda}{D_s} = 0.085 \frac{(D_s G_0/\mu)^{0.43}}{1 - \epsilon} \tag{24}$$

and, for $\epsilon = 0.4$,

$$\frac{\lambda}{D_s} = 0.141 \left(\frac{D_s G_0}{\mu}\right)^{0.43} \tag{25}$$

Thus, for flow in the bed producing a Reynolds number of 225, $\lambda/D_s = 1.45$. In this case the minimum length of bed would be 3λ, or $4.35\ D_s$. For a D_s of 1.375 in, the minimum bed length would be about 6 in, a relatively short bed. Each circumstance will be different, but the selection of an appropriate length to the bed should not be difficult. The criterion of $L > 3\lambda$ ought to be satisfied easily in most cases.

5. Air-Solid Heat Transfer Coefficient

The transfer of heat within the bed is by a convective process and is described in terms of the heat transfer coefficient h, in Btu/(h·ft²·°F) [W/(m²·°C)]. Correlations for this quantity have been the subject of many investigations during the past 50 years. Barker [18] has conducted a comprehensive review of this topic and has summarized the results of about 30 investigations. Other studies include those in Refs. 19 to 25, the results of which are shown in Fig. 6. As may be seen, a fairly wide variation exists. This is largely a result of inherent differences in the structure of the physical systems, the flow distribution, and the difficulty in describing the processes in a simple form. However, to provide guidance in determining the heat transfer coefficients for bed design three correlations are cited.

For Reynolds numbers in the range 100 to 1000, a large body of correlated data are bracketed by the results of Gamson et al. [23] (lower bound) and Eq. (34) of Chap. 6 of this handbook (upper bound), as shown in Fig. 6. These equations are:

1. Upper bound: Re $= D_s G_0/\mu = 100$ to 1000 [Chap. 6, Eq. (34)]

$$j_h = \frac{h}{-c_{p_i}G_0} \text{Pr}^{2/3} = 2.0\text{Re}^{-1}\text{Pr}^{-1/3} + 1.8\ (\epsilon\ \text{Re})^{-1/2} \tag{26}$$

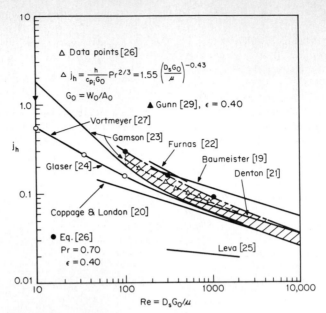

FIG. 6 Heat transfer data.

2. Lower bound: $\mathrm{Re} = D_sG_0/\mu = 100$ to 1000

$$j_h = \frac{h}{c_{p_i}G_0}\,\mathrm{Pr}^{2/3} = 1.064\mathrm{Re}^{-0.41} \tag{27}$$

A correlation which falls intermediate to those in Eqs. (26) and (27) is that of Van den Broek and Clark [26], also shown in Fig. 6, as

3. Intermediate: $\mathrm{Re} = D_sG_0/\mu = 100$ to 1000

$$j_h = \frac{h}{c_{p_i}G_0}\,\mathrm{Pr}^{2/3} = 1.55\mathrm{Re}^{-0.43} \tag{28}$$

It is important to note that the heat transfer coefficient computed by these equations is the average value over the area of the bed, A_s. Because of variations in void volume, flow, and other factors, local values of h can be expected to vary significantly from the average values.

At low Reynolds numbers ($\mathrm{Re} < 100$) there is significant variation also in the correlated data. The results of Gamson et al. [23] and Vortmeyer and Schaefer [27] appear to bracket the data in this region, as indicated in Fig. 6. These results are:

4. Upper bound: $\mathrm{Re} = D_sG_0/\mu \leq 100$

$$j_h = \frac{h}{c_{p_i}G_0}\,\mathrm{Pr}^{2/3} = 18.1\ \mathrm{Re}^{-1} \tag{29}$$

5. Lower bound: $\mathrm{Re} = D_sG_0/\mu \leq 100$

$$j_h = \frac{h}{c_{p_i}G_0}\,\mathrm{Pr}^{2/3} = 1.75\mathrm{Re}^{-0.51} \tag{30}$$

Martin [28] has shown that at very low Reynolds number flow (beds having fine particles), the heat transfer coefficient falls below that of a single sphere (corresponding to $hD_s/k = 2.0$) by several orders of magnitude. A correlation is proposed which takes into account the effect of variation in the void fraction within the bed. Gunn [29] considers the same problem and concludes that at very low Reynolds numbers, the Nusselt number approaches a constant value of $7 - 5\epsilon(2 - \epsilon)$. Gunn also provides a correlation for the Nusselt number in terms of the Reynolds and Prandtl numbers and void fraction. At a Reynolds number of 10 this correlation gives a value of j_h (Fig. 6) of 1.126 for a void fraction of 0.4 and roughly follows the results of Gamson to Re equal to 100.

6. Bed Pressure Drop and Pumping Power

An important design parameter for a packed bed is the pressure drop ΔP across the bed and the power required to pump the air through the bed. Among the various correlations available for pressure drop that of Ergun [30] is recommended. This correlation is

$$\Delta P = f\frac{L}{D_s}\frac{\rho V_0^2}{g_c}\frac{1 - \epsilon}{\epsilon^3} \tag{31}$$

where

$$f = \frac{150(1 - \epsilon)}{\text{Re}} + 1.75 \tag{32}$$

$$\text{Re} = \frac{D_s G_0}{\mu} \qquad G_0 = \frac{W_0}{A_0} \tag{33}$$

The relationship of this correlation to some experimental measurements [26] is shown in Fig. 7.

The pumping power for the bed is written

$$P = 0.0771\dot{Q}\,\Delta P \qquad \text{Btu/h} \tag{34}$$

or

$$P = 0.0226\dot{Q}\,\Delta P \qquad \text{W} \tag{35}$$

where $\dot{Q}$ — volumetric flow rate through bed, ft³/min
ΔP = bed pressure drop, lb_f/ft²

To avoid excessive flow channeling it is recommended that the bed pressure drop be in the range of 0.15 to 0.30 inches of water (37 to 75 Pa, or 0.78 to 1.56 lb_f/ft²) [1]. The Solaron Corporation [15] recommends a minimum pressure drop of 0.15 in of water (37 Pa). As a guideline the volume flow rate $\dot{Q}$ would be $2.0A_c$ for the "standard" flow of 2 ft³/min per square foot of collector. Under this condition the bed pumping power per square foot of collector for the range of ΔP of 0.15 to 0.30 inches of water would be

$$\frac{P}{A_c} = 0.12 \text{ to } 0.24 \qquad \text{Btu/(h·ft}^2)$$
$$= 0.035 \text{ to } 0.070 \qquad \text{W/ft}^2 \tag{36}$$

The actual power input to the motor driving the fan or blower will be greater by the efficiency of the unit. In general, the pumping power should be very small in a properly designed bed in relation to the rate of energy collection.

A convenient set of design curves for a rock bed is shown in Fig. 8, taken from Ref. 1. In this figure the pressure gradient in the bed, $\Delta P/L$, is given as a function of face

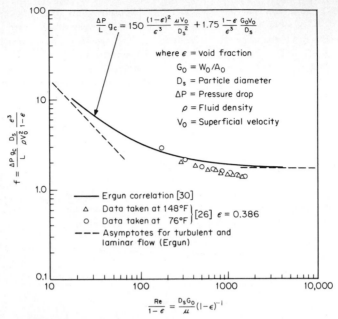

FIG. 7 Packed-bed pressure drop [26,30].

velocity V_0 for various effective rock diameters. The data are constructed similar to Eqs. (31) and (32). For the range of face velocities from 20 to 30 ft/min and a range of pressure gradients in a bed of 0.03 to 0.06 inches of water per foot (7.4 to 14.8 Pa/ft), corresponding to the range of pressure drops of 0.15 to 0.30 inches of water, in a bed having a length of 5 ft, the design range is shown by the crosshatched box. This design range indicates an approximate range of effective rock diameters D_s to be from 0.50 to 1.0 in (1.27 to 2.54 cm), or, for a shape factor of 0.80, the range of equivalent spherical diameters D_e to be from 0.625 to 1.25 in (1.59 to 3.18 cm), approximately the range indicated earlier as recommended.

7. Temperature Stratification Calculations

The temperature distribution within a packed bed for both the air (fluid) and rock (solid) has been the subject of many studies. Of importance are the time and spatial variations of these temperatures for a time-varying inlet air temperature and for a spatial variation in initial temperatures. Physical factors that influence this thermal response are void fraction distribution, size of particle or size distribution of particle, flow rate distribution, and heat transfer coefficient and its spatial distribution as related to spatial variations in void, flow, and particle size, among others. Temperature stratification calculations allow estimates to be made of the thermal response to be expected from the bed during heat recovery, the quantity of energy stored, and temperature levels for materials selection. Certain analytical results are available but as a practical matter numerical results using a computer are generally the most useful. The significant literature, including some results, will be cited.

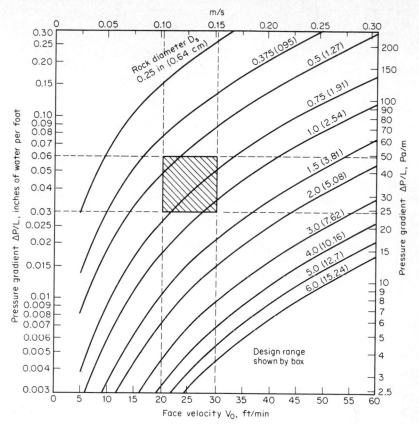

FIG. 8 Rock bed performance map [1]. Crosshatched area is for $\Delta P = 0.15$ to 0.30 inches of water, length $= 5$ ft and $V_0 = 20$ to 30 ft/min.

A detailed description of these studies is not within the scope of this chapter and, in many cases, probably not of first importance anyway to a bed designer whose objective is to construct a practical working bed. For purposes of bed optimization and design improvement, however, a detailed knowledge of this subject is essential. Analytical studies are given by Schumann [31], Burch et al. [32], Hughes et al. [33], Riaz [34, 35], Clark and Arpaci [36], and Spiga and Spiga [37], among others. Numerical solutions (computer models) have been published by Clark et al. [38], Von Fuchs [39], Mumma and Marvin [40], Gross et al. [41], and Beasley and Clark [42].

Typical results from a numerical solution for the rock bed shown in Fig. 5 are given in Fig. 9 [26]. These results show the stratification in air temperature within the bed during a 6-h charging period resulting from a time-dependent inlet air temperature. A comparison of computer predictions with experimental measurements is shown in Fig. 10 [42], indicating the importance of the void fraction to the performance. Comparisons of the results of various numerical models are summarized in Ref. [42] and show essential agreement among the one-dimensional models [38, 41]. A two-dimensional (axial-radial) dynamic model is described in Ref. 42.

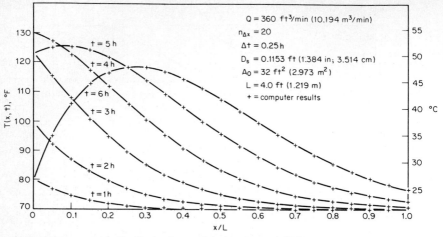

FIG. 9 Air temperature distribution in a rock bed storage unit [26].

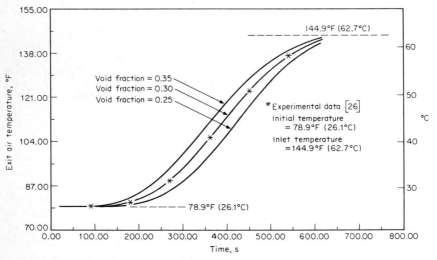

FIG. 10 Comparison of computer results and experimental data [42].

8. Costs

The cost of natural rock, usually in the form of screened gravel, will vary from region to region but will probably fall in the range of $3 to $10 per ton (1980 dollars) at the supplier. When an order is to be placed it is recommended that the purchaser visit the supplier, inspect the rock, and give very specific instructions as to the requirements. Some variation in size and type of rock should be expected, especially from quarries and gravel pits.

Rock bed containers will vary in cost depending on the region and design. In general, wood containers are the least expensive, cinder blocks somewhat greater in cost, and poured concrete the most costly. Costs per unit volume will decrease as the volume increases.

F. LIQUID STORAGE DESIGN

1. Sizing and Stratification

The sizing of liquid storage tanks for long-term applications in solar systems can be done using Fig. 3 and Eq. (2). These results assume the liquid volume to be well mixed, i.e., to have a uniform temperature at any given time. Under this condition the volume of storage to satisfy a given heating load and hot water load can be computed from Eq. (7). The thermal loss from well-mixed tanks (nonstratified) is determined from Eq. (12), (13), or (15) and Tables 3 and 4.

In principle, a liquid storage tank will always have some degree of temperature stratification because of thermal buoyancy that exists from nonuniform densities within the tank. The higher-temperature fluid usually enters and leaves the tank in the upper third of its volume and the colder, makeup fluid is introduced into the tank at its bottom. However, conditions of operation are variable, and whenever an intermediate density exists between that of an entering or leaving fluid and a local density, both diffusion and thermal convection occur. These effects tend to destroy temperature stratification. Also, any fluid flowing into (or from) a tank possesses a momentum that causes fluid entrainment and mixing, thus acting to reduce thermal stratification.

Unlike packed beds, where stratification is inherent and reasonably good models exist for its determination, a liquid tank tends to become mixed readily, and well-established models for its stratification are few. Duffie and Beckman [2, Chap. 9] give a numerical model for this calculation. Clark and Barakat [45], Clark [46], and Sha et al. [48] describe and summarize the effects of natural convection in tanks on stratification. Computer simulations of solar systems having both stratified and mixed storage indicate a small (5 to 10 percent) improvement in performance with stratified storage [2, Chap. 10]. This is a consequence of improved collector performance stemming from lower thermal losses resulting from lower collector inlet temperatures. This improvement is probably the most that can be expected, as in practical systems it is difficult to obtain and maintain the stratification predicted by the models. Considering the various uncertainties in the input parameters for the simulation of solar system performance for the long term, a mixed tank model will probably be satisfactory in most cases [47, 52]. Stratification modeling is useful, however, for short-term performance studies where control strategies are being examined and a detailed analysis of system dynamics is required. In general, a stratified condition is desirable, and this is promoted by introducing the heated fluid into the upper regions of a tank at as low a velocity (inlet Reynolds number) and as low a maximum dispersion as possible with mixing with *in situ* fluid having a density at or close to that of the incoming fluid. The use of baffles or perforated tubes and plates will assist in achieving this condition [49, 50]. Internal heat exchangers and timing of the load will disturb the stratification. The influence of mixing in the upper, stratified regions of a tank is modeled by Han and Wu [51].

The actual choice of storage tank configuration and placement is usually governed by architectural and retrofit considerations. For these reasons present modeling techniques, which may represent the performance of laboratory models, will not necessarily be appropriate for full-sized systems.

Kennedy et al. [52] have examined the thermal behavior of three full-sized operating solar systems: (1) for a multiple dwelling, (2) for a restaurant, and (3) for a vocational technical school, all in the Washington, D.C., region. The schematic diagram for the system in the multiple dwelling (188 apartments) is shown in Fig. 11, and its solar input–demand data are shown in Fig. 12. Data were collected during a 24-h period, spanning two days, February 18 to 19, 1980, Monday to Tuesday. The temperature distribution

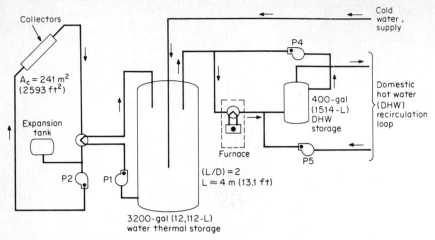

FIG. 11 System 1 schematic (multiple-dwelling, 188 apartments).

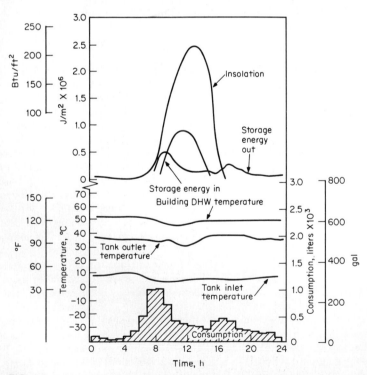

FIG. 12 System 1 average hourly performance.

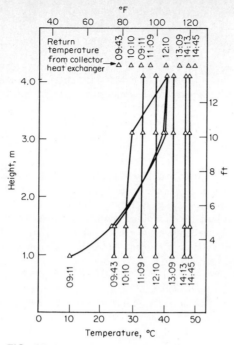

FIG. 13 System 1 temperature profiles in storage during collection—February 18, 1980.

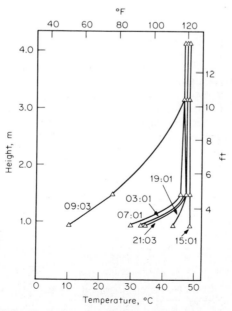

FIG. 14 System 1 temperature profiles in storage between collections—February 18 to February 19, 1980.

in the 3200-gal (12,112-liter) storage tank is given in Fig. 13 (during daylight charging) and Fig. 14 (evening, between collection periods). During most of the periods the tank is quite well mixed, especially during and at the end of the charge cycle, despite an initial stratified condition. This is believed to be due to the effect of the nonbaffled vertical inlet [52], something that may be expected from most flow arrangements.

2. Tank Types

The selection of the type of storage tank will depend primarily on its service requirements. As a guide the following may be used. Space heating applications will require maximum storage temperatures of approximately 160°F (71°C), although some systems may go higher. Generally, this application will not require pressurized tanks. Solar-assisted heat pump systems will require temperatures no greater than about 100°F (38°C), which reduces the amount of insulation needed. Absorption air conditioning systems use hot storage at temperatures above 170°F (77°C) and cold storage temperatures below 55°F (13°C) with pressurization a possibility. Direct heating of potable water usually requires temperatures not greater than 140°F (60°C), but a pressurized tank is needed. The preheating of water can be done at lower temperatures, about 120°F (49°C).

Materials commonly used for these tanks are steel, fiberglass, concrete, or wood with plastic lining. The advantages and disadvantages of each of these materials are summarized in Table 7, taken from Ref. 1.

3. Corrosion

The corrosion of metal tanks, especially steel, is a primary factor in design and operation of a system which uses water as the storage medium. Rates of corrosion increase rapidly with level of temperature. For steel tanks there are four types of corrosion that can cause problems: electrochemical corrosion, oxidation (rusting), galvanic corrosion, and pitting. Methods for corrosion prevention include sealing the system, adding chemical corrosion inhibitors, using protective coatings or liners, using cathodic protection, increasing metal thickness, and using noncorroding alloys [1]. Sealing, coating, and the use of sodium sulfite as an oxygen scavenger are commonly used. Any inhibitor must be tested periodically and replenished. Hot water systems cannot be sealed, and the use of inhibitors is inconvenient. Glass or stone lining and the use of a sacrificial anode (magnesium) are common in this case. Galvanic corrosion occurs whenever dissimilar metals (copper and steel, for example) form an electrical contact in the presence of water. This can be prevented by avoiding the use of dissimilar metals or by electrically insulating them. Cole et al. [1] recommend the following for steel tanks:

1. Use four coats of baked-on phenolic epoxy on the inside of the tank.
2. Use a magnesium bar as cathodic protection.
3. Electrically insulate dissimilar metals, except the magnesium anode, with dielectric bushings, gaskets, or pads.
4. Use sodium sulfite to scavenge oxygen.
5. Do not use chromate-type corrosion inhibitors. They are highly toxic, and carcinogenic, cause damage to a common type of pump seal, and are difficult to dispose of

TABLE 7 Advantages and Disadvantages of Tank Types [1]

Steel tank	Fiberglass tank	Concrete tank	Wooden tank with liner
		Advantages	
Steel tanks can be designed to withstand pressure.	Factory-insulated tanks are available.	Cost is moderate.	Cost is low.
Much field experience is available.	Considerable field experience is available.	Concrete tanks may be cast in place or may be precast.	Indoor installation is easy.
Connections to plumbing are easy to make.	Some tanks are designed specifically for solar energy storage.		
Some steel tanks are designed specifically for solar energy storage.	Fiberglass does not rust or corrode.		
		Disadvantages	
Complete tanks are difficult to install indoors.	Maximum temperature is limited, even with special resins.	Careful design is required to avoid cracks, leaks, and excessive cost.	Maximum temperature is limited.
Steel tanks are subject to rust and corrosion.	Fiberglass tanks are relatively expensive.	Concrete tanks must not be pressurized.	Wooden tanks must not be pressurized.
Steel tanks are relatively expensive.	Complete tanks are difficult to install indoors.	Connections to plumbing are difficult to make leaktight.	Wooden tanks are not suitable for underground installation.
	Fiberglass tanks must not be pressurized.		

properly. Since chromates act by passivating the steel surface, their effectiveness is much reduced when combined with cathodic protection.

6. If possible, limit the maximum tank temperature to 160°F (71°C) or less.

7. Protect the outside of a buried tank with two coats of coal tar epoxy and a magnesium anode. Use a coat of primer and two coats of enamel to protect the outside of above-ground tanks.

8. If the tank will be pressurized, it must comply with Section VIII of the ASME Boiler and Pressure Vessel Code or local codes.

9. Do not install large steel tanks in basements, crawl spaces, or other locations where building modifications would be necessary to replace them.

4. Costs

The costs of new tanks without volume discounts are summarized by Pickering [53]. Transportation, installation, and insulation costs are included. Cost estimates are given in Tables 8, 9, and 10 for concrete, fiberglass-reinforced plastic, and miscellaneous tanks,

TABLE 8 Cost Estimates, Concrete Tanks [53]*

Type and size	Dimensions†	Capacity, gal	Cost, $				$/gal
			FOB plant	Transportation and installation‡	Insulation	Total	
1. Precast concrete utility vaults, 7 ft deep, buried with 4 in polyurethane board insulation plus asphalt waterproofing.	6 ft × 8 ft	2500	999	200	300	1499	0.60
	6 ft × 10 ft	3000	1050	250	350	1650	0.55
	6 ft × 12 ft	3800	1105	300	400	1805	0.48
2. Precast concrete septic tanks in ground, insulation same as vaults.	Various	1000	190	150	150	490	0.49
		1300	220	175	200	595	0.46
		1500	260	200	250	710	0.47
3. Precase concrete drainpipe sections in ground, insulation same as vaults.	4 ft × 12 ft	1000	254	200	200	654	0.65
	5 ft × 16 ft	2000	534	275	350	1159	0.58
	6 ft × 16 ft	3200	1227	350	500	2077	0.65
4. Reinforced concrete block, built in place in basement, 6 ft high. 6 in batt insulation plus lagging. Butyl lining.	4 ft 0 in × 5 ft 8 in	1000		590	75	665	0.67
	5 ft 4 in × 7 ft 4 in	1700		800	100	900	0.53
	6 ft 0 in × 8 ft 0 in	2000		950	115	1065	0.53
5. Reinforced concrete cast in place in basement, 6 ft high. Insulation same as block. No lining.	4 ft × 6 ft	1000		600	75	675	0.68
	5 ft × 7 ft	1500		700	100	800	0.53
	6 ft × 8 ft	2000		800	115	915	0.46

*Spring 1975 prices, new construction, no volume discount.

†Dimensions for precast concrete drainpipe sections are diameter × length; all others are width × length.

‡Includes lining where specified.

TABLE 9 Cost Estimates, Fiberglass-Reinforced Plastic Tanks [53]*

Type and size	Dimensions, D × L	Capacity, gal	Cost, $				$/gal
			FOB plant	Transportation and installation	Insulation	Total	
1. Underground service station type, insulated on job with 3 in polyurethane foam plus asphalt waterproofing.	4 ft 3 in × 6 ft 6 in	550	545	310	120	975	1.77
	4 ft 3 in × 11 ft 0 in	1000	725	430	180	1335	1.34
	6 ft 6 in × 10 ft 10 in	2000	1640	600	300	2540	1.27
	8 ft 0 in × 14 ft 7 in	4000	2160	1100	480	3740	0.94
2. Vertical, above ground, closed top, insulated on job with 6 in batt plus lagging.	6 ft 6 in × 5 ft 10 in	1000	1065	200	90	1355	1.36
	6 ft 7 in × 8 ft 4 in	1500	1305	250	100	1655	1.10
	6 ft 7 in × 10 ft 1 in	2000	1450	300	125	1875	0.94
	6 ft 7 in × 15 ft 6 in	3000	2150	400	190	2740	0.91
	12 ft 0 in × 10 ft 4 in	5800	2645	500	255	3400	0.59
3. Vertical, above ground, closed top, insulated at plant with 2 in polyurethane foam covered with fiberglass.	6 ft 10 in × 6 ft 0 in	1000	2000	200		2200	2.20
	6 ft 11 in × 8 ft 5 in	1500	2385	250		2635	1.76
	6 ft 11 in × 10 ft 3 in	2000	2600	300		2900	1.45
	6 ft 11 in × 15 ft 8 in	3000	3850	400		4250	1.42
	12 ft 3 in × 10 ft 6 in	5800	5030	500		5530	0.95

*Spring 1975 prices, new construction, no volume discount.

TABLE 10 Cost Estimates, Miscellaneous Tanks [53]*

Type and size	Dimensions, D × L	Capacity, gal	Cost, $				$/gal
			FOB plant	Transportation and installation†	Insulation	Total	
1. Steel shells lined with butyl rubber. Installed in basement. 6 in batt insulation plus lagging.	5 ft × 5 ft	700	250	180	50	480	0.69
	6 ft × 6 ft	1200	400	235	75	710	0.60
	6 ft × 8 ft	1500	500	270	90	860	0.57
	7 ft × 8 ft	2000	600	315	110	1025	0.51
2. Wood stave tank lined with butyl rubber. Installed in basement. Top insulation only of 6 in batt.	4 ft × 4 ft	350	180	250	25	455	1.30
	6 ft × 6 ft	1200	400	300	40	740	0.62
	8 ft × 8 ft	2500	650	400	50	1100	0.44
3. Vertical pipe type, loose fill insulation:							
● Filament wound fiberglass	4 ft × 10 ft	900	480	155	25	660	0.73
	4 ft × 16 ft	1500	765	200	40	1005	0.67
● Reinforced plastic mortar	4 ft × 10 ft	900	480	155	25	660	0.73
	4 ft × 16 ft	1500	765	200	40	1005	0.67
● Asbestos cement, sewer grade	3 ft × 10 ft	500	200	155	20	375	0.75
	3 ft × 16 ft	800	320	200	35	555	0.69
● High-density extruded PVC	3 ft × 10 ft	500	450	155	20	625	1.25
	3 ft × 16 ft	800	720	200	35	955	1.19

*Spring 1975 prices, new construction, no volume discount.
†Includes lining where specified.

including steel, in terms of spring 1975 dollars [53]. Pickering [53] indicates that volume production could result in some cost reduction, as follows:

1. Steel tanks: 20 percent reduction for 500- to 2000-gal tanks to 40 to 55 cents per gallon
2. Concrete tanks: 25 percent reduction to about 37 to 45 cents per gallon
3. Fiberglass tanks: For a 2000-gal size to about 75 cents per gallon
4. Pipe-type tanks: To about 50 cents per gallon

G. PHASE-CHANGE (LATENT HEAT) STORAGE DESIGN

1. Background

When a substance undergoes a change in its physical structure, such as a transition between liquid and solid states or an alteration in its crystalline form, much larger quantities of energy are transferred than when the same mass of the substance changes its temperature a few degrees in either the solid or liquid state. The process of changing its physical structure is known as a phase change, and the quantity of energy transferred during the process is called the latent heat, or heat of fusion in the case of solid-liquid transformations. Further, an important characteristic of this process is that the temperature of the substance remains relatively constant during the phase change. Thus, the process of phase change is one in which significantly large quantities of energy can be absorbed or released at reasonably constant temperatures, a process that has obviously important applications to thermal storage in residential, commercial, and industrial systems.

A large number of materials have been identified that possess phase-change properties in the range of temperatures from 0 to 1400°C (32 to 2552°F). Lane et al. [55] have identified and classified 210 organic, inorganic, and eutectic materials having congruent melting (phase change without component separation) in the range of 10 to 90°C (50 to 194°F), which would be suitable for space heating and cooling applications. Schröder [56] has identified nine eutectic fluoride mixtures suitable for high-temperature storage applications in the range 449 to 832°C (840 to 1530°F) and several other substances whose melting temperatures range from 180 to 1400°C (356 to 2552°F), all candidates for thermal storage in power cycles. The application of thermal (including cooling) storage devices to permit electrical load shifting to off-peak demand periods is being given increased attention [57, 58].

2. Properties of Phase-Change Materials

Although hundreds of substances have been identified as candidates for heat or cooling storage, practical considerations such as the magnitude of latent heat, level of fusion temperature, corrosion and materials compatibility, subcooling, volume change, stability, flammability, toxicity, availability, manufacturing problems, and cost have limited the choice to about a dozen for storage temperatures below 117°C (243°F). A number of these phase-change materials (PCMs) are listed in Table 2, and a compilation of others is given in Table 11. The data are taken from Refs. 1, 59, 60, and 61. Basically, the

TABLE 11 Properties of Phase-Change Materials (PCM)

Melting temperature, °F (°C)	Material	Latent heat		Solid thermal conductivity, Btu/(h·ft·°F) [W/(m·°C)]	Density, solid, lbm/ft³ (kg/m³)	Specific heat, Btu/(lbm·°F) [kJ/(kg·°C)]	
		Btu/lbm (kJ/kg)	Btu/ft³ (kJ/m³)			Solid	Liquid
32 (0)	Water	143.1 (333.4)	8191 (0.306 × 10⁶)	1.28 (2.21)	57.24 (916.8)	0.487 (2040)	1.00 (4210)
38 (3.3)	Na$_2$SO$_4$, KCl, additives	50 (116.5)	4600 (0.172 × 10⁶)		92 (1473)		
45 (7.2)	Na$_2$SO$_4$, NH$_4$Cl, KCl, additives	50 (116.5)	4600 (0.172 × 10⁶)		92 (1473)		
55 (13)	Na$_2$SO$_4$, NaCl, NH$_4$Cl (eutectic)	78 (181.3)	7150 (0.267 × 10⁶)		92 (1473)		
84 (29)	CaCl$_2$·6H$_2$O	82 (190.8)	8364 (0.311 × 10⁶)	0.629 (1.088)	102 (1630)	0.320 (1340)	0.552 (2310)
90.3 (32.4)	Na$_2$SO$_4$·10H$_2$O	109 (253)	9940 (0.358 × 10⁶)		91.2 (1416)	0.459 (1920)	0.779 (3260)
97 (36)	Na$_2$S$_2$O$_3$·12H$_2$O	114 (280)	10830 (0.426 × 10⁶)		95 (1520)	0.404 (1690)	0.464 (1940)
120.0 (49)	Na$_2$S$_2$O$_3$·5H$_2$O ("hypo")	86 (200)	9116 (0.338 × 10⁶)		106 (1690)	0.346 (1450)	0.570 (2389)
135 (57.2)	Mg(NO$_3$)$_2$·6H$_2$O, NH$_4$NO$_3$	47 (109.3)	5000 (0.185 × 10⁶)	0.392 (0.678)	106 (1690)		
136 (58)	Mg(NO$_3$)$_2$·6H$_2$O, MgCl$_2$·6H$_2$O	56.8 (132.2)	5783 (0.215 × 10⁶)		102 (1630)		
192 (89)	Mg(NO$_3$)$_2$·6H$_2$O	70.0 (162.8)	7140 (0.265 × 10⁶)	0.353 (0.611)	102 (1630)		
243 (117)	MgCl$_2$·6H$_2$O	72.5 (168.6)	7105 (0.264 × 10⁶)	0.401 (0.694)	98 (1569)		

present choices in this temperature range are water (ice), a number of salt hydrates and their mixtures, and paraffin wax. For cooling storage, ice has been used for centuries and is still a valid, practical choice for this application. To provide for heating storage a range of higher temperatures is necessary, and this application is presently being served largely by the salt hydrates and their various mixtures, as shown in Table 11.

3. Commercial PCM Units

Commercial thermal PCM storage units usually consist of the PCM either encapsulated in tubes, spheres, or trays, with the heat transfer fluid external to the PCM, or in bulk with the heat transfer fluid pumped through tubular coils imbedded in the PCM. For air systems the encapsulated arrangement is generally used and for liquid systems a bulk storage design is convenient. An encapsulated design of a PCM in which $CaCl_2 \cdot 6H_2O$ is contained within 2-in cylinders is shown in Fig. 15. This is a design tested by the Dow Chemical Company, Midland, Michigan, the manufacturer of the PCM. The tubes are made of high-density polyethelyne (HDPE). Other PCMs manufactured by Dow are reviewed by Lane [60]. Four materials are now commercially available which are free from the problem of component separation (incongruency during melting or freezing) and are characterized by negligible subcooling. These important characteristics have been

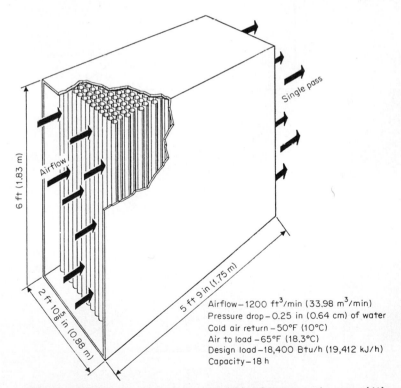

6 ft (1.83 m)

Airflow

Single pass

2 ft 10$\frac{5}{8}$ in (0.88 m)

5 ft 9 in (1.75 m)

Airflow—1200 ft³/min (33.98 m³/min)
Pressure drop—0.25 in (0.64 cm) of water
Cold air return—50°F (10°C)
Air to load—65°F (18.3°C)
Design load—18,400 Btu/h (19,412 kJ/h)
Capacity—18 h

FIG. 15 Closet storage unit for nighttime setback or heat pump solar system [61]: calcium chloride hexahydrate encapsulated in 2-in cylinders.

obtained by formulating the bulk chemicals with nucleating additives [60]. The four PCMs are: $CaCl_2 \cdot 6H_2O$ (melting point 29°C, or 84°F), 58.7% $Mg(NO_3)_2 \cdot 6H_2O$ + 41.3% $MgCl_2 \cdot 6H_2O$ (melting point 58°C, or 136°F), $Mg(NO_3)_2 \cdot 6H_2O$ (melting point 89°C, or 192°F), and $MgCl_2 \cdot 6H_2O$ (melting point 117°C, or 243°F). Freezing curves provided by the Dow Chemical Company for $MgCl_2 \cdot 6H_2O$ and $Mg(NO_3)_2 \cdot 6H_2O$ are shown in Figs. 16 and 17, which demonstrate the phase change without subcooling at temperatures of 117°C (243°F) and 89°C (192°F).

Bulk PCM storage devices have been developed by the CALMAC Manufacturing Corporation, Englewood, N.J., for three PCM salt hydrates at 7.2°C (45°F), 46.1°C (115°F), and 113°C (235°F) and for ice at 0°C (32°F). Separation problems in the

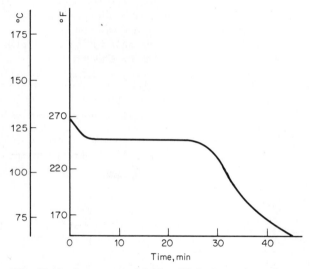

FIG. 16 Cooling curve for $MgCl_2 \cdot 6H_2O$ plus nucleator (courtesy Dr. G. A. Lane, Dow Chemical Co.).

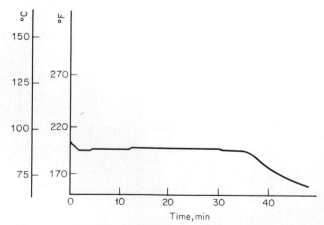

FIG. 17 Cooling curve for $Mg(NO_3)_2 \cdot 6H_2O$ plus nucleator (courtesy Dr. G. A. Lane, Dow Chemical Co.).

PCMs have been overcome. Known as HEATBANKs®, these units consist of insulated polyethylene tanks containing the PCM and an internal coiled plastic-tube heat exchanger [57, 58, 59]. No loss in performance is reported for as many as 10,000 cycles [62]. Discharge-charge (freezing-melting) curves from a commercially available storage unit for $Na_2SO_4 + NH_4CL +$ additives [eutectic 7.2°C (45°F)] and (impure) $Na_2S_2O_3 \cdot 5H_2O$ ("hypo") [46.1°C (115°F)] are given in Figs. 18 and 19.

Other manufactures of PCM thermal storage units, as of 1979, are listed by Cole [1, p. 135].

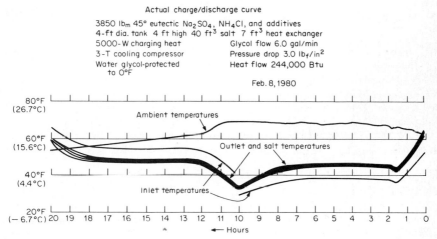

FIG. 18 Heating-cooling curves for Na_2SO_4, NH_4Cl, and additivies; 45°F (7.2°C) eutectic.

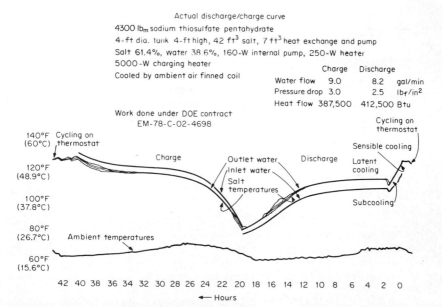

FIG. 19 Heating-cooling curves for $Na_2S_2O_3 \cdot 5H_2O$.

4. Costs

The installed cost of a PCM storage unit will vary according to its design, size, and type of installation (i.e., new or building retrofit). Cole [1, p. 140] reports that a survey of several manufacturers indicated a range of costs to be $10 to $12 per million Btu ($9.50 to $11.40 per million joules), but that the overall range is still large. The costs are in 1979 dollars. MacCracken [59] indicates that costs range from $6 to $27 per million Btu in 1980 dollars.

H. TESTING OF STORAGE UNITS

The thermal performance of sensible heat energy storage devices used in heating, air conditioning, and service hot water systems can be evaluated using a standard test procedure adopted by the American Society of Heating, Refrigerating and Air-Conditioning Engineers (ASHRAE). This standard is ASHRAE 94-77 (ANSI B 199.1–1977), "Methods of Testing Thermal Storage Devices Based on Thermal Performance" [63]. Although originally intended to be used also in testing PCM units, the ASHRAE standard 94-77 was found to have shortcomings for this application. A new test standard is being formulated for these units which is expected to be available before the end of 1985.

REFERENCES

1. R. L. Cole, K. J. Nield, R. R. Rohde, and R. M. Wolosewicz, "Design and Installation Manual for Thermal Energy Storage," 2d ed., *ANL-79-15*, Argonne National Laboratory, Argonne, Ill., January 1980. (Available from National Technical Information Service, U.S. Dept. of Commerce, 5285 Port Royal Road, Springfield, VA 22161.)

2. J. A. Duffie and W. A. Beckman, *Solar Engineering of Thermal Processes,* Wiley, New York, 1980.

3. D. Balcomb, J. C. Hedstrom, S. W. Moore and B. T. Rogers, *Solar Heating Handbook for Los Alamos,* Los Alamos Scientific Laboratory *LA-5967-MS,* Los Alamos, N. Mex., May 1975.

4. *Pacific Regional Solar Heating Handbook,* San Francisco Operations Office, ERDA, November 1976.

5. "Interim Performance Criteria for Solar Heating and Combined Heating/Cooling Systems and Dwellings," U.S. Dept. of Housing and Urban Development, *NP-20551,* 1975.

6. *Heating and Air Conditioning Systems Installation Standards for One and Two Family Dwellings and Multifamily Housing, Including Solar,* Sheet Metal and Air Conditioning Contractors' National Association (SMACNA), 8224 Old Courthouse Road, Tyson's Corner, Vienna, VA 22810, 1977.

7. E. Mazria, *The Passive Solar Energy Book,* Rodale Press, 1979.

8. R. Judkoff and F. Sokol, "Performance of a Slective-Surface Trombe Wall in a Small Commercial Building," *SERI/TP-721-1158,* Solar Energy Research Institute, Golden, Colo., March 1981.

9. U.S. Department of Energy, *Passive Solar Design Handbook,* Vols. I and II, January 1980, National Technical Information Center, Springfield, Va.

10. B. T. Tamblyn, "Maximizing Solar Collection with Thermal Storage," *ASHRAE J.,* pp. 72–75, November 1980.

11. L. Wen and H. Steele, "Thermal Storage Requirements for Parabolic Dish Solar Power Plants," *ASME Paper 80-WA/Sol-21,* 1980.

12. S. Sillman, "The Trade-off between Collector Area, Storage Volume, and Building Conservation in Annual Storage Solar Heating Systems," *SERI/TR-721-907,* Solar Energy Research Institute, Golden, Colo., April 1981.

13. F. Baylin, R. Monte, and S. Sillman, "Annual-Cycle Thermal Energy Storage for a Community Solar System: Details of a Sensitivity Analysis," *SERI/TR-721-575,* Solar Energy Research Institute, Golden, Colo., July 1980.

14. F. Baylin and S. Sillman, "System Analysis Techniques for Annual-Cycle Thermal Energy Storage Solar Systems," *SERI/RR-721-676,* Solar Energy Research Institute, Golden, Colo., July 1980.

15. *Application Engineering Manual,* Solaron Corporation, Englewood, Colo., 1977.

16. J. D. Gabor and J. S. M. Botterill, "Heat Transfer in Fluidized and Packed Beds," Chap. 6 of this handbook.

17. D. P. Haughey and G. S. G. Beveridge, "Structural Properties of Packed Beds—A Review," *Can. J. Chem. Eng.,* Vol. 47, pp. 130–140, April 1969.

18. J. J. Barker, "Heat Transfer in Packed Beds," *Ind. Eng. Chem.,* Vol. 57 pp. 43–51, 1965.

19. E. B. Baumeister and C. C. Bennett, "Fluid-Particle Heat Transfer in Packed Beds," *AIChE J.,* Vol. 4, pp. 69–74, 1958.

20. J. Coppage and A. London, "Heat Transfer and Flow Friction Characteristics of Porous Media," *Chem. Eng. Prog.,* Vol. 52, p. 57, 1956.

21. W. H. Denton, "The Heat Transfer and Flow Resistance for Fluid Flow through Randomly Packed Spheres," *Proc. Inst. Mech. Eng. Gen. Discussion Heat Transfer,* Sept. 11–13, 1951.

22. C. C. Furnas, "Heat Transfer from a Gas Stream to a Bed of Broken Solids," *Ind. Eng. Chem.,* Vol. 22, pp. 26 31, 1930.

23. B. W. Gamson, G. Thodos, and O. A. Hougen, "Heat, Mass, and Momentum Transfer in the Flow of Gases through Granular Solids," *Trans. Am. Inst. Chem. Eng.,* Vol. 39, pp. 1–35, 1943.

24. M. B. Glaser and G. Thodos, "Heat and Momentum Transfer in the Flow of Gases through Packed Beds," *AIChE J.,* Vol. 4, pp. 63–68, 1958.

25. M. Leva, "Heat Transfer to Gases through Packed Tubes," *Ind. Eng. Chem.,* Vol. 39, pp. 657–662, 1947.

26. C. D. Van Den Broek and J. A. Clark, "Heat Transfer in a Column Packed with Spheres," Von Karman Institute of Fluid Dynamics, Lecture Series, Brussels, Feb. 12–16, 1979.

27. D. Vortmeyer and R. J. Schaefer, "Equivalence of One- and Two-Phase Models for Heat Transfer Processes in Packed Beds: One Dimensional Theory," *Chem. Eng. Sci.,* Vol. 29, p. 485–491, 1974.

28. H. Martin, "Low Peclet Number Particle to Fluid Heat and Mass Transfer in Packed Beds," *Chem. Eng. Sci.,* Vol. 33, p. 913, 1978.

29. D. J. Gunn, "Transfer of Heat or Mass to Particles in Fixed and Fluidized Beds," *Int. J. Heat Mass Transfer,* Vol. 21, No. 4, pp. 467–476, 1978.

30. S. Ergun, "Fluid Flow through Packed Columns," *Chem. Eng. Prog.,* Vol. 48, No. 2, p. 89–94, 1952.

31. T. E. W. Schumann, "Heat Transfer: A Liquid Flowing through a Porous Prism," *J. Franklin Inst.,* Vol. 28, p. 405, 1929.

32. D. M. Burch, R. W. Allen, and B. A. Peavy, "Transient Temperature Distributions within Porous Slabs Subjected to Sudden Transpiration Heating," *J. Heat Transfer,* Vol. 98, No. 2, pp. 221–226, 1976.

33. P. J. Hughes, S. A. Klein, and D. J. Close, "Packed Bed Thermal Storage Models for Solar Air Heating and Cooling Systems," *J. Heat Transfer,* Vol. 98, No. 2, pp. 336–339, May 1976.

34. M. Riaz, "Transient Analysis of Packed-Bed Thermal Storage Systems," *Sol. Energy,* Vol. 21, No. 2, pp. 123–129, 1978.

35. M. Riaz, "Analytical Solutions for Single and Two-Phase Models of Packed-Bed Thermal Storage Systems," *J. Heat Transfer,* Vol. 99, No. 3, pp. 489–492, August 1977.

36. J. A. Clark and V. S. Arpaci, "Dynamic Response of a Packed-Bed Energy Storage System to a Time Varying Inlet Temperature," *Proc. Solar Heating and Cooling Forum,* Clean Energy Institute, University of Miami, Miami, Fla., December 1976.

37. G. Spiga and M. Spiga, "A Rigorous Solution to a Heat Transfer Two-Phase Model in Porous Media and Packed-Bed," *Int. J. Heat and Mass Transfer,* Vol. 24, No. 2, p. 355, 1981.

38. J. A. Clark, R. L. Nabozny, and J. H. Heetderks, "ROCKBED, A Computer Program for Thermal Storage," *Proc. Ann. Mtg. Am. Sec. Int. Sol. Energy Soc.,* Orlando, Fla., 1977.

39. G. F. Von Fuchs, "Validation of a Rockbed Thermal Storage Model," Solar Engineering— 1981, *Proc., 3d Annual Conf. Systems Simulation Economic Analysis/Solar Heating and Cooling Operational Results,* ed. R. L. Reid, L. M. Murphy, and D. S. Wald, ASME, Reno, Nev., April 1981.

40. S. A. Mumma and W. C. Marvin, "A Method of Simulating the Performance of a Pebble-Bed Thermal Energy Storage and Recovery System," *ASME Paper 76-HT-73,* ASME-AIChE Heat Transfer Conference, August 1976.

41. R. J. Gross, C. E. Hickox, and C. E. Hackett, "Numerical Simulation of Dual-Media Thermal Energy Storage Systems," *J. Sol. Energy Eng.* (ASME), Vol. 102, p. 287, November 1980.

42. D. E. Beasley and J. A. Clark, "Transient Response of a Packed Bed for Thermal Energy Storage," *Int. J. of Heat and Mass Transfer,* Vol. 27, No. 9, pp. 1659–1669, 1984.

43. F. W. Schmidt and J. A. Willmott, *Thermal Energy Storage and Regeneration,* McGraw-Hill, New York, 1981.

44. E. C. Guyer, J. G. Bourne, L. Paglia, and D. H. Walker, "Rockbed Storage for Cooling," *COO-4481-1,* Final Report, September 1978. Available from National Technical Information Service, Springfield, Va., Distribution Category VC-59. (See App. J, "Microbial Activity in Rockbed Thermal Storage Devices," p. 158.)

45. J. A. Clark and H. Z. Barakat, "Analytical and Experimental Study of the Transient Laminar Natural Convection Flow in Partially Filled Liquid Containers," *Proc. 3d Int. Heat Transfer Conf.,* Vol. II, Chicago, 1966.

46. J. A. Clark, "A Review of Pressurization, Stratification and Interfacial Phenomena," *International Advances in Cryogenic Engineering,* Vol. 10, Plenum, New York, 1965.

47. P. I. Cooper, S. A. Klein, and C. W. S. Dixon, "Experimental and Simulated Performance of a Closed Loop Solar Water Heating System," *Extended Abstracts 1975 Int. Sol. Energy Cong. Exposition,* Los Angeles, July 28 to August 1, 1975.

48. W. T. Sha, E. I. H. Lin, et al., "COMMIX-SA-1: A Three Dimensional Thermohydrodynamic Computer Program for Solar Applications," *ANL-80-8,* December 1979, Components Technology Division, Argonne National Laboratory, Argonne, Ill.

49. R. I. Loehrice, M. K. Sharp, H. N. Gari, and R. D. Haberstroh, "Thermal Stratification Enhancement for Solar Energy Applications," Civil Engineering Laboratory, *CR-78.011,* Naval Construction Battalion Center, Port Hueneme, Calif., July 1977.

50. E. S. Davis and R. Bartera, "Stratification in Solar Water Heater Storage Tanks," *Proc. Sol. Energy Storage Sub-Systems Heating and Cooling of Buildings, NSF RA-N-75-041,* National Science Foundation, April 1975.

51. S. M. Han and S. T. Wu, "Experimental and Numerical Study of Liquid Thermal Storage Tank Models," *Proc. Am. Sec. Int. Sol. Energy Soc.,* Vol. 2, No. 1, pp. 265–271, 1978.

52. M. J. Kennedy, S. J. Sersen, and S. M. Rossi, "Comparisons of Liquid Solar Thermal Storage Subsystems in the National Solar Data Network," *ASME Paper 80-WA/SOL-36,* American Society of Mechanical Engineers, 345 E. 47th St., New York, NY 10017.

53. E. E. Pickering, "Residential Hot Water Solar Energy Storage," *Proc. Sol. Energy Storage Sub-Systems Heating and Cooling of Buildings, NSF RA-N-75-041,* National Science Foundation, April 1975.

54. B. T. Rogers, "Rockbed Storage Systems Group Report" (Remarks by F. G. Hogg, p. 167), *Proc. Sol. Energy Storage Subsystems Heating and Cooling of Buildings, NSF RA-N-75-041,* April 1975.

55. G. A. lane, D. N. Glew, E. C. Clarke, et al., "Heat of Fusion Systems for Solar Energy Storage," *Proc. Sol. Energy Storage Sub-Systems Heating and Cooling of Buildings, NSF-RA-N-75-041,* April 1975.

56. J. Schröder, "Thermal Energy Storage and Control," *J. Eng. for Industry,* pp. 893–896, August 1975.

57. C. D. MacCracken, "Refrigeration, Cooling Storage and Load Management," *ASHRAE J.,* p. 35, December 1979.

58. C. D. MacCracken, "Ice Storage Systems and Eutectic Salt Concepts," *Int. Conf. Thermal Storage in Buildings,* Toronto, Nov. 13, 1980.

59. C. D. MacCracken, "Salt Hydrate Thermal Energy Storage System for Space Heating and Air Conditioning," CALMAC Manufacturing Co., Englewood, N.J., *EM-78-C-02-4698,* December 1977 to June 1980.

60. G. A. lane, "Congruent-Melting Phase-Change Heat Storage Materials," Dow Chemical Co., Midland, Mich., *Publ. Release B-600-203-81,* 1981.

61. G. A. Lane, "Low Temperature Heat Storage with Phase-Change Materials," *Int. J. Ambient Energy,* Vol. 1, No. 3, pp. 155–168, July 1980.

62. C. D. MacCracken, "PCM Bulk Storage," *Int. Conf. Energy Storage,* Paper H3, Brighton, England, April 29 to May 1, 1981.

63. *Methods of Testing Thermal Storage Devices Based on Thermal Performance, Std. 94-77,* American Society of Heating, Refrigerating and Air Conditioning Engineers, Inc., 345 E. 47th St., New York, NY 10017.

NOMENCLATURE

Symbol, Definition, SI Units, English Units

A_c	collector area: m^2, ft^2
A_s	storage container area: m^2, ft^2
A_0	bed cross-sectional area (Fig. 5): m^2, ft^2
c_{p_i}	heat capacity of air: $J/(kg \cdot K)$, $Btu/(lb_m \cdot °F)$
c_{pl}	heat capacity of liquid: $J/(kg \cdot K)$, $Btu/(lb_m \cdot °F)$
c_{p_s}	heat capacity of solid: $J/(kg \cdot K)$, $Btu/(lb_m \cdot °F)$
D_e	equivalent spherical diameter, Eq. (16): m, ft
D_s	effective particle diameter, Eq. (17): m, ft
F	fraction of month
f	friction factor, Eq. (32)
f_0	fraction of specified heat loss, Eq. (12)
G_0	superficial mass velocity: $kg/(h \cdot m^2)$, $lb_m/(h \cdot ft^2)$
g_c	conversion factor: $lb_m \cdot ft/(lb_f \cdot s^2)$; 1 in SI units
h	heat transfer coefficient: $W/(m^2 \cdot K)$, $Btu/(h \cdot ft^2 \cdot °F)$
i_{sl}	solid-liquid latent heat: kJ/kg, Btu/lb_m
Δi	enthalpy change: kJ/kg, Btu/lb_m
j_h	j factor, Eqs. (26) to (30)
k	thermal conductivity of fluid, $W/m \cdot K$, $Btu/(h \cdot ft \cdot °F)$
L	length: m, ft
m_s	storage mass: kg, lb_m
n	sample number
$n_{\Delta x}$	number of divisions in bed
P	pressure: Pa, lb_f/ft^2
P	pumping power: W, Btu/h
PCM	phase-change material
Pr	Prandtl number
Q	$m_s c_{p_s}(T_0 - T_{min})$: kJ, Btu

Q_w water heating load: kJ, Btu

$\dot{Q}$ volumetric flow rate: m^3/min, ft^3/min

Re   Reynolds number, Eq. (19)

T temperature: K, °F

T_a ambient temperature: K, °F

$T_{\min}$ minimum cooling temperature: K, °F

T_m melting temperature: K, °F

T_0 maximum storage temperature: K, °F

t time: s, h

t_0 time for heat loss: s, h

U_s storage heat transfer coefficient: W/(m$^2 \cdot$K), Btu/(h$\cdot$ft$^2 \cdot$°F)

UA   structure load factor: kJ/(K$\cdot$day), Btu/(°F$\cdot$day)

$(UA)_s$ storage load factor: kJ/(K$\cdot$day), Btu/(°F$\cdot$day)

V_R rock sample volume: m^3, ft^3

V_s storage volume: m^3, ft^3

V_0 superficial (face) velocity: m/s, ft/s

$W_i c_{p_i}$ infiltration capacity rate: kJ/(K$\cdot$day), Btu/(°F$\cdot$day)

W_0 air mass flow rate per unit area of bed cross section: kg/s, lb$_m$/h

X distance along bed: m, ft

Greek Symbols

ϵ void fraction

θ degree-days per month: °C$\cdot$days/month, °F$\cdot$days/month

λ relaxation length: m, ft

μ dynamic viscosity: Pa$\cdot$s, lb$_m$/(ft$\cdot$h)

ρ density: kg/m^3, lb$_m$/ft^3

ρ_s solid density: kg/m^3, lb$_m$/ft^3

ρ_0 inlet density: kg/m^3, lb$_m$/ft^3

ϕ shape factor

CHAPTER 9

Heat Transfer in Buildings

By T. Kusuda

National Bureau of Standards
Washington, D.C.

A. INTRODUCTION

Building heat transfer calculations are performed for different applications such as:

1. Heat loss and heat gain through the exterior envelope:
   - Conduction through the exterior envelope
   - Conduction heat transfer through basement walls and slab-on-grade floor (to semi-infinite space)
   - Short-wavelength (or solar heat) transmission, absorption, and reflection for fenestration
   - Air leakage through exterior envelopes as well as the interior partition walls, ceilings and floors
   - Thermal storage in exterior masses of buildings

2. Interior environment analyses:
   - Radiant heat exchange among interior surfaces and heat sinks or heat sources
   - Convective heat exchange between room air and room interior surfaces
   - Room air convection: inter- and intraroom convective motion
   - Convective and radiative heat transfer of internal heat sources such as heaters, coolers, and occupants
   - Thermal storage in interior masses

3. Material or building element–related problems:
   - Thermal-bridge effect
   - Convection within porous insulation
   - Moisture condensation due to simultaneous flow of air, moisture, and heat

Problems associated with building heat transfer are unique and differ from those associated with industrial processes or the design and performance of various heat exchangers. This is because the latter are usually governed by better defined boundary conditions than the former. Moreover, most heat transfer equipment is of relatively simple geometry and is composed of homogeneous material, much more so than most buildings. In addition,

1. Buildings are exposed to environmental conditions that are constantly and randomly changing with respect to time, and these changing environmental conditions consist of many parameters that affect the heat transfer process: temperature, solar radiation, infrared radiation (including heat loss to the sky), wind velocity, relative humidity, and soil temperature.

2. The heat transfer process for a building includes a combination of conductive and radiative processes which affect each other. Even the indoor environment is dependent upon the performance of the building heating and cooling system, and occupancy schedules.

3. Thermal property values and heat transfer coefficients are not well defined, changing constantly not only with respect to time and space but also with respect to temperature and humidity. In other words, the heat transfer process is nonlinear.

4. A building is made up of many different components of differing thermal mass, all of which react to the change in external and internal environmental conditions with different thermal time constants. In other words, in a dynamic simulation analysis a building is an extremely stiff system in which the response equations of different thermal time constants have to be solved simultaneously.

Most textbook solutions, developed for heat exchangers and industrial processes of simple and well-defined boundary conditions and geometries, are not applicable for the analysis of building heat transfer. This is the reason that many of the existing handbook data for building heat transfer are based upon empirical relationships that are derived from many years of experience with practical considerations rather than rigorous experimental and mathematical analyses.

Most building heat transfer calculation procedures in the United States have been developed by the American Society for Heating, Refrigerating and Air Conditioning Engineers (ASHRAE) and summarized in the *Handbook of Fundamentals* [1] and other special publications. The majority of calculations are directed toward developing design and evaluation data for the following subjects:

1. Energy consumption
2. Sizing of heating and cooling equipment
3. Moisture condensation
4. Material deterioration due to temperature cycling
5. Comfort requirements and determination

While comfort determination requires accurate prediction of indoor temperature, radiation field, air velocity, and humidity with respect to outdoor environmental conditions for different envelope designs, calculations for energy analysis and equipment sizing are usually performed for the prescribed indoor conditions which meet comfort and/or habitability criteria (see Fanger [2]). Moisture condensation in the building requires a predictive methodology to analyze the simultaneous gradient of the temperature and water vapor pressure through the envelope component, as well as the interior surface temperature and room dew-point temperature, as discussed in Sec. D. The material degradation problem is usually associated with moisture condensation and extreme temperature cycling in the building.

The five subjects mentioned above have been treated in most traditional heat transfer books, such as those by Van Stratten [3], Kimura [4], Muncey [5], and McIntyre [6]. There are a growing number of new problems to be addressed in advanced heat transfer calculations; some examples are:

1. Convective heat transfer within a room and between rooms due to temperature as well as wind pressure effects
2. Interaction between the operation of the heating and cooling systems and building thermal response
3. Moisture absorption by interior surfaces concurrent with heat transfer processes

While most of the traditional calculations can be handled by a relatively simple empirical formula, advanced topics such as those mentioned here require comprehensive computer simulation, introductory materials for which are also included in this chapter.

B. OUTDOOR WEATHER DATA

The external boundary conditions for building heat transfer calculations are weather data, including temperature, humidity, barometric pressure, wind speed, wind direction, total and diffuse solar radiation, and ground temperature. These data must be available for various localities for different seasons as well as for different hours of the day.

Two kinds of weather data are generally used for building heat transfer calculations. The first are those for relatively extreme conditions in summer or winter, generally used for design and selection of heating and cooling systems; these data are available from the *ASHRAE Handbook,* as discussed below. The second kind of weather data needed is for annual energy consumption calculations, which require information beyond that available through ASHRAE.

1. ASHRAE Weather Data

Relatively extreme conditions are represented by very hot and humid hours of a summer day, or by a very cold spell during typical winter nights. These extreme values are generally used for the design and selection of heating and cooling systems and equipment. This is to ensure satisfactory environmental control for most of the summer and winter hours. The *ASHRAE Handbook* [1] tabulates, for more than 750 U.S. cities and 300 foreign cities, dry- and wet-bulb temperatures equaled and exceeded by 1, 2.5, and 5 percent of total summer hours (June through September). Also presented in the *Handbook,* together with these percentile dry-bulb temperatures, are coincident wet-bulb temperatures. For the design of winter heating equipment, the *Handbook* lists 99 and 97.5 percentile values representing the temperatures which equal or exceed these portions of the total winter hours (December through February).

Solar radiation data for equipment design and selection are provided for clear sky conditions calculated at 8° intervals from 0 to 64° north latitude for eight compass orientations, as well as for horizontal surfaces. The data are, however, for the twenty-first day of each of the 12 months. The standard wind data used for the design and selection calculations are 15 mi/h (6.7 m/s) for winter and 7½ mi/h (3.35 m/s) for summer.

2. Weather Data for Energy Analysis

The second kind of weather data needed is for annual energy consumption calculations. For the purpose of estimating the annual energy consumption for the heating and cooling of buildings, these ASHRAE percentile data and clear sky solar radiation data are inadequate.

The most commonly used data for energy analysis are degree-days and monthly normal temperatures, which are available for the major cities in the NOAA publications [7,8]. The daily average as well as the seasonal average heating energy consumption of homes is usually found to be directly proportional to average outdoor temperatures, and is the basis for the well-accepted degree-day calculation method. This dependency of building energy consumption upon outdoor temperature becomes more complicated for nonresidential buildings. Modern energy analysis procedures developed for nonresidential buildings such as high-rise office buildings, hospitals, schools, etc., are able to account for the hour-by-hour variation of the coincident nature of temperature, humidity, solar radiation, and wind velocity. This is because the dynamic response of a building and its heating and cooling system is influenced by the changing nature of these outdoor factors. Most of the existing hourly simulation procedures employ weather data tapes containing these hourly weather parameters for at least one year. There are several sources of hourly weather data available in magnetic tapes through the National Climatic Center in Asheville, North Carolina.

The Series 1440 tape contains hourly observations of major weather parameters such as temperature, humidity, wind data, barometric pressure, and precipitation, but does not contain solar radiation data. The data are available for over 300 Weather Bureau stations and 213 Air Force stations in the United States, covering as many as 30 years of historical data. The data are available only at three-hour intervals after 1965, however. Since Series 1440 was the only weather tape suitable for the hourly energy analysis until 1977, these tapes have been widely used by hourly simulation energy analysis programs.

In an effort to expedite the use of standard reference years for the comparative energy analysis based upon hourly simulation energy consumption, ASHRAE, in 1977, developed in conjunction with the National Climatic Center the so-called Test Reference Year, TRY (see Stamper [9]). TRY is not, however, considered typical enough to give an accurate estimate of the long-time average annual energy consumption of a building. The selection procedures for determining a TRY for a specific location are based on the elimination of years that contain extremely hot or cold months. Monthly normal values (based upon at least 30 years' weather data) of average temperatures published in the Climatological Data Summary by the National Climatic Center publication were used for the elimination process.

Another important form of hourly weather data for energy analysis developed by Crow [11] is called WYEC, or weather year for energy calculation. It is available for 51 (as of March 1984) North American cities and was synthesized by using the best judgment of a professional meteorologist to yield a long-time average annual energy consumption estimate.

In 1978, the Department of Energy developed the SOLMET tape (see Ref. 10), which is compatible with the hour-by-hour simulation annual energy analysis and contains not only the solar radiation data (global radiation on a horizontal surface) but also coincident meteorological data from the 1440 tapes and estimates of direct normal solar radiation. SOLMET tapes are available for 26 stations and most of them cover 25 years of data. The SOLMET tapes have been widely used by the designers of solar heating and cooling units.

In order to select typical design year data from the SOLMET tapes, Hall et al. [12] developed an empirical procedure based upon the cumulative distribution function (CDF). For each of the 12 calendar months over a 23-year period, they statistically selected a typical month by comparing the CDF for each month with that of the long-term average CDF of the 23 years for 13 statistics computed from the dry-bulb, dew-point, wind-speed, and solar radiation data. The final selection of a typical meteorological year (TMY) was made by comparing the sum of the weighted deviation of each of the 13 statistics' CDFs from their long-term average CDFs. The TMY is available for 26 original SOLMET stations and 222 rehabilitated SOLMET stations, for which hourly solar data were synthesized using the cloud cover data.

3. Degree-Day Data and Bin Data

The degree-day data have been widely used for energy calculations because of their simplicity. They are generated from the daily mean temperature (arithmetic average of daily maximum and daily minimum temperatures) from the past weather records of many years, using the following mathematical relationship:

$$H_{DD} = \sum_{i=1}^{N_{DH}} (T_B - T_{di})\delta_i$$

where H_{DD} = heating degree-days
T_B = balance temperature
T_{di} = daily mean temperature of day i
$\delta_i = 1$ if $T_B > T_{di}$
$\delta_i = 0$ if $T_B \leq T_{di}$
N_{DH} = total number of days in heating season

and

$$C_{DD} = \sum_{i=1}^{N_{DC}} (T_{di} - T_B)\delta_i$$

where C_{DD} = cooling degree-day
$\delta_i = 1$ if $T_{di} \geq T_B$
$\delta_i = 0$ if $T_{di} < T_B$
N_{DC} = total number of days in cooling season

The currently available degree-day data [1] are based upon $T_B = 65°F$ (18.3°C). The balance point T_B is defined as the temperature above which the building heat loss is offset by the heat gain from solar energy and internal heat sources. A recent study by Kusuda et al. [13] showed that the accuracy of the building energy analysis by the use of the degree-day method depends heavily upon the accuracy of the balance point temperature. The annual energy estimated by the degree-day method can be made very close to that determined by more comprehensive hourly computer simulation calculations, provided that the balance point temperature, instead of the standard 65°F (18.3°C), has been determined by accurate building heat transfer calculations. Heating and cooling degree-day data for several selected balance temperatures other than 65°F (18.3°C) have been developed by using Thom's statistical technique [14] for major cities in the United States. The data are available in microfiche form from the National Climatic Center [15].

Many simplified energy analysis procedures including nonresidential applications employ the bin temperature data. The bin data are nothing but a frequency of hourly temperature data for a series of 5°F (2.8°C) bins. Air Force Manual 88-29 [16] contains these bin data, together with the coincident average wet-bulb temperature. These data are available on a monthly basis for three time periods of the day for all of the U.S. Air Force bases throughout the world.

4. Soil Temperature Data

Since most buildings are erected on the ground and exchange heat with the ground through the basement walls and slab-on-grade basement floors, ground temperature is extremely important for the assessment of building heat transfer, especially for the sizing of the summer air conditioning system and for the annual energy consumption analysis. Moreover, there are many buildings which are completely underground, such as underground survival shelters, missile silos, earth berms, and passive dwellings. Kusuda and Achenbach [17] made an extensive survey of earth temperatures throughout the United States covering more than 100 ground temperature station data. They found that the monthly average temperature of the earth at various depths z can be estimated by the following equation, with thermal diffusivity α approximately 0.056 m²/day (0.025 ft²/h), and with the annual average $\overline{T_A}$ and amplitude B of the monthly normal temperature cycle:

$$T_G = \overline{T_A} + B \exp\left(-\sqrt{\frac{\pi}{\alpha \tilde{P}} }\, z\right) \cos\left(\frac{2\pi}{\tilde{P}}\, t - \sqrt{\frac{\pi}{\alpha \tilde{P}} }\, z - \phi\right) \qquad (1)$$

A recommended value of ϕ for the United States in this equation is 0.6 rad.

Using these survey results, Kusuda and Achenbach [17] developed a recommended set of seasonal design temperatures based upon 10-ft-depth average data. It should be noted that ground temperature around the building varies from one corner of the building to another, as well as with the distance from the building, depending upon the type of moisture content and ground cover. Kusuda [18], for instance, found the ground temperature under a paved surface to be approximately 15°F (8.3°C) higher throughout the summer, even at the 10-ft (3-m) depth.

5. Solar Radiation Data

Solar radiation data under clear sky conditions for the design and selection of building heating and cooling equipment are found in the *ASHRAE Handbook of Fundamentals* [1977]. Section C.6 of this chapter discusses the basic solar heat gain equations. For the annual energy analysis, solar radiation data for cloudless sky conditions, as published in the *ASHRAE Handbook*, are not particularly useful because there are many non-clear-sky days in the year. More appropriate data would be the monthly normal-day hourly solar radiation data. Kusuda and Ishii [19] applied and modified the Liu and Jordan method [20] to generate the average-day hourly solar radiation data for horizontal as well as vertical surfaces of various orientations for 80 cities in the United States and Canada.

C. ENVELOPE HEAT TRANSFER

1. Steady-State Heat Conduction

Steady-state conduction heat loss or gain q through walls, roofs, and floor slabs is determined by a well-known equation,

$$q = UA(T_i - T_o) \tag{2}$$

where T_i and T_o are interior and exterior air temperatures, respectively. The value of U is in turn determined by another well-known equation,

$$\frac{1}{U} = \frac{1}{h_o} + \Sigma \frac{l_i}{k_i} + \frac{1}{C_i} + \frac{1}{h_i} \tag{3}$$

where h_o = external surface heat transfer coefficient
$\quad\quad h_i$ = interior surface heat transfer coefficient
$\quad\quad C_i$ = air-cavity thermal conductance
$\quad\quad \Sigma \dfrac{l_i}{h_i}$ = sum of the thermal resistances of the various composite layers that make up the wall

Tables 1, 2, and 3 show data for this equation commonly used for typical building heat transfer calculations.

TABLE 1 Surface Heat Transfer Coefficients of Different Surfaces

Position of surface	Direction of heat flow	Surface heat transfer coefficient	
		$W/(m^2 \cdot K)$	$Btu/(h \cdot ft^2 \cdot °F)$
Vertical	Horizontal	8.34	1.47
Horizontal	Up	11.0	1.94
Horizontal	Down	6.6	1.20
Horizontal with underside having emissivity = 0.05*	Down	1.3	0.23
Horizontal with underside having emissivity = 0.05*	Up	4.3	0.75
Vertical, with emissivity = 0.05*	Horizontal	3.2	0.56
External surfaces	All directions	20.0	3.50

*Emissivity of a surface is a measure of its ability to radiate heat. Emissivities at room temperatures for most building materials, irrespective of color, are of the order of 0.9, while those for bright metal surfaces can generally be taken at 0.05.

2. Thermal Bridges

Equation (3), however, is not applicable when there are parallel heat flow paths such as wall studs, floor or ceiling joists, window frames, and metallic tie rods. These parallel elements normal to the planes of heat transfer surfaces are commonly known as "thermal bridges" or "cold bridges." When a wall with thermal bridges is exposed to cold outdoor air, the interior surface temperature over the bridge region is colder than the rest of the insulated region; this is the source of thermal discomfort and undesirable moisture condensation. If the warm room-side surface temperature over the cold bridge region is

TABLE 2 Thermal Conductance of Air Cavity

Position of air space	Direction of heat flow	Conductance, C_i	
		W/(m²·K)	Btu/(h·ft²·°F)
Vertical	Horizontal	6.2	1.10
Horizontal	Up	6.8–7.4	1.2~1.30
Horizontal	Down	5.2–5.3	0.91~0.93
45° angle	Up	6.2	1.10
45° angle	Down	5.7	1.0
Horizontal, underside of upper surface having emissivity = 0.05			
Thickness of air layer 100 mm and over	Down	0.68	0.12
Thickness of air layer 40 mm	Down	1.1	0.19
Thickness of air alyer 20 mm	Down	2.1	0.37
Horizontal, underside of upper boundary surface having emissivity = 0.05			
Thickness of air layer 20 mm and over	Up	2.6	0.46

denoted by T_c, the surface temperature remote from the cold bridge zone by T_{wi}, and room air temperature by T_i, the cold bridge effect may be evaluated by

$$\rho = \frac{T_i - T_c}{T_i - T_{wi}} \tag{4}$$

where ρ is called by Berthier [21] the "thermal bridge effect."

TABLE 3 Thermal Conductivities of Different Building Materials at Room Temperatures

Type of material	Density		Thermal conductivity	
	kg/m³	lb_m/ft³	W/(m·K)	Btu·in/(h·ft²·°F)
Dense concrete	2323	145	1.5	10.4
Cellular concrete	961	60	0.26	1.80
Brickwork·	1826	114	0.82	5.69
Asbestos cement	1602	100	0.65	4.51
Burnt-clay tiles	1922	120	0.84	5.83
Slate	2563~2723	160~170	1.4–1.9	9.7~13.19
Mineral wool	12~144	0.7~9	0.033–0.042	0.23~0.29
Steel	7850	490	53	368
Asphalt tile	2243	140	1.3	9.02
Wood	368~801	23~50	0.10–0.22	0.69~1.53
Soft fiberboard	264	16.5	0.058	0.40
Hardboard	1121	70	0.20	1.39
Compressed cork	136~176	8.5~11	0.042–0.046	0.29~0.32
Window glass	2483	155	0.75	5.21
Gypsum plasterboard	993	62	0.17	1.18
Expanded polystyrene	24	1.5	0.033–0.042	0.23~0.29
Expanded polyurethane	24~48	1.5~3.0	0.025	0.17

If there is no two- or three-dimensional heat conduction effect, or there is no lateral heat conduction among parallel heat flow elements, the overall heat transfer coefficient of the parallel heat flow system can be determined by

$$U = \frac{A_c U_c + A_p U_p}{A_c + A_p} \tag{5}$$

and the value of ρ over the thermal bridge should be

$$\rho_c = \frac{U_p}{U_c} \tag{6}$$

where U_p and U_c and A_p and A_c are overall heat transfer coefficients and areas of the thermal bridge zone and of the zone remote from the thermal bridge, respectively.

The deviation of Eq. (6) from Eq. (4), therefore, represents the multidimensional heat conduction effect. Figure 1 shows temperature profiles of a simple thermal bridge system where wood [$k = 0.173$ W/(m·K), or 1.2 Btu·in/(h·ft²·°F)] and pressed fiberboard frames [$k = 0.0697$ W/m·K), or 0.480 Btu·in/(h·ft²·°F)] of different thicknesses were imbedded in a 3-cm (1.18-in) panel of expanded polystyrene insulation board [$k = 0.0348$ W/(m·K), or 0.24 Btu·in/(h·ft²·°F)].

Shown in this figure is ρ_m, which is determined by the minimum surface temperature T_{cm} over the thermal bridge using the relation

$$\rho_m = \frac{T_i - T_{cm}}{T_i - T_{wi}} \tag{7}$$

Calculated ρ_c's based upon U_p/U_c for the system are 2.8 for the wood frame and 1.7 for the pressed-wood frame. The figure demonstrates strong two-dimensional heat conduction effects for thin and less conductive bridges than for the wider, more conductive bridges.

The thermal bridge effects of different configurations, measured by Berthier [21], are

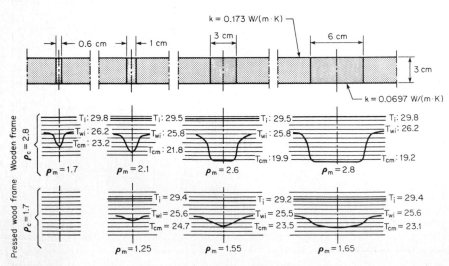

FIG. 1 Temperature profile over the thermal bridge (SI units are used).

shown in Fig. 2, in which 4-mm (0.15 in) thick iron frames [$k = 58 \text{ W/(m·K)}$, or 403 Btu·in/(h·ft²·°F)] as well as 4.6-mm (0.18-in) tie rods constitute parallel heat flow paths. The theoretical bridge effect for the system is 6.0, and most of the systems show the measured maximum bridge effect to be lower than ρ_c, except for the frame with the cold-side flange.

The value of θ^+ shown in Fig. 2 is calculated by

$$\theta^+ = \frac{T_i - T_{cm}}{T_i - T_o} \qquad (8)$$

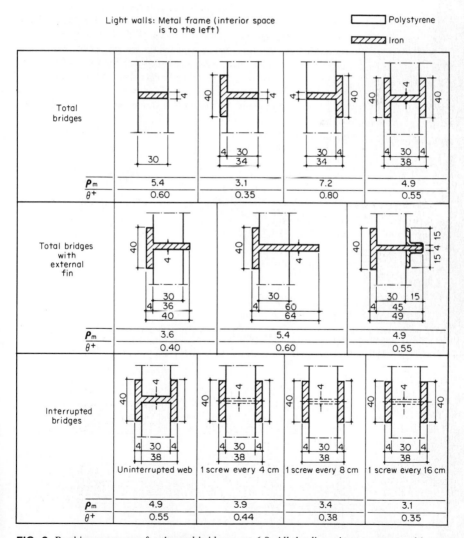

FIG. 2 Berthier parameters for thermal bridge $\rho_c = 6.0$. All the dimensions are expressed in mm.

where T_i and T_o are inside and outside temperature, respectively. For the perfect thermal bridge with no two-dimensional heat flow, θ^+ should assume a value calculated by

$$\theta_c^+ = \frac{U_p}{h_i} \tag{9}$$

where h_i is the inside surface heat transfer coefficient and is expected to be 7.00 W/(m²·K), or 1.23 Btu/(h·ft²·°F), for this particular system.

Again the measured θ^+ exceeded the θ_c^+ for the frame with the cold-side flange.

While previous discussions centered around the surface temperature over the thermal bridge, Berthier also investigated the heat transfer rate influenced by the thermal bridge as shown in Fig. 3. Compared side by side in the figure are measured and calculated overall heat transfer coefficients for a 3-cm (1.18-in) polystyrene insulation board sandwiched between 5-mm (0.197-in) insulating fiberboard.

While the calculated and measured values agree well for the systems without parallel heat flow paths or thermal bridges, calculated U values that ignore the two-dimensional heat conduction are considerably different from the measured values for the panel with an iron frame of 5-mm (0.197-in) thickness with 20-cm (7.87-in) pitch. Also shown is the fact that calculated U values (ignoring the two-dimensional effect) are practically identical for frames with and without flanges, yet the measured U values for those with flanges are much higher than for flangeless frames.

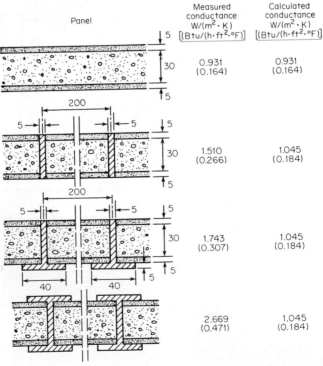

Panel	Measured conductance W/(m²·K) [(Btu/(h·ft²·°F)]	Calculated conductance W/(m²·K) [(Btu/(h·ft²·°F)]
	0.931 (0.164)	0.931 (0.164)
	1.510 (0.266)	1.045 (0.184)
	1.743 (0.307)	1.045 (0.184)
	2.669 (0.471)	1.045 (0.184)

FIG. 3 Heat transfer through thermal bridge (dimensions are expressed in mm).

Unfortunately data such as presented above are all specific to particular wall designs and no generalized formulas are available for estimating the ρ_m, θ', and U values due to thermal bridges.

Hoffman and Schwartz [22] conducted numerical calculations using finite difference methods to analyze the Berthier data. Such a procedure is needed to evaluate any specific thermal bridge design, inclusive of wall or roof systems consisting of concrete blocks of various cavity designs.

The calculated values, no matter how elaborately done, however, are subject to errors because of uncertainties in surface heat transfer coefficients and air-cavity thermal resistances, which vary from location to location due to the surface convection flow characteristics.

3. Surface Heat Transfer Coefficients

Although there are numerous data available for the surface heat transfer coefficients (see Refs. 50 to 52), for the purpose of consistency, most American building heat transfer analysts use the data presented in the *ASHRAE Handbook of Fundamentals* [1]. The ASHRAE data may be represented by the equations in the following section.

a. Exterior-Surface Heat Transfer Coefficient h_o

$$h_o = aV^2 + bV + c \qquad \text{W}/(\text{m}^2 \cdot \text{K})*$$
(10)

where V is wind speed in miles per hour and a, b, and c are empirical constants depending upon the type of surface, as shown below:

Type of surface	a	b	c
Stucco	0.0	2.634	11.583
Brick and rough plaster	0.006	1.817	12.492
Concrete	0.0	1.874	10.788
Clear pine	−0.006	1.789	8.233
Smooth plaster	0.0	1.386	10.221
Glass, white paint on pine	0.0	1.488	8.233

This ASHRAE equation is, however, a combined radiative and convective surface heat transfer coefficient, and not conducive to realistic surface temperature determination. Ito and Kimura [23] obtained the convective component of the building surface heat transfer coefficient from an actual building; some of their results are shown in Fig. 4. The convective surface heat transfer coefficient is strongly dependent upon the wind direction and whether the subject surface is on the windward or leeward side, and can be expressed by the following relations:

$$h_{oc} = 18.63 V_c^{0.605} \qquad \text{W}/(\text{m}^2 \cdot \text{K})$$
(11)

where
$$V_c = \begin{cases} 0.25V & \text{if } V > 2 \text{ m/s} \\ 0.50V & \text{if } V < 2 \text{ m/s} \end{cases}$$

for the windward surface, while $V_c = 0.3 + 0.05V$ for the leeward surface.

*Conversion multiplier to Btu/(h·ft²·°F) is 0.176.

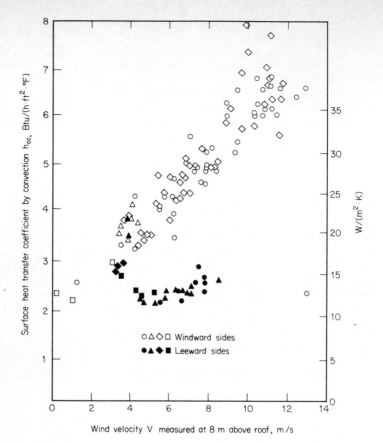

FIG. 4 Correlation between wind velocity above roof and convective heat transfer coefficient for the surface at the center part of a mid-floor of a high-rise building.

b. Inside-Surface Heat Transfer Coefficient h_i

The ASHRAE *Handbook* [1] lists recommended sets of data for inside surfaces at different tilt angles and with different surface emissivity. The results can be represented by the following equation. In still air,

$$h_i = h_c + 5.72 \, \epsilon \, W/(m^2 \cdot K) \tag{12}$$

where h_c is dependent on the direction of heat flow so that

$$h_c = \begin{cases} 4.04 & \text{upward} \\ 3.87 & 45° \text{ upward} \\ 3.08 & \text{horizontal} \\ 2.28 & 45° \text{ downward} \\ 0.920 & \text{downward} \end{cases}$$

In moving air,

$$h_i = 11.36 \qquad W/(m^2 \cdot K)$$

The above equation is derived from more general expressions such as

$$h_i = h_c + h_r \tag{13}$$

where $h_c = C(\Delta T)^{1/3}$, the convective heat transfer coefficient
ΔT = temperature difference between the surface and the surrounding air
C = constant depending upon the direction of the heat flow
$h_r = 4\sigma\Sigma F_{i-k}\ (\epsilon_i, \epsilon_k)\ \overline{T}^3_{i-k}$, the radiative heat transfer coefficient
σ = Stefan-Boltzmann constant
F_{i-k} = radiation exchange factor between surfaces i and k
ϵ_i, ϵ_k = surface emittances of surfaces i and k
$\overline{T}_{i-k}$ = average absolute temperature of the surfaces i and k

4. Air-Cavity Heat Transfer Coefficient

There are many air spaces in a building envelope such as those between the panes of multiple-glazed windows, between a ceiling and the floor above it, and within the wall. The convective heat transfer coefficient h_c across the air spaces of various thicknesses and orientations was extensively measured by Robinson and Powell [24] using a guarded-hot-box apparatus. While the downward convective heat flow is very closely approximated by the purely conductive process, upward flow is very much affected by the convection within the space (Fig. 5). For the horizontal heat flow through vertical air spaces, however Fig. 6 reveals an optimum air space thickness of approximately 0.75 in (1.9 cm) where the thermal resistance is highest. A generalized representation of the Robinson-

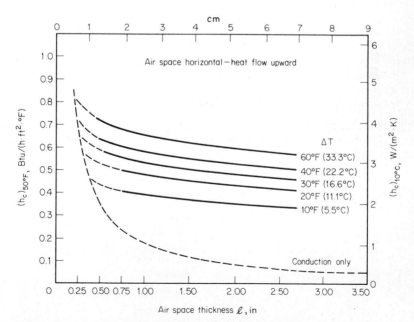

FIG. 5 Natural convection heat transfer coefficient evaluated at the mean air space temperature of 50°F (10°C) through horizontal air space with upward heat flow at temperature differences of 10, 20, 30, 40, and 60°F.

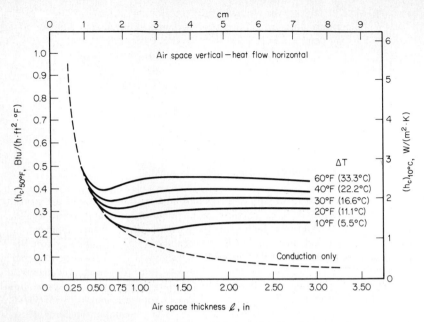

FIG. 6 Natural convection heat transfer coefficient evaluated at the mean air space temperature of 50°F (10°C) through vertical air space with horizontal heat flow at temperature differences of 10, 20, 30, 40, and 60°F.

Powell results is presented in Fig. 7 in terms of $h_c\ell$ at an average cavity temperature of 50°F (10°C), as a function of $\Delta T \cdot \ell^3$, where ΔT is the temperature difference across the air cavity and ℓ is the air-cavity thickness.

The convective heat transfer coefficient h_c at a cavity temperature other than 50°F is determined by

$$h_c = [1 - (\overline{T}_m - 50)f_c]h_{c,50°F} \qquad \text{Btu}/(\text{h}\cdot\text{ft}^2\cdot°\text{F})^* \qquad (14)$$

where $f_c = -0.001$ if $(h_c\ell)_{50} > 0.3$
       $= 0.00035$ if $0.2 < (h_c\ell)_{50} \leq 0.3$
       $= 0.0017$ if $(h_c\ell)_{50} \leq 0.2$
$\overline{T}_m$ = average temperature of the cavity, °F

Robinson and Powell suggest that the radiative heat transfer coefficient h_r given by the following expression be used in conjunction with the convective heat transfer coefficient h_c:

$$h_r = 0.00686 \left[\frac{460 + \overline{T}_m}{100} \right]^3 \epsilon \qquad \text{Btu}/(\text{h}\cdot\text{ft}^2\cdot°\text{F}) \qquad (15)$$

where

$$\epsilon = \frac{1}{1/\epsilon_1 + 1/\epsilon_2 - 1}$$

*Conversion to W/(m²·K): multiply by 5.6785.

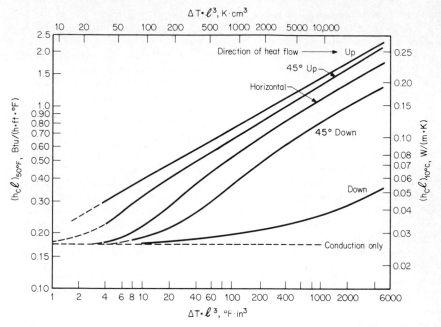

FIG. 7 Convection-conduction coefficient evaluated at the mean air space temperature of 50°F for heat transfer across an air space for five orientations of the air space and directions.

and ϵ_1 and ϵ_2 are emissivities of the surfaces separated by the cavity.

Perhaps the most important application of the air space heat transfer coefficient is for the evaluation of double-glazed windows.

A number of studies have been reported in recent years in which numerical solutions of the Navier-Stokes fluid dynamics equation and the energy equation were obtained for two-dimensional airflow within a confined space. The solutions are for the laminar-flow regime, such as obtained by Bankvall [25] and Tien [26]. Tien extended the solution to partially filled insulation cavities, as well as to permeable-boundary surface problems to simulate air leakage through insulated walls.

5. Air Leakage Calculation

Air infiltration in buildings is defined as the uncontrolled leakage of air through cracks and openings in the building envelope. This mass transfer of air is driven by weather (in particular, indoor-outdoor temperature differences and wind) and is a function of the building architecture, occupant activity, and topological location (siting). Air infiltration accounts for 20 to 50 percent of the heating load of residential structures. By maintaining intake air in excess of exhaust air, commercial buildings are usually maintained at a positive pressure relative to the outdoors, which reduces infiltration to a great extent but increases energy consumption and cost.

In the past 30 years, international research has been conducted primarily on residences to understand air infiltration, and especially to model or predict infiltration rates in order to improve the design of buildings and to size properly mechanical equipment. Although the research is extensive, exterior wall leakage characteristics are very poorly understood; dynamic effects, such as pulsating flow, eddies, and turbulent flows through multiple cracks, have only recently undergone any examination; and the effect of moisture on crack size is not understood well enough to be incorporated into any mathematical model.

Traditionally, models of residential buildings have taken the form

$$Q = A + B \Delta T^m + C V^n \tag{16}$$

where Q is the infiltration rate expressed in air changes per hour; ΔT is the indoor-outdoor temperature difference; V is wind speed; A, B, and C are regression coefficients; and m and n are empirically determined exponents. The best known example is the Achenbach and Coblentz equation [27] where $m = n = 1$, $A = 0.15$, $B = 0.005$, and $C = 0.013$ when ΔT is expressed in °F and V in miles per hour. Because these regression coefficients reflect structural characteristics as well as shielding effects, occupant behavior, and heating and cooling systems, the respective values of A, B, and C may vary by 20 to 1 for similar residences.

Grimsrud, Sherman, and Sonderegger [28] developed a reasonably accurate procedure applicable to predicting air infiltration for houses based upon numerous test results for pressurization and tracer gas measurements. Their suggested formula takes the following form:

$$Q = \sqrt{Q_{stack}^2 + Q_{wind}^2} \tag{17}$$

where $Q_{stack} = \tilde{L} F_s^* \sqrt{\Delta T}$
$Q_{wind} = \tilde{L} F_w^* V$

In these expressions, Q, Q_{stack}, and Q_{wind} are total, stack-dominated, and wind-dominated infiltration expressed in m^3/s, and

$$F_s^* = \frac{1 + R^+/2}{3} \left[1 - \frac{(X^+)^2}{(2 - R^+)^2} \right]^{3/2} \sqrt{\frac{gH}{T}} \tag{18}$$

$$F_w^* = C'(1 - R^+)^{1/3} \left[\frac{\tilde{a}'(H'/10)^{\tilde{\gamma}'}}{\tilde{\alpha}\,(H/10)^{\tilde{\gamma}}} \right] \tag{19}$$

$$X^+ = \frac{\tilde{L}_{ceil} - \tilde{L}_{floor}}{\tilde{L}} \tag{20}$$

$$R^+ = \frac{\tilde{L}_{ceil} + \tilde{L}_{floor}}{\tilde{L}} \tag{21}$$

where ΔT = indoor-outdoor temperature difference, °C
   H = height of the house from grade to the top of the living space, m
   H' = height of the wind measurement site, m
   g = gravitational constant, 9.8 m/s^2
   $\tilde{L}$ = effective leakage area of the house, m^2 (deduced from the pressurization test)
   $\tilde{L}_{ceil}$ = effective leakage area of the ceiling, m^2 (deduced from the pressurization test)
   $\tilde{L}_{floor}$ = effective leakage area of the floor, m^2 (deduced from the pressurization test)
   C' = shielding coefficient (Table 4)
   $\tilde{\alpha}, \tilde{\gamma}$ = terrain constants for the house (Table 5)
   $\tilde{\alpha}', \tilde{\gamma}'$ = terrain constants for the wind measuring site (Table 5)

TABLE 4 Shielding Coefficients for Standard Shielding Classes

Class	C'	Description
I	0.34	No obstruction or local shielding
II	0.30	Few obstructions; a few trees or a small shed
III	0.25	Some obstructions within two house heights; thick hedge, solid fence, neighboring house
IV	0.19	Heavy shielding; buildings and trees within 30 ft (10 m) in most directions
V	0.11	Very heavy shielding; typical downtown shielding

TABLE 5 Terrain Parameters for Standard Terrain Classes

Class	$\tilde{\alpha}, \tilde{\alpha}'$	$\tilde{\gamma}, \tilde{\gamma}'$	Description
I	1.30	0.1	Ocean or unrestricted expanse
II	1.00	0.15	Flat terrain with some isolated obstructions
III	0.85	0.20	Rural areas with low buildings and trees
IV	0.67	0.25	Urban, industrial, or forest area
V	0.47	0.35	Center of large city

There are numerous yet similar approaches suggested by several researchers which are found in the comprehensive literature survey on air leakage prepared by Ross and Grimsrud [29].

6. Window Solar Heat Gain Calculation

a. Solar Radiation Data under Clear Sky Conditions

It has been customary for building engineers to use solar radiation data under clear sky conditions for sizing of the cooling equipment. In this section the method developed by the ASHRAE Task Group on Energy Requirements [30] for the clear-day solar radiation calculation will be given first, followed by modifications for nonclear sky conditions.

For solar radiation heat transfer problems, the first thing to know is whether the sun is above the horizon or not. The sunrise and sunset time in hours can be determined by the latitude angle **L**, the sun's declination angle δ_s, and its longitude angle l' as follows:

$$\text{Sunrise time} = 12 - Y - \text{ET} - \text{TZN} + \frac{l'}{15}$$
$$\text{Sunset time} = 24 - \text{sunrise time}$$

(22)

where $Y = \dfrac{12}{\pi} \cos^{-1}(-\tan L \tan \delta_s)$

ET = equation of time (see Table 6)

TZN = time zone index

    Atlantic zone: 4

    Eastern zone: 5

    Central zone: 6

    Mountain zone: 7

    Pacific zone: 8

Also it is customary to use solar time or solar hour angle ω for the determination of the solar altitude angle as well as the solar azimuth angle. The hour angle ω and the direction cosines of the direct solar beam are determined by

$$\omega = 15 \,(\text{standard time} - 12 + \text{TZN} + \text{ET}) - l' \tag{23}$$

$$
\begin{aligned}
\cos z &= \sin L \sin \delta_s + \cos L \cos \delta_s \cos \omega \\
\cos w &= \cos \delta_s \sin \omega \\
\cos s &= -\sin \delta_s \cos L + \cos \delta_s \sin L \cos \omega
\end{aligned}
\tag{24}
$$

whereby $\cos s > 0$ if

$$\cos \omega > \frac{\tan \delta_s}{\tan L} \qquad \text{in northern hemisphere}$$

$$\cos \omega < \frac{\tan \delta_s}{\tan L} \qquad \text{in southern hemisphere}$$

The solar altitude angle ψ is then (see Fig. 8)

$$\psi = \sin^{-1}(\cos z) \tag{25}$$

TABLE 6 Values of δ_s, ET, $\overline{A}$, $\overline{B}$, and $\overline{C}$ for Northern Hemisphere*

Date	δ_s, deg	ET, h	$\overline{A}$ W/m^2	$\overline{A}$ Btu/(h·ft^2)	$\overline{B}$, air mass^{-1}	$\overline{C}$
Jan. 21	−20.0	−0.190	1230	390	0.142	0.058
Feb. 21	−10.8	−0.230	1214	385	0.144	0.060
Mar. 21	0.0	−0.123	1186	376	0.156	0.071
Apr. 21	11.6	0.020	1135	360	0.180	0.097
May 21	20.0	0.060	1104	350	0.196	0.121
June 21	23.45	−0.025	1088	345	0.205	0.134
July 21	20.6	−0.103	1085	344	0.207	0.136
Aug. 21	12.3	−0.051	1107	351	0.201	0.122
Sept. 21	0.0	0.113	1151	365	0.177	0.092
Oct. 21	−10.5	0.255	1186	378	0.160	0.073
Nov. 21	−19.8	0.235	1220	387	0.149	0.063
Dec. 21	−23.45	0.033	1233	391	0.142	0.057

*Derived from the 1972 *ASHRAE Handbook of Fundamentals*, Table 1, p. 387, Chap. 22. For southern hemisphere $\overline{B}$ and $\overline{C}$ should be shifted by six months; thus Jan. 21 data take those of July 21.

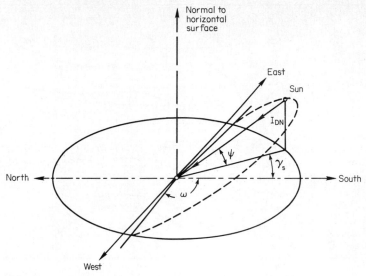

FIG. 8 A schematic showing apparent path of sun and hour angle.

and the solar azimuth angle γ_s is

$$
\gamma_s = \begin{cases} \sin^{-1} \dfrac{\cos w}{\cos \psi} & \text{if } \cos s > 0 \\[2ex] \pi - \sin^{-1} \dfrac{\cos w}{\cos \psi} & \text{if } \cos s < 0 \end{cases} \tag{26}
$$

The intensity of the direct normal radiation under a clear sky is calculated by

$$
I_{DN} = \bar{A} N_c e^{-\bar{B}/(\cos z)} \tag{27}
$$

where the values of $\bar{A}$ and $\bar{B}$ are shown in Table 6 and N_c is the clearness number derived by Threlkeld [31] shown in Fig. 9.

On the other hand the direction cosines of the normal to the surface having the azimuth angle γ and tilt angle β are (see Fig. 10)

$$
\begin{align}
\bar{a} &= \cos \beta \\
\bar{b} &= \sin \gamma \sin \beta \\
\bar{c} &= \cos \gamma \sin \beta
\end{align} \tag{28}
$$

The incident angle of the direct solar radiation upon a given surface of direction cosines $\bar{a}$, $\bar{b}$, and $\bar{c}$ is calculated by

$$
\cos \eta = \bar{a} \cos z + \bar{b} \cos w + \bar{c} \cos s \tag{29}
$$

where $\cos z$, $\cos w$, and $\cos s$ are determined by Eq. (24).

The direct radiation on a surface under consideration for cloudless sky conditions is

$$
I_D = \begin{cases} I_{DN} \cos \eta & \text{as long as } \cos \eta > 0 \\ 0 & \text{when } \cos \eta \leq 0 \end{cases} \tag{30}
$$

The diffuse sky radiation (sky brightness) for a cloudless condition is

$$
B_S = \bar{C} \frac{I_{DN}}{N_c^2} \tag{31}
$$

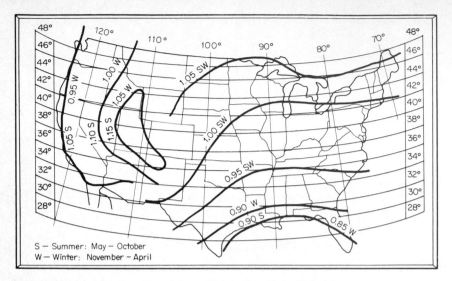

FIG. 9 Clearness numbers of nonindustrial atmosphere in the United States.

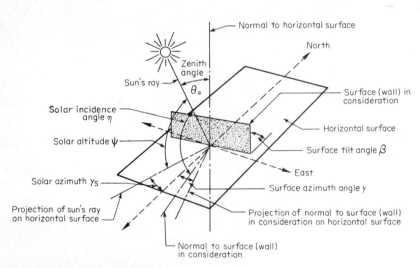

FIG. 10 Solar angles for tilted and horizontal surfaces.

where $\overline{C}$ is a constant from Table 6.

The ground reflected radiation for a cloudless condition (ground brightness) is

$$B_G = r_g(B_S + I_{DN} \cos z) \tag{32}$$

where r_g is the reflectivity of the ground. The diffuse radiation on a horizontal surface under a cloudless sky is

$$I_{dH} = B_S$$

The total radiation on a horizontal surface under a cloudless sky is

$$I_{TH} = I_{DH} + I_{dH} \tag{34}$$

b. Solar Radiation under Nonclear Sky Conditions

The non-clear-day solar radiation can be estimated by the use of the cloud cover factor F_{cc} to modify the total solar radiation on a horizontal surface by means of the observed cloud cover data for a cloudy sky condition. The cloud cover observations are made every hour at major weather stations by experienced observers who estimate the amount of cloud on a scale of 0 to 10 and indicate the type of cloud in four different layers. Kimura and Stephenson [32] analyzed Canadian data for observed solar radiation with respect to the cloud cover data, type of cloud, and the calculated solar radiation under a cloudless condition at the same solar time. Based upon their analysis, a comprehensive methodology was developed for calculating the cloud-day solar radiation. The value of F_{cc} (cloud cover factor) is first defined as

$$F_{cc} = \frac{I_{THC}}{I_{TH}} \tag{35}$$

where I_{THC} = total solar radiation on a horizontal surface under a cloudy sky of given cloud amount and types of clouds

I_{TH} = total solar radiation calculated for a horizontal surface under a cloudless sky at the same solar hour as I_{THC}

and F_{cc} may be calculated by the cloud amount data of four cloud layers in the sky as well as by the total cloud amount as follows:

$$F_{cc} = \overline{P} + \overline{Q}\,(\text{CC}) + \overline{R}\,(\text{CC})^2 \tag{36}$$

where CC = TCA $- 0.5X$ where $0 \leq$ CC ≤ 10

$X = (\Sigma\, \text{CA}_j)_{\text{cirrus}} + (\Sigma\, \text{CA}_j)_{\text{cirrostratus}} + (\Sigma\, \text{CA}_j)_{\text{cirrocumulus}}$

$\Sigma\, \text{CA}_j$ = cloud amount at the jth layer, j = 1, 2, 3, or 4

TCA = total cloud amount

and $\overline{P}$, $\overline{Q}$, and $\overline{R}$ are found in Table 7.

TABLE 7 Cloud Cover Constants

Season	$\overline{P}$	$\overline{Q}$	$\overline{R}$
Spring	1.06	0.012	−0.0084
Summer	0.96	0.033	−0.0106
Autumn	0.95	0.030	−0.0108
Winter	1.14	0.003	−0.0082

Kimura and Stephenson [32] found a general relationship among direct and diffuse solar radiation and CC as shown in Fig. 11.

Using the Kimura-Stephenson data, direct radiation on a horizontal surface under a cloudy sky is determined by

$$I_{DHC} = (I_{DN} \cos z + I_{dH})K\left(1 - \frac{\text{CC}}{10}\right) \tag{37}$$

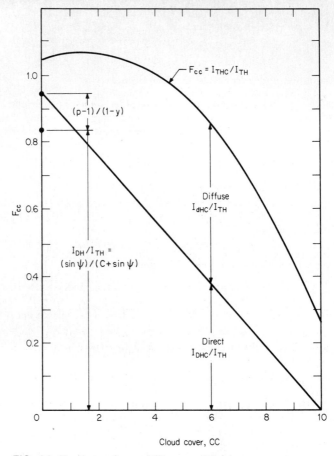

FIG. 11 Cloud cover factor of Kimura and Stephenson.

where

$$K = \frac{\sin \psi}{\overline{C} + \sin \psi} + \frac{\overline{P} - 1}{1 - y} \tag{38}$$

Where $\overline{C}$ and $\overline{P}$ are found in Tables 6 and 7, respectively and

$$y = 0.309 - 0.137 \sin \psi + 0.394 \sin^2 \psi$$

Direct radiation under a cloudy sky is then generally

$$I_{DC} = I_D \frac{I_{DHC}}{I_{DH} \cos z} \tag{39}$$

Diffuse radiation for a cloudless sky is

$$I_d = \begin{cases} B_S & \text{for horizontal surface} \\ B_s Y + \dfrac{B_G}{2} & \text{for vertical surfaces} \end{cases} \tag{40}$$

Where

$$Y = 0.55 + 0.437U + 0.313U^2$$

and

$$U = \cos \eta$$

However, if $U < -0.2$, then $Y = 0.45$.

Diffuse radiation upon a horizontal surface under a cloudy sky is

$$I_{dHC} = I_{TH}\left[F_{cc} - K\left(1 - \frac{CC}{10}\right)\right] \tag{41}$$

Diffuse radiation on a surface under consideration is

$$I_{dC} = I_d \frac{I_{dHC}}{I_{dH}} \tag{42}$$

Total radiation upon a surface under a cloudy sky is

$$I_{TC} = I_{DC} + I_{dC} \tag{43}$$

When the cloud cover CC is zero,

$$I_{TC} = I_T = I_D + I_d \tag{44}$$

c. Solar Optical Performance of Window Systems

One of the most important aspects of solar heat transfer in buildings is the window heat gain problem, which requires calculations for solar heat absorptivity, transmissivity, and reflectivity of various glazings.

The cosine of the refraction angle of a solar ray incident upon single-pane glass with an incident angle η is

$$\cos \xi = \sqrt{1 - \frac{(1 - \cos^2\eta)}{n^2}} \tag{45}$$

where n = index of refraction = 1.52 for ordinary glass.

The fraction of radiation that is absorbed in a single pass through a sheet of glass is

$$a = 1 - e^{-\kappa l/(\cos \xi)} \tag{46}$$

where κ = extinction factor, cm^{-1}
l = thickness of glass, cm

The single-glaze air-glass interface reflectivity r is found by the Fresnel formula for vibration parallel to the plane of the glass,

$$r = \frac{\tan^2(\eta - \xi)}{\tan^2(\eta + \xi)} \tag{47}$$

and for vibration normal to the plane of glass is

$$r' = \frac{\sin^2(\eta - \xi)}{\sin^2(\eta + \xi)} \tag{48}$$

The absorptivity for direct radiation (equal parallel and normal components) is

$$\alpha_\eta = \frac{x + x'}{2} \tag{49}$$

where $x = \dfrac{a(1 - r)[1 + r(1 - a)]}{1 - r^2(1 - a)^2}$

$x' = \dfrac{a(1 - r')[1 + r'(1 - a)]}{1 - r'^2(1 - a)^2}$

The transmissivity for direct radiation is

$$\tau_\eta = \frac{y + y'}{2} \tag{50}$$

where $y = \dfrac{(1 - r)^2(1 - a)}{1 - r^2(1 - a)^2}$

$y' = \dfrac{(1 - r')^2(1 - a)}{1 - r'^2(1 - a)^2}$

The absorptivity and transmissivity for diffuse radiation are

$$\alpha_d = \int_0^{\pi/2} \alpha_\eta \sin 2\eta \, d\eta \tag{51}$$

$$\tau_d = \int_0^{\pi/2} \tau_\eta \sin 2\eta \, d\eta \tag{52}$$

Stephenson [33] developed polynomial expressions for α_η, τ_η, α_d, and τ_d with respect to $\cos \eta$ as follows:

$$\begin{aligned}
\alpha_\eta &= \sum_{j=0}^{5} a_j (\cos \eta)^j \\
\tau_\eta &= \sum_{j=0}^{5} t_j (\cos \eta)^j \\
\alpha_d &= 2 \sum_{j=0}^{5} \frac{a_j}{j + 2} \\
\tau_d &= 2 \sum_{j=0}^{5} \frac{t_j}{j + 2}
\end{aligned} \tag{53}$$

The polynomial coefficients a_j and t_j are found in Table 8.

d. Solar Heat Gain Calculation Procedure

Using these single-sheet absorptivity and transmissivity data, solar heat absorptivity and transmissivity of multiple-glazing windows can be calculated as follows. Let, for example,

$\alpha_{1\eta}$ = absorptivity of inner pane for direct radiation

$\alpha_{2\eta}$ = absorptivity of outer pane for direct radiation

$\tau_{1\eta}$ = transmissivity of inner pane for direct radiation

$\tau_{2\eta}$ = transmissivity of outer pane for direct radiation

α_{1d} = absorptivity of inner pane for diffuse radiation

α_{2d} = absorptivity of outer pane for diffuse radiation

TABLE 8 Coefficients for Use in Calculation of Transmittance and Absorptance of Glass

		Single glazing		Double glazing		
κl	j	a_j	t_j	a_j (outer)	a_j (inner)	t_j
0.05	0	0.01154	−0.00885	0.01407	0.00228	−0.00401
	1	0.77674	2.71235	1.06226	0.34559	0.74050
	2	−3.94657	−0.62062	−5.59131	−1.19908	7.20350
	3	8.57881	−7.07329	12.15034	2.22366	−20.11763
	4	−8.38135	9.75995	−11.78092	−2.05287	19.68824
	5	3.01188	−3.89922	4.20070	0.72376	−6.74585
0.10	0	0.01636	−0.01114	0.01819	0.00123	−0.00438
	1	1.40783	2.39371	1.86277	0.29788	0.57818
	2	−6.79030	0.42978	−9.24831	−0.92256	7.42065
	3	14.37378	−8.98262	19.49443	1.58171	−20.26848
	4	−13.83357	11.51798	−18.56094	−1.40040	19.79706
	5	4.92439	−4.52064	6.53940	0.48316	−6.79619
0.15	0	0.01837	−0.01200	0.01905	0.00067	−0.00428
	1	1.92497	2.13036	2.47900	0.26017	0.45797
	2	−8.89134	1.13833	−11.74266	−0.72713	7.41367
	3	18.40197	−10.07925	24.14037	1.14950	−19.92004
	4	−17.48648	12.44161	−22.64299	−0.97138	19.40969
	5	6.17544	−4.83285	7.89954	0.32705	−6.66603
0.20	0	0.01902	−0.01218	0.01862	0.00035	−0.00401
	1	2.35417	1.90950	2.96400	0.22974	0.36698
	2	−10.47151	1.61391	−13.48701	−0.58381	7.27324
	3	21.24322	−10.64872	27.13020	0.84626	−19.29364
	4	−19.95978	12.83698	−25.11877	−0.67666	18.75408
	5	6.99964	−4.95199	8.68895	0.22102	−6.43968
0.40	0	0.01712	−0.01056	0.01423	−0.00009	−0.00279
	1	3.50839	1.29711	4.14384	0.15049	0.16468
	2	−13.86390	2.28615	−16.66709	−0.27590	6.17715
	3	26.34330	−10.37132	31.30484	0.25618	−15.84811
	4	−23.84846	11.95884	−27.81955	−0.12919	15.28302
	5	8.17372	−4.54880	9.36959	0.02859	−5.23666
0.60	0	0.01406	−0.00835	0.01056	−0.00016	−0.00192
	1	4.15958	0.92766	4.71447	0.10579	0.08180
	2	−15.06279	2.15721	−17.33454	−0.15035	4.94753
	3	27.18492	−8.71429	30.91781	0.06487	−12.43481
	4	−23.88518	9.87152	−26.63898	0.02759	11.92495
	5	8.03650	−3.73328	8,79495	−0.02317	−4.07787
0.80	0	0.01153	−0.00646	0.00819	−0.00015	−0.00136
	1	4.55946	0.68256	5.01768	0.07717	0.04419
	2	−15.43294	1.82499	−17.21228	−0.09059	3.87529
	3	26.70568	−6.95325	29.46388	0.00050	−9.59069
	4	−22.87993	7.80647	−24.76915	0.06711	9.16022
	5	7.57795	−2.94454	8.05040	−0.03394	−3.12776
1.00	0	0.00962	−0.00496	0.00670	−0.00012	−0.00098
	1	4.81911	0.51043	5.18781	0.05746	0.02576
	2	−15.47137	1.47607	−16.84820	−0.05878	3.00400
	3	25.86516	−5.41985	27.90292	−0.01855	−7.33834
	4	−21.69106	6.05546	−22.99619	0.06837	6.98747
	5	7.08714	−2.28162	7.38140	−0.03191	−2.38328

τ_{1d} = transmissivity of inner pane for diffuse radiation

τ_{2d} = transmissivity of outer pane for diffuse radiation

Then the reflectivity of the inner and outer panes for direct and diffuse radiation are, respectively,

$$r_{1\eta} = 1 - \alpha_{1\eta} - \tau_{1\eta}$$
$$r_{2\eta} = 1 - \alpha_{2\eta} - \tau_{2\eta}$$
$$r_{1d} = 1 - \alpha_{1d} - \tau_{1d}$$
$$r_{2d} = 1 - \alpha_{2d} - \tau_{2d}$$

(54)

The direct radiation absorptivity of the double-glazed system is

$$\alpha_{\eta,\text{outer}} = \alpha_{2\eta}\left(\frac{1 + r_{1\eta}\tau_{2\eta}}{1 - r_{1\eta}r_{2\eta}}\right) \tag{55}$$

$$\alpha_{\eta,\text{inner}} = \alpha_{1\eta}\frac{\tau_{2\eta}}{1 - r_{1\eta}r_{2\eta}} \tag{56}$$

For the diffuse radiation,

$$\alpha_{d,\text{outer}} = \alpha_{2d}\left(1 + \frac{r_{1d}\tau_{2d}}{1 - r_{1d}r_{2d}}\right) \tag{57}$$

$$\alpha_{d,\text{inner}} = \frac{\alpha_{1d}\,\tau_{2d}}{1 - r_{1d}r_{2d}} \tag{58}$$

The direct radiation transmissivity of the double glazed system is

$$\tau_\eta = \frac{\tau_{1\eta}\,\tau_{2\eta}}{1 - r_{1\eta}r_{2\eta}} \tag{59}$$

For the diffuse radiation,

$$\tau_d = \frac{\tau_{1d}\,\tau_{2d}}{1 - r_{1d}r_{2d}} \tag{60}$$

Using these overall absorptance and transmittance data, solar heat gain through the window is calculated by the following equation:

Direct solar heat gain:

$$q_D'' = S_f I_{DN} \cos\eta \, (\tau_\eta + N_o\alpha_{\eta,\text{outer}} + N_i\alpha_{\eta,\text{inner}}) \tag{61}$$

Diffuse solar heat gain:

$$q_d'' = (B_S F_{WS} + B_G F_{WG})(\tau_d + N_o\alpha_{d,\text{outer}} + N_i\alpha_{d,\text{inner}}) \tag{62}$$

Total solar heat gain:

$$q'' = q_D'' + q_d''$$

where S_f = sunlight factor due to shadow of the overhangs, side fins, and external obstructions, which can be determined by a computer program such as by Sun [34]

F_{WS} = window view factor of the sky
F_{WG} = window view factor of the ground

$$N_o = \frac{R_o + R_a}{R_o + R_a + R_i}$$

$$N_i = \frac{R_o}{R_o + R_a + R_i}$$

R_o, R_a, R_i = thermal resistances of exterior surface, air space, and interior surface

Similar equations can be derived for triple-glaze and quadruple-glaze window systems. A generalized solar heat gain formula for the n-glaze system may be expressed as follows:

$$q'' = I\left(\bar{\tau} + \sum_{j=1}^{n} N_j\alpha_j\right) \tag{63}$$

where I = incident solar radiation
$\bar{\tau}$ = overall transmittance
α_j = absorptance of the jth glazing
N_j = fraction of absorbed solar radiation that is transmitted toward room by the jth glazing

Figure 12 shows the formula for calculating $\bar{\tau}$ and $\Sigma\ N_j\alpha_j$ for single, double, triple, and quadruple glazing systems.

In Eq. (61), the sunlight factor S_f is a function of the sun's altitude angle ψ and the wall-solar angle γ. The values for ψ and γ are combined to yield a profile angle δ such that

$$\tan \delta = \frac{\tan \psi}{\cos \gamma} \tag{64}$$

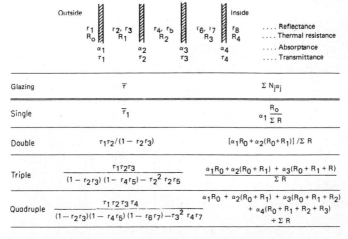

FIG. 12 Formulas for the solar heat gain calculations of multiple-glaze systems.

If a window of height a and width c has a setback of b from the outside plane of a building, the sunlit factor S_f is calculated by the following equation:

$$S_f = 1 - \frac{b}{a} \tan \delta - \frac{b}{c} \tan \gamma + \frac{b^2}{ac} \tan \delta \tan \gamma \qquad (65)$$

There are several computer programs (see for example Sun [34]) that can simulate shadow patterns and calculate S_f for more complex shading geometries of various architectural designs.

When the window system (fenestration) has various shading devices such as venetian blinds, draperies, curtains, and louvers, it is customary to use shading coefficients, which are experimentally determined. The shading coefficient SC is defined relative to the solar heat gain through standard double-strength (⅛-in thick) clear glass (DSA):

$$SC = \frac{q'' \text{ of fenestration}}{q'' \text{ of DSA}} \qquad (66)$$

The total solar heat gain is then evaluated by

$$q'' = (SC) \cdot (SHGF) + U(T_o - T_i) \qquad (67)$$

where SHGF, the solar heat gain factor, is nothing but the q'' of DSA and is found in the *ASHRAE Handbook of Fundamentals* [1].

D. INTERIOR HEAT AND MASS TRANSFER

Since conduction and convection heat transfer data applicable for interior heat transfer are found in the previous section, only the radiative heat transfer and moisture transfer aspects will be discussed in this section.

1. Radiant Heat Exchange Angle Factors

Although the fundamental theory of radiation heat exchange and an extensive list of angle factors are given in Ref. 53, some of the radiant heat exchange factors unique to building heat transfer problems are given in NBSLD [35]. The radiation exchange among wall and window surfaces is an important parameter for determining room heating and cooling loads based upon the accurate evaluation of room surface temperatures. In most of the current calculation methods the room surfaces are treated as having equal thermal emissivities with relatively good agreement with the measured surface temperatures. More exact treatment using radiosity-type calculations [53] will be needed for rooms having heat mirror–type windows. Other important radiation heat exchange factors are those involving the comfort of the human occupant with the cold wall surfaces. Fanger [2] developed radiation heat exchange factor data useful for such calculations.

2. Moisture Transfer

Although it is well understood that airborne moisture will condense wherever it finds a surface whose temperature is below the dew-point temperature, the rate of moisture absorption and release by the building structure, especially within the porous insulation, has not been clear. A classical treatment of the water vapor transfer through building materials is to use Fick's law, which states that water vapor flow per unit area $\dot{W}_v$ through a wall of permeability μ and thickness l is a function of vapor pressure difference, such that

$$\dot{W}_v = \mu \frac{P_{vi} - P_{vo}}{l} \tag{68}$$

Extensive tables are available in which the permeability μ of various building materials is tabulated [1]. In order to prevent condensation in the structure, it is customary to install vapor barriers—films of extremely low permeability—at the higher–vapor pressure side of the walls and ceilings.

Actual water vapor transfer, however, has been found to be very different from that predicted by Eq. (68). Moreover, the equation does not explain the situation commonly found in building walls. While condensation, in spite of vapor barriers, is found in the upper parts of walls near the wall-ceiling corners, the lower parts of walls are commonly found to be dry even when the vapor barrier is damaged. Since vapor pressure differences across the wall from inside to outside are usually similar between the upper and lower part of the wall, it is obvious that the water vapor flow is not governed by Eq. (68).

A crucial factor that controls the wall moisture condensation has been recognized to be the penetration of moist air through minute cracks. Thus the place as well as rate of wall moisture condensation can be determined by locating the air leakage and air leakage rate. If the volume flow rate of air is estimated to be Q and the vapor pressures of indoor and outdoor air across the wall are P_{vi} and P_{vo}, the water vapor flow should be determined by

$$\dot{W}_v = \rho Q (\chi_i - \chi_o) \tag{69}$$

where ρ = density of air

$\chi_i = \dfrac{M_v}{M_a} \dfrac{P_{vi}}{P_t - P_{vi}}$

$\chi_o = \dfrac{M_v}{M_a} \dfrac{P_{vo}}{P_t - P_{vo}}$

P_t = total barometric pressure
M_v = molecular weight of water vapor
M_a = molecular weight of air

This equation reflects realistic conditions much more accurately than the conventional equation (68).

A simple equation such as Eq. (69) has been found vital to the prediction as well as cure and prevention of moisture condensation problems not only in the wall insulation, but also in attics and interstitial ceiling cavities. While undesirable air leakage is a major cause of moisture condensation, it is also important to realize that well-controlled ventilation is the best cure for drying out the condensation. Ventilation control can be obtained by locating small weep holes (vent holes) so that the water vapor condensed during the cold period can be evaporated and escape during the warm spell.

It is also well known that humidity is one of the major elements in the realistic assessment of the indoor environment, is the cause of condensation and premature decay of building materials, is responsible for mildew growth, and is the source of bacteria. The indoor humidity has generally been calculated by a simple mass balance equation between the ventilated moisture and moisture generation. This calculation usually ignores the moisture absorbed by the room surfaces and yields very erroneous results. Recently, however, Tsuchiya [36] reported a method for analyzing the room absorption capability of moisture. He simulated a room air moisture transfer process by the following three equations:

$$M \frac{d\chi_i}{dt} = W_G - \frac{dw_i}{dt} + \dot{V}(\chi_o - \chi_i) - K_g A_g (\chi_i - \chi_g) \tag{70}$$

$$\frac{dw_i}{dt} = K A_i (\chi_i - \chi_s) \tag{71}$$

$$\chi_i = \frac{\chi_s^*}{41} \left(\frac{w_i}{C} T - 5.46 \right) \tag{72}$$

where M = dry air weight in the room, kg
  W_G = moisture generation rate, g/h
  w_i = moisture absorbed by the room surface, g
  $\dot{V}$ = ventilation rate, kg/h
  K_g = vapor condensation rate to glass surface, $g/[m^2 h \cdot (g/kg)]$
  A_g = glass surface area whose temperature is lower than the room dew-point temperature, m^2
  χ_g = humidity ratio of the wet glass surface air, g/kg
  K = moisture transfer factor for the room surface, $g/[m^2 h \cdot (g/kg)]$
  A_i = room surface area, m^2 (other than the condensing glass)
  χ_s = humidity ratio of the room surface air, g/kg
  T = room air temperature, K
  χ_s^* = humidity ratio of the air saturated at temperature T, g/kg
  χ_i = humidity ratio of room air, g/kg
  χ_o = humidity ratio of outdoor air, g/kg
  t = time, h
  C = equivalent dry weight of room surfaces, g

Tsuchiya speculates that the absorption of moisture by the room is limited to a thin film over the solid surface (excepting the window glass). Moreover, he assumes that this thin film attains instantaneous moisture equilibrium with respect to the room air moisture content. By applying his three equations to experimental rooms with carefully monitored and calibrated operating conditions, Tsuchiya [36] was able to determine typical values for the rate of moisture absorption by the room surface as well as the room surface average moisture content. Using the values found by the experiment, Tsuchiya showed that the calculated pattern of the hour-by-hour fluctuation of the room humidity ratio agreed very closely with that observed for rooms having different moisture-generating activities such as shown in Fig. 13. It has been found that approximately one-third of the moisture generated in the room was absorbed by the room surface, resulting in room air humidity much lower than calculated by the conventional method that ignores the moisture absorption. Tsuchiya also found that ventilation or air leakage dramatically affects the room humidity as long as it is less than one air change per hour, which is the case for typical American homes.

3. Room Air Convection

There have been in past years concerted efforts to simulate room air convection by solving momentum and energy balance equations [50–54]. However, most of the past efforts have been limited to laminar flow, two-dimensional problems. Recently, however, Yoshikawa

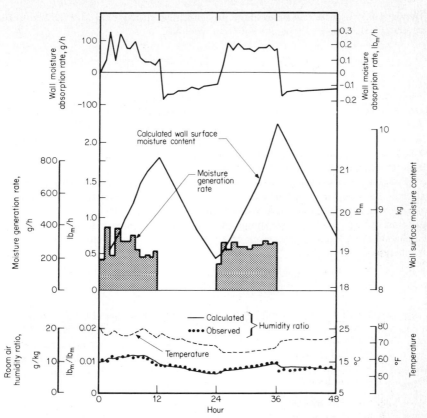

FIG. 13 Calculated wall moisture content and air humidity ratio performance in the living room with a given moisture generation pattern under two-cycle test.

and Yamaguchi [37] were able to solve three-dimensional, turbulent-flow room problems and to compare the solutions with the visualized flow patterns. They have developed extensive sets of formulations for the three-dimensional, turbulent-flow problems.

E. HEATING AND COOLING LOAD CALCULATIONS

1. Transient Heat Conduction Using Conduction Transfer Functions

Conduction transfer functions are used widely and are considered a convenient and effective tool for the evaluation of transient heat transfer through building construction components.

An expression for conduction heat transfer from a room to randomly fluctuating outdoor conditions is to use the hourly history of temperatures in conjunction with conduction transfer functions (CTF). For example, in calculating the heat transfer from an inside surface of a building envelope, the equation would be:

$$q''_t = \sum_{j=0}^{N} X_j T_{a,t-j} - \sum_{j=0}^{N} Y_j T_{o,t-j} + R q''_{t-1} \tag{73}$$

where X_j, Y_j (for $j = 0, 1, 2, \ldots, N$), and R are CTFs. In Eq. (73), $T_{o,t-j}$ and $T_{a,t-j}$ represent outdoor and room air temperatures, respectively, at the jth hour prior to the time for which the value of q''_t is needed; and q''_{t-1} is the heat loss q''_t from the room at the previous hour. By having a record of $T_{o,t-1}$, $T_{o,t-2}$, $T_{o,t-3}$, $\ldots$, $T_{o,t-j}$ and $T_{a,t-1}$, $\ldots$, $T_{a,t-j}$, it is possible to determine the instantaneous conduction heat transfer, provided that the values for X_j, Y_j, and R are available.

The number of terms involved for the calculation of q''_t [or the value of N in Eq. (73)] depends upon the type of roof or wall construction. Generally heavy constructions require a large value, although for most conventional constructions it seldom exceeds 20. Stephenson and Mitalas [38] have shown that the value of N can be further decreased by employing more than one past record of q''_t (or q''_{t-j} with j more than 1) in the following manner:

$$q''_{i,t} = \sum_{j=0}^{N'} A_j T_{a,t-j} - \sum_{j=0}^{N'} B_j T_{o,t-j} - \sum_{j=1}^{N} D_j q''_{i,t-j} \tag{74}$$

where A_j, B_j, and D_j are the modified CTFs. Table 9 gives values of CTF calculated for a brick wall having an overall heat transfer coefficient of 2.38 W/(m²·K) [0.419 Btu/(h·ft²·°F)]. In Table 9, the factors designated by Z_j and C_j are the additional transfer functions to be used for evaluating the instantaneous heat loss $q''_{o,t}$ at the exterior side of the structure. The applicable equations for this side of the structure are then:

$$q''_{o,t} = \sum_{j=0}^{N} Y_j T_{a,t-j} - \sum_{j=0}^{N} Z_j T_{o,t-j} + R q''_{o,t-1} \tag{75}$$

or

$$q''_{o,t} = \sum_{j=0}^{N'} B_j T_{a,t-j} - \sum_{j=0}^{N'} C_j T_{o,t-j} - \sum_{j=1}^{N'} D_j q''_{o,t-j} \tag{76}$$

The table also shows that for the calculation of $q''_{o,t}$ the use of Stephenson-type transfer functions would permit the reduction of N from 10 to $N' = 4$ with a corresponding increase of three terms in the past record of $q''_{o,t}$.

Equations (73) and (74), however, must be valid also for steady-state heat transfer problems. One of the best ways to check the consistency of the CTFs is to use them in the solution of steady-state problems and see if the following criteria are met:

$$\sum_{}^{N} X_j = \sum_{}^{N} Y_j = \sum_{}^{N} Z_j = U(1 - R) \tag{77}$$

In fact, the CTF of the sample wall shown in Table 9 can be shown to satisfy this requirement.

Using the data in Table 9, it can be shown that

$$\Sigma\, X_j = \Sigma\, Y_j = \Sigma\, Z_j = 0.379$$

and

$$U(1 - R) = 2.373(1 - 0.8398) = 0.380 \text{ W/(m}^2\cdot\text{K)}$$

TABLE 9 Sample Conduction Transfer Functions

Construction data	Thickness		Thermal conductivity		Density		Specific heat		Thermal resistance	
	m	ft	$W/(m \cdot K)$	$Btu/(h \cdot ft \cdot °F)$	kg/m^3	lb_m/ft^3	$J/(kg \cdot K)$	$Btu/(lb_m \cdot °F)$	$m^2 \cdot K/W$	$h \cdot ft^2 \cdot °F/Btu$
Inside surface									0.146	0.830
4-in common brick	0.10	0.333	0.72	0.420	1602	100	921	0.220	0.140	0.793
4-in face brick	0.10	0.333	1.33	0.770	2003	125	921	0.220	0.076	0.432
Outside surface								*	0.058	0.330
									0.420	2.385

Time increment: 1 h.

Overall heat transfer coefficient $U = 2.38 \ W/(m^2 \cdot K) \ [0.419 \ Btu/(h \cdot ft^2 \cdot °F)]$.

TABLE 9 Sample Conduction Transfer Functions (*Continued*)

Conduction transfer functions (CTF)

j	X_j W/(m²·K)	X_j Btu/(h·ft²·°F)	Y_j W/m²·K	Y_j Btu/(h·ft²·°F)	Z_j W/(m²·K)	Z_j Btu/(h·ft²·°F)	R (dimensionless)
0	5.2204	0.9194	0.0006	0.0001	11.2612	1.9833	0.8398
1	−5.3339	−0.9391	0.0454	0.0080	−12.3695	−2.1785	
2	0.3440	0.0606	0.1380	0.0243	1.1271	0.1983	
3	0.0868	0.0153	0.1506	0.0186	0.2197	0.0387	
4	0.0346	0.0061	0.0511	0.0090	0.0818	0.0144	
5	0.0148	0.0026	0.0221	0.0039	0.0340	0.0060	
6	0.0062	0.0011	0.0097	0.0017	0.0148	0.0026	
7	0.0028	0.0005	0.0040	0.0007	0.0062	0.0011	
8	0.0011	0.0002	0.0017	0.0003	0.0028	0.0005	
9	0.0006	0.0001	0.0006	0.0001	0.0011	0.0002	
10	0.0000	0.0000	0.0005	0.0001	0.0006	0.0001	

Stephenson-type transfer functions

j	A_j W/(m²·K)	A_j Btu/(h·ft²·°F)	B_j W/(m²·K)	B_j Btu/(h·ft²·°F)	C_j W/(m²·K)	C_j Btu/(h·ft²·°F)	D_j (dimensionless)
0	5.2204	0.9194	0.0006	0.0001	11.2612	1.9833	1.000
1	−8.0219	−1.4128	0.0448	0.0079	−18.1707	−3.2002	−1.3552
2	3.2847	0.5785	0.1147	0.0202	7.9163	1.3942	0.4699
3	−0.2901	−0.0511	0.0363	0.0064	−0.8222	−0.1448	0.0315
4	0.0004	0.0007	0.0011	0.0002	0.01363	0.0024	0.0002

An algorithm for the generation of the CTF for various building constructions will not be given here, since it involves lengthy mathematical solutions to the standard transient heat conduction differential equation. A paper by Kusuda [39] provides an excellent background for this calculation. Several computer programs (Kusuda [39], Mitalas and Arseneault [40], and Peavy [41]) are available for the calculation of CTF for multilayer wall, roof, and floor constructions.

The program of the National Bureau of Standards, for example, requires the thermal property data for each of the layers to be placed successively from the inside layer to the outside, and calculates the conduction transfer functions of multilayered plane, cylindrical, and spherical walls. It also calculates the transfer functions for solid objects of plane, cylindrical, and spherical shapes as well as the heat conduction systems involving semi-infinite solids, approximated by basement floors and underground constructions.

2. Application of the Conduction Transfer Function (CTF)

a. Exterior Walls and Roof

The heat balance equation at the exterior surface at times t is given by

$$q''_{R,t} + q''_{A,t} + q''_{o,t} - q''_{s,t} = 0 \tag{78}$$

where $q''_{R,t}$ = incident solar radiation absorbed by the surface at time t

$q''_{A,t} = h_o(T_{DB,t} - T_{os,t})$, convection heat transfer from the outdoor air

$$q''_{o,t} = \sum_{j=0}^{N} Y_j(T_{is,t-j} - T_M) - \sum_{j=0}^{N} Z_j(T_{os,t-j} - T_M) + R \cdot q''_{o,t-1}, \quad \text{conduction}$$

heat flow from the inside surface†

$q''_{s,t}$ = heat loss to the sky

By letting

$$\text{SUM1} = \sum_{j=0}^{N} Y_j(T_{is,t-j} - T_M) + Rq''_{o,t-1}$$

$$\text{SUM2} = \sum_{j=1}^{N} Z_j(T_{os,t-j} - T_M)$$

the outside surface temperature can be determined by

$$T_{os,t} = \frac{q''_{R,t} - q''_{s,t} + h_{o,t}T_{DB,t} + \text{SUM1} - \text{SUM2} + Z_0 T_M}{h_{o,t} + Z_0} \tag{79}$$

Using this new $T_{os,t}$, the heat loss at the interior surface is then determined as follows:

$$q''_t = \sum_{j=0}^{N} X_j(T_{is,t-j} - T_M) - \sum_{j=0}^{N} Y_j(T_{os,t-j} - T_M) + Rq''_{t-1} \tag{80}$$

†In the use of the CTF it is customary for a value of T_M, an arbitrary datum value such as a mean temperature, to be subtracted from the interior surface and exterior surface temperatures. This subtraction usually helps to minimize the digital errors which occur and are sometimes significant when a large number of numerical data are multiplied and added. The CTF should be determined in such a manner that surface thermal resistance layers are not included.

Since $\sum_{j=0}^{N} X_j = \sum_{j=0}^{N} Y_j$, the net effect of the subtraction is zero.

b. Interior Walls and Floor or Ceiling Sandwich

The calculation sequence for the partition wall and floor or ceiling sandwich is different from that of the exterior wall or roof. The difference is due to the fact that the air temperature at the exterior side of the construction can be assumed the same as at the interior side, at least for a climate-controlled building. In order to take advantage of this fact, CTF should be determined with the surface thermal resistance layer added at the exterior side of the structure. The heat loss through a partition wall is then calculated by

$$q''_t = \sum_{j=0}^{N} \mathsf{X}_j \, (T_{is,t-j} - T_M) - \sum_{j=0}^{N} \mathsf{Y}_j(T_{A,t-j} - T_M) + \mathsf{R}q''_{t-1} \tag{81}$$

where $T_{is,t-j}$ = inside surface temperature at the $(t-j)$th hour and where $T_{A,t-j}$ = room temperature at the $(t-j)$th hour.

If the room temperature were maintained constant at T_M, which is usually the case, the terms involving the Y_j's would then drop out of the equation.

c. Slab-on-Grade Floor

The transient heat loss to the ground through a slab-on-grade floor may be calculated by using conduction transfer functions determined on the basis of flooring, concrete, and 0.30 m (12 in) of ground layer, such that

$$q''_t = \sum_{j=0}^{N} \mathsf{X}_j(T_{is,t-j} - T_M) - \sum_{j=0}^{N} \mathsf{Y}_j(T_G - T_M) + \mathsf{R}q''_{t-1} \tag{82}$$

Here T_{is} represents the floor temperatures, T_G the ground temperature below the floor slab, and T_M the arbitrary mean temperature discussed above. Since T_G usually remains constant, at least during the diurnal heat transfer calculation, one can write

$$q''_t = \sum_{j=0}^{N} \mathsf{X}_j(T_{is,t-j} - T_M) - U_G(1 - \mathsf{R})(T_G - T_M) + \mathsf{R}q''_{t-1} \tag{83}$$

where U_G = overall floor slab heat conductance. The value of the ground temperature T_G must be determined separately by using an appropriate ground heat transfer algorithm. A recommended procedure is given in Sec. F.4.

The same method is applicable to a floor with a crawl space as long as the space is not vented. The conduction transfer functions for the floor with an unvented crawl space may be obtained by assuming an additional air film resistance layer to account for the dead air space between the floor and the ground. In many cases it is assumed that $T_G = T_M$, and then the term involving U_G drops out of Eq. (83).

d. Floor over Vented Crawl Space

The floor over a well-vented crawl space may be treated in the same manner as an exterior wall or roof except that the solar radiation and sky radiation terms would not be included in the energy balance and that the outside-surface heat transfer coefficient is replaced by a value similar in magnitude to the inside-surface heat transfer coefficient. If the conduction transfer functions include the outside-surface heat transfer resistance, the calculation is simply

$$q''_t = \sum_{j=0}^{N} \mathsf{X}_j(T_{is,t-j} - T_M) - \sum_{j=0}^{N} \mathsf{Y}_j(T_{DB,t-j} - T_M) + \mathsf{R}q''_{t-1} \tag{84}$$

If the crawl space temperature is expected to be considerably different from the outdoor dry-bulb temperature T_{DB}, a separate heat balance calculation is required among heat gains from the crawl space walls, floor above ground surface, and air leakage.

3. Room Temperature and Cooling (Heating) Load Calculations

The basic heat transfer process that occurs in a room can best be illustrated by an electrical circuit network as shown in Fig. 14. The figure represents a situation in a typical room (see Fig. 15) having two exterior walls, each of which contains a window, and two interior partition walls, in addition to the roof and floor. In Fig. 14, heat conduction paths through the walls, roof, and floor are depicted by resistance and capacitance circuits, and those through windows are represented by resistance circuits, implying that the windows do not have significant thermal mass. Points T_{s1} through T_{s8} in Fig. 14 indicate interior

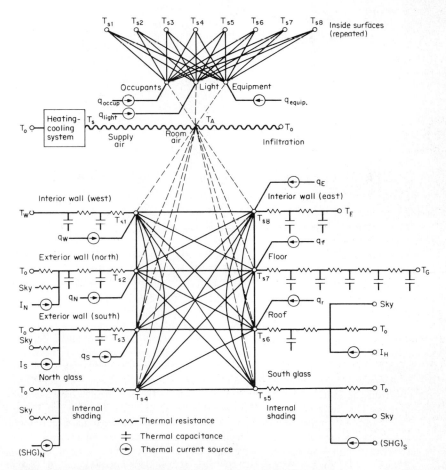

FIG. 14 Analogous electric circuit for heat exchange process in a room.

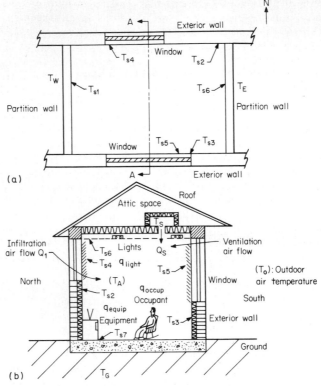

FIG. 15 Physical model of a typical room (a) used in Figure 14. Cross section A-A (b).

surfaces of the walls, roof, floor, and windows, all of which receive conduction heat through solid material, solar radiation (represented by ⊝q) through transparent surfaces, and long-wavelength radiation from other solid surfaces indicated by solid lines connecting the surface nodes; and they lose heat to the room air (represented by a point called T_A) by the convection process (dashed lines).

At the top of Fig. 14, the radiation heat exchange among the room surfaces, the surfaces of lighting fixtures, equipment such as business machines, and occupants is depicted. Also indicated in this same location is the convective heat exchange between these items and the room air. Actual heat or power input to these internal heat sources is indicated by ⊝q. Although not indicated in this figure, it is possible to represent the conduction heat gain from the inner core of lighting fixtures and equipment if they have sufficient thermal mass. The equipment could of course include the unit heaters or air conditioners if they are a part of the room heat exchange system. The room air exchanges energy with outdoor air or with the conditioned air from the central climate control system (or with the forced ventilation system) and is depicted in this figure by wavy lines $\overline{T_s T_A}$ and $\overline{T_o T_A}$. The heat exchange at the exterior surfaces with outdoor air, sky, and sun (except for the partition walls and floor on grade) are also included using a normal design calculation procedure. The temperature and/or heat flow at the exterior sides of the room surfaces is usually available either by calculation or as input data. The exception occurs when two or more rooms adjacent to one another are treated simultaneously. This latter case is commonly referred to as a "multiroom problem" and is very complex. Kusuda [42] describes a recommended procedure for calculating the multiroom heat transfer problem.

The room heat transfer system depicted in Figs. 14 and 15 can be solved in two different modes: the room temperature calculation mode or the room load calculation mode. The room temperature calculation mode requires the simultaneous solution of heat balance equations in order to determine all the surface temperatures together with air temperature. On the other hand, the room cooling load calculation mode requires that the room temperature be prescribed and only the room surface temperatures be solved for. The convective heat exchange between the room air and the heat-emitting surfaces is then the cooling load (or the heating load if the heat is lost from the air to the surfaces). These two modes of computation can be combined to simulate the actual thermal behavior of a room and its environment where the temperature fluctuates. The floating temperature would be calculated as long as it remained between prescribed limits. A heat load would be computed when the room temperature fell to the lower limit and a cooling load would be calculated when the temperature rose to the upper limit. In this manner, the calculation of load and subsequent energy requirement would more closely correspond to actual building and system operation.

In order to formulate the room heat balance process, special additional symbols are introduced as follows:

N_s — Total number of heat transfer surfaces contributing to the room heat balance.

A_{ij} — Area of ith heat transfer surface, where $i = 1, 2, 3, \ldots, N_s$ $\mathbf{X}_{ij}$ $\mathbf{Y}_{ij}$, $\mathbf{Z}_{ij}$. Conduction transfer functions of the ith surface where $i = 1, 2, 3, \ldots,$ N_s and $j = 1, 2, \ldots, N_i$. These conduction transfer functions are usually evaluated without the interior or room-side surface thermal resistance. The thermal resistance layer of the exterior surface is also omitted if the exterior surface temperature is to be computed as a result of a heat balance involving solar radiation, sky radiation, and convective loss to the outdoor air.

N_i — Number of conduction transfer function terms to be used for the calculation of the ith surface conduction heat gain.

$\mathbf{R}_i$ — Common ratio for the conduction transfer function of the ith surface.

$T_{os,i,t-j}$ — For $i = 1, 2, 3, \ldots, N_s$ and $j = 0, 1, 2, \ldots, N_i$. Outside surface temperature history from present hour to that of N_i hours earlier for the ith surface. Subscript t refers to the present time and $t - j$ refers to the present time minus j hours.

$T_{is,i,t-j}$ — For $i = 1, 2, 3, \ldots, N_s$ and $j = 1, 2, 3, \ldots, N_i$. Inside surface temperature history from one hour earlier to N_i hours earlier for the ith surface. The present value (for $j = 0$) will be computed in this routine and stored for future use.

$T_{A,t}$ — Air temperature of the room at time t.

$T_{DB,t}$ — Outdoor air dry-bulb temperature at time t.

$T_{S,t}$ — Supply air temperature from the central system at time t.

$\overline{T}_{i-k}$ — Average absolute temperature of surfaces i and k.

h_i — Inside surface convection heat transfer coefficient for the ith surface.

F_{i-k} Radiation heat exchange view factor between the ith surface and the kth surface to be determined by appropriate equations found in Ref. 53.

ϵ_i Emissivity of the ith surface.

$q''_{R,i,t}$ Radiant heat flux impinging upon the ith surface at time t from various sources, which includes solar radiation and radiation from lights, occupants, and equipment.

$q''_{i,t}$ Heat flux conducted into the ith surface at time t.

$G_{L,t}$ Mass air flow rate due to air leakage at time t.

$G_{S,t}$ Mass air flow rate of the supply air from the central system at time t.

q_{equip} Internal heat generated from equipment such as business machines and computers.

q_{occup} Internal heat (sensible) generated from occupants (a function of room air temperature).

q_{light} Heat from lights.

RE   Fraction of internal heat gain from equipment that can be assumed to be convective.

RO   Fraction of internal heat gain from occupants that can be assumed to be convective.

RL   Fraction of heat gain from lights that can be assumed to be convective.

$I_{i,t}$ Solar incident radiation on the ith surface at time t.

4. Calculation Sequence

a. Heat Balance Equation at the ith Surface at Time t

$$q''_{i,t} = \sum_{j=0}^{N_i} \mathsf{X}_{i,j} T_{is,i,t-j} - \sum_{j=0}^{N_i} \mathsf{Y}_{i,j} T_{os,i,t-j} + \mathsf{R}_i q''_{i,t-1}$$

$$= h_i(T_{A,t} - T_{is,i,t}) + \sum_{k=1}^{N_i} G_{i,k}(T_{is,k,t} - T_{is,i,t}) \qquad (85)$$

$$+ q''_{R,i,t}$$

where $G_{i-k} = 4\sigma\epsilon_i F_{i-k}\overline{T}^3_{i-k}$†

$$q''_{R,i,t} = I_{i,t} + \frac{(1 - RE)Q_{equip} + (1 - RO)Q_{occup} + (1 - RL)Q_{light}}{\sum\limits_{i=1}^{N_s} A_i}$$

b. Heat Balance for the Room Air

$$\sum_{i=1}^{N_s} h_i A_i(T_{is,i,t} - T_{A,t}) + G_{L,t}c(T_{DB} - T_{A,t}) + G_{S,t}c(T_{S,t} - T_{A,t})$$

$$+ q_{equip}\, RE + q_{occup}\, RO + q_{light}\, RL = 0 \qquad (86)$$

where c is the specific heat of air.

---

†This formulation is valid as long as the emissivity values of room surfaces do not vary from surface to surface.

The values of $G_{S,t}$ and $T_{S,t}$, supply air flow rate and its temperature, are the link between the load calculation and the heating and cooling system and equipment simulation, details of which are beyond the scope of this section.

c. Definition of Matrix Elements

Matrix elements (for i = 1, 2, 3, ... , N_s and for k = 1, 2, 3, ...) may be defined as follows:

$$A_{i,i} = X_{i,1} + H_i + \sum_{k=1}^{N_s} G_{i,k}$$

$$A_{i,k} = -G_{i,k}$$

$$A_{k,i} = -G_{k,i}$$

$$A_{i,Ns} = h_i$$

$$B_i = -\sum_{j=1}^{N_i} X_{i,j} T_{is,i,t-j} + \sum_{j=0}^{N_i} Y_{i,j} T_{os,i,t-j} - Rq''_{i,t-1}$$
$$+ q''_{R,i,t} \tag{87}$$

$$A_{Ns+1,k} = A_k h_k$$

$$A_{Ns+1,Ns+1} = -(G_{L,t} + G_{S,t})c - \sum_{k=1}^{N_s} h_k A_k$$

$$B_{Ns+1} = q_{\text{equip}} \, RE + q_{\text{occup}} \, RO + q_{\text{light}} \, RL + G_{L,t} c T_{DB,t}$$
$$+ G_{S,t} c T_{S,t}$$

d. Solution for $T_{is,i,t}$ and $T_{A,t}$

Using these matrix elements, the following $N_s + 1$ equation should be solved simultaneously for $T_{is,i,t}$ (i = 1, 2, ... , N_s) and for $T_{A,t}$:

$$\begin{bmatrix} A_{1,1} & \cdots & A_{1,Ns+1} \\ A_{2,1} & \cdots & A_{2,Ns+1} \\ \cdots \cdots \cdots \\ A_{Ns,1} & \cdots & A_{Ns,Ns+1} \\ A_{Ns+1,1} & \cdots & A_{Ns+1,Ns+1} \end{bmatrix} \cdot \begin{bmatrix} T_{is,1,t} \\ T_{is,2,t} \\ \cdots \\ T_{is,Ns,t} \\ T_{A,t} \end{bmatrix} = \begin{bmatrix} B_1 \\ B_2 \\ \cdots \\ B_{Ns} \\ B_{Ns+1} \end{bmatrix} \tag{88}$$

When the value of $T_{A,t}$ has been specified, as in the case of a controlled enviornmental condition, the following N_s equations should be solved instead of the $N_s + 1$ equations given above:

$$\begin{bmatrix} A_{1,1} & \cdots & A_{1,Ns} \\ A_{2,1} & \cdots & A_{2,Ns} \\ \cdots \cdots \cdots \\ A_{Ns,1} & \cdots & A_{Ns,Ns} \end{bmatrix} \cdot \begin{bmatrix} T_{is,1,t} \\ T_{is,2,t} \\ \cdots \\ T_{is,Ns,t} \end{bmatrix} = \begin{bmatrix} B'_1 \\ B'_2 \\ \cdots \\ B'_{Ns} \end{bmatrix} \tag{89}$$

where $B'_i = B_i - A_{i,Ns+1} T_{A,t}$

This matrix-type solution is not required if it is assumed that

$$\sum_{k=1}^{N_s} G_{i,k}(T_{is,k,t} - T_{is,i,t}) = h_{r,i}(T_{A,t} - T_{is,i,t})$$

where $h_{r,i} = 4\epsilon\sigma \overline{T}^3_{A,t}$, or $A_{i,k} = 0$ when $i \neq k$.

e. Room Sensible Thermal Load

The room sensible thermal load is then determined by

$$q_{LS,t} = \sum_{i=1}^{N_s} A_i h_i (T_{is,i,t} - T_{A,t}) + G_{L,t} c (T_{DB,t} - T_{A,t})$$
$$+ q_{\text{equip}} \text{RE} + q_{\text{occup}} \text{RO} + q_{\text{light}} \text{RL} \tag{90}$$

In this expression, q_{LS} is a cooling load if positive and it is a heating load if negative. This is the heat picked up by the room air (or that lost by the room air) which has to be removed (or added) by the central air conditioning system.

It should be noted that for ordinary load calculations, $G_{S,t}$ and $T_{S,t}$ are not used as long as the following condition is satisfied:

$$|q_{LS,t}| \leq G_{S,t} c (T_{A,t} - T_{S,t})|, \text{ maximum capacity of the heating}$$

$$\text{or cooling system}$$

In other words, the desired or prescribed room temperature can be maintained as long as the calculated load is less than the maximum capacity of the central system. When the above condition is not satisfied because of inadequate values for either the air supply or the supply air temperature, the room temperature used for the load calculation is no longer valid. The calculation must then be repeated, first calculating the room temperature as outlined in previous sections.

F. EARTH-CONTACT HEAT TRANSFER

1. Slab-on-Grade Floor and Basement Wall Problems

One of the most challenging aspects of building heat transfer problems is to determine heat loss from earth-contact surfaces; this involves slab-on-grade floors, basement walls, basement floors, and underground rooms. While the heat flow near the center of the floor slab is nearly a one-dimensional semi-infinite problem, three-dimensional effects become significant around the floor perimeter. The real situation is further complicated by the fact that:

1. The earth temperature and thermal properties vary with respect to the distance from the building, with depth, and with time.
2. The soil near the building usually is found to contain a high moisture content due to drainage problems and capillary action, all of which defy conventional heat conduction analysis.

Although several attempts have been made in the past to apply numerical calculation techniques [54], no satisfactory analytical solutions have been developed for the realistic earth-contact system. Kusuda and Achenbach [43], for example, used three-dimensional and transient finite difference models for the prediction of temperature and humidity in underground survival shelters. Several authors, such as Meixel and Shipp [44], Adamson [45], and Akasaka [46], reported results of numerical calculations using two-dimensional heat conduction analysis for slab-on-grade as well as basement wall problems. Latta and Boileau [47] suggested an approximate method based on circular temperature profiles such as shown in Fig. 16 to determine a steady-state heat loss from basement walls and floors. Their simplistic solution found, however, a surprisingly good agreement with the finite difference calculation conducted by Meixel and Shipp, as shown in Fig. 17. The

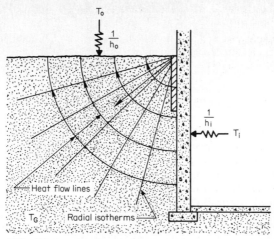

FIG. 16 Simplified heat transfer model for basement wall and floor system suggested by Latta and Boileau [1968].

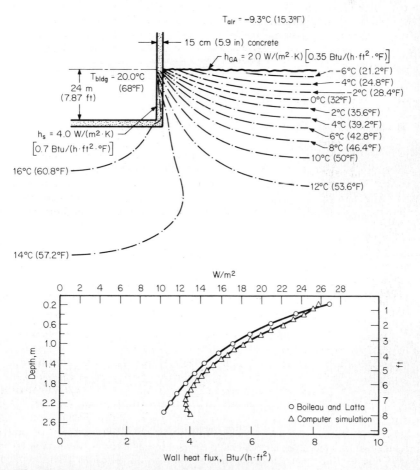

FIG. 17 Heat conduction through uninsulated basement wall.

Latta-Boileau temperature and heat flow path deviate from the finite difference solution when the basement wall is insulated, as shown in Fig. 18.

2. Muncey-Spencer Solution

An elegant analytical solution for the slab-on-grade floor problem has been reported by Muncey and Spencer [48]. The results of their calculations, which applied to a 10-m × 10-m (32.8-ft × 32.8-ft) square slab on the soil of 1 W/(m·K)[6.94 Btu·in/(h·ft²·°F)] thermal conductivity, are shown in Fig. 19. Based upon extensive calculations, they have developed ground thermal resistance values for selected values of the floor surface resistance and thickness of the wall, ℓ. A general procedure for determining the thermal resistance R of slabs of different shapes, when resting on soils of different thermal conductivity, can be obtained by using the standard ground thermal resistance data R_S presented in Fig. 19 by using a relationship such that

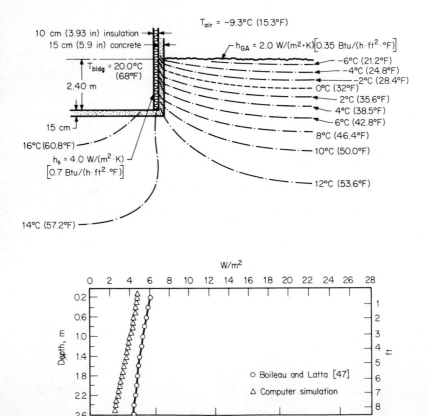

FIG. 18 Heat conduction through insulated basement wall.

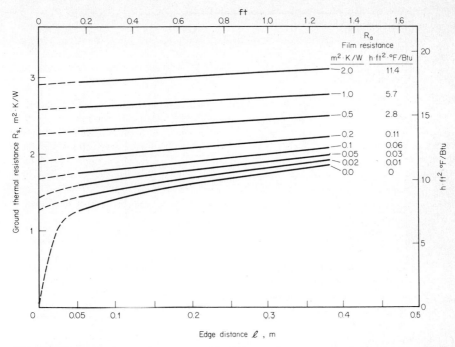

FIG. 19 Thermal resistance for a square slab of 40 m perimeter for a ground conductivity of 1.0 W/m·K for various values of edge distance ℓ and air film resistance R_a.

by using a relationship such that

$$R(k^+k,\ \mathsf{P}^+\mathsf{P},\ \ell,\ R_a) = F_s \frac{\mathsf{P}^+}{k^+}\ R_s\left(k,\ \mathsf{P},\ \frac{\ell}{\mathsf{P}^+},\ \frac{k^+R_a}{\mathsf{P}^+}\right) \qquad (91)$$

where k^+ = ratio of thermal conductivity with respect to the standard k [= 1 W/(m²·K)]

P^+ = ratio of slab perimeter with respect to the standard P (= 40 m)

ℓ = wall thickness

R_a = slab-surface film resistance

F_s = shape correction factor

The shape correction factor F_s as a function of $A/(\mathsf{P}/4)^2$, where A and P are slab area and perimeter length, respectively, is shown in Fig. 20.

a. Sample Calculation

1. Determine R for a slab of the shape shown in Fig. 21. For the slab shown the perimeter is 50 m (164 ft), so P^+ = 1.25, and for a ground conductivity of 0.8 W/(m·K) [5.55 Btu·in/(h·ft²·°F)], so k^+ = 0.8. Assuming the wall thickness to be 0.2 m (0.656 ft) and the surface thermal resistance to be 0.5 m²·K/W (2.84 h·ft²·°F/Btu), and substituting these values in Eq. (91) the value of R_s (1·40·0.2/1.25·0.8 × 0.5/1.25) is found from Fig. 19. Enter the figure along the abscissa at the modified value of the

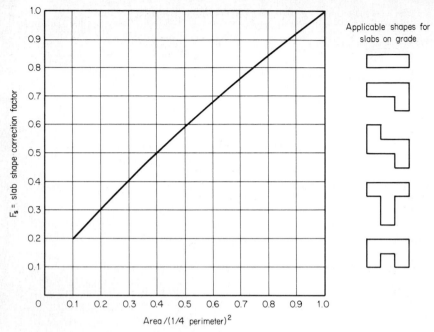

FIG. 20 Correction factor for the slab shape.

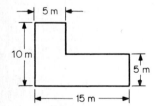

FIG. 21 Example slab used for showing Muncey/Spencer calculation method.

edge distance ($l/k^+ = 0.16$) and intersect at a modified value of the film resistance ($k^+\mathrm{Ra}/\mathbf{P}^+ = 0.32$) and read off the value of $2.16\ \mathrm{m^2 \cdot K/W}$ ($12.26\ \mathrm{h \cdot ft^2/Btu}$).

2. Using Eq. (91), the thermal resistance for a square of the same perimeter is $(\mathbf{P}^+/k^+)R_s = (1.25/0.8) \times 2.16 = 3.37\ \mathrm{m^2 \cdot K/W}$ ($19.13\ \mathrm{h \cdot ft^2 \cdot °F/Btu}$).

3. For the slab shown, the area A is $100\ \mathrm{m^2}$ ($1076\ \mathrm{ft^2}$), so that $A/(\mathbf{P}/4)^2 = 0.64$. Reference to Fig. 20 then gives $F_s = 0.72$ and hence the thermal resistance of the ground for the case illustrated is $3.37\ 0.72 = 2.4\ \mathrm{m^2 \cdot K/W}$ ($13.62\ \mathrm{h \cdot ft^{2°}F/Btu}$).

3. Lachenbruch Solution

In addition to the Muncey-Spencer solution, a more generalized solution applicable to evaluating the periodic heat transfer from heated slabs of arbitrary shapes has been developed by Lachenbruch [49] who used the Green's function approach.

Figures 22 and 23 show the January and July earth temperature below a heated slab of 20 ft × 20 ft (6.1 m × 6.1 m) calculated by the Lachenbruch method, while Fig. 24 is the annual average or steady-state temperature distribution below the same slab.

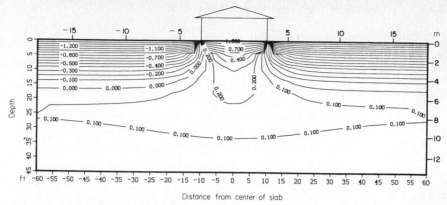

FIG. 22 Dimensionless January earth temperature under a heated house having 20′ × 20′ (6.1m × 6.1m) slab floor. The temperature is expressed in $(T - \overline{T}_A)/(T_R - \overline{T}_A)$.

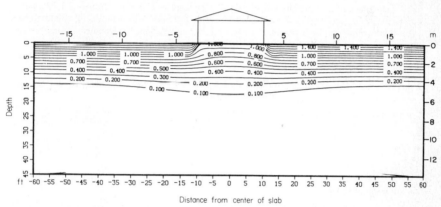

FIG. 23 Dimensionless July temperature under a house having 20′ × 20′ (6.1m × 6.1m) slab floor. The temperature is expressed in $(T - \overline{T}_A)/(T_R - \overline{T}_A)$.

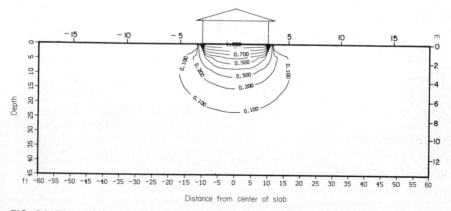

FIG. 24 Dimensionless annual; average earth temperature under a house having 20′ × 20′ (6.1m × 6.1m) slab floor. The temperature is expressed in $(T - \overline{T}_A)/(T_R - \overline{T}_A)$.

REFERENCES

1. *ASHRAE Handbook of Fundamentals,* Chaps. 21–26, American Society for Heating, Refrigerating and Air Conditioning Engineers, 1981.
2. P. O. Fanger, *Thermal Comfort Analysis and Applications in Environment Engineering,* 1st ed., pp. 170–198, McGraw-Hill, New York, 1970.
3. J. F. Van Straaten, *Thermal Performance of Buildings,* 1st ed., Elsevier, Amsterdam, 1967.
4. K. Kimura, *Science of Air Conditioning,* Applied Science Publishers, London, 1978.
5. R. W. Muncey, *Heat Transfer Calculations for Buildings,* Applied Science Publishers, London, 1979.
6. D. A. McIntyre, *Indoor Climate,* Applied Science Publishers, London, 1980.
7. *NOAA Comparative Climatic Data through 1976,* National Climatic Center, Asheville, NC 28801, 1976.
8. *NOAA* Climatic Atlas of the United States, U.S. Department of Commerce, Environmental Science Administration Environmental Data Service, 1968.
9. E. Stamper, Weather Data, *ASHRAE J.,* p. 47, February 1977.
10. U.S. Department of Energy, Division of Solar Technology, *SOLMET—Hourly Solar Radiation Surface Meterological Observations—TD9724,* Vol. 1: Users' Manual; Vol. 2: Final Report; published by National Oceanic and Atmospheric Administration, Environmental Data Service, National Climatic Center, Asheville, NC 28801, 1979.
11. L. W. Crow, Development of Hourly Data for Weather Year for Energy Calculations (WYEC), *ASHRAE J.,* pp. 37–41, October 1981.
12. I. J. Hall et al., Generation of Typical Meterological Years (TMY) for 26 SOLMET Stations, *Sandia Lab. Rep.,* Albuquerque, NM 81785, 1979.
13. T. Kusuda, T. Alereza, and I. Sud, Comparison of DoE-2 Generated Residential Design Energy Budgets with Those Calculated by the Degree-Day Method, ASHRAE *Trans.,* Vol. 81, Pt. 1, pp. 49–506, 1975.
14. H. C. Thom, Normal Degree Days below Any Base, *Monthly Weather Rev.,* Vol. 82, No. 5, Aug. 15, 1954.
15. *NOAA Degree Days to Selected Bases for First Order Tape Station,* Microfiche, NOAA, National Climatic Center, Asheville, NC 28801, 1979.
16. *Engineering Weather Data, AFM 88-29 Tm 5-785, NAVFAC P-89,* Department of the Air Force, the Army and the Navy, July 1978.
17. T. Kusuda and P. R. Achenbach, Earth Temperature and Thermal Diffusivity at Selected Stations in the United States, *ASHRAE Trans.,* Vol. 71, Pt. I, pp. 66–75, 1965.
18. T. Kusuda, Earth Temperature beneath Five Different Surfaces, *N. B. S. Rep. 10373,* National Bureau of Standards, 1971.
19. T. Kusuda and K. Ishii, Hourly Solar Radiation Data for Vertical and Horizontal Surfaces on Average Days in the United States and Canada, *NBS Building Sci. Ser. 96,* April 1976.
20. B. Y. H. Liu and R. C. Jordan, The Interrelationship and Characteristic Distribution of Direct, Diffuse and Total Solar Radiation, *Sol. Energy,* Vol. 4, No. 3, pp. 1–19, 1960.
21. J. Berthier, Weak Thermal Points on Thermal Bridges, *NBS Tech. Note 710-7,* National Bureau of Standards, May 1973.
22. M. Hoffman and B. Schwartz, Comparison of Steady State and Time-Dependent Temperature Distribution for Building Elements, *Building and Environment,* Vol. 15, pp. 63–72, Pergamon, New York, 1980.
23. N. Ito, K. Kimura, and J. Oka, A Field Experiment Study on the Convective Heat Transfer Coefficient on the Exterior Surface of a Building, *ASHRAE Trans.,* Vol. 78, Pt. 1, January 1972.
24. H. E. Robinson and F. J. Powell, Thermal Resistance of Air Spaces and Fibrous Insulation Bounded by Reflective Surfaces, *Building Material and Structures Rep. 151,* National Bureau of Standards, 1957.

25. C. G. Bankvall, Natural Convective Heat Transfer in Permeable Insulation, Thermal Transmission Measurements of Insulation, *STP60 ASTM,* pp. 70–81, 1977.

26. C. L. Tien and P. J. Burns, Convection in a Vertical Slot Filled with Porous Insulation, *Proc. Int. Center Heat and Mass Transfer,* Dubrovnik, Yugoslavia, 1977.

27. P. R. Achenbach and C. W. Coblentz, Field Measurement of Air Infiltration in Ten Electrically Heated House, *ASHRAE Trans.,* Vol. 69, pp. 358–365, 1963.

28. D. T. Grimsrud, M. H. Sherman, and R. C. Sonderregger, Calculating Infiltration; Implication for a Construction Quality Standards, *Proc. ASHRAE/DoE Conf. Thermal Performance Exterior Envelopes of Buildings II, 7P422–452,* Las Vegas, Nev., Dec. 6–9, 1982.

29. H. Ross and D. Grimsrud, Air Infiltration in Buildings: Literature Survey and Proposed Research Agenda, *Lawrence Berkeley Lab. Rep. LBL-W7822,* UC-95d, 1978.

30. ASHRAE Task Group on Energy Requirements, *Procedure to Determine Heating and Cooling Loads for Computerized Energy Calculations for Buildings,* special ASHRAE bulletin, 1974.

31. J. L. Threlkeld, *Thermal Environmental Engineering,* 2d ed., pp. 346–347, Prentice-Hall, Englewood Cliffs, N.J., 1970.

32. K. Kimura and D. G. Stephenson, Solar Radiation on Cloudy Days, *ASHRAE Trans.,* Vol. 75, Pt. 1, pp. 227–233, 1969.

33. D. G. Stephenson, Tables of Solar Altitude, Azimuth, Intensity, and Heat Gain Factors for Latitudes from 43 to 55 Degrees North, National Research Council of Canada, Division of Building Research, *Tech. Paper No. 243,* April 1967;.

34. T. -Y. Sun, Shadow Area Equations for Window Overhangs and Side Fins and Their Application in Computer Calculation, *ASHRAE Trans.,* Vol. 74, Pt. 1, 1968.

35. T. Kusuda, NBSLD, The Computer Program for Heating and Cooling Loads in Buildings, *NBS Building Sci. Ser. 69,* July 1976.

36. T. Tsuchiya, Infiltration and Indoor Air-Temperature and Moisture Variation in a Detached Residence, *J. SHASE,* Vol. 54, No. 11, pp. 13–19, 1980.

37. A. Yoshikawa and K. Yamaguchi, Numerical Solution of Room Air Distribution, *Trans. Soc. Heating, Air-Conditioning and Sanitary Eng. Jpn.,* Vol. 1–7, 1978.

38. D. G. Stephenson and G. P. Mitalas, Calculation of Heat Conduction Transfer Functions of Multi-layer Slabs, *ASHRAE Trans.,* Vol. 77, Pt. 2, 1971.

39. T. Kusuda, Thermal Response Factors for Multi-layer Structures of Various Heat Conduction Systems, *ASHRAE Trans.,* Vol. 75, Pt. 1, pp. 250–269, 1969.

40. G. P. Mitalas and J. G. Arseneault, Fortran IV Program to Calculate Z-Transfer Functions for the Calculation of Transient Heat Transfer through Walls and Roofs, Use of Computers for Environmental Engineering Related to Buildings, *NBS Building Sci. Ser. 39,* pp. 663–668, 1971.

41. B. A. Peavy, A Note on Response Factors and Conduction Transfer Functions, *ASHRAE Trans.,* Vol. 84, P. 1, pp. 688–690, 1978.

42. T. Kusuda, Fundamentals of Building Heat Transfer, *J. Res. N. B. S.,* Vol. 82, No. 2, pp. 97–106, 1977.

43. T. Kusuda and P. R. Achenbach, Numerical Analysis of the Thermal Environment of Occupied Underground Spaces with Finite Cover Using Digital Computer, *ASHRAE Trans.,* 1963.

44. G. O. Meixel and P. M. Shipp, The Importance of Insulation Placement on the Seasonal Heat Loss through Basement and Earth Sheltered Walls, *DoE/ASHRAE Conf. Thermal Performance of Exterior Envelope of Buildings,* Orlando, Fla. Oct. 3–5, 1979.

45. B. Adamson et al., Floor System on Ground Foundation, *Rep. BKL 1972: 8 (E),* Lund Institute of Technology (Sweden), 1972.

46. H. Akasaka, Calculation Method of Heat Loss through Floor and Basement Wall, *Trans. Soc. Heating, Air-Conditioning Eng. Jpn.,* No. 7, June 1978.

47. J. K. Latta and G. G. Boileau, Calculation of Basement Heat Losses, National Research Council of Canada, Division of Building Research *NRC 10477,* December 1968.

48. R. W. Muncey and J. W. Spencer, Heat Flow into the Ground under a House, *Energy Con-*

servation in Heating, Cooling and Ventilating Buildings, Vol. 2, pp. 649–660, Hemisphere, Washington, D.C., 1978.

49. A. H. Lachenbruch, Three Dimensional Heat Conduction in Permafrost beneath Heated Buildings, *Geological Survey Bulletin 1052-B,* U.S. Government Printing Office, Washington, D.C., 1957.

50. G. D. Raithby and K. G. T. Hollands, Natural Convection, *Handbook of Heat Transfer Fundamentals,* W. M. Rohsenow, J. P. Hartnett, and E. N. Ganić, eds., Chap. 6, McGraw-Hill, New York, 1985.

51. W. M. Kays and H. C. Perkins, Forced Convection, Internal Flow in Ducts, *Handbook of Heat Transfer Fundamentals,* W. M. Rohsenow, J. P. Hartnett, E. N. Ganić, eds., Chap. 7, McGraw-Hill, New York, 1985.

52. M. W. Rubesin, M. Inouye, and P. G. Parikh, Forced Convection, External Flows, *Handbook of Heat Transfer Fundamentals,* W. M. Rohsenow, J. P. Hartnett, and E. N. Ganić, eds., Chap. 8, McGraw-Hill, New York, 1985.

53. E. R. G. Eckert, C. L. Tien, and D. K. Edwards, Radiation *Handbook of Heat Transfer Fundamentals,* W. M. Rohsenow, J. P. Hartnett, and E. N. Ganić, eds., Chap. 14, McGraw-Hill, New York, 1985.

54. K. E. Torrance, Numerical Methods in Heat Transfer, *Handbook of Heat Transfer Fundamentals,* W. M. Rohsenow, J. P. Hartnett, and E. N. Ganić, eds., Chap. 5, McGraw-Hill, New York, 1985.

NOMENCLATURE

Symbol, Definition, SI Units, English Units

Unless specifically indicated in the text, the following general nomenclature is used throughout this chapter.

A area:m^2, ft^2

A_g glass area: m^2, ft^2

B temperature amplitude of monthly normal cycle: °C, °F

B_G ground-reflected solar radiation under cloudless conditions: W/m^2, Btu/(h·ft^2)

B_S diffuse sky radiation under cloudless conditions: W/m^2, Btu/(h·ft^2)

C surface moisture content constant

C' shielding coefficient

C_{DD} cooling degree-days: °C·days, °F·days

C_i air-cavity thermal conductance per unit area: W/(m^2·K), Btu/(h·ft^2·°F)

c specific heat: J/(kg·°C), Btu/(lb$_m$·°F)

E energy consumption rate: W°h/yr, Btu/yr

F_{cc} cloud cover factor

f_c temperature correction factor

F_{i-j} radiation heat exchange shape factor between surfaces i and j

F_s shape correction factor

G mass velocity of dry air: kg/(h·kg/h), lb$_m$/(h·lb$_m$/h)

H height: m, ft

H_{DD} heating degree-days: °C·days, °F·days

h heat transfer coefficient: W/(m^2·K), Btu/(h·ft^2·F)

$\tilde{h}$ height; m, ft

h_c convection heat transfer coefficient: $W/(m^2 \cdot K)$, $Btu/(h \cdot ft^2 \cdot {}^\circ F)$

h_i inside-surface heat transfer coefficient: $W/(m^2 \cdot K)$, $Btu/(h \cdot ft^2 \cdot {}^\circ F)$

h_o outside-surface heat transfer coefficient: $W/(m^2 \cdot K)$, $Btu/(h \cdot ft^2 \cdot {}^\circ F)$

h_r radiative heat transfer coefficient: $W/(m^2 \cdot K)$, $Btu/(h \cdot ft^2 \cdot {}^\circ F)$

I solar radiation: W/m^2, $Btu/(h \cdot ft^2)$

I_D direct solar radiation: W/m^2, $Btu/(h \cdot ft^2)$

I_{DC} direct solar radiation under cloudy conditions: W/m^2, $Btu/(h \cdot ft^2)$

I_{DH} direct radiation on a horizontal surface: W/m^2, $Btu/(h \cdot ft^2)$

I_{DHC} direct solar radiation on a horizontal surface under cloudy conditions: W/m^2, $Btu/(h \cdot ft^2)$

I_{DN} direct normal solar radiation for a cloudless sky: W/m^2, $Btu/(h \cdot ft^2)$

I_d diffuse solar radiation: W/m^2, $Btu/(h \cdot ft^2)$

I_T total solar radiation: W/m^2, $Btu/(h \cdot ft^2)$

K moisture transfer coefficient: $kg/[h \cdot m^2(kg/kg)]$, $lb_m/[h \cdot ft^2(lb_m/lb_m)]$

K_g moisture transfer coefficent to glass surface: $kg/(h \cdot m^2(kg/kg)]$, $lb_m/[h \cdot ft^2(lb_m/lb_m)]$

k thermal conductivity: $W/(m \cdot K)$, $Btu \cdot in/(h \cdot ft^2 \cdot {}^\circ F)$

k^+ dimensionless thermal conductivity

L length: m, ft

L' latitude: deg

$\tilde{L}$ air leakage area: m^2, ft^2

l thickness: m, ft

l' longitude: deg

M molecular weight: kg/mol, lb_m/mol

N_c clearness number

N_s total number of heat transfer surfaces

n index of refraction

P pressure: Pa, in Hg

$\hat{P}$ period: h

P perimeter: m, ft

P^+ dimensionless perimeter

P_v vapor pressure: Pa, in Hg

Q air leakage rate: m^3/h, ft^3/min

q heat flow rate: W, Btu/h

q'' heat flux: W/m^2, $Btu/(h \cdot ft^2)$

R thermal resistance: $m^2 \cdot K/W$, $h \cdot ft^2 \cdot {}^\circ F/Btu$

R common ratio of thermal response factors

r reflectance or reflectivity

T temperature: $^\circ C$, $^\circ F$

$\overline{T}_A$ annual average outdoor temprature: $^\circ C$, $^\circ F$

T_B balance-point temperature (usually $65^\circ F$ or $18.3^\circ C$): $^\circ C$, $^\circ F$

T_{DB} outdoor air dry-bulb temperature: $^\circ C$, $^\circ F$

T_{DP} outdoor air dew-point temperature: $^\circ C$, $^\circ F$

T_{di} mean temperature of day i: $^\circ C$, $^\circ F$

T_{is} inside-surface temperature: °C, °F

T_{os} outside-surface temperature: °C, °F

T_R annual average indoor temperature: °C, °F

T absolute temperature

t time: h or day

U overall heat transfer coefficient: $W/(m^2 \cdot K)$, $Btu/(h \cdot ft^2 \cdot °F)$

V wind speed: m/s, mi/h

V volume: m^3, ft^3

V_c corrected wind speed: m/s, mi/h

$\dot{W}_G$ moisture generation rate: kg/h, lb_m/h

$\dot{W}_v$ mass flow rate of water vapor: $kg/(h \cdot m^2)$, $lb_m/(h \cdot ft^2)$

wi moisture content of room surface: g/kg, lb_m/lb_m

X thermal response factor: $W/(m^2 \cdot K)$, $Btu/(h \cdot ft^2 \cdot °F)$

x distance variable: m, ft

Y thermal response factor: $W/(m^2 \cdot K)$, $Btu/(h \cdot ft^2 \cdot °F)$

Z thermal response factor: $W/(m^2 \cdot K)$, $Btu/(h \cdot ft^2 \cdot °F)$

z distance variable: m, ft

Greek Symbols

α absorptivity

$\tilde{\alpha}$ terrain constant

α thermal diffusivity: m^2/day, ft^2/h

β surface tilt angle: rad

γ solar azimuth angle: rad

$\tilde{\gamma}$ terrain constant

δ_s solar declination angle: rad

ϵ radiation emittance

ζ function

η solar incidence angle: rad

θ^+ dimensionless temperature

θ_c^+ minimum value of θ^+

κ extinction coefficient: m^{-1}, ft^{-1}

μ permeability: $kg/(m \cdot h \cdot Pa)$, $lb_m/[ft^2 \cdot h(in\ Hg/in)]$

ρ density: kg/m^3, lb_m/ft^3

ρ thermal bridge effect

σ Stefan-Boltzmann constant: $W/(m^2 \cdot K^4)$, $Btu/(h^2 \cdot ft \cdot °R^4)$

σ_z zenith angle: rad

τ transmissivity

ϕ phase angle: rad

χ_o humidity ratio of outdoor air: g/kg, lb_m/lb_m

χ_g humidity ratio of the wet glass surface air: g/kg, lb_m/lb_m

χ_i humidity ratio of room air: g/kg, lb_m/lb_m

χ_s humidity ratio of surface air: g/kg, lb_m/lb_m

χ_s^* humidity ratio of surface air saturated at room air temperature: g/kg, lb_m/lb_m

ψ solar altitude angle: rad

ψ function

ω hour angle: rad

CHAPTER 10

Cooling Towers and Cooling Ponds

By J. S. Maulbetsch
J. A. Bartz

Electric Power Research Institute
Palo Alto, California

A. INTRODUCTION

1. The Use of Closed-Cycle Cooling

Most industrial power generating and space conditioning processes require rejection of heat to the environment. In particular, steam electric power plants reject heat at approximately twice the rate at which electricity is generated. For a long time, industrial process and building air conditioning system designers have found the use of evaporative cooling towers, often in the form of roof-top cooling towers, to be an efficient and convenient alternative to simple water cooling. On the other hand, central-station power plants have normally opted for once-through cooling. Prior to 1970, over 75 percent of power plants used once-through systems. Water for power plant cooling accounted for nearly 10 percent of the total water withdrawals from the nation's surface waters. In the past decade, in response both to a shortage of sites where large withdrawals could be supported and to specific environmental regulations from the Clean Water Act, the use of closed-cycle cooling at new plants has drastically increased. The following discussion will be cast primarily in the context of power plant cooling.

2. Types of Cooling Systems

Closed-cycle cooling systems circulate heated water from the plant condenser to a device in which cooling is achieved either by heat transfer to the air or by evaporation of a small fraction of the total flow. The types of cooling devices in most common use are cooling towers, cooling ponds, and spray systems.

a. Cooling Towers

Wet Towers.   Cooling towers may be both wet and dry. In wet cooling towers, water is pumped to the top of the cooling section and allowed to flow and splash down through the fill or packing region while air flows counter to or crosswise to the water flow. As a fraction of the water evaporates, the air is heated and humidified while the remainder of the water is cooled. The airflow may be created by buoyancy (natural-draft) or by fans (mechanical-draft).

Dry Towers.   In dry towers, the cooling is accomplished in an air-cooled heat exchanger. In a *direct* system, the turbine steam is ducted directly to the tower and is condensed there; in an *indirect* system, a secondary circulating loop is cooled by the tower, while the steam is condensed in a conventional condenser. Both natural- and mechanical-draft towers have been used.

Wet-Dry Systems.   Combined wet-dry systems are primarily dry systems which utilize supplemental wet cooling during periods of high ambient temperature. Several arrangements are possible including separate towers and a combined or hybrid structure.

b. Cooling Ponds

Cooling ponds are artificial impoundments into which the heated cooling water is discharged and from which cooled water is withdrawn. Heat is rejected from the pond surface to the atmosphere by conduction, radiation, and evaporation.

c. Spray Ponds and Spray Canals

Spray ponds are cooling ponds in which the transfer processes to the atmosphere are enhanced by spraying the water into the air to form a large interfacial area for heat and mass transfer. Spray canals consist of a series of spray modules distributed along a canal which connects the plant intake and discharge. Spray systems are similar to towers in that a dispersed water phase is created but the contact region is unenclosed and no positive control of airflow is possible.

d. Mixed Systems

Any of the above devices can be used in a "mixed" system. These are once-through systems in which the heated discharge flow is partially cooled before being returned to the natural receiving water body in order to maintain a predetermined temperature limit.

3. Selection Criteria

Heat rejection systems are chosen on the basis of achieving the required cooling with acceptable environmental effects at minimum cost.

a. Cost-Performance Trade-Offs

The cooling system is a significant cost item at a power plant. Not only are the capital costs high, but the performance of the cooling system has a dominant effect on plant heat rate. If the cooling system does not provide adequate cooling, the turbine back pressure rises. Therefore, a complete cost analysis must include penalties for reduced plant efficiency and reduced plant capacity associated with cooling system performance limits on plant operation. Larger, and more costly, systems will result in reduced penalties. Therefore, an optimum exists and comparisons among alternative systems should be made on the basis of comparing optimized systems.

b. Environmental Considerations

Much of the motivation for the increased use of closed-cycle cooling has been environmental protection. The primary environmental benefits, in comparison to once-through cooling, are reduced water withdrawals and reduced thermal discharge. Withdrawal rates are 2 to 4 percent of those required for once-through systems, and such discharges as are required may be made from the cold side of the system.

Other environmental considerations include water consumption, blowdown, plumes and drift, land use, noise, and aesthetics. Water consumption is a factor in any wet cooling system. Evaporative systems typically dissipate 1100 to 1300 Btu/lb_m (2.6 to 3.0 J/kg) of water evaporated. Even once-through systems increase the local evaporation rates from the surface of the warmed receiving waters. The relative consumption rates of the different closed-cycle systems have been the subject of several investigations; they are strongly site-dependent.

Blowdown or liquid discharge from closed-cycle cooling systems is required to control the buildup of dissolved and suspended solids in the circulating water. Typical blowdown rates range from 0.5 to 5 (gal/min)/MW, or 3.1×10^{-5} to 3.1×10^{-4} $(m^3/s)/MW$. Elevated salinity or the presence of maintenance chemicals in the blowdown may pose environmental problems.

Vapor plumes and drift are atmospheric effects of closed-cycle systems, primarily towers and sprays. These are primarily "nuisance" effects but may result in reduced visibility, potentially dangerous icing, or damage to vegetation from saline drift under particular, site-specific conditions.

Land use, noise, and aesthetics are secondary effects which, in general, favor once-through cooling.

B. EVAPORATIVE COOLING TOWERS

1. General Description

In an evaporative cooling tower, reject heat is dissipated to the atmosphere by bringing the circulating cooling water into contact with air. Cooling is accomplished by evaporation of a small portion of the water as well as sensible heat transfer from the water to the air. Water that is heated in the steam condenser is pumped to a distribution section at the top of the tower. The water then falls by gravity through the fill to the tower basin as air flows through the fill section. This results in cooling of the water toward the ambient wet-bulb temperature and a transfer of latent heat to the air, which rapidly approaches saturation.

There are two basic fill configurations. In the cross flow, the air flows horizontally, or perpendicular to the downward flow of water (Fig. 1a). In the counterflow, Fig. 1b, the air flows upward, or parallel to the water flow. The counterflow design is theoretically

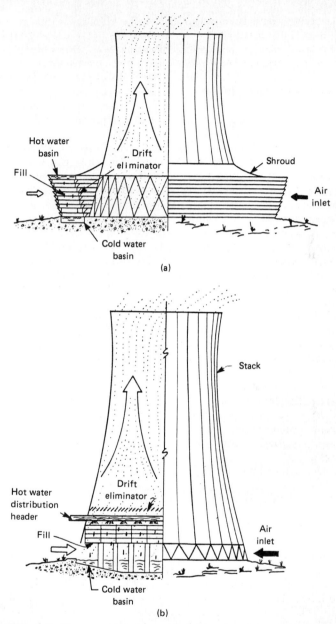

FIG. 1 Cooling tower fill configurations. (*a*) Cross-flow tower (natural draft). (*b*) Counterflow tower (natural draft).

more efficient, but the cross-flow configuration lends itself to simpler designs for good air distribution to the fill.

Fill types fall into two categories. Splash fills break the water into droplets, promoting evaporation. Film packings produce a thin film of water on extended surfaces to enhance evaporation.

Three different designs are used to supply air to the fill, resulting in three basic tower types [1]: mechanical-draft, natural-draft, and fan-assisted natural-draft cooling towers. In the mechanical-draft cooling tower, fans are used to provide an air flow rate that is nearly independent of ambient conditions.

There are two basic fan-draft configurations. In the induced-draft design, air is pulled through the fill, while in the forced-draft configuration, air is pushed through the fill. Most mechanical-draft cooling towers use induced-draft designs in power plant applications. The induced-draft design generally provides a more uniform airflow, but requires mounting fans and motors on top of the structure rather than at grade level.

Air flow rate through the fill is controlled by varying the number of fans in operation, the motor speed, or the fan blade pitch.

In the natural-draft tower, air flows through the fill as the result of buoyancy of the warm, moist air leaving the tower stack, illustrated in Fig. 1. The cost of the fan installation and the cost of power to operate the fans are eliminated in a natural-draft tower, but the tower is large and expensive because of the stack. Air flow rate through the fill, and consequently natural-draft cooling tower performance, depend more strongly on ambient weather conditions.

In the fan-assisted natural-draft tower, the chimney effect of the tower stack is supplemented with a fan. This design reduces cooling tower size and improves airflow reliability, but at the cost of the fan installation and operation.

2. Design Procedure

The common practice in the design and performance evaluation of cooling towers is based on mathematical models which depend on one or more of the following assumptions [2]:

- The hydrodynamic and thermal processes are not coupled.
- There are no spatial variations in the flow and heat transfer processes (the "point" or "zero-dimensional" model).
- There are no spatial variations in the flow process, but one- (or two-) dimensional variations in the heat transfer process (the one- (or two-) dimensional model).

The introduction of additional dimensions in the models provides for increased accuracy of representation, but at the expense of added complexity and increased computation cost.

In the point models, the governing equations for the thermal processes, written for a control volume, are the heat and mass transfer equation

$$G \, di_G = KA \, dV(i_l - i_G) \qquad (1)$$

and the energy conservation equation*

$$G \, di_G = L \, di_l \qquad (2)$$

*L and G are the symbols commonly used for the mass flow rate of water and air, respectively.

Elimination of G from Eqs. (1) and (2) yields Merkel's equation

$$\frac{di_l}{i_l - i_G} = \frac{KA\, dV}{L} \tag{3}$$

A detailed derivation is provided in Ref. 1.

Equation (3) is a fundamental design equation because it relates variables associated with thermal performance of the tower (i_l, i_G) with variables associated with the heat transfer performance of the fill (K, A, V, L).

The solution of Eq. (3) is obtained graphically after integrating the equation across the fill:

$$c_{p,w} \int \frac{dT_l}{i_l - i_G} = \int \frac{KA}{L}\, dV \tag{4}$$

or

$$I_L = I_R \tag{5}$$

The graphical solution of Eq. (5) is shown in Fig. 2a. The value of L/G at the point of intersection of the curves I_L and I_R is the required solution for the ratio of water and air flow rates.

Figure 2b from Ref. 1 shows the enthalpy and temperature relationships for water and air and the driving potential for heat transfer in an evaporative counterflow cooling tower. The water operating line, AB, is determined by the tower inlet and outlet water temperatures. Line AB corresponds to the conditions of the air adjacent to the falling water surface. The line is labeled "saturation curve" because it is assumed that the air next to the water surface is saturated at the water temperature. The air operating line, CD, corresponds to the bulk air conditions as the air flows through the tower. The value of the ordinate between lines AB and CD represents the enthalpy driving force. The slope of line CD is the value L/G. The cooling range and approach are also illustrated.

Figure 2b together with Eqs. (1) to (5) summarizes the governing parameters of the cooling tower:

- L, mass flow rate of water
- G, mass flow rate of air
- K, fill performance characteristic
- Cooling range
- Cooling approach
- Ambient wet-bulb temperature

An example of cooling tower design curves is given in Fig. 3a, from Ref. 3. The Merkel equation can be written in the empirical form

$$\frac{KAV}{L} = c_0 \left(\frac{L}{G}\right)^{-n_0} \tag{6}$$

The constants c_0 and n_0, which depend on the packing design, define the intercept and slope of EF in Fig. 3a. Values of n_0 range from about ¼ to 1. The lower values are associated with packings in which the water splashes over the film, and the larger values correspond to packings in which the water forms a surface film.

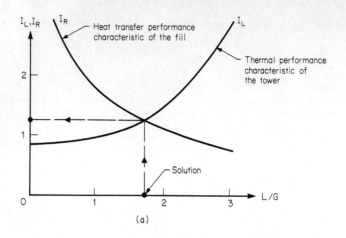

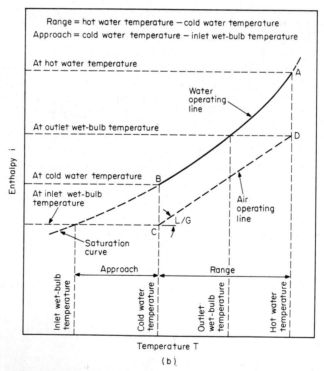

FIG. 2 Governing parameters of cooling tower performance. (*a*) Graphical solution of the Merkel equation. (*b*) Driving potential for heat transfer.

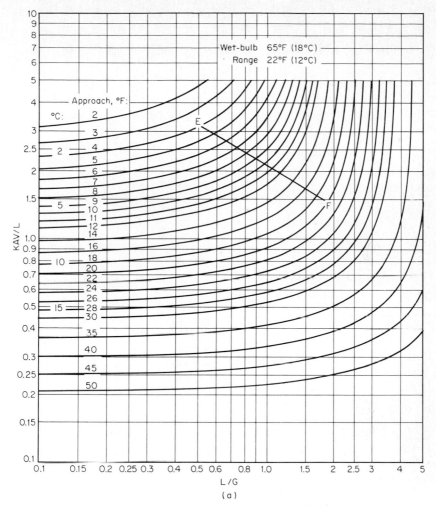

FIG. 3 Evaporative cooling tower design. (*a*) Design curves. (*b*) Fill configurations.

To design a tower, a plot of the type shown in Fig. 3*a* is entered with the appropriate information on

- Wet-bulb temperature
- Range
- L/G
- c_0, n_0

The solution to Merkel's equation is then obtained as shown in Fig. 2*a*.

Cooling tower design handbooks provide the values of I_L versus L/G as a function of range, approach, and wet-bulb temperature, and experimental data provide values of I_R

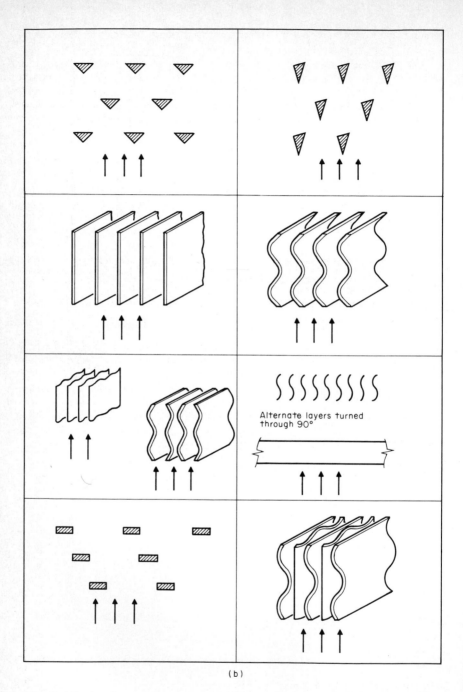

Alternate layers turned
through 90°

(b)

FIG. 3 (*Continued*)

versus L/G [4–7]. The only required empirical data are the performance characteristics of the fill. Some types of fill geometry are shown in Fig. 3b.

The preceding thermal analysis for point models is identical for natural-draft and mechanical-draft towers, but the airflow characteristics of these two tower types differ. In the natural-draft tower, upward airflow results from buoyant forces of the heated effluent, which are balanced by the forces that oppose flow resistance. In the point model, the hydrodynamic equation for the natural-draft tower can be expressed as

$$H \, \Delta\rho = N_v \rho \frac{v_a^2}{2g} \tag{7}$$

Values of N_v for internal flow resistances such as fill, struts, etc., are empirical.

In the mechanical-draft tower, upward flow results from the inertial forces produced by the fan, which are balanced by flow resistance. The required fan power is calculated from an equation of the type

$$P_f = K_p Q^3 \tag{8}$$

One-dimensional methods have been developed [8, 9] for solving the governing equations for air and water mixing in the fill of a counterflow tower. Merkel's equation is integrated numerically in an iterative process over a number of vertical cells in the flow direction. A coupled iterative analysis has also been developed [10] in which the air flow rate is adjusted to satisfy Eq. (7).

Two-dimensional methods are required in order to model flows in a cross-flow configuration. Several investigators [9, 11, 12] have developed two-dimensional grid methods to calculate thermal performance. Kelly [4] used the procedure of Ref. 11 to calculate performance of a cross-flow tower, with a "point" model for the hydrodynamics.

Several two-dimensional models have been developed [13–16] that couple the thermodynamics and hydrodynamics, and are capable of simulating the performance of natural- and mechanical-draft cooling towers in both cross-flow and counterflow configurations. The following governing equations [14] (written here in radial coordinates) are solved using finite difference techniques in the computer code:

a. Mass of air-vapor and of vapor:

$$\frac{\partial}{\partial x}(\rho u) + \frac{1}{r}\frac{\partial}{\partial r}(\rho r v) = 0$$

b. Vertical momentum:

$$\frac{\partial}{\partial x}(\rho u^2) + \frac{1}{r}\frac{\partial}{\partial r}(\rho r v u) - \frac{\partial}{\partial x}\left(\mu_{eff}\frac{\partial u}{\partial x}\right) - \frac{1}{r}\frac{\partial}{\partial r}\left(r\mu_{eff}\frac{\partial u}{\partial r}\right) = -\frac{\partial P}{\partial x} - f_x - \rho g$$

c. Radial (or horizontal) momentum:

$$\frac{\partial}{\partial x}(\rho uv) + \frac{1}{r}\frac{\partial}{\partial r}(\rho r v^2) - \frac{\partial}{\partial x}\left(\mu_{eff}\frac{\partial v}{\partial x}\right) - \frac{1}{r}\frac{\partial}{\partial r}\left(r\mu_{eff}\frac{\partial v}{\partial r}\right) = \frac{\partial P}{\partial r} - f_y$$

$$\tag{9}$$

d. Air enthalpy:

$$\frac{\partial}{\partial x}(\rho u i_G) + \frac{1}{r}\frac{\partial}{\partial r}(\rho r v i_G) - \frac{\partial}{\partial x}\left(\Gamma_G\frac{\partial v}{\partial x}\right) - \frac{1}{r}\frac{\partial}{\partial r}\left(r\Gamma_G\frac{\partial v}{\partial x}\right)$$

$$-\frac{1}{r}\frac{\partial}{\partial r}\left(r\Gamma_G\frac{\partial v}{\partial r}\right) = m'' i_G$$

e. Water enthalpy:

$$\frac{\partial}{\partial x}(\rho u_l i_l) = -m'' i_g$$

f. Moisture fraction:

$$\frac{\partial}{\partial x}(\rho u f_G) + \frac{1}{r}\frac{\partial}{\partial r}(\rho r v f_G) - \frac{\partial}{\partial x}\left(\Gamma_G \frac{\partial u}{\partial x}\right) - \frac{1}{r}\frac{\partial}{\partial r}\left(r\Gamma_f \frac{\partial v}{\partial r}\right) = m''$$

where $f_x = N_v \dfrac{\rho u^2}{2}$

$$f_y = N_v \frac{\rho v^2}{2}$$

C. DRY AND WET-DRY COOLING TOWERS

1. General Description

Dry cooling towers use heat exchangers with extended surfaces to reject heat to the atmosphere by circulating fluid, either water or steam, inside the tubes. Ambient air flows over the outside of the tubes and extended surfaces, which improve the convective heat transfer as a result of increased contact area.

There are two basic cooling system designs. In the direct system, steam from the low-pressure turbine flows in ducts to the heat exchangers and is condensed by the cooling air. This design eliminates an intermediate loop and additional heat exchanger, with resulting savings in thermal resistance and capital costs. The density of steam at the vacuum conditions of the turbine exhaust is very low, however, which results in the need for large ducts to transport the steam. For this reason, there is a practical upper limit on the size of direct condensation systems.

In the indirect system, an intermediate loop is used to avoid transporting steam in large ducts. There are two indirect system designs, as illustrated in Fig. 4.

The Heller system (Fig. 4a) uses a direct-contact condenser to condense the steam leaving the turbine by mixing it with cooling water that is circulated through the cooling tower. In Fig. 4b, a surface condenser is used to separate the steam cycle from the cooling water loop. Turbine steam is condensed by the cooling water in the surface condenser, and the circulating water is cooled by air in the cooling coils.

The Heller system eliminates the thermal resistance of the boundary that separates the fluid cycles in the surface condenser, leading to a smaller and less expensive condenser. However, the Heller system requires cooling water of a high quality, because the cooling water is used for boiler feed water. This is a disadvantage, because the required cooling water flow rate is much greater than the feed water flow rate, which means that a large inventory of high-quality water is needed for cooling.

Wet-dry cooling towers combine the advantages of both evaporative and dry systems. The dry tower provides cooling much of the time with no water consumption. During periods of high ambient temperature, the dry cooling is supplemented with evaporative cooling. Many configurations are possible: separate wet and dry towers, combined towers with parallel or series airflow through the dry and wet sections, etc. Figure 5 illustrates a configuration in which the airflow is in parallel and the water flow is in series [17, 18].

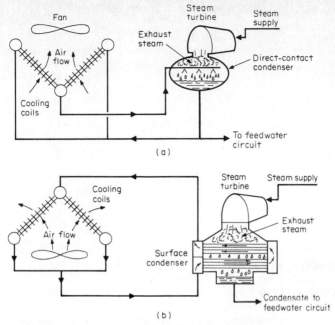

FIG. 4 Dry cooling systems. (*a*) Indirect dry cooling system with direct-contact condenser. (*b*) Indirect dry cooling system with surface condenser.

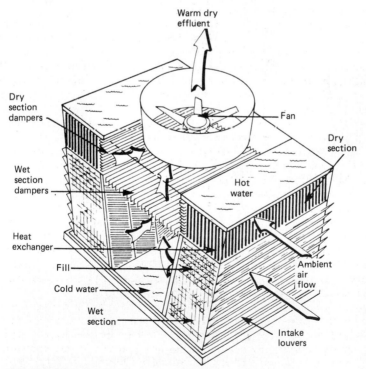

FIG. 5 Wet-dry cooling tower.

In addition to conventional evaporative cooling designs, supplemental cooling is provided in some systems by evaporating the water in the inlet air stream before the dry section, and in others by deluging the heat exchangers with water.

2. Design Procedure

Heat transfer from the extended surface of the heat exchanger in a dry cooling tower occurs primarily by forced convection. The heat rejected is given by

$$q = U \Delta T_{lm} A_s F \tag{10}$$

where F is a function of the geometrical flow arrangement [19] and

$$\Delta T_{lm} = \frac{\Delta T_G - \Delta T_L}{\ln (\Delta T_G / \Delta T_L)} \tag{10a}$$

The temperature variations in the heat rejection system are illustrated in Fig. 6. Heat transfer in the air-cooled heat exchanger can also be analyzed in terms of heat transfer effectiveness and the number of heat transfer units [1].

When the extended surface is deluged with water, the heat rejection capability of the system is enhanced considerably, typically 2 to 5 times [20]. It has been shown [21] that the driving potential for heat transfer is the difference between the saturation enthalpy of air at the tube-side temperature minus the enthalpy of the coolant air stream, which is similar to the Merkel approach, Eq. (1). The heat rejected on a deluged surface can be expressed in direct analogy to Eq. (10) as

$$q^* = \frac{U}{c_{p,a}^*} \Delta i_{lm} A_s F \tag{11}$$

Reference 21 provides simple techniques for estimating the benefit of heat transfer augmentation by deluge.

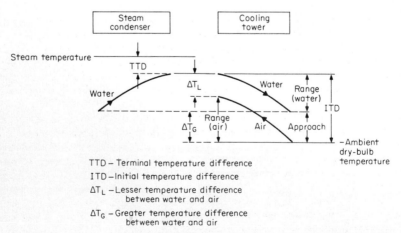

TTD – Terminal temperature difference
ITD – Initial temperature difference
ΔT_L – Lesser temperature difference between water and air
ΔT_G – Greater temperature difference between water and air

FIG. 6 Temperature variations in a heat rejection system (dry cooling, indirect).

3. Advanced Systems

In addition to the direct and indirect systems that are commercially available, there are a number of promising concepts for water-conserving cooling systems under development or becoming commercially available. A few will be described to indicate the wide range of possibilities, rather than trying to provide a comprehensive survey.

In one process [22], reject heat is extracted to evaporate wastewater (brine) recirculating through a falling-film heat exchanger. The wastewater evaporates into an air stream that flows across the falling water film, as shown in Fig. 7a. Evaporation increases the concentration of dissolved solids in the wastewater, and conventional water quality control techniques are used to remove potential scaling constituents. Only about 5 percent of the wastewater is removed to maintain desired levels of dissolved solids.

In another development [23], a direct condensation system is combined with a conventional evaporative tower to provide supplemental cooling. In this system, illustrated in Fig. 7b, steam is condensed either in the direct condenser during colder ambient conditions or in both the direct and surface condenser during warmer periods. In the all-dry mode the circulating water in the evaporative system is shut off and all steam is condensed in the direct condenser. In the wet-dry mode, heat rejection is modulated between the two condensing systems by means of airflow control. Advantages of this system include small dry tower, evaporative tower, and turbine building, plus the fact that its components are commercially available.

A third advanced wet-dry system [24] uses ammonia to transport reject heat from the steam condenser to the cooling tower, as shown in Fig. 7c. Liquid ammonia is boiled in the tubes of the surface condenser. The ammonia vapor that is produced is condensed in the tower, using ambient air as the coolant. A prime advantage of the system is the elimination of the sensible heating range (see Fig. 6) as a temperature increment between the ambient dry-bulb and steam condensing temperature, which improves the thermodynamic efficiency of the steam cycle. Other advantages include reduced pumping power of the transport loop and elimination of concern over freeze-up of the transport fluid. A disadvantage of the system is the possibility of leakage of the ammonia, which is toxic and corrosive.

Two methods of providing water augmentation to the ammonia system are attractive. In one design, the air-cooled surfaces are deluged with water to provide supplemental cooling. This results in a compact, uncomplicated design, but intermittent wetting of the surfaces causes concerns of corrosion or scale buildup. In another design, supplemental cooling is provided by a conventional evaporative tower. This eliminates concerns associated with wetting the air-cooled surfaces, but a second heat exchanger is needed in this design.

Variations in the ammonia heat rejection system that are under development offer added advantages. In one dry cooling variation [25], an ammonia turbine is added to the ammonia loop, resulting in a binary steam-ammonia cycle. Additional work is extracted from the cycle during periods of low ambient temperatures, resulting in improved system economics in cases where peak electrical demand occurs in winter months, as in Europe and parts of the United States.

In another variation [26], a thermal capacitor is added to the system. During periods of high ambient temperature, some of the reject energy is used to heat a quantity of water, which is stored until nighttime. The heat is then rejected to the atmosphere from the cooling tower. This system offers a means to supplement cooling during periods of high ambient temperatures without evaporative consumption of water. In addition, the prob-

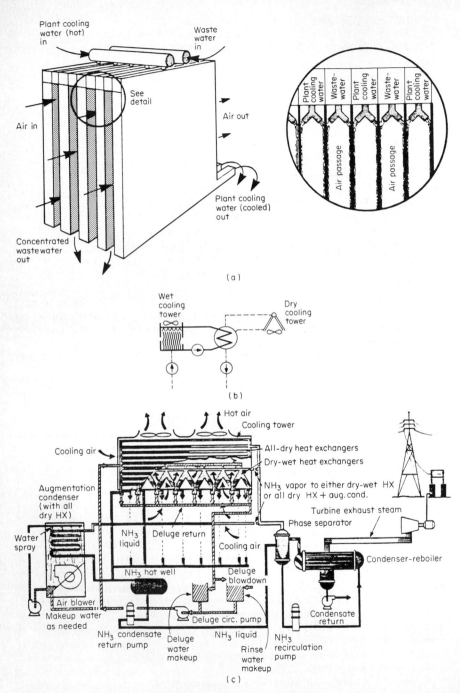

FIG. 7 Examples of advanced water-conserving cooling systems. (*a*) Brine concentrator. (*b*) Direct condensation plus evaporative tower. (*c*) Ammonia phase change.

lems associated with water treatment for evaporative cooling and the disposal of blow-down are eliminated. The economics of the thermal capacitor system are most favorable when peak electrical demand occurs in summer months, as in much of the United States.

D. COOLING PONDS

1. General Description

Cooling ponds are artificial bodies of water for the dissipation of heat to the atmosphere. By convention, cooling *ponds* are defined as artificial water impoundments which do not impede the flow of a natural stream, while cooling *lakes* are usually formed by impounding a natural stream. Hence, cooling lakes typically conform to the natural contours of the local topography, are irregular in shape, may have extensive dead-end zones, and, in general, have uncontrolled internal flow patterns. Cooling ponds may be constructed with a preferred, regular shape or equipped with baffles for control of flow patterns, and thus may be more heavily loaded (greater heat transfer per unit of surface area). Typical thermal loadings in power plant applications require 0.6 to 1.6 hectares per megawatt of plant capacity for cooling lakes and 0.2 to 0.6 hectares per megawatt for cooling ponds.

The primary use of cooling ponds is for heat rejection from steam-electric power plants. As of mid-1981, approximately 126 plants, representing 78,943 MW of generating capacity (15 percent of the total U.S. capacity) used cooling ponds. The advantages and disadvantages of cooling ponds vis-à-vis other power plant cooling systems are summarized in Table 1. This combination of characteristics has made cooling ponds the system of choice primarily in the midwestern and southeastern United States, where land is readily available at modest cost.

TABLE 1 Characteristics of Cooling Ponds

	Advantages	Disadvantages

Performance	Low pumping power
	High thermal inertia
	Compatible with other plant water systems; i.e.,
	  Provides storage
	  Requires no makeup for extended periods
	  Can serve as water treatment (settling) basin
		Low heat transfer rate (large land area required)
		Water consumption difficult to predict or control; may be more or less than cooling towers

Cost	Low maintenance Low capital cost (in comparison to cooling towers)	Costs increase if pond liners are required for seepage control.

Environmental	Adaptable for multiple use:
	  Recreation
	  Wildlife management
	  Flood control
	Minimum blowdown requirements
		High land use requirements
		Low-level fogging or icing
		Potential groundwater contamination

The primary technical issues for cooling ponds are design of the pond to enhance its thermal performance and the accurate prediction of required size and water consumption. Recommended methods and correlations for design and performance calculations are provided in the following sections. More detailed discussions on the general subject of cooling ponds are found in Refs. 1 and 27.

2. Classification of Ponds

The heat transfer performance of a pond is ultimately set by the transfer mechanisms of conduction, evaporation, and radiation at the pond surface, which are determined by the surface temperature distribution. A precise calculation of the surface temperature would depend on the details of the three-dimensional transient velocity and temperature fields throughout the pond and would be prohibitively complex. Tractable design and performance computations require approximations based on spatial and temporal averaging. To this end, a classification of ponds by heat loading, by depth, by aspect ratio, and by retention time (ratio of pond volume to through-flow rate) is required. A convenient classification scheme was developed by Jirka et al. [28].

a. Thermal Loading

Ponds are classified as "lightly" or "heavily" loaded on the basis of the artificially imposed heat load per unit area, q'', in MW(th)/hectare. Lightly loaded ponds [$q'' < 0.5$ MW(th)/hectare] act as "natural impoundments." Their vertical temperature variation and flow structure are dominated by natural meteorological and hydrological processes. They are characterized by little horizontal variability.

Heavily loaded ponds [$q'' > 0.5$ MW(th)/hectare] may exhibit significant horizontal variability, which may be confined to a thin surface layer or extend throughout the depth of the pond. The heat load, the pumped through flow, and the design and location of the inlet and outlet structures can dominate the pond behavior.

b. Pond "Depth"

Heavily loaded ponds are further classified as "deep" or "shallow" ponds. The distinction refers to the degree of vertical stratification and is controlled by flow, temperature difference, and aspect ratio in addition to simply pond depth. "Deep" ponds are strongly stratified with a thin, heated surface layer on top of a horizontally uniform subsurface region. "Shallow" ponds are only partially stratified or vertically fully mixed. A dimensionless "pond number" has been suggested by Jirka et al. [28]:

$$\mathrm{IP} \equiv \left\{ \frac{f_i}{4} \left(\frac{Q_0^2}{\beta \, \Delta T_0 \, ga^3 b^2} \right) D_v^3 \frac{c}{a} \right\}^{1/4} \tag{12}$$

for categorizing ponds as follows:

$$\mathrm{IP} < 0.3: \quad \text{deep ponds}$$
$$0.3 < \mathrm{IP} < 1.0: \quad \text{shallow–partially mixed}$$
$$\mathrm{IP} > 1.0: \quad \text{shallow–fully mixed}$$

Examination of Eq. (12) shows that low flows, long retention times, and low aspect ratios lead to strongly stratified or "deep" ponds.

c. Flow Structure

A further classification of shallow ponds, in which the stratification and hence the density currents are relatively weak, is used to account for the influence of through flow and the inlet-outlet configuration or flow structure. A length-to-width criterion ($c/b \lesssim 3$ to 5) may be used for subcategorizing shallow ponds into recirculating ($c/b < 3$ to 5) and dispersive ($c/b > 3$ to 5) types. This categorization determines the choice of computational model for design and performance predictions.

3. Pond Performance

In order to completely predict or describe the performance of a cooling pond, it is necessary to compute the three-dimensional time-varying velocity and temperature profiles throughout the pond. In principle, this requires the satisfaction of (1) conservation relations throughout the water body; (2) heat and mass transfer boundary conditions at the pond surface; and (3) imposed boundary conditions regarding through flow and heat load from the cooling operation. The situation is inherently unsteady; the three-dimensional nature of the problem is, in general, important; and the spatial boundaries are complex and irregular. As a result, a complete solution for either design or performance calculations is intractable and prohibitively costly. The following sections describe two intermediate levels at which useful calculations can be made:

- Transient, hydrothermal models based on simplifications appropriate to the categorizations proposed in Sec. D.2
- Simplified, steady-state models useful for design approximations or preliminary estimates

a. Transient, Hydrothermal Models

Three classes of models were proposed [28]:

- Natural impoundment model
- Shallow pond models (dispersive and recirculating)
- Deep, stratified pond model

Natural Impoundments.   Ponds which are lightly loaded [$q'' < 0.5$ MW(th)/hectare] are dominated in their hydrothermal behavior by a natural, meteorological process with little horizontal variability. They are modeled in a manner similar to that for natural lakes and reservoirs, which is a one-dimensional, time-variant, variable-area model which includes the processes of absorption and transmission of ambient radiation, conduction, and evaporation from the pond surface, vertical advection due to through flows, and turbulent diffusion and mixing due to buoyancy and wind-driven flows. Descriptions of the mathematical formulation and solution procedure are given in Refs. 28, 29, and 30.

Shallow, Dispersive Pond.   Ponds which are more heavily thermally loaded [$q'' > 0.5$ MW(th)/hectare], are "shallow" (IP > 0.5), and are of high aspect ratio ($c/b > 3$

to 5) may be characterized by an average width and depth, no vertical or transverse horizontal gradients of temperature or velocity, and a negligibly small entrance mixing zone.
The describing equation is

$$\frac{\partial T_m}{\partial t} + \overline{U}\frac{\partial T_m}{\partial x_s} = E_L\frac{\partial^2 T_m}{\partial x_s^2} - \frac{q_n''}{\rho c_p a} \tag{13}$$

Boundary conditions at the inlet and outlet are

$$T_m\bigg|_{x_s=0} - \frac{E_L}{\overline{U}}\frac{\partial T_m}{\partial x_s}\bigg|_{x_s=0} = T_{in} \tag{13a}$$

and
$$\frac{E_L}{\overline{U}}\frac{\partial T_m}{\partial x_s}\bigg|_{x_s=c} = 0 \tag{13b}$$

The value E_L is estimated, according to Ref. 31, as

$$E_L = 0.6\frac{U^*l^2}{K_0^2 D_h} \tag{13c}$$

where
$$U^* \equiv \sqrt{\frac{f}{8}}\,\overline{U}$$

The value q_n'' is determined in accordance with formulas of Sec. D.3.b. Numerical solution is achieved through conventional finite difference methods as described in Ref. 28.

Extension of the solution to the case where b, a, and E_L vary with x_s is straightforward, but is readily applicable only to situations where the channel geometry changes
gradually in the streamwise direction. A complete pond can be divided into several com-
partments, each represented as a single compartment and linked by boundary conditions
of the type in Eq. (13).

Shallow, Recirculating Pond.   Many ponds may be of the heavily loaded, shallow
type discussed in the previous section but have a lower aspect ratio $(c/b < 3)$. This shape
gives rise to recirculating flows which may be represented by a two-zone model: a forward
flow zone and a return flow zone, as shown in Fig. 8. The describing equation in each

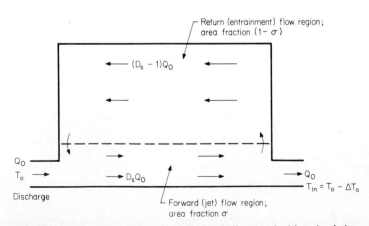

FIG. 8 Schematic of mathematical model for shallow pond with recirculating
flow [28].

zone is the same as for the shallow dispersive model, Eq. (13). The two zones are linked by a dilution factor D_s, defined in Fig. 8. The relative size of the two zones is given by the area fraction σ, such that

$$\frac{A_f}{A_p} = \sigma \tag{14}$$

These parameters must be determined independently. The area fraction is a function of pond shape but is typically estimated as $\sigma \approx 0.3$ on the basis of laboratory studies [32, 33]. The dilution factor estimates are based on studies of jet behavior [34] as

$$D_s \approx 0.62 \sqrt{\frac{L_j}{2b_0}} \tag{15}$$

Typical values are 1.5 to 3.

The dispersion coefficient E_L can be estimated as for the dispersive pond case, Eq. (13c), for each region. The solution is very insensitive to the choice of E_L in this case, as it is dominated by the recirculating behavior. The numerical solution is performed as for the previous case and applied repetitively to the two zones.

Deep, Stratified Pond. The final case is that of a heavily loaded [$q'' > 0.5$ MW(th)/hectare] deep (IP < 0.3) pond. For computational purposes, Ryan and Harleman [35] have suggested a three-zone model as shown in Fig. 9: (1) an entrance mixing zone, (2) a surface layer, and (3) a subsurface region.

The entrance mixing region, in which the entering flow is rapidly diluted by vertical entrainment of the surrounding water, is relatively small in relation to the pond surface.

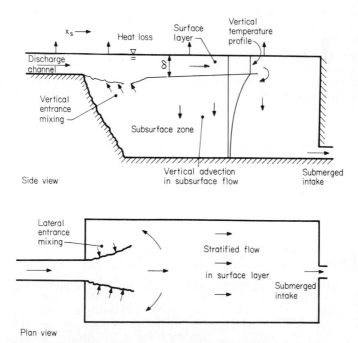

FIG. 9 Schematic view of deep stratified cooling pond [28].

The two describing variables are the surface area A_m and the mixed temperature T_{mix}, calculated as:

$$A_m = 550(1 + Fr^2)a_o b' \tag{16}$$

where Fr is a modified discharge Froude number

$$Fr = \frac{U_o}{\sqrt{\beta \, \Delta T_o \, g(a_o b')^{1/2}}} \tag{16a}$$

The mixed temperature T_{mix} is based on the dilution factor D_v, defined in Eq. (18), and the temperature at the top of the subsurface region.

The surface layer is characterized by a constant thickness δ, complete vertical mixing, and an absence of diffusion or entrainment across the interface with the subsurface region due to the presence of a strong thermocline. The flow in the surface layer is dominated by density flows, while through flow and horizontal dispersion are of minor importance. As a result, the temperature distribution is independent of pond shape and $T = T(A_x)$ where A_x is the surface area fraction.

The describing equation in the surface layer is one-dimensional and transient and accounts for the radiation heat transfer into and through the layer as:

$$\frac{\partial T}{\partial t} + \frac{1}{\delta}\frac{\partial}{\partial A_x}(D_v Q_0 T) = \frac{\partial}{\partial A_x}\left(E_L b^2 \frac{\partial T}{\partial A_x}\right) + \frac{1}{\rho c_p \delta}[q''_n - q''_{sn}(1 - \beta^*)e^{-\eta\delta}] \tag{17}$$

The boundary conditions are set at the end of the mixing region and at the end of the pond:

Upstream:
$$T\bigg|_{A_x=0} + \frac{E_L b}{\overline{U}}\frac{\partial T}{\partial A_x}\bigg|_{A_x=0} = T_{mix} \tag{17a}$$

Downstream:
$$\frac{E_L b}{\overline{U}}\frac{\partial T}{\partial A_x}\bigg|_{A_x=A_p} = 0 \tag{17b}$$

The subsurface region is characterized by a horizontally uniform temperature distribution and is treated as a natural impoundment using the relationships and methods discussed in Sec. D.3.a.

The vertical mixing in the entrance region is characterized by the dilution factor D_v as

$$D_v = 1 + 1.2(Fr - 1) \tag{18}$$

At entrance Froude numbers less than or close to unity, this correlation gives physically unrealistic results. Ryan and Harleman [35] suggest minimum dilution coefficients of

$$D_{v \, min} = 1.5 \quad \text{(for rectangular inlets)} \tag{18a}$$

and
$$D_{v \, min} = 1.2 \quad \text{(for radial inflow)} \tag{18b}$$

b. Steady-State, Simplified Models.

The models described in the previous section were transient formulations, as required to account properly for the effect of the pond's thermal inertia in responding to diurnal and seasonal changes in natural and plant-imposed thermal loads. However, for preliminary design or estimating purposes, simplified, steady-state approximations can be useful. Typically a monthly average meteorology for a summer month might be chosen to estimate the most demanding conditions for the pond.

Surface Heat Transfer. The surface heat transfer mechanism is linearized on the basis of an equilibrium temperature T_E as follows:

$$q_n'' = h(T_E - T_s) \tag{19}$$

For the three categories of pond defined in Sec. D.2, the simplified describing equations are:

Shallow-dispersive:

$$\rho c_p Q_0 \frac{dT}{dx_s} = \rho c_p b a E_L \frac{d^2 T}{dx_s^2} + h(T_E - T)b \tag{20}$$

Shallow-recirculating:

$$D_s Q_0 \rho c_p \frac{dT}{dx_s} = h(T_E - T)\sigma b \qquad \text{forward flow zone}$$

$$(D_s - 1)Q_0 \rho c_p \frac{dT}{dx_s} = h(T_E - T)(1 - \sigma)b \qquad \text{return flow zone} \tag{21}$$

Deep-stratified:

$$D_v Q_0 \rho c_p \frac{dT}{dx_s} = h(T_E - T)b \tag{22}$$

These equations are readily solved for several limiting cases:

- Single-pass: where the inlet temperature is independent of the discharge temperature
- Continuous operation: where the difference between the inlet and discharge temperature is equal to the plant condenser temperature rise

In addition, the effect of mixing can be investigated by examining the extreme cases of no mixing (plug flow) and of complete mixing. The solutions are summarized in Table 2.

c. Surface Heat Transfer

The ultimate heat rejection from a pond takes place from the pond surface to the atmosphere. The net interchange is made up of seven separate components including

- Incident solar radiation q_s
- Incident atmospheric radiation q_a
- Emitted long-wave back radiation q_{br}
- Reflected solar radiation q_{sr}
- Reflected atmospheric radiation q_{ar}
- Evaporation heat loss q_e
- Sensible heat loss to atmosphere q_c

The net heat transfer from the pond is given by

$$q_n = (q_{br} + q_e + q_c) - (q_s + q_a - q_{sr} - q_{ar}) \tag{23}$$

TABLE 2 Summary of Simplified, Steady-State Solutions for Cooling Ponds

$r' = hA_p/\rho c_p Q_0$; $E_t^* \equiv E_L/\overline{U}c$; $a' \equiv \sqrt{1 + 4r'E_t^*}$

	Operating mode	
Pond type	Single-pass	Continuous
Shallow, dispersive	$$\frac{T_{in} - T_E}{T_o - T_E} = \frac{4a' \exp(1/2E_t^*)}{(1+a')^2 \exp(a'/2E_t^*) - (1-a')^2 \exp(-a'/2E_t^*)}$$ $E_t^* = 0$ (plug): $$\frac{T_{in} - T_E}{T_o - T_E} = e^{-r'}$$ $E_t^* \to \infty$ (fully mixed): $$\frac{T_{in} - T_E}{T_o - T_E} = \frac{1}{1 + r'}$$	$$\frac{T_{in} - T_E}{\Delta T_0} = \frac{4a' \exp(1/2E_t^*)}{(1+a')^2 \exp(a'/2E_t^*) - (1-a')^2 \exp(-a'/2E_t^*) - 4a' \exp(1/2E_t^*)}$$ $E_t^* = 0$: $$\frac{T_{in} - T_E}{\Delta T_0} = \frac{e^{-r'}}{1 - e^{-r'}}$$ $E_t^* \to \infty$: $$\frac{T_{in} - T_E}{\Delta T_0} = \frac{1}{r'}$$
Shallow, recirculating	$$\frac{T_{in} - T_E}{T_o - T_E} = \frac{e^{-r'\sigma/D_s}}{D_s - (D_s - 1)\exp[-r'(1 - \sigma/D_s)/(D_s - 1)]}$$	$$\frac{T_{in} - T_E}{\Delta T_0} = \frac{e^{r'\sigma/D_s}}{D_s - e^{r'\sigma/D_s} - (D_s - 1)\exp[-r'(1 - \sigma/D_s)/(D_s - 1)]}$$
Deep, stratified	$$\frac{T_{in} - T_E}{T_o - T_E} = \frac{e^{-r'/D_v}}{D_v - (D_v - 1)e^{-r'/D_v}}$$ $D_v = 0$ (plug): $$\frac{T_{in} - T_E}{T_o - T_E} = e^{-r'}$$ $D_v \to \infty$ (fully mixed): $$\frac{T_{in} - T_E}{T_o - T_E} = \frac{1}{1 + r'}$$	$$\frac{T_{in} - T_E}{T_o - T_E} = \frac{1}{D_v} \frac{e^{-r'/D_v}}{1 - e^{-r'/D_v}}$$ $D_v = 0$: $$\frac{T_{in} - T_E}{\Delta T_0} = \frac{e^{-r'}}{1 - e^{-r'}}$$ $D_v \to \infty$: $$\frac{T_{in} - T_E}{\Delta T_0} = \frac{1}{r'}$$

and q_n must balance the thermal loading on the pond. All of these terms, particularly the natural radiation components, are functions of location, season, and local meteorological variables such as air temperature and humidity, wind speed, and cloud cover. Empirical correlations are available in several references [36–39].

A linearized simplification suggested by Edinger and Geyer [36] is based on the concept of an equilibrium temperature T_E which results from natural environmental conditioning when there is no plant-imposed loading and takes the form

$$q_n = hA_s(T_E - T_s)$$

where h is a function primarily of wind speed as well as air and surface temperature. This approximation is the basis for the simplified solutions in Table 2. Correlations for the determination of h are given by Brady [37].

$$h = 15.7 + (\gamma + 0.26)\phi \qquad \text{(English units)} \qquad (24)$$

where

$$\gamma = 0.255 - 0.0085 T_{\text{avg}} + 0.00204 T_{\text{avg}}^2 \qquad (24a)$$

$$T_{\text{avg}} = \frac{T_s + T_d}{2} \qquad (24b)$$

$$\phi = 70 + 0.7V_w^2 \qquad (24c)$$

and

$$T_E = T_d + \frac{q_s''}{h} \qquad (25)$$

Charts to facilitate the computation are given in Fig. 10.

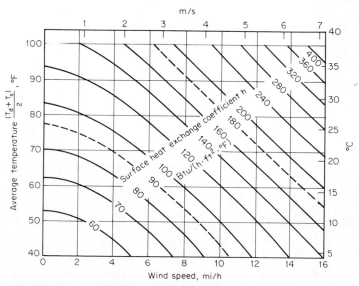

FIG. 10 Design surface heat exchange coefficient for cooling ponds [37].

E. SPRAYS

1. General Description

Spray systems are essentially enhanced cooling ponds in which the thermal performance is augmented, at the expense of increased pumping power requirements, by spraying the water into the air. The resultant formation of droplets vastly increases the air-water interfacial area available for heat and mass transfer. In this sense, the principle of operation is similar to that of a cooling tower, except that no fill or packing is present. In addition, no independent provision of or control over airflow is possible, and the proper distribution of fresh dry air throughout the spray zone for cooling is the major uncertainty associated with the design and performance of spray systems.

In comparison to ponds, the heat rejection, per unit of *pond* surface area, is significantly increased and less land is required. In comparison to towers, *more* land is required and performance is less predictable and lower both per unit volume of spray zone and per unit of pumping power. Control of drift is more difficult in the unenclosed configuration.

Spray systems are the least common of the major classes of cooling systems for main plant cooling at power plants. As of mid-1981, only three plants representing 2000 MW(e) of generating capacity (less than 1 percent of the total U.S. capacity) used sprays.

Additional use of sprays has been made for mixed or "trimming" systems where the discharge temperature of a once-through system is reduced slightly just prior to return to the receiving water for those periods of the year when a discharge limit might be exceeded. Sprays have also been chosen for ultimate heat sink application at nuclear plants on the basis that they provide more positive controllable cooling than ponds and more thermal inertia than towers.

2. Classification of Sprays

Spray systems can be classified both on the basis of equipment type—floating-module versus fixed-piping—and on the basis of system configuration—pond versus canal. The larger systems required for main plant cooling are normally of the floating-module type arranged in a cooling canal which connects the plant discharge and intake as illustrated in Fig. 11. The module itself consists of a flotation platform supporting a pump and motor, a submerged intake, manifold piping, and spray nozzles. Each module in a large spray field is individually supplied with electric power and, in principle, may be individually turned on or off to match the cooling requirements with minimum power consumption. A typical module capacity is 10,000 gal/min (0.7 m^3/s).

The smaller systems, as for nuclear ultimate heat sinks, are typically fixed piping systems arranged in ponds. A typical installation is that at the Rancho Seco Nuclear Plant and is described in detail in Refs. 41 and 42.

3. Spray System Performance

a. General Approach

The performance of a spray system is determined both by the performance of the individual modules and by the interaction among them when they are grouped together to

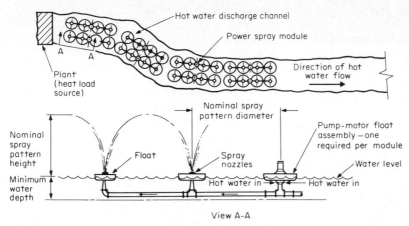

FIG. 11 Typical power spray canal system with power spray module details [40].

form a large spray field. Individual module performance is set by the local values of water temperature, atmospheric wet-bulb temperature, and wind speed. The effects of interacting sprays are attributable primarily to modification of airflow patterns and wet-bulb temperature distributions by neighboring modules.

The usual approach to computing spray field performance is based on an arrangement of modules illustrated schematically in Fig. 12. The water flow in the canal is divided into channels each corresponding to a single row of modules. At each pass some fraction of the channel flow is sprayed into the air and cooled according to the performance of an individual module and local temperature. The cooled flow is returned to the channel and the mixed channel flow temperature is determined by the ratio of the channel flow to the pumped spray flow. The row-to-row variation in wet-bulb temperature is accounted for by empirical correction factors. The channel flows are assumed to be fully mixed at the end of each pass. This assumption is consistent with field observations that the canal temperature is essentially uniform throughout.

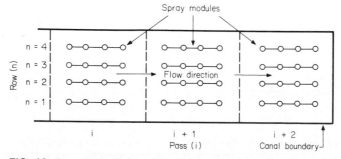

FIG. 12 Control volume for sizing spray canal systems.

b. Module Performance

A thorough treatment of single-module performance is given by Porter et al. [43]. In an analysis identical to that for cooling tower performance, a Merkel-type equation is derived:

$$c_{p,w} \int_{T_{sp}}^{T} \frac{dT'}{Q_c(T') - Q_c(T_{wb})} = \text{NTU} \tag{26}$$

where

$$\text{NTU} = \frac{\xi A_t}{\dot{m}_s} \tag{26a}$$

$$\xi = \frac{Q_D}{1 + \omega} \tag{26b}$$

and

$$Q_c(T_{as}) = i_a(T_{db}, \omega) - \omega i_f(T_{as}) \tag{26c}$$

An energy balance on the air and water streams would yield, as for cooling towers,

$$\frac{dQ_c(T_{wb})}{dT'} = -c_{p,w} \frac{L}{G} \tag{27}$$

but for sprays, G is not determinable, and Eq. (26) must be integrated using an approximation of an average local wet-bulb temperature. A convenient assumption is

$$\frac{dQ_c(T)}{dT'} \equiv b_f = \text{constant at } T_f \tag{28}$$

where

$$T_f \equiv \frac{T_{wb,amb} + T_{hot}}{2}$$

On this basis, the cooling range for a single module is given by

$$\frac{T - T_s}{T - T_{wb}} = 1 - \exp\left(-\text{NTU} \frac{b_f}{c_{p,w}}\right) \tag{29}$$

Values of b_f are given in Table 3. The value of NTU can be estimated from single-module data or back-calculated from field data.

TABLE 3 Thermodynamic Data for Eq. (28)

Value of $\dfrac{b_f}{c_{p,w}} = \dfrac{1}{c_{p,w}} \dfrac{dQ_c(T)}{dT}$

T, °F	T, °C	$b_f / c_{p,w}$
50	10	0.545
59	15	0.650
68	20	0.785
77	25	0.960
86	30	1.185
95	35	1.470
104	40	1.845

c. Spray Field Performance

Using the modular arrangement and nomenclature of Fig. 12, consider the canal temperature change across the module in the nth row of the ith pass.

The inlet temperature to all rows in the ith pass is given by T_i, since the canal is well mixed between passes. Therefore, from Eq. (29), the cooled spray temperature ($T_{s_{i,n}}$) is given by:

$$\frac{T_i - T_{s_{i,n}}}{T_i - T_{wb_n}} = 1 - \exp\left(-\text{NTU}\,\frac{b_f}{c_{p,w}}\right) \tag{30}$$

where

$$T_{wb_n} = T_{wb,\infty} + f_n(T_i - T_{wb,\infty}) \tag{30a}$$

and f_n is based on the hottest local water temperature and the ambient wet bulb.

The mixed channel temperature in the nth row is

$$T_{i+1,n} = T_i - \frac{r_0}{m}(T_i - T_{s_{i,n}}) \tag{31}$$

Averaging over the n rows yields

$$T_{i+1} - T_i = \frac{r_0}{m}\sum_{n=1}^{m}(T_i - T_{s_{i,n}}) \tag{32}$$

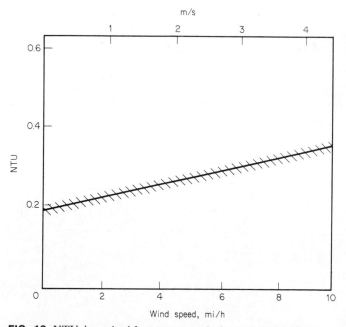

FIG. 13 NTU determined from tests on a single spray module [46].

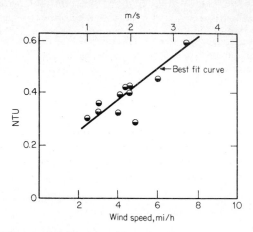

FIG. 14 NTU determined from system performance tests [46].

Proceeding along the entire canal,

$$\frac{T_c - T_{wb,\infty}}{T_h - T_{wb,\infty}} = \exp\left\{-Nr_0(1 - \bar{f})\left[1 - \exp\left(-\text{NTU}\,\frac{b_f}{c_{p,w}}\right)\right]\right\} \qquad (33)$$

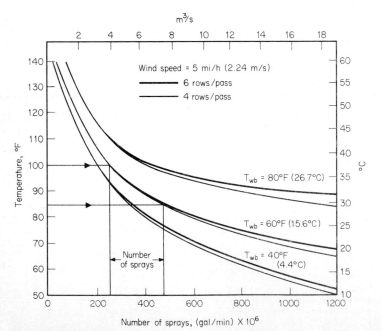

FIG. 15 Design curves for sizing spray canal systems [46].

The NTU approximations are available from data. Each data set is tied to a particular wet-bulb correction factor. Hoffman [45] reports NTU values determined from tests on a single module shown in Fig. 13. A wet-bulb correction factor of 1°F (0.56°C) for each downwind row is recommended.

Alternatively, Porter and Chen [46] give NTU values based on large field performance tests as shown in Fig. 14 coupled with an interference factor correlation of

$$\bar{f} = \frac{1}{n}[0.18 + 0.26(n-2)] \qquad (34)$$

An alternative graphical method has been developed and reported in Ref. 45 in which manufacturer-supplied curves are used to determine the total number of modules required for a given wind-speed range of wet-bulb temperatures and module arrangements. Figure 15 illustrates the use of the curves.

d. Design Calculations

Ryan and Myers [46] performed a series of design calculations for a single case using the several recommended approaches. The results are summarized in Table 4.

It can be seen that considerable uncertainty exists with the calculated number of modules required differing by a factor of nearly 2.5. The experience with full scale spray installations suggests that design calculations for large spray fields are difficult to make and that extreme caution should be exercised in designing for any specific use. Adequate space should be provided to expand the field if it should be necessary.

Water consumption in spray systems is given by using Bowen's ratio of sensible to evaporative heat transfer as demonstrated in Ref. [48]. The fraction of the sprayed flow evaporated is

$$\alpha = \frac{c_{p,w}}{i_{lg}} \frac{T_a - T_{sp}}{1 + B} \qquad (35)$$

TABLE 4 Comparison of Spray Canal Design Methods

Method	Performance correlation		Number of modules required
	NTU	Wet-bulb correction	
Tennessee Valley Authority [46]	(From Fig. 15)		104
Hoffman [45]	NTU = 0.255 (Fig. 13)	1°F per row	112
Porter [47]	NTU = 0.42 (Fig. 14), "best fit"	$\bar{f}$ = 0.175 (f_n = 0.18, 0.26, 0.26)	84
Porter [47]	NTU = 0.29 (Fig. 14), "conservative"	$\bar{f}$ = 0.175 (f_n = 0.18, 0.26, 0.26)	112

Design conditions: Flow = 500,000 gal/min (32 m³/s); T_h = 100°F (38°C); T_c = 85°F (29°C); wind speed = 5 mi/h (2.2 m/s); $T_{wb,\infty}$ = 60°F (16°C).

where $\overline{B} \equiv$ Bowen's ratio:

$$\overline{B} = \frac{P'c_{p,a}}{i_{lg}} \frac{\overline{T} - T_{db}}{[\omega_s(T) - \omega]} \tag{35a}$$

The maximum consumption is estimated by setting $\overline{B} = 0$ and corresponds to 1 to 2 percent.

Secondary variables affecting design calculations include canal bank height and wind angle to the canal. Intuitively, improved performance is expected with minimum bank height and wind normal to the canal, but data are inadequate to support reliable corrections to the basic correlations. The effects of the several thermodynamic simplifications leading to Eqs. (26) and (33) are well within the uncertainty of the data for design purposes. A discussion of these approximations is given by Porter et al. [43].

F. PERFORMANCE TESTING OF COOLING SYSTEMS

1. Cooling Towers

The thermal and hydraulic performance of cooling towers is tested mainly with conventional industrial-type instrumenation [49, 50]. Typical measurements performed on a mechanical-draft wet cooling tower [51] and representative instrumentation are given in Table 5.

The most difficult measurements are wet-bulb temperature and water flow rate. Wet-bulb temperature measurements are affected by various factors, such as solar radiation, improper wetting of the wick surrounding the sensor, and airflow maldistribution to the wick. As a further complication, small errors in the measurement of wet-bulb temperature are magnified as large uncertainties in the thermal performance. Water flow rate measurements with Pitot tubes in large pipes are complicated by flow maldistributions due to elbows, uncertainties in sensor calibration, sensor vibration and misalignment, etc. [52]. Alternate measurement techniques include dye-tracer and ultrasonic methods.

TABLE 5 Measurements and Instrumentation for a Performance Test on an Evaporative Mechanical Draft Cooling Tower

Measurement	Instrument
Ambient wet-bulb temperature	Psychrometer with platinum RTD (resistance temperature device)
Inlet wet-bulb temperature	Psychrometer with platinum RTD
Hot water temperature	Platinum RTD
Cold water temperature	Platinum RTD
Electrical power to the fans	Electric power meter
Water flow rate	Pitot tubes
Wind speed and wind direction	Vane and cup anemometer
Exhaust air dry-bulb temperature	Platinum RTD or mercury-in-glass thermometer

Instrumentation requirements for performance measurements on dry and wet-dry towers are similar. Figure 16 illustrates the arrangement of instruments in a wet-dry tower test [17].

There are a number of factors which can lead to performance deficiencies in cooling towers and complicate the measurement of performance. These include recirculation of the heated effluent and its ingestion by the cooling air inlet, interference between towers in close proximity, maldistribution of air and water flows within the tower, effects of ambient wind on tower performance, and fouling of heat exchanger surfaces.

Recirculation and interference can substantially reduce the heat rejection capability of cooling towers, typically by 10 percent, with severe resultant economic penalties. It has been shown [51] that significant recirculation of the cooling tower plume can occur at ambient wind speeds that are less than those specified in the current test codes [49, 50]. This is because the testing techniques that are used at present do not satisfactorily account

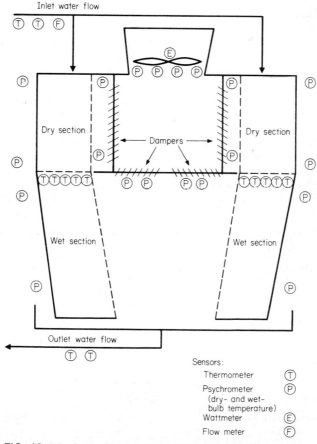

FIG. 16 Wet-dry cooling tower instrumentation arrangement.

for increases in the wet-bulb temperature that are caused by recirculation on the leeward face of the tower. Use of additional wet-bulb sensors is required to obtain an accurate measure of performance [51].

Use of mathematical models for prediction of recirculation effects is limited at this time because of the complexity and cost of solving the governing equations. Physical modeling of recirculation effects in the laboratory in water flumes [53] holds promise because of relatively low costs.

Maldistribution of fluid flows within the cooling tower can also lead to significant performance losses. Ambient wind effects, both steady and unsteady, in general contribute to the problem of maldistribution. In wet towers, a common problem is nonuniform wetting of the fill as the result of nonuniform airflow distribution through the fill, end effects, or improper distribution of hot water to the fill section. In dry towers, airflow nonuniformities result from wind effects [54] and heat exchangers that are not aerodynamically designed [55].

Serious performance degradations can result from fouling or corrosion of heat exchanger surfaces [56]. This problem is more critical in dry towers, because the controlling thermal resistance is on the air side. In one test [17], air-side fouling of a heat exchanger with airborne dust decreased the dry tower performance capability by about a factor of 2 prior to cleaning of the surfaces.

2. Cooling Ponds and Spray Systems

The thermal performance of a cooling pond or spray system is measured by how much the temperature of the hot water is reduced before it reenters the cooling system. This cooling is strongly dependent on the amount of water evaporated by the pond, because the evaporation component is normally the largest term in the governing conservation equations. The spray system merely augments the capability of the cooling pond in evaporating water.

There are several techniques to determine evaporative loss [57]:

- Energy balance
- Water balance
- Vapor budget
- Pan measurement
- Eddy correlation
- Bulk aerodynamic method

In the energy balance method, all energy fluxes entering and leaving the pond, with the exception of evaporation and conduction, are measured or calculated over a period of time. From the energy conservation equation, the evaporative energy loss is given by

$$q_E = \frac{q_A + q_R - \Delta q_s}{(1 + \overline{B})} \tag{36}$$

In the water balance approach, all mass flows of water entering and leaving the pond, with the exception of evaporation, are measured or calculated over a period of time. From the mass conservation equation, the evaporative mass loss is given by

$$W_e = W_i + W_p - W_o - W_r - W_g - \rho \, \Delta V' \tag{37}$$

In the vapor budget technique [58], measurements of water vapor flux are performed at the leeward and windward ends of a pond, and the evaporation rate is given by

$$W_e = W_l - W_w \tag{38}$$

Another technique [59] uses evaporation pans placed in or near the pond. The measured evaporation rate from the pan is used to infer the pond evaporation rate by means of an empirical correlation:

$$W_e = C_0 W_{pm} \tag{39}$$

Evaporation rate can be determined locally from the eddy correlation method by means of

$$W_e = \rho_a A_s \overline{\theta' \psi'} \tag{40}$$

where θ' and ψ' are measured directly [58].

In summary, the energy budget, water balance, and vapor budget methods apply to the whole water body, while the pan measurement, eddy correlation, and bulk aerodynamic techniques are point measurements. The energy budget has the potential for accuracy, and allows the treatment of nonuniform surface temperatures with a suitable mathematical model. The water balance technique requires estimation of groundwater losses and runoff, which may compromise accuracy. In addition, nonuniform surface temperatures cannot be treated with a model. The vapor budget method provides a direct measurement of fluxes, but may be inaccurate if the formation of vapor is not steady. The pan measurement is simple, but proper design of a pan experiment is not. Furthermore, extrapolation of the point measurement to the entire pond is not straightforward. The eddy correlation technique provides a direct measurement of fluxes, but instrumentation is expensive. Finally, the bulk aerodynamic technique is potentially accurate, but data interpretation in an unstable atmosphere is difficult.

G. COOLING SYSTEM SELECTION AND OPTIMIZATION

In addition to the design and optimization of each cooling system component, the system itself must be optimized. This is complex because, for power plant applications, there are major costs over and above the capital and installation costs of the system itself which dominate the optimization. These stem from the coupling between cooling system performance and plant heat rate and capacity. They include:

• Capacity reduction at high ambient temperatures (capital and operating costs of replacement capacity)
• Increased heat rate and energy costs
• Cooling system auxiliary power requirements (fan and pumping power)
• Cooling system maintenance costs

In general, the optimum is set by a trade-off between cooling system capital costs and plant capacity replacement costs.

1. General Approach

A complete analysis requires:

- Specification of design conditions
- Determination of a base case against which penalties are evaluated
- Determination of how penalty costs are computed
- Hour-by-hour cost analysis of plant and system operation for a design year
- Selection of "optimum" system size for each type of cooling system
- Comparison among designs

a. Design Conditions

The necessary design parameters fall into two categories: environmental parameters and plant-related variables. The environmental parameters include dry- and wet-bulb temperature, wind velocity and direction, and height above sea level. In order to assess the performance penalties correctly over the course of a year, it is necessary to know the variation in atmospheric conditions over time. A typical representation is shown in Fig. 17.

b. Base-Case Determination

The penalty costs are determined both by how the capacity and energy losses are evaluated and by how their replacement is provided. Computationally, the output of the plant at any hour is calculated as a function of ambient conditions and cooling system and plant performance and then compared to the output of a reference base plant. Since the design optimization is a strong function of the penalty terms, the selection of the reference plant

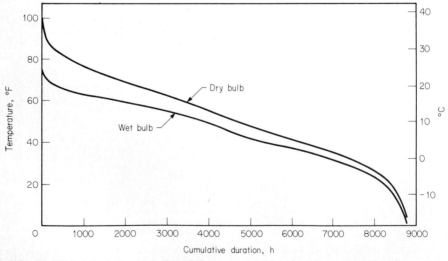

FIG. 17 Typical dry bulb–wet bulb variations.

can influence the result. In a survey of selection and optimization methods, Fryer [60] identifies three methods for defining the baseline:

- Fixed demand with fixed heat source
- Fixed demand with scalable steam supply and scalable plant
- Negotiable demand with fixed heat source

The fixed-demand analyses, illustrated in Fig. 18, impose a fixed power output requirement on the reference plant against which the design plant output is measured. For the fixed heat source analysis the reference plant is taken as having a nameplate rating equal to the fixed demand at a specified turbine back pressure; the design plant is assumed to have the identical plant design, heat rate, and output at the same back pressure. The output throughout the year is then computed as a function of ambient conditions for a range of cooling system designs. Figure 19a, b, and c illustrates the several aspects of the analysis. Figure 19a shows the output shortfall or excess during the year. The deficiency is divided into two portions: that due to back pressure differences imposed by the cooling system and that due to cooling system auxiliary power requirements (pumps and fans). Figure 19b shows the variation in the magnitude of the penalty with cooling system size; Fig. 19c illustrates the existence of a minimum cost and hence an optimum design.

Analyses based on other reference plant selections are conceptually the same. In the fixed demand–scalable supply approach, the capacity deficit is made up in whole or in part by assuming a plant larger than the reference plant. The incremental cost of the enlarged plant is then added to the cost of the cooling system as an equivalent capital cost. Similarly, the reference plant can be derated to correspond to the output of the design plant at the hottest hour. Conceptually this is equivalent to scaling up the design plant. In both cases, a large amount of "excess" capacity is available throughout the year and the cost of the system is strongly dependent on how credit is given for this excess capacity.

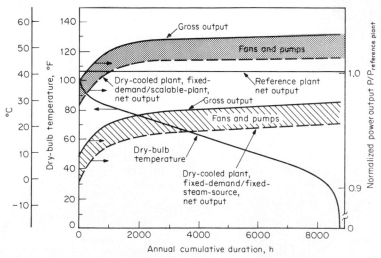

FIG. 18 Relative performance of a dry-cooled plant utilizing a high–back pressure turbine under the fixed demand–scalable steam source–scalable plant approach [1].

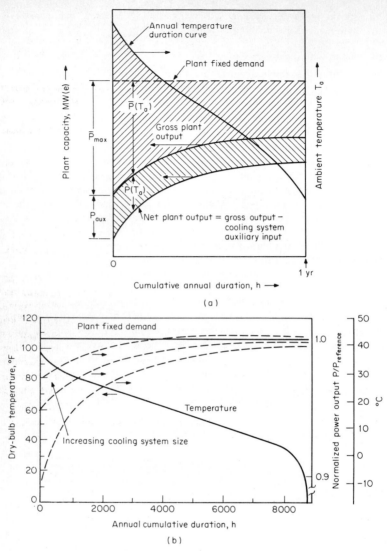

FIG. 19 Cooling system optimization methodology. (*a*) Ambient temperature duration and corresponding plant performance for fixed demand–fixed heat source approach. (*b*) Relative performance of different-size cooling systems. (*c*) Schematic diagram of economic trade-offs and optimum selection of cooling systems. From Ref. 1.

c. Cost Assessment

In order to develop a systematic comparison of competing systems, all the components of the cost must be put on a common basis. A procedure for doing this is discussed in detail by Senges et al. [1].

The penalty and operating costs are evaluated on an annual basis, capitalized over the

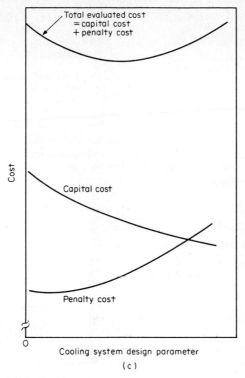

Cooling system design parameter

(c)

FIG. 19 (*Continued*)

plant lifetime through the use of an annual fixed charge rate, and added to the cooling system capital cost to arrive at a "total evaluated cost":

$$C_t = C + \frac{1}{X_{af}} \sum_{j=1}^{N_p} P_J \qquad (41)$$

The four cost components are given by

Capacity penalty (P_1):

$$P_1 = X_{af} K_c \overline{P}_{\max} \qquad (41a)$$

Replacement energy penalty (P_2):

$$P_2 = p^* \int_{t=0}^{8760 \text{ h}} [P_o + F'Q^*(T_a)]\overline{P}(T_a)\, dt \qquad (41b)$$

Cooling system auxiliary power (P_3):

$$P_3 = X_{af} K_c P_{\text{aux}} \qquad (41c)$$

Cooling system auxiliary energy (P_4):

$$P_4 = p^* \int_{t=0}^{8760 \text{ h}} [P_0 + F'Q^*(T_a)]P(T_a)\, dt \qquad (41d)$$

where $\overline{P}_{\max}$, $\overline{P}(T_a)$, P_{aux}, and $P(T_a)$ are shown in Fig. 19a.

Trade-offs are available depending upon how the capacity and energy deficits are made up. These choices, which may include enlarged base units (high capital cost, low operating cost), peaking units such as gas turbines (low capital cost, high energy cost), purchased power from a power pool, or some combination of the three, will affect the magnitude of K_c, F', and P_o in Eq. (41).

2. Example Cost Comparisons

Several studies [61–64] have performed economic comparisons based on the foregoing analyses. The results from Ref. 61, using a fixed demand with fixed supply and comparing seven alternative cooling systems, are given in Tables 6 and 7.
The cooling systems are

1. Mechanical-draft wet-tower cooling system
2. Fan-assisted natural-draft wet-tower cooling system
3. Natural-draft wet-tower cooling system
4. Constructed-pond cooling system
5. Power spray module canal cooling system
6. Mechanical-draft dry-tower cooling system
7. Natural-draft dry-tower cooling system

TABLE 6   Economic Factors

Cost year	1978
Plant capacity factor	75%
Annual fixed charge rate	18%
Plant life	40 years
Capacity penalty charge rate (for capacity loss at the peak ambient temperature and auxiliary power)	$563/kW (nuclear)
	$450/kW (fossil)
Energy cost	10 mills/kWh (nuclear)
	15 mills/kWh (fossil)
Escalation rate for material and labor costs	7% per year
Cooling system maintenance charge	0.5% of direct capital cost

TABLE 7   Costs of Typical Conventional Cooling Systems for Fossil Power Plants (1978 Dollars)

Cooling system	Capital cost, $/kW	Penalty cost, $/kW	Total evaluated cost	
			$/kW	Mills/kWh
Once-through	15.16	6.36	21.52	0.60
Mechanical-draft wet	21.57	27.72	49.29	1.35
Natural-draft wet	26.96	21.02	47.98	1.31
Fan-assisted wet	27.77	22.73	50.50	1.38
Pond	38.50	32.74	71.24	1.95
Spray canal	23.99	25.63	49.62	1.36
Mechanical-draft dry	34.29	125.04	159.33	4.37
Natural-draft dry	37.87	115.98	153.85	4.22

In addition to the cooling unit itself the system includes the steam condenser and the circulating water system. Engineering, purchase, and installation costs are included; a uniform 25 percent contingency is imposed.

The economic factors corresponding to the quantities identified in Eq. (41) are tabulated in Table 6.

The environmental and reference plant information is as follows:

Environmental specifications, based on meteorological conditions for Boston, Massachusetts:

Design dry-bulb temperature: 93°F   (34°C)

Design wet-bulb temperature: 74°F   (23°C)

Wind speed: 5 mi/h   (2.2 m/s)

Base plant:

Nominal: 1000-MW(e) Fossil plant

Gross output: 1043 MW(e)

Turbine heat rate: 7365 Btu/kWh (7.8 $\times$ 10^6 J/kWh)

Exhaust pressure: 1.5 in Hg abs (5.0 $\times$ 10^3 Pa)

These costs are summarized in Table 7.

REFERENCES

1. D. C. Senges et al., "Closed-Cycle Cooling Systems for Steam-Electric Power Plants: A State-of-the-Art Manual," *U.S. Environmental Protection Agency Rep. No. EPA-600/7-79-001,* January 1979.
2. A. K. Majumdar et al., "Numerical Modeling of Wet Cooling Towers," *J. Heat Transfer,* Vol. 105, p. 728, 1983.
3. Cooling Tower Institute, *Cooling Tower Performance Curves,* Millican Press, Houston, 1967.
4. N. W. Kelly, *Kelly's Handbook of Crossflow Cooling Tower Performance,* Neil W. Kelly and Associates, Kansas City, Mo., 1976.
5. *Counterflow Cooling Tower Performance,* J. F. Pritchard and Co. of California, 1957.
6. *Design Handbook of Fill Characteristics,* The Munters Corporation, Fort Myers, Fla., September 1973.
7. H. J. Lowe and D. G. Christie, "Heat Transfer and Pressure Drop Data on Cooling Tower Packings and Model Studies of the Resistance of Natural-Draft Towers to Airflow," *Proc. Int. Heat Transfer Conf.* Pt. V, pp. 933–950, ASME, New York, 1962.
8. M. M. Mesarovic, "Ararat-A Computer Code for Thermal Design of Cooling Towers," *Nucl. Eng. Des.,* Vol. 24, pp. 57–70, 1973.
9. K. E. Cross et al., "Theory and Application of Engineering Models for Crossflow and Counterflow Induced Draft Cooling Towers," *Union Carbide Rep. K/CSD-1,* May 1976.
10. L. D. Winniarski and B. A. Tichenor, "Model of Natural Draft Cooling Tower Performance," Journal of the Sanitary Engineering Division, *Proc. Am. Soc. Civil Eng.,* August 1970.
11. S. M. Zivi and B. B. Brand, "An Analysis of the Cross-Flow Cooling Tower," *Refrigeration Eng.,* Vol. 64, pp. 31–34, 1956.
12. D. R. Baker and H. A. Shryock, "A Comprehensive Approach to the Analysis of Cooling Tower Performance," *J. Heat Transfer,* Vol. 83, pp. 339–349, 1961.
13. T. R. Penney and D. B. Spalding, "Validation of Cooling Tower Analyzer (VERA)," *EPRI FP-1279,* December 1979.

14. A. K. Majumdar and A. K. Singhal, "VERA2D: Program for 2-D Analysis of Flow, Heat, and Mass Transfer in Evaporative Cooling Towers," *EPRI CS-2923,* March 1983.

15. C. Bourillot, "TEFERI, Numerical Model for the Calculation of Performance of an Evaporative Cooling Tower." *EPRI CS-3212-5R,* August 1983.

16. D. Benton, "A Numerical Simulation of Heat Transfer in Evaporative Cooling Tower," *TVA Rep. WR28-1-900-110,* September 1983.

17. D. M. Burkhart, and J. A. Bartz, "Wet/Dry Cooling Tower Field Tests," *Proc. Am. Power Conf.,* Vol. 42, 1980.

18. D. M. Burkhart, "Test Report: Wet/Dry Cooling Tower Test Module," *EPRI CS-1565,* October 1980.

19. R. D. Woodson, "Cooling Towers for Large Steam-Electric Generating Units," in *Electric Power and Thermal Discharges,* pp. 351–378, Gordon and Breach, New York, 1969.

20. R. T. Allemann et al., "Development of an Advanced Concept of Dry/Wet Cooling of Power Generating Plants," *EPRI CS-1668,* February 1981.

21. D. K. Kreid et al., "Approximate Analysis of Evaporative Heat Transfer from a Finned Heat Exchanger," *ASME Paper No. 78-HT-26,* Battelle Pacific Northwest Laboratories, 1978.

22. C. T. Zeien and R. J. York, "Engineering Evaluation of Magma Cooling-Tower Demonstration at Nevada Power Company's Sunrise Station," *EPRI CS-1626,* November 1980.

23. H. P. Fay et al., "Improved Wet/Dry Heat Rejection by Means of Direct Condensation in Combination with Evaporative Cooling," *Proc. IAHR/EPRI Cooling Tower Workshop,* San Francisco, September 1980.

24. B. M. Johnson et al., "Development of an Advanced Concept of Dry/Wet Cooling for Power Plants," *Proc. Am. Power Conf.,* Vol. 43, 1981.

25. J. Fleury, "Ammonia Bottoming Cycle Development at Electricité de France for Nuclear Power Plants," *1981 Int. ASME Gas Turbine Conf.,* Houston, March 1981.

26. R. J. Naegelen et al., "Detailed Design for Incorporating CBI Capacitive Cooling System in the ACT Facility in Bakersfield, California," *EPRI CS-2463,* July 1982.

27. J. C. Sonnichsen et al., "Cooling Ponds—A Survey of the State-of-the-Art," Hanford Engineering Development Laboratory, Richland, Wash., *HED-TME 72-101,* 1972.

28. G. H. Jirka et al., "Mathematical Predictive Models for Cooling Ponds and Lakes, Part A: Model Development and Design Considerations," *R. M. Parsons Lab. Rep. No. 238,* Massachusetts Institute of Technology, Cambridge, Mass., December 1978.

29. K. H. Octavio et al., "Vertical Heat Transport Mechanisms in Lakes and Reservoirs," *R. M. Parsons Lab. Rep. No. 227,* Massachusetts Institute of Technology, Dept. of Civil Engineering, Cambridge, Mass., August 1977.

30. P. J. Ryan and D. R. F. Harleman, "Prediction of the Annual Cycle of Temperature Changes in a Stratified Lake or Reservoir: Mathematical Model and User's Manual," *R. M. Parsons Lab. Rep. No. 137,* Massachusetts Institute of Technology, Dept. of Civil Engineering, Cambridge, Mass., April 1971.

31. H. B. Fisher, "The Mechanics of Dispersion in Natural Streams," *Proc. ASCE, J. Hydraulic Div.,* Vol. 93, No. HY6, November 1967.

32. C. Iamandi, and H. Rouse, "Jet-Induced Circulation and Diffusion," *Proc. ASCE, J. Hydraulic Div.,* No. HY2, March 1969.

33. G. N. Abramovich, *The Theory of Turbulent Jets,* MIT Press, Cambridge, Mass., 1963.

34. M. L. Albertson, et al., "Diffusion of Submerged Jets," *Trans. ASCE,* Vol. 115, 1950.

35. P. T. Ryan and D. R. F. Harleman, "An Analytical and Experimental Study of Transient Cooling Pond Behavior," *R. M. Parsons Lab. Rep. No. 161,* Massachusetts Institute of Technology, Dept. of Civil Engineering, Cambridge, Mass., January 1973.

36. J. E. Edinger and J. C. Geyer, "Heat Exchange in the Environment," *EEI Publ. No. 65-902,* Edison Electric Institute, New York, 3d printing, 1971.

37. D. K. Brady et al., "Surface Heat Exchange at Power Plant Cooling Lakes," *EEI Publ. No. 69-901,* Edison Electric Institute, New York, 1969.

38. W. T. Hogan et al., "An Engineering-Economic Study of Cooling Pond Performance," U.S. Environmental Protection Agency, *1613130 DFX 05/70,* 1970.

39. E. L. Thackston and F. L. Parker, "Effect of Geographical Location on Cooling Pond Requirements and Performance," U.S. Environmental Protection Agency, *16130 FDQ 03/71,* 1971.

40. P. A. Frohwerk, "Power Spray Module: A New Concept in Evaporative Water Cooling Systems," Ceramic Cooling Tower Company, Fort Worth, Texas. Presented at Spring Conference, Southeastern Electric Exchange, 1971.

41. W. E. Hebden and A. M. Shah, "Effects of Nozzle Performance on Spray Ponds," *Proc. Am. Power Conf.,* Vol. 38, pp. 1449–1457, 1976.

42. V. E. Schrock et al., "Performance of a Spray Pond for Nuclear Power Plant Ultimate Heat Sink," *ASME Paper No. 75-WA/HT-41,* New York, 1975.

43. R. W. Porter et al., "Thermal Performance of Spray-Cooling System," *Proc. Am. Power Conf.,* Vol. 38, pp. 1458–1472, 1976.

44. C. H. Berry, "Mixtures of Gases and Vapors," in *Mechanical Engineers' Handbook,* T. Baumeister, ed., 6th ed., p. 4-82, McGraw-Hill, New York, 1958.

45. D. P. Hoffman, "Spray Cooling for Power Plants," *Proc. Am. Power Conf.,* Vol. 35, pp. 702–712, 1973.

46. P. J. Ryan and D. M. Myers, "Spray Cooling: A Review of Thermal Performance Models," *Proc. Am. Power Conf.,* Vol. 38, pp. 1473–1481, 1976.

47. R. W. Porter and K. H. Chen, "Heat and Mass Transfer of Spray Canals," *J. Heat Transfer,* Vol. 96, pp. 286–291, August 1974.

48. R. W. Porter, "Analytical Solution for Canal Heat and Mass Transfer," *ASME Paper No. 74-HF-58,* New York, 1974.

49. American Society of Mechanical Engineers, "Atmospheric Water Cooling Equipment," *Power Test Code No. 23,* 1958.

50. The Cooling Tower Institute, "Acceptance Test Code for Water Cooling Towers," *ATC-105,* February 1975.

51. K. R. Wilber, "Examination of Specific Aspects of Cooling Tower Testing Methodology," *EPRI FP-953,* December 1978.

52. R. E. Moore et al., "Investigation of Pitot Tube Water Flow Rate Measurements in Large Pipes," Environmental Systems Corporation Report (unnumbered), February 1980.

53. S. C. Jain and J. F. Kennedy, "Development and Verification for Prediction of Near-Field Behavior of Cooling-Tower Plumes," *EPRI CS-1370,* March 1980.

54. G. Dibelius and A. Ederhof, "The Influence of Wind on the Updraft Flow of Natural Draft Cooling Towers," *Proc. IAHR/EPRI Cooling Tower Workshop,* September 1980.

55. F. K. Moore, "Aerodynamic Design Problems of Dry Cooling Systems," *Proc. IAHR/EPRI Cooling Tower Workshop,* September 1980.

56. J. G. De Steese and K. Simhan, "European Dry Cooling Tower Operating Experience," *Battelle Pacific Northwest Labs. Rep. No. BNWL-1995,* March 1976.

57. K. R. Helfrich et al., "Evaluation of Models for Predicting Evaporative Water Loss in Cooling Impoundments," *EPRI CS-2325,* March 1982.

58. R. E. West, "Field Investigation of Cooling Tower and Cooling Pond Plumes," *U.S. Environmental Protection Agency Rep. No. EPA-600-17-78-059,* 1978.

59. J. B. Nystron and G. E. Hecker, "Experimental Evaluation of Water to Atmosphere Heat Transfer Equations," paper presented at ASCE Annual Convention, Denver, 1975.

60. B. C. Fryer, "A Review and Assessment of Engineering Economic Studies of Dry Cooling Electric Generating Plants," *Battelle Pacific Northwest Labs., Rep. No. BNWL-1976,* Richland, Wash., 1976.

61. United Engineers & Constructors, Inc., "Heat Sink Design and Cost Study for Fossil and Nuclear Power Plants," *UE&C-AEC-740401,* Philadelphia, 1974. (Available from National Technical Information Service, Springfield, Virginia, *WASH-1360*)

62. R. D. Mitchell, "Methods for Optimizing and Evaluating Indirect Dry-Type Cooling Systems for Large Electric Generating Plants," *ERDA-74*, R. W. Beck and Associates, Denver, 1975.

63. D. J. Braun et al., "A User's Manual for the BNW-I Optimization Code for Dry Cooled Power Plants," *Battelle Pacific Northwest Labs. Rep. No. BNWL-2180*, Richland, Wash., 1977.

64. L. G. Hauser et al., "An Advanced Optimization Technique for Turbine, Condenser, Cooling System Combinations," *Proc. Am. Power Conf.*, Vol. 33, pp. 427–445, 1971.

NOMENCLATURE

Symbol, Definition, SI Units, English Units

A area of water-air interface per unit fill volume, Eq. (1): m^{-1}, ft^{-1}

A_f area of pond forward flow zone, Eq. (14): m^2, ft^2

A_m mixing zone surface area, Eq. (16): m^2, ft^2

A_p pond surface area, Eq. (14): m^2, ft^2

A_s surface area, Eq. (10): m^2, ft^2

A_t arbitrary transfer area: m^2, ft^2

A_x surface area fraction (as a function of longitudinal distance), Eq. (17): m^2, ft^2

a pond depth, Eq. (12): m, ft

a_0 discharge jet depth, Eq. (16): m, ft

a' parameter in Table 2 ($= \sqrt{1 + 4r'E_L^*}$)

$\overline{B}$ Bowen's ratio, Eqs. (35a) and (36)

b pond width, Eq. (12): m, ft

b_f constant, Eq. (28): $J/(kg \cdot °C)$, $Btu/(lb_m \cdot °F)$

b_0 inlet jet half-width $= \frac{1}{2}\sigma b$, Eq. (15): m, ft

b' width of discharge channel (half-width for symmetric discharge; otherwise full width), Eq. (16): m, ft

C capital cost of cooling system, Eq. (41): \$

C_t total evaluated cost, Eq. (41): \$

C_0 empirical pan coefficient, Eq. (39)

c pond length, Eq. (12): m, ft

c_p specific heat, Eq. (13): $J/(kg \cdot °C)$, $Btu/(lb_m \cdot °F)$

$c_{p,a}$ specific heat of dry air, Eq. (35a): $J/(kg \cdot °C)$, $Btu/(lb_m \cdot °F)$

$c_{p,w}$ specific heat of water, Eqs. (4) and (20): $J/(kg \cdot °C)$, $Btu/(lb_m \cdot °F)$

c_0 constant, Eq. (6)

$c_{p,a}^*$ specific heat of moist air (for deluged heat exchangers), Eq. (11): $J/(kg \cdot °C)$, $Btu/(lb_m \cdot °F)$

D_h hydraulic diameter, Eq. (13c): m, ft

D_s dilution factor, Fig. 8

D_v dilution coefficient, Eq. (18)

E_L longitudinal dispersion coefficient, Eq. (13): m^2/s, ft^2/s

E_L^* dimensionless dispersion coefficient (defined as $E_L/\overline{U}c$), Table 2

F log-mean temperature difference correction factor, Eq. (10)

F'	fuel cost for the generating unit to make up the loss of energy, Eq. (41b): \$/J, \$/Btu
Fr	discharge Froude number, Eq. (16)
f	friction factor
f_G	mass fraction humidity, Eq. (4)
f_i	interfacial friction factor, Eq. (12)
f_n	interference factor for the nth spray row
f_x	vertical component of internal flow resistance force, Eq. (9): N/m^3, lb_f/ft^3
f_y	radial component of internal flow resistance force, Eq. (9): N/m^3, lb_f/ft^3
$\bar{f}$	average interference factor for all rows, Eq. (34)
G	mass flow rate of air, Eqs. (1) and (27): kg/s, lb_m/s
g	acceleration of gravity, Eqs. (7), (9), and (12): m/s^2, ft/s^2
H	natural draft tower height, Eq. (7): m, ft
h	surface heat transfer coefficient, Eq. (19): $\text{W}/(\text{s}\cdot\text{m}^2\cdot{}^\circ\text{C})$, $\text{Btu}/(\text{s}\cdot\text{ft}^2\cdot{}^\circ\text{F})$
h_D	mass transfer coefficient: $\text{kg}/(\text{m}^2\cdot\text{s})$, $\text{lb}_m/(\text{ft}^2\cdot\text{s})$
I_L	integral of left-hand side of Eq. (4)
I_R	integral of right-hand side of Eq. (4)
IP	pond number, Eq. (12)
i_a	enthalpy of dry air: J/kg, Btu/lb_m
i_f	enthalpy of saturated water: J/kg, Btu/lb_m
i_G	enthalpy of air, Eqs. (1) and (9): J/kg, Btu/lb_m
i_g	enthalpy of vapor, Eq. (9): J/kg, Btu/lb_m
i_l	enthalpy of water, Eqs. (1) and (9): J/kg, Btu/lb_m
i_{lg}	heat of vaporization, Eq. (35): J/kg, Btu/lb_m
Δi_{lm}	log mean enthalpy difference, Eq. (11): J/kg, Btu/lb_m
j	index for penalty cost component, Eq. (41)
K	fill performance characteristic, Eq. (1): $\text{kg}/(\text{m}^2\cdot\text{s})$, $\text{lb}_m/(\text{ft}^2\cdot\text{s})$
K_c	capacity penalty charge rate, Eq. (41a): \$/kW
K_p	fan characteristic constant, Eq. (8): $\text{W}\cdot\text{s}^3/\text{m}^9$, $\text{W}\cdot\text{s}^3/\text{ft}^9$
K_0	von Karman constant (≈ 0.4)
L	mass flow rate of water, Eqs. (2) and (27): kg/s, lb_m/s
L_d	Secchi disk depth, Eq. (17): m, ft
L_j	jet length (distance to where jet stagnates or encounters a boundary), Eq. (15): m, ft
l	characteristic pond dispersion length ($\approx b/2$), Eq. (13c): m, ft
m	number of rows of modules per pass, Eq. (31)
m''	mass rate of vapor generation per unit volume, Eq. (9): $\text{kg}/(\text{m}^3\cdot\text{s})$, $\text{lb}_m/(\text{ft}^3\cdot\text{s})$
$\dot{m}_s$	spray mass flow rate, Eq. (26a): kg/s, lb_m/s
N	total number of modules in spray canal, Eq. (33)
N_p	total number of penalty cost components, Eq. (41)
N_v	number of velocity heads of flow resistance, Eq. (7)
NTU	number of transfer units, Eq. (26)
n	number of rows, Eq. (34)

n_0	constant, Eq. (6)
P	hydrostatic pressure, Eq. (9): Pa, lb$_f$/in^2
$P(T_a)$	cooling system auxiliary power requirement at ambient temperature, Eq. (41d): kW
P_{aux}	cooling system auxiliary power requirement at maximum ambient temperature, Fig. 19a: kW
P_f	fan power, Eq. (8): W
P_j	annual economic penalty for the jth component, Eq. (41): $
P_0	operating and maintenance cost for the generating unit, Eq. (41b): $/kWh
P_1	capacity penalty, Eq. (41a): $
P_2	replacement energy penalty, Eq. (41b): $
P_3	cooling system auxiliary power cost, Eq. (41c): $
P_4	cooling system auxiliary energy cost, Eq. (41d): $
P'	psychrometric ratio ($= h/c_{p,a}K$), Eq. (35a)
$\overline{P}$	loss of capacity at ambient temperature T_a, Fig. 19a and Eq. (41b): kW
$\overline{P}_{max}$	maximum loss of capacity at maximum ambient temperature, Fig. 19a and Eq. (41a): kW
p^*	average capacity factor of the plant, Eq. (41b): %/100
Q	volumetric flow rate of air, Eq. (8): m^3/s, ft^3/s
Q_c	Carrier's (total heat) sigma function: J/kg, Btu/lb$_m$
Q_0	pond through flow, Eqs. (12) and (17): m^3/s, ft^3/s
Q^*	heat rate for the generating unit to make up the loss of energy, Eq. (41b): J/kWh, Btu/kWh
q	heat transfer rate, Eq. (10): W, Btu/s
q_A	advected energy, Eq. (36): W, Btu/s
q_a	incident atmospheric radiation: W, Btu/s
q_{ar}	reflected atmospheric radiation: W, Btu/s
q_{br}	emitted back radiation: W, Btu/s
q_c	sensible loss: W, Btu/s
q_E	evaporative energy loss, Eq. (36): W, Btu/s
q_e	evaporative loss: W, Btu/s
q_n	net heat transfer: W, Btu/s
q_R	net incoming radiation $q_s + q_a - q_{br}$, Eq. (36): W, Btu/s
q_s	incident solar radiation: W, Btu/s
q_{sr}	reflected solar radiation: W, Btu/s
Δq_s	change in stored energy, Eq. (36): W, Btu/s
q''	pond thermal loading per unit area: MW(th)/hectare, MW(th)/acre
q''_n	net surface heat flux, Eq. (13): W/m^2, Btu/(s·ft^2)
q''_{sn}	short-wave solar radiation per unit area (incident − reflected), Eq. (17): W/m^2, Btu/(s·ft^2)
q''_s	incident solar radiation flux: W/m^2, Btu/(s·ft^2)
q^*	heat transfer rate from deluged surface, Eq. (11): W, Btu/s
r	radial coordinate, Eq. (9): m, ft
r_0	ratio of pumped flow per row of module to total channel flow
r'	cooling capacity parameter (defined as $hA_p/\rho c_p Q_0$), Table 2

T	temperature: °C, °F
T_a	ambient temperature, Eq. (41b): °F
T_{as}	adiabatic saturation temperature: °C, °F
T_{avg}	average temperature, Eq. (24a): °F
T_c	exit water temperature for the entire spray canal, Eq. (33): °C, °F
T_d	dew point, Eq. (24): °F
T_{db}	dry-bulb temperature, Eq. (35a): °F
T_E	equilibrium temperature, Eq. (19): °C, °F
T_f	film temperature: °C, °F
T_h	inlet water temperature for the entire spray canal, Eq. (33): °C, °F
T_{hot}	hot water temperature: °C, °F
T_i	inlet temperature to all spray rows in the ith pass: °C, °F
T_{in}	pond inlet temperature, Eq. (13a) and Table 2: °C, °F
T_l	temperature of water, Eq. (14): °C, °F
T_m	cross-sectional mean pond temperature, Eq. (13): °C, °F
T_{mix}	mixing zone mixed temperature, Eq. (16): °C, °F
T_o	exit pond temperature, Table 2: °C, °F
T_s	surface temperature, Eq. (19); cooled spray temperature, Eq (29): °C, °F
$T_{s_{i,n}}$	cooled spray temperature: °C, °F
T_{sp}	sprayed water temperature, Eq. (26): °C, °F
T_{wb}	local wet-bulb temperature, Eq. (26): °C, °F
$T_{wb,\,amb}$	ambient wet-bulb temperature: °C, °F
T_{wb_n}	local wet-bulb temperature at nth spray row, Eq. (30): °C, °F
$T_{wb,\infty}$	ambient wet-bulb temperature, Eq. (30a): °C, °F
ΔT_G	greater temperature difference, Eq. (10a): °C, °F
ΔT_L	lesser temperature difference, Eq. (10a): °C, °F
ΔT_{lm}	log mean temperature difference, Eq. (10a): °C, °F
ΔT_0	pond temperature difference, $T_{inlet} - T_{discharge}$, Eq. (12): °C, °F
T'	local spray temperature, Eq. (26): °C, °F
$\overline{T}$	average water temperature, Eq. (35a): °C, °F
t	time, Eq. (13): s; Eq. (41b): h
U	overall heat transfer coefficient, Eq. (10): W/(m²·°C), Btu/(s·ft²·°F)
U_o	discharge velocity, Eq. (16): m/s, ft/s
$\overline{U}$	cross-sectional mean velocity, Eq. (13): m/s, ft/s
U^*	shear velocity, Eq. (13c): m/s, ft/s
u	vertical velocity component, Eq. (9): m/s, ft/s
u_l	liquid velocity, Eq. (9): m/s, ft/s
$\overline{u}$	mean horizontal air velocity: m/s, ft/s
V	active cooling volume, Eq. (1): m³, ft³
V_w	wind speed, Eq. (24): mi/h
V'	pond volume, Eq. (37): m³, ft³
v	radial velocity component, Eq. (9): m/s, ft/s
v_a	velocity of air, Eq. (7): m/s, ft/s
W_e	evaporative mass loss, Eq. (37): kg/s, lb$_m$/s

W_g groundwater flow, Eq. (37): kg/s, lb_m/s
W_i mass inflows, Eq. (37): kg/s, lb_m/s
W_l mass flow at leeward end, Eq. (38): kg/s, lb_m/s
W_o mass outflows, Eq. (37): kg/s, lb_m/s
W_p precipitation, Eq. (37): kg/s, lb_m/s
W_{pm} evaporation from pan, Eq. (39): kg/s, lb_m/s
W_r overland runoff, Eq. (37): kg/s, lb_m/s
W_w mass flow at windward end, Eq. (38): kg/s, lb_m/s
X_{af} annual fixed charge rate, Eq. (41): %/100
x vertical coordinate, Eq. (9): m, ft
x_s streamwise directional coordinate, Eq. (13): m, ft

Greek Symbols

α fraction of sprayed flow evaporated, Eq. (35)
β volumetric expansion coefficient, Eq. (12): $°C^{-1}$, $°F^{-1}$
β^* fraction of q_{sn} penetrating pond surface (typically 0.4 to 0.5), Eq. (17)
Γ_G generalized exchange coefficient of air, Eq. (9): for Eq. (9d), J/m^2, Btu/ft^2; for Eq. (9f), kg/m^2, lb_m/ft^2
Γ_l generalized exchange coefficient of water, Eq. (9): for Eq. (9d), J/m^2, Btu/ft^2; for Eq. (9f), kg/m^2, lb_m/ft^2
γ slope of saturated vapor pressure curve, Eq. (24): mmHg/$°F$
Δ finite difference, Eq. (7)
δ surface layer thickness: m, ft
η extinction coefficient = $1.7/L_d$, Eq. (17): m^{-1}, ft^{-1}
θ' turbulent vertical velocity fluctuation, Eq. (40): m/s, ft/s
μ_{eff} effective viscosity of air-water mixture, Eq. (9): Pa·s, $lb_m/(h·ft)$
ξ $h_D/(1 + \omega)$, defined by Eq. (26a): $kg/(m^2·s)$, $lb_m/(ft^2·s)$
ρ density, Eqs. (7), (9), (13), and (37): kg/m^3, lb_m/ft^3
ρ_a air density, Eq. (40): kg/m^3, lb_m/ft^3
σ area ratio, Eq. (14)
ϕ wind speed function, Eq. (24c): $Btu/(ft^2·day·mmHg)$
ψ' turbulent specific humidity fluctuation, Eq. (40): kg/kg, lb_m/lb_m
$\overline{\psi}$ mean specific humidity: kg/kg, lb_m/lb_m
ω specific humidity
ω_s specific humidity of air saturated at spray temperature, Eq. (35a)

CHAPTER 11

Geothermal Heat Transfer

======= By Ping Cheng

Department of Mechanical Engineering
University of Hawaii at Manoa, Honolulu

A. INTRODUCTION

Broadly speaking, all of the thermal energy that is trapped in the earth's interior can be regarded as a geothermal resource. Geothermal systems can be classified as hydrothermal convective, geopressured, dry hot rock, and magma systems, depending upon geological, hydrothermal, and thermodynamic conditions. To date, only hydrothermal convective systems have been developed for commercial use. Moderate- to high-temperature hydrothermal convective systems have been developed for power generation, while those at low temperature have been utilized for space or industrial heating, desalination of water, and mineral recovery.

A hydrothermal convective system can be further classified as a hot-water, steam-water two-phase, or dry-steam system. Hot-water systems are more abundant than dry-steam systems, which produce superheated steam with little or no associated liquid water [1, 2]. Of about one hundred hot-water systems throughout the world that have been explored by drilling, there are only a few which exceed 250°C in temperature [1].

In general, the higher the temperature of the geothermal field, the greater the concentration of dissolved minerals in the geothermal water. Thus, the salinity of water is usually very high in high-temperature hot-water systems such as, for example, those at the Salton Sea geothermal field in the Imperial Valley, California, and at the Cerro Prieto field in Mexico, where maximum temperatures as high as 236 and 388°C, respectively, have been reported. Commercial development of hot-water systems in many parts of the world is often handicapped by the problems of corrosion, as well as by plugging and scaling due to mineral depositions as temperature and pressure are reduced in the formation, well bore, or heat transfer equipment [2]. On the other hand, dry-steam systems are free of many of these troubles and are therefore the most sought after.

It is generally agreed that the heat source of a geothermal field is intimately associated with the occurrence of recent volcanism or intense tectonic movements [3]. As a result of these activities, magma from the earth's interior can seep or well up through cracks, faults, or fractures to the earth's crust. The geothermal fluid, mostly of meteoric origin, serves as the medium by which heat is transferred from a deep igneous source to shallow depths where it can be tapped by drilled holes. For power generation, the hot geothermal fluids are piped to a geothermal power plant to drive a turbine directly or indirectly, depending on whether the plant is operating on a flashed-steam or binary system. The residual water discharged from the power plant is usually disposed of through reinjection wells into the formation for the purposes of reducing chemical and thermal pollution, prolonging the lifespan of the reservoir by replenishing the groundwater, and controlling ground surface subsidence by maintaining the pore pressure of formation [4–6].

An understanding of the heat transfer processes in geothermal systems is important for the assessment of thermal data obtained during exploration, for innovative design of apparatus for the *in situ* measurements of physical properties, and for improved methodology for estimating reservoir capacity. Moreover, numerical and physical models can be used to forecast reservoir behavior under production and reinjection. These studies can provide important information on the amount of energy recoverable, the optimum well spacing and withdrawal rates, the cooling effects of injected fluids, and the adverse effects of ground surface subsidence due to excessive fluid withdrawal. Many of these problems will be discussed in the subsequent paragraphs.

B. FUNDAMENTAL CONCEPTS AND DEFINITIONS

Although geothermal systems are characterized by many near-vertical faults and relatively impermeable intrusives interspersed in the reservoir, a qualitative understanding of the large-scale convection processes in a geothermal system can be obtained by modeling the reservoir as a fluid-filled porous medium. In this section, some of the fundamental concepts and definitions of flow through a porous medium will be discussed.

1. Porosity and Darcy's Velocity

The volumetric porosity ϕ is defined as the ratio of the volume of void space V_p to the total volume V of the medium; i.e.,

$$\phi = \frac{V_p}{V} \tag{1}$$

where ϕ is dimensionless and has magnitude less than 1. In general, the porosity of a porous medium is a function of position and pressure. For an idealized porous medium which is nondeformable, homogeneous, and isotropic, the porosity is assumed to be constant. However, measurements [7] have shown that it is inevitable that porosity is drastically increased near an impermeable wall even for a homogeneous porous medium such as a packed-sphere bed.

For flow through a column filled with a porous medium with interconnected pores, Darcy's velocity is defined as

$$v = \frac{Q}{A} \tag{2}$$

where Q is the volume flow rate and A is the total cross-sectional area of the column. On the other hand, the actual velocity in the pores (or the pore velocity) v_p is given by

$$v_p = \frac{Q}{A_p} \tag{3}$$

where A_p is the total pore area of the cross section.

The areal porosity A_p^+ is defined as the ratio of the area of the pores to the total cross-sectional area; i.e.,

$$A_p^+ = \frac{A_p}{A} \tag{4}$$

It follows from Eqs. (1) and (4) that

$$\phi = A_p^+ \tag{5}$$

for an idealized porous medium. Moreover, Eqs. (2) to (5) give

$$\frac{v}{v_p} = \phi \tag{6}$$

which shows that the pore velocity is larger than Darcy's velocity.

2. Darcy's Law and Its Modifications for Single-Phase Flow

The empirical law governing the isothermal vertical flow of water at low flow rate through a column of unconsolidated sands was first stated by Darcy [8], who found that the volume flow rate Q is proportional to the difference in hydraulic heads ΔH_d and the cross-sectional area of the column A, and inversely proportional to the height of the sand column Δs; i.e.,

$$Q = \frac{\mathsf{K}_c A\, \Delta H_d}{\Delta s} \tag{7}$$

where K_c is a proportionality constant called the hydraulic conductivity and $H_d = z + P/\rho g$, with z denoting the elevation, P the pressure, and ρ the density of the fluid. Substituting Eq. (7) into Eq. (2) yields

$$v = -\mathsf{K}_c \frac{d}{ds}\left(z + \frac{P}{\rho g}\right) \tag{8}$$

where the negative sign has been added to indicate that the direction of the flow is opposite to the gradient of the hydraulic head.

Later experiments have shown that the hydraulic conductivity is proportional to the density of the fluid, and inversely proportional to the viscosity of the fluid μ. If the intrinsic permeability of a porous medium is defined as $K = \mu \mathsf{K}_c/\rho g$, Eq. (8) can be rewritten as

$$v = -\frac{K}{\mu}\frac{d}{ds}(P + \rho g z) \tag{9}$$

where the intrinsic permeability K depends only on the microstructure of the medium.

The relationship between the intrinsic permeability and the porosity for a porous matrix composed of solid spheres is deduced theoretically by Kozeny [9] to be of the form

$$K \propto \frac{d^2 \phi^3}{(1 - \phi)^2} \tag{10}$$

where d is the diameter of the spheres.

Equation (9) is essentially a balance of viscous force, gravitational force, and pressure gradient with inertia effects neglected, and is therefore valid only at low velocity. For horizontal flow at high velocity, Forchheimer [10] proposed that Darcy's law be modified by the addition of a quadratic term in velocity to give

$$v + \frac{\chi \rho v^2}{\mu} = - \frac{K}{\mu} \frac{dP}{ds} \tag{11a}$$

where χ is Forchheimer's coefficient. Note that the ratio K/μ has been referred to as *mobility* in the petroleum engineering literature. Equation (11a) can be rewritten in the following dimensionless form:

$$f_K = \frac{1}{\mathrm{Re}_K} + \frac{\chi}{\sqrt{K}} \tag{11b}$$

where f_K and Re_K are the friction factor and Reynolds number based on the square root of permeability as a characteristic length; that is, $f_K \equiv (-dP/ds) \sqrt{K}/\rho v^2$ and $\mathrm{Re}_K = \rho v \sqrt{K}/\mu$. Based on experimental data for isothermal flow of water through twenty different porous media with Reynolds numbers Re_K up to 18.1, Ward [11] determined that

$$\frac{\chi}{\sqrt{K}} = 0.55 \tag{11c}$$

Equations (11b) and (11c) together with experimental data are plotted in Fig. 1, where it is shown that Darcy's law is valid for $\mathrm{Re}_K < 1$.

On the other hand, Ergun [12] proposed the following correlation for pressure drop through packed columns based on 640 experimental data points with various-sized spheres, sand, pulverized coke, and different gases:

$$- \frac{dP}{ds} = 150 \frac{(1 - \phi)^2}{\phi^3} \frac{\mu v}{d^2} + 1.75 \frac{(1 - \phi)}{\phi^3} \frac{\rho v^2}{d} \tag{12a}$$

Equation (12a) can be rewritten in dimensionless form as

$$f_d = 150 \frac{(1 - \phi)}{\mathrm{Re}_d} + 1.75 \tag{12b}$$

where f_d and Re_d are the friction factor and Reynolds number based on the grain diameter d which are defined as $f_d = [d\phi^3/(1 - \phi)\rho v^2](-dP/ds)$ and $\mathrm{Re}_d = \rho v d/\mu$. Ergun [12] noted that Eq. (12a) reduces to Burke and Plummer's correlation [13] at high speeds. Comparing Eqs. (11a) and (12a) gives

$$K = \frac{d^2 \phi^3}{150(1 - \phi)^2} \tag{13a}$$

$$\chi = \frac{1.75d}{150(1 - \phi)} \tag{13b}$$

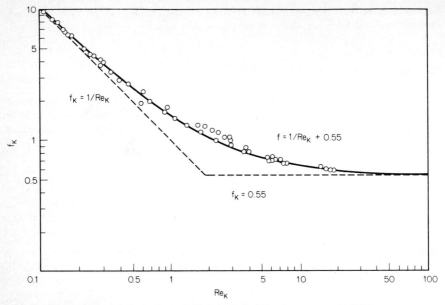

FIG. 1 f_K versus Re_K for single-phase isothermal flow through porous media [11].

For a three-dimensional flow in a gravitational field, the Forchheimer equation can be generalized [14] to give

$$\bar{v} + \frac{\rho \chi \bar{v}}{\mu} |\bar{v}| = -\frac{K}{\mu} (\nabla P - \rho \bar{g}) \tag{14}$$

which can be reduced to the following generalized Darcy's law if $\chi = 0$:

$$\bar{v} = -\frac{K}{\mu} (\nabla P - \rho \bar{g}) \tag{15}$$

Here $\bar{g}$ is the gravitational acceleration vector.

For flow through an isotropic porous medium with a high permeability, Brinkman [15] as well as Chan et al. [16] argued that the momentum equation for flow in a porous medium must be reduced to the viscous flow limit and proposed that Eq. (15) be modified to

$$-(\nabla P - \rho \bar{g}) = \frac{\mu \bar{v}}{K} - \mu \nabla^2 \bar{v} \tag{16}$$

Equation (16) was later derived by Tam [17] and Lundgren [18] using a more rigorous mathematical approach. For forced convection in a porous medium, Vafai and Tien [19] propose the following equation as the momentum equation:

$$-\nabla P = \frac{\mu \bar{v}}{K} + \frac{\rho \chi \bar{v}}{K} |\bar{v}| - \frac{\mu}{\phi} \nabla^2 \bar{v} \tag{17}$$

Equations (16) and (17) with no-slip boundary conditions imply the existence of a thin fluid sublayer adjacent to the wall, which is known as the momentum boundary layer [19]. As will be discussed in Sec. D.3, the effect of the momentum boundary layer tends to decrease the heat transfer rate. Its effect, however, is small for low-speed flows in geothermal systems [20].

3. Darcy's Law for Two-Phase Flow

When a two-phase flow passes through a porous medium, one phase will preferentially adhere to the solid surface. The one that adheres to the solid phase more strongly is called the wetting phase, and the other is called the nonwetting phase. The volume fraction of the total pore volume occupied by each phase is called the saturation of that phase. Thus, the wetting and nonwetting fluid saturation are defined as

$$S_l = \frac{V_l}{V_p} \tag{18a}$$

$$S_g = \frac{V_g}{V_p} \tag{18b}$$

where the subscripts l and g denote the quantities associated with the wetting and the nonwetting phase and V_p is the total pore volume. Since $V_l + V_g = V_p$, it follows from Eqs. (18a) and (18b) that

$$S_l + S_g = 1 \tag{19}$$

The value of the effective permeability for each phase is reduced as compared to the intrinsic permeability as defined for single-phase flow. This is because part of the pore space is filled with the wetting phase and the remaining pore space is occupied by the nonwetting phase. The relative permeability is defined as the ratio of the effective permeability to the intrinsic permeability. Thus, the relative permeabilities of the wetting phase (R_l) and the nonwetting phase (R_g) are defined as

$$R_l = \frac{K_{el}}{K} \quad \text{and} \quad R_g = \frac{K_{eg}}{K} \tag{20}$$

where K_{el} and K_{eg} are the effective permeabilities for the wetting and the nonwetting phase, respectively. With the concept of relative permeability introduced, Darcy's law for two-phase flow in a three-dimensional isotropic porous medium in a gravitational field is

$$\bar{v}_l = -\frac{KR_l}{\mu_l}(\nabla P_l - \rho_l \bar{g}) \tag{21a}$$

$$\bar{v}_g = -\frac{KR_g}{\mu_g}(\nabla P_g - \rho_g \bar{g}) \tag{21b}$$

where $\bar{v}_l$ and $\bar{v}_g$ are the Darcian velocity vectors of the wetting and nonwetting phases. Note that because of surface tension and the curvature of the interfaces between the wetting and the nonwetting phases, the pressure in the nonwetting phase (P_g) is higher than the pressure in the wetting phase (P_l). The difference between these two pressures is defined as the capillary pressure P_c; i.e.,

$$P_c = P_g - P_l \tag{22}$$

where P_c is a function of S_l; that is, $P_c = P_c(S_l)$.

It has been determined experimentally that the relative permeability is a function of saturation. A typical relative permeability curve is sketched in Fig. 2, which has the fol-

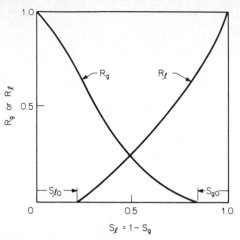

FIG. 2 Typical relative permeability curves for two-phase flow through porous media.

lowing characteristics: (1) the sum of R_l and R_g is always less than 1 and (2) $R_l = 0$ at $S_l \le S_{l0}$ (with S_{l0} denoting the irreducible wetting fluid saturation) and $R_g = 0$ at $S_g \le S_{g0}$ (with S_{g0} denoting the residual nonwetting fluid saturation). According to the two parts of Eq. (21), this implies that the wetting phase becomes immobile for $S_l \le S_{l0}$ (because of viscosity) while the nonwetting phase becomes immobile for $S_g \le S_{g0}$ (because the nonwetting phase begins to break up into bubbles).

Corey [21, 22] proposed the following correlations for the relative permeabilities of oil (wetting phase) and gas (nonwetting phase) based on a semiempirical method:

$$R_l = \hat{S}_l^4 \quad \text{and} \quad R_g = (1 - \hat{S}_l)^2(1 - \hat{S}_l^2) \tag{23a}$$

where

$$\hat{S}_l = \frac{S_l - S_{l0}}{1 - S_{l0} - S_{g0}} \tag{23b}$$

The above equations give $R_l = 0$ at $S_l = S_{l0}$ and $R_g = 0$ at $S_g = S_{g0}$. The two parts of Eq. (23) have also been used as the relative permeabilities of water and steam, where water is the wetting phase and steam is the nonwetting phase. Another simplified equation for the relative permeabilities of a steam and water mixture is given by Sondergeld and Turcotte [23] as

$$R_l = S_l \quad \text{and} \quad R_g = S_g = 1 - S_l \tag{24}$$

Equation (24) ignores the fact that water and steam become immobile at small S_l and S_g, respectively.

4. Diffusion and Dispersion Processes in Porous Media

Fluid flow in a porous medium takes on a tortuous path. The macroscopic mixing of fluid elements due to the presence of the solid matrix is referred to as mechanical dispersion. This process differs from molecular diffusion, which is microscopic in nature. Thus, the dispersion process in a porous medium consists of both molecular diffusion and mechanical dispersion. For a nonisothermal flow in a porous medium with concentration gradients, the effective hydrodynamic dispersion coefficient D_e and the effective thermal dispersion coefficient k_e are given by

$$D_e = D_0 + D' \quad \text{and} \quad k_e = k + k' \tag{25}$$

where D_0 and D' are the mass diffusion coefficient and mass dispersion coefficient and k and k' are the stagnation thermal conductivity and the thermal dispersion coefficient of the fluid-filled porous medium, respectively. It has been established that the dispersion processes are anisotropic in nature, with the longitudinal dispersion coefficient usually larger than the transverse dispersion coefficient [24, 25].

A considerable amount of work has been devoted to the determination of mass dispersion coefficients [24–26]. The correlation equation for the longitudinal mass dispersion coefficient D'_L is given by

$$\frac{D'_L}{D_0} = C_1(\hat{Pe})^m \tag{26}$$

with $C_1 = 0.5$ and $1 < m < 1.2$ for $\hat{Pe} < 10^2$, and $C_1 = 1.8$ and $m = 1$ for $10^2 < \hat{Pe} < 10^5$, where $\hat{Pe} = vd/D_0$ is the molecular Peclet number for mass transfer, d is the diameter of the grain, D_0 is the mass diffusion coefficient of the fluid, and v is the Darcy velocity. A compilation of experimental results for the transverse mass dispersion coefficient D'_T is given by List and Brooks [27], who plotted the data in terms of the inverse of the dispersion Peclet number $(1/\hat{Pe}_T)$ against the molecular Peclet number $(\hat{Pe})$ at different values of the Schmidt number Sc; i.e.,

$$\frac{1}{\hat{Pe}_T} = f(\text{Sc}, \hat{Pe}) \tag{27a}$$

where $\text{Sc} = \nu/D_0$ and $\hat{Pe}_T$ is the transverse dispersion Peclet number for mass transfer, which is defined as $\hat{Pe}_T = vd/D_T$. List and Brooks [27] concluded that at small values of $\hat{Pe}$, $\hat{Pe}_T$ is independent of Sc, i.e.,

$$\frac{1}{\hat{Pe}_T} = f_1(\hat{Pe}) \tag{27b}$$

and that at large values of $\hat{Pe}$, $\hat{Pe}_T$ is independent of both $\hat{Pe}$ and Sc, i.e.,

$$\frac{1}{\hat{Pe}_T} = \text{constant} \tag{27c}$$

The thermal dispersion process differs from the mass dispersion process in two different aspects: (1) heat does transfer through the solid phase while solute cannot, and (2) the thermal diffusion coefficient is about three orders of magnitude larger than the mass diffusion coefficient [24]. As a result, while mass dispersion effects must be considered even at very low speeds, thermal dispersion effects may be negligible at low speeds. For this reason, most of the analyses of free convection in porous media have so far neglected thermal dispersion effects.

Some experiments have been performed for the measurements of longitudinal and transverse thermal dispersion coefficients [28–31]. For air flowing in a tube filled with glass beads, Yagi et al. [28] obtain the following correlation equation for the longitudinal thermal dispersion coefficient k'_L:

$$\frac{k'_L}{k_g} = 0.8 \, \text{Pe}_d \tag{28}$$

where k_g is the thermal conductivity of the fluid and $\text{Pe}_d = vd/\alpha$, where α is the thermal diffusivity of the fluid.

Yagi et al. [29] obtained the following correlation for the transverse thermal dispersion coefficient k_T' for air flowing in a tube filled with glass beads

$$\frac{k_T'}{k_g} = 0.1 \text{ Pe}_d \tag{29}$$

C. LARGE-SCALE CONVECTION IN HOT-WATER GEOTHERMAL SYSTEMS

In this section, the heat transfer processes in a hot-water geothermal system consisting of compressed water at elevated temperature will be considered.

1. Governing Equations

For theoretical studies of convective heat transfer in hot-water systems, the following assumptions are often made:

1. The convective fluid remains a single-phase compressed liquid. For water in a subsurface formation, this assumption is valid for a wide range of temperature. For example, at a depth of 3000 ft ($\sim$900 m) where the hydrostatic pressure is approximately 1300 psi ($\sim 8.96 \times 10^6$ N/m^2), geothermal fluids remain a compressed liquid at a temperature up to 580°F ($\sim$300°C).

2. The rock formation is nondeformable. Except for the studies of ground surface subsidence (see, for example, Lippmann et al. [32]), this assumption has been used in most of the convective heat transfer problems in geothermal systems.

3. The convective fluid and the rock formation are in local thermal equilibrium everywhere. As the velocity of the convective fluid in a geothermal system is generally low, this assumption is valid for most of the situations except for the problems of a hot spring [33] or near an injection well.

4. Compression work is negligible.

5. Chemical reactions in the formation are negligible.

6. Physical properties of the formation and the convective fluid are assumed to be constant in most of the analytical studies. In reality, the permeability of the rock formation is anisotropic and is a function of pressure and temperature. For the temperature range typically from 25 to 250°C (77 to 482°F) in a geothermal system, the viscosity of water decreases by over one order of magnitude while the thermal expansion coefficient increases by two orders of magnitude [34]. The specific heats, however, vary only slightly with temperature. The assumption of constant physical properties may, however, be relaxed in a numerical study.

7. The Boussinesq approximation is valid; i.e., the density of water may be considered to be constant except in the body force term.

8. In considering the body force term, the density of water is assumed to be a linear function of temperature and concentration; i.e., the effect of pressure on density of water is negligible. This approximation has been confirmed by a number of theoretical studies [34, 35].

9. Darcy's law is applicable.

10. Thermal dispersion is negligible.

11. Viscous dissipation is negligible.

Recent studies [19, 36–39] have shown that assumptions 9 and 10 are valid only for low-speed flows in porous media.

   With these assumptions, the macroscopic conservation equations for single-phase flow in a hot-water geothermal reservoir can be obtained from the microscopic conservation equations by a volumetric averaging process [40–45]. The foundation of this approach is a theorem relating the volume average of a space derivative to the space derivative of the volume average [42, 43]. Whitaker [43] used this approach to derive Darcy's law for slow flows from the Navier-Stokes equations, while Slattery [44, 45] used the theorem to obtain the following conservation equations of mass, momentum, and energy for a nonisothermal flow in a porous medium:

$$\nabla \cdot \bar{v} = 0 \tag{30}$$

$$\bar{v} = -\frac{K}{\mu}(\nabla P - \rho\bar{g}) \tag{31}$$

$$\sigma\frac{\partial T}{\partial t} + \bar{v} \cdot \nabla T = \alpha\nabla^2 T \tag{32}$$

where σ it the ratio of the heat capacity of the fluid-filled rock formation to that of water defined as $\sigma \equiv [\phi(\rho c_p)_f + (\rho c_p)_m(1 - \phi)]/(\rho c_p)_f$, with $(\rho c_p)_f$ and $(\rho c_p)_m$ denoting the heat capacity of water and the solid matrix, respectively; and where T is the temperature of the fluid and the rock. Equations (30) to (32) can also be obtained directly from simple macroscopic balance considerations without resorting to the averaging procedures [46].

   If solute gradients exist in the reservoir, the macroscopic equation for the conservation of solute is

$$\frac{\partial c}{\partial t} + \bar{v} \cdot \nabla c = \nabla \cdot (\mathbf{D}_e\nabla c) \tag{33}$$

where $\mathbf{D}_e$ is the effective hydrodynamic dispersion coefficient. To complete the mathematical formulation of the problem, the equation of state is needed. According to assumption 8, it is given by

$$\rho = \rho_r[1 - \beta(T - T_r) + \beta_s(c - c_r)] \tag{34}$$

where β and β_s are respectively the thermal and solute coefficients of volume expansion of water and the subscript r indicates the reference condition. The linearized versions of Eqs. (30) to (34) will be used to study the growth of infinitesimal disturbances and to obtain the conditions for the onset of convection in Secs. C.2 and Sec. C.3. The nonlinear equations, Eqs. (30) to (34), will be used to study finite amplitude convection in Sec. C.4.

2. Onset of Thermally Induced Convection

Consider first the most simple problem of the onset of natural convection in a confined geothermal reservoir covered with cap rock at the top and heated by bedrock from below. As a first approximation, the problem can be idealized as a fluid-filled porous medium bounded by two infinite parallel horizontal impermeable surfaces heated from below. This is the classical "Benard problem" in porous media. A linear stability analysis for this problem was performed by Horton and Rogers [47] and independently by Lapwood

[48]. They [47, 48] found that the condition for the onset of free convection for this problem is

$$\text{Ra}^* = 4\pi^2 \tag{35}$$

where Ra is the Rayleigh number for the parallel-plate geometry defined as Ra $= K\rho_0 g\beta(T_0 - T_1)H/\mu\alpha$, with μ and β denoting the viscosity and the thermal expansion coefficient of the fluid, T_0 and T_1 the temperatures of the lower (heated) and upper surfaces respectively, ρ_0 the density of the fluid at T_0, H the distance between the parallel plates, and α the equivalent thermal diffusivity of the fluid-filled porous medium. The asterisk with Ra in Eq. (35) indicates its critical value.

The horizontal wave number at the onset of natural convection in a porous medium between parallel plates is

$$a = \sqrt{\Delta^2 + \gamma^2} = \pi \tag{36}$$

where γ and Δ are the wave numbers in the two horizontal directions. Equation (36) indicates that infinitely many convective patterns are possible at the onset of natural convection. Consider the case of two-dimensional rolls for which either $\Delta = 0$ or $\gamma = 0$. To be specific, let $\Delta = 0$ so that Eq. (36) gives $a = \gamma = \pi$. Consequently, the wavelength of the two-dimensional rolls at the onset of natural convection is unity. Another possible convection pattern at the onset of instability is that of square cells for which $\Delta = \gamma$. It follows from Eq. (36) that the wavelength for this convection pattern is $\sqrt{2}$. Other patterns are also possible.

The onset of natural convection in an unconfined geothermal reservoir with a permeable upper boundary was also studied by Lapwood [48], who used the approximate boundary condition with constant pressure for the upper boundary. Lapwood found that the critical Rayleigh number for these boundary conditions is 27.1 and the associated wave number is 2.33. Other boundary conditions were investigated by Nield [49], Ribando and Torrance [50], and Elder [51]. A summary of results for the critical Rayleigh numbers and the associated wave numbers for different boundary conditions is presented in Table 1, where the Rayleigh number is defined as $\text{Ra}_q = K\rho_0\beta g H^2 q''/\mu\alpha k$ for problems with prescribed surface heat flux q''. Table 1 indicates that convection at zero wave number occurs if the heat flux is kept constant at each boundary. In this case, instead of a horizontal array of convective cells, there is only a single cell, whose size is limited only by the presence of lateral walls. From Table 1 it can be concluded that the case of a confined geothermal reservoir with upper and lower boundaries at constant temperatures and heated from below has the largest critical Rayleigh number, and thus is most stable in the presence of disturbances.

The Benard problem was investigated theoretically from the limit of a viscous fluid to that of a Darcian fluid by Katto and Masuoka [52] as well as by Walker and Homsy [53], using Brinkman's model, Eq. (16), as the momentum equation. It was found that the critical parameter for the onset of natural convection is $\tilde{\text{Ra}}$, defined as $\tilde{\text{Ra}} = \rho_0 g\beta(T_0 - T_1)H^3/\alpha\mu$, and its critical value depends on the Darcy number Da_H, defined as $\text{Da}_H = K/H^2$. Walker and Homsy [53] concluded from this analysis that the Darcian limit is reached when $\text{Da}_H < 10^{-3}$, whereas the viscous fluid limit is reached when $\text{Da}_H > 10$.

The effect of lateral insulated walls on the onset of natural convection in an enclosed three-dimensional porous medium heated from below was studied by Beck [54]. He found that whenever the height is not the smallest dimension, two-dimensional rolls invariably occur; the axes and the direction of the rolls are usually such that each roll has the closest possible approximation to a square cross section. The value of the critical Rayleigh num-

TABLE 1 Values of the Critical Rayleigh Numbers and the Associated Wave Numbers for Onset of Free Convection in a Porous Medium with Heating from Below

Upper boundary conditions				Lower boundary conditions				Ra, $\dfrac{K\rho_0 g\beta\,\Delta T\,H}{\mu\alpha}$	Ra_φ, $\dfrac{K\rho_0 g\beta q'' H^2}{\mu\alpha k}$	a	References
Thermal		Hydrodynamic		Thermal		Hydrodynamic					
Constant temperature	Constant heat flux	Impermeable wall	Constant pressure	Constant temperature	Constant heat flux	Impermeable wall	Constant pressure				
X		X		X		X		39.48 ($4\pi^2$)		3.14 (π)	Lapwood [48]
	X	X		X		X			27.10	2.33	Nield [49]
	X	X	'		X	X			12.0	0	Nield [49]
X			X	X			X	27.10		2.33	Lapwood [48]
X			X		X	X			17.65	1.75	Nield [49]; Ribando and Torrance [50]
X			X		X		X		9.87 (π^2)	1.57 ($\pi/2$)	Elder [51]
	X		X		X		X		3.0	0	Nield [49]

ber Ra* quickly approaches $4\pi^2$ as either side of the parallelepiped becomes larger than its height. Thus, the lateral walls appear to have little effect on the critical Rayleigh number except for very narrow tall boxes where the critical Rayleigh numbers are larger than $4\pi^2$. Similarly, for a cylindrical reservoir heated from below, the critical Rayleigh number also approaches the value of $4\pi^2$ when the ratio of the radius to the height of the reservoir approaches unit [55].

The effect of inclination angle on the onset of cellular convection in a porous medium bounded by infinite parallel plates heated from below was studied by Bories and Combarnous [56]. The result of a linear stability analysis shows that the critical Rayleigh number is also equal to $4\pi^2$, if the Rayleigh number is based on the component of the gravitational acceleration parallel to the inclined surface.

The effect of a prescribed horizontal temperature gradient on the top and the bottom impermeable surfaces on the onset of natural convection in a porous medium was investigated by Weber [57], who found that the critical Rayleigh number for this problem is

$$Ra^* = 4\pi^2(1 + \hat{\Gamma}^2 + 1.73\hat{\Gamma}^4 + \cdots) \qquad (37)$$

where $\hat{\Gamma} = \Gamma H/\Delta T$, with Γ denoting the prescribed horizontal temperature gradient and ΔT the temperature difference between lower and upper wall at the origin. Equation (37) shows that the existence of the horizontal temperature gradient delays the onset of cellular convection.

The effect of a uniform horizontal flow on the onset of cellular convection in a horizontal porous layer confined by parallel plates heated from below was investigated theoretically by Prats [58], who found that the critical Rayleigh number of $4\pi^2$ is unaffected by the horizontal flow. An extension of Prats' analysis to include the non-Darcian and thermal dispersion effects has been performed by Rubin [59, 60]. He shows that the preferred mode at the onset of the thermally induced flow is two-dimensional rolls with axes parallel to the unperturbed horizontal flow, and that the value of the critical Rayleigh number is unchanged although the definition of the Rayleigh number is modified. On the other hand, Homsy and Sherwood [61] found that the effect of vertical through flow (such as in the case of a fault zone) is to delay the onset of cellular convection as the flow rate is increased; however, the direction of the through flow relative to gravity has no effect on the critical value. Homsy and Sherwood's analysis is an extension of the earlier work by Wooding [62], who had considered a similar problem for a semi-infinite porous medium with a permeable top.

The effect of vertical through flow with small dimensionless flow strength on the onset of cellular convection in a two-dimensional horizontal layer bounded laterally by vertical insulated walls was studied theoretically by Sutton [63]. It was found that the critical Rayleigh number Ra* decreases as the width to thickness ratio is increased. In the limit of small flow strength and large aspect ratio (width to thickness ratio), the critical Rayleigh number approaches the classical value of $4\pi^2$.

A linear stability analysis to study the effect of layered structure on the onset of natural convection in a geothermal reservoir was performed by McKibbin and O'Sullivan [64]. It was shown that large permeability differences between the horizontal layers are required so that the onset modes are substantially different from that of a homogeneous system. The effect of anisotropic permeability (i.e., with different values of vertical permeability K_v and horizontal permeability K_h) on the onset of natural convection in a geothermal reservoir was studied by Wooding [65] and by Castinel and Combarnous [66]. They found that both the values of the critical Rayleigh number (based on K_v) and the horizontal wave number are reduced as the permeability ratio K_v/K_h decreases from unity.

The effect of temperature dependence of viscosity on the onset of natural convection in a porous medium between parallel plates and inside a box (heated from below) was studied theoretically by Kassoy and Zebib [67, 68]. The critical Rayleigh number Ra* (with viscosity evaluated at the upper surface temperature) for this case depends on an additional dimensionless parameter relating to the temperature difference between the upper and lower walls. The values of the critical Rayleigh numbers are substantially reduced from the constant-viscosity analysis, although the associated wave numbers change only slightly. The streamlines and isotherms at the onset of natural convection are asymmetric with respect to the vertical direction. This is as expected since the hotter, less viscous fluid near the bottom (heated) surface moves faster than the colder fluid near the top and thus convects relatively more energy away, which, in turn, yields steeper velocity and temperature gradients near the bottom heated surface.

In addition to the temperature-dependent viscosity, Straus and Schubert [37] have taken into consideration the temperature dependence of other properties such as thermal expansion coefficient and density of water in their analysis of onset of natural convection in a confined geothermal reservoir with infinite horizontal extent. They found that the critical Rayleigh number Ra*, with fluid properties evaluated at the upper wall temperature, is a function of the thickness of the reservoir and the imposed vertical temperature gradient. Moreover, the critical Rayleigh number based on the variable-properties analysis was shown to be reduced considerably compared with the constant-property analysis. A similar analysis was also carried out by Morland et al. [69].

The first experimental confirmation of the critical Rayleigh number Ra* for the onset of natural convection in a porous layer between parallel plates heated from below is given by Katto and Masuoka [52], who used nitrogen as the saturating fluid with glass, steel, or aluminum balls as the porous medium. Satisfactory agreement between theory and measurements for the critical Rayleigh number is achieved if the diffusivity in the Rayleigh number is taken as the conductivity of the medium divided by the heat capacity of the *fluid* (not of the medium, as taken by earlier investigators). Similar experimental results were obtained by Yen [70] as well as by Buretta and Berman [71] for glass beads saturated with water. In other experiments, however, it has been observed that the critical Rayleigh number depends on the thermal characteristics of the porous medium and the saturating fluid. For example, Kaneko et al. [72] found that the critical Rayleigh number for a heptane-sand system was approximately 40, while that for ethanol-sand systems was only 28, which is considerably less than the theoretical value.

3. Onset of Thermohaline Convection

The onset of thermohaline convection in a porous medium has important applications to saline geothermal fields such as the one located in the Imperial Valley, California, where salinity as high as 2.5×10^5 ppm has been reported [73]. A linear stability analysis for the onset of thermohaline convection in a porous medium between parallel horizontal plates separated by a distance H with an unstable vertical temperature gradient and stable vertical solute gradient was performed by Nield [49], who found that the critical Rayleigh number is

$$Ra^* = Ra_s + 4\pi^2 \tag{38}$$

where $Ra_s = \rho_0 g \beta_s (c_0 - c_1) K H / \mu D_0$ is the salinity Rayleigh number with c_0 and c_1 denoting the salt concentration at the lower and upper walls. Equation (38) shows that the critical Rayleigh number increases for $\text{Ra}_s > 0$; i.e., stabilizing solute gradients inhibit cellular convection. In addition, Nield [49] found that the presence of a stabilizing solute gradient introduces the possibility of oscillatory motion, the so-called overstability motion.

An extension of Nield's analysis for nonlinear stabilizing salinity profiles has been performed by Rubin [74]. The effects of the anisotropic solute dispersion as well as a non-Darcian horizontal flow at the onset of thermohaline convection in a porous medium were also investigated by Rubin [75-77]. He found that (1) the transverse dispersion has a stabilizing effect; (2) the critical Rayleigh number is independent of the longitudinal dispersion coefficient; and (3) the preferred mode of instability is stationary two-dimensional rolls. A similar analysis for the onset of thermohaline convection in a thermally driven shear flow was performed by Weber [78].

The case of a strong stabilizing temperature gradient ($\text{Ra} < 0$) and a destabilizing solute gradient ($\text{Ra}_s < 0$) resulting in a mean stable density gradient is of special interest. This occurs in geothermal systems whenever relatively hot water with a high solute concentration is injected into a geothermal reservoir with a small solute concentration. In this case, salt fingers will develop even though the fluid is still statically stable. The condition for the onset of salt fingers in an isotropic porous medium was investigated by Taunton and Lightfoot [79]. An extension of the analysis to include the effects of anisotropic permeability and thermal and hydrodynamic dispersion was performed by Tyvand [80].

4. Finite-Amplitude Convection

In the preceding sections, linear theories were used to investigate the conditions for onset of cellular convection due to infinitesimal disturbances. These theories fail to predict the convective patterns and the heat transfer rate at a Rayleigh number (Ra) higher than the critical value. For these situations, other approaches such as finite-amplitude analyses, perturbation methods, or numerical solutions must be used instead. Earlier efforts have been devoted mostly to the finite-amplitude analysis for natural convection in an enclosed two-dimensional rectangular porous medium with isothermal heating from below. For more complicated geometry or boundary conditions, numerical methods have been used.

For a two-dimensional nonisothermal flow with no concentration difference, it is convenient to express Eqs. (30) to (34) in the following dimensionless form:

$$\frac{\partial^2 \Psi}{\partial X^2} + \frac{\partial^2 \Psi}{\partial Y^2} = -\text{Ra}_H \frac{\partial \Theta}{\partial X} \tag{39}$$

$$\frac{\partial \Theta}{\partial \tau} + \left(\frac{\partial \Psi}{\partial Y} \frac{\partial \Theta}{\partial X} - \frac{\partial \Psi}{\partial X} \frac{\partial \Theta}{\partial Y} \right) = \frac{\partial^2 \Theta}{\partial X^2} + \frac{\partial^2 \Theta}{\partial Y^2} \tag{40}$$

where $\Theta \equiv (T - T_r)/(T_m - T_r)$ and $\Psi \equiv \psi/\alpha$ are dimensionless temperature and stream function, with T_m denoting the maximum temperature of the heated surface; and where $\text{Ra}_H \equiv K\rho_r \beta g(T_m - T_r)H/\mu\alpha$, with H denoting the depth of the reservoir and the subscript r denoting a reference condition. The dimensionless independent variables are $X = x/H$, $Y = y/H$, and $\tau = \alpha t/\sigma H^2$, with x denoting the coordinates placed on the lower (heated) horizontal surface and y the coordinate perpendicular to the heated surface pointing toward the porous medium. For problems involving the withdrawal and injection of fluids, it is sometimes convenient to express Eqs. (30) to (34) in terms of pressure and temperature [81]. With appropriate boundary and initial conditions, Eqs.

(39) and (40) are solved for the study of finite amplitude convection in geothermal reservoirs in the subsequent paragraphs.

a. Natural Convection in a Confined Geothermal Reservoir Heated from Below

The most simple model of cellular convection in a geothermal reservoir is that of steady natural convection in a porous medium between parallel horizontal plates or in an enclosed porous medium heated from below. Using a perturbation method, Palm et al. [82] have performed a finite-amplitude analysis of steady two-dimensional cellular convection in an enclosed porous medium with aspect ratio α^+ equal to 1, where the aspect ratio is defined as the width to height ratio, i.e., $\alpha^+ = L/H$. This approach results in a series of linear inhomogeneous subproblems that were solved analytically up to the sixth order to obtain an analytical expression for the Nusselt number, which for a parallel-plate configuration is defined as

$$\text{Nu} = \frac{\overline{q}'' H}{k(T_0 - T_1)} = \frac{H}{(T_0 - T_1)} \int_0^L \left(\frac{\partial T}{\partial y}\right)_{y=0} dx \tag{41}$$

where $\overline{q}''$ is the average heat flux from the bottom (heated) surface and L is the length of the heated surface. The series expression for the Nusselt number as obtained by Palm et al. [82] is valid for Rayleigh numbers slightly larger than the critical Rayleigh number. Using the same approach, Zebib and Kassoy [83] obtained a two-term expansion of the Nusselt number for the three-dimensional convection patterns in a porous medium inside a rectangular parallelepiped with horizontal dimensions larger than the vertical dimension. They found that the two-dimensional rolls transfer more heat than three-dimensional convective cells when the Rayleigh number is just above the critical value. An extension of the two-dimensional analysis of Palm et al. [82] to include the effects of the temperature dependence of viscosity is given by Zebib and Kassoy [84]. The same problem was also considered by Straus [85], who used the Galerkin method to obtain an analytical expression for the Nusselt number which is valid for a Rayleigh number 10 times larger than the critical value. In addition, Straus [85] shows that for Ra < 380, there is a limited band of wave number for which two-dimensional rolls are stable. On the other hand, Busse and Joseph [86] as well as Gupta and Joseph [87] used a variational method to derive an upper bound of the Nusselt number as a function of the Rayleigh number for the three-dimensional convection in an enclosed porous medium heated from below.

Finite difference solutions were obtained by Donaldson [88] to study steady free convection in a confined geothermal reservoir with an aspect ratio equal to 1 ($\alpha^+ = 1$) and with the Rayleigh number (Ra) up to 165. For a reservoir with a single porous layer, the mushroom shape of the isotherms became apparent at Ra = 96. Donaldson also considered a two-layer system consisting of an upper permeable layer and an underlying impermeable basement rock. The numerical results show that the mushroom shape of the isotherms in the two-layer system at the same Rayleigh number are less pronounced than those of a single-layer, indicating the conducting layer tends to retard convection. The two-layer model was later reexamined by Sorey [38], who solved the governing equations numerically on the basis of the integrated finite difference method [89].

Steady natural convection in a three-dimensional geothermal reservoir of arbitrary geometry was considered by Chen and Conel [90] using the finite element method. In their analysis, the governing equations are expressed in terms of a vector potential and temperature. If the buoyancy force terms in the vector potential equations and the convective terms in the energy equations are considered as source terms, the governing equations are of a Laplacian form that can be solved numerically by iteration. The numerical solution of a sample two-dimensional problem with Ra = 200 and α^+ = 0.6 was executed using NASTRAN, a NASA-developed general-purpose computer code. Numerical results for temperature and pressure distribution as well as Nusselt numbers were found in good agreement with finite difference solutions performed earlier by Ribando and Torrance [50].

Elder [51] considered the roles of end effects and nonuniform heating of the bottom (heated) surface on steady natural convection in an enclosed two-dimensional geothermal reservoir. Numerical solutions for isotherms Θ and streamlines Ψ [both Θ and Ψ are defined in Eqs. (39) and (40)] at Rayleigh numbers (Ra) up to 120 were obtained for different boundary conditions and aspect ratios. The numerical results for isotherms and streamlines at Ra = 80 and α^+ = 10 with different dimensionless heating length l^+ (ratio of the length of the heated lower boundary to the depth of the reservoir) are plotted in Fig. 3, where it is shown that the number of convective cells depends on the value of l^+.

The first transient numerical solution for transient two-dimensional natural convection in a confined geothermal reservoir heated from below is given by Elder [91]. A similar study taking into consideration the temperature dependence of fluid properties is given by Holst and Aziz [92]. Elder's problem was later reexamined by Horne and O'Sullivan [93], who used a more sophisticated numerical method. Computations were carried out for Rayleigh numbers (Ra) up to 1250. Both uniformly heated and nonuniformly heated lower boundaries were considered. For the uniformly heated case, Horne and O'Sullivan [93] found that two distinct convective patterns are possible depending on the rate of

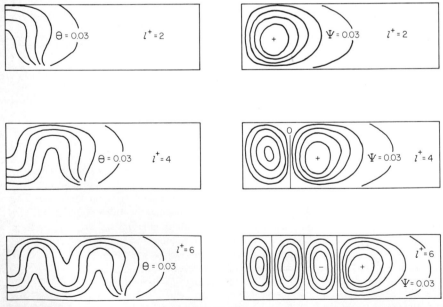

FIG. 3 Effects of heating length l^+ on isotherms (left) and streamlines (right) in an enclosed geothermal reservoir at Ra = 80 and α^+ = 10 [51].

heating: (1) when the lower boundary is suddenly heated, a steady multicellular motion is developed; and (2) when the lower boundary is heated slowly, a unicellular motion is developed which is unsteady (or fluctuating) at a Rayleigh number larger than 280—this is in agreement with earlier experimental observations by Combarnous and Le Fur [94] (see discussion below). For the nonuniformly heated case, the system is self-restricting to a unique mode which can either be a steady multicellular motion or a periodic oscillatory motion, depending on how much the lower boundary is heated.

Caltagirone [95] found that, apart from the Rayleigh number (Ra), the aspect ratio is the most important parameter for the occurrence of fluctuating convection in an enclosed porous medium heated from below. Figure 4 is a plot of numerical results showing conduction and stable and fluctuating convection regions as a function of Rayleigh number Ra and cell aspect ratio α^+. The neutral stability curve obtained numerically was found to be in agreement with that obtained from the linear stability analysis. It is interesting to note that at Ra = 1000 the cell aspect ratio is about 0.1 at the onset of natural convection. At a Rayleigh number between 40 and 250, the cell aspect ratio determines

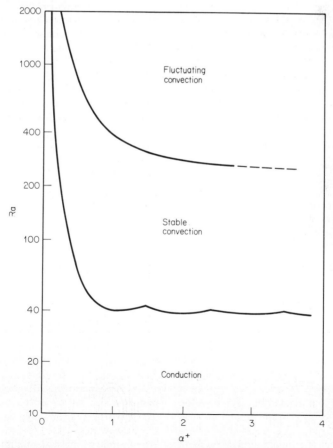

FIG. 4 Rayleigh number versus aspect ratio showing three types of heat transfer in an enclosed porous medium heated from below [95].

whether heat is transferred by conduction or stable convection. For a Rayleigh number greater than 250, conduction and stable or fluctuating convection are all possible, depending on the aspect ratio. The numerical results show that the Nusselt number increases to a maximum and then decreases as cell aspect ratio is increased, which is in agreement with the earlier study by Combarnous and Bia [96].

Finite difference solutions for transient three-dimensional natural convection in an enclosed porous medium heated from below have been obtained by Holst and Aziz [97]. The governing equations are expressed in terms of temperature and a vector potential which may be regarded as the three-dimensional counterpart of a stream function. Holst and Aziz found that while two-dimensional rolls transfer more energy than three-dimensional flows at low Rayleigh numbers (Ra), the reverse is true at higher Rayleigh numbers. Using a more sophisticated numerical technique, Horne [98] considered the same three-dimensional problem based on both the pressure and the vector potential formulation. Numerical solutions were carried out for Rayleigh numbers up to 300. He concluded that even though the pressure formulation with its two equations may seem more manageable than the vector potential formulation with three equations, the numerical solution of the latter formulation is much faster and has greater accuracy. Horne [98–100] found that there exists more than one mode of convection for any particular physical configuration and Rayleigh number. For a cubic region and a Rayleigh number (Ra) less than 300, the flow pattern may be two- or three-dimensional, and steady or unsteady, depending on its initial conditions and subsequent history. The same problem was also considered by Straus and Schubert [101–104], who arrive at similar conclusions based on Galerkin's method.

Experimental studies of steady natural convection in a porous medium bounded by parallel plates heated from below have been performed by a number of investigators. Using glass spheres (ranging from 3 to 18 mm in diameter) in water, Elder [51] obtained the following empirical formula for the Nusselt number:

$$\text{Nu} = \frac{\text{Ra}}{40} \tag{42}$$

which was correlated with experimental data for a Rayleigh number (Ra) slightly higher than the onset of natural convection.

An extensive experimental investigation for the same configuration using glass beads, lead spheres, and sands as porous media and silicone oil and water as the saturating fluids was performed by Combarnous and Le Fur [94]. They found that when the Rayleigh number (Ra) of the porous layer is about 240 to 280, the convective motion changes in character. A plot of the Nusselt number (Nu) versus the Rayleigh number (Ra) shows that there is a change in slope in this range of Rayleigh number. Furthermore, the data of Combarnous and Le Fur [94] confirmed the earlier findings by Schneider [105], who had observed that at a given Rayleigh number, the Nusselt number depends on the thermal characteristics of the porous medium and the saturating fluid. A similar experiment was carried out by Kaneko et al. [72], who used heptane and ethanol as the saturating fluids and sand as the porous medium. They found that the heat transfer results for the heptane-sand system are in good agreement with previous experiments using oil and sand [95] and oil and glass [105]. The data for ethanol-sand systems are in close agreement with water-sand, water-lead, and oil-steel systems used by previous investigators [94, 105]. Combarnous and Bories [106] have attributed the scatter of the experimental data to the fact that the fluid and the solid matrix were not in local thermal equilibrium, and that different porous media and saturating fluids have different values of heat transfer coefficient for the fluid-solid interface. Based on a two-temperature nonlocal thermal equilibrium model, Combarnous and Bories [106] found that it is possible to choose a correct value of interface heat transfer coefficient such that the numerical results for the Nu-Ra relationship coincide with experimental data.

In another experiment with glass beads and demineralized water, Buretta and Berman [71] obtained heat transfer data for Rayleigh numbers (Ra) up to 10^4. Their data were found in excellent agreement with Gupta and Joseph's theoretical results based on upper-bound analysis [87]. Buretta and Berman [71] obtained the following correlation equations:

$$\log (\text{Nu} + 0.076) = 1.154 \log \text{Ra} - 1.823 \qquad 40 < \text{Ra} < 100 \qquad (43a)$$

$$\log (\text{Nu} + 0.35) = 0.835 \log \text{Ra} - 1.214 \qquad 100 < \text{Ra} < 1000 \qquad (43b)$$

Another experiment of similar nature at a lower temperature range was performed by Yen [70].

More recently, Seki et al. [107] found that heat transfer in a horizontal porous layer heated from below depends not only on the Rayleigh number but also on the Prandtl number as well as the ratio of the beads' diameter to the porous bed's depth (d/H). They found the following correlation equations:

$$\text{Nu} = 0.1 \ \text{Pr}^{0.132} \left(\frac{d}{H}\right)^{-0.655} \text{Ra}^{0.5} \qquad 200 < \text{Ra} < 1400 \qquad (44a)$$

$$\text{Nu} = 0.88 \ \text{Pr}^{0.132} \left(\frac{d}{H}\right)^{-0.655} \text{Ra}^{0.2} \qquad 1400 < \text{Ra} < 40000 \qquad (44b)$$

These correlations were obtained for glass beads with diameters ranging from 3.02 to 16.4 mm with the depth of the bed varying from 16.4 to 103 mm, and with water, ethyl alcohol, fluorocarbon R-11, and transformer oil as the saturating fluids. The Prandtl number for this experiment ranged from 1.1 to 7.3.

A partial compilation of analytical, numerical, and experimental results for heat transfer in a horizontal porous layer heated from below is shown in Fig. 5. It is seen that there are substantial discrepancies between experimental data obtained by different investigators as well as between theory and experiments. This is probably due to the nonlocal thermal equilibrium effect [106] as discussed earlier, as well as to the fact that the data presented in Fig. 5 do not take into consideration the ratio of the grain diameter to the depth of the porous bed effect (d/H). This dimensionless parameter is related to the Darcy number (Da_H), which is a measure of the boundary and inertia effects. As will be discussed in Sec. D.3, the boundary and inertia effects become important at high Rayleigh numbers.

b. Natural Convection in an Unconfined Geothermal Reservoir Heated from Below

In the absence of cap rock, a geothermal reservoir may recharge or discharge through the upper boundary. Numerical solutions for steady two-dimensional natural convection in such a geothermal reservoir were obtained by Elder [51]. It was assumed that pressure was constant along the permeable upper boundary and that the upwelling of the water table was ignored. Ribando et al. [50, 108] carried out a similar numerical study to explain the spatially periodic heat flux data obtained adjacent to the Galapagos spreading center. Consideration was given to two cases, of (1) a porous layer having finite depth with constant mobility and (2) a semi-infinite porous layer with mobility decreasing exponentially with depth. Numerical solutions at different Rayleigh numbers (Ra_q) were obtained for the constant-mobility case with the aspect ratio equal to unity and for the

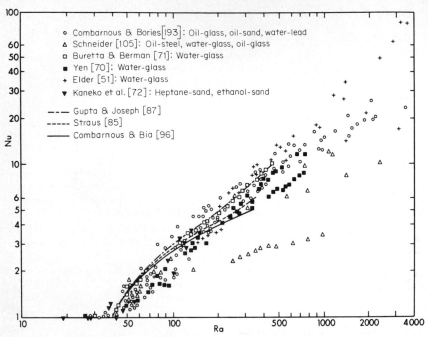

FIG. 5 A compilation of some experimental, analytical, and numerical results of Nusselt number versus Rayleigh number for natural convection in a horizontal porous layer heated from below [194].

variable mobility case with the aspect ratio approximately equal to 5.64. The numerical results for the ratio of maximum surface heat flux (occurring at the top of a rising hot plume) to the minimum surface heat flux (occurring at the top of a descending cold plume) as a function of the Rayleigh number were plotted. Of particular interest is the case where this ratio is approximately 6.1, which is considered to be typical of the Galapagos spreading center. This heat flux corresponds to $Ra_q = 40$ for the case of constant mobility and to $Ra_q = 49$ for the case of variable mobility. With these values of the Rayleigh number, an independent estimate of the permeability is possible if the depth of circulation and the mean heat flux are known. Using this method, the permeability of the ocean crust was estimated to be 4.5×10^{-12} cm^2 (4.84×10^{-13} ft^2).

Rana et al. [109] studied transient two-dimensional natural convection in a layered geothermal reservoir having a permeable boundary at the top and heated from below. The reservoir consists of three horizontal permeable layers of different thickness, with the middle layer the least permeable. The vertical boundaries of the reservoir are impermeable surfaces where prescribed temperature increases linearly with depth. Since the layers have different physical properties, the governing equations in terms of stream function and temperature are applied separately to each of the layers. The boundary conditions at the interfaces are such that dimensional temperature and normal flow rates are continuous. Steady state isotherms Θ and streamlines Ψ [both defined in Eqs. (39) and (40)] in a reservoir consisting of three layers with Rayleigh numbers (Ra) equal to 300, 120, and 750 and aspect ratio of 2 are presented in Fig. 6. In this figure only half of the reservoir is shown with the dashed lines indicating the interfaces of the horizontal layers. It is seen that the least permeable (middle) layer allows some of the groundwater in the top layer to penetrate into the permeable bottom layer, where closed-streamlined convective cells are formed.

A two-temperature model was used by Turcotte et al. [36] to simulate steady natural convection at Steamboat Springs, Nevada, which was assumed to be an unconfined geothermal reservoir. Fluid enters the upper permeable boundary at the ambient temperature, while that leaving the layers is at a temperature greater than ambient temperature, thereby resulting in hot springs. At large Rayleigh numbers, it was found that significant temperature differences between the fluid and the formation are restricted to a thin layer near the permeable upper boundary.

Horne and O'Sullivan [110] made a numerical study of transient natural convection in a two-dimensional geothermal reservoir with nonuniform heating from below and with recharge and discharge from the top. They found that oscillatory convection starts at a lower Rayleigh number (Ra) than in the case of an enclosed reservoir. The lowering of the Rayleigh number at which oscillation occurs is attributed to the steeper thermal gradients in the nonrecirculating case. In another paper [111], these authors have extended their analysis to include the effects of the temperature dependence of viscosity and thermal expansion coefficient.

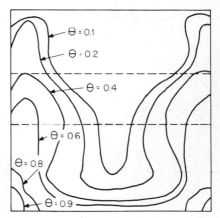

c. Natural Convection in a Geothermal Reservoir with Magmatic Intrusions

Norton and coworkers [112, 113] have performed a numerical study of transient cooling of a massive pluton after its emplacement in a two-dimensional geothermal

FIG. 6 Steady-state streamlines (top) and isotherms (bottom) for natural convection in a multilayer geothermal reservoir, dashed lines indicating the interfaces of the horizontal layers [109].

reservoir. The intrusive was idealized as an impermeable block initially at a high temperature. Variable fluid properties were taken into consideration in the numerical model. The surface heat flux directly above the pluton was found to increase tenfold after 10^4 years following its emplacement. A more refined numerical model taking into consideration thermal crackings in the solidified intrusion was used by Torrance and Sheu [114].

Convection in a rift zone can be modeled as convection in an enclosed porous medium heated by vertical walls at T_0 and T_1. Using the boundary-layer approximations together with Gill's approximate method [115], Weber [116] has determined the Nusselt number for steady natural convection in such a two-dimensional reservoir to be

$$Nu = 0.58 \left(\frac{H}{L}\right)^{1/2} Ra^{1/2} \qquad (45)$$

where L is the depth of the reservoir and H is the distance between the vertical walls. Weber's approximate analysis was later refined by Walker and Homsy [117], who developed a self-consistent boundary-layer theory by taking into consideration the thick horizontal boundary layers. The refined analysis gives the following expression for the Nusselt number:

$$Nu = (0.51 \pm 0.01) \left(\frac{H}{L}\right)^{1/2} Ra^{1/2} \qquad (46)$$

An experimental investigation of natural convection in an enclosed porous medium heated by two vertical walls at different temperatures was performed by Bories and Combarnous [56], who obtained the following correlation equation:

$$Nu = 0.245 \left(\frac{H}{L}\right)^{0.397} Ra^{0.625} \qquad (47)$$

Equation (47) was obtained based on the data for glass beads saturated with water for $100 < Ra < 1000$ and with an aspect ratio (L/H) of the layer ranging from 0.05 to 15. Similar experiments have been performed by Seki et al. [118] using two kinds of solid particles (glass beads and iron balls) and three types of saturating fluids (water, transformer oil, and ethyl alcohol). They obtained the following correlation equation

$$Nu = 0.627 \, Pr^{0.13} \left(\frac{H}{L}\right)^{0.527} Ra^{0.463} \qquad (48)$$

Equation (48) was obtained based on data for Prandtl numbers between 1 and 200 and aspect ratios (L/H) between 5 and 26, and Ra between 10^2 and 10^4.

d. Natural Convection in an Island Geothermal Reservoir

Natural convection in a volcanic island geothermal reservoir with recharge from the ocean through its vertical boundaries was considered in a series of papers by Cheng and coworkers [119–122]. In all of these analyses, the variations of the salt concentration were neglected. A perturbation analysis [119, 120] was used to study the upwelling of the water table at small and moderate Rayleigh numbers (Ra); this resulted in a series of linear subproblems which were solved numerically. In another paper [121], the finite difference method was used to study steady natural convection in an axisymmetric island reservoir confined by cap rock from the top and heated from below. The numerical solutions show that cold seawater moves inland from the lower portion of the reservoir, rises up as a thermal plume after sufficient heat is absorbed, and spreads around under the cap rock, and that finally part of the warm water discharges to the ocean and the remainder recirculates in the reservoir. The amount of recirculation is small at large Rayleigh numbers. Vertical temperature profiles show that steep temperature gradients exist adjacent to the heated surface at large Rayleigh numbers.

To study the formation of an island geothermal reservoir after a magma chamber was intruded from below, Cheng and Teckchadani [122] obtained a finite difference solution for transient two-dimensional natural convection in a porous medium with an impermeable upper boundary and with recharge and discharge through vertical boundaries. For an island geothermal reservoir 6000 m ($\sim$2000 ft) wide and 1500 m ($\sim$5000 ft) thick at Ra = 200, a steady-state condition is reached approximately 7000 years after a sudden heating from below. Oscillatory convection becomes noticeable for Ra > 200, with a higher frequency of oscillation as the value of Ra is increased.

e. Mixed Convection in a Geothermal Reservoir

Externally imposed pressure gradients may exist in a geothermal reservoir due to the difference in hydrostatic heads generated either by natural recharge or discharge of meteoric water, or by artificial withdrawal and injection of geothermal fluids. In a numerical study, Elder [51] showed that cellular convection may be annihilated by a sufficiently strong recharge due to variations of water table heads.

The transient behavior of isotherms in a confined geothermal reservoir resulting from a simultaneous withdrawal and reinjection of fluids was studied numerically by Lippmann et al. [34]. The reservoir was considered to be deformable and two production-injection schemes were considered. In the first case, water was removed at the top of the ascending convective current while injection took place at the bottom of the ascending current. In the second case, the relative positions of production and injection wells were reversed. In both cases it was shown that a cold-water zone was generated near the injection wells, and a general cooling of the reservoir was observed. In another paper [123], the same model was used to study temperature recovery of a production well after the cooler fluid was injected. It was concluded that if injection of cold water must be carried out, it would be advantageous to inject at the lower portion of the reservoir and produce at the upper portion of the reservoir. In this way, the colder fluid with a higher density and a larger viscosity would slow down its migration rate toward the warmer fluid above so that water at a higher temperature could be produced.

The effect of fluid withdrawal on heat transfer from the bottom (heated) surface in a two-dimensional geothermal reservoir with recharge from the top was investigated numerically by Horne and O'Sullivan [124]. The numerical results show that fluid withdrawal can increase or decrease the heat transfer rate from the bottom (heated) surface depending on its position relative to the heat source. In another paper [125], the same authors considered the contraction of isotherms and the drawdown of the water table resulting from fluid withdrawal. The numerical results show that the drawdown does not continue indefinitely and that the pressure decline at a location midway between the sink and the water table quickly levels off.

The problem of mixed convection in an island geothermal reservoir resulting from fluid withdrawal was investigated by Cheng and coworkers [81, 122]. The production wells were realized either as a point sink (for a cylindrical reservoir), a horizontal-line sink, or a vertical-plane sink (for a two-dimensional reservoir). The problem was formulated in terms of pressure and temperature, where either the flow rate or the pressure at the sink may be specified. The numerical results show that, while the fluid below the well experiences a favorable pressure gradient, which tends to accelerate its vertical upward motion driven by the buoyancy force, the fluid above the production well experiences an adverse pressure gradient, which tends to retard its upward motion due to the buoyancy force. If the withdrawal rate is sufficiently strong that the negative pressure gradient can overcome the buoyancy force, the isotherms in the reservoir will begin to contract. The contraction of isotherms in the reservoir will eventually lead to the decrease in temperature of hot water produced from the well, and consequently the decrease in the amount of recoverable energy.

Mixed convection in a saline coastal geothermal reservoir resulting from injection of fluid was studied numerically and experimentally by Henry and Kohout [126]. The movement of groundwater in this problem was controlled by the hydraulic pressure gradient, the geothermal gradient, and the salt concentration gradient. It was assumed that the reservoir was bounded by impermeable surfaces from the bottom as well as from the top, except near the center of the upper boundary, where freshwater was recharged continuously. Both the numerical and experimental results show that the reservoir can be divided into three zones: a zone of essentially fresh water flowing seaward, a convective zone consisting mostly of seawater in the lower portion of the reservoir near the coast, and a transition zone in between. It is interesting to note that the effect of geothermal heating from below tends to increase seawater intrusion.

f. Convection in a Fault-Controlled Zone

Donaldson [127, 128] proposed a two-dimensional model to simulate steady hydrothermal convection in fracture zones at Wairakei, New Zealand. The model consists of a cold permeable reservoir (open to the ground surface to permit recharge) next to a vertical permeable column heated from below. The reservoir and column are connected by a permeable horizontal channel and all surrounding media are assumed to be impermeable. As a result of density unbalance between the column and the reservoir, the heated fluid will rise up through the column to the ground surface to become hot springs and fumaroles. Numerical solutions for isotherms and streamlines in a column with a small aspect ratio for Rayleigh numbers (Ra) up to 800 were obtained by a finite difference method. As predicted by the linear-stability analysis [63], local cellular motion in the fault will be initiated when the Rayleigh number exceeds the critical value. For a small upward flow rate, it was found that the local circulatory flow substantially increases the heat transfer rate.

Kassoy and Zebib [129] have modeled the fracture zone as a two-dimensional vertical channel of a porous medium bounded by vertical impermeable walls with temperature increasing linearly with depth. The channel is charged with heated liquid from below which rises due to buoyancy effects. Solutions to the entry and fully developed flow at large Rayleigh numbers (based on the width of the channel and temperature difference between top and bottom of the channel) were obtained by a singular perturbation technique. It was shown that steady fully developed two-dimensional flow is not possible for certain ranges of the Rayleigh number and channel width. In a subsequent paper, Kassoy and Goyal [130] considered the problem of convection in a fault zone and the adjacent reservoir. It was postulated that the fault is charged from below by liquid which has been heated in a basement fracture system. Because of the presence of cap rock, water is pushed out of the fault and flows horizontally into the adjacent reservoir with zero vertical permeability. A quasi-analytical theory was developed for convection in the fault zone and the reservoir at high Rayleigh numbers. A similar model [131] was used to study the forced convection in the Monroe hot springs geothermal system in central Utah.

g. Numerical Simulation of Real Geothermal Systems

A three-dimensional transient model based on the integrated finite difference method was used by Sorey [132, 133] to simulate the natural hot springs discharge at Long Valley Caldera, California. The model consists of five horizontal layers: two permeable layers with permeabilities of 0.03 to 0.35 $\times$ 10^{-12} m^2 (0.32 to 3.78 $\times$ 10^{-12} ft^2) and with different porosities (0.1 and 0.05), bounded by cap rock at the top, and by two impermeable layers from below. Heating is provided by a magma chamber deep in the western part of the caldera. The cap rock is considered to be impermeable except along parts of the caldera rim where recharging groundwater moves downward by way of fractures in the west and northwest. Hot water flows upward along faults to discharge at lower altitudes in the gorge springs. The density difference between the recharging cold water and discharging warm water provides the additional driving force for the movement of groundwater. The discharge rate is estimated to be 0.25 to 0.3 m^3/s (8.8 to 10.6 ft^3/s) at 210°C (410°F) with a total heat discharge of 3×10^8 W ($\sim 10^9$ Btu/h). The numerical simulation shows that in order to sustain these discharge characteristics after a period of 30,000 years, a magma chamber at a depth of 6 km (3.72 mi) with a permeable layer 1.5 to 2.5 km (0.92 to 1.55 mi) thick is required.

Numerical simulation of the Wairakei geothermal field for the period from 1953 to 1963 was performed by Mercer and coworkers [134, 135]. The model assumes the existence of a principal production zone, the Wairoa aquifer, overlaid by the less permeable Huka Falls formation and underlain by the impermeable Wairakei ignimbrites. Fluid is allowed to be transmitted vertically between the Wairoa aquifer and the overlying Wairakei breccia through the intervening Huka Falls formation. As a first approximation the governing transient three-dimensional equations were integrated vertically, resulting in a two-dimensional areal model that neglects the vertical convective movement. The resulting equations are thus in terms of vertically averaged quantities. The boundary conditions for the areal model are assumed to be impermeable surfaces at constant temperature. The areal model was solved numerically on the basis of the finite element method together with Galerkin's procedures. A trial and error method was used to adjust the unknown bottom heat flux distribution so that the calculated steady-state temperature distribution and the heat flux leaving the top surface matched those of the observed data in 1958 prior to exploitation. The steady-state distribution was then used as the initial conditions for the simulation of the Wairakei geothermal reservoir under production.

Both the observed and the calculated temperature distribution in 1962 were found to differ only slightly from those in 1958. The calculated mass discharge rate is much higher than the observed value, while the pressure drop history displays only a qualitative agreement with field data. The calculated results do not match with field data after 1963, presumably due to a considerable amount of steam that had formed in the Wairakei geothermal field since that time.

In a recent paper, Mercer and Faust [136] have revised their model by (1) allowing mass recharge through the underlying Wairakei ignimbrites, (2) increasing the vertical permeability of the Huka Falls formation, and (3) allowing two-phase flow in the formation. The model was used to simulate the Wairakei geothermal field from 1953 to 1973. The numerical results for temperature, pressure, and mass discharge versus time obtained by the revised model are shown in better agreement with field data.

A transient three-dimensional radial numerical model was used by Tansev and Wasserman [137] to simulate the Heber geothermal field under production and reinjection. The reservoir was assumed to be a nearly circular area of 7500 acres which extended from 2000 to 6000 ft (~ 600 to 1800 m) in depth. The initial temperature in the reservoir was assumed to be a function of the radial coordinate and independent of the vertical coordinate. The reservoir consists of three horizontal layers where 100 MW of energy is produced from each layer. The produced fluid is initially at 360°F (182°C) and the injected fluid at 200°F (93°C). Water at 200°F (93°C) is recharged continuously at the outer boundary of the reservoir. The numerical results show that the bottom hole temperature drops from 360°F (182°C) to about 320°F (160°C) after 40 years of production. During this period of time, 30 percent of heat in place is recovered.

D. CONVECTIVE COOLING OF INTRUSIVES OR HEATED BODIES IN HOT-WATER GEOTHERMAL SYSTEMS

In this section, analytical solutions for single-phase convective cooling of intrusives or heated bodies in hot-water geothermal systems with infinite or semi-infinite extent will be discussed. An exact solution for natural convection about a concentrated heat source will first be presented; this will be followed by the discussion of approximate solutions based on perturbation methods and boundary-layer approximations.

1. Exact Solutions for Natural Convection about a Point Source of Heat

A small intruded hot body in a large geothermal reservoir can be considered to be a point source of heat in a fluid-filled porous medium. Hickox and Watts [138] have considered the problems of natural convection about a point source in an infinite porous medium as well as a point source located on the lower, insulated boundary of a semi-infinite region. The governing partial differential equations in spherical polar coordinates (for the former case) and cylindrical polar coordinates (for the latter case) are transformed into two sets of ordinary differential equations by similarity transformations. Exact solutions for the axisymmetric flow and temperature fields are obtained by solving the sets of ordinary differential equations numerically. For a Rayleigh number ($\mathrm{Ra}_p = qg\beta K/2\pi\alpha^2\mu c_p$, where q is the heat generated by the point source per unit time) of 0.1, the temperature distribution is found to be closely approximated by the steady-state heat conduction solution with isotherms approximately of hemispherical or spherical shape. When $\mathrm{Ra}_p = 100$, the solutions exhibit a plumelike behavior in agreement with the boundary-layer model proposed by Wooding [139]. Streamlines (solid lines) and isotherms (dashed lines) about a point source in an infinite geothermal reservoir at $\mathrm{Ra}_p = 10$ are shown in Fig. 7.

2. Perturbation Solutions for Small Rayleigh Numbers

The problem of transient and steady-state temperature distribution and flow pattern in natural convection around a concentrated heat source in an infinite porous medium at low Rayleigh numbers (Ra_p) was considered by Bejan [140]. The dependent variables were expanded in power series in terms of the Rayleigh number, resulting in a series of linear subproblems. Analytic solutions up to 0 (Ra_p) were obtained for the transient case and up to 0 (Ra_p^2) for the steady-state case. Using this approach, Hickox [141] showed that the solutions of transient and steady-state convection about a concentrated heat source and a finite vertical line source in a semi-infinite porous medium can be obtained by superposition.

3. Boundary-Layer Approximations for High Rayleigh Numbers (for Natural Convection) and High Peclet Numbers (for Mixed Convection)

The numerical results of natural convection in an island geothermal reservoir led Cheng [121] to observe that steep temperature gradients exist near the heated surface when the

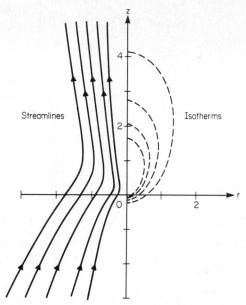

FIG. 7 Streamlines (solid lines) and isotherms (dashed lines) about a point source (at the origin) in an infinite geothermal reservoir [138].

Rayleigh number is large. In other words, the thermal boundary layer is thin under this situation. Thus, for natural convection in a porous medium at large Rayleigh numbers, the boundary-layer approximations analogous to the classical boundary-layer theory can be applied. Using an order-of-magnitude estimate, Cheng [142] has shown that the boundary-layer equations in terms of stream function ψ and temperature T for steady two-dimensional Darcian flow adjacent to a heated vertical surface are

$$\frac{\partial^2 \psi}{\partial y^2} = \frac{\rho_\infty g \beta K}{\mu} \frac{\partial T}{\partial y} \tag{49}$$

$$\frac{\partial^2 T}{\partial y^2} = \frac{1}{\alpha} \left(\frac{\partial \psi}{\partial y} \frac{\partial T}{\partial x} - \frac{\partial \psi}{\partial x} \frac{\partial T}{\partial y} \right) \tag{50}$$

where the stream function ψ is defined as

$$u = \frac{\partial \psi}{\partial y} \qquad v = - \frac{\partial \psi}{\partial x} \tag{51}$$

with the x axis placed on the heated surface and the y axis pointing toward the porous medium.

Similarly, for steady two-dimensional Darcian flow adjacent to a horizontal surface with the x axis placed on the heated surface and the y axis pointing vertically upward toward the porous medium, the boundary-layer equations are

$$\frac{\partial^2 \psi}{\partial y^2} = - \frac{K \rho_\infty g \beta}{\mu} \frac{\partial T}{\partial x} \tag{52}$$

$$\frac{\partial^2 T}{\partial y^2} = \frac{1}{\alpha} \left(\frac{\partial \psi}{\partial y} \frac{\partial T}{\partial x} - \frac{\partial \psi}{\partial x} \frac{\partial T}{\partial y} \right) \tag{53}$$

where the stream function is defined by Eq. (51). Equations (52) and (53) could have been obtained directly from Eqs. (39) and (40) by assuming that $(\partial^2/\partial X^2) \ll (\partial^2/\partial Y^2)$ and neglecting the unsteady term.

a. Similarity Solutions

Equations (49) and (50) or Eqs. (52) and (53) with appropriate boundary conditions can sometimes lead to closed-form solutions for convective heat transfer problems in a porous medium. In this section similarity solutions for steady natural and mixed convection about intrusives such as dikes or sills in a reservoir with infinite extent will be discussed. It will be assumed that the wall temperature of the heated surface is a power function of distance, that is, $T_w = T_\infty + \tilde{A}x^\lambda$, where x is the coordinate on the heated surface and $\tilde{A}$ and λ are positive constants and T_∞ is the temperature at infinity. Solutions to problems of prescribed surface heat flux can be obtained by a simple transformation of variables [143]. Solutions for plume rise above a horizontal-line source and injection of hot water along wells and vertical fissures can be obtained by a modification of boundary conditions for the problem of natural convection about a heated vertical surface [145]. Similarity solutions for other prescribed wall temperature and unsteady problems can be found in the work by Johnson and Cheng [144].

Natural Convection about a Dike. A dike is a vertical or nearly vertical sheetlike intrusive whose surface becomes relatively impermeable after it has been cooled by groundwater or country rock. Dike complexes in a volcanic formation often provide the heat source for the heating of geothermal fluids. The problem of natural convection about a dike can be idealized as a vertical heated plate embedded in a water-saturated porous medium. A similarity solution of the problem has been obtained by Cheng and Minkowycz [145], who show that the thermal boundary-layer thickness δ_T and the local surface heat flux q'' are given by

$$\frac{\delta_T}{x} = \frac{\eta_T}{\sqrt{\mathrm{Ra}_x}} \tag{54}$$

and
$$q'' = -k\left(\frac{\partial T}{\partial y}\right)_{y=0} = k\tilde{A}^{3/2}\left(\frac{\rho_\infty g \beta K}{\mu \alpha}\right)^{1/2} x^{(3\lambda-1)/2}[-\theta'(0)] \tag{55}$$

$$= k(T_w - T_\infty)^{3/2} \sqrt{\frac{K\rho_\infty \beta g}{\mu \alpha x}}[-\theta'(0)]$$

where $\mathrm{Ra}_x = \rho_\infty g \beta K (T_w - T_\infty)x/\mu\alpha$ is the local Rayleigh number, η_T the boundary thickness parameter, and $\theta'(0)$ the dimensionless temperature gradient at the wall where the dimensionless temperature θ is defined as $\theta \equiv (T - T_\infty)/(T_w - T_\infty)$. The values of η_T and $[-\theta'(0)]$ for different values of λ are given in Table 2. Equation (55) shows that $\lambda = \frac{1}{3}$ is for the case of uniform prescribed heat flux. The average surface heat flux can be obtained by an integration of Eq. (55) to give

$$\bar{q}'' = \frac{1}{L}\int_0^L q''(x)\,dx = \frac{2k\,(1+\lambda)^{3/2}}{(3\lambda+1)}(\overline{T_w - T_\infty})^{3/2}\sqrt{\frac{K\rho_\infty \beta g}{\mu \alpha L}}[-\theta'(0)] \tag{56}$$

where
$$(\overline{T_w - T_\infty}) = (1/L)\int_0^L (T_w - T_\infty)\,dx$$

$= \tilde{A}L^\lambda/(1+\lambda)$ and L is the height of the dike. Equations (55) and (56) can be rewritten

TABLE 2 Values of η_T and $-\theta'(0)$ for Different λ [145]

λ	η_T	$-\theta'(0)$
$-\frac{1}{3}$	7.2	0
$-\frac{1}{4}$	6.9	0.1621
0	6.3	0.4440
$\frac{1}{4}$	5.7	0.6303
$\frac{1}{3}$	5.5	0.6788
$\frac{1}{2}$	5.3	0.7615
$\frac{3}{4}$	4.9	0.8926
1	4.6	1.001

in the following dimensionless form:

$$\frac{\mathrm{Nu}_x}{\sqrt{\mathrm{Ra}_x}} = -\theta'(0) \tag{57}$$

$$\frac{\overline{\mathrm{Nu}_L}}{\mathrm{Ra}_L^{1/2}} = \frac{2(1 + \lambda)^{3/2}}{(1 + 3\lambda)}[-\theta'(0)] \tag{58}$$

where $\mathrm{Nu}_x = q''x/k(T_w - T_\infty)$ and $\overline{\mathrm{Nu}_L} = \bar{q}''L/k(T_w - T_\infty)$ are the local and average Nusselt number, respectively, and $\overline{\mathrm{Ra}_L} = \rho_\infty g\beta KL(T_w - T_\infty)/\mu\alpha$.

Equations (54) to (58) are based on Darcy's law with non-Darcy, boundary, and thermal dispersion effects neglected. Recently, Plumb and Huenefeld [36] have extended Cheng and Minkowycz's analysis [145] to include the non-Darcy effect in the natural-convection porous layer adjacent to a heated vertical plate. They use the generalized Forchheimer equation, Eq. (14), as the momentum equation and obtain a similarity for the problem. They found that the inertia effects tend to increase the thermal boundary-layer thickness while decreasing the heat transfer rate.

To assess the boundary effect in natural-convection porous layers, Evans and Plumb [20] have reexamined the problem considered by Cheng and Minkowycz [145]. Evans and Plumb formulated the problem based on Eq. (16) plus convective terms with no-slip boundary conditions imposed. A comparison of their numerical results with the similarity solution by Cheng and Minkowycz [145] reveals that the boundary effect tends to decrease the local heat transfer rate for $\mathrm{Da}_x > 10^{-7}$, where $\mathrm{Da}_x = K/x^2$, with x denoting the location at the vertical heated plate.

Experimental investigation of natural convection in porous media adjacent to a vertical plate has been performed by Evans and Plumb [20], while that for an inclined plate has been performed by Cheng and coworkers [37, 39]. Evans and Plumb [21] observed that their data of the dimensionless temperature profiles θ and the local Nusselt number Nu_x at $\mathrm{Ra}_x < 400$ are in good agreement with Cheng and Minkowycz's similarity solution. Moreover, they observed that temperature fluctuations began to appear at a local Rayleigh number higher than 400, which lead to the scattering of data at higher values of the Rayleigh number. Evans and Plumb's data for the dimensionless temperature profile θ [where $\theta = (T - T_\infty)/(T_w - T_\infty)$] versus the similarity variable η (where $\eta = \sqrt{\mathrm{Ra}_x}\, y/x$) together with Cheng and Minkowycz's similarity solution are presented in Fig. 8, which shows that they are in good agreement.

Similarly, Cheng and coworkers' experimental data [37, 39] at moderate values of the local Rayleigh number (Ra_x) for both the vertical and inclined heated plate in water-saturated glass beads compare well with the similarity solution. However, they found that

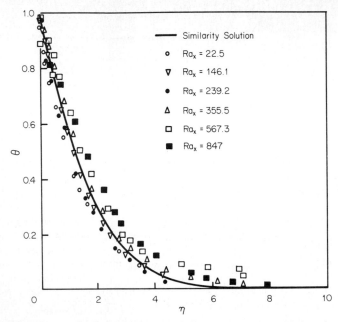

FIG. 8 Comparison of similarity solution with experimental data for dimensionless temperature distribution in natural-convective flow about a heated vertical surface in a porous medium [20].

the local Nusselt number (Nu_x) data are lower than those predicted by the theory at high local Rayleigh numbers. The value of the local Rayleigh number at which the data of the local Nusselt number begins to deviate from the similarity solution depends on the local Darcy number Da_x (where $Da_x = K/x^2$). The decrease in the value of the local Nusselt number data at high local Rayleigh numbers is probably due to the boundary and non-Darcy effects and the onset of secondary flow.

Natural Convection about a Stock.   An intrusive with cylindrical shape is called a *stock*. The problem of natural convection about a stock (modeled as a vertical heated cylinder in a saturated porous medium) was considered by Minkowycz and Cheng [146]. For a given temperature distribution, they found that the ratio of the local surface heat flux of a cylinder (q_c'') to that of a flat plate (q'') is a nearly linear function of the curvature parameter $\xi = 2x/(r_0 Ra_x^{1/2})$ (with r_0 denoting the radius of the cylinder); i.e.,

$$\frac{q_c''}{q''} = 1 + C_2\xi \qquad (59)$$

where q'' is given by Eq. (55) and the constant C_2 for different values of λ is tabulated in Table 3. The values of C_2 were obtained by fitting Eq. (59) to the numerical results given by Minkowycz and Cheng [146]. Similarly, the numerical results for the average heat flux ratio $\overline{q}_c''/\overline{q}''$ can be approximated by

$$\frac{\overline{q}_c''}{\overline{q}''} = 1 + 0.26\xi_L \qquad (60)$$

TABLE 3 The Values of C_2 for Different λ

λ	C_2
0	0.30
¼	0.23
⅓	0.21
½	0.20
¾	0.17
1	0.15

where $\xi_L \equiv 2L/(r_0 \overline{Ra}_L^{1/2})$ and $\overline{q}''$ is given by Eq. (56). Note that both $\xi \rightarrow 0$ and $\xi_L \rightarrow 0$ for a flat plate where $r_0 \rightarrow \infty$. Equations (59) and (60) show that for the same surface area, a stock transfers more heat than a dike.

Plume Rise from a Horizontal Line Source. Natural convection about a horizontal intrusive with a horizontal length considerably larger than other dimensions can be approximated as a horizontal-line source of heat in a fluid-filled porous medium. Similarity solutions for this problem at large Rayleigh numbers based on the boundary-layer approximations were obtained by Wooding [139] as well as Yih [147]. Cheng and Minkowycz [145] have pointed out that this problem is actually a special case of natural convection about a vertical heated surface with $\lambda = -\frac{1}{3}$.

Injection of Hot Water along Wells or Vertical Fissures. Wastewater from geothermal power plants is often reinjected through injection wells into the formation. If the temperature and the horizontal discharge velocity of the injected water along the well or fissure are of the form $T_w = T_\infty + Ax^\lambda$ and $v = \tilde{a}x^N$, Cheng [148] has shown that similarity solutions for the problem exist if $N = (\lambda - 1)/2$. The expression for thermal boundary-layer thickness is also given by Eq. (54) with the value η_T increasing with the injection parameter $f_w = -2\tilde{a}/(\alpha\rho_\infty g\beta KA/\mu)^{1/2}(1 + \lambda)$. The more realistic case of a constant discharge velocity at a uniform temperature was considered by Merkin [149].

Natural Convection about a Sill or Bedrock. A sill is a horizontal intrusive which can be modeled as a horizontal heated surface in a porous medium [150]. For natural convection about a sill or bedrock, Cheng and Chang [150] have obtained the following expressions for the thermal boundary-layer thickness and the local and average heat transfer rates:

$$\frac{\delta_T}{x} = \frac{\eta_T}{Ra_x^{1/3}} \tag{61}$$

$$q'' = k\tilde{A}^{4/3}\left(\frac{K\rho_\infty g\beta}{\mu\alpha}\right)^{1/3} x^{(4\lambda-2)/3}[-\theta'(0)] \tag{62}$$

$$= k(T_w - T_\infty)^{4/3}\left(\frac{K\rho_\infty g\beta}{\mu\alpha x^2}\right)^{1/3}[-\theta'(0)]$$

$$\overline{q}'' = \frac{1}{L}\int_0^L q''(x)\,dx$$

$$= \frac{3k}{4\lambda + 1}(T_w - T_\infty)^{4/3}(1 + \lambda)^{4/3}\left(\frac{K\rho_\infty\beta g}{\mu\alpha L^2}\right)^{1/3}[-\theta'(0)] \tag{63}$$

TABLE 4 Values of η_T and $-\theta'(0)$ for Different λ [150]

λ	η_T	$-\theta'(0)$
0	5.5	0.420
0.5	5.0	0.8164
1.0	4.5	1.099
1.5	4.0	1.351
2.0	3.7	1.571

where the values of η_T and $-\theta'(0)$ are tabulated in Table 4. Equation (62) shows that $\lambda = \frac{1}{2}$ corresponds to the case of uniform surface heat flux. Equations (62) and (63) can be rewritten in the following dimensionless form:

$$\frac{\mathrm{Nu}_x}{\mathrm{Ra}_x^{1/3}} = -\theta'(0) \tag{64}$$

$$\frac{\overline{\mathrm{Nu}_L}}{\overline{\mathrm{Ra}_L^{1/3}}} = \frac{3(1 + \lambda)^{4/3}}{(1 + 4\lambda)} [-\theta'(0)] \tag{65}$$

where $\overline{\mathrm{Nu}_L}$ and $\overline{\mathrm{Ra}_L}$ are defined in Eq. (58). Similar analyses for natural convection above a heated circular horizontal surface in a porous medium were considered by McNabb [151] and by Cheng and Chau [152].

Natural Convection about a Cylindrical and Spherical Intrusive.   Steady natural convection about two-dimensional and axisymmetric isothermal intrusives of arbitrary shape was considered by Merkin [153]. In Merkin's analysis, a curvilinear orthogonal coordinate was used. With the boundary-layer approximation and the component of the buoyancy force normal to the impermeable heated surface neglected, a similarity solution was obtained. The solution contains a local configuration parameter which can be obtained once the shape of the heated surface is specified. Applying Merkin's solution to the case of an isothermal horizontal cylinder and an isothermal sphere, Cheng [154] has obtained the following expressions for the local Nusselt number (Nu_Φ) and the average Nusselt number (Nu_D) for these cases:

$$\frac{\mathrm{Nu}_\Phi}{\sqrt{\mathrm{Ra}_D}} = \begin{cases} 0.628 \dfrac{\sin \Phi}{\sqrt{1 - \cos \Phi}} & \text{for a horizontal cylinder} \\[4mm] 0.628 \dfrac{\sin^2\Phi}{\sqrt{(\cos^3\Phi)/3 - \cos \Phi + \frac{2}{3}}} & \text{for a sphere} \end{cases} \tag{66a} \tag{66b}$$

$$\frac{\mathrm{Nu}_D}{\sqrt{\mathrm{Ra}_D}} = \begin{cases} 0.565 & \text{for a horizontal cylinder} \\ 0.362 & \text{for a sphere} \end{cases} \tag{67a} \tag{67b}$$

where $\mathrm{Nu}_\Phi = q''D/k(T_w - T_\infty)$, $\mathrm{Nu}_D = \bar{q}''D/k(T_w - T_\infty)$, and $\mathrm{Ra}_D = \rho_\infty g\beta K(T_w - T_\infty)D/\mu\alpha$, and where Φ is the angle measured from the lower stagnation point and D is the diameter of the cylinder and the sphere.

Mixed Convection about a Dike.   Similarity solutions for mixed convection in a porous medium adjacent to an inclined heated surface with inclination angle α_0 (with respect to the vertical) and with external free-stream velocity $u_\infty = Bx^n$ were obtained by Cheng [155]. The value of n is related to the angle of incidence β_0 by $n = \beta_0/(\pi -$

β_0), where $0 < \beta_0 < \pi/2$. It turns out that similarity solutions exist for the problem if the wall temperature parameter λ is equal to the free-stream parameter n. The governing parameter of the problem is $\mathrm{Gr}_{x,\alpha}/\mathrm{Re}_x = g \cos \alpha_0 \tilde{A}\beta K/B\nu$, where $\mathrm{Gr}_{x,\alpha} \equiv K\beta g \cos \alpha_0 (T_w - T_\infty)x/\nu^2$ and $\mathrm{Re}_x \equiv u_\infty x/\nu$. The thermal boundary-layer thickness and the local Nusselt number are of the form

$$\frac{\delta_T}{x} = \frac{\eta_T}{\mathrm{Pe}_x^{1/2}} \tag{68}$$

$$\frac{\mathrm{Nu}_x}{\mathrm{Pe}_x^{1/2}} = -\theta'(0) \tag{69}$$

where $\mathrm{Pe}_x \equiv u_\infty x/\alpha$ is the Peclet number. The values of η_T and $-\theta'(0)$ as a function of $\mathrm{Gr}_{x,\alpha}/\mathrm{Re}_x$ for aiding flows (i.e., the buoyancy force has a component in the direction of the free-stream velocity) of the two special cases are given in Ref. 155. It is relevant to note that the boundary layer approximation is applicable to mixed convection problems with high Peclet numbers.

Mixed Convection about a Sill or Bedrock.   Cheng [156] has considered the problem of mixed convection in parallel flow as well as stagnation-point flow about a heated horizontal surface embedded in a porous medium. The former is applicable to the mixed convective cooling of a sill or bedrock, while the latter is applicable to cold-water injection near the heated bedrock in a geothermal reservoir. For both cases, the velocity at infinity can be represented by $u_\infty = Bx^n$, where $n = 0$ is for parallel flow and $n = 1$ is for the stagnation-point flow. It was found that similarity solutions exist if $\lambda = (3n + 1)/2$ and that the governing parameter for the problem is $\mathrm{Ra}_x/\mathrm{Pe}_x^{3/2} = (\rho_\infty g\beta K\tilde{A}/\mu B)(\alpha/B)^{1/2}$, which is independent of x. The thermal boundary-layer thickness and the local Nusselt number are also given by Eqs. (68) and (69) where the values η_T and $-\theta'(0)$ as a function of $\mathrm{Ra}_x/\mathrm{Pe}_x^{3/2}$ for the two special cases are presented in Ref. 156.

b. Integral Methods for Convective Heat Transfer about Intrusives

As in the classical heat transfer theory, integral methods can provide approximate solutions to problems where similarity solutions do not exist. To assess the accuracy of the integral methods for porous-layer flow, Cheng [142] has obtained approximate solutions for free and mixed convection in porous layers adjacent to vertical and horizontal heated surfaces. A comparison of these results with those based on the similarity solution (see Table 5) shows that the integral method is accurate for the predictions of the Nusselt number for all of the cases considered. Hardee [157] has considered the problems of steady natural convection about a heated horizontal cylinder and a heated sphere embed-

TABLE 5 Comparison of Nusselt Numbers Based on Similarity Solutions [145, 150] and Integral Method [142] for Free Convection in Porous Layers Adjacent to Vertical and Horizontal Isothermal Heated Surfaces

Temperature profiles	Vertical heated surfaces, $\mathrm{Nu}_x/(\mathrm{Ra}_x)^{1/2}$	Horizontal heated surfaces, $\mathrm{Nu}_x/(\mathrm{Ra}_x)^{1/3}$
Exact	0.444	0.420
$\theta = (1 - \eta)^2$	0.447	0.390
$\theta = (1 - \eta)^3$	0.463	0.415
$\theta = (1 - 1/2\eta)e^{-\eta}$	0.451	0.408

ded in a fluid-saturated porous medium, using the integral method. Hardee's approximate solution agrees well with those given by Eqs. (66) and (67).

c. Refinements of the Boundary-Layer Approximations

Boundary-layer approximations are accurate only for thin thermal boundary layers at high Rayleigh numbers (for natural convection) or high Peclet numbers (for mixed convection). Since there are many situations where the Rayleigh or Peclet numbers are of moderate values, it is of interest to refine the boundary-layer analyses so that results can be applicable to these situations. This can be achieved by the method of matched asymptotic expansions as discussed by Cheng and Chang [158]. For free convection in a porous medium, the small parameter in the perturbation series is $1/\mathrm{Ra}_r^{1/3}$ for a horizontal plate, and $1/\mathrm{Ra}_r^{1/2}$ for a vertical plate, where $\mathrm{Ra}_r = \rho_\infty g \beta K (T_r - T_\infty) L / \mu \alpha$, with L denoting the length of the plate. For the first-order inner problem, the governing equations reduced to the boundary-layer approximations [145, 150]. The second-order corrections are mainly due to fluid entrainment along the edge of the thermal boundary layer. Figure 9 is a comparison of the local Nusselt number (Nu_x) based on the boundary-layer theory and the second-order theory for natural convection about a vertical heated surface in a porous medium. It is shown that the second-order corrections are zero for isothermal ($\lambda = 0$) and constant heat flux ($\lambda = \frac{1}{2}$) surfaces. For all other temperature distributions, the second-order corrections for the local Nusselt number are all positive. The streamwise

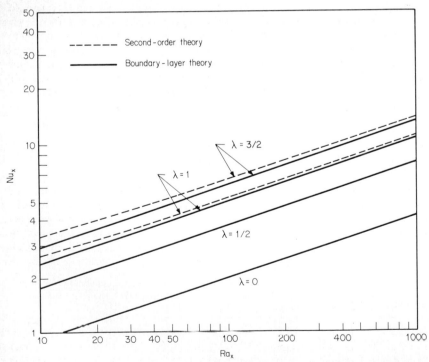

FIG. 9 Comparison of boundary-layer theory and the second-order theory for local Nusselt number versus local Rayleigh number in natural-convective flow about a heated vertical surface in a porous medium [158].

heat conduction and the upward drift–induced friction were taken into consideration in the third-order theory [159, 160], where it is shown that the third-order corrections to the local Nusselt number are also positive for most of the cases. The higher-order theories have a profound effect on the velocity profiles both inside and outside the boundary layer, but have a relatively small effect on temperature distribution and consequently the heat transfer rate.

d. Vortex Instability of Thermal Boundary Layers

Recent experiments by Cheng and Ali [39] show that large amplitude fluctuations occur downstream of the flow adjacent to an upward-facing inclined heat surface in a porous medium. The fluctuations appear to be periodic, with amplitudes increasing with the angle of inclination α_0 (with respect to the vertical). Based on a linear stability theory, Cheng and coworkers [161–164] have investigated theoretically the conditions for the onset of vortex instability in natural and mixed convection in a porous medium adjacent to a heated surface with a power-law variation of wall temperature. In their analyses, it was assumed that the basic undisturbed states are the steady two-dimensional boundary-layer flow discussed in Sec. D.3. They found that the critical value for the onset of vortex instability in natural convection about an inclined isothermal surface in a porous medium is $\mathrm{Ra}_{x,\alpha}^{*} \tan^2 \alpha_0 = 120.7$ [where $\mathrm{Ra}_{x,\alpha} = K\rho_\infty \beta g \cos \alpha_0 (T_w - T_\infty)x/\mu\alpha$ with $\alpha_0 \neq \pi/2$], while that for an isothermal horizontal surface is $\mathrm{Ra}_x^* = 33.4$, where $\mathrm{Ra}_x = K\rho_\infty \beta g (T_w - T_\infty)x/\mu\alpha$. It follows that the larger the inclination angle α_0, the more susceptible is the flow for the vortex mode of disturbances, and in the limit of zero inclination (i.e., a vertical heated surface) the flow is stable with respect to this form of disturbances. On the other hand, Hsu and Cheng [163, 164] found that the critical parameter for the onset of vortex instability in mixed convection in a porous medium adjacent to an inclined surface is $\mathrm{Pe}_x^* \tan^2 \alpha_0$, while that for a horizontal surface is Pe_x^*. The critical values, however, depend on the mixed-convection parameters, which, in turn, depend on the free-stream velocity.

E. HEAT TRANSFER IN TWO-PHASE AND DRY-STEAM GEOTHERMAL SYSTEMS

The heat transfer processes in steam-water two-phase geothermal systems as well as dry-steam geothermal systems will be discussed in this section. A two-phase geothermal system may be classified further as vapor-dominated (with water saturation < 0.5) or liquid-dominated (with water saturation > 0.5). A hot-water or a liquid-dominated geothermal system may eventually evolve into a dry-steam or a vapor-dominated geothermal system if the withdrawal rate is higher than the recharge rate; the reverse may also occur.

1. Governing Equations

To study the transport processes in a two-phase geothermal system, it is generally assumed that (1) local thermal equilibrium prevails, (2) capillary pressure is negligible, (3) the rock formation is nondeformable, (4) Darcy's law is applicable to both phases provided that the relative permeability concept is introduced, and (5) pressure work and viscous dissipation are negligible. The macroscopic equations of mass and momentum for two-phase flow in a porous medium have been obtained by Whitaker [165] and by Gray et al. [166] using the method of volumetric averaging. Gray et al. [166] showed that the general macroscopic momentum equations reduce to Darcy's law when the convective and inertia terms are neglected. The macroscopic energy equation for two-phase flow in a porous medium in terms of internal energy was obtained by Witherspoon et al. [167], while that in terms of enthalpy was obtained by Faust [168] and by Faust and Mercer [169]. In addition to the macroscopic conservation equations, the state equation and the auxiliary relations between the relative permeabilities of stream and water as a function of saturation must be known for the complete formulation of the problem.

Truesdell and White [170] have speculated that most of the liquid water in a vapor-dominated geothermal system is more or less immobilized in small pores (due to surface tension and adsorption), and that steam is the pressure-controlling phase which dominates the larger fractures and voids of the reservoir. Based on this conceptual model, Moench [171] has derived a set of simplified equations for vapor-dominated geothermal systems where water saturation is treated as a mass source or sink that does not move in response to the pressure gradient.

2. Two-Phase Flow in a Fault Zone

A steady one-dimensional model was used by Donaldson [172] to study two-phase upward flow in fissures or fault zones where hot water enters at depth and rises upward when boiling commences, due to the pressure drop. Computations were carried out for different prescribed values of the mass flow rate to permeability ratio G/K, which is the controlling parameter of the problem. It was found that at the point where boiling commences, there was a change in slope in the vertical temperature profiles and a sudden jump in steam saturation. The latter is due to the relative permeability effects. The effect of a step increase in the vertical permeability was also investigated. It was found that a jump in water saturation occurs whenever there is a sudden change in permeability. Donaldson's analysis was later extended by Sheu et al. [173], who modified the model to include a surface water zone overlying the two-phase zone, and also to include the pressure work term in the energy equation. The condition for which a two-phase zone exists was obtained in terms of G/K, of the permeability–thermal conductivity ratio K/k, and of the energy flux–thermal conductivity ratio q''/k.

3. Heat Transfer and Fluid Flow Characteristics in a Two-Phase Geothermal Reservoir before Production

a. One-Dimensional Two-Phase Counterflow

When the temperature of the bedrock exceeds the saturation temperature, boiling heat transfer will initiate. However, if the permeability of the reservoir is low enough that the Rayleigh number (Ra) is sufficiently below the critical Rayleigh number for the onset of natural convection, one-dimensional two-phase counterflow of saturated steam and liquid may occur in a geothermal reservoir [174]. Such a model was used by Sondergeld and Turcotte [23] to study two-phase flows in a geothermal reservoir where steam was generated at the bottom (heated) surface and condensed at the top (cooled) surface. The necessary condition for the existence of a two-phase zone (with isothermal temperature) in between was investigated.

Another steady one-dimensional model of a counterflowing steam-water mixture above quiescent saturated water at a given pressure in a geothermal reservoir was considered by Schubert and Straus [175]. It was found that counterflowing of the steam-water mixture exists only if the temperature gradient in the underlying quiescent saturated water lies between specified limits. Outside these limits, there is an abrupt jump in the steam fraction and the temperature gradient at the interface between the two-phase region and the underlying water. When the permeability–thermal conductivity ratio K/k is small, the thickness of the two-phase region is essentially independent of K/k; for large values of K/k the thickness of the region increases dramatically with K/k. For fixed values of K/k the thickness of the steam-water region increases substantially with a decreasing temperature gradient of the underlying water. A similar model was used by Martin et al. [176] to study the temperature and pressure distributions in a two-phase geothermal reservoir. Numerical solutions were carried out for the case where the temperature of the cap rock as well as conduction heat flux from below were specified. Martin et al. [176] found that there were two different numerical solutions that satisfied these boundary conditions for a reservoir bounded by impermeable surfaces at the top and bottom. One solution is a high hot-water saturation in which hot water is the principal mobile phase with the pressure gradient approximately that of hot water. The other solution is a high steam saturation case in which steam is the principal mobile phase, characterized by a small pressure gradient approximately equal to that of steam with a nearly isothermal temperature distribution. However, if a steam-water interface exists near the bottom (heated) surface, only one solution is possible when the temperature of the cap rock and the total energy transfer rate are specified. For comparison purposes, computations were carried out for two different relative permeability curves. The results show that the effects of different relative permeability curves on temperature and pressure distributions were small, although their effects on water saturation were pronounced.

b. Convection in a Two-Phase Geothermal Reservoir

Two-phase convection will occur in a geothermal reservoir when the temperature of the bedrock exceeds the saturation temperature and the Rayleigh number (Ra) is higher than the critical Rayleigh number for the onset of natural convection [174]. Schubert and Straus [177] applied a homogeneous model (valid only for a mixture with small amounts of steam) to study the onset of natural convection in a two-phase geothermal reservoir. They found that the two-phase geothermal reservoirs were more susceptible to convection than those of single-phase. Experimental investigation of two-phase convection in a porous medium heated from below and cooled from above has been reported by Sondergeld and Turcotte [23]. It was observed that polyhedral convective cells exist in the two-phase zone in a porous layer heated from below. This suggests that ore mineralization under the conditions of boiling may appear as periodic zones of concentration and deposition in the formation.

The problem of transient-boiling heat transfer around a pluton in a two-dimensional geothermal reservoir has been studied numerically by Cathles [178]. His mathematical model assumes an abrupt interface between the liquid and dry steam, and thus ignores the existence of a two-phase zone. The numerical results show that a self-supported vapor-dominated steam zone forms briefly above the intrusive. The occurrence of the vapor region is found long before the intrusive event can be detected by any surface heat flux anomaly. For example, numerical solutions show that the maximum surface heat flux occurs 10^4 years after the intrusion has taken place.

4. Heat Transfer and Fluid Flow Characteristics in a Two-Phase Geothermal Reservoir under Production

a. One-Dimensional Models

A few one-dimensional rectilinear or radial flow models with gravity neglected have been used to study two-phase nonisothermal flow in geothermal reservoirs under production. A transient finite element solution for one-dimensional horizontal two-phase flow was obtained by Faust and Mercer [179] to simulate the evolution of an initially hot-water reservoir to a two-phase reservoir resulting from fluid withdrawal. It is shown that pressure drops rapidly everywhere in the reservoir during the early stages of production. When vaporization takes place, pressure continues to drop rapidly in the rest of the reservoir, but drops only slightly in the two-phase zone. This is because, once vaporization takes place, the pressure is maintained by the formation of steam. The effect of permeability on water saturation distribution was also investigated. It was shown that water saturation decreases rapidly near the sink for reservoirs with low permeability, indicating that mass is being removed from the vicinity of the sink instead of from the entire reservoir. Experimental investigation of steady and transient one-dimensional horizontal two-phase flow in a porous medium with cold and hot water injection were performed by Arihara [180]. The numerical simulation of Arihara's depletion experiment was later carried out by Garg et al. [181] and by Thomas and Pierson [182]. Their calculated temperature and pressure distributions are in good agreement with experimental data.

Moench [171] has used a simplified two-phase model to study transient one-dimensional vertical nonisothermal flow in a vapor-dominated geothermal system under production. In this model, the reservoir initially has a small degree of water saturation. Steam is being produced at constant pressure at the top of the reservoir. The two-phase zone in the reservoir is confined by an adiabatic cap rock from the top and a liquid-vapor interface at the bottom, where it is assumed that vaporization at the interface will not occur until the liquid water overlying the interface has been depleted. Numerical results show that there is a change in slope on the vertical temperature profiles at the point where boiling commences. Moreover, it is shown that compressible work cannot be neglected in a vapor-dominated geothermal reservoir.

b. Multidimensional Models

A numerical study of nonisothermal two-phase flow in a two-dimensional geothermal reservoir under production is given by Toronyi and Farouq Ali [183]. Since the model is applicable only when steam and water coexist, they found it convenient to express the governing equations in terms of water saturation and pressure. Finite difference solutions for a cross-sectional model (including the gravitational force) and an areal model (omitting the gravitational force) were obtained to study the effects of porosity, permeability, gravity, initial pressure, and liquid-phase saturation on bottom-hole pressure and quality. It was shown that smaller initial liquid saturation or smaller permeability leads to a smaller pressure decline. In the cross-sectional formulation, the bottom-hole pressure decline curves are all concave upward. On the other hand, the pressure decline curves in the areal model may be concave upward or downward depending on the initial liquid saturation. Comparison of results from the cross-sectional and the areal models shows that the effect of the gravitational force tends to accelerate the development of a superheated region.

The first transient two-phase two-dimensional model applicable to both hot-water and vapor-dominated geothermal reservoirs was presented by Mercer and Faust [184]. The governing three-dimensional equations were integrated vertically, resulting in a two-dimensional areal model which was solved numerically by a Galerkin finite element method. To check the validity of their model, Mercer and Faust [184] applied the model to Toronyi and Farouq Ali's two-dimensional problem (with gravity neglected) and found that their results were in good agreement. The model was then used to study the development of a two-phase zone in the vicinity of a sink in a two-dimensional reservoir initially filled with hot water.

Another numerical model applicable to both hot-water and two-phase two-dimensional geothermal reservoirs is given by Lasseter [185]. This model was applied to an axisymmetric vapor-dominated geothermal reservoir, 2000 m ($\sim$6500 ft) in radius and 3000 m ($\sim$10,000 ft) in depth. The initial temperature and pressure in the reservoir were 250°C (482°F) and 35 bars ($\sim$35 atm), respectively, and the system consisted of vapor except for the bottom 250 m ($\sim$820 ft), which was liquid. The bottom impermeable surface was maintained at 250°C (482°F) and other boundaries were impermeable and adiabatic. Fluid was withdrawn at a rate of 3×10^7 kg/day (6.61×10^7 lb$_m$/day) from a well which penetrated to a depth of 1100 m (3600 ft) along the axis of the reservoir. The numerical results show that the temperature at the well drops from 250 to 220°C (from 482 to 428°F) after 1000 days and a low-pressure zone is generated near the well. Calculations indicate that after 5000 days of operation, the liquid phase would completely vaporize.

Garg et al. [181] studied both the withdrawal and injection of fluids in a circular two-phase geothermal reservoir using an areal model with gravity neglected. The reservoir was initially filled with a compressed liquid. When fluid was withdrawn at a pressure less than the saturated pressure from a central crack, *in situ* flashing flow resulted near the crack. At the same time, injection of cold water took place in two cracks of the same length located symmetrically along the reservoir boundary and in the general direction perpendicular to the central crack. It was shown that after one year of cold water injection, the injected cold water was beginning to invade the production zone. Continuous injection would adversely affect the temperature of the production fluids at a later time. The numerical results also show that the effect of injection will result in higher gross power output. This is because that injection helps to maintain the reservoir pressure and thereby sustains the production rate.

A transient three-dimensional finite difference simulator for two-phase flow in a geothermal reservoir has been developed by Thomas and Pierson [182]. Computations were carried out for a hypothetical reservoir with an area of 6000 ft $\times$ 1250 ft ($\sim$1800 m $\times$ 380 m) and 500 ft ($\sim$150 m) thick. The reservoir, having a permeability of 150 millidarcy (md) (1.593×10^{-12} ft^2) and a porosity of 0.2, was filled with a compressed liquid initially at 500°F (260°C) and 1200 psia (8.27×10^6 N/m^2). Production from the partially penetrated well was set equivalent to a constant surface steam rate of 4×10^3 lb$_m$/day ($\sim$1800 kg/day). It was found that the amount of water produced at the bottom hole decreased rapidly while that of steam increased steadily. The numerical results are presented in Fig. 10, which shows the formation of a small region of superheated steam below the cap rock near the well and the effect of water coning into the partially penetrating well during the production period.

The numerical simulation of a real two-phase geothermal system is given by Mercer and Faust [136] and by Pritchett et al. [186] for the Wairakei geothermal field. The former is a vertically integrated areal model, which has been discussed in Sec. C.4.g, while the latter is a cross-sectional model. Pritchett et al. [186] assumed that the Wairakei geothermal reservoir was 3 km (1.86 mi) thick and 5 km (3.1 mi) in length and saturated with compressed liquid initially. Figure 11 shows the vapor saturation (S_g) contours in the vertical cross section of the Wairakei geothermal field at the beginning of the years

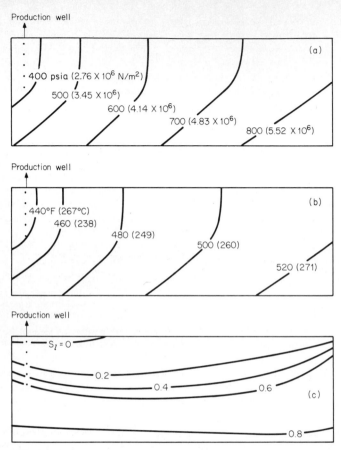

FIG. 10 Calculated pressure (*a*), temperature (*b*), and water saturation (*c*), distribution in a three-dimensional steam-water two-phase geothermal reservoir under production at the end of 12 years [182].

1955, 1960, and 1968. It may be seen that boiling started in the region directly above the production zone shortly after production began in 1955. From 1955 to 1960, the growth of the two-phase zone helped to maintain the reservoir pressure, thereby resulting in a relatively small pressure drop during this period of time. As the two-phase flow began to invade the production zone, the pressure began to drop rapidly due to the relative permeability effect. The entire production zone finally became two-phase in 1964 and 1965, resulting in a relatively small pressure drop during this period of time.

5. Boiling Heat Transfer about Intrusives

It is known that the maximum temperature of an intrusive can be as high as 1200°C (~ 2200°F) [187]. Thus, depending on pressure in the reservoir, boiling heat transfer may

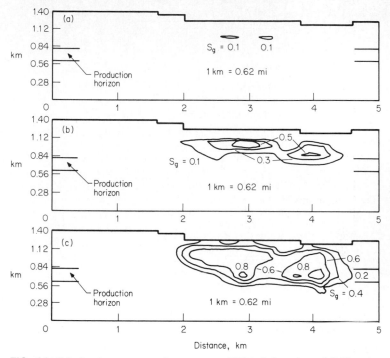

FIG. 11 Calculated vapor saturation contours of Wairakei geothermal field at the beginning of (*a*) 1955, (*b*) 1960, and (*c*) 1968 [186].

be the mechanism for the cooling of intrusives during the initial stage after their emplacement. Parmentier [188] has postulated that when boiling occurs adjacent to the dike, a thin vapor film will form in direct contact with the dike. Based on an argument similar to that given by Cathles [178], Parmentier [188] assumed that the vapor film and the subcooled water was separated by a distinct interface with no two-phase zone in between. Parmentier then obtained an approximate solution for steady film boiling about an isothermal dike. The solution of this problem was later refined by Cheng and Verma [189]. They obtained an exact solution based on a two-phase boundary-layer theory where the density of the subcooled liquid is assumed to be linearly proportional to temperature— i.e., the Boussinesq approximation is applicable—and the density of the vapor phase is assumed to be constant. According to their theory, the local Nusselt number is given by

$$\mathrm{Nu}_x = \frac{[K(\rho_\infty - \rho_g)gx/\pi\mu_g\alpha_g]^{1/2}}{\mathrm{erf}\,(\eta_{g\delta}/2)} \tag{70}$$

where the subscript g denotes the quantities associated with the vapor phase while the subscript ∞ denotes the quantities associated with the subcooled liquid at infinity and $\eta_{g\delta}$ is the film thickness parameter, which is an implicit function of the degree of super-heating of the wall $\Delta T_{\mathrm{sup}}^+$, the extent of the subcooling of the surrounding liquid $\Delta T_{\mathrm{sub}}^+$, and the property ratio of the vapor and the liquid phases $\hat{R}$—i.e.,

$$\eta_{g\delta} = f(\Delta T_{\mathrm{sup}}^+, \Delta T_{\mathrm{sub}}^+, \hat{R}) \tag{71}$$

where $\Delta T_{\text{sup}}^{+} \equiv \dfrac{c_{pg}(T_w - T_s)}{i_{lg}}$; $\Delta T_{\text{sub}}^{+} \equiv \dfrac{c_{pl}(T_s - T_\infty)}{i_{lg}}$; and

$$\hat{R} \equiv \frac{\rho_g}{\rho_\infty} \left[\frac{\mu_l \alpha_g (\rho_\infty - \rho_g) c_{pl}}{\mu_g \alpha_l \rho_\infty \beta_l i_{lg}} \right]^{1/2}.$$

and where the subscript l represents the quantities associated with the liquid phase, T_w and T_s are the wall temperature and the saturation respectively, and i_{lg} is the latent heat of vaporization at T_s. The functional relationship given by Eq.(71) is presented graphically in Ref. 189. Figure 12 shows the boundary-layer thickness of the vapor film (δ_g) and the size of the hot water zone (δ_l) for a dike located along the positive x axis. The dike is 100 m (328 ft) in length with an average surface temperature of 400°C (752°F) embedded in an aquifer with $K = 10^{-12} \text{ m}^2$ ($\sim 10^{-11} \text{ ft}^2$) at 20°C (68°F) at a mean static pressure of 10 atm (10,133 N/m²) and with $T_s = 180$°C (356°F). It is seen that the vapor film is thin compared to the thickness of the hot water zone.

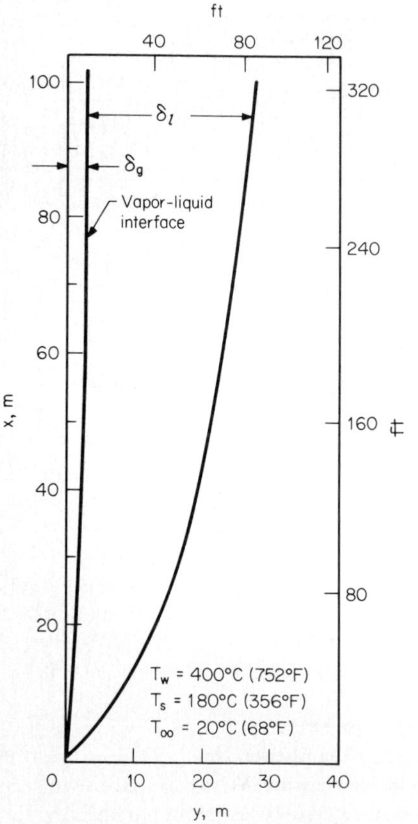

FIG. 12 Vapor film and the hot-water zone along a dike [190].

For the special case of zero subcooling ($\Delta T_{\text{sub}}^{+} = 0$), Cheng [190] found that $\eta_{g\delta}$ is a function of $\Delta T_{\text{sup}}^{+}$ only and independent of $\hat{R}$. For this case, Eq. (70) can be fitted approximately by the following expression:

$$\text{Nu}_x = \left(\frac{1}{2\Delta T_{\text{sup}}^{+}} + \frac{1}{\pi} \right)^{1/2} \left[\frac{K(\rho_\infty - \rho_g)gx}{\mu_g \alpha_g} \right]^{1/2} \tag{72}$$

which can be rewritten as

$$\text{Nu}_x = \left[\frac{i'_{lg}}{2c_{pl}(T_s - T_w)} \right]^{1/2} \left[\frac{K(\rho_\infty - \rho_g)gx}{\mu_g \alpha_g} \right]^{1/2} \tag{73a}$$

where

$$i'_{lg} = i_{lg}(1 + 0.636 \, \Delta T_{\text{sup}}^{+}) \tag{73b}$$

For $\Delta T_{\text{sup}}^{+} \to \infty$, Eq. (72) gives

$$\text{Nu}_x = 0.564 \left[\frac{K(\rho_\infty - \rho_g)gx}{\mu_g \alpha_g} \right]^{1/2} \tag{74}$$

which is the heat transfer rate about a dike in a dry-stream geothermal reservoir, with a high degree of wall superheating.

F. CONCLUDING REMARKS

This chapter has focused attention on heat transfer in hydrothermal convective geothermal systems. The results presented here cover the literature up to 1981 and are in condensed and useful form for quick reference. More detailed reviews on the subject are available elsewhere [167, 174, 193–196]. Although this chapter discusses heat transfer in geothermal systems, many of the analyses are actually applicable to a wide range of engineering problems whenever the system can be idealized as a porous medium. To name but a few, these include the problems of the secondary recovery of oil by thermal methods, the use of fibrous materials for thermal insulation, the design of underground energy storage systems, and the deposition of mineral ore in the subsurface formation.

REFERENCES

1. D. E. White, *U.N. Symp. Dev. Utilization Geotherm. Resources Rapp. Rep.,* Sec. V, Pisa, 1970.
2. D. E. White, L. J. P. Muffler, and A. H. Truesdell, *Econ. Geol.,* Vol. 66, pp. 75–97, 1971.
3. D. E. White, in *Geothermal Energy: Resources, Production, Stimulation,* ed. P. Kruger and C. Otte, pp. 70–94, Stanford University Press, Stanford, Calif., 1973.
4. A. J. Chasteen, *Proc. 2d U.N. Symp Dev. Use Geotherm. Resources,* Vol. 2, pp. 1335–1336, 1976.
5. S. S. Einarsson, *Proc. 2d U.N. Symp. Dev. Use Geotherm. Resources,* Vol. 2, pp. 1349–1363, 1976.
6. C. A. Swanberg, *Proc. 2d U.N. Symp. Dev. Use Geotherm. Resources,* Vol. 2, pp. 1435–1443, 1976.
7. R. F. Benenati and C. B. Brosilow, *AIChE J.,* Vol. 8, No. 3, pp. 359–361, 1962.
8. H. Darcy, *Les fontaines publiques de la Ville de Dijon,* Victor Dalmont, Paris, 1856.
9. J. Kozeny, *Ber. Wien Akad.,* Vol. 136a, p. 271, 1927.
10. P. H. Forchheimer, *Z. Vereines Deutscher Ing.,* Vol. 45, pp. 1782–1788, 1901.
11. C. J. Ward, *J. Hydraul. Div., ASCE,* Vol. 90, pp. 1–12, 1964.
12. S. Ergun, Chem. Eng. Prog., Vol. 48, pp. 89–94, 1952.
13. S. P. Bruke and W. B. Plummer, *Ind. Eng. Chem.,* Vol. 20, pp. 1196–1200, 1928.
14. V. Stanek and J. Szekely, *AIChE J.,* Vol. 20, pp. 974–980, 1974.
15. H. C. Brinkman, *Appl. Sci. Res.,* Sec. A., Vol. 1, pp. 27–34, 1947.
16. B. K. C. Chan, C. M. Ivey, and J. M. Barry, *J. Heat Transfer,* Vol. 92, pp. 21–27, 1970.
17. C. K. W. Tam, *J. Fluid Mech.,* Vol. 38, pp. 537–546, 1969.
18. T. S. Lundgren, *J. Fluid Mech.,* Vol. 51, pp. 273–299, 1972.
19. K. Vafai and C. L. Tien, *Int. J. Heat Mass Transfer,* Vol. 24, pp. 195–202, 1981.
20. G. H. Evans and O. A. Plumb, paper presented at the 2d AIAA/ASME Thermophys. & Heat Transfer Conf., *No. 78-HT-55,* 1978.
21. A. T. Corey, *Producers Monthly,* Vol. 19, pp. 38–41, 1954.
22. A. T. Corey, C. H. Tathjens, J. H. Henderson, and M. R. J. Wyllie, *Trans. AIME,* Vol. 207, pp. 349–351, 1956.
23. C. H. Sondergeld and D. L. Turcotte, *J. Geophys. Research,* Vol. 82, pp. 2045–2053, 1977.
24. J. Bear, *Dynamics of Fluids in Porous Media,* American Elsevier, New York, 1972.
25. J. J. Fried and M. A. Combarnous, *Advances in Hydroscience,* Vol. 11, pp. 169–282, 1976.
26. D. R. F. Harleman and R. R. Rumer, *J. Fluid Mech.,* Vol. 16, pp. 385–394, 1963.
27. E. J. List and N. H. Brooks, *J. Geophys Research,* Vol. 72, pp. 2531–2541, 1967.
28. S. Yagi, D. Kunii, and N. Wako, *AIChE J.,* Vol. 6, pp. 543–546, 1960.
29. S. Yagi, D. Kunii, and N. Wakao, *Int. Heat Transfer Conf.,* Boulder, Colo., pp. 742–749, 1960.
30. G. P. Willhite, J. S. Dranoff, and J. M. Smith, *Soc. Pet. Eng. J.,* Vol. 3, pp. 185–188, 1963.
31. D. J. Gunn and M. Khalid, *Chem. Eng. Sci.,* Vol. 30, pp. 261–267, 1975.

32. M. J. Lippmann, T. N. Narasimhan, and P. A. Witherspoon, paper presented at the 2d International Symposium on Land Subsidence, Anaheim, Calif., Dec. 10–17, 1976.

33. D. L. Turcotte, R. J. Ribando, and K. E. Torrance, in *The Earth's Crust,* ed. J. G. Heacock et al., *AGU Monogr. No. 20,* pp. 722–736, American Geophysical Union, Washington, D.C., 1977.

34. J. M. Straus and G. Schubert, *J. Geophys. Research,* Vol. 82, pp. 325–333, 1977.

35. L. M. Sorey, *U.S. Geological Survey Open-File Rep. 75-613,* 1975.

36. O. A. Plumb and J. S. Huenefeld, *Int. J. Heat Mass Transfer,* Vol. 24, pp. 765–768, 1981.

37. P. Cheng, C. L. Ali, and A. K. Verma, *Lett. Heat Mass Transfer,* Vol. 8, No. 4, pp. 261–266, 1981.

38. P. Cheng, *Lett. Heat Mass Transfer,* Vol. 8, No. 4, pp. 267–270, 1981.

39. P. Cheng and C. L. Ali, *Paper #81-HT-85,* presented at the 20th National Heat Transfer Conference, August 1981.

40. S. Whitaker, *Chem. Eng. Sci.,* Vol. 21, pp. 291–300, 1966.

41. S. Whitaker, *AIChE J.,* Vol. 13, pp. 420–427, 1967.

42. J. C. Slattery, *AIChE J.,* Vol. 13, pp. 1066–1071, 1967.

43. S. Whitaker, *Ind. Eng. Chem.,* Vol. 61, No. 12, pp. 15–28, 1969.

44. J. C. Slattery, *AIChE J.,* Vol. 15, No. 6, pp. 866–872, 1969.

45. J. C. Slattery, *Momentum, Energy, and Mass Transfer in Continua,* McGraw-Hill, New York, 1972.

46. W. B. Fulks, R. B. Guenther and E. L. Roetman, *Acta Mech.,* Vol. 12, pp. 121–129, 1971.

47. C. W. Horton and F. T. Rogers, Jr., *J. Appl. Phys.,* Vol. 16, pp. 367–370, 1945.

48. E. R. Lapwood, *Proc. Cambridge Philos. Soc.,* Vol. 44, pp. 508–521, 1948.

49. D. A. Nield, *Water Resources Res.,* Vol. 4, pp. 553–560, 1967.

50. R. J. Ribando and K. E. Torrance, *J. Heat Transfer,* Vol. 98, pp. 42–48, 1976.

51. J. W. Elder, *J. Fluid Mech.,* Vol. 27, pp. 29–48, 1967.

52. Y. Katto and T. Masuoda, *Int. J. Heat Mass Transfer,* Vol. 10, pp. 297–309, 1967.

53. K. Walker and G. M. Homsy, *J. Heat Transfer,* Vol. 99, pp. 338–339, 1977.

54. J. L. Beck, *Phys. Fluids,* Vol. 15, pp. 1377–1383, 1972.

55. A. Zebib, *Phys. Fluids,* Vol. 21, pp. 699–700, 1978.

56. S. A. Bories and M. A. Combarnous, *J. Fluid Mech.,* Vol. 57, pp. 63–70, 1973.

57. J. E. Weber, *Int. J. Heat Mass Transfer,* Vol. 17, pp. 241–248, 1974.

58. M. Pratt, *J. Geophys. Research,* Vol. 71, pp. 4835–4838, 1966.

59. H. Rubin, *J. Hydrol.,* Vol. 21, pp. 173–185, 1974.

60. H. Rubin, *Water Resources Res.,* Vol. 13, pp. 34–40, 1977.

61. G. M. Homsy and A. E. Sherwood, *AIChE J.,* Vol. 22, pp. 168–174, 1976.

62. R. A. Wooding, *J. Fluid Mech.,* Vol. 9, pp. 183–192, 1960.

63. F. M. Sutton, *Phys. Fluids,* Vol. 13, pp. 1931–1934, 1970.

64. R. McKibbin and M. J. O'Sullivan, *J. Fluid Mech.,* Vol. 96, pp. 375–393, 1980.

65. R. A. Wooding, *2d Workshop Geotherm. Reservoir Eng.,* pp. 339–345, Stanford University, Dec. 1–3, 1976.

66. G. Castinel and M. Combarnous, *Int. Chem. Eng.,* Vol. 17, No. 4, pp. 605–614, 1977.

67. D. R. Kassoy and A. Zebib, *Phys. Fluids,* Vol. 18, pp. 1649–1651, 1975.

68. A. Zebib and D. R. Kassoy, *Phys. Fluids,* Vol. 20, pp. 4–9, 1977.

69. L. W. Morland, A. Zebib, and D. R. Kassoy, *Phys. Fluids,* Vol. 20, pp. 1255–1259, 1977.

70. Y. C. Yen, *Int. J. Heat Mass Transfer,* Vol. 17, pp. 1349–1356, 1974.

71. R. J. Buretta and A. S. Berman, *J. Appl. Mech.,* Vol. 43, pp. 249–253, 1976.

72. T. Kaneko, M. F. Mohtadi, and K. Aziz, *Int. J. Heat Mass Transfer,* Vol. 17, pp. 485–496, 1974.

73. H. C. Helgeson, *Am. J. Sci.,* Vol. 266, pp. 129–166, 1968.
74. H. Rubin, *Water Resources Res.,* Vol. 9, pp. 211–221, 1973.
75. H. Rubin, *Water Resources Res.,* Vol. 9, pp. 968–974, 1973.
76. H. Rubin, *Water Resources Res.,* Vol. 11, pp. 154–158, 1975.
77. H. Rubin, *Water Resources Res.,* Vol. 12, pp. 141–147, 1976.
78. J. E. Weber, *J. Hydrol.,* Vol. 25, pp. 59–70, 1975.
79. J. W. Taunton and E. N. Lightfoot, *Phys. Fluids,* Vol. 15, pp. 748–753, 1972.
80. P. A. Tyvand, *Water Resources Res.,* Vol. 16, No. 2, pp. 325–330, 1980.
81. P. Cheng and K. H. Lau, *Proc. 2d U.N. Symp. Dev. Use Geotherm. Resources, 1975,* Vol. 3, pp. 1591–1598, 1976.
82. E. Palm. J. E. Weber, and O. Kvernvold, *J. Fluid Mech.,* Vol. 54, pp. 153–161, 1972.
83. A. Zebib and R. Kassoy, *Phys. Fluids,* Vol. 21, pp. 1–3, 1978.
84. A. Zebib and D. R. Kassoy, *Phys. Fluids,* Vol. 20, pp. 4–9, 1978.
85. J. M. Straus, *J. Fluid Mech.,* Vol. 64, pp. 51–63, 1974.
86. F. H. Busse and D. D. Joseph, *J. Fluid Mech.,* Vol. 54, pp. 521–543, 1972.
87. V. P. Gupta and D. D. Joseph, *J. Fluid Mech.,* Vol. 57, pp. 491–514, 1973.
88. I. G. Donaldson, *J. Geophys. Research,* Vol. 67, pp. 3449–3459, 1962.
89. T. N. Narasimhan and P. A. Witherspoon, *Water Resources Res.,* Vol. 12, pp. 57–64, 1976.
90. J. Chen and J. E. Conel, *J. Energy,* Vol. 1, No. 6, pp. 364–369, 1977.
91. J. W. Elder, *J. Fluid Mech.,* Vol. 27, pp. 609–623, 1967.
92. P. H. Holst and K. Aziz, *Can. J. Chem. Eng.,* Vol. 50, pp. 232–241, 1972.
93. R. N. Horne and M. J. O'Sullivan, *J. Fluid Mech.,* Vol. 66, pp. 339–352, 1974.
94. M. A. Combarnous and B. Le Fur, *C.R. Hood. Seamces Acad. Sci.,* Ser. B., pp. 1009–1012, 1969.
95. J. P. Caltagirone, *J. Fluid Mech.,* Vol. 72, pp. 269–287, 1975.
96. M. A. Combarnous and P. Bia, *Soc. Pet. Eng. J.,* Vol. 4, pp. 399–405, 1971.
97. P. H. Holst and K. Aziz, *Int. J. Heat Mass Transfer,* Vol. 15, pp. 73–90, 1972.
98. R. N. Horne, *J.F.M.,* Vol. 92, pp. 751–766, 1979.
99. R. N. Horne and J. Caltagirone, *J. Fluid Mech.,* Vol. 100, pp. 385–395, 1980.
100. R. N. Horne, *ASME Paper #80-HT-85,* 1980.
101. J. M. Straus and G. Schubert, *J. Fluid Mech.,* Vol. 87, pp. 385–394, 1978.
102. J. M. Straus and G. Schubert, *J. Fluid Mech.,* Vol. 91, pp. 155–165, 1979.
103. G. Schubert and J. M. Straus, *J. Fluid Mech.,* Vol. 94, pp. 25–38, 1979.
104. J. M. Straus and G. Schubert, *J. Fluid Mech.,* Vol. 103, pp. 23–32, 1981.
105. K. J. Schneider, *Proc. 11th Int. Cong. Refrig., Paper 11-4,* 1963.
106. M. A. Combarnous and S. Bories, *Int. J. Heat Mass Transfer,* Vol. 17, pp. 505–515, 1974.
107. N. Seki, S. Fukusako and Y. Ariake, *Wärme Stoffübertrag.,* Vol. 13, pp. 61–71, 1980.
108. R. J. Ribando, K. E. Torrance, and D. L. Turcotte, *J. Geophys. Research,* Vol. 81, pp. 3007–3012, 1976.
109. R. Rana, R. Horne, and P. Cheng, *J. Heat Transfer,* Vol. 101, pp. 411–416, 1979.
110. R. N. Horne and M. J. O'Sullivan, *Phys. Fluids,* Vol. 21, pp. 1260–1265, 1978.
111. R. N. Horne and M. J. O'Sullivan, *J. Heat Transfer,* Vol. 100, pp. 448–552, 1978.
112. D. Norton and J. Knight, *Am. J. Sci.,* Vol. 277, pp. 937–981, 1977.
113. D. Norton, in *The Earth's Crust,* ed. J. G. Heacock et al., *AGU Monogr. #20,* pp. 693–704, American Geophysical Union, Washington, D.C., 1977.
114. K. E. Torrance and J. P. Sheu, *Num. Heat Transfer,* Vol. 1, pp. 147–161, 1978.
115. A. E. Gill, *J. Fluid Mech.,* Vol. 26, pp. 515–536, 1966.
116. J. E. Weber, *Int. J. Heat Mass Transfer,* Vol. 18, pp. 569–573, 1975.

117. K. L. Walker and G. M. Homsy, *J. Fluid Mech.*, Vol. 87, pp. 449–474, 1978.

118. N. Seki, S. Fukusako, and H. Inaba, *Int. J. Heat Mass Transfer,* Vol. 21, pp. 985–989, 1978.

119. P. Cheng and K. H. Lau, *J. Geophys. Research,* Vol. 79, pp. 4425–4431, 1974.

120. K. H. Lau and P. Cheng, *Int. J. Heat Mass Transfer,* Vol. 20, pp. 1205–1210, 1977.

121. P. Cheng, K. C. Yeung, and K. H. Lau, in *Future Energy Production Systems,* ed. J. C. Denton and N. Afgan, pp. 429–448, Hemisphere, Washington, D.C., 1976.

122. P. Cheng and L. Teckchandani, in *The Earth's Crust,* ed. J. G. Heacock et al., *AGU Monogr. #20,* pp. 705–721, American Geophysical Union, Washington, D.C., 1977.

123. M. J. Lippmann, C. F. Tsang, and P. A. Witherspoon, *47th Annual Calif. Regional Mtg. Soc. Pet. Eng. AIME, Paper No. SPE 6537,* Bakersfield, Calif., April 13–15, 1977.

124. R. N. Horne and M. J. O'Sullivan, *5th Australasian Conf. Hydraul. Fluid Mech.,* University of Canterbury, Christchurch, N.Z., Dec. 9–13, 1974.

125. R. N. Horne and M. J. O'Sullivan, *47th Annual Calif. Regional Mtg. Soc. Pet. Eng. AIME, Paper No. SPE 6536,* Bakersfield, Calif., April 13–15, 1977.

126. H. R. Henry and F. A. Kohout, *Underground Waste Mgmt. Environmental Implications,* Vol. 18, pp. 202–221, 1973.

127. I. G. Donaldson, *Proc. 3d Australasian Conf. Hydraul. Fluid Mech., Paper No. 2580,* pp. 200–204, 1968.

128. I. G. Donaldson, *Geothermics,* Vol. 2, pp. 649–654, 1970.

129. D. R. Kassoy and A. Zebib, *J. Fluid Mech.,* Vol. 88, pp. 769–792, 1978.

130. D. R. Kassoy and K. P. Goyal, *Lawrence Berkeley Laboratory Rept. 8784, GREMP-3, UC-66a,* 1979.

131. K. Kilty, D. S. Chapman, and C. W. Mase, *J. Volcanol. Geotherm. Res.,* Vol. 6, pp. 257–277, 1979.

132. M. L. Sorey and R. E. Lewis, *J. Geophys. Research,* Vol. 81, pp. 785–791, 1976.

133. M. Sorey, *Proc. 2d Workshop Geotherm. Reservoir Eng.,* pp. 324–338, Stanford University, Stanford, Calif., Dec. 1–3, 1976.

134. J. W. Mercer, Jr., and G. F. Pinder, *Geothermics,* Vol. 2, pp. 81–89, 1973.

135. J. W. Mercer, G. F. Pinder, and I. G. Donaldson, *J. Geophys. Research,* Vol. 80, pp. 2608–2621, 1975.

136. J. W. Mercer and C. R. Faust, *Water Resources Res.,* Vol. 15, pp. 653–671, 1979.

137. E. O. Tansev and M. L. Wasserman, *Proc. 3d Geotherm. Reservoir Eng.,* pp. 107–115, Stanford University, Stanford, Calif., 1977.

138. C. E. Hickox and H. A. Watts, *J. Heat Transfer,* Vol. 102, pp. 248–253, 1980.

139. R. A. Wooding, *J. Fluid Mech.,* Vol. 15, pp. 527–544, 1963.

140. A. Bejan, *J. Fluid Mech.,* Vol. 89, pp. 97–107, 1978.

141. C. E. Hickox, *J. Heat Transfer,* Vol. 103, pp. 232–236, 1981.

142. P. Cheng, *Lett. Heat Mass Transfer,* Vol. 5, pp. 243–252, 1978.

143. P. Cheng, *Lett. Heat Mass Transfer,* Vol. 4, pp. 119–128, 1977.

144. C. H. Johnson and P. Cheng, *Int. J. Heat Mass Transfer,* Vol. 21, pp. 709–718, 1978.

145. P. Cheng and W. J. Minkowycz, *J. Geophys. Research,* Vol. 82, pp. 2040–2044, 1977.

146. W. J. Minkowycz and P. Cheng, *Int. J. Heat Mass Transfer,* Vol. 19, pp. 805–813, 1976.

147. C. S. Yih, *Dynamics of Nonhomogeneous Fluids,* Macmillan, New York, 1965.

148. P. Cheng, *Int. J. Heat Mass Transfer,* Vol. 20, pp. 201–206, 1977.

149. J. H. Merkin, *Int. J. Heat Mass Transfer,* Vol. 21, pp. 1499–1504, 1978.

150. P. Cheng and I. D. Chang, *Int. J. Heat Mass Transfer,* Vol. 19, pp. 1267–1272, 1976.

151. A. McNabb, *Proc. 2d Australasian Conf. Hydraul. Fluid Mech.,* pp. C161–C171, 1965.

152. P. Cheng and W. C. Chau, *Water Resources Res.,* Vol. 13, No. 4, pp. 768–772, 1977.

153. J. H. Merkin, *Int. J. Heat Mass Transfer,* Vol. 22, pp. 1461–1462, 1979.

154. P. Cheng, class notes of M.E. 627, Heat Transfer in Porous Media, University of Hawaii at Manoa, Honolulu, 1980.

155. P. Cheng, *Int. J. Heat Mass Transfer,* Vol. 20, pp. 804–814, 1977.

156. P. Cheng, *Int. J. Heat Mass Transfer,* Vol. 20, pp. 893–898, 1977.

157. H. C. Hardee, *SAND 76-0075,* Sandia Laboratories, Albuquerque, N.M., 1976.

158. P. Cheng and I. D. Chang, *Lett. Heat Mass Transfer,* Vol. 6, pp. 253–258, 1979.

159. I. D. Chang, P. Cheng, and C. T. Hsu, *Int. J. Heat Mass Transfer,* Vol. 26, pp. 163–174, 1983.

160. P. Cheng and C. T. Hsu, *J. Heat Transfer,* Vol. 106, pp. 143–151, 1984.

161. C. T. Hsu, P. Cheng, and G. M. Homsy, *Int. J. Heat Mass Transfer,* Vol. 21, pp. 1221–1228, 1978.

162. C. H. Hsu and P. Cheng, *J. Heat Transfer,* Vol. 101, pp. 660–665, 1979.

163. C. H. Hsu and P. Cheng, *J. Heat Mass Transfer,* Vol. 102, pp. 544–549, 1980.

164. C. T. Hsu and P. Cheng, *Int. J. Heat Mass Transfer,* Vol. 23, pp. 789–798, 1980.

165. S. Whitaker, *Chem. Eng. Sci.,* Vol. 28, pp. 139–147, 1973.

166. W. G. Gray and K. O'Neill, *Water Resources Res.,* Vol. 12, pp. 148–154, 1976.

167. P. A. Witherspoon, S. P. Neuman, and M. L. Sorey, Lawrence Berkeley Laboratory *Rep. No. 3263,* 1975.

168. C. R. Faust, Ph.D. thesis, Geosciences Dept., Pennsylvania State University, Philadelphia, 1976.

169. C. R. Faust and J. W. Mercer, *U.S.G.S. Open-File Rep. 77-60,* 1977.

170. A. H. Truesdell and D. E. White, *Geothermics,* Vol. 2, pp. 154–173, 1973.

171. A. F. Moench, *U.S.G.S. Open-File Rep. 76-607,* 1976.

172. I. G. Donalson, *N.Z. J. Sci.,* Vol. 11, pp. 3–23, 1968.

173. J. P. Sheu, K. E. Torrance, and D. L. Turcotte, *J. Geophys. Research,* Vol. 84, pp. 7524–7532, 1979.

174. K. E. Torrance, *ASME-JSME Therm. Eng. Joint Conf.,* Vol. 2, pp. 593–606, 1983.

175. G. Schubert and J. M. Straus, *J. Geophys. Research,* Vol. 84, pp. 1621–1628, 1979.

176. J. C. Martin, R. E. Wegner, and F. J. Kelsey, *Proc. 2d Workshop Geotherm. Reservoir Eng.,* pp. 251–262, Stanford University, Dec. 1–3, 1976.

177. G. Schubert and J. M. Straus, *J. Geophys. Research,* Vol. 82, No. 23, pp. 3411–3421, 1977.

178. L. M. Cathles, *Econ. Geol.,* Vol. 72, pp. 804–826, 1977.

179. C. R. Faust and J. W. Mercer, *Proc. 2d U.N. Symp. Dev. Use Geotherm. Resources,* Vol. 3, pp. 1635–1641, 1976.

180. N. Arihara, Ph.D. thesis, Department of Petroleum Engineering, Stanford University, 1974.

181. S. K. Garg, J. W. Pritchett, and D. H. Brownell, *Proc. 2d U.N. Symp. Dev. Use Geotherm. Resources,* pp. 1651–1656, 1976.

182. L. K. Thomas and R. Pierson, *Soc. Pet. Eng. J.,* Vol. 18, pp. 151–161, 1978.

183. R. M. Toronyi and S. M. Farouq Ali, *Soc. Pet. Eng. J.,* Vol. 17, pp. 171–183, 1977.

184. J. W. Mercer and C. R. Faust, *Soc. Pet. Eng. AIME, Paper No. SPE 5520,* 1975.

185. T. J. Lasseter, *ASME Paper No. 75-WA/HT-71.*

186. J. W. Pritchett, S. K. Garg, and D. H. Brownell, *Proc. 2d Workshop Geotherm. Reservoir Eng.,* pp. 310–323, 1976.

187. G. A. MacDonald and A. T. Abbott, *Volcanoes in the Sea,* University of Hawaii Press, Honolulu, 1970.

188. E. M. Parmentier, *Int. J. Heat Mass Transfer,* Vol. 22, pp. 849–855, 1979.

189. P. Cheng and A. K. Verma, *Int. J. Heat Transfer,* Vol. 24, pp. 1151–1160, 1981.

190. P. Cheng, *Int. J. Heat Mass Transfer,* Vol. 24, pp. 983–990, 1981.

191. J. W. Elder, Bulletin 169, Department of Scientific & Industrial Research, Wellington, N.Z., 1966.

192. J. W. Elder, in *Terrestrial Heat Flow*, ed. W. H. K. Lee, *AGU Monogr. #8*, pp. 211–239, American Geophysical Union, Washington, D.C., 1965.

193. M. A. Combarnous and S. A. Bories, *Advances in Hydroscience*, Vol. 10, pp. 231–307, 1975.

194. P. Cheng, pp. 1–105 in *Advances in Heat Transfer*, Vol. 14, ed. T. F. Irvine, Jr., and J. P. Hartnett, Academic, New York, 1978.

195. S. K. Garg and D. R. Kassoy, Chap. 2, *Geothermal Systems: Principles and Case Histories*, ed. P. Muttler and L. Rybach, J. Wiley, New York, 1981.

196. G. A. Sudol, R. F. Harrison, and H. J. Ramey, Jr., *Lawrence Berkeley Lab. Rept. #LBL-8664, GREMP-1, UC-66a*, August 1979.

NOMENCLATURE

Symbol, Definition, SI Units, English Units

A area: m^2, ft^2

A_p total pore area of the cross section: m^2, ft^2

A_p^+ areal porosity $= A_p/A$

$\tilde{A}$ constant in the prescribed wall temperature: K/m^λ, $°F/ft^\lambda$

a horizontal wave number $= \sqrt{\Delta^2 + \gamma^2}$

$\tilde{a}$ horizontal injection velocity parameter: $m^{(1-N)}/s$, $ft^{(1-N)}/s$

B constant in the free-stream velocity: $m^{(1-n)}/s$, $ft^{(1-n)}/s$

C_1 constant defined in Eq. (26)

C_2 constant defined in Eq. (59)

c concentration

c_r concentration evaluated at a reference condition

c_p scientific heat at constant pressure: $J/(kg \cdot K)$, $Btu/(lb_m \cdot °F)$

c_{pg} specific heat of vapor at constant pressure: $J/(kg \cdot K)$, $Btu/(lb_m \cdot °F)$

c_{pl} specific heat of liquid water at constant pressure: $J/(kg \cdot K)$, $Btu/(lb_m \cdot °F)$

c_r concentration at a reference condition

D diameter of a cylinder or sphere: m, ft

D_e effective mass dispersion coefficient: m^2/s, ft^2/s

D_0 mass diffusivity: m^2/s, ft^2/s

D' mass dispersion coefficient: m^2/s, ft^2/s

D'_L longitudinal mass dispersion coefficient: m^2/s, ft^2/s

D'_T transverse mass dispersion coefficient: m^2/s, ft^2/s

Da_H Darcy number $= K/H^2$

Da_L Darcy number $= K/L^2$

Da_x local Darcy number $= K/x^2$

d diameter of the grains: m, ft

f_d friction factor based on the diameter of the grains $= [d\phi^3/(1 - \phi)\rho v^2] \times (-dP/ds)$

f_k friction factor based on the permeability of the grains $= (\sqrt{K}/\rho v^2) (-dP/ds)$

f_w injection parameter $= -2\tilde{a}/(\alpha\rho_\infty g\beta K\tilde{A}/\mu)^{1/2}(1 + \lambda)$

G	mass flow rate: $\text{kg}/(\text{m}^2 \cdot \text{s})$, $\text{lb}_\text{m}/(\text{h} \cdot \text{ft}^2)$
Gr_x	Grashof number $= K\beta g(T_w - T_\infty)x/\nu^2$
$\text{Gr}_{x,\alpha}$	Grashof number of an inclined plate $= K\beta g \cos \alpha_0(T_w - T_\infty)x/\nu^2$
g	gravitational acceleration: m/s^2, ft/s^2
H	distance between parallel plates: m, ft
H_d	hydraulic head: m, ft
h	heat transfer coefficient: $\text{W}/(\text{m}^2 \cdot \text{K})$, $\text{Btu}/(\text{h} \cdot \text{ft}^2 \cdot {}^\circ\text{F})$
$\overline{h}$	average heat transfer coefficient: $\text{W}/(\text{m}^2 \cdot \text{K})$, $\text{Btu}/(\text{h} \cdot \text{ft}^2 \cdot {}^\circ\text{F})$
i_{lg}	latent heat of evaporation: J/kg, $\text{Btu}/\text{lb}_\text{m}$
i'_{lg}	modified latent heat of evaporation $= i_{lg}(1 + 0.636 \, \Delta T^+_\text{sup})$: J/kg, $\text{Btu}/\text{lb}_\text{m}$
K	intrinsic permeability: m^2, ft^2
K_{eg}	effective permeability for the nonwetting phase: m^2, ft^2
K_{el}	effective permeability for the wetting phase: m^2, ft^2
K_h	permeability in the horizontal direction: m^2, ft^2
K_v	permeability in the vertical direction: m^2, ft^2
K_c	hydraulic conductivity: m/s, ft/s
k	stagnant thermal conductivity of the saturated porous medium: $\text{W}/(\text{m} \cdot \text{K})$, $\text{Btu}/(\text{h} \cdot \text{ft} \cdot {}^\circ\text{F})$
k_e	effective thermal conductivity: $\text{W}/(\text{m} \cdot \text{K})$, $\text{Btu}/(\text{h} \cdot \text{ft} \cdot {}^\circ\text{F})$
k_g	thermal conductivity of the fluid: $\text{W}/(\text{m} \cdot \text{K})$, $\text{Btu}/(\text{h} \cdot \text{ft} \cdot {}^\circ\text{F})$
k'	thermal dispersion coefficient: $\text{W}/(\text{m} \cdot \text{K})$, $\text{Btu}/(\text{h} \cdot \text{ft} \cdot {}^\circ\text{F})$
k'_L	longitudinal thermal dispersion coefficient: $\text{W}/(\text{m} \cdot \text{K})$, $\text{Btu}/(\text{h} \cdot \text{ft} \cdot {}^\circ\text{F})$
k'_T	transverse thermal dispersion coefficient: $(\text{W}/(\text{m} \cdot \text{K})$, $\text{Btu}/(\text{h} \cdot \text{ft} \cdot {}^\circ\text{F})$
L	length: m, ft
l^+	dimensionless heating length
M_w	molecular weight: kg/mol, $\text{lb}_\text{m}/\text{mol}$
m	exponent in Eq. (26)
N	exponent in the injection velocity
Nu	average Nusselt number for the parallel-plate geometry with prescribed wall temperatures $\overline{q}''H/k(T_0 - T_1)$
Nu_D	average Nusselt number for a horizontal cylinder and a sphere $= q''D/k(T_w - T_\infty)$
Nu_L	average Nusselt number for an isothermal flat plate $= \overline{q}''L/k(T_w - T_\infty)$
$\overline{\text{Nu}_L}$	average Nusselt number for a nonisothermal flat plate $= \overline{q}''L/k(\overline{T_w - T_\infty})$
Nu_x	local Nusselt number for a flat plate $= q''x/k(T_w - T_\infty)$
Nu_Φ	local Nusselt number for a cylinder or sphere $= q''D/k(T_w - T_\infty)$
n	exponent in external free-stream velocity
P	pressure: N/m^2, $\text{lb}_\text{f}/\text{ft}^2$
P_c	capillary pressure: N/m^2, $\text{lb}_\text{f}/\text{ft}^2$
P_g	pressure in the nonwetting phase: N/m^2, $\text{lb}_\text{f}/\text{ft}^2$
P_l	pressure in the wetting phase: N/m^2, $\text{lb}_\text{f}/\text{ft}^2$
Pe_d	Peclet number $= vd/\alpha$
Pe_x	local Peclet number $= u_\infty x/\alpha$

$\hat{P}e$ molecular Peclet number for mass transfer $= vd/\mathbf{D}_0$

$\hat{P}e_T$ transverse dispersion Peclet number for mass transfer $= vd/\mathbf{D}_T$

Pr Prandtl number

Q volume flow rate: m^3/s, ft^3/s

q heat transfer rate: W, Btu/h

q'' local surface heat flux: W/m^2, $Btu/(h \cdot ft^2)$

q''_c local surface heat flux from a cylinder: W/m^2, $Btu/(h \cdot ft^2)$

$\overline{q}''$ average surface heat flux: W/m^2, $Btu/(h \cdot ft^2)$

R_g relative permeability of the nonwetting phase $= K_{eg}/K$

R_l relative permeability of the wetting phase $= K_{el}/K$

$\hat{R}$ property ratio $= (\rho_g/\rho_\infty)[\mu_l\alpha_g(\rho_\infty - \rho_g)c_{pl}/\mu_g\alpha_l\rho_\infty\beta_l i_{lg}]^{1/2}$

Ra Rayleigh number for the parallel-plate configuration with prescribed wall temperatures $= K\rho_0\beta g(T_0 - T_1)H/\mu\alpha$

Ra_D Rayleigh number based on the diameter of an isothermal cylinder or sphere $= K\rho_\infty g\beta(T_w - T_\infty)D/\mu\alpha$

Ra_H Rayleigh number based on the length H, $= K\rho_r\beta g(T_w - T_r)H/\mu\alpha$

Ra_L Rayleigh number based on the length L, $= K\rho_\infty\beta g(T_w - T_\infty)L/\mu\alpha$

Ra_p Rayleigh number for a point source, $= Kqg\beta/2\pi\alpha^2\mu c_p$

Ra_q Rayleigh number for the parallel-plate configuration with constant heat flux, $Kq''\rho_0 g\beta H^2/\mu\alpha k$

Ra_r Rayleigh number based on T_r, the reference temperature $= \rho_\infty g\beta K (T_r - T_\infty)L/\mu\alpha$

Ra_s salinity Rayleigh number, $= \rho_0 g\beta_s(c_0 - c_1)KH/\mu\mathbf{D}_0$

Ra_x local Rayleigh number, $= K\rho_\infty\beta g(T_w - T_\infty)x/\mu\alpha$

$Ra_{x,\alpha}$ local Rayleigh number for an inclined plate $= K\rho_\infty\beta g\cos\alpha_0(T_w - T_\infty) x/\mu\alpha$

$\overline{Ra}_L$ average Rayleigh number for nonisothermal plate $= K\rho_\infty\beta g(\overline{T_w - T_\infty})L/\mu\alpha$

Ra^* critical Rayleigh number for the parallel-plate configuration

$\tilde{R}a$ Rayleigh number of a viscous fluid, $= \rho_0 g\beta(T_0 - T_1)H^3/\alpha\mu$

Re_d Reynolds number based on the diameter of the grains, $= \rho vd/\mu$

Re_K Reynolds number based on the square root of the permeability, $= \rho v\sqrt{K}/\mu$

Re_x local Reynolds number, $= u_\infty x/\nu$

r_w radius of well: m, ft

r_0 radius of a cylinder or a sphere: m, ft

S_g nonwetting fluid saturation, $= V_g/V_p$

S_{g0} residual nonwetting fluid saturation

S_l wetting fluid saturation, $= V_l/V_p$

S_{l0} irreducible wetting fluid saturation

$\hat{S}_l$ $(S_l - S_{l0})/(1 - S_{l0} - S_{g0})$

Sc Schmidt number, $\nu/\mathbf{D}_0$

s distance along an arbitrary direction: m, ft

T temperature: °C, K; °F, °R

T_m maximum temperature of heated surface: °C, K; °F, °R

T_r reference temperature: °C, K; °F, °R

T_s	saturation temperature: °C, K; °F, °R
T_w	wall temperature: °C, K; °F, °R
T_0	temperature of the hotter surface of the two parallel plates: °C, K; °F, °R
T_1	temperature of the colder surface of the two parallel plates: °C, K; °F, °R
T_∞	temperature at infinity: °C, K; °F, °R
ΔT_{sub}^+	degree of subcooled liquid, $= c_{pl}(T_s - T_\infty)/i_{lg}$
ΔT_{sup}^+	degree of superheated steam, $= c_{pg}(T_w - T_s)/i_{lg}$
t	time: s
u_∞	external free-stream velocity: m/s, ft/s
V	total volume: m³, ft³
V_g	volume occupied by the nonwetting phase: m³, ft³
V_l	volume occupied by the wetting phase: m³, ft³
V_p	total pore volume: m³, ft³
v	Darcy velocity: m/s, ft/s
v_p	pore velocity: m/s, ft/s
X	dimensionless distance along the heated surface, $= x/H$
x	distance along the heated surface: m, ft
Y	dimensionless coordinate perpendicular to the heated surface, $= y/H$
y	distance normal to the heated surface: m, ft
Z	compressibility factor
z	coordinate in the vertical direction: m, ft

Greek Symbols

α_g	thermal diffusivity of the vapor phase: m²/s, ft²/s
α	equivalent thermal diffusivity of the saturated porous medium: m²/s, ft²/s
α_0	inclination angle with respect to the vertical: rad, deg
α^+	aspect ratio $= L/H$
β	thermal expansion coefficient: K^{-1}, $°R^{-1}$
β_l	thermal expansion coefficient of liquid phase K^{-1}, $°R^{-1}$
β_s	solute expansion coefficient
β_0	angle of incidence: rad, deg
Γ	temperature gradient °C/m, °F/ft
$\hat{\Gamma}$	dimensionless temperature gradient
γ	wave number in horizontal direction
Δ	wave number in horizontal direction normal to direction of γ
δ_g	boundary layer thickness of the vapor film: m, ft
δ_l	thickness of the liquid zone: m, ft
δ_T	thermal boundary-layer thickness: m, ft
η	similarity variable
$\eta_{g\delta}$	dimensionless vapor film thickness
η_T	value of η at the edge of the thermal boundary layer
Θ	dimensionless temperature $= (T - T_r)/(T_m - T_r)$
θ	dimensionless temperature $= (T - T_\infty)/(T_w - T_\infty)$

λ	exponent in the prescribed wall temperature
μ	dynamic viscosity: Pa·s, $lb_m/h \cdot ft$
μ_g	dynamic viscosity of gas phase: Pa·s, $lb_m/(h \cdot ft)$
μ_l	dynamic viscosity of liquid phase: Pa·s, $lb_m/(h \cdot ft)$
ν	kinematic viscosity: m^2/s, ft^2/s
ξ	local curvature parameter $= 2x/(r_0 Ra_x^{1/2})$
ξ_L	curvature parameter $= 2L/(r_0 \overline{Ra}_L^{1/2})$
ρ	density of fluid: kg/m^3, lb_m/ft^3
ρ_g	density of the vapor: kg/m^3, lb_m/ft^3
ρ_l	density of the liquid: kg/m^3, lb_m/ft^3
ρ_o	density of the fluid at T_o: kg/m^3, lb_m/ft^3
ρ_r	density of the fluid at T_r: kg/m^3, lb_m/ft^3
ρ_∞	density of the fluid at the edge of boundary layer: kg/m^3, lb_m/ft^3
σ	heat capacity ratio $[\phi(\rho c_p)_f + (1 - \phi)(\rho c_p)_m]/(\rho c_p)_f$
τ	dimensionless time, $\alpha t/\sigma H^2$
Φ	angle measured from the bottom stagnation point: rad, deg
ϕ	porosity
χ	Forchheimer coefficient: m, ft
Ψ	dimensionless stream function $= \psi/\alpha$
ψ	stream function: m^2/s, ft^2/s

CHAPTER 12

Measurement of Temperature and Heat Transfer

By R. J. Goldstein
H. D. Chiang

Department of Mechanical Engineering
University of Minnesota, Minneapolis

A. INTRODUCTION

In heat transfer studies, many different properties or quantities may be measured. These include thermodynamic properties, transport properties, velocities, mass flow, concentration, etc. There are, however, two key quantities that must be measured in almost every heat transfer experiment or project, be it for research purposes or for a specific application. These are temperature and the quantity of heat transferred. The present chapter will discuss the means of measuring these quantities. First, we review definitions of temperature and temperature scales and then discuss specific instruments that are used to measure temperature. The next section covers heat flux measurement techniques. We close the chapter with a discussion of mass transfer analogies that are used in experiments to simulate heat transfer.

B. TEMPERATURE MEASUREMENT

1. Basic Concepts and Definitions

Temperature measurement is important in virtually every heat transfer study. The temperature difference is, of course, the driving force that causes heat to be transferred from one system to another. In addition, knowledge of temperature is often required for determining the properties of systems in equilibrium.

Temperature is a thermodynamic property and is defined for systems in equilibrium. Even with nonequilibrium systems—e.g., in a medium in which there is a temperature gradient—local measurements are generally assumed to give a temperature which can help determine local thermodynamic properties by assuming local quasi equilibrium.

From the Zeroth Law of Thermodynamics we know that two systems which are in thermal equilibrium with a third system are in thermal equilibrium with one another and, by definition, have the same temperature. This Zeroth Law not only is important in defining systems that have the same temperature, but it is also a basic principle in thermometry; one measures temperatures of different systems by thermometers which are, in turn, compared to some standard temperature systems or standard thermometers.

Initial concepts of temperature came from the physical sensation of the relative hotness or coldness of bodies. This sensation of warmth or cold is so subjective relative to our immediate prior exposure that it is difficult to use for anything but simple qualitative comparison. The scientific or thermodynamic definition of temperature comes from Kelvin, who defined the ratio of the thermodynamic or absolute temperatures of two systems as being equal to the ratio of the heat added to the heat rejected for a reversible heat engine operated between the systems. The use of an idealized (Carnot) engine for measuring temperature or the ratio of two temperatures is not practical for most systems though such an engine is approximated in some devices at very low temperatures. The definition of temperature from a Carnot cycle leads to the concept of ideal gas systems for thermodynamic temperature measurement. These are approached by special low-pressure high-precision gas thermometers. The thermodynamic (or ideal-gas) temperature scale is elegant in its definition, but its implementation and transfer in general use is too complex and costly. Hence, an alternative temperature scale which is unique in definition and simpler to implement and transfer is desirable.

The definition of a temperature scale requires one or more fixed points (i.e., quantitatively defined and easily reproduced temperatures), and a device (thermometer) which is sensitive primarily to temperature and not other variables. Such a device would permit comparing different systems with fixed reference (temperature) points. In addition, an interpolation equation is required to measure temperatures between the fixed points.

Fixed points are temperatures which are fairly easy to reproduce in any laboratory given a reasonable apparatus and sufficient care. They usually involve two- (or three-) phase thermodynamic equilibrium points of pure substances such as the freezing point of a liquid, the boiling point of a liquid or a triple point where solid, liquid, and vapor are in equilibrium. For a pure substance, the triple point is a uniquely fixed temperature, while the freezing point usually has only a slight dependence on pressure. Liquid-vapor equilibrium temperatures are dependent on pressure and thus are not as convenient except at temperatures where other equilibrium points are not readily available.

The Kelvin or thermodynamic temperature scale uses a single fixed point to define the size of a degree—the triple point of water at 273.16 Kelvin (K). Special low-pressure constant-volume gas thermometers are used in a handful of laboratories to measure the thermodynamic temperatures of other fixed points.

Many different thermometers have been used in temperature measurements. The most common types are those in which the temperature-dependent measured variable is: (1) volume or length of a system, as with liquid-in-glass thermometers, (2) electrical resistance (platinum and other resistance thermometers, including thermistors); (3) electromotive force (emf)—particularly as used in thermocouples; and (4) radiation emitted by a surface, as in various types of pyrometers which are used primarily with high-temperature systems. All of these thermometers and some others will be described below.

In principle, any device which has one or more physical properties that are uniquely related to temperature in a reproducible way can be used as a thermometer. Such a device is usually classified as either a primary or a secondary thermometer. If the relation between the temperature and the measured physical quantity is described by a physical law which is exact, the thermometer is referred to as a primary thermometer; otherwise, it is called a secondary thermometer. Examples of primary thermometers include special low-pressure gas thermometers which behave according to the ideal gas law and some radiation-sensitive thermometers which are based upon the Planck radiation law. Resistance thermometers, thermocouples, and liquid-in-glass thermometers all belong to the category of secondary thermometers. A primary thermometer ideally is capable of measuring the thermodynamic temperature directly, whereas a secondary thermometer requires a calibration prior to use; furthermore, the temperature measured by a secondary thermometer is not the thermodynamic temperature, even with an "exact" calibration at fixed points.

Low-pressure gas thermometers have the accuracy required for the determination of thermodynamic temperature. However, from a practical standpoint, where precision and simplicity in the implementation and transfer are the major considerations, secondary thermometers were chosen as the defining standard thermometers for a practical temperature scale. This scale was defined by the use of fixed reference points whose thermodynamic temperatures were determined from gas thermometer measurements. The International Committee of Weights and Measures is responsible for the development and refinement of such a scale.

National laboratories, such as the National Bureau of Standards (NBS) in the United States, implement and maintain the practical temperature scale for their respective countries. They also help in the transfer of the scale by calibrating the defining standard thermometers. These defining standard thermometers are costly to maintain and are primarily used in standards laboratories in industries or universities. They are directly or indirectly used for calibration of thermometers used in actual applications.

2. Standards and Temperature Scales

The standardized scale now used in temperature measurement is the International Practical Temperature Scale of 1968, IPTS-68—last amended in 1975 [1]. IPTS-68 has been designed to give values numerically close to the corresponding thermodynamic temperatures. It covers the range of temperatures above 13.81 K. For temperatures lower than this there is no internationally accepted standard as yet [2]. Temperature in the IPTS-68 scale is presented in International Practical Kelvin Temperature, T_{68}, and International Practical Celsius Temperature, t_{68}, with units kelvin (K) and degree Celsius (°C) respectively. The temperatures are related by

$$t_{68} = T_{68} - 273.15 \text{ K} \tag{1}$$

Note that a temperature difference has the same numerical value when expressed in either of these units; i.e.,

$$\Delta t_{68} \, (°\text{C}) = \Delta T_{68} \, (\text{K}) \tag{2}$$

The IPTS scale includes 13 defining fixed points ranging from the triple point of equilibrium hydrogen (13.81K) to the normal freezing point of gold (1337.58 K) (see Table 1A). Four temperature ranges are defined: for three of these a defining standard thermometer and a specific form of interpolation or calibration equation are specified [1]. In brief, a standard platinum resistance thermometer is used from the lowest temperature covered (13.81 K) up to 630.74°C. This temperature span is divided into two ranges, one up to 0°C and the second from 0°C to 630.74°C; different interpolation equations are used for these two ranges. In range 3, from 630.74°C to 1064.43°C, a standard platinum–10% rhodium vs. platinum thermocouple is used with a specific form of calibration equation. The constants of the interpolation equations are determined by calibration at specific defining fixed points [1]. Minimum specifications on the properties of the resistance sensor or thermocouple are also included in the IPTS-68. In the fourth range, above 1064.43°C, IPTS-68 temperatures are based on the radiant energy emitted at a given wavelength as given by the Planck equation and compared to similar radiance at the gold point (1064.43°C). No standard temperature sensor is recommended for this range, but optical pyrometers are often used by the NBS and other standards laboratories as the defining standard thermometer for this temperature range.

IPTS-68 is an improvement over previous practical temperature scales, but it is far from an ideal one. Areas targeted for future improvement [2, 3, and 4] include:

1. Stability and accuracy. The stability and accuracy provided by the standard thermocouple in range 3 is less than desirable. Deviation from the true thermodynamic tem-

TABLE 1A International Practical Temperature Scale IPTS-68: Defining Fixed Points [1]

Equilibrium state	Assigned Value of International Practical Temperature[a]	
	T_{68} (K)	t_{68} (°C)
Equilibrium between the solid, liquid, and vapor phases of equilibrium hydrogen (triple point of equilibrium hydrogen)[b]	13.81	−259.34
Equilibrium between the liquid and vapor phases of equilibrium hydrogen at a pressure of 33,330.6 Pa (25/76 standard atmosphere)[b,c]	17.042	−256.108
Equilibrium between the liquid and vapor phases of equilibrium hydrogen (boiling point of equilibrium hydrogen)[b,c]	20.28	−252.87
Equilibrium between the liquid and vapor phases of neon (boiling point of neon)[c]	27.102	−246.048
Equilibrium between the solid, liquid, and vapor phases of oxygen (triple point of oxygen)	54.361	−218.789
Equilibrium between the solid, liquid, and vapor phases of argon (triple point of argon)[d]	83.798	−189.352
Equilibrium between the liquid and vapor phases of oxygen (condensation point of oxygen)[c,d]	90.188	−182.962
Equilibrium between the solid, liquid, and vapor phases of water (triple point of water)	273.16	0.01
Equilibrium between the liquid and vapor phases of water (boiling point of water)[e]	373.15	100
Equilibrium between the solid and liquid phases of tin (freezing point of tin)[e]	505.1181	231.9681
Equilibrium between the solid and liquid phases of zinc (freezing point of zinc)	692.73	419.58
Equilibrium between the solid and liquid phases of silver (freezing point of silver)	1235.08	961.93
Equilibrium between the solid and liquid phases of gold (freezing point of gold)	1337.58	1064.43

[a]Except for the triple points and one equilibrium hydrogen point (17.042 K) the assigned values of temperature are for equilibrium states at a pressure P_0 = 101,325 Pa (1 standard atmosphere). The effects of small deviations from this pressure are shown in Ref. 1. In those cases where differing isotopic abundances could significantly affect the fixed-point temperature, the abundances specified in Ref. 1 must be used.

[b]The term equilibrium hydrogen is defined in Sec. III, 5 of Ref. 1.

[c]Fractionation of isotopes or impurities dictates the use of boiling points (vanishingly small vapor fraction) for hydrogen and neon, and condensation point (vanishingly small liquid fraction) for oxygen [1].

[d]The triple point of argon may be used as an alternative to the condensation point of oxygen.

[e]The freezing point of tin may be used as an alternative to the boiling point of water.

perature of 0.5°C around 800°C and calibration uncertainties of 0.2°C or more in that temperature range suggest eventual replacement of thermocouples by platinum resistance thermometers as the defining standard.

2. Extension of the IPTS scale to temperatures below 13.81 K.

3. The possible elimination of boiling points from the defining fixed-point list to avoid the need for pressure measurement.

4. Simplification of the interpolation formulas for temperatures below 0°C.

TABLE 1B International Practical Temperature Scale IPTS-68: Secondary Reference Points [1]

Equilibrium state[b]	International Practical Temperature[a]	
	T_{68} (K)	t_{68} (°C)
Equilibrium between the solid, liquid, and vapor phases of normal hydrogen (triple point of normal hydrogen)	13.956	−259.194
Equilibrium between the liquid and vapor phases of normal hydrogen (boiling point of normal hydrogen)	20.397	−252.753

$$\log \frac{P}{P_0} = A + \frac{B}{T_{68}} + CT_{68} + DT_{68}^2$$

$A = 1.734791$, $B = -44.62368$ K, $C = 0.0231869$ K^{-1},
   $D = -0.000048017$ K^{-2}
for the temperature range from 13.956 K to 30 K.

Equilibrium between the solid, liquid, and vapor phases of neon (triple point of neon)	24.561	−248.589
Equilibrium between the liquid and vapor phases of neon		

$$\log \frac{P}{P_0} = A + \frac{B}{T_{68}} + CT_{68} + DT_{68}^2$$

$A = 4.61152$, $B = -106.3851$ K, $C = -0.0368331$ K^{-1},
   $D = 4.24892 \times 10^{-4}$ K^{-2}
for the temperature range from 24.561 K to 40 K

Equilibrium between the solid, liquid, and vapor phases of nitrogen (triple point of nitrogen)	63.146	−210.004
Equilibrium between the liquid and vapor phases of nitrogen (boiling point of nitrogen)	77.344	−195.806

$$\log \frac{P}{P_0} = A + \frac{B}{T_{68}} + C \log \frac{T_{68}}{T_0} + DT_{68} + ET_{68}^2$$

$A = 5.893271$, $B = -403.96046$ K, $C = -2.3668$,
   $D = -0.0142815$ K^{-1}, $E = 72.5872 \times 10^{-6}$ K^{-2}, $T_0 =$
   77.344 K
for the temperature range from 63.146 K to 84 K

Equilibrium between the liquid and vapor phases of argon (boiling point of argon)	87.294	−185.856
Equilibrium between the liquid and vapor phases of oxygen		

$$\log \frac{P}{P_0} = A + \frac{B}{T_{68}} + C \log \frac{T_{68}}{T_0} + DT_{68} + ET_{68}^2$$

$A = 5.961546$, $B = -467.45576$ K, $C = -1.664512$,
   $D = -0.01321301$ K^{-1}, $E = 50.8041 \times 10^{-6}$ K^{-2}, $T_0 =$
   90.188 K
for the temperature range from 54.361 K to 94 K

Equilibrium between the solid and vapor phases of carbon dioxide (sublimation point of carbon dioxide)	194.674	−78.476

$$T_{68} = 194.674 + 12.264 \left(\frac{P}{P_0} - 1 \right) - 9.15 \left(\frac{P}{P_0} - 1 \right)^2 \quad \text{K}$$

for the temperature range from 194 K to 195 K

TABLE 1B International Practical Temperature Scale IPTS-68: Secondary Reference Points [1] (*Continued*)

Equilibrium state[b]	International Practical Temperature[a]	
	T_{68} (K)	t_{68} (°C)
Equilibrium between the solid and liquid phases of mercury (freezing point of mercury)[c]	234.314	−38.836
Equilibrium between ice and air-saturated water (ice point)[d]	273.15	0
Equilibrium between the solid, liquid, and vapor phases of phenoxybenzene (diphenyl ether)(triple point of phenoxybenzene)	300.02	26.87
Equilibrium between the solid, liquid, and vapor phases of benzoic acid (triple point of benzoic acid)	395.52	122.37
Equilibrium between the solid and liquid phases of indium (freezing point of indium)[c]	429.784	156.634
Equilibrium between the solid and liquid phases of bismuth (freezing point of bismuth)[c]	544.592	271.442
Equilibrium between the solid and liquid phases of cadmium (freezing point of cadmium)[c]	594.258	321.108
Equilibrium between the solid and liquid phases of lead (freezing point of lead)[c]	600.652	327.502
Equilibrium between the liquid and vapor phases of mercury (boiling point of mercury)	629.81	356.66

$$t_{68} = 356.66 + 55.552 \left(\frac{P}{P_0} - 1 \right) - 23.03 \left(\frac{P}{P_0} - 1 \right)^2$$
$$+ 14.0 \left(\frac{P}{P_0} - 1 \right)^3 \quad °C$$

for P = 90 kPa to 104 kPa

Equilibrium between the liquid and vapor phases of sulfur (boiling point of sulfur)	717.824	444.674

$$t_{68} = 444.674 + 69.010 \left(\frac{P}{P_0} - 1 \right) - 27.48 \left(\frac{P}{P_0} - 1 \right)^2$$
$$+ 19.14 \left(\frac{P}{P_0} - 1 \right)^3 \quad °C$$

for P = 90 kPa to 104 kPa.

Equilibrium between the solid and liquid phases of the copper-aluminum eutectic	821.41	548.26
Equilibrium between the solid and liquid phases of antimony (freezing point of antimony)[c]	903.905	630.755
Equilibrium between the solid and liquid phases of aluminum (freezing point of aluminum)	933.61	660.46
Equilibrium between the solid and liquid phases of copper (freezing point of copper)	1358.03	1084.88
Equilibrium between the solid and liquid phases of nickel (freezing point of nickel)	1728	1455
Equilibrium between the solid and liquid phases of cobalt (freezing point of cobalt)	1768	1495
Equilibrium between the solid and liquid phases of palladium (freezing point of palladium)	1827	1554
Equilibrium between the solid and liquid phases of platinum (freezing point of platinum)	2042	1769

TABLE 1B International Practical Temperature Scale IPTS-68: Secondary Reference Points [1] (*Continued*)

Equilibrium state[b]	International Practical Temperature[a]	
	T_{68} (K)	t_{68} (°C)
Equilibrium between the solid and liquid phases of rhodium (freezing point of rhodium)	2236	1963
Equilibrium between the solid and liquid phases of aluminum oxide, Al_2O_3 (temperature of melting aluminum oxide)	2327	2054
Equilibrium between the solid and liquid phases of iridium (freezing point of iridium)	2720	2447
Equilibrium between the solid and liquid phases of niobium (temperature of melting niobium)	2750	2477
Equilibrium between the solid and liquid phases of molybdenum (temperature of melting molybdenum)	2896	2623
Equilibrium between the solid and liquid phases of tungsten (temperature of melting tungsten)	3695	3422

[a]The temperatures listed in this table are the best available at the time of compilation. At present it is not possible to assign levels of accuracy to these temperatures; errors may be as high as several units of the last digit shown. Assessments of accuracies of these temperatures will be made and are expected to be published from time to time under the auspices of the Comité International des Poids et Mesures.

[b]The equilibrium states in this table are at a pressure $P_0 = 101,325$ Pa (one standard atmosphere) except for the triple points and in those cases where a range of pressures is explicitly allowed.

[c]See Ref. 1 for the effect of pressure variations on these freezing points.

[d]The ice point is a very close approximation to the temperature defined as being 10 mK below the triple point of water.

Recent improvements in constant-volume gas thermometers are expected to provide thermodynamic temperature values that would eventually result in the update of fixed-point temperatures [2, 5]. An improved IPTS scale is expected in the early 1990s.

Two common temperature scales having different-size degrees are widely used: the Celsius and Fahrenheit scales. The Celsius scale is essentially described by the IPTS-68 and Eq. (1), as noted above. This closely agrees with its historical definition that the normal freezing point of water be 0°C and the boiling point of pure water at standard atmospheric pressure be 100°C.

The Fahrenheit scale (°F) has an analogous absolute temperature scale, the Rankine scale (°R) with a value of zero at absolute zero. Temperatures in the Rankine scale can be determined from the Kelvin temperature by

$$T \ (°R) = 1.8\,T \ (K) \tag{3}$$

The Fahrenheit scale has the same-size degree as the Rankine scale and can be determined from

$$t \ (°F) = T \ (°R) - 459.67 \tag{4}$$

In accordance with the historical definitions, this leads to a value of 32°F at the freezing point of water and 212°F at the boiling point of water at standard atmospheric pressure. Temperatures in °C and °F are related by

$$t \ (°F) = \tfrac{9}{5}t \ (°C) + 32 \tag{5}$$

It should be noted that the Fahrenheit and Rankine scales are, with few exceptions, now used only in the United States.

3. Sensors

a. Introduction

Historically, low-pressure gas thermometers were the only primary thermometers that had the accuracy required in determining the temperatures of defining fixed points. In the last decade, advances in thermometry resulted in the development of several types of primary thermometers capable of accurate thermodynamic temperature measurements [2]. A list of present-day primary thermometers used for thermodynamic temperature measurement includes:

1. Gamma-ray anisotropy thermometers—applicable over the range 10 μK to 1 K
2. Melting curve of ^{3}He—applicable from 1 mK to 1 K or higher, still under development
3. Noise thermometers
    a. Josephson-junction type—1 mK to 1 K, under development
    b. "Normal" type—2 K to 2000 K
4. Acoustic thermometers—1 K to 273 K
5. Constant-volume gas thermometers—1 K to 1400 K
6. Dielectric-constant gas thermometers—under development
7. Total radiation thermometers—100 K to 700 K

The total radiation thermometers are not really primary thermometers since a one-point (or more) calibration is needed for thermodynamic temperature measurements. For that matter, vapor-pressure thermometers, magnetic thermometers, and various radiation thermometers all belong to this class—pseudo-primary thermometers.

Almost all thermometers used in actual applications belong to the class of secondary thermometers. They can be categorized according to their temperature-dependent function, represented by the type of transducer or output they employ. They include systems whose temperature is indicated by:

1. **Changes in Dimension.**   With such devices, a change in physical dimension occurs with a change in temperature. In this category are liquid-in-glass or other fluid-expansion thermometers, bimetallic strips, and others.

2. **Electrical Effects.**   Electrical methods are convenient because an electrical signal can be easily processed. Resistance thermometers (including thermistors) and thermocouples are the most widely used. Other electrical methods include: noise thermometers using the Johnson noise as a temperature indicator; resonant-frequency thermometers, which rely on the temperature dependence of the resonant frequency of a medium, including the nuclear quadrupole resonance thermometers, ultrasonic thermometers, and quartz thermometers; and semiconductor-diode thermometers, where the relation between temperature and junction voltage at constant current is used.

3. **Radiation.**   The thermal radiation emitted by a body is a function of the temperature of the body; hence, measurement of the radiation can be used to indicate the temperature. Commonly employed in this category are optical pyrometers, infrared scanners, spectroscopic techniques, and total-radiation calorimeters.

4. **Other Methods.**   In addition to the systems mentioned above, there are many others.

Commonly encountered ones include: optical methods in which the index-of-refraction variation is determined and from this, the temperature; and liquid-crystal or other contact thermographic methods in which the color changes of cholesteric liquid crystals or thermally sensitive materials are determined by the temperature.

Different devices have their own temperature range of operation—some suitable for low-temperature measurements, some only applicable at high temperature, and others in between. No single device is used over the entire range of temperatures in the physical environment. Even IPTS-68 requires three different defining standard thermometers: the standard platinum resistance thermometer; the platinum–10% rhodium vs. platinum thermocouple; and a radiant energy sensor. Figure 1 gives an approximate range of operation for some commonly used practical thermometers, and Table 2 summarizes the relative merits of four common insertion-type thermometers.

The accuracy of a thermometer in reproducing the IPTS-68 temperature is achieved through calibration. Calibration of a thermometer is usually done directly or indirectly against a defining standard thermometer. For example, many laboratories maintain standard platinum resistance thermometers (SPRT) and standard platinum platinum-rhodium type thermocouples calibrated according to the specifications of IPTS-68. Calibration of these defining standard thermometers is normally done at a government standards laboratory. Other temperature sensors are calibrated by comparison against one of these defining standard thermometers in a constant-temperature environment—usually a well-stirred constant-temperature liquid bath or special furnace. Comparisons are usually made at a modest number of temperatures with intermediate values described by a cor-

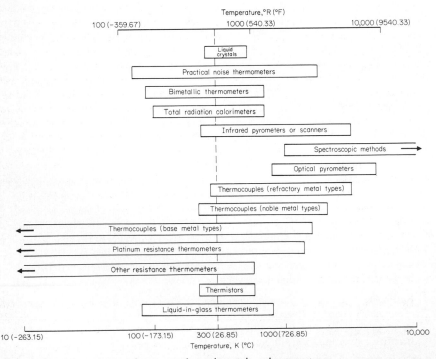

FIG. 1 Temperature ranges for commonly used secondary thermometers.

TABLE 2 Advantages and Disadvantages of Various Insertion- or Probe-Type Sensors

Sensor	Advantages	Disadvantages
Mercury-in-glass thermometers	1. Stable 2. Cheap 3. Good accuracy	1. Slow time response 2. Automation and remote sensing impractical
Thermocouples	1. Wide temperature range 2. Fast response 3. Ease of remote sensing 4. Reasonable cost 5. Rugged 6. Small	1. Need reference junction 2. Do not have extreme sensitivity and stability 3. Fair accuracy
Thermistors	1. Fast response 2. Ease of remote sensing 3. Low cost 4. Rugged	1. Highly nonlinear 2. Small temperature range 3. Limited interchangeability
Platinum resistance thermometers	1. Good linearity 2. Wide temperature range 3. Very good stability even at high temperature 4. Very good precision and accuracy	1. Slow response 2. Vibration and shock fragility 3. High cost

relating equation for the specific thermometer being used. Easily established fixed points, such as the water-ice equilibrium point or water triple point, are used for precision thermometers to see if there has been any drift in the correlation. Because of the fragility of SPRTs and other defining standard thermometers, thermometers for many applications are often directly calibrated against working standard thermometers which have been previously calibrated against a transfer standard thermometer or a defining standard thermometer which is traceable to NBS. A working standard thermometer is usually of the same type as the thermometers to be calibrated to reduce the instrumentation requirements; a transfer standard thermometer serves as an intermediate standard to minimize the use and drift of the defining standard thermometer. All standard thermometers should be calibrated periodically to ensure the best accuracy in the realization of IPTS-68.

In the following sections, some of the commonly used thermometers are reviewed. For applications at extremes of temperatures, readers are referred to Refs. 6 and 7 (cryogenic thermometry), and to Ref. 8 (applications at very high temperature). Also included are some statements about the reliability (accuracy, precision, stability, etc.) of individual sensing methods. The number quoted as the accuracy is the "ideal," which is the accuracy a thermometer is capable of producing but is not the same as the accuracy of the actual measurement, which involves more than just the thermometer.

b. Liquid-in-Glass Thermometers

Definitions and Principles.   A typical liquid-in-glass thermometer consists of a liquid in a thin-wall glass bulb attached to a glass stem, with the bulb and stem system sealed against the environment. The volume of the liquid in the bulb depends on the liquid temperature, and with almost all real thermometers both the volume of the liquid and the cavity itself increase with increasing temperature. The increase in the volume of the cavity is a secondary effect in the measurement. The volume of the liquid is indicated by the height of liquid in the stem. A scale is provided on the stem to facilitate the reading of the temperature of the bulb. The volume above the liquid into which it can expand is either evacuated or filled with a dry inert gas. An expansion chamber is provided at the top end of the stem to protect the thermometer in case of overheating. Liquid-in-glass thermometers for use at high temperatures have an auxiliary scale for calibration at the ice-point temperature and a contraction chamber to reduce the stem length. A schematic of a high-temperature thermometer is shown in Fig. 2.

Liquid-in-glass thermometers are almost exclusively used to determine the temperature of fluids which are relatively uniform—i.e., which contain no large temperature gradients. Conduction along the glass stem can affect the temperature of the glass as well as that of the liquid. Therefore, thermometers are usually calibrated for a specified depth of immersion. They should then show the correct temperature when inserted to that level in the fluid whose temperature is to be measured. When a thermometer is used in situations where the immersion is other than that for which it was designed and calibrated, a stem correction [9] should be applied.

Three immersion types are available: (1) partial immersion—the bulb and a specified portion of the stem are exposed to the fluid whose temperature is being measured; (2) total immersion—the bulb and all of the liquid column are in the fluid; and (3) complete immersion—the entire thermometer is immersed in the fluid whose temperature is being measured.

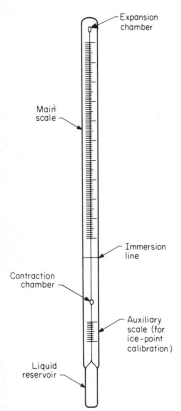

FIG. 2 High-temperature liquid-in-glass thermometer.

The most widely used liquid-in-glass thermometer is the mercury-in-glass thermometer, which can be used between $-38.9°C$ and about $300°C$ ($-38°F$ to $570°F$). The upper limit can be extended to about $540°C$ ($1000°F$) by filling the space above the mercury with nitrogen. The lower limit can be extended to $-56°C$ ($-69°F$) by alloying the mercury with thallium.

For lower temperatures, down to $-200°C$ ($-330°F$), organic liquids (e.g., alcohol) can be used as the working fluid. Thermometers filled with organic liquids are not considered as reliable as mercury-filled thermometers.

Accuracy. Mercury-in-glass thermometers are relatively inexpensive and can be obtained in a wide variety of accuracies and temperature ranges. For example, between 0 and $100°C$ (32 and $212°F$), thermometers with a $0.1°C$ graduation interval (0.2 or $0.1°F$ graduation) are readily obtained. With proper calibration by the NBS [9, 10], an accuracy of from 0.01 to $0.03°C$ (0.02 to $0.05°F$) can be achieved. Table 3 summarizes

TABLE 3A Expected Accuracies of Various Mercury-in-Glass Thermometers: Total-Immersion Thermometers (°C) [9]

Temperature range, °C	Graduation interval	Tolerance	Accuracy
Thermometer graduated under 150°C			
0 up to 150	1.0 or 0.5	0.5	0.1 to 0.2
0 up to 150	0.2	0.4	0.02 to 0.05
0 up to 100	0.1	0.3	0.01 to 0.03
Thermometers graduated under 300°C			
0 up to 100	1.0 or 0.5	0.5	0.1 to 0.2
Above 100 up to 300		1.0	0.2 to 0.3
0 up to 100	0.2	0.4	0.02 to 0.05
Above 100 up to 200		0.5	0.05 to 0.1
Thermometers graduated above 300°C			
0 up to 300	2.0	2.0	0.2 to 0.5
Above 300 up to 500		4.0	0.5 to 1.0
0 up to 300	1.0 or 0.5	2.0	0.1 to 0.5
Above 300 up to 500		4.0	0.2 to 0.5

TABLE 3B Expected Accuracies of Various Mercury-in-Glass Thermometers: Total-Immersion Thermometers (°F) [9]

Temperature range, °F	Graduation interval	Tolerance	Accuracy
Thermometers graduated under 300°F			
32 up to 300	2.0	1.0	0.2 to 0.5
32 up to 300	1.0 or 0.5	1.0	0.1 to 0.2
32 up to 212	0.2 or 0.1	0.5	0.02 to 0.05
Thermometers graduated under 600°F			
32 up to 212	2 or 1	1.0	0.2 to 0.5
Above 212 up to 600		2.0	0.5
Thermometers graduated above 600°F			
32 up to 600	5	4.0	0.5 to 1.0
Above 600 up to 950		7.0	1.0 to 2.0
32 up to 600	2 or 1	3.0	0.2 to 1.0
Above 600 up to 950		6.0	0.5 to 1.0

the tolerances and the expected accuracies for various mercury thermometers accepted for calibration by the NBS. Guidelines for acceptance for calibration by the NBS can be found in Ref. 9. Total-immersion types are more accurate than partial-immersion types. For accurate measurement, thermometers should be checked periodically at a reference point (normally, the ice point) for changes in calibration. In general, stem correction is not significant for applications below 100°C. At higher temperatures, accuracy can only be assured with a proper stem correction [9].

Factors that affect the accuracy of the thermometer reading include changes in volume

TABLE 3C Expected Accuracies of Various Mercury-in-Glass Thermometers: Partial-Immersion Thermometers (°C) [9]

Temperature range, °C	Graduation interval*	Tolerance	Accuracy†
Thermometers not graduated above 150°C			
0 up to 100	1.0 or 0.5	1.0	0.1 to 0.3
0 up to 150	1.0 or 0.5	1.0	0.1 to 0.5
Thermometers not graduated above 300°C			
0 up to 100	1.0	1.0	0.1 to 0.3
Above 100 up to 300	1.0	1.5	0.5 to 1.0
Thermometers graduated above 300°C			
0 up to 300	2.0 or 1.0	2.5	0.5 to 1.0
Above 300 up to 500		5.0	1.0 to 2.0

*Partial-immersion thermometers are sometimes graduated in smaller intervals than shown in these tables, but this in no way improves the performance of the thermometers, and the listed tolerances and accuracies still apply.

†The accuracies shown are attainable only if emergent stem temperatures are closely known and accounted for.

TABLE 3D Expected Accuracies of Various Mercury-in-Glass Thermometers: Partial-Immersion Thermometers (°F) [9]

Temperature range, °F	Graduation interval*	Tolerance	Accuracy†
Thermometers not graduated above 300°F			
32 up to 212	2.0 or 1.0	2.0	0.2 to 0.5
32 up to 300	2.0 or 1.0	2.0	0.2 to 1.0
Thermometers not graduated above 600°F			
32 up to 212	2.0 or 1.0	2.0	0.2 to 0.5
Above 212 up to 600	2.0 or 1.0	3.0	1.0 to 2.0
Thermometers graduated above 600°F			
32 up to 600	5.0 or 2.0	5.0	1.0 to 2.0
Above 600 up to 950		10.0	2.0 to 3.0

*Partial-immersion thermometers are sometimes graduated in smaller intervals than shown in these tables, but this in no way improves the performance of the thermometers, and the listed tolerances and accuracies still apply.

†The accuracies shown are attainable only if emergent stem temperatures are closely known and accounted for.

of the glass bulb under thermal stress, pressure effects, and response lag. A source for more information on liquid-in-glass thermometry is Ref. 11.

c. Resistance Thermometers

Principles and Sensors.   The electrical resistivity of a material is a function of temperature. This functional dependence is the basis for the operation of resistance thermometers. A resistance thermometer includes a properly mounted and protected resistor and a resistance measuring instrument. The resistance values are then converted to temperature readings.

Resistive materials used in thermometry include platinum, copper, nickel, carbon, germanium, and certain semiconductors known as thermistors. Sensors made from platinum wires are called platinum resistance thermometers (PRT) and, though expensive, are widely used. They have excellent stability and have the potential for high-precision measurements. The temperature range of operation is from $-260°C$ to $1100°C$ ($-436°F$ to $2000°F$). Other resistance thermometers are less expensive than a PRT and are useful in certain situations. Copper has a fairly linear resistance-temperature relationship, but its upper temperature limit is only about $150°C$ ($300°F$), and because of its low resistivity, special measurements may be required. Nickel has an upper temperature limit of about $300°C$ ($570°F$), but it oxidizes easily at high temperature and is quite nonlinear [12]. Carbon and germanium are used in cryogenic temperature measurements for temperatures below those defined in IPTS-68 [7]. Generally these instruments (except thermistors) have a positive temperature coefficient of resistance—the resistance increases with temperature.

Thermistors are usually made from ceramic metal oxide semiconductors which have a large negative temperature coefficient of electrical resistance. Thermistor is a contraction of *thermal-sensitive-resistor*. The recommended temperature range of operation is from -55 to $300°C$ (-70 to $570°F$). Their popularity has grown rapidly in recent years. Special thermistors for cryogenic applications are also available [13].

Among the different resistance-type thermometers, we shall consider only PRTs and thermistors in greater detail.

PRTs.   PRTs are the most widely used temperature measuring devices for high-precision applications. The precision of PRTs is due to: (1) the very high degree of purity of platinum which can be obtained in the manufacture of platinum wires, (2) the reproducibility of that purity from batch to batch, and (3) the inertness of pure platinum. These characteristics allow for a high degree of stability and repeatability in PRTs.

As mentioned in Sec. B.2, a special PRT which satisfies the specifications of IPTS-68 is the defining standard thermometer (called the standard platinum resistance thermometer, SPRT) for the range of temperatures from 13.81 K to 903.89 K (630.74°C). IPTS-68 in this temperature range as maintained by the NBS is described in Refs. 14 and 15.

The choice of an SPRT as the standard thermometer stems from the extremely high stability, repeatability, and accuracy that can be achieved by an SPRT in a strain-free annealed state. The high cost of calibrating and maintaining an SPRT, the care needed in proper handling, the slow time response, and the small change in resistance per degree are factors that make it a standard device rather than one to be used in most measurements. Figure 3a and b shows the construction of the sensing element for an SPRT and a high-temperature PRT standard, respectively. Figure 3c shows the sensing element for a PRT standard which has a more rugged design. However, its stability and precision are not as good as the units shown in Fig. 3a and in Fig. 3b; thus, it is only recommended as a transfer or working standard. Further information on SPRTs is available in Ref. 16.

Besides SPRTs, platinum resistance thermometers are available in many different forms and sizes for a variety of applications. The basic design of a PRT involves a length of platinum wire wound on an inert supporting material with proper insulation to prevent shorting. Figure 4 shows the construction of some industrial PRTs.

The sensitivity of a PRT depends on its nominal resistance—the resistance measured at the ice point, R_0. The sensitivity is measured by the change in resistance per degree

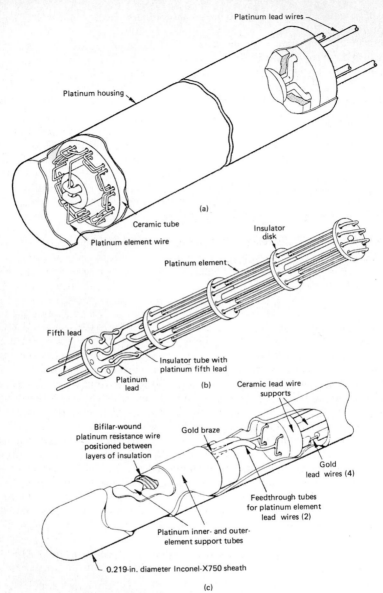

FIG. 3 Sensing elements for platinum resistance thermometers: (*a*) Standard thermometer (SPRT). (*b*) High-temperature standard thermometer. (*c*) Transfer or working standard thermometer. (Courtesy of Rosemount, Inc.)

change in temperature. Pure platinum has a change in resistivity of approximately 0.4 percent per °C near the ice point [see Eq. (6), below]. A typical industrial PRT having an R_0 of 100 Ω has a sensitivity of about 0.4 Ω/°C; SPRTs usually have an R_0 of 25 Ω with a sensitivity of only 0.1 Ω/°C. Industrial PRTs have a range of R_0 from 10 Ω to 2000 Ω. For a custom-made sensor, an R_0 of 10,000 Ω or greater can be achieved. The

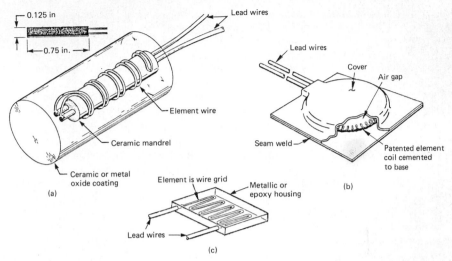

FIG. 4 Industrial platinum resistance thermometers: (*a*) Rod type. (*b*) Disk type. (*c*) Strain-gauge type. (Courtesy of Rosemount, Inc.)

choice of R_0 normally depends on the operating temperature of the sensor; a PRT with a low R_0 would be used in high-temperature applications to avoid shorting through the insulation, while a large R_0 is normally chosen for cryogenic applications to increase the sensitivity.

The platinum wire used in PRTs generally has a diameter between 7.6 μm and 76 μm (0.0003 in and 0.003 in). The wire resistance is about 4 Ω/cm for 17-μm wire (10 Ω/in for 0.0007-in wire). The choice of wire diameter would depend on the sensitivity and physical size of the sensor.

The resistance of a wire is a function of temperature, deformation, and impurities in the wire. The temperature effect is the one sought; the other two introduce uncertainties (errors). The effect of deformation is minimized in an SPRT by a strain-free mounting (thus requiring care in handling to avoid straining the wire) and annealing (to remove the strains). PRTs with compact insulation and support have some inherent uncertainty in repeatability due to mechanical strain. Additional deformation can be caused by vibration and mechanical or thermal shock. These should be avoided in applications if the stability of the PRT is important.

A measure of the impurities in the wire is given by a material constant α, which is an averaged temperature coefficient of resistivity defined by

$$\alpha = \frac{R_{100} - R_0}{100 R_0} \frac{1}{{}^\circ\text{C}} \tag{6}$$

The higher the value for α, the better the wire purity. IPTS-68 requires a minimum of $\alpha = 0.003925 {}^\circ\text{C}^{-1}$ for the platinum wires used in SPRTs. In industrial applications, a value of $0.00385 {}^\circ\text{C}^{-1}$ is used as a standard for PRTs.

PRTs are used to measure temperatures of gases, liquids, granular solids, etc. They are usually limited to steady-state or relatively slow transient measurements (compared to thermistors or thermocouples) as a result of constraints imposed by the construction, mainly of physical size and mass. For the same reason, PRTs are not generally used to measure local values in systems with significant temperature gradients. Commercially available PRTs with a small measuring volume have typical sizes of 1.5 mm (0.06 in) in diameter by 15.0 mm (0.6 in) in length for a rod-type sensor, and 0.76 mm (0.03 in) in thickness with a square surface of side 6.4 mm (0.25 in) long for a disk-type sensor.

The common types of insulation used in PRTs include mica, ceramic, glass, epoxy resin, and aluminum oxide (Al_2O_3). Mica and ceramic are used in SPRTs as both insulation and support. In other PRTs, ceramic glazes or glass or epoxy coatings, as well as Al_2O_3 packing, are frequently used with metal supports. In situations where better response time is required, an open-end type construction is used with the platinum wire wound on top of an insulation coating and with no insulation on the outside other than a thin coating on the wire.

The two ends of the platinum wire in a PRT are connected to lead wires. Two-, three-, or even four-lead configurations are available. The choice would depend on the accuracy required, which is discussed in the subsection on resistance measurement, pp. 12-18 to 12-22.

Thermistors.   Thermistors [17] are made from semiconductors which have   large temperature coefficients (change in resistance with temperature) as compared to metal-wire resistance sensors. Although positive and negative temperature coefficient thermistors are available, only the latter are normally used as temperature sensors. The most widely used thermistor sensors are made of metal oxides including oxides of manganese, nickel, cobalt, copper, iron, and titanium. Their high resistance allows for remote sensing capability as the lead resistances can be safely ignored.

Commercial thermistors are available in two major classes: (1) embedded-lead types, including (glass-coated or uncoated) bead, glass-probe, and glass-rod thermistors; and (2) metal-contact types, including thermistor disks, chips, rods, washers, etc. The above division is based on the means of attaching the lead wires. Thermistors of the first type use direct sintering of metal oxide onto the lead wires. They exhibit high stability, are available in very small-size units with fast response, and can be used at temperatures up to 300°C. Those of the other type have the lead wires soldered on after the thermistor is fabricated. They have close tolerances and good interchangeability but lack the stability of the first type. They are generally not usable at temperatures above 125°C (260°F) and have slower response time than the first type. Table 4 gives a summary of the characteristics of thermistors.

Figure 5 shows the schematic of some thermistor sensors. More information on thermistors can be obtained from Refs. 18 to 21. The characteristic rating for a thermistor is given by its resistance at 25°C. Unlike PRTs, which have a rather limited range of resistance values, commercially available thermistor probes have a large range of characteristic resistances, from as low as 30 Ω to as high as 20 MΩ. The physical size of a thermistor could be as small as 0.075 mm (0.003 in) in diameter on lead wires of 0.018 mm (0.0007 in) diameter for a bead thermistor. Metal-contact thermistors are generally bigger, starting from 0.25 mm (0.01 in) in thickness and 1.0 mm (0.04 in) in diameter for disk types and from 0.5 mm (0.02 in) in diameter and 5 mm (0.2 in) in length for rod-type thermistors.

Resistance Measurement.   The common methods of resistance measurement in resistance thermometry are the bridge method and the potentiometric method. Basically, the bridge method uses the resistance sensor together with a variable resistor and two fixed resistors to form the four legs of a conventional Wheatstone bridge circuit. On the

TABLE 4A Characteristics of Thermistors: Embedded-Lead Thermistors (Beads, Probes) [18]

Advantages	Disadvantages
1. High stability, reliability 2. Available in very small (0.1 mm) diameter 3. Fast response 4. Can be used at high temperatures (300°C)	1. Relatively high cost for close tolerances and interchangeability. 2. Interchangeability requires resistor network padding or use of matched pairs. 3. Comparatively low dissipation constants.

TABLE 4B Characteristics of Thermistors: Metal-Contact Thermistors (Disks, Wafers, Chips) [18]

Advantages	Disadvantages
1. Close tolerances and interchangeability at relatively low cost 2. Comparatively high dissipation constants 3. Reasonably fast response times for uncoated units—particularly for 1 mm × 1 mm × ⅛ mm thick units	1. Difficult to obtain reliable electrodes. 2. Reliability and stability leave much to be desired. 3. Size limited to about 1 mm diameter (or 1 mm × 1 mm) uncoated; about 1.5 mm diameter coated. 4. Cannot be used at high temperatures—requires cold sterilization. 5. Comparatively long response times for coated units.

other hand, the potentiometric method, also called a half bridge, connects the resistance sensor in series with a known resistor.

The basic difference between the bridge method and the potentiometric method is that in the bridge method the resistance is measured directly by comparing the sensor's resistance with that of a known resistor, while in the potentiometric method the measurement of the IR voltage drop is used to determine the sensor's resistance. The bridge method is inherently tedious and is not easily automated. In the potentiometric circuit, a known fixed resistor and a constant voltage source are used. From a measurement of the IR

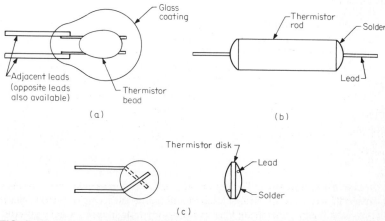

FIG. 5 Thermistors: (*a*) Glass-coated bead type. (*b*) Rod type. (*c*) Disk type.

voltage drop across the fixed resistor, the sensor's resistance can be determined. This system is simple and versatile, and is widely used.

Figure 6 shows a simple circuit used to determine resistance of a thermistor using a constant voltage source. Variations to this basic circuit can be found in Ref. 22.

In precision measurements, for SPRTs in particular, the conventional Wheatstone bridge (not shown) and the simple potentiometric circuit shown in Fig. 6 are not sufficient. With the bridge method, the resistance of the lead wires to the sensor is not accounted for; with the potentiometric method, a problem arises from the difficulty in maintaining a constant voltage source and measuring it to an acceptable accuracy (and the inability to account for lead-wire resistance with a two-lead sensor). Thus, for high-precision measurements, more sophisticated circuits are used.

To overcome the problems of lead resistances in the bridge method, a Mueller bridge [23, 24] can be used with an SPRT. A schematic of the bridge is shown in Fig. 7. Its advantage over the conventional Wheatstone bridge is the provision for interchanging the leads (c and C with t and T) so that the average of two readings is independent of the lead resistances. Referring to Fig. 7, at null condition, the equation of balance is

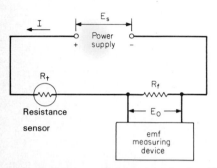

FIG. 6 Resistance thermometer circuit with constant voltage source.

$$R_{D1} + R_C = R_X + R_T \tag{7}$$

where R_C and R_T are the resistances of leads C and T, respectively, and R_{D1} is the value of R_D required for zero current in the galvanometer with these connections to the SPRT. After interchanging the leads c and C with t and T (reverse connection, see Fig. 7) and rebalancing (changing only R_D)

$$R_{D2} + R_T = R_X + R_C \tag{8}$$

From Eqs. (7) and (8), the unknown resistance of the SPRT, R_X, can be found:

$$R_X = \frac{R_{D1} + R_{D2}}{2} \tag{9}$$

To eliminate the error caused by spurious emf's, the bridge current is reversed (not shown on Fig. 7)—i.e., four bridge balances are made in the order NRRN (where N = normal and R = reverse connections of the bridge) for each determination of R_X.

Conventional potentiometric methods for precision thermometry may utilize a potentiometer [25] in measuring the resistance. The procedure is tedious, requiring four balancings for each reading. A modification using an isolating potential comparator requires only two balancings [25]. A schematic of a potentiometric circuit with an isolating potential comparator (for an SPRT or a thermistor standard, both with four lead wires) is shown in Fig. 8. A variable resistor (precalibrated so that its resistance is known "exactly") is used. The IR voltage drop across this resistor is compared to that across the SPRT by a potential comparator, with provision for current reversal. The potential comparator basically consists of a high-quality capacitor that is successively connected to the leads of the SPRT and those of the known resistor by a high-speed double-pole double-throw switch (commonly called a chopper). If the resistance of the known resistor is not

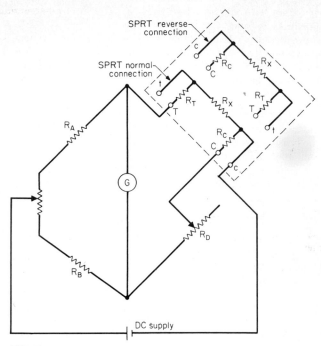

FIG. 7 Mueller bridge circuit for precision resistance measurement.

equal to the resistance of the SPRT, the capacitor will experience a process of successive charging and discharging, thus causing a current to flow through the galvanometer (or null detector). The resistance of the SPRT is given by the resistance of the known resistance under null condition (no current through the galvanometer). A provision for current reversal (not shown on Fig. 8) is used to eliminate the error caused by spurious emf's.

For the methods described above, which use a dc power source, two to four balancings are required for each reading and automation is difficult. The development of ac bridges eliminates the need of current reversal, and permits automation in precision thermometry [26, 27].

PRTs can have either two, three, or four lead wires. Besides the Mueller bridge setup, two-wire and three-wire bridges are available [28]. A two-wire (Wheatstone) bridge is the least accurate because the lead-wire resistances are not accounted for. A three-wire bridge allows for the compensation of lead-wire resistances by adding the resistance of the third lead wire to the leg of the bridge where the known variable resistor resides. Hence, if the resistances of the two

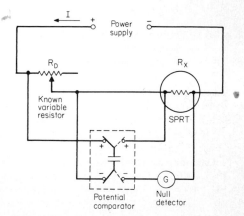

FIG. 8 Potentiometric circuit for resistance measurement with comparator.

sation of lead-wire resistances by adding the resistance of the third lead wire to the leg of the bridge where the known variable resistor resides. Hence, if the resistances of the two lead wires (one connected to one leg and the second to the other leg of the bridge) are the same, they will effectively cancel each other out. However, the resistances of the two lead wires would normally not be equal, if for no other reason than the presence of the temperature gradient along the lead wires. With potentiometric circuits, only two– or four–lead-wire sensors are used. A two–lead-wire sensor is less accurate because of the unknown lead-wire resistances.

Resistance-Temperature Conversion. A measuring circuit (bridge or potentiometer) can provide the resistance of the sensor in a resistance thermometer. The next step is to convert the resistance reading into a value of sensor temperature.

In the definition of IPTS-68, interpolation formulas are provided for the calibration of SPRTs. For temperatures between 0 and 630.74°C (32 and 1167.33°F), the interpolation formula is a quadratic expression relating the resistance ratio R/R_0 to temperature.

$$\frac{R(t)}{R_0} = 1 + At + Bt^2 \tag{10a}$$

The three unknowns in the formula (resistance at 0°C, and the two coefficients A and B) are determined from calibration at the triple point of water, the boiling point of water (or the freezing point of tin), and the freezing point of zinc.

The temperature determined from Eq. (10a) is not quite equal to the IPTS-68 temperature. A slight correction (maximum less than 0.05°C) given by Eq. (10b) is still needed if high accuracy is required.

$$t_{68} = t + 0.045 \left(\frac{t}{100°C}\right)\left(\frac{t}{100°C} - 1\right)\left(\frac{t}{419.58°C} - 1\right)\left(\frac{t}{630.74°C} - 1\right) \tag{10b}$$

For IPTS temperatures below 0°C, the interpolation formulas are complex and will not be covered here. More details on the interpolation formulas and procedures are available in Refs. 1, 16, and 29.

An industrial standard of $\alpha = 0.00385°C^{-1}$ is assumed for PRTs. A standard R_0 of 100 Ω is also adopted. This enables the specification of the standard resistance versus temperature curve as well as the interchangeability tolerances. Two standards are available today: the German Specification DIN 43760, and the British Specification BS 1904. An international standard may be adopted [29] from a specification of the International Electrotechnical Commission (IEC). The IEC values of R at different t are given in Table 5. Figure 9 shows the expected maximum deviations from the standard R versus t relationship (Table 5) for different classes (e.g., A, B) of PRTs. Individual manufacturers often provide their own specifications for PRTs. Smaller tolerances can usually be achieved with higher cost.

The resistance of a thermistor varies approximately exponentially with temperature. Sample resistance-temperature characteristics of thermistors are shown in Fig. 10. The relationship for an idealized thermistor can be written as

$$R = a \exp \frac{b}{t} \tag{11}$$

The existence of impurities causes a deviation from the simple exponential relation given in Eq. (11). Accurate representation is given by addition of higher powers of $1/t$ to the exponential term in Eq. (11).

TABLE 5 Resistance versus Temperature for Standard Industrial Platinum Resistance Thermometers ($\alpha = 0.00385°C^{-1}$) [29]

Temperature, °C	Resistance, Ω	Temperature, °C	Resistance, Ω
−200	18.49	150	157.31
−180	27.08	200	175.84
−160	35.53	250	194.07
−140	43.67	300	212.02
−120	52.11	350	229.67
−100	60.25	400	247.04
−80	68.33	450	264.11
−60	76.33	500	280.90
−40	84.27	550	297.39
−20	92.16	600	313.59
0	100.00	650	329.51
20	107.79	700	345.13
40	115.54	750	360.47
60	123.24	800	375.51
80	130.89	850	390.26
100	138.50		

For most applications, an alternative is employed. Recall that in measuring the resistance of a thermistor a fixed resistor is normally connected in series with the sensor. If a constant-voltage source is used, the circuit current is inversely proportional to the total resistance. Then the relationship between the measured IR voltage drop across the fixed resistor and the thermistor temperature can be almost linear over a range of temperature while the overall functional relationship between the two exhibits an S-shaped behavior. The linear part of this S curve can be shifted along the temperature scale by changing the value of the fixed resistor.

$$E_0 = IR_f = \frac{E_s}{R_t + R_f} R_f \qquad (12)$$

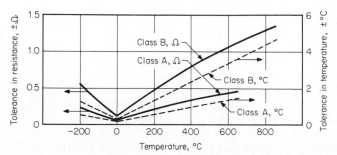

FIG. 9 Interchangeability tolerances for industrial platinum resistance thermometer standards ($\alpha = 0.00385°C^{-1}$) [29].

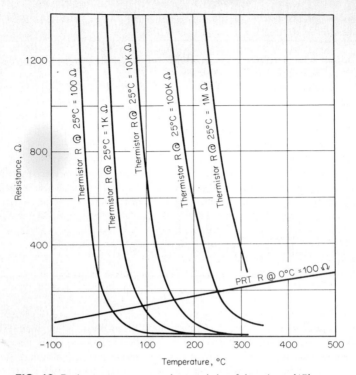

FIG. 10 Resistance-temperature characteristics of thermistors [17].

The resistance of the thermistor is normally expressed in terms of its resistance at a standard reference temperature of 25°C, R_{25}, by introducing a ratio γ called the resistance-ratio characteristic:

$$\gamma = \frac{R_t}{R_{25}} \qquad (13)$$

The dependence of γ on t would come from Eq. (11), or some variation of it, for the particular thermistor used. Since R_{25} is a known fixed value, the ratio R_{25}/R_f is a circuit constant and is designated by s. With the introduction of γ and s, Eq. (12) becomes

$$\frac{E_0}{E_s} = \frac{R_f}{R_t + R_f} = \frac{1}{1 + R_t/R_f} = \frac{1}{1 + s\gamma} = F(t) \qquad (14)$$

The function $F(t)$ defines the S-shaped curve discussed above. By varying the parameter s, a family of such curves can be generated. They can be used to define the standard curves for different thermistor materials.

The relation between E_0 and t [Eq. (14)] can be closely linear over a range of 40 to 60°C at temperatures above 0°C, and over a range of 30°C for temperatures below 0°C. The choice of the value for the fixed resistor R_f is such that the operating temperature range of the thermistor would coincide with the linear portion of $F(t)$, thus simplifying the data reduction process. Actual $F(t)$ curves for different thermistor materials are given in Ref. 22.

Reliability.   A discussion of the reliability of a sensor requires the use of terms like precision, accuracy, and stability. These are popular but not always well defined. Here we shall give brief descriptions of these terms. A more detailed discussion on reliability will be covered in Sec. B.4, on calibration, together with calibration procedures.

Within its specified temperature range of operation, an SPRT has very good *precision* (used here interchangeably with repeatability—the ability to reproduce the same reading for the same conditions) as a temperature sensor. Thus it is used as the defining standard thermometer for IPTS. The precision of an SPRT is of the order of 0.001°C (1 mK). The ability of an SPRT to realize IPTS-68 temperature (its *accuracy*) at the calibration points (defining fixed points) can be better than 5 mK. At temperatures other than the calibration points there is an additional error due to the interpolation process.

When using an SPRT, high precision and accuracy are difficult to maintain. The delicate construction of an SPRT is vulnerable to mechanical strains. An audible tapping of the sheath of an SPRT could produce an increase as large as 100 $\mu\Omega$ in R_0 (corresponding to a 1-mK change in temperature if R_0 is 25 Ω). The cost and fragility of a calibrated SPRT plus the need to have high-precision auxiliary equipment for resistance measurement makes the SPRT a standard thermometer rather than a normal laboratory sensor.

For many years thermistors were not popular because of limitations of *stability* (ability to maintain a reading over a period of time). Even today the stability of metal-contact-type thermistors could still pose a problem for high-accuracy measurement. However, the development of bead-type (embedded-lead) thermistors makes possible a low-drift, fast-response resistance sensor. The stability of bead-type thermistors was investigated by the NBS in a recent study [30] and found to be very good (a mean drift of less than 4 mK over a 100-day period). In the same study, temperature drift of the order of 0.1°C over 100 days was recorded for disk-type (metal-contact-type) thermistors.

Recently [31], a standard thermistor was developed for use at temperatures between 0 and 60°C (32 and 140°F). It is a bead-in-glass probe-type thermistor assembled into a thin-wall stainless steel housing with R_{25} = 1000 Ω. The combination of the stability of bead-type thermistors with the rugged design (good vibration and mechanical shock resistance) yields a good transfer (or working) standard thermometer.

d. Thermoelectric Thermometers

Basic Principles.   Thermocouples are widely used for temperature measurement. The advantages of a thermocouple include simplicity in construction (they are formed by only two wires joined at their ends), ease of remote measurement, flexibility in construction (the size of the measuring junction can be made large or small according to the needs in life expectancy, drift, local measurement, and response time), simplicity in operation and signal processing (there is negligible self-heating, and the electrical signal output can be recorded directly), and the availability of thermocouple wires at a nominal cost.

The principle of operation of a thermocouple is described by the Seebeck effect: when two dissimilar materials are joined together at two junctions and these junctions are maintained at different temperatures an emf exists across the two junctions.

An elementary thermocouple circuit is shown in Fig. 11. The emf generated in this circuit is a function of the materials used and the temperatures of the junctions. It is useful to describe briefly the basic thermoelectric phenomena or effects that are related to the Seebeck effect and are present in thermocouple measurements. They include two well-known irreversible phenomena—Joule heating and thermal conduction—and two reversible phenomena—the Peltier effect and the Thompson effect.

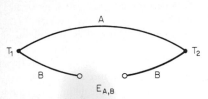

FIG. 11 Simple thermoelectric circuit.

Joule heating is the energy dissipation that occurs with an electric current flowing through a resistor and has magnitude I^2R. Thermal conduction is often quantitatively determined from Fourier's law: the heat conduction in a material is proportional to the temperature gradient.

The Peltier effect relates to reversible heat absorption or heat rejection at the junction between two dissimilar materials through which an electric current flows. The quantity of heat rejected or received is proportional to the electric current:

$$q_p = \pi_{AB} I \tag{15}$$

where π_{AB}, the Peltier coefficient for the junction, is a function of the temperature and the materials (A and B) used. Note that if the current were reversed, the sign of the heat transfer would also change.

The Thompson effect refers to a heat addition or rejection per unit length of a conductor. The heat addition is proportional to the product of the electric current and the temperature gradient along the conductor. The rate of transfer of Thompson heat per unit length of a wire (conductor) is given by $\sigma_T I \, dT/dx$, where σ_T is the Thompson coefficient. Integration over the entire length of a wire gives the Thompson heat:

$$q_T = \int_{\text{along wire}} \left(\sigma_T I \frac{dT}{dx} \right) dx = I \cdot \int_{\text{along wire}} \sigma_T \, dT \tag{16}$$

The direction or sign of this heat flow depends on the direction of the current relative to the temperature gradient. The magnitude of the Thompson coefficient is a function of the material (assumed homogeneous) in the conductor and the temperature level.

A simple and convenient means of determining the emf generated in a complex thermocouple circuit can be derived from the relations of irreversible thermodynamics. The result of this analysis [32] is that the zero-current emf for a single homogeneous wire of length dx is

$$-\left(\frac{dE}{dx} \right)_{I=0} = S^* \frac{dT}{dx} \tag{17}$$

and over a finite-length wire

$$\Delta E = -\int_0^L S^* \frac{dT}{dx} \, dx = -\int_{\text{along wire length}} S^* \, dT \tag{18}$$

S^*, the entropy transfer parameter, is the ratio of the electric current–driven entropy transport to the electric current transport itself in a conductor. It should be noted that this ratio exists even when the current is zero. The key point is that S^* is a function solely of the material composition and the temperature. Equation (18) can be integrated over any circuit to give the net emf. In doing this individual wires are generally assumed to be homogeneous.

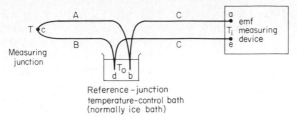

FIG. 12 Basic circuit for single thermocouple. C represents connector (often copper) wires.

Consider a basic thermocouple circuit with one measuring junction as shown in Fig. 12. The two thermoelements A and B are jointed at point c to form the measuring junction at temperature T. The thermoelements are connected to wires C at points b and d, both immersed in an ice bath (liquid water and ice in equilibrium) at T_0. The wires C are connected to the input of an emf measuring device. The input ports a and e are maintained at temperature T_i. Applying Eq. (18) over the various legs of the circuit gives

$$E_a - E_b = \int_{T_i}^{T_0} S_C^* \, dT \qquad E_c - E_d = \int_{T}^{T_0} S_B^* \, dT$$

$$E_b - E_c = \int_{T_0}^{T} S_A^* \, dT \qquad F_d - E_e = \int_{T_0}^{T_i} S_C^* \, dT$$

Adding the four equations together, we can see that the thermoelectric properties of the (similar) connecting wires C do not influence the overall emf:

$$E_a - E_e = \int_{T_0}^{T} (S_A^* - S_B^*) \, dT = E_{AB} \tag{19}$$

$S_A^* - S_B^*$ is called the Seebeck coefficient, α_{AB}, and is a function of material A, material B, and temperature. Equation (19) is often presented in the form

$$E_{AB} = \int_{T_0}^{T} \alpha_{AB} \, dT \tag{20}$$

Equation (18) is particularly useful in deriving the net emf of a complex thermoelectric circuit. In many reference works, three laws—the law of homogeneous materials, the law of intermediate materials, and the law of intermediate temperature—are used to show how the emf at an emf measuring device is affected by various lead wires from a thermocouple junction (See Ref. 33). These laws can also be derived from Eq. (18). They are often difficult to apply in a complex circuit.

Thermoelectric Circuits. A typical circuit for a single thermocouple of materials A and B is shown in Fig. 12. The reference temperature (at which junctions b and d are maintained) is usually the ice point, 0°C (32°F). The connecting wires C are usually copper wires. Note that according to Eq. (19), the connecting (copper) wires C should not affect the emf E_{AB}, which, for given materials A and B, is just a function of the temperature T.

With more than one thermocouple, the circuit given in Fig. 13a can be used for precision work—with each reference junction at the ice point. To avoid using multiple ref-

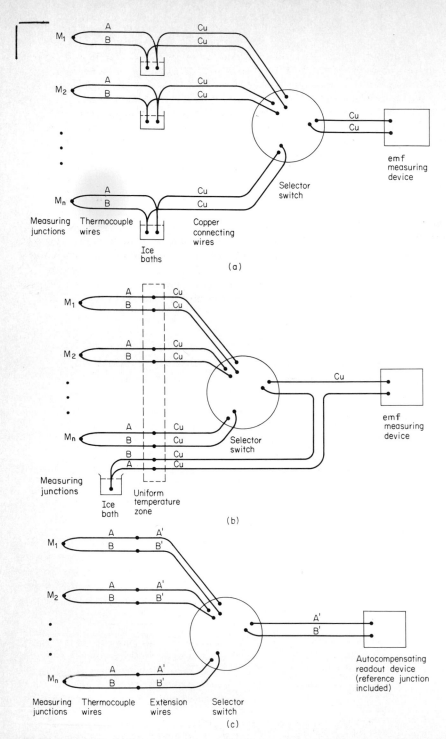

FIG. 13 Measuring circuits for thermocouples: (*a*) Ideal circuit for multiple thermocouples. (*b*) Zone-box circuit with only one reference junction. (*c*) Typical circuit for industrial use—no ice bath.

erence junctions, a circuit with only one reference junction and a uniform temperature zone box can be used (see Fig. 13b). Alternatively, if high precision is required and the calibrations of the individual thermocouples are not identical, a variant of the apparatus in Fig. 13b can be used. With this, the junction in the ice bath is treated as a regular junction, and in this way is used to determine the temperature of the selector switch which serves as the uniform temperature zone. Then the temperature (not the equivalent emf) of the selector switch can be subtracted (added) to the temperature readings of the individual junctions. A typical industrial thermocouple circuit using extension wires which approximate the composition of the thermocouple wires is shown in Fig. 13c.

Some different thermocouple circuits are shown in Figs. 14a and b. Figure 14a shows a thermopile, which has a number of thermocouples in series. This circuit amplifies the output by the number of thermocouples in series. The amplification permits detection of small temperature differences, and such a circuit is often used in heat flux gauges. Figure 14b shows a parallel arrangement for sensing average temperature level; one reading gives the arithmetic mean of the temperatures sensed by a number of individual junctions.

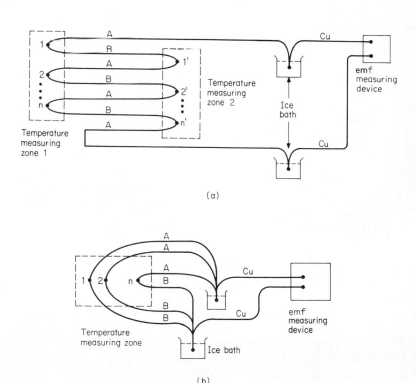

FIG. 14 Thermocouples in series and parallel circuits: (a) Thermopile for magnifying emf when measuring small temperature differences. (b) Parallel arrangement for sensing average temperature level.

Thermocouple Materials.   Thermoelectric properties of different materials can be represented by their respective Seebeck coefficients. Usually this is done with reference to Eq. (20) assuming a standard reference material (material B) and a standard reference temperature T_0 (usually 0°C; see Fig. 12). The thermoelectric standard reference material in use today is a special form of platinum referred to as Pt-67 [35].

In principle, any two different materials can be used in a thermocouple combination. Of the enormous number of possible combinations, only a few are commonly used. Based on NBS recognition and general usage, a standard was established which designates a number of standard thermocouples [34, 35, and 36]. Table 6 gives the designations, popular names, materials (with color codes), temperature ranges, and Seebeck coefficients for seven of the standard thermocouples. Three of these (types S, R, and B) are referred to as noble metal thermocouples since they are platinum-rhodium vs. platinum combinations; the other four (types E, J, K, and T) are called base metal thermocouples.

The letter designations for thermocouples were originally assigned by the Instrument Society of America and accepted as an American Standard in ANSI-C96.1-1964. They are not used to specify the precise chemical composition of the materials but only the thermoelectric properties. Hence the composition of a given type of thermocouple may be different and still be acceptable by the standard as long as the combined thermoelectric properties remain within specified error limits (see Table 7).

Some wire sizes and their corresponding upper temperature limits for each type of standard thermocouple as recommended by the American Society for Testing and Materials (ASTM) are summarized in Table 8. These limits are for protected thermocouples in conventional closed-end protecting tubes. When a standard type of thermocouple is used beyond the ASTM recommended ranges given in Table 8, accuracy and reliability may be compromised. Commercially available thermocouple wires come in many more gauge sizes than those indicated in Table 8.

Environmental limitations of standard thermocouple materials compiled by ASTM [34] are reproduced in Table 9. The thermoelectric emf of the standard thermoelements relative to platinum is presented in Fig. 15 [34]. Seebeck coefficients (first derivative of thermoelectric emf with respect to temperature) for each of the standard thermocouples as a function of temperature are tabulated in Table 10.

Type S† thermocouples (platinum–10% rhodium vs. platinum) were introduced in 1886 by Le Châtelier. Because of their stability and reproducibility they became an early standard and are still used as the defining instrument for IPTS-68 between 630.74 and 1064.43°C (1167.33 and 1947.97°F).

Early SP‡ thermoelements were found to contain iron impurities from impure rhodium. Later, when pure rhodium was available, 13 percent rhodium instead of the usual 10 percent had to be used in the platinum-rhodium alloy to match the previous standard thermoelectric emf—hence the development of the type R thermocouple.

Type B thermocouples have better stability and greater mechanical strength, and can be used at higher temperatures than types R and S. The small Seebeck coefficient for the noble metal thermocouples makes them unattractive for low-temperature measurements. With type B thermocouples the particularly small Seebeck coefficient at low temperatures means the actual temperature of the reference junction does not greatly affect the emf as long as it is between 0 and 50°C (32 and 120°F).

---

†Hereafter, standard thermocouples will be referred to by their type designation—see Table 6 for their compositions.

‡The second capital letter (P) designates the polarity of the thermoelement: P = positive, N = negative.

TABLE 6 Standard Thermocouple Types

SLD[1]	Popular name	Materials (color code)[2] (positive material appears first)	Typical temperature range[3]	Seebeck coefficient at 100°C (212°F)[4], $\mu V/°C$
S	—	Platinum–10% rhodium vs. platinum	−50 to 1767°C (−60 to 3200°F)	7.3
R	—	Platinum–13% rhodium vs. platinum	−50 to 1767°C (−60 to 3200°F)	7.5
B	—	Platinum–30% rhodium vs. platinum–6% rhodium	0 to 1820°C (32 to 3300°F)	0.9
T	Copper-constantan	Copper (blue) vs. a copper-nickel alloy[5] (red)	−270 to 400°C (−450 to 750°F)	46.8
J	Iron-constantan	Iron (white) vs. a slightly different copper-nickel alloy[6] (red)	−210 to 760°C[9] (−350 to 1400°F)	54.3
E	Chromel-constantan	Nickel-chromium alloy[7] (purple) vs. a copper-nickel alloy[5] (red)	−270 to 1000°C (−450 to 1830°F)	67.5
K	Chromel-Alumel	Nickel-chromium alloy[7] (yellow) vs. nickel-aluminum alloy[8] (red)	−270 to 1372°C (−450 to 2500°F)	41.4

[1]SLD stands for Standardized Letter Designation. The letter designation is for the combined thermocouple with each individual thermoelement designated by P or N for positive or negative legs, respectively—for example, SN stands for platinum, TP stands for copper, etc.

[2]The color codes given in parentheses are the color of the duplex-insulated wires [36]. Color codes are not available for the noble metal types (S, R, and B). For the base metal types (T, J, E, and K), the overall insulation color is brown.

[3]These temperature ranges are taken from Ref. 35.

[4]Conversion to $\mu V/°F$ is by dividing the figures by 1.8.

[5]This copper-nickel alloy is the same for both EN and TN, often referred to as Adams constantan or, sometimes, constantan.

[6]This copper-nickel alloy is used in JN. It is similar to, but not always interchangeable with, EN and JN; SAMA specification, often referred to as SAMA constantan but also loosely called constantan.

[7]EP and KP is a nickel-chromium alloy which is usually referred to by its trade name, Chromel (Hoskins Manufacturing Co.).

[8]KN is a nickel-aluminum alloy usually referred to by its trade name, Alumel (Hoskins Manufacturing Co.).

[9]Even though emf-temperature values are available up to 1200°C (2200°F) [35], thermophysical properties of type J thermocouples are not stable above 760°C (1400°F).

In order to reduce cost and to obtain higher emf's, base metal thermocouples were introduced. Iron and nickel were good candidates. However, pure nickel was found to be very brittle after oxidation; to avoid this a copper-nickel alloy commonly referred to as constantan was introduced. The iron-constantan combination is designated type J. Later, type K was developed for use at higher temperatures than type J, and type T was introduced particularly for measurements below 0°C. Unlike the other three, the type E ther-

TABLE 7 Limits of Error for Thermocouples* (Deviation from Standard Tables)

Thermocouple		Temperature range		Limits of error†			
				Standard		Special	
Type	Materials	°F	°C	°F	°C‡	°F	°C‡
B	Pt–30% Rh vs. Pt–6% Rh	1600 to 3100	871 to 1705	±½ percent			
E	Chromel-constantan	32 to 600	0 to 316	±3°F		±2¼°F	
		600 to 1600	316 to 871	±½ percent		±⅜ percent	
J	Iron-constantan	32 to 530	0 to 277	±4°F		±2°F	
		530 to 1400	277 to 760	±¾ percent		±⅜ percent	
K	Chromel-Alumel	32 to 530	0 to 277	±4°F		±2°F	
		530 to 2300	277 to 1260	±¾ percent		±⅜ percent	
R or S	Pt–13% Rh vs. Pt, or Pt–10% Rh vs. Pt	32 or 1000	0 to 538	±2½°F			
		1000 to 2700	538 to 1482	±¼ percent			
T§	Copper-constantan	−300 to −150	−184 to −101			±1 percent	
		−150 to −75	−101 to −59	±2 percent		±1 percent	
		−75 to 200	−59 to 93	±1½°F		±¾°F	
		200 to 700	93 to 371	±¾ percent		±⅜ percent	

*In this table the limits of error for each type of thermocouple apply only over the temperature range for which the wire size in question is recommended (see Table 8). These limits of error should be applied only to standard wire sizes. The same limits may not be obtainable in special sizes.

†Limits of error apply to thermocouples as supplied by the manufacturer. The calibration of a thermocouple may change during use. The magnitude of the change depends upon such factors as temperature, the length of time, and the conditions under which it was used.

‡Where limits of error are given in percent, the percentage applies to the temperature being measured when expressed in degrees Fahrenheit. To determine the limit of error in degrees Celsius, multiply the limit of error in degrees Fahrenheit by 5/9.

§Type T wire cannot be expected to meet the limits of error at temperatures below the ice point unless so specified at time of purchase. Selection is usually required.

Copyright, American Society for Testing and Materials, 1916 Race Street, Philadelphia, PA 19103. Reprinted with permission.

TABLE 8 Recommended Upper Temperature Limits for Protected Thermocouples

Thermocouple		Upper temperature limit for various wire sizes (AWG), °C (°F)					
Type	Materials	8 gauge, 3.25 mm (0.128 in)	14 gauge, 1.63 mm (0.064 in)	20 gauge, 0.81 mm (0.032 in)	24 gauge, 0.51 mm (0.020 in)	28 gauge, 0.33 mm (0.013 in)	30 gauge, 0.25 mm (0.010 in)
T	Copper-constantan		370 (700)	260 (500)	200 (400)	200 (400)	150 (300)
J	Iron-constantan	760 (1400)	590 (1100)	480 (900)	370 (700)	370 (700)	320 (600)
E	Chromel-constantan	870 (1600)	650 (1200)	540 (1000)	430 (800)	430 (800)	370 (700)
K	Chromel-Alumel	1260 (2300)	1090 (2000)	980 (1800)	870 (1600)	870 (1600)	760 (1400)
S	Pt–10% Rh vs Pt				1480 (2700)		
R	Pt–13% Rh vs. Pt				1480 (2700)		
B	Pt–30% Rh vs. Pt–6% Rh				1700 (3100)		

NOTE: This table gives the recommended upper temperature limits for the various thermocouples and wire sizes. These limits apply to protected thermocouples, i.e., thermocouples in conventional closed-end protecting tubes. They do not apply to sheathed thermocouples having compacted mineral oxide insulation. Properly designed and applied sheathed thermocouples may be used at temperatures above those shown in the tables. Other literature sources should be consulted.

Copyright, American Society for Testing and Materials, 1916 Race Street, Philadelphia, PA 19103. Reprinted with permission.

TABLE 9 Environmental Limitations of Thermoelements

Thermoelement	Environmental recommendations and limitations (see notes)
JP (iron)	For use in oxidizing, reducing, or inert atmospheres or in vacuum. Oxidizes rapidly above 540°C (1000°F). Will rust in moist atmospheres as in subzero applications.
	Stable to neutron radiation transmutation. Change in composition is only 0.5 percent (increase in manganese) in 20-year period.
JN, TN, EN (constantan)	Suitable for use in oxidizing, reducing, and inert atmospheres or in vacuum. Should not be used unprotected in sulfurous atmospheres above 540°C (1000°F).
	Composition changes under neutron radiation since copper content is converted to nickel and zinc. Nickel content increases 5 percent in 20-year period.
TP (copper)	Can be used in vacuum or in oxidizing, reducing, or inert atmospheres. Oxidizes rapidly above 370°C (700°F). Preferred to type JP element for subzero use because of its superior corrosion resistance in moist atmospheres.
	Radiation transmutation causes significant changes in composition.
	Nickel and zinc grow into the material in amounts of 10 percent each in a 20-year period.
KP, EP (Chromel)	For use in oxidizing or inert atmospheres. Can be used in hydrogen or cracked ammonia atmospheres if dew point is below −40°C (−40°F). Do not use unprotected in sulfurous atmospheres above 540°C (1000°F).
	(Not recommended for service in vacuum at high temperatures except for short time periods because preferential vaporization of chromium will alter calibration. Large negative calibration shifts will occur if exposed to marginally oxidizing atmospheres in temperature range 815 to 1040°C (1500 to 1900°F).
	Quite stable to radiation transmutation. Composition change is less than 1 percent in 20-year period.
KN (Alumel)	Can be used in oxidizing or inert atmospheres. Do not use unprotected in sulfurous atmospheres as intergranular corrosion will cause severe embrittlement.
	Relatively stable to radiation transmutation. In 20-year period, iron content will increase approximately 2 percent. The manganese and cobalt contents will decrease slightly.
RP, SP, SN, RN, BP, BN (platinum and platinum-rhodium alloys)	For use in oxidizing or inert atmospheres. Do not use unprotected in reducing atmospheres in the presence of easily reduced oxides, atmospheres containing metallic vapors such as lead or zinc, or those containing nonmetallic vapors such as arsenic, phosphorus, or sulfur. Do not insert directly into metallic protecting tubes. Not recommended for service in vacuum at high temperatures except for short time periods.

TABLE 9 Environmental Limitations of Thermoelements (*Continued*)

Thermoelement	Environmental recommendations and limitations (see notes)
	Types RN and SN elements are relatively stable to radiation transmutation. Types BP, BN, RP, and SP elements are unstable because of the rapid depletion of rhodium. Essentially, all the rhodium will be converted to palladium in a 10-year period.

NOTE 1: Refer to Table 8 for recommended upper temperature limits.

NOTE 2: Stability under neutron radiation refers to chemical composition of thermoelement, not to stability of thermal emf.

NOTE 3: Radiation transmutation rates are based on exposure to a thermal neutron flux of 1×10^{14} neutrons/($cm^2 \cdot s$). See Browning, W. E., Jr., and Miller, C. E., Jr., "Calculated Radiation Induced Changes in Thermocouple Composition," *Temperature, Its Measurement and Control in Science and Industry*. Part 2, Reinhold, New York, Vol. C, 1962, p. 271.

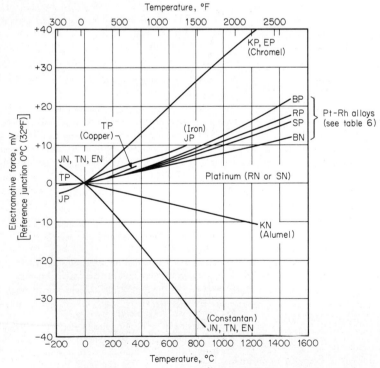

FIG. 15 Electromotive force of thermoelements relative to platinum.

TABLE 10 Nominal Seebeck Coefficients (Thermoelectric Power), α_{AB}: $\mu V/°C$

Temperature, °C	Thermocouple						
	E Chromel-constantan	J Iron-constantan	K Chromel-Alumel	R Pt–13% Rh vs. Pt	S Pt–10% Rh vs. Pt	T Copper-constantan	B Pt–30% Rh vs. Pt–6% Rh
−190	27.3	24.2	17.1			17.1	
−100	44.8	41.4	30.6			28.4	
0	58.5	50.2	39.4			38.0	
200	74.5	55.8	40.0	8.8	8.5	53.0	2.0
400	80.0	55.3	42.3	10.5	9.5		4.0
600	81.0	58.5	42.6	11.5	10.3		6.0
800	78.5	64.3	41.0	12.3	11.0		7.7
1000			39.0	13.0	11.5		9.2
1200			36.5	13.8	12.0		10.3
1400				13.8	12.0		11.3
1600					11.8		11.6

mocouple does not have a well-documented history. It was formed by a combination of KP and TN.

Though not recommended for subzero temperature applications by the ASTM standard [34], the type E thermocouple has been found [43] to be a very useful type-designated thermocouple combination for subfreezing applications. Type E thermocouples have the largest Seebeck coefficient among the letter-designated thermocouples, which can be important especially in differential temperature measurements.

In situations when mechanical strength between measuring junction and reference junction is required (e.g., when very fine wires are used as the measuring junction for high response time or small conductance) or when cost reduction is a significant factor (e.g., when using noble metal thermocouples), extension wires can be used between the measuring and reference junctions. Extension wires are wires having similar or compensating thermoelectric properties to the thermocouple wires to which they are attached. However, they can introduce additional sources of errors due to improper combinations of thermoelements and extension wires. Errors as great as ±4 to 5°C (±8°F)—twice the commercial limit shown in Table 7—can be introduced. Additional errors can also arise if the temperature of the two thermoelement–extension wire junctions are different. One should avoid extension wires unless absolutely necessary; if used, care should be exercised in the selection of extension wires to ensure proper matching and the equalization of thermoelement–extension wire junction temperatures. Detailed information on extension wires can be found in Refs. 34 and 36. Once the leads have been brought out to the reference junction, copper or other homogeneous wires can be used for connection to the emf-measuring device.

In addition to the seven designated thermocouple combinations, other, nonstandard combinations are also available. Over the years, hundreds of combinations had been investigated for various applications, many of which never went beyond the research stage. Reference 37 reviews well over a hundred combinations. Other reference sources on thermocouple materials are available in Refs. 38 and 39.

Thermocouple materials can be divided into four categories: (1) noble metals, (2) base metals, (3) refractory metals, and (4) nonmetal types. Of the seven letter-designated thermocouples, three are from the first category and four from the second. Refractory metal thermocouples are normally used for high-temperature applications (e.g., tungsten-rhenium alloys [40, 41]). Carbides, graphites, and ceramics are the main nonmetal thermoelements [39]; they are not as popular, because they are bulky and have poor reproducibility.

Of the hundreds of combinations of thermocouple materials, only a few (besides the seven standard types) are used to any significant extent and/or are commercially available. Table 11 summarizes the characteristics of several combinations that might be useful for special applications. More information is available in Refs. 34 and 38.

Thermocouple Components and Fabrication. A thermocouple measurement assembly includes a sensing element assembly, extension wires (when used), reference junction, connecting wires, emf-measuring device (possibly with signal-processing equipment), and other hardware needed for applications in adverse environments such as protection tubes, connectors, adaptors, etc. Each of the above components will be discussed in the following paragraphs.

The sensing element assembly includes two thermoelements properly joined together to form the measuring junction with appropriate electrical insulation to avoid shorting the two thermoelements except at the measuring junction.

Commercial off-the-shelf thermocouple wires are available in different forms: bare wires, insulated wires, and sheathed wires. Bare wires come on spools in a wide range of gauge sizes. They have to be individually matched, fabricated, and insulated before use.

Insulated wires come as single insulated thermoelements or duplex (two insulated) wires. Duplex wires can be obtained with a stainless steel overbraid for wear and abrasion protection. Table 12 lists characteristics of insulations used with thermocouple wires.

Sheathed wires have two bare wires embedded within a sheath packed with ceramic insulation. The crushed ceramic insulation rapidly absorbs moisture. To avoid deterioration due to moisture and contamination, the wires should be properly sealed at both ends at all times and clean tools should be used during fabrication. If the wires have to be left unsealed temporarily, they should be kept in an oven at about 100°C (200°F) or higher.

Fabrication of a thermocouple [36] requires some skill and familiarity, especially when using small-diameter wires. The measuring junction should be a joint of good thermal and electrical contacts without destroying the thermoelectric properties of the wires at the junction. For applications below 500°C (1000°F), silver solder with borax flux is sufficient for most base metal types, whereas welded junctions are recommended for use above 500°C (1000°F). Common welding methods include gas, electric arc, resistance, butt, tungsten–inert gas, and plasma arc welding. Table 13 summarizes fabrication methods for bare wires. If present, the insulation and/or sheath should first be removed and the wires properly cleaned before the joining process. For the removal of sheath materials, commercially available stripping tools can be used. Figure 16 shows pairs of wires prepared for different means of welding to form junctions.

Bare wires or insulated wires are normally fabricated into butt-welded or beaded (welded or soldered) junctions. Sheathed wires can be fabricated into these exposed types as well as into protected (grounded or ungrounded) types. Figure 17 shows different thermocouple junctions.

At the reference junction, each thermoelement (or extension wire) is soldered to a separate copper connecting wire. These copper wires are then connected to the input panel of an emf-measuring device. Depending on the application and accuracy required, different emf-measuring devices are available: (1) direct emf readout, normally a preci-

TABLE 11 Characteristics of Some Nonstandard Thermocouples

Materials	Typical temperature range	Seebeck coefficient, μV/°C (at temperature, °C)[1]	Characteristics
KP (Chromel) vs. Au–0.07 atomic % Fe	−270 to 0°C (−450 to 32°F)	16.8 (−259.34°C)	Large Seebeck coefficient, recommended for applications below 50 K.
Nicrosil vs. Nisil[2]	−270 to 1300°C (−450 to 2370°F)	29.6 (100°C) 38.6 (1000°C)	Excellent resistance to preferential oxidation in oxidizing and reducing atmosphere at temperature above 1000°C (1830°F) (preferable over type K at high temperatures). Not recommended under vacuum or wet hydrogen atmosphere.
Platinel 1813 vs. Platinel 1503[3]	−100 to 1360°C (−150 to 2480°F)	~42 (100 to 1000°C) ~33 (1000 to 1300°C)	A noble metal type thermocouple having a Seebeck coefficient approaching that of the type K thermocouple. Not recommended in vacuum, reducing atmosphere (nonhydrogen), and wet hydrogen atmosphere.
Ir–40% Rh vs. Ir[4]	0 to 2100°C (32 to 3810°F)	5.6 (2100°C)	Can be used in vacuum and inert atmosphere. May be used in oxidizing atmosphere with shortened life, but not recommended for use in reducing atmosphere.
W–5% Re vs. W–26% Re[5]	0 to 2760°C (32 to 5000°F)	12.1 (2316°C)	Applicable in vacuum, inert atmosphere, or high-purity hydrogen atmosphere. Not recommended for oxidizing or reducing atmospheres.

[1]The Seebeck coefficient at the temperature given in the bracket is listed. Conversion to μV/°F is by dividing the numbers by 1.8.

[2]This is a new thermocouple combination. See Ref 42 for details.

[3]Trademark of Engelhard Industries. The negative thermoelement is a 65% gold–35% palladium alloy. Two combinations are available for the positive thermoelement. Platinel 1813 is also called Platinel II (55% palladium–31% platinum–14% gold); the other one is Platinel I (or Platinel 1786, 83% palladium–14% platinum–3% gold). Platinel II is preferred due to its better mechanical fatigue properties.

[4]Iridium-rhodium types. Three different alloys are commonly used for the positive thermoelement: Ir–40% Rh, Ir–50% Rh, and Ir–60% Rh. The first one has the highest maximum operating temperature of the three.

[5]Tungsten-rhenium types. Three different combinations are commonly used: W–5% Re vs. W–26% Re, W–3% Re vs. W–25% Re, and W vs. W–26% Re. The first one is the most widely used of the three, comprising about 70 percent of the market for high-temperature applications. See Refs. 40 and 41 for more details.

sion digital voltmeter or potentiometer; and (2) automatic compensating readout, giving a reading directly in temperature, available in both analog and digital off-the-shelf devices with built-in reference junctions to be used with a specified type of thermocouple. The choice between the two depends primarily on the required accuracy in the measurements. The automatic compensator is convenient to use but accuracy beyond the commercial limits should not be expected. On the other hand, the better accuracy obtainable in the direct emf readout requires calibration and calculation of temperature from emf.

TABLE 12 Thermocouple Insulation Characteristics

Insulation	Continuous use temperature limit, °C (°F)	Single exposure temperature limit, °C (°F)	Moisture resistance	Abrasion resistance
Cotton	95 (200)	95 (200)	poor	fair
Polyvinyl	105 (220)	105 (220)	excellent	good
Enamel and cotton	95 (200)	95 (200)	fair	fair
Nylon	125 (260)	125 (260)	poor	good
Teflon[a]	205 (400)	315 (600)	excellent	good
Polyimide	315 (600)	400 (750)	excellent	good
Teflon and fiber glass[b]	315 (600)	370 to 540 (700 to 1000)	excellent to 600°F	good
Fiber glass—varnish or silicone impregnation E[c]	480 (900)	540 (1000)	fair to 400°F, poor above 400°F	fair to 400°F, poor above 400°F
Fiber glass, nonimpregnated S[d]	540 (1000)	650 (1200)	poor	fair
Asbestos and fiber glass with silicone[e]	480 (900)	650 (1200)	fair to 400°F	fair to 400°F, poor above 400°F
Felted asbestos	540 (1000)	650 (1200)	poor	poor
Asbestos over asbestos	540 (1000)	650 (1200)	poor	poor
Refrasil[f]	870 (1600)	1100 (2000)	very poor	very poor
Ceramic fibers (for example, Nextel[g])	1000 (1830)	1370 (2500)	very poor	poor

[a]Trademark of E. I. duPont de Nemours & Co.

[b]The Teflon vaporizes at 315°C (600°F) with possible toxic effects.

[c]E = electrical grade fiber glass.

[d]S = structural grade fiber glass.

[e]Individual wires are asbestos and overbraid is fiber glass.

[f]Trademark of the H. I. Thompson Co.

[g]Trademark of the 3M Corp.

Copyright, American Society for Testing and Materials, 1916 Race Street, Philadelphia, PA 19103. Reprinted with permission.

The emf measured by an emf-readout device is the difference between the emf's of the measuring junction and the reference junction; thus the temperature at the reference junction should be properly controlled. It is usually maintained at the ice point (0°C or 32°F) by using an ice bath (Fig. 18). The ice bath should be a mixture of crushed ice and liquid water in a Dewar flash with considerably more ice than liquid. It can be very accurate but not too convenient in some applications since frequent replacement of melted ice is required. With a thermoelectric refrigeration system [34], ice can be maintained by cooling the bath of water with thermoelectric cooling elements.

In contrast to these systems, electrical compensation can be used [34]. The ice bath is replaced by an electrical circuit that has a temperature-sensitive resistor. The circuit is preset to compensate for variation in the reference junction temperature.

Another method to control the reference junction temperature is by the use of a constant-temperature oven. The oven will maintain the reference junction at a fixed temperature which is above the highest anticipated ambient temperature. Corrections to the standard reference emf values must be made when using this method.

For applications in adverse environments, such as measurement in high-temperature furnaces, combustion chambers, nuclear reaction cores, etc., a protection tube is normally

TABLE 13 Summary of Methods for Joining of Bare-Wire Thermocouples

Type	Materials	Lower melting	Silver brazing*	Welding			Resistance welding	Plasma arc or tungsten inert gas	Butt welding‡
				Flux	Gas	Arc			
T	Copper-constantan	TP	✓	Borax	✓	✓	N.R.	✓	N.R.
J	Iron-constantan	JN	✓	Broax	✓	✓	✓	✓	✓
E	Chromel-constantan	EN	✓	Fluorspar	✓	✓	N.R.	✓	✓
K	Chromel-Alumel	KN	✓	Fluorspar	✓	✓	✓	✓	✓
R	Pt–13% Rh vs. Pt	RN	N.R.	None	✓	✓	✓	✓	—
S	Pt–10% Rh vs. Pt	SN	N.R.	None	✓	✓	✓	✓	—
B	Pt–30% Rh vs. Pt–6% Rh	BN	N.R.	None	✓	✓	✓	✓	—

*Only recommended for use below 500°C for base metal thermocouple.

†Boric acid also recommended for types J, E, and K.

‡Recommended for 8 through 20 AWG wires.

N.R.—Not recommended.

Reprinted from ANSI MC96.1-1982 with the permission of the copyright holder. © Instrument Society of America. Users should refer to a complete copy of the standard available from the publisher. ISA, 67 Alexander Drive, P.O. Box 12277, Research Triangle Park, NC 27709

used with the thermocouple. Detailed descriptions on the selection of protection tubes and typical industrial high-temperature thermocouple assemblies are available [34].

Reference Tables and Reliability. The National Bureau of Standards reference tables and equations [35] are now widely accepted for the seven letter-designated ther-

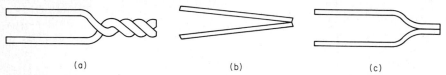

(a)   (b)   (c)

FIG. 16 Preparation of thermocouple wires: (*a*) Wire placement for gas and arc welding of base metal thermocouples. (*b*) Wire placement for arc welding of all standardized thermocouples. (*c*) Wire placement for resistance welding.

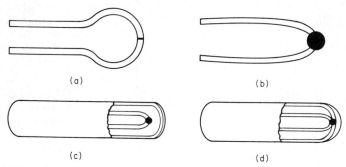

(a)   (b)

(c)   (d)

FIG. 17 Thermocouple junctions: (*a*) Butt-welded junction. (*b*) Beaded junction. (*c*) Protected junction, ungrounded. (*d*) Protected junction, grounded.

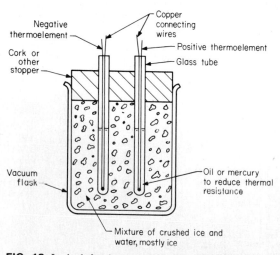

FIG. 18 Ice bath for thermocouple reference junction.

mocouples: S, R, B, T, J, E, and K. These tables and the appropriate limits of error (Table 7) serve as the standard for industry in manufacturing thermocouple wires. Tables 14 to 20 list emf's at different temperatures for each of the seven thermocouple types. These tables are generated from equations given in Ref. 35. The tables can be used directly using linear interpolation. However, it is often more convenient, once the temperature range and thermocouple materials are known, to fit a polynomial (usually a cubic or a quadratic) to the emf values, in this temperature range, from the appropriate table.

Where accuracy of measurement is not vital, thermocouple reference tables (Tables 14 to 20) are sufficient for conversion of emf's to temperatures. Accuracy of at least ± 2.2°C (±4°F) or 0.75 percent for standard wires and ±1.1°C (±2°F) or 0.4 percent for special wires can be achieved without calibration (see Table 7). However, the precision or reproducibility of a single thermocouple is much better than these limits. Automatic compensators are available commercially for conversion of emf's to temperatures in these applications.

When accuracy greater than the commercial limit of error is required, the wires should be calibrated. Normally this requires comparison calibration of sample thermocouples made from each spool to account for spool variability. Typically, two thermocouples—one fabricated from the beginning of the spool and one from the end of the spool—are calibrated to determine the average calibration of the entire spool. If the agreement between the two calibrations is not satisfactory, a third thermocouple fabricated from the center of the spool can be used.

Transfer or working standard thermocouples (including connecting and/or extension wires— see Fig. 12) are individually calibrated by comparison calibration against a defining standard or another transfer standard thermometer (possibly thermocouple). Fixed-

TABLE 14A EMF-Temperature for Type S (Pt-10% Rh vs. Pt) Thermocouples

Temperature in °C   Reference junction at 0°C

°C	0	10	20	30	40	50	60	70	80	90	100	°C
					Thermoelectric voltage in absolute millivolts							
−0	0.000	−0.053	−0.103	−0.150	−0.194	−0.236						−0
0	0.000	0.055	0.113	0.173	0.235	0.299	0.365	0.432	0.502	0.573	0.645	0
100	0.645	0.719	0.795	0.872	0.950	1.029	1.109	1.190	1.273	1.356	1.440	100
200	1.440	1.525	1.611	1.698	1.785	1.873	1.962	2.051	2.141	2.232	2.323	200
300	2.323	2.414	2.506	2.599	2.692	2.786	2.880	2.974	3.069	3.164	3.260	300
400	3.260	3.356	3.452	3.549	3.645	3.743	3.840	3.938	4.036	4.135	4.234	400
500	4.234	4.333	4.432	4.532	4.632	4.732	4.832	4.933	5.034	5.136	5.237	500
600	5.237	5.339	5.442	5.544	5.648	5.751	5.855	5.960	6.064	6.169	6.274	600
700	6.274	6.380	6.486	6.592	6.699	6.805	6.913	7.020	7.128	7.236	7.345	700
800	7.345	7.454	7.563	7.672	7.782	7.892	8.003	8.114	8.225	8.336	8.448	800
900	8.448	8.560	8.673	8.786	8.899	9.012	9.126	9.240	9.355	9.470	9.585	900
1000	9.585	9.700	9.816	9.932	10.048	10.165	10.282	10.400	10.517	10.635	10.754	1000
1100	10.754	10.872	10.991	11.110	11.229	11.348	11.467	11.587	11.707	11.827	11.947	1100
1200	11.947	12.067	12.188	12.308	12.429	12.550	12.671	12.792	12.913	13.034	13.155	1200
1300	13.155	13.276	13.397	13.519	13.640	13.761	13.883	14.004	14.125	14.247	14.368	1300
1400	14.368	14.489	14.610	14.731	14.852	14.973	15.094	15.215	15.336	15.456	15.576	1400
1500	15.576	15.697	15.817	15.937	16.057	16.176	16.296	16.415	16.534	16.653	16.771	1500
1600	16.771	16.890	17.008	17.125	17.243	17.360	17.477	17.594	17.711	17.826	17.942	1600
1700	17.942	18.056	18.170	18.282	18.394	18.504	18.612					1700

TABLE 14B EMF-Temperature for Type S (Pt–10% Rh vs. Pt) Thermocouples

Temperature in °F   Reference Junction at 32°F

°F	0	10	20	30	40	50	60	70	80	90	100	°F
					Thermoelectric voltage in absolute millivolts							
−0	−0.092	−0.119	−0.145	−0.170	−0.194	−0.218						−0
0	−0.092	−0.064	−0.035	−0.006	0.024	0.055	0.087	0.119	0.152	0.186	0.221	0
100	0.221	0.256	0.291	0.328	0.365	0.402	0.440	0.478	0.517	0.557	0.597	100
200	0.597	0.637	0.678	0.719	0.761	0.803	0.846	0.889	0.932	0.976	1.020	200
300	1.020	1.064	1.109	1.154	1.199	1.245	1.291	1.337	1.384	1.431	1.478	300
400	1.478	1.525	1.573	1.620	1.669	1.717	1.765	1.814	1.863	1.912	1.962	400
500	1.962	2.011	2.061	2.111	2.161	2.211	2.262	2.313	2.363	2.414	2.465	500
600	2.465	2.517	2.568	2.620	2.672	2.723	2.775	2.828	2.880	2.932	2.985	600
700	2.985	3.037	3.090	3.143	3.196	3.249	3.302	3.356	3.409	3.463	3.516	700
800	3.516	3.570	3.624	3.678	3.732	3.786	3.840	3.895	3.949	4.004	4.058	800
900	4.058	4.113	4.168	4.223	4.278	4.333	4.388	4.443	4.498	4.554	4.609	900
1000	4.609	4.665	4.721	4.776	4.832	4.888	4.944	5.000	5.057	5.113	5.169	1000
1100	5.169	5.226	5.283	5.339	5.396	5.453	5.510	5.567	5.625	5.682	5.740	1100
1200	5.740	5.797	5.855	5.913	5.971	6.029	6.087	6.146	6.204	6.263	6.321	1200
1300	6.321	6.380	6.439	6.498	6.557	6.616	6.675	6.734	6.794	6.853	6.913	1300
1400	6.913	6.972	7.032	7.092	7.152	7.212	7.272	7.333	7.393	7.454	7.514	1400
1500	7.514	7.575	7.636	7.697	7.758	7.819	7.880	7.942	8.003	8.065	8.126	1500
1600	8.126	8.188	8.250	8.312	8.374	8.436	8.498	8.560	8.623	8.685	8.748	1600
1700	8.748	8.811	8.874	8.937	9.000	9.063	9.126	9.190	9.253	9.317	9.380	1700
1800	9.380	9.444	9.508	9.572	9.636	9.700	9.764	9.829	9.893	9.958	10.023	1800
1900	10.023	10.087	10.152	10.217	10.282	10.348	10.413	10.478	10.544	10.609	10.675	1900
2000	10.675	10.740	10.806	10.872	10.938	11.004	11.070	11.136	11.202	11.268	11.335	2000
2100	11.335	11.401	11.467	11.534	11.600	11.667	11.734	11.800	11.867	11.934	12.001	2100
2200	12.001	12.067	12.134	12.201	12.268	12.335	12.402	12.469	12.536	12.604	12.671	2200
2300	12.671	12.738	12.805	12.872	12.940	13.007	13.074	13.142	13.209	13.276	13.344	2300
2400	13.344	13.411	13.478	13.546	13.613	13.681	13.748	13.815	13.883	13.950	14.018	2400
2500	14.018	14.085	14.152	14.220	14.287	14.354	14.422	14.489	14.556	14.624	14.691	2500
2600	14.691	14.758	14.826	14.893	14.960	15.027	15.094	15.161	15.228	15.295	15.362	2600
2700	15.362	15.429	15.496	15.563	15.630	15.697	15.763	15.830	15.897	15.963	16.030	2700
2800	16.030	16.096	16.163	16.229	16.296	16.362	16.428	16.494	16.560	16.626	16.692	2800
2900	16.692	16.758	16.824	16.890	16.955	17.021	17.086	17.152	17.217	17.282	17.347	2900
3000	17.347	17.412	17.477	17.542	17.607	17.672	17.736	17.801	17.865	17.929	17.993	3000
3100	17.993	18.056	18.119	18.182	18.245	18.307	18.369	18.431	18.492	18.552	18.612	3100
3200	18.612	18.672										3200

point calibrations are normally used only for calibration of a defining standard thermocouple (type S). Procedures and care required in a fixed-point calibration (at the temperature range defined by the type S thermocouple) are such that even type S thermocouples are usually calibrated by comparison against another type S standard thermocouple at a fraction of the cost.

Type S thermocouples are used as the defining standard thermometer for IPTS-68 at temperatures between 630.74 and 1064.43°C (1167.33 and 1947.97°F). The precision of a type S thermocouple is about 0.02°C (0.04°F). Its ability to reproduce the IPTS-68 defining fixed points (its accuracy) is 0.2°C (0.4°F). The accuracy between fixed points is about 0.2 to 0.3°C (about 0.5°F). At lower temperatures (between about 0 and 200°C), a base metal–type thermocouple, e.g., type T, is capable of a precision of about 0.01°C

TABLE 15A EMF-Temperature for Type R (Pt–13% Rh vs. Pt) Thermocouples

Temperature in °C   Reference junction at 0°C

°C	0	10	20	30	40	50	60	70	80	90	100	°C
				Thermoelectric voltage in absolute millivolts							
−0	0.000	−0.051	−0.100	−0.145	−0.188	−0.226						− 0
0	0.000	0.054	0.111	0.171	0.232	0.296	0.363	0.431	0.501	0.573	0.647	0
100	0.647	0.723	0.800	0.879	0.959	1.041	1.124	1.208	1.294	1.380	1.468	100
200	1.468	1.557	1.647	1.738	1.830	1.923	2.017	2.111	2.207	2.303	2.400	200
300	2.400	2.498	2.596	2.695	2.795	2.896	2.997	3.099	3.201	3.304	3.407	300
400	3.407	3.511	3.616	3.721	3.826	3.933	4.039	4.146	4.254	4.362	4.471	400
500	4.471	4.580	4.689	4.799	4.910	5.021	5.132	5.244	5.356	5.469	5.582	500
600	5.582	5.696	5.810	5.925	6.040	6.155	6.272	6.388	6.505	6.623	6.741	600
700	6.741	6.860	6.979	7.098	7.218	7.339	7.460	7.582	7.703	7.826	7.949	700
800	7.949	8.072	8.196	8.320	8.445	8.570	8.696	8.822	8.949	9.076	9.203	800
900	9.203	9.331	9.460	9.589	9.718	9.848	9.978	10.109	10.240	10.371	10.503	900
1000	10.503	10.636	10.768	10.902	11.035	11.170	11.304	11.439	11.574	11.710	11.846	1000
1100	11.846	11.983	12.119	12.257	12.394	12.532	12.669	12.808	12.946	13.085	13.224	1100
1200	13.224	13.363	13.502	13.642	13.782	13.922	14.062	14.202	14.343	14.483	14.624	1200
1300	14.624	14.765	14.906	15.047	15.188	15.329	15.470	15.611	15.752	15.893	16.035	1300
1400	16.035	16.176	16.317	16.458	16.599	16.741	16.882	17.022	17.163	17.304	17.445	1400
1500	17.445	17.585	17.726	17.866	18.006	18.146	18.286	18.425	18.564	18.703	18.842	1500
1600	18.842	18.981	19.119	19.257	19.395	19.533	19.670	19.807	19.944	20.080	20.215	1600
1700	20.215	20.350	20.483	20.616	20.748	20.878	21.006					1700

(0.02°F) and an accuracy of 0.1°C (0.2°F). An accuracy of better than 0.1°C is almost impossible to attain with a thermocouple.

Factors that can affect thermocouple reliability include wire inhomogeneity and mechanical strain. Wire inhomogeneity (short-range) can be detected by observing whether a spurious emf is generated when a local heat source is applied to a thermocouple wire [44]. Short-range inhomogeneities are small and usually not significant in most wires manufactured today. Medium- and longer-range inhomogeneities—changes in calibration from thermocouple to thermocouple or from one end of a spool of wire to the other— are normally small and gradual. They are compensated by the calibration method outlined above (for spool variability). Calibration of base metal thermocouples at temperatures above about 300°C (600°F) will generate inhomogeneities in the wires which can change the calibration itself [45]. The usual practice to overcome this dilemma for application at high temperature is to calibrate sample thermocouples to obtain the calibration for the rest of the spool and discard those thermocouples themselves. If possible, the thermal environment found in the actual application should be simulated during calibration to reduce the shift in calibration due to differences in the thermal gradient along thermocouple wires. This procedure cannot account for the occurrence, albeit seldom, of a localized defect (short-range inhomogeneity) in the wire which can result in erroneous results.

Mechanical strains in the wires can be removed by annealing. However, commercial base metal thermocouple wire is already annealed by the manufacturer. Further annealing is seldom needed. Some noble metal thermocouple wires can also be obtained in an annealed form. Annealing is needed only when high accuracy is required [46].

The inability to control precisely the reference junction temperature or inaccurate

TABLE 15B EMF-Temperature for Type R (Pt–13% Rh vs. Pt) Thermocouples

Temperature in °F   Reference junction at 32°F

°F	0	10	20	30	40	50	60	70	80	90	100	°F
				Thermoelectric voltage in absolute millivolts								
−0	−0.089	−0.116	−0.141	−0.165	−0.188	−0.210						−0
0	−0.089	−0.063	−0.035	−0.006	0.024	0.054	0.086	0.118	0.150	0.184	0.218	0
100	0.218	0.253	0.289	0.326	0.363	0.400	0.439	0.478	0.517	0.557	0.598	100
200	0.598	0.639	0.681	0.723	0.766	0.809	0.852	0.897	0.941	0.986	1.032	200
300	1.032	1.077	1.124	1.170	1.217	1.265	1.313	1.361	1.409	1.458	1.508	300
400	1.508	1.557	1.607	1.657	1.708	1.758	1.810	1.861	1.913	1.964	2.017	400
500	2.017	2.069	2.122	2.175	2.228	2.282	2.335	2.389	2.443	2.498	2.552	500
600	2.552	2.607	2.662	2.718	2.773	2.829	2.885	2.941	2.997	3.053	3.110	600
700	3.110	3.167	3.224	3.281	3.338	3.396	3.453	3.511	3.569	3.627	3.686	700
800	3.686	3.744	3.803	3.862	3.921	3.980	4.039	4.099	4.158	4.218	4.278	800
900	4.278	4.338	4.398	4.458	4.519	4.580	4.640	4.701	4.762	4.824	4.885	900
1000	4.885	4.947	5.008	5.070	5.132	5.194	5.256	5.319	5.381	5.444	5.507	1000
1100	5.507	5.570	5.633	5.696	5.759	5.823	5.886	5.950	6.014	6.078	6.143	1100
1200	6.143	6.207	6.272	6.336	6.401	6.466	6.532	6.597	6.662	6.728	6.794	1200
1300	6.794	6.860	6.926	6.992	7.059	7.125	7.192	7.259	7.326	7.393	7.460	1300
1400	7.460	7.527	7.595	7.663	7.731	7.799	7.867	7.935	8.004	8.072	8.141	1400
1500	8.141	8.210	8.279	8.348	8.417	8.487	8.556	8.626	8.696	8.766	8.836	1500
1600	8.836	8.907	8.977	9.048	9.118	9.189	9.260	9.331	9.403	9.474	9.546	1600
1700	9.546	9.617	9.689	9.761	9.833	9.906	9.978	10.050	10.123	10.196	10.269	1700
1800	10.269	10.342	10.415	10.488	10.562	10.636	10.709	10.783	10.857	10.931	11.006	1800
1900	11.006	11.080	11.155	11.229	11.304	11.379	11.454	11.529	11.605	11.680	11.756	1900
2000	11.756	11.831	11.907	11.983	12.059	12.135	12.211	12.287	12.363	12.440	12.516	2000
2100	12.516	12.593	12.669	12.746	12.823	12.900	12.977	13.054	13.131	13.208	13.286	2100
2200	13.286	13.363	13.440	13.518	13.595	13.673	13.751	13.828	13.906	13.984	14.062	2200
2300	14.062	14.140	14.218	14.296	14.374	14.452	14.530	14.608	14.686	14.765	14.843	2300
2400	14.843	14.921	15.000	15.078	15.156	15.235	15.313	15.391	15.470	15.548	15.627	2400
2500	15.627	15.705	15.784	15.862	15.941	16.019	16.097	16.176	16.254	16.333	16.411	2500
2600	16.411	16.490	16.568	16.646	16.725	16.803	16.882	16.960	17.038	17.116	17.195	2600
2700	17.195	17.273	17.351	17.429	17.507	17.585	17.663	17.741	17.819	17.897	17.975	2700
2800	17.975	18.053	18.130	18.208	18.286	18.363	18.441	18.518	18.595	18.673	18.750	2800
2900	18.750	18.827	18.904	18.981	19.058	19.135	19.211	19.288	19.365	19.441	19.518	2900
3000	19.518	19.594	19.670	19.746	19.822	19.898	19.974	20.050	20.125	20.200	20.275	3000
3100	20.275	20.350	20.424	20.498	20.572	20.645	20.718	20.791	20.863	20.935	21.006	3100
3200	21.006	21.077										3200

compensation would also affect thermocouple reliability. In the case of an ice bath, factors that can cause the junction temperature to depart from 0°C (32°F) include nonuniformity in the temperature of the ice-water mixture, small depth of immersion, insufficient ice, and large wire sizes (conduction effects). Reference 47 describes various sources of errors in an ice bath.

e. Radiation Thermometers

Introduction. The relationship between the radiant energy emitted by an ideal (black) body and its temperature is described by the Planck radiation law. Radiation thermometers measure the radiation emitted by a body to determine its temperature.

TABLE 16A   EMF-Temperature for Type B (Pt–30% Rh vs. Pt–6% Rh) Thermocouples

Temperature in °C   Reference juntion at °C

°C	0	10	20	30	40	50	60	70	80	90	100	°C
					Thermoelectric voltage in absolute millivolts							
0	0.000	−0.002	−0.003	−0.002	−0.000	0.002	0.006	0.011	0.017	0.025	0.033	0
100	0.033	0.043	0.053	0.065	0.078	0.092	0.107	0.123	0.140	0.159	0.178	100
200	0.178	0.199	0.220	0.243	0.266	0.291	0.317	0.344	0.372	0.401	0.431	200
300	0.431	0.462	0.494	0.527	0.561	0.596	0.632	0.669	0.707	0.746	0.786	300
400	0.786	0.827	0.870	0.913	0.957	1.002	1.048	1.095	1.143	1.192	1.241	400
500	1.241	1.292	1.344	1.397	1.450	1.505	1.560	1.617	1.674	1.732	1.791	500
600	1.791	1.851	1.912	1.974	2.036	2.100	2.164	2.230	2.296	2.363	2.430	600
700	2.430	2.499	2.569	2.639	2.710	2.782	2.855	2.928	3.003	3.078	3.154	700
800	3.154	3.231	3.308	3.387	3.466	3.546	3.626	3.708	3.790	3.873	3.957	800
900	3.957	4.041	4.126	4.212	4.298	4.386	4.474	4.562	4.652	4.742	4.833	900
1000	4.833	4.924	5.016	5.109	5.202	5.297	5.391	5.487	5.583	5.680	5.777	1000
1100	5.777	5.875	5.973	6.073	6.172	6.273	6.374	6.475	6.577	6.680	6.783	1100
1200	6.783	6.887	6.991	7.096	7.202	7.308	7.414	7.521	7.628	7.736	7.845	1200
1300	7.845	7.953	8.063	8.172	8.283	8.393	8.504	8.616	8.727	8.839	8.952	1300
1400	8.952	9.065	9.178	9.291	9.405	9.519	9.634	9.748	9.863	9.979	10.094	1400
1500	10.094	10.210	10.325	10.441	10.558	10.674	10.790	10.907	11.024	11.141	11.257	1500
1600	11.257	11.374	11.491	11.608	11.725	11.842	11.959	12.076	12.193	12.310	12.426	1600
1700	12.426	12.543	12.659	12.776	12.892	13.008	13.124	13.239	13.354	13.470	13.585	1700
1800	13.585	13.699	13.814									1800

Radiation thermometers can be sensitive to radiation in all wavelengths (total-radiation thermometers) or only to radiation in a band of wavelengths (spectral-radiation thermometers). Thermocouple and thermopile junctions or a calorimeter are the usual detectors in a total-radiation thermometer. For spectral systems, the classification is normally based on the effective wavelength or wavelength band used—as determined, for example, by a filter, which allows only monochromatic radiation to reach the detector, or by the use of a detector sensitive only to radiation in a specific wavelength band. Optical pyrometers utilize the visible portion of the radiation spectrum, infrared pyrometers or scanners measure infrared radiation, and spectroscopic thermometers operate with radiation which is normally of shorter wavelength than the other two methods.

Thermal Radiation.   The wavelength distribution of an ideal thermal radiation emitter (a blackbody) in a vacuum is given by the Planck distribution function

$$e_{b\lambda} = \frac{c_1 \lambda^{-5}}{\exp(c_2/\lambda T) - 1} \tag{21}$$

where $e_{b\lambda}$ = monochromatic blackbody emissive power
= energy/(time · area · wavelength interval)
λ = wavelength in μm
T = absolute temperature
c_1 = 1.1872×10^8 Btu·μm^4/(h·ft^2)
= $3.741832 \pm 0.000020 \times 10^8$ W·μm^4/m^2
c_2 = 2.5898×10^4 μm·°R
= $1.438786 \pm 0.000045 \times 10^4$ μm·K

Integration of Eq. (21) over all wavelengths gives the total thermal radiation emitted by a blackbody:

$$e_b = \int_0^\infty e_{b\lambda}\, d\lambda = \sigma T^4 \tag{22}$$

where e_b = emissive power of a blackbody at temperature T, W/m^2

σ = Stefan-Boltzmann constant

= 0.1716×10^{-8} Btu/$(h \cdot ft^2 \cdot {}^\circ R^4)$ = $5.67032 \pm 0.00071 \times 10^{-8}$ W/$(m^2 \cdot K^4)$

TABLE 16B EMF-Temperature for Type B (Pt–30% Rh vs. Pt–6% Rh) Thermocouples

Temperature in °F   Reference junction at 32°F

°F	0	10	20	30	40	50	60	70	80	90	100	°F
					Thermoelectric voltage in absolute millivolts							
0					−0.001	−0.002	−0.002	−0.003	−0.002	−0.002	−0.001	0
100	−0.001	0.000	0.002	0.004	0.006	0.009	0.012	0.015	0.019	0.023	0.027	100
200	0.027	0.032	0.037	0.043	0.049	0.055	0.061	0.068	0.075	0.083	0.090	200
300	0.090	0.099	0.107	0.116	0.125	0.135	0.144	0.155	0.165	0.176	0.187	300
400	0.187	0.199	0.210	0.223	0.235	0.248	0.261	0.275	0.288	0.303	0.317	400
500	0.317	0.332	0.347	0.362	0.378	0.394	0.410	0.427	0.444	0.462	0.479	500
600	0.479	0.497	0.515	0.534	0.553	0.572	0.592	0.612	0.632	0.652	0.673	600
700	0.673	0.694	0.716	0.737	0.759	0.782	0.804	0.827	0.851	0.874	0.898	700
800	0.898	0.922	0.947	0.972	0.997	1.022	1.048	1.074	1.100	1.127	1.153	800
900	1.153	1.181	1.208	1.236	1.264	1.292	1.321	1.350	1.379	1.409	1.438	900
1000	1.438	1.468	1.499	1.529	1.560	1.591	1.623	1.655	1.687	1.719	1.752	1000
1100	1.752	1.785	1.818	1.851	1.885	1.919	1.953	1.988	2.022	2.058	2.093	1100
1200	2.093	2.128	2.164	2.201	2.237	2.274	2.311	2.348	2.385	2.423	2.461	1200
1300	2.461	2.499	2.538	2.576	2.615	2.655	2.694	2.734	2.774	2.814	2.855	1300
1400	2.855	2.896	2.937	2.978	3.019	3.061	3.103	3.146	3.188	3.231	3.274	1400
1500	3.274	3.317	3.361	3.404	3.448	3.492	3.537	3.581	3.626	3.672	3.717	1500
1600	3.717	3.762	3.808	3.854	3.901	3.947	3.994	4.041	4.088	4.136	4.183	1600
1700	4.183	4.231	4.279	4.327	4.376	4.425	4.474	4.523	4.572	4.622	4.672	1700
1800	4.672	4.722	4.772	4.823	4.873	4.924	4.975	5.027	5.078	5.130	5.182	1800
1900	5.182	5.234	5.286	5.339	5.391	5.444	5.497	5.551	5.604	5.658	5.712	1900
2000	5.712	5.766	5.820	5.875	5.930	5.984	6.039	6.095	6.150	6.206	6.262	2000
2100	6.262	6.318	6.374	6.430	6.487	6.543	6.600	6.657	6.714	6.772	6.829	2100
2200	6.829	6.887	6.945	7.003	7.061	7.120	7.178	7.237	7.296	7.355	7.414	2200
2300	7.414	7.473	7.533	7.592	7.652	7.712	7.772	7.833	7.893	7.953	8.014	2300
2400	8.014	8.075	8.136	8.197	8.258	8.319	8.381	8.442	8.504	8.566	8.628	2400
2500	8.628	8.690	8.752	8.814	8.877	8.939	9.002	9.065	9.128	9.191	9.254	2500
2600	9.254	9.317	9.380	9.443	9.507	9.570	9.634	9.697	9.761	9.825	9.889	2600
2700	9.889	9.953	10.017	10.081	10.145	10.210	10.274	10.338	10.403	10.467	10.532	2700
2800	10.532	10.596	10.661	10.726	10.790	10.855	10.920	10.985	11.050	11.115	11.179	2800
2900	11.179	11.244	11.309	11.374	11.439	11.504	11.569	11.634	11.699	11.764	11.829	2900
3000	11.829	11.894	11.959	12.024	12.089	12.154	12.219	12.284	12.349	12.413	12.478	3000
3100	12.478	12.543	12.608	12.672	12.737	12.801	12.866	12.930	12.995	13.059	13.124	3100
3200	13.124	13.188	13.252	13.316	13.380	13.444	13.508	13.572	13.635	13.699	13.763	3200
3300	13.763											3300

TABLE 17A EMF-Temperature for Type T (Copper-Constantan) Thermocouples

Temperature in °C

Reference junction at 0°C

°C	0	1	2	3	4	5	6	7	8	9	10	°C
					Thermoelectric voltage in absolute millivolts							
−40	−1.475	−1.510	−1.544	−1.579	−1.614	−1.648	−1.682	−1.717	−1.751	−1.785	−1.819	−40
−30	−1.121	−1.157	−1.192	−1.228	−1.263	−1.299	−1.334	−1.370	−1.405	−1.440	−1.475	−30
−20	−0.757	−0.794	−0.830	−0.867	−0.903	−0.940	−0.976	−1.013	−1.049	−1.085	−1.121	−20
−10	−0.383	−0.421	−0.458	−0.496	−0.534	−0.571	−0.608	−0.646	−0.683	−0.720	−0.757	−10
−0	0.000	−0.039	−0.077	−0.116	−0.154	−0.193	−0.231	−0.269	−0.307	−0.345	−0.383	−0
0	0.000	0.039	0.078	0.117	0.156	0.195	0.234	0.273	0.312	0.351	0.391	0
10	0.391	0.430	0.470	0.510	0.549	0.589	0.629	0.669	0.709	0.749	0.789	10
20	0.789	0.830	0.870	0.911	0.951	0.992	1.032	1.073	1.114	1.155	1.196	20
30	1.196	1.237	1.279	1.320	1.361	1.403	1.444	1.486	1.528	1.569	1.611	30
40	1.611	1.653	1.695	1.738	1.780	1.822	1.865	1.907	1.950	1.992	2.035	40
50	2.035	2.078	2.121	2.164	2.207	2.250	2.294	2.337	2.380	2.424	2.467	50
60	2.467	2.511	2.555	2.599	2.643	2.687	2.731	2.775	2.819	2.864	2.908	60
70	2.908	2.953	2.997	3.042	3.087	3.131	3.176	3.221	3.266	3.312	3.357	70
80	3.357	3.402	3.447	3.493	3.538	3.584	3.630	3.676	3.721	3.767	3.813	80
90	3.813	3.859	3.906	3.952	3.998	4.044	4.091	4.137	4.184	4.231	4.277	90
100	4.277	4.324	4.371	4.418	4.465	4.512	4.559	4.607	4.654	4.701	4.749	100
110	4.749	4.796	4.844	4.891	4.939	4.987	5.035	5.083	5.131	5.179	5.227	110
120	5.227											120

TABLE 17B EMF-Temperature for Type T (Copper-Constantan) Thermocouples

Temperature in °C

Reference junction at 0°C

°C	0	10	20	30	40	50	60	70	80	90	100	°C
						Thermoelectric voltage in absolute millivolts						
−200	−5.603	−5.753	−5.889	−6.007	−6.105	−6.181	−6.232	−6.258				−200
−100	−3.378	−3.656	−3.923	−4.177	−4.419	−4.648	−4.865	−5.069	−5.261	−5.439	−5.603	−100
−0	0.000	−0.383	−0.757	−1.121	−1.475	−1.819	−2.152	−2.475	−2.788	−3.089	−3.378	−0
0	0.000	0.391	0.789	1.196	1.611	2.035	2.467	2.908	3.357	3.813	4.277	0
100	4.277	4.749	5.227	5.712	6.204	6.702	7.207	7.718	8.235	8.757	9.286	100
200	9.286	9.820	10.360	10.905	11.456	12.011	12.572	13.137	13.707	14.281	14.860	200
300	14.860	15.443	16.030	16.621	17.217	17.816	18.420	19.027	19.638	20.252	20.869	300
400	20.869											400

TABLE 17C EMF-Temperature for Type T (Copper-Constantan) Thermocouples

Temperature in °F

Reference junction at 32°F

°F	0	1	2	3	4	5	6	7	8	9	10	°F
				Thermoelectric voltage in absolute millivolts								
−40	−1.475	−1.494	−1.513	−1.533	−1.552	−1.571	−1.591	−1.610	−1.629	−1.648	−1.667	−40
−30	−1.279	−1.299	−1.319	−1.338	−1.358	−1.377	−1.397	−1.416	−1.436	−1.455	−1.475	−30
−20	−1.081	−1.101	−1.121	−1.141	−1.160	−1.180	−1.200	−1.220	−1.240	−1.260	−1.279	−20
−10	−0.879	−0.899	−0.920	−0.940	−0.960	−0.980	−1.000	−1.021	−1.041	−1.061	−1.081	−10
−0	−0.674	−0.695	−0.716	−0.736	−0.757	−0.777	−0.798	−0.818	−0.838	−0.859	−0.879	−0
0	−0.674	−0.654	−0.633	−0.613	−0.592	−0.571	−0.550	−0.529	−0.509	−0.488	−0.467	0
10	−0.467	−0.446	−0.425	−0.404	−0.383	−0.362	−0.341	−0.320	−0.299	−0.277	−0.256	10
20	−0.256	−0.235	−0.214	−0.193	−0.171	−0.150	−0.129	−0.107	−0.086	−0.064	−0.043	20
30	−0.043	−0.022	0.000	0.022	0.043	0.065	0.086	0.108	0.130	0.151	0.173	30
40	0.173	0.195	0.216	0.238	0.260	0.282	0.303	0.325	0.347	0.369	0.391	40
50	0.391	0.413	0.435	0.457	0.479	0.501	0.523	0.545	0.567	0.589	0.611	50
60	0.611	0.634	0.656	0.678	0.700	0.722	0.745	0.767	0.789	0.812	0.834	60
70	0.834	0.857	0.879	0.902	0.924	0.947	0.969	0.992	1.014	1.037	1.060	70
80	1.060	1.082	1.105	1.128	1.151	1.173	1.196	1.219	1.242	1.265	1.288	80
90	1.288	1.311	1.334	1.357	1.380	1.403	1.426	1.449	1.472	1.495	1.518	90
100	1.518	1.542	1.565	1.588	1.611	1.635	1.658	1.681	1.705	1.728	1.752	100
110	1.752	1.775	1.799	1.822	1.846	1.869	1.893	1.917	1.940	1.964	1.988	110
120	1.988	2.011	2.035	2.059	2.083	2.107	2.131	2.154	2.178	2.202	2.226	120
130	2.226	2.250	2.274	2.298	2.322	2.347	2.371	2.395	2.419	2.443	2.467	130
140	2.467	2.492	2.516	2.540	2.565	2.589	2.613	2.638	2.662	2.687	2.711	140

TABLE 17C EMF-Temperature for Type T (Copper-Constantan) Thermocouples (*Continued*)

Temperature in °F

Reference junction at 32°F

°F	0	1	2	3	4	5	6	7	8	9	10	°F
					Thermoelectric voltage in absolute millivolts							
150	2.711	2.736	2.760	2.785	2.809	2.834	2.859	2.883	2.908	2.933	2.958	150
160	2.958	2.982	3.007	3.032	3.057	3.082	3.107	3.131	3.156	3.181	3.206	160
170	3.206	3.231	3.256	3.281	3.307	3.332	3.357	3.382	3.407	3.432	3.458	170
180	3.458	3.483	3.508	3.533	3.559	3.584	3.609	3.635	3.660	3.686	3.711	180
190	3.711	3.737	3.762	3.788	3.813	3.839	3.864	3.890	3.916	3.941	3.967	190
200	3.967	3.993	4.019	4.044	4.070	4.096	4.122	4.148	4.174	4.199	4.225	200
210	4.225	4.251	4.277	4.303	4.329	4.355	4.381	4.408	4.434	4.460	4.486	210
220	4.486	4.512	4.538	4.565	4.591	4.617	4.643	4.670	4.696	4.722	4.749	220
230	4.749	4.775	4.801	4.828	4.854	4.881	4.907	4.934	4.960	4.987	5.014	230
240	5.014	5.040	5.067	5.093	5.120	5.147	5.174	5.200	5.227	5.254	5.281	240

TABLE 17D EMF-Temperature for Type T (Copper-Constantan) Thermocouples

Temperature in °F

Reference junction at 32°F

°F	0	10	20	30	40	50	60	70	80	90	100	°F
					Thermoelectric voltage in absolute millivolts							
−400	−6.105	−6.150	−6.187	−6.217	−6.240	−6.254						−400
−300	−5.341	−5.439	−5.532	−5.620	−5.705	−5.785	−5.860	−5.930	−5.995	−6.053	−6.105	−300
−200	−4.149	−4.286	−4.419	−4.548	−4.673	−4.794	−4.912	−5.025	−5.135	−5.240	−5.341	−200
−100	−2.581	−2.753	−2.923	−3.089	−3.251	−3.410	−3.565	−3.717	−3.864	−4.009	−4.149	−100
−0	−0.674	−0.879	−1.081	−1.279	−1.475	−1.667	−1.856	−2.042	−2.225	−2.405	−2.581	−0
0	−0.674	−0.467	−0.256	−0.043	0.173	0.391	0.611	0.834	1.060	1.288	1.518	0
100	1.518	1.752	1.988	2.226	2.467	2.711	2.958	3.206	3.458	3.711	3.967	100
200	3.967	4.225	4.486	4.749	5.014	5.281	5.550	5.821	6.094	6.369	6.647	200
300	6.647	6.926	7.207	7.490	7.775	8.062	8.350	8.641	8.933	9.227	9.523	300
400	9.523	9.820	10.120	10.420	10.723	11.027	11.333	11.640	11.949	12.260	12.572	400
500	12.572	12.885	13.200	13.516	13.834	14.153	14.474	14.795	15.118	15.443	15.769	500
600	15.769	16.096	16.424	16.753	17.084	17.416	17.750	18.084	18.420	18.757	19.095	600
700	19.095	19.434	19.774	20.116	20.458	20.801						700

TABLE 18A EMF-Temperature for Type J (Iron-Constantan) Thermocouples

Temperature in °C

Reference junction at 0°C

°C	0	1	2	3	4	5	6	7	8	9	10	°C
					Thermoelectric voltage in absolute millivolts							
−40	−1.960	−2.008	−2.055	−2.102	−2.150	−2.197	−2.244	−2.291	−2.338	−2.384	−2.431	−40
−30	−1.481	−1.530	−1.578	−1.626	−1.674	−1.722	−1.770	−1.818	−1.865	−1.913	−1.960	−30
−20	−0.995	−1.044	−1.093	−1.141	−1.190	−1.239	−1.288	−1.336	−1.385	−1.433	−1.481	−20
−10	−0.501	−0.550	−0.600	−0.650	−0.699	−0.748	−0.798	−0.847	−0.896	−0.945	−0.995	−10
−0	0.000	−0.050	−0.101	−0.151	−0.201	−0.251	−0.301	−0.351	−0.401	−0.451	−0.501	−0
0	0.000	0.050	0.101	0.151	0.202	0.253	0.303	0.354	0.405	0.456	0.507	0
10	0.507	0.558	0.609	0.660	0.711	0.762	0.813	0.865	0.916	0.967	1.019	10
20	1.019	1.070	1.122	1.174	1.225	1.277	1.329	1.381	1.432	1.484	1.536	20
30	1.536	1.588	1.640	1.693	1.745	1.797	1.849	1.901	1.954	2.006	2.058	30
40	2.058	2.111	2.163	2.216	2.268	2.321	2.374	2.426	2.479	2.532	2.585	40
50	2.585	2.638	2.691	2.743	2.796	2.849	2.902	2.956	3.009	3.062	3.115	50
60	3.115	3.168	3.221	3.275	3.328	3.381	3.435	3.488	3.542	3.595	3.649	60
70	3.649	3.702	3.756	3.809	3.863	3.917	3.971	4.024	4.078	4.132	4.186	70
80	4.186	4.239	4.293	4.347	4.401	4.455	4.509	4.563	4.617	4.671	4.725	80
90	4.725	4.780	4.834	4.888	4.942	4.996	5.050	5.105	5.159	5.213	5.268	90
100	5.268	5.322	5.376	5.431	5.485	5.540	5.594	5.649	5.703	5.758	5.812	100
110	5.812	5.867	5.921	5.976	6.031	6.085	6.140	6.195	6.249	6.304	6.359	110
120	6.359											120

TABLE 18B EMF-Temperature for Type J (Iron-Constantan) Thermocouples

Temperature in °C

Reference junction at 0°C

°C	0	10	20	30	40	50	60	70	80	90	100	°C
				Thermoelectric voltage in absolute millivolts								
−200	−7.890	−8.096										−200
−100	−4.632	−5.036	−5.426	−5.801	−6.159	−6.499	−6.821	−7.122	−7.402	−7.659	−7.890	−100
−0	0.000	−0.501	−0.995	−1.481	−1.960	−2.431	−2.892	−3.344	−3.785	−4.215	−4.632	−0
0	0.000	0.507	1.019	1.536	2.058	2.585	3.115	3.649	4.186	4.725	5.268	0
100	5.268	5.812	6.359	6.907	7.457	8.008	8.560	9.113	9.667	10.222	10.777	100
200	10.777	11.332	11.887	12.442	12.998	13.553	14.108	14.663	15.217	15.771	16.325	200
300	16.325	16.879	17.432	17.984	18.537	19.089	19.640	20.192	20.743	21.295	21.846	300
400	21.846	22.397	22.949	23.501	24.054	24.607	25.161	25.716	26.272	26.829	27.388	400
500	27.388	27.949	28.511	29.075	29.642	30.210	30.782	31.356	31.933	32.513	33.096	500
600	33.096	33.683	34.273	34.867	35.464	36.066	36.671	37.280	37.893	38.510	39.130	600
700	39.130	39.754	40.382	41.013	41.647	42.283	42.922					700

TABLE 18C EMF-Temperature for Type J (Iron-Constantan) Thermocouples

Temperature in °F

Reference junction at 32°F

°F	0	1	2	3	4	5	6	7	8	9	10	°F
					Thermoelectric voltage in absolute millivolts							
−40	−1.960	−1.987	−2.013	−2.039	−2.066	−2.092	−2.118	−2.144	−2.171	−2.197	−2.223	−40
−30	−1.695	−1.722	−1.748	−1.775	−1.802	−1.828	−1.855	−1.881	−1.908	−1.934	−1.960	−30
−20	−1.428	−1.455	−1.481	−1.508	−1.535	−1.562	−1.589	−1.615	−1.642	−1.669	−1.695	−20
−10	−1.158	−1.185	−1.212	−1.239	−1.266	−1.293	−1.320	−1.347	−1.374	−1.401	−1.428	−10
−0	−0.885	−0.913	−0.940	−0.967	−0.995	−1.022	−1.049	−1.076	−1.103	−1.131	−1.158	−0
0	−0.885	−0.858	−0.831	−0.803	−0.776	−0.748	−0.721	−0.694	−0.666	−0.639	−0.611	0
10	−0.611	−0.583	−0.556	−0.528	−0.501	−0.473	−0.445	−0.418	−0.390	−0.362	−0.334	10
20	−0.334	−0.307	−0.279	−0.251	−0.223	−0.195	−0.168	−0.140	−0.112	−0.084	−0.056	20
30	−0.056	−0.028	0.000	0.028	0.056	0.084	0.112	0.140	0.168	0.196	0.224	30
40	0.224	0.253	0.281	0.309	0.337	0.365	0.394	0.422	0.450	0.478	0.507	40
50	0.507	0.535	0.563	0.592	0.620	0.648	0.677	0.705	0.734	0.762	0.791	50
60	0.791	0.819	0.848	0.876	0.905	0.933	0.962	0.990	1.019	1.048	1.076	60

70	1.076	1.105	1.134	1.162	1.191	1.220	1.248	1.277	1.306	1.335	1.363
80	1.363	1.392	1.421	1.450	1.479	1.507	1.536	1.565	1.594	1.623	1.652
90	1.652	1.681	1.710	1.739	1.768	1.797	1.826	1.855	1.884	1.913	1.942
100	1.942	1.971	2.000	2.029	2.058	2.088	2.117	2.146	2.175	2.204	2.233
110	2.233	2.263	2.292	2.321	2.350	2.380	2.409	2.438	2.467	2.497	2.526
120	2.526	2.555	2.585	2.614	2.644	2.673	2.702	2.732	2.761	2.791	2.820
130	2.820	2.849	2.879	2.908	2.938	2.967	2.997	3.026	3.056	3.085	3.115
140	3.115	3.145	3.174	3.204	3.233	3.263	3.293	3.322	3.352	3.381	3.411
150	3.411	3.441	3.470	3.500	3.530	3.560	3.589	3.619	3.649	3.678	3.708
160	3.708	3.738	3.768	3.798	3.827	3.857	3.887	3.917	3.947	3.976	4.006
170	4.006	4.036	4.066	4.096	4.126	4.156	4.186	4.216	4.245	4.275	4.305
180	4.305	4.335	4.365	4.395	4.425	4.455	4.485	4.515	4.545	4.575	4.605
190	4.605	4.635	4.665	4.695	4.725	4.755	4.786	4.816	4.846	4.876	4.906
200	4.906	4.936	4.966	4.996	5.026	5.057	5.087	5.117	5.147	5.177	5.207
210	5.207	5.238	5.268	5.298	5.328	5.358	5.389	5.419	5.449	5.479	5.509
220	5.509	5.540	5.570	5.600	5.630	5.661	5.691	5.721	5.752	5.782	5.812
230	5.812	5.843	5.873	5.903	5.934	5.964	5.994	6.025	6.055	6.085	6.116
240	6.116	6.146	6.176	6.207	6.237	6.268	6.298	6.328	6.359	6.389	6.420

TABLE 18D EMF-Temperature for Type J (Iron-Constantan) Thermocouples

Temperature in °F

Reference junction at 32°F

°F	0	10	20	30	40	50	60	70	80	90	100	°F
					Thermoelectric voltage in absolute millivolts							
-300	-7.519										-7.519	-300
-200	-5.760	-5.962	-6.159	-6.350	-6.536	-6.716	-6.890	-7.057	-7.218	-7.372	-7.519	-200
-100	-3.492	-3.737	-3.978	-4.215	-4.448	-4.678	-4.903	-5.124	-5.341	-5.553	-5.760	-100
-0	-0.885	-1.158	-1.428	-1.695	-1.960	-2.223	-2.483	-2.740	-2.994	-3.245	-3.492	-0
0	-0.885	-0.611	-0.334	-0.056	0.224	0.507	0.791	1.076	1.363	1.652	1.942	0
100	1.942	2.333	2.526	2.820	3.115	3.411	3.708	4.006	4.305	4.605	4.906	100
200	4.906	5.207	5.509	5.812	6.116	6.420	6.724	7.029	7.335	7.641	7.947	200
300	7.947	8.253	8.560	8.867	9.175	9.483	9.790	10.098	10.407	10.715	11.023	300
400	11.023	11.332	11.640	11.949	12.257	12.566	12.874	13.183	13.491	13.800	14.108	400
500	14.108	14.416	14.724	15.032	15.340	15.648	15.956	16.264	16.571	16.879	17.186	500
600	17.186	17.493	17.800	18.107	18.414	18.721	19.027	19.334	19.640	19.947	20.253	600
700	20.253	20.559	20.866	21.172	21.478	21.785	22.091	22.397	22.704	23.010	23.317	700
800	23.317	23.624	23.931	24.238	24.546	24.853	25.161	25.469	25.778	26.087	26.396	800
900	26.396	26.705	27.016	27.326	27.637	27.949	28.261	28.573	28.887	29.201	29.515	900
1000	29.515	29.831	30.147	30.464	30.782	31.100	31.420	31.740	32.061	32.384	32.707	1000
1100	32.707	33.031	33.356	33.683	34.010	34.339	34.668	34.999	35.331	35.664	35.999	1100
1200	35.999	36.334	36.671	37.009	37.348	37.688	38.030	38.372	38.716	39.061	39.407	1200
1300	39.407	39.754	40.103	40.452	40.802	41.154	41.506	41.859	42.212	42.567	42.922	1300
1400	42.922											1400

TABLE 19A EMF-Temperature for Type E (Chromel-Constantan) Thermocouples

Temperature in °C

Reference junction at 0°C

°C	0	1	2	3	4	5	6	7	8	9	10	°C
					Thermoelectric voltage in absolute millivolts							
−40	−2.254	−2.308	−2.362	−2.416	−2.469	−2.522	−2.575	−2.628	−2.681	−2.734	−2.787	−40
−30	−1.709	−1.764	−1.819	−1.874	−1.929	−1.983	−2.038	−2.092	−2.146	−2.200	−2.254	−30
−20	−1.151	−1.208	−1.264	−1.320	−1.376	−1.432	−1.487	−1.543	−1.599	−1.654	−1.709	−20
−10	−0.581	−0.639	−0.696	−0.754	−0.811	−0.868	−0.925	−0.982	−1.038	−1.095	−1.151	−10
− 0	0.000	−0.059	−0.117	−0.176	−0.234	−0.292	−0.350	−0.408	−0.466	−0.524	−0.581	− 0
0	0.000	0.059	0.118	0.175	0.235	0.295	0.354	0.413	0.472	0.532	0.591	0
10	0.591	0.651	0.711	0.770	0.830	0.890	0.950	1.011	1.071	1.131	1.192	10
20	1.192	1.252	1.313	1.373	1.434	1.495	1.556	1.617	1.678	1.739	1.801	20
30	1.801	1.862	1.924	1.985	2.047	2.109	2.171	2.233	2.295	2.357	2.419	30
40	2.419	2.482	2.544	2.607	2.669	2.732	2.795	2.858	2.921	2.984	3.047	40
50	3.047	3.110	3.173	3.237	3.300	3.364	3.428	3.491	3.555	3.619	3.683	50
60	3.683	3.748	3.812	3.876	3.941	4.005	4.070	4.134	4.199	4.264	4.329	60
70	4.329	4.394	4.459	4.524	4.590	4.655	4.720	4.786	4.852	4.917	4.983	70
80	4.983	5.049	5.115	5.181	5.247	5.314	5.380	5.446	5.513	5.579	5.646	80
90	5.646	5.713	5.780	5.846	5.913	5.981	6.048	6.115	6.182	6.250	6.317	90
100	6.317	6.385	6.452	6.520	6.588	6.656	6.724	6.792	6.860	6.928	6.996	100
110	6.996	7.064	7.133	7.201	7.270	7.339	7.407	7.476	7.545	7.614	7.683	110
120	7.683											120

TABLE 19B EMF-Temperature for Type E (Chromel-Constantan) Thermocouples

Temperature in °C

Reference junction at 0°C

°C	0	10	20	30	40	50	60	70	80	90	100	°C
					Thermoelectric voltage in absolute millivolts							
−200	−8.824	−9.063	−9.274	−9.455	−9.604	−9.719	−9.797	−9.835				−200
−100	−5.237	−5.680	−6.107	−6.516	−6.907	−7.279	−7.631	−7.963	−8.273	−8.561	−8.824	−100
−0	0.000	−0.581	−1.151	−1.709	−2.254	−2.787	−3.306	−3.811	−4.301	−4.777	−5.237	−0
0	0.000	0.591	1.192	1.801	2.419	3.047	3.683	4.329	4.983	5.646	6.317	0
100	6.317	6.996	7.683	8.377	9.078	9.787	10.501	11.222	11.949	12.681	13.419	100
200	13.419	14.161	14.909	15.661	16.417	17.178	17.942	18.710	19.481	20.256	21.033	200
300	21.033	21.814	22.597	23.383	24.171	24.961	25.754	26.549	27.345	28.143	28.943	300
400	28.943	29.744	30.546	31.350	32.155	32.960	33.767	34.574	35.382	36.190	36.999	400
500	36.999	37.808	38.617	39.426	40.236	41.045	41.853	42.662	43.470	44.278	45.085	500
600	45.085	45.891	46.697	47.502	48.306	49.109	49.911	50.713	51.513	52.312	53.110	600
700	53.110	53.907	54.703	55.498	56.291	57.083	57.873	58.663	59.451	60.237	61.022	700
800	61.022	61.806	62.588	63.368	64.147	64.924	65.700	66.473	67.245	68.015	68.783	800
900	68.783	69.549	70.313	71.075	71.835	72.594	73.350	74.104	74.857	75.608	76.358	900
1000	76.358											1000

TABLE 19C EMF-Temperature for Type E (Chromel-Constantan) Thermocouples

Temperature in °F

Reference junction at 32°F

°F	0	1	2	3	4	5	6	7	8	9	10	°F
					Thermoelectric voltage in absolute millivolts							
−40	−2.254	−2.284	−2.314	−2.344	−2.374	−2.404	−2.433	−2.463	−2.493	−2.522	−2.552	−40
−30	−1.953	−1.983	−2.014	−2.044	−2.074	−2.104	−2.134	−2.164	−2.194	−2.224	−2.254	−30
−20	−1.648	−1.678	−1.709	−1.740	−1.770	−1.801	−1.831	−1.862	−1.892	−1.923	−1.953	−20
−10	−1.339	−1.370	−1.401	−1.432	−1.463	−1.494	−1.525	−1.555	−1.586	−1.617	−1.648	−10
−0	−1.026	−1.057	−1.089	−1.120	−1.151	−1.183	−1.214	−1.245	−1.276	−1.308	−1.339	−0
0	−1.026	−0.994	−0.963	−0.931	−0.900	−0.868	−0.836	−0.805	−0.773	−0.741	−0.709	0
10	−0.709	−0.677	−0.645	−0.613	−0.581	−0.549	−0.517	−0.485	−0.453	−0.421	−0.389	10
20	−0.389	−0.357	−0.324	−0.292	−0.260	−0.227	−0.195	−0.163	−0.130	−0.098	−0.065	20
30	−0.065	−0.033	0.000	0.033	0.065	0.098	0.131	0.163	0.196	0.229	0.262	30
40	0.262	0.295	0.327	0.360	0.393	0.426	0.459	0.492	0.525	0.558	0.591	40
50	0.591	0.624	0.658	0.691	0.724	0.757	0.790	0.824	0.857	0.890	0.924	50
60	0.924	0.957	0.990	1.024	1.057	1.091	1.124	1.158	1.192	1.225	1.259	60
70	1.259	1.292	1.326	1.360	1.394	1.427	1.461	1.495	1.529	1.563	1.597	70
80	1.597	1.631	1.665	1.699	1.733	1.767	1.801	1.835	1.869	1.903	1.936	80
90	1.937	1.972	2.006	2.040	2.075	2.109	2.143	2.178	2.212	2.247	2.281	90
100	2.281	2.316	2.350	2.385	2.419	2.454	2.489	2.523	2.558	2.593	2.627	100
110	2.627	2.662	2.697	2.732	2.767	2.802	2.837	2.872	2.907	2.942	2.977	110
120	2.977	3.012	3.047	3.082	3.117	3.152	3.187	3.223	3.258	3.293	3.329	120
130	3.329	3.364	3.399	3.435	3.470	3.506	3.541	3.577	3.612	3.648	3.683	130
140	3.683	3.719	3.755	3.790	3.826	3.862	3.898	3.933	3.969	4.005	4.041	140

TABLE 19C EMF-Temperature for Type E (Chromel-Constantan) Thermocouples (*Continued*)

Temperature in °F

Reference junction at 32°F

°F	0	1	2	3	4	5	6	7	8	9	10	°F
150	4.041	4.077	4.113	4.149	4.185	4.221	4.257	4.293	4.329	4.365	4.401	150
160	4.401	4.437	4.474	4.510	4.546	4.582	4.619	4.655	4.691	4.728	4.764	160
170	4.764	4.801	4.837	4.874	4.910	4.947	4.983	5.020	5.056	5.093	5.130	170
180	5.130	5.166	5.203	5.240	5.277	5.314	5.350	5.387	5.424	5.461	5.498	180
190	5.498	5.535	5.572	5.609	5.646	5.683	5.720	5.757	5.794	5.832	5.869	190
200	5.869	5.906	5.943	5.981	6.018	6.055	6.092	6.130	6.167	6.205	6.242	200
210	6.242	6.280	6.317	6.355	6.392	6.430	6.467	6.505	6.543	6.580	6.618	210
220	6.618	6.656	6.693	6.731	6.769	6.807	6.845	6.882	6.920	6.958	6.996	220
230	6.996	7.034	7.072	7.110	7.148	7.186	7.224	7.262	7.300	7.339	7.377	230
240	7.377	7.415	7.453	7.491	7.530	7.568	7.606	7.645	7.683	7.721	7.760	240

TABLE 19D EMF-Temperature for Type E (Chromel-Constantan) Thermocouples

Temperature in °F

Reference junction at 32°F

°F	0	10	20	30	40	50	60	70	80	90	100	°F
					Thermoelectric voltage in absolute millivolts							
-400	-9.604	-9.672	-9,729	-9.774	-9.809	-9.830	-9.837	-9.825	-9.783	-9.696	-9.543	-400
-300	-8.404	-8.561	-8.710	-8.852	-8.986	-9.112	-9.229	-9.338	-9.437	-9.525	-9.604	-300
-200	-6.471	-6.692	-6.907	-7.116	-7.319	-7.516	-7.707	-7.891	-8.069	-8.240	-8.404	-200
-100	-3.976	-4.248	-4.515	-4.777	-5.034	-5.287	-5.534	-5.776	-6.013	-6.245	-6.471	-100
-0	-1.026	-1.339	-1.648	-1.953	-2.254	-2.552	-2.845	-3.134	-3.419	-3.700	-3.976	-0
0	-1.026	-0.709	-0.389	-0.065	0.262	0.591	0.924	1.259	1.597	1.937	2.281	0
100	2.281	2.627	2.977	3.329	3.683	4.041	4.401	4.764	5.130	5.498	5.869	100
200	5.869	6.242	6.618	6.996	7.377	7.760	8.145	8.532	8.922	9.314	9.708	200
300	9.708	10.103	10.501	10.901	11.302	11.706	12.111	12.518	12.926	13.336	13.748	300
400	13.748	14.161	14.576	14.992	15.410	15.829	16.249	16.670	17.093	17.517	17.942	400
500	17.942	18.368	18.795	19.223	19.643	20.083	20.514	20.947	21.380	21.814	22.248	500
600	22.248	22.684	23.120	23.558	23.996	24.434	24.873	25.313	25.754	26.195	26.637	600
700	26.637	27.079	27.522	27.966	28.409	28.854	29.299	29.744	30.189	30.636	31.082	700
800	31.082	31.529	31.976	32.423	32.871	33.319	33.767	34.215	34.664	35.113	35.562	800
900	35.562	36.011	36.460	36.909	37.358	37.808	38.257	38.707	39.157	39.606	40.056	900
1000	40.056	40.505	40.955	41.404	41.853	42.303	42.752	43.201	43.650	44.098	44.547	1000
1100	44.547	44.995	45.443	45.891	46.339	46.786	47.234	47.681	48.127	48.574	49.020	1100
1200	49.020	49.466	49.911	50.357	50.802	51.246	51.691	52.135	52.578	53.022	53.465	1200
1300	53.465	53.907	54.349	54.791	55.233	55.674	56.115	56.555	56.995	57.434	57.873	1300
1400	57.873	58.312	58.750	59.188	59.626	60.063	60.499	60.935	61.371	61.806	62.240	1400
1500	62.240	62.675	63.108	63.542	63.974	64.406	64.838	65.269	65.700	66.130	66.559	1500
1600	66.559	66.988	67.416	67.844	68.271	68.698	69.124	69.549	69.974	70.398	70.822	1600
1700	70.822	71.244	71.667	72.088	72.509	72.930	73.350	73.769	74.188	74.606	75.024	1700
1800	75.024	75.441	75.858	76.274								1800

TABLE 20A EMF-Temperature for Type K (Chromel-Alumel) Thermocouples

Temperature in °C

Reference junction at 0°C

°C	0	1	2	3	4	5	6	7	8	9	10	°C
						Thermoelectric voltage in absolute millivolts						
−40	−1.527	−1.563	−1.600	−1.636	−1.673	−1.709	−1.745	−1.781	−1.817	−1.853	−1.889	−40
−30	−1.156	−1.193	−1.231	−1.268	−1.305	−1.342	−1.379	−1.416	−1.453	−1.490	−1.527	−30
−20	−0.777	−0.816	−0.854	−0.892	−0.930	−0.968	−1.005	−1.043	−1.081	−1.118	−1.156	−20
−10	−0.392	−0.431	−0.469	−0.508	−0.547	−0.585	−0.624	−0.662	−0.701	−0.739	−0.777	−10
− 0	0.000	−0.039	−0.079	−0.118	−0.157	−0.197	−0.236	−0.275	−0.314	−0.353	−0.392	− 0
0	0.000	0.039	0.079	0.119	0.158	0.198	0.238	0.277	0.317	0.357	0.397	0
10	0.397	0.437	0.477	0.517	0.557	0.597	0.637	0.677	0.718	0.758	0.798	10
20	0.798	0.838	0.879	0.919	0.960	1.000	1.041	1.081	1.122	1.162	1.203	20
30	1.203	1.244	1.285	1.325	1.366	1.407	1.448	1.489	1.529	1.570	1.611	30
40	1.611	1.652	1.693	1.734	1.776	1.817	1.858	1.899	1.940	1.981	2.022	40
50	2.022	2.064	2.105	2.146	2.188	2.229	2.270	2.312	2.353	2.394	2.436	50
60	2.436	2.477	2.519	2.560	2.601	2.643	2.684	2.726	2.767	2.809	2.850	60
70	2.850	2.892	2.933	2.975	3.016	3.058	3.100	3.141	3.183	3.224	3.266	70
80	3.266	3.307	3.349	3.390	3.432	3.473	3.515	3.556	3.598	3.639	3.681	80
90	3.681	3.722	3.764	3.805	3.847	3.888	3.930	3.971	4.012	4.054	4.095	90
100	4.095	4.137	4.178	4.219	4.261	4.302	4.343	4.384	4.426	4.467	4.508	100
110	4.508	4.549	4.590	4.632	4.673	4.714	4.755	4.796	4.837	4.878	4.919	110
120	4.919											120

TABLE 20B EMF-Temperature for Type K (Chromel-Alumel) Thermocouples

Temperature in °C

Reference junction at 0°C

°C	0	10	20	30	40	50	60	70	80	90	100	°C
					Thermoelectric voltage in absolute millivolts							
−200	−5.891	−6.035	−6.158	−6.262	−6.344	−6.404	−6.441	−6.458				−200
−100	−3.553	−3.852	−4.138	−4.410	−4.669	−4.912	−5.141	−5.354	−5.550	−5.730	−5.891	−100
−0	0.000	−0.392	−0.777	−1.156	−1.527	−1.889	−2.243	−2.586	−2.920	−3.242	−3.553	−0
0	0.000	0.397	0.798	1.203	1.611	2.022	2.436	2.850	3.266	3.681	4.095	0
100	4.095	4.508	4.919	5.327	5.733	6.137	6.439	6.939	7.338	7.737	8.137	100
200	8.137	8.537	8.938	9.341	9.745	10.151	10.560	10.969	11.381	11.793	12.207	200
300	12.207	12.623	13.039	13.456	13.874	14.292	14.712	15.132	15.552	15.974	16.395	300
400	16.395	16.818	17.241	17.664	18.088	18.513	18.938	19.363	19.788	20.214	20.640	400
500	20.640	21.066	21.493	21.919	22.346	22.772	23.198	23.624	24.050	24.476	24.902	500
600	24.902	25.327	25.751	26.176	26.599	27.022	27.445	27.967	28.288	28.709	29.128	600
700	29.128	29.547	29.965	30.383	30.799	31.214	31.629	32.042	32.455	32.866	33.277	700
800	33.277	33.686	34.095	34.502	34.909	35.314	35.718	36.121	36.524	36.925	37.325	800
900	37.325	37.724	38.122	38.519	38.915	39.310	39.703	40.096	40.488	40.879	41.269	900
1000	41.269	41.657	42.045	42.432	42.817	43.202	43.585	43.968	44.349	44.729	45.108	1000
1100	45.108	45.486	45.863	46.238	46.612	46.985	47.356	47.726	48.095	48.462	48.828	1100
1200	48.828	49.192	49.555	49.916	50.276	50.633	50.990	51.344	51.697	52.049	52.399	1200
1300	52.399	52.747	53.093	53.439	53.782	54.125	54.466	54.807				1300

TABLE 20C EMF-Temperature for Type K (Chromel-Alumel) Thermocouples

Temperature in °F

Reference junction at 32°F

°F	0	1	2	3	4	5	6	7	8	9	10	°F
					Thermoelectric voltage in absolute millivolts							
−40	−1.527	−1.547	−1.567	−1.588	−1.608	−1.628	−1.648	−1.669	−1.689	−1.709	−1.729	−40
−30	−1.322	−1.342	−1.363	−1.383	−1.404	−1.424	−1.445	−1.465	−1.486	−1.506	−1.527	−30
−20	−1.114	−1.135	−1.156	−1.177	−1.197	−1.218	−1.239	−1.260	−1.280	−1.301	−1.322	−20
−10	−0.904	−0.925	−0.946	−0.968	−0.989	−1.010	−1.031	−1.051	−1.072	−1.093	−1.114	−10
− 0	−0.692	−0.714	−0.735	−0.756	−0.777	−0.799	−0.820	−0.841	−0.862	−0.883	−0.904	− 0
0	−0.692	−0.671	−0.650	−0.628	−0.607	−0.585	−0.564	−0.543	−0.521	−0.500	−0.478	0
10	−0.478	−0.457	−0.435	−0.413	−0.392	−0.370	−0.349	−0.327	−0.305	−0.284	−0.262	10
20	−0.262	−0.240	−0.218	−0.197	−0.175	−0.153	−0.131	−0.109	−0.088	−0.066	−0.044	20
30	−0.044	−0.022	0.000	0.022	0.044	0.066	0.088	0.110	0.132	0.154	0.176	30
40	0.176	0.198	0.220	0.242	0.264	0.286	0.308	0.331	0.353	0.375	0.397	40
50	0.397	0.419	0.441	0.464	0.486	0.508	0.530	0.553	0.575	0.597	0.619	50
60	0.619	0.642	0.664	0.686	0.709	0.731	0.753	0.776	0.798	0.821	0.843	60

70	0.843	0.865	0.888	0.910	0.933	0.955	0.978	1.000	1.023	1.045	1.068	70
80	1.068	1.090	1.113	1.135	1.158	1.181	1.203	1.226	1.248	1.271	1.294	80
90	1.294	1.316	1.339	1.362	1.384	1.407	1.430	1.452	1.475	1.498	1.520	90
100	1.520	1.543	1.566	1.589	1.611	1.634	1.657	1.680	1.703	1.725	1.748	100
110	1.748	1.771	1.794	1.817	1.839	1.862	1.885	1.908	1.931	1.954	1.977	110
120	1.977	2.000	2.022	2.045	2.068	2.091	2.114	2.137	2.160	2.183	2.206	120
130	2.206	2.229	2.252	2.275	2.298	2.321	2.344	2.367	2.390	2.413	2.436	130
140	2.436	2.459	2.482	2.505	2.528	2.551	2.574	2.597	2.620	2.643	2.666	140
150	2.666	2.689	2.712	3.735	2.758	2.781	2.804	2.827	2.850	2.873	2.896	150
160	2.896	2.920	2.943	2.966	2.989	3.012	3.035	3.058	3.081	3.104	3.127	160
170	3.127	3.150	3.173	3.196	3.220	3.243	3.266	3.289	3.312	3.335	3.358	170
180	3.358	3.381	3.404	3.427	3.450	3.473	3.496	3.519	3.543	3.566	3.589	180
190	3.589	3.612	3.635	3.658	3.681	3.704	3.727	3.750	3.773	3.796	3.819	190
200	3.819	3.842	3.865	3.888	3.911	3.934	3.957	3.980	4.003	4.026	4.049	200
210	4.049	4.072	4.095	4.118	4.141	4.164	4.187	4.210	4.233	4.256	4.279	210
220	4.279	4.302	4.325	4.348	4.371	4.394	4.417	4.439	4.462	4.485	4.508	220
230	4.508	4.531	4.554	4.577	4.600	4.622	4.645	4.668	4.691	4.714	4.737	230
240	4.737	4.759	4.782	4.805	4.828	4.851	4.873	4.896	4.919	4.942	4.964	240

TABLE 20D EMF-Temperature for Type K (Chromel-Alumel) Thermocouples

Temperature in °F

Reference junction at 32°F

°F	0	10	20	30	40	50	60	70	80	90	100	°F
					Thermoelectric voltage in absolute millivolts							
-400	-6.344	-6.380	-6.409	-6.431	-6.447	-6.456						-400
-300	-5.632	-5.730	-5.822	-5.908	-5.989	-6.064	-6.133	-6.195	-6.251	-6.301	-6.344	-300
-200	-4.381	-4.527	-4.669	-4.806	-4.939	-5.067	-5.190	-5.308	-5.421	-5.529	-5.632	-200
-100	-2.699	-2.883	-3.065	-3.242	-3.417	-3.587	-3.754	-3.917	-4.075	-4.230	-4.381	-100
-0	-0.692	-0.904	-1.114	-1.322	-1.527	-1.729	-1.929	-2.126	-2.320	-2.511	-2.699	-0
0	-0.692	-0.478	-0.262	-0.044	0.176	0.397	0.619	0.843	1.068	1.294	1.520	0
100	1.520	1.748	1.977	2.206	2.436	2.666	2.896	3.127	3.358	3.589	3.819	100
200	3.819	4.049	4.279	4.508	4.767	4.964	5.192	5.418	5.643	5.868	6.092	200
300	6.092	6.316	6.539	6.761	6.984	7.205	7.427	7.649	7.870	8.092	8.314	300
400	8.314	8.537	8.759	8.983	9.206	9.430	9.655	9.880	10.106	10.333	10.560	400
500	10.560	10.787	11.015	11.243	11.472	11.702	11.931	12.161	12.392	12.623	12.854	500
600	12.854	13.085	13.317	13.549	13.781	14.013	14.246	14.479	14.712	14.945	15.178	600
700	15.178	15.412	15.646	15.880	16.114	16.349	16.583	16.818	17.053	17.288	17.523	700

800	17.523	17.759	17.994	18.230	18.466	18.702	18.938	19.174	19.410	19.646	19.883
900	19.883	20.120	20.356	20.593	20.830	21.066	21.303	21.540	21.777	22.014	22.251
1000	22.251	22.488	22.725	22.961	23.198	23.435	23.672	23.908	24.145	24.382	24.618
1100	24.618	24.854	25.091	25.327	25.563	25.799	26.034	26.270	26.505	26.740	26.975
1200	26.975	27.210	27.445	27.679	27.914	28.148	28.382	28.615	28.849	29.082	29.315
1300	29.315	29.547	29.780	30.012	30.244	30.475	30.706	30.937	31.168	31.399	31.629
1400	31.629	31.859	32.088	32.317	32.546	32.775	33.003	33.231	33.459	33.686	33.913
1500	33.913	34.140	34.366	34.593	34.818	35.044	35.269	35.494	35.718	35.942	36.166
1600	36.166	36.390	36.613	36.836	37.058	37.280	37.502	37.724	37.945	38.166	38.387
1700	38.387	38.607	38.827	39.046	39.266	39.485	39.703	39.922	40.140	40.358	40.575
1800	40.575	40.792	41.009	41.225	41.442	41.657	41.873	42.088	42.303	42.518	42.732
1900	42.732	42.946	43.159	43.373	43.585	43.798	44.010	44.222	44.434	44.645	44.856
2000	44.856	45.066	45.276	45.486	45.695	45.904	46.113	46.321	46.529	46.736	46.944
2100	46.944	47.150	47.356	47.562	47.767	47.972	48.177	48.381	48.584	48.787	48.990
2200	48.990	49.192	49.394	49.595	49.796	49.996	50.196	50.395	50.594	50.792	50.990
2300	50.990	51.187	51.384	51.530	51.776	51.971	52.165	52.360	52.553	52.747	52.939
2400	52.939	53.132	53.324	53.515	53.706	53.897	54.087	54.277	54.466	54.656	54.845
2500	54.845										

Thermal radiation emitted by a real body has an irregular wavelength dependence (see Ref. 99). The emissivity is defined to relate the emissive power of a real body, e_λ or e, to that of a blackbody:

$$e_\lambda = \epsilon_\lambda e_{b\lambda} \tag{23a}$$

$$e = \epsilon e_b \tag{23b}$$

where ϵ_λ and ϵ are the monochromatic emissivity and the total emissivity of the body, respectively. A gray body is defined as one for which ϵ_λ is independent of wavelength; then $\epsilon_\lambda = \epsilon$.

Principles of Optical Pyrometers.   Optical pyrometers are used by the NBS to maintain IPTS-68 above the gold point [48]. A typical optical pyrometer is shown in Fig. 19. Radiation from the object whose temperature is to be measured is focused on a standard adjustable-brightness lamp. The power supply to the lamp filament can be adjusted to change the brightness of the filament. This brightness is then compared with the brightness of the incoming radiation.

The wavelength at which $e_{b\lambda}$ is a maximum, λ_{max}, can be determined from Wien's displacement law, which can be derived from Eq. (21):

$$\lambda_{max} T = 2897.8 \ \mu m \ K \tag{24}$$

As the gold point, 1337.58 K,

$$\lambda_{max} = 2.17 \ \mu m$$

Although this is well into the infrared region, a substantial fraction of the emitted radiation at the gold point is in the visible region. The effective wavelength of an optical pyrometer, determined by the filter (usually red), is often taken to be 0.655 μm.

The emissive power is determined by comparing the incoming radiation from the object with the radiation from the standard lamp. The filament current is adjusted until the filament image just disappears in the image of the test body; such a condition is similar to a null condition in resistance measurement with a bridge. Prior calibration of the optical pyrometer to determine the filament current setting and the corresponding filament temperature is used to calculate the temperature of the test body from the current setting. As a real body emits less thermal radiation than a blackbody at the same temperature, the temperature reading of an optical pyrometer must be corrected. The measured emissive power can be represented by a blackbody at temperature T_m corresponding to the

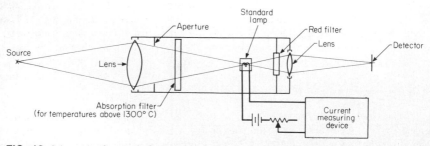

FIG. 19 Schematic of an optical pyrometer.

calibrated-filament lamp reading. For a monochromatic measurement, from Eqs. (21) and (23a),

$$\exp \frac{c_2}{\lambda T} - 1 = \epsilon_\lambda \left(\exp \frac{c_2}{\lambda T_m} - 1 \right)$$

Rearranging,

$$\exp \frac{c_2}{\lambda T} = \epsilon_\lambda \exp \frac{c_2}{\lambda T_m} + (1 - \epsilon_\lambda)$$

Neglecting $1 - \epsilon_\lambda$ and taking the natural logarithm of both sides,

$$\frac{1}{T} - \frac{1}{T_m} = \frac{\lambda}{c_2} \ln \epsilon_\lambda$$

$$\approx 4.55 \times 10^{-5} \ln \epsilon_\lambda \qquad \text{for } \lambda = 0.655 \ \mu m \tag{25}$$

With ϵ_λ known and T_m having been measured, Eq. (25) can be solved for the unknown temperature, T [49]. Note that $T \geq T_m$, and the smaller the value of ϵ_λ, the greater the temperature correction. This correction is small unless the emissivity is considerably less than 1.0.

If the total emissive power were used in the pyrometer, then

$$e_b(T_m) = e(T) = \epsilon e_b(T)$$

$$\sigma T_m^4 = \epsilon \sigma T^4 \tag{26}$$

$$T = T_m \left(\frac{1}{\epsilon} \right)^{1/4}$$

Similarly, as an approximation to Eq. (25), the variation of $e_{b\lambda}$ with T can be assumed to follow a power law at least over a limited temperature range:

$$e_{b\lambda} = c T^n \tag{27}$$

Then,

$$e_{b\lambda}(T_m) = e_\lambda(T) = \epsilon_\lambda e_{b\lambda}(T)$$

$$c T_m^n = \epsilon_\lambda c T^n \tag{28}$$

$$T = T_m \left(\frac{1}{\epsilon_\lambda} \right)^{1/n}$$

Note that for $\lambda \approx 0.655 \ \mu m$ and $T \approx 1340$ K, n is about 16.

Table 21 [50] shows the variation in ϵ_λ for different materials. For oxides, ϵ_λ has a strong dependence on surface roughness, which leads to the large uncertainty shown. Care should be exercised to ensure that the surface condition of the object whose temperature is being measured is such that the emissivity is known with a reasonable certainty. Sometimes a small hole with a depth of about five diameters is made in the object being studied and the pyrometer is focused on this hole. Internal reflections in the hole cause it to approximate a black surface.

Reliability of Optical Pyrometers. Calibration of pyrometers at the NBS is accomplished by balancing the standard lamp brightness against a blackbody furnace at the gold point [51]. To increase the stability and life of the standard lamp, the filament is normally

TABLE 21A Spectral Emissivity of Materials ($\lambda = 0.65 \ \mu m$): Unoxidized Surfaces [50]

Element	Solid	Liquid	Element	Solid	Liquid
Beryllium	0.61	0.61	Thorium	0.36	0.40
Carbon	0.80–0.93		Titanium	0.63	0.65
Chromium	0.34	0.39	Tungsten	0.43	
Cobalt	0.36	0.37	Uranium	0.54	0.34
Columbium	0.37	0.40	Vanadium	0.35	0.32
Copper	0.10	0.15	Yttrium	0.35	0.35
Erbium	0.55	0.38	Zirconium	0.32	0.30
Gold	0.14	0.22	Steel	0.35	0.37
Iridium	0.30		Cast iron	0.37	0.40
Iron	0.35	0.37	Constantan	0.35	
Manganese	0.59	0.59	Monel	0.37	
Molybdenum	0.37	0.40	Chromel P		
Nickel	0.36	0.37	(90Ni–10Cr)	0.35	
Palladium	0.33	0.37	80Ni–20Cr	0.35	
Platinum	0.30	0.38	60Ni–24Fe–16Cr	0.36	
Rhodium	0.24	0.30	Alumel		
Silver	0.07	0.07	(95Ni; bal. Al, Mn, Si)	0.37	
Tantalum	0.49		90Pt–10Rh	0.27	

not operated above a temperature of 1350°C (2460°F). Measurement at higher temperatures is achieved through the use of absorbing glass filters [51] to reduce the brightness of the incoming radiation (see Fig. 19). Instruments for laboratory use can be compared to an NBS-calibrated pyrometer.

With calibration performed at the NBS, optical pyrometers can realize IPTS to within 0.4°C (0.8°F) at the gold point, within about 2°C at 2200°C (3.5°F at 4000°F), and within about 10°C at 4000°C (about 20°F at 7250°F).

Other Radiation Thermometers. Optical pyrometers use a red filter to achieve a monochromatic comparison between the incoming radiation and that from a standard lamp. The comparison can be performed by visual inspection, but this is subject to human error. With the development of photomultipliers, better precision and automation are possible including direct detection—eliminating the use of a standard lamp.

Infrared pyrometers are similar to optical pyrometers, the main difference being that an infrared pyrometer uses a detector sensitive to a band of radiation in the infrared region. A band filter is often used. An infrared scanner is an infrared pyrometer coupled with a scanning mechanism to produce a two-dimensional output display. In order to eliminate the errors due to thermal noise (infrared radiation from the casing, etc.), the detector has to be cooled and properly sealed. Many infrared scanners used in airborne measuring systems for thermal mapping are sensitive to a portion of the spectrum in which the atmosphere has low absorption, 8 to 14 μm. Other commonly found units for ground-level application—often smaller in physical size—are sensitive in the 3 to 5 μm band of radiation. More information on infrared scanners and remote sensing is available in Refs. 52 and 53.

Spectroscopic methods [8] depend on the spectral line intensity emitted by the media of interest. These techniques have been used for temperature measurement in high-temperature gases. The wavelength involved is generally shorter than those in the infrared band.

TABLE 21B Spectral Emissivity of Materials ($\lambda = 0.65 \ \mu$m): Oxides [50]

Material*	Range of observed values	Probable value for the oxide formed on smooth metal
Aluminum oxide	0.22 to 0.40	0.30
Beryllium oxide	0.07 to 0.37	0.35
Cerium oxide	0.58 to 0.80	
Chromium oxide	0.60 to 0.80	0.70
Cobalt oxide		0.75
Columbium oxide	0.55 to 0.71	0.70
Copper oxide	0.60 to 0.80	0.70
Iron oxide	0.63 to 0.98	0.70
Magnesium oxide	0.10 to 0.43	0.20
Nickel oxide	0.85 to 0.96	0.90
Thorium oxide	0.20 to 0.57	0.50
Tin oxide	0.32 to 0.60	
Titanium oxide		0.50
Uranium oxide		0.30
Vanadium oxide		0.70
Yttrium oxide		0.60
Zirconium oxide	0.18 to 0.43	0.40
Alumel (oxidized)		0.87
Cast iron (oxidized)		0.70
Chromel P (90Ni–10Cr) (oxidized)		0.87
80Ni–20Cr (oxidized)		0.90
60Ni–24Fe–16Cr (oxidized)		0.83
55Fe–37.5Cr–7.5Al (oxidized)		0.78
70Fe–23Cr–5Al–2Co (oxidized)		0.75
Constantan (55Cu–45Ni) (oxidized)		0.84
Carbon steel (oxidized)		0.80
Stainless steel (18-8) (oxidized)		0.85
Porcelain	0.25 to 0.50	

*The emissivity of oxides and oxidized metals depends to a large extent upon the roughness of the surface. In general, higher values of emissivity are obtained on the rougher surfaces.

f. Other Thermometers

Among the many other types of thermometers, we will briefly discuss the following: optical methods (shadowgraph, schlieren, and interferometer), bimetallic thermometers, liquid-crystal and other contact thermographic methods, noise thermometers, and semiconductor diode thermometers.

Shadowgraph, schlieren, and interferometric systems [54] measure variations in the index of refraction or one of its spatial derivatives in a transparent medium. Shadowgraph systems are used to indicate the variation of the second derivative (normal to the light beam) of the index of refraction. Schlieren systems are sensitive to the first derivative of the index of refraction in a direction normal to the light beam. Interferometers measure the index of refraction field directly. The index of refraction of a fluid is closely related to the density of the fluid. In many cases the fluid density may be a function only of the temperature; this often holds for homogeneous liquids and gases at constant pressure.

Then from the measured variation in the index of refraction, the temperature field can be inferred.

Being optical techniques, these systems essentially have zero inertia. As no probe is used, they would not influence a flowing fluid whose temperature is being studied. However, they do require a transparent medium—almost always a fluid—with windows to permit a light beam or beams to enter and pass through the medium. In addition, since the light beam essentially integrates the variation in index of refraction or its gradients along the path of the beam, these techniques are usually not useful for local measurements in a three-dimensional temperature field. The development of holographic systems allows for the possibility of three-dimensional temperature measurement (local measurement) rather than the averaging process along the light beam direction in traditional interferometers [55].

Bimetallic thermometers which measure temperature by the change in physical dimension of the sensor have often been used [56]. The sensor consists of a composite strip of material, normally in a helical shape, formed by two different metals. Differences in the thermal expansion of the two metals cause the curvature of the strip to be a function of temperature. The strip is used as a temperature indicator with a self-contained scale. Bimetallic thermometers have been used at temperature from -185 to $425°C$ (-300 to $800°F$). The precision and accuracy of such thermometers are described in Ref. 57.

Liquid-crystal thermography depends on the color change of cholesteric liquid crystals with temperature. From black at low temperature they progress through a whole spectrum of colors before turning black again as the temperature is increased over a distinct reproducible range. With proper calibration, a sheet of liquid crystal can be used for quantitative as well as qualitative determination of constant temperature contours on the crystal [58]. Other contact thermographic methods using various temperature-sensitive materials (paints, coatings, etc.) are also available [53].

With noise thermometers, the Johnson noise fluctuation generated in a resistor is the temperature indicator [59]. The random noise fluctuation in a frequency band is related to the temperature and resistance of the resistor, thus ideally enabling the noise thermometer to measure absolute temperature. Noise thermometers have a wide temperature range, from cryogenic temperatures up to about $1500°C$ ($2700°F$). They have a long integration time—between 1 and 10 h for thermodynamic temperature measurements and of the order of seconds for an industrial sensor. Problems encountered in their use include the existence of noise of nonthermal origin and the stability of the measuring instrumentation during integration.

In a resonant-frequency thermometer the resonance frequency of the medium serves as the temperature indicator. Included in this category are nuclear quadrupole resonance thermometers, quartz thermometers, and ultrasonic thermometers. These thermometers usually come complete with their frequency measuring (counting) device. The improvement in modern frequency counting allows high precision measurements.

Nuclear quadrupole resonance thermometers [60] can be used between 20 and 400 K. A precision and accuracy of $\pm$ 1 mK can be achieved. However, they are quite expansive, generally limiting them to being good transfer standards. A quartz thermometer developed for use between -80 and 250 °C [100] has a resolution of 0.1 mK. If used at the same temperature, a comparable precision can be achieved. However, with temperature cycling, hysteresis can reduce its repeatability. An accuracy of 0.05 °C can be achieved with calibration. Ultrasonic thermometers [101] measure the temperature from its dependence on sound velocity. They are mainly used for measurements at high temperatures (300 up to 2000 K).

When a pn-junction diode is subjected to a constant forward bias current, its junction voltage is inversely proportional to the absolute temperature [61]. This is the physical basis for semiconductor diode thermometers. They can be used from liquid helium temperature to 200°C (480°F). Germanium, silicon, and gallium arsenide diodes have been used as thermometers. Due to the difficulty in maintaining a uniform manufacturing process, the interchangeability of diode thermometers is a significant problem. This variation from unit to unit can be accommodated by adjustment in the signal conditioning circuit which normally comes as part of the thermometer.

4. Calibration—Reliability of Temperature Sensors

a. Introduction

The task of relating a thermometer's output (i.e., magnitude of the variable dependent on temperature) to its temperature is achieved through calibration. Two general means of calibration are available: (1) fixed-point calibration and (2) comparison calibration. IPTS-68 provides 13 defining fixed points to which unique values of temperature are assigned (Table 1a). These fixed points are used for fixed-point calibration of defining standard thermometers and are also used, albeit seldom, in calibration of other standard thermometers. A number of secondary reference points are also provided by IPTS-68 (Table 1b).

The second means of calibration compares the output of the sensor to be calibrated to that of a standard thermometer, and it is called comparison calibration. The temperature of the sensor and of the standard thermometer at the time of calibration should be identical. This temperature, called the calibration temperature, does not have to be a fixed point; it can be any temperature as long as it can be determined by the standard thermometer.

Any experimental procedure contains uncertainties, and an error analysis is essential to attach significance to the results. The calibration process is no exception, and the reliability of a particular temperature sensor is only assured after the calibration and accompanying error estimate are completed.

b. Calibration Systems and Procedures

In calibrating a temperature sensor, the sensor should be maintained at a known temperature while its output signal (e.g., emf or resistance) is measured. The following description is only meant to be a brief introduction. Interested readers are referred to Refs. 29, 33, 45, 51, and 62.

Fixed-point calibration has the highest accuracy, but to achieve such accuracy requires great precautions and is time-consuming. Most freezing- and triple-point cells are difficult to maintain and cannot accommodate more than one sensor at a time. The maintenance of boiling points is even more complex, often requiring very precise pressure measurement. Thus, fixed-point calibrations are usually carried out only for high-precision thermometers and usually only at national laboratories. However, easily maintained water triple-point (or ice-point) systems are used in many laboratories to correct thermometer drift, particularly for PRTs.

Constant-temperature baths (or furnaces) and defining standard thermometers (such as SPRTs or type S thermocouples previously calibrated by a national laboratory) or a transfer standard thermometer generally are used when calibrating temperature measuring devices for laboratory experiments. In an industrial setting, when production calibration is required, comparison calibration against a working standard thermometer is used.

The working standard is usually the same type of thermometer as the thermometers to be calibrated to reduce the instrumentation requirements; it should have been previously calibrated against a defining or transfer standard thermometer.

The defining standard thermometers are the thermometers specified for IPTS-68. They are calibrated by the NBS at regular time intervals and are then used to calibrate transfer or working standard thermometers. Transfer standard thermometers serve as intermediate standards to reduce the use and drift of the defining standard thermometer.

Constant-temperature baths or furnaces are needed to maintain the uniform temperature environment for comparison calibration. Stirred-liquid baths and temperature controllers have good characteristics as a constant-temperature medium due to their ability to maintain temperature uniformity [63]. The liquids used include refrigerants, water, oils, molten tin, and molten salt. Water can be used for temperatures between 0 and 100°C (32 and 212°F) and oils from 100 to 315°C (212 to 600°F). Refrigeration units are available for commercial constant-temperature baths with alcohol as the working fluid down to −80° (−110°F). At the NBS, special cryostats [9, 16] with liquid nitrogen or liquid helium as the coolant are used for comparison calibration at low temperatures. A molten-tin bath is used at the NBS for calibration between 315 and 540°C (600 and 1000°F). A molten-salt bath [62] can be used for calibration in a temperature range similar to that for the tin baths. Air–fluidized solid baths for calibration up to 1100°C (2000°F) have been developed [64]. Gas furnaces are used for high-temperature calibrations. A furnace has been used for calibration of high-temperature thermocouples up to 2200°C (4000°F) [65]. However, the temperature uniformity in a furnace is not as good as that in a liquid bath because of the poor heat transfer characteristics of gases. To ensure uniformity, a copper block is sometimes used as a holder for the thermometer to be calibrated and the standard thermometer. For best accuracy, the copper block is partially insulated from the surroundings to reduce the effect of small fluctuation in the temperature of the surroundings, and the positions of the two thermometers are interchanged to eliminate the influence of spatial variation in temperature.

A typical procedure for comparison calibration of a thermometer is as follows. First, the temperature range of operation for the thermometer is decided. Then, an interpolation formula is selected, usually a polynomial. Third, a number of calibration temperatures are chosen; these are distributed over the desired temperature range of calibration. For comparison calibration, the number of calibration points, usually at regular temperature intervals, is greater than the unknown coefficients in the interpolation formula to enable least-squares determination. The fourth step is to measure the thermometer output while it is maintained at each of the calibration temperatures and while the actual temperature is determined with a standard thermometer maintained at the same temperature. From the thermometer outputs and the corresponding temperatures, the unknown coefficents in the correlation equation are determined.

The thermometer can be calibrated separately from the measuring instrument (e.g., a digital voltmeter) that is used to read the thermometer output in actual application. Sometimes, the measuring instrument and thermometer are a unit and are calibrated together.

c. Error Analysis

The terms accuracy, precision, repeatability, stability, and interchangeability all relate to the reliability of the measurements of a thermometer. *Accuracy* indicates the closeness of a sensor's reading to the actual IPTS-68 value. The terms *precision* and *repeatability* indicate the ability of the sensor to reproduce the same readings at the same temperature or the ability to reproduce the same calibration [66]. *Stability* is an indication of the ability of the sensor to maintain the same reading for a given condition over a period of time. *Interchangeability* is a measure of how closely replacement thermometer readings would be to similar thermometers without recalibration.

The terms error and uncertainty quantify the accuracy and precision of a sensor. *Error* indicates the difference between the measured and the actual values. *Uncertainty* normally refers to the inability to pinpoint the value; it is the range of readings (precision) obtained when repeatedly measuring the same temperature.

The stability of a sensor and its measuring instrumentation are not determined in calibration. The precision of a sensor can be found by repeated measurements, while determining or ensuring the accuracy can be very difficult. A 1-mK precision for a good SPRT is not unusual; however, to achieve a 10-mK accuracy, even at the calibration points, is beyond the capability of many calibration laboratories.

Sources of error in calibration include (1) difficulty in maintaining the fixed points, (2) accuracy of the standard thermometer, (3) the uniformity of the constant-temperature medium, (4) accuracy in the signal-reading instrumentation used, (5) stability of each of the components, (6) operator error, (7) hysteresis effect, and (8) interpolation uncertainty. Techniques for error analysis are described in a number of books on experimental measurement.

Reducing errors to attain very high accuracy is difficult. An accuracy of better than about 0.05°C (0.1°F) at temperatures less than 600°C (1100°F) and 1°C (2°F) for temperatures between 600°C (1100°F) and the gold point is usually attainable only in a few standards laboratories. Today high accuracy is maintained in some laboratories by the Measurement Assurance Program. Through this program, SPRTs circulate among a number of standards laboratories each of which recalibrates the thermometers and then returns them to the NBS for another recalibration. This procedure permits the detection of any drift in the overall calibration system at the participating laboratories [67].

5. Local Temperature Measurement

a. Introduction

In choosing a thermometer, consideration must be given to the specific environment in which the temperature is being measured (see Ref. 68). Many measurements are taken in systems in which the temperature varies with position and perhaps with time as well. The presence of a sensor and/or its connecting leads can change the temperature field being studied. Factors that affect the suitability and accuracy of a particular thermometer for making a temperature measurement include the size and the physical characteristics (thermal properties) of the sensing element.

Small size is important if there are steep gradients of temperature in the medium being studied. Even with a small sensor such as a thermocouple junction, conduction along the lead wires can affect the temperature of the junction. With a thermometer well, conduction along the well should also be considered. The leads near a sensor should pass when possible through an isothermal region.

b. Steady Temperature Measurement in Solids

In measuring internal temperatures of a solid the leads to the temperature sensor should follow an isothermal path. In addition, there should be good thermal contact between the sensor and the surrounding solid. This is usually provided by a thermally conducting paste or, in a thermometer well, a liquid.

Often knowledge of a solid's surface temperature is required. Figure 20 [33] shows common methods of thermocouple attachment to surfaces. The junction can be held close to the surface by soldering, welding, using insulating cement, or simply by pressure. The lead wires should be held in good contact with an isothermal portion of the surface for the order of 20 wire diameters to avoid conduction errors. Temperatures can also be measured at different locations below the surface and extrapolation used to determine the temperature at the surface.

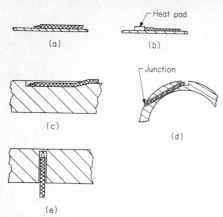

c. Temperature Measurement in Fluids

The definitions of temperature and temperature scale come from thermodynamic considerations; the meaning of temperature is strictly valid only for a state of thermodynamic equilibrium. In most real situations, nonuniformities are present. We generally assume the fluid is in local thermodynamic equilibrium and the properties of the fluid are interrelated through the material's equilibrium equation of state.

FIG. 20 Common methods of thermocouple attachment to surfaces [33]. (*a*) Junction mounted directly on surface. (*b*) Junction in heat collecting pad. (*c*) Junction mounted in groove. (*d*) Junction mounted in chordal hole in tube wall. (*e*) Junction mounted from rear of surface.

The local temperature can be determined from a probe that is small compared to the region over which the temperature varies by any significant amount. Problems arise in special cases such as rarefied gases, as well as regions with very large gradients, as in shock waves.

A temperature-sensing element measures its own temperature—not that of the surrounding fluid. Ideally, it is in thermal equilibrium with the surrounding region and does not affect the local temperature because of its presence. Often this ideal is not met and factors that must be considered include heat transfer to or from the sensor by radiation and conduction, conversion of kinetic energy into internal energy in the boundary layer surrounding the probe, and convective heat transfer from the surrounding medium, including that following a transient change in temperature.

An energy balance can be made on the temperature sensor to take into account the energy flows to and from it. This could indicate that its temperature is the same as that of the surrounding medium or that corrections should be made. For example, the principal heat transfer mechanisms might be convection from the adjacent fluid, conduction along the leads, and thermal radiation transport. If the conduction and radiation heat flows are small and/or the sensor is in good thermal contact (high heat transfer coefficient) with the immediately surrounding fluid, the sensor's temperature can closely approximate that of the adjacent fluid.

Radiation shields [69] can be used to isolate the probe from a distant medium so that there will be relatively little radiation heat transfer to it; at the same time they do not interfere with good thermal contact between the probe and the surrounding fluid. Designs of thermometer probes for gas temperature measurement are described in Refs. 70 and 71. Analyses to account for some uncertainties in probe measurement are in Ref. 72.

When a probe is immersed in a flowing fluid, the flow comes to rest in the immediate vicinity of the probe. In this deceleration kinetic energy is converted into internal energy, which can significantly increase the fluid temperature. Although this change in temperature is generally small in liquid flows, it can be significant in gas flows. The total and static temperatures of a gas with velocity V and (constant) specific heat c_p are related by the equation

$$T_t = T_{st} + \frac{V^2}{2c_p} \tag{29}$$

The static temperature T_{st} would be observed by a thermometer moving along with the flow, while the total temperature T_t would be attained by the fluid following adiabatic conversion of the kinetic energy into internal energy. As a flow is brought to rest at a real probe, the temperature generally is not equal to the total temperature other than in an idealized case or in specially designed probes [69, 70]. Often, as a result of dissipative processes (conduction, viscosity), the temperature is some value less than T_t. The temperature at an adiabatic surface is called the recovery temperature T_r,

$$T_r = T_{st} + rT_d \tag{30}$$

where r is the recovery factor and T_d is the dynamic temperature, $T_t - T_{st}$. With a laminar boundary layer flowing along a flat surface parallel to the flow, the recovery factor is equal to the square root of the Prandtl number [69]. In general, the recovery factor around a temperature probe is not uniform and often must be measured if accurate measurements are required.

Specific values of the recovery (dynamic correction) factors for the three probe geometries indicated in Fig. 21 have been measured [73]. The result is reproduced in Fig. 22. Note that the "half-shielded" probe has a relatively constant recovery factor of about 0.96 over the range of conditions studied.

d. Transient Temperature Measurements

The response of a thermometer to a change in the temperature of its environment depends on the physical properties of the sensor and the dynamic properties of the surrounding environment. A standard approach is to determine a time constant for the probe, assuming the probe is "zero-dimensional" (i.e., assuming the probe is small enough or its conductivity is high enough that as a first approximation the temperature within the probe is

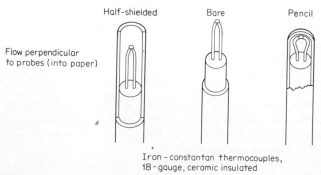

Iron-constantan thermocouples,
18-gauge, ceramic insulated

FIG. 21 Thermocouple probes for measuring temperatures of flowing fluids.

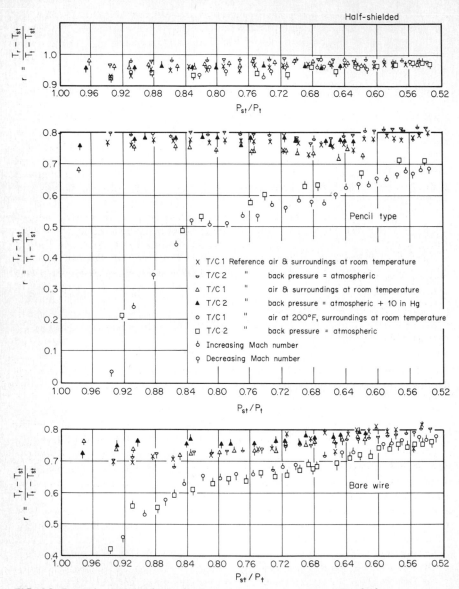

FIG. 22 Dynamic correction factors for temperature measurement in airflow [73].

uniform). Neglecting radiation and lead-wire conduction, the increase in energy stored within the probe would be equal to the net heat transfer convected into it:

$$mc\frac{dT}{d\theta} = hA(T_f - T) \qquad (31)$$

where m, c, and A are mass, specific heat, and surface area of the sensor, respectively; T and T_f are the temperatures of the sensor and the surrounding material (often a fluid), respectively; θ is the time; and h is the heat transfer coefficient between sensor and surrounding. A time constant, τ, can be defined in the form

$$\tau = \frac{mc}{hA} = \frac{\text{thermal capacitance of the sensor}}{\text{thermal conductance of the fluid}} \qquad (32)$$

The response of such a sensor has been analyzed for a number of fluid temperature transients—in particular, a step change, a ramp change (linear increase with time), and a periodic change. For a step change in fluid temperature, τ is the time for the sensor to have changed its temperature reading T so that $(T_f - T)_{\theta=\tau}$ is equal to $e^{-1}(T_f - T)_{\theta=0}$. For ramp and periodic changes, the time constant is (after the initial transient has faded) the time the sensor lags its environment. Examples of error estimates for transient temperature measurement in solids can be found in Refs. 74 to 77. For measurements in unsteady fluid flows, analysis of sensor response to temperature transients can be found in Ref. 72.

The above analyses provide error estimates in measurement of time-varying temperatures. However, the best practice is to reduce the potential lag by minimizing the time constant, usually by having a small sensor. One can also sometimes reduce the thermal resistance (i.e., increase h) to the surroundings. In a solid, for example, a good conductor such as copper oxide paste can be used to provide good thermal contact between the sensor and its surroundings.

C. HEAT FLUX MEASUREMENT

1. Basic Principles

Unlike temperature, heat is not a thermodynamic property. The definition of heat comes from the first law of thermodynamics, which, for a closed, fixed-mass system can be written

$$\mathbf{Q} = \mathbf{W} + \Delta\mathbf{U} \qquad (33)$$

The term $\mathbf{Q}$ has to be introduced into this energy balance when considering the interaction of a closed system with a surrounding which is not in thermal equilibrium (not at the same temperature) with the system itself. Heat can thus be defined as energy transported between two systems, and it is transported due to the temperature difference between the two systems. Note that in Eq. (33), $\mathbf{Q}$ is the heat added to the system and $\mathbf{W}$ is the work done by the system. From the second law of thermodynamics we know that the heat transferred between two systems goes from the one with the higher temperature to the one with the lower temperature.

Measurements of heat transfer are usually based on one or more of a small number of physical laws. For many measurements, the first law, Eq. (33), is used directly. This is true for some heat flux gauges, such as transient ones which normally have $\mathbf{W} = 0$, where the measurement of $\Delta\mathbf{U}$ would then give $\mathbf{Q}$. Some experimental systems achieve a constant heat flux boundary condition by electric heating. Then the energy input is a work term in the form of electrical energy. If a steady state is achieved, there is no change in internal energy and the work (electrical energy) added is equal to the heat transferred from the electric element.

With an open system, the first law for steady flow through a control volume with one exit and one entrance can be written in simplified form to obtain the net heat flow through the boundaries of the control volume:

$$q = \dot{\mathbf{Q}} = \dot{\mathbf{W}} + W(i_{\text{exit}} - i_{\text{ent}}) \qquad (34)$$

where i is enthalpy per unit mass. This equation is often used to determine the net heat (or energy) flow to a fluid passing through some system, such as a heat exchanger, nuclear reactor, or test apparatus.

Some heat flow measuring instruments are based on Fourier's law. For a one-dimensional system Fourier's law in rectangular coordinates is

$$q'' = \frac{q}{A} = -k \frac{\partial T}{\partial x} \tag{35}$$

Often, k is difficult to measure directly and a thermal resistance R_{th} of a finite-thickness material across which the temperature difference is measured is determined from a calibration of the heat flux gauge.

Thus, the first law of thermodynamics provides the definition of heat flow and it, along with Fourier's law, provides the basis for most instrumentation used to measure heat transfer or heat flux. Specific systems using these laws are described below.

2. Methods

a. Introduction

Most heat flux measuring devices can be placed into one of two categories: (1) the thermal resistance type, where measurement of temperature drop gives an indication of the heat flux from Fourier's law (e.g., the sandwich type, or Gardon type); and (2) the thermal capacitance or calorimetric type, where a heat balance on the device yields the desired heat flux (e.g., the wall-heating (cooling) type or temperature-transient type). For reliable results, heat flux measuring systems should be calibrated before use.

With a thermal resistance heat flux sensor, the presence of the instrument in the environment will disturb the temperature field somewhat and introduce an error in the measurement. Wall-heating systems require a heat source (or sink) and an appropriate heat balance equation to determine the heat flux. The temperature-transient types require a measurement of the temperature variation with time. A brief discussion of the various types of heat flux sensors is given below.

b. Thermal Resistance Gauges

Sandwich Type.   When heat is conducted through a thin slab of material (a thermal resistance), there is a temperature drop across the material. A measurement of the temperature difference across the slab can be used as a direct indication of the heat flux [78]. A schematic of a sandwich-type gauge is shown in Fig. 23. The steady one-dimensional heat flux through the material is related to the temperature difference ΔT by

$$q'' = \frac{k}{\delta} \Delta T \tag{36}$$

or

$$q'' = \frac{\Delta T}{R_{th}} \tag{37}$$

where k and δ are the thermal conductivity and thickness of the material, respectively, and R_{th} is the thermal resistance, δ/k. Even for a finite-size system where the temperature and/or R_{th} are not perfectly uniform along the gauge, Eq. (37) can still be used with R_{th} determined by calibration.

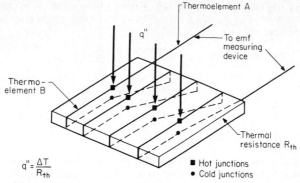

$$q'' = \frac{\Delta T}{R_{th}}$$

■ Hot junctions
● Cold junctions

FIG. 23 Sandwich-type heat flux gauge.

The temperature drop across the thermal resistance is usually measured with a multijunction thermocouple (thermopile) to increase the sensitivity of the device. The sensitivity of a heat flux sensor depends on the slab material and slab thickness, which essentially determine R_{th} and the number of junctions in the thermopile, which determines the output emf as a function of ΔT.

Heat flux sensors having a large sensitivity (output per unit heat flux) generally have a low maximum allowable heat flux, and a relatively large physical size (thus slower response time). For example, one commercial unit with a sensitivity of 60 μV per W/m^2, or 200 μV per Btu/(h·ft^2), has a maximum allowable heat flux of 1.6×10^4 W/m^2, or 5000 Btu/(h·ft^2), with a response time of a few seconds [79]; whereas a unit that can accommodate heat fluxes up to 6×10^5 W/m^2, or 2×10^5 Btu/(h·ft^2), has a sensitivity of 0.006 μV per W/m^2, or 0.02 μV per Btu/(h·ft^2), and a response time of less than one second [80].

Gardon Type.   The Gardon heat flux gauge [81] shown in Fig. 24 is also based on Fourier's law. A copper heat sink is installed in the wall of the measuring site with a thin constantan disk mounted over it. A small copper wire is attached to the center of the constantan disk. Another copper wire attached to the copper heat sink completes a thermocouple circuit. Heat flow to the constantan disk is conducted radially outward to the

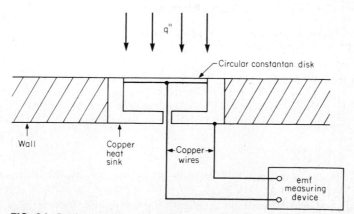

FIG. 24 Gardon-type heat flux gauge.

copper heat sink, creating a temperature difference between the center and the edge of the disk. The copper and constantan (other materials could be used) act as a thermocouple pair to measure this temperature difference.

The Gardon heat flux gauge is also called a circular foil heat flux gauge. It is used for measurement of convective and/or radiative heat fluxes. Although ideally the relationship between the heat flux and sensor temperature difference can be analyzed, calibration of the gauge is almost always necessary. More information is available in Refs. 82 and 83.

c. Calorimetric Gauges

Wall-Heating (Cooling) Type.   A wall-heating system employing a heat source (or heat sink) can be used not only to measure the heat flow from a surface but also to control the thermal boundary conditions. Two distinct types of systems can be considered: one, where the heat source or sink is behind the wall to which heat is being transferred to or from its front face; and the other, where the system is placed directly on the wall itself. The former type has often been used to obtain the heat flux. The rear of a wall can be sectioned off into individual regions where a forced liquid flow [84] or condensing vapor is used to measure the local heat flow. The mass flow and change of enthalpy of the fluid flowing to the rear face of the wall are used [Eq. (34)] to determine the heat input to each individual region of the front surface. Other systems measure the total heat loss from the surface by having a single fluid region behind the wall and measuring the mass flow and enthalpy change to that fluid. Sometimes the temperature drop across the wall and a calculated thermal resistance can be used as a check on the total heat flow.

For some systems, the heat input to different regions of the surface is adjusted to obtain a specific boundary condition—often to approximate an isothermal wall. This is most easily done with a number of small heating elements which can be individually set to maintain a constant wall temperature. The local heat flux or at least the heat flux averaged over the size of each individual heater can be determined from the power input to the heaters. These heaters can be placed quite close to the surface.

If a constant heat flux boundary condition is required, an electrical heating element, often a thin, metallic foil, can be stretched over an insulated wall. The uniform heat flux is obtained by Joule heating. If the wall is well insulated, then under steady-state conditions all of the energy input to the foil goes to the fluid flowing over the wall. Thermocouples attached to the wall beneath the heater can be used to measure local surface temperature. From the energy dissipation per unit time and area, the local surface temperature, and the fluid temperature, the convective heat transfer coefficient can be determined. Corrections to the total heat flow—e.g., due to radiation heat transfer or wall conduction—may have to be made.

Temperature-Transient Type.   With a temperature-transient gauge, the time history of temperature (an indicator of the change in internal energy) is used to determine the heat flux. Assuming one-dimensional (rectangular coordinates) heat flow, the governing equation for the temperature within a homogeneous gauge is

$$k \frac{\partial^2 T}{\partial x^2} = \rho c \frac{\partial T}{\partial \theta} \qquad (38)$$

Equation (38) can readily be solved to give a relationship between the temperature and the heat flux to the surface.

Various temperature-transient type gauges are available [85]. They can be categorized by the boundary conditions used: (1) the semi-infinite type, where the heat flux is derived from the solution of Eq. (38) for heat flow into a semi-infinite conductor; and (2) the finite-thickness type, where the wall beneath the gauge is assumed to be insulated and the heat flux is determined by the temporal variation of the (assumed uniform) sensor temperature.

The solution to Eq. (38) for transient conduction in a semi-infinite solid is utilized for the design of the thin-film gauge. Such a gauge consists of a thin—usually metallic, often platinum—film attached on the wall to (from) which the heat flows. The film is assumed to have no thickness and the wall beneath it is semi-infinite in depth. Then, the film temperature will be the surface temperature of the semi-infinite solid and the heat flux is related to the film temperature by the solution to Eq. (38). The surface temperature is usually measured by measuring the electrical resistance of the film gauge, though thermocouples near the surface of a homogeneous wall have also been used.

Finite-thickness type gauges include slug (or plug) calorimeters and thin-wall (or thin-skin) calorimeters. They assume the gauge is exposed to a heat flux on the front surface and is insulated on the back. The slug calorimeter consists of a small mass of high-conductivity material inserted into the insulated wall. A thin-wall calorimeter covers a large (or the entire) surface of a well-insulated wall. Both calorimeters assume the temperature within the gauge is uniform; thus the energy balance equations for a slug (plug) calorimeter (Fig. 25a) and a thin-wall calorimeter (Fig. 25b) are, respectively,

$$\frac{q}{A} = \frac{mc}{A}\frac{dT_s}{d\theta} \tag{39a}$$

or

$$q'' = \rho\delta c\,\frac{dT_s}{d\theta} \tag{39b}$$

where m and A are the mass and surface area of the slug (plug); ρ and δ are the density and thickness of the wall; and c and T_s are the specific heat and mean temperature of the slug or wall.

The time derivative of the mean temperature is needed to determine the heat flux.

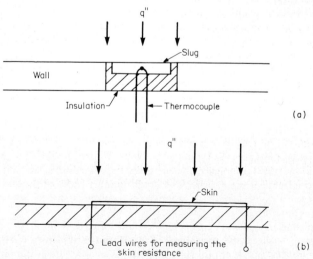

FIG. 25 Calorimeters: (a) Slug (plug) calorimeter. (b) Thin-wall (thin-skin) calorimeter.

This is obtained either by using a thermocouple to measure the back surface temperature of the sensor or by using the sensor as a resistance thermometer to measure its average temperature [85–87]. Since the mean temperature increases as long as the heat flow is positive, these sensors generally are limited to short-duration measurement of transient heat flux, as is also true for the thin-film calorimeter.

Uncertainties in the use of these sensors include the temperature nonuniformity across the sensors' thickness at high heat flux, the edge correction for localized gauges, and the disturbance of the temperature field caused by the presence of the sensor [85, 88].

D. ANALOGY

1. Introduction

Analogies have been widely used in studying heat transfer. An analog system is often simpler to construct than a heat transfer test apparatus. In addition, analog systems can often be set up to avoid secondary effects (e.g., conduction) that tend to introduce errors in temperature and heat transfer measurements.

Electric networks have been used to describe radiation heat transfer. Such an analog is useful because of the commonly available solutions for electric networks. It also permits use of an analog computer for solution of complex problems. Similarly, conduction systems have been studied using small analog models made of various materials, including conducting paper.

Numerical analysis using high-speed digital computers has taken the place of the above analogies in many situations requiring accurate analysis. Many conduction and radiation problems with known physical properties are amenable to computer modeling and solution. For this reason, the analogies for conduction and radiation heat transfer, though still used as teaching tools, will not be discussed here.

Computer modeling of convection has had mixed success. Many convection problems, particularly those involving laminar flow, can already be solved by special computer programs. However, in situations where turbulence and complex geometries are involved, computer analysis is still under development. Mass transfer analogies can play a key role in the study of convective heat transfer processes. Two mass transfer systems, the sublimation technique and the electrochemical technique, are of particular interest because of their convenience and advantages relative to direct heat transfer measurements.

The principal governing equations in convective heat transfer are the continuity equation, the momentum equations, and the energy equation. In a mass transfer system involving a two-component single-phase medium, the energy equation is replaced by the species diffusion equation. For the analogy between heat and mass transfer to be valid, the energy and species diffusion equations have to be similar in both form and boundary conditions. The conditions for similarity can be readily derived [89]. For laminar flow, the Prandtl number must be equal to the Schmidt number and there must be similarity in the boundary conditions. With turbulence, the energy and species diffusion equations both have an additional term involving a turbulent Prandtl number and a turbulent Schmidt number, respectively. Fortunately, experimental evidence suggests an equality between these two turbulent quantities in many flows.

In most heat transfer systems, the component of velocity normal to the active boundaries is zero, while for the corresponding mass transfer system this may not be the case. However, the magnitude of this normal velocity is usually sufficiently small that the analogy is not affected.

The advantages of using a mass transfer system to simulate a heat transfer system include the potential for improved accuracy of measurement and control of boundary conditions. For example, electric current and mass changes can generally be measured with greater accuracy than heat flux. Also while adiabatic walls are an ideal that, at best, we can only approach, impermeable walls are an everyday reality. Thus, mass transfer systems are gaining popularity in precision experimental studies.

In convective heat transfer, knowledge of the heat transfer coefficient h is often required:

$$h = \frac{q''}{\Delta T} \tag{40}$$

where ΔT is the driving force for heat transfer. For mass transfer (of component i), a mass transfer coefficient h_{Dj} can be defined:

$$h_{Dj} = \frac{j_i}{\Delta C_i} \tag{41}$$

where j_i is the mass flux of component i and the concentration difference ΔC_i is the driving force for mass transfer.

The dimensionless forms of the transfer coefficients are the Nusselt number, Nu, and the Sherwood number, Sh, for the heat and mass transfer processes, respectively:

$$\text{Nu} = \frac{hL}{k} \qquad \text{Sh} = \frac{h_{Dj}L}{D} \tag{42}$$

Each of these is a ratio of a convective transfer rate to the corresponding diffusion rate of transfer. Dimensionless analysis indicates that for fixed geometry and constant properties, the Nusselt number and the Sherwood number depend on the Reynolds number (forced convection), Rayleigh number (natural convection), position, Prandtl number (heat transfer), and Schmidt number (mass transfer).

2. Sublimation Technique

Sublimation of naphthalene is a frequently used mass transfer technique because of its simplicity when compared to heat transfer experiments. Early users of this mass transfer system include Jakob and coworkers [90, 91].

The molecular weight of naphthalene ($C_{10}H_8$) is 128.2 and its melting point is about $80°C$. At room temperature, solid naphthalene sublimes into vapor with a rate controlled by its vapor pressure and the convective process. Hence, it can be used to obtain information on average and local mass transfer coefficients (and heat transfer coefficients by analogy) along a surface exposed to the flow of a gas (usually air). The vapor pressure of naphthalene at $20°C$ ($68°F$) is about 0.05 mmHg. It has a strong, approximately exponential, variation with temperature—near $20°C$ an increase of $1°C$ will increase the vapor pressure by about 10 percent [92]. Hence control to ensure isothermal conditions during naphthalene experiments is essential.

In normal use naphthalene is cast in a mold so it will have the desired shape. For average or total measurements on a surface, the naphthalene section can be weighed before and after exposure to airflow to determine the sublimation rate. Once the vapor pressure of the naphthalene is known, the mass transfer coefficient can then be determined. Local mass transfer coefficients can be determined by profilometer measurements which measure the change in naphthalene thickness from the change in the surface profile before and after each test run [93]. Great accuracy and ease of use can be attained using a computer-controlled data acquisition system to measure the profile.

3. Electrochemical Technique

The sublimation method is used for mass transfer measurements in airflows. For measurement in liquids, the electrochemical technique can be used.

Early systematic study on the mass transfer process in an electrochemical system date back to the 1940s [94, 95]. Later investigators extended the use of the method to both natural- and forced-convection flows. Extensive bibliographies of natural- and forced-convection studies using the electrochemical technique are available [96]. A review of the technique [97] and a book [98] are convenient sources of information on the general treatment of electrochemical transport phenomena.

The working fluid in an electrochemical system is the electrolyte. When an electric potential is applied between two electrodes in the system, the positive ions of the electrolyte will move toward the cathode while the negative ions move toward the anode. The movement of the ions is controlled by (1) migration due to the electric field, (2) diffusion because of the ion density gradient, and (3) convection if the fluid is in motion. Fluid motion can be driven by the pressure drop in forced flow or by the density gradient in natural convection.

With heat transfer, convection and diffusion processes are present, but there is no equivalent to migration. In order to use ion transport as an analog to the heat transfer process, the ion migration has to be made negligible. This is done by introducing a second electrolyte—the so-called supporting electrolyte. It is normally in the form of an acid or base with a concentration of the order of 30 times that of the active electrolyte, and selected so that its ions do not react at the electrodes at the potential used in the experiment. This supporting electrolyte tends to neutralize the charge in the bulk of the fluid. The addition of a supporting electrolyte does not significantly affect the transport phenomena in forced flow. However, it introduces an additional density gradient into the buoyancy force term for natural convection. Therefore, the analogy between heat and mass transfer in natural-convection flow does not rigorously apply, and the effect of this additional gradient must be considered in applying the results of a mass transfer study.

Among the more commonly used electrolytic solutions are:

1. Copper sulfate–sulfuric acid solution: $CuSO_4–H_2SO_4–H_2O$
2. Potassium ferrocyanide–potassium ferricyanide–sodium hydroxide solution: $K_3[Fe(CN)_6]–K_4[Fe(CN)_6]–NaOH–H_2O$

With a copper sulfate solution, copper is dissolved from the anode and deposited on the cathode. For the other solution (also known as redox couples solution), only current transfer occurs at the electrodes. The respective reactions at the electrodes are:

1.
$$Cu^{++} + 2e^- \underset{\text{anode}}{\overset{\text{cathode}}{\rightleftarrows}} Cu$$

2.
$$[Fe(CN)_6]^{3-} + e^- \underset{\text{anode}}{\overset{\text{cathode}}{\rightleftarrows}} [Fe(CN)_6]^{4-}$$

Typical transport properties for the above systems are listed in Table 22. Note that they are high–Schmidt number (analogous to Prandtl number) fluids.

TABLE 22 Transport Properties of Typical Electrolyte Solutions

Solution*	Density ρ, g/cm^3	Viscosity μ, N·s/m^2	Diffusion coefficient of active species, $D \times 10^5$, cm^2/s		Schmidt number Sc = $\mu/\rho D$
A	1.095	0.0124	CuSO$_4$	0.648	1750
B	1.095	0.0139	K$_3$[Fe(CN)$_6$]	0.537	~2500
			K$_4$[Fe(CN)$_6$]	0.460	

Solution A:	CuSO$_4$:	0.05 gmol	H$_2$SO$_4$	1.5 gmol	
Solution B:	K$_3$[Fe(CN)$_6$]:	0.05 gmol	NaOH	1.9 gmol	
	K$_4$[Fe(CN)$_6$]:	0.05 gmol			

The mass transfer coefficient for species i, $h_{D,i}$, is defined in the usual manner,

$$h_{D_i} = \frac{(\dot{N}_i'')_{DC}}{\Delta C_i} \tag{43}$$

where $(\dot{N}_i'')_{DC}$ is the transfer flux of species i in kg mol/(s·m^2) due to diffusion and convection and ΔC_i is the concentration difference of species i in kg mol/m^3 across the region of interest.

The total mass flux can be determined from the electric current using the basic electrochemical relations. With the introduction of the supporting electrolyte, diffusion and convection are the prime contributors to the total mass flux, while the migration effect is accounted for as a correction. The concentration difference ΔC_i is determined from the bulk and surface concentrations. The bulk concentration is determined through chemical analysis of the solution; however, the surface concentration is an unknown. This is resolved by the use of the limiting current condition [98]. As the voltage across the system is increased, the current increases monotonically until a plateau—in the graph of current versus voltage—occurs. At this limiting current, the surface concentration of the active species at one electrode is zero and ΔC_i is just the bulk concentration.

APPENDIX

Definitions

Primary thermometer	One which uses a physical law to define an exact relation between the temperature and the measured physical quantity.
Secondary thermometer	One which does not have an exact relationship between the temperature and measured quantity; calibration is required.
Defining standard thermometer	A thermometer used in the definition of IPTS-68; calibration performed at defining fixed points or against another defining standard thermometer.
Transfer standard thermometer	An intermediate standard used to minimize the use and drift of a defining standard thermometer.

Working standard thermometer — A standard used in calibration which is the same type of thermometer as the thermometers to be calibrated. Calibration against a defining standard thermometer or a transfer standard thermometer is required at periodic intervals to ensure accuracy.

Thermometer Types

PRT	Platinum resistance thermometer
SPRT	Standard platinum resistance thermometer
Type S thermocouple	Platinum–10% rhodium vs. platinum thermocouple
Type R thermocouple	Platinum–13% rhodium vs. platinum thermocouple
Type B thermocouple	Platinum–30% rhodium vs. platinum–6% rhodium thermocouple
Type T thermocouple	Copper-constantan
Type J thermocouple	Iron-constantan
Type E thermocouple	Chromel-constantan
Type K thermocouple	Chromel-Alumel

Abbreviations

ANSI	American National Standard Institute
ASTM	American Society for Testing and Materials
IEC	International Electrotechnical Commission
IPTS	International Practical Temperature Scale
IPTS-68	International Practical Temperature Scale of 1968
NBS	National Bureau of Standards

REFERENCES

1. The International Practical Temperature Scale of 1968. Amended Edition of 1975, *Metrologia*, Vol. 12, pp. 7–17, 1976.
2. R. P. Hudson, Measurement of Temperature, *Rev. Sci. Instrum.*, Vol. 51, pp. 871–881, 1980.
3. J. F. Schooley, Toward a New Scale of Temperature, *Dimension/NBS*, pp. 20–25, 1978.
4. J. F. Schooley, Progress toward a New Scale of Temperature, *Dimension/NBS*, pp. 10–15, 1980.
5. L. A. Guildner and R. E. Edsinger, Progress in NBS Gas Thermometry above 500°C, in *Temperature, Its Measurement and Control in Science and Industry*, Vol. 5, Pt. 1, pp. 43–48, American Institute of Physics, New York, 1982.
6. L. G. Rubin, Cryogenic Thermometry: A Review of Recent Progress, *Cryogenics*, Vol. 10, pp. 14–22, 1970.
7. J. A. Clark, Temperature Measurement in Cryogenics, in *Measurement in Heat Transfer*, E. R. G. Eckert and R. J. Goldstein (eds.), 2d ed., pp. 163–240, McGraw-Hill, New York, 1976.
8. E. Pfender, Spectroscopic Temperature Determination in High Temperature Gases, in *Measurement in Heat Transfer*, E. R. G. Eckert and R. J. Goldstein (eds.), 2d ed., pp. 295–336, McGraw-Hill, New York, 1976.
9. J. A. Wise, Liquid-in-Glass Thermometry, *NBS Monograph 150*, National Bureau of Standards, 1976.

10. Calibration and Related Measurement Services of National Bureau of Standards, *NBS Spec. Publ. 250,* 1980 Edition and June 1981 Appendix, 1981.

11. R. D. Thompson, Recent Developments in Liquid-in-Glass Thermometry, in *Temperature, Its Measurement and Control in Science and Industry,* Vol. 3, Pt. 1, pp. 201–218, Reinhold, New York, 1962.

12. D. A. Grant and W. F. Hickes, Industrial Temperature Measurement with Nickel Resistance Thermometers, in *Temperature, Its Measurement and Control in Science and Industry,* Vol. 3, Pt. 2, pp. 305–315, Reinhold, New York, 1962.

13. W. F. Schlosser and R. H. Munnings, Thermistors as Cryogenic Thermometers, in *Temperature, Its Measurement and Control in Science and Industry,* Vol. 4, Pt. 2, pp. 795–801, Instrument Society of America, Pittsburgh, 1972.

14. G. T. Furukawa, J. L. Riddle, and W. R. Bigge, The International Practical Temperature Scale of 1968 in the Region 13.81K to 90.188K as Maintained at the National Bureau of Standards, *J. Res. (NBS),* Vol. 77A, pp. 309–332, 1973.

15. G. T. Furukawa, J. L. Riddle, and W. R. Bigge, The International Practical Temperature Scale of 1968 in the Region 90.188K to 903.89K as Maintained at the National Bureau of Standards. *J. Res. (NBS),* Vol. 80A, pp. 477–504, 1976.

16. J. L. Riddle, G. T. Furukawa, and H. H. Plumb, Platinum Resistance Thermometry, *NBS Monograph 126,* 1973; also in *Measurement in Heat Transfer,* E. R. G. Eckert and R. J. Goldstein (eds.), 2d ed., pp. 25–104, McGraw-Hill, New York, 1976.

17. M. Sapoff, Thermistors for Resistance Thermometry, *Measurements and Control,* Vol. 14, No. 2, pp. 110–121, 1980.

18. M. Sapoff, Thermistors—2: Manufacturing Techniques, *Measurements and Control,* Vol. 14, No. 3, pp. 112–117, 1980.

19. Thermistor Definitions and Test Methods, *EIA Std. RS-275-A,* 1971.

20. General Specification for Thermistors, Insulated and Non-insulated, *EIA Std. RS-309,* 1965.

21. General Specification for Glass Coated Thermistor Beads and Thermistor Beads in Glass Probes and Glass Rods, *EIA Std. RS-337,* 1967.

22. M. Sapoff, Thermistors: Part 4—Optimum Linearity Techniques, *Measurements and Control,* Vol. 14, No. 5, pp. 112–119, 1980.

23. E. F. Mueller, Precision Resistance Thermometry, in *Temperature, Its Measurement and Control in Science and Industry,* Vol. 1, pp. 162–179, Reinhold, New York, 1941.

24. J. P. Evans, An Improved Resistance Thermometer Bridge, in *Temperature, Its Measurement and Control in Science and Industry,* Vol. 3, Pt. 1, pp. 285–289, Reinhold, New York, 1962.

25. T. M. Dauphinee, Potentiometric Methods of Resistance Measurement, in *Temperature, Its Measurement and Control in Science and Industry,* Vol. 3, Pt. 1, pp. 269–283, Reinhold, New York, 1962.

26. R. D. Cutkosky, Automatic Resistance Thermometer Bridges for New and Special Applications, in *Temperature, Its Measurement and Control in Science and Industry,* Vol. 5, Pt. 2, pp. 711–712, American Institute of Physics, New York, 1982.

27. N. L. Brown, A. J. Fougere, J. W. McLeod, and R. J. Robbins, An Automatic Resistance Thermometer Bridge, in *Temperature, Its Measurement and Control in Science and Industry,* Vol. 5, Pt. 2, pp. 719–727, American Institute of Physics, New York, 1982.

28. S. Anderson and D. Myhre, Resistance Temperature Detectors—A Practical Approach to Application Analysis, *Rosemount Rep. 108123,* 1981.

29. D. J. Curtis, Temperature Calibration and Interpolation Methods for Platinum Resistance Thermometers, *Rosemount Rep. 68023F,* 1980.

30. S. D. Wood, B. W. Mangum, J. J. Filliben, and S. B. Tillett, An Investigation of the Stability of Thermistors, *J. Res. (NBS),* Vol. 83, pp. 247–263, 1978.

31. M. Sapoff and H. Broitman, Thermistors—Temperature Standards for Laboratory Use, *Measurements & Data,* Vol. 10, pp. 100–103, 1976.

32. M. W. Zemansky, *Heat and Thermodynamics,* 4th ed., pp. 298–309, McGraw-Hill, New York, 1957.

33. W. F. Roeser, Thermoelectric Thermometry, in *Temperature, Its Measurement and Control in Science and Industry,* Vol. 1, pp. 180–205, Reinhold, New York, 1941.

34. Manual on the Use of Thermocouples in Temperature Measurement, *ASTM Spec. Tech. Publ. STP-470B,* American Society for Testing and Materials, 1981.

35. R. L. Powell, W. J. Hall, C. H. Hyink, L. L. Sparks, G. W. Burns, M. C. Scroger, and H. H. Plumb, Thermocouple Reference Tables Based on the IPTS-68, *NBS Monograph 125,* National Bureau of Standards, 1974.

36. American National Standard for Temperature Measurement Thermocouples, *ANSI-MC96.1-1982,* Instrument Society of America (sponsor), 1982.

37. P. A. Kinzie, *Thermocouple Temperature Measurement,* Wiley, New York, 1973.

38. R. F. Caldwell, Thermocouple Materials, *NBS Monograph 40,* National Bureau of Standards, 1962.

39. E. D. Zysk and A. R. Robertson, Newer Thermocouple Materials, in *Temperature, Its Measurement and Control in Science and Industry,* Vol. 4, Pt. 3, pp. 1697–1734, Instrument Society of America, Pittsburgh, 1972.

40. T. M. Anderson and P. Bliss, Tungsten-Rhenium Thermocouples Summary Report, in *Temperature, Its Measurement and Control in Science and Industry,* Vol. 4, Pt. 3, pp. 1735–1746, Instrument Society of America, Pittsburgh, 1972.

41. G. W. Burns and W. S. Hurst, Studies of the Performance of W-Re Type Thermocouples, in *Temperature, Its Measurement and Control in Science and Industry,* Vol. 4, Pt. 3, pp. 1751–1766, Instrument Society of America, Pittsburgh, 1972.

42. N. A. Burley, R. L. Powell, G. W. Burns, and M. G. Scroger, The Nicrosil vs. Nisil Thermocouple: Properties and Thermoelectric Reference Data, *NBS Monograph 161,* National Bureau of Standards, 1978.

43. L. L. Sparks, R. L. Powell, and W. J. Hall, Reference Tables for Low-Temperature Thermocouples, *NBS Monograph 124,* National Bureau of Standards, 1972.

44. R. L. Powell, Thermocouple Thermometry, in *Measurement in Heat Transfer,* E. R. G. Eckert and R. J. Goldstein (eds.), 2d ed., pp. 112–115, McGraw-Hill, New York, 1971.

45. C. A. Mossman, J. L. Horton, and R. L. Anderson, Testing of Thermocouples for Inhomogeneities,: A Review of Theory, with Examples, in *Temperature, Its Measurement and Control in Science and Industry,* Vol. 5, Pt. 2, pp. 923–929, American Institute of Physics, New York, 1982.

46. W. F. Roeser and S. T. Lonberger, Methods of Testing Thermocouples and Thermocouple Materials, *NBS Circ. 590,* National Bureau of Standards, 1958.

47. F. R. Caldwell, Temperature of Thermocouple Reference Junctions in an Ice Bath, *J. Res. (NBS),* Vol. 69c, pp. 256–262, 1965.

48. R. D. Lee, The NBS Photoelectric Pyrometer and Its Use in Realizing the International Practical Temperature Scale above 1063°C, *Metrologia* Vol. 2, pp. 150–162, 1966.

49. R. P. Benedict, *Fundamentals of Temperature: Pressure and Flow Measurements,* 2d ed., pp. 144–146, 265–273, Wiley, New York, 1977.

50. W. F. Roeser and H. T. Wensel, Appendix to *Temperature, Its Measurement and Control in Science and Industry,* Vol. 1, p. 1313, Reinhold, New York, 1941.

51. H. J. Kostkowski and R. D. Lee, Theory and Methods of Optical Pyrometry, *NBS Monograph 41,* National Bureau of Standards, 1962.

52. R. D. Hudson, Jr., *Infrared System Engineering,* Wiley, New York, 1969.

53. R. E. Engelhardt and W. A. Hewgley, Thermal and Infrared Testing, in *Non-destructive Testing, NASA Rep. SP-5113,* Chap. 6, pp. 119–140, 1973.

54. R. J. Goldstein, Optical Techniques for Temperature Measurement, in *Measurement in Heat Transfer,* E. R. G. Eckert and R. J. Goldstein (eds.), 2d ed., pp. 241–293, McGraw-Hill, New York, 1976.

55. C. M. Vest and P. T. Radulovic, Measurement of Three-Dimensional Temperature Field by Holographic Interferometry, in *Applications of Holography and Optical Data Processing,* E. Marom, A. A. Friesem, and E. Wiener-Avnear (eds.), pp. 241–249, Pergamon, London, 1977.

56. Bimetallic Thermometers, *SAMA Std. PMC-4-1-1962,* 1962.

57. W. D. Huston, The Accuracy and Reliability of Bimetallic Temperature Measuring Elements, in *Temperature, Its Measurement and Control in Science and Industry,* Vol. 3, Pt. 2, pp. 949–957, Reinhold, New York, 1962.

58. J. L. Fergason, Liquid Crystals in Nondestructive Testing, *Appl. Optics,* Vol. 7, pp. 1729–1737, 1968.

59. T. V. Blalock and R. L. Shepard, A Decade of Progress in High Temperature Johnson Noise Thermometry, in *Temperature, Its Measurement and Control in Science and Industry,* Vol. 5, Pt. 2, American Institute of Physics, New York, 1982.

60. A. Ohte and H. Iwaoka, A New Nuclear Quadrupole Resonance Standard Thermometer, in *Temperature, Its Measurement and Control in Science and Industry,* Vol. 5, Pt. 2, pp. 1173–1180, American Institute of Physics, New York, 1982.

61. R. W. Treharne and J. A. Riley, A Linear-Response Diode Temperature Sensor, *Instrum. Technol.,* Vol. 25, No. 6, pp. 59–61, 1978.

62. Temperature Measurement Instruments and Apparatus, *ASME-PTC 19.3-1974,* supplement to ASME performance test codes, 1974.

63. G. N. Gray and H. C. Chandon, Development of a Comparison Temperature Calibration Capability, in *Temperature, Its Measurement and Control in Science and Industry,* Vol. 4, Pt. 2, pp. 1369–1378, Instrument Society of America, Pittsburgh, 1972.

64. H. K. Staffin and C. Rim, Calibration of Temperature Sensors between 538°C (1000°F) and 1092°C (2000°F) in Air Fluidized Solids, in *Temperature, Its Measurement and Control in Science and Industry,* Vol. 4, Pt. 2, pp. 1359–1368, Instrument Society of America, Pittsburgh, 1972.

65. D. B. Thomas, A Furnace for Thermocouple Calibrations to 2200°C, *J. Res. (NBS),* Vol. 66C, pp. 255–260, 1962.

66. Use of the Terms Precision and Accuracy as Applied to Measurement of a Property of a Material, *ASTM Std. E-177,* Annual Standard of ASTM, P. 41, 1981.

67. G. T. Furukawa, and W. R. Bigge, A Measurement Assurance Program—Thermometer Calibration, National Conference on Testing Laboratory Performance Evaluation and Accreditation, Sept. 25–26, 1979.

68. H. D. Baker, E. A. Ryder, and N. H. Baker, *Temperature Measurement in Engineering,* Vol. II, Omega Press, Stamford, Conn., 1975.

69. E. R. G. Eckert and R. M. Drake, Jr., *Analysis of Heat and Mass Transfer,* pp. 417–422, 694–695, McGraw-Hill, New York, 1972.

70. R. J. Moffat, Gas Temperature Measurements, in *Temperature, Its Measurement and Control in Science and Industry,* Vol. 3, Pt. 2, pp. 553–571, Reinhold, New York, 1962.

71. S. J. Green and T. W. Hunt, Accuracy and Response of Thermocouples for Surface and Fluid Temperature Measurement, in *Temperature, Its Measurement and Control in Science and Industry,* Vol. 3, Pt. 2, pp. 695–722, Reinhold, New York, 1962.

72. E. M. Sparrow, Error Estimates in Temperature Measurements, in *Measurement in Heat Transfer,* E. R. G. Eckert and R. J. Goldstein (eds.), 2d ed., pp. 1–24, McGraw-Hill, New York, 1976.

73. R. P. Benedict, Temperature Measurement in Moving Fluids, *ASME Paper 59A-257,* 1959.

74. C. D. Henning and R. Parker, Transient Reponse of an Intrinsic Thermo-couple, *J. Heat Transfer,* Vol. 89, pp. 146–154, 1967.

75. J. V. Beck and H. Hurwicz, Effect of Thermocouple Cavity on Heat Sink Temperature, *J. Heat Transfer,* Vol. 82, pp. 27–36, 1960.

76. J. V. Beck, Thermocouple Temperature Disturbances in Low Conductivity Materials, *J. Heat Transfer,* Vol. 84, pp. 124–132, 1962.

77. R. C. Pfahl, Jr., and D. Dropkin, Thermocouple Temperature Perturbations in Low Conductivity Materials, *ASME Paper 66-WA/HT-8,* 1966.

78. N. E. Hager, Jr., Thin Foil Heat Meter, *Rev. Sci. Instrum.,* Vol. 36, pp. 1564–1570, 1965.

79. Hy-Cal Engineering, sale brochure, Santa Fe, Calif.

80. RdF Corporation, sale brochure, Hudson, N.H.

81. R. Gardon, An Instrument for the Direct Measurement of Thermal Radiation, *Rev. Sci. Instrum.,* Vol. 24, pp. 366–370, 1953.

82. Standard Method for Measurement of Heat Flux Using a Copper-Constantan Circular Foil Heat Flux Gauge, *ASTM Std. E-511,* Annual Standard of ASTM, Pt. 41, 1981.

83. N. R. Keltner and M. W. Wildin, Transient Response of Circular Foil Heat-Flux Gauges to Radiative Fluxes, *Rev. Sci. Instrum.,* Vol. 46, pp. 1161–1166, 1975.

84. Standard Method for Measuring Heat Flux Using a Water-Cooled Calorimeter, *ASTM Std. E-422,* Annual Standard of ASTM, Pt. 41, 1981.

85. D. L. Schultz and T. V. Jones, Heat Transfer Measurements in Short-Duration Hypersonic Facilities, *AGARDograph No. 165,* 1973.

86. Standard Method for Measuring Heat Transfer Rate Using a Thermal Capacitance (Slug) Calorimeter, *ASTM Std. E-457,* Annual Standard of ASTM, Pt. 41, 1981.

87. Standard Method for Design and Use of a Thin-Skin Calorimeter for Measuring Heat Transfer Rate, *ASTM Std. E-459,* Annual Standard of ASTM, Pt. 41, 1981.

88. C. S. Landram, Transient Flow Heat Transfer Measurements Using the Thin-Skin Method, *J. Heat Transfer,* Vol. 96, pp. 425–426, 1974.

89. E. R. G. Eckert, Analogies to Heat Transfer Processes, in *Measurement in Heat Transfer,* E. R. G. Eckert and R. J. Goldstein (eds.), 2d ed., pp. 397–423, McGraw-Hill, New York, 1976.

90. H. H. Sogin and M. Jakob, Heat and Mass Transfer from Slender Cylinders to Air Streams in Axisymmetrical Flow, *Heat Transfer Fluid Mech. Inst.,* Preprint of Papers, p. 5, Stanford University Press, Stanford, Calif., 1953.

91. W. J. Christian and S. P. Kezios, Experimental Investigation of Mass Transfer by Sublimation from Sharp-Edged Cylinders in Axisymmetric Flow with Laminar Boundary Layer, *1957 Heat Transfer Fluid Mech. Inst.,* p. 359, Stanford University Press, Stanford, Calif., 1957.

92. H. H. Sogin, Sublimation from Disks to Air Streams Flowing Normal to Their Surfaces, *Trans. ASME,* Vol. 80, pp. 61–71, 1958.

93. R. J. Goldstein, M. K. Chyu, and R. C. Hain, Measurement of Local Mass Transfer on a Surface in the Region of the Base of a Protruding Cylinder with a Computer-Controlled Data Acquisition System, *Int. J. Heat Mass Transfer,* Vol. 28, pp. 977–985, 1985.

94. J. N. Agar, Diffusion and Convection at Electrodes, *Discussion Faraday Soc.,* Vol. 1, pp. 26–37, 1947.

95. C. Wagner, The Role of Natural Convection in Electrolytic Processes, *Trans. Electrochem. Soc.,* vol. 95, pp. 161–173, 1949.

96. A. A. Wragg, Applications of the Limiting Diffusion Current Technique in Chemical Engineering, *Chem. Eng.* (London), January 1977, pp. 39–44.

97. J. R. Selman and C. W. Tobias, Mass-Transfer Measurements by the Limiting-Current Technique, *Advances in Chemical Engineering,* Vol. 10, pp. 211–318, 1978.

98. J. Newman, *Electrochemical Systems,* 1st ed., Prentice-Hall, Englewood Cliffs, N.J., 1973.

99. E. R. G. Eckert, C. L. Tien, and D. K. Edwards, Radiation, in *Handbook of Heat Transfer Fundamentals,* W. M. Rohsenow, J. P. Hartnett, and E. N. Ganić (eds.), Chap. 14, McGraw-Hill, New York, 1985.

100. A. Benjaminson and F. Rowland, The Development of the Quartz Resonator as a Digital Temperature Sensor with a Precision of 1×10^{-4}, in *Temperature, Its Measurement and Control in Science and Industry,* Vol. 3, Pt. 1, pp. 701–708, Reinhold, New York, 1962.

101. L. C. Lynnworth, Temperature Profiling using Multizone Ultrasonic Waveguides, in *Temperature, Its Measurement and Control in Science and Industry,* Vol. 5, Pt. 2, pp. 1181–1190, American Institute of Physics, New York, 1982.

NOMENCLATURE

Symbols, Definitions, SI Units, English Units

A	area m^2, ft^2
a, b	constants used in Eq. (10)
C_i	mass concentration of species i: kg/m^3, lb_m/ft^3

C_i	molar concentration of species i in an electrochemical system: kg mol/m^3, lb mol/ft^3
c	specific heat: J/(kg·K), Btu/(lb$_m$·°F)
c_p	specific heat at constant pressure: J/(kg·K), Btu/(lb$_m$·°F)
c_1; c_2	constants, defined in Eq. (21)
D	diffusion coefficient: m^2/s, ft^2/s
E	electric potential, or thermoelectric emf: V
E_{AB}	electric potential of thermocouple circuit with materials A and B: V
E_a to E_e	electric potentials at junctions a to e, respectively (Fig. 8): V
E_0, E_s	electric potential used in Eq. (11): V
e	emissive power: W/m^2, Btu/(h·ft^2)
e_b	blackbody emissive power: W/m^2, Btu/(h·ft^2)
exp	exponential function
F	function defined in Eq. (14)
h	heat transfer coefficient: W/(m^2·K), Btu/(h·ft^2·°F)
$h_{D,i}$	mass transfer coefficient for species i: m/s, ft/s
I	electric current: A
i_{exit}, i_{ent}	enthalpy per unit mass at exit and entrance, respectively: J/kg, Btu/lb$_m$
j_i	mass flux of species i: kg/(s·m^2), lb$_m$/(s·ft^2)
k	thermal conductivity: W/(m·K), Btu/(h·ft·°F)
L	length: m, ft
m	mass: kg, lb$_m$
$(\dot{N}_i'')_{DC}$	diffusion and convection flux for species i in an electrochemical system: kg mol/(s·m^2), lb mol/(s·ft^2)
Nu	Nusselt number
n	exponential defined in Eq. (27)
P	pressure: Pa (N/m^2), lb$_f$/ft^2
Pr	Prandtl number
Q	heat: J, Btu
$\dot{Q}$	heat flow rate: W, Btu/h
q	heat transfer rate: W, Btu/h
q_P	Peltier heat: W, Btu/h
q_T	Thompson heat: W, Btu/h
q''	heat flux: W/m^2, Btu/(h·ft^2)
q_e'', q_r''	external energy source input and radiative flux input: W/m^2, Btu/(h·ft^2)
R	electrical resistance: Ω
R_C, R_D, R_T, R_X	resistances used in Eqs. (7) to (9): Ω
R_f, R_t	resistance of a fixed resistor and of a thermistor, respectively: Ω
R_{th}	thermal resistance: °C/W, °F·h/Btu
R_0, R_{25}, R_{100}	electrical resistance at 0°C, 25°C, and 100°C, respectively: Ω
Re	Reynolds number
r	recovery factor
S^*	entropy transfer parameter: V/°C, V/°F

S_A^* to S_D^* entropy transfer parameters for materials A to D, respectively: V/°C, V/°F

Sh   Sherwood number

s circuit constant defined in Eq. (13)

T absolute temperature: K, °R

T_f fluid temperature: K, °C; °R, °F

T_m blackbody temperature corresponding to the pyrometer-measured radiant energy: K, °R

T_r, T_{st}, T_t recovery temperature, static temperature, and total temperature, respectively Eq. (29) and (30): K, °C; °R, °F

T_s, T_w temperature of a heat flux sensor and that of the wall behind the sensor: K, °C; °R, °F

T_0, T_e, T_i junction temperatures (fig. 8): K, °C; °R, °F

t   temperature: °C, °F

U overall heat transfer coefficient: W/(m²·°C), Btu/(h·ft²·°F)

U   internal energy: J, Btu

V velocity: m/s, ft/s

W mass flow rate: kg/s, lb$_m$/h

W   work: J, Btu

Ẇ   rate of work: W, Btu/h

x rectangular coordinate: m, ft

Greek Symbols

α material constant for platinum, Eq. (6): °C⁻¹, °F⁻¹

α_{AB} Seebeck coefficient: V/°C, V/°F

γ resistance-ratio defined in Eq. (12)

Δ finite increment

δ thickness: m, ft

ϵ emissivity

θ time: s

λ wavelength: m, ft

λ_{max} wavelength at which $e_{b\lambda}$ is maximum: m, ft

μ dynamic viscosity: Pa·s, lb$_m$/(h·ft)

π_{AB} Peltier coefficient: W/A, Btu/(h·A)

ρ density: kg/m³, lb$_m$/ft³

σ Stefan-Boltzmann constant: W/(m²·K⁴), Btu/(h·ft²·°R⁴)

σ_T Thompson coefficient: W/(A·°C), Btu/(h·A·°F)

τ time constant, Eq. (32): s

Subscripts

st   static

t total

λ monochromatic value, evaluated at wavelength λ

Index